ANNUAL REVIEW OF BIOCHEMISTRY

ANNUAL REVIEW OF BIOCHEMISTRY

ESMOND E. SNELL, *Editor*
University of Texas at Austin

PAUL D. BOYER, *Associate Editor*
University of California, Los Angeles

ALTON MEISTER, *Associate Editor*
Cornell University Medical College

CHARLES C. RICHARDSON, *Associate Editor*
Harvard Medical School

VOLUME 48

1979

ANNUAL REVIEWS INC.   4139 EL CAMINO WAY   PALO ALTO, CALIFORNIA 94306

ANNUAL REVIEWS INC.
Palo Alto, California, USA

REPRINTS   The conspicuous number aligned in the margin with the title
of each article in this volume is a key for use in ordering reprints. Available
reprints are priced at the uniform rate of $1.00 each postpaid. The minimum
acceptable reprint order is 5 reprints and/or $5.00 prepaid. A quantity
discount is available.

International Standard Serial Number: 0066-4154
International Standard Book Number: 0-8243-0848-4
Library of Congress Catalog Card Number: 32-25093

Annual Reviews Inc. and the Editors of its publications assume no
responsibility for the statements expressed by the contributors to this
Review.

PREFACE

Often as I approach one of our large cities by air, I reflect on the complexity of our society and the many support systems that must function for cities to survive. The complexity of a cell is far greater. The variation of chemical structures, the myriad of reactions, and the subtlety of controls in cells suggest a final organism that should not work. But it does. The chemistry that occurs in such living cells is the subject of this volume, another in the influential series—the *Annual Review of Biochemistry.*

One of the pleasures of serving with the Editorial Committee for the *Annual Review of Biochemistry* is the perspective it gives of the breadth and nature of present-day biochemistry and molecular biology. Such perspective is extended to the user of this volume. Another pleasure is the excellent response we receive from invited authors, who, in most instances, accept our invitation to contribute a chapter. Another pleasure is that the nonprofit status of Annual Reviews Inc. allows this volume to be reasonably priced.

While assuming responsibility for the words above, I also speak on behalf of the authors, the other Editorial Committee Members, and the staff members of Annual Reviews Inc. We are pleased and proud to present the current volume as an additional chronicle for our remarkable science.

PAUL D. BOYER

Annual Review of Biochemistry
Volume 48, 1979

CONTENTS

SOME RELATED ARTICLES IN OTHER *ANNUAL REVIEWS*

From the *Annual Review of Biophysics and Bioengineering,* Volume 8, 1979

Analysis of Intact Tissue with ^{31}P NMR, C. Tyler Burt, Sheila M. Cohen, and Michael Bárány

Molecular Structure Determination by High Resolution Electron Microscopy, F. P. Ottensmeyer

Kinetic Analysis of Myosin and Actomyosin ATPase, Richard W. Lymn

Chemical Cross-Linking in Biology, Manjusri Das and C. Fred Fox

From the *Annual Review of Genetics,* Volume 12, 1978

Environmental Mutagens and Carcinogens, Minako Nagao, Takashi Sugimura, and Taijiro Matsushima

From the *Annual Review of Medicine,* Volume 30, 1979

Biological and Enzymatic Events in Chemical Carcinogenesis, Henry C. Pitot

From the *Annual Review of Microbiology,* Volume 33, 1979

Arginine Catabolism by Microorganisms, Ahmed T. Abdelal

Cellular and Molecular Mechanisms of Action of Bacterial Endotoxins, S. G. Bradley

Biosynthesis of Polysaccharides by Prokaryotes, Sheri J. Tonn and J. E. Gander

The Chemistry and Biosynthesis of Selected Bacterial Capsular Polymers, Frederic A. Troy II

Lipopolysaccharides of Photosynthetic Prokaryotes, Jürgen Weckesser, Gerhart Drews and Hubert Mayer

From the *Annual Review of Physical Chemistry,* Volume 30, 1979

Photosynthesis—The Light Reactions, Kenneth Sauer

Developments in Extended X-ray Absorption Fine Structure Applied to Chemical Systems, Donald R. Sandstrom and Farrel W. Lytle

Homogeneous Catalysis in Solution, Y. Pocker

Hemoglobin and Myoglobin Ligand Kinetics, Lawrence J. Parkhurst

Annual Reviews are published in the following sciences: Anthropology, Astronomy and Astrophysics, Biochemistry, Biophysics and Bioengineering, Earth and Planetary Sciences, Ecology and Systematics, Energy, Entomology, Fluid Mechanics, Genetics, Materials Science, Medicine, Microbiology, Neuroscience, Nuclear and Particle Science, Pharmacology and Toxicology, Physical Chemistry, Physiology, Phytopathology, Plant Physiology, Psychology, and Sociology. The *Annual Review of Public Health* will begin publication in 1980. In addition, four special volumes have been published by Annual Reviews Inc.: *History of Entomology* (1973), *The Excitement and Fascination of Science* (1965), *The Excitement and Fascination of Science, Volume Two* (1978), and *Annual Reviews Reprints: Cell Membranes, 1975–1977* (published 1978). For the convenience of readers, a detachable order form/envelope is bound into the back of this volume.

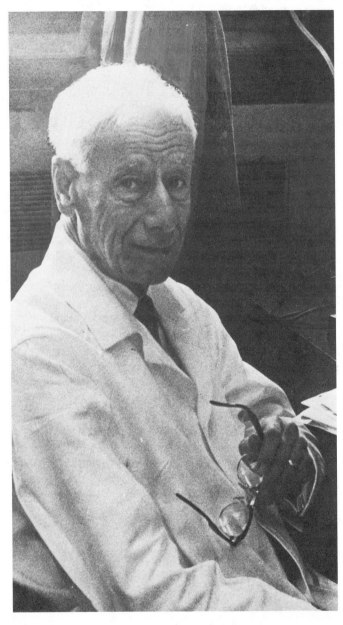

Michael Heidelberger

Ann. Rev. Biochem. 1979. 48:1–21
Copyright © 1979 by Annual Reviews Inc. All rights reserved

A "PURE" ORGANIC CHEMIST'S ♦12000
DOWNWARD PATH: CHAPTER 2—
THE YEARS AT P. AND S.*

Michael Heidelberger

Emeritus Professor of Immunochemistry, College of Physicians and Surgeons,
Columbia University, New York, New York 10032 and Adjunct Professor
of Pathology (Immunology), New York University School of Medicine,
New York, New York 10016

> The World of Science . . . is the purest democracy on earth, a brotherhood from which
> poverty bars no one—neither the color of skin nor religious belief. Intelligent devotion
> to the pursuit of scientific truth, and competent effort, or support of effort, toward this
> end, automatically enrolls one in the great company of the "Fifth Estate." One may find
> in this real "League of Nations" some of the most inspiring of human associations . . .
> Here may be encountered some of life's greatest opportunities for service to humanity and
> certainly some of its greatest compensations. Happy are its devotees!
>
> Charles Fuller Baker,
> Dean of the College of Agriculture,
> University of the Philippines
> In *The Philippine Agriculturalist,* 14:455 (1926)

In 1928 the Department of Practice of Medicine of the College of Physicians
and Surgeons of Columbia University moved with the Presbyterian Hospital from 70th Street and Park Avenue to the newly built Medical Center
on the site of the Yankees' baseball park at 168th Street and Broadway.
Walter W. Palmer was, at that time and for many years later, chairman of
the department. One of his close friends and advisors was Henry D. Dakin,
an ex-English chemist of "chloramine T" and Dakin pad fame, and they
had collaborated in a study of thyroidal function. Partly as a result of

---

*Chapter 1 of this account appeared in the *Annual Review of Microbiology,* Volume 31,
1977.

1

Dakin's influence and accomplishments, Palmer was determined to have a chemist in his newly expanded department and succeeded in obtaining one-half million dollars for that purpose from Edward S. Harkness, a principal benefactor of the Center. Thus was the Harkness Research Fund established, and my lucky earlier acquaintance with "Bill" Palmer in Van Slyke's laboratory at the Rockefeller Institute led to my choice as the first full-time chemist in a Department of Medicine. With the appointment as Associate Professor of Medicine went the title, Chemist to the Presbyterian Hospital, with purely consultative duties.

Medicine at the Center was taught and practised by a closely knit group of clinicians with widely divergent interests in research. Its leaders were Palmer himself, who studied the thyroid gland with Alexander Gutman and Ethel Benedict (later, Mrs. Gutman); Alphonse R. Dochez, concentrating at that time on streptococcal infections with Alvin F. Coburn, Franklin M. Hanger, and David Seegal; Robert F. Loeb, interested in a wide range of metabolic disorders; and Franklin M. Stevens, an allergist. Every Tuesday, after lunch, there was a departmental meeting in Palmer's spacious office at which he opened the proceedings with a discussion of problems of general interest or with medical gossip when there were no problems. Visitors were presented and anyone was free to voice suggestions or complaints or to bring up a topic of interest. Interactions with other departments were frequent; for example, Fordyce (Johnny) St. John, a Professor of Surgery, came in one day to protest the burden placed on his department by operations which could have been prevented by earlier medical intervention. The weekly meetings were continued by Robert Loeb when he succeeded Palmer on the latter's mandatory "retirement" to become Director of the Public Health Research Institute of the City of New York. I stress these meetings because, in another department at the Center, there was only one departmental meeting in the entire twenty-seven years I spent at P. and S. We also had afternoon teas in Dochez's and Hanger's offices, during which stimulating ideas were tossed about. For these teas we contributed small sums or a monthly cake.

I was given two laboratories at the northwestern end of the department's (8th) floor. Offered an office in the main stem of the hospital, near Dr. Palmer's, I chose instead a narrow architectural afterthought leading off from my personal laboratory past the powerhouse stack and designed to bring that feature out to the main building line. Although intolerably hot in summer, it was cozy in winter and saved long walks back and forth. I was soon joined by Forrest E. Kendall, who had just received his PhD in organic chemistry at the University of Illinois with Professor W. A. Noyes. It was a surprise, when Kendall arrived, to find that he had only one hand. The other had been cut off by a threshing machine on the family farm, but in spite of his handicap he had earned enough to put his younger brothers and sisters through college. In the eight years that we worked together, the absent hand was never missed. I had brought from Mt. Sinai Hospital, as technician, Check M. Soo Hoo, a California-trained Chinese who served faithfully and admirably with me for thirty-five years at P. and S. and at the Institute of Microbiology, until a series of heart attacks ended his life. The Harkness Fund also provided us with a laboratory helper and assistance for washing glassware.

During the last years at the Rockefeller Institute and the year at Mt. Sinai Hospital I had been reading Jules Bordet's huge, novelesque *Traité d'Immunologie* and Arrhenius and Madsen's *Immunochemistry,* a brave attempt in 1907 to apply principles of physical chemistry to the immunology of the time. It appeared to me that there was a crying need to determine the true nature of antibodies and that until this was done there could be no end to the polemics and uncertainties that were plaguing immunology. It was evident, also, that the purely relative methods in use, giving titers or their reciprocals, were inaccurate, often misleading, and incapable of permitting a decision as to whether antibodies were really globulins or substances of unknown nature adsorbed to these proteins. To a chemist, it appeared reasonable to assume that the principles of analytical microchemistry could be applied to the estimation of antibodies in units of weight, a prerequisite to the determination of their nature. This had been attempted by Hsien Wu et al (1a) but not carried to a conclusion.

Actually, the time was propitious for the development of the absolute method we envisaged. Felton had just devised a simple procedure for partial purification of antibodies in antipneumococcal horse sera by precipitation with the so-called euglobulins on dilution with large volumes of water and solution of the sediment in physiological saline (2). And if antibodies should indeed be proteins, the ideal initial antigen for their estimation would be the nitrogen-free, specifically precipitating capsular polysaccharide of type II or type III pneumococcus, which I had isolated with Oswald T. Avery at the Rockefeller Institute several years before (1b). Any nitrogen in a specific precipitate, therefore, would be derived from the antibody if conditions could be found to eliminate nonspecific, adsorbed proteins.

Realization of such a quantitative analytical micromethod became the immediate task for Kendall and me. I felt, though, that I had to justify Palmer's insistence on having a chemist in his department, so that we were soon collaborating with Bill on a study of thyroglobulin (3) and with Dochez's group on fractionation of the antigens of hemolytic streptococci (4, 5). In addition, the novel presence of organic chemists in a medical department often brought microbiologists, surgeons, and others to our laboratory to borrow chemicals or obtain suggestions on overcoming unforeseen difficulties. Our horizons were immeasurably widened and there was scarcely a day without excitement of one kind or another.

Antibodies to type III Pneumococcus, purified according to Felton, gave precipitates with the nitrogen-free type III pneumococcal capsular polysaccharide which were centrifuged down. The supernatants were analyzed for nitrogen, and this value, subtracted from the original amount of total nitrogen taken, gave a measure of antibody nitrogen by difference. Varying proportions of antigen and antibody were studied and we attempted to explain the results according to the law of mass action, much as other chemical precipitations could be accounted for quantitatively. Shortly after our publication (6), John R. Marrack, whose *Chemistry of Antigens and Antibodies* (7) soon attracted wide notice, came to the laboratory, having made the trip from England as ship's physician on a small Cunard liner. He objected to our assumptions that there were only two antigen-antibody complexes and that equilibria existed, because we had made no study of the kinetics. This resulted in an enduring friendship and caused Kendall and me to shift our emphasis, without abandoning the mass law, in the final development of a quantitative theory of the precipitin reaction (8).

Realizing that analyses by difference were applicable only to solutions of purified antibody, we added normal serum and began analyzing the precipitates. When well-washed in the cold with 0.9% saline, these precipitates yielded results identical to those previously obtained. We could then use whole sera and determine their content of precipitable antibody in units of weight.

This was all very well for antibodies to polysaccharides, but what about those elicited by the vast numbers of protein antigens? To solve that problem we had to distinguish between antigen nitrogen and antibody nitrogen. ^{15}N had not yet been discovered and radioactive isotopes had not been tamed for use in biology. We therefore had recourse to dyes. By coupling tetrazotized benzidine on one side with R-salt, a naphtholdisulfonic acid, and on the other side with crystalline hen's egg albumin, we produced an Easter-egg-like purplish protein antigen. This was treated rather drastically until it no longer precipitated anti-egg albumin. Injected into rabbits, it gave rise to antibodies that formed pink to red precipitates with the antigen, depending upon the proportions used. Washed precipitates were dissolved with alkali, rinsed into a colorimeter cup and compared, visually, with alkaline solutions of the dye. Photoelectric colorimeters were not yet developed. The solutions of the precipitates were rinsed quantitatively into micro-Kjeldahl flasks. Total nitrogen minus antigen-dye N gave antibody N (9). When we had worked out the course of this reaction, we were ready to study that of a colorless protein, crystalline egg albumin (10). Later, with H. P. Treffers, now Professor of Pathology at Yale, and a young B. Davis, whose short career was ended by poliomyelitis, we went on to study the immunochemical effects of phosphorylating the protein. The denaturation of egg albumin was also investigated with Catherine F. C. MacPherson (11) (who helped develop the method for estimation of antibodies in human sera, then studied the polysaccharides of several types of *Hemophilus influenzae* at Babies Hospital, went on to E. G. D. Murray's department at McGill, and is now an authority on proteins of the brain at the University of Western Ontario), the deamination of egg albumin with Paul H. Maurer (12a, 12b) (now for many years Professor of Biochemistry at Jefferson Medical), and the cross-reaction between hen's egg and duck's egg albumins (13) with Abraham G. Osler (who became Professor of Microbiology at Johns Hopkins, spent many years in research at the Public Health Research Institute of the City of New York, and has just gone to the University of California at San Diego).

Quite reasonably, some of our colleagues were worried about soluble antigen-antibody complexes in the egg albumin system, but as we had shown that they did not occur measurably and as we claimed only the estimation of precipitable antibody, our data appeared valid. Most welcome confirmation of our analyses came by an entirely independent physical-chemical method (14). Even this did not convince all "hard-core" immunologists, for as late as 1939 a new edition of a standard text insisted that we were "doubtless measuring non-specific nitrogen" instead.

Development of a quantitative theory of the precipitin reaction taxed our ingenuity, but the experiments we had done led inevitably and inexorably toward the interaction of multivalent antigen with multivalent antibody (8). We had partially hydrolyzed the type III polysaccharide and shown that the larger fragments could precipitate portions of the antibodies, so that the intact substance was obviously multivalent. From the proportions of antibody to antigen in precipitates from protein systems, it seemed reasonable to assume that a protein such as egg albumin must have at least five or six combining groups, so that our theory was not limited to polysaccharide antigens. As for the antibody, we knew from the beginning that precipitates in the region of antibody excess could combine with more antigen; therefore, two or more combining groups did not seem unreasonable. My friend, Felix Haurowitz, however, with true scientific skepticism, insisted for many years that antibodies need not be more than univalent, but he finally capitulated. We always remained friends in spite of this disagreement as to theory, and we argued in our correspondence and at meetings with as much relish as we experienced when playing sonatas together, for Felix was both a superb scientist and an excellent pianist. For many years, Waldo E. Cohn organized and played cello at evening sessions of chamber music at the annual meetings of the Federation of American Societies for Experimental

Biology. Felix and I frequently took part in these. Another critical friend was Linus Pauling, who believed that the valence of antibodies could not exceed two. However, I attended a lecture of his at Brooklyn Polytechnic and, with great glee, detected a trivalent antibody on one of his lantern slides, inserted to give a specific precipitate the proper degree of compactness. Linus also, with Dan Campbell and David Pressman, criticized our assumptions as "arbitrary and unlikely," which surely was their privilege. With a different set of assumptions, however, they arrived at an equation practically identical with ours (15). This, of course, pleased us immensely. Later, our simple equation for a quantitative theory was shown to be a special case of a more general theory (16) and this, in turn, was found to be a special case of a still more general theory requiring calculations by a computer (17). This was progress, but for immunologists needing simple, quantitative explanations and calculations in many, but not all, precipitating systems, our formulation and equation remained practical and useful. Moreover, the theory permitted a number of valid predictions, as appears later.

While Forrest and I were polishing up the precipitin reaction, our first graduate student, Elvin A. Kabat, arrived. He had received his BS *cum laude* at City College and was to work as a laboratory helper while studying for the PhD degree. Within a few weeks, this young whirlwind had read H. Gideon Wells's and Arrhenius's books and all of our papers and wanted to know why we hadn't adapted our quantitative method to bacterial agglutination. "We've been too busy," we said: "You go ahead with it," and he did.

Studies of the agglutination of microorganisms had been hampered, not only by the sole use of titers for measurement, but also by overemphasis of the effects of electrical charge and by lack of recognition that the combining ratios of bacteria and antibodies could vary greatly in different systems. Because of such differences, standard textbooks contained statements like "anti-*Salmonella* sera (titers up to 1:1,000,000) are stronger than anti-pneumococcal (titers up to several thousand)." Had this comparison been valid, most anti-*Salmonella* animals would have died of heart failure owing to the viscosity of their blood. Our plan was to wash killed type-specific pneumococci until the washings no longer gave tests for capsular polysaccharide. This was to avoid confusion of agglutinins with precipitins. A measured volume of the washed cells could then be added to a known volume of homologous antiserum. Agglutination occurred, as sufficient type-specific polysaccharide remained on the cells, and the added antibody nitrogen could be measured. Elvin worked out the proper conditions for accuracy, and bacterial agglutination was found to be a precipitin reaction at the cellular surface and subject to the same theory and similar quantitative expression (18).

An immediate practical result was the termination of the still ongoing dispute as to whether agglutinins and precipitins were identical or different. Though not as virulent as that over unitarian or pantheism, the ineffectual thought and effort given to it retarded progress. With our genuinely quantitative methods it was easy to show that, with respect to a single antigen and its homologous antibodies, precipitation and agglutination were functions of the same fraction of antibody (19). As protection had already been found to parallel the antipolysaccharide, three independent manifestations of the action of antibodies were shown to be due to the same protein molecules.

There was another intensely practical result when Hattie E. Alexander, microbiologist of the Babies Hospital at the Center, asked for help to improve the potency of the rabbit anti-*H. influenzae b* sera with which she wished to treat influenzal meningitis in infants. Quantitative agglutinin estimations that precisely measured the results of various dosages and routes of injection of the bacilli into rabbits soon resulted in sera with up to ten times the former content of agglutinins and precipitins for the polysaccharide of type *b*. Injection of the new sera into infants with influenzal meningitis cured most of them (20).

Confident of our methodology, we turned to our original quest: to find out whether or not antibodies were actually modified globulins. Again, luck favored us, for we were joined for several months by Torsten Teorell, presently (1978) Emeritus Professor of Physiology at the University of Uppsala. Torsten had been working with W. J. V. Osterhout at the Rockefeller Institute and had a consuming interest in the biological functions of electrolytes. We put him to work on the effects of varying concentrations of salts on the precipitin reaction with the aid of his own modification of the micro-Kjeldahl technique. An increase in sodium chloride diminished precipitation, but all added polysaccharide was still precipitated in the region of antibody excess. This, we soon realized, offered an approach to the isolation of pure antibody: a thoroughly washed precipitate formed in the region of antibody excess in 0.9% saline should liberate only antibody when equilibrated in 0.5M or M NaCl. This was all very well on a micro-Kjeldahl scale, at which nonspecific proteins could easily be washed out within the limit of error of the method. Our first large-scale preparation, however, was only about 70% pure, and it took Kendall, Kabat, and me two years before the first batch of analytically pure antibody was obtained (21a, 21b). This was typical globulin and provided the last link in the chain of evidence that antibodies were actually globulins. With this secure knowledge of the chemical nature of antibodies, one could now proceed to the study of their biosynthesis, cellular origin, and control. Visiting Denmark the following summer, I tried to interest August Krogh, a leading physiologist, in undertaking such a study, but he was fully occupied with other problems. Other Scandinavians, notably Mogens Björneboe and Astrid Fagraeus, however, stimulated by the new findings, soon implicated the plasma cell in the formation of antibodies (22a, 22b). These workers not only started cellular immunology on its prodigious expansion but also contributed much to its development.

While the quantitative studies were in progress, Florence Rena Sabin came to the Rockefeller Institute with radical ideas and experiments on the nature of tuberculosis. A warm friendship soon developed, and Florence was often at our home, particularly when we also invited Oswald T. Avery. My first wife, Nina Tachau, and I also went to the Cosmopolitan Club as her guests or to her home, where Florence was almost as proud of her skill in broiling steaks as of her scientific accomplishments. At her urging and with a grant from the National Tuberculosis Association, the complex mixtures of polysaccharides and proteins of various strains of mycobacteria were studied by means of fractionations meticulously carried out by Arthur E. O. Menzel (23a, 23b) and later by Sulo A. Karjala.

In 1934 and 1936 I was given partial John Simon Guggenheim Memorial Fellowships to work with The Svedberg's ultracentrifuge in Uppsala for periods of about six weeks, as I did not wish to be absent longer from the laboratory. The first time, I took samples of purified thyroglobulin along and prepared others with Swedish material. To our great surprise, this protein secreted by the thyroid gland was found to be of high molecular weight (24). On the second trip, purified equine antibodies to pneumococcal polysaccharides were centrifuged and also found to have high molecular weight (25), confirming Kendall's and my findings with the Northrop cell and in accord with independently arrived at results by ultrafiltration (26a, 26b). This time Nina and I were there in May and part of June, when we could enjoy the explosive Swedish Spring and its twenty-four hour shades of brightness. When there were not too many calculations to be done in the evening we would drive along an east-west road, enjoying colorful sunsets toward the north and watching them merge into equally beautiful sunrises. Svedberg had assigned his codeveloper of the ultracentrifuge, Kai O. Pedersen, to most visitors, to teach them the many precautions, controls, and actual use of the complicated, oil-driven machine with its terrifyingly numerous gauges and its optical system perfected by Ole Lamm. Kai, a gentle, self-effacing man of immense competence, had been nursing numerous visiting scientists along for some years, to the detriment of his own research. The many papers published with his help contained only brief acknowledgments for his efforts. As my time was limited, he practically gave up his work while I was there. We and our wives became fast friends and have often exchanged visits on both sides of the Atlantic. For his essential part in the work, I naturally made him a co-author of the resulting publications, a precedent which happily encouraged subsequent visitors to do the same.

By 1937 Elvin Kabat had completed two projects worthy of the PhD awarded for one of them. Thanks to a fellowship from the Rockefeller Foundation, he, too, went to Uppsala to work with Arne Tiselius and Kai Pedersen on the electrophoretic and ultracentrifugal properties of antibodies, including purified samples from various species of animals[1] which we sent over (27a, 27b). This research confirmed our analytical data on antibody content, as already noted, and even more firmly established the protein nature of antibodies and demonstrated their variations in molecular size.

Goodner & Horsfall had also shown that antipolysaccharide in rabbit antipneumococcal sera was of relatively low molecular weight, and had reported good clinical results with these sera (28). This led us to immunize rabbits with type I, II, or III pneumococci, and, with Joseph C. Turner of our department as clinician, to devise a simple method of partial purification of the antibodies for use in the Presbyterian Hospital (29). Cases of pneumonia due to I, II, or III received an intravenous infusion of 400–600 mg of the appropriate antibody. Nurses were instructed to bring me a few cc of the patient's blood half an hour later. Serum was drawn off and tested with homologous capsular polysaccharide. A positive precipitin reaction showing excess antibody usually resulted, but one severely infected type III patient required five bottles of antibody solution before an excess was established and recovery ensued. Under this regime deaths from these types of pneumonia were practically abolished at the Hospital as was also serum sickness. Soon afterward, the sulfa-drugs and penicillin came into use and typing and serum therapy were abandoned.

Having quantitated specific precipitation and agglutination, we took on the problem of complement (alexin), so important for diagnosis and immune lytic action. Was it a substance, as Ehrlich and his followers maintained, or was it merely a colloidal state of freshly drawn serum, as the school of Bordet insisted? Interest in this question was stimulated by the arrival of Alfred J. Weil from Germany and Otto G. Bier from Brazil, both of whom were familiar with the peculiarities of complement. An adequate series of controls soon made it possible to show that complement-containing sera added appreciable weight to specific precipitates formed with rabbit antipolysaccharide or antiprotein (30). Complement was therefore a substance or a series of substances (four components were then known) and Ehrlich's ideas were vindicated. Simultaneously, Pillemer, Ecker, Oncley, and Cohn purified the first of the four components (31), and arrived at the same conclusion. Further quantitative studies in the laboratory at P. and

---

[1] A turkey that had received injections of formalinized pneumococci was ultimately roasted by a Hospital chef and became the main dish of a laboratory luncheon. The formalin and pneumococci did not affect its flavor.

S., carried out with Manfred M. Mayer, Graciela Leyton of Chile, and Abraham Osler, established optimal conditions for the estimation and more efficient use of complement, and showed that, in many immune systems, its "fixation" ran parallel to specific precipitation.

Quantitative immunochemical methods also provided a check on metabolic studies with heavy nitrogen. Because of the rapid entry of ^{15}N into proteins in intact animals, Schoenheimer concluded that preformed peptide bonds opened and closed easily (32). Collaborating with his group, Treffers and I injected an ^{15}N-fed rabbit with killed type III pneumococci and antibodies to type I polysaccharide from another rabbit. Quantitative analyses of the circulating actively forming and passive antibodies at varying intervals showed that only the anti-III molecules being synthesized were capable of taking up ^{15}N. The simultaneous presence, with ^{15}N, of independent, immunologically specific markers thus settled an important aspect of the metabolism of proteins in the living animal (33). The central idea for this work, also quantitative studies of hemolysins, of antibodies to proteins raised in horses (in part with Jules Freund and Robert C. Krueger), and friendly assistance to foreign visitors, were contributions made by Henry P. Treffers during four postdoctoral years in the laboratory.

A foreign visitor not previously mentioned was Pierre Grabar, who arrived in 1937 as a Rockefeller Foundation Fellow, at the suggestion of André Boivin of Strasbourg. Pierre, a physical chemist, had only shortly before become interested in immunochemistry but developed into its leading European exponent on returning to the Pasteur Institute in Paris. Then there was Bertil Josephson, of Stockholm, whose interest in immunochemistry grew out of his specialization in the functions of the kidney. Still another was Sverre Dick Henriksen, a microbiologist from the laboratory of T. Thjötta in Oslo. Sverre could not return to the Rikshospital because of the invasion of Norway by the Germans, but was helped by a Rockefeller Foundation fellowship until he and his wife, Aase, departed for medical service in the Little Norway camp in Canada. Lifelong friendships developed with all of these visitors and their families.

Meanwhile, in part with D. L. Shrivastava, another Rockefeller Foundation Fellow and later Assistant Director of the Central Drug Research Laboratory at Lucknow and an expert on cholera, we did a quantitative study of homologous and cross-precipitation in the pneumococcal type III and type VIII systems (34a, 34b). This not only disclosed five different kinds of heterogeneity in the equine antibodies used, but also enabled us to predict the most likely arrangement of the sugars in the type VIII substance. There was also a practical sequel. A paper on oxidized cotton made it obvious to me that this linear polymer of cellobiose had been converted into analogs of types III and VIII polysaccharides by conversion of many $-CH_2OH$

groups to –COOH. One could predict from our quantitative theory that substances with multiple identically or similarly linked sugars or amino acids should give cross-precipitation in appropriate antisera. I wrote for samples of oxidized cotton and found at once that their neutralized solutions precipitated antipneumococcal III and VIII sera heavily (35). Our surgeons were then using oxidized cotton pads as a hemostatic and as a packing for wounds, for they could safely be left inside and would slowly disappear. By means of a rapid precipitin test of a patient's serum or urine with equine antipneumococcal type III serum, I was able to tell the surgeons, almost to the hour, how much time elapsed before the pads were wholly absorbed.

Then came the Second World War and theoretical problems were slowed or shelved and immediate concerns of the military were given priority. Two of our projects were highly secret and both illustrated the damaging effects of scientific secrecy. The first, with Forrest Kendall and L. A. Julianelle, was to protect and cure animals infected with anthrax, an agent our enemies were supposed to be planning to use. Julianelle found that for mice, penicillin, just then available, was the answer. Secrecy prevented Julianelle from publishing this and also lost him the benefits of priority when microbiologists at the Mayo Clinic, working outside of governmental auspices, announced the same finding in *The New York Times*. The second secret project concerned ricin, the enormously toxic principle of the castor bean which, it was feared, might also be used as a weapon. Elvin Kabat and I further purified samples of ricin, and colleagues in two other laboratories actually crystallized it. Suddenly, secrecy as to the chemical studies was lifted because of an incautious general's speech. I telephoned my friends in the project, proposing that we all publish our data in the same number of the same journal. They agreed, so we quickly wrote our paper, in which their independent work was mentioned, and sent it to Washington. Clearance was speedily granted, but alarming reports caused secrecy to be clamped down again before our colleagues' papers were written. This created the ridiculous situation that our publication (36) disclosed results that those who had obtained them were forbidden to mention.

Two nonsecret projects, however, proceeded normally. Because of a continuing large series of pneumococcal pneumonias during several years at a training camp for aviators at Sioux Falls, South Dakota, it became advisable to find out whether or not human subjects could be protected by immunization with purified capsular polysaccharides of the causative types. Felton had attempted this during the Great Depression by immunizing many thousands in camps of the Civilian Conservation Corps. His massive experiment was ill-fated, however, because healthful, outdoor living conditions led to almost no pneumonia in any of the camps, unimmunized or immunized, and mouse protection tests of sera of vaccinated persons varied from titers (alas!) of 0–2,000,000, an uninterpretable spread. I asked for volunteers from the entering classes at P. and S. in 1942 and 1943. Most of the students agreed to be injected with the polysaccharides of types I, II, and V and were wonderfully faithful in appearing for injections and bleedings, even though the latter often meant going without lunch during the hectic wartime compression of the medical course into three years. For a few exquisitely sensitive individuals it was fortunate that we first injected only about one third (15–20 μg) of the total dose, waiting 24–48 hours before adding the remainder if no reaction, or only a slight one, occurred. Strangely enough, in no case was a marked reaction due to known previous infection with pneumococci of types I, II, or V. Maximal values of precipitin in serum were reached in two to six weeks and a good response to one type did not guarantee an equal result with the others. The antibodies, unlike those to diphtheria toxoid, for example, were evoked at remarkably permanent levels, diminishing only slightly during months and even years (37). But were the volunteers protected against types I, II, and V? Our busy students would not have enjoyed being challenged with virulent pneumococci, but luck favored us. A paper by Barry Wood had appeared opportunely (38), in which rats sprayed 12 hr before with virulent type I pneumococci were saved by 0.02 cc of a rabbit antipneumococcal I serum but not by 0.002 cc. We analyzed this antiserum and calculated that 0.2 cc gave a concentration of antibody in the blood of Wood's rats about three times that of the average of our volunteers. Assuming that if three times our average could cure, one-third as much should protect, I recommended to the Surgeon General that vaccination be tried at Sioux Falls. We decided to inject polysaccharides of types I, II, V, and VII, as the resident microbiologists had found these types in about 60% of the pneumonias. The camp's population was randomly marshaled into two lines of 8,500 each. Those in one line were injected with 1 cc of saline as control subjects. Those in the other line were given 1 cc containing 50–70 μg of each of the polysaccharides. These had been prepared by us, by E. R. Squibb and Sons under Tillman D. Gerlough's direction, and by Augustus C. Wadsworth and Rachel Brown. As seen in Table 1, within two weeks there were no more pneumonias caused by the four types among those vaccinated. Even the nonvaccinated were partially protected because, as the microbiologists found, the vaccinated personnel no longer carried these pneumococci in their noses and throats. Pneumonias due to pneumococcal types not in the vaccine remained equal in both groups, which showed the strict specificity of the protection (39). The entire study, so beautifully organized and monitored in the field under Colin M. MacLeod's direction showed that epidemics of pneumococcal pneumonia in closed populations could be ter-

Table 1   Interval between injection and the development of pneumonia in immunized and nonimmunized subjects[a]

Interval Weeks	Number of cases of pneumonia			
	Types I, II, V, VII in immunized subjects	Types I, II, V, VII in nonimmunized subjects	All other types in immunized subjects	All other types in nonimmunized subjects
1	2	0	1	1
2	2	3	5	3
3	0	3	7	5
4	0	2	8	12
6	0	2	6	7
8	0	2	3	4
10	0	1	4	4
12	0	0	2	4
14	0	2	2	1
16	0	3	2	4
16+	0	8	16	14
Total	4	26	56	59

[a]Reprinted from *J. Exp. Med.* 1945. 82:453.

minated within two weeks after vaccination with the polysaccharides of the causative types. As many of the protected subjects had shown levels of antibody too low for our relatively insensitive precipitin technique to detect, it was evident that very little antibody could prevent the droplet infection by a few pneumococci that presumably initiates the disease.

Our fourth wartime project was carried out with Manfred M. Mayer and Myron A. Leon. The latter came from the Bronx High School of Science at the age of 16, was drafted into the Navy at 18, was reassigned to the project through the good offices of Admiral H. W. Smith, and is now Professor of Immunology and Microbiology at Wayne State University in Detroit. We had earlier prepared malarial vaccine from the hemolyzed red cells of volunteers among the malaria-infected sailors who had been admitted to the Marine Hospital on Staten Island. Manfred's wife, Elinor, a pianist, suggested that we try curing victims of malaria between relapses with vaccines made from their own parasites. This was done in about fifteen patients with mildly encouraging results. As the hospitals of the armed forces were filling up with malarial patients from the Solomon Islands and elsewhere in the Pacific, I proposed a trial of our vaccine to the Surgeon General of the Army. Better proof of efficiency was demanded, however, so I went to the Bureau of Medicine of the Navy. An entire ward of about 200 patients was quickly assigned to the project at St. Albans Hospital in Queens, New York, under the medical supervision of Commander W. A.

Coates. He explained the need for, and use of, blood from the most highly parasitized cases during relapses, and these men gave their blood liberally in spite of their discomfort. During the six months of the test our laboratory did little else than prepare malarial vaccine, and at the end the vaccine-treated and two control groups showed exactly the same rate of relapses (40). By-products of scientific value, however, were an electromagnetic method of concentrating the relatively few parasitized red cells in infections due to *Plasmodium malariae* (41) and an understanding of the triple nature of complement-fixation in malarial blood (42).

Postwar inflation soon made the Harkness Research Fund inadequate and recourse was had to the Rockefeller Foundation and the Office of Naval Research (ONR), precursor of the National Science Foundation (NSF). When I asked how the Navy could possibly be interested in one of our projects on an ONR grant, the reply was: "Do the research as you think best and the Navy will look out for its own interests." This liberal attitude was continued in the NSF when Alan Waterman left ONR to be Director of NSF.

During all these years I had been making sporadic attempts to learn more about the capsular polysaccharides of pneumococci: for example, that of type IV with Kendall, and types I, III, V, and XVIII with Harold Marko-witz, now with the Mayo Clinic. The polysaccharide of type I, Neufeld's "typical" pneumococcus, was taken up several times, but the instability of its hydrolytic products proved baffling. Happily, I have interested Bengt Lindberg in this unusual amphoteric polygalacturonate. Now that commercial quantities are being prepared one may expect that he will solve its fine structure, as he has with types II and XIV.

Unlike the problem with type I, another major investigation had been slowly expanding: the use of cross-reactions to clarify relations between chemical structure and immunological specificity. This had begun (see 1b) with Avery's intuitive feeling that analogs of pneumococcal polysaccharides must exist "free in Nature." His hunch was quickly justified by the cross-reaction of gum arabic and its products of partial hydrolysis in anti-pneumococcal type II serum and the analogous precipitation by the polysaccharide of Friedlander's bacillus B (*Klebsiella* K2). The cross-reaction of pneumoccal types III and VIII has already been mentioned. Also, during a consultantship with Merck and Co., of which Per K. Frohlich was Director of Research, I had looked into their synthetic polyglucoses as possible blood-extenders and found that, like the dextrans, certain fractions precipitated heavily with antipneumococcal sera of several types. Then glycogens of various origins were found to react in this way (43). To our astonishment, the paper describing this was rejected by the *Journal of Biological Chemistry* as not being biochemical! Since many gums of plants and other polysaccharides with nonreducing lateral end-groups of D-galactose precipitated antipneumococcal type XIV serum, one could predict that the capsular polysaccharide of this type would also contain this structural feature. I spent a summmer in the organic chemical laboratories at the University of Birmingham, England, learning techniques of chromatography and methylation from Maurice Stacey (who had been sent on a wedding trip mini-sabbatical to my laboratory in 1936 by Sir Norman Haworth) and S. Alan Barker. We verified the prediction (44). Other early uses of cross-reactions were the demonstration that an acidic impurity in the galactan of bovine lungs contained D-glucuronic acid (45) and that gum arabic was a mixture from which a fraction of quite different composition could be precipitated by antipneumococcal type II serum (46). Samples began to come in from many countries and it was often possible, by means of rapid cross-precipitation, in which the initial qualitative tests were frequently followed by quantitative estimations, to inform the sender of the nature, position, and/or linkage of one or more sugars in the product—information which might require weeks or months by purely chemical means.

In 1938 I first met Selman A. Waksman. We served on a planning committee for the International Congress of Biochemistry held in New York in 1939. My first wife, Nina, mother of my son, Charles, was on the Women's and Hospitality Committees, and many lasting friendships, especially with the foreign visitors, resulted from her activities as well. The hurried departure of the English, French, and German biochemists on the outbreak of World War II cast a heavy pall over the last days of the Congress.

From 1943 to 1946 Warren Weaver was organizing some eighty scientific talks sponsored by the US Rubber Company and transmitted over the Columbia Broadcasting System during the intermissions of Sunday concerts of the Philharmonic Orchestra of New York. It was, I am sure, Weaver's insistence that resulted in the rare circumstance of scientists being paid entertainers' fees—$500 for an eleven-minute talk. My subject was Resistance to Infectious Disease. Although my wife, Nina, was dying of cancer, as a writer she worked with me over every word, insisting on simplicity and clarity and helping shorten long sentences. The vivid memories of her aid required intense self-control when I gave the talk several months later, nervous as I also was with the realization that a million or more listeners might hear me. Perhaps they did, for this last message with her help, so frequent with earlier papers as well, brought letters from as far away as Los Angeles and Saskatchewan.

In December 1946, I was invited to the first postwar international scientific symposium, to be held in Paris to commemorate the 50th anniversary of the death of Pasteur. The twenty-one hour flight in a nonpressurized DC-4 across the Atlantic, via Goose Bay and Shannon, was a true adventure. The hotel in Paris was allowed to have heat for only an hour in the morning and another in the evening, but we were given an adequate supply of food stamps. The resurgent spirit of the French, after their enormous losses, was inspiring, and when a minister of education spoke proudly of French science, he was rebuked by Jules Bordet in an impassioned address stressing its international rather than national quality. Another early postwar international congress on microbiology was held in Copenhagen in the summer of 1947 and I was on our delegation to it. Selman Waksman was given an award and donated the quite large monetary portion for a Danish-American exchange fellowship. At the "banquet," by chance I sat next to Gabrielle Hoerner, a statuesque Alsatian pathologist who had come along as secretary to J. E. Morin, Professor of Microbiology at Laval University in Quebec. (I was much touched when, the following winter, Professor Morin and a carful of students drove through a heavy snowstorm to attend my French lecture in Professor Armand Frappier's Department at the University of Montreal.) The cursory table talk with Gabrielle eventually narrowed down to music and my impending visit to Paris, where Gaby was intending to visit a "musical sister." An invitation resulted, and I was embarrassed, being only an amateur, to find that the "musical sister" was the leading Wagnerian soprano of the Paris opera, Germaine Hoerner. The three of us, and Gaby's eventual husband, Robert Vogt, became good friends and I owed many fine sessions at the Opéra and participation in chamber music with French musicians to these accidentally acquired friends.

Through pure chance I was twice President of the American Society of Immunologists. The society's original constitution strangely provided that the elected president assume the vice-presidency *after* his presidential year. Elected president in 1947, I was vice-president in 1948 when the membership decided to change to the more usual progression. This automatically made me president again in 1949. My presidential addresses were: "Science, Freedom, and Peace," and its variant, "Ivory Pawn in the Ivory Tower," stimulated by activities novel for me up to that time.

Nina had been Chairman on Foreign Policy for the New York City League of Women Voters and active in the Speakers' Bureau of the American Association for the United Nations. After her death, her friends in the latter, lacking funds to hire experts and knowing I was going to Europe, asked me to try to carry on her work in the US delegation to the World Federation of United Nations Associations. There were extraordinary people in our group: among them, its leader, Clark Eichelberger, the UNA's executive secretary, Charles Marburg of Baltimore, who had been in the State Department, and Cyril Bath of Cleveland, who had begun as a laborer, had reorganized the Mexican railroads into a viable system, and in 1946 was a manufacturer of machine tools with an intensely loyal force of skilled workers. At the meeting in Prague, I acquired two warm friends in the Norwegian group, and because I knew French fairly well I was able, at a meeting of the French, English, and US delegations, to eliminate differences in meaning that had prevented agreement on two resolutions. These actually were voted on by the General Assembly of the United Nations that winter, and although they were too international to get enough votes from the nationalistic delegates, it was thrilling to know that my first dive into international politics had created even a ripple. Another year at the plenary sessions, I sat next to a pharmacist, Asare, one of two Ewe tribesmen from Togoland who had traveled through the native villages, talking about the United Nations and obtaining small sums for the journey to Geneva. They told an affecting story of a village that had spent years collecting money to build a bridge, but gave them the entire amount, considering their mission even more useful. Asare became an important official after Ghana's independence and sent a young student, Samuel Essandoh, to the United States for premedical studies and the medical course at P. and S., during which he was president of his class for one year. At the same meeting, in Geneva, were Evelyn Fox and Mary Dingman, who had been active there in the YWCA during the war, and we became fast friends. And later still, in order to finance a trip to the first meeting of the UNAs in Asia, I lectured in Hong Kong, Tokyo, Osaka, Kyoto, Hokkaido, and Bangkok, aided by the Rockefeller Foundation. Eleanor Roosevelt was a member of our delegation at the meeting in Bangkok, and I learned from her how to prepare a resolution on world peace that would be acceptable to our UNA. It was a joy for me and for my second wife, Charlotte Rosen, to be accepted by her as friends.

In the meantime, Duncan A. MacIness, of the Rockefeller Institute, unable to get the New York Academy of Sciences to agree to convene small meetings of experts to consider unsolved problems in the sciences, had been holding several highly successful midweek meetings of that type at an inn on Long Island under the auspices of the National Academy of Sciences. As the knowledge of complement was in a rather primitive state, Duncan was willing to add a conference on it to his series, with about twenty participants. All of us contributing to the study of complement were there in the autumn of 1950. We spend the first evening telling the others, among them physical chemists and enzymologists who had scarcely heard the word, what we knew, or thought we knew, of complement. The remaining two days were devoted to discussing the things that needed to be known and done, with highly stimulating and useful results (47).

In 1951 Linus Pauling, presently still very much alive, was forbidden by his physician to go to the Indian Science Congress to which he had been invited. As his substitute, I departed for India by air on Christmas eve and spent six busy weeks in that fascinating country as part of an international group that also included Erwin Brand, a biochemist of our Medical Center who was wise in the ways of amino acids and proteins and who stimulated the development of a photoelectric polarimeter for their study. Another invited foreigner was Arthur Stoll, of Basel, who had been Willstätter's first assistant when I studied in Zurich, and whose company, Sandoz, was equipping a new, air-conditioned animal house at the Central Drug Research Laboratory in Lucknow. In Calcutta, where the congress was held, and wherever we went afterward on our tour through India, Stoll was met by a Swiss consul general or commercial attaché and was usually housed in a governor's mansion. By contrast, our government left the amenities of hospitality entirely to our hosts, who did their very best under the direction of a young man named Gonçalves from the Ministry of Science who hailed from Goa, looked very Indian, and said his family had been Christians for 800 years. The general lectures of the Congress were held in an enormous open "pandal," or tent, which seated 8000 because of complaints that the general public was excluded in the original plans. As a result, entire families, from grandparents to small children, listened patiently to our technical talks. These were made difficult by the belated reverberation of our words from loud speakers at the back of the tent. I was greatly impressed by the fine research our Indian colleagues were doing, often with not-too-good facilities, but with a slowly increasing supply of excellent locally manufactured apparatus. Impressive, too, was a hospital in Bombay run by Col. A. A. Bhatnagar, with its high standards of sanitation and care of patients. In Calcutta we foreigners were given a dinner in the governor's palace and the privilege of individual talks with Nehru, who had actively furthered the development of the sciences in India. He and I, however, did not talk science but found a congenial topic in criticism of the policies of John Foster Dulles. During our tour we were kept so busy giving lectures that when we reached Delhi I refused to give any more until I had seen the Taj Mahal. It was arranged that the J. B. S. Haldanes, Brand, and I would be taken early in the morning to Agra by car, but at the appointed time the Haldanes had gone off birding and were left behind after much delay. The sad story of Shah Jahan and his beloved Mumtaz, and the glorious beauty of the monument to her were a tremendous emotional experience.

After the Congress I stayed on to complete a pet project. In the old days, Avery would say: "If we only had an elephant instead of rabbits, we'd have plenty of antiserum." I did not want to miss this chance to study the immunological properties of elephants, potential suppliers of huge quantities of antisera. But first, at Lucknow, my friend, D. L. Shrivastava, was having trouble with the supply of guinea pig complement for diagnostic tests. I suggested elephant complement instead, which apparently had never been tried. Serum was obtained from the nearest available elephant, a day's trip away, and we set up tests with great excitement, only to be disappointed. As for elephants and antibodies, chance and luck had intervened again. Back in New York, the same Mrs. Vally Weigl, through whom I met my present wife (see 1b), had been giving piano lessons to the wife of a former maharajah. A young friend of the maharani, sister of the first secretary of the Maharajah of Mysore, came to hear us play, and we became friends through our interest in the United Nations. When the younger lady learned of my approaching visit to India and my desire to immunize elephants, she notified her brother. The result was that in Mysore and Bangalore I was quartered in the Maharajah's No. 1 guest house and given an appointment to play one of the clarinet sonatas of Brahms with the Maharajah, an excellent pianist. However, only a brief, cordial interview resulted, owing to the Maharajah's sudden violent toothache. When he heard that I could not stay for the usual weekend illumination of the fountains in the beautiful park below the massive dam that was bringing electricity to the farms of Mysore, he ordered a special midweek illumination and throngs came for the unusual event. It was difficult to imagine that Thomas E. Dewey, then Governor of New York State, would do the same for a troublesome visiting scientist. The Maharajah also made two animals available, one of which, a work elephant at the edge of the jungle, was willing to lie down on command of his mahout. After several attempts the elephant was injected through its massive ear vein and the thin skin behind the ear with a few grams of human gamma globulin which I had brought along, thinking an antiserum to it might be of use in India. When we went for a bucket of blood some days later, the mahout and forester in charge would only let us draw about 5 cc, fearing we might kill the animal and the mahout would be jobless! This, although it was customary to calm obstreperous bull elephants by taking ten liters of blood. Anyhow, the 5 cc sufficed to show that one elephant, at least, was a good producer of anti-human globulin. Whether or not this has ever been followed up I do not know. What did follow, however, was a practical joke played by my young Indian friend who had been so helpful. Because of the limited success of the project, she said, she had arranged to have three elephants sent from Mysore. Seriously alarmed because of her previous efficiency, I was able to get the New York Zoo in the Bronx to agree to receive them. When the three elephants arrived, it was in a cardboard box, and they were exquisitely carved out of rosewood!

This brings me nearly to the end of my twenty-seven happy and busy years at P. and S. Mandatory retirement from Columbia in 1956 was looming ahead, and for some time Selman Waksman, who had founded and funded the Institute of Microbiology at Rutgers University with his royalties from streptomycin, had been asking me to start an immunochemical group in his institute. From 1954 on, my associate, Dr. Otto J. Plescia, and I had been giving occasional lectures at the Institute of Microbiology. "Retirement" to the laboratories there was humanely facilitated by Columbia in 1955, at Professor Elvin Kabat's suggestion, by giving me a year's terminal leave, with salary, so that I would not have to resign and could receive the proud title of Emeritus Professor of Immunochemistry. As no one wanted our specialized equipment or antigens and antibodies, we assembled a truckload and with it my small staff and I transferred to New Brunswick. Thus ends Chapter 2; within two weeks we were working again as smoothly as if we had never moved.

Literature Cited

1a. Wu, H., Cheng, L. H., Li, C. P. 1927–28. *Proc. Soc. Exp. Biol. Med.* 25:853–55; Wu, H., Sah, P. P. T., Li, C. P. 1928–29. *Ibid* 26:737–38

1b. Heidelberger, M. 1977. *Ann. Rev. Microbiol.* 31:1–12

2. Felton, L. D. 1926. *Bull. Johns Hopkins Hosp.* 38:33–60

3. Heidelberger, M., Palmer, W. W. 1933. *J. Biol. Chem.* 101:433–39

4. Heidelberger, M., Kendall, F. E. 1931. *J. Exp. Med.* 54:515–31

5. Seegal, D., Heidelberger, M., Jost, E. 1931–32. *Proc. Soc. Exp. Biol. Med.* 29:939–42

6. Heidelberger, M., Kendall, F. E. 1929. *J. Exp. Med.* 50:809–23

7. Marrack, J. R. 1934. *Chemistry of Antigens and Antibodies.* London: HMSO. (2nd ed. 1938.)

8. Heidelberger, M., Kendall, F. E. 1935. *J. Exp. Med.* 61:563–91

9. Heidelberger, M., Kendall, F. E. 1935. *J. Exp. Med.* 62:467–83

10. Heidelberger, M., Kendall, F. E. 1935. *J. Exp. Med.* 62:697–720

11. MacPherson, C. F. C., Heidelberger, M. 1945. *J. Am. Chem. Soc.* 67:574–77; 585–91; with Moore, D. H. 578–85

12a. Maurer, P. H., Heidelberger, M. 1951. *J. Am. Chem. Soc.* 73:2070–80

12b. Maurer, P. H., Heidelberger, M. 1952. *J. Am. Chem. Soc.* 74:1089–90

13. Osler, A. G., Heidelberger, M. 1948. *J. Immunol.* 60:327–37

14. Tiselius, A., Kabat, E. A. 1939. *J. Exp. Med.* 69:119–31

15. Pauling, L., Campbell, D. H., Pressman, D. 1943. *Physiol. Rev.* 23:203–19

16. Goldberg, R. J. 1952. *J. Am. Chem. Soc.* 74:5715–25

17. Aladjem, F., Palmiter, M. T., Chang, F-W. 1966. *Immunochemistry* 3:419–24

18. Heidelberger, M., Kabat, E. A. 1937. *J. Exp. Med.* 65:885–902

19. Heidelberger, M., Kabat, E. A. 1936. *J. Exp. Med.* 63:737–46

20. Alexander, H. E. 1965. In *Bacterial and Mycotic Infections in Man,* ed. R. J. Dubos, J. G. Hirsch, p. 738. Philadelphia: Lippincott. 4th ed.

21a. Heidelberger, M., Kendall, F. E. 1936. *J. Exp. Med.* 64:161–72

21b. Heidelberger, M., Kabat, E. A. 1938. *J. Exp. Med.* 67:181–99

22a. Björneboe, M., Gormsen, H. 1941. *Nord. Med.* 9:891–94; with Lundquist, F. 1947. *J. Immunol.* 55:125–29

22b. Bing, J., Fagraeus, A., Thorell, B. 1945. *Acta Physiol. Scand.* 10:282–94

23a. Menzel, A. E. O., Heidelberger, M. 1938. *J. Biol. Chem.* 124:89–101; 301–7

23b. Menzel, A. E. O., Heidelberger, M. 1939. *J. Biol. Chem.* 127:221–36. (See also earlier papers)

24. Heidelberger, M., Pedersen, K. O. 1935. *J. Gen. Physiol.* 19:95–108

25. Heidelberger, M., Pedersen, K. O. 1937. *J. Exp. Med.* 65:393–414

26a. Elford, W. J., Grabar, P., Fischer, W. 1936. *Biochem. J.* 30:92–99

26b. Goodner, K., Horsfall, F. L. Jr., Bauer, J. H. 1936. *Proc. Soc. Exp. Biol. Med.* 34:617–19

27a. Kabat, E. A. 1939. *J. Exp. Med.* 69:103–18;

27b. Tiselius, A., Kabat, E. A. 1939. *J. Exp. Med.* 69:119–31

28. Horsfall, F. L., Jr., Goodner, K., MacLeod, C. M. 1936. *Science* 84:579–81

29. Heidelberger, M., Turner, J. C., Soo Hoo, C. M. 1938. *Proc. Soc. Exp. Biol. Med.* 37:734–36

30. Heidelberger, M. 1941. *J. Exp. Med.* 73:681–94
31. Pillemer, L., Ecker, E. E., Oncley, J. L., Cohn, E. 1941. *J. Exp. Med.* 74:297–308
32. Schoenheimer, R. 1949. *The Dynamic State of Body Constituents.* Cambridge, Mass: Harvard Univ. Press
33. Heidelberger, M., Treffers, H. P., Schoenheimer, R., Ratner, S., Rittenberg, D. 1942. *J. Biol. Chem.* 144:555–62
34a. Heidelberger, M., Kabat, E. A., Shrivastava, D. L. 1937. *J. Exp. Med.* 65:487–96
34b. Heidelberger, M., Kabat, E. A., Mayer, M. M. 1942. *J. Exp. Med.* 75:35–47
35. Heidelberger, M., Hobby, G. L. 1942. *Proc. Natl. Acad. Sci. USA* 28:516–18
36. Kabat, E. A., Heidelberger, M., Bezer, A. E. 1947. *J. Biol. Chem.* 168:629–39
37. Heidelberger, M., MacLeod, C. M., Kaiser, S. J., Robinson, B. 1946. *J. Exp. Med.* 83:303–20
38. Wood, W. B., Jr. 1941. *J. Exp. Med.* 73:201–22
39. MacLeod, C. M., Hodges, R. G., Heidelberger, M., Bernhard, W. G. 1945. *J. Exp. Med.* 82:445–65
40. Heidelberger, M., Coates, W. A., Mayer, M. M. 1946. *J. Immunol.* 53:101–7
41. Heidelberger, M., Mayer, M. M., Demarest, C. R. 1946. *J. Immunol.* 52:325–30
42. Heidelberger, M., Mayer, M. M. 1946. *J. Immunol.* 54:89–102
43. Heidelberger, M., Aisenberg, A. C., Hassid, W. Z. 1954. *J. Exp. Med.* 99:343–53
44. Barker, S. A., Heidelberger, M., Stacey, M., Tipper, D. J. 1958. *J. Chem. Soc.* pp. 3468–74
45. Heidelberger, M., Dische, Z., Neely, W. B., Wolfrom, M. L. 1955. *J. Am. Chem. Soc.* 77:3511–14
46. Heidelberger, M., Adams, J., Dische, Z. 1956. *J. Am. Chem. Soc.* 78:2853–55.
47. Heidelberger, M. 1951.. *Proc. Natl. Acad. Sci. USA* 37:185–89

Ann. Rev. Biochem. 1979. 48:23–45

THE ASSEMBLY OF PROTEINS INTO BIOLOGICAL MEMBRANES: THE MEMBRANE TRIGGER HYPOTHESIS

♦12001

William Wickner

Molecular Biology Institute and Department of Biological Chemistry,
University of California at Los Angeles, 405 Hilgard Avenue,
Los Angeles, California 90024

CONTENTS

23

0066-4154/79/0701-0023$01.00

PERSPECTIVES AND SUMMARY

In addition to the unsolved problem of how a protein folds into its native conformation, there are several major problems in the proper assembly of an integral membrane protein into the lipid bilayer: (a) How are hydrophobic membrane proteins made by the water-soluble protein synthetic machinery without aggregating? (b) How are the polar regions of proteins that span the membrane transferred across the hydrocarbon core of the bilayer? (c) How is membrane protein asymmetry with respect to the plane of the bilayer established? (d) How are the lateral motions of proteins in the bilayer plane controlled? (e) How do newly synthesized proteins make the correct choice from among the numerous intracellular membranes? While there are almost no data that bear on this last question, two theories are reviewed that propose solutions for the others. One, an adaptation of the signal hypothesis (1, 2) to membrane assembly (3), emphasizes the role of catalysis in membrane assembly. In this model, a protein destined to span the bilayer has a hydrophobic N-terminal signal sequence that causes it and the ribosome on which it is synthesized to bind to a specific protein transport channel. It is proposed that the force of polypeptide chain elongation then drives the chain through the bilayer. When a hydrophobic sequence enters the protein pore, it is thought to signal the release of the peptide from this pore into the bilayer. Although leader (signal) peptides and membrane-bound ribosomes have been identified, there is no direct evidence for the proposed protein pore.

A second model of the assembly of proteins into membranes, termed the membrane trigger hypothesis, is presented. This model minimizes the role of catalysis in assembly and emphasizes the ability of a membrane lipid bilayer to trigger the folding of a polypeptide into a conformation that spans the bilayer or is at least integrally associated with it. The N-terminal leader peptide is thought to activate the protein for membrane assembly by altering the folding pathway.

Studies of the assembly of a wide variety of membrane proteins are presented that support the following conclusions: (a) Hydrophobic leader peptides are widespread if not ubiquitous and are normally removed during or shortly after membrane assembly. (b) Proteins can be specifically induced to assume alternative conformations in which their apolar residues are exposed to solvent or buried within. (c) A number of water-soluble proteins (complement, melittin, a-toxin) will spontaneously assemble into membranes. (d) The synthesis of membrane proteins such as those of the mitochondria and the major capsid protein of bacteriophage M13 occurs prior to assembly into the membrane. Coat protein with a hydrophobic N-terminal leader peptide is made on free polysomes as a water-soluble pro-protein. It then assembles into the membrane and is cleaved.

These two models are compared in the light of these data and some further experiments are proposed.

INTRODUCTION

Recent advances in understanding the structure of biological membranes have focused interest on their modes of growth. In this short review, questions will be posed about the insertion of integral proteins into membranes. A brief examination of what is known of the assembly of a few selected proteins will then permit a critical appraisal of two current theories of membrane assembly. This discussion focuses on the relative roles of enzymic catalysis and of the free energy of lipid-polypeptide interactions in guiding proteins into cellular membranes. Catalysis is taken here to include topographic catalysis, where there is only a spatial change distinguishing substrate from product.

Pivotal Facts

Several facts about membrane proteins are central to the assembly questions: (a) Membrane proteins are synthesized by the same (water-soluble) enzymic machinery used by soluble proteins. However, their synthesis is distinguished by unusual sensitivities to several antibiotics known to act on ribosomes (4–7), which suggests that membrane protein synthesis may occur on specialized ribosomes or at specific sites in the cell. (b) Integral membrane proteins are distinguished from soluble proteins by their hydrophobic surfaces which will promote their aggregation in aqueous buffers (8–10). Although they do not always have an unusually high content of hydrophobic amino acids (11), their amino acid sequence and folding expose their hydrophobic aminoacyl side chains rather than bury them in the interior as with soluble proteins (12). (c) In general, species of protein associate with a single cellular membrane. This may be a binary choice, as between the inner and outer membranes of *Escherichia coli,* or a choice between the multiple distinct membranes of a eukaryotic cell (cytoplasmic, nuclear, endoplasmic reticulum, golgi, mitochondrial, lysosomal, etc). (d) Each membrane protein has its own characteristic orientation, termed transverse asymmetry, with respect to the plane of the bilayer (3, 13). Small proteins such as glycophorin (14) or the major M13 capsid protein (15, 16) may span the bilayer once, while larger proteins such as bacteriorhodopsin (17–19) and the erythrocyte band III (20–22) span it several times. (e) Proteins can have specific lateral distributions on the surface of the bilayer. Examples of this form of organization include patching and capping of specific cell surface receptors (23, 24), mesosomes in the plasma membranes of prokaryotic cells (25, 26), and viral capsids formed during extrusion or budding processes (27–29). With the advent of cleavable bifunctional cross-linking reagents, surface topology maps are being drawn with increasing sophistication (30).

Major Problems

Each of these facts raises important questions for membrane assembly.

1. How are hydrophobic proteins synthesized in the hydrophilic milieu of the protein synthetic machinery without aggregating or improperly folding? Proposed solutions to this problem currently rely on special sites of synthesis or on special roles for the N-terminal portion of nascent membrane polypeptides.

2. How do membrane proteins make the correct membrane choice? Three facts underline the difficulty of this choice. First, there are cytologically only two protein synthetic compartments expressing nuclear genes, namely soluble and membrane-bound polysomes (rough endoplasmic reticulum in eukaryotes), yet there are a much larger number of membrane choices. If membrane proteins are made exclusively on the membrane-bound polysomes of the rough endoplasmic reticulum, then there must be an additional mechanism for directing these proteins to the appropriate membrane. Second, proper segregation of proteins among membranes occurs in cultured cells with varying lipid composition (31–36) and proportions of different membrane proteins (37). It is thus quite unlikely that membrane proteins form one extremely large multisubunit protein in the bilayer. Third, proteins are capable of crossing one membrane, a possible intermembrane compartment, and specifically assembling into a second membrane. This behavior, seen for example in bacteria (38, 39) and in mitochondria (40), is inconsistent with membrane proteins assembling with stability into the first bilayer that they encounter. Regions of apparent membrane fusion (41–43) may have a role in membrane protein segregation.

3. How are polar regions of membrane proteins transferred across the hydrocarbon center of the bilayer? Most water-soluble proteins are clearly incapable of spontaneously wriggling across membranes, although exceptions to this rule are presented later in this review. How, then, do water-soluble domains of membrane proteins, such as those exposed on the outer surface of cytoplasmic membranes, perform this feat? Once accomplished, it appears to be irreversible, that is, polar domains of integral membrane proteins do not rapidly rotate across the bilayer (13, 44).

4. Where is the information that specifies each protein's transverse asymmetry? While the primary protein sequence is surely of paramount importance, roles have also been proposed for the site of synthesis, for polypeptide transport elements, and for the preexisting asymmetries of lipid polar head groups, fatty acyl chains, and other membrane proteins.

5. What determines the lateral organization of membrane proteins? Peripheral membrane elements, those that are not directly bound to the hydro-

carbon core of the bilayer by hydrophobic interactions, are thought to play a major role in organizing the integral proteins' distributions (23, 24, 45).

Current Models

Two basic hypotheses have been presented to guide experimentation on these questions. The first, termed the signal hypothesis by Milstein et al (46) and by Blobel and colleagues (1, 2), was proposed for secreted proteins and extended by Rothman & Lenard (3) to membrane proteins. It gives emphasis to the potential role of protein catalysis in the complete transport of proteins across membranes (secretion) and in membrane protein assembly. It was suggested by the cytological observation that cells that export proteins have more abundant rough endoplasmic reticulum than those that do not, and by the biochemical observation that soluble polysomes seem to make largely soluble proteins, while polysomes isolated in association with the endoplasmic reticulum synthesize proteins that are destined for export (47–54). It seeks to explain the widespread occurrence of hydrophobic N-terminal leader peptides on membrane proteins and secreted proteins. These peptides are 15–30 residues long and are most readily detected during cell-free protein synthesis (1, 2, 46, 55–64). They are removed rapidly during, or shortly after, the peptides' passage across the membrane. The signal hypothesis (1, 2) suggests that this N-terminal region binds to a receptor in the bilayer and is recognized by a membrane-bound peptide transport system. The ribosome catalyzing the peptide's synthesis then binds to this transport system and the force of polypeptide chain elongation drives the peptide through this specific pore. On the opposite face of the bilayer, the emerging leader peptide is removed by a specific protease. Secreted proteins are proposed to entirely cross the membrane by this means, while integral membrane proteins are presumably released from the peptide transport system at an earlier stage of transport (3).

I present a second model, termed the membrane-triggered folding hypothesis (membrane trigger hypothesis), which places less emphasis upon catalysis in membrane assembly. This model rests on data from several laboratories, as is discussed below, and has in some aspects undoubtedly occurred to others [see, for example, the particularly cogent statement by Bretscher in (13)]. In this model, the leader peptide allows the growing chain to fold in a manner compatible with the aqueous environment. Membrane recognition is proposed to usually involve more than the leader peptide portion of the new protein, and the membrane element recognized may be either protein, lipids, or particular physical properties of the lipid phase. Upon binding to the appropriate membrane, the protein interacts with lipid components to fold into a conformation that exposes hydrophobic residues to the bilayer's fatty acyl chains. This interaction may begin before synthesis of the peptide chain is complete. Finally, the N-terminal leader oligopeptide is in many cases removed proteolytically, rendering the process irreversible. The central distinctions between this model and the signal hypothesis are the absence of a peptide transport system and the proposed role of the N-terminal leader peptide in facilitating the proper protein folding as it encounters a lipid bilayer. This model does not call for specific ribosome-membrane interactions.

In the following section, studies of several membrane proteins are examined from the perspectives of these two hypotheses. It is of course probable that neither model will prove to be completely correct for all proteins and that different proteins will integrate into membranes by different pathways. Rather, their utility as models will be measured by the degree to which they suggest useful experiments.

STUDIES OF SPECIFIC PROTEINS

Bacteriorhodopsin

Under anaerobic growth conditions, *Halobacterium halobium* produces a purple region of membrane bearing a single protein, termed bacteriorhodopsin, for its similarity to the visual pigment. Bacteriorhodopsin has a single 26,000 dalton polypeptide chain, one covalently bound retinaldehyde, and is 70–80% α-helical. It has been shown to function as a light-driven proton pump (65). Fragments of purple membrane have been isolated in pure form by virtue of their unusual density. Electron microscopy and electron diffraction (17–19) have revealed a regular two-dimensional array of this protein, which allows the determination of a 7Å resolution, three-dimensional structure (17). Bacteriorhodopsin spans the bilayer seven times in a largely α-helical conformation. Both diffraction and proteolysis studies (66) show that very little of the protein projects beyond the membrane into either aqueous compartment. In the absence of three-dimensional crystals of any known integral membrane protein, bacteriorhodopsin offers a unique insight into the structure of a membrane protein. It is difficult to imagine that any peptide transport system has the capacity to specifically catalyze the transmembrane orientation of the seven α-helical segments of this protein. The implications of this structure for assembly are further discussed below.

Erythrocyte Band III Glycoprotein

The best studied membrane, from a structural perspective, is that of the human erythrocyte (for review see 13, 67). The great advantages of the erythrocyte are that it has only one cellular membrane, that it can readily be obtained in large quantity, and that both right-side-out and inside out hemoglobin-free ghosts can be readily prepared. Proteins of the erythrocyte are named according to their order of migration on SDS-acrylamide gels. Band III, a 90,000 dalton glycoprotein, has been extensively characterized by Steck and colleagues (22, 68, 69). It spans the bilayer at least once, with its N-terminus exposed to the cytoplasm and its C-terminus on the cell exterior. This orientation differs from that of many membrane proteins, such as M13 coat protein, erythrocyte glycophorin, VSV G protein, rhodopsin, and bacteriorhodopsin, which have their N-termini "out" and their C-termini "in" (3). This striking result, which has also been found by Drickamer (20) and by Marchesi and colleagues (21) is not readily explained by models in which the protein assembles into the membrane by threading its N-terminus through the bilayer.

Mitochondrial Membranes

While the preponderance of mitochondrial proteins are made on cytoplasmic polysomes, a few are specified by the mitochondrial genome itself and are made within the mitochondrion. Several proteins of the mitochondrial inner membrane are comprised of polypeptides of cytoplasmic origin as well as others of mitochondrial origin. In yeast, these include cytochrome oxidase (70–76), cytochrome b (77), and the ATPase (71) with 3/7, 1/2 and 4/9 of the subunits of mitochondrial origin, respectively. This mosaicism poses several striking assembly questions: How are cytoplasmic proteins transported across the mitochondrial membranes? What determines whether a given polypeptide will reside in the outer membrane, the intermembrane space, the inner membrane, or the matrix? What coordinates the synthesis of cytoplasmic and mitochondrial subunits of each of these proteins to the proper levels?

Several interesting attempts have been made to answer these questions. Butow et al (78) have detected polysomes bound to the surface of yeast mitochondria in both thin sections and in the isolated organelle. Binding appears to be largely at sites of adhesion of the inner and outer membranes and is only disrupted by EDTA or by both puromycin and high salt. However, it is not clear from these studies whether these are the polysomes that make proteins that segregate into the membrane. Schatz and colleagues (75) have examined cytochrome c oxidase and cytochrome c_1 assembly in a deprived δ-aminolevulinic acid auxotroph. Cytochrome c oxidase has 2 heme groups and seven polypeptide subunits, I to III coded by the mitochondrial DNA and IV to VII by nuclear genes. When cells were grown in the absence of heme, only subunits II, III, and VI were found in the mitochondria and these were found by immunological tests to be abnormally assembled. In another study (79), the assembly of cytochrome c_1 into the inner membrane was examined. Cytochrome c_1, a 31,000 dalton peptide with a convalently bound heme, is made in the cytoplasm. Under anaerobic growth conditions or when heme synthesis was blocked, no cytochrome c_1 was found in the mitochondria but a cross-reacting peptide of the appropriate molecular weight was found in the cytosol. It was suggested that this might represent the primary translation product which, in the absence of heme, could not properly assemble into the membrane.

Insight into the coordination of the synthesis of a single enzyme's subunits by both cytoplasmic and mitochondrial machinery came from the observation that purified mitochondria require a soluble cytoplasmic fraction to carry on protein synthesis (70). Treatment of this fraction with antiserum to cytochrome oxidase or to the cytoplasmic subunits IV or VI abolished the extract-stimulated synthesis of mitochondrial-coded subunits I, II, and III. This treatment had no effect on the overall stimulation of mitochondrial protein synthesis, and control serum was without effect in any regard. The interpretation of these findings was that certain soluble proteins entered the mitochondria and that the synthesis of subunits I, II, and III within the mitochondrial matrix required a supply of the four other subunits from the cytoplasm. In this interpretation, the cytoplasmic proteins that entered the mitochondria did so after their synthesis was complete.

This interpretation has recently been confirmed and extended in several respects (R. Poyton, personal communication). Pulse-labeling yeast with radioactive amino acids has shown that subunits IV to VII are initially made as one polypeptide precursor, termed Pr_{IV-VII}. Pr_{IV-VII} is made on unattached polysomes, is probably the initial translation product, and has been purified. The presence of subunit IV to VII sequences in Pr_{IV-VII} is shown by the fact that antibody to each of the four subunits precipitates Pr_{IV-VII} and that tryptic peptides of each subunit are found in the set of tryptic peptides from Pr_{IV-VII}. Pr_{IV-VII} can be added to isolated mitochondria and will integrate into the inner membrane. This assembly requires both an energized inner membrane and concomitant mitochondrial synthesis of subunits I, II, and III. Finally, specific proteolysis of Pr_{IV-VII} yields the cytochrome oxidase subunits.

5'-Nucleotidase

Farquar, Bergeron, & Palade (80) have reported experiments that trace the topology of 5'-nucleotidase from its synthesis on the endoplasmic reticulum to its final location on the external face of the plasma membrane. Cytological assays of this enzyme indicated that it was located on the cytoplasmic face of endoplasmic reticulum vesicles and of those Golgi vesicles near the endoplasmic reticulum. Transfer of the catalytic site across the bilayer to the noncytoplasmic face only occurred in those secretion vesicles that were budding from the Golgi, i.e., distinctly posttranslationally. These results were confirmed and extended by Little & Widnell (81). They assayed rough endoplasmic reticulum, Golgi, and secretion vesicles for 5'-nucleotidase in the presence or absence of a permeabilizing detergent, for sidedness of lead phosphate deposits in the cytological assay, and for accessibility of the enzyme to concanavalin A or to specific antibody. Each of these tests confirmed that the enzyme's active site is on the cytoplasmic face of endoplasmic reticulum and Golgi vesicles and the noncytoplasmic face of secretion granules. The temporal separation of this enzyme's synthesis and the establishment of its final orientation raises important questions about its mode of assembly into the bilayer, questions that are not clearly answered by either the signal or membrane trigger hypotheses.

E. coli Lipoprotein

Of the many N-terminal leader peptides that have been identified, only that of the *E. coli* lipoprotein has a defined mutation. The outer membrane of *E. coli* has a short lipoprotein (82–85) of 58 amino acid residues with an N-terminal glycerylcysteine. Two fatty acids are bound to the N-terminus by ester linkages and one by an amide bond. At the C-terminus, approximately one third of the lipoprotein molecules are covalently bound to the peptidoglycan. Inouye has studied this protein in some detail and has proposed a model (86) in which its largely α-helical structure spans the bilayer. Protein synthesis in toluene-treated cells or in cell-free reactions yields a prolipoprotein with a hydrophobic N-terminal leader peptide of 20 residues (87, 88). This prolipoprotein is presumed to be the immediate biosynthetic product in vivo and must be quite rapidly proteolyzed to the mature protein. Recently, a point mutant in the prolipoprotein gene has been described (89) in which an aspartyl residue replaces glycine at position 14 in the leader region. The mutation prevents the processing of prolipoprotein to lipoprotein in vivo. The prolipoprotein was nevertheless still found predominantly in the outer membrane, which suggests that processing is not necessary for this protein to partition to the proper membrane.

VSV Spike (G) Protein

Vesicular stomatitis virus (VSV) buds from the plasma membrane of infected cells with a membrane envelope of cellular lipid and virus-specific protein. The virion contains a single transmembrane glycoprotein, the G protein, which is synthesized on the rough endoplasmic reticulum of infected cells and transported to the plasma membrane. Cell-free synthesis of the G protein has been achieved (90–92). These elegant studies have shown that addition of microsomal membrane to the synthetic reaction at or near the time when synthesis is begun causes proper proteolytic processing, glycosylation, and sequestration of all but the extreme C-terminus. Rothman & Lodish (91) have shown that none of these three criteria of correct spike protein assembly is seen if the membranes are added to the synthetic reaction at a time when only one fifth of the peptide chain is completed. This striking result indicates that the nascent chain will fold into a conformation that does not enter the assembly pathway unless it interacts with membranes at an early stage of its synthesis.

M13 Coat Protein

The (gene 8) capsid protein of the filamentous coliphage M13 is a 5,260 dalton, 50 amino acid peptide with an acidic N-terminus, a basic C-terminus, and a hydrophobic center (93, 94). Although the virus itself has no lipid, its 2,000 copies of coat protein are inserted asymmetrically into the host plasma membrane during infection (15, 16, 27, 95, 96). Later in the infectious process, newly synthesized coat protein also rapidly assembles into the plasma membrane. It has the same asymmetry as the "parental" coat from the virus, with the acidic N-terminus exposed on the cell exterior, and the basic C-terminus exposed on the inner membrane surface where it can interact with the M13 DNA during virus assembly. Both the coat protein from infecting virus and that synthesized de novo within the cell are utilized for progeny virus assembly (97), a physiologic confirmation of their common orientation. The fact that coat proteins that assemble into the membrane by two quite distinct paths achieve the same orientation is most consistent with protein sequence determining orientation.

Coat protein is one of the major products of M13 DNA-directed cell-free protein synthesis (98, 99). The coat protein made in such reactions has an extra 23 amino acids on its N-terminus and is referred to as procoat (99–102). Procoat is processed to coat protein and leader peptide by this same cell-free extract when a nonionic detergent is present during synthesis (101). Recently, this proteolytic activity has been localized to the membrane fraction (G. Mandel and W. Wickner, in press). Processing of procoat to coat protein can occur after procoat has been completely synthesized *if* detergent is present during the synthetic reaction (G. Mandel and W. Wickner, in press).

Procoat that is synthesized in a membrane-free reaction has been shown to be in a soluble form that will irreversibly bind to added lipid vesicles (102). Chymotrypsin trapped within such liposomes has access to the bound procoat, which indicates that the procoat spans the bilayer in the absence of any other membrane proteins.

Pulse-chase labeling of M13-infected cells (K. Ito and W. Wickner, in press) has established that procoat synthesis is complete prior to its conversion into membrane-bound coat protein. These results have been strengthened by the finding that the polysomes from M13-infected cells that are *not* membrane associated support vigorous procoat synthesis, while those that are membrane bound do not (W. Wickner, in press).

The simplest interpretation of these data is that coat protein is made in a precursor form on the soluble polysomes of M13-infected cells. The N-terminal leader peptide may play a crucial role in the ability of this protein to fold into a water-soluble conformation. When the procoat reaches the cell membrane, it assembles into a conformation that spans the bilayer and the leader peptide is removed proteolytically. The ability to mimic this reaction in vitro with lipid vesicles suggests that peptide transport may occur without catalysis.

Secreted Proteins

It cannot be overemphasized that the fundamental problems in secreting proteins are quite different from those involved in assembling proteins into membranes. Referring back to the list of central problems in membrane assembly, it is clear that: (*a*) Secreted proteins have only one orientation relative to the membrane, namely completely outside it. (*b*) In their final conformation, secreted proteins have a predominantly polar surface, in contrast to the apolar surfaces which anchor membrane proteins to the bilayer. (*c*) It is sufficient for all the proteins that are secreted from a cell to enter the microsomal lumen, which is topologically outside the cell. If all the new membrane proteins were to integrate into the microsomal membrane, a mechanism would still be needed to partition them to the membranes of the various organelles. Nevertheless, secretion is briefly considered here.

Palade and colleagues (47, 48, 52, 104, 105) have studied the synthesis of secreted proteins by cytochemistry, by pulse labeling, and by isolation of the relevant subcellular fractions. They have shown that secreted proteins are synthesized on the rough endoplasmic reticulum and discharged into its lumen, then are found successively in the lumen of the Golgi vesicles and in the secretory granules or vesicles. Finally, they are discharged from the cell when the granule membrane fuses with the plasma membrane. This pathway of secretion has been questioned by Rothman (106). In the late 1960s, Ganoza & Williams (49), Redman (51), and Hicks et al (50) separated hepatic membrane-bound polysomes (microsomes) and unattached polysomes and allowed completion of polypeptide synthesis in a cell-free extract. They found, in agreement with Palade's studies, that the microsomes synthesized serum proteins such as albumin while the unattached polysomes made hepatic cellular proteins such as transferrin. In 1972,

Milstein et al (46) reported that free mRNA from a mouse myeloma directed synthesis in a reticulocyte extract of a polypeptide that was approximately 1500 daltons larger than an immunoglobulin light chain but that had an essentially identical peptide map. When myeloma microsomes were used instead of the mRNA, light chains of mature length were produced. Polysomes derived from the microsomes by detergent extraction produced the longer form of light chain. They suggested that the larger size material was a precursor with approximately 15 additional amino acid residues on its N-terminus which might function as a signal to direct ribosomes with nascent light chain to the microsomal membrane. This signaling concept was combined with the concept of a specific peptide transport system by Blobel & Dobberstein (1), who termed this the signal hypothesis. They proposed that the leader (signal) peptide guided its ribosome to the peptide transport system or channel, that a specific membrane-ribosome bond was formed, and that subsequent chain elongation extruded the peptide through the protein channel. Blobel and colleagues (56, 58–60, 90) and others (46, 55, 57, 61–64, 91, 92) have since identified a wide variety of secreted proteins that are synthesized in vitro with an additional 15–30 hydrophobic residues at their N-terminus. Microsomal membranes have been found to remove leader peptides if they are present from the start of the synthesis. The proteins synthesized in the presence of microsomal membranes are protected from added protease, which reflects their transit into the microsomal lumen.

While many studies support this model, certain aspects have been questioned. Davis and co-workers (107–109) have demonstrated that membrane-bound polysomes from *E. coli* which synthesize secreted proteins such as alkaline phosphatase are only attached via their growing peptide chain. Highfield & Ellis (110) have demonstrated that translocation of secreted protein across a membrane need not always occur during the protein's synthesis. They studied ribulose bisphosphate carboxylase, an abundant soluble enzyme in chloroplasts. This enzyme has two subunits, a large one coded by chloroplast DNA and a small one made in the cytosol from message of nuclear origin. Cytoplasmic mRNA from light-induced peas directed synthesis of a small subunit precursor peptide in a wheat germ extract. After translation was complete, chloroplasts were added. The completed pre-small subunit was taken up by the chloroplasts (as assayed with protease) and was processed to the mature small subunit size. This translocation and processing were unaffected by chloramphenicol or cycloheximide and were therefore independent of protein synthesis. This data is similar to Poyton & Kavanagh's studies (70) of mitochondrial cytochrome oxidase, described above. Finally, ovalbumin has been found to be synthesized and secreted without the aid of a transient leader sequence (111).

A CRITICAL APPRAISAL

This background of information on a wide spectrum of proteins permits consideration of the strengths and weaknesses of the proposed mechanisms of membrane assembly. These mechanisms, the signal hypothesis and the membrane trigger hypothesis, were presented earlier in this review and are compared for convenience in Table 1 and Figure 1.

Signal Hypothesis

If a specific protein transport pore recognizes an N-terminal "signal" peptide, then it follows that proteins that span the bilayer should do so with their N-terminus on the exterior surface and their C-terminus facing the cytoplasm. This prediction has been substantiated for a remarkably wide variety of membrane proteins (3) but is clearly not true for the band III protein of the human erythrocyte, which is oriented with the opposite polarity. A more telling problem is that a growing number of membrane proteins span the bilayer more than once. Transport proteins will undoubtedly often fall into this class. One such protein, bacteriorhodopsin, has been discussed earlier in this review. It is difficult to imagine how a protein

Table 1   A comparison of two models for membrane assembly

Stage of synthesis	Signal hypothesis	Membrane trigger hypothesis
Site of initiation	soluble polysomes	soluble polysomes
Role of the leader peptide	recognition by the protein transport channel	to alter the folding pathway
Association of the new protein with membrane	time: when leader peptide is complete place: protein transport channel	time: during or after protein synthesis place: receptor protein or lipid portions of the bilayer
Specific ribosome associations	with the protein transport channel	none
Catalysis for assembly	a specific pore	the effect of the leader peptide on conformation
Driving force for assembly	polypeptide chain elongation	protein-protein and protein-lipid associations: self-assembly
Removal of leader peptide	during polypeptide extrusion	during or after polypeptide assembly into bilayer
Segregation of proteins into different cellular membranes	not clearly addressed	not clearly addressed
Final orientation	C-terminus in, N-terminus out	specified by the primary sequence

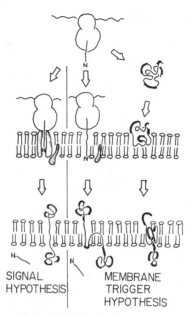

Figure 1 A schematic comparison of the signal hypothesis and the membrane trigger hypothesis. Thick lines are polar regions, thin lines represent hydrophobic regions. A soluble complex of nascent peptide, tRNA, ribosome, and mRNA is shown at the top of the figure. The shaded integral membrane proteins are the preprotein protease and the peptide transport pore.

transport pore could catalyze the ordered weaving of this 27,000 dalton polypeptide back and forth across the bilayer seven times. If, to account for transport of polar peptide regions to the exterior, the interior of the pore is polar, it would be hard for the pore to transport the seven apolar segments of bacteriorhodopsin.

A second prediction of the signal hypothesis is that proteins can only insert into or cross membranes during their synthesis. This is clearly not substantiated in several cases. M13 coat protein is synthesized in vivo as a precursor on soluble polysomes. It then enters the membrane and is converted to coat protein. In vitro, procoat has been shown to assemble into protein-free lipid vesicles post-translationally. Finally, procoat can be processed to coat protein post-translationally. A second clear example is the post-translational processing and entry into the chloroplast of the small subunit of ribulose bisphosphate carboxylase. Finally, many mitochondrial proteins that are made in the cytosol enter the organelle after their synthesis is complete. Even clearer examples of proteins that cross membranes in the absence of synthesis, namely melittin and complement, are discussed below.

Since much of the experimental evidence is consistent with the signal hypothesis without establishing it, it is important to consider what sort of experimental evidence would prove it. This might include the purification and reconstitution of the membrane components necessary for protein insertion and processing and the demonstration that a particular protein can recognize the leader (signal) peptides, bind the large ribosomal subunit, and conduct peptides across the bilayer. A genetic approach might also be possible, i.e., the isolation of a conditionally lethal mutant which, under nonpermissive conditions in vivo (and in vitro), failed to assemble new proteins into its membrane(s) while both continuing the functions of those previously present and exhibiting normal synthesis of soluble proteins. Biochemical studies would still be necessary to establish that such a lesion was indeed in a protein pore rather than, for example, in an enzyme of lipid metabolism.

Membrane Trigger Hypothesis

The essence of this hypothesis is that the thermodynamics of protein folding governs membrane assembly with little intervention of catalysis. The leader peptide is proposed to function by modifying the folding pathway where necessary. It may be proteolytically removed when assembly is complete to permit the protein to function or to drive the assembly reaction. In the latter function, the leader peptide may be viewed as a means of activating a protein for assembly; this mechanism has ample precedent in virus assembly (112). Specific features of the hypothesis are presented above and are summarized in Table 1. Experiments are discussed below that bear on three issues that are central to the membrane trigger hypothesis: (a) Integral membrane proteins are capable of more than one functional conformation. (b) Individual proteins can assume an asymmetric transmembrane conformation without catalysis of this assembly, and the information for a protein's asymmetry lies in its structure. (c) Isolated, water-soluble proteins can enter bilayers as integral membrane proteins. This process is accompanied by membrane-triggered conformational change and often by proteolysis.

CONFORMATIONAL FLEXIBILITY   While the membrane trigger hypothesis does not necessarily require a temporal and spatial dissociation of protein synthesis from membrane assembly, it is most easily visualized for proteins where the synthesis is quite separate from the assembly. This would necessitate the protein being able to assume at least two conformations, one in which its hydrophobic groups are buried in the interior such that it is water-soluble, and another in which it has an exposed hydrophobic face that anchors it to the bilayer. Clark (113, 114) has demonstrated that fumarase, an enzyme of the mitochondrial matrix that is synthesized in the cytoplasm, can exist in two stable conformations. The isolated soluble fumarase binds no Triton X-100, which indicates that it has an entirely polar surface. After denaturation with SDS, sedimentation through a Triton solution displaces the SDS and leaves the fumarase with an equal weight of bound Triton, which indicates that its apolar groups are now exposed to the solvent. Unlike fumarase, bovin serum albumin binds the same low level of Triton after treatment with SDS as before. Boquet et al (115) used the same technique to probe the structure of diphtheria toxin. This 62,000 dalton polypeptide binds to cell surface receptors and, by an unknown path, transfers its N-terminal region to the cytoplasm. Like fumarase, diphtheria toxin did not bind Triton unless it had been first denatured in SDS. It was shown that the central portion of the toxin molecule had the masked hydrophobic segment and that this segment was exposed when either the N-terminal or C-terminal portions of the protein were removed or when the protein was first denatured by SDS. Haber and colleagues (116, 117) found similar conformational changes for adenyl cyclase after dodecyl sepharose chromatography. Dog heart adenyl cyclase required detergent for solubilization and aggregated when the detergent was removed. Once it had been chromatographed on dodecyl sepharose, it no longer required detergent for its solubility. Other techniques have been used to switch the conformation of a protein into a second stable state. Henning and co-workers (118) converted protein II* of the *E. coli* outer membrane to a water-soluble form by addition of urea and dialysis into aqueous buffers. Although native protein II* has a phage receptor role and is selectively cleaved by trypsin, the water-soluble form is inactive and is degraded to oligopeptides by trypsin. If the water-soluble II* is incubated with a tenfold molar excess of lipopolysaccharide (or the lipid A portion of lipopolysaccharide), it becomes fully active as a phage receptor and regains the exact native pattern of trypsin cleavage. Neither total *E. coli* lipid nor dimyristoyl lecithin will replace lipopolysaccharide in this reconstitution. These data show that the water-soluble form of protein II* can properly assemble into a protein-free bilayer without additional catalysis. Similar assembly events have been documented for proteins that are usually isolated in a soluble form, such as melittin, α-toxin, and complement, and are discussed below. The selectivity of outer membrane protein II* assembly for a lipid that is exclusively found in the outer membrane might be the basis for its proper membrane choice in vivo.

Two additional examples of conformational flexibility bear review. Kimura & Futai (119) have shown that *E. coli* L-lactate dehydrogenase that was purified in the presence of detergent will, upon detergent removal, oligomerize and (*a*) have altered α-helical content, (*b*) lose half of its catalytic activity, (c) become insensitive to antibody, and (d) become heat-labile. Each of these properties is reversed upon incubation with E. coli lipid, which indicates a return to the native conformation. Schrock & Gennis (120) have studied a particularly instructive conformational change in E. coli pyruvate oxidase. This enzyme was extracted from the membrane fraction by salt and purified without detergents. Nevertheless, it bound detergents such as SDS or cetyl trimethylammonium bromide at levels well below their critical micelle concentrations. This binding was only seen in the presence of each of this enzyme's cofactors (magnesium, pyruvate, and thiamine pyrophosphate) and was accompanied by a dramatic increase in the enzymatic activity. With this bound detergent and in the presence of its cofactors, the pyruvate oxidase had a strong tendency to aggregate. This elegant study indicates that pyruvate oxidase can exist in two distinct conformations. In one, the enzyme is inactive but water-soluble. In the presence of its enzymatic cofactors and detergent (or, presumably, lipid) it is activated and exposes hydrophobic segments. This not only clearly demonstrates the ability of this protein to refold but raises an important caution in the classification of membrane proteins as integral or peripheral (8, 9).

These examples serve to show that membrane proteins can exist in alternate conformations that are compatible with aqueous or apolar environments. In the case of protein II* and pyruvate oxidase, transformation between these states is triggered by specific lipids or specific cofactors. In the following section, we discuss proteins that are water-soluble and whose function requires integration into biomembranes. The conformational changes just reviewed which are seen with purified membrane proteins also accompany the insertion of purified soluble proteins into membranes.

MEMBRANE-TRIGGERED ASSEMBLY   While the membrane trigger hypothesis does not require that the synthesis of a protein be complete before its insertion into a membrane begins, this hypothesis is clearly applicable in such cases. A number of soluble secreted proteins must assemble into membranes to carry out their biological functions. Complement is a group of nine serum proteins (121–126) involved in a proteolytic activation cascade that is itself triggered by clusters of antibody. Complement components 5b, 6, and 7 form a complex that binds firmly to natural or synthetic bilayers. Component 8 binds to this complex, and this causes a slow leak of cellular or liposomal contents. This is greatly accelerated by the binding of component 9; a pore is formed in the bilayer by the complex of complement components 5–9. This reaction has been duplicated with purified complement components acting on synthetic liposomes, and the pore has been demonstrated biochemically and visualized in the electron microscope. Such pores cause target cell lysis.

Small soluble proteins can also assemble into membranes. Melittin (127–135) is a soluble protein that is 50% of the dry weight of bee venom. It is a tetramer of identical 2840 dalton subunits. The amino acid sequence of the melittin peptide (127) is striking in that 19 of the N-terminal 20 residues are hydrophobic and the other 6 C-terminal residues are polar. Melittin has been shown by a variety of techniques to insert into liposomes or cell membranes and cause lysis. Sessa et al (129) showed that levels of melittin as low as 10^{-6} M caused radioactive chromate or glucose release from liposomes. Melittin was also shown to insert into a lipid monolayer with concomitant increase in the surface pressure. Williams & Bell (131) showed by electron spin resonance that melittin disrupted the hydrocarbon chain matrix. Electron micrographs (129) showed extensive liposome disruption by this protein. These data add up to a striking picture of a detergent-like peptide which, despite its heavily apolar character, can fold to form a water-soluble tetramer. Upon meeting a cell or a liposome, the peptide inserts into the lipid bilayer, presumably via its hydrophobic N-terminus. Studies of melittin-liposome interactions hold great promise for illuminating the chemistry of membrane-triggered refolding.

Other examples of highly purified soluble proteins inserting into membranes include staphylococcal α-toxin (136, 137) and streptolysin S. Two instances of proteins that post-translationally assemble into membranes are the M13 coat protein and the cytoplasmically synthesized mitochondrial proteins. M13 coat protein is made on the soluble polysome fraction in a precursor form with a leader sequence. The initial translation product is not membrane associated but, in vivo and in vitro, it rapidly associates with the membrane and spans it in an asymmetric manner. The evidence for this and for the mitochondrial proteins being first synthesized in the cytoplasm and then assembled into the organelle is discussed above.

MEMBRANE ASYMMETRY   A corollary of the membrane trigger hypothesis is that proteins can assume an asymmetric orientation with respect to the plane of the bilayer without catalysis of their assembly into the bilayer. This has been established during the past four years; four clear examples are reviewed. The initial indication of spontaneous asymmetric assembly came from the work of Racker and colleagues (reviewed in 138) on reconstitution of oxidative phosphorylation. This multienzyme process is widely believed to depend on an asymmetric, transmembrane arrangement of the proteins of the electron transport chain. Functional reconstitutions of resolved proteins with lipid were achieved by sonication, addition of cholate and its removal by dialysis or dilution, exposure of mixtures of preformed vesicles and proteins to low levels of lysolecithins, or by calcium-induced fusion of liposomes. Knowles & Racker (139) and Goldin (140) showed the physical asymmetry of reconstituted proteins. Wickner & Zwizinski (16, 141, 142) studied the parameters affecting asymmetric incorporation of M13 coat protein into synthetic lecithin vesicles and found that the physical state of the lipids' fatty acyl chains during assembly was crucial. Eytan & Broza (143) achieved a particularly elegant reconstitution of cytochrome oxidase. This enzyme, discussed above, was purified and freed of lipid. Incubation with low levels of liposomes allowed the enzyme to bind one fifth its weight of lipid and to regain activity. This activity was not under respiratory control, i.e. uncouplers were without effect. At higher concentrations of liposomes, the protein assembled into the bilayer and became dependent on uncoupler for activity, which indicates that it spanned the bilayer. Labeling studies with [^{35}S]-diazonium benzene sulfonate demonstrated the physical asymmetry of cytochrome oxidase in these vesicles. It is noteworthy that this reconstitution was performed with presealed vesicles and did not involve detergent.

The essence of the membrane trigger hypothesis is that the interaction of a protein with the lipid bilayer triggers spontaneous folding of the protein into the membrane without additional catalysis. In the preceding sections, it has been shown that proteins have the capacity to assume alternative conformations with their hydrophobic regions either exposed or buried. Examples were cited of proteins in their soluble forms that can be triggered by membranes to expose their apolar face and insert asymmetrically into the bilayer. It should again be emphasized that many membrane proteins might not be capable of a stable, water-soluble conformation. Rather, as their synthesis proceeds in the N- to C- terminal direction, their folding may be critically dependent on the availability of an appropriate lipid bilayer and may lead to their insertion into this membrane. Despite the variety of evidence presented that such assembly can occur, it is difficult, if not impossible, to prove that a protein pore is not involved in vivo. The issue will undoubtedly be decided either by continued accumulation of understanding of how proteins assemble into lipid bilayers or by convincing genetic and biochemical identification of a protein pore that recognizes leader peptides and conducts them through a membrane.

FUTURE DIRECTIONS

It is undoubtedly clear to the reader that the author favors the membrane trigger hypothesis as a mechanism for assembly of integral membrane proteins. It is also clear that a great deal of further work is needed to clarify the mechanisms of protein assembly. The following approaches seem of particular interest:

1. A search should be made for pleiotropic, conditionally lethal mutants which, at the nonpermissive condition, make soluble proteins but fail to assemble membrane proteins. Mutants in the protease(s) that remove leader

peptides might be conditional lethals and will, of course, always be of great interest.

2. A large collection of mutants should be made in genes for membrane proteins such as the M13 coat protein and *E. coli* lipoprotein. Such mutations, particularly in the leader peptide region, could then be correlated with the fate of the protein in vivo. Elegant studies of fusions between soluble and membrane proteins (144, 145) have already been reported.

3. Membrane proteins *with* their N-terminal leader sequences need to be isolated and to have studies made of the chemistry of their folding, their association with membranes, and their proteolytic processing.

4. Prepeptide proteases need to be isolated and studied to determine the basis of their substrate specificities. Their asymmetric reconstitution into liposomes may, in conjunction with in vitro protein synthesis extracts, provide an assay for other components (such as a protein pore) involved in assembly.

ACKNOWLEDGMENTS

Stimulating conversations with Randy Schekman, Jim Rothman, Koreaki Ito, and Gail Mandel are gratefully acknowledged.

Literature Cited

1. Blobel, G., Dobberstein, B. 1975. *J. Cell Biol.* 67:835–51
2. Blobel, G., Dobberstein, B. 1975. *J. Cell Biol.* 67:852–62
3. Rothman, J. E., Lenard, J. 1977. *Science* 195:743–53
4. Vambutas, V. K., Salton, M. R. J. 1970. *Biochim. Biophys. Acta* 203:83–89
5. Glew, R. H., Heath, E. C. 1971. *J. Biol. Chem.* 246:1566–74
6. Hirashima, A., Childs, G., Inouye, M. 1973. *J. Mol. Biol.* 79:373–84
7. Halegoua, S., Hirashima, A., Inouye, M. 1976. *J. Bacteriol.* 126:183–91
8. Yu, J., Fischman, D. A., Steck, T. L. 1973. *J. Supramol. Struct.* 1:233–48
9. Steck, T. L., Yu, J. 1973. *J. Supramol. Struct.* 1:220–32
10. Spatz, L., Strittmatter, P. 1971. *Proc. Natl. Acad. Sci. USA* 68:1042–46
11. Capaldi, R. A., Vanderkooi, G. 1972. *Proc. Natl. Acad. Sci. USA* 69:930–32
12. Nobbs, C. L., Watson, H. C. and Kendrew, J. C. 1966. *Nature* 209:339–41
13. Bretscher, M. S. 1973. *Science* 181:622–29
14. Tomita, M., Marchesi, V. T. 1975. *Proc. Natl. Acad. Sci. USA* 72:2964–68
15. Wickner, W. 1975. *Proc. Natl. Acad. Sci. USA* 72:4749–53
16. Wickner, W. 1976. *Proc. Natl. Acad. Sci. USA* 73:1159–63
17. Henderson, R., Unwin, P. 1975. *Nature* 257:28–32
18. Henderson, R. 1975. *J. Mol. Biol.* 93:123–38
19. Unwin, P., Henderson, R. 1975. *J. Mol. Biol.* 94:425–40
20. Drickamer, L. K. 1977. *J. Biol. Chem.* 252:6909–17
21. Fukuda, M., Eshdat, Y., Tarone, G., Marchesi, V. T. 1978. *J. Biol. Chem.* 253:2419–28
22. Steck, T. L., Koziarz, J. J., Singh, M. K., Reddy, G., Köhler, H. 1978. *Biochemistry* 17:1216–22
23. Edidin, M., Weiss, A. 1972. *Proc. Natl. Acad. Sci. USA* 69:2456–59
24. Yahara, I., Edelman, G. M. 1972. *Proc. Natl. Acad. Sci. USA* 69:608–12
25. Reusch, V. M., Jr., Burger, M. M. 1974. *J. Biol. Chem.* 249:5337–45
26. Beining, P. R., Huff, E., Prescott, B., Theodore, T. S. 1975. *J. Bacteriol.* 121:137–43
27. Smilowitz, H., Carson, J., Robbins, P. W. 1972. *J. Supramol. Struct.* 1:8–18
28. Sefton, S. M., Gaffney, B. J. 1974. *J. Mol. Biol.* 90:343–58

29. Compans, R. W., Meier-Ewert, H., Palese, P. 1974. *J. Supramol. Struct.* 2:496–511
30. Peters, K., Richards, F. M. 1977. *Ann. Rev. Biochem.* 46:523–52
31. Mindich, L. 1971. *Proc. Natl. Acad. Sci. USA* 68:420–24
32. Nunn, W. D., Cronan, J. E., Jr. 1974. *J. Biol. Chem.* 249:724–31
33. McIntyre, T. M., Bell, R. M. 1975. *J. Biol. Chem.* 250:9053–59
34. Randall, L. L. 1975. *J. Bacteriol.* 122:347–51
35. Weisberg, L. J., Cronan, J. E., Jr., Dunn, W. E. 1975. *J. Bacteriol.* 123:492–96
36. Mindich, L. 1970. *J. Mol. Biol.* 49:433–39
37. Henning, U., Haller, I. 1975. *FEBS Lett.* 55:161–64
38. Osborn, M. J., Gander, J. E., Parisi, E., Carson, J. 1972. *J. Biol. Chem.* 247:3962–72
39. Ito, K., Sato, T., Yura, T. 1977. *Cell* 11:551–59
40. Mahler, H. R. 1973. *Critical Reviews in Biochemistry* 1:381–460
41. Bayer, M. E. 1968. *J. Gen. Microbiol.* 53:395–404
42. Bayer, M. E. 1968. *J. Virol.* 2:346–56
43. Bayer, M. E., Starkey, T. W. 1972. *Virology* 49:236–56
44. Singer, S. J. 1974. *Ann. Rev. Biochem.* 43:805–833
45. Yu, J., Branton, D. 1976. *Proc. Natl. Acad. Sci. USA* 73:3891–95
46. Milstein, C., Brownlee, G. G., Harrison, T. M., Mathews, M. B. 1972. *Nature New Biol.* 239:117–20
47. Siekevitz, P., Palade, G. E. 1960. *J. Biophys. Biochem. Cytol.* 7:619–30
48. Redman, C. M., Siekevitz, P., Palade, G. E. 1966. *J. Biol. Chem.* 241:1150–58
49. Ganoza, M. C., Williams, C. A. 1969. *Proc. Natl. Acad. Sci. USA* 63:1370–85
50. Hicks, S. J., Drysdale, J. W., Munro, H. N. 1969. *Science* 164:584–85
51. Redman, C. M. 1969. *J. Biol. Chem.* 244:4308–15
52. Andrews, T. M., Tata, J. R. 1971. *Biochem. J.* 121:683–94
53. Cancedda, R., Schlesinger, M. J. 1974. *J. Bacteriol.* 117:290–301
54. Randall, U. L., Hardy, S. J. S. 1977. *Eur. J. Biochem.* 75:43–53
55. Lodish, H. F. 1973. *Proc. Natl. Acad. Sci. USA* 70:1526–30
56. Devillers-Thiery, A., Kindt, T., Scheele, G., Blobel, G. 1976. *Proc. Natl. Acad. Sci. USA* 72:5016–20
57. Lomedico, P. T., Chan, S. J., Steiner, D. F., Saunders, G. F. 1977. *J. Biol. Chem.* 252:7971–78
58. Lingappa, V. R., Devillers-Thiery, A., Blobel, G. 1977. *Proc. Natl. Acad. Sci. USA* 74:2432–36
59. Jackson, R. C., Blobel, G. 1977. *Proc. Natl. Acad. Sci. USA* 74:5598–5602
60. Shields, D., Blobel, G. 1977. *Proc. Natl. Acad. Sci. USA* 74:2059–63
61. Palmiter, R. D., Gagnon, J., Ericsson, L. H., Walsh, K. A. 1977. *J. Biol. Chem.* 252:6386–93
62. Inouye, H., Beckwith, J. 1977. *Proc. Natl. Acad. Sci. USA* 74:1440–44
63. Inouye, S., Wang, S., Sekizawa, J., Halegoua, S., Inouye, M. 1977. *Proc. Natl. Acad. Sci. USA* 74:1004–8
64. Randall, L. L., Hardy, J. S., Josefsson, L. G. 1978. *Proc. Natl. Acad. Sci. USA* 75:1209–12
65. Danon, A., Stoeckenius, W. 1974. *Proc. Natl. Acad. Sci. USA* 71:1234–38
66. Gerber, G. E., Gray, C. P., Wildenauer, D., Khorana, H. G. 1977. *Proc. Natl. Acad. Sci. USA* 74:5426–30
67. Marchesi, V. T. 1976. In *Biochemistry of Cell Walls and Membranes,* ed. C. F. Fox, pp. 123–54. London: Butterworth
68. Yu, J., Steck, T. L. 1975. *J. Biol. Chem.* 250:9170–75
69. Yu, J., Steck, T. L. 1975. *J. Biol. Chem.* 250:9176–84
70. Poyton, R. O., Kavanagh, J. 1976. *Proc. Natl. Acad. Sci. USA* 73:3947–51
71. Koch, G. 1976. *J. Biol. Chem.* 251:6097–6107
72. Harmey, M. A., Hallermayer, G., Korb, H., Neupert, W. 1977. *Eur. J. Biochem.* 81:533–44
73. Frey, T. G., Chan, S. H. P., Schatz, G. 1978. *J. Biol. Chem.* 253:4389–95
74. Cabral, F., Schatz, G. 1978. *J. Biol. Chem.* 253:4396–4401
75. Saltzgaber-Müller, J., Schatz, G. 1978. *J. Biol. Chem.* 253:305–10
76. Cabral, F., Solioz, M., Rudin, Y., Schatz, G., Clavilier, L., Slonimski, P. P. 1978. *J. Biol. Chem.* 253:297–304
77. Weiss, H., Schwab, A. J., Werner, S. 1975. In *Membrane Biogenesis,* ed. A. Tzagoloff, pp. 125–53. New York: Plenum
78. Butow, R. A., Bennett, W. F., Finkelstein, D. B., Kellems, R. E. 1975. In *Membrane Biogenesis,* ed. A. Tzagoloff, pp. 155–199. New York: Plenum
79. Ross, E., Schatz, G. 1976. *J. Biol. Chem.* 251:1997–2004
80. Farquhar, M. G., Bergeron, J. J. M., Palade, G. E. 1974. *J. Cell Biol.* 60:8–25
81. Little, J. S., Widnell, C. C. 1975. *Proc. Natl. Acad. Sci. USA* 72:4013–17

82. Braun, V., Sieglin, J. 1970. *Eur. J. Biochem.* 13:336–46
83. Braun, V., Bosch, V. 1972. *Proc. Natl. Acad. Sci. USA* 69:970–74
84. Braun, V., Bosch, V. 1972. *Eur. J. Biochem.* 28:51–69
85. Hantke, K., Braun, V. 1973. *Eur. J. Biochem.* 34:284–96
86. Inouye, M. 1974. *Proc. Natl. Acad. Sci. USA* 71:2369–2400
87. Sekizawa, J., Inouye, S., Halegoua, S., Inouye, M. 1977. *Biochem. Biophys. Res. Commun.* 77:1126–33
88. Inouye, S., Wang, S., Sekizawa, J., Halegoua, S., Inouye, M. 1977. *Proc. Natl. Acad. Sci. USA* 74:1004–8
89. Lin, J. J. C., Kanazawa, H., Wu, H. C. 1978. *Fed. Proc.* 37:1691 (Abstr.)
90. Katz, F. N., Rothman, J. E., Lingappa, V. R., Blobel, G., Lodish, H. F. 1977. *Proc. Natl. Acad. Sci. USA* 74:3278–82
91. Rothman, J. E., Lodish, H. F. 1977. *Nature* 269:775–80
92. Toneguzzo, F., Ghosh, H. P. 1978. *Proc. Natl. Acad. Sci. USA* 75:715–19
93. Asbeck, V. F., Beyreuther, K., Köhler, H., von Wettstein, G., Braunitzer, G. 1969. *Hoppe-Seyler's Z. Physiol. Chem.* 350:1047–66
94. Nakashima, Y., Konigsberg, W. 1974. *J. Mol. Biol.* 88:598–600
95. Trenkner, E., Bonhoeffer, F., Gierer, A. 1967. *Biochem. Biophys. Res. Commun.* 28:932–39
96. Marco, R., Jazwinski, S. M., Kornberg, A. 1974. *Virology* 62:209–23
97. Smilowitz, H. 1974. *J. Virol.* 13:94–99
98. Model, P., Zinder, N. D. 1974. *J. Mol. Biol.* 83:231–51
99. Konings, R. N. H., Hulsebos, T., van den Hondel, C. A. 1975. *J. Virol.* 15:570–84
100. Sugimoto, K., Sugisaki, H., Takanami, M. 1977. *J. Mol. Biol.* 111:487–507
101. Chang, C. N., Blobel, G., Model, P. 1978. *Proc. Natl. Acad. Sci. USA* 75:361–65
102. Wickner, W., Mandel, G., Zwizinski, C., Bates, M., Killick, T. 1978. *Proc. Natl. Acad. Sci. USA* 75:1754–58
103. Deleted in proof
104. Redman, C. M., Sabatini, D. D. 1966. *Proc. Natl. Acad. Sci. USA* 56:608–15
105. Sabatini, D. D., Blobel, G. 1970. *J. Cell Biol.* 45:146–57
106. Rothman, S. S. 1975. *Science* 190:747–53
107. Ron, E. Z., Kohler, R. E., Davis, B. D. 1966. *Science* 153:1119–20
108. Smith, W. P., Tai, P.-C., Thompson, R. D., Davis, B. D. 1977. *Proc. Natl. Acad. Sci. USA* 74:2830–34
109. Smith, W. P., Tai, P.-C., Davis, B. D. 1978. *Proc. Natl. Acad. Sci. USA* 75:814–17
110. Highfield, P. E., Ellis, J. 1978. *Nature* 271:420–24
111. Palmiter, R. D., Gagnon, J., Walsh, K. A. 1978. *Proc. Natl. Acad. Sci. USA* 75:94–98
112. Casjens, S., King, J. 1975. *Ann. Rev. Biochem.* 44:555–611
113. Clarke, S. 1975. *J. Biol. Chem.* 250:5459–69
114. Clarke, S. 1977. *Biochem. Biophys. Res. Commun.* 79:46–52
115. Boquet, P., Silverman, M. S., Pappenheimer, A. M., Jr., Vernon, W. B. 1976. *Proc. Natl. Acad. Sci. USA* 73:4449–53
116. Homcy, C. J., Wrenn, S. M., Haber, E. 1977. *J. Biol. Chem.* 252:8957–64
117. Homcy, C. J., Wrenn, S. M., Haber, E. 1978. *Proc. Natl. Acad. Sci. USA* 75:59–63
118. Schweizer, M., Hindennach, I., Garten, W., Henning, U. 1978. *Eur. J. Biochem.* 82:211–17
119. Kimura, H., Futai, M. 1978. *J. Biol. Chem.* 253:1095–1100
120. Schrock, H. L., Gennis, R. B. 1977. *J. Biol. Chem.* 252:5990–95
121. Müller-Eberhard, H. J. 1975. *Ann. Rev. Biochem.* 44:697–724
122. Stolfi, R. L. 1968. *J. Immunol.* 100:46–54
123. Kolb, W. P., Müller-Eberhard, H. J. 1975. *Proc. Natl. Acad. Sci. USA* 72:1687–89
124. Hammer, C. H., Nicholson, A., Mayer, M. M. 1975. *Proc. Natl. Acad. Sci. USA* 72:5076–80
125. Hammer, C. H., Shin, M. L., Abramovitz, A. S., Mayer, M. M. 1977. *J. Immunol.* 119:1–8
126. Michaels, D. W., Abramovitz, A. S., Hammer, C. H., Mayer, M. M. 1978. *Proc. Natl. Acad. Sci. USA* 73:2852–56
127. Habermann, V. E., Jentsch, J. 1967. *Hoppe-Seyler's Z. Physiol. Chem.* 348:37–50
128. Kreil, G., Kreil-Kiss, G. 1967. *Biochem. Biophys. Res. Commun.* 27:275–80
129. Sessa, G., Freer, J. H., Colacicco, G., Weissmann, G. 1969. *J. Biol. Chem.* 244:3575–82
130. Kreil, G., Bachmayer, H. 1971. *Eur. J. Biochem.* 20:344–50
131. Williams, J. C., Bell, R. M. 1972. *Biochim. Biophys. Acta* 288:255–62
132. Kreil, G. 1973. *Eur. J. Biochem.* 33:558–66
133. Suchanek, G., Kindås-Mugge, I., Kreil,

G., Schreier, H. 1975. *Eur. J. Biochem.* 60:309–15
134. Suchanek, G., Kreil, G. 1977. *Proc. Natl. Acad. Sci. USA* 74:975–78
135. Suchanek, G., Kreil, G., Hermodson, M. A. 1978. *Proc. Natl. Acad. Sci. USA* 75:701–4
136. Weissman, G., Sessa, G., Bernheimer, A. W. 1966. *Science* 154:772–74
137. Freer, J. H., Arbuthnott, J. P., Bernheimer, A. W. 1968. *J. Bacteriol.* 95:1153–68
138. Racker, E. 1975. *Proc. FEBS Meet.,* 10th, pp. 25–34
139. Knowles, A. F., Racker, E. 1075. *J. Biol. Chem.* 250:3538–44
140. Goldin, S. M. 1977. *J. Biol. Chem.* 252:5630–42
141. Wickner, W. 1977. *Biochemistry* 16:254–58
142. Zwizinski, C., Wickner, W. 1977. *Biochim. Biophys. Acta* 471:169–76
143. Eytan, G. D., Broza, R. 1978. *J. Biol. Chem.* 253:3196–3202
144. Silhavy, T. J., Casadaban, M. H., Shuman, H. A., Beckwith, J. A. 1976. *Proc. Natl. Acad. Sci. USA* 73:3423–27
145. Silhavy, T. J., Shuman, H. A., Beckwith, J., Schwartz, M. 1977. *Proc. Natl. Acad. Sci. USA* 74:5411–15

Ann. Rev. Biochem. 1979. 48:47–71

LIPID ASYMMETRY
IN MEMBRANES

♦12002

Jos A. F. Op den Kamp

Laboratory of Biochemistry, University of Utrecht, Transitorium 3,
University Centre "De Uithof", Utrecht, The Netherlands

CONTENTS

PERSPECTIVES AND SUMMARY

Nearly half a century after the presentation of a model in which the erythrocyte membrane was depicted as a lipid bilayer (1) evidence started to accumulate to suggest that this structure is asymmetric. The specific and asymmetric arrangement of membrane proteins was the first to be recognized (again the erythrocyte membrane has been studied in most detail). The asymmetry is twofold. Proteins that are attached to, or partly penetrate, the lipid bilayer are located specifically at one side. The interior of the membrane contains those proteins that are involved in intracellular events and the outside contains proteins involved in the defense mechanism of the

47

cell. In addition, the membrane-spanning proteins have a defined orientation. Part of these proteins is outside the cell and usually contains sugar residues, a small portion, consisting mainly of apolar amino acids, spans the lipid bilayer, and the residual, polar, part of these proteins is inside the cell. These studies on erythrocytes and on a variety of other membranes have led to the general conclusion that membranes have an absolute protein asymmetry, a conclusion which is not surprising considering the differences in function between the outer and inner surface of a membrane.[1]

The functional asymmetry provides the first indication that membrane lipids are also localized in a specific and maybe asymmetric way, because membrane-bound enzymes can be surrounded by specific phospholipids (9, 16). Further argument favouring lipid asymmetry comes from the fact that the chemical environment of both layers is, in most cases, completely different and may subsequently induce or maintain a specific localization of the various (phospho) lipids. This argument is substantiated by the fact that most membranes contain a large variety of lipids differing not only in their apolar parts, but also in the size, charge, and chemical reactivity of their polar headgroups.

When it was realized that labelling techniques, which were developed for proteins, could also be applied to localize at least some of the lipids, experiments were initiated to investigate the concept of lipid asymmetry (17). Considering the evidence already in existence concerning protein asymmetry in erythrocytes, the choice of this membrane as the first target for studies on lipid localization was a logical one, but as it stands now, also a very fortunate one. The erythrocyte membrane is, until now (July 1978) the only biological membrane in which lipid asymmetry has been demonstrated in a convincing and conclusive way, although the evidence for asymmetry in several virus membranes is fairly compelling. The other results, however, obtained with various membranes, indicate at best that lipid asymmetry does occur in these membranes, but more frequently lead to contradictory conclusions about lipid topology. The inconsistent and controversial data cannot be readily explained. It has to be realized, however, that many problems are inherent to the most commonly used techniques. A short description of techniques and specific objections against each method therefore precedes the compilation and discussion of the data. In addition, it is noteworthy that the interpretation of data is often based on the assumption that the original lipid localization is not altered during (as a consequence of) the experiment. This review emphasizes that this assumption is questionable and in some experimental approaches even unlikely, mainly because

---

[1]The reader is referred to references (2–15) for recent reviews covering this field and closely related topics.

transbilayer movements of lipids do occur more frequently and, under special conditions, much faster than was realized before (2, 18).

It has been suggested in one of the first papers on this subject that "... in erythrocytes, and by extension in other mammalian cells, the two phospholipids, phosphatidylcholine and sphingomyelin are located chiefly in the outer half of the lipid bilayer ..." (17). Five years later it was mentioned in a review that "these results firmly establish lipid asymmetry as a general property of biological membranes" (13). A critical appraisal of data, techniques, and interpretations shows that the concept of lipid asymmetry in biological membranes is presumably correct. It is too early, however, to declare it as a firmly established, general phenomenon.

METHODOLOGY

General Remarks

Lipid localization in biological membranes has been carried out primarily via chemical or enzymatic modification, via exchange techniques, and in some cases by immunochemical procedures. Physicochemical approaches have been used mainly in studies on model membrane systems. Before describing in more detail these methods, two general remarks have to be made concerning lipid localization studies.

If one tries to localize a lipid it is necessary to check that: (a) the reagent (enzyme, chemical probe, exchange protein, etc) facing the outer layer reacts only with the lipid in the outer layer and with none of the inner layer (or vice versa) and (b) the reagent does not induce alterations in the original lipid distribution. This means that the reagent should not lyse or penetrate the membrane (effects that can be easily controlled), and also that the transbilayer movement of lipids should be slow in comparison with the time required for the experiment. This condition can be readily met in model systems. With a variety of techniques it is shown that in artificial bilayers in which the lipids are in equilibrium distribution, flip-flop is an extremely slow process with half-time values of several days or longer (18–25). However, in artificial membranes, flip-flop rates can be increased considerably by the presence of intrinsic membrane proteins or by the disturbance of the equilibrium in lipid distribution that exists between inner and outer layer. In the presence of glycophorin, the flip-flop of 1-palmitoyl lysolecithin and dioleoylphosphatidylcholine in unilamellar lecithin vesicles is two orders of magnitude faster than in the absence of this membrane-spanning protein (26, 27). Alterations in the equilibrium distribution of lipids have been induced in several systems. Using a phospholipid exchange protein, dioleoyl lecithin was introduced into the outer layer of dimyristoyl lecithin vesicles (23); phospholipase D was used to form phosphatidic acid in the outer layer of dimyristoyl lecithin vesicles (28); and cholesterol was depleted from cholesterol-lecithin vesicles using red cell ghosts (29). In the first two experiments the rate of lecithin flip-flop was increased from several days to half-times of 8 hr and 30 min, respectively. In the latter experiment cholesterol from the inner layer of the vesicle became available for exchange at the outside. These data indicate that unilateral modifications in the amount and properties of lipids (surface pressure, charge, degree of unsaturation), may induce a rapid transbilayer movement. The question arises, are such unilateral alterations also involved in the observed transbilayer movements in natural membranes? In the relatively inert membrane of the influenza virus, transbilayer movements are negligible (30, 31); however, lecithin (32–35) and cholesterol (36–38) translocations have been detected in erythrocytes. Half-times of a few hours have been reported, depending on the fatty acid composition of the lecithin involved and the temperature (34, 35). In highly active membrane systems these half-times are decreased to a few minutes. This is the case for phosphatidylethanolamine in the growing membrane of *Bacillus megaterium* (39) and for dioleoylphosphatidyltempocholine, a spin-labelled lecithin analogue, when present in excitable membrane vesicles (40). Altogether these data imply that transbilayer movements of lipids may occur during localization experiments, especially when techniques are applied that require long incubation times (exchange, phospholipases) or induce alterations in the lipid distribution (phospholipases). It cannot be excluded that lipids that were originally present in the inner layer migrate and become exposed to the reagent facing the outer layer.

For complete localization of lipids, which in most cases is necessary to distinguish between a symmetric or an asymmetric distribution, all of the lipid has to be available for the reagent. Several objections can be raised against conclusions that are drawn from studies in which only initial velocities of the reaction with membrane lipids are measured. A biological membrane is not a homogenous dispersion of lipid molecules and the reactivity of lipids can differ greatly within one membrane leaflet, as well as between inner and outer layer. Differences in reactivity of lipids can be caused by lipid-lipid or lipid-protein interactions, lipid-phase transitions or phase separation, and environmental conditions such as differences in pH. The following examples illustrate this phenomenon. Evidence for a decreased reactivity due to lipid-lipid and lipid-protein interaction is presented for the erythrocyte membrane in which part of the phospholipids was accessible for phospholipase C only after sphingomyelin hydrolysis or disruption of lipid-protein interaction by ATP depletion (41). Protective effects of proteins have been observed in *Mycoplasma hominis* (169) and *Acholeplasma laidlawii* (42) membranes in which all respectively part of the phosphatidylglycerol is shielded against phospholipase attack. Furthermore, in rat liver microsomal membranes part of the phosphatidylinositol is inaccessible for exchange (43). It has also been demonstrated that lipid phase transitions and phase separations influence phospholipase activity. When lecithin molecules coexist in a liposome in the expanded liquid crystalline configuration and the condensed gel state, the more fluid species is hydrolyzed much faster by pancreatic phospholipase A_2 (44). When used on membranes of *A. laidlawii,* the same enzyme was found to discriminate between phosphatidylglycerol which was in a liquid crystalline state and therefore hydrolyzed, and that fraction of phosphatidylglycerol which was in a crystalline state and so not accessible to enzymic attack (45). The reaction of lipids in the membrane with covalent lipid reagents also depends on parameters such as packing density of the lipids, local pH, presence of neighbouring charged groups, and steric accessibility (46), and it is unlikely that these paramaters are similar throughout the entire surface of the membrane.

Enzymatic Modification of Lipids

The digestion of phospholipids in the outer monolayer of a membrane by exogenous phospholipases may reveal the distribution of phospholipids between the two membrane layers [see (47) for an extensive description of methods involved]. Phospholipases A_2 and C from various sources as well as sphingomyelinase and lysophospholipase have been used successfully in this type of experiment. The application of phospholipase D has been rather limited because this enzyme is not available in large quantities in a purified form. Extensive purification is necessary because crude phospholipase preparations frequently contain nonenzymatic direct lytic factors or other enzymatic activities which may alter the membrane structure (48, 49). On the other hand, phospholipase treatment itself may induce lysis or severe perturbations of membrane structure. It is well known that hydrolysis products such as lysophospholipids, free fatty acids, and diglycerides are not suitable compounds to form stable bilayers. Thus, when these compounds are formed in an already existing bilayer the best one can hope is that the original structure is not disturbed too much and that other membrane constituents are able to stabilize the bilayer. Lysophospholipids in particular tend to disrupt bilayer structures and for this reason the application of phospholipase A_2 as a localizing reagent has been limited to a very small number of membranes so far. Lysis by phospholipase A_2 treatment has been observed with several viruses (50, 51), *Escherichia coli* (52) and *Bacillus subtilis* (53, 54). Various membranous organelles from rat liver also lysed following treatment with phospholipase A_2 (55). In this case, however, lysis could be prevented by the addition of high amounts of bovine serum albumin which binds the released lysophospholipids and free fatty acids. The lytic effect of released lysophospholipids can also be overcome by an intrinsic membrane bound lysophospholipase. Such an enzyme is present in the membrane of *A. laidlawii* (56) and is capable of degrading instantaneously the lysophosphatidylglycerol that is produced by phospholipase A_2 treatment of these cells (57). Erythrocytes stay intact during treatment with phospholipase A_2 from several sources (4, 58–60). In this case no clearcut explanation is available. It can be speculated that the stabilization is due to a tight association of the residual phospholipids in the inner monolayer with the spectrin-actin complex (61), and that the high content of cholesterol eliminates the lytic effect of the lysolecithin. However, the situation seems to be more complex because it has been reported that: hemolysis of human erythrocytes does occur with a basic phospholipase A_2 from Agkistrodon halys blomhofii (62); that hemolysis occurs in the presence of albumin (60, 63); and that ATP depletion of Avian red cells enhanced hemolysis (64).

The formation of diglycerides in membranes by phospholipase C does not easily result in lysis. However, perturbation of the membrane structure by these compounds cannot be excluded. This is indicated by experiments on erythrocyte membranes which show that diglycerides tend to aggregate into discrete pools (65–67). Also ceramides, following treatment with sphingomyelinase, show this behaviour. The latter enzymes alter the characteristic freeze fracture appearance of erythrocyte membranes. The combination of phospholipase C and sphingomyelinase treatment of erythrocytes may result in lysis (67). Due to these complicating factors relatively few types of membrane appear to be suitable for the phospholipase approach of determining lipid asymmetry.

Apart from the phospholipases mentioned above, a highly purified lysophospholipase has been applied to localize lysophosphatidylcholine in unilamellar vesicles (24, 68), and a neuraminidase was used to localize the ganglioside hematoside in the membrane of vesicular stomatitis virus (69, 70). Galactose oxidase treatment followed by reduction with radioactive borohydride showed that glycolipids are located at the external surface of erythrocyte membranes (71, 72).

Chemical Modification of Lipids

A nonenzymatic modification of phospholipids can be easily obtained with those phospholipids containing a reactive amino group. The reagents required for the modification of the other polar headgroups are in general too harsh and either lead to disruption of the membrane or penetrate readily into the cell. An illustration of this is found in the modification of phosphatidylglycerol by successive treatment with sulfanilic acid diazonium salt, and NaB^3H_4. This approach was successful in the case of PM_2 virus (73) but not in localization studies with *A. laidlawii* because the reagents readily penetrated the membrane of this organism (74). To label amino groups, much milder reagents can be applied which leave the membrane intact. In most cases, however, these reagents permeate the membrane so it is essential to establish conditions under which this permeability is restricted as much as possible.

To find such conditions it is not sufficient to refer to previously published reports in which the same probe was used because different membrane systems may react completely differently with respect to their permeability properties. The widely used reagent trinitrobenzene sulfonic acid offers a good example. Erythrocyte (75) and vesicular stomatitis virus (76) membranes are impermeable to this probe, at room temperature. In contrast the membrane of *B. megaterium* was found to be permeable to trinitrobenzene sulfonic acid at 15°C (77).

A further drawback of this technique may be found in the perturbation of membrane structure which arises from the chemical modification although one can expect that these perturbations are perhaps less drastic than in the case of phospholipase treatment. Trinitrobenzene sulfonic acid treatment of membranes introduces a bulky electronegative group both in membrane lipids and proteins. It can be envisaged that this causes spatial restrictions at the membrane surface and prevents the reaction from going to completion. Indeed incomplete labelling of phosphatidylethanolamine and phosphatidylserine has been reported in several membrane systems (78–81), and monolayer experiments indicated that an increase in surface pressure of the lipids may be responsible for this incomplete reaction (82). The reaction rate is greatly retarded with more saturated phosphatidylethanolamine and in the presence of negatively charged lipids. For this reason it appears dangerous to assign an asymmetric or symmetric distribution of lipids in the membrane solely on the basis of values for initial rates of the reaction, which are measured at both sides of the membrane. Complete modification of the lipid being investigated, if achievable without change in membrane structure, is preferable for its complete localization.

Recognition of lipids is also possible using compounds that bind specifically, but noncovalently, to the polar head-group of lipids, and this method is particularly applicable to the localization of sugar-containing lipids. Glycolipids in several mycoplasma species have been localized with the use of lectins (83, 84) or antibodies directed against, presumably, the sugar moiety of these lipids (85). A similar approach has been reported for the localization of lipopolysaccharides in the outer membrane of gram-negative bacteria (86–88), phosphatidylglycerol in mycoplasma membranes (89), and phosphatidylinositol in myelin and microsomal membranes from rat liver (123).

Exchange of Lipids

Lipids can exchange from one membrane to another, a process which, in the case of phospholipids, can be enhanced by specific phospholipid exchange proteins. These water soluble proteins transport individual phospholipid molecules and exchange the molecules at the membrane surface with one of the membrane phospholipids. In this way, phosphatidylcholine, phosphatidylinositol, sphingomyelin, phosphatidylserine, and phosphatidylethanolamine can be exchanged (for review see 90). In addition it has been reported that phosphatidylglycerol and cardiolipin can be exchanged via a crude supernatant of a rat liver homogenate (91). The exchange is possible only with phospholipids in the outer layer of a membrane (19, 21, 92, 93) and the process is unlikely to perturb the membrane structure to any great extent. Therefore, under conditions where the membrane vesicle or cell stays intact and the rate of transbilayer movement of phospholipids is low, these exchange proteins are suitable tools with which to study the localizations of phospholipids.

The exchange of lipids between membranes in the absence of specific exchange proteins has been used to study the localization of cholesterol. Phosphatidylcholine-cholesterol vesicles or plasma lipoproteins can serve as a source for the exchangeable cholesterol. In these experiments localization is possible only in the absence of a transbilayer movement of cholesterol.

Physicochemical Techniques

Among the physicochemical techniques that are available, NMR appears to be the most powerful for the study of lipid localization in sonicated vesicles. Both ^{1}H and ^{31}P NMR have been applied to localize phosphatidylcholine, phosphatidylethanolamine, phosphatidylserine, phosphatidylglycerol, phosphatidic acid, and sphingomyelin in various mixtures, and have provided data, which, indirectly, have enabled the localization of cholesterol. (For a compilation of the data obtained so far, see Table 1). The lipid localization studies are based upon the use of paramagnetic ions as chemical shift or broadening reagents. The signal from the lipids in the membrane layer which is in contact with these paramagnetic ions is broadened or shifted away thus revealing the signal from the lipids in the other layer. The merits and disadvantages of this technique have been discussed excellently elsewhere (14). It has to be emphasized here that in lipid localization studies NMR has only been applied to the observation of small lipid vesicles. The reasons are obvious. The NMR spectra of biological membranes are in general very broad and unresolved which is caused by both the complexity of the structure as well as their large size. The large dimensions of biological membranes results in a broadening of the resonances as a consequence of a slow tumbling rate. With small unilamellar vesicles the

Table 1 Lipid localization in viral, bacterial and model membranes[a]

Membrane	Lipids localized			Method	References
	Preferential outside	Preferential inside	Equally distributed		
Viral membranes					
PM$_2$ phage	PG	PE		Sulfanilic acid dia-zonium salt	73
Vesicular stomatitis virus		PE (unsaturated)		TNBS	76, 125
	hematoside			Neuraminidase	69, 70
Influenza virus	PC, Sph	PS, PE		Pl'ase C	124
	PC, PI	Sph	PS, PE cho-lesterol	Pl'ase C, PLEP exchange	30 / 31
Bacterial membranes					
S. typhimurium[b]	LPS			Immunochemical	86–88
		PE		Cyanogen Br-dextran	126
E. coli[b]		PE		Pl'ase	52, 127
C. vinosium, A. vinelandii[c]		PE		TNBS	128
M. lysodeikticus	PG	PI	CL	Pl'ase A$_2$, C, PLEP	91
B. megaterium		PE		TNBS, IAI	77
	3'GlucNH$_2$PG		PE	Pl'ase C, TNBS	129
B. subtilis	PE, LysPG			Pl'ase C, TNBS	81
Vesicle membranes					
PC	unsaturated fatty acids			NMR	130
Unsaturated PC/cholesterol	PC	cholesterol[d]		NMR	131–133
Saturated PC/cholesterol			cholesterol	NMR	133
PC/Chol/LysoPC	LysoPC			NMR, LysoPl'ase	24, 25, 68
PC/PE		PE		TNBS	134, 135
PC/PG	PG			NMR	136
PC/PE; PC/PS; PC/PI; PC/PA		PE, PS, PI, PA		NMR	137
PC/Sph	Sph			NMR	137

[a] The abbreviations used are as follows: PE: phosphatidylethanolamine; PC: phosphatidylcholine; PS: phosphatidyl-serine; PI: phosphatidylinositol; Sph: sphingomyelin; CL: cardiolipin; lysoPC: 1-acyl-lysophosphatidylcholine; PG: phosphatidylglycerol; PA: phosphatidic acid; lysPG: lysylphosphatidylglycerol; 3'GlucNH$_2$PG: 3'glucosaminylphos-phatidylglycerol; IAI: isethionylacetimidate; TNBS: trinitrobenzene sulfonic acid; Pl'ase: phospholipase; lysoPl'ase: 1-acyl-lysophospholipase; Sph'ase: sphinogomyelinase; PLEP: phospholipid exchange protein.
[b] Data refer to outer membrane only.
[c] Data obtained with intracytoplasmic membranes.
[d] Observed only at cholesterol concentration above 30 mole %.

technique is extremely useful: with ^{31}P-NMR because almost each class of phospholipid can be localized separately, and with ^{13}C-NMR because the fate of individual ^{13}C-labelled lipids can be followed.

In contrast to NMR, electron spin resonance (ESR) studies have not provided direct information about the localization of lipids in membranes because naturally occurring lipids do not contain unpaired electrons. Still this technique has provided useful information on two aspects of lipid asymmetry: transbilayer movements of lipids (18, 40) and overall fluidity of the two membrane layers (94–96). Synthetic spin-labelled phospholipids have been introduced into artificial and biological membranes. In the latter approach the phospholipid exchange proteins were found to be suitable tools with which the labelled lipids could be introduced into the membrane (97). The transbilayer movements of the spin-labelled molecules can be measured by selective reduction at one side of the membrane. Furthermore, the temperature dependence of the distribution of spin-labelled molecules between hydrocarbon and aqueous domains of membrane preparations may indicate differences in overall fluidity of membrane lipids within the two layers.

A few other techniques have to be mentioned here although the data obtained are rather limited. X-ray analysis indicated a nonequal distribution of cholesterol in myelin membranes (98) and an asymmetry with respect to the total amount of lipids on both layers of the myelin membrane (99). A promising procedure with which to study the composition of the two layers of a membrane is based on the nonrandom fracturing of membranes bound to polylysine treated glass (100). Inner and outer layer of the erythrocyte membrane have been separated this way and comparison of the lipids in both fractions showed an enrichment of cholesterol in the exterior plane of the membrane (101).

RESULTS AND DISCUSSION

A summary of the data is presented in Tables 1 and 2; a discussion of the results is given below.

Table 2   Lipid localization in membranes from eukaryotic cells[a]

| | Lipids localized | | | | |
Membrane	Preferential outside	Preferential inside	Equally distributed	Methods	References
Erythrocytes		PE, PS		FMMP	17, 102
human, rat, chicken,		PE, PS		IAI, EAI	103
pig, sheep		PE, PS		TNBS, FDNB	75, 78, 79, 104, 105, 106
	PC, Sph	PE, PS, PI		Pl'ase A$_2$, C, Sph'ase	32, 41, 62, 66, 67, 107, 112
	glycolipids			galactose oxidase	71, 72
	cholesterol			freeze fracturing	101
	PC			ESR	95
Blood platelets	Sph	PC, PE, PS, PI		Pl'ase A$_2$, C, Sph'ase	113
human, rat	PE	PC, PS, Sph		Pl'ase C, Sph'ase	114
		PE, PS		TNBS	115
Rat liver organelles[b]	PS, PE	PI, Sph, CL	PC	Pl'ase A$_2$	9, 55, 116, 117
			PC, PE, PI	Pl'ase A$_2$	118
	PC, Sph	PE, PS, PI		Pl'ase C	119, 111
	PE			TNBS	120
Sarcoplasmic reticulum	PE	PS		TNBS, FDNB	80
rabbit white muscle	PE	PS		fluorescamine	122
			PC	NMR	
			LysoPC	LysoPl'ase	154
Plasma membrane	Sph	PE	PC	PC-PLEP, TNBS	121
mouse LM cell	unsaturated fatty acid				
Myelin	cholesterol			X ray	98

[a] For abbreviations used see footnote a to Table 1. In addition: FMMP: formylmethionyl (sulfonyl) methylphosphate; EAI: ethylacetimidate, FDNB: fluorodinitrobenzene.
[b] The membrane preparations summarized here are microsomes (9, 55, 116–119); Golgi vesicles (55, 118); inner mitochondrial membranes (55, 120); lysosomes (55), and nuclear membranes (55).

Erythrocytes and Platelets

The most convincing evidence for an asymmetric phospholipid distribution has been obtained for the membrane of erythrocytes. All the techniques that have been used provide similar results: the majority of the choline-containing lecithin and sphingomyelin is found in the outer layer, whereas phosphatidylethanolamine and phosphatidylserine are mainly located inside. The first experiments that indicated this distribution were obtained with the probe formylmethionyl (sulfonyl) methylphosphate which reacted much more readily with red cell ghosts than with intact cells (17, 102). Criticism was raised against this concept mainly because it was based on initial rate studies, but the objections were partly overcome by experiments showing that other reagents such as trinitrobenzene sulfonic acid (78) and the combination of ethyl acetimidate and isethionylacetimidate labelled many more of the amino groups on the inside than in the outer layer. Furthermore, independent evidence for the asymmetric phospholipid distribution was obtained with phospholipases (67). The phospholipids in the outer layer of the membrane could be hydrolyzed with exogenous phospholipases under nonlytic conditions. In addition the inner membrane layer has been investigated by entrapping enzymes within resealed ghosts (108) or by using inside out sealed ghosts (107). Phospholipases have been applied to show that, contrary to the polar headgroup distribution, the fatty acid composition of lecithin in the inner and outer layers is similar (35).

A large number of control experiments have accompanied these studies. The chemical probes did not permeate (17, 75, 78, 102); phospholipases could be used under nonlytic conditions (67, 108); and transbilayer movement of phospholipids, at least lecithin, is a relatively slow process in these membranes (33, 34). Nevertheless, the proposed phospholipid distribution can be criticized. The conclusions made are based in part on the assumption that the localization of phospholipids in each of the two layers is fixed and cannot be altered by exposing the membrane to variations in its environment, such as the preparation of ghosts, resealed ghosts and inside out vesicles, the physiological condition of the cell interior, and the formation of nonmembrane lipids such as diglycerides and ceramides. First of all, it is evident that alterations occur during the preparation of red cell ghosts. This is most convincingly shown by the ability of phospholipid exchange protein to exchange lecithin from resealed ghosts whereas in intact cells, exchange is negligible (33). Phospholipases also react differently when added to intact cells and resealed ghosts; the polar headgroups of the phospholipids are more rapidly degraded by phospholipase C after the hypotonic lysis procedure (112, 138, 139). In addition, anionic detergents attack resealed ghosts more readily than intact cells (139). Although the extent of the alterations is a matter of debate (140–143) the fact that the membrane structure is changed during the preparation of ghosts and subsequent resealing has to be considered because the alterations may lead to a difference in the lipid disposition. In this respect, questions can be raised about the localization of the phosphatidylethanolamine and phosphatidylserine. That these phospholipids are localized in the interior layer is based merely on experiments obtained with such membrane derivatives as ghosts and resealed ghosts, and even then, only part of this lipid is assayed directly with various techniques. Covalent lipid probes do not label completely (78, 79) and phospholipases entrapped in resealed ghosts hydrolyze 50–65% of these phospholipids before lysis starts (108).

In conclusion the assignment of 80% of phosphatidylethanolamine and all of the phosphatidylserine at the membrane interior is partially based on the evidence derived from negative results (there is either limited or no reaction of these lipids at the outside of intact cells) in combination with the assumption that the phospholipids are distributed in a 1:1 ratio over inner and outer membrane. As long as no data are available about the exact localization of cholesterol, the area occupied by proteins in inner and outer membrane layer, and the average surface pressure in each layer, the latter assumption is merely speculative. It cannot be excluded therefore that more phosphatidylethanolamine and phosphatidylserine, or even more total lipid, are present in the outer layer but protected against reaction. Indeed evidence has been presented to substantiate this idea. In human erythrocytes, 76% of the lecithin and 20% of the phosphatidylethanolamine is assigned to the outer layer. Additional hydrolysis of phosphatidylcholine and phosphatidylethanolamine, prior to hemolysis, has been observed with phospholipase A_2 at slightly elevated pH and higher Ca^{2+} concentrations (62).

Furthermore, it is evident that the energetic condition of the cell interior affects the membrane structure in such a way that it can be measured at the outside. Glucose suppresses the pH/Ca^{2+} effect mentioned above (62). Depletion of ATP in rat, chicken, or sheep erythrocytes alters the susceptibility of the exterior phospholipids to chemical and enzymatic modifications. Under these conditions more phospholipids are available for hydrolysis (41, 110), trinitrobenzene sulfonic acid labelling (144), and extraction with ether (145). Energy depletion due to aging was found to increase the rate (not the extent) at which phosphatidylethanolamine was labelled with trinitrobenzene sulfonic acid (105), and the amount that could be hydrolyzed with phospholipase A_2 from *Naja naja* (109). Recently it has been found that when ATP is entrapped in resealed ghosts the opposite effect is obtained. An extensive reduction in the hydrolysis of phosphatidylethanolamine and sphingomyelin at the outside was found to be due to the internal presence of ATP (139). Several explanations have been offered for these effects. ATP depletion has been correlated with dephosphorylation of membrane proteins, particle aggregation, and increase in lipid bilayer area and thus less protection of lipids by proteins (144). It is feasible also that transbilayer movements may contribute to these effects although no direct evidence is presented that under mild conditions of energy starvation and ATP depletion transbilayer movement is enhanced. On the other hand a more extensive damage of the erythrocyte membrane by treatment with SH-oxidizing reagents leads to the accessibility of 50% of the phosphatidylethanolamine and even 30% of phosphatidylserine without hemolysis (109). The latter event could be correlated with a significant cross-linking of spectrin at the inner surface of the membrane and therefore a loss of matrix which is able to maintain the specific localization of phospholipids in the interior side of the membrane (61). The localization of cholesterol using exchange techniques has been hampered by the fast transbilayer movement of this compound (37, 38, 92). More success has been obtained with the freeze fracturing approach which indicated that the outer layer contains relatively more cholesterol than the inner layer (101).

In summary, the erythrocyte membrane offers the most convincing example of asymmetric lipid distribution. However, part of the phospholipids, especially phosphatidylethanolamine and phosphatidylserine, have been localized in an indirect way and alterations in environmental conditions may well induce alterations in membrane structure and possibly in lipid localization. Another type of blood cell, the platelet, has been investigated with techniques similar to those described above. The data, however, are fewer and partly contradictory. None of the phosphatidylserine and only 12–18% of the total phosphatidylethanolamine of the human platelets react with exogenous trinitrobenzene sulfonic acid (115). Treatment with phospholipases A₂ and C and sphingomyelinase also showed a limited accessibility of both these phospholipids, whereas sphingomyelin and nearly half of the plasma membrane lecithin could be hydrolyzed (113). A possible function of such an asymmetric distribution has been indicated by experiments that show that phosphatidylserine and phosphatidylethanolamine, which are assigned to the inner layer of the platelet plasma membrane, reduce blood coagulation times considerably, in contrast to lipids that are present in the outer layer. Lipid asymmetry in blood cells may contribute to the processes that regulate hemostasis (146). Different results have been obtained in an independent study in which phospholipases were also applied to localize lipids in the human platelet plasma membrane (114). The assignment of phosphatidylethanolamine and sphinogomyelin is completely opposite to that concluded from the other studies (113, 115). Such discrepancies are not unusual in studies on lipid localization. The following sections amplify this. Today one can only speculate about the reasons for the observed differences.

One clear conclusion from a comparative study of all these reports is that the data obtained with phospholipases, especially that on complex membrane systems, have to be handled very carefully.

Subcellular Membranes

In contrast to the numerous studies on erythrocyte membranes, all of which demonstrate phospholipid asymmetry, attempts to determine the localization of phospholipids in the various membranes of more complex cells are confusing because each provides a different answer. Agreement appears to exist only about the site of phospholipid synthesis that occurs in the endoplasmic reticulum. The enzymes involved in phosphatidylcholine, phosphatidylethanolamine, phosphatidylinositol, and triglyceride biosynthesis are all active on the cytoplasmic side of the microsomal vesicles (43, 147, 148). But whether this unilateral synthesis results in asymmetric membranes, especially in microsomal membranes, remains unclear. Studies with various phospholipases have indicated that phosphatidylethanolamine and phosphatidylserine are located preferentially in the outer (cytoplasmic) layer (9, 55, 117), in the inner (luminal) layer (119), or are distributed equally over the two layers (118). Similar contradictory conclusions have been drawn for the phosphatidylcholine and sphingomyelin localization. A reviewer can simply acknowledge the conflicting data and conclude that the field is highly controversial or try to explain the observed discrepancies. The latter course should be facilitated by the fact that in this case similar approaches (i.e. hydrolysis by phospholipases A_2 and C) have been applied. However, although in some experiments the same enzymes have been used, the incubation conditions vary greatly or are not completely described which makes a direct comparison impossible. When phospholipases are involved the incubation conditions are extremely important because external parameters such as pH, temperature, and ion concentration do influence the lipid organization in membranes and the activity of phospholipases is dependent on the quality of the interface (149, 150). It is unlikely, therefore, that conclusions about lipid localization can be based on comparative studies between microsomal lipids present in Triton X-100 micelles, sonicated vesicles, and microsomal membranes. The activities of phospholipases toward the individual lipids at these three different surfaces are not necessarily identical, particularly when incubations are carried out under conditions that differ for each system (118). Studies with phospholipases can be misleading when transbilayer movements of lipids are possible. Studies with exchange proteins have shown that these processes may proceed relatively fast in microsomes (151, 152). An upper limit of 45 min or less for the half-time of lecithin translocation has been estimated (151), and this process can be enhanced considerably by unilateral alterations in the membrane.

Alterations in the charge of lipids (28), increases in the amount of lipid due to biosynthesis (39), or depletion via phospholipase treatment (81) may result in very fast and possibly also specific transbilayer movements. It is feasible that treatment of microsomes with phospholipase C (119), which results in the loss of complete polar headgroups, enhances the already fast translocation of those lipids that are hydrolyzed (Figure 1). Phospholipase A_2 treatment may induce a similar event especially when albumin is present to remove the split products from the membrane thus creating a lipid deficiency at one side (55, 116, 117). Such considerations, plus the fact that the specificity of phospholipases varies greatly depending on the organization of lipids in the membrane (which is not known for microsomes) may offer an explanation for the contradictory results so far obtained (55, 116–119). When a particular lipid is hydrolyzed faster than the other lipids, transbilayer movements may supply the outer layer with this particular lipid from the inside. As a result the data do not reflect lipid localization but lipid accessibility and mobility. Lipid mobility has to be considered very carefully in lipid localization experiments, particularly those using microsomal membranes, because recent evidence shows that the phospholipids of this membrane are not exclusively organized in classical bilayer structure. ^{31}P-NMR provided evidence that a sizeable fraction of the phospholipids experience isotropic motion, which not only allows a fast transbilayer movement of lipids, but also makes studies on lipid localization, as a static phenomenon, irrelevant (153).

The phospholipids that are synthesized in the endoplasmic reticulum are transported toward the other intracellular membranes and the plasma membrane. These different membrane structures have defined specific func-

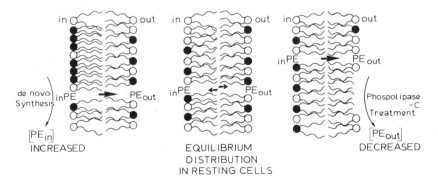

Figure 1 Hypothetical model showing the induction of phospholipid transbilayer movements during de novo lipid synthesis [(39) left] and following phospholipid degradation (right). Both processes disturb the existing equilibrium distribution of phospholipids over the two membrane layers and, perhaps, the driving force behind the transbilayer movements observed.

tions in the cell, contain specific proteins, and have a specific phospholipid composition, and therefore probably each has a specific phospholipid distribution. Treatment of microsomes, golgi membranes, inner mitochondrial membranes, lysosomal, and nuclear membranes by phospholipase A_2 reveals a similar asymmetry (55). In contrast to the conflicting data on most (intra)cellular membranes, nearly identical results have been obtained in two studies with sarcoplasmic reticulum membranes from rabbit white muscles (80, 122). Although different chemical probes have been used, both studies demonstrate the preferential localization of phosphatidylethanolamine in the outer and phosphatidylserine in the inner layer of this membrane. Phosphatidylcholine as well as its lysoderivative are equally distributed over the two membrane layers. The latter observations were made using a combined approach of NMR, exchange of lecithin, and lysophospholipase treatment of these membranes (154).

Plasma Membranes

The transbilayer distribution of phospholipids in plasma membranes has been studied via two indirect techniques. The first approach is based upon the determination of phospholipid distribution in a virus that obtains its membrane lipids from the host plasma membrane by a budding process (30). The second approach is based on the opposite mechanism. Phagolysosomes, resulting from the phagocytosis of latex beads by mouse LM cells, are assumed to be inside out plasma membrane derivatives (121). Convincing evidence has been presented concerning the distribution of phosphatidylcholine, phosphatidylethanolamine, and sphingomyelin in the virus (Table 1) and the phagolysosomes (Table 2) despite the fact that the results are completely opposite in the two systems. A direct comparison is difficult because different cells have been used as starting material for the plasma membrane derivatives. In addition, it is feasible that the derivatives do not give a correct reflection of the overall plasma membrane. Phagocytosis as well as budding may be processes confined to restricted (specific) areas of the plasma membrane and may induce alterations in membrane composition (155, 156). A more extensive discussion of the data obtained with virus membranes is given below.

Viruses

A virus membrane is supposed to arise from the host cell plasma membrane by a budding process in which the host cell lipids are not actively involved but merely form the glue between the specific virus membrane proteins (157). One should therefore expect that viruses grown on a similar cell type, in this case BHK-25 cells, should have identical phospholipid distributions. However, treatment with trinitrobenzene sulfonic acid showed that phosphatidylethanolamine is preferentially localized in the inside layer of vesicular stomatitis virus (76). A similar preference of phosphatidylethanolamine for the inner layer has been found in studies with two phospholipases C on influenza virus (124). Lecithin and sphingomyelin are nearly completely hydrolyzed in these cells and therefore assigned to the outer layer. Cholesterol was found to be present in equal amounts in both layers (31). In a later study, completely the opposite distribution of phosphatidylethanolamine and sphingomyelin was found (30). These data were obtained in a very extensive study using two phospholipases C and two different exchange proteins. It was found that phosphatidylcholine and phosphatidylinositol are preferentially located in the outer layer, sphingomyelin is in the inner layer, and phosphatidylethanolamine and phosphatidylserine are present, in similar proportions, at each surface. In addition it was observed that the absolute amount of phospholipid in the two layers differed.

The shortage of phospholipid in the outer layer can be compensated for by a specific localization of all the glycolipids at the outside (69, 70). The differences between these data and the earlier results obtained with the same virus grown on the same BHK cell line (124), have been ascribed to an abrupt change in the phospholipid composition but cannot be explained further (30). This summary of the published data on lipid distribution in viruses emphasizes the heterogeneity and contradictory nature of the results, possibly due to variations in lipid composition of host cells and viruses and to variations in experimental approaches and conditions. Alternatively, the data may illustrate that the virus membrane is not a direct derivative of the host plasma membrane.

One of the early studies on lipid localization was carried out with the bacteriophage PM2 (73), which is impermeable to sulfanilic acid diazonium salt below 0.75 M LiCl. Under these conditions the reagent preferentially oxidizes phosphatidylglycerol, a reaction that can be followed by a reduction with NaB^3H_4. At higher salt concentrations the probe is permeable and also modifies the internal phosphatidylethanolamine. Although the labelling of both lipids is incomplete, the distribution is clearly asymmetric. A model is presented in which phosphatidylethanolamine interacts in the inner layer with the lipophilic nucleocapsid core, whereas the negatively charged phosphatidylglycerol interacts with the positively charged protein II which covers nearly the entire surface of the phage (73). The large differences in phospholipid composition between host cell and phage membranes[2] have led to the conclusion (158) that active and selective processes are involved in the formation of phage membrane.

[2]The ratios of the phospholipids, phosphatidylethanolamine, and phosphatidylglycerol in the host cell are inversely proportional to those of the virus.

Bacteria

Various techniques have been used to localize lipids in bacterial membranes. The limited data do not provide a coherent picture of lipid distribution in bacterial membranes and probably a general model does not exist. Bacterial membranes are very flexible, particularly the phospholipid composition, which differs considerably, not only between the various organisms but also between organisms cultivated under different conditions (159, 160). The variations in lipid composition are not restricted to the fatty acid side chains, as is usual in eukaroytic organisms, but alterations in the polar headgroups can also be induced (161). Furthermore, it is easy to modify membrane functions and, to some extent, membrane structures, by mutagenic treatments. The latter effect, as well as a good example of membrane asymmetry, is found in the outer membrane of *E. coli* and *Salmonella typhimurium*. Recent evidence suggests that these outer membranes are highly asymmetric, and that they contain most or all of the lipopolysaccharide but only a small amount of the phosphatidylethanolamine which is the major phospholipid in these organisms. This model is based upon evidence obtained from: (*a*) X-ray experiments showing that the phospholipid is partly present in a monolayer and partly in a bilayer configuration (162); (*b*) localization of lipopolysaccharides with ferritin-labelled antibodies (86–88); (*c*) phospholipases, which shows that phospholipids are inaccessible in intact cells (52, 126, 127); and (*d*) studies with cyanogen bromide–activated dextran which gave similar results to those obtained using the phospholipases (126). Mutants, however, lacking part of the lipopolysaccharide and/or part of the outer membrane proteins showed an altered freeze fracture appearance (163–165) and furthermore an increased accessibility of phospholipids to both exogenous probes (126, 127). The latter observation has been interpreted as a loss in protection of phospholipids (127) as well as an absolute increase in the amount of phosphatidylethanolamine in the outer layer of the outer membrane (126).

Studies on gram-positive organisms have provided interesting information not so much on the lipid distribution as such, but more especially on the maintenance of a specific lipid distribution in membranes. Using the probes trinitrobenzene sulfonic acid and isethionyl acetimidate, 31–35% of the phosphatidylethanolamine in *B. megaterium* could be assigned to the outer layer of the membrane (77). That the residual phosphatidylethanolamine is all localized in the inner layer is less convincingly demonstrated because complete labelling of phosphatidylethanolamine is not shown, and because the preparations, especially the characterization of the inside out vesicle preparation, can be criticized. Nevertheless additional experiments have elegantly shown that newly synthesized phosphatidylethanolamine is found first in the inner layer of the membrane but a few minutes after synthesis can be found in the outer layer (39). The transbilayer movement of the newly synthesized phospholipid is four orders of magnitude faster than in artificial membrane systems in which the phospholipid distribution is in equilibrium. A completely different approach substantiates this observation that transbilayer movements can be very fast in natural membranes. When the equilibrium distribution of specific phospholipids over the two membrane layers is disturbed by selective removal of phospholipid at one side, transbilayer movement is enhanced (Figure 1). A comparison was made between the extent of trinitrobenzene sulfonic acid labelling of phosphatidylethanolamine in *B. subtilis* protoplasts and the accessibility of this lipid towards phospholipase C (81). The results show that at least 60% of this phospholipid is in the outer layer but that the residual phosphatidylethanolamine can also be hydrolyzed under nonlytic conditions, which indicates that the transbilayer movement is stimulated by the phospholipid hydrolysis at the outside. Similar results have been obtained in studies on the phosphatidylethanolamine distribution in *B. megaterium* (129) and phosphatidylglycerol distribution in *A. laidlawii* membranes (57). In addition, it was found that in the latter organism phospholipid localization and/or transbilayer movement may depend on the physiological condition of the cells, as in the erythrocytes. The accessibility of phosphatidylglycerol for phospholipase A_2 is complete in resting cells but decreases to about 50% when glucose is actively transported and fermented (166).

Altogether, these data do not allow general conclusions to be made about lipid asymmetry but indicate that the bacterial membranes can be suitable systems for studying processes that are intimately associated with the formation and maintenance of specific lipid distributions.

Model Membranes

To solve the problem of how phospholipid localization is established and maintained in biological membranes one may study the arrangements and behavior of well-defined phospholipids in artificial bilayers. With simple phospholipid mixtures one can find out if the fatty acid composition and the size and charge of the polar headgroups are involved in determining a specific preferential localization. By increasing the complexity of the systems (e.g. by the introduction of proteins) an approximation of the biological membrane can be obtained. Today, however, most studies are carried out with the simplest system, i.e. the single bilayer vesicle, the drawback of which is that the lipid localization is largely determined by the high curvature of the bilayer, a parameter that is not of major significance in most biological membranes. On the other hand, it is relatively easy to obtain vesicle preparations that are well-defined and accessible to a variety of techniques. As mentioned earlier, ^{13}C- and ^{31}P-NMR in particular have been applied to study the lipid localization and transbilayer movements in these vesicles (171). Due to the high curvature, the vesicle membrane is already asymmetric with respect to the amount of phospholipids. The outside/inside ratio of about 2 in vesicles containing pure lecithin molecules depends on the fatty acid composition. The ratio is lowered upon an increase in the hydrocarbon chain length and to a smaller extent also by the presence of double bonds (167).

^{13}C-NMR has shown that the lecithin molecules, containing unsaturated fatty acids, are preferentially located in the outer layer (130). In most cases mixing lecithin with other phospholipids results in asymmetric vesicles. In mixed vesicles of lecithin, with small amounts of lysolecithin, more than 80% of the latter lipid is located in the outer layer. Such a distribution has also been reported for larger single bilayer vesicles (68). This high outside/inside ratio is substantially lowered by the presence of cholesterol (24, 25). Sphingomyelin (137) and phosphatidylglycerol (136) also prefer the outside surface, when present in lecithin vesicles. However, other negatively charged phospholipids such as phosphatidylinositol, phosphatidylserine, and phosphatidic acid prefer the inner layer (137). Phosphatidylethanolamine is preferentially located inside, an observation that was made with NMR (137) and with trinitrobenzene sulfonic acid treatment of vesicles (134). The latter approach demonstrated that the polar headgroup is largely responsible for the localization and not the fatty acid composition of the phosphatidylethanolamine (135), a conclusion that seems valid for the other lipid mixtures mentioned above. Much attention has been focused on the behavior of cholesterol in these vesicles. In a systematic study of the localization of cholesterol and various lecithin species in vesicles it was found that the cholesterol is equally distributed below 30 mole %, but shows a preference for the inner layer at higher concentrations. Under these conditions the lecithin distribution was found to change also, especially for those species containing *cis* unsaturated fatty acids (131, 132).

The localization of cholesterol was carried out in these studies in an indirect way. A more direct approach based on cholesterol exchange, indicates that the cholesterol distribution follows closely the phospholipid distribution in these small sonicated vesicles (170). Twice as much cholesterol is located in the outer layer as in the inside layer and this ratio is little influenced by the nature of the lecithin present (29). Transbilayer movements of cholesterol do not seem to occur in such systems (29, 170). It has, however, been indicated that the slow transbilayer movement is found only under equilibrium conditions. Depletion of cholesterol at the outside of the vesicle enhances the transbilayer movement considerably (29). The latter observation might explain an earlier report of complete exchangeability and fast flip-flop times for cholesterol in sonicated vesicles (168).

CONCLUDING REMARKS

Phospholipid asymmetry is clearly established in erythrocyte membranes. The choline-containing phospholipids are located mainly in the outer layer and the amino phospholipids preferentially in the inner layer. Evidence has been presented that phospholipid asymmetry also occurs in other mammalian membranes; the data, however, are incomplete and in many cases controversial. Similar conclusions can be made for membranes of viruses and bacteria. Phage PM_2 and influenza virus are surrounded by an asymmetric membrane but the data obtained from studies on membranes from bacteria at best indicate that lipid asymmetry may exist.

It is obvious that further studies on lipid asymmetry in biological membranes requires the development of new techniques or a better understanding of the present methods. In particular it is important to know if reagents that modify lipids cause alterations in membrane organization, including lipid localization. Improved understanding of localization techniques should lead to better insight into the naturally occurring flip-flop movement and the transbilayer movements induced by the lipid localization experiments.

Furthermore, one can envisage that a specific lipid asymmetry may be a transient feature of many membranes and that alterations in environment or metabolic activities may be coupled to alterations in the distribution of lipids between the two layers of a membrane. As stated in the introduction, the recognition of an asymmetric distribution of membrane proteins coupled to the different functions that are expressed at the two sides of a membrane initiated the search for lipid asymmetry. The next step should be the search for a correlation between lipid and protein asymmetry. This should lead to a better understanding of the role of lipids in membrane function.

ACKNOWLEDGMENTS

The author is indebted to Drs. H. van den Bosch, L. L. M. van Deenen, B. de Kruyff, and G. van Meer for critical reading of the manuscript and helpful discussions. I wish to thank Dr. M. Hope for adapting the manuscript to the standard rules of English grammar.

Literature Cited

1. Gorter, E., Grendel, F. 1925. *J. Exp. Med.* 41:439–43
2. Singer, S. J., Nicholson, G. L. 1972. *Science* 175:720–31
3. Bretscher, M. S. 1973. *Science* 181:622–29
4. Zwaal, R. F. A., Roelofsen, B., Colley, C. M. 1973. *Biochim. Biophys. Acta* 300:159–82
5. Juliano, R. L. 1973. *Biochim. Biophys. Acta* 300:341–78
6. Steck, T. L. 1974. *J. Cell. Biol.* 62:1–19
7. Singer, S. J. 1974. *Ann. Rev. Biochem.* 43:805–33
8. Bretscher, M. S., Raff, M. C. 1975. *Nature* 258:43–49
9. DePierre, J. W., Dallner, G. 1975. *Biochim. Biophys. Acta* 415:411–72
10. Marchesi, V. T., Furthmayr, H., Tomita, M. 1976. *Ann Rev. Biochem.* 45:667–98
11. Lenaz, G., Sechi, A. M. 1976. *Ital. J. Biochem.* 25:427–510
12. DePierre, J. W., Ernster, L. 1977. *Ann. Rev. Biochem.* 46:201–62
13. Rothman, J. E., Lenard, J. 1977. *Science* 195:743–53
14. Bergelson, L. D., Barsukov, L. I. 1977. *Science* 197:224–30
15. Singer, S. J. 1977. *J. Colloid Interface Sci.* 58:452–58
16. Warren, G. B., Birdsall, N. J. M., Lee, A. G., Metcalfe, J. C. 1974. In *Membrane Proteins in Transport and Phosphorylation*, ed. G. F. Azzone, M. E. Klingenberg, E. Quagliarello, N. Silliprandi, pp. 1–12. Amsterdam: North Holland. 293 pp.
17. Bretscher, M. S. 1972. *Nature New Biol.* 236:11–12
18. Kornberg, R. D., McConnell, H. M. 1971. *Biochemistry* 10:1111–20
19. Rothman, J. E., Dawidowicz, E. A. 1975. *Biochemistry* 14:2809–16
20. Johnson, L. W., Hughes, M. E., Zilversmit, D. B. 1975. *Biochim. Biophys. Acta* 375:176–85
21. Dawidowicz, E. A., Rothman, J. E. 1976. *Biochim. Biophys. Acta* 455:621–30
22. Roseman, M., Litman, B. J., Thompson, T. E. 1975. *Biochemistry* 14:4826–30
23. de Kruijff, B., Wirtz, K. W. A. 1977. *Biochim. Biophys. Acta* 468:318–25
24. van den Besselaar, A. M. H. P., van den Bosch, H., van Deenen, L. L. M. 1977. *Biochim. Biophys. Acta* 465:454–65
25. de Kruijff, B., van den Besselaar, A. M. H. P., van Deenen, L. L. M. 1977. *Biochim. Biophys. Acta* 465:443–53
26. de Kruijff, B., van Zoelen, E. J. J., van Deenen, L. L. M. 1978. *Biochim. Biophys. Acta* 509:537–42
27. van Zoelen, E. J. J., de Kruijff, B., van Deenen, L. L. M. 1978. *Biochim. Biophys. Acta* In press
28. de Kruijff, B., Baken, P. 1978. *Biochim. Biophys. Acta* 507:38–47
29. Poznansky, M. J., Lange, Y. 1978. *Biochim. Biophys. Acta* 506:256–64
30. Rothman, J. E., Tsai, D. K., Dawidowicz, E. A., Lenard, J. 1976. *Biochemistry* 15:2361–70
31. Lenard, J., Rothman, J. E. 1976. *Proc. Natl. Acad. Sci. USA* 73:391–95
32. Renooij, W., van Golde, L. M. G., Zwaal, R. F. A., van Deenen, L. L. M. 1976. *Eur. J. Biochem.* 61:53–58
33. Bloj, B., Zilversmit, D. B. 1976. *Biochemistry* 15:1277–83
34. Renooij, W., van Golde, L. M. G. 1976. *FEBS Lett.* 71:321–24
35. Renooij, W., van Golde, L. M. G. 1977. *Biochim. Biophys. Acta* 470:465–74
36. Bloj, B., Zilversmit, D. B. 1977. *Proc. Soc. Exp. Biol. Med.* 156:539–43
37. Kirby, C. J., Green, C. 1977. *Biochem. J.* 168:575–77
38. Lange, Y., Cohen, C. M., Poznansky, M. J. 1977. *Proc. Natl. Acad. Sci. USA* 74:1538–42
39. Rothman, J. E., Kennedy, E. P. 1977. *Proc. Natl. Acad. Sci. USA* 74:1821–25
40. McNamee, M. G., McConnell, H. M. 1973. *Biochemistry* 12:2951–58
41. Gazitt, Y., Loyter, A., Reichler, Y., Ohad, I. 1976. *Biochim. Biophys. Acta* 419:479–92
42. Bevers, E. M. 1978. *Phosphatidylglycerol in the membrane of Acholeplasma laidlawii.* PhD thesis. Univ. Utrecht, Utrecht. 53 pp.
43. Brophy, P. J., Burbach, P., Nelemans, A. S., Westerman, J., Wirtz, K. W. A., van Deenen, L. L. M. 1978. *Biochem. J.* 174:413–20
44. Op den Kamp, J. A. F., Kauerz, M. Th., van Deenen, L. L. M. 1975. *Biochim. Biophys. Acta* 406:169–77
45. Bevers, E. M., Op den Kamp, J. A. F., van Deenen, L. L. M. 1977. *Eur. J. Biochem.* 84:35–42
46. Carraway, K. L. 1975. *Biochim. Biophys. Acta* 415:379–410
47. Roelofsen, B., Zwaal, R. F. A. 1976. *Methods in Membrane Biology*, ed. E. D. Korn, 7:147–77. New York: Plenum. 267 pp.
48. Karlsson, E. 1973. *Experientia* 29:1319–27

49. Condrea, E., de Vries, A., Mager, J. 1964. *Biochim. Biophys. Acta* 84:60–73
50. Simpson, R., Hauser, R. 1966. *Virology* 30:684–97
51. Wahlström, A. 1971. *Toxicon* 9:45–56
52. Duckworth, D. H., Bevers, E. M., Verkleij, A. J., Op den Kamp, J. A. F., van Deenen, L. L. M. 1974. *Arch. Biochem. Biophys.* 165:379–87
53. Op den Kamp, J. A. F., Kauerz, M. Th., van Deenen, L. L. M. 1972. *J. Bacteriol.* 112:1090–98
54. Nanninga, N., Tijssen, F. C., Op den Kamp, J. A. F. 1973. *Biochim. Biophys. Acta* 298:184–94
55. Nilsson, O. S., Dallner, G. 1977. *Biochim. Biophys. Acta* 464:453–58
56. van Golde, L. M. G., McElhaney, R. N., van Deenen, L. L. M. 1971. *Biochim. Biophys. Acta* 231:245–49
57. Bevers, E. M., Singal, S. A., Op den Kamp, J. A. F., van Deenen, L. L. M. 1977. *Biochemistry* 16:1290–95
58. Ibrahim, S. A., Thompson, R. H. S. 1965. *Biochim. Biophys. Acta* 99:331–41
59. Condrea, E., Barzilay, M., Mager, J. 1970. *Biochim. Biophys. Acta* 210:65–73
60. Gul, S., Smith, A. D. 1972. *Biochim. Biophys. Acta* 288:237–40
61. Haest, C. W. M., Plasa, G., Kamp, D., Deuticke, B. 1978. *Biochim. Biophys. Acta* 509:21–32
62. Martin, J. K., Luthra, M. G., Wells, M. A., Watts, R. P., Hanahan, D. J. 1975. *Biochemistry* 14:5400–8
63. Gul, S., Smith, A. D. 1974. *Biochim. Biophys. Acta* 367:271–81
64. Gazitt, Y., Ohad, I., Loyter, A. 1975. *Biochim. Biophys. Acta* 382:65–72
65. Coleman, R., Finean, J. B., Knutton, S., Limbrick, A. R. 1970. *Biochim. Biophys. Acta* 219:81–92
66. Colley, C. M., Zwaal, R. F. A., Roelofsen, B., van Deenen, L. L. M. 1973. *Biochim. Biophys. Acta* 241:925–29
67. Verkleij, A. J., Zwaal, R. F. A., Roelofsen, B., Comfurius, P., Kastelijn, D., van Deenen, L. L. M. 1973. *Biochim. Biophys. Acta* 323:178–93
68. De Oliveira Filgueiras, O. M., van den Besselaar, A. M. H. P., van den Bosch, H. 1977. *Biochim. Biophys. Acta* 471:391–400
69. Stoffel, W., Anderson, R., Stahl, J. 1975. *Hoppe-Seylers Z. Physiol. Chem.* 356:1123–29
70. Stoffel, W., Sorgo, W. 1976. *Chem. Phys. Lipids* 17:324–35
71. Gahmberg, C. G., Hakomori, S. 1973. *J. Biol. Chem.* 248:4311–17
72. Steck, T. L., Dawson, G. 1974. *J. Biol. Chem.* 249:2135–42
73. Schäfer, R., Hinnen, R., Franklin, R. M. 1974. *Eur. J. Biochem.* 50:15–27
74. Amar, A., Rottem, S., Razin, S. 1974. *Biochim. Biophys. Acta* 352:228–44
75. Gordesky, S. E., Marinetti, G. V., Love, R. 1975. *J. Membrane Biol.* 20:111–32
76. Fong, B. S., Hunt, R. C., Brown, J. C. 1976. *J. Virology* 20:658–63
77. Rothman, J. E., Kennedy, E. P. 1977. *J. Mol. Biol.* 110:603–18
78. Gordesky, S. E., Marinetti, G. V. 1973. *Biochem. Biophys. Res. Commun.* 50:1027–31
79. Marinetti, G. V., Love, R. 1976. *Chem. Phys. Lipids* 16:239–54
80. Vale, M. G. P. 1977. *Biochim. Biophys. Acta* 471:39–48
81. Bishop, D. G., Op den Kamp, J. A. F., van Deenen, L. L. M. 1977. *Eur. J. Biochem.* 80:381–91
82. Bishop, D. G., Bevers, E. M., van Meer, G., Op den Kamp, J. A. F., van Deenen, L. L. M. 1978. *Biochim. Biophys. Acta.* In press
83. Schiefer, H. G., Gerhardt, U., Brunner, H., Krüpe, M. 1974. *J. Bacteriol.* 120:81–88
84. Schiefer, H. G., Krauss, H., Brunner, H., Gerhardt, U. 1975. *J. Bacteriol.* 124:1598–1600
85. Schiefer, H. G., Gerhardt, U., Brunner, H. 1977. *Zentralbl. Bakteriol. Parasitenkd. Infektionskr. Hyg. Abt. 1: Orig. Reihe A* 239:262–69
86. Mühlradt, P. F., Menzel, J., Golecki, J. R., Speth, V. 1973. *Eur. J. Biochem.* 35:471–81
87. Mühlradt, P. F., Menzel, J., Golecki, J. R., Speth, V. 1974. *Eur. J. Biochem.* 43:533–39
88. Mühlradt, P. F., Golecki, J. R. 1975. *Eur. J. Biochem.* 51:343–52
89. Schiefer, H. G., Gerhardt, U., Brunner, H. 1975. *Hoppe-Seylers Z. Physiol. Chem.* 356:559–65
90. Wirtz, K. W. A. 1974. *Biochim. Biophys. Acta* 344:95–117
91. Barsukov, L. I., Kulikov, V. I., Bergelson, L. D. 1976. *Biochem. Biophys. Res. Commun.* 71:704–11
92. Johnson, L. W., Hughes, M. E., Zilversmit, D. B. 1975. *Biochim. Biophys. Acta* 375:176–85
93. Rousselet, A., Colbeau, A., Vignais, P. M., Deveaux, P. F. 1976. *Biochim. Biophys. Acta* 426:372–84
94. Rottem, S. 1975. *Biochem. Biophys. Res. Commun.* 64:7–12
95. Tanaka, K., Ohnishi, S. 1976. *Biochim. Biophys. Acta* 426:218–31
96. Wisnieski, B. J., Iwata, K. K. 1977. *Biochemistry* 16:1321–26

97. Rousselet, A., Guthmann, C., Matricon, J., Bienvenue, A., Deveaux, P. F. 1976. *Biochim. Biophys. Acta* 426: 357–71

98. Caspar, D. L. D., Kirscher, D. A. 1971. *Nature New Biol.* 231:46–49

99. Blaurock, A. E., King, G. I. 1977. *Science* 196:1101–4

100. Fisher, K. A. 1975. *Science* 190:983–85

101. Fisher, K. A. 1976. *Proc. Natl. Acad. Sci. USA* 73:173–77

102. Bretscher, M. S. 1972. *J. Mol. Biol.* 71:523–28

103. Whiteley, N. M., Berg, H. C. 1974. *J. Mol. Biol.* 87:541–61

104. Gordesky, S. E., Marinetti, G. V., Segel, G. B. 1972. *Biochem. Biophys. Res. Commun.* 5:1004–9

105. Haest, C. W. M., Deuticke, B. 1975. *Biochim. Biophys. Acta* 401:468–80

106. Marinetti, G. V. 1977. *Biochim. Biophys. Acta* 465:198–209

107. Kahlenberg, A., Walker, C., Rohrlick, R. 1974. *Can. J. Biochem.* 52:803–6

108. Zwaal, R. F. A., Roelofsen, B., Comfurius, P., van Deenen, L. L. M. 1975. *Biochim. Biophys. Acta* 406:83–96

109. Haest, C. W. M., Deuticke, B. 1976. *Biochim. Biophys. Acta* 436:353–65

110. Gazitt, Y., Loyter, A., Ohad, I. 1977. *Biochim. Biophys. Acta* 471:361–71

111. Low, M. G., Finean, J. B. 1977. *Biochem. J.* 162:235–40

112. Barzilay, M., Kaminsky, E., Condrea, E. 1978. *Toxicon* 16:153–61

113. Chap, H. J., Zwaal, R. F. A., van Deenen, L. L. M. 1977. *Biochim. Biophys. Acta* 467:146–64

114. Otnaess, A. B., Holm, T. 1976. *J. Clin. Invest.* 57:1419–25

115. Schick, P. K., Kurica, K. B., Chacko, G. K. 1976. *J. Clin. Invest.* 57:1221–26

116. Nilsson, O. S., Dallner, G. 1975. *FEBS Lett.* 58:190–93

117. Nilsson, O. S., Dallner, G. 1977. *J. Cell Biol.* 72:568–83

118. Sundler, R., Sarcione, S. L., Alberts, A. W., Vagelos, P. R. 1977: *Proc. Natl. Acad. Sci. USA* 74:3350–54

119. Higgins, J. A., Dawson, R. M. C. 1977. *Biochim. Biophys. Acta* 470:342–56

120. Marinetti, G. V., Senior, A. E., Love, R., Broadhurst, C. I. 1976. *Chem. Phys. Lipids* 17:353–62

121. Sandra, A., Pagano, R. E. 1978. *Biochemistry* 17:332–38

122. Hidalgo, C., Ikemoto, N. 1977. *J. Biol. Chem.* 252:8446–54

123. Guarnieri, M. 1975. *Lipids* 10:294–98

124. Tsai, K. H., Lenard, J. 1975. *Nature* 253:554–55

125. Fong, B. S., Brown, J. C. 1978. *Biochim. Biophys. Acta* 510:230–41

126. Kamio, Y., Nikaido, H. 1976. *Biochemistry* 15:2561–70

127. van Alphen, L., Lugtenberg, B., van Boxtel, R., Verhoef, K. 1977. *Biochim. Biophys. Acta* 466:257–68

128. Shimada, K., Murata, N. 1976. *Biochim. Biophys. Acta* 455:605–20

129. Demant, E. J. F., Op den Kamp, J. A. F., van Deenen, L. L. M. 1978. *Biochim. Biophys. Acta* Submitted for publication

130. Yeagle, P. L., Hutton, W. C., Martin, R. B. 1976. *J. Biol. Chem.* 251:2110–12

131. Huang, C. H., Sipe, J. P., Chow, S. T., Martin, R. B. 1974. *Proc. Natl. Acad. Sci. USA* 71:359–62

132. de Kruijff, B., Cullis, P. R., Radda, G. K. 1976. *Biochim. Biophys. Acta* 436:729–40

133. Sears, B., Hutton, W. C., Thompson, T. E. 1976. *Biochemistry* 15:1635–39

134. Litman, B. J. 1974. *Biochemistry* 13:2844–48

135. Litman, B. J. 1975. *Biochim. Biophys. Acta* 413:157–62

136. Michaelson, D. M., Horwitz, A. F., Klein, M. P. 1973. *Biochemistry* 12:2637–45

137. Berden, J. A., Barker, R. W., Radda, G. K. 1975. *Biochim. Biophys. Acta* 375:186–208

138. Woodward, C. B., Zwaal, R. F. A. 1972. *Biochim. Biophys. Acta* 274: 272–78

139. Shukla, S. D., Billah, M. M., Coleman, R., Finean, J. B., Michell, R. H. 1978. *Biochim. Biophys. Acta* 509:48–57

140. Bretscher, M. S. 1973. *Nature New Biol.* 245:116–17

141. Staros, J. V., Haley, B. E., Richards, F. M. 1974. *J. Biol. Chem.* 249:5004–07

142. Wallach, D. F. H., Schmidt-Ullrich, R., Knüfermann, H. 1974. *Nature* 248: 623–24

143. Cabantchik, Z. I., Balshin, M., Breuer, W., Markus, H., Rothstein, A. 1975. *Biochim. Biophys. Acta* 382:621–33

144. Gazitt, Y., Ohad, I., Loyter, A. 1976. *Biochim. Biophys. Acta* 436:1–14

145. Gazitt, Y., Ohad, I., Loyter, A. 1976. *Biochem. Biophys. Res. Commun.* 72:1359–66

146. Zwaal, R. F. A., Comfurius, P., van Deenen, L. L. M. 1977. *Nature* 268:358–60

147. Vance, D. E., Choy, P. C., Farren, S. B., Lim, P. H., Schneider, W. J. 1977. *Nature* 270:268–69

148. Coleman, R., Bell, R. M. 1978. *J. Cell Biol.* 76:245–53

149. Verger, R., de Haas, G. H. 1976. *Ann. Rev. Biophys. Bioeng.* 5:77–117
150. Roberts, M. F., Otnaess, A., Kensil, C. A., Dennis, E. A. 1978. *J. Biol. Chem.* 253:1252–57
151. Zilversmit, D. B., Hughes, M. E. 1977. *Biochim. Biophys. Acta* 469:99–110
152. van den Besselaar, A. M. H. P., de Kruijff, B., van den Bosch, H., van Deenen, L. L. M. 1978. *Biochim. Biophys. Acta* 510:242–55
153. de Kruijff, B., van den Besselaar, A. M. H. P., Cullis, P. R., van den Bosch, H., van Deenen, L. L. M. 1978. *Biochim. Biophys. Acta.* In press
154. van den Besselaar, A. M. H. P. 1978. *Aspects of phospholipid dynamics in membranes.* PhD thesis. University of Utrecht, Utrecht. 100 pp.
155. Lenard, J., Compans, R. W. 1974. *Biochim. Biophys. Acta* 344:51–94
156. Berlin, R. D., Fera, J. P. 1977. *Proc. Natl. Acad. Sci. USA* 74:1072–76
157. Simons, K., Garoff, H., Helenius, A., Ziemiecki, A. 1978. In *Frontiers of Physicochemical Biology,* ed. B. Pullman. London: Academic. In press
158. Tsukagoshi, N., Franklin, R. M. 1974. *Virology* 59:408–17
159. Cronan, J. E., Gelmann, E. P. 1975. *Bacteriol. Rev.* 39:232–56
160. Salton, M. R. J., Owen, P. 1976. *Ann. Rev. Microbiol.* 30:451–82

161. Op den Kamp, J. A. F., van Deenen, L. L. M., Tomasi, V. 1969. In *Structural and Functional Aspects of Lipoproteins in Living Systems,* ed. E. Tria, A. M. Scanu, pp. 227–328. London: Academic. 661 pp.
162. Overath, P., Brenner, M., Gulik-Krzywicki, T., Shechter, E., Letellier, L. 1975. *Biochim. Biophys. Acta* 389:358–69
163. Verkleij, A. J., Lugtenberg, E. J. J., Ververgaert, P. H. J. Th. 1976. *Biochim. Biophys. Acta* 426:581–86
164. Verkleij, A. J., van Alphen, L., Bijvelt, J., Lugtenberg, B. 1977. *Biochim. Biophys. Acta* 466:269–82
165. Smit, J., Kamio, Y., Nikaido, H. 1975. *J. Bacteriol.* 124:942–58
166. Bevers, E. M., Leblanc, C., Le Grimellec, C., Op den Kamp, J. A. F., van Deenen, L. L. M. 1978. *FEBS Lett.* 87:49–51
167. de Kruijff, B., Cullis, P. R., Radda, G. K. 1975. *Biochim. Biophys. Acta* 406:6–20
168. Bloj, B., Zilversmit, D. B. 1977. *Biochemistry* 16:3943–48
169. Rottem, S., Hasin, M., Razin, S. 1973. *Biochim. Biophys. Acta* 323:520–31
170. Poznansky, M. J., Lange, Y. 1976. *Nature* 259:420–21
171. Bystrov, V. F., Shapiro, Y. E., Viktorov, A. V., Barsukov, L. I., Bergelson, L. D. 1972. *FEBS Lett.* 25:337–38

Ann. Rev. Biochem. 1979. 48:73–101
Copyright © 1979 by Annual Reviews Inc. All rights reserved

USE OF MODEL ENZYMES IN THE DETERMINATION OF THE MODE OF ACTION OF PENICILLINS AND Δ³-CEPHALOSPORINS[1]

❖12003

Jean-Marie Ghuysen, Jean-Marie Frère, Mélina Leyh-Bouille, Jacques Coyette, Jean Dusart, and Martine Nguyen-Distèche

Service de Microbiologie, Faculté de Médecine, Institut de Botanique, Université de Liège, 4000 Sart Tilman, Liège, Belgium

CONTENTS

PERSPECTIVES AND SUMMARY

Studies carried out with exocellular penicillin-sensitive enzymes (PSEs) from various strains of actinomycetes (the only organisms known to excrete such enzymes during growth), show that:

[1]Abbreviations used are: PBP, penicillin-binding protein; PSE, penicillin-sensitive enzyme.

73

0066-4154/79/0701-0073$01.00

(*a*) The penam or 3-cephem nuclei are the main portion of the antibiotic molecules concerned with the initial reversible binding to the enzyme. Binding to the enzyme active center is neither very selective nor very efficient but it has specific effects on the enzyme conformation. The sites thus induced in the enzyme react with suitably oriented and structured N_{14} substituents and force the β-lactam ring to gain a much increased chemical reactivity. As the β-lactam ring opens (with the help of electron attracting groups) and achieves acylation of the enzyme, the thiazolidine or dihydrothiazine ring interacts with a nearby site thus greatly increasing the stability of the acyl-enzyme complex.

(*b*) Elimination of the bound residue and regeneration of a free active enzyme by deacylation in the presence of a suitable exogenous nucleophile may proceed according to two different pathways. PSEs slowly overcome the stabilizing effect of the bound penicilloyl molecule by C_5–C_6 cleavage; deacylation of the N-acylglycyl-enzyme complex thus formed is then immediate whereas the other fragmentation product is further processed and eventually released as N-formyl-D-penicillamine. Alternatively, fragmentation does not occur and the bound residue is slowly eliminated as a whole. Irrespective of the pathway, the overall reactions between PSEs and β-lactams are characterized by very low turnover numbers.

(*c*) In some respects the above mechanism resembles that through which analogues of the natural substrates, L-R-D-Ala-D-Ala-terminated peptides (where R is an amino acid residue), interact with the PSEs. The D-Ala-D-Ala portion is mainly responsible for the initial binding whereas a suitable side chain in the preceding L residue is required to induce enzyme action on the bound peptide, which forces the D-Ala-D-Ala amide linkage to lose all double bond character. However, whether the enzyme functions as a catalyst template or forms a transitory acyl intermediate, the transfer of the L-R-D-Ala moiety to the exogenous nucleophile is rapid, and therefore the reactions are characterized by high turnover numbers.

(*d*) Depending upon the PSEs, the independence between the penicillin and L-R-D-Ala-D-Ala active centers is more or less pronounced. Indirect but strong evidence suggests that the physiological function of the β-lactam center in the PSEs may be regulatory rather than catalytic. By interacting with it, the β-lactam inactivates the specific catalytic center of the enzyme thus preventing action on natural substrate analogues.

Exocellular PSEs are probably not physiologically important whereas the membrane-bound components (or at least some of them) are likely to be crucial in mediating the antibacterial β-lactam effect. The relevance of the exocellular model systems to the actual problem of the mechanism of penicillin action has been examined by extending the above studies to several membrane-bound PSEs originating from taxonomically unrelated bacteria. The data suggest that they react with β-lactam antibiotics according to the same basic mechanism as that described above.

SCOPE AND LIMITATION

The damages caused by penicillins and Δ^3-cephalosporins to growing bacteria are related to the ability of these antibiotics to bind to several proteins (penicillin-binding proteins or PBPs) and to inactivate several enzymes (penicillin-sensitive enzymes or PSEs). With few exceptions, PBPs and PSEs are membrane-bound and in some cases, PBPs and PSEs are known to be synonymous. The understanding of the mechanism through which the β-lactams exert their antibacterial effects requires a combination of studies. They include: (a) the pore properties of those proteins (porins) which facilitate the permeation of the β-lactams through the outer membrane of the gram-negative bacteria (1–3); (b) the identification and genetic analysis of the PBPs and PSEs and the roles that they fulfil in vivo (4–7); (c) the physiological and biochemical events that follow the inactivation of the PBPs/PSEs and frequently culminate in cell death and lysis (8); and (d) the interaction at the molecular level between the β-lactams and the PBPs/PSEs (9–11).

Penicillins and Δ^3-cephalosporins (Figure 1) are stereochemically similar, not because of their detailed dimensions and conformations, but because the N–CO bond of the β-lactam ring does not have the normal character of a free amide. On the basis of X-ray data in the solid state, the dihedral angles about the pyramidal β-lactam nitrogen (ω_2) are 135° for penicillins and 155° for Δ^3-cephalosporins (16). Among the two sets of dihedral angles (ϕ_2, ψ_2) and (ϕ_3, ψ_3), ψ_2 and ϕ_3 are fixed to specific values by the lactam and the thiazolidine or dihydrothiazine rings. Hence the conformation depends on the two variable angles ϕ_2 and ψ_3 which determine the orientations of the aminoacyl substituent and the carboxyl grouping, respectively. A review on the functional modifications and nuclear analogues of β-lactam antibiotics can be found in (17, 18). Antibacterial activity (a) requires a β-lactam ring of sufficient strain and possibilities for electron delocalization outside the lactam ring; (b) appears to be associated with ϕ_2 rotational angles ranging, in solution, between $-180°$ and $-160°$, i.e. with a ϕ_2 region termed region A (12); and (c) is considerably enhanced by the occurrence of suitable substituents on N_{14} and the presence of a carbonyl group in α position to the nitrogen atom of the lactam amide bond.

The present work is restricted to the roles played by the various portions of penicillins and Δ^3-cephalosporins and by the various residues of well-defined peptides used as natural substrate analogues, in their interactions with several model PBPs/PSEs.

1

2

3

MODEL ENZYMES

PSEs and PBPs are members of the bacterial wall peptidoglycan cross-linking system which performs attachment of the nascent peptidoglycan strands to the preexisting wall peptidoglycan, controls the final extent of cross-linkages, and imparts to the bacterial wall its shape and physical strength (19). The main known reactions catalyzed by the PSEs are nucleophilic attacks of glycan-substituted pentapeptides.

$$\text{L-Ala-D-Glu} \underset{\rule{0.8cm}{0.4pt}}{\rule{0.4pt}{0.3cm}} \text{L-X-D-Ala-D-Ala}$$
$$| \, \omega$$
$$\text{(H)}$$

where X is a diamino acid residue whose lateral chain is either free or extended by one or several additional amino acids. Attacks occur on the carbonyl carbon of the penultimate D-Ala. The nucleophile may be either water causing the formation of tetrapeptides (DD-carboxypeptidase activity; reaction 1 in Figure 2), or an amino compound (NH_2-R) causing the formation of

$$\text{L-Ala-D-Glu} \underset{\rule{0.8cm}{0.4pt}}{\rule{0.4pt}{0.3cm}} \text{L-X-D-Ala-CONH-R}$$
$$| \, \omega$$
$$\text{(H)}$$

products (DD-transpeptidase activity; reaction 2 in Figure 2). Free D-Ala is produced by either pathway. DD-transpeptidase action on the same pep-

---

Figure 1   (*1*) Structure of penicillins (*2*) Δ^3-cephalosporins; and (*3*) X-D-Ala-D-Ala showing the backbone dihedral angles (ω_i, ϕ_i, ψ_i). The β-lactam ring is fused to another ring, either thiazolidine forming the penam nucleus of penicillins, or dihydrothiazine forming the 3-cephem nucleus of Δ^3-cephalosporins. With penicillins, $\omega_2 = 0$ when the bond C_7-N_4 eclipses the bond C_6-N_{14}, and $\phi_2 = 0$ when the bond C_6-C_7 eclipses the bond N_{14}-C_{15}; clockwise rotation is considered as positive (12). All penicillins are derivatives of 6-aminopenicillanic acid (a condensed L-cysteinyl-D-valine dipeptide); R_{17} is C_6H_5-CH_2- in benzylpenicillin, C_6H_5-O-CH_2- in phenoxymethylpenicillin, C_6H_5-CH(NH_2)- in ampicillin, and C_6H_5-CH(COOH)- in carbenicillin. Cephalosporins with a 3'-acetoxy side chain are derivatives of 7-aminocephalosporinic acid; R_{17} is C_6H_5-CH(NH_2)- in cephaloglycine,

$$\begin{array}{c}\text{HOOC} \\ \text{H}_2\text{N}\end{array}\!\!>\!\!\text{CH-(CH}_2)_3\text{-}$$

(i.e. D-α-aminoadipic acid) in cephalosporin C, and

$$\underset{S}{\boxed{}}\text{-CH}_2\text{-}$$

in cephalothin. Cephalexin is a Δ^3-cephalosporin with a 3-methyl side chain (instead of an acetoxy) and with $R_{17} = C_6H_5$-CH(NH_2)-. Antibacterial β-lactams are known that either do not contain a sulphur atom in their ring system, such as thienamycine (13) and clavulanic acid (14), or are not fused to another ring, such as the monocyclic nocardicins (15). [Adapted from (12) with the permission of Munksgaard, Copenhagen.]

tide acting both as carbonyl donor and as nucleophilic acceptor leads to peptide dimer formation (reaction 2 in Figure 2, middle part). DD-carboxypeptidases can act on interpeptide bonds extending between D-Ala and another D-center in α-position to a free carboxyl group (reaction 5 in Figure 2); such DD-carboxypeptidases behave as endopeptidases, performing hydrolysis of those peptide dimers previously formed by DD-transpeptidase action. Finally, the peptidoglycan cross-linking system also performs LD-carboxypeptidase (reaction 3 in Figure 2) and LD-transpeptidase (reaction 4 in Figure 2) activities. Model PSEs catalyzing these latter reactions are not discussed in this work.

Three exocellular PSEs have been used. The R39 (M_r 53,000), R61 (M_r 38,000), and G (M_r 20,000) enzymes are excreted by *Actinomadura* R39, *Streptomyces* R61, and *Streptomyces albus* G, respectively. They have been purified to homogeneity (20–22) and their chemical, physical, and immunological properties have been studied extensively (20–25). The three enzymes give rise to different peptide maps (after proteolysis with trypsin; C. Duez and J. M. Frere, unpublished). Both R39 and R61 enzymes probably consist of a single polypeptide chain but the G enzyme might consist of two chains

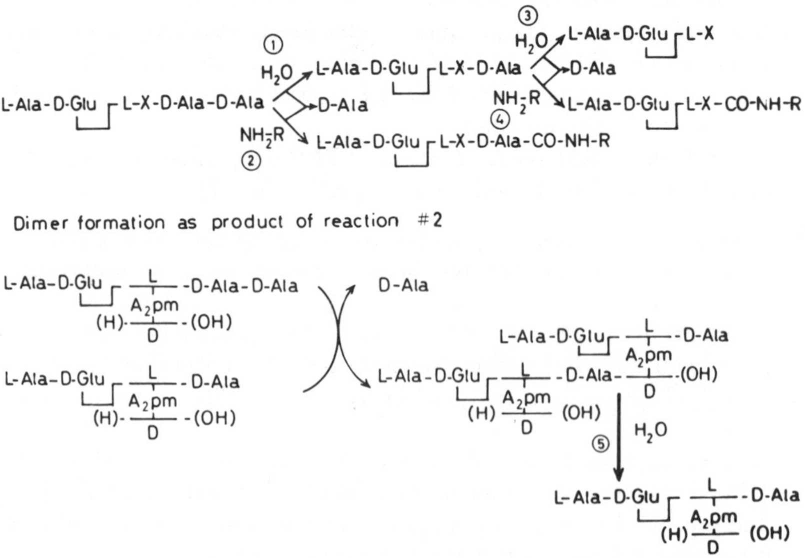

Figure 2 Schematic representation of the enzyme activities performed by the peptidoglycan cross-linking enzyme system. (*1*) DD-carboxypeptidase activity; (*2*) DD-transpeptidase activity; (*3*) LD-carboxypeptidase activity; (*4*) LD-transpeptidase activity; (*5*) DD-endopeptidase activity. In the natural substrates, the L-Ala residues substitute the glycan strands through *N*-acetylmuramyl-L-Ala linkages.

of similar size. Ac_2-L-Lys-D-Ala-D-Ala is an analogue of the natural carbonyl donor substrate. The general equation for the enzyme-catalyzed nucleophilic attacks (HY = the nucleophile) is Ac_2-L-Lys-D-Ala-D-Ala + HY → Ac_2-L-Lys-D-Ala-Y + D-Ala. With HY = H_2O (hydrolysis pathway), the reaction products are Ac_2-L-Lys-D-Ala and free D-Ala. With HY = NH_2-R (transpeptidation pathway), the reaction products are the transpeptidated derivative Ac_2-L-Lys-D-Ala-CONH-R and free D-Ala. In DD-carboxypeptidase assays, the turnover number of the R61 and R39 enzymes are high (3300 and 1050, respectively). In the presence of suitable amino nucleophiles, both enzymes also function as DD-transpeptidases; they catalyze transpeptidation reactions occurring concomitantly with the hydrolysis of the peptide donor (26–30). The G enzyme is a DD-carboxypeptidase that has a relatively low turnover number of 150. It does not function as a transpeptidase (31) but it is the only one of the three enzymes under consideration that exhibits a high endopeptidase activity and solubilizes isolated bacterial walls where the peptidoglycan subunits are cross-linked through C-terminal N^α-(D-Ala-D) linkages (32, 33) (reaction 5 in Figure 2).

Bacteria possess from three to eight or more membrane-bound PBPs. PBPs of low molecular weight (20,000–50,000) have been identified as PSEs. Like the exocellular PSEs, they act on simple carbonyl donor-nucleophilic acceptor systems. The possible enzymic functions of the PBPs of high molecular weight (50,000 to 100,000 or more) is unknown. The three following membrane-bound PBPs/PSEs are especially relevant to this work but others also receive attention.

1. *Streptomyces* R61 possesses a major PBP (or perhaps a triplet; M_r about 22,000) that has the following properties (34–37):

(*a*) It catalyzes nucleophilic attacks on Ac_2-L-Lys-D-Ala-D-Ala of suitable amino nucleophiles and functions primarily as a DD-transpeptidase.

(*b*) A temperature as low as – 30°C is necessary to prevent the membranes from catalyzing the reaction, which suggests that the enzyme functions in a lipid environment that remains remarkably fluid at low temperature.

(*c*) The enzyme can be solubilized with N-acetyl- or N-dodecyl-N,N,N-trimethyl-ammonium bromide or chloride. The solubilized and partially purified enzyme continues to function almost exclusively as a DD-transpeptidase but it is inactive in the frozen state.

(*d*) The enzyme is probably the lethal target of penicillins and cephalosporins in this bacterium.

2. *Streptococcus faecalis* ATCC 9790 possesses at least six PBPs (38). The 43,000 dalton PBP is a DD-carboxypeptidase-transpeptidase but its

ability to perform transpeptidation reactions in vitro (with Ac_2-L-Lys-D-Ala-D-Ala) is limited to simple amino nucleophiles such as D-amino acids and Gly-Gly (39). The enzyme can be solubilized with Genapol X-100 and has been highly purified (38). The isolated membranes also perform an LD-transpeptidase activity (Ac_2-L-Lys-D-Ala + NH_2-R → Ac_2-L-Lys-CONH-R + D-Ala) which is sensitive to various penicillins and Δ^3-cephalosporins (40, 41). This latter enzyme has not been isolated and its possible identification with one of the PBPs is unknown.

3. *Proteus mirabilis* strain 19 possesses six PBPs. It can grow as an unstable L-form in the presence of benzylpenicillin; the wall peptidoglycan made under these conditions is a round and fragile structure of apparently normal composition (42). Binding of [^{14}C]benzylpenicillin to the membranes isolated from these L-forms reveals the presence of PBPs 5 and 6. PBP5 (M_r: about 40,000) has been highly purified (43, 44). It is a DD-carboxypeptidase, active on L-Ala-D-Glu$_\Gamma$(L)-*meso*-A_2pm-(L)-D-Ala-D-Ala. Whether or not it also functions as a DD-transpeptidase remains to be established.

INTERACTION BETWEEN PSEs AND β-LACTAM ANTIBIOTICS

The three exocellular R61, R39, and G enzymes have been of particular interest in this work. They react with β-lactams according to the same scheme (24, 45, 46)

$$E + I \underset{}{\overset{K}{\rightleftharpoons}} EI \xrightarrow{k_3} EI^* \xrightarrow{k_4} E + \text{degraded antibiotic}$$

In the above equation E = enzyme and I = antibiotic. K is the dissociation constant of the stoichiometric complex EI; k_3 and k_4 are first order rate constants. The second and third steps are irreversible and the third step regenerates the free, active enzyme. Complex EI^* exhibits a rather high stability (low k_4); hence, enzyme regeneration is a slow or very slow process.

Formation of complex EI^* as a function of time is given by:

$$[EI^*]/[E_0] = k_a(1-e^{-(k_4 + k_a)t})/(k_4 + k_a) \text{ or}$$

$$t = -\ln(1-(k_4 + k_a)[EI^*]/k_a \cdot E_0)/(k_4 + k_a)$$

where $k_a = k_3/(1 + K/[I])$.

The time at which $[EI^*]/E_0$ is 95% of the value at the steady state is $t_{0.95} = 3/(k_4 + k_a)$. At the steady state (ss) $[EI^*]_{ss}/E_0 = k_a/(k_4 + k_a) = 1/(1 + (k_4/k_3) + k_4K/k_3[I])$ and the fraction of enzyme that remains functional under these conditions is given by $[E]_{ss}/E_0 = 1/(1 + ([I]/K) + k_3[I]/Kk_4)$.

Initial Binding and Activation

Table 1 gives the K, k_3 and k_4 values for the interactions between the three exocellular model enzymes and several penicillins and Δ^3-cephalosporins. (Because of technical difficulties, only the k_3/K ratios could be determined in some cases.) In no case is breakdown of the EI^* complex a rapid process.

The analysis of the K and k_3 values that govern the formation of complex EI^*, shows that:

(a) Whether the β-lactam belongs to the penam or 3-cephem series, and irrespective of the nature of the substituents on N_{14} or of the enzyme used, the K term is never low. The relative dimensions and conformations of the various groupings influence the strength of the initial binding.

(b) The side chains on N_{14} exert great influences on the k_3 term. Thus, for example, k_3 for the interaction between the R61 enzyme and benzylpenicillin is 180 sec^{-1} whereas the corresponding k_3 for 6-aminopenicillanic acid is 2×10^{-4} sec^{-1}; hence, removal of the N_{14} substituent results in a 10^6-fold decreased k_3 value.

(c) The effects of a given side chain vary greatly depending upon the enzyme.

The best interpretation of the above observations is that: (a) the enzymes appear to have a binding surface that is neither very efficient nor very

Table 1 Values of the constants involved in the interaction between the exocellular model PSEs and β-lactam antibiotics (at 37°C unless otherwise stated)

Enzyme	Antibiotic[a]	Formation of complex EI^*			Breakdown of complex EI^*	
		k_3/K (M^{-1} sec^{-1})	K (mM)	k_3 (sec^{-1})	k_4 (sec^{-1})	Half-life (min)
G (45)	phenoxymethylpenicillin	0.005	150	0.0008	9×10^{-5}	130
	cephalothin	0.06	9.5	0.0005	3.3×10^{-5}	350
	cephalosporin C	0.06	1.6	0.0001	8×10^{-5}	145
R61 (44)	6-aminopenicillanic acid[b]	0.2	1	0.0002		
	cephaloglycine	22	0.4	0.009	3×10^{-6}	3,800
	ampicillin	107	7.2	0.77	1.4×10^{-4}	82
	carbenicillin	820	0.11	0.09	1.4×10^{-4}	82
	cephalosporin C	1,150	>1	>1	1×10^{-6}	10,000
	phenoxymethylpenicillin	1,500	>1	>1	2.8×10^{-4}	40
	benzylpenicillin	13,700 (25°C)	13 (25°C)	180 (25°C)	1.4×10^{-4}	82
R39 (25)	cephalosporin C	67,000 (20°C)	0.19 (20°C)	12.5 (20°C)	0.3×10^{-6}	38,000
	carbenicillin	2,920 (20°C)			5.4×10^{-6}	2,125
	cephalexin	3,000 (20°C)			2.4×10^{-6}	4,800
	ampicillin	74,000 (20°C)			4.4×10^{-6}	2,600
	cephaloglycine	74,000 (20°C)			0.8×10^{-6}	14,000
	benzylpenicillin	>90,000 (20°C)			2.8×10^{-6}	4,100
	cephalosporin 87–312	3,000,000 (10°C)			1.5×10^{-6}	7,700

[a] For the structure of the antibiotics, see Figure 1. Cephalosporin 87–312 is [3(3, 4 dinitrostyryl)-(6R-7R)-7-(2-thienylacetamido)-cephem-3-em-4-carboxylic acid, E-isomer (47).
[b] Unpublished data.

selective; (b) the penam or 3-cephem nuclei are the main portion of the β-lactam antibiotics concerned with the initial binding; and (c) both efficiency and selectivity of β-lactam action are, primarily, the consequence of efficiency and selectivity in the k_3 term.

The sequence of events may be visualized as that suggested by Rando (48). Once the β-lactam is bound to the enzyme, the substituent on a correctly positioned N_{14} (so that ϕ_2 falls inside region A) functions as the handle through which the enzyme distorts the antibiotic from its already strained ground state structure, greatly increasing the chemical reactivity of the β-lactam amide bond. As the β-lactam ring rapidly opens, the strain is relieved and acylation of a nearby enzyme side chain is completed. The free COOH grouping of the fused thiazolidine or dihydrothiazine rings may also be part of the handle or may function as an electron attracting group or both. According to this model, the k_3 term expresses the efficiency with which the β-lactam is converted by the enzyme target into a very reactive form which in turn rapidly acylates the enzyme.

Enzyme Acylation

Serine is one of the enzyme residues with which benzylpenicillin collides as a result of its binding to and activation by the R61 enzyme (49). (This enzyme possesses 29 serine residues.) Nucleophilic attack occurs on C_7 of the bound antibiotic molecule and a benzylpenicilloyl-serine ester linkage is formed (Figure 3). The process (i.e. formation of complex EI^*) causes conformational changes to the enzyme as shown by a decrease of its fluorescence and by an extensive alteration of its CD spectrum in the near ultra violet (23). Phenylmethanesulfonyl fluoride has no effect on the ability of the R61 enzyme to bind penicillin (49) but diisopropylphosphofluoridate prevents benzylpenicillin binding; (P. Charlier and J. M. Frere, unpublished). Moreover, on the basis of the effects of 2,4-dinitrofluorobenzene, O-methylisourea, and the fused ring trimer of 2,3-butanedione, benzylpenicillin binding also requires one free ϵ-amino group of lysine, (P. Charlier and J.-M. Frère, unpublished). (The R61 enzyme possesses 6 lysine residues.)

When bound to the R61 or R39 enzymes in the form of their EI^* complexes, cephalexin, cephalosporin C, and cephaloglycine have their molar extinction coefficients at 260 nm decreased to an extent identical to that obtained after β-lactamase action [(50) and J.-M. Frère, unpublished]. When bound to the R39 enzyme, the chromogenic cephalosporin 87–312 has a $\epsilon_{482}/\epsilon_{386}$ ratio of 2.40 which is identical to that obtained after β-lactamase action (51). With the R61 enzyme, however, the $\epsilon_{482}/\epsilon_{386}$ ratio value is only 1.20 (50) (it becomes 2.40 through breakdown of complex EI^*). This blue shift may indicate differences between the two EI^* complexes with regard to the electrostatic and/or the conformational changes involved.

With those Δ^3-cephalosporins that have a 3'-acetoxy side chain (Figure 1), opening of the β-lactam ring may cause further alterations of the bound molecule. Migration of the electron density to the ester oxygen of the CH_2OAc on C_3 may result in the release of acetate and formation of an exocyclic methylene group. The tendency of the 3 substituents to attract or accept electrons from the 3-cephem nucleus should make the β-lactam carbonyl a stronger acylating agent (52). Cephalothin or cephaloglycine which have a 3'-acetoxy side chain might function this way. However, cephalexin (Figure 1) with its acaudal 3-methyl side chain cannot. Hence, both the chemical reactivity of the β-lactam ring in its ground state conformation, and the increased reactivity gained by the β-lactam as a result of

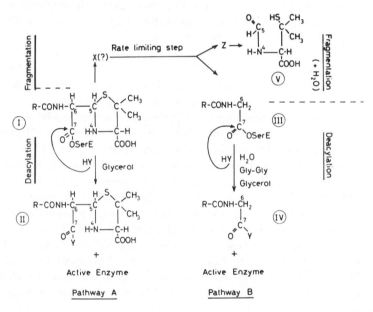

Figure 3   Breakdown of complex *EI** formed between penicillin and the R61 enzyme showing direct nucleophilic attack of the penicilloyl-ester linkage (*Pathway A*) and fragmentation of the penicilloyl moiety and attack of the acylglycyl ester linkage (*Pathway B*). (*I*) penicilloyl-enzyme (or *EI**) complex with the serine ester linkage on C_7 (R = C_6H_5-CH_2 or C_6H_5-O-CH_2); (*II*) the reaction product resulting from the direct nucleophilic attack of *I*. With the R61 enzyme, glycerol is the only functional reagent; hence *II* is the penicilloyl ester of glycerol; (*III*) acylglycyl-enzyme complex; (*IV*) the reaction product resulting from the nucleophilic attack of *III*, i.e. acylglycine if HY is H_2O; acyl-Gly-Gly-Gly if HY is Gly-Gly, and the acylglycyl ester of glycerol if HY is glycerol; (*X*) a hypothetical modified complex *EI** (see text); (*Z*) the primary product that arises from thiazolidine (see text and Figure 5); (*V*) *N*-formyl-D-penicillamine.

its binding to the enzyme, dictate the effectiveness of the acylation step. The virtual inactivity of the Δ^2-cephalosporins, which have less ring strain than the Δ^3-cephalosporins, probably results from the height of the barrier to reaction with the enzyme (53).

Nucleophilic Attack

Deacylation of the benzylpenicilloyl and phenoxymethylpenicilloyl-R61 enzyme complexes has received much attention. To reject the bound penicilloyl moiety and consequently to recover its initial conformation and activities, the R61 enzyme has developed two possible mechanisms (Figure 3). Neither of them is a rapid process (low k_4).

In Figure 3, *Pathway A* is a direct nucleophilic attack of the serine ester linkage resulting in the transfer of the penicilloyl moiety to the nucleophile HY and the regeneration of a free active enzyme. In *Pathway B,* the penicilloyl moiety is first transformed by C_5–C_6 cleavage into an N-acylglycyl residue (54–56) which remains linked to the enzyme, and a compound Z which in turn gives rise to free N-formyl-D-penicillamine. Z has a half-life of 10–15 min at 37°C and a neutral pH. As shown by isotopic studies with D_2O (56), the fragmentation reaction involves a rate-limiting step that is immediately followed by C_5–C_6 cleavage, protonation (or deuteration) of C_6, transfer of the acylglycyl moiety to the nucleophile, and regeneration of an active enzyme. Since the formation of free N-formyl-D-penicillamine is delayed when compared with the release of acylglycine, it follows that the processing of Z to N-formyl-D-penicillamine is irrelevant to the regeneration of the enzyme activity.

The various sites that in their ground state conformations form the β-lactam active center provide the R61 enzyme with a rather loose binding surface (Figure 4, site I). Sites II–VI in Figure 4, fulfill more selective functions. A revealing feature of *Pathway B* in Figure 3 is that C_5–C_6 cleavage is immediately followed by the release of the acylglycyl moiety. Thus, the stability of the ester bond between the penicilloyl moiety and the enzyme fixation site III (Figure 4) can be attributed to the monocyclic thiazolidine part of the molecule which by interacting with some stabilizing site IV probably confers to the penicilloyl-enzyme complex a conformation that is not favorable to nucleophilic attack. As the fragmentation step catalyzed by site V is completed, the catalytic groups of the releasing site VI undergo the correct alignment, and catalyze the immediate attack of the serine ester bond.

Figure 5 shows a hypothetical mechanism for the fragmentation step. The delayed release of N-formyl-D-penicillamine is tentatively attributed to the interaction of the primary fragmentation product Z with an enzyme amino group.

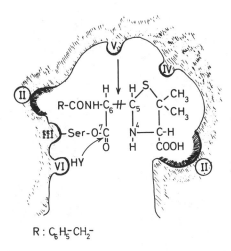

$R : C_6H_5\text{-}CH_2\text{-}$

Figure 4 The penicillin-active center of the R61 enzyme. The enzyme binding surface (site *1*; not shown in the figure) is supposed to be a mosaic work of sites *II–VI*. Whether binding involves the α or β phases of the β-lactam fused ring system is not known. (*II*): activation sites : interaction with the substituents on N_{14} and the carboxyl grouping; (*III*): fixation or acylation site : nucleophilic attack of C_7 by a serine residue; (*IV*): stabilization site : interaction with the monocyclic thiazolidine part of the molecule; (*V*): fragmentation site catalyzing C_5–C_6 cleavage; (*VI*): releasing site catalyzing the attack of the serine ester linkage by a suitable nucleophile HY. Depending upon the PSEs and the nucleophile HY, *Pathways A* and/or *B* may occur (Figure 3). Although *A* has fewer steps than *B*, the overall rate of the former, where fragmentation does not occur, need not be faster.

Channelling of the enzyme-bound penicilloyl moiety through *Pathways A* and/or *B* (Figure 3) depends upon the exogenously available nucleophile. With the R61 enzyme and in water, *B* only is operational (54–56): phenoxymethyl- and benzylpenicillin are quantitatively fragmented into acylglycine and *N*-formyl-D-penicillamine. The effects of nucleophiles other than water are necessarily studied in aqueous media, i.e. under competition conditions (59). With Gly-Gly and other NH$_2$-R nucleophiles, the reaction also proceeds exclusively through *Pathway B:* acylglycine (from H$_2$O), acyl-Gly-Gly-Gly (or the corresponding acyl-Gly-Gly-CONH-R derivatives) and *N*-formyl-D-penicillamine are the end products (59). In a 20% solution of glycerol both *A* and *B* (Figure 3) occur. The reaction products are the penicilloyl ester of glycerol from *A,* and acylglycine, the acylglycyl ester of glycerol, and *N*-formyl-D-penicillamine from *B*. Attack on the penicilloyl moiety by glycerol through *Pathway A* is rather rapid but ceases after a short time (10–15 min). This observation suggests that the rate-limiting step of penicillin fragmentation is preceded by a faster reaction through which an intermediate (*X* in Figure 3) is formed which escapes

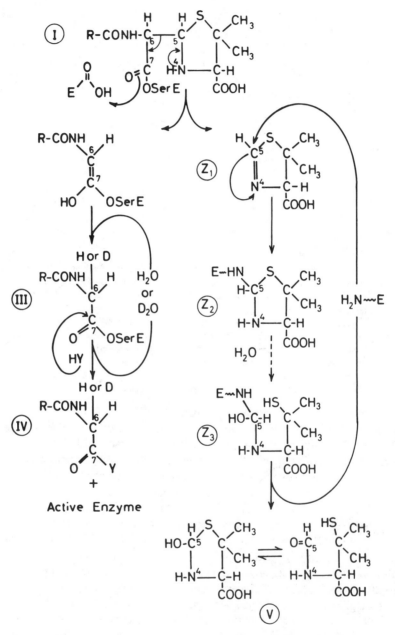

Figure 5 Hypothetical mechanism for the fragmentation of the penicilloyl-R61 enzyme complex *EI** (*Pathway B* in Figure 3). (*I*), (*III*), (*IV*), and (*V*) see Figure 3 except that here N-formyl-D-penicillamine, i.e. *V*, is represented in equilibrium with hydroxythiazolidine.

nucleophilic attack by glycerol. When fragmentation is achieved, attack of the acylglycyl moiety by both H_2O and glycerol occurs in a competitive manner and the partitioning between the two nucleophiles present does not alter the total rate of transfer.

The attacks on the enzyme-bound penicilloyl or enzyme-bound acylglycyl moieties by the exogenous nucleophile HY are directed against the same C_7 of the original penicillin molecule (Figures 3 and 4). The two processes exhibit great differences in their specificity profiles, which suggests that they are catalyzed by distinct enzyme sites. However, these sites may be different conformations containing the same grouping of amino acid residues, the change in conformation being induced by the fragmentation process (59). Finally, denatured complex EI^* also breaks down but an active enzyme is not regenerated, and in water, penicilloate is very slowly released (59).

The two mechanisms described above for the deacylation of complex EI^* formed between the R61 enzyme and benzylpenicillin are probably general but the channelling through either *Pathway A or B* (or both) appears to be the result of a complex interplay between the three reagents involved: the enzyme, the nucleophile and, most likely, the β-lactam itself. Thus in water, the R39 enzyme (54) fragments penicillin with formation of acylglycine (as observed with the R61 enzyme) but the G enzyme (46) catalyzes the hydrolysis of penicillin to penicilloate.

Membrane-Bound PSEs

Although it is not impossible that a suitable amino acid other than serine (for example, cysteine) might be involved in the initial attack of the β-lactam, membrane-bound PSEs essentially behave like the exocellular enzymes and give rise to rather stable EI^* complexes. Depending upon the β-lactam antibiotics, the k_3/K values range from 3–400 M^{-1} sec $^{-1}$ with the

---

$Z_1 \rightarrow V$ represents the enzyme-catalyzed processing of 5,5-dimethyl-Δ^2-thiazoline-4-carboxylate. Pathway $Z_1 \rightarrow V$ would proceed with a half-life of about 10–15 min (at 37°C and neutral pH). The proposed mechanism rests upon the following observations : (*a*) Isotopic studies with D_2O show that the hydrogen or deuterium atom that is fixed on C_6 comes from H_2O or D_2O (56); (*b*) Benzylpenicillin methylester can be degraded nonenzymatically to methyl 5,5,-dimethyl-Δ^2-thiazoline-4-carboxylate in trifluoroacetic acid. The phenylacetylglycyl fragment was isolated by conversion to its *N*-benzyl amide (57). Enzymatically, however, free 5,5-dimethyl-Δ^2-thiazoline-4-carboxylate is not the primary product released from the thiazolidine part of penicillin (58). Evidence suggests the existence of an enzyme site that interacts with the monocyclic thiazolidine part of the bound benzylpenicilloyl molecule (see text). However, the proposed mechanism implies that the processing $Z_1 \rightarrow V$ occurs on enzyme sites unrelated to those involved in DD-carboxypeptidase-transpeptidase activity.

Streptomyces membrane-bound DD-transpeptidase (34, 36) and from 1–1200 M^{-1} sec $^{-1}$ with the purified *S. faecalis* enzyme (38, 39). With this latter enzyme and benzylpenicillin, *EI** formation is a two-step process with K = 0.025 mM and k_3 = 0.025 sec $^{-1}$ (k_4 = 2.8 × 10^{-5} sec $^{-1}$) (38). A similar mechanism has been proposed for the DD-carboxypeptidase of *B. subtilis* (60); with all the antibiotics tested, the K values are high, ranging from 0.1 mM to more than 10 mM. Finally, PBPs of high molecular weight and of unknown enzyme activity also form with [^{14}C]benzylpenicillin radioactive complexes that subsequently decay more or less rapidly (5, 61, 62).

With the *S. faecalis* enzyme (39) and the DD-carboxypeptidases from various bacilli (63, 64), breakdown of the complex *EI** formed with penicillin proceeds in water via the fragmentation pathway and acylglycine is produced. Incidentally, fixation of penicillin to the *B. subtilis* enzyme involves the formation of a penicilloyl-ester bond and the hydroxylaminolysis of the complex is catalyzed enzymatically (65). In contrast to the above enzymes, the 46,000 dalton PBP (a DD-carboxypeptidase-transpeptidase) of *S. aureus* (66) and the enzyme of the L-form of *Proteus* (67) catalyze in water the hydrolysis of penicillin to penicilloate. They are thus β-lactamases of low efficiency. PSEs of this type may have served as protoenzymes in the evolution of true β-lactamases. True β-lactamases have tremendously increased turnover numbers on β-lactam antibiotics but they have lost the ability to react with DD-carboxypeptidase-transpeptidase substrates. The *Streptomyces* DD-transpeptidase, when it is membrane-bound, is also a penicillinase of this type (34) but once it has been solubilized with the help of the cationic detergent, the same enzyme catalyzes penicillin fragmentation (36).

INTERACTION BETWEEN PSEs AND L-R-D-ALA-D-ALA-TERMINATED SUBSTRATE ANALOGUES

The natural substrates that undergo the nucleophilic attacks catalyzed by the PSEs do not end in an N-substituted β-lactam-fused ring system but in an N-substituted L-R-D-Ala-D-Ala sequence. Tipper & Strominger (68) had proposed that penicillin would be isosteric with the C-terminal D-Ala-D-Ala portion of the natural substrate and thus would function as an affinity labelling agent. However, the dihedral angles ω_2 in penicillins and Δ^3-cephalosporins (135 and 155°) in their ground state conformations are considerably smaller than that about the peptide bond (180°), and, in addition, the

$$O = C \overset{N\,=}{\underset{C\,\leq}{\diagup}}$$

bond angle of the β-lactam is 90.5°, whereas the corresponding bond angle in the dipeptide is 117° (Figure 6). Alternatively, it was proposed (68, 69) that penicillin would be isosteric with the transition state structure of the D-Ala-D-Ala portion of the natural substrate. On the basis of steric maps of D-Ala-D-Ala generated according to the ω_2 angle of penicillins and Δ^3-cephalosporins, β-lactams have conformations similar to the distorted dipeptide (12). Hence, according to this modified model, enzyme action on the natural substrate or its analogues should involve the distortion of the

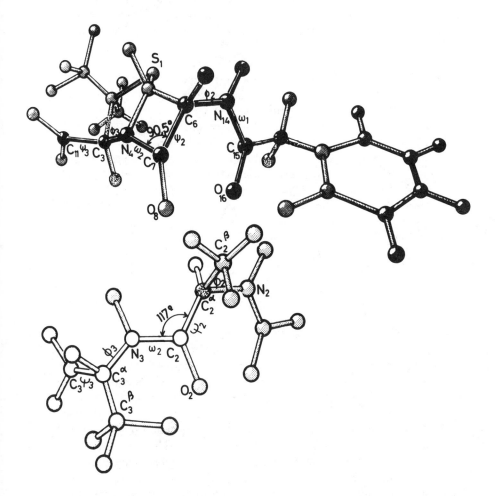

Figure 6   Ground state conformation of benzylpenicillin and its comparison with D-alanyl-D-alanine. [Adapted from (48) with the permission of Pergamon, Oxford.]

D-Ala-D-Ala peptide bond to a pyramidal shape. Such a mechanism seems to be very unlikely (48). Finally, to add confusion to the picture, it is worth mentioning that the LD-transpeptidase of *S. faecalis* (40) and the LD-carboxypeptidase of *Gaffkya homari* (70) are penicillin-sensitive.

Hydrolysis Pathway

The substrate requirements of the R39, R61, and G enzymes have been characterized by measuring the effects that each residue in the standard tripeptide Ac$_2$-L-Lys-D-Ala-D-Ala exerts on the K_m and V_{max} parameters, respectively, for the release of the C-terminal residue in water (hydrolysis pathways) (33, 71–73). These observations were made (Table 2). (*a*) The occurrence of L-Ala instead of D-Ala at the C-terminal or the penultimate positions and the occurrence of D-amino acids other than D-Ala at the penultimate position change the peptide into a nonsubstrate. The enzymes have a less strict specificity for a D-Ala residue at the C-terminal position where it can be replaced by other D-amino acids, although most often at the expense of substrate activity. Substitution of D-Ala by Gly at either one of the two positions under consideration also causes a more or less pronounced decrease of substrate activity; (*b*) Shortening the neutral side chain of the L-residue that precedes D-Ala-D-Ala (i.e. replacement of Ac$_2$-L-Lys by Ac$_2$-L-A$_2$bu, Ac-L-Hse, Ac-L-Ala, and Ac-Gly) causes a drastic decrease of substrate activity. Succinylation instead of acetylation of the L-lysine tripeptide, lack of substitution of the ε-amino group, or substitution with other charged groups may increase, decrease, or suppress substrate activity; (*c*) K_m values of good substrates are not much different and are often as high as those of poor substrates but good substrates have much higher V_{max} values. Hence, efficiency in substrate activity is primarily the consequence of efficiency in the V_{max} term; (*d*) Nonsubstrate peptides or peptides exhibiting very low substrate activity may act as inhibitors of the G enzyme (as measured on Ac$_2$-L-Lys-D-Ala-D-Ala). Ac-D-Ala-D-Ala is a weak inhibitor but Ac-Gly-D-Ala-D-Glu is a good inhibitor. Since Ac$_2$-L-Lys-D-Ala-D-Glu is a fair substrate, it follows that lack of the acetylated side chain of L-lysine can change a good substrate into a nonsubstrate inhibitory peptide. Ac$_2$-L-Lys-D-Glu-D-Ala is also a good inhibitor, whereas Ac$_2$-L-Lys-D-Ala-D-Ala is, of course, an excellent substrate; hence, replacement of the penultimate D-Ala whose carbonyl carbon undergoes nucleophilic attack, by another D-amino acid residue such as D-Glu can also change a substrate into an inhibitor. Inhibition of the R61 enzyme is usually smaller than that observed with the G enzyme, and none of the peptide inhibitors of the G and R61 enzymes have any effect on the R39 enzyme (in fact some of them are fair substrates of this latter enzyme). Obviously, the substrate, nonsub-

Table 2 Substrate activity of LDD terminated peptides for the excellular model PSEs, (hydrolysis pathway)[a]

Peptide	G enzyme K_m[b]	G V_{max}[c]	G Eff.[d]	G Act.[e] (%)	R61 K_m[b]	R61 V_{max}[c]	R61 Eff.[d]	R61 Act.[e] (%)	R39 K_m[b]	R39 V_{max}[c]	R39 Eff.[d]	R39 Act.[e] (%)
1. Ac_2-L-Lys-D-Ala-D-Ala	0.33	100	300	100	12	890	72	100	0.8	330	410	100
Ac_2-L-Lys-D-Ala-D-Lys	0.80	85	106		13	90	7					
Ac_2-L-Lys-D-Ala-D-Leu	0.33	33	100		10	50	5	10	0.7	230	320	
Ac_2-L-Lys-D-Ala-D-Glu				100	36	200	6	0				
Ac_2-L-Lys-D-Ala-Gly	2.5	60	24						2.5	100	40	
Ac_2-L-Lys-D-Ala-L-Ala				0				0				0
2. Ac_2-L-Lys-Gly-D-Ala	15.0	107	7		15	1.7	0.1					
Ac_2-L-Lys-D-Leu-D-Ala				0	10	10	1					
Ac_2-L-Lys-D-Glu-D-Ala				0				0				0
Ac-L-Lys-L-Ala-D-Ala				0				0				0
3. Ac-L-A_2bu-D-Ala-D-Ala	0.6	42	70					85				16
R_1-L-Hse-D-Ala-D-Ala[f]	1.0	16	16					3				7
Ac-L-Ala-D-Ala-D-Ala	3.3	0.7	0.2					1.4				0.2
Ac-Gly-D-Ala-D-Ala	1.1	1.9	2									
Ac-D-Ala-D-Ala	—							1				5
4. N^α-Ac-L-Lys-D-Ala-D-Ala	6	20	3	0	15	4	0.3		0.2	600	3,000	
R_2-(L)-meso-A_2pm-(L)-D-Ala-D-Ala[f]												
R_3-L-Lys-D-Ala-D-Ala[f]	0.4	10	25		11	8	0.3		0.25	400	1,600	
$(Gly)_5$	0.28	9	32		14	800	57		0.30	420	1,400	
Suc_2-L-Lys-D-Ala-D-Ala				5				16				36

a From (33, also 71–73). The experiments reported here were carried out at a time when none of the enzymes had been purified to homogeneity.
b K_m values are expressed in mM.
c V_{max} values are in μmoles of peptide hydrolyzed/mg protein/h.
d Eff. (efficiency) = V_{max}/K_m.
e Act. = substrate activity, simply expressed as a percentage of that of the standard tripeptide Ac_2-L-Lys-D-Ala-D-Ala.
f R_1: UDP-MurNAc-L-Ala-Gly-D-Glu⌐; R_2: UDP-MurNAc-D-Glu⌐; R_3: β-1,4 GlcNAc-MurNAc-L-Ala-D-Glu(amide)⌐.

strate, and inhibitor activities of LDD-terminated peptides result from a complex interplay between the enzyme and each of the three amino acid residues under consideration.

The most general equation for the interaction between PSEs and L-R-D-Ala-D-Ala-terminated substrates is

$$E + D \underset{k_2}{\overset{k_1}{\rightleftharpoons}} ED \xrightarrow{k_3} EP \xrightarrow{k_4} E + P,$$

where D is the carbonyl donor and P is the reaction product. If k_4 is very high (a likely assumption at least with the exocellular PSEs and good substrates as shown by the high turnover numbers), then $V_{max} = k_3 \times [E_0]$; if, in addition, $k_2 \gg k_3$, then $K_m = (k_3 + k_2)/k_1$ reduces to $K_m = k_2/k_1$ and becomes equal to the dissociation constant K of the reversible complex ED. Hence, as a first approximation the above observations suggest that (a) the enzymes have a binding surface that is neither very efficient nor very selective (high K_m or K); (b) the C-terminal D-Ala-D-Ala dipeptide is the main portion of the molecule concerned with the initial binding; and (c) a side chain of very definite molecular characteristics is required in the L residue that precedes D-Ala-D-Ala to induce enzymic action on the bound peptide (high V_{max} or k_3). In other words, an L residue whose lateral chain has the appropriate size, shape, and charge may be regarded as the handle through which the peptide, once bound to the enzyme, induces a correct alignment of the catalytic groups which in turn forces the amide bond of the D-Ala-D-Ala portion to adopt a configuration intermediate between *cis* and *trans* and to undergo rapid nucleophilic attack. With poor substrates, the induced change in the enzyme would be unfavorable or incorrect, markedly decreasing the V_{max} (or k_3) value or even preventing the enzyme action (73).

The above model is analogous to that proposed for the interaction with β-lactam antibiotics; (a) the D-Ala-D-Ala and the penam or 3-cephem nuclei, respectively, would be mainly involved in the initial binding. However, the corresponding ground state conformations exhibit little isosterism; and (b) the L residue that precedes D-Ala-D-Ala and the substituents on N_{14} of the β-lactam ring, respectively, would be mainly responsible for the activation of the relevant bound molecules. These side chains, however, are not structurally related.

The membrane-bound DD-carboxypeptidase of *S. faecalis* (40) and the enzyme that in the membrane-wall system of *Gaffkya homari* (74) catalyzes the cross-linking stage of peptidoglycan synthesis (presumably the transpeptidase) also discriminate between the residues in the C-terminal position of the peptide donor with a high specificity for a D-Ala-D-Ala sequence.

Whether the exocellular model enzymes function as catalyst templates or form Ac$_2$-L-Lys-D-Ala-enzyme intermediates remains to be established. PSEs isolated from membranes of various bacilli, *Escherichia coli,* and *S. aureus,* are transitorily acylated by the Ac$_2$-L-Lys-D-Ala moiety (66, 75, 76). By using trapping procedures, acyl enzyme intermediates can be isolated by Sephadex filtration (76) and/or detected after SDS polyacrylamide gel electrophoresis and fluorography (66). Enzyme acylation is probably mediated through an ester linkage (76). On the basis of these observations, Strominger and his colleagues have proposed a Ping Pong Bi Bi mechanism. An Ac$_2$-L-Lys-D-Ala-enzyme intermediate (*ED**) would be formed in the first step of the reaction and the Ac$_2$-L-Lys-D-Ala moiety (*D**) would be transferred to a suitable nucleophile in the second step:

$$E + D \xrightleftharpoons{K} ED \xrightarrow{k_3} \begin{array}{l} ED^* + HY \xrightarrow{k_4} D^*Y + E \\ \text{D-Ala} \end{array}$$

The release of D-Ala precedes the nucleophilic attack. With two competing nucleophiles (such as H$_2$O and NH$_2$-R), partitioning occurs at the level of complex *ED**.

Transpeptidation Pathway

Some PSEs utilize either H$_2$O or amino compounds as nucleophilic reagents. They behave exclusively as DD-carboxypeptidases or DD-transpeptidases. Many PSEs, however, catalyze concomitant hydrolysis and transpeptidation reactions, the two pathways competing with each other. Their transpeptidation abilities may be limited to simple amino nucleophiles but several PSEs are known that catalyze much more complex reactions. The DD-carboxypeptidases of *E. coli* (77–79), the PBPs 4 and 5 of *Salmonella typhimurium* (61), and both R61 (26, 27, 30, 80) and R39 (26, 28, 29) enzymes perform cross-linking reactions leading to the formation of peptide dimers. Such PSEs must possess sites especially devised to operate with complex amino nucleophiles.

The exocellular model enzymes have revealed interesting features:

1. The profiles of the R61 and R39 enzymes for amino nucleophiles closely reflect the type of cross-linkage that exists in the peptidoglycan of the strain that produces the enzyme. In strain R39, the cross-link is from D-Ala to the amino group at the D-center of *meso*-diaminopimelic acid (28). Correspondingly, suitable amino groups must be in α-position to the carboxyl group of glycine or a D-amino acid (26). In strain R61, on the other hand, the cross-link is between D-Ala and a glycine residue that is attached to the ε-amino group of LL-diamino pimelic acid (81). In this case, glycine

and various peptides with an N-terminal glycine residue are efficient nucleophiles, although other amino compounds such as ω-amino acids, amino-
hexuronic acids, D-cycloserine, and 6-aminopenicillanic acid can also
function (26). In addition, the R39 enzyme catalyzes dimerization reactions
such as that shown in Figure 2 (29), and the R61 enzyme transforms the
tetrapeptide monomer

Ac-L-Lys-D-Ala-D-Ala
Gly ⏌

into the hexa- and heptapeptides,

Ac-L-Lys-D-Ala-(D-Ala)
Ac-L-Lys-D-Ala-Gly⏌
Gly⏌

(30, 80). The dimers thus made in vitro are identical or very similar to those
formed in vivo during peptidoglycan synthesis in the corresponding bacteria.

2. The amino nucleophiles act as modulators of the enzyme activities.
The effects of Ac$_2$-L-Lys-D-Ala-D-Ala and amino nucleophile concentrations on the rates of hydrolysis and transpeptidation suggest, for the
transpeptidation reaction, an ordered pathway mechanism in which the
amino nucleophile NH$_2$-R binds first to the R61 enzyme (27, 82)

$$E \overset{NH_2-R}{\underset{H_2O}{<}} \quad \begin{matrix} E \cdot NH_2\text{-}R \overset{D}{\longrightarrow} E \cdot NH_2\text{-}R \cdot D & \rightarrow \text{transpeptidation} \\ E \cdot H_2O \overset{D}{\longrightarrow} E \cdot H_2O \cdot D \longrightarrow \text{hydrolysis} \end{matrix}$$

In such a mechanism, hydrolysis is decreased (by competition) whereas the
overall reaction (hydrolysis + transpeptidation) may be unchanged, increased, or decreased. With the R61 enzyme and simple amino nucleophiles
(such as *meso*-diaminopimelic acid), the overall reaction is unchanged and
the transpeptidation/hydrolysis ratio is proportional to the nucleophile
concentration. However, at high concentrations of complex nucleophiles
(related to peptidoglycan structure), transpeptidation is also inhibited and
the transpeptidation/hydrolysis ratio is smaller than expected, which suggests that a second molecule of the amino nucleophile is fixed on the
enzyme, yielding a nonproductive quaternary complex E. D. (NH$_2$-R)$_2$.
Features of complex amino nucleophiles also exert profound effects on the
relative amounts of transpeptidation and hydrolysis catalyzed by the R39
enzyme (28, 29). Thus, when increasing concentrations of the amidated
tetrapeptide L-Ala-D-Glu (amide)─(L)-*meso*-A$_2$pm-(L)-D-Ala are provided as nucleophile to the R39 enzyme with Ac$_2$-L-Lys-D-Ala-D-Ala,

hydrolysis is progressively inhibited and transpeptidation rises to a maximum at a definite tetrapeptide concentration; at higher concentrations, both transpeptidation and hydrolysis are progressively inhibited until eventually the tripeptide donor Ac_2-L-Lys-D-Ala-D-Ala remains unused (Figure 7). Excess of amidated tetrapeptide can thus freeze the enzyme. The phenomenon is dependent on the α-amide group of the D-Glu residue, since high concentrations of the same nonamidated L-Ala-D-Glu┌-(L)-*meso*-A_2pm-

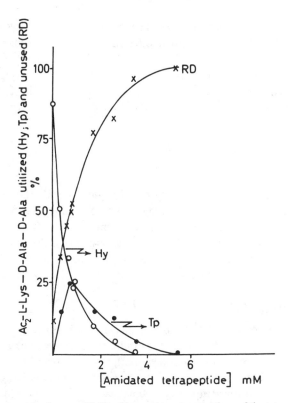

Figure 7 Competition between H_2O and increasing concentrations of the tetrapeptide L-Ala-D-Glu(amide)┐-(L)-A_2pm-(L)-D-Ala for the nucleophilic attack of Ac_2-L-Lys-D-Ala-D-Ala (0.27 mM). Hy : hydrolysis (formation of Ac_2-L-Lys-D-Ala); Tp : transpeptidation, i.e. formation of

$$\text{L-Ala-D-Glu(amide)} \underline{} \begin{array}{c} \text{L} \\ \overline{} \\ A_2\text{pm} \\ \end{array} \text{-D-Ala}$$
$$Ac_2\text{-L-Lys-D-Ala-} \overline{\underset{D}{}}$$

RD : residual Ac_2-L-Lys-D-Ala-D-Ala. Note that at 5.4 mM of amidated tetrapeptide, the enzyme is completely inactive. [Adapted from (29)].

(L)-D-Ala tetrapeptide limits the hydrolysis of Ac$_2$-L-Lys-DAla-D-Ala but does not decrease the amount of transpeptidation product formed. The phenomenon is also dependent on the degree of saturation of the enzyme by the carbonyl donor peptide; the less saturated the enzyme, the lower is the concentration of amidated tetrapeptide required to inhibit transpeptidation as well as hydrolysis. The complex interplay between enzyme, D-Ala-D-Ala-terminated substrates and nucleophilic acceptor that the above experiments reveal is not accounted for by a simple acylation and deacylation model.

3. Whether the substrate is a β-lactam or an L-R-D-Ala-D-Ala terminated peptide, PSEs exhibit great differences in their specificities for the nucleophilic reagents. Thus, glycerol (a 20% solution) is an excellent nucleophilic reagent for the attack of the penicilloyl- and acylglycyl- R61 enzyme complexes, but it is a very poor acceptor for the R61 enzyme-catalyzed transfer of Ac$_2$-L-Lys-D-Ala from Ac$_2$-L-Lys-D-Ala-D-Ala (59). The tetrapeptide

Ac-L-Lys-D-Ala-D-Ala
Gly—⌐

is not utilized as a nucleophile by the R61 enzyme for the attack of the penicilloyl or acylglycyl ester linkages. The *Streptomyces* DD-transpeptidase utilizes water to degrade penicillin to penicilloate or to fragment it with formation of free acylglycine but the same enzyme utilizes preferentially, if not exclusively, amino compounds for the attack of Ac$_2$-L-Lys-D-Ala-D-Ala.

β-Lactams as Inhibitors of the PSEs

L-R-D-Ala-D-Ala-terminated substrates (D) inhibit the formation of complex EI^*; reciprocally, β-lactams (I) inhibit the processing of D both in the carboxypeptidase and transpeptidase pathways. Mathematical treatments giving the concentration of complex EI^* at time t in the presence of D, and the amount of D utilized at time t in the presence of I, can be found in (83). For very large values of t, the phenomenon looks like a competition between two substrates for the same enzyme. The question therefore arises whether D and I are processed on the same enzyme center (in which case the inhibition is competitive and the ternary complex EID is not formed) or on two distinct enzyme centers (in which case the inhibition is noncompetitive and the ternary complex EID is formed). In many cases, kinetics suggest a competitive inhibition. However, because of the very rapid formation of a stable complex EI^*, the concentration of EI, and of EID, may be so small that a seemingly competitive inhibition may be obtained even if the interaction is truly noncompetitive (83). However, with the enzyme of the L-form of *Proteus,* in spite of a very rapid formation of complex EI^* ($k_3/K = 2$–$8 \times 10^4 M^{-1} sec^{-1}$) benzylpenicillin inhibits the hydrolysis of UDP-N-acetylmuramyl-pentapeptide in a noncompetitive manner, with a dissociation constant K'

$$EI + D \overset{K'}{\rightleftharpoons} EID$$

lower than 3 μM (67). The interactions between the R61 enzyme and 6-aminopenicillanic acid or between the G enzyme and cephalosporin C are characterized by a very low k_3 value (10^{-4} sec $^{-1}$; Table 1) so that the inhibition of the enzyme activity by the β-lactams can be studied under conditions where it is due almost exclusively to the formation of complex EI, since complex EI^* is virtually not formed. With the R61 enzyme, the inhibition is competitive (the K value is about 1 mM; J. M. Frere, unpublished); with the G enzyme, the inhibition is noncompetitive (49) and the dissociation constants of the various complexes formed ($E + D \rightleftharpoons ED$; $E + I \rightleftharpoons EI$; $EI + D \rightleftharpoons EID$; $ED + I \rightleftharpoons EID$) exhibit similar values of about 0.5 mM. The ternary complexes EID which occur with the G and *Proteus* enzymes are dead ends at least with regard to substrate D.

To be a good inhibitor of the enzyme activity on L-R-D-Ala-D-Ala-terminated substrates, a β-lactam must be able to immobilize a large proportion of the enzyme in the form of complex EI (very low K) or of complex EI^* (high k_3 and low k_4). Although the first alternative may occur, with all those systems where inhibition is observed at low [I] values, in fact [I] is much smaller than K and therefore inhibition is mainly due to the immobilization of the enzyme in the form of complex EI^*. Under these conditions, the concentration of free enzyme at the steady state is a function of [I], k_3/K, and k_4. The higher the k_3/K and the lower the k_4, the better is the β-lactam as an inhibitor. For a given k_3/K, the higher the k_4, the better is the antibiotic as a substrate and the poorer it is as an inhibitor. At equal values of K, intrinsic resistance to β-lactams may be caused by different mechanisms. The main cause of resistance of the G enzyme is a very low k_3 value (Table 1) whereas the resistance of the *Proteus* enzyme to benzylpenicillin (44, 67) is due to a relatively high k_4 value of 3.3 $\times$ $10^{-3} sec^{-1}$ (half life of complex EI^*: 3.5 min).

CONCLUDING REMARKS

The mechanisms of the PSEs-catalyzed reactions on β-lactams and L-R-D-Ala-D-Ala-terminated substrates, respectively, differ in many respects:

1. The stabilizing site *IV* and the fragmentation site *V* in Figure 4 which play important roles in the processing of the β-lactams, are not involved in the processing of the L-R-D-Ala-D-Ala substrates.

2. The penicillin sensitivity of the PSEs and their ability to degrade the β-lactams according to either one of the two pathways described above are not related to the ability of the same enzymes to function as carboxypeptidases, transpeptidases, or endopeptidases.

3. The DD-carboxypeptidase-transpeptidase activities of the PSEs can be regulated by changing the carbonyl donor and amino acceptor concentrations or by altering some particular structural features of these compounds. Conversely, the β-lactam-degrading activity of the same PSEs is not subject to such precise regulations.

4. The corresponding functional portions of the β-lactam and L-R-D-Ala-D-Ala molecules exhibit very little isosterism or no isosterism at all.

5. The binding surface of at least some PSEs seems to be able to accomodate β-lactam antibiotics and L-R-D-Ala-D-Ala substrates at the same time. All these observations lead to the two following proposals:

1. The β-lactam and the L-R-D-Ala-D-Ala active centers are essentially distinct entities although, depending upon the PSEs, the independence between them may be more or less pronounced (84), in which case the two centers may share one or more active sites.

2. The normal physiological function of the β-lactam center may be regulatory rather than catalytic. Such a model has the following additional implications: (a) Activation of one center by its relevant substrate freezes the other center as a result of the conformational changes; (b) mutations affecting one center may exist that have no effect on the functioning of the other center; and (c) loss of the L-R-D-Ala-D-Ala center and further alterations of the β-lactam center might have been important steps in the evolution of the PSEs to true β-lactamases. The R61 enzyme (now available in large quantities) (85) which is a highly penicillin-sensitive DD-carboxypeptidase-transpeptidase, and the G enzyme, which is a highly penicillin-resistant DD-carboxypeptidase-endopeptidase have been crystallized (86) (O. Dideberg, J. M. Frère, and J. M. Ghuysen, in preparation). X-ray crystallographic and sequencing studies should provide more information on the exact geometries of the relevant active centers.

The complex series of reactions catalyzed by the peptidoglycan cross-linking enzyme system during bacterial wall synthesis have to be regulated. Mechanisms similar to the modulating effects that the carbonyl donor and the nucleophilic acceptor exert on the DD-carboxypeptidase-transpeptidase activities of some model PSEs could well be involved in vivo. As suggested above, the β-lactam center of the PSEs probably fulfills a regulatory function; however, the natural effector thus postulated is unknown. By interacting with this regulator center, penicillins and Δ^3-cephalosporins impart to the catalytic center a conformation that is incorrect for enzyme action. Inactivation of the wall peptidoglycan cross-linking system as a whole or of some of its essential components, causes inhibition of wall synthesis and cessation of cell growth. With the continuous action of the autolytic system (or perhaps its triggering) (8), the process eventually leads to cell death and cell lysis.

ACKNOWLEDGMENTS

This work was supported in part by the National Institutes of Health, Washington, DC (contract 1 RO1 AI 13364-02 MBC). We thank Drs R. Huls, J. Knox, D. Mirelman, M. J. Osborn, H. R. Perkins, M. Richmond, G. D. Shockman, A. Vasella, and A. R. Zeiger for useful discussions and comments.

Literature Cited

1. Nakae, T. 1976. *J. Biol. Chem.* 251: 2176–78
2. Nikaido, H. 1976. *Biochim. Biophys. Acta* 433:118–32
3. Van Alpen, W., Van Boxtel, R., Van Selmi, N., Lutgenberg, B. 1978. *FEMS Microbiol. Lett.* 3:103–6
4. Blumberg, P. M., Strominger, J. L. 1974. *Bacteriol. Rev.* 38:291–335
5. Spratt, B. G. 1977. *Eur. J. Biochem.* 72:341–352
6. Tamaki, S., Nakajima, S., Matsuhashi, M. 1977. *Proc. Natl. Acad. Sci. USA* 74:5472–76
7. Suzuki, H., Nishimura, Y., Hirota, Y. 1978. *Proc. Natl. Acad. Sci.* 75:664–68
8. Tomasz, A., Höltje, J. 1977. *Microbiology,* 209–16
9. Ghuysen, J. M. 1977. *The Bacterial DD-Carboxypeptidase-Transpeptidase Enzyme System. A New Insight Into the Mode of Action of Penicillin.* E. R. Squibb Lectures on Chemistry of Microbial Products. Series Ed: W. E. Brown. Tokyo: Univ. Tokyo Press. 162 pp.
10. Ghuysen, J. M. 1977. *J. Gen. Microbiol.* 101:13–33
11. Frère, J. M. 1977. *Biochem. Pharmacol.* 26:2203–10
12. Virudachalam, R., Rao, V. S. R. 1977. *Int. J. Peptide Protein Res.* 10:51–59
13. Kropp, J., Kahan, S., Kahan, F. M., Sundelof, J., Darland, G., Birnbaum, J. 1976. *Program Abstr. Intersci. Conf. Antimicrob. Agents Chemother. 16th, Chicago* (Abstr. 228)
14. Brown, A. G., Butterworth, D., Cole, M., Hanscomb, G., Hood, J. D., Reading, C., Robinson, G. N. 1976. *J. Antibiot.* 29:668–69
15. Hashimoto, M., Kamori, J., Kamiya, T. 1976. *J. Antibiot.* 29:890–901
16. Sweet, R. M., Dahl, L. F. 1970. *J. Am. Chem. Soc.* 92:5489–5507
17. Jaszbirenyi, J. Cs., Gunda, E. T. 1975. *Progress in Medicinal Chemistry* 12: 395–477
18. Gunda, E. T., Jaszberenyi, J. Cs. 1977. *Progress in Medicinal Chemistry* 14:182–248
19. Ghuysen, J. M. 1977. In *The Synthesis, Assembly and Turnover of Cell Surface Components,* Cell Surface Reviews, ed. G. Poste, G. L. Nicolson. 4:463–595 Amsterdam: North-Holland
20. Frère, J. M., Moreno, R., Ghuysen, J. M., Perkins, H. R., Dierickx, L., Delcambe, L. 1974. *Biochem. J.* 143: 233–40
21. Frère, J. M., Ghuysen, J. M., Perkins, H. R., Nieto, M. 1973. *Biochem. J.* 135: 463–68
22. Duez, C., Frère, J. M., Geurts, F., Ghuysen, J. M., Dierickx, L., Delcambe, L. 1978. *Biochem. J.* 175:793–800
23. Nieto, M., Perkins, H. R., Frère, J. M., Ghuysen, J. M. 1973. *Biochem. J.* 135:493–505
24. Fuad, N., Frère, J. M., Ghuysen, J. M., Duez, C., Iwatsubo, M. 1976. *Biochem. J.* 155:623–29
25. Nguyen-Distèche, M., Frère, J. M., Dusart, J., Leyh-Bouille, M., Ghuysen, J. M., Pollock, J. J., Iacono, V. J. 1977. *Eur. J. Biochem.* 81:29–32
26. Perkins, H. R., Nieto, M., Frère, J. M., Leyh-Bouille, M., Ghuysen, J. M. 1973. *Biochem. J.* 131:707–18
27. Frère, J. M., Ghuysen, J. M., Perkins, H. R., Nieto, M. 1973. *Biochem. J.* 135:483–92

28. Ghuysen, J. M., Leyh-Bouille, M., Campbell, J. N., Moreno, R., Frère, J. M., Duez, C., Nieto, M., Perkins, H. R. 1973. *Biochemistry* 12:1243–51
29. Ghuysen, J. M., Reynolds, P. E., Perkins, H. R., Frère, J. M., Moreno, R. 1974. *Biochemistry* 13:2539–47
30. Zeiger, A. R., Frère, J. M., Ghuysen, J. M., Perkins, H. R. 1975. *FEBS Lett.* 52:221–25
31. Pollock, J. J., Ghuysen, J. M., Linder, R., Salton, M. R. J., Perkins, H. R., Nieto, M., Leyh-Bouille, M., Frère, J. M., Johnson, K. 1972. *Proc. Natl. Acad. Sci. USA* 69:662–66
32. Ghuysen, J. M., Leyh-Bouille, M., Bonaly, R., Nieto, M., Perkins, H. R., Schleifer, K. H., Kandler, O. 1970. *Biochemistry* 9:2955–61
33. Leyh-Bouille, M., Ghuysen, J. M., Bonaly, R., Nieto, M., Perkins, H. R., Schleifer, K. H., Kandler, O. 1970. *Biochemistry* 9:2961–71
34. Marquet, A., Dusart, J., Ghuysen, J. M., Perkins, H. R. 1974. *Eur. J. Biochem.* 46:515–23
35. Dusart, J., Marquet, A., Ghuysen, J. M., Perkins, H. R. 1975. *Eur. J. Biochem.* 56:57–65
36. Dusart, J., Leyh-Bouille, M., Ghuysen, J. M. 1977. *Eur. J. Biochem.* 81:33–44
37. Leyh-Bouille, M., Dusart, J., Nguyen-Distèche, M., Ghuysen, J. M., Reynolds, P. E., Perkins, H. R. 1977. *Eur. J. Biochem.* 81:19–28
38. Coyette, J., Ghuysen, J. M., Fontana, R. 1978. *Eur. J. Biochem.* 88:297–305
39. Coyette, J., Ghuysen, J. M., Binot, F., Adriaens, P., Meesschaert, B., Vanderhaeghe, H. 1977. *Eur. J. Biochem.* 75:231–39
40. Coyette, J., Perkins, H. R., Polacheck, I., Shockman, G. D., Ghuysen, J. M. 1974. *Eur. J. Biochem.* 44:459–68
41. Coyette, J., Ghuysen, J. M., Perkins, H. R. 1977. *Eur. J. Biochem.* 75:225–29
42. Katz, W., Martin, H. H. 1970. *Biochem. Biophys. Res. Commun.* 39:744–49
43. Martin, H. H., Maskos, C., Burger, R. 1975. *Eur. J. Biochem.* 55:465–73
44. Martin, H. H., Schilf, W., Maskos, C. 1976. *Eur. J. Biochem.* 71:585–93
45. Frère, J. M., Ghuysen, J. M., Iwatsubo, M. 1975. *Eur. J. Biochem.* 57:343–351
46. Frère, J. M., Geurts, F., Ghuysen, J. M. 1978. *Biochem. J.* 175:801–5
47. O'Callaghan, C., Morris, A., Kirby, S. A., Shingler, H. A. 1972. *Antimicrob. Agents Chemother.* 1:283–88
48. Rando, R. R. 1975. *Biochem. Pharmacol.* 24:1153–60

49. Frère, J. M., Duez, C., Ghuysen, J. M., Vandekerckhove, J. 1976. *FEBS Lett.* 70:257–60
50. Frère, J. M., Leyh-Bouille, M., Ghuysen, J. M., Perkins, H. R. 1974. *Eur. J. Biochem.* 50:203–14
51. Frère, J. M., Ghuysen, J. M., Reynolds, P. E., Moreno, R., Perkins, H. R. 1974. *Biochem. J.* 143:241–49
52. Boyd, D. B., Hermann, R. B., Presti, D. E., Marsh, M. M. 1975. *J. Med. Chem.* 18:408–17
53. Topp, W. C., Christenson, B. G. 1974. *J. Med. Chem.* 17:342–47
54. Frère, J. M., Ghuysen, J. M., Degelaen, J., Loffet, A., Perkins, H. R. 1975. *Nature* 258:168–70
55. Frère, J. M., Ghuysen, J. M., Vanderhaeghe, H., Adriaens, P., Degelaen, J., De Graeve, J. 1976. *Nature* 260:451–54
56. Frère, J. M., Ghuysen, J. M., De Graeve, J. 1978. *FEBS Lett.* 88:147–50
57. Bell, M. R., Carlson, J. A., Oesterlin, M. 1972. *J. Org. Chem.* 37:2733–35
58. Adriaens, P., Meesschaert, B., Frère, J. M., Vanderhaeghe, H., Degelaen, J., Ghuysen, J. M., Eyssen, H. 1978. *J. Biol. Chem.* 253:3660–64
59. Marquet, A., Frère, J. M., Ghuysen, J. M., Loffet, A. 1979. *Biochem. J.* In press
60. Umbreit, J. N., Strominger, J. L. 1973. *J. Biol. Chem.* 248:6767–71
61. Shepherd, S. T., Chase, H. A., Reynolds, P. E. 1977. *Eur. J. Biochem.* 78:521–32
62. Chase, H. A., Reynolds, P. E. 1978. *Eur. J. Biochem.* 88:275–85
63. Hammarström, S., Strominger, J. L. 1975. *Proc. Natl. Acad. Sci. USA* 72:3463–67
64. Georgopapadakou, N., Hammarström, S., Strominger, J. L. 1977. *Proc. Natl. Acad. Sci. USA* 74:1009–12
65. Kozarich, J. W., Nishino, T., Willoughby, E., Strominger, J. L. 1977. *J. Biol. Chem.* 252:7525–29
66. Kozarich, J. W., Strominger, J. L. 1978. *J. Biol. Chem.* 253:1272–78
67. Schilf, W., Frère, Ph., Frère, J. M., Martin, H. H., Ghuysen, J. M., Adriaens, P., Meeschaert, B. 1978. *Eur. J. Biochem.* 85:325–30
68. Tipper, D. J., Strominger, J. L. 1965. *Proc. Natl. Acad. Sci. USA* 54:1133–36
69. Lee, B. 1971. *J. Mol. Biol.* 61:463–69
70. Hammes, W. P., Seidel, H. 1978. *Eur. J. Biochem.* 84:141–47
71. Leyh-Bouille, M., Nakel, M., Frère, J. M., Johnson, K., Ghuysen, J. M., Nieto, M., Perkins, H. R. 1972. *Biochemistry* 11:1290–98

72. Leyh-Bouille, M., Coyette, J., Ghuysen, J. M., Idczak, J., Perkins, H. R., Nieto, M. 1971. *Biochemistry* 10:2163–70

73. Nieto, M., Perkins, H. R., Leyh-Bouille, M., Frère, J. M., Ghuysen, J. M. 1973. *Biochem. J.* 131:163–71

74. Carpenter, C. V., Goyer, S., Neuhaus, F. C. 1976. *Biochemistry* 15:3146–52

75. Nishino, T., Kozarich, J. W., Strominger, J. L. 1977. *J. Biol. Chem.* 252:2934–39

76. Rasmussen, J. R., Strominger, J. L. 1978. *Proc. Natl. Acad. Sci. USA* 75:84–88

77. Pollock, J. J., Nguyen-Distèche, M., Ghuysen, J. M., Coyette, J., Linder, R., Salton, M. R. J., Kim, K. S., Perkins, H. R., Reynolds, P. E. 1974. *Eur. J. Biochem.* 41:439–46

78. Nguyen-Distèche, M., Ghuysen, J. M., Pollock, J. J., Reynolds, P. E., Perkins, H. R., Coyette, J., Salton, M. R. J. 1974. *Eur. J. Biochem.* 41:447–455

79. Nguyen-Distèche, M., Pollock, J. J., Ghuysen, J. M., Puig, J., Reynolds, P.

E., Perkins, H. R., Coyette, J., Salton, M. R. J. 1974. *Eur. J. Biochem.* 41:457–63

80. Frère, J. M., Ghuysen, J. M., Zeiger, A. R., Perkins, H. R. 1976. *FEBS Lett.* 63:112–16

81. Leyh-Bouille, M., Bonaly, R., Ghuysen, J. M., Tinelli, R., Tipper, D. J. 1970. *Biochemistry* 9:2944–52

82. Frère, J. M. 1973. *Biochem. J.* 135:469–81

83. Frère, J. M., Ghuysen, J. M., Perkins, H. R. 1975. *Eur. J. Biochem.* 57:353–59

84. Ghuysen, J. M., Leyh-Bouille, M., Frère, J. M., Dusart, J., Marquet, A., Perkins, H. R., Nieto, M. 1974. *Ann. NY Acad. Sci.* 235:236–66

85. Fossati, P., Saint-Ghislain, M., Sicard, P. J., Frère, J. M., Dusart, J., Klein, D., Ghuysen, J. M. 1978. *Biotechnol. Bioeng.* XX:577–87

86. Moews, P. C., Kelly, J. A., Delucia, M. L., Knox, J. R. 1977. *Eur. Crystallogr. Meet. 4th Oxford,* (Abstr. PI 119)

Ann Rev. Biochem. 1979. 48:103–31
Copyright © 1979 by Annual Reviews Inc. All rights reserved

MEMBRANE ADENOSINE TRIPHOSPHATASES OF PROKARYOTIC CELLS[1]

♦12004

J. Allan Downie, Frank Gibson, and Graeme B. Cox

Department of Biochemistry, The John Curtin School of Medical Research, Australian National University, Canberra, A.C.T. 2601

CONTENTS

[1]The following abbreviations are used: ANS, 1-anilinonaphthalene-8-sulfonic acid; CCCP, carbonyl cyanide *m*-chlorophenylhydrazone; DCCD, N,N'-dicyclohexylcarbodiimide.

PERSPECTIVES AND SUMMARY

A membrane-bound ATPase concerned with energy metabolism appears to be associated with all living cells. In the mitochondria and chloroplasts of eukaryotes the ATPase is required for the formation of ATP from ADP by a mechanism that is still obscure. In prokaryotes the ATPase is located in the plasma membranes of both aerobic and anaerobic species. In addition to being involved in ATP synthesis the ATPase may also play a role in coupling ATP, produced by substrate-level phosphorylation, to the energization of the membrane. The ATPase complex has the same general structure in all organisms and consists of an aggregate of at least eight polypeptides. Five of these polypeptides (α, β, γ, δ, ϵ) constitute a complex (F_1) that can be readily solubilized from the membranes of prokaryotes and retains ATPase activity. The ATPase components remaining in the membrane (F_0) after solubilization of the F_1 have been less well characterized.

Much of the earlier work on the membrane-bound ATPases from prokaryotes has concentrated on the characterization of the enzyme from different organisms. The combination of the complexity of this membrane-bound enzyme, the inactivity of separated subunits, and their possible loss or alteration during purification, has made difficult its analysis. Even the number and stoichiometry of the subunits are not yet known with certainty.

Recent work on prokaryotic ATPases has been concerned with the dissociation of F_1 into its individual subunits, their purification, and attempts to reconstitute a functional complex. Most success in this area has been achieved by using ATPase preparations from a thermophilic bacterium in which the subunits are particularly resistant to denaturation. Using various combinations of purified subunits of the F_1 from this organism it has been possible to gain some insight into the functions of the individual subunits. Some progress has also been achieved along these lines using F_1 from *Escherichia coli.*

The demonstration that it is possible to obtain mutants, particularly of *E. coli,* in which oxidative phosphorylation is uncoupled from electron transport, has allowed the application of the techniques of bacterial genetics to the study of the ATPase. Such *unc* mutants have now been isolated in many laboratories and used in a variety of studies on energy-requiring processes, ranging from active transport across membranes to bacterial motility. Genetic characterization of *unc* mutants is in progress and to date five genes concerned with oxidative phosphorylation, *uncA, uncB, uncC, uncD,* and *uncE,* have been distinguished. These genes have been shown to form part of an operon on the *E. coli* chromosome and there is no reason to suppose at present that all the *unc* genes will not eventually be found to be localized in this operon. Gene-polypeptide relationships have been established for two of the genes: the *uncA* gene codes for the α-subunit of F_1 and the *uncD* gene codes for the β-subunit of F_1. It appears likely that the *uncB* and *uncE* genes code for components of the F_0 portion of the ATPase, while the location of the defect in the membranes resulting from a mutation in the *uncC* gene is still not clear. It would seem that further development of the biochemical-genetic approach with respect to prokaryotic ATPases will make significant contributions to our understanding of energy transformations in living cells, a problem that has proved one of the most intractable in biochemistry.

INTRODUCTION

Although studies of oxidative phosphorylation in prokaryotic cells by the fractionation and examination of heat labile "coupling factors" were reported over 25 years ago (1, 2), it was not until 1960 that a membrane-bound ATPase was shown to be associated with bacterial cell membranes (3). Furthermore it was only during the last decade that it was generally realized that the bacterial cytoplasmic membrane has many features in common with the membranes of mitochondria and chloroplasts of eukaryotic organisms. This realization, together with the demonstration that the techniques of biochemical genetics can be used to study energy transduction mechanisms, has resulted in the bacterial ATPases receiving greatly increased attention. ATPase has been demonstrated in the membranes of prokaryotic cells from many genera, including both aerobic and anaerobic species.

Because of limitation of space it is not possible to deal with all aspects of prokaryotic membrane ATPases. There are other recent reviews (4–7) that deal with bacterial membrane ATPases. However, we hope that the present review provides a general key to the literature regarding ATPases, while dealing in some detail with two important aspects of the current studies on the energy-transducing ATPase. These are the fractionation and reconstitution of energy transducing systems, and the study of the biochemical genetics of the system. Both of these approaches promise to play an important role in the attack on the general problems relating to bioenergetics. Studies on the physiological properties of whole cells carrying a mutation affecting one polypeptide of the complex energy-transducing systems provide a powerful addition to the armory for the attack on the difficult problems of oxidative phosphorylation and related phenomena.

The polypeptide complex that can be readily separated from membranes and carries the ATP hydrolytic activity is commonly referred to as F_1, a convention that is adopted for this review. The term F_0 is used to describe the section of the membrane, other than F_1, specifically concerned with

ATP synthesis, and the F_1-F_0 complex is referred to as the ATPase complex. The F_1 derived from various bacterial sources has been often designated with a prefix related to the source, e.g. BCF_1, bacterial coupling factor; ECF_1, from *E. coli;* TF_1, from the thermophilic bacterium PS3, and so on.

GENERAL

Although it is evident that mutant strains of some bacteria lacking an active ATPase complex can grow on fermentable substrates [(8) and see below] no naturally occurring bacterium yet studied appears to lack a membrane-bound ATPase complex. From the literature on membrane-bound ATPases from prokaryotic microorganisms it is clear that, although there are several specific differences, the ATPases from different microorganisms are generally very similar. Thus, in order to examine the development of current ideas about the structure and function of bacterial ATPases, this section gives an overall picture rather than separate reports on individual bacteria.

The first tentative report of oxidative phosphorylation in a fractionated bacterial system was by Pinchot & Racker in 1951 (1) using cell free extracts of *E. coli.* Subsequently, Pinchot extended this work using *Alcaligenes faecalis* and showed that the oxidative phosphorylation system could be fractionated into membranes and soluble fractions (2) which could be recombined to form a coupled phosphorylating system (10).

Abrams et al (3) first demonstrated in 1960 that ATPase activity was localized in the plasma membrane of a bacterium, when they showed that a high level of ATPase activity was associated with washed membranes prepared from *Streptococcus faecalis.* Using a cytochemical stain specific for ATPase activity on thin sections of cells of *E. coli* or *Bacillus cereus,* Voelz (11) demonstrated the presence of ATPase activity on membranes. Negative staining of membrane preparations of several bacteria revealed the presence of "knobs" (12–15) protruding from the membrane. Using ferritin-labelled antibody to purified F_1, Oppenheim & Salton (12), confirmed the observation previously made with mitochondria that these knobs are the ATPase and that they are localized on the inner face of the plasma membrane.

Solubilization and Purification of F_1

Ishikawa & Lehninger (16) in 1962 showed that the membrane-bound ATPase activity could be released from membranes of *Micrococcus lysodeikticus* by a "shock wash" using cold water, and that the solubilized fraction could be used to reconstitute oxidative phosphorylation in the shock washed membranes. It is now clear that a similar procedure, involving washing bacterial membranes with buffers of low ionic strength in the presence of EDTA, can be used to solubilize the membrane-bound ATPase from the membranes of many bacterial species, including *Strep. faecalis* (17), *B. stearothermophilus* (18), *B. megaterium* (9, 19), *Azotobacter vinelandii* (20), *M. lysodeikticus* (21), *E. coli* (22, 23), *Rhodopseudomonas sphaeroides* (24), *Alc. faecalis* (25), and a Proteus sp. (26). However, in some microorganisms such as *Rhodospirillum rubrum* (27) or *Mycoplasma laidlawii* (28), it appears that, as with chloroplasts or mitochondria, the ATPase cannot be solubilized by this procedure.

Recently it has been shown that the protease inhibitor *p*-aminobenzamidine inhibits the solubilization of the F_1 from membranes of *E. coli* although over 60% of the membrane-associated proteins are solubilized (29). Moreover, it appears that this effect is not due to the inhibition of a protease since the solubilization of F_1, which has been rebound to depleted membranes, is still inhibited by the presence of *p*-aminobenzamidine (30).

ATPase of high specific activity can be solubilized from membranes of *M. lysodeikticus* (31, 32) or *E. coli* (29) by extracting the membranes with *n*-butanol or chloroform. Although the F_1 solubilized by these procedures is of much higher specific activity than that released by the low-ionic-strength wash procedure (29, 32), the resulting ATPase fraction cannot be rebound to membranes or reconstitute energy-linked activities in membranes depleted of F_1. In general, F_1 solubilized by the low-ionic-strength wash procedure can reconstitute ATP-dependent membrane energization or oxidative phosphorylation (9, 16, 25, 29, 33–36) and this washing procedure is often used as a primary step in the purification of F_1. Alternative methods of solubilization are treatment of membranes with Triton X-100 (28, 37, 38), passing washed membranes through a French pressure cell (27), or sonication (39).

The method of choice for purifying F_1 from membranes of *E. coli* appears to depend upon the growth conditions and strains of *E. coli* used (40, 41). In general, purification steps include precipitation with ammonium sulfate (40, 42) or polyethylene glycol (41, 43), ion exchange chromatography on DEAE cellulose (35, 40, 41, 44, 45) or DEAE Sephadex (37), and gel filtration using Sepharose 6B (35, 41, 43) or Biogel A (40, 42, 44). The methods devised by Bragg & Hou (35, 45), Futai et al (40), Giordano et al (44), Vogel et al (43, 46) and Senior et al (41), yield F_1 preparations that can rebind to F_1-depleted membranes and reconstitute energy-linked reactions. However some preparations do not rebind to membranes (22, 37, 42, 47) and have presumably been altered during preparation.

An apparent 200-fold purification of the F_1 from *Mycobacterium phlei* was achieved in a single step by chromatography on a Sepharose-ADP affinity column; this purified F_1 had latent ATPase activity unmasked by trypsin treatment (48, 49).

Partial purification of the ATPase activity from *Strep. faecalis* by chromatography on Biogel A 1.5 M gave a preparation incapable of binding to F_1-depleted membranes (50). Binding of the ATPase activity required the addition of a fraction from the Biogel column, subsequently shown to contain the δ-subunit of F_1 (34, 50). Purification of the F_1 was achieved by preparative gel electrophoresis in the presence of Mg^{2+} (34).

The purification of F_1 from membranes of *Micrococcus* species has been reported by several workers (21, 32, 51–55) who used, in addition to the techniques described for the purification of the *E. coli* F_1, preparative gel-electrophoresis (56) or affinity chromatography using an ATP analogue (54). The F_1 purified by the method of Munoz et al (21) was found to rebind to F_1-depleted membranes (31).

Partially purified F_1 preparations that rebind to F_1-depleted membranes have also been reported for *Alc. faecalis* (25) and *Salmonella typhimurium* (45). Although the ATPase is more difficult to solubilize from membranes of photosynthetic bacteria, preparations that reconstitute ATP-dependent membrane energization (27, 57) have been purified from *Rhodosp. rubrum* (27, 58) and *Rhodopseudomonas capsulata* (59).

The F_1 from the thermophilic bacterium PS3 has been purified by Kagawa and his colleagues and is discussed in detail below.

Solubilization and Purification of the ATPase Complex

In contrast to the F_1 preparations described above, the ATPase activity of the membrane-bound form is sensitive to inhibition by DCCD. Thus DCCD-sensitivity is used to check for the integrity of the ATPase complex during a purification procedure.

An ATPase complex, sensitive to DCCD, has been solubilized from *E. coli* using cholate or deoxycholate (60). Using deoxycholate to solubilize the ATPase complex from the membranes of *E. coli*, Hare (61) purified a DCCD-sensitive ATPase complex about 20-fold by sucrose density-gradient centrifugation. Added phospholipids were required for maximal ATPase activity and DCCD sensitivity. Further purification of the ATPase complex by DEAE-cellulose chromatography and sucrose gradient centrifugation, gave a 50-fold purification (62). A DCCD-sensitive ATPase complex was solubilized from membranes of *Clostridium pasteurianum* using Triton X-100 and purified using a Sepharose-4B column to give about a sixfold purification with respect to ATPase activity (63).

An ATPase complex was solubilized from *Myco. phlei* membranes by Triton X-100 and the complex purified by affinity chromatography on a

Sepharose-ADP column. The purified complex could be incorporated into Triton-extracted membranes to reconstitute oxidative phosphorylation (64).

Kagawa and his colleagues (see below) have also described the purification of the ATPase complex from the thermophilic bacterium, PS3.

Molecular Weights of F_1

The reported values for the molecular weights of the purified F_1 from bacterial membranes lie within the range 250×10^3 (48) to 400×10^3 (22; see also 4, 7). However the reported molecular weight of the purified F_1 from *E. coli* alone varies over the range 340×10^3 up to 400×10^3, and therefore the differences reported for the molecular weights of the F_1 from different species may not be significant.

Subunit Composition and Arrangement of F_1

Although analyses of some early preparations of F_1 from different bacteria suggested the presence of only two subunits (65, 66) it is probable that at least five subunits are present in preparations of F_1 that can be used to reconstitute energy-linked reactions. These subunits have generally been designated α, β, γ, δ, and ϵ in order of decreasing molecular weights. The reported molecular weights for the subunits of F_1 from *E. coli* fall within the following ranges: α-subunit, 54×10^3 (22) – 60×10^3 (37); β-subunit, 52×10^3 (43, 45) – 56×10^3 (37); γ-subunit, 32×10^3 (43, 45) – 35×10^3 (37); δ-subunit about 21×10^3 (43, 45); ϵ-subunit, 11×10^3 (22) – 13×10^3 (37, 45). Further information about subunit molecular weights is given in previous reviews (4, 7).

Analysis of the subunit composition of the F_1 from *M. lysodeikticus* is complicated by instability of purified preparations of the enzyme (53, 55, 67, 68) and the use of different substrains (55, 68). Andreu et al (71–73) have recently shown that there are sugars associated with F_1 purified from *M. lysodeikticus.* It was estimated (72) that, on a weight basis, about 10% of the F_1 is carbohydrate and it was concluded that it is a glycoprotein in which the α, β and γ-subunits contain sugar moities. There also seems to be heterogeneity in F_1 preparations from *Alc. faecalis* (69, 70). However, it is probable that the F_1 from each of these species consists of five subunits (25, 55, 70, 71). Five-subunit F_1 preparations have also been described for *Strep. faecalis* (34), *Salm. typhimurium* (45), *Rhodosp. rubrum* (27, 74) and *Myco. phlei* (49).

The stoichiometry of the different subunits is still unclear. Bragg & Hou (45) suggested a stoichiometry for the subunits of the F_1 from both *E. coli* and *Salm. typhimurium* as being a_3, β_3, γ_1, δ_1, and ϵ_1 based on the relative amounts of ^{14}C incorporated into the subunits during growth of the organisms in a medium containing a ^{14}C-labelled protein hydrolysate. Similarly, Kagawa et al (75) postulated a subunit ratio of a_3, β_3, γ_1, δ_1, and ϵ_1. Vogel & Steinhart (43) have concluded, on the basis of the titration of aggregates of subunits required to reconstitute ATPase activity, that the subunit stoichiometry of *E. coli* F_1 is a_2, β_2, γ_2, δ_{1-2}, and ϵ_2.

In an attempt to assess which subunits of the F_1 are adjacent, Bragg & Hou (45, 76) used the bifunctional cross-linking reagent dithiobis (succinimidyl propionate) to cross-link subunits of the purified F_1 from *E. coli.* It was concluded from the experimental results that the a-subunit was cross-linked to the β- and δ-subunits, while the β-subunit was cross-linked also to the γ- and ϵ-subunits. However, as the authors point out, the results do not exclude the possibility of other subunits being in close proximity.

Kinetic Properties of ATPase

The reported specific activities of various F_1 preparations vary greatly depending upon the preparative method used, assay conditions, the organism used, and the latency and method of activation of the ATPase. It is difficult to compare kinetic data obtained using F_1 preparations with widely differing specific activities and with the added complication of different forms of the enzyme in a single preparation [see (68, 70, 77) for examples]. The activity of the ATPase varies as a function of both the ATP and Mg^{2+} concentrations; the optimal ATP:Mg^{2+} ratio has been reported as about 2:1 for the F_1 from *E. coli* (42). Using this ratio of ATP:Mg^{2+}, Nelson et al (42) reported a K_m for ATP of 0.25 mM, although higher values have been reported in *E. coli* and other bacteria (37, 51, 78, 79).

The ATPase activity of membrane-bound and soluble F_1 from a variety of organisms is dependent on Ca^{2+} or Mg^{2+} (9, 21, 23, 37, 42, 51, 58, 78, 80–84), although it has been reported that Mg^{2+} but not Ca^{2+} stimulates the ATPase from *Strep. faecalis* (17) and *Cl. pasteurianum* (85). Conversely Ca^{2+}, but not Mg^{2+}, has been reported to stimulate the ATPase from *B. subtilis* (86).

The range of nucleotides hydrolyzed by F_1 has been shown to include ATP, GTP, CTP, UTP, and ITP (37, 78).

Tightly-Bound Nucleotides

In parallel with the studies of the role of the tightly-bound nucleotides of mitochondrial F_1 (see 87) several workers have also shown that tightly-bound nucleotides are present in the bacterial F_1 (88–93). A mutant strain that has an inactive ATPase has also been shown to contain tightly-bound nucleotides (94, 95). After measuring the rate of turnover of ^{32}P, using

F_1 labelled with ^{32}P-ATP, Maeda et al (88) concluded that the tightly-bound ATP was not a direct intermediate in oxidative phosphorylation in *E. coli.*

Inhibitors of ATPase Activity

There are a number of inhibitors that affect the ATPases of prokaryotic microorganisms and, of these, DCCD has proven to be the most useful and most widely used. DCCD inhibits the ATPase activity of membrane-bound, but not of soluble F_1 (63, 78, 81, 96–101, 122) and it has been shown to react with the F_0 portion of the ATPase since (*a*) the hydrophobic DCCD was a much more efficient ATPase inhibitor than other hydrophilic carbodi-mides (103) and (*b*) ATPase-depleted membranes from a mutant strain of *Strep. faecalis* resistant to DCCD, when reconstituted with normal F_1, had DCCD-resistant ATPase activity (104). In addition, membranes of *E. coli* that had been reacted with DCCD and then depleted of F_1 were found to inhibit the ATPase activity of added F_1 prepared from untreated membranes (81).

Using ^{14}C -labelled DCCD, Fillingame (105) showed with *E. coli* that a protein of molecular weight 9000 was specifically labelled. This protein was not significantly labelled in a DCCD-resistant mutant of *E. coli.* Altendorf & Zitzmann (106) also showed that a low molecular weight protein bound DCCD. The DCCD-binding protein was extracted from the membranes (107) using chloroform-methanol as described by Cattell et al (108) for mitochondrial membranes. The purified protein contained a very low amount (16%) of polar amino acids and was completely lacking in histidine, serine, cysteine, or tryptophan (107). It was suggested that more than one copy of the DCCD-binding protein is likely to be present since only about 30% of the membrane DCCD-binding protein was labelled by ^{14}C-DCCD when maximal ATPase inhibition occurred (107). The DCCD-binding protein appears to be involved with proton movements through the membranes since DCCD can reconstitute electron transport-dependent membrane energization in membranes lacking F_1 (109–111).

DCCD does not inhibit membrane-bound ATPase in membranes from strains carrying the *uncB402* allele (112) and it has been shown that depleted membranes of a strain carrying this allele do not become proton permeable (113).

Bathophenanthroline, like DCCD, inhibits the membrane-bound AT-Pase of *E. coli* but not the soluble ATPase (114). This inhibition is unusual in that it is reversed by the uncoupler CCCP (114). However uncoupler-sensitive inhibition of soluble mitochondrial F_1 has been observed (115) and it has been pointed out that the reversal of bathophenanthroline inhibition may be due to a chemical interaction between bathophenanthroline and the uncoupler (116). Another inhibitor that inhibits membrane bound F_1 from mitochondria is oligomycin, but this does not affect the ATPase activities of most prokaryotes. The ATPases of photosynthetic bacteria differ from those of other bacteria since oligomycin does inhibit the membrane-bound ATPase of *Rhodosp. rubrum* (27, 58, 117) and *Rhodops. capsulata* (118).

A number of inhibitors are known that affect the solubilized ATPase activity of bacteria and among these are azide (21, 58, 78, 80, 119), the antibiotic dio-9 (78, 120–122), quercetin (40), and the zinc chelator, Zincon (123). Aurovertin also inhibits soluble ATPase activity (70, 89, 124–126) but the enhancement of aurovertin fluorescence associated with its binding to mitochondrial F_1 (127) is not observed with prokaryotic F_1 (see 128). Aurovertin-resistant mutants of *E. coli* have recently been isolated (126) and these mutations, which confer resistance, map in the region of the *unc* genes (see below) and do not confer resistance to DCCD, quercetin, azide, or 7-chloro-4-nitrobenzo-2-oxa-1,3-diazole chloride (NBDCl). The latter compound covalently modifies a tyrosine residue in mitochondrial F_1, which results in loss of ATPase activity (129) and inhibition of the ATPase activity of bacterial F_1 (70, 130). In experiments with *Micrococcus denitrificans* F_1, spectrophotometric analyses could not be used to distinguish whether the sulfhydryl groups of cysteine or the phenolic oxygen groups of tyrosine were being covalently modified by the inhibitor NBDCl (130).

RECONSTITUTION OF ATPase FROM ITS SUBUNITS

Thermophilic Bacterium, Strain PS3

The dissociation of the purified ATPase complex into its component subunits, the purification of these subunits, and their reassociation into a functional complex has not yet been achieved. Most progress in this direction has been made by Kagawa and his colleagues using a thermophilic bacterium, strain PS3. This particular strain was selected from a number of thermophiles, isolated from hot springs at various locations in Japan, for its high membrane ATPase activity, high sensitivity to DCCD, and rapid growth (doubling time, 35 min at 65°C) (131).

The F_1 from strain PS3, which will be referred to as TF_1, was solubilized from the membranes by incubation at 25°C in 0.2 mM EDTA, and purified by DEAE-cellulose and Sepharose-6B chromatography (132). The molecular weight of TF_1 as determined by Sepharose chromatography was 380,000 (132). Acrylamide gel electrophoresis of TF_1 in the presence of SDS indicated that there were five dissimilar subunits of molecular weight 56,000, 53,000, 32,000, 16,000, and 11,000 (132). These subunits were subsequently designated α, β, γ, δ and ϵ in order of decreasing molecular weight in accordance with the usual nomenclature for subunits of F_1.

The properties of TF_1 differed markedly from those of other F_1 preparations in that it was generally more stable to dissociating agents, to organic solvents, and to extremes of temperature. The maximum velocity of ATPase activity occurred at 75°C and was not cold-sensitive, being stable for several months at 2–4°C (132). The concentrations of various chemicals giving 50% inhibition of activity was considerably greater for TF_1 as compared to the F_1 prepared from ox heart, e.g. LiCl 3.95 M compared to 0.43 M, KSCN 0.53 M compared to 0.07 M, and acetone 63% compared to 13% (132).

The ATPase activity of TF_1 has an absolute metal requirement, and the range of metal ions giving activity was greater than that for the F_1 from ox heart, with Cd^{2+}, Ca^{2+}, Co^{2+}, Mn^{2+}, and Mg^{2+} giving about equal activity (132). It was also possible to completely dissociate TF_1 in the presence of SDS and to regain about 60% of the original activity following the removal of SDS by dialysis. Similarly, urea and guanidine treatment resulted in reversible inactivation, although the recovery of activity was less at 46% and 38% respectively (132).

The ATPase complex ($TF_0 \cdot F_1$) was solubilized and purified to give a 39% yield with an increase in specific activity from 1.2 to 17.6. The solubilization and purification of $TF_0 \cdot F_1$ involved washing the membranes with 1% cholate then extracting the $TF_0 \cdot F_1$ with 1% Triton. The Triton extract was then chromatographed on DEAE cellulose and Sepharose-6B, and the active fractions concentrated by $(NH_4)_2SO_4$ precipitation (133). Acrylamide gel electrophoresis in the presence of SDS indicated the presence of eight polypeptides which included the five subunits of TF_1 (133). The three subunits attributed to TF_0 had molecular weights of 19,000, 14,000, and 5,500. The densitometric tracings of the gels stained with amido black indicated the presence of only a low level of a number of impurities. The molecular weight of the $TF_0 \cdot F_1$ was difficult to determine from the Sepharose chromatography, due to aggregation, and it was deduced from the relative staining intensities on SDS-acrylamide gels that the molecular weight was about 550,000. The ATPase activity of the purified $TF_0 \cdot F_1$ was stimulated fivefold by the addition of phospholipid prepared from either strain PS3 or soybean. The effects of metal ions, temperature, dissociating agents, and organic solvents were the same as for the purified TF_1 and the ATPase activity of the $TF_0 \cdot F_1$ was 75% inhibited by 10^{-5}M DCCD and 10^{-7}M trialkyltin (133).

Radioactive-labelled TF_1 and $TF_0 \cdot F_1$ were prepared from strain PS3 grown in the presence of an L-[U-^{14}C]-amino acid mixture (75). The stoichiometries of the subunits were calculated, on the basis of the molecular weights determined by SDS-acrylamide gel electrophoresis, to be 3:3:1:1 :1 for the α-, β-, γ-, δ-, and ϵ-subunits, respectively, of TF_1, and 1:2:5 for the 19,000, 14,000, and 5,500 subunits respectively of the TF_0. However, the actual figures are sufficiently different from these stoichiometries for there to remain some uncertainty as to the ratios of the subunits present.

The ATP-dependent enhancement of ANS fluorescence was used as an assay to determine the optimum conditions for reconstitution of vesicles from purified $TF_0 \cdot F_1$ and phospholipids (134). The optimum ratio of phospholipid to $TF_0 \cdot F_1$ protein was 40:1 and the optimum ratio of deoxycholate to cholate in the reconstitution mixture was 1:2. The reconstitution mixture was dialyzed against Tricine-NaOH buffer, pH 8.0, (containing 2.5 mM $MgSO_4$) for 20 hr at 30°C, when soybean phospholipid was used, or 45°C, when the phospholipid from strain PS3 was used. The ATPase activity was constant during the formation of the reconstituted vesicles, and was not increased by the addition of Tween-80, which indicated that the F_1 was on the outside of the vesicles (134).

ATP synthesis was demonstrated by these reconstituted vesicles in response to an electrochemical proton gradient but only if phospholipids from strain PS3, and not soybean phospholipids, were used (135). The electrochemical proton gradient was generated by first incubating the vesicles in malonate buffer (pH 5.5) and valinomycin, and then rapidly adding glycylglycine buffer (final pH 8.33) containing 0.15 M KCl. ATP synthesis was determined by measuring the esterification of $^{32}P_i$; the reaction was essentially complete in 25 sec with a total ATP formation of about 100 n mole/mg $TF_0 \cdot F_1$ protein. The pH gradient was essential for ATP synthesis but if the electrogenic K^+ gradient was not formed, then ATP synthesis was 25% of that under optimal conditions.

The remarkable stability of TF_1 and its component subunits was demonstrated when it was found possible to purify each of the five subunits of TF_1 and then to reconstitute ATPase activity and, in combination with TF_0, ATP-dependent ANS-fluorescence enhancement (136). TF_1 was dissociated by 8M guanidine HCl and the guanidine replaced by 8M urea after passage through a Bio-gel P-6 column. The γ-subunit was retained on a CM-cellulose column and eluted with about 0.06 M sodium chloride. The unadsorbed material from the CM-cellulose column was then applied to DEAE-cellulose and the α-, β-, and δ-subunits were retained on the column. The ϵ-subunit was isolated from the unadsorbed material. The α-, β-, and δ-subunits, plus two unidentified proteins, were eluted with a linear gradient of 0–0.1 M NaCl. The purity of each subunit was checked by acrylamide gel electrophoresis in the presence of SDS or urea. However the protein profile obtained for TF_1 on the urea gel was not the same as the sum of the five subunits (136).

Reconstitution of ATPase activity could be obtained by mixing the subunits in the presence of Mg^{2+} at temperatures between 20°C and 45°C and at a pH between 6.3 and 7.0 for 2–3 hr, and better reconstitution was obtained with higher protein concentrations (136). Combinations of subunits including β and γ, or α, β, and δ, had ATPase activity, whereas combinations without β were inactive. Acrylamide gel electrophoresis under nondissociating conditions indicated that combinations of subunits not giving ATPase activity did not aggregate (136). It would appear, then, that complex formation is only initiated in the presence of the β-subunit. The presence of the γ-subunit on the other hand conferred the characteristic properties of TF_1 on combinations of subunits with ATPase activity. The pH profile, temperature profile, and sensitivity to NaN_3 of the ATPase activity of combinations of subunits including γ were all similar to TF_1, but those combinations lacking γ had different properties. No evidence could be obtained for an inhibitory role for either the δ or ϵ-subunits (136).

The ATPase activity obtained with a combination of β- and γ-subunits was surprising in view of the ATPase activity of trypsin-treated F_1 from other bacteria, which apparently contained only α- and β-subunits (42, 137). However, recent work has shown that, at least for the *E. coli* enzyme, a portion of the γ-subunit remains after trypsin treatment (138).

The TF_0 was prepared from the purified $TF_0 \cdot F_1$ complex after incubation in 7 M urea at 4°C for 5 hr. The TF_0 was pelleted by centrifugation at 144,000 g for 45 min, and acrylamide gel electrophoresis in the presence of SDS gave about eleven bands, three of which predominated (139). These three polypeptides had previously been characterised as TF_0 components (see above) and have molecular weights of 19,000, 14,000, and 5,500.

Vesicles were prepared from purified TF_0 and soybean phospholipids by the dialysis method, and permeability of the vesicles to protons was tested (139). The vesicles were either K^+-loaded or K^+-deficient and the pH changes were recorded in response to electrogenic K^+-movement initiated by the addition of valinomycin. Fluorescence-quenching of 9-aminoacridine was also used with the K^+-loaded vesicles to monitor proton movements. Using these assays a proportionality was demonstrated between the TF_0/phospholipid ratio and the proton permeability of the resultant vesicles. However, only 15% of the TF_0 incorporated into the vesicles was able to bind TF_1. Both added TF_1 and DCCD blocked the proton permeability of the TF_0 vesicles. The uptake of protons by TF_0 vesicles was dependent on temperature, and definite transition temperatures were evident in the Arrhenius plot. The authors speculate that this might indicate that TF_0 functions as a mobile carrier rather than as an immobile channel (139).

The binding of ATPase activity to TF_0 vesicles required the presence of all 5 subunits of TF_1 (140). However δ- and/or ϵ-subunits could be bound to the TF_0 vesicles, and the presence of these subunits did not prevent TF_1 from subsequently binding. There was no preferred sequence in the binding of the δ- or ϵ-subunits. The ability of various combinations of TF_1 subunits to prevent proton permeability of TF_0 vesicles was tested using K^+-loaded vesicles, in which electrogenic K^+-movement was initiated with valinomycin, and 9-aminoacridine fluorescence-quenching used to monitor proton movements (140). TF_1 and the combination of α-, β-, γ-, δ- and ϵ-subunits were both effective, as was the combination of γ-, δ-, and ϵ-subunits. ATP-dependent ANS fluorescence-enhancement could only be reconstituted using TF_1 or the combination of α-, β-, γ, δ-, and ϵ-subunits. The $^{32}P_i$-ATP exchange activity could also be reconstituted using the TF_0 vesicles plus TF_1, and to a lesser extent with the combination of α-, β-, γ-, δ-, and ϵ-subunits. Surprisingly, the combination of β-, γ-, δ-, and ϵ-subunits gave about 15% of $^{32}P_i$-ATP exchange activity compared with that obtained using TF_1. The authors proposed a model for the subunit structure of TF_1 (140).

Escherichia coli

An analogous study to that using TF_1 has not yet been reported using the F_1 from any other bacterial source, although progress has been made with the F_1 from *E. coli* (which will be referred to as ECF_1). The separation of the δ- and ϵ-subunits from the complex was achieved by treating ECF_1 with pyridine by the method developed by Nelson et al (142) for chloroplast F_1. After centrifugation at 8000 g for 10 min, δ- and ϵ-subunits were found in the supernatant, and the α-, β-, and γ-subunits, and some δ- and ϵ-subunits, were found in the pellet (141). The δ- and ϵ-subunits in the supernatant were purified by repeated gel filtration through Sephadex G-75 (143, 144). The four subunit ATPase (α, β, γ, ϵ) was prepared by gel filtration of ECF_1 at pH 9.4 and a three subunit ATPase (α, β, γ) was prepared by passing the 4-subunit complex through a Sepharose 6B column carrying immobilised antibodies to the ϵ-subunit. Using combinations of the 3- and 4-subunit enzymes and the purified δ- and ϵ-subunits, it was shown that both δ- and ϵ-subunits were required to bind the ATPase to ECF_1-depleted membranes with concomitant reconstitution of energy-linked reactions, and that only one molecule of δ and of ϵ per ATPase molecule were necessary (143, 144). Neither the δ- nor the ϵ-subunits by themselves bound significantly to ECF_1-depleted membranes and no evidence was obtained that these subunits directly interact with the membrane (143, 144). Addition of purified ϵ-subunit to ECF_1 caused inhibition of ATPase activity by up to 65% (145). However the ϵ-subunit is required for the energy-linked activities involving the membrane-bound ATPase (145), and whether or not the ϵ-subunit functions in vivo as an ATPase inhibitor remains to be determined.

Vogel & Steinhart (43) reported that preparations of ECF_1 that had been inactivated by repeated freezing and thawing in high concentration salt solution regained ATPase activity when the salt concentration was decreased and the temperature raised above 22°C in the presence of ATP and $MgCl_2$. The cold-inactivated enzyme was fractionated into three components (IA, IB, and II) by chromatography on DEAE-cellulose. Fraction IA contained α-, γ-, and ϵ-subunits, fraction IB contained α-, γ-, δ-, and ϵ-subunits, and fraction II contained the β-subunit. Reconstitution of ATPase activity required combination of fractions IA and II, but reconstitution of ATP-dependent acridine fluorescence-quenching required the combination of all three fractions. Futai (146) has been able to improve the fractionation of the cold-dissociated ECF_1 and to purify, by hydrophobic column chromatography, each of the α-, β-, and γ-subunits. Reconstitution of ATPase activity from the isolated subunits required α-, β-, and γ-subunits, a result which is in contrast to that obtained with TF_1 where only the β- and γ-subunits were required (see above). Trypsin treatment of ECF_1 had previously been reported (42) as producing a protein with ATPase activity consisting of only the α- and β-subunits. However, recent work (138) has indicated that a fragment of the γ-subunit remains tightly bound to the α- and β-subunits and is essential for ATPase activity. The α-, β-, and γ-subunits from cold-dissociated ECF_1 have also been purified by chromatography on hydroxylapatite and DEAE-Sepharose (147). The purified α-subunit has been used to reconstitute cold-dissociated inactive F_1 from a mutant strain carrying the *uncA401* allele (see below).

BIOCHEMICAL GENETICS

The use of biochemical genetics in the study of oxidative phosphorylation and related phenomena is of relatively recent origin. In 1971 the first mutant of *E. coli* in which oxidative phosphorylation was uncoupled from electron transport (*unc* mutant) was described (8). Since that time uncoupled mutants have been isolated in a number of laboratories (148–160) and used in the study of a number of energy-requiring processes including oxidative phosphorylation (8, 151, 153, 158, 160, 161); ATP-dependent transhydrogenase activity (151, 153, 155–158, 160, 162, 163, 165); electron transportdependent transhydrogenase activity (151, 153, 156, 157, 160, 163); active transport and proton translocation (113, 148, 149, 152, 153, 154, 157, 158, 164, 166–179, 204); membrane energization (109, 110, 113, 164, 165, 176, 180–182); DNA synthesis (183); adsorption of bacteriophage (190); and motility and chemotaxis (184–186). The effect of *unc* mutations on aerobic growth yield (8, 149, 152, 155, 165, 187), anaerobic growth (152, 160, 188), nucleotide binding (94, 95, 189), and ATP-$^{32}P_i$ exchange (95, 158) have also been measured. An uncoupled mutant of *B. megaterium* has also been isolated and it was found that the mutant strain did not sporulate in glucose-limited medium (203).

Methods of Isolation of Uncoupled (unc) Mutants

E. coli is a facultative anaerobe and mutant strains unable to carry out oxidative phosphorylation cannot grow on nonfermentable substrates such as acetate, malate, or succinate (8, 148), but can grow on fermentable substrates that allow sufficient substrate-level phosphorylation to provide ATP for growth. Thus, uncoupled mutant strains of E. coli were isolated after mutagenesis using such mutagens as N-methyl-N'-nitro-N-nitroso-guanidine (8, 148, 153, 158, 160, 191), hydroxylamine (156, 157), ethyl methylsulfonate (149), and ultra-violet irradiation (155), and screened for the uncoupled phenotype as described in detail by Cox & Downie (191). Such uncoupled mutant strains have low aerobic growth yields on limiting glucose medium and lack ATP-dependent energization of membrane functions (191).

Once the phenotype of unc mutants had been established, alternative methods were used to isolate additional mutants. About 3% of neomycin-resistant strains were found to have an uncoupled phenotype (150–152, 154) although the reason for this is not clear (see below). Since the unc mutants that have so far been mapped all occur in the region of the ilv genes (see below), it has been possible to use the procedure of localized mutagenesis (156, 157). In this method, a transducing phage lysate is mutagenized using hydroxylamine and ilv+ transductants selected using an appropriate recipient. These transductants are then screened for the unc mutant phenotype.

Genetic Characterisation of unc Mutants

All of the known mutations that uncouple oxidative phosphorylation map at about 83 min on the E. coli chromosome and are about 40% cotransducible with the ilv genes (8, 109, 150, 153, 155–160), and about 70% cotransducible with asn (153, 155, 158). Additionally the unc genes are more than 80% cotransducible with the bglR gene, the mutant form of which permits strains of E. coli to grow on salicin as a carbon source (192). Mutations that give rise to resistance to DCCD (105, 192) or aurovertin (126) have been shown to be very closely linked to the unc genes.

The location of all the known unc mutations at 83 min on the E. coli chromosome, together with the overall similarity of properties of various unc mutants, has made the characterization of the number of genes concerned with oxidative phosphorylation difficult. However unc mutant strains could be broadly classified into two groups, those that retain ATPase activity (e.g. uncB402) and those that lose ATPase activity (e.g. uncA401). The uncA401 and uncB402 alleles were further characterized as affecting the F_1 and F_0 portions of the ATPase, respectively, from the results of membrane fractionation and reconstitution experiments (161). These two alleles were therefore used as reference alleles to develop a system for genetically characterizing new *unc* mutant strains. This involved the construction of partial diploids containing two different *unc* alleles, one on the chromosome and one on a plasmid. The dominance of *unc*$^+$ over *unc*$^-$ alleles was shown (193) using a plasmid covering the *ilv-unc* region of the *E. coli* chromosome. A procedure was then devised (159) in which mutant *unc* alleles could be introduced into a similar plasmid. Thus, a plasmid was identified that covered the *pyrE, unc, ilv,* and *argH* region of the chromosome, but that contained a deletion of the *unc-ilv* region. If this plasmid was introduced into a strain carrying a mutant *unc* allele, the deleted region of the plasmid could be repaired due to a double cross-over that resulted in the formation of a new plasmid that now contained the *ilv*$^+$ and *unc* genes. (The latter included the mutant *unc* allele.) Such recombinant plasmids were selected by mating into a *recA*$^-$ strain carrying mutations in the *pyrE, ilvC,* and *argH* genes. The plasmids cannot undergo further recombination with the chromosome in the *recA*$^-$ strain, and the partial diploid thus formed could then be used as a donor in genetic complementation tests.

Partial diploids carrying combinations of mutant *unc* alleles on either the plasmid or the chromosome were isolated and tested for genetic complementation by examining for growth on succinate and growth yields on limiting amounts of glucose. Presence or absence of genetic complementation was confirmed by biochemical tests. Using these genetic procedures (159) it was possible to confirm the previous biochemical complementation studies (161) which showed that the *uncA401* and *uncB402* mutations affected different genes. In addition, three further genes, *uncC* (159), *uncD* (194) and *uncE* (195) were identified (see below). The level of complementation of ATPase activity and energy-linked reactions in membranes prepared from partial diploid strains carrying mutant *uncA* and *uncD* alleles are generally low due to the assembly of F$_1$ complexes containing normal and mutant α- and β-subunits (194).

The unc Operon

Since mutations affecting five different *unc* genes were closely linked, it was possible that the *unc* genes formed an operon. The bacteriophage Mu, when used as a mutagen, inserts into the *E. coli* chromosome, causing mutations that have strong polar effects; such polar mutations can be used to define which genes occur in a particular transcriptional unit, and the order of those genes. Thus, a series of uncoupled mutants were isolated using bacteriophage Mu as a mutagen. On the basis of complementation between strains carrying these Mu-induced *unc* alleles and plasmids carrying point mutations in known *unc* genes, it was concluded that the *uncA, uncB, uncC, uncD,* and *uncE* genes were in an operon, and that the order of genes was *uncBEADC* (165, 195). No uncoupled mutants yet tested have been found to map outside this operon. The operon arrangement of the *unc* genes might suggest that equimolar amounts of the gene products are formed and hence that the ATPase complex might consist of equimolar amounts of the component subunits.

The orientation of the *unc* operon was established on the basis of transduction tests, using a P_1 preparation made on a Mu-induced *unc* strain in which the Mu was inserted between the *uncB* and *uncE* genes. A normal cotransduction frequency (40%) between *ilv* and *uncB* was found but only a very low cotransduction frequency ($<1\%$) was found between the *uncD* and *ilv* genes (F. Gibson and C. Yanofsky, unpublished). The reason for this low cotransduction frequency was due to the large insert of Mu DNA in the *unc* operon. Thus, it can be concluded (as shown in Figure 1) that in the *unc* operon the *uncB* gene is proximal to the *ilv* gene cluster.

Gene-Polypeptide Relationships

Gene-polypeptide relationships have thus far been established for two *unc* genes, and in both cases the mutant alleles have caused a loss of ATPase activity—the *uncA* gene codes for the α-subunit of the F_1 and the *uncD* gene codes for the β-subunit of the F_1.

It was shown that F_1-depleted membranes from a mutant strain carrying the *uncD409* allele could not be reconstituted for ATP-dependent reactions using a normal F_1 preparation, and that the lack of reconstitution was due to an altered β-subunit that remained bound to membranes even after washing with low-ionic-strength buffer [(196) and see below]. The altered β-subunit was identified using the two-dimensional polyacrylamide gel-electrophoresis method of O'Farrell [isoelectric focusing followed by sodium dodecyl sulphate electrophoresis, (see 197)]. This observation was confirmed using a partial diploid strain carrying both *unc*⁺ and *uncD409* alleles: both normal and altered β-subunits were found on the membranes

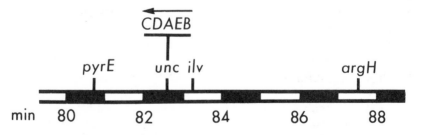

Figure 1 The location and polarity of the *unc* operon on the chromosome of *E. coli*. The conventions of genetic nomenclature and calibration of the chromosome in minutes are as in Bachmann et al (202).

of the partial diploid strain, and in the ATPase solubilized from the membranes (196).

The observation that the α-subunit of the F_1 is coded for by the *uncA* gene was made independently, using two different approaches. In one approach, partially purified inactive F_1 from a strain carrying the *uncA401* allele was cold-dissociated. It was found that ATPase activity could be reconstituted if an excess of α subunit, purified from a normal F_1, was added during the reconstitution (198, 199). No complementation of ATPase activity was observed after addition of purified β- or γ-subunits to the cold-dissociated mutant F_1.

Purified F_1 from three strains of *E. coli* carrying different *uncA* alleles (see below) was examined by two-dimensional gel-electrophoresis and it was found that the α-subunit in the F_1 from one of the mutant strains had an altered isoelectric point (41). In addition, in an analogous study to that described above for the *uncD409* allele, an altered α-subunit was identified bound to membranes prepared from strains carrying a fourth *uncA* allele: both an altered and a normal α-subunit were found to be present in F_1 purified from a partial diploid strain that carried both the mutant *uncA* and normal *unc*$^+$ alleles (41).

Gene-polypeptide relationships have not yet been established for any other subunits of the ATPase complex. A mutation causing DCCD resistance has been used to identify the DCCD-binding protein (105), but the mutation has not yet been allocated to a particular *unc* gene.

Characteristics of Mutations in Different unc Genes

It has become apparent (see below) that different mutations affecting the same *unc* gene can result in different biochemical characteristics. This confirms the suggestion made above that the only way in which a particular mutation can be ascribed to a particular gene is by genetic complementation tests or identification of an altered polypeptide. The characteristics of genetically characterized mutations affecting different *unc* genes are summarized in Table 1.

uncA MUTANTS   The *uncA* gene codes for the α-subunit of the F_1 (see above). There have been four *uncA* mutants described and these can be divided into two classes according to their biochemical characteristics (41). The *uncA401, uncA447,* and *uncA453* alleles form one class, while the *uncA450* allele is in a different class.

Membranes prepared from strains of *E. coli* carrying the *uncA401* allele lack ATPase activity, ATP-dependent membrane energization, and oxidative phosphorylation (8, 94). An ATPase complex is formed on the membrane and can be washed off using the same procedure as is used for normal F_1 (8, 41, 94, 95). The F_1-depleted membranes then become permeable to

Table 1   Some effects[a] of mutations in various *unc* genes

Mutant alleles	ATPase activity	Other characteristics	Component of ATPase affected (polypeptide)
uncA401, uncA447, uncA453	−	inactive F_1 on membranes, and F_1-depleted membranes can be reconstituted with normal F_1	F_1 (α-subunit)
uncA450	−	low-ionic-strength washed membranes cannot be reconstituted with normal F_1; mutant α-subunit remains on membranes	
uncB402	+	normal F_1 on membranes. F_1-depleted membranes bind normal F_1 but energy-linked reactions are not reconstituted. ATPase activity in cytoplasm	F_0
uncC424	+	membrane-bound ATPase activity sensitive to DCCD and bathophenanthroline. Cytoplasmic ATPase activity insensitive to DCCD and stimulated by bathophenanthroline	F_1 and/or F_0
uncD405, uncD406, uncD409, uncD427	−	β-subunit with abnormal net charge present on low-ionic strength washed membranes. These membranes bind differing amounts of normal F_1	F_1 (β-subunit)
uncD412	(−)[b]	F_1 with low ATPase activity on membranes. F_1-depleted membranes can be fully reconstituted with normal F_1	
uncE408, uncE429	+	ATPase activity found only in cytoplasm. Low-ionic-strength washed membranes do not bind normal F_1	F_0

[a] All strains carrying the mutant alleles listed show little or no growth on nonfermentable carbohydrates, have low aerobic growth yields, and cannot carry out oxidative phosphorylation.
[b] F_1 prepared from a mutant strain carrying the *uncD412* allele has about 10% of normal specific activity.

protons as judged by the lack of NADH-dependent acridine dye fluorescence-quenching (41, 113); the solubilized F_1 from *uncA401* can be used to reseal F_1-depleted membranes, but ATP-dependent reactions are not reconstituted. Alternatively, the addition of normal F_1 to F_1-depleted membranes from an *uncA401* mutant will reconstitute energy-linked activities (161).

Using F_1 purified from a strain carrying the *uncA401* allele it has been shown that all five of the subunits of the F_1 are present (41, 94, 95). In addition, a normal amount of tightly-bound nucleotide was found to be associated with the *uncA401* F_1 (94, 189) although the dissociation constant of the tightly-bound nucleotide was somewhat lower than normal (94).

Membranes prepared from strains of *E. coli* carrying the *uncA450* allele are different from those prepared from the other *uncA* mutants. The membranes do not become proton permeable after washing in low-ionic-strength buffer, nor can energy-linked activities be reconstituted by the addition of normal F_1 (41). An α-subunit with an abnormal isoelectric point was identified by two-dimensional gel electrophoresis as being present on washed membrane preparations from the *uncA450* mutant strain.

uncB MUTANTS   Strains carrying the *uncB402* allele [originally called *uncB401*, (see 193)] possess membrane-bound ATPase activity that cannot be coupled to membrane energization or oxidative phosphorylation (113, 159, 161) and, in addition, cytoplasmic ATPase activity. The membrane-bound ATPase activity is resistant to inhibition by DCCD or bathophenan-throline (112, 113). Lack of inhibition of ATPase activity by DCCD is not due to lack of the DCCD-binding protein (105).

The F_1 can be solubilized from the membranes by low-ionic-strength washing and can subsequently be used to reconstitute energy-dependent reactions in F_1-depleted membranes from normal or *uncA* mutant strains (161). However energy-linked reactions cannot be reconstituted by the addition of normal F_1 to F_1-depleted membranes from the *uncB* mutant strain which indicates that the *uncB402* allele affects an F_0 component (193). Unlike normal membranes, low-ionic-strength washing of membranes from an *uncB* mutant strain does not make them become permeable to protons or cause a loss of respiratory-driven functions, even though a normal F_1 has been solubilized (113).

uncC MUTANTS   Only one point mutation affecting the *uncC* gene has thus far been identified. Mutant strains carrying this allele, *uncC424,* have both membrane-bound and cytoplasmic ATPase activities (112, 159), but in contrast to the *uncB402* mutant strain, the membrane-bound ATPase activity is sensitive to inhibition by DCCD (112). Although the ATPase activities of the cytoplasmic F_1 found in strains carrying either the *uncB402* or *uncC424* allele are not inhibited by DCCD, the cytoplasmic ATPase activity is inhibited by bathophenanthroline in the *uncC* mutant compared with a stimulation in the *uncB* mutant (112).

On the basis of biochemical complementation between two strains carrying the *uncC424* allele or *uncD405* allele [(159) and see below] after using the techniques of disrupting a mixture of cells from the two mutant strains in a French pressure cell (180), it was concluded that the *uncC424* allele affected the F_0. However, subsequent work (J. A. Downie and G. B. Cox, unpublished) has shown that both the F_1 and F_0 appear to be affected by the *uncC424* allele and it is not yet known which peptide is affected.

uncD MUTANTS   The *uncD* gene codes for the β-subunit of the F_1, and the first mutation that was identified as occurring in the *uncD* gene was the *uncD409* allele (194).

Membranes prepared from a strain carrying the *uncD409* allele lacked ATPase activity and did not become proton permeable even after extensive washing with low-ionic-strength buffer (196). This observation suggested that the membranes had been altered, and two-dimensional gel electrophoresis of low-ionic-strength washed membranes showed that an altered β-subunit remained bound to the membranes (196).

Subsequently, four new mutant *uncD* alleles, *uncD405, uncD406, un-cD412*, and *uncD427* were identified and characterized. Two-dimensional gel-electrophoretic analysis of membranes prepared from strains carrying each of the *uncD* alleles revealed that β-subunits were present and altered in all strains. Thus, altered β-subunits were identified that have either higher isoelectric points (*uncD409, uncD427, uncD406*) or lower isoelectric points (*uncD412, uncD405*) than that of the normal β-subunit (200).

Unlike membranes from normal strains or *uncD409* mutant strains, it was found that membranes prepared from strains carrying the *uncD405, uncD406, uncD412*, or *uncD427* alleles were proton-permeable, as judged by the low levels of NADH-dependent atebrin fluorescence-quenching (200). Low-ionic-strength washing of the membranes from strains carrying the *uncD405, uncD427*, or *uncD406* alleles did not significantly increase the proton permeability of the membranes, whereas the membranes from the strain carrying the *uncD412* allele became as proton-permeable as normal membranes washed under the same conditions.

Washed membranes from the five *uncD* mutant strains bound different amounts of normal F_1; membranes from a strain carrying the *uncD412* allele bound normal amounts of F_1, whereas membranes from a strain carrying the *uncD409* allele bound the least F_1 (200). Different levels of reconstitution of ATP-dependent atebrin fluorescence-quenching were observed after the addition of normal F_1 to low-ionic-strength washed membranes from the various *uncD* mutants. Complete reconstitution was observed with the *uncD412* mutant, an intermediate level with *uncD405* membranes, and only a low level with the *uncD409, uncD427*, or *uncD406* membranes (200).

An F_1 aggregate was purified from the *uncD412* mutant using the same procedure as had been used for normal F_1, and it was found to contain an abnormal β-subunit and about 10% of normal activity. This purified F_1 could then be rebound to washed membranes from an *uncA* mutant strain and reconstitute NADH-dependent membrane energization and a low level of ATP-dependent energization (200).

uncE MUTANTS   Three mutant alleles *uncE408, uncE429*, and *unc-410* have been identified (195) that affect the *uncE* gene. Strains carrying each of these mutant alleles retained ATPase activity; in strains carrying the *uncE408* and *uncE429* alleles no ATPase activity was found bound to membranes, whereas the ATPase activity in strains carrying the *unc-410* allele was present both on the membranes and in the cytoplasm (195).

Membrane-bound ATPase solubilized from the *unc-410* mutant strain could be used to reconstitute ATP-dependent reactions in F_1-depleted membranes from a normal or an *uncA401* mutant strain (195) which indicates that the *unc-410* allele affects an F_0 component. Normal F_1 did not bind to membranes from *uncE408* or *uncE429* mutant strains, but using F_1-depleted *unc-410* membranes, normal F_1 bound to a level of about 30% compared with normal membranes, without reconstituting ATP-dependent energy-linked activities. The rebound ATPase activity of the *unc-410* membranes was not sensitive to inhibition by DCCD (195).

On the basis of the pattern of complementation in partial diploids carrying the *unc-410* allele and mutator (Mu) phage-induced polar mutants (see above), it was concluded that the *unc-410* allele may be a deletion affecting the *uncE* gene and another as yet unidentified gene. Thus, the *unc-410* allele has not yet been given a gene designation (195).

OTHER UNC MUTANTS: ATPASE⁻ A variety of mutants of *E. coli* lacking ATPase activity have been isolated in different laboratories. Some of these mutants, such as *uncA103c* (148), MDA1, MDA2, MDA3 (155), DL54 (149), DG20/1, BH212, AS69/1, BH273, and AS12/25 (158) were isolated as being able to grow on glucose but not on nonfermentable substrates. Another group of mutants, including NI44 (150, 151, 162), NR70 (154, 167), NR76 (176), and *unc17* (183, 152) were first selected as neomycin-resistant strains and subsequently screened for lack of growth on nonfermentable substrates.

It appears that, in general, those strains that were first selected on the basis of neomycin-resistance have membranes that are leaky to protons. With these strains, DCCD increases respiration-dependent membrane energization (109, 164, 176) and active transport (154, 166, 167). Membranes from strain NI44 are less proton-permeable than those from NR70, NR76, or DL54 because some respiration-dependent membrane energization was observed (109) and this low membrane energization appeared to be sufficient to drive the transhydrogenase reaction in membranes from strain NI44 (163). Using an isogenic series of partial diploid strains that have proton-permeable membranes, a correlation has been observed between the degree of leakiness and the level of neomycin-resistance (F. Gibson, unpublished). It should be noted that not all *unc* mutant strains are neomycin-resistant (158).

Among those strains selected on the basis of their inability to grow on nonfermentable substrates, there appear to be two different groups: those, such as DL54 (163, 168, 179), that have proton-permeable membranes (see also *uncD* mutants above) and those, such as *uncA103c* (148), DG20/1, BH212, AS12/25, BH273, and AS69/1 (158), like *uncA401* and *uncD409* membranes (see above), that appear to have normal respiration-dependent membrane energization. Apparently, the proton permeability of the leaky mutants, NR70 and DL54, can be suppressed by growing the mutants anaerobically, although the reason for this is not clear (179). In a study of binding of normal F_1 to the membranes of ATPase-deficient mutants, three classes of mutant membranes were identified: MDA1, which could bind F_1 only if membranes were first treated by low-ionic-strength washing; MDA2, which could bind F_1 even before the washing procedure; and MDA3 which could not bind F_1, even after the membranes were washed (155).

Although it is difficult to draw comparisons between different mutant strains that are in widely varying genetic backgrounds, it is possible that the differences in the phenotypes of various ATPase-deficient *unc* mutants may even result from mutations affecting a single gene (see section on *uncA* or *uncD* mutants, above).

OTHER UNC MUTANTS: ATPASE+   Of those *unc* mutants that retain ATPase activity, some, such as BV4, K11, and A144 (180), were isolated using neomycin-resistance as a primary selection factor, whereas others, such as *unc253* (153), BG31 (157), MDB (155), *unc373* (156), DG15/10, DG26/4, DG7/10, and DG31/3 (158) were selected because of their inability to grow on nonfermentable substrates. Where tested, the ATPase activities of these mutants were insensitive to inhibition by DCCD. Membranes prepared from mutants *unc253* (153), BG31 (157), DG15/10, DG26/4, DG7/10, and DG31/3 (158) were not proton-permeable even after washing with low-ionic-strength buffer.

In strain BV_4, the mutation affects an F_1 component, since energy-linked reactions in washed membranes from this strain can be reconstituted using F_1 from a normal strain or the mutant strains K11 or A144 (180). Likewise, the addition of normal F_1 to washed membranes of strain DG25/9 reconstituted some ATP-dependent membrane energization (158). The unwashed membranes from this strain had a low level of oxidative phosphorylation and ATP-dependent membrane energization; this latter activity could be increased by the addition of normal F_1 (158). No reconstitution of membrane energization has been reported with any of the other *unc* mutants that retain ATPase activity.

It appears that with membranes from K11, A144 (180), BG31 (157), *unc-373* (156), DG15/10, DG26/4, DG7/10, DG31/3, and DG25/9, (158) but not MDB (155), that ATPase remains bound to the membranes. F_1 from K11, A144 (180), BG31 (157), DG15/10, DG7/10, DG31/3, and DG25/9 (158) could be used to reconstitute ATP-dependent energy-linked reactions.

DCCD-RESISTANT MUTANTS   A mutant strain that has a membrane-bound ATPase resistant to inhibition by DCCD, was isolated after selection of DCCD-resistant mutants on succinate medium (105). These mutants are normal with respect to ATPase activity, ATP-linked reactions, and oxidative phosphorylation. The mutation conferring DCCD resistance affects an F_0 component of the ATPase complex that has a molecular weight of about

9000. It is not yet clear whether or not this mutation affects a previously identified gene. Surprisingly, if these DCCD-resistant mutants are grown on glucose, the DCCD resistance is suppressed (201). It is possible that the mutant lesion affecting strain DG7/10 affects the DCCD-binding protein, since this protein was not labelled by ^{14}C-DCCD in membranes from that strain (106).

CONCLUSIONS

As with the ATPase from the mitochondria of eukaryotic cells, there is still a great deal to be learnt about the prokaryotic energy-transducing ATPases. Even information about the number of polypeptides in the complex and the way in which these polypeptides are assembled and arranged is still far from complete. The molecular mechanisms of energy transductions in which the ATPase is involved are still matters of hypothesis and are likely to remain so until more factual information is available regarding the ATPase complex. Recent experiments on the reconstitution of the ATPase complex by assembly of purified subunits and the application of the techniques of molecular biology are yielding interesting information. It is apparent that single amino-acid substitutions in the ATPase subunits lead to profound phenotypic changes. Further study of these changes, together with studies on the protein chemistry of normal and mutant ATPases, should contribute very significantly to our understanding of energy transductions.

ACKNOWLEDGMENTS

We wish to express our thanks to the following for making available unpublished manuscripts; A. Abrams, P. D. Bragg, S. D. Dunn, M. Futai, Y. Kagawa, E. Munoz, and A. E. Senior. We are grateful for the expert secretarial assistance and patience of Mrs. D. Grimshaw during the preparation of the manuscript.

Literature Cited

1. Pinchot, G. B., Racker, E. 1951. In *Phosphorus Metabolism, Vol. 1,* ed. W. D. McElroy, B. Glass, p. 366. Baltimore: Johns Hopkins Univ. Press
2. Pinchot, G. B. 1953. *J. Biol. Chem.* 205:65–74
3. Abrams, A., McNamara, P., Johnson, F. B. 1960. *J. Biol. Chem.* 235:3659–62
4. Abrams, A., Smith, J. B. 1974. In *The Enzymes* ed. P. D. Boyer, pp. 395–429. New York/London: Academic. 886 pp.
5. Simoni, R. D., Postma, P. W. 1975. *Ann. Rev. Biochem.* 44:523–54

6. Melandri, B. A., Baccarini-Melandri, A. 1976. *J. Bioenerg.* 8:109–19
7. Haddock, B. A., Jones, C. W. 1977. *Bacteriol. Rev.* 41:47–99
8. Butlin, J. D., Cox, G. B., Gibson, F. 1971. *Biochem. J.* 124:75–81
9. Mirsky, R., Barlow, V. 1971. *Biochim. Biophys. Acta* 241:835–45
10. Pinchot, G. B. 1957. *J. Biol. Chem.* 229:1–9
11. Voelz, H. 1964. *J. Bacteriol.* 86:1196–98
12. Oppenheim, J. D., Salton, M. R. J.

1973. *Biochim. Biophys. Acta* 298:297–322
13. Abram, D. 1965. *J. Bacteriol.* 89:855–73
14. Asano, A., Cohen, N., Baker, R. F., Brodie, A. F. 1973. *J. Biol. Chem.* 248:3386–97
15. Jones, C. W., Redfearn, E. R. 1967. *Biochim. Biophys. Acta* 143:354–62
16. Ishikawa, S., Lehninger, A. L. 1962. *J. Biol. Chem.* 237:2401–8
17. Abrams, A. 1965. *J. Biol. Chem.* 240:3675–81
18. Hachimori, A., Murumatsu, N., Nosoh, Y. 1970. *Biochim. Biophys. Acta* 206:426–37
19. Ishida, M., Mizushima, S. 1969. *J. Biochem.* 66:33–43
20. Pandit-Hovenkamp, H. G. 1967. *Methods Enzymol.* 10:152–57
21. Munoz, E., Salton, M. R. J., Ng, M. H., Schor, M. T. 1969. *Eur. J. Biochem.* 7:490–501
22. Kobayashi, H., Anraku, Y. 1972. *J. Biochem.* 71:387–99
23. Davies, P. L., Bragg, P. D. 1972. *Biochim. Biophys. Acta* 266:273–84
24. Melandri, B. A., Baccarini-Melandri, A., Fabbri, E. 1972. *Biochim. Biophys. Acta* 275:383–94
25. Adolfsen, R., Moudrianakis, E. N. 1971. *Biochemistry* 10:2247–53
26. Monteil, H., Schoun, J., Guinard, M. 1974. *Eur. J. Biochem.* 41:525–32
27. Lücke, F. K., Klemme, J. H. 1976. *Z. Naturforsch.* 31:272–79
28. Neeman, Z., Kahane, I., Razin, S. 1971. *Biochim. Biophys. Acta* 249:169–76
29. Cox, G. B., Downie, J. A., Fayle, D. R. H., Gibson, F., Radik, J. 1978. *J. Bacteriol.* 133:287–92
30. Downie, J. A., Senior, A. E., Cox, G. B., Gibson, F. 1979. *J. Bacteriol.* In press
31. Salton, M. R. J., Schor, M. T. 1972. *Biochem. Biophys. Res. Commun.* 49:350–57
32. Salton, M. R. J., Schor, M. T. 1974. *Biochim. Biophys. Acta* 345:74–82
33. Ishikawa, S. 1970. *J. Biochem.* 67:297–312
34. Abrams, A., Jensen, C., Morris, D. H. 1976. *Biochem. Biophys. Res. Commun.* 69:804–11
35. Bragg, P. D., Hou, C. 1972. *FEBS Lett.* 28:309–12
36. Cox, G. B., Gibson, F., McCann, L., Butlin, J. D., Crane, F. L. 1973. *Biochem. J.* 132:689–95
37. Hanson, R. L., Kennedy, E. P. 1973. *J. Bacteriol.* 114:772–81

38. Reed, D. W., Raveed, D. 1972. *Biochim. Biophys. Acta* 283:79–91
39. Baccarini-Melandri, A., Gest, H., San Pietro, A. 1970. *J. Biol. Chem.* 245:1224–26
40. Futai, M., Sternweis, P. C., Heppel, L. A. 1974. *Proc. Natl. Acad. Sci. USA* 71:2725–29
41. Senior, A. E., Downie, J. A., Cox, G. B., Gibson, F., Langman, L., Fayle, D. R. H. 1979. *Biochem. J.* In press
42. Nelson, N., Kanner, B. I., Gutnick, D. L. 1974. *Proc. Natl. Acad. Sci. USA* 71:2720–24
43. Vogel, G., Steinhart, R. 1976. *Biochemistry* 15:208–16
44. Giordano, G., Riviere, C., Azoulay, E. 1975. *Biochim. Biophys. Acta* 389:203–18
45. Bragg, P. D., Hou, C. 1975. *Arch. Biochem. Biophys.* 167:311–21
46. Vogel, G., Schairer, H. U., Steinhart, R. 1978. *Eur. J. Biochem.* 87:155–60
47. Kobayashi, H., Anraku, Y. 1974. *J. Biochem.* 76:1175–82
48. Higashi, T., Kalra, V. K., Lee, S. H., Bogin, E., Brodie, A. F. 1975. *J. Biol. Chem.* 250:6541–48
49. Kalra, V. K., Lee, S. H., Ritz, C. J., Brodie, A. F. 1975. *J. Supramol. Struct.* 3:231–41
50. Baron, C., Abrams, A. 1971. *J. Biol. Chem.* 246:1542–44
51. Mileikovskaya, E. I., Kozlov, I. A., Tikhonova, G. V. 1975. *Biokimiya.* 40:846–51
52. Ishikawa, S. 1967. *Methods Enzymol.* 10:169–75
53. Andreu, J. M., Albendea, J. A., Munoz, E. 1973. *Eur. J. Biochem.* 37:505–15
54. Hulla, F. W., Höckel, M., Risi, S., Dose, K. 1976. *Eur. J. Biochem.* 67:469–76
55. Risi, S., Höckel, M., Hulla, F. W., Dose, K. 1977. *Eur. J. Biochem.* 81:103–9
56. Andreu, J. M., Munoz, E. 1975. *Biochim. Biophys. Acta* 387:228–33
57. Melandri, A., Baccarini-Melandri, A., Crofts, A. R., Codgell, R. 1972. *FEBS Lett.* 24:141–45
58. Johansson, B. C., Baltscheffsky, M., Baltscheffsky, H., Baccarini-Melandri, A., Melandri, M. 1973. *Eur. J. Biochem.* 40:109–17
59. Baccarini-Melandri, A., Melandri, B. A. 1971. *Methods Enzymol.* 23:556–61
60. Nieuwenhuis, F. J. R. M., Thomas, A. A. M., Van Dam, K. 1974. *Biochem. Soc. Trans.* 2:512–13

61. Hare, J. F. 1975. *Biochem. Biophys. Res. Commun.* 66:1329–37
62. Foster, D. L., Fillingame, R. H. 1978. *Fed. Proc.* 37(6):1520 (Abstr.)
63. Clarke, D. J., Morris, J. G. 1976. *Biochem. J.* 154:725–29
64. Lee, S. H., Cohen, N. S., Brodie, A. F. 1976. *Proc. Natl. Acad. Sci. USA* 73:3050–53
65. Schnebli, H. P., Vatter, A. E., Abrams, A. 1970. *J. Biol. Chem.* 245:1122–27
66. Mirsky, R., Barlow, V. 1973. *Biochim. Biophys. Acta* 291:480–88
67. Carriera, J., Andreu, J. M., Munoz, E. 1977. *Biochim. Biophys. Acta* 492:387–98
68. Carriera, J., Andreu, J. M., Nieto, M., Munoz, E. 1976. *Mol. Cell. Biochem.* 10:67–76
69. Adolfsen, R., McClung, J. A., Moudrianakis, E. N. 1975. *Biochemistry* 14:1727–35
70. Adolfsen, R., Moudrianakis, E. N. 1976. *Biochemistry* 15:4163–70
71. Andreu, J. M., Carriera, J., Munoz, E. 1976. *FEBS Lett.* 65:198–203
72. Andreu, J. M., Larraga, V., Munoz, E. 1977. *Bioenergetics of Membranes* ed. L. Packer, G. C. Popageorgiou, A. Trebst, pp. 529–38. New York: Academic
73. Andreu, J. M., Warth, R., Munoz, E. 1978. *FEBS Lett.* 86:1–5
74. Johansson, B. C., Baltscheffsky, M. 1975. *FEBS Lett.* 53:221–24
75. Kagawa, Y., Sone, N., Yoshida, M., Hirata, H., Okamoto, H. 1976. *J. Biochem.* 80:141–51
76. Bragg, P. D., Hou, C. 1976. *Biochem. Biophys. Res. Commun.* 72:1042–48
77. Azocar, O., Munoz, E. 1977. *Biochim. Biophys. Acta* 482:438–52
78. Roisin, M-P., Kepes, A. 1972. *Biochim. Biophys. Acta* 275:333–46
79. Carriera, J., Munoz, E. 1975. *Mol. Cell. Biochem.* 9:85–95
80. Evans, D. J. 1969. *J. Bacteriol.* 100:914–22
81. Roisin, M-P., Kepes, A. 1973. *Biochim. Biophys. Acta* 305:249–59
82. Günther, T., Pellnitz, W., Mariss, G. 1974. *Z. Naturforsch.* 29c:54–59
83. Lastras, M., Munoz, E. 1974. *J. Bacteriol.* 119:593–601
84. Adolfsen, R., Moudrianakis, E. N. 1973. *Biochemistry* 12:2926–33
85. Riebeling, V., Jungerman, K. 1976. *Biochim. Biophys. Acta* 430:434–44
86. Serrahima-Zieger, M., Monteil, H. 1978. *Biochim. Biophys. Acta* 502:445–57
87. Senior, A. E. 1978. In *Membrane Proteins in Energy Transduction* ed. R. A. Capaldi. New York: Dekker. In press
88. Maeda, M., Kobayashi, H., Futai, M., Anraku, Y. 1977. *J. Biochem.* 82:311–14
89. Harris, D. A., John, P., Radda, G. K. 1977. *Biochim. Biophys. Acta* 459:546–59
90. Abrams, A., Nolan, E. A., 1972. *Biochem. Biophys. Res. Commun.* 48:982–89
91. Abrams, A., Nolan, E. A., Jensen, C., Smith, J. B. 1973. *Biochem. Biophys. Res. Commun.* 55:22–29
92. Abrams, A., Jensen, C., Morris, D. 1975. *J. Supramol. Struct.* 3:261–74
93. Adolfsen, R., Moudrianakis, E. N. 1976. *Arch. Biochem. Biophys.* 172:425–33
94. Bragg, P. D., Hou, C. 1977. *Arch. Biochem. Biophys.* 178:486–94
95. Maeda, M., Futai, M., Anraku, Y. 1977. *Biochem. Biophys. Res. Commun.* 76:331–38
96. Harold, F. M., Baarda, J. R., Baron, C., Abrams, A. 1969. *J. Biol. Chem.* 244:2261–68
97. Franklin, R. M., Datta, A., Dahlberg, J. E., Braunstein, S. N. 1971. *Biochim. Biophys. Acta* 233:521–37
98. Mileikovskaya, E. I., Tikhonova, G. V., Kondrashin, A. A., Kozlov, I. A. 1976. *Eur. J. Biochem.* 62:613–17
99. Evans, D. J. 1970. *J. Bacteriol.* 104:1203–12
100. Fisher, R. J., Sanadi, D. R. 1971. *Biochim. Biophys. Acta* 245:34–41
101. Kalra, V. K., Brodie, A. F. 1971. *Arch. Biochem. Biophys.* 147:653–59
102. Tsuchiya, T. 1977. *J. Bacteriol.* 129:763–69
103. Abrams, A., Baron, C. 1970. *Biochem. Biophys. Res. Commun.* 41:858–63
104. Abrams, A., Smith, J. B., Baron, C. 1972. *J. Biol. Chem.* 247:1484–88
105. Fillingame, R. H. 1975. *J. Bacteriol.* 124:870–83
106. Altendorf, K., Zitzmann, W. 1975. *FEBS Lett.* 59:268–72
107. Fillingame, R. H. 1976. *J. Biol. Chem.* 251:6630–37
108. Cattell, K. J., Lindop, C. R., Knight, I. G., Beechey, R. B. 1971. *Biochem. J.* 125:169–77
109. Nieuwenhuis, F. J. R. M., Kanner, B. I., Gutnick, D. L., Postma, P. W., Van Dam, K. 1973. *Biochim. Biophys. Acta* 325:62–71
110. Hasan, S. M., Rosen, B. P. 1977. *Biochim. Biophys. Acta* 459:225–40

111. Patel, L., Kaback, H. R. 1976. *Biochemistry* 15:2741–46
112. Cox, G. B., Crane, F. L., Downie, J. A., Radik, J. 1977. *Biochim. Biophys. Acta* 462:113–20
113. Hasan, S. M., Tsuchiya, T., Rosen, B. P. 1978. *J. Bacteriol.* 133:108–13
114. Sun, I. L., Phelps, D. C., Crane, F. L. 1975. *FEBS Lett.* 54:253–58
115. Phelps, D. C., Nordenbrand, K., Nelson, B. D., Ernster, L. 1975. *Biochem. Biophys. Res. Commun.* 63:1005–12
116. Phelps, D. C., Nordenbrand, K., Hundal, T., Carlsson, C., Nelson, B. D., Ernster, L. 1975. *Electron Transfer Chains and Oxidative Phosphorylation* ed. E. Quagliariello, S. Papa, F. Palmieri, E. C. Slater, N. Siliprandi, pp. 385–400. Amsterdam: North Holland. 450 pp.
117. Oren, R., Gromet-Elhanan. Z. 1977. *FEBS Lett.* 79:147–50
118. Melandri, B. A., Fabbri, E., Firstarer, E., Baccarini-Melandri, A. 1974. *Membrane Proteins in Transport and Phosphorylation* ed. G. F. Azzone, M. E. Klingenberg, E. Quagliariello, N. Siliprandi, pp. 55–60. Amsterdam: Elsevier
119. Kobayashi, H., Maeda, M., Anraku, Y. 1977. *J. Biochem.* 81:1071–77
120. Harold, F. M., Baarda, J. R., Baron, C., Abrams, A. 1969. *Biochim. Biophys. Acta* 183:129–36
121. Eilermann, L. J. M., Pandit-Hovenkamp, H. G., Van Der Meer-Van Buren, M., Kolk, A. H. J., Feenstra, M. 1971. *Biochim. Biophys. Acta* 245:305–12
122. Bragg, P. D., Hou, C. 1975. *Arch. Biochem. Biophys.* 174:553–61
123. Sun, I. L., Crane, F. L. 1975. *Biochem. Biophys. Res. Commun.* 65:1334–42
124. Ravizzini, R. A., Lescano, W. I. M., Vallejos, R. H. 1975. *FEBS Lett.* 58:285–88
125. Baccarini-Melandri, A., Fabbri, E., Melandri, B. A. 1975. *Biochim. Biophys. Acta* 376:82–88
126. Satre, M., Klein, G., Vignais, P. V. 1978. *J. Bacteriol.* 134:17–23
127. Chang, T. M., Penefsky, H. S. 1973. *J. Biol. Chem.* 248:2746–54
128. Van De Stadt, R. J., Van Dam, K., Slater, E. C. 1974. *Biochim. Biophys. Acta* 347:224–39
129. Ferguson, S. J., Lloyd, W. J., Radda, G. K., 1974. *FEBS Lett.* 38:234–36
130. Ferguson, S. J., Lloyd, W. J., Radda, G. K., Whatley, F. R. 1974. *Biochim. Biophys. Acta* 357:457–61
131. Kagawa, Y. 1976. *J. Cell Physiol.* 89:569–74
132. Yoshida, M., Sone, N., Hirata, H., Kagawa, Y. 1975. *J. Biol. Chem.* 250:7910–16
133. Sone, N., Yoshida, M., Hirata, H., Kagawa, Y. 1975. *J. Biol. Chem.* 250:7917–23
134. Sone, N., Yoshida, M., Hirata, H., Kagawa, Y. 1977. *J. Biochem.* 81:519–28
135. Sone, N., Yoshida, M., Hirata, H., Kagawa, Y. 1977. *J. Biol. Chem.* 252:2956–60
136. Yoshida, M., Sone, N., Hirata, H., Kagawa, Y. 1977. *J. Biol. Chem.* 252:3480–85
137. Höckel, M., Hulla, F. W., Risi, S., Dose, K. 1976. *Biochim. Biophys. Acta* 429:1020–28
138. Smith, J. B., Wilkowski, C. 1978. *Fed. Proc.* 37(6):1521 (Abstr.)
139. Okamoto, H., Sone, N., Hirata, H., Yoshida, M., Kagawa, Y. 1977. *J. Biol. Chem.* 252:6125–31
140. Yoshida, M., Okamoto, H., Sone, N., Hirata, H., Kagawa, Y. 1977. *Proc. Natl. Acad. Sci. USA* 74:936–40
141. Smith, J. B., Sternweis, P. C. 1975. *Biochem. Biophys. Res. Commun.* 62:764–71
142. Nelson, N., Deters, D. W., Nelson, H., Racker, E. 1973. *J. Biol. Chem.* 248:2049–55
143. Sternweis, P. C., Smith, J. B. 1977. *Biochemistry* 16:4020–25
144. Sternweis, P. C. 1978. *J. Biol. Chem.* 253:3123–28
145. Smith, J. B., Sternweis, P. C. 1977. *Biochemistry* 16:306–11
146. Futai, M. 1977. *Biochem. Biophys. Res. Commun.* 79:1231–37
147. Dunn, S. D., Futai, M. 1978. *Fed. Proc.* 37(6):1518 (Abstr.)
148. Schairer, H. U., Haddock, B. A. 1972. *Biochem. Biophys. Res. Commun.* 48:544–51
149. Simoni, R. D., Shallenberger, M. K. 1972. *Proc. Natl. Acad. Sci. USA* 69:2663–67
150. Kanner, B. I., Gutnick, D. L. 1972. *J. Bacteriol.* 111:287–89
151. Gutnick, D. L., Kanner, B. I., Postma, P. W. 1972. *Biochim. Biophys. Acta* 283:217–22
152. Yamamoto, T. H., Mével-Ninio, M., Valentine, R. C. 1973. *Biochim. Biophys. Acta* 314:267–75
153. Schairer, H. U., Gruber, D. 1973. *Eur. J. Biochem.* 37:282–86
154. Rosen, B. P. 1973. *J. Bacteriol.* 116:1124–29
155. Daniel, J., Roisin, M-P, Burstein, C.,

Kepes, A. 1975. *Biochim. Biophys. Acta* 376:195–209

156. Thipayathasana, P. 1975. *Biochim. Biophys. Acta* 408:47–57
157. Simoni, R. D., Shandell, A. 1975. *J. Biol. Chem.* 250:9421–27
158. Schairer, H. U., Friedl, P., Schmid, B. I., Vogel, G. 1976. *Eur. J. Biochem.* 66:257–68
159. Gibson, F., Cox, G. B., Downie, J. A., Radik, J. 1977. *Biochem. J.* 164:193–98
160. Butlin, J. D., Cox, G. B., Gibson, F. 1973. *Biochim. Biophys. Acta* 292:366–75
161. Cox, G. B., Gibson, F., McCann, L. 1973. *Biochem. J.* 134:1015–21
162. Kanner, B. I., Gutnick, D. L. 1972. *FEBS Lett.* 22:197–99
163. Bragg, P. D., Hou, C. 1973. *Biochem. Biophys. Res. Commun.* 50:729–36
164. Tsuchiya, T., Rosen, B. P. 1975. *J. Biol. Chem.* 250:8409–15
165. Gibson, F., Downie, J. A., Cox, G. B., Radik, J. 1978. *J. Bacteriol.* 134:728–36
166. Van Thienen, G., Postma, P. W. 1973. *Biochim. Biophys. Acta* 323:429–40
167. Rosen, B. P. 1973. *Biochem. Biophys. Res. Commun.* 53:1289–96
168. Altendorf, K., Harold, F. M., Simoni, R. D. 1974. *J. Biol. Chem.* 249:4587–93
169. Kobayashi, H., Kin, E., Anraku, Y. 1974. *J. Biochem.* 76:251–61
170. Curtis, S. J. 1974. *J. Bacteriol.* 120:295–303
171. Wilson, D. B. 1974. *J. Bacteriol.* 120:866–71
172. Cowell, J. L. 1974. *J. Bacteriol.* 120:139–46
173. Berger, E. A., Heppel, L. A. 1974. *J. Biol. Chem.* 249:7747–55
174. Rosenberg, H., Cox, G. B., Butlin, J. D., Gutowski, S. J. 1975. *Biochem. J.* 146:417–23
175. Maloney, P. C., Kashket, E. R., Wilson, T. H. 1974. *Proc. Natl. Acad. Sci. USA* 71:3896–3900
176. Rosen, B. P., Adler, L. W. 1975. *Biochim. Biophys. Acta* 387:23–36
177. del Campo, F. F., Hernández-Asensio, M., Ramírez, J. M. 1975. *Biochem. Biophys. Res Commun.* 63:1099–1105
178. Tsuchiya, T., Rosen, B. P. 1975. *Biochem. Biophys. Res. Commun.* 63:832–38
179. Boonstra, J., Gutnick, D. L., Kaback, H. R. 1975. *J. Bacteriol.* 124:1248–55

180. Kanner, B. I., Nelson, N., Gutnick, D. L. 1975. *Biochim. Biophys. Acta* 396:347–59
181. Grinius, L., Brazenaite, J. 1976. *FEBS Lett.* 62:186–89
182. Wilson, D. M., Alderete, J. F., Maloney, P. C., Wilson, T. H. 1976. *J. Bacteriol.* 126:327–37
183. Mével-Ninio, M. T., Valentine, R. C. 1975. *Biochim. Biophys. Acta* 376:485–91
184. Larsen, S. H., Adler, J., Gargus, J. J., Hogg, R. W. 1974. *Proc. Natl. Acad. Sci. USA* 71:1239–43
185. Thipayathasana, P., Valentine, R. C. 1974. *Biochim. Biophys. Acta* 347:464–68
186. Zukin, R. S., Koshland, D. E. 1976. *Science* 193:405–8
187. Stouthamer, A. H., Bettenhaussen, C. W. 1977. *Arch. Microbiol.* 113:185–89
188. Miki, K., Lin, E. C. C. 1975. *J. Bacteriol.* 124:1282–87
189. Maeda, M., Kobayashi, H., Futai, M., Anraku, Y. 1976. *Biochem. Biophys. Res. Commun.* 70:228–34
190. Hancock, R. E. W., Braun, V. 1976. *J. Bacteriol.* 125:409–15
191. Cox, G. B., Downie, J. A. 1978. *Methods Enzymol.* 56:106–17
192. Friedl, P., Schmid, B. I., Schairer, H. U. 1977. *Eur. J. Biochem.* 73:461–68
193. Gibson, F., Cox, G. B., Downie, J. A., Radik, J. 1977. *Biochem. J.* 162:665–70
194. Cox, G. B., Downie, J. A., Gibson, F., Radik, J. 1978. *Biochem. J.* 170:593–98
195. Downie, J. A., Senior, A. E., Gibson, F., Cox, G. B. 1979. *J. Bacteriol.* In press
196. Fayle, D. R. H., Downie, J. A., Cox, G. B., Gibson, F., Radik, J. 1978. *Biochem. J.* 172:523–31
197. O'Farrell, P. H. 1975. *J. Biol. Chem.* 250:4007–21
198. Kanazawa, H., Saito, S., Futai, M. 1978. *J. Biochem.* In press
199. Dunn, S. D. 1978. *Biochem. Biophys. Res. Commun.* 82:596–602
200. Senior, A. E., Fayle, D. R. H., Downie, J. A., Gibson, F., Cox, G. B. 1979. *Biochem. J.* In press
201. Fillingame, R. H., Wopat, A. E. 1978. *J. Bacteriol.* 134:687–89
202. Bachmann, B. J., Low, K. B., Taylor, A. L. 1976. *Bacteriol. Rev.* 40:116–67
203. Decker, S. J., Lang, D. R. 1977. *J. Bacteriol.* 131:98–104

Ann. Rev. Biochem. 1979. 48:133–58

REDUCTION OF RIBONUCLEOTIDES

♦12005

Lars Thelander and Peter Reichard

Medical Nobel Institute, Department of Biochemistry, Karolinska Institute, S-104 01 Stockholm, Sweden

CONTENTS

PERSPECTIVES AND SUMMARY

In the early 1950s in vivo experiments with rats (1, 2) first suggested that deoxyribonucleotides are formed by a direct reduction of ribonucleotides. Later, the enzyme involved in this process was found in all procaryotic and eucaryotic cells that synthesize DNA. Ribonucleotide reductase catalyzes the first unique step in DNA synthesis and the enzyme provides the cell with

133

0066-4154/79/0701-0133$01.00

a balanced supply of the four deoxyribonucleotides. A close correlation of the regulation of DNA synthesis and ribonucleotide reduction can be expected and has also been observed in many systems. However, there is no evidence that the general supply of deoxynucleotides, i.e. the overall activity of ribonucleotide reductase, by itself regulates DNA synthesis. Instead, more sophisticated regulatory interactions have been invoked but remain so far largely speculative.

All ribonucleotide reductases catalyze the substitution of the OH-group at position 2' of ribose by a hydrogen, with NADPH as the ultimate hydrogen donor:

$$\text{NADPH} + \text{H}^+ + \text{ribonucleotide} \rightarrow \text{NADP}^+$$
$$+ \text{deoxyribonucleotide} + \text{H}_2\text{O}$$

Two distinct classes of reductases have been described. One class is represented by the enzyme isolated from *Escherichia coli,* the other by the enzyme from *Lactobacillus leichmannii.* The latter is a monomeric enzyme and uses adenosylcobalamin as a dissociable cofactor. The *E. coli* enzyme lacks this requirement. It consists of two nonidentical subunits, one of which contains two bound iron atoms and a tyrosyl radical as part of the polypeptide chain. This subunit may be thought of as the functional counterpart of cobalamin.

In spite of these structural dissimilarities, the mechanism of reduction at the "substrate level" appears to be similar and involves a direct replacement of the OH-group by a hydrogen with retention of configuration at the carbon atom. At the "protein level" a radical-dithiol system appears to be the reducing agent in both cases. Elucidation of the intermediate steps may help to understand why widely different structural elements are used for the same chemical purpose.

In both cases small proteins participate as hydrogen carriers. The first such protein described, *E. coli* thioredoxin, contains an oxidation-reduction active disulfide in a protrusion of its three-dimensional structure. The disulfide is reduced with NADPH by a specific flavoprotein (thioredoxin reductase) and the dithiol then acts as hydrogen donor for the reduction of ribose. Similar thioredoxin systems have been found in *L. leichmannii* and other types of cells. They participate as hydrogen carriers in a variety of reductive processes. One common denominator for their function could be the ability of reduced thioredoxin to function as a powerful protein disulfide reductase.

This simple picture has recently been upset by two discoveries: (*a*) prototrophic mutants of *E. coli* lacking either thioredoxin or thioredoxin reductase show no decreased ability to reduce ribonucleotides and (*b*) most thioredoxin in wild type *E. coli* occurs as phosphothioredoxin, containing a phosphothiol linkage. Further work is required to appreciate the full implications of these results which, nevertheless, clearly demonstrate that thioredoxin is not an obligatory intermediate in ribonucleotide reduction. Another small protein (glutaredoxin) containing an oxidation-reduction active disulfide and active in ribonucleotide reduction was recently identified in *E. coli*. For its function, glutaredoxin requires the presence of glutathione, NADPH, and glutathione reductase. Figure 1 schematically summarizes the properties of the two classes of ribonucleotide reductases.

Ribonucleotide reductases are allosteric enzymes, i.e. they contain one or several effector-binding sites that are distinct from the catalytic site. One remarkable feature of all enzymes is that the same catalytic site reduces all four ribonucleotides but that the specificity of this site is modulated by binding of nucleoside triphosphates to allosteric sites. This effect is developed in a rather rudimentary fashion for the *Lactobacillus* enzyme where effectors, in addition, enhance binding of cobalamine. It is already found as a highly sophisticated mechanism for the *E. coli* enzyme and appears in all its beauty with the mammalian enzymes.

Our knowledge of the properties of ribonucleotide reductases from mammalian sources has been lagging. In most respects the properties of highly purified preparations are similar to those of the *E. coli* enzyme, and iron, not cobalamine, is involved. One enzyme reduces all four ribonucleotides, and earlier conflicting data concerning this point can now be explained by the allosteric regulation of substrate specificity. Recently, the allosteric behavior of the mammalian enzyme has also provided an interesting explanation for hereditary immune-deficiency diseases caused by a lack of adenosine deaminase or purine nucleoside phosphorylase.

Little is known about the organization of ribonucleotide reductase within the cell and its possible spatial or functional relation to the DNA synthesizing machinery. Does the enzyme occur as part of a multienzyme complex, together with other precursor enzymes and possibly with DNA replication enzymes? Is it located in the cytoplasm or in the cell nucleus in eucaryotic cells?

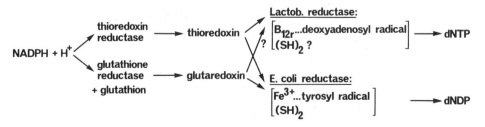

Figure 1   Alternative modes of ribonucleotide reduction.

During recent years two excellent reviews (3, 4) have appeared which particularly deal with the mechanism of action of the *Lactobacillus* enzyme.

RIBONUCLEOTIDE REDUCTASE FROM *E. COLI*

Structural Aspects

The enzyme consists of two nonidentical subunits (5) that were named protein B1 and B2. These are defined by two closely linked structural genes (*nrdA* and *nrdB*), located at 48 min between *nalA* and *glpT* (6). So far only one mutant in each gene has been described (7–9). The two genes appear to be part of one operon and show a coordinate regulation (10). A strain of *E. coli,* lysogenic for a defective λ-phage carrying both *nrdA* and *nrdB* was constructed (11). On induction, extracts contained up to 10% of the soluble protein as B1+B2, in a roughly 1:1 ratio. By a new purification procedure, involving affinity chromatography on dATP-Sepharose (12), large amounts of highly active enzyme can now be obtained from such extracts (11).

The schematic structure of *E. coli* ribonucleotide reductase is given in Figure 2. The active enzyme contains B1 and B2 in a 1:1 stoichiometry (13, 14), and each subunit by itself is completely inactive. Binding of subunits requires Mg^{2+} and is weak (13). During purification, the two subunits dissociate easily and were purified as separate entities (5).

Protein B1 has a molecular weight of 160,000 and, as isolated, is a dimer of the general structure $\alpha\alpha'$ (14). The two polypeptide chains are of similar or identical size with identical COOH termini but different NH_2 termini. It is not clear whether the difference is caused by a preparation artefact or

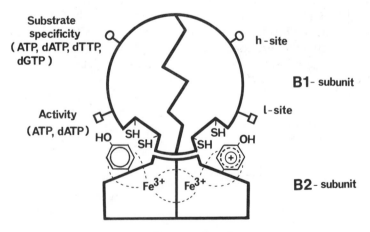

Figure 2   Model of ribonucleoside diphosphate reductase from *E. coli.*

whether the two peptides are products of separate structural genes. In Figure 2 they are assumed to have identical functions. B1 contains binding sites both for ribonucleoside diphosphate substrates (15) and for the nucleoside triphosphates which act as allosteric effectors (16). Equilibrium dialysis experiments demonstrated the presence of two substrate-binding sites and suggest that the four diphosphate substrates are bound to identical sites (15). B1 also contains oxidation-reduction active sulfhydryl groups which reduce a stoichiometric amount of substrate in the absence of an external hydrogen donor and therefore are part of the active site (17). No binding site for thioredoxin has been found on either B1 or B2 (glutaredoxin has as yet not been investigated), but it may be assumed that during the catalytic turnover of the enzyme, reduced thioredoxin (or glutaredoxin) reacts with the oxidation-reduction active disulfides.

Ribonucleotide reductase is an allosteric enzyme. Protein B1 contains two classes of effector-binding sites (h- and l-sites, Figure 2), each class consisting of two subsites (16). h-Sites are defined by their $h$igh affinity ($K_d = 0.03$ μM) for dATP, l-sites by their $l$ow affinity ($K_d = 0.1$–0.5 μM). From competition experiments it is apparent that h-sites also bind the allosteric effectors ATP, dTTP, and dGTP while l-sites only bind ATP in addition to dATP. As discussed below, effector-binding to the two classes results in distinct allosteric effects. In Figure 2 we arbitrarily place one site from each class onto one polypeptide chain. An asymmetric distribution is, however, not excluded and may in fact be related to the $\alpha\alpha'$ asymmetry of B1.

Protein B2 has a molecular weight of 78,000 (14) and consists of two apparently identical polypeptide chains (general structure β_2). Each molecule contains two atoms of iron (18), presumably one per polypeptide chain. Iron is required for enzyme activity; its removal by dialysis against 8-hydroxyquinoline gives an inactive apoprotein B2 which can be reactivated to more than 100% by reconstitution with Fe(II) ascorbate. Mössbauer spectroscopy of protein B2 enriched in ^{57}Fe suggested that B2 contains two nonidentical high spin Fe(III) ions in an antiferromagnetically coupled binuclear complex (19). This is supported by the electronic spectrum of B2 which closely resembles that of methydroxohemerythrin (20). In the latter case there is strong evidence for binuclear iron centers. B2 also contains a free radical which is characterized by a sharp peak at 410 nm in the optical spectrum and a doublet EPR signal centered around $g = 2.0047$ (21). From isotope substitution experiments the radical was assigned to a tyrosine residue in the protein (22). Radical spin density is delocalized over the aromatic ring (23) and not, as suggested originally (22), to the β-position of tyrosine. The presence of the radical is closely linked to the presence of iron. Radical is lost on removal of iron and reformed on reconstitution. On the other hand, the radical can be destroyed selectively by aging of B2 or by treatment with chemicals, e.g. hydroxyurea or hydroxylamine, without loss of iron. Enzyme activity is destroyed irreversibly, however, and is regained only after removal and reintroduction of iron. The function of iron as it binds to the protein is both to generate the tyrosyl radical, probably by an iron-catalyzed 1-electron aerobic oxidation, and to stabilize the radical in the metalloprotein by some continued interaction. It has so far not been possible to obtain preparations of B2 with more than one radical equivalent per two iron atoms, even though there is no apparent loss of radical during the preparation (11). This does, of course, not exclude the existence of B2 with even higher radical content. If it represents a maximal attainable value, however, it might imply that the radical participates only in one of the two active sites at a time, such as depicted in Figure 2, and point to a reaction mechanism with alternative active sites (cf 24).

Reaction Mechanism

Ribonucleotide reduction involves the replacement of an OH-group by a hydrogen. The closest identified possible hydrogen donors are the dithiol groups of B1, thus, an understanding of the mechanism should demonstrate how a dithiol can reduce a secondary alcohol. Studies concerning the stereochemistry of the reaction have been reviewed in detail earlier (3, 4). In summary, they demonstrate that only one new hydrogen enters at position 2' and that configuration at this carbon atom is retained. These results excluded formation of intermediates with C=C or C=O-containing structures. Identical results were found with the *E. coli* and *L. leichmannii* enzymes. In the *Lactobacillus* system deoxyadenosyl-cobalamin functions as intermediate hydrogen carrier with formation of free radicals by homolytic cleavage of the carbon-cobalt bond (see below). A deoxyadenosyl radical together with a dithiol apparently forms the agent used for the reduction of the OH-group. In the *E. coli* system, the presence of the tyrosyl radical in B2 also points to a radical mechanism. There is ample evidence for a correlation between the amount of free radical in a given B2 preparation and its specific enzyme activity (21). The presence of the radical in the active site of the reductase was demonstrated by experiments with the substrate analogue 2'-azidoCDP (25). With the complete B1:B2 enzyme this nucleotide is reduced and irreversibly destroys the tyrosyl radical. With B2 alone, the analogue is inert. Comparable results were found concerning the oxidation-reduction active SH-groups of protein B1 with a different nucleotide analogue, 2'-chloroCDP (25). On reduction of this nucleotide by the complete enzyme the active site SH-groups were inactivated. Taken together, the results with analogues demonstrate the presence of structural elements from both B1 and B2 in the active site (Figure 2).

Allosteric Control

The substrate specificity of all ribonucleotide reductases is controlled by allosteric mechanisms (26–29). The *E. coli* enzyme represents the best studied case and it has been possible to arrive at a plausible model linking the results of effector-binding studies to the expression of enzyme activity (16). Binding studies revealed the existence of the two classes of sites (*h* and *l*) on B1, depicted in Figure 2. This model was recently confirmed by the chromatographic behavior of B1 on affinity columns (30). Binding of effectors (ATP or dATP) to *l*-sites (= activity sites) regulates the general level of activity, with dATP acting as negative and ATP as positive effector. Inhibitory effects are accompanied by a tendency of the enzyme to sediment as dimers (13) or higher oligomers (14) during centrifugation. These results resemble similar regulatory effects observed with many other allosteric enzymes. The uniqueness of the control of the reductase derives from the binding of effectors to *h*-sites (= specificity sites). This results in conformational changes at the active site, which induce preferential binding of one or the other substrate. The changes are manifested both by a decrease in the dissociation constant of ribonucleoside diphosphates to the isolated B1 subunit (15), as well as by decreased K_m and increased K_{cat} values of the complete enzyme for a given substrate (26, 27). At low substrate concentration, binding of a positive effector may result in a 50- to 100-fold increase in the rate of the enzyme reaction. The multiplicity of effectors and effector-binding sites makes possible a large number of different states of the enzyme with different substrate specificities (16), a few of which are tabulated in Table 1. There, we distinguish between states containing ATP (or dATP) bound to *l*-sites and those with "naked" *l*-sites. While the latter cases may be of considerable interest to the biochemist working with purified enzymes, the enzyme in the cell in all probability has its *l*-sites occupied. It is then apparent that essentially three positive and one negative state exist: the

Table 1   Allosteric regulation of ribonucleotide reductase from *E. coli*[a]

Effector binding to		Reduction of			
l-sites	*h*-sites	CDP	UDP	GDP	ADP
0	ATP	+	+	0	0
0	dTTP	+	+	+	+
0	dGTP	0	0	+	+
ATP	ATP or dATP	+	+	0	0
ATP	dTTP	–	–	+	(+)
ATP	dGTP	nd	nd	(+)	+
dATP	any effector	–	–	–	–

[a] 0 = no effect; + = stimulation; – = inhibition; and nd = not determined

positive states contain ATP bound to l-sites and will reduce CDP or UDP with ATP (or dATP) at the h-sites, reduce GDP (and ADP) with dTTP at the h-sites, and reduce ADP (and GDP) with dGTP at the h-sites. All combinations containing dATP at l-sites are inactive. These results are, in the last section, integrated into a scheme linking the action of the reductase to the requirements of DNA synthesis for a supply of all four deoxyribonu-cleotides.

Hydrogen Transport System

Purification of ribonucleotide reductase became feasible when it was discov-ered that dithiols, such as reduced lipoate or dithiothreitol, could act as direct hydrogen donors (31). Monothiols, e.g. mercaptoethanol or gluta-thione, did not support the reaction. This led to the discovery of the thio-redoxin system of *E. coli* (32, 33), and later of similar systems in *L. leichmannii* (34) and eucaryotes (35–38). The key compound, thioredoxin, is a small protein (108 amino acids in *E. coli*) containing an oxidation-reduction active disulfide made up of the sequence (-Cys-Gly-Pro-Cys-) (39). This sequence represents the active center of the molecule, and in the three-dimensional structure (40) is found as an exposed protrusion. With one exception [T4-induced thioredoxin (41)] thioredoxin from all sources investigated so far shows the same sequence at the active center. The $(SH)_2$-form of *E. coli* thioredoxin serves as an efficient hydrogen donor for both the *E. coli* ribonucleotide reductase (32) and the enzymes from *L. leichmannii* (42) or mammalian sources (43). The oxidized form of thiore-doxin is reduced in the cell at the expense of NADPH by a specific FAD-protein, thioredoxin reductase (33):

$$\text{thioredoxin-S}_2 + \text{NADPH} + \text{H}^+ \xrightarrow[\text{reductase}]{\text{thioredoxin}} \text{thioredoxin-(SH)}_2 + \text{NADP}^+$$

In the test tube, also, low concentrations of artificial dithiols such as dithiothreitol, efficiently reduce thioredoxin. The FAD-containing thiore-doxin reductases appear to be ubiquitous enzymes. The enzyme from *E. coli* which has been studied in great detail (44–46) has a molecular weight of 66,000 with the general structure $(FAD)_2a_2$. Each polypeptide chain (45) contains one oxidation-reduction active disulfide (-Cys-Ala-Thr-Cys-) (46). The reaction mechanism (45, 47) involves a stepwise reduction of FAD and the oxidation-reduction active disulfides, and shows some differences from that of the two similar S_2-reducing flavoproteins, lipoyl dehydrogenase (48) and glutathione reductase (49). Since oxidation-reduction active disulfides also appear on ribonucleotide reductase the overall sequence of electron transfer from NADPH to ribonucleotides involves a shuttle of $S_2/(SH)_2$ interchanges as described in Figure 3 (17).

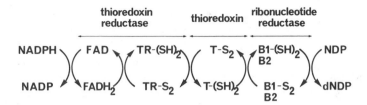

Figure 3   Involvement of oxidation-reduction active disulfides in ribonucleotide reduction in *E. coli.*

E. coli contains 10,000–20,000 molecules of thioredoxin per cell (50), more than sufficient to provide the reducing power for the 1500–3000 molecules of ribonucleotide reductase (11). It seemed possible that thioredoxin participates in other types of reductions, and the *E. coli* protein was found to be an efficient protein disulfide reductase (33). In yeast, participation of thioredoxin in sulfate and methionine sulfoxide reduction was implicated (51). More recent experiments support these suggestions (52, 53). In spite of clear demonstration that the thioredoxin system in the test tube efficiently participates in all these enzyme reactions, its true physiological role is presently unclear and, at least in *E. coli,* it does not seem to have any obligatory function. This became apparent from unrelated studies of prototrophic *E. coli* mutants that had lost the ability to support the growth of bacteriophage T7 (54). One class of such mutants (*tsnC*) was defective in thioredoxin (55) and one of these mutants (*tsnC* 7004) appears to be a nonsense or deletion mutant, completely devoid of any thioredoxin activity as measured by very sensitive enzymatic and immunological assays (50). In addition, a mutant of *E. coli* lacking thioredoxin reductase has also been described (56). Neither mutant shows a decreased capacity to reduce ribonucleotides. The *tsnC* mutants do not support the growth of bacteriophage T7 since *E. coli* thioredoxin together with a phage-induced protein forms the active phage DNA-polymerase (55). At the present time this is the only firmly established physiological function of thioredoxin.

A new development in the thioredoxin story is the discovery of phosphothioredoxin, with the phosphate bound in a phosphothiol linkage (57). Careful studies showed that as much as 94% of the thioredoxin of *E. coli* may be recovered in the phosphorylated form, with phosphate predominately located at Cys_{32} (58). The functional significance of these results is not yet clear.

In the absence of thioredoxin, ribonucleotide reduction in extracts of *tsnC* 7004 is mediated by a newly discovered hydrogen transport system that involves NADPH, glutathione reductase, glutathione, and a small protein, named glutaredoxin (59). This glutaredoxin system exists side by side with the thioredoxin system in wild type *E. coli.* Glutaredoxin and thioredoxin are clearly two different proteins, coded by two separate genes (60). Like thioredoxin, glutaredoxin is a small acidic protein with an oxidation-reduction active disulfide made up of two half-cystines. The disulfide is reduced by glutathione, glutathione reductase, and NADPH. It is also reduced chemically by dithiothreitol and then acts as a direct hydrogen donor in ribonucleotide reduction. The glutaredoxin system is also a powerful disulfide reductase (60). The major difference between thioredoxin and glutaredoxin lies in the reduction of their disulfide groups by NADPH: thioredoxin is reduced directly by the specific flavoprotein thioredoxin reductase, glutaredoxin via glutathione and glutathione reductase. The reduced disulfide groups interact, probably in both cases with the oxidation-reduction active disulfide of the B1 subunit of ribonucleotide reductase. Thus, glutaredoxin has the special capacity to mediate the utilization of the monothiols of glutathione for the reduction of disulfide bridges.

Is glutaredoxin rather than thioredoxin the obligatory intermediate in ribonucleotide reduction? As yet no glutaredoxin mutants have been described, but mutants defective in glutathione synthesis (61, 62) apparently show an unabated reduction of ribonucleotides. In this case, the thioredoxin system may provide the required reducing power. The glutathione mutants described so far may, however, retain a small amount of glutathione synthesis, sufficient for the reduction of glutaredoxin. Since pure glutaredoxin and antibodies to the protein are now available this problem should soon be solved.

RIBONUCLEOTIDE REDUCING SYSTEMS INDUCED BY BACTERIOPHAGES

Phage infection of *E. coli* results, in several cases, in induction of new virus-coded ribonucleotide reductases and thioredoxins. It was first noted that infection with phages T6 (63) or T2 (64) resulted in increased reductase activity. That a similar increase after infection with T4 depended on the induction of phage-coded ribonucleotide reductase and thioredoxin was demonstrated by the isolation and characterization of these proteins from extracts of infected bacteria (65, 66) as well as by identification of the genes [*nrdA* and *nrdB* for the reductase (67), *nrdC* for thioredoxin (68)] coding for their synthesis. The T4-coded reductase shows the same general construction as the *E. coli* enzyme, with a molecular weight of 225,000 and $\alpha_2\beta_2$ structure (69). The binding of the nonidentical subunits is much tighter, however, and does not require Mg^{2+}. α_2 has a molecular weight of 160,000, contains binding sites for dATP, and thus corresponds to the B1 subunit of the *E. coli* enzyme. β_2 contains 2 moles of iron as well as a paramagnetic species with a signal similar to that found in the *E. coli* enzyme, and corresponds to the B2 subunit. The modulations of the phage enzyme by positive effectors closely mimics those of the bacterial enzyme, and the major difference is the absence of inhibition by dATP, since this nucleotide only acts as positive effector for the reduction of CDP and UDP (70). All data can be explained by assuming that the phage enzyme lacks activity sites (cf Figure 2), is modulated only via substrate specificity sites, and therefore is not turned off by accumulation of dNTPs.

T4 also induces a thioredoxin (65). The location of $nrdC$ on the T4 map is not linked to $nrdA$ and B (68). T4 thioredoxin contains 87 amino acids (41) and shows no primary sequence homology with $E.$ $coli$ thioredoxin. The protein contains one oxidation-reduction active disulfide (-Cys-Val-Tyr-Cys-) that is reduced by the bacterial thioredoxin reductase (71) in spite of the nonhomology of the primary sequences of the two thioredoxins. The three-dimensional structure of T4 thioredoxin shows, however, large structural similarities to that of the bacterial protein (72). In contrast to bacterial thioredoxin, T4 thioredoxin is also reduced by glutathione, glutathione reductase, and NADPH and thus appears to be a functional hybrid between bacterial thioredoxin and glutaredoxin (73). The redox potential of T4 thioredoxin is –0.23 V as compared to a value of –0.26 V for the $E.$ $coli$ protein which makes it possible for thioredoxin reductase to catalyze the reduction of T4-thioredoxin-S_2 by bacterial thioredoxin-$(SH)_2$ (74). The metabolism of the T4-infected cells thus appears to be geared to specifically favor the reduction of the phage-induced thioredoxin. This should be related to the specificity requirements of the bacterial and phage-induced ribonucleotide reductases. Each enzyme interacts exclusively with its homologous thioredoxin. This is quite unusual, since, in most other cases, ribonucleotide reductases, e.g. from $E.$ $coli,$ $L.$ $leichmanni,$ and mammalian sources, are readily reduced by heterologous thioredoxins. In summary, both the properties of the T4 thioredoxin and the allosteric properties of the T4 reductase lead to a preferential use of the phage-induced system after virus-infection.

Induction of new ribonucleotide reductases and thioredoxins has also been demonstrated after infection with bacteriophage T6 and T5, but not T7 or λ (75). The T6-induced proteins cross-react functionally and immunologically with the T4-induced proteins. The T5-induced proteins show a more specific behavior. T5-thioredoxin is reduced by the bacterial thioredoxin reductase but cannot interact with the $E.$ $coli$ ribonucleotide reductase and thus, in this respect, resembles T4-thioredoxin. On the other hand, T5-ribonucleotide reductase accepts bacterial thioredoxin as hydrogen donor. The partially purified enzyme appears to use triphosphates as substrates and showed an allosteric behavior different from both the $E.$ $coli$- and T4-induced enzyme.

RIBONUCLEOTIDE REDUCTASE
FROM LACTOBACILLUS LEICHMANNII

Structural Aspects

The enzyme was purified to homogeneity both by conventional methods (76, 77) and by affinity chromatography on dGTP-Sepharose (78, 79). The active form of the enzyme is a monomer with a molecular weight of 76,000 and a $s^0_{20,w}$ of 5.13S. It does not aggregate over a wide range of protein concentration (76, 77). The single polypeptide chain has COOH-terminal lysine and the NH$_2$-terminal sequence

$$\text{Ser-Glu-Ile-}\begin{matrix}\text{Ser}\\\text{Cys}\end{matrix}\text{ Leu-}\begin{matrix}\text{Ser}\\\text{Cys}\end{matrix}\text{ -Ala- (76, 79).}$$

The activity of the enzyme is completely dependent on the presence of 5'-deoxy-5'-adenosyl-cobalamin (80).

In Figure 4 we make an attempt to summarize the knowledge of the structure of the *Lactobacillus* enzyme. Equilibrium dialysis experiments demonstrated that the enzyme contains one common binding site for deoxyribonucleoside triphosphates with K_d ranging from 9–80 μM. This site can also bind ribonucleoside triphosphates, but with a 100- to 1000-fold lower affinity (77, 79). It is depicted as the regulatory site in Figure 4. The existence of a substrate-binding site could not be demonstrated directly by binding of ribonucleoside triphosphate substrates (77). Kinetic experiments showed that the apparent K_m for GTP was as high as 0.24 mM which might impede the demonstration of a site by equilibrium dialysis experiments. Deoxyribonucleoside triphosphates showed little product inhibition, which supports the existence of two separate sites as shown in Figure 4. Binding of the cobalamin coenzyme could be demonstrated directly and required effector-binding to the regulatory site (79). No experiment demonstrating

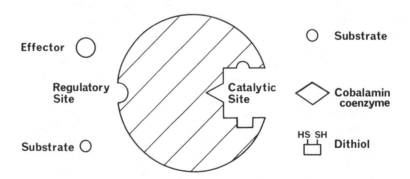

Figure 4   Model of ribonucleoside triphosphate reductase from *L. leichmannii.*

binding of thioredoxin has been reported but in Figure 4 thioredoxin is assumed to interact with the catalytic site.

Reaction Mechanism

In adenosylcobalamin-dependent enzyme reactions the coenzyme has been implicated as an intermediate hydrogen carrier (81), and in the ribonucleotide reductase reaction it assists in the transfer of hydrogen from a dithiol substrate to the reduction of the ribonucleoside triphosphate. The enzyme catalyzes an isotope exchange between the hydrogens at the 5'-methylene group of the coenzyme and water (82, 83). This reaction requires the presence of a dithiol and an allosteric effector, necessary for binding of the coenzyme (see below). The hydrogen exchange presumably occurs via the dithiols, and the reaction between cobalamin and the thiols is considered to be an integral part of the overall reaction. Both hydrogens at the 5'-methylene carbon are involved, in spite of the fact that they are stereochemically nonequivalent. This was explained by postulating the formation of an enzyme-bound intermediate, containing *three* equivalent hydrogens, one derived from the reducing thiol (83).

The nature of such an intermediate was investigated by spectrophotometric stopped flow studies and EPR measurements (84, 85). Both techniques demonstrated that a paramagnetic species was formed rapidly in the presence of dithiol and an allosteric effector, and disappeared on addition of substrate ribonucleoside triphosphate. The kinetic properties of this species were compatible with it being an intermediate in the overall reaction. Experiments with analogues of the coenzyme also support this contention (86). The properties of the intermediate were interpreted to be similar to cob(II)alamin (B_{12r}), which suggests that the paramagnetic intermediate was formed by a homolytic cleavage of the carbon-cobalt bond. However, no evidence of a deoxyadenosyl radical, or other organic radical, was obtained (85, 87). Such a radical has, nevertheless, to be postulated as a necessary complement to the Co(II) complexes of B_{12r}.

A second paramagnetic species is formed slowly by the enzyme in the presence of dithiol and effector, and has been identified as cob(II)alamin. Together with 5'-deoxyadenosine, it is the product of a slow irreversible degradation of adenosylcobalamin. Its formation is considered to be a side reaction (86, 88).

A third paramagnetic species, referred to as the "doublet" EPR spectrum, is formed in the presence of the complete ribonucleotide reductase system, including ribonucleoside triphosphate substrate. Even though this species only accumulates slowly during the reaction, its function as an intermediate appears possible (89). Similar doublets have been observed in other adenosylcobalamin-dependent reactions, and a seemingly satisfactory explanation has been given for these spectra: they arise from the interaction between an organic radical, probably on the substrate, and a low spin [Co(II)] complex (B_{12r}) (90, 91).

Two of the paramagnetic species described above qualify as intermediates in the overall reaction as parts of a radical-dithiol system. The dithiol involved in this process could either be an oxidation-reduction active disulfide on the enzyme—as appears to be the case with the *E. coli* enzyme —or the reduced form of thioredoxin (or glutaredoxin). Studies to differentiate between these alternatives do not appear conclusive (92, 93).

Allosteric Regulation

L. leichmannii enzyme also can regulate its substrate specificity by allosteric mechanisms. However, the most important consequence of effector binding to the regulatory site is an increased affinity of the enzyme for cobalamin coenzyme (79). Deoxynucleoside triphosphates are the most efficient effectors but ribonucleoside triphosphates also bind to the regulatory site and therefore can act both as substrates and effectors (77, 79). Thus, in kinetic studies, where the reduction of GTP was measured at low concentrations of adenosylcobalamin in the absence of deoxynucleoside triphosphates, a plot of $1/v$ versus $1/(GTP)$ demonstrated substrate activation. This effect disappeared when either the concentration of coenzyme was increased or small amounts of deoxynucleotides were added during incubation (79).

Similarly, formation of the radical intermediate, or tritium exchange between cobalamin and water in the presence of dithiol, required the presence of either deoxynucleoside or ribonucleoside triphosphates. These results correlate well with binding studies that demonstrated that insignificant amounts of coenzyme are bound to the enzyme when the regulatory site is not occupied (79).

The stimulation of cobalamin-binding shows little specificity for the nature of the base of the effector. In contrast, the allosteric control of the substrate specificity of the enzyme is base-dependent even though the results are not as clearcut as for the *E. coli* or mammalian enzymes and the effects are less pronounced. Nevertheless, the various deoxynucleoside triphosphates could be ordered into a scheme ranging from moderate activation to weak inhibition (80, 94, 95). The most positive effectors were dATP for the reduction of CTP, dCTP for UTP reduction, dTTP for GTP reduction, and dGTP for ATP reduction. In general, these positive effects show a qualitative similarity to those observed in other systems. No strong negative effector has been found for the *L. leichmannii* enzyme.

Allosteric effects are usually believed to involve subunit interactions. Here we are dealing with a monomeric enzyme and the interactions must involve the tertiary structure of the polypeptide chain. Evidence for conformational changes of the protein that depend on effector-binding were demonstrated by the findings that addition of dATP lowered the rate constant for alkylation of reduced enzyme by N-ethylmaleimide (93), and binding of dGTP caused significant changes in the aromatic region of the circular dichroism spectrum and gave a significant fractional change in the sedimentation coefficient (79).

OTHER B_{12}-DEPENDENT REDUCTASES

The distribution of adenosylcobalamin-dependent ribonucleotide reductases in different organisms has been surveyed by application of the tritium exchange reaction between $(5'-^3H_2)$ 5'-deoxyadenosylcobalamin and water (96). The enzyme was reported to be more common among procaryotes than the iron-containing reductase. No general rules were found, and even closely related species could contain either reductase. Among eucaryotes, only *Euglenophyta* and the fungus *Phitomyces chartarum* (97) were found to contain the B_{12}-requiring enzyme. No activity was found in different extracts from mammalian sources (96). Freshwater blue-green algae had an adenosylcobalamin-dependent enzyme, whereas extracts of the marine species *Agmenellum quadruplicatum* showed no activity in the exchange reaction (98). Instead, the activity of this organism was inhibited by hydroxyurea (99), a phenomenon typical of iron-containing reductases. Also, green algae, yeast, and higher plants have enzymes that are not dependent on adenosylcobalamin (100).

Aside from the enzyme from *L. leichmannii,* only the B_{12}-dependent enzyme from *Euglena gracilis* has been purified to homogeneity (101). This enzyme was completely dependent on adenosylcobalamin and was also in other respects similar to the *Lactobacillus* reductase. A high molecular weight of 440,000 was reported which suggests that the enzyme may occur as a tetramer. However, it cannot be excluded that this value was due to aggregation and redetermination of it appears to be required.

The *Lactobacillus* and *Euglena* reductases use ribonucleoside triphosphates as substrates and it has been usually assumed that this is characteristic of all cobalamin-dependent reductases. However, cobalamin-dependent enzymes from a few microorganisms (*Rhizobium meliloti* and *Corynebacterium nephridii*) appear to use ribonucleoside diphosphates as substrates (102, 103).

MAMMALIAN RIBONUCLEOTIDE REDUCTASE

Structural Aspects and Reaction Mechanism
No mammalian reductase has as yet been obtained as a homogeneous preparation. Enzyme activity has been demonstrated in extracts from many

growing organs, both normal (104–107) and malignant (43, 108–110), as well as from cells in tissue culture (111, 112). Highly purified preparations were obtained from Novikoff hepatoma of the rat (109) and from calf thymus (107). The latter enzyme was purified more than 3000-fold, obtained in good yield, and had the highest specific activity as yet reported.

Mammalian reductases contain two separable subunits, somewhat comparable to those from *E. coli*. Partially purified enzymes from rabbit bone marrow (106), Novikoff hepatoma (109), and Ehrlich ascites tumor (110) could be separated into two protein fractions that were required together for enzyme activity. Reconstitution was, however, difficult, resulted only in partial reactivation, and each fraction proved to be very labile on further purification. By combining results obtained with reductases from different sources, one may tentatively conclude that one of the subunits binds to dATP-Sepharose, and migrates, during electrophoresis on denaturing gels, with a molecular weight of 80,000–90,000, while the other fraction requires iron (107, 109).

During preparation of the thymus enzyme an attempt was made to avoid separation of subunits (107). The highly purified enzyme showed only one slightly asymmetric band on nondenaturing gels and one main protein peak at about 9S in glycerol gradients. Addition of dATP resulted in aggregation of the enzyme with a shift to a position at 16S. On denaturing gels one major band was found, corresponding to a molecular weight of 84,000, together with several minor components of lower molecular weight. Among those, a band with a molecular weight of 55,000 was the most prominent, but occurred in far from stoichiometric amounts. EDTA inactivated the enzyme and activity could be restored to 90–170% by addition of iron (or manganese). ^{59}Fe-reactivated enzyme showed cosedimentation of enzyme activity and radioactivity in glycerol gradients. While these data strongly argue for iron being an essential component of the enzyme, the amount of iron found either by direct analysis of the enzyme or by incorporation of ^{59}Fe after reconstitution only amounted to 0.03–0.1 g-atoms of iron per 250.000 g of protein (107). EPR spectroscopy could not demonstrate the presence of a free radical similar to that found in the B2 protein of the *E. coli* enzyme (L. Thelander, unpublished).

The results with the thymus reductase are feasible if the bulk of the protein in the final preparation corresponded to one subunit of the enzyme, while the other subunit was present in substoichiometric amounts. One can not disregard, however, that the preparation is still far from pure, in spite of the close to homogeneous behavior during electrophoresis and on centrifugation.

Mammalian reductases are inhibited by hydroxyurea (113–115). However, with the purified thymus enzyme inhibition was reversible, since removal of the drug by gel filtration resulted in a fully active enzyme (107).

This is in contrast to the irreversible inhibition of the B2 subunit of the *E. coli* reductase. Similarly, inhibition of the thymus enzyme by 2' -azidoCDP was reversible, in contrast to the inhibition of the *E. coli* enzyme (107). Both hydroxyurea and the azido-analogue act by scavenging the tyrosyl radical of the B2 subunit (19, 25) and their effect with the mammalian enzyme may suggest participation of a free radical in this case also. However, the reversibility of the process then might indicate a somewhat different catalytic process, with the radical possibly being formed during catalysis rather than being present in the isolated enzyme.

A difference between the two classes of enzymes was also observed with respect to their behavior toward substituted thiosemicarbazones (116). These drugs are strong iron chelators but do not act by removing iron from the enzyme, since Fe-thiosemicarbazone chelates are better inhibitors than the nonchelated forms (117, 118). The reductase from Novikoff hepatoma was inhibited to 50% by inhibitor concentrations of about 0.1 μM, while no inhibition of the *E. coli* enzyme was found (E. C. Moore, personal communication).

Much further work is clearly required in order to reach a more detailed understanding of the structure and catalytic mechanism of mammalian reductases. It is well established that these enzymes, like the *E. coli* enzyme, do not require cobalamin coenzymes (96, 105, 106, 108), and in all probability they contain iron (105–107, 109). One major obstacle is the lability of the presumed iron-containing subunit during purification (109).

Allosteric Regulation

The similarily between the mammalian enzymes and the enzyme from *E. coli* is particularly apparent in their allosteric control. The first description of allosteric behavior of ribonucleotide reductase actually concerned an enzyme present in extracts from chick embryos (119), and not until later were the details of these effects outlined with partially purified preparations from *E. coli* (26, 27) and Novikoff hepatoma (29). The results with the Novikoff hepatoma demonstrated a complicated pattern of activations and inhibitions of the reduction of all four ribonucleoside diphosphates by nucleoside triphosphates, which suggested an allosteric regulation of substrate specificity quite similar to that observed in *E. coli*. However, the enzyme preparation was still very crude, and in particular, the question whether one or several reductases were involved in the reduction of the four substrates was left open. Since then several published articles have described the existence of separable reductases specific for the reduction of various substrates (120–122).

Results with the thymus reductase now definitely establish that one enzyme catalyzes the reduction of all four ribonucleoside diphosphates and that substrate specificity is regulated by allosteric effects (123). Thus, during the 3000-fold purification of the enzyme, the activities towards CDP and GDP were purified together, provided the assays were carried out under optimal allosteric conditions. With the highly purified enzyme, all four substrates were reduced at similar rates, and furthermore, CDP and GDP completed for the same catalytic site.

In the absence of positive effectors the enzyme was inactive with any ribonucleoside diphosphate. Reduction of CDP and UDP was stimulated by ATP, reduction of GDP by dTTP (or better still dTTP + ATP), and reduction of ADP by dGTP (or dGTP + ATP). dATP served as a general inhibitor whose negative effects could be reversed by high concentrations of ATP (123). The general pattern of effects is thus similar to that of the *E. coli* enzyme. However, with the mammalian enzyme there is an absolute requirement for effectors and each effector shows a more distinct effect. Thus, dTTP stimulates only the reduction of GDP, and dATP is only a negative effector. While binding studies have as yet not been carried out with any mammalian enzyme, the data fit a model containing two classes of effector sites, with one class regulating substrate specificity via binding of ATP, dTTP, and dGTP, and the other class regulating overall activity via binding of ATP and dATP (123).

Hydrogen Donor System

It was first shown for the Novikoff hepatoma enzyme that a chemical dithiol (reduced lipoate) or the thioredoxin system from *E. coli* can function as hydrogen donors (43). This led to the isolation of thioredoxins (35, 124, 125) and thioredoxin reductases (126–128) from different mammalian sources. All mammalian ribonucleotide reductases studied so far can use chemical dithiols (usually dithiothreitol) and, whenever tested, reduced thioredoxin from homologous or heterologous sources (36, 43, 107, 109, 129). While the K_m for thioredoxin is much lower than that for the chemical dithiols, V_{max} is actually slightly lower (129). This result is not observed with the *E. coli* reductase, probably because the bacterial enzyme has a higher K_m for dithiothreitol than the mammalian enzyme and inhibition occurs at high concentrations (5).

Homogeneous preparations of thioredoxins have been prepared from Novikoff hepatoma (124) and from calf liver (125). Both proteins have a molecular weight of close to 12,000, contain NH_2-terminal valine and one oxidation-reduction active disulfide, but show some differences in amino acid composition. The sequence (-Cys-Gly-Pro-Cys-) of calf liver thioredoxin at the active center is identical to that of the thioredoxins from yeast (130) and *E. coli* (39).

Mammalian thioredoxins are reduced by mammalian thioredoxin reductases and not by the *E. coli* enzyme. Highly purified mammalian thioredoxin reductases were obtained from calf liver, calf thymus (128), and Novikoff hepatoma (127). The latter was reported to be about 95% pure. This enzyme contains two subunits of approximately 58,000 daltons, with one FAD per subunit. All mammalian enzymes directly reduce 5,5'-dithi-obis(2-nitrobenzoic acid) with NADPH, and this reaction can be used as a convenient assay during their purification.

The presence of thioredoxin systems in cells with little or no DNA synthesis (131, 132) point to a function besides ribonucleotide reduction. During liver regeneration, the activity of ribonucleotide reductase rises 10- to 20-fold without concomitant increase in thioredoxin activity (126). In view of the results obtained with the *tsnC* mutants of *E. coli* (50, 55, 59) one may ask whether the thioredoxin system is the only physiological hydrogen transport system for ribotide reduction in mammalian cells, whether it shares such a function with a glutaredoxin system, or whether, in fact, it may not be involved at all. These questions cannot be answered at the present time, but evidence has recently been obtained for the existence of a glutaredoxin system in calf thymus also (129).

RIBONUCLEOTIDE REDUCTION AND DNA SYNTHESIS

Correlation of In Vitro and In Vivo Activities

The significance of ribonucleotide reductase as *the* enzyme that provides building blocks for DNA synthesis was in doubt for a long time because of the low enzyme activity in cell extracts. For *E. coli* such doubts were effectively dispelled by the discovery of temperature-sensitive *nrd* mutants (7–9). In eucaryotic systems, isotope experiments with regenerating rat liver (133) also strongly suggested that all DNA was derived via the reductase pathway. Furthermore, deoxyadenosine (134), thymidine (135), or hydroxyurea (113, 136) completely depress DNA synthesis. These compounds interact with the mammalian reductase: the deoxyribosides, after phospho-rylation to the triphosphates, by upsetting the allosteric regulation of the enzyme (29, 123), and hydroxyurea by reversibly inactivating the enzyme (107, 115), possibly by reacting with a free radical. Cell lines selected for resistance toward inhibition by hydroxyurea (112, 122) or deoxyadenosine (137), overproduce the reductase, and, in the case of deoxyadenosine, also show altered allosteric properties.

Why then do crude cell extracts contain such low activities? The assay presents considerable difficulties. Enzyme concentration curves are highly sigmoidal, and it is difficult to keep the substrate at the diphosphate level and to avoid the influence of negative effectors. But also, when the amount of enzyme in *E. coli* extracts was quantitized by an immunological method, and enzyme activity was computed from idealized optimal conditions involving, among other things, an excess of the second subunit, the calculated activity did not suffice to support all DNA synthesis (11). Such results show that the enzyme in the cell performs its function more efficiently and suggest that the isolated proteins represent a degenerate species. Evidence for this derives also from the finding that permeabilized cells of *E. coli* (138) and cell lysates on cellophane discs (139) show up to 50-fold higher activities than extracts prepared by alumina grinding. Similarly, freshly prepared "gentle lysates" give a high activity that rapidly decays on aging (139). Such bacterial extracts, as well as the cellophane system, show much less sigmoidicity in the assay, and activity is not further stimulated by addition of an excess of complementary subunit. This suggests a stronger B1:B2 interaction and the possibility that ribonucleotide reductase is present as part of a multienzyme complex catalyzing synthesis of deoxynucleotides. Purification of the highly active form of the reductase is difficult because of its instability. Progress was made recently when the enzyme, together with thymidylate synthetase, was obtained in a "membrane" fraction of *E. coli* (V. Pigiet, personal communication). Similar considerations may apply to the mammalian enzyme which also shows a high degree of sigmoidicity when assayed in extracts (105–107, 109) but gives a linear enzyme concentration curve in permeabilized cells (122).

A multienzyme aggregate containing several (i.e. dNMP kinase, dCTPase, dCMP-hydroxymethylase, dTMP-synthetase, and dUTPase) but not all enzymes involved in deoxynucleotide synthesis induced by T4 infection, was observed after infection of *E. coli* with the bacteriophage (140). T4-induced ribonucleotide reductase could not be demonstrated directly in this aggregate but indirect evidence for its association within the cells was cited (141). Indirect evidence for a complex containing enzymes of both deoxynucleotide and DNA synthesis has also been presented (142–144).

In vitro, the activity and substrate specificity of the reductases from *E. coli* and mammalian sources behave similar towards allosteric effectors. The data can be integrated into a scheme (145) that links ribonucleotide reduction to DNA synthesis (Figure 5). Deoxynucleotide synthesis begins with the reduction of CDP and UDP by an ATP-activated enzyme, proceeds to GDP reduction via a dTTP-enzyme, and, finally, reaches ADP reduction by a dGTP-activated enzyme. Accumulation of dATP (e.g. in the absence of DNA synthesis) completely turns off the activity of the reductase. Accumulation of dTTP shuts off the reduction of pyrimidine substrates; accumulation of dGTP (at least with eucaryotic enzymes) also turns off GDP reduction. The scheme represents, of course, a formal concept of a dynamic situation, and its operation in vivo is indicated by the finding that deoxyadenosine and thymidine added to cells in culture inhibits DNA synthesis. As predicted from the scheme of Figure 5, addition of thymidine gives a specific depletion of the dCTP pool, with increased dGTP and dATP pools (146). Furthermore, mutant cell lines with increased resistance towards deoxyadenosine contained an altered reductase with decreased sensitivity towards inhibition by dATP (137). These cells also contained larger dNTP pools (147) in accordance with the postulate that the dATP pool normally regulates the size of the other dNTP pools (Figure 5).

A hereditary immunodeficiency disease in man (148) may also ultimately be caused by an imbalance of the allosteric control of the reductase. The patients show a deficiency of the enzyme adenosine deaminase which in normal individuals rapidly catabolizes both adenosine and deoxyadenosine. In patients lacking the enzyme, deoxyadenosine can instead be phosphorylated to dATP and erythrocytes of such individuals were shown to contain a 50-fold increased pool of dATP (149, 150). The consequence of such an accumulation of dATP are illuminated by early experiments with a DNA-synthesizing enzyme system from chick embryos (119). In these experiments DNA synthesis was coupled to ribonucleotide reduction so that one of the four deoxynucleoside triphosphates was derived from the reduction of a ribonucleotide (CDP or GDP). Addition of increasing amounts of dATP (or dGTP) then strongly inhibited DNA synthesis and inhibition was shown to occur at the level of ribonucleotide reductase. The tissue-specific effect in the patients, which results in a disease of the immune

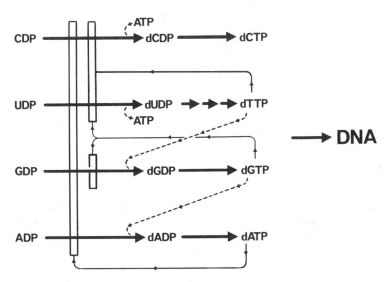

Figure 5 Scheme of the physiological regulation of deoxyribonucleotide synthesis. The broken arrows stand for positive effects, the open bars for negative effects.

system, may be explained by the finding that lymphoid tissues are particularly active in the phosphorylation of deoxyadenosine (151). A similar explanation could also apply to the immuno-deficiency associated with an inherited deficiency in the enzyme purine nucleoside phosphorylase (152).

Regulation of Enzyme Synthesis

The activity of ribonucleotide reductase is correlated to the extent of DNA synthesis. In eucaryotes, the reduction was difficult to find in tissues with low DNA synthesis and the highest activities were reported in rapidly growing malignant cells (153). Cells induced to synthesize DNA, e.g. by liver regeneration (105) or by virus infection (154), rapidly increase reductase activity. In E. coli, enzyme activity is also correlated to growth rate. When DNA synthesis was blocked by treatment with certain chemicals (naladixic acid, bleomycin), by thymine starvation of thymine auxotrophs, or by shifting cultures of temperature-sensitive dna mutants to nonpermissive conditions large increases in reductase activity were observed (155, 156). Immunological methods showed that thymine starvation causes a fivefold increase in the amount of enzyme (11). It was suggested (155) that under these conditions the cell accumulates an unknown compound that both stimulates initiation of DNA synthesis and induces ribonucleotide reductase.

Deoxyribonucleotide Pools

The E. coli experiments invoke a common factor regulating both ribonucleotide reduction and DNA synthesis (155). For mammalian cells the suggestion was made that a product synthesized by the reductase, dCTP or a nucleotide related to it, may have a specific regulatory function in DNA synthesis (145). This hypothesis arose from measurements of deoxynucleoside triphosphate pools in cultured cells (157). Such pools reflect the balance between deoxynucleotide synthesis by the reductase and deoxynucleotide consumption by DNA synthesis or other synthetic or degradative pathways. Studies of pools of several cell lines in cultures under different growth conditions demonstrated that the dCTP pool showed a close correlation to the rate of DNA synthesis, while the coupling was much weaker for the other pools. Also, in experiments where cells were blocked with thymidine, which acts by depleting the dCTP pool, inhibition of DNA synthesis occurred when the size of the dCTP pool was still quite large compared to other pools (146).

Pool measurements also demonstrated (157) that (a) resting cells contain small but definite amounts of all four triphosphates; (b) much larger pools are found in cells in S-phase but that the dGTP pool only suffices for about 15 sec of DNA synthesis; and (c) very large differences in pool size exist during S-phase for the four deoxynucleotides, with dGTP always being the smallest and dCTP usually the largest pool, the difference in some cases being 100-fold. Pool measurements were also used to differentiate between inhibitors affecting precursor synthesis and DNA synthesis proper (158). The pool measurements discussed so far were made with whole cells. When cytoplasmic and nuclear pools were separated (159) the cell nucleus accumulated deoxynucleotides during S-phase and this effect showed different magnitudes for the four deoxynucleotides. It was most pronounced for dTTP and least pronounced for dATP.

ACKNOWLEDGMENTS

Work by the authors of this review was supported by the Swedish Natural Science Research Council and the Swedish Medical Research Council.

Literature Cited

1. Hammarsten, E., Reichard, P., Saluste, E. 1950. *J. Biol. Chem.* 183:105–9
2. Rose, I. A., Schweigert, B. S. 1953. *J. Biol. Chem.* 202:635–45
3. Hogenkamp, H. P. C., Sando, G. N. 1974. *Struct. Bonding (Berlin)* 20:23–58
4. Follmann, H. 1974. *Angew. Chem. Int. Ed. Engl.* 13:569–79
5. Brown, N. C., Canellakis, Z. N., Lundin, B., Reichard, P., Thelander, L. 1969. *Eur. J. Biochem.* 9:561–73
6. Bachmann, B. J., Low, K. B., Taylor, A. L. 1976. *Bacteriol. Rev.* 40:116–67
7. Fuchs, J. A., Karlström, H. O., Warner, H. R., Reichard, P. 1972. *Nature New Biol.* 238:69–71
8. Fuchs, J. A., Neuhard, J. E. 1973. *Eur. J. Biochem.* 32:451–56
9. Fuchs, J. A., Karlström, H. O. 1973. *Eur. J. Biochem.* 32:457–62
10. Fuchs, J. A. 1977. *J. Bacteriol.* 130:957–59
11. Eriksson, S., Sjöberg, B.-M., Hahne, S., Karlström, O. 1977. *J. Biol. Chem.* 252:6132–38
12. Berglund, O., Eckstein, F. 1974. *Methods Enzymol.* 34B:253–61
13. Brown, N. C., Reichard, P. 1969. *J. Mol. Biol.* 46:25–38
14. Thelander, L. 1973. *J. Biol. Chem.* 248:4591–4601
15. von Döbeln, U., Reichard, P. 1976. *J. Biol. Chem.* 251:3616–22
16. Brown, N. C., Reichard, P. 1969. *J. Mol. Biol.* 46:39–55
17. Thelander, L. 1974. *J. Biol. Chem.* 249:4858–62
18. Brown, N. C., Eliasson, R., Reichard, P., Thelander, L. 1969. *Eur. J. Biochem.* 9:512–18
19. Atkin, C. L., Thelander, L., Reichard, P., Lang, G. 1973. *J. Biol. Chem.* 248:7464–72
20. Garbett, K., Darnall, D. W., Klotz, I. M., Williams, R. J. P. 1969. *Arch. Biochem. Biophys.* 103:419–34
21. Ehrenberg, A., Reichard, P. 1972. *J. Biol. Chem.* 247:3485–88
22. Sjöberg, B.-M., Reichard, P., Gräslund, A., Ehrenberg, A. 1977. *J. Biol. Chem.* 252:536–41
23. Sjöberg, B.-M., Reichard, P., Gräslund, A., Ehrenberg, A. 1978. *J. Biol. Chem.* 253:6863–65
24. Smith, D. J., Boyer, P. D. 1975. *Proc. Natl. Acad. Sci. USA* 73:4314–18
25. Thelander, L., Larsson, B., Hobbs, J., Eckstein, F. 1976. *J. Biol. Chem.* 251:1398–1405
26. Larsson, A., Reichard, P. 1966. *J. Biol. Chem.* 241:2533–39
27. Larsson, A., Reichard, P. 1966. *J. Biol. Chem.* 241:2540–49
28. Goulian, M., Beck, W. S. 1966. *J. Biol. Chem.* 241:4233–42
29. Moore, E. C., Hurlbert, R. B. 1966. *J. Biol. Chem.* 241:4802–9
30. von Döbeln, U. 1977. *Biochemistry* 16:4368–71
31. Reichard, P. 1962. *J. Biol. Chem.* 237:3513–19
32. Laurent, T. C., Moore, E. C., Reichard, P. 1964. *J. Biol. Chem.* 239:3436–44
33. Moore, E. C., Reichard, P., Thelander, L. 1964. *J. Biol. Chem.* 239:3445–52
34. Orr, M. D., Vitols, E. 1966. *Biochem. Biophys. Res. Commun.* 25:109–15

35. Moore, E. C. 1967. *Biochem. Biophys. Res. Commun.* 29:264–68
36. Larson, G., Larsson, A. 1972. *Eur. J. Biochem.* 26:119–24
37. Porqué, G. P., Baldesten, A., Reichard, P. 1970. *J. Biol. Chem.* 245:2363–70
38. Wagner, W., Follmann, H. 1977. *Biochem. Biophys. Res. Commun.* 77: 1044–51
39. Holmgren, A. 1968. *Eur. J. Biochem.* 6:475–84
40. Holmgren, A., Söderberg, B.-O., Eklund, H., Brändén, C.-I. 1975. *Proc. Natl. Acad. Sci. USA* 72:2305–9
41. Sjöberg, B.-M., Holmgren, A. 1972. *J. Biol. Chem.* 247:8063–68
42. Beck, W. S., Goulian, M., Larsson, A., Reichard, P. 1966. *J. Biol. Chem.* 241:2177–79
43. Moore, E. C., Reichard, P. 1964. *J. Biol. Chem.* 239:3453–56
44. Thelander, L. 1967. *J. Biol. Chem.* 242:852–59
45. Thelander, L. 1968. *Eur. J. Biochem.* 4:407–22
46. Thelander, L. 1970. *J. Biol. Chem.* 245:6026–29
47. Zanetti, G., Williams, C. H. Jr. 1967. *J. Biol. Chem.* 242:5232–36
48. Massey, V., Hofmann, T., Palmer, G. 1962. *J. Biol. Chem.* 237:3820–28
49. Massey, V., Williams, C. H. Jr. 1965. *J. Biol. Chem.* 240:4470–80
50. Holmgren, A., Ohlsson, I., Grankvist, M.-L. 1978. *J. Biol. Chem.* 253:430–36
51. Porqué, G. P., Baldesten, A., Reichard, P. 1970. *J. Biol. Chem.* 245:2371–74
52. Holmgren, A. 1977. *J. Biol. Chem.* 252:4600–6
53. Tsang, M. L.-S., Schiff, J. A. 1976. *J. Bacteriol.* 125:923–33
54. Chamberlin, M. 1974. *J. Virol.* 14: 509–16
55. Mark, D. F., Richardson, C. C. 1976. *Proc. Natl. Acad. Sci. USA* 73:780–84
56. Fuchs, J. A. 1977. *J. Bacteriol.* 129:967–72
57. Pigiet, V., Conley, R. R. 1978. *J. Biol. Chem.* 253:1910–20
58. Conley, R. R., Pigiet, V. 1978. *J. Biol. Chem.* 253:5568–72
59. Holmgren, A. 1976. *Proc. Natl. Acad. Sci. USA* 73:2275–79
60. Holmgren, A. 1979. *J. Biol. Chem.* In press
61. Apontoweil, P., Berends, W. 1975. *Biochim. Biophys. Acta* 399:10–22
62. Fuchs, J. A., Warner, H. R. 1975. *J. Bacteriol.* 124:140–48
63. Cohen, S. S., Barner, H. D. 1962. *J. Biol. Chem.* 237:PC1376–78

64. Biswas, C., Hardy, J., Beck, W. S. 1965. *J. Biol. Chem.* 240:3631–40
65. Berglund, O., Karlström, O., Reichard, P. 1969. *Proc. Natl. Acad. Sci. USA* 62:829–35
66. Berglund, O. 1969. *J. Biol. Chem.* 244:6306–8
67. Yeh, Y. C., Dubovi, E. J., Tessman, I. 1969. *Virology* 37:615–23
68. Yeh, Y.-C., Tessman, I. 1972. *Virology* 47:767–72
69. Berglund, O. 1975. *J. Biol. Chem.* 250: 7450–55
70. Berglund, O. 1972. *J. Biol. Chem.* 247: 7276–81
71. Berglund, O., Sjöberg, B.-M. 1970. *J. Biol. Chem.* 245:6030–35
72. Söderberg, B.-O., Sjöberg, B.-M., Sonnerstam, U., Brändén, C.-I. 1978. *Proc. Natl. Acad. Sci. USA.* In press
73. Holmgren, A. 1978. *J. Biol. Chem.* 253: 7424–30
74. Berglund, O., Holmgren, A. 1975. *J. Biol. Chem.* 250:2778–82
75. Eriksson, S., Berglund, O. 1974. *Eur. J. Biochem.* 46:271–78
76. Panagou, D., Orr, M. D., Dunstone, J. R., Blakley, R. L. 1972. *Biochemistry* 11:2378–88
77. Chen, A. K., Bhan, A., Hopper, S., Abrams, R., Franzen, J. S. 1974. *Biochemistry* 13:654–61
78. Hoffmann, P. J., Blakley, R. L. 1975. *Biochemistry* 14:4804–12
79. Singh, D., Tamao, Y., Blakley, R. L. 1977. *Adv. Enzyme Regul.* 15:81–100
80. Vitols, E., Brownson, C., Gardiner, W., Blakley, R. L. 1967. *J. Biol. Chem.* 242:3035–41
81. Abeles, R. H. 1971. *The Enzymes.* 5:481–97. 3rd ed.
82. Abeles, R. H., Beck, W. S. 1967. *J. Biol. Chem.* 242:3589–93
83. Hogenkamp, H. P. C., Ghambeer, R. K., Brownson, C., Blakley, R. L., Vitols, E. 1968. *J. Biol. Chem.* 243:799–808
84. Tamao, Y., Blakley, R. L. 1973. *Biochemistry* 12:24–34
85. Orme-Johnson, W. H., Beinert, H., Blakley, R. L. 1974. *J. Biol. Chem.* 249:2338–43
86. Sando, G. N., Blakley, R. L., Hogenkamp, H. P. C., Hoffmann, P. J. 1975. *J. Biol. Chem.* 250:8774–79
87. Coffman, R. E., Ishikawa, Y., Blakley, R. L., Beinert, H., Orme-Johnson, W. H. 1976. *Biochim. Biophys. Acta* 444:307–12
88. Yamada, R., Tamao, Y., Blakley, R. L. 1971. *Biochemistry* 10:3959–68

89. Hamilton, J. A., Tamao, Y., Blakley, R. L., Coffman, R. E. 1972. *Biochemistry* 11:4696–4705
90. Schepler, K. L., Dunham, W. R., Sands, R. H., Fee, J. A., Abeles, R. H. 1975. *Biochim. Biophys. Acta* 397: 510–18
91. Buettner, G. R., Coffman, R. E. 1977. *Biochim. Biophys. Acta* 480:495–505
92. Vitols, E., Hogenkamp, H. P. C., Brownson, C., Blakley, R. L., Connellan, J. 1967. *Biochem. J.* 104:58C–60C
93. Kim, J. J., Abrams, R., Franzen, J. S. 1977. *Arch. Biochem. Biophys.* 182: 674–82
94. Beck, W. S. 1967. *J. Biol. Chem.* 242:3148–58
95. Ludwig, W., Follmann, H. 1978. *Eur. J. Biochem.* 82:393–403
96. Gleason, F. K., Hogenkamp, H. P. C. 1972. *Biochim. Biophys. Acta* 277: 466–70
97. Stutzenberger, F. 1974. *J. Gen. Microbiol.* 81:501–3
98. Gleason, F. K., Wood, J. M. 1976. *Science* 192:1343–44
99. Gleason, F. K., Wood, J. M. 1976. *J. Bacteriol.* 128:673–76
100. Feller, W., Follmann, H. 1976. *Biochem. Biophys. Res. Commun.* 70: 752–58
101. Hamilton, F. D. 1974. *J. Biol. Chem.* 249:4428–34
102. Cowles, J. R., Evans, H. J. 1968. *Arch. Biochem. Biophys.* 127:770–78
103. Tsai, P. K., Hogenkamp, H. P. C. 1978. *Fed. Proc.* 37:1803(2927)
104. Murphree, S., Moore, E. C., Beall, P. T. 1968. *Cancer Res.* 28:860–63
105. Larsson, A. 1969. *Eur. J. Biochem.* 11:113–21
106. Hopper, S. 1972. *J. Biol. Chem.* 247: 3336–40
107. Engström, Y., Eriksson, S., Thelander, L., Åkerman, M. 1979. *Biochemistry.* Submitted for publication
108. Fujioka, S., Silber, R. 1970. *J. Biol. Chem.* 245:1688–93
109. Moore, E. C. 1977. *Adv. Enzyme Regul.* 15:101–14
110. Cory, J. G., Fleischer, A. E., Munro, J. B. III. 1978. *J. Biol. Chem.* 253:2898–2901
111. Nordenskjöld, B. A., Skoog, L., Brown, N. C., Reichard, P. 1970. *J. Biol. Chem.* 245:5360–68
112. Lewis, W. H., Wright, J. A. 1974. *Biochem. Biophys. Res. Commun.* 60: 926–33
113. Young, C. W., Hodas, S. 1964. *Science* 146:1172–74
114. Turner, M. K., Abrams, R., Lieberman, I., 1966. *J. Biol. Chem.* 241:5777–80
115. Moore, E. C. 1969. *Cancer Res.* 29: 291–95
116. Moore, E. C., Zedeck, M. S., Agrawal, K. C., Sartorelli, A. C. 1970. *Biochemistry* 9:4492–98
117. Agrawal, K. C., Schenkman, J. B., Denk, H., Mooney, P. D., Moore, E. C., Wodinsky, I., Sartorelli, A. C. 1977. *Cancer Res.* 37:1692–96
118. Sartorelli, A. C., Agrawal, K. C., Tsiftsoglou, A. S., Moore, E. C. 1977. *Adv. Enzyme Regul.* 15:117–139
119. Reichard, P., Canellakis, Z. N., Canellakis, E. S. 1961. *J. Biol. Chem.* 236:2514–19
120. Cory, J. G., Mansell, M. M., Whitford, T. W. Jr. 1976. *Adv. Enzyme Regul.* 14:45–62
121. Peterson, D. M., Moore, E. C. 1976. *Biochim. Biophys. Acta* 432:80–91
122. Lewis, W. H., Kuzik, B. A., Wright, J. A. 1978. *J. Cell. Physiol.* 94:287–98
123. Eriksson, S., Thelander, L., Åkerman, M. 1979. *Biochemistry.* Submitted for publication
124. Herrmann, E. C., Moore, E. C. 1973. *J. Biol. Chem.* 248:1219–23
125. Engström, N.-E., Holmgren, A., Larsson, A., Söderhäll, S. 1974. *J. Biol. Chem.* 249:205–10
126. Larsson, A. 1973. *Eur. J. Biochem.* 35:346–49
127. Chen, C.-C., Borns McCall, B. L., Moore, E. C. 1977. *Prep. Biochem.* 7(2): 165–77
128. Holmgren, A. 1977. *J. Biol. Chem.* 252:4600–6
129. Luthman, M., Eriksson, S., Holmgren, A., Thelander, L. 1979. *Proc. Natl. Acad. Sci. USA.* Submitted for publication
130. Hall, D. E., Baldesten, A., Holmgren, A., Reichard, P. 1971. *Eur. J. Biochem.* 23:328–35
131. Chen, C.-C., Moore, E. C. 1977. *Anal. Biochem.* 83:609–14
132. Holmgren, A., Luthman, M. 1978. *Biochemistry.* 17:407–77
133. Larsson, A., Neilands, J. B. 1966. *Biochem. Biophys. Res. Commun.* 25: 222–26
134. Klenow, H. 1959. *Biochim. Biophys. Acta* 35:412–21
135. Morris, N. R., Fischer, G. A. 1960. *Biochim. Biophys. Acta* 42:183–84
136. Castillot, J. J., Miller, M. R., Pardee, A. B. 1978. *Proc. Natl. Acad. Sci. USA* 75:351–55
137. Meuth, M., Green, H. 1974. *Cell* 3:367–74

138. Warner, H. R. 1973. *J. Bacteriol.* 115:18–22
139. Eriksson, S. 1975. *Eur. J. Biochem.* 56:289–94
140. Reddy, G. P. V., Singh, A., Stafford, M. E., Mathews, C. K. 1977. *Proc. Natl. Acad. Sci. USA* 74:3152–56
141. Reddy, G. P. V., Mathews, C. K. 1978. *J. Biol. Chem.* 253:3461–67
142. Chiu, C.-S., Greenberg, G. R. 1973. *J. Virol.* 12:199–201
143. North, T. W., Stafford, M. E., Mathews, C. K. 1976. *J. Virology* 17:973–82
144. Flanegan, J. B., Greenberg, G. R. 1977. *J. Biol. Chem.* 252:3019–27
145. Reichard, P. 1978. *Fed. Proc.* 37:9–14
146. Bjursell, G., Reichard, P. 1973. *J. Biol. Chem.* 248:3904–9
147. Meuth, M., Aufreiter, E., Reichard, P. 1976. *Eur. J. Biochem.* 71:39–43
148. Giblett, E. R., Anderson, J. E., Cohen, F., Pollara, B., Meuwissen, H. J. 1972. *Lancet* 2:1067–69
149. Cohen, A., Hirschhorn, R., Horowitz, S. D., Rubinstein, A., Polmar, S. H., Hong, R., Martin, D. W. Jr. 1978. *Proc. Natl. Acad. Sci. USA* 75:472–76
150. Coleman, M. S., Donofrio, J., Hutton, J. J., Hahn, L., Daoud, A., Lampkin, B., Dyminski, J., 1978. *J. Biol. Chem.* 253:1619–26
151. Carson, D. A., Kaye, J., Seegmiller, J. E. 1977. *Proc. Natl. Acad. Sci. USA* 74:5677–81
152. Giblett, E. R., Ammann, A. J., Wara, D. W., Sandman, R., Diamond, L. K. 1975. *Lancet* 1:1010–13
153. Elford, H. L., Freese, M., Passamani, E., Morris, H. P. 1970. *J. Biol. Chem.* 245:5228–33
154. Lindberg, U., Nordenskjöld, B. A., Reichard, P., Skoog, L. 1969. *Cancer Res.* 29:1498–1506
155. Filpula, D., Fuchs, J. A. 1977. *J. Bacteriol.* 130:107–13
156. Filpula, D., Fuchs, J. A. 1978. *J. Bacteriol.* 135:429–35
157. Skoog, L., Bjursell, G., Nordenskjöld, B. 1974. *Adv. Enzyme Regul.* 12:345–54
158. Skoog, L., Nordenskjöld, B. 1971. *Eur. J. Biochem.* 19:81–89
159. Skoog, L., Bjursell, G. 1974. *J. Biol. Chem.* 249:6434–38

Ann. Rev. Biochem. 1979. 48:159–91

HISTONES

❖12006

Irvin Isenberg

Department of Biochemistry and Biophysics, Oregon State University,
Corvallis, Oregon 97331

CONTENTS

PERSPECTIVES AND SUMMARY

Chromatin consists of two parts—the nucleosomal particle, and the linker region joining the particles. The core of the nucleosomal particle consists of two each of the inner histones, H2a, H2b, H3, and H4, plus a 145 base pair stretch of DNA wrapped around the inner histones. The H1 class of histones is associated with the linker region.

0066-4154/79/0701-0159$01.00

The inner histones show variability. H2b is an evolutionary hybrid: approximately the first third of the molecule is variable; the rest is very highly conserved. H2a may also be a hybrid, but there are not yet enough sequences known to test this. Work on H3 from yeast and from *Tetrahymena* shows that in both species H3 has diverged from calf or plant H3. *Tetrahymena* H4 has diverged from calf H4. Nevertheless, these divergences are much smaller than those shown by most proteins and H3 and H4 must still be regarded as highly conserved proteins. Since H3 and H4 have two different domains, a basic amino part and a carboxyl globular part, the evolutionary pressure maintaining each region must be different. The reason for the conservation of the basic region is completely unknown. The pressure conserving the globular region may arise, in part, from the necessity for conserving histone-histone interactions, since these interactions occur between globular regions, and appear to be highly conserved.

Subtypes of histones H1, H2a, H2b, and H3 exist, which differ in amino acid sequence and arise either during embryogenesis, or during the maturation of certain specialized cells.

Histones may be modified postsynthetically in a variety of ways. Phosphorylation has been correlated with chromosome condensation. Acetylation has been associated with regions of the genome that are structurally active. A modification of H2a, named A24, appears to be preferentially located in inactive regions of the genome. Histones may also be methylated and ribosylated, but there is as yet no clue as to any correlation of these modifications with any structural or functional aspects of chromatin.

INTRODUCTION

Only a short time ago, most people studying histones had but a simplistic idea regarding the conservation of the inner histones, H2a, H2b, H3, and H4; they were simply "highly conserved." This attitude has changed recently.

Within each of the five classes of histones, H1, H2a, H2b, H3, and H4, the molecules may vary. The variability of H1 has long been recognized, as have changes of the histones due to postsynthetic modifications. However, recently it has become clear that there are two additional types of diversity. Within the inner histones there are singificant evolutionary divergencies, and within a particular class of an inner histone of a single species there are subtypes that have different primary structures.

If there is a unifying theme to this review, it is a discussion of what varies, and what remains constant, in the properties of the histones. The review tries, from time to time, to relate these properties of the histones to our knowledge of chromatin, but this is not a review of chromatin, and many important features of chromatin are not even mentioned. On the other hand, an attempt is made to cover what appear to be the most important things that have been learned about histones during the last few years.

PRIMARY STRUCTURES

Histone H4

The well-known result that histone H4 of pea plants differs from H4 of calf thymus by only two conservative replacements established H4 as a highly conserved molecule (1). There are, however, four eukaryotic kingdoms: animals, plants, fungi, and protista (2), and it would be interesting to have complete sequences of representative examples of H4 from fungi and protista. Unfortunately, only a partial sequence is known at this time.

Sixty-five residues of H4 of the protista, *Tetrahymena thermophila,* have recently been sequenced. Fifteen of these are different from those in calif (3), and H4 thus appears to show some evolutionary divergence. Nevertheless, compared to either globular proteins or other histones, H4 still appears as a highly conserved molecule. In fact, what must now be emphasized is not just that H4 is conserved, but that it is conserved in both the amino and carboxyl halves of the molecule. There is a simple reason for such an emphasis.

As illustrated in Figure 1, the amino terminal region is highly basic, whereas the carboxyl region has an amino acid distribution characteristic of globular proteins.

As discussed later, histone-histone complexing occurs between the globular, carboxyl regions of the inner histones, and the strengths of the interactions between the histones appear to have been conserved. The evolutionary pressure maintaining the strengths of the interactions might therefore be considered responsible, at least in part, for the sequence conservation of the globular region. On the other hand, the conservation of the basic amino terminal half of H4 must be due to quite a different evolutionary pressure, and we have as yet no clue as to what this might be.

There are no primary sequences known for H4 from fungi. We know that the gel mobility of *Saccharomyces* H4 is similar to that of calf thymus H4 (4), and there are as yet no reasons to suspect that fungal and calf H4 might show significant divergencies. On the other hand, arguments based on similar gel mobilities could be misleading, and a direct sequence determination of yeast H4 is needed.

Histone H3

The primary structure of H3 of pea differs from that of calf thymus in only four places (5). H3, like H4, is therefore a highly conserved histone in both

```
                    10                                              20
Ac-Ser-Gly-Arg-Gly-Lys-Gly-Gly-Lys-Gly-Leu-Gly-Lys-Gly-Gly-Ala-Lys-Arg-His-Arg-Lys-

                    30                                              40
Val-Leu-Arg-Asp-Asn-Ile-Gln-Gly-Ile-Thr-Lys-Pro-Ala-Ile-Arg-Arg-Leu-Ala-Arg-Arg-

                    50                                              60
Gly-Gly-Val-Lys-Arg-Ile-Ser-Gly-Leu-Ile-Tyr-Glu-Glu-Thr-Arg-Gly-Val-Leu-Lys-Val-

                    70                                              80
Phe-Leu-Glu-Asn-Val-Ile-Arg-Asp-Ala-Val-Thr-Tyr-Thr-Glu-His-Ala-Lys-Arg-Lys-Thr-

                    90                                              100
Val-Thr-Ala-Met-Asp-Val-Val-Tyr-Ala-Leu-Lys-Arg-Gln-Gly-Arg-Thr-Leu-Tyr-Gly-Phe-Gly-Gly
```

Figure 1 (*a*) Calf H4 sequence (72, 73). Also known are H4 from pea (1), the sea urchin, *Parechinus angulosus* (199), the sea urchin, *Psammechimus miliaris* (200), the rat (201), and the pig (202). (*b*) Calf H4 distribution of basic, acidic, hydrophobic, and uncharged hydrophilic residues. An upward line indicates Arg, Lys, or His, a downward line shows Glu or Asp, and a dot represents Val, Met, Leu, Ile, Tyr, Phe, Pro, or Ala.

the amino and carboxyl regions. However, the gel mobilities of both yeast H3 (4, 6) and *Tetrahymena* H3 (7) are distinctly different from that of calf thymus H3. One suspects, therefore, that the primary structures are different. Complete fungal histone amino acid sequences are now lacking; the first 15 residues of yeast H3 (6) are known, however, and as these are identical to their calf counterpart, differences must appear further down the polypeptide chain.

Histone H2b

It has become clear recently that H2b is an evolutionary hybrid: the carboxyl two thirds is highly conserved, and the amino one third is poorly conserved. Figure 3a shows the sequence of those H2b's whose primary structure is now known. The alignments in the amino part are somewhat arbitrary. For the most part the residues have been positioned for maximum homology, but this has been violated to permit a striking pentapeptide repeat, Pro-Thr-Lys-Arg-Ser, near the beginning of the H2b (1) of *S. angulosis* (8). The arbitrariness of alignment in the amino terminal region itself emphasizes the poor conservation of that part of the molecule, for strong conservation would impose an alignment.

Pea H2b has been cleaved by BrCN and trypsin and the cleaved fragments compared with the sequence of calf H2b (9). It is interesting that it is possible to align the fragments of the carboxyl region with the calf sequence, but the amino terminal portions cannot be so aligned. This indicates that the general conservation pattern of Figure 3 is followed by pea H2b.

Histone H2a

Not much is as yet known about the variable and constant regions of H2a. The data of Figure 4 are for animal H2a's only; no others have as yet been sequenced. It is known, however, that pea H2a differs from mammalian H2a in gel mobility and in amino acid composition (10), and it would be interesting to have H2a sequences of these or, for that matter, of any species other than animals.

Histone H1

Histone H1 is the most variable of the histones (11–15). The number of subfractions varies from tissue to tissue in a given species, and, for a given tissue, it varies from one species to another. The molecular weight varies as well as the sequence. Yet H1, like H2b, has strongly conserved stretches and must be regarded as an evolutionary hybrid.

Figure 5 shows that there are three, or perhaps more, structurally distinct domains in H1. For about the first 40 residues, there is a variable region,

10
20
Ala-Arg-Thr-Lys-Gln-Thr-Ala-Arg-Lys-Ser-Thr-Gly-Gly-Lys-Ala-Pro-Arg-Lys-Gln-Leu-

30
40
Ala-Thr-Lys-Ala-Ala-Arg-Lys-Ser-Ala-Pro-Ala-Thr-Gly-Gly-Val-Lys-Lys-Pro-His-Arg-

50
60
Tyr-Arg-Pro-Gly-Thr-Val-Ala-Leu-Arg-Glu-Ile-Arg-Arg-Tyr-Gln-Lys-Ser-Thr-Glu-Leu-

70
80
Leu-Ile-Arg-Lys-Leu-Pro-Phe-Gln-Arg-Leu-Val-Arg-Glu-Ile-Ala-Gln-Asp-Phe-Lys-Thr-

90
100
Asp-Leu-Arg-Phe-Gln-Ser-Ser-Ala-Val-Met-Ala-Leu-Gln-Glu-Ala-Cys-Glu-Ala-Tyr-Leu-

110
120
Val-Gly-Leu-Phe-Glu-Asp-Thr-Asn-Leu-Cys-Ala-Ile-His-Ala-Lys-Arg-Val-Thr-Ile-Met-

130
Pro-Lys-Asp-Ile-Glu-Leu-Ala-Arg-Arg-Ile-Arg-Gly-Glu-Arg-Ala

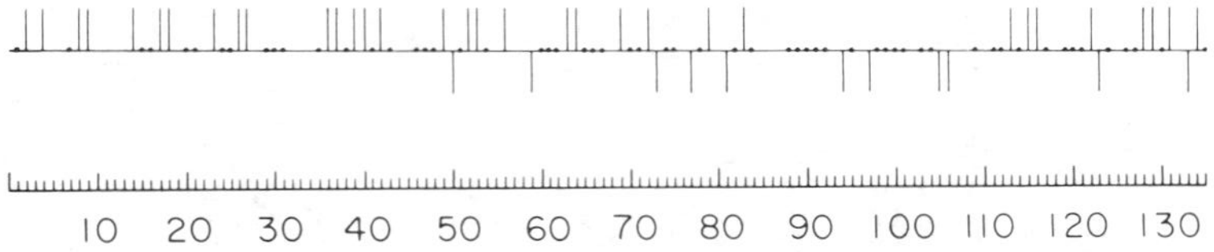

Figure 2 (*a*) Calf H3 sequence (203, 204). Also known are H3 from pea (5), chicken (205), shark (206), sea urchin (207), and carp (208). (*b*) Calf H3 distribution of residue types. See Figure 1*b* for meaning of symbols.

```
                         10                              20
P. Ang 1  Pro-Ser-Gln-Lys-Ser-Pro-Thr-Lys-Arg-Ser-Pro-Thr-Lys-Arg-Ser-Pro-Thr-Lys-Arg-Ser—
P. Ang 2  Pro————————Lys-Ser-Pro-Thr-Lys-Arg-Ser-Pro-Arg-Ser-Pro-Arg-Lys-Gly-Ser—
Calf      Pro————Glu————————Pro-Ala-Lys-Ser-Ala-Pro-Ala-Pro-Lys——————
Trout     Pro————Glu————————Pro-Ala-Lys-Ser-Ala-Pro——————Lys——————
Dros      Pro————————Pro————Lys-Thr-Ser-Gly-Lys-Ala-Ala——————Lys——————

                         30                              40
          Pro-Gln-Lys-Gly-Gly-——————————Lys-Gly-Gly-Lys-Gly-Ala-Lys-Arg-Gly-Gly-Lys-Ala—
          Pro-Ser-Arg-Lys-Ala-Ser-Pro-Lys-Arg-Gly-Gly-Lys-Gly-Ala-Lys-Arg-Ala-Gly-Lys-Gly—
          ——————————————Lys——————Gly-Ser-Lys-Lys—
          ——————————————Lys——————Gly-Ser-Lys-Lys—
          ——————————————Lys-Ala-Gly——————Lys-Ala—

                         50                              60
          ——————————Gly-Lys-Arg-Arg-Arg-Gly-Val-Gln-Val-Lys-Arg-Arg-Arg-Arg-Arg-Arg—
          ——————————Gly-Arg-Arg-Arg-Arg-Arg——————Val——————Val-Lys-Arg-Arg-Arg-Arg-Arg-Arg—
          Ala——————————Val-Thr-Lys-Ala-Gln-Lys-Lys-Asp-Gly-Lys-Lys-Arg-Lys-Arg-Ser-Arg—
          Ala——————————Val-Thr-Lys-Thr-Ala-Gly-Lys-Gly-Lys-Gly-Lys-Lys-Arg-Ser-Arg-Ser-Arg—
          Gln-Lys-Asn-Ile——————Thr-Lys-Thr——————Asp-Lys——————Lys-Lys-Lys-Arg-Lys-Arg—
```

80

70

——Glu-Ser-Tyr-Gly-Ile-Tyr-Ile-Tyr-Lys-Val-Leu-Lys-Gln-Val-His-Pro-**Asp-Thr**-Gly-

——Glu-Ser-Tyr-Gly-Ile-Tyr-Ile-Tyr-Lys-Val-Leu-Lys-Gln-Val-His-Pro-Asp-Thr-Gly-

Lys-Glu-Ser-Tyr-Gly-Ile-Tyr-Val-Tyr-Lys-Val-Leu-Lys-Gln-Val-His-Pro-Asp-Thr-Gly-

Lys-Glu-Ser-Tyr-Ala-Ile-Tyr-Val-Tyr-Lys-Val-Leu-Lys-Gln-Val-His-Pro-Asp-Thr-Gly-

Lys-Glu-Ser-Tyr-Ala-Ile-Tyr-Val-Tyr-Lys-Val-Leu-Lys-Gln-Val-His-Pro-Asp-Thr-Gly-

100

90

Ile-Ser-Ser-Arg-Ala-Met-Ser-Val-Met-Asn-Ser-Phe-Val-Asn-Asp-Val-Phe-Glu-Arg-Ile-

Ile-Ser-Ser-Arg-Ala-Met-Ser-Val-Met-Asn-Ser-Phe-Val-Asn-Asp-Val-Phe-Glu-Arg-Ile-

Ile-Ser-Ser-Lys-Ala-Met-Gly-Ile-Met-Asn-Ser-Phe-Val-Asn-Asp-Ile-Phe-Glu-Arg-Ile-

Ile-Ser-Ser-Lys-Ala-Met-Gly-Ile-Met-Asn-Ser-Phe-Val-Asn-Asp-Ile-Phe-**Glu**-Arg-Ile-

Ile-Ser-Ser-Lys-Ala-Met-Ser-Ile-Met-Asn-Ser-Phe-Val-Asn-Asp-Ile-Phe-Glu-Arg-Ile-

120

110

Ala-Ala-Glu-Ala-Gly-Arg-Leu-Thr-Thr-Tyr-Asn-Arg-Arg-Ser-Thr-Val-Ser-Ser-Arg-Glu-

Ala-Gly-Glu-Ala-Ser-Arg-Leu-Thr-Ser-Ala-Asn-Arg-Arg-Ser-Thr-Val-Ser-Ser-Arg-Glu-

Ala-Gly-Glu-Ala-Ser-Arg-Leu-Ala-His-Tyr-Asn-Lys-Arg-Ser-Thr-**Ile-Thr**-Ser-Arg-Glu-

Ala-Gly-Glu-Ser-Ser-Arg-Leu-Ala-His-Tyr-Asn-Lys-Arg-Ser-Thr-Ile-Thr-Ser-Arg-Glu-

Ala-Ala-Glu-Ala-Ser-Arg-Leu-Ala-His-Tyr-Asn-Lys-Arg-Ser-Thr-Ile-Thr-Ser-Arg-Glu-

130   140

Val-Gln-Thr-Ala-Val-Arg-Leu-Leu-Leu-Pro-Gly-Glu-Leu-Ala-Lys-His-Ala-Val-Ser-Glu–

Ile-Gln-Thr-Ala-Val-Arg-Leu-Leu-Leu-Pro-Gly-Glu-Leu-Ala-Lys-His-Ala-Val-Ser-Glu–

Ile-Gln-Thr-Ala-Val-Arg-Leu-Leu-Leu-Pro-Gly-Glu-Leu-Ala-Lys-His-Ala-Val-Ser-Glu–

Ile-Gln-Thr-Ala-Val-Arg-Leu-Leu-Leu-Pro-Gly-Glu-Leu-Ala-Lys-His-Ala-Val-Ser-Glu–

Ile-Gln-Thr-Ala-Val-Arg-Leu-Leu-Leu-Pro-Gly-Glu-Leu-Ala-Lys-His-Ala-Val-Ser-Glu–

150

Gly-Thr-Lys-Ala-Val-Thr-Lys-Tyr-Thr-Thr-Ser-Arg

Gly-Thr-Lys-Ala-Val-Thr-Lys-Tyr-Thr-Thr-Ser-Arg

Gly-Thr-Lys-Ala-Val-Thr-Lys-Tyr-Thr-Thr-Ser-Ser-Lys

Gly-Thr-Lys-Ala-Val-Thr-Lys-Tyr-Thr-Thr-Ser-Ser-Lys

Gly-Thr-Lys-Ala-Val-Thr-Lys-Tyr-Thr-Thr-Ser-Ser-X

Figure 3 (*a*) Calf H2b sequence (209). Also known are sequences of H2b from trout (210), two sea urchin sperm H2b's (8, 54), and *Drosophila* (211). (*b*) The top portion shows the distribution of residue types for one of the sea urchin sperm H2b's. See Figure 1*b* for meaning of symbols. A diagonal slash means a deletion in the sequence. In the bottom portion a black square indicates that the residues for all five H2b's are identical.

```
                                              10                                                      20
Calf       Ac - Ser-Gly-Arg-Gly-Lys-Gln-Gly-Gly-Lys-Ala-Arg-Ala-Lys-Ala-Lys-Thr-Arg-Ser-Ser-Arg-

Trout      Ac - Ser-Gly-Arg-Gly-Lys-Thr-Gly-Gly-Lys-Ala-Arg-Ala-Lys-Ala-Lys-Thr-Arg-Ser-Ser-Arg-

P. Miliaris Ac- Ser-Gly-Arg-Gly-Lys——-Gly-Ala-Lys-Gly-Lys-Ala-Lys-Ala-Lys-Ser-Arg-Ser-Ser-Arg-


                       30                                                      40
           Ala-Gly-Leu-Gln-Phe-Pro-Val-Gly-Arg-Val-His-Arg-Leu-Leu-Arg-Lys-Gly-Asn-Tyr-Ala-

           Ala-Gly-Leu-Gln-Phe-Pro-Val-Gly-Arg-Val-His-Arg-Leu-Leu-Arg-Lys-Gly-Asn-Tyr-Ala-

           Ala-Gly-Leu-Gln-Phe-Pro-Val-Gly-Arg-Val-His-Arg-Phe-Leu-Arg-Lys-Gly-Asn-Tyr-Ala-


                       50                                                      60
           Glu-Arg-Val-Gly-Ala-Gly-Ala-Pro-Val-Tyr-Leu-Ala-Ala-Val-Leu-Glu-Tyr-Leu-Thr-Ala-

           Glu-Arg-Val-Gly-Ala-Gly-Ala-Pro-Val-Tyr-Leu-Ala-Val-Ala-Val-Leu-Glu——-Leu-Thr-Ala-

           Asn-Arg-Val-Gly-Ala-Gly-Ala-Pro-Val-Tyr-Leu-Ala-Ala-Val-Leu-Glu-Tyr-Leu-Ala-Ala-


                       70                                                      80
           Glu-Ile-Leu-Glu-Leu-Ala-Gly-Asn-Ala-Ala——-——-Arg-Asp-Asn-Lys-Lys-Thr-Arg-

           Glu-Ile-Leu-Glu-Leu-Ala-Gly-Asx-Ala-Ala-Arg-Ile-Pro-Arg-Asx-Asx-Lys-Lys-Thr-Arg-

           Glu-Ile-Leu-Glu-Leu-Ala-Gly-Asn-Ala-Ala——-——-Arg-Asp-Asn-Lys-Lys-Thr-Arg-
```

```
                                     90                                    100
Ile-Ile-Pro-Arg-His-Leu-Gln-Leu-Ala-Ile-Arg-Asn-Asp-Glu-Glu-Leu-Asn-Lys-Leu-Leu-
--Ile-Pro-Arg-His-Leu-Gln-Leu-Ala-Val-Arg-Asn-Asp-Glu-Glu-Leu-Asx-Lys-Leu-Leu-
Ile-Ile-Pro-Arg-His-Leu-Gln-Leu-Ala-Ile-Arg-Asn-Asp-Glu-Glu-Leu-Asn-Lys-Leu-Leu-

                                    110                                    120
Gly-Lys-Val-Thr-Ile-Ala-Gln-Gly-Gly-Val-Leu-Pro-Asn-Ile-Gln-Ala-Val-Leu-Leu-Pro-
Gly-Gly-Val-Thr-Ile-Ala-Glx-Gly-Gly-Val-Leu-Pro-Asx-Ile-Glx-Ala-Val-Leu-Leu-Pro-
Gly-Gly-Val-Thr-Ile-Ala-Gln-Gly-Gly-Val-Leu-Pro-Asn-Ile-Gln-Ala-Val-Leu-Leu-Pro-

                                    130
Lys-Lys-Thr-Glu-Ser-His-His-Lys-Ala-Lys-Gly-Lys
Lys-Lys-Thr-Glu--------------Lys-Ala-Lys-Val-Ala-Lys
Lys-Lys-Thr-Gly-Ser------Lys-Ser-Ser------Lys
```

Figure 4 (a) H2a sequence of calf (212, 213), trout (216) and sea urchin (217). Also known are sequences of H2a from chicken erythrocytes (214) and a rat chloroleukemia (215). (b) Residue type distribution and residue identity for calf, trout, and sea urchin H2a's. See Figure 3b for meaning of symbols.

```
                     10                        20
RTL-3   Ac -Ser-Glu-Ala-Pro-Ala-Glu-Thr-Ala-Ala-Pro-Ala-Glu-Lys-Ser-Pro-Ala-Lys- —— -Lys-
Trout   Ac -Ala-Glu-Ala-Pro-Ala-Glu-Val-Ala- —— -Pro-Ala-Pro-Ala-Ala-Pro-Ala-Ala-Lys-Ala-Pro-Lys-

                     30                        40
RTL-3   Lys-Lys-Ala-Ala-Lys-Lys-Pro-Gly- —— -Ala-Gly-Ala-Ala-Lys-Ala-Ala-Gly-Pro-Pro-Val-
Trout   Lys-Lys-Ala-Ala-Lys-Lys-Pro-Lys-Lys-Ala-Gly- —— - —— - —— -Gly-Pro-Ala-Val-

                     50                        60
RTL-3   Ser-Glu-Leu-Ile-Thr-Lys-Ala-Val-Ala-Ala-Ser-Lys-Glu-Arg-Asn-Gly-Leu-Ser-Leu-Ala-Ala-Leu-
Trout   Gly-Glu-Leu-Ile-Gly-Lys-Ala-Val-Ala-Ala-Ser-Lys-Glu-Arg-Ser-Gly-Val-Ser-Leu-Ala-Ala-Leu-

                     70                        80
RTL-3   Lys-Lys-Ala-Leu-Ala-Ala-Gly-Gly-Tyr-Asp-Val-Glu-Lys-Asn-Asn-Ser-Arg-Ile-Lys-Leu-Gly-Leu-
Trout   Lys-Lys-Ser-Leu-Ala-Ala-Gly-Gly-Tyr-Asp-Val-Glu-Lys-Asn-Asn-Ser-Arg-Val-Lys-Ile-Ala-Val-

                     90                 100                 110
RTL-3   Lys-Ser-Leu-Val-Ser-Lys-Gly-Thr-Leu-Val-Glu-Thr-Lys-Gly-Thr-Gly-Ala-Ser-Gly-Ser-Phe-Lys-
Trout   Lys-Ser-Leu-Val-Thr-Lys-Gly-Thr-Leu-Val-Glu-Thr-Lys-Gly-Thr-Gly-Ala-Ser-Gly-Ser-Phe-Lys-

                    120                       130
RTL-3   Leu-Asn-Lys-Lys-Ala-Ala-Ser-Gly-Glu-Ala-Lys-Pro-Lys-Pro- —— -Lys-Lys-Ala-Gly-Ala-Ala-Lys-
Trout   Leu-Asn-Lys-Lys-Ala- —— - —— -Val-Glu-Ala-Lys-Lys- —— -Lys-Pro-Ala-Lys-Lys-Ala-Ala-Ala-Pro-Lys-

                    140                       150
RTL-3   Pro-Lys-Lys-Pro-Ala-Gly- —— - —— - —— -Ala-Thr-Pro-Lys-Lys-Pro-Lys-Lys-Ala-Gly-Ala-Lys-
Trout   Ala-Lys-Lys-Val-Ala-Lys-Ala-Lys-Lys-Pro-Ala-Ala-Ala-Lys-Lys-Pro-Lys-Lys-Val-Ala- —— -Ala-Lys.

                    160                       170
RTL-3   Lys-Ala-Val- —— - —— -Lys-Lys-Thr-Pro-Lys-Ala-Pro-Lys-Ala-Ala-Ala-Lys-Pro-Lys- ——
Trout   Lys-Ala-Val-Ala-Ala-Lys-Lys-Ser-Pro-Lys-Lys-Ala- —— - —— -Lys- —— - —— -Lys-Pro- ——
```

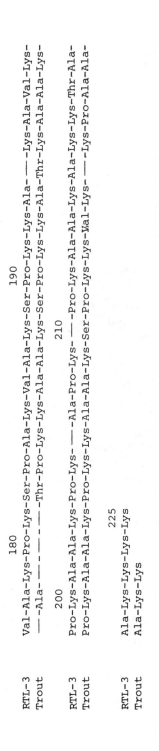

180
190

RTL-3 Val-Ala-Lys-Pro-Lys-Ser-Pro-Ala-Lys-Val-Ala-Lys-Ser-Pro-Lys-Lys-Ala— ——Lys-Ala-Val-Lys-
Trout ——Ala— — —— ——Thr-Pro-Lys-Lys-Ala-Ala-Lys-Ser-Pro-Lys-Ala-Thr-Lys-Ala-Ala-Lys-

200
210

RTL-3 Pro-Lys-Ala-Ala-Lys-Pro-Lys— ——Ala-Pro-Lys— ——Pro-Lys-Ala-Ala-Lys-Ala-Lys-Lys-Thr-Ala-
Trout Pro-Lys-Ala-Ala-Lys-Pro-Lys-Lys-Ala-Ala-Lys-Ser-Pro-Lys-Val-Lys— ——Lys-Pro-Ala-Ala-

225

RTL-3 Ala-Lys-Lys-Lys
Trout Ala-Lys-Lys

Figure 5 (*a*) Sequence of an H1 from rabbit thymus (15) and trout (218). (*b*) Residue type distribution of the rabbit thymus H1 of 5a and identity of residues between this H1 and the H1 of trout. See Figure 3b for meaning of symbols.

very basic and rich in alanines and prolines. This is followed by a region of about 75 residues with quite a different character. The residue distribution pattern here is typical of globular proteins, and the region has long stretches of sequences that are identical in the trout and rabbit proteins. The rest of the molecule shows a high degree of variability. The carboxyl terminal half of H1 is highly basic. It contains only a few amino acids, mainly lysine, alanine, and proline, and it is possible, therefore, that some of the short stretches of homology that appear there are accidental.

H1 has this in common with H2b. The regions with the globular-like residue distributions are the ones that are highly conserved. The others, with important exceptions, are variable.

HISTONES OF THE FUNGI

Histones have been found in a number of fungi: *Physarum polycephalum* (16), *Aspergillus nidulans* (17), and *Saccharomyces cerevisiae* (18).

The most extensive studies of fungal histones have been made on those from *S. cerevisiae*. It is clear that this yeast has four distinct inner histones (4, 19–22), and all four have now been isolated and identified (4). It is not yet known if yeast contains H1.

Yeast H2a, H2b, and H3 have all diverged from their plant and animal counterparts. Since plant and animal H2a and H2b differ from one another, it was not surprising to find that yeast and animal H2a and H2b also differed. However, yeast H3 also differs from plant and animal H3 in that it has a higher mobility on acid-urea gels than calf H3 and has neither cysteine nor methionine. Furthermore, when yeast H3 complexes with yeast H4, the α-helical content does not rise (4), although there is a marked increase in α-helical content when the calf counterparts complex with one another (23).

HISTONES OF PROTISTA

Histones have been reported in *Euglena gracilis* (24), *Stylonychia mytilus* (25), *Oxytricha* sp. (26), *Tetrahymena* (7, 27), and *Peridinium balticum* (28). The most extensive investigations have been made on *Tetrahymena* by Gorovsky and his co-workers.

Tetrahymena, like most ciliated protozoans, exhibit nuclear dimorphism. During the vegetative life cycle, the macronuclei divide amitotically and are transcriptionally active; the micronuclei divide mitotically and show little RNA synthesis. The macronuclei contain a full complement of all five histones, H1 plus the four inner histones.

The micronuclei of *Tetrahymena* appear to have only histones H2a, H2b, and H4, and apparently lack both H3 and H1 (27, 29).

Recent complexing studies have shown that the protein originally called HX is really H2a (3). This histone has two genetically distinct subspecies.

As already discussed, the *Tetrahymena* inner histones show significant divergencies from their calf counterparts.

PRIMARY STRUCTURE SUBTYPES

During embryogenesis, or during the development or maturation of specific cell types, a particular histone may no longer be synthesized, or a new histone of the same class may begin to be synthesized. Although of the same class, the histones differ in primary structure. They are referred to as subtypes or variants.

Sequence differences between subtypes may range from a change of only a few residues to massive substitutions. There are undoubtedly important differences between many of these cases, but they are discussed together here, since we do not yet understand any of the biological mechanisms involved, and it is not yet possible to make meaningful distinctions. In fact, the juxtaposition of what are often regarded as quite different phenomena may have some virtue, since it will call attention to certain common features. In a similar way, both the H1 and the inner histone subtypes are discussed in this section, even though there must obviously be profound differences between them.

The H1 subtypes were the first to be discovered since they appear as H1 subfractions in histone preparations from a huge variety of tissues. For over a decade it has been known that the number and relative amounts of these subfractions vary from tissue to tissue, and species to species (11–13, 30). Since the original reports, the heterogeneity of H1 has been shown repeatedly (31–44) and is clearly a widespread phenomenon.

Sea urchins have often been organisms of choice to examine subtypes, and it is known that the number of H1 subfractions increases from the morula to the gastrula stage of the sea urchin (35, 45). The increase is due to a switch from the synthesis of one type to that of another (39) with the earlier H1 persisting for a long time after its synthesis has terminated (39, 46).

The development of Triton gels (47, 48) has been an important technical advance, since subfractions of the inner histones may be resolved on such gels. By their use, it has been shown that the existence of subtypes is by no means limited to the H1 histones. H2a, H2b, and H3 also have variants (47–50).

The syntheses of the inner histone variants, as well as those of H1, are turned on or off at specific times during embryogenesis (44, 46, 49). Furthermore, the suggestion has been made that the switching is intimately related to the chromatin remodeling that occurs in the development of new cell types (44, 51).

During spermatogenesis the syntheses of specific histones are switched on or off. In teleosts, histones are replaced by protamines (52). In echinoderms, the sperm still have histones, but these are generally different from the somatic ones (53). Sperm of the sea urchin, *Parechinus angelosus* has two H2b's, both of which have been sequenced (8, 54) (see above).

Extensive switching also occurs during mammalian spermatogenesis. At a stage before the formation of spermatids there are specific meiotic histones of classes H1, H2b, and possibly H3 (55–58). In the maturation of the spermatid, new histones (and other basic proteins) become associated with the DNA (59–61).

Franklin & Zweidler (50) have identified the amino acid differences between a number of mammalian variants; these are shown in Table 1. (The second subtype of H2b in Table 1 has been found only in the mouse, but all of the others have been found in a wide variety of species and tissues.)

Careful examination of the variants shows that most, and perhaps all, of the switches could result in marked conformational changes of the nucleosome, drastically altered interactions of the nucleosome, or both. For example, Figure 3 shows that Glu 103 of H2b is stringently conserved, which suggests that it engages in a critically important, though as yet unknown, interaction. If so then the variant shift, Glu → Gln, would be a drastic one indeed.

Consider the H3 variants. These occur in the globular regions of the molecule, in a sequence that is rich in hydrophobic and acidic residues. Hydrophobics are often found in the interior of globular proteins. Alterations of these might lead to sharply changed conformations, which could,

Table 1   Known mammalian inner histone variants[a]

Variant	Residues substituted		
		H2a	
1	Thr 16		Leu 51
2	Ser 16		Met 51
		H2b	
1	Gly 102		Glu 103
2	Ser 102		Glu 103
3	Gly 102		Gln 103
		H3	
1	Val 89	Met 90	Cys 96
2	Val 89	Met 90	Ser 96
3	(Ile 89	Gly 90)	Ser 96

[a]The numbering is according to Figures 1–4. The data is taken from (50). There is some uncertainty regarding variant 3 of H3.

of course, also change the interactions of the negatively charged residues that are nearby.

That such speculations may have some merit is indicated by an examination of the H2a variants. Position 51 differs from one subtype to another. It is also in the midst of a hydrophobic region that contains acidics.

It is true that only a few specific residue changes of inner histones have so far been identified. It is nevertheless striking that in these cases, either an otherwise invariant negative residue is switched, or switches occur in hydrophobic regions that are negatively charged or that are near an acidic residue. It will be interesting to see if this feature occurs in other subtypes when these are identified.

Goose and pigeon H5 shows extensive homologies to H1 (62) so that H5 and H1 may be considered as variants of the same histone class, even though the differences are extensive. Viewed in this light, H5 becomes another example of a new variant occurring during a change in cell type, since H5 accumulates in the erythroid cells of developing chick embryos, as the cells mature (63, 64). Furthermore, H5 exists in the nucleated erythrocytes of a variety of fish (65, 66); this also fits into the pattern in which the synthesis of a new variant accompanies the formation of a specialized cell.

As indicated, there are reasons for believing that different histone variants will show markedly different interactions with various cellular constituents, but almost no actual cases are known. Only two possibilities have been demonstrated. Different H1 subfractions bind to different proteins of the High Mobility Group with considerable specificity (67, 68), and various H1 subfractions can bind differently to DNA (69). It is not yet known if either of these specificities has any physiological significance.

POSTSYNTHETIC MODIFICATIONS

Five types of postsynthetic modifications are now known: acetylation, phosphorylation, methylation, poly(ADP) ribosylation, and the formation of A24.

Acetylation

For almost all cases that have been studied to date, amino-terminal acetylation, in the form of acetylserine, occurs in histones H1, H2a, and H4. *Tetrahymena* H4 is not so acetylated, however (3), and presumably other H4's of *Protista* will also be found to have unblocked amino terminii. Amino-terminal acetylation occurs in the cytoplasm before the histones are transported to the nucleus (70).

The acetylation of residues internal to the polypeptide chain occurs on lysines in the amino-terminal region; any of the inner histones may be so acetylated. Acetylation sites in H4 are lysines at positions 5, 8, 12, and 16; in H3 at positions 9, 14, 18, and 23; in H2b at positions 8, 35, 39, and 47; and in H2a at position 5 (52, 71–79).

There is, of course, some arbitrariness in aligning histones of different species, but despite this arbitrariness, it is still of great interest that the sequences may be aligned so that the sites of acetylation are conserved. The acetylation sites may be invariant.

Consider, for example, H2b (Figure 3). Lysines at 8, 35, and 39 may be acetylated. These are invariant. The fourth lysine that may be acetylated is at position 47 and the three somatic histones of Figure 3 all have lysines at this position.

There is abundant evidence that ϵ-N-lysyl acetylation is correlated with active regions of the genome. The acetylation of histones has been reviewed recently [see for example, (78, 79)], and no attempt to summarize the field is made here. However, the recent finding that n-butyrate induces the formation of multi-acetylated forms of histones of HeLa cells and Friend erythroleukemia cells (80) has stimulated rapid and exciting progress. This superacetylation arises because butyrate suppresses the deacetylation of the histones without affecting the rate of acetylation (81, 82). The acetylation leads to increased DNase I sensitivity of the DNA (82), and DNase I preferentially digests regions of the chromatin that are active in some structural sense (83–87). (Structurally active regions are not necessarily being transcribed. What appears to be true, however, is that sensitivity to DNase I can characterize two different types of chromatin, one that may be transcribed and one that may not be transcribed.)

It has also been recently shown (88) that DNase I preferentially degrades the chromatin of hepatoma tissue culture cells in regions whose histones are acetylated. In these studies the cells had not been treated with butyrate.

Phosphorylation

The phosphorylation of histones has been reviewed a number of times in recent years (78, 89–91). Only recent findings are emphasized here, and early work is mentioned only as it appears important to the understanding of later results.

Phosphorylation occurs during S phase, during mitosis, and as a response to hormonal stimulation. In each case, different and highly specific serines and threonines are phosphorylated.

Bradbury and his co-workers (92–94) have studied the phosphorylation of H1 of the slime mold *Physarum polycephalum* during the mitotic cycle. Studying mitosis in this species is advantageous because the nuclei divide synchronously, and thereby enable the phase relationships of phosphorylation to be observed. It was reported (93) that the phosphorylation of H1 occurred just before mitosis, when the chromatin condensed and was much reduced in metaphase. Furthermore, if a preparation of phosphokinase from Ehrlich ascites tumor cells was added to the plasmodia, mitosis was shifted to earlier times (94). These observations led to the suggestion that phosphorylation might be a trigger for mitosis.

All classes of histones are phosphorylated in developing trout testis (78, 95–97) and a summary of the identified sites is given in (78). All classes of histones are phosphorylated in avian erythroid cells (98, 99) and the phosphorylation of various histones has also been reported in a variety of other tissues (100–104).

Langan (91) has identified the sites of phosphorylation of mammalian H1 for two different types of phosphorylation. One is a cyclic AMP-dependent phosphorylation which occurs on a serine at position 37 in mammalian sequences, but would be aligned at position 40 in the numbering used in Figure 5. In the rabbit subfraction shown in Figure 5, the residue at 40 is an alanine, which illustrates the interesting point that not all subfractions have a serine at this position to be phosphorylated.

The other type of phosphorylation has been designated as growth-associated. The phosphorylations associated with growth are presumably the same as some of the phosphorylations found in Chinese hamster CHO cells, which is discussed below. The sites of the growth-associated phosphorylations are in both the amino and carboxyl regions of the protein and, in the alignment of Figure 5, are Thr 143, Thr 162, Ser 189, and Thr 16. Serine 108 is phosphorylated enzymatically in vitro, but has not yet been observed in vivo.

A number of laboratories have studied phosphorylation during mitosis of synchronized cultures of a variety of mammalian cells: Chinese hamster cells (105–108), HeLa cells (106, 108, 109), liver hepatoma cells (110), rat nephroma cells (106), and Chinese hamster CHO cells (37, 38, 91, 111, 112).

In mammalian cells H1, H3, and H2a are all phosphorylated; the phosphorylation of H2a is constant throughout the cell cycle, but that of H3 and H1 varies with the stage.

Peptide mapping studies show that H1 phosphorylations are of two distinct and separate types (90, 112). One type occurs during interphase and has been designated as $H1_I$; another occurs during mitosis and has been named $H1_M$. The phosphorylation of H3 also occurs during mitosis. More specifically, $H1_M$ and phosphorylated H3 appear together during prophase, metaphase, and anaphase; dephosphorylation of these molecules occurs as the cells leave anaphase (90).

$H1_I$ differs from $H1_M$ in many ways. An $H1_I$ molecule contains only 1–3 phosphates, all on serine and all located on the carboxyl side of Tyr 73. $H1_M$ has 3–6 phosphates, both serines and threonines are phosphorylated, and these are on both sides of Tyr 73. Furthermore, only some of the H1 molecules undergo phosphorylation to $H1_I$, while all, or almost all H1 molecules are phosphorylated to $H1_M$ during mitosis. During mitosis, both $H1_I$ and previously unphosphorylated H1 are phosphorylated to the $H1_M$ level. When cells are arrested in early G1, neither $H1_I$ or $H1_M$ phosphorylation is observed.

Studies of Chinese hamster cells have shown that it is essential to include sodium bisulfite (or some other equivalent phosphatase inhibitor) in buffers used for isolating chromatin and histones. If not, H1 and H3 will be rapidly dephosphorylated (90, 113). Even metaphase chromosomes are highly phosphorylated, a finding that is missed if a phosphatase inhibitor is not used (113).

It remains to be seen if the reported low phosphorylation in *Physarum* at metaphase is due to the absence of a phosphatase inhibitor during the preparation of *Physarum* histones or is perhaps due to a real difference between *Physarum* and CHO cells.

Gurley et al (111) have compared the phosphorylation in cultured cell lines of two species of deer mice, *Peromyscus crinitus* and *Peromyscus eremicus*. These have about the same euchromatic content, but *P. eremicus* has about 36% more heterochromatin. The steady state H1 phosphorylation of the two cells was the same, but the H2a of *P. eremicus* was phosphorylated 58% more than the H2a of *P. crinitus*.

It has been suggested (111) that different levels of chromatin organization might be correlated with different types of phosphorylation: $H1_I$ phosphorylation might be correlated with a change in organization at a molecular level, H2a phosphorylation with heterochromatin condensation, and $H1_M$ and H3 phosphorylation with condensation to chromosomes. This, however, is only a correlation; cause and effect relationships, and structural data, are still to be determined.

All of the phosphates that have been discussed so far are acid stable. In addition, however, it has been reported (114, 115) that, in regenerating rat liver, and in other actively growing tissues, the histidines of H4 are modified to 3-phosphohistidine, and that phosphoryllysines occur in H1. Very little is known about such phosphorylations but, in contrast to serine and threonine phosphates, they are alkali-stable and acid-labile, at least in hot acid.

Methylation

Methylation of lysines has been seen only in H3 and H4, but histidines may also be methylated in H1 and H5 of duck erythroid cells (116).

In H3, lysines 9 and 27 may be methylated; in H4 methylation is at lysine 10. Sometimes a lysine may have more than one methyl group (5, 72, 73, 76, 77).

Methylation occurs late in S phase or in G_2, in both HeLa cells (117), and in Ehrlich ascites tumor cells (118). It occurs after DNA synthesis (and presumably well after histone synthesis) in regenerating liver (119).

Methylation appears to be irreversible (118, 120, 121); once a histone is methylated, it remains methylated. Since nucleosomes apparently form as DNA replication proceeds (122), probably methylation occurs after nucleosome formation.

The role of methylation is unknown although Byvoet & Baxter (123) have suggested that methylation might alter DNA-histone interactions.

ADP Ribosylation

Poly(ADP-ribose) and ADP-ribosylation have been reviewed recently (124), and no attempt is made here to repeat the material covered in that review.

As long ago as 1968, Nishizuka et al (125) suggested that H1, H2a, H2b, and H3 of rat liver could all be covalently linked to poly(ADP-ribose). Ribosylation of histones has since been reported not only in mammalian nuclei (126–128), but in the sea urchin, *Echinus esculentus* (129), and in trout testis (130). In the trout, only H1 and not the inner histones are reported to be ribosylated. Poly(ADP-ribose) polymerase (131) and the glycohydrolase (132) have been found in *Physarum polycephalum.* Presumably, therefore, the ribosylation of histones is widespread.

It appears reasonable that ADP-ribosylation might modify dramatically the interactions of histones, and the state of the chromatin might be affected drastically, but nothing is yet known about either the physical or physiological consequences of such an alteration.

Protein A24

Protein A24 was so designated because of its position in the two-dimensional gel in which it was first seen. It appeared as the twenty-fourth spot of a gel area that had been named region A (133).

A24 is a modification of H2a (134). The ϵ-amino of lysine 119 of the H2a is conjugated, by an isopeptide linkage, to the carboxyl terminus of another peptide chain. The amino portion of that chain is the protein, ubiquitin (135).

Ubiquitin is a very highly conserved protein that has been found in a variety of animal cells, bacteria, yeast, and higher plants (136). The complete sequence is known (137).

A24 has been found in nucleoli of rat liver and Novikoff hepatoma ascites cells (133), calf thymus chromatin (138), and chromatin from human lymphocytes (139). A24 is a constitutent of nucleosomes (140). It has been estimated that about 10% of calf thymus H2a is modified to A24 (141). This means that about 1 in 10 to 1 in 5 nucleosomes are so altered, depending upon whether one or both H2a's of a particular nucleosome are in the A24 form.

The amount of A24 decreases during liver regeneration (142, 143), and is markedly reduced in acute lymphocytic leukemic cells compared to normal lymphocytes (139). If we can extrapolate these results to more general possibilities, it appears that the amount of A24 has an inverse correlation with activity.

HISTONE CONFORMATION CHANGES

Inner Histone Refolding

To prepare pure histones it is necessary to use denaturing solvents. The resultant preparations are consequently in a nonphysiological state.

Pure histones dissolved in water with no added salt are in an extended loose form (144–151), and have been denatured. To perform meaningful studies it is necessary to refold the molecules, which occurs when salt is added.

However, one cannot simply add salt, for then not only do the molecules fold, but they may also aggregate to large structures containing hundreds of protein molecules. It is necessary to fold the molecule and simultaneously block the formation of high molecular weight aggregates.

Fortunately, at low protein concentrations, the refolding and the formation of the large aggregates occur on time scales that differ by many orders of magnitude (146–151). Refolding is fast; aggregation is slow. This permits the two phenomena to be separated from one another, and thus allows a study of histone-histone complexing to be made, for the histones will complex to one another in the folded state, but not in the high molecular weight aggregated form.

The refolding is a highly cooperative conformational change. It is similar to the transition that is observed in the renaturation of a denatured globular protein (151).

Aggregation of the Inner Histones

In the remainder of this review the formation of high molecular weight aggregates is referred to simply as aggregation, and the formation of oligomers is called complexing.

The literature on the aggregation of histones extends back many decades. The earlier papers are today mainly of historical interest; they have been reviewed recently (151), and are not discussed here. It must be emphasized, however, that aggregation is not a simple juxtaposition of many folded molecules. When histones aggregate, their secondary structure changes, and such a change implies that a large change in tertiary structure occurs as well.

The folded form of H4 has no detectible β-sheet, while aggregated H4 has about 33% of the residues in β-sheet (146). H3 also has about 33% of the residues in β-sheet in the aggregated state, but none in the folded form (149). Thus an enormous change in secondary structure accompanies aggregation.

The histone aggregates are 80 Å wide ribbons twisted at intervals of about 330 Å, or bundles of such ribbons (152–155). The fibers show diffraction patterns with layer lines at axial spacings of 27, 37, and 55 Å. These spacings are similar to those found in diffraction patterns of chromatin (155) and there may be structural similarities between the fibers and chromatin. However, since histone aggregates contain large amounts of β-sheet and chromatin has little or none (156) chromatin cannot, therefore, contain the fibers as such.

Aggregation is blocked by lowering the temperature (157–159) and at 0°C, no aggregation is detectible for over 80 hours.

The Folding of H1

Like the inner histones, H1 also appears in a loose, random conformation when dissolved in water, and folds as salt is added. The fluorescence emission of the lone tyrosine increases (160, 161), the anisotropy increases (160), and UV absorbance of the tyrosine red shifts, and the resolution of the fine structure is increased. All these changes are typical of the transfer of tyrosine from an aqueous to a nonpolar environment (160). The circular dichroic spectra changes, and a small amount (6–10%) of α-helix formation occurs (160). Bradbury et al (162) reported that no β-sheet formation occurs upon H1 folding; Smerdon & Isenberg (160) found a small amount, of the order of 5%. NMR data (90, 162) indicate that only part of the H1 molecule becomes folded as salt is added; the folded region extends from residues 40–115. This region has a residue distribution typical of globular proteins (Figure 5), and is the part of the molecule that is highly conserved.

The folding is highly cooperative (160) and obeys a two-state model. In this model, one state is a completely unfolded molecule; the other is one in which part of the molecule is folded. The equilibrium constant between the states is a function of salt concentration, and the transition from predominantly unfolded molecules to predominantly folded ones occurs in the range of 75–100 mM NaCl. It is of some interest, therefore, that the binding properties of H1 are dramatically different below and above this range.

At low salt concentrations, H1 binds to polynucleosomes noncooperatively; at high salt concentrations it binds cooperatively. At the high salt concentrations, the binding is stronger with longer lengths of the polynucleosomal chain, saturating at about 7 nucleosomes per H1 molecule (163). This suggests that the globular region of H1 is involved in H1-nucleosome binding. Bonner & Stedman (164) have obtained the interesting result that, in mouse L1210 cell nuclei, H1 is linked to H2a by carbodiimide.

It has also been reported (165) that the type of binding of H1 to closed circular DNA changes as the salt concentration traverses the above range. Again, at low salt concentrations the binding is noncooperative; at high concentrations it is cooperative.

The globular region of chicken erythrocyte H5, in contrast to that of H1, appears to start at the amino terminus, and NMR evidence suggests that it extends from residue 1 to beyond residue 99 (166).

HISTONE-HISTONE INTERACTIONS

The inner histones interact with one another in specific ways (23, 167–177). There are three strong interactions leading to the formation of two dimers, (H2a, H2b) and (H2b, H4), and one tetramer (H3$_2$H4$_2$). There are two weak interactions, H2a-H4 and H2b-H3, and one intermediate interaction, H2a-H3 (23). The intermediate interaction is not much weaker than the strong ones.

In addition, individual inner histones may form low molecular weight homo-oligomers (146, 159, 177). The binding constant has been measured for only one of these, the H4 dimer (146). Here, at least, the interaction between subunits is weak.

Cross-linking of chromatin by tetranitromethane results in the formation of H2b-H4 (178), UV irradiation leads to H2a-H2b linking (179), and H3 is linked to H4 by carbodiimide (180). All of these cross-linkers are of zero length, and such cross-linking implies that the histones that are linked had previously been in contact, and hence presumably interacting. It is interesting that the partners of all of the strong interactions have been linked in chromatin by zero length linkers and no other pairs have been so linked. It seems, therefore, that the histones that interact strongly in vitro also interact within the nucleosome.

It should be emphasized that cross-linking and complexing studies measure two essentially different things. Cross-linking experiments measure a geometrical property—which residues of what histones are within a linking distance of one another. Complexing experiments study thermodynamic properties—which histones interact with one another, and the strengths of the interactions.

Cross-linking agents of various lengths may be used to probe the relative positions of parts of different histones, but the use of long cross-linking agents, in contrast to zero length linkers, can yield no information about histone interactions. Attempts to draw conclusions on histone interactions from experiments using long cross-linking agents can be misleading. It is particularly dangerous to try to get information on histone-histone interactions by using long linkers that join two lysines. NMR evidence (181, 182) and the inter-kingdom complexing studies, described below, indicate that it is the carboxyl regions of the inner histones that complex with one another, not the lysine-rich arms. Long lysine-lysine linking studies, if properly interpreted, might give some information on how the lysines are distributed with respect to one another, but they can give no information about histone-histone interactions.

As already indicated, the inner histones complex with one another only in the folded state; they will not complex with one another either in the extended unfolded (denatured) form, or in the form in which the histones are aggregated to high molecular weight species.

A simple trick allows us simultaneously to obtain the folded form for complexing experiments and to study the complexes themselves. The histones are first mixed together in water with no added salt. Then when salt is added the histones fold and complex with one another. For the strong complexes, the aggregation is blocked.

The pattern of histone interactions of calf thymus histones (23) is repeated by pea histones (10), yeast histones (4), and those of *Tetrahymena thermophila* (3). The pattern of binding has been evolutionarily conserved.

Complexes may form between histones of species from different kingdoms (3, 4, 183). A study of such complexes permits one to determine whether the surfaces between interacting histones or only the binding interaction has been conserved. It is easy to imagine that the binding surfaces could vary even though the complexing itself was conserved. However, the complexing experiments between histones of different kingdoms suggest otherwise. The binding surfaces themselves appear to have been highly conserved.

Data for a number of H2a-H2b complexes appear in Table 2. Such data contain considerable experimental error but, even so, permit significant conclusions to be drawn. We first note that, with one possible exception, all of the binding constants are close to one another. The inter-kingdom constants are not weaker than the intraspecies ones and, in fact, the binding constant for yeast H2a-calf H2b may be larger than the others.

The standard free energy of binding is related to the association constant by $\Delta G^0 = -RT\ln K$. Klotz et al (184) have analyzed the implications of this logarithmic relationship for protein-protein binding studies. Small changes in a binding surface lead to small changes in ΔG^0, but large changes in K. A variation in K by an order of magnitude can result from a small change in the binding surface.

Table 2   Summary of association constants for the H2a–H2b complex

Complex	$K \times 10^{-6}$ (M^{-1})	$\Delta G^0 - \Delta G^0$ (Pea–Pea) (kcal/mole)[a]
Calf H2a – Calf H2b[b]	2	0.4
Yeast H2a – Yeast H2b[c]	4	0.8
Pea H2a – Pea H2b[d]	1	0
Calf H2a – Yeast H2b[c]	5	1
Yeast H2a – Calf H2b[c]	14	1.6
Pea H2a – Calf H2b[e]	2	0.4
Calf H2a – Pea H2b[e]	3	0.6

[a] This assumes that $RT = 0.6$ kcal/mole. The values are nominal and, for most of them, a value of zero would be within experimental error.
[b] (169).
[c] (4).
[d] (10).
[e] (183).

A single residue change in the region of contact between subunits of hemoglobin may change the binding energy between hemoglobin dimers of 1–8 kcal/mole (184, 185). A change of Try 37 to Ser 37 on the β chain of hemoglobin results in a free energy change of 8 kcal/mole, with the tetramer-dimer dissociation constant increasing by a factor of a million (185). Table 2 shows fluctuations of only about 1 kcal/mole as we go from one complex to another. Such a small value suggests that the binding surfaces have been very highly conserved (4, 183).

The existence of strong inter-kingdom complexes, plus the marked variability of the amino-terminal region of H2b (Figure 3), suggest that the amino-terminal regions of the histones play no role in histone-histone interactions. This agrees with NMR data which indicate that the regions important for H3-H4 interaction are residues 42–120 in H3 and 38–102 in H4 (182), and the regions important for H2a-H2b interaction are residues 31–95 in H2a and 37–114 in H2b (181). [Similar NMR studies for the H2b-H4 complex are still lacking. These would be of great interest in view of the observation that the H2b site for binding to H2a is different from the H2b site for binding to H4, so that both interactions can occur simultaneously (179).]

In high salt, 2M NaCl or above, it is now generally agreed that the inner histones may form an octamer (186–189). However, there is sharp controversy over whether or not the heterotypic tetramer (H2a, H2b, H3, H4) also exists. Data obtained by a variety of techniques have been interpreted as showing that the heterotypic tetramer exists (173, 188, 190–192) and other data have been interpreted as showing that it does not (186, 187, 189).

DO ACIDS DENATURE HISTONES IRREVERSIBLY?

Histones have been reported to contain lysyl and histidyl phosphates (114, 115); such linkages are disrupted in hot acids. It is therefore possible, in principle at least, that the use of acids in the customary preparative procedures could alter histones irreversibly. However, reported instances of such occurrences (171, 172, 193, 194) leading, for example, to an inability of the histones to form complexes, may more reasonably be attributed to the formation of high molecular weight histone aggregates since, in each of the cases cited, the solution conditions favored such an aggregation, and high molecular weight histone aggregates do not complex with one another (151). Histones are indeed denatured by acid, but there is now abundant evidence that such denaturation is reversible, at least as far as the ability to form complexes, core protein, and nucleosomal particles is concerned.

The complexing of acid-extracted histones has already been discussed, and Lewis (175) has cycled the H3-H4 tetramer through low pH values, where they denature, back to neutral pH, where the tetramer reforms. In addition, Laskey et al (195) have shown that acid-extracted histones may be assembled into nucleosomes by a tissue extract of Xenopus eggs.

All of this demonstrates that complexing and assembly is not blocked by acid treatment of the inner histones. It is possible, of course, that other histone properties are altered, but no evidence for this has yet been presented.

CLASSIFICATION OF THE HISTONES

Historically, fractionation procedures and classification methods were developed first for animal histones. The various histone classes were first defined by operational criteria, based mainly on solubility characteristics and gel mobilities. These were then taken as definitions of the histone classes. It is now clear that such definitions are inadequate for any histones other than those from animals, and perhaps not even for all animal histones. Thus, for example, the solubility in 5% perchloric acid was long regarded as a distinctive characteristic of H1, but wheat germ H2b is also soluble in that medium (196). Pea plant H2a and H2b migrate differently from their calf counterparts on SDS gels (10), the inner histones from yeast (4), and *Tetrahymena* (3) also migrate differently from calf inner histones. It is thus impossible to use either gel mobility or specific solubility characteristics to identify the histones, and histone preparations not previously studied cannot be classified by such means.

The fact that the pattern of strong complexes follows a linear array, H2a-H2b-H4-H3, makes it possible to use tests of strong complexing to identify the histones, and such tests have been used to identify the inner histones from pea plants (10), yeast (4), and *Tetrahymena* (3). Since it now appears that the pattern of complexing is evolutionarily conserved, cross-complexing may indeed offer definitive identifications of the inner histones.

If enough protein has been purified, sequencing can identify histones, since each class has conserved regions. However, large amounts of purified histones are not always available, but it may still be important to obtain at least a tentative idea of the classes to which the various histones belong. Several recent observations have proven helpful in this respect.

H2a has the lowest mobility of any of the histones on gels containing Triton X-100 (48) and this was the property that was first used to provide a tentative identification of pea H2a (197). It has also been observed that gel bands of H2b (as well as of H1) destain much more rapidly in ferric chloride than bands of the other histones, and such destaining can be used to tentatively identify H2b (197). Furthermore, after staining with amido black, following extensive destaining, gel bands of H2a and H3 appear green and H1 and H2b appear purple (198). However, it should be emphasized that the slow mobility of H2a in Triton gels, the rapid destaining of H2b bands by ferric chloride, and the characteristic staining colors are not yet understood, and such tentative tests must therefore be used cautiously.

It is noted above that H1 can no longer be identified solely by 5% perchloric acid solubility, but such a solubility characteristic is still useful if used in conjunction with other criteria. The best of the present day criteria for H1, aside from extensive sequence homology in the central globular region, appears to be that it is a very lysine-rich, highly basic histone, but until we obtain additional structural information a cautious approach might dictate that we regard even this as only a tentative classification criterion. It must be recognized, for example, that no one to date has identified H1 in yeast, and we still do not know if this is because yeast really has none, or whether H1 in yeast is rapidly degraded during preparative procedures, or whether we simply do not know how to recognize H1 in yeast.

ACKNOWLEDGMENTS.

The work of the author's laboratory has been supported by Grant CA 10872 awarded by the National Cancer Institute, Department of Health, Education and Welfare. I should like to thank the many investigators who have provided manuscripts prior to publication.

Literature Cited

1. DeLange, R. J., Fambrough, D. M., Smith, E. L., Bonner, J. 1969. *J. Biol. Chem.* 244:5669–79
2. Whittaker, R. H. 1969. *Science* 163:150–60
3. Glover, C. L. C., Gorovsky, M. A. 1978. *Biochemistry.* In press
4. Mardian, J. K. W., Isenberg, I. 1978. *Biochemistry.* 17:3825–33
5. Patthy, L., Smith, E. L., Johnson, J. 1973. *J. Biol. Chem.* 248:6834–40
6. Brandt, W. F., von Holt, C. 1976. *FEBS Lett.* 65:386–90
7. Johmann, C. A., Gorovsky, M. A. 1976. *Biochemistry* 15:1249–56
8. Strickland, M., Strickland, W. N., Brandt, W. F., von Holt, C. 1977. *Eur. J. Biochem.* 77:263–75
9. Hayashi, H., Iwai, K., Johnson, J. D., Bonner, J. 1977. *J. Biochem.* 82:503–10
10. Spiker, S., Isenberg, I. 1977. *Biochemistry* 16:1819–26
11. Kinkade, J. M., Cole, R. D. 1966. *J. Biol. Chem.* 241:5790–97
12. Kinkade, J. M., Cole, R. D. 1966. *J. Biol. Chem.* 241:5798
13. Bustin, M., Cole, R. D. 1969. *J. Biol. Chem.* 244:5286–90
14. Rall, S. C., Cole, R. D. 1971. *J. Biol. Chem.* 246:7175–90
15. Cole, R. D. 1977. In *The Molecular Biology of the Mammalian Genetic Apparatus,* Vol. 1, ed. P. O. P. Ts'o, pp. 93–104, Amsterdam: North-Holland
16. Mohberg, J., Rusch, H. P. 1969. *Arch. Biochem. Biophys.* 134:577–89
17. Felden, R. A., Sanders, M. M., Morris, R. N. 1976. *J. Cell Biol.* 68:430–39
18. Tonino, G. J. M., Rozijn, Th. H. 1966. In *The Cell Nucleus—Metabolism and Radiosensitivity.* pp. 125–133, London: Taylor & Francis
19. Moll, R., Wintersberger, E. 1976. *Proc. Natl. Acad. Sci. USA* 73:1863–67
20. Thomas, J. O., Furber, V. 1976. *FEBS Lett.* 66:274–80
21. Brandt, W. F., von Holt, C. 1976. *FEBS Lett.* 65:386–90
22. Nelson, D. A., Beltz, W. R., Rill, R. L. 1977. *Proc. Natl. Acad. Sci. USA* 74:1343–47
23. D'Anna, J. A., Jr., Isenberg, I. 1974. *Biochemistry* 13:4992–97
24. Jardine, N. J., Leaver, J. L. 1975. *Biochem. Soc. Trans.* 3:1012–14
25. Lipps, H. J., Hantke, K. G. 1974. *Chromosoma* (Berlin) 49:309–20
26. Caplan, E. B. 1975. *Biochim. Biophys. Acta* 407:109–13
27. Gorovsky, M. A., Keevert, J. B. 1975. *Proc. Natl. Acad. Sci. USA* 72:2672–76
28. Rizzo, P. J., Cox, E. R. 1977. *Science* 198:1258–60
29. Johmann, C. A., Gorovsky, M. A. 1976. *J. Cell. Biol.* 71:89–95
30. Bustin, M., Cole, R. D. 1968. *J. Biol. Chem.* 243:4500–5
31. Fambrough, D. M., Bonner, J. 1969. *Biochim. Biophys. Acta* 175:113–22
32. Kinkade, J. M. 1969. *J. Biol. Chem.* 244:3375–86
33. Panyim, S., Chalkley, R. 1969. *Biochem. Biophys. Res. Commun.* 37:1042–49
34. Langan, T. A., Rall, S. C., Cole, R. D. 1971. *J. Biol. Chem.* 246:1942–48
35. Seale, R. L., Aronson, A. I. 1973. *J. Mol. Biol.* 75:647–58
36. Stout, J. T., Phillips, R. L. 1973. *Proc. Natl. Acad. Sci. USA* 70:3043–47
37. Gurley, L. R., Walters, R. A., Tobey, R. A. 1974. *J. Cell Biol.* 60:356–64
38. Gurley, L. R., Walters, R. A., Tobey, R. A. 1975. *J. Biol. Chem.* 250:3936–44
39. Ruderman, J. V., Gross, P. R. 1974. *Dev. Biol.* 36:286
40. Ruderman, J. V., Baglioni, C., Gross, P. R. 1974. *Nature* 247:36–38
41. Sherod, D., Johnson, G., Chalkley, R. 1974. *J. Biol. Chem.* 249:3923–31
42. Smerdon, M. J., Isenberg, I. 1976. *Biochemistry* 15:4233–42
43. Spiker, S. 1976. *J. Chromatogr.* 128:244–48
44. Newrock, K. M., Alfageme, C. R., Nardi, R. V., Cohen, L. H. 1978. *Cold Spring Harbor Symp. Quant. Biol.* 42:421–31
45. Hill, R. J., Poccia, D. L., Doty, P. 1971. *J. Mol. Biol.* 61:445–62
46. Poccia, D. L., Hinegardner, R. T. 1975. *Dev. Biol.* 45:81–89
47. Zweidler, A., Cohen, L. H. 1972. *Fed. Proc.* 31:926
48. Alfageme, C. R., Zweidler, A., Mahowald, A., Cohen, L. H. 1974. *J. Biol. Chem.* 249:3729–36
49. Cohen, L. H., Newrock, K. M., Zweidler, A. 1975. *Science* 190:994–97
50. Franklin, S. G., Zweidler, A. 1977. *Nature* 266:273–75
51. Weintraub, H., Flint, S. J., Leffak, I. M., Groudine, M., Grainger, R. M. 1978. *Cold Spring Harbor Symp. Quant. Biol.* 42:401–7
52. Sung, M. T., Dixon, G. H. 1970. *Proc. Natl. Acad. Sci. USA* 67:1616–23
53. Easton, D., Chalkley, R. 1972. *Exp. Cell Res.* 72:502–6
54. Strickland, W. N., Strickland, M., Brandt, W. F., von Holt, C. 1977. *Eur. J. Biochem.* 77:277–86

55. Shires, A., Carpenter, M. P., Chalkley, R. 1975. *Proc. Natl. Acad. Sci. USA* 72:2714–18
56. Shires, A., Carpenter, M. P., Chalkley, R. 1976. *J. Biol. Chem.* 251:4155–58
57. Branson, R. E., Grimes, S. R., Jr., Yonuschot, G., Irvin, J. L. 1975. *Arch. Biochem, Biophys.* 168:403–12
58. Kistler, W. S., Geroch, M. E. 1975. *Biochem. Biophys. Res. Commun.* 63:378–84
59. Kistler, W. S., Geroch, M. E., Williams-Ashman, H. G. 1973. *J. Biol. Chem.* 248:4532–43
60. Marushige, Y., Marushige, K. 1975. *J. Biol. Chem.* 250:39–45
61. Balhorn, R., Gledhill, B. L., Wyrobek, A. J. 1977. *Biochemistry* 16:4074–80
62. Yaguchi, M., Roy, C., Dove, M., Seligy, M. 1977. *Biochem. Biophys. Res. Commun.* 76:100–6
63. Moss, B. A., Joyce, W. G., Ingram, V. M. 1973. *J. Biol. Chem.* 248:1025–31
64. Mura, C., Huang, P. C., Craig, S. W. 1978. *Mech. Ageing Dev.* 7:109–22
65. Goetz, G., Esmailzadeh, A. K., Huang, P. C. 1978. *Biochim. Biophys. Acta* 517:236–45
66. Miki, B. L. A., Neelin, J. M. 1975. *Can. J. Biochem.* 53:1158–69
67. Smerdon, M., Isenberg, I. 1976. *Biochemistry* 15:4242–47
68. Yu, S. H., Spring, T. G. 1977. *Biochim. Biophys. Acta* 492:20–28
69. Welch, S., Cole, R. D. 1978. *J. Biol. Chem.* In press
70. Liew, C. C., Haslett, G. W., Allfrey, V. G. 1970. *Nature* 226:414–17
71. Louie, A. J., Candido, E. P. M., Dixon, G. H. 1974. *Cold Spring Harbor Symp. Quant. Biol.* 38:803
72. DeLange, R. J., Fambrough, D. M., Smith, E. L., Bonner, J. 1969. *J. Biol. Chem.* 244:319–34
73. Ogawa, Y., Quagliarotti, G., Jordan, J., Taylor, C. W., Starbuck, W. C., Busch, H. 1969. *J. Biol. Chem.* 244:4387–92
74. Candido, E. P. M., Dixon, G. H. 1972. *Proc. Natl. Acad. Sci. USA* 69:2015–19
75. Candido, E. P. M., Dixon, G. H. 1972. *J. Biol. Chem.* 247:3868–73
76. DeLange, R. J., Hooper, J. A., Smith, E. L. 1973 *J. Biol. Chem.* 248:3261–74
77. Hooper, J. A., Smith, E. L., Sommer, K. R., Chalkley, R. 1973. *J. Biol. Chem.* 248:3275–79
78. Dixon, G. H., Candido, E. P. M., Honda, B. M., Louie, A. J., MacLeod, A. R., Sung, M. T. 1975. *The Structure and Function of Chromatin.* Presented at Ciba Found. Symp. 28, London, pp. 229–258

79. Allfrey, V. G. 1977. In *Chromatin and Chromosome Structure,* ed. H. J. Li, R. A. Eckhardt. p. 167. New York: Academic
80. Riggs, M. G., Whittaker, R. G., Neumann, J. R., Ingram, V. M. 1977. *Nature* 268:462–64
81. Boffa, L. C., Vidali, G., Mann, R. S., Allfrey, V. G. 1978. *J. Biol. Chem.* 253:3364–66
82. Vidali, G., Boffa, L. C., Bradbury, E. M., Allfrey, V. G. 1978. *Proc. Natl. Acad. Sci. USA* 75:2239–43
83. Weintraub, H., Groudine, M. 1976. *Science* 193:848–56
84. Garel, A., Axel, R. 1976. *Proc. Natl. Acad. Sci. USA* 73:3966–70
85. Flint, S. J., Weintraub, H. 1977. *Cell* 12:783–94
86. Garel, A., Zolan, M., Axel, R. 1977. *Proc. Natl. Acad. Sci. USA* 74:4867–71
87. Levy, W. B., Dixon, G. H. 1977. *Nucleic Acids Res.* 4:883–98
88. Nelson, D. A., Perry, W. M., Chalkley, R. 1978. *Biochem. Biophys. Res. Comm.* 82:356–63
89. Elgin, S. C. R., Weintraub, H. 1975. *Ann. Rev. Biochem.* 44:725–74
90. Gurley, L. R., D'Anna, J. A., Barkham, S. S., Deaven, L. L., Tobey, R. A. 1978. *Eur. J. Biochem.* 84:1–15
91. Langan, T. A. 1978. In *Methods Cell Biol.* 19:127–42
92. Bradbury, E. M., Inglis, R. J., Matthews, H. R., Sarner, N. 1973. *Eur. J. Biochem.* 33:131–39
93. Bradbury, E. M., Inglis, R. J., Matthews, H. R. 1974. *Nature* 247:257–61
94. Bradbury, E. M., Inglis, R. J., Matthews, H. R., Langan, T. A. 1974. *Nature* 249:553–56
95. Sung, M. T., Dixon, G. H. 1970. *Proc. Natl. Acad. Sci. USA* 67:1616–23
96. Louie, A. J., Dixon, G. H. 1972. *Proc. Natl. Acad. Sci. USA* 69:1975–79
97. Louie, A. J., Dixon, G. H. 1972. *J. Biol. Chem.* 247:5498–5505
98. Tobin, R. S., Seligy, V. L. 1975. *J. Biol. Chem.* 250:358–64
99. Sung, M. T., Dixon, G. H., Smithies, O. 1971. *J. Biol. Chem.* 246:1358–64
100. Shlyapnikov, S. V., Arutyunyan, A. A., Kurochkin, S., Memelova, L. V., Nesterova, M. V., Sashchenko, L. P., Severin, E. S. 1975. *FEBS Lett.* 53:316–19
101. Langan, T. A. 1969. *Proc. Natl. Acad. Sci. USA* 64:1276–83
102. Langan, T. A., Rall, S. C., Cole, R. D. 1971. *J. Biol. Chem.* 246:1942–44
103. Farago, A., Rombanyi, T., Antoni, F., Takato, A., Fabian, F. 1975. *Nature* 254:88

104. Hashimoto, E., Takeda, N., Nishizuka, Y., Hamana, K., Iwai, K. 1975. *Biochem. Biophys. Res. Comm.* 66:547–55
105. Lake, R. S., Salzman, N. P. 1972. *Biochemistry* 11:4817–26
106. Lake, R. S., Goidl, J. A., Salzman, N. P. 1972. *Exp. Cell Res.* 73:113–21
107. Lake, R. S. 1973. *Nature New Biol.* 242:145–46
108. Lake, R. S. 1973. *J. Cell Biol.* 58:317–31
109. Marks, D. B., Park, W. K., Borun, T. W. 1973. *J. Biol. Chem.* 248:5660–67
110. Balhorn, R., Jackson, V., Granner, D., Chalkley, R. 1975. *Biochemistry* 14:2504–11
111. Gurley, L. R., Walters, R. A., Barham, S. S., Deaven, L. L. 1978. *Exp. Cell Res.* 111:373–83
112. Hohmann, P., Tobey, R. A., Gurley, L. R. 1976. *J. Biol. Chem.* 251:3685–92
113. D'Anna, J. A., Jr., Gurley, L. R., Deaven, L. L. 1978. *Nucl. Acids Res.* 5:3195–3207
114. Chen, C. C., Smith, D. L., Bruegger, B. B., Halpern, R. M., Smith, R. A. 1974. *Biochemistry* 13:3785–89
115. Smith, D. L., Chen, C. C., Bruegger, B. B., Holtz, S. L., Halpern, R. M., Smith, R. A. 1974. *Biochemistry* 13:3780–85
116. Gershey, E. L., Haslett, G. W., Vidali, G., Allfrey, V. G. 1969. *J. Biol. Chem.* 244:4871–77
117. Borun, T., Pearson, D., Paik, W. K. 1973. *J. Biol. Chem.* 247:4288
118. Thomas, G., Lange, H. W., Hempel, K. 1975. *Eur. J. Biochem.* 51:609–15
119. Tidwell, T., Allfrey, V. G., Mirsky, A. E. 1968. *J. Biol. Chem.* 243:707–15
120. Byvoet, P., Shepherd, G. R., Hardin, J. M., Noland, B. J. 1972. *Arch. Biochem. Biophys.* 148:558–567
121. Honda, B. M., Candido, E. P. M., Dixon, G. H. 1975. *J. Biol. Chem.* 250:8686–89
122. McKnight, S. L., Miller, O. L. Jr. 1977. *Cell* 12:795
123. Byvoet, P., Baxter, C. S. 1975. In *Chromosomal Proteins and Their Role in the Regulation of Gene Expression,* ed. G. S. Stein, L. J. Kleinsmith, pp. 127–51. New York: Academic
124. Hayaishi, O., Ueda, K. 1977. *Ann. Rev. Biochem.* 46:95
125. Nishizuka, Y., Ueda, K., Honjo, T., Hayaishi, O. 1968. *J. Biol. Chem.* 243:3765–67
126. Smith, J. A., Stocken, L. A. 1973. *Biochem. Biophys. Res. Commun.* 54:297–300
127. Ord, M. G., Stocken, L. A. 1975. See Ref. 78, pp. 259–67
128. Ueda, K., Omachi, A., Kawaichi, M., Hayaishi, O. 1975. *Proc. Natl. Acad. Sci. USA* 72:205–9
129. Ord, M. G., Stocken, L. A. 1977. *Biochem. J.* 161:583–92
130. Wong, N. C. W., Poirier, G. G., Dixon, G. H. 1977. *Eur. J. Biochem.* 77:11–21
131. Brightwell, M. D., Leech, C. E., O'Farrell, M. K., Whish, W. J. D., Shall, S. 1975. *Biochem. J.* 147:119–29
132. Tanaka, M., Miwa, M., Matsushima, T., Sugimara, T., Shall, S. 1976. *Arch. Biochem. Biophys.* 172:224–29
133. Orrick, L. R., Olson, M. O. J., Busch, H. 1973 *Proc. Natl. Acad. Sci. USA* 70:1316
134. Goldknopf, I. L., Busch, H. 1977. *Proc. Natl. Acad. Sci. USA* 74:864–68
135. Hunt, L. T., Dayhoff, M. O. 1977. *Biochem. Biophys. Res. Comm.* 74:650–55
136. Goldstein, G., Scheid, H., Hammerling, U., Boyse, E. A., Schlesinger, D. H., Niall, H. D. 1975. *Proc. Natl. Acad. Sci. USA* 72:11–15
137. Schlesinger, D. H., Goldstein, G., Niall, H. D. 1975. *Biochemistry* 14:2214–18
138. Goldknopf, I. L., Taylor, C. W., Baum, R. M., Yeoman, L. C., Olson, O. J., Prestayko, A. W., Busch, H. 1975. *J. Biol. Chem.* 250:7182–87
139. Yeoman, L. C., Seeber, S., Taylor, C. W., Fernbach, D. J., Falletta, J. M., Jordan, J. J., Busch, H. 1976. *Exp. Cell Res.* 100:47–55
140. Goldknopf, I. L., French, M. F., Musso, R., Busch, H. 1977. *Proc. Natl. Acad. Sci. USA* 74:5492–95
141. Busch, H., Ballal, N. R., Busch, R. K., Choi, Y. C., Davis, F., Goldknopf, I. L., Matsui, S. I., Rao, M. S., Rothblum, L. I. 1978. *Cold Spring Harbor Symp. Quant. Biol.* 42:665–83
142. Ballal, N. R., Goldknopf, I. L., Goldberg, D. A., Busch, H. 1974. *Life Sci.* 14:1835–45
143. Ballal, N. R., Kang, Y. J., Olson, M. O. J., Busch, H. 1975. *J. Biol. Chem.* 250:5921–25
144. Boublik, M., Bradbury, E. M., Crane-Robinson, C. 1970. *Eur. J. Biochem.* 14:486–97
145. Boublik, M., Bradbury, E. M., Crane-Robinson, C., Johns, E. W. 1970. *Eur. J. Biochem.* 17:151–59
146. Li, H. J., Wickett, R., Craig, A. M., Isenberg, I. 1972. *Biopolymers* 11:375–97
147. D'Anna, J. A., Jr., Isenberg, I. 1972. *Biochemistry* 11:4017–25
148. D'Anna, J. A., Jr., Isenberg, I. 1974. *Biochemistry* 13:2093–97

149. D'Anna, J. A., Jr., Isenberg, I. 1974. *Biochemistry* 13:4987–92
150. van Holde, K. E., Isenberg, I. 1975. *Accoutns of Chem. Res.* 8:327–35
151. Isenberg, I. 1977. In *Search and Discovery, A Tribute to Albert-Szent-Gyorgyi,* ed. B. Kaminer, pp. 195–215. New York: Academic
152. Sperling, R., Bustin, M. 1974. *Proc. Natl. Acad. Sci. USA* 71:4625–29
153. Sperling, R., Bustin, M. 1975. *Biochemistry* 14:3322–31
154. Sperling, R., Bustin, M. 1976. *Nucleic Acids Res.* 3:1263–75
155. Sperling, R., Amos, L. A., 1977. *Proc. Natl. Acad. Sci. USA* 74:3772–76
156. Thomas, G. J., Jr., Prescott, B., Olins, D. E. 1977. *Science* 197:385–88
157. Small, E. W., Craig, A. M., Isenberg, I. 1973. *Biopolymers* 12:1149–60
158. Smerdon, M. J., Isenberg, I. 1973. *Biochem. Biophys. Res. Comm.* 55:1029–34
159. Smerdon, M. J., Isenberg, I. 1974. *Biochemistry* 13:4046–49
160. Smerdon, M. J., Isenberg, I. 1976. *Biochemistry* 15:4233–42
161. Giancotti, V., Fonda, M., Crane-Robinson, C. 1977. *Biophys. Chem.* 6:379–83
162. Bradbury, E. M., Cary, P. D., Chapman, G. E., Crane-Robinson, C., Danby, S. E., Rattle, H. W. E. 1975. *Eur. J. Biochem.* 52:605–13
163. Renz, M., Nehls, P., Hozier, J. 1977. *Proc. Natl. Acad. Sci. USA* 74:1879–83
164. Bonner, W. M., Stedman, J. 1978. In press
165. Singer, D. S., Singer, M. F. 1978. *Biochemistry* 17:2086–95
166. Crane-Robinson, C., Briand, G., Sautiere, P., Champagne, M. 1977. *Biochim. Biophys. Acta* 443:283–92
167. Skandrani, E., Mizon, J., Sautiere, P., Biserte, G. 1972. *Biochimie* 54:1267–72
168. D'Anna, J. A. Jr., Isenberg, I. 1973. *Biochemistry* 12:1035–43
169. D'Anna, J. A. Jr., Isenberg, I. 1974. *Biochemistry* 13:2098–2104
170. D'Anna, J. A. Jr., Isenberg, I. 1974. *Biochem. Biophys. Res. Commun.* 61:343–47
171. Kornberg, R. D., Thomas, J. O. 1974. *Science* 184:865–68
172. Roark, D. E., Geoghegan, T. E., Keller, G. H. 1974. *Biochem. Biophys. Res. Commun.* 59:542–47
173. Weintraub, H., Paiter, K., van Lente, F. 1975. *Cell* 6:85–110
174. Lewis, P. N. 1976. *Can. J. Biochem.* 54:641–49
175. Lewis, P. N. 1976. *Can. J. Biochem.* 54:963–70
176. Lewis, P. N. 1976. *Biochem. Biophys. Res. Commun.* 68:329–35
177. Nicola, N. A., Fulmer, A. W., Schwartz, A. M., Fasman, G. D. 1978. *Biochemistry* 17:1779–85
178. Martinson, H. G., McCarthy, B. J. 1975. *Biochemistry* 14:1073–78
179. Martinson, H. G., Shetlar, M. D., McCarthy, B. J. 1976. *Biochemistry* 15:2002–7
180. Bonner, W. M., Pollard, H. B. 1975. *Biochem. Biophys. Res. Commun.* 64:282–88
181. Moss, T., Cary, P. D., Abercrombie, B. D., Crane-Robinson, C., Bradbury, E. M. 1976. *Eur. J. Biochem.* 71:337–50
182. Böhm, L., Hayashi, H., Cary, P. D., Moss, T., Bradbury, E. M. 1977. *Eur. J. Biochem.* 77:487–93
183. Spiker, S., Isenberg, I. 1977. *Cold Spring Harbor Symp. Quant. Biol.* 42:157–63
184. Klotz, I. M., Darnall, D. W., Langerman, N. R. 1975. In *The Proteins,* Vol. 1, ed. H. Neurath, R. L. Hill, pp. 293–411. New York: Academic
185. Sasaki, J., Imamura, T., Yanase, T. 1978. *J. Biol. Chem.* 253:87–94
186. Thomas, J. O., Kornberg, R. D. 1975. *Proc. Natl. Acad. Sci. USA* 72:2626–30
187. Thomas, J. O., Butler, P. J. G. 1977. *J. Mol. Biol.* 116:769–81
188. Chung, S. Y., Hill, W. E., Doty, P. 1978. *Proc. Natl. Acad. Sci. USA* 75:1680–84
189. Eickbush, T. H., Moudrianakis, E. M. 1978. *Biochemistry* 17:4955–64
190. Campbell, A. M., Cotter, R. I. 1976. *FEBS Lett.* 70:209–11
191. Pardon, J. F., Worcester, D. L., Wooley, J. C., Cotter, R. I., Lilley, D. M., Richards, B. M. 1977. *Nucl. Acids Res.* 9:3199–3214
192. Wooley, J. C., Pardon, J. F., Richards, B. M., Worcester, D. L., Campbell, A. M. 1977. *Fed. Proc.* 36:810
193. Roark, D. E., Geoghegan, T. E., Keller, G. H., Matter, K. V., Engle, R. L. 1976. *Biochemistry* 15:3019–25
194. Feldman, L., Stollar, B. D. 1977. *Biochemistry* 16:2767–71
195. Laskey, R. A., Mills, A. D., Morris, N. R. 1977. *Cell* 10:237–43
196. Fazal, M., Cole, R. D. 1977. *J. Biol. Chem.* 252:4068–72
197. Spiker, S., Key, J. L., Wakim, B. 1976. *Arch. Biochem. Biophys.* 176:510–18
198. Zweidler, A. 1978. *Methods Cell Biol.* 17:223–33
199. Strickland, M., Strickland, W. N., Brandt, W. F., von Holt, C. 1974. *FEBS Lett.* 40:346–48

200. Wouters-Tyrou, D., Sautiere, P., Biserte, G. 1976. *FEBS Lett.* 65:225–28
201. Sautiere, P., Tyrou, D., Moschetto, Y., Biserte, G. 1971. *Biochimie* 53:479–83
202. Sautiere, P., Lamberlin-Brernaert, M. D., Moschetto, Y., Biserte, G. 1971. *Biochimie* 53:711–15
203. DeLange, R. J., Hooper, J. A., Smith, E. L. 1972. *Proc. Natl. Acad. Sci. USA* 69:882–84
204. Olson, M. O. J., Jordan, J., Busch, H. 1972. *Biochem. Biophys. Res. Commun.* 46:50–55
205. Brandt, W. F., von Holt, C. 1974. *Eur. J. Biochem.* 46:407–17
206. Brandt, W. F., Strickland, W. N., von Holt, C. 1974. *FEBS Lett.* 40:349–52
207. Brandt, W. F., Strickland, W. N., Morgan, M., von Holt, C. 1974. *FEBS Lett.* 40:167–72
208. Hooper, J. A., Smith, E. L., Sommer, K. R., Chalkley, R. 1973. *J. Biol. Chem.* 248:3275–79
209. Iwai, K., Hayashi, H., Ishikawa, K. 1972. *J. Biochem.* 72:357–67
210. Koostra, A., Bailey, G. S. 1976. *FEBS Lett.* 68:76–78
211. Elgin, S. C. R., Schilling, J. Hood, L. E. 1978. In press.
212. Yeoman, L. C., Olson, W. O., Sugano, N., Jordan, J. J., Taylor, C. W., Starbuck, W. C., Busch, H. 1972. *J. Biol. Chem.* 247:6018–23
213. Sautiere, P., Tyrou, D., Laine, B., Mizon, J., Ruffin, P., Biserte, G. 1974. *Eur. J. Biochem.* 41:563–76
214. Laine, B., Kmiecik, D., Sautiere, P., Biserte, G. 1978. *Biochimie* 60:147–50
215. Laine, B., Sautiere, P., Biserte, G. 1976. *Biochimie* 15:1640–45
216. Koostra, A., Bailey, G. S. 1978. *Biochemistry* 17:2504–10
217. Wouters-Tyrou, D. 1977 *Les histones de gonades d'oursin Psammechinus Miliaris.* PhD thesis. Univ. des Sciences et Techniques de Lille, Lille. 189 pp.
218. Macleod, A. R., Wong, N. C. W., Dixon, G. H. 1977. *Eur. J. Biochem.* 78:281–91

Ann. Rev. Biochem. 1979. 48:193–216

EPIDERMAL GROWTH FACTOR

♦12007

Graham Carpenter

Department of Biochemistry and Division of Dermatology, Department of Medicine, Vanderbilt University School of Medicine, Nashville, Tennessee 37232

Stanley Cohen

Department of Biochemistry, Vanderbilt University School of Medicine, Nashville, Tennessee 37232

CONTENTS

0066-4154/79/0701-0193$01.00

PERSPECTIVES AND SUMMARY

Among the basic challenges confronting modern biological research is the understanding of the biochemical processes that regulate or otherwise influence the proliferation of mammalian cells. While classical endocrinological studies have established the importance of circulating hormones, such as growth hormone and thyroxine, in the control of growth, development, and differentiation, evidence has accumulated which suggests that many more extracellular signals are involved in these highly complex processes.

One experimental approach to understanding the control of cell proliferation and/or differentiation has been the development of culture techniques and media for the propagation of eukaryotic cells in vitro. The results have demonstrated the dependence of cell proliferation on the presence of macromolecular growth factors in serum, some of which have been purified (for review see 1–4). Although many growth factors are present in serum and body fluids such as urine (5), the site(s) of synthesis of these factors is obscure. A number of mitogenic polypeptides have been successfully isolated from sources other than blood serum: nerve growth factor from the mouse submaxillary gland (6) and from snake venom (7), epidermal growth factor from mouse submaxillary gland (8) and from human urine (9), and fibroblast growth factor from bovine pituitary and brain (10, 11). Growth factors have also been isolated from media "conditioned" by the growth of cultured cells: multiplication-stimulating activity from media "conditioned" by rat liver cells (12), and macrophage growth factor and colony-stimulating factor from L-cell "conditioned" media (13).

The existence in both mouse and man of epidermal growth factors that have similar chemical characteristics and exhibit identical biological activities indicates that these polypeptides have been retained throughout a long evolutionary process and probably have a function of general significance in nature. The conservation of specific receptors for epidermal growth factors in a wide variety of cells from diverse species supports this idea. Of the growth factors that have been purified to date, epidermal growth factor (EGF) is one of the most biologically potent and best characterized as to its physical, chemical, and biological properties.

EGF clearly stimulates the proliferation of epidermal and epithelial tissues in animals, and cell culture studies suggest that additional cell types might be responsive in vivo. Current research concerning epidermal growth factor is concentrated mainly on (*a*) describing the growth factor's interaction with plasma membrane receptors and its subsequent internalization and degradation, (*b*) analyzing the enhanced phosphorylation of membrane proteins induced by EGF in vitro, (*c*) determining the sequence of biochemical events in intact cells that lead to cell proliferation, and (*d*) defining the physiological role of endogenous EGF in mammalian biology.

INTRODUCTION

The existence of an "epidermal growth factor" was first detected some 20 years ago as an incidental observation during the course of studies on the nerve growth factor. Extracts of the submaxillary gland of the mouse, when injected into newborn animals, induced precocious eyelid opening and incisor eruption (7, 8) due to a direct stimulation of epidermal growth and keratinization (14, 15). The factor responsible for these effects was isolated (8) and found to be a low molecular weight, heat stable, nondialyzable polypeptide that accounted for approximately 0.5% of the protein content of the submaxillary gland. A rapid two-step procedure for the purification of EGF from this source has been reported (16). An homologous polypeptide, human epidermal growth factor (hEGF) was detected (17) and isolated (9) from human urine.

A new aspect of the biology of EGF emerged with the report by Gregory (18) that urogastrone, a gastric antisecretory hormone isolated from human urine, is probably identical to human EGF. Several reviews are available dealing with various aspects of the biology and chemistry of EGF (19–22).

CHEMICAL AND PHYSICAL PROPERTIES OF EPIDERMAL GROWTH FACTOR

Mouse EGF

Taylor, et al (23) reported that EGF is a single polypeptide chain of 53 amino acid residues, devoid of alanine, phenylalanine, and lysine, that has an isoelectric point at pH 4.6 and an extinction coefficient ($E_{1cm}^{1\%}$ at 280 nm) of 30.9. The primary amino acid sequence (24) of mouse EGF (mEGF) and the location of the three intramolecular disulfide bonds (25) are shown in Figure 1.

The disulfide bonds in mEGF are required for biological activity (23). Extensive reduction of the disulfides in the presence of mercaptoethanol and urea yielded an inactive polypeptide. The biological activity, however, was completely restored by removal of the mercaptoethanol and urea by dialysis and subsequent reoxidation by air.

EGF lacking the carboxy-terminal Leu-Arg dipeptide also may be isolated from the submaxillary gland and is biologically active (16). Exposure of intact mEGF to mild tryptic digestion resulted in cleavage of the peptide bond between residues 48 and 49, and produced a pentapeptide containing residues 49–53 and a derivative of mEGF composed of residues 1–48 (24) that retained full biological activity when assayed in vivo, but only about 10% of its activity when assayed with cultured fibroblasts (26). It is possible that when injected subcutaneously, EGF_{1-53} is converted to EGF_{1-48} or

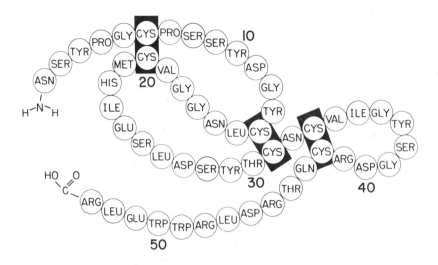

Figure 1   Amino acid sequence of mEGF with placement of disulfide bonds (25).

that receptors on fibroblasts and epidermal cells differ in their affinities for EGF and its derivatives.

The major physiochemical properties of EGF have been investigated. Circular dichroic (CD) examination of the far-ultraviolet spectrum of mEGF indicated the absence of significant α-helical structure, the presence of approximately 25% β-helix, and a random coil content of 75% (23). These results have been confirmed by Holladay, et al (27), who extended CD studies of mEGF to include the near-ultraviolet spectra. These workers reported that while intact mEGF has 22% β-structure, derivates of the native molecule, namely $mEGF_{1-48}$ and cyanogen bromide–treated mEGF, have approximately 10% and 12% β-structure, respectively. Thus, the carboxyl terminal residues 49–53 and the peptide bond between residues 21 and 22, which is cleaved by cyanogen bromide, appear to contribute to ordered structure in the intact molecule.

A space filling model of mEGF has been constructed (27) based on CD spectra and the application of a set of predictive rules for secondary structure and bends to the sequence of EGF. The model indicates that mEGF exists as a very compact structure, due mainly to many β-bends and the three disulfide bonds. A sedimentation constant of 1.25S (8) and a frictional ratio (f/f_o) of 1.12 (L. Holladay, personal communication) for mEGF is consistent with a relatively compact and globular structure. Interestingly, one face of the space filling model of EGF is composed of a three-tiered β-sheet region. At present, it is not possible to describe the relationship of this area of β-structure to the formation of hormone : receptor complexes or to the biological activity of mEGF.

Holladay et al (27) have studied the unfolding thermodynamics of EGF and derivatives. The reversible unfolding of mEGF in the presence of guanidinium hydrochloride showed that the transition midpoint is quite high (6.89 M guanidinium hydrochloride at 25°C) compared to other proteins. By extrapolation, an estimate of 16 kcal/mole was made for the free energy of unfolding (ΔG^0) in the absence of denaturant. The thermal stability of mEGF also was examined and various thermodynamic properties were estimated for mEGF at 40°C in the absence of denaturant. The data indicated that the apparent enthalpy of unfolding (ΔH_{app}) is 20.4 cal mol^{-1} deg^{-1}. The absolute magnitude of the apparent heat capacity (ΔCp) was estimated at less than 0.5 kcal/(mole deg^{-1}), substantially lower than values reported for other proteins. It was suggested that the small ΔCp value for mEGF is due in part to the relatively high degree of exposure to the solvent of hydrophobic side chains in native EGF as suggested by CD spectra analysis and the space filling model. In the absence of completely buried aromatic residues, it is not clear what forces beyond the three disulfide bonds produce such a highly stable native conformation of mEGF, although the β-structure may play a large role. These data indicate that mEGF is one of the most energetically stable proteins that has been described.

High-Molecular-Weight form of mEGF

Taylor, et al (28, 29) reported that in crude homogenates of the mouse submaxillary gland prepared at neutral pH, EGF was isolated as a component of a high-molecular-weight (HMW) complex. The complex has a molecular weight of approximately 74,000, as judged by sedimentation equilibrium, and is composed of 2 molecules of mEGF ($M_r = 6045$) and 2 molecules of binding protein ($M_r = 29,300$). The HMW complex is stable over the pH range 5–8, has an isoelectric point of 5.4, a sedimentation value of 4.81S, an extinction coefficient ($E_{1cm}^{1\%}$ at 280 nm) of 19.1, and a diffusion constant ($D°_{20,w}$) of 6.7 X 10^{-7} cm^2/second.

The HMW complex can be dissociated into subunits by ion-exchange chromatography at pH 7.5 or gel filtration at pH values below 5 or above 8. Low-molecular-weight (LMW) mEGF released by dissociation of the complex is identical to the mEGF described in the preceding sections. The dissociated mEGF and binding protein reassociate at a neutral pH to form a complex of approximately the same molecular weight as the native complex. The capacity of mEGF to reassociate with its binding protein is dependent on the presence of the carboxyl-terminal arginine residue (30). Mouse EGF$_{1-51}$, lacking the carboxyl-terminal Arg-Leu residues, does not recombine with the binding protein. The importance of the carboxyl-terminal arginine residue was demonstrated by treating native mEGF with carboxypeptidase B to produce a derivative lacking the carboxyl-terminal arginine residue. The derivative showed no capacity to recombine with the binding protein.

The biological significance of the carboxyl-terminal arginine residue of mEGF was indicated by the identification of the binding protein's enzymatic activity toward arginine esters (31). A similar binding protein associated with nerve growth factor (NGF) has been previously isolated from the mouse submaxillary gland and shown to possess arginine esterase activity (32). Although the arginine esteropeptidases associated with NGF and mEGF have similar molecular weights, amino acid compositions, and substrate specificities, they differ in their electrophoretic properties. Also, while they cross-react immunologically, they are not identical antigenically (31), and the mEGF-binding protein does not substitute for the binding protein associated with the formation of 7S NGF (33). The results indicate that specific associations occur between each of these growth factors and their respective arginine esteropeptidases. Since both mEGF and NGF have carboxyl-terminal arginine residues and are associated with arginine esteropeptidases, it is possible that these LMW polypeptides are produced from precursor proteins by the action of these peptidases (28, 34).

Human EGF (Urogastrone)

Although human EGF has not been as well characterized as the mouse-derived polypeptide, the available data indicate that the two polypeptides are very similar but not identical with respect to their chemical and physical properties (9). Because mEGF and hEGF exhibit identical biological activities and have some common antigenic sites, it was suggested that the human growth factor is an evolved form of the mouse polypeptide.

The completely unexpected finding by Gregory (18) that a relationship exists between mEGF and the gastric antisecretory hormone urogastrone was based on the comparison of the amino acid sequences of the two polypeptides. Of the 53 amino acid residues comprising each of the two polypeptides, 37 are common to both molecules, and the 3 disulfide bonds are formed in the same relative positions. Further, mEGF and urogastrone elicited nearly identical biological responses; mEGF elicited a gastric antisecretory response, and urogastrone produced the biological effects of EGF, as judged by its ability to induce precocious eyelid opening in the newborn mouse. Also, urogastrone and hEGF are the only polypeptides that compete with ^{125}I-mEGF in radioreceptor assays (9, 35). The available data strongly suggest that human EGF and human urogastrone are identical.

Moore (58) has reported the isolation and characterization of EGF from rat submaxillary glands. The amino acid composition of rat EGF resembles the compositions of both mouse and human EGF, but is not identical to either.

BINDING AND METABOLISM OF EGF

Several rather basic questions may be formulated regarding the interaction of any polypeptide hormone with its target cell: 1. Does the hormone enter the cell? 2. If the hormone enters the cell, what is its destination and is it accompanied by its plasma membrane receptor? 3. If the hormone penetrates the interior of the cell, is the internalization a necessary step for the evocation of at least some of the cellular responses? To obtain answers to these questions, let alone elucidate the biochemical mechanisms involved in growth control, is a formidable task. From the published experiments with EGF we have reached the following conclusions regarding the above questions: 1. It has been shown unambiguously that EGF is internalized subsequent to its binding to the plasma membrane. 2. A major portion of the internalized EGF ultimately is metabolized in lysosomes. 3. It is probable that the receptor for EGF is internalized as an EGF: receptor complex. 4. It is not known whether internalization is required for the expression of the mitogenic effect of EGF.

Receptors for EGF

Specific, saturable receptors for EGF have been demonstrated using ^{125}I-labeled mEGF or hEGF (urogastrone) and a wide variety of cultured cells including corneal cells (36–38), human fibroblasts (26, 39–42), lens cells (43), human glial cells (44), human epidermoid carcinoma cells (45, 46), 3T3 cells (47), granulosa cells (48), human vascular endothelial cells (49), human choriocarcinoma cells (123), and a number of other cell types (19, 45).

For different strains of human fibroblasts it has been estimated that each cell contains 40,000–100,000 binding sites for EGF, and apparent dissociation constants of 2–4×10^{-10} M have been calculated (40, 41).

Receptors for EGF were detected by O'Keefe et al (50) with crude membrane fractions prepared from a variety of mammalian tissues. These authors reported that placental and liver membranes have a high capacity to bind EGF. Specific binding of EGF also has been detected in liver membrane fractions of evolutionarily distant organisms, such as certain teleosts and the dogfish shark (51). The specificity and high affinity of cells and membrane preparations for ^{125}I-EGF may be employed as the basis for competitive radioreceptor assays for EGF.

Internalization and Degradation of EGF

As a step toward understanding the biochemical mechanisms involved in the growth-stimulating effects of EGF, the fate of the cell-bound ^{125}I-EGF was examined using cultured human diploid fibroblasts. Carpenter & Cohen (42) postulated that subsequent to the initial binding of ^{125}I-EGF to specific plasma membrane receptors, the EGF : receptor complex is internalized and the hormone is ultimately degraded in lysosomes. These conclusions were drawn from the following series of observations: 1. Cell-bound ^{125}I-EGF is rapidly degraded to mono [^{125}I]iodotyrosine at 37°. 2. At 0° cell-bound ^{125}I-EGF is not degraded but slowly dissociates from the cell. 3. When the binding of ^{125}I-EGF was carried out at 37° and the cells then incubated at 0°, almost no release of cell-bound radioactivity was detected. 4. The degradation, but not the binding, required metabolic energy. 5. The degradation was inhibited by drugs that inhibit lysosomal function, such as chloroquine and ammonium chloride. 6. When ^{125}I-EGF was bound to cells at 0°, the hormone was much more accessible to surface reactive agents, such as trypsin and antibodies to EGF, than when the hormone was bound at 37°. 7. Exposure of fibroblasts to EGF resulted in an apparent loss of plasma membrane receptors for EGF, which suggests that the receptor also is internalized.

Internalization and degradation of ^{125}I-EGF has been confirmed with granulosa cells (48), 3T3 and SV40-transformed 3T3 cells (52), human choriocarcinoma cells (K. V. Speeg, et al, manuscript in preparation), epidermoid carcinoma cells (46), and BSC-1 cells (54) using similar methodologies.

The problem of direct visualization of the internalization of EGF has been approached in a number of laboratories using three procedures: the preparation and tracing of fluorescent derivatives of EGF, the tracing of ^{125}I-EGF by electron microscope autoradiography, and the preparation and tracing of EGF-ferritin conjugates by electron microscopy.

Schlessinger et al (55, 128) prepared a highly fluorescent rhodamine derivative of EGF and examined the binding and internalization of this derivative to 3T3 cells using image-intensified video fluorescent microscopy. Haigler et al (46) prepared fluorescein-conjugated EGF and examined the binding and internalization of this derivative by direct fluorescence microscopy using human epidermoid carcinoma cells (A-431) which are capable of binding much larger quantities of EGF than fibroblasts, thus rendering visualization possible. Both laboratories found that the initial binding of the derivatives to the cell surface was diffuse (except for a concentration of staining at the cell borders with the A-431 cells). Within 10–30 min at 37° the labeled hormone was found within the cells in endocytotic vesicles. An intermediary patching stage was seen in the 3T3 cells but was not detected in the A-431 cells; a "microclustering" of receptors was suggested for the latter cells.

Gorden et al (56, 57) used quantitative electron microscopic autoradiography to localize ^{125}I-EGF in human fibroblasts. The initial binding of the labeled hormone was localized to the plasma membrane with some preference to coated pit regions. The membrane-bound ^{125}I-EGF was internalized by the cell in a time- and temperature-dependent fashion. The internalized grains were almost exclusively related to lysosomal structures.

Cohen and co-workers (59), and Haigler et al (manuscript in preparation) have prepared a biologically active EGF-ferritin conjugate and have observed that at 4° the EGF-ferritin binds to the plasma membrane of A-431 cells. Within one minute at 37° the ferritin-EGF redistributed on the surface of the membrane to form an increased number of small clusters that were subsequently internalized into pinocytotic vesicles. Within 30 min approximately 90% of the ferritin was seen in lysosomes. All of these data directly confirm the original kinetic studies with ^{125}I-EGF (42).

MODULATION OF RECEPTOR ACTIVITY

As plasma membrane receptors are crucial in determining the capacity of a cell to respond to a particular extracellular signal, an increasing area of investigation has been the study of mechanisms that control the level of receptors in the plasma membrane.

The most interesting and perhaps significant means of lowering EGF receptor activity is by incubation of cells with the growth factor. This phenomenon has been observed with many other polypeptide hormones and has been termed, for better or worse, "down regulation", and is quite similar to other receptor-macromolecule interactions, such as antigenic modulation. Carpenter & Cohen (42) demonstrated that incubation of human fibroblasts in serum-free media with EGF for a short period of time (1 hr) produced a rapid decrease in receptor binding activity that remained at a low level (20% of initial binding activity) for prolonged periods of time in the absence of the growth factor. They suggested that the loss of receptor activity was due to the internalization of EGF : receptor complexes, without the concomitant production of new receptors. The down regulation of EGF receptors has subsequently been demonstrated in other cell types—3T3 (52), granulosa and luteal cells (48), and choriocarcinoma cells (K. V. Speeg et al, manuscript in preparation). Two studies (48, 52) have reported that incubation of cells with subsaturating concentrations of EGF produces the down regulation of unoccupied receptors which may suggest cooperative interactions between occupied and unoccupied receptors. Scatchard analysis carried out prior to and after down regulation indicated no alteration of binding affinity by this process (52). There is no clear evidence to indicate what relationship down regulation may have to the biologic effects produced by EGF.

Carpenter & Cohen (42) showed that upon removal of the growth factor and addition of serum to down regulated cells, ^{125}I-EGF-binding activity was quickly recovered—100% within 9 hr. This recovery process was inhibited by cycloheximide or Actinomycin D which indicates a requirement for concomitant macromolecular synthesis. They suggested that the recovery process most likely involved the de novo synthesis of receptor molecules, although recycling of receptors also was considered as a possible explanation.

Two reports have shown that the inhibition of protein synthesis decreases EGF receptor binding activity. Aharonov et al (52) estimated the half-life of the EGF receptor in 3T3 cells to be 6 hr in the presence of cycloheximide. Carpenter (60) measured the capacity of human fibroblasts to bind ^{125}I-EGF following the inhibition of protein synthesis with histidinol and the reinitiation of synthesis by the addition of histidine. Under these conditions the half-life of the receptor was estimated to be 14.5 hr. ^{125}I-EGF-binding activity could be recovered upon the reinitiation of protein synthesis by the addition of L-histidine; 70% of the original binding capacity was recovered in 10 hr. The recovery of ^{125}I-EGF binding capacity was stimulated by serum, but was not blocked by Actinomycin D which indicates that RNA for the EGF receptor may accumulate during the period of histidinol-mediated inhibition of protein synthesis. The time course of recovery of receptors for EGF in fibroblasts following reinitiation of protein synthesis was very similar to that observed following down regulation of EGF receptor activity (42). Recovery from down regulation, as mentioned previously, is markedly sensitive to Actinomycin D as well as cycloheximide.

Density-dependent regulation of the binding of ^{125}I-EGF has been noted in BSC-1 cells (54). The number of available EGF receptors decreased drastically as the cell density increased and it was suggested that the accompanying decrease in sensitivity of the cells to EGF contributes to the density-dependent regulation of growth of these cells. However, no alteration of the number of EGF binding sites per cell in dense and sparse glial cell cultures was detected (44) and an increase in the number of binding sites per cell with increasing cell density was noted for 3T3 cells (61). Modulation of the binding and action of EGF on human fibroblasts by glucocorticoids has been reported (62).

The decreased ability of mouse sarcoma virus–transformed cells to bind ^{125}I-EGF was first noted by Carpenter et al (41). In detailed studies by Todaro, et al (63, 64) it was found that transformation of either epithelial or fibroblastic cells by murine or feline sarcoma virus greatly reduced, and in some cases, completely eliminated the available membrane EGF receptors. This phenomenon was specific in that transformation by DNA viruses did not alter EGF receptor binding activity, and sarcoma virus–transformed cells that had lost their capacity to bind ^{125}I-EGF had no alteration in their capacity to bind other ligands. Fibroblastic and epithelioid clones have been isolated from a normal rat kidney (NRK) line. It was found that the epithelioid clones had many more receptors for EGF than the fibroblastic clones—1.3 X 10^5 binding sites per cell respectively for the epithelioid cells, versus 1.3 X 10^4 binding sites per cell for the fibroblastic cells (65). When the clones were transformed by Moloney murine sarcoma virus the EGF binding decreased. Whether the low binding was due to the production of an EGF-like factor by these cells (63) or to defective glycoprotein synthesis (61), is not known. Pruss & Herschman (47) have isolated variants of 3T3 cells that are unable to bind ^{125}I-EGF. These variants, however, can be transformed by the Kirsten sarcoma virus and exhibit unregulated growth properties characteristic of transformed 3T3 cells (66). They suggest that the basis for uncontrolled growth in sarcoma virus–transformed cells, therefore, cannot be the action of a sarcoma gene product on the EGF receptor.

Lee & Weinstein (53) have reported that certain tumor promoters, such as phorbol esters and related plant macrocyclic diterpenes, block the binding of ^{125}I-EGF to cultured cells. They suggest that these agents, which produce mitogenic activities similar to those provoked by EGF, may exert their biological activity by interaction with EGF receptors.

NATURE OF THE RECEPTOR

Carpenter & Cohen suggested (67) that the membrane receptor for EGF is a glycoprotein based on the ability of various lectins to reversibly inhibit the binding of ^{125}I-EGF to human fibroblasts. This is supported by data that demonstrate a lowered EGF-binding capacity of mutant 3T3 cells that have a markedly decreased content of cell surface carbohydrate due to a block in the acetylation of N-acetylglucosamine-6-phosphate (61).

Two attempts have been made to crosslink EGF and its putative membrane receptor. Sahyoun et al (68) covalently linked ^{125}I-labeled human EGF to rat liver membrane binding sites with glutaraldehyde followed by sodium borohydride reduction. Upon solubilization one major and two minor ^{125}I-labeled components were detected by chromatography on Sepharose. It was suggested that the major binding component is a glycoprotein subunit of approximately 100,000 daltons. Das et al (69, 70) have synthesized a photoreactive derivative of ^{125}I-EGF and devised a procedure for labeling the EGF receptor on 3T3 cells. Although in very low yield, a single cross-linked complex, with a molecular weight of approximately 190,000, was detected by electrophoretic analysis. Incubation of the photoaffinity-labeled cells at 37° resulted in a time-dependent loss of radioactivity from the EGF : receptor complex and the appearance of three LMW radioactive components, which appeared in the lysosomal fraction of the cell. These data provide more direct experimental support for the proposal (42) that EGF-induced down regulation of plasma membrane receptors for this hormone involves the internalization of the EGF : receptor complex. In conjunction with the data previously discussed on the internalization of EGF it may be concluded that the EGF : receptor complex is internalized in pinocytotic vesicles and probably reaches the lysosome as an intact unit prior to extensive degradation of either component.

BIOLOGICAL EFFECTS OF EPIDERMAL GROWTH FACTOR

EGF has been demonstrated to elicit significant biological responses in intact animals, organ cultures, and, most recently, in cell culture systems. These responses are summarized in Table 1. Since the biological effects of EGF in whole animals and in organ cultures have been reviewed several times previously (19–22), only studies performed in cell culture systems are reviewed at this time. A second reason for concentrating on cell culture investigations is that they best afford an opportunity to examine the cascade of events that are part of the mitogenic response produced by EGF.

Stimulation of Transport

Increases in the active transport of LMW compounds from the extracellular milieu have been, in most experimental systems, one of the first detectable alterations of cellular biochemistry in mitogen-stimulated cells. The significance of transport changes to the control of cell proliferation has been discussed elsewhere (3, 94, 95).

Using $^{86}Rb^+$ as a tracer for K^+ transport, Rozengurt & Heppel (96) demonstrated that the influx of $^{86}Rb^+$ was increased twofold within 20 min after the addition of EGF to quiescent 3T3 cells. Barnes & Colowick (97) reported that increased uptake of $[^3H]$-deoxyglucose occurred 15 min after the addition of EGF to 3T3 cells. Maximal enhancement of sugar transport (2-fold in 3T3 cells and 1.5-fold in human fibroblasts) was achieved 2 hr after the addition of EGF. That transport per se was stimulated by EGF was indicated by the enhanced (two- to three fold) uptake of $[^{14}C]$-3-O-methylglucose in both 3T3 and SV40-3T3 cells. In contrast Hollenberg & Cuatrecasas (40) reported that incubation of human fibroblasts with EGF did not affect the transport of $[^{14}C]$3-O-methylglucose, but did enhance (1.2- to 2-fold) the rate of uptake of radioactive α-aminoisobutyrate. Since these observed increases in the active transport of nutrients occur quite soon (15–30 min) after the addition of EGF to cell cultures, it is unlikely that intracellular degradation of the growth factor is required to elicit these biochemical activities. This conclusion is supported by the fact that EGF

Table 1 Biological effects of EGF

Biological effect	Reference
In vivo	
Accelerated proliferation/differentiation	
of skin tissue	14, 71
of corneal epithelial tissue	72–74
of lung and trachae epithelia	75, 76
Potentiation of methylcholanthrene carcinogenesis	77, 78
Inhibition of gastric acid secretion	79
Increased activity of ornithine decarboxylase and accumulation	
of putrescine	80, 81
Formation of fatty liver	82
Increase of disulfide group content in skin	83
Hepatic hypertrophy and hyperplasia	84
Potentiation of cleft palate	85
Organ cultures	
Accelerated proliferation/differentiation	
of skin tissue	15, 86
of corneal epithelial tissue	72
of mammary gland epithelial tissue	87
Induction of ornithine decarboxylase and accumulation of	
putrescine	88
Enhanced protein synthesis, RNA synthesis	89–91
Inhibition of palate fusion	92, 93
Cell cultures	
Increased transport	
α-aminoisobutyrate	40
deoxyglucose	97
K^+	96
Activation of glycolysis	98, 102
Stimulation of macromolecular synthesis	
hyaluronic acid	104
RNA	39, 91, 107
protein	89, 90, 91, 107
DNA	26, 38, 39, 43, 49, 54, 107–114
Enhanced cell multiplication	26, 39, 108–112, 114, 118, 119
Alteration of membrane properties	120–122
Stimulated hCG secretion	123
Increased biogenesis of LETS protein	124
Enhanced prostaglandin biosynthesis	125
Alteration of viral growth	126
Increased squame production by keratinocytes	127

stimulated deoxyglucose uptake in the presence of ammonium chloride, a potent inhibitor of EGF degradation (97). At present it is not possible to ascertain whether the phosphorylation of membrane proteins by EGF (see below) and/or the internalization of EGF : receptor complexes are parts of the mechanism whereby mitogens produce increased rates of nutrient transport in target cells.

Activation of Glycolysis

Activation of quiescent 3T3 cells by EGF produces a rapid increase in glycolytic activity as shown by enhanced production of lactic acid (98). The level of lactate in the media increased very rapidly (fourfold within 2 hr) after addition of EGF, insulin, or serum to quiescent cells. The increases in lactic acid production during the first 3 hr of exposure to EGF were not affected by cycloheximide which indicates that the initial activation of glycolysis did not require new protein synthesis. The addition of dibutyryl cyclic AMP, dibutyryl cyclic GMP, or theophylline alone or in combination with EGF did not alter the production of lactic acid. The authors also report that ouabain did not block the activation of glycolysis by EGF which suggests that the Na^+ pump is not crucial, although others (99, 100) have implicated the Na^+ pump in the control of glycolysis. The ability of EGF to stimulate lactate production was markedly decreased by omission of Ca^{2+} from the media. This result is of interest in view of the proposed role of Ca^{2+} as a hormone second messenger (101).

The stimulatory effect of EGF on glycolysis could be demonstrated in cell-free homogenates prepared from cells activated with EGF (102). Addition of EGF to homogenates prepared from quiescent cells did not affect glycolysis. These studies have further shown that the specific activity of phosphofructokinase is significantly increased in cell-free homogenates prepared from quiescent cells exposed to EGF for 3 hr. The ability of EGF to stimulate phosphofructokinase activity was dependent on the presence of Ca^{2+} in the medium and was not affected by the presence of cycloheximide. The results of mixing experiments, and the persistence of EGF-activated phosphofructokinase after gel chromatography and at alkaline pH, where the influence of many allosteric effectors of the enzyme is reduced, indicate that diffusible activators or inhibitors are probably not involved in this response to EGF.

Activation of the Synthesis of Extracellular Macromolecules

When human fibroblasts are grown for an extended period of time in the presence of EGF, the cells form multilayers 3–4 cells thick (103). Dissociation of these multilayered cells into single cells cannot be achieved by trypsinization, but does occur if the cells are subsequently treated with collagenase and hyaluronidase which suggests that increased synthesis and excretion of extracellular materials are associated with extended cell multiplication in the presence of EGF (G. Carpenter, unreported observation). Recently, Lembach (104) has demonstrated that the addition of EGF to cultured human fibroblasts increases the incorporation of ^{3}H-glucosamine into both cellular and extracellular glycosaminoglycans and glycoproteins. Incorporation of $^{35}SO_4^{2-}$ into these cells was not altered by the presence of EGF. EGF stimulated the incorporation of ^{3}H-glucosamine into cellular glycosaminoglycans within 4 hr and into extracellular material within 8–12 hr. The labeled extracellular material was identified as mostly hyaluronic acid. Increased production of hyaluronic acid (105) and elevated activity of hyaluronic acid synthetase (106) have been reported in other cells stimulated to proliferate. EGF stimulated incorporation of ^{3}H-glucosamine into both cellular and extracellular glycoproteins within 4 hr (104).

Activation of RNA and Protein Synthesis

Relatively few studies have been reported recently of alterations in protein and RNA synthesis following the activation of cells by EGF. Hollenberg & Cuatrecasas (39) have presented data that show increased incorporation of [^{3}H]-uridine into RNA in human fibroblasts approximately 15–20 hr after the addition of EGF. This experiment was performed in serum-deficient media which may account for the rather slow response—serum stimulation by itself produced an increased ^{3}H-uridine incorporation in 5 hr. Experiments with HeLa cells (107) and chick embryo epidermis (91) have shown increased (two- to fivefold) ^{3}H-uridine incorporation within 5 hr following addition of EGF.

Experiments with chick embryo epidermis in organ culture have shown a rapid increase in the rate of protein synthesis following the addition of EGF (89). Increased polysome formation (91) and increased activity of ribosomes in vitro (90) were evident within 5 hr after activation by EGF. Increased incorporation of labeled amino acids at approximately 10 hr after addition of EGF to HeLa cells has been noted (107). There are no reports to date indicating what effects EGF may have on the process of protein degradation which plays an important role in determining the rate of net protein synthesis.

Initiation of DNA Synthesis

Armelin (108) and Hollenberg & Cautrecasas (39) first reported that EGF stimulated DNA synthesis in cultured fibroblasts. Other workers have since shown that EGF enhances DNA synthesis in a variety of cell types: mouse (108) and human (39, 26) fibroblasts, glia (109), rabbit lens epithelial cells (43), rabbit chondrocytes (110), mammary epithelial cells (111), bovine corneal endothelial cells (112), human endothelial cells (49), monkey kidney cells (54), rat hepatocytes (113), HeLa (107), and bovine granulosa cells (114). In most cases the level of EGF-mediated stimulation of DNA synthesis is dependent on small amounts (1–2%) of serum in the medium (26, 103). Depending on the cell type under study and the presence or absence of serum various agents enhance the EGF-stimulation of DNA synthesis. These agents include ascorbic acid (103, 115), insulin (113, 116), B_{12} (116), glucagon (113), thrombin (49, 117), and EGF-binding arginine esterase (115). One study (39) has reported that cholera toxin, theophylline, and theophylline plus dibutyryl cyclic AMP inhibited the stimulation of DNA synthesis by EGF. In most cell culture systems increased DNA synthesis begins approximately 15 hr after the addition of EGF and reaches a maximum at about 22 hr. Removal of EGF from the media by the addition of rabbit anti-EGF (103) or by washing (52) indicates that the growth factor must be continuously present in the media for about 5 hr in order to initiate even a small level of DNA synthesis. Removal of EGF from the media 7½ hr after its addition reduces stimulated DNA synthesis to 50% of that achieved in the continuous presence of the hormone. Removal of the growth factor after DNA synthesis begins (12–15 hr) has very little effect on the rate of DNA synthesis measured at 20–24 hr. Several studies have shown that maximal stimulation of DNA synthesis is achieved at a concentration of EGF that yields 25% of the maximal binding (40, 42, 48, 52). Thus, only a fraction of the available receptors need to be occupied by the growth factor to elicit a maximal biological activity.

Increased Cell Multiplication

In cell culture systems EGF has been shown to enhance cell proliferation, as measured by increased cell numbers, in a wide variety of nontransformed cell types: mouse fibroblasts (108); human fibroblasts (26, 39); glia (109); mammary epithelial (111); keratinocytes (118); vascular endothelial (38); rabbit chondrocytes (110); bovine granulosa (114); and corneal endothelial (112). The growth factor also increases the multiplication of transformed cells such as HeLa (119) and SV40–3T3 (G. Carpenter, unpublished data) when they are maintained in a serum-free or serum-limited media.

Characterization of the EGF-stimulation of cell multiplication has been most extensively studied with cultured human fibroblasts (26, 39, 40, 102). The growth of human fibroblasts in vitro is characterized by three parameters that regulate the growth of most "normal" cells: 1. density-dependent inhibition of growth (contact inhibition), 2. a high serum requirement, and 3. anchorage-dependent growth.

When human fibroblasts are cultured in the continuous presence of EGF (4 ng/ml), the first two of these growth-controlling mechanisms are modified (103). Cells grown in the presence of EGF continue to proliferate when the culture becomes confluent, and form a multilayered cell population not restricted by density dependent inhibition of growth, that reaches population densities four- to sixfold higher than the controls. Also, when human fibroblasts are incubated in low or deficient serum (1% calf serum or 10% γ-globulin free serum or plasma) cell proliferation is normally restricted, but supplementation of these media with EGF yields cell growth equal to that obtained in the presence of optimal serum concentrations (10% calf serum). The apparent loss of sensitivity to these growth-regulating mechanisms by cells cultured in the presence of EGF bears some resemblance to the behavior of transformed cells in vitro. Transformed cells are frequently characterized by their insensitivity to normal growth-controlling mechanisms. Thus, EGF, while not transforming cells, does have the effect, under cell culture conditions, of creating an imbalance of homeostatic signals favoring cell proliferation.

Miscellaneous Biological Effects of EGF

Morphological studies have shown that membrane ruffling and macropinocytosis are enhanced within 4–6 hr after the addition of EGF to quiescent glial cells (120). Also, the addition of EGF to 3T3 cells maintained in serum-free media was reported to strengthen cell attachment and produce marked increases in the number of microvilli and the mobility of Con A receptors (121). A study of the increased binding of EGF-treated 3T3 cells to Con A–coated nylon fibers also indicates that the growth factor enhances the lateral mobility of Con A receptors in the plasma membrane (122).

Several reports have appeared concerning effects of EGF which do not, at this time, appear to be directly related to mitogenesis. These biological activities may be indirect consequences of the mitogenic signal or they may in fact demonstrate that EGF, like other hormones, such as insulin, has multiple biological activities. The ability of EGF to inhibit gastric acid secretion (79) is one example and is noted previously. The addition of EGF to cultured human choriocarcinoma cells stimulated the secretion of biologically active human chorionic gonadotropin (hCG) and to a lesser extent the secretion of free hCG-α units (123). Subsequent studies (K. Y. Speeg et al, manuscript in preparation) have shown that EGF-stimulated secretion of hCG begins approximately 20 hr after addition of the growth factor and is threefold higher than controls after 44 hr. Addition of rabbit anti-EGF to these cells at varying times after the addition of EGF indicates that the growth factor must be in contact with the cells for 6 hr to achieve maximal stimulation of hCG secretion. Chen et al (124) have reported that when 3T3 cells are maintained in low serum, the fibrous network of LETS cell surface protein is lost. Subsequent addition of EGF resulted in the reappearance of the LETS protein network. Levine & Hassid (125) have reported that EGF stimulated prostaglandin $F_{2\alpha}$ and E_2 by 40% and 25%, respectively, in cultured canine kidney cells. EGF did not affect prostaglandin biosynthesis in other cell lines. Knox et al (126) reported that pretreatment of human fibroblasts for 12–48 hr with EGF suppressed the infectivity of cytomegalovirus and enhanced the infectivity of herpes simplex type 1 virus in vitro. Green (127) has shown that EGF increased the production of squames two- to fivefold in cultured human keratinocytes.

PHYSIOLOGICAL ASPECTS

Mouse EGF

Although mEGF is synthesized in the submaxillary gland of the mouse (129) and stored in granules of the tubular duct cells (22, 130–133), plasma levels of the growth factor, determined by a radioimmunoassay, are not decreased by excision of the submaxillary glands (134). It appears, therefore, that there is a secondary site(s) of EGF synthesis in the mouse.

The morphology and EGF content of tubular duct cells of the rodent submaxillary gland are dependent on the hormonal status (mainly androgen levels) of the animal. The cells are developed fully in the male only after puberty; castration results in the atrophy of the tubular portion of the gland, and injection of testosterone into female mice results in hypertrophy and hyperplasia of these cells (reviewed in 136). The quantity of EGF present in the submaxillary gland closely parallels the development of the tubular system (137). In the immature 15-day-old male mouse, the amount of submaxillary gland EGF is very low, about 0.02 ng/mg wet tissue. The concentration increases rapidly with age until maximum levels of about 1000 ng/mg wet tissue are reached in adult male mice aged 50 days or more (138). In contrast to the adult male mouse, the concentration of EGF in the submaxillary gland of the adult female is only 70 ng/mg wet tissue. These results have been confirmed using an immunocytochemical procedure (133). As expected, the tfm/y mouse, with a genetic defect that results in target organ insensitivity to androgens, has very low levels of EGF in the submaxillary gland (139). The relative potencies of a variety of androgens and progestins to induce the accumulation of EGF in the submaxillary gland have been examined (139, 140) and the results reviewed (19).

The androgen-dependent levels of EGF in the submaxillary gland do not reflect levels in plasma and other body fluids (134). While there are 1000 ng of EGF/mg wet submaxillary gland tissue in the mouse, the concentration in plasma is 1 ng/ml; in milk, approximately 300 ng/ml; and in saliva or urine, about 1000 ng/ml. The plasma levels of immunoactive EGF are not significantly different in adult male and female mice. Testosterone treatment of females or castrated males does not alter EGF plasma levels. Also, EGF concentrations in the plasma of immature male mice are not detectably different from adult levels.

The release of submaxillary EGF into the lumen of the gland and the circulatory system is controlled, either directly or indirectly, by α-adrenergic agents (134, 141) which are known to bring about the degranulation of the peritubular cells in the submaxillary gland (136). The intravenous administration of an α-adrenergic agent, such as phenylephrine, into adult male mice, results in a 65% decrease in submaxillary gland EGF within 60 min and a concomitant increase in plasma EGF levels from 1.2 to 152 ng/ml (134). The rise in plasma EGF after administration of an α-adrenergic agent was blocked by the prior injection of α-adrenergic blocking agents. Similar results were obtained in vitro using minced submaxillary glands (142, 143). A circadian periodicity in the submaxillary gland concentration of EGF, which was abolished by superior cervical ganglionectomy, has been noted (144).

The administration of isoprenaline (isoproterenol), a stimulant of exocrine secretion by the submaxillary glands of mice, reduces the content of amylase in the gland by 90% but does not significantly decrease the amount of glandular EGF (145) or increase plasma EGF levels (134).

Human EGF (Urogastrone)

With the isolation of human EGF (9) and urogastrone (18, 146), and the realization that the two polypeptides are biologically equivalent and probably chemically identical, much recent effort has been directed toward determining the distribution and concentration of EGF in various tissue and body fluids and correlating these data with normal or abnormal physiological states.

Urogastrone (hEGF) has been detected in the salivary gland and in the gland of Brunner by immunofluorescent staining (135). Whether these tissues are the sites of synthesis is not yet proven.

Both heterologous and homologous radioimmunoassays for hEGF have been designed by Orth and co-workers (147, 148) Twenty-four-hour urinary excretion of hEGF by normal adult males and females was 63.0 ± 3.0 and 52 ± 3.5 (mean ± SE) μg per total volume. No diurnal or postprandial variation was detected; however, an excellent linear correlation between urinary hEGF and creatinine concentration in each urine sample was noted. The concentration of hEGF in human saliva ranged from 6–17 ng/ml, was about 80 ng/ml in human milk, ranged from 29–272 ng/ml in human urine, and was undetectable in human amniotic fluid. Preliminary data indicated

2–4 ng/ml in human plasma. As yet no particular disease state has been clearly defined in which urinary EGF levels are abnormally high or low; it was noted, however, that excretion by females taking oral contraceptives was significantly greater than that by females who were not.

PHOSPHORYLATION AND MOLECULAR MECHANISMS

We have reviewed the evidence that EGF forms a complex with plasma membrane receptors, initiating an intricate series of biochemical and morphological events within the cell that include internalization and degradation of the hormone:receptor complex. These events ultimately result in cell growth. It is reasonable to assume that the observed biochemical and morphological alterations induced by EGF result from the generation, amplification, and propagation of a series of "signals." To understand the nature of these signals it would appear necessary to obtain cell-free systems that are responsive in vitro to the addition of either EGF, a fragment of EGF, or an EGF-induced cellular product.

Since phosphorylation and dephosphorylation reactions participate in the regulation of many metabolic pathways (149) and membranes contain endogenous protein kinases, a study was initiated on the possible role of EGF as a control element in these processes. Carpenter et al (150) have reported that membranes may be prepared from A-431 cells which retain the ability to bind ^{125}I-EGF in vitro and that following the formation of EGF:receptor complexes the capacity of these membranes to phosphorylate endogenous proteins in the presence of γ-labeled [^{32}P]ATP is markedly enhanced. The A-431 membrane preparation appeared to have EGF-stimulated protein kinase activity toward exogenous substrates (histone) as well as toward membrane associated proteins.

Both the endogenous phosphorylation and EGF-stimulated phosphorylation were dependent on the presence of Mg^{2+} or Mn^{2+}; Ca^{2+} was ineffective. Partial acid hydrolysis and electrophoresis of the phosphorylated membranes showed that the major phosphorylated product, in both the presence and absence of EGF, was phosphothreonine. SDS polyacrylamide gel electrophoresis and autoradiography indicated that the phosphorylation of a number of membrane associated proteins, including components with molecular weights of approximately 160,000, 80,000 and 22,500, were enhanced in the presence of EGF.

The mechanism by which EGF enhances phosphorylation is not known; it is possible that the receptor is a protein kinase or a regulatory subunit of a kinase. Since neither the identity nor function of the EGF-stimulated phosphorylated membrane components have been determined, it is currently not possible, however tempting, to relate these findings to specific metabolic alterations induced in intact cells by EGF. Nevertheless, these studies demonstrate for the first time a functional alteration of membranes in a cell-free preparation as a consequence of the binding of EGF.

CONCLUDING REMARKS

Among the many alternative hypotheses that may be considered in attempting to understand the mechanisms by which EGF exerts its effect are:

1. All of the biological effects of EGF are exerted at the plasma membrane and the internalization process merely serves to remove the stimulus.
2. Membrane binding is necessary for some of the observed effects (such as transport) whereas others (such as DNA synthesis) require internalization of the cell-bound hormone.
3. Phosphorylation of membrane components induced by EGF is related to altered transport of ions and nutrients or internalization of the cell-bound growth factor.
4. A degradation product of EGF is biologically active within the cell.
5. Internalization of the receptor, with or without degradation, is related to the mitogenic signal.
6. Activation and internalization of the membrane phosphorylating systems (receptor related?) is required for mitogenicity. This could serve to phosphorylate specific membrane-associated cytoplasm proteins which in turn could act as metabolic signals.

We hope that the biochemical information available concerning the structure of epidermal growth factor and the variety of systems affected by this hormone will facilitate further exploration of the molecular mechanisms by which mitogens and polypeptide hormones regulate cell growth.

ACKNOWLEDGMENTS

The authors thank Drs. Leon Cunningham, Harry Haigler, and William Mitchell for critical reading of this manuscript. The support of Biomedical Research Support Grant RR–05424–16 and US Public Health Service Grant HD–00700 is acknowledged. S. Cohen is an American Cancer Society Research Professor.

214 CARPENTER & COHEN

Literature Cited

1. Temin, H. M., Pierson, R. W., Jr., Dulak, N. C. 1972. In *Growth, Nutrition and Metabolism of Cells in Culture,* ed. G. H. Rothblat, V. J. Cristofalo, Vol. 1, pp. 50–81, New York: Academic
2. Gospodarowicz, D., Moran, J. S. 1976. *Ann. Rev. Biochem.* 45:531–58
3. Pardee, A. B., Dubrow, R., Hamlin, J. L., Kletzien, R. F. 1978. *Ann. Rev. Biochem.* 47:715–50
4. Sato, G. H. 1975. In *Biochemical Actions of Hormones,* ed. G. Litwack, Vol. 3 pp. 391–96. New York: Academic
5. Holley, R. W., Kiernan, J. A. 1968. *Proc. Natl. Acad. Sci. USA* 60:300–4
6. Cohen, S. 1960. *Proc. Natl. Acad. Sci. USA* 46:302–11
7. Cohen, S. 1959. *J. Biol. Chem.* 234:1129–37
8. Cohen, S. 1962. *J. Biol. Chem.* 237:1555–62
9. Cohen, S., Carpenter, G. 1975. *Proc. Natl. Acad. Sci. USA.* 72:1317–21
10. Gospodarowicz, D. 1975. *J. Biol. Chem.* 250:2515–20
11. Gospodarowicz, D., Bialecki, H., Greenberg, G. 1978. *J. Biol. Chem.* 253:3736–43
12. Smith, G. L., Temin, H. 1974. *J. Cell Physiol.* 84:181–92
13. Stanley, E. R., Cifone, M., Heard, P. M., Defendi, V. 1976. *J. Exp. Med.* 143:631–47
14. Cohen, S., Elliott, G. A. 1963. *J. Invest. Dermatol.* 40:1–5
15. Cohen, S. 1965. *Dev. Biol.* 12:394–7
16. Savage, C. R. Jr., Cohen, S. 1972. *J. Biol. Chem.* 247:7609–11
17. Starkey, R. H., Cohen, S., Orth, D. N. 1975. *Science* 189:800–2
18. Gregory, H. 1975. *Nature* 257:325–27
19. Carpenter, G., Cohen, S. 1978. In *Biochemical Actions of Hormones* ed. G. Litwack, Vol. 5 New York: Academic. pp. 203–47
20. Carpenter, G. 1978. *J. Invest. Dermatol.* 71:283–87
21. Cohen, S., Taylor, J. M. 1974. *Recent Prog. Horm. Res.* 30:533–50
22. Cohen, S., Savage, C. R., Jr. 1974. *Recent Prog. Horm. Res.* 30:551–74
23. Taylor, J. M., Mitchell, W. M., Cohen, S. 1972. *J. Biol. Chem.* 247:5928–34
24. Savage, C. R., Jr., Inagami, T., Cohen, S. 1972. *J. Biol. Chem.* 247:7612–21
25. Savage, C. R., Jr., Hash, J. H., Cohen, S. 1973. *J. Biol. Chem.* 248:7669–72
26. Cohen, S., Carpenter, G., Lembach, K. J. 1975. *Adv. Metab. Disord.* 8:265–84
27. Holladay, L. A., Savage, C. R., Jr.,

Cohen, S., Puett, D. 1976. *Biochemistry* 15:2624–33
28. Taylor, J. M., Cohen, S., Mitchell, W. M. 1970. *Proc. Natl. Acad. Sci. USA* 67:164–71
29. Taylor, J. M., Mitchell, W. M., Cohen, S. 1974. *J. Biol. Chem.* 249:3198–3
30. Server, A. C., Sutter, A., Shooter, E. M. 1976. *J. Biol. Chem.* 251:1188–96
31. Taylor, J. M., Mitchell, W. M., Cohen, S. 1974. *J. Biol. Chem.* 249:2188–94
32. Greene, L. A., Shooter, E. M., Varon, S. 1969. *Biochemistry* 8:3735–41
33. Server, A. C., Shooter, E. M. 1976. *J. Biol. Chem.* 251:165–73
34. Angeletti, R. H., Bradshaw, R. A. 1971. *Proc. Natl. Acad. Sci. USA* 68:2417–20
35. Hollenberg, M. D., Gregory, H. 1976. *Life Sci.* 20:267–74
36. Covelli, I., Rossi, R., Mozzi, R., and Frati, L. 1972. *Eur. J. Biochem.* 27:225–30
37. Frati, L., Daniele, S., Delogu, A., Covelli, I. 1972. *Exp. Eye Res.* 14:135–41
38. Gospodarowicz, D., Mescher, A. L., Brown, K. D., Birdwell, C. R. 1977. *Exp. Eye Res.* 25:631–49
39. Hollenberg, M. D., Cuatrecasas, P. 1973. *Proc. Natl. Acad. Acad. Sci. USA* 70:2964–68
40. Hollenberg, M. D., Cuatrecasas, P. 1975. *J. Biol. Chem.* 250:3845–53
41. Carpenter, G., Lembach, K. J., Morrison, M., Cohen, S. 1975. *J. Biol. Chem.* 250:4297–4
42. Carpenter, G., Cohen, S. 1976. *J. Cell Biol.* 71:159–71
43. Hollenberg, M. D. 1975. *Arch. Biochem. Biophys.* 171:371–77
44. Westermark, B. 1977. *Proc. Natl. Acad. Sci. USA* 74:1619–21
45. Fabricant, R. N., DeLarco, J. E., Todaro, G. J. 1977. *Proc. Natl. Acad. Sci. USA* 74:565–69
46. Haigler, H., Ash, J. F., Singer, S. J., Cohen, S. 1978. *Proc. Natl. Acad. Sci. USA* 75:3317–21
47. Pruss, R. M., Herschman, H. R. 1977. *Proc. Natl. Acad. Sci. USA* 74:3918–21
48. Vlodavsky, I., Brown, K. D., Gospodarowicz, D. 1978. *J. Biol. Chem.* 253:3744–50
49. Gospodarowicz, D., Brown, K. D., Birdwell, C. R., Zetter, B. R. 1978. *J. Cell Biol.* 77:774–88
50. O'Keefe, E., Hollenberg, M. D., Cuatrecasas, P. 1974. *Arch. Biochem. Biophys.* 164:518–26
51. Naftel, J., Cohen, S. 1978. *J. SC Med. Assoc.* 74:53

52. Aharonov, A., Pruss, R. M., Herschman, H. R. 1978. *J. Biol. Chem.* 253:3970–77
53. Lee, L. S., Weinstein, I. B. 1978. *Science.* 202:313–15
54. Holley, R. W., Armour, R., Baldwin, J. H., Brown, K. D., Yeh, Y. 1977. *Proc. Natl. Acad. Sci. USA* 74:5046–50
55. Schlessinger, J., Shechter, Y., Willingham, M. C., Pastan, I. 1978. *Proc. Natl. Acad. Sci. USA* 75:2659–63
56. Gorden, P., Carpentier, J., Cohen, S., Orci, L. 1978. *C. R. Acad. Sci.* 286:1471–74
57. Gorden, P., Carpentier, J., Cohen, S., Orci, L., 1978. *Proc. Natl. Acad. Sci. USA.* 75:5025–29
58. Moore, J. B., Jr. 1978. *Arch. Biochem. Biophys.* 189:1–7
59. Cohen, S., Haigler, H., McKanna, J., Carpenter, G., King, L. Jr. 1979. *Cold Spring Harbor Conf. Cell Proliferation, Vol. 6.* In press
60. Carpenter, G. 1979. *J. Cell Physiol.* In press
61. Pratt, R. M., Pastan, I. 1978. *Nature* 272:68–70
62. Baker, J. B., Barsh, G. S., Carney, D. H., Cunningham, D. D. 1978. *Proc. Natl. Acad. Sci. USA* 75:1882–86
63. Todaro, G. J., DeLarco, J. E., Cohen, S. 1976. *Nature* 246:26–31
64. Todaro, G. J., DeLarco, J. E., Nissley, S. P., Rechler, M. M. 1977. *Nature* 267:526–28
65. DeLarco, J. E., Todaro, G. J. 1978. *J. Cell. Physiol.* 94:335–42
66. Pruss, R. M., Herschman, H. R., Klement, V. 1978. *Nature* 274:272–74
67. Carpenter, G., Cohen, S. 1977. *Biochem. Biophys. Res. Commun.* 79:545–52
68. Sahyoun, N., Hock, R. A., Hollenberg, M. D. 1978. *Proc. Natl. Acad. Sci. USA* 75:1675–79
69. Das, M., Miyakawa, T., Fox, C. F., Pruss, R. M., Aharonov, A., Herschman, H. 1977. *Proc. Natl. Acad. Sci. USA* 74:2790–94
70. Das, M., Fox, F. C. 1978. *Proc. Natl. Acad. Sci. USA* 75:2644–48
71. Birnbaum, J. E., Sapp, T. M., Moore, J. B. 1976. *J. Invest. Dermatol.* 66:313–18
72. Savage, C. R., Jr., Cohen, S. 1973. *Exp. Eye Res.* 15:361–66
73. Frati, L., Daniele, S., Delogu, A., Covelli, I. 1972. *Exp. Eye Res.* 14:135–41
74. Ho, P. C., Davis, W. H., Elliott, J. H. 1974. *Invest. Ophthalmol.* 13:804–9
75. Sundell, H., Serenius, F. S., Barthe, P., Friedman, Z., Kanarek, K. S., Escobedo, M. B., Orth, D. N., Stahlman, M. T. 1975. *Pediatr. Res.* 9:371
76. Catterton, W. Z., Escobedo, M. B., Sexson, W. R., Gray, M. E., Sundell, H. W., Stahlman, M. T. 1978. *Pediatr. Res.* In press
77. Reynolds, V. H., Boehm, F. H., Cohen, S. 1965. *Surg. Forum* 16:108–9
78. Rose, S. P., Stahn, R., Passovoy, D. S., Herschman, H. 1976. *Experientia* 31:913–15
79. Bower, J. M., Camble, R., Gregory, H., Gerring, E. L., Willshire, I. R. 1975. *Experientia* 32:825–26
80. Stastny, M., Cohen, S. 1972. *Biochim. Biophys. Acta* 261:177–80
81. Blosse, P. T., Fenton, E. L., Henningsson, S., Kahlson, G., Rosengren, E. 1974. *Experientia* 30:22–23
82. Heimberg, M., Weinstein, I., LeQuire, V. S., Cohen, S. 1965. *Life Sci.* 4:1625–33
83. Frati, C., Covelli, I., Mozzi, R., Frati, L. 1972. *Cell Differ.* 1:239–44
84. Bucher, N. L. R., Patel, V., Cohen, S. 1978. *Adv. Enzyme Regul.* 16:205–13
85. Bedrick, A. D., Ladda, R. L. 1978. *Teratology* 17:13–18
86. Bertsch, S., Marks, F. 1974. *Nature* 251:517–19
87. Turkington, R. W. 1969. *Exp. Cell Res.* 57:79–85
88. Stastny, M., Cohen, S. 1970. *Biochim. Biophys. Acta* 204:578–89
89. Hoober, J. K., Cohen, S. 1967. *Biochim. Biophys. Acta* 138:347–56
90. Hoober, J. K., Cohen, S. 1967. *Biochim. Biophys. Acta* 138:357–68
91. Cohen, S., Stastny, M. 1968. *Biochim. Biophys. Acta* 166:427–37
92. Hassell, J. R. 1975. *Develop. Biol.* 45:90–2
93. Hassell, J. R., Pratt, R. M. 1977. *Exp. Cell Res.* 106:55–62
94. Kaplan, J. G. 1978. *Ann. Rev. Physiol.* 40:19–41
95. Holley, R. W. 1972. *Proc. Natl. Acad. Sci. USA* 69:2840–41
96. Rozengurt, E., Heppel, L. A. 1975. *Proc. Natl. Acad. Sci. USA* 72:4492–95
97. Barnes, D., Colowick, S. P. 1976. *J. Cell. Physiol.* 89:633–40
98. Diamond, I., Legg, A., Schneider, J. A., Rozengurt, E. 1978. *J. Biol. Chem.* 253:866–71
99. Scholnick, P., Land, D., Racker, E. 1973. *J. Biol. Chem.* 248:5175–82
100. Racker, E. 1976. *J. Cell. Physiol.* 89:697–700
101. Rasmussen, H. 1970. *Science* 170:404–12

102. Schneider, J. A., Diamond, I., Rozengurt, E. 1978. *J. Biol. Chem.* 253: 872–77
103. Carpenter, G., Cohen, S. 1976. *J. Cell. Physiol.* 88:227–38
104. Lembach, K. J. 1976. *J. Cell. Physiol.* 89:277–88
105. Moscatelli, D., Rubin, H. 1975. *Nature* 254:65–66
106. Tomida, M., Koyama, H., Ono, T. 1975. *J. Cell. Physiol.* 86:121–30
107. Covelli, I., Mozzi, R., Rossi, R., Frati, L. 1972. *Hormones* 3:183–91
108. Armelin, H. 1973. *Proc. Natl. Acad. Sci. USA* 70:2702–6
109. Westermark, B. 1976. *Biochem. Biophys. Res. Commun.* 69:304–10
110. Gospodarowicz, D., Mescher, A. L. 1977. *J. Cell. Physiol.* 93:117–28
111. Stoker, M. G. P., Pigott, D., Taylor-Papadimitriou, J. 1976. *Nature* 264: 764–67
112. Gospodarowicz, D., Mescher, A. L., Birdwell, C. R. 1977. *Exp. Eye Res.* 25:75–89
113. Richman, R. A., Claus, T. H., Pilkis, S. J., Friedman, D. L. 1976. *Proc. Natl. Acad. Sci. USA* 73:3589–93
114. Gospodarowicz, D., Ill, C. R., Birdwell, C. R. 1977. *Endocrinology* 100:1108–20
115. Lembach, K. J. 1976. *Proc. Natl. Acad. Sci. USA* 73:183–87
116. Mierzejewski, K., Rosengurt, E. 1976. *Biochem. Biophys. Res. Commun.* 73:271–78
117. Zetter, B. R., Sun, T. T., Chen, L. B., Buchanan, J. M. 1977. *J. Cell. Physiol.* ?2:233–40
118. Rheinwald, J. G., Green, H. 1977. *Nature* 265:421–24
119. Hutchings, S. E., Sato, G. 1978. *Proc. Natl. Acad. Sci. USA* 75:901–4
120. Brunk, U., Schellens, J., Westermark, B. 1976. *Exp. Cell Res.* 103:295–2
121. Berliner, J. A. 1977. *J. Cell Biol.* 75:219a
122. Aharonov, A., Vlodavsky, I., Pruss, R. M., Fox, C. F., Herschman, H. 1978. *J. Cell. Physiol.* 95:195–202
123. Benveniste, R., Speeg, K. V., Jr., Carpenter, G., Cohen, S., Lindner, J., Rabinowitz, D. 1978. *J. Clin. Endocrinol. Metab.* 46:169–72
124. Chen, L. B., Gudor, R. C., Sun, T. T., Chen, A. B., Mosesson, M. W. 1977. *Science* 197:776–78
125. Levine, L., Hassid, A. 1977. *Biochem. Biophys. Res. Commun.* 76:1181–87
126. Knox, G. E., Reynolds, D. W., Cohen, S., Alford, C. A. 1978. *J. Clin. Invest.* 61:1635–44
127. Green, H. 1977. *Cell* 11:405–16
128. Shechter, Y., Schlessinger, J., Jacobs, S., Chang, K., and Cuatrecasas, P. 1978. *Proc. Natl. Acad. Sci. USA* 75:2135–39
129. Turkington, R. W., Males, J. L., Cohen, S. 1971. *Cancer Res.* 31:252–56
130. Pasquini, F., Petris, A., Sbaraglia, G., Scopelliti, R., Cenci, G., Frati, L. 1974. *Exp. Cell Res.* 86:233–36
131. Gresik, E., Barka, T. 1977. *J. Histochem. Cytochem.* 25:1027–35
132. Van Noorden, S., Heitz, P., Kasper, M., Pearse, A. G. E. 1977. *Histochem.* 52:329–40
133. Gresik, E., Barka, T. 1978. *Am. J. Anat.* 151:1–10
134. Byyny, R. L., Orth, D. N., Cohen, S., Doyne, E. S. 1974. *Endocrinology* 95:776–82
135. Elder, J. B., Williams, G., Lacy, B., Gregory, H. 1978. *Nature* 271:466–467
136. Sreebny, L. M., Meyer, J. 1964. *Salivary Glands and Their Secretion.* New York: Pergamon 380 pp.
137. Cohen, S. 1965. In *M. D. Anderson Tumor Inst. Symp. Developmental and Metabolic Control Mechanisms and Neoplasia,* pp. 251–72. Baltimore: Williams & Wilkins
138. Byyny, R. L., Orth, D. N., Cohen, S. 1972. *Endocrinology* 90:1261–66
139. Barthe, P. L., Bullock, L. P., Mowszowicz, I., Bardin, C. W., Orth, D. N. 1974. *Endocrinology* 95:1019–25
140. Bullock, L. P., Bardin, P. L., Mowszowicz, I., Orth, D. N., Barthe, C. W. 1975. *Endocrinology* 97:189–95
141. Barka, T., Gresik, E. W., van der Hoen, H. 1978. *Cell Tiss. Res.* 186:269–78
142. Roberts, M. L. 1977. *Arch. Pharmacol.* 296:301–5
143. Roberts, M. L. 1978. *Biochim. Biophys. Acta* 540:246–52
144. Krieger, D. T., Hauser, H., Liotta, A., Zelenetz, A. 1976. *Endocrinology* 99:1589–96
145. Roberts, M. L., Reade, P. C. 1975. *Arch. Oral Biol.* 20:693–94
146. Gregory, H., Willshire, I. R. 1975. *Z. Physiol. Chem.* 356:1765–74
147. Starkey, R. H., Orth, D. N. 1977. *J. Clin. Endocrinol. Metab.* 45:1144–53
148. Dailey, G. E., Kraus, J. W., Orth, D. N. 1978. *J. Clin. Endocrinol. Metab.* 46:929–36
149. Nimmo, H. G., Cohen, P. 1977. *Adv. Cyclic Nucleotide Res.* 8:145–266
150. Carpenter, G., King, L. Jr., Cohen, S. 1978. *Nature.* 276:409–10

Ann. Rev. Biochem. 1979. 48:217–50

PROBES OF SUBUNIT ASSEMBLY AND RECONSTITUTION PATHWAYS IN MULTISUBUNIT PROTEINS

◆12008

Fred K. Friedman and Sherman Beychok

Departments of Chemistry and Biological Sciences, Columbia University, New York, New York 10027

CONTENTS

217

PERSPECTIVES AND SUMMARY

Single polypeptide chains assemble into complexes that range in complexity from the simple oligomeric proteins with molecular weights of less than 100,000 to supramolecular complexes with molecular weights in the millions. Paralleling the magnitude of this variation in structure is the diversity of function. The largest complexes include the filamentous and fibrous proteins that are responsible for structural integrity and motility, such as collagen, actomyosin, and microtubules. Intermediate in size are the largest assemblies of globular proteins, such as multienzyme complexes, ribosomal subunits, and viral coats. At a more fundamental level and more numerous are the smaller multisubunit proteins considered in this review, that play a central role in a wide array of cellular activities.

The factors dictating the biological activity of a completely synthesized and folded polypeptide chain are its fine structure and molecular state, both of which are in turn influenced by environmental conditions. By manipulating various environmental parameters, such as concentration of an appropriate ligand, the cell regulates molecular activity. Since many globular proteins are oligomeric, knowledge of protein-protein interactions is essential for the understanding of cellular function at the molecular level. To this end, there has been considerable research on the characterization of subunits in both monomeric and polymeric states, and on correlation of state of aggregation with function. Recently, studies have focused on the mechanism of assembly of several well-defined multisubunit proteins. The identification of assembly intermediates, defined here as species lacking the full complement of subunits present in the native form, without regard to minor changes in tertiary or quaternary interactions, may demonstrate the existence of specific, but not necessarily exclusive, assembly pathways. In a few cases, these assembly intermediates have been isolated, and limited characterization achieved. In other studies, the existence of intermediates can only be inferred from kinetic analysis. Studies on the dynamics of protein-protein interaction complement the much more extensive work pertaining to the thermodynamic parameters of assembly. Knowledge of both these aspects of protein chemistry is essential for a proper assessment and evaluation of the forces responsible for the manner in which a polypeptide chain expresses itself through interaction with other chains and with its environment.

Assembly of multisubunit proteins must be viewed in the broader context of the general problem of protein folding. Identical physical principles govern the chemistry of single and multichain proteins, permitting a common methodology in their study. It is sometimes difficult to determine whether an alteration in some property of a subunit originates from a change in the environment, or from interaction with neighboring subunits.

Renaturation is often monitored by observing the rate of recovery of a physical parameter of the system. Unfortunately, evaluation of the (protein) concentration dependence of the rates is frequently omitted despite the fact that such measurements may reveal the refolding and reassembly processes responsible for the observed change. Admittedly, such simple solutions are often precluded by the complexity of the system, in which the numerous interacting components can alter structure at different levels. Still, such work is necessary for the establishment of the molecular basis of the structure-function relationship.

Related to the topic under review are the studies on the pathway and kinetics of recovery of a specific tertiary structure of single chain proteins, and the search for intermediates in their folding. These studies, as well as determinations of structure at an atomic level, are now extensive and numerous enough for the emergence of theories of folding and prediction of structure by empirical methods, both of which should be applicable to reconstitution of subunits. The concept of interacting nucleation centers in a single polypeptide chain is analogous to the association of two prefolded chains; the difference between the two is the covalent linkage in the former case. The relative ease with which oligomers can be disassembled and the constituent subunits manipulated prior to reassociation, provides the protein chemist with a valuable tool and model with which to investigate interactions between structured regions within a single chain. This will aid in the pursuit of the ultimate goal of defining the pathway by which disordered polypeptide chains are transformed to active oligomers.

SCOPE

Research related to our topic has been reviewed from various viewpoints. The coverage includes protein folding (1–6), prediction of structure by empirical methods (7), and determination of structure by X-ray diffraction studies (8, 9). The oligomeric proteins and their subunit structure have been tabulated (10, 11). Reviews on various aspects of the structure and function of multisubunit proteins are available (11–18).

We review the multisubunit assembly problem from a different perspective than previous reviewers. Most research has focused on recovery of biological and/or physical properties as probes of refolding and reassembly. As pointed out by Wetlaufer & Ristow (1), the first of these is inherently more informative and should be the preferred method. We believe that for multisubunit proteins, the very identification of an intermediate species by both techniques, as well as the evaluation of its structural and functional properties relative to the initial and final states of the system, is in itself an informative probe in the determination of assembly pathways. Hence we consider multisubunit proteins in which specific intermediates have been isolated or inferred. The thermodynamic stability of an intermediate is determined by the rates of its association from, and dissociation to, lower molecular weight species. We therefore consider both assembly and disassembly processes. Examples of current research in the field are presented. Our approach is that of a selective rather than exhaustive survey of the literature.

The scope of this article is limited to the simplest oligomers. A recent review on the assembly of multisubunit respiratory proteins is available (19). Larger assemblies that have been reviewed and are not covered in this article are glutamate dehydrogenase (20), microtubules (21), contractile proteins (22), and viruses (23, 24).

In this article, monomer is defined as a single polypeptide chain; oligomer as a species consisting of two or more monomers; and subunit as a separable, observable component of the native oligomer, which may be either a monomer or oligomer.

GENERAL CONSIDERATIONS

Multisubunit proteins, generally, are functionally more versatile than single chain proteins. When the individual subunits catalyze distinct consecutive reactions, such as occurs within the multienzyme complexes, the very proximity of the reaction sites tends to enhance activity by providing higher local concentrations of substrate about the active site of the subunit catalyzing the second reaction. Also significant are the effects on activity induced through intersubunit contacts, especially the highly important allosteric interactions. Before studies of the assembly of oligomers are initiated, the constituent subunits and their individual polypeptide chains must first be separated and characterized, physically and functionally. Careful attention must be paid to the establishment of conditions where the subunits are sufficiently stable, since a medium that maintains the integrity of the oligomer might not suffice for the isolated subunits. Dissociation of an oligomer often yields oligomeric species, the very existence of which suggests a degree of stability that may confer on them the role of assembly intermediate.

The difference between the oligomer and the sum of the isolated subunits with respect to suitable observable properties must be determined. Standard methods are then employed to monitor assembly after mixing of the different subunits. In some cases, noncovalent association, at the protein concentrations frequently used, is so fast that stopped flow or relaxation methods are required. However, conventional techniques utilizing such sensitive monitoring techniques as enzymic activity can sometimes be used if the rate is reduced by lowering the protein concentration.

The identification of an intermediate species suggests, but is not evidence for, a particular pathway(s) of assembly. The simplest interpretation is that it lies directly on the pathway from initial to final state, but it may also be an intermediate not directly on the assembly pathway. These two alternatives can be kinetically distinguished from one another (25, 26). Further complicating the interpretation is the possibility that intermediates that lie on the major assembly pathway may rapidly convert to the product and escape detection. In summary, observation of particular intermediates is not sufficient, but must be coupled with kinetic data, for the evaluation of alternative pathways.

NONCOVALENTLY ASSOCIATED MULTISUBUNIT PROTEINS

Proteins Composed of Nonidentical Subunits

Interactions between the subunits in the active oligomer are elucidated by studies on the individual, isolated components, and on multisubunit intermediates, prepared either by dissociation of the whole molecule or by mixing various combinations of subunits.

ASPARTATE TRANSCARBAMOYLASE   Aspartate transcarbamoylase (ATCase) regulates pyrimidine biosynthesis in *Escherichia coli* by catalyzing the formation of carbamoylaspartate from carbamoyl phosphate and aspartate. The enzyme exerts a fine control on this important metabolic pathway, with its allosteric behavior, as evidenced by its homotropic interaction with aspartate, and heterotropic interaction with ATP and CTP. It consists of six chains that are regulatory (r), and six that are catalytic (c) in function. Their molecular weights are respectively 17,000 and 33,000. Structural studies on ATCase have determined the spatial arrangement of the chains (27, 28): the c chains are organized as a pair of trimers oriented parallel to one another and bonded through three r dimers.

Treatment of ATCase with mercurials dissociates the multimer into its c_3 and r_2 subunits. Catalytic species smaller than trimer have not been detected under nondenaturing conditions. Although the regulatory subunit is a dimer (29), exchange experiments with its succinylated derivative indicate a rapidly reversible equilibrium with the monomer (30). The catalytic subunit contains the active site and exhibits a simple hyperbolic saturation curve that is insensitive to ATP or CTP. The regulatory subunit is inactive but binds heterotropic effectors. Recombination of the separated subunits results in recovery of quaternary structure and function (31).

One method used to produce intermediates is to react one subunit with a large excess of the other (32–34). When excess r_2 is added to a dilute solution of c_3, a complex with greater activity than c_3 is formed (32, 33).

At lower r_2 concentrations the activity of the mixture eventually decreases with time to that corresponding to ATCase. An enzyme lacking one catalytic trimer subunit, c_3r_6, was identified as the complex, from consideration of the maximum number of binding sites for r_2 on c_3, and from electrophoretic (34) and sedimentation studies (33–34). The enhanced activity of this intermediate serves as a useful probe for further studies. Like the catalytic subunit, but unlike ATCase, the complex displays Michaelis-Menten kinetics with aspartate (32, 33). Addition of CTP or ATP has no effect on activity. From the increase in activity upon addition of a specified amount of r_2, the association constant between r_2 and its catalytic bonding partner was determined in the presence and absence of several effectors (35). Neither CTP nor ATP alters the association constant.

The aforementioned studies involve relatively low concentrations of subunits. At higher concentrations and nonsaturating levels of r_2, conversion of c_3r_6 to c_6r_6 and c_3 is detected on sucrose gradients, and their identities verified from the enzymic behavior in the presence of the substrate aspartate or the inhibitor CTP (33). c_3r_6 and its conversion to c_6r_6 were also monitored electrophoretically with a technique involving radioactively labeled subunits (34). The activity decrease accompanying reconstitution of ATCase from c_3 and r_2 follows second order kinetics. The presence of either ATP or CTP decreases the rate. Furthermore, the rate increases with c_3, but decreases with r_2 concentration at the concentrations studied. A more recent investigation of the assembly demonstrated a more complex relationship between the rate and r_2 concentration (35). These observations were rationalized by a proposed mechanism that involves an equilibrium mixture of species with general formula c_3r_{2n} (where $n = 0, 1, 2,$ or 3) that reacts with c_3r_{6-2n} to yield c_6r_6. At high r_2 levels, the equilibrium shifts to high c_3r_6, which is presumably sterically inhibited from combining with any r_2-containing species.

With catalytic subunit in excess during recombination, a c_6r_4 complex was identified on the basis of molecular weight, electrophoresis, and hybridization with succinylated regulatory subunits (36, 37). The complex can also be obtained by partial dissociation of the whole enzyme (28, 38). This species displays the homotropic and heterotropic interactions of whole ATCase, although to a lesser extent (28, 36–38). It reacts with r_2 to produce c_6r_6 and lacks the stability of the native enzyme, slowly converting to c_6r_6 and c_3 (37, 39). This process is accelerated by the bisubstrate analogue, N-(phosphonacetyl)-L-aspartate, and was proposed to proceed by a pathway involving disproportionation of c_6r_4 to individual subunits which subsequently assemble into c_6r_6 (39). The reaction of c_6r_4 with r_2 to yield c_6r_6 was monitored electrophoretically and obeyed second order kinetics (34).

The subunit-deficient species that have been detected behave similarly in that partial disassembly occurs prior to formation of c_6r_6. Both intermediates exist due to kinetic factors. Since they do not lie directly on the pathway from separated subunits to c_6r_6, they are analogous to the incorrectly folded intermediates in the refolding of single chain polypeptides (25, 26).

HEMOGLOBIN   The hemoglobin (Hb) molecule is best described by the formula $(\alpha\beta)_2$. Studies of the tetramer, dimer, and separated chains have helped elucidate the relationship between structure and allosteric behavior (40). The structure and state of association are linked to pH, the presence of heme ligands, and organic phosphates. The unliganded deoxy-Hb is in a constrained state in which the tetramer is favored over dimer much more than in liganded hemoglobin, which is in a relatively unconstrained conformation. Hemoglobin reactions can be monitored by taking advantage of the sensitivity of the heme absorption to binding of ligands, conformation of subunits, iron oxidation state, and environmental factors.

The large difference in the tetramer stabilities of the deoxy and liganded hemoglobins allows study of association-dissociation reactions by perturbation of an equilibrium with the rapid addition or removal of a ligand. When HbO_2 is deoxygenated with dithionite, the deoxy dimers associate, resulting in a second-order spectral change in the Soret region (41–44). Kinetic parameters can also be determined by perturbing the dimer-tetramer equilibrium with a pH change, and monitoring the resulting spectral changes (45). Dissociation of tetramer may be followed by trapping released dimers with haptoglobin, which exclusively and irreversibly binds dimer, and monitoring absorption (42–44) or fluorescence quenching (46, 47). The rate of tetramer dissociation has also been determined by hybridization studies in which hybrid dimer formation is followed electrophoretically (48, 49).

Organic phosphates influence the structure and function of hemoglobin (40, 50). In the presence of 2,3-diphosphoglycerate, which stabilizes the tetrameric state, an accelerated rate of association of deoxygenated dimers is observed (51). However, the stabilization of tetramer by the phosphate was attributed primarily to its inhibitory effect on the dissociation rate. A related study focused on the effect of inositol hexaphosphate on the rate of association of deoxy Hb dimers after a pH drop (52). As ligand concentration increases, the rate initially increases, reaches a maximum, and then declines to the value observed in the absence of ligand. The proposed mechanism assigns a greater reaction rate for the combination of a liganded with an unliganded dimer than for combination of two identical dimers. The molecular interpretation ascribes the rate difference to electrostatic interactions: a bound phosphate converts a positively charged region to a negatively charged one, which facilitates the approach of surfaces of unlike charge.

Although most of the assembly work has involved dimers, similar studies have been performed with the individual chains. These are obtained by treating the whole molecule with p-mercuribenzoate (PMB), which is subsequently removed. The alpha chains exist in a monomer-dimer equilibrium (53, 54), and the beta chains in a monomer-tetramer equilibrium (54). The initial nonmonomeric states increase the complexity of the assembly mechanism, as evidenced by the Soret spectral changes resulting from mixing of deoxygenated α- and β- chains (55, 56). The half time is concentration-dependent, but a lag phase is observed. A more recent study of deoxy-Hb assembly demonstrated similar kinetics (57). At very low concentrations, the data are consistent with a two step mechanism involving formation of $\alpha\beta$-dimers from the separated chains, followed by association of dimer to tetramer. At higher concentrations, beta tetramers predominate and their dissociation to monomers become the rate-determining step in assembly. Preincubation of beta chains with moderate concentrations of inositol hexophosphate decreases the rate, an effect presumably due to stabilization of the tetramer by the phosphate. High phosphate levels lead to an increased rate, which suggests that the phosphate binds monomers and thereby makes the tetramer less energetically favorable. Modification of free sulfhydryls with PMB provides another means of resolving assembly into its component steps. With α-PMB, in which the lone sulfhydryl at the $\alpha_1\beta_1$ interface is blocked, the rate of $\alpha\beta$ dimer formation is retarded. Tetramer formation, which involves formation of the unmodified $\alpha_1\beta_2$-interface, proceeds with the same velocity. In β-PMB, sulfhydryls in both contact regions with α chains are blocked, resulting in reduced rate for both dimer and tetramer formation.

The influence of the heme group in the refolding and reassembly processes is currently being assessed. The individual heme-free α- and β-globin chains differ structurally from the native chains (58). The globin chains do not recombine at neutral pH, whereas apohemoglobin is a dimer, a kinetic barrier apparently inhibiting interconversion between the dimer and separated subunits. However, association can be induced at lower pH (Y. K. Yip, J. I. Luchins, and S. Beychok, manuscript in preparation). Mixing of either heme-containing chain with its corresponding heme-deficient globin chain yields a hybrid dimer (59). Addition of heme to this half-filled intermediate, or to a mixture of both globin chains, reconstitutes native Hb, as judged by a series of rigorous functional and structural tests (60). The rates of the various recombination reactions can be followed through spectral changes. In particular, the resulting large changes in secondary structure make the processes amenable to monitoring by stopped-flow far-UV circular dichroism (61).

RNA POLYMERASE DNA-dependent RNA polymerase catalyzes the formation of RNA from ribonucleoside triphosphates. The *E. coli* enzyme consists of the four polypeptide chains α, β, β', and σ, with molecular weights of 40,000, 155,000, 165,000, and 90,000 respectively. The $\alpha_2\beta\beta'\sigma$ holoenzyme, the $\alpha_2\beta\beta'$ core enzyme, and the individual subunits have been partially characterized (62). Tentatively, certain functional attributes of the chains have been inferred from the observations that the β' subunit binds DNA, β binds several transcription-inhibiting antibiotics, and σ is required for proper initiation of transcription. In vivo as well as in vitro assembly have been studied (63).

Although treatment with PMB or low concentrations of urea partially dissociates the enzyme (64), assembly studies are typically carried out by denaturing the protein with urea or guanidinium hydrochloride, removing denaturant, and monitoring reactivation and recovery of physical parameters. When the denaturant is rapidly removed by dilution, recovery of protein fluorescence and circular dichroism precedes reactivation (65). When guanidine is removed from core enzyme at low temperature, the renatured product is structured but inactive (66). Sedimentation and molecular sieve chromatography experiments indicate that the preparation may be a mixture of subassemblies. The inactive preparation is partly functional, retaining ATP (67), DNA (67); P. D. Orphanos, B. J. Levine, B. Fischman, and S. Beychok (manuscript in preparation), and rifampicin binding ability (67, 68). Differences between the renatured-inactive and native enzymes are further confirmed by their different structural parameters (66) and resistance to proteolysis (68). Raising the temperature results in a time-dependent full reactivation and recovery of structure (66). When enzyme with β subunit proteolytically cleaved into two fragments is subjected to this denaturation-renaturation procedure, both fragments are incorporated into the reassembled enzyme (69, 70).

To identify intermediates in reconstitution, the stabilities of various subassemblies must first be evaluated. The isolated subunits are mixed in different proportions, and examined for evidence of recombination. When glycerol (71) or sucrose gradient (72) centrifugation is performed on various combinations of subunits, only an $\alpha_2\beta$-assembly is observed. An $\alpha\beta$-complex is independently confirmed by proteolytic digestion experiments in which preincubation with α-subunit protects β from cleavage (70). β' is digested to a similar extent in the presence or absence of α-subunit (70), which is consistent with the lack of complex formation between β' and

α. Addition of β' to $\alpha_2\beta$ reconstitutes core enzyme. Treatment of the enzyme with cyanate renders β', but neither α nor β, incompetent in reassembly (73). Therefore, when unreactive β' is used, the $\alpha_2\beta$-intermediate accumulates in reconstitution mixtures.

The kinetics and extent of reactivation of holoenzyme have been monitored after initiation by addition of β to a solution containing the remaining subunits (72). Aliquots at different times are withdrawn, further reconstitution inhibited by addition of heparin, and enzymatic assays carried out. The profound influence of temperature on reactivation of previously dissociated but nonisolated core subunits (renatured-inactive enzyme) is also apparent here: both rate and yield increase with temperature. Association of subunits is not the rate-limiting step in reactivation (72, 74). However, an initial lag phase at lower enzyme concentrations suggests a rapid recombination prior to recovery of activity (72). The assembly of $\alpha_2\beta$ at 0°C was monitored by a competition experiment: aliquots from a mixture of α- and rifampicin-resistant or -sensitive β were added to a mixture of β-subunit of opposite susceptibility, β' and σ; after incubation at higher temperature and enzymic assay in the presence of rifampicin, the extent of recombination was calculated. Assembly of $\alpha_2\beta$ was complete in two minutes. When the aliquot was mixed with competing β-subunit, and β' and σ added afterward, the activity level indicated a similar extent of association of α with the two types of β. These observations led to a proposed mechanism in which α and β are in rapid equilibrium with $\alpha_2\beta$, and the complex is stabilized by association with β'. The resulting core then incorporates σ to yield holoenzyme. Complete assembly only occurs at sufficiently high temperature. At low temperature, sedimentation analysis of the reconstituted enzyme reveals a peak sedimenting more slowly than native enzyme but composed of stoichiometric amounts of the subunits. Although these observations were attributed to conformational change, they are also consistent with a mixture of various subassemblies (67).

Full reactivation of core enzyme has been achieved in the absence of σ factor in some studies. Other investigations have suggested that σ stimulates full recovery of activity. This discrepancy may arise from the use of different procedures for denaturation or renaturation (75) but the evidence is still incomplete.

The smallest subassembly considered thus far is $\alpha_2\beta$. It can be formed in a one-step combination with α_2 or by sequential addition of monomer to β. The molecular state of the purified α subunit depends on its concentration. Because it associates with β at a concentration where the dimer predominates, it has been assumed that the dimer is the immediate precursor (75). However, another plausible mechanism involves dissociation of nonreactive dimers to reactive monomers. These alternative mechanisms might be kinetically distinguishable.

The σ subunit is in a dynamic equilibrium with core enzyme (76). The association constant has been estimated by light scattering experiments (77). The kinetics of association of core enzyme with σ subunit labeled with the fluorescent probe N-(1-pyrene) maleimide was determined (78), and the data shown to be consistent with a mechanism involving a rapid combination and a subsequent conformation change. The temperature dependence of the second process is complex, exhibiting a nonlinear Arrhenius plot.

TRYPTOPHAN SYNTHETASE   This complex catalyzes the final reactions of tryptophan biosynthesis. The $\alpha_2\beta_2$-enzyme is in equilibrium with its α- and β-subunits (79). α is normally present as the monomer. A dimer can be produced by a denaturation-renaturation procedure (80, 81) and can bind to β_2 in this state (80). β is usually present as a dimer, and denoted by β_2, although monomers can be detected by sedimentation analysis (82). The subunits are distinct in function: α catalyzes the conversion of indoleglycerol phosphate to indole, and β_2 catalyzes the formation of tryptophan from indole. The complex exhibits a great enhancement in these activities over that of the isolated subunits. Using level of activity to assess the extent of recombination, the equilibrium constant was determined from static measurements and kinetic data (79). Dissociation is readily monitored by displacement of a mutant inactive α-subunit by an active subunit. The data are consistent with identical and independent binding sites on the β_2-subunit. Independent catalytic sites on β_2 are also suggested by a hybridization study in which a dimer of active β and a modified inactive β exhibits half the activity of the native dimer (83). However, differences between β and β_2 do exist, as indicated by differential binding of pyridoxal phosphate (82). α reacts with excess β_2 to produce a stable $\alpha\beta_2$-species (84), which may therefore be an assembly intermediate. Assembly apparently proceeds by the sequential addition of α to β_2.

To further probe the intersubunit interactions in the tetramer, β_2 was proteolyzed to two major fragments and their properties determined (85). The structural integrity and spectral properties of the dimer are retained, although the number of coenzyme binding sites is slightly reduced. Moreover, denatured and separated fragments recombine upon renaturation. The nicked dimer is inactive and does not bind α.

OTHER PROTEINS   The cyclic AMP-dependent protein kinase from bovine heart muscle consists of two regulatory (R) and two catalytic (C) subunits. Cyclic AMP activates the enzyme by binding to R and promoting dissociation to R_2 dimers and C monomers; its removal results in reassociation. In addition to cyclic AMP itself, phosphorylation of R inhibits recombination (86–88), as does ATP (88). The holoenzyme and its individual subunits can be rapidly resolved by chromatographic means, which provides a method for following the kinetics of assembly (87, 88). The identification of an R_2C intermediate (88) indicates that assembly proceeds by sequential addition of C to R_2.

Phage Qβ RNA replicase consists of the following four subunits, in order of decreasing molecular weight: *E. coli* 30S ribosomal protein S1, viral protein, protein biosynthesis elongation factors EF-Tu, and EF-Ts. Dimers of the two larger and two smaller chains are separable (89), and have been shown to be in equilibrium with the whole molecule by a cross-linking study (90a). Maximal reactivation is obtained when the separated complexes are denatured in urea prior to reconstitution in a nondenaturing medium (90b). The protein concentration has only a minor effect on the renaturation kinetics (91). Renaturation of EF-Tu is apparently the rate-limiting step in reassembly, as evidenced by similar rates of renaturation for the replicase and EF-Tu, and by the observed enhancement in polymerase activity when undenatured EF-Tu is present in the renaturing replicase mixture (91).

Proteins Composed of Identical Subunits

In oligomeric proteins consisting of more than one type of subunit, the isolated subunits may be partially renatured and structured prior to recombination. When a protein contains identical subunits, interpretation of assembly data is often more complicated because maintenance of the nonassociated state requires the same denaturing conditions that perturb secondary and tertiary structure. These added effects are often reflected in complex renaturation kinetics, which can in principle be resolved into elementary first and second order processes if carried out over a sufficiently wide concentration range. Failure to do so may result in discrepancies between different studies with respect to the kinetic order of a renaturation process: identical denaturation-renaturation procedures performed on a particular protein may exhibit either first or second order renaturation kinetics at sufficiently high or low protein concentrations. Investigations of several oligomeric enzymes have demonstrated a general feature of renaturation of this kind: major recovery of structure does not occur simultaneously with reactivation (92, 93).

LACTATE DEHYDROGENASE Lactate dehydrogenase (LDH) is a nonallosteric enzyme composed of four identical subunits of molecular weight 36,000. The rate-limiting step in the renaturation of the acid-denatured beef enzyme depends on time of exposure to low pH. Brief exposure results in second order kinetics, as measured by ability to bind co-enzyme (94) and recovery of activity (95). The first-order kinetics of reactivation that are observed following prolonged incubation at low pH has been attributed to the acid-induced formation of unfolded species which either irreversibly denature, or refold slowly, upon restoration of neutral pH (96).

The renaturation of the acid-denatured pig heart (H_4) and muscle (M_4) isozymes is frequently monitored through the recovery of activity and fluorescence emission properties. The renaturation process includes both refolding and reassembly steps. The kinetics of renaturation may be strictly first order (95, 97, 98), second order (99, 100), or a combination of first order and second order steps (100, 101), depending on the isozyme under study and such experimental variables as protein concentration. The kinetic parameters are the same for enzyme previously dissociated using different denaturant systems, which suggests common intermediates in the rate-determining renaturation steps (101).

Hybridization studies constitute a valuable tool for the identification of assembly intermediates. Assembly of subunits of an LDH isozyme is followed by withdrawing aliquots at appropriate times, quenching further association by addition of an excess of the other isozyme, and identifying the oligomeric structure electrophoretically (95, 98). Dimers are found to be the predominant intermediate, with only small quantities of trimers observed.

ALDOLASE   This nonallosteric tetrameric enzyme from rabbit muscle has a monomeric molecular weight of 40,000. Dissociation requires denaturing conditions with concomitant perturbation of secondary and tertiary structure. Renaturation, accordingly, is a multistep process, involving first and second order kinetics.

When guanidine-denatured aldolase is transferred to a nondenaturing medium, activity reappears over a longer time period than reformation of secondary structure, which is complete within one minute, as determined by optical rotatory dispersion (92, 93, 102). Light scattering and sedimentation results indicate rapidly formed refolded monomers and dimers as intermediates (102). The rate of the subsequent reassociation to tetramer parallels recovery of activity and follows first order kinetics, which implicates an obligatory unimolecular isomerization prior to rapid recombination of chains. A structured monomeric intermediate is also observed in renaturation of acid-denatured aldolase (103).

The difference in stability between the tetramer and smaller species in 2.3M urea, and of folded and unfolded species in the presence of trypsin provide useful probes of assembly of guanidine-denatured aldolase (104,

105). By simultaneously monitoring recovery of activity with assay systems containing or lacking urea, the levels of intermediate species and reconstituted tetramer can be calculated. Rate constants corresponding to the refolding and reassociation processes are obtained by fitting the kinetic data to postulated models (104, 105).

Reconstitution of aldolase denatured at pH 2.3 yields a product indistinguishable from the native enzyme in activity and using a number of physiochemical criteria (106). The renaturation process has been followed through protein fluorescence and activity (107). First- and second-order processes are observed, and a similar reactivation mechanism is observed after denaturation by guanidine (105).

GLYCERALDEHYDE-3-PHOSPHATE DEHYDROGENASE   The interactions between the yeast (Y_4) and rabbit (R_4) glyceraldehyde-3-phosphate dehydrogenases (GPD) have been studied. The predominant hybrid formed upon mixing the two is R_2Y_2 (108). Analysis of the rate and extent of reaction leads to a mechanism involving a rate-limiting dissociation to a dimer intermediate (109) which is influenced by NAD^+ and $NADH$ (110). When a mixture of the enzymes in the monomeric state is transferred to an associating environment, the same symmetrical hybrid is formed, which suggests thermodynamic control. R_2Y_2 would also be expected to be energetically favored over R_3Y or RY_3 because of the greater number of interactions between identical subunits (109). That dimer is an intermediate in rabbit GPD assembly is consistent with an ultracentrifuge study, which indicates a dimer-tetramer equilibrium (111). Moreover, the kinetics of inactivation of yeast GPD are consistent with a model that incorporates a dimer intermediate (112). A combination of sedimentation experiments and hybridization with succinylated derivatives of rabbit GPD indicates isologous association of dimers (113).

To investigate the complete reassembly of yeast GPD, the subunits were dissociated with urea (114) or guanidine (115); upon dilution a concentration-dependent process is observed. Study of reassociation in the absence of extensive refolding is made feasible by utilizing ATP, low temperature, and appropriate solvent conditions to induce dissociation of tetramer to structured monomers (116, 117) or dimers (118a). Recombination is effected by warming, and is concentration-dependent. The reactivation kinetics are similar to that observed in renaturation of guanidine denatured enzyme, which suggests common intermediates (117). A similar procedure applied to the rabbit muscle enzyme yielded similar results, although a rapidly equilibrating monomer-dimer mixture was observed (118b).

PHOSPHOFRUCTOKINASE   This important regulator of glycolysis is an allosteric tetrameric enzyme composed of four identical polypeptide chains of molecular weight 80,000. The extent of association is influenced by pH (119–121), temperature (122), and various activators and inhibitors (123). Species smaller than tetramer do not exhibit significant activity (119, 120, 122).

A structural study has been carried out by treatment of dilute solutions of the protein with the bifunctional cross linking reagent, dimethyl suberimidate, and analysis by sodium dodecyl sulfate gel electrophoresis (124). Only even numbered oligomers are observed, which implicates the dimer as the basic polymerizing unit that mediates the equilibrium with larger aggregates.

Treatment of phosphofructokinase with guanidine has multiple consequences, as determined by fluorescence, enzymatic activity, and light scattering measurements (125). In addition to disproportionation, prolonged exposure to denaturant leads to irreversible formation of an unfolded monomer, which is reminiscent of the situation observed with acid-denatured lactate dehydrogenase (96). An extension of this work explored the kinetics of reassembly (126). Steps corresponding to the dissociations of tetramer to dimer and of dimer to monomer were resolved fluorometrically (125). The latter process exhibits a negative temperature dependence, which suggests hydrophobic interactions in the stabilization of dimer. The reassembly follows second order kinetics and is also consistent with a scheme involving a dimer intermediate (126).

The cold lability of the enzyme also suggests a role for hydrophobic interactions (122). At low temperature, a pH drop promotes inactivation and dissociation. The inactivation data are also consistent with dissociation to dimer intermediate species.

HEMOCYANIN   The association-dissociation behavior of hemocyanin from the snail *Helix pomatia* has been studied under a variety of conditions. The whole molecule has a molecular weight of 9×10^6 and can be dissociated into half, tenth, or twentieth size molecules by appropriate alterations of pH, ionic strength, oxygen pressure, or concentration of Ca^{2+} (127). Owing to its large size, direct observation of the molecule's various conformational and oligomeric states is feasible with electron microscopy (128).

Light scattering and sedimentation experiments demonstrate the efficacy of NaCl in promoting dissociation of hemocyanin to half molecules (129). Several observations, however, indicate that a simple equilibrium is not operative under conditions of partial dissociation: the light scattering data obtained at different protein concentrations cannot be explained by the law of mass action, and the sedimentation patterns exhibit two discrete peaks instead of the single one expected of a rapidly equilibrating system. The rapid reversibility of association and dissociation suggests that the observed behavior is not due to metastable states. Microheterogeneity of the hemocyanin preparation was invoked to account for the results. A subsequent analysis of the data (130) led to a proposed mechanism in which some of the molecules are unreactive while the remainder participate in a reversible equilibrium. Relaxation experiments with the concentration and pressure jump techniques (131) yielded values for the forward and reverse rate constants that are consistent with the association constant calculated from the static experimental data.

Dissociation of hemocyanin to half and tenth size molecules by high pH also indicates microheterogeneity (132). The relative proportions of the various species in the dissociation mixture are independent of concentration. When isolated from one another they do not reequilibrate and differ from the whole molecule in susceptibility to pH-induced dissociation. Furthermore, the pH-dependent association-dissociation to tenth size molecules displays hysteresis. The existence of subpopulations of molecules with different pH sensitivities was proposed to account for these results. Highly cooperative interconversions between states render the concentration dependence negligible.

Assembly of hemocyanin is induced by the establishment of a suitable pH for reassociation of tenth size molecules (133). The accompanying molecular weight changes are large enough to permit following the reaction by turbidity changes. The first phase of the observed biphasic reaction is too rapid for detailed analysis, but the second, slower step can be monitored in the analytical ultracentrifuge. The sedimentation patterns indicate that half molecules as well as two other species, probably dimers and tetramers of the tenth size molecules, are assembly intermediates. The data were incorporated into a proposed scheme in which tenth size molecules rapidly associate to the intermediate species, which in turn undergo a slower combination to yield the whole molecule.

OTHER PROTEINS   When urea-dissociated *E. coli* L-asparaginase subunits are allowed to renature, rapid refolding is followed by recovery of the tetrameric structure. Dissociation (134) and reassembly (135) are monitored by changes in sedimentation, activity, and absorption and emission properties. The latter process is rapid, but its study is facilitated by reducing the rate via renaturation by dilution into solutions of various urea concentrations. Variable rates and extent of assembly are obtained, and reassociation is concentration-dependent. One notable observation is that monomer and tetramer, but no intermediate species, are observed in the sedimentation velocity profiles, which indicates that associations of chains is a highly cooperative process.

Renaturation of guanidine-treated triose phosphate isomerase dimer at different concentrations was followed by direct dilution into the enzyme assay mixture (136). Presence of the inhibitor phosphoglycolate at the higher enzyme concentrations slows the reaction sufficiently for activity measurements to be performed over a wider concentration range. A mechanism consisting of a unimolecular change, followed by a bimolecular step that is associated with activity, was proposed to account for the rate data. When the isomerase is renatured in the presence of subtilisin, which hydrolyzes the unfolded, but not the native, enzyme, an enhanced reactivation rate results. A possible mechanism was suggested in which a dimer composed of one refolded and one incorrectly folded monomer readily dissociates after hydrolysis of the latter, thus permitting a more rapid reassociation rate between correctly folded monomers.

The influence of kinetic factors on refolding and reassociation is exemplified by the renaturation of the tryptophanase tetramer after denaturation by urea (137). The extent of reactivation is concentration-dependent, higher concentrations resulting in lower recovery of activity and higher levels of aggregated species. Since the presence of foreign proteins does not influence the extent of aggregation, a model that involves competitive inter- and intrachain associations between structured segments of the polypeptide chain was proposed. To follow association independent of extensive renaturation, advantage was taken of the stabilizing effect of pyridoxal phosphate on the quaternary structure (138). The apoenzyme consists of inactive dimers and the holoenzyme is an active tetramer. Association of dimers is induced by added coenzyme, and the progress of the reaction is readily monitored by continuous assay of activity.

Another example of ligand-induced association is the binding of magnesium to enolase (139). Since the ion binds the dimer in preference to monomer, addition or removal of magnesium promotes association or dissociation, respectively. The reaction is monitored by absorbance changes, and exhibits both first- and second-order kinetic steps. Moreover, the dissociation process is not simply the reverse of association, but is more complex.

The monomers of formyltetrahydrofolate synthetase associate to the active tetramer upon introduction of monovalent cations (140, 141). In the case of the enzymes from *Clostridium cylindrosporum* and *Clostridium acidi-urici*, assembly exhibits bimolecular reactivation kinetics under appropriate conditions (141). Neither dimer nor trimer intermediates are detected by sedimentation analysis (140), which suggests a scheme in which a rate-limiting dimerization is followed by rapid formation of tetramer by either sequential aggregation of monomers or by condensation of dimers.

Monomer-dimer equilibria and kinetics of conversion of insulin, β-lactoglobulin, and α-chymotrypsin have been investigated utilizing release or uptake of protons (142). Monitoring was achieved by coupling the reaction to that of a pH indicator and following the spectral changes using rapid techniques. In this way, the forward and reverse rate constants for the dimerizations of these three proteins were determined. The kinetic parameters of α-chymotrypsin dimerization were also determined by monitoring the loss of active sites that occurs upon dimer formation, with proflavin dye as a spectral probe of availability of active sites (143). The agreement between the equilibrium constants obtained by this technique and by direct methods demonstrates the suitability of this probe because of its noninterference with the reaction of interest.

TABLE OF INTERACTION PARAMETERS OF MULTISUBUNIT PROTEINS

Table 1 lists the forward and reverse rate constants (k_1 and k_{-1}, respectively) and the association constant (K_{assoc}) where directly determined, for the listed reaction. Where the parameter has been determined by a number of investigators, or has been measured under a wide range of conditions of pH, temperature, ligand concentration, etc, the values presented are those obtained under conditions most nearly approximating physiological.

COVALENTLY ASSOCIATED PROTEIN SUBUNITS

A number of studies on single-chain proteins has focused on disulfide bond formation and its relation to the overall folding process (1, 3). Owing to the more limited conformational fluctuations accompanying covalent assembly of prefolded chains, a less complex process is expected. However, research on reoxidation of immunoglobulins reveals that disulfide bond formation is not a simple process, but involves subtle changes in conformation of nonadjacent regions of the molecule. The function of the disulfide is often evaluated by comparing the properties of the native and reduced proteins. The native protein is usually regenerated either by oxidation with oxygen, or through disulfide exchange with an added reagent.

Immunoglobulins

The vast literature on immunoglobulins attest to their significance and availability for study. Extensive research on their biosynthesis (144–146) complements and stimulates much of the in vitro work. Their various properties are well-defined (147, 148), so that they may serve alongside hemoglobin as a prototype for studies of multisubunit proteins. Immunoglobulin G (IgG) is the simplest member of this group of proteins. Other classes are defined according to the nature of one kind of chain in the tetramer subunit, and may include polymers of its basic structure. The mode of intersubunit bonding may also vary. We deal primarily with the in vitro assembly of human immunoglobulins. Research on the assembly of mouse myeloma proteins, both intracellular and in vitro, is presented only for comparative purposes.

IMMUNOGLOBULIN G   The IgG molecule may be visualized as a dimer of heavy (H) chains, with a light (L) chain bound to the amino-terminal half of each H chain, which yields a symmetrical oligomer with the formula H_2L_2. Bonding appears to be primarily noncovalent, although interchain disulfides confer additional stability on the tetramer. After reduction of these disulfides, structural integrity is maintained and the noncovalently bound tetramer is indistinguishable from native IgG by several physio-chemical criteria, thus demonstrating that noncovalent interactions are the dominant stabilizing forces. There is evidence however that the disulfides influence molecular flexibility and interaction of IgG with the complement system. The noncovalent and covalent components of in vitro assembly are conceptually and temporally largely independent processes and are treated separately.

Noncovalent assembly   To separate the chains, mild reduction of the inter-chain disulfides is followed by treatment with reagents that disrupt non-covalent interactions, such as aliphatic acids and/or urea. When not essential for subsequent bonding, the free sulfhydryls are inactivated by reaction with alkylating agents. This procedure contributes greatly to the stability of the separated chains. Heavy chain poses the greater storage problem because it spontaneously aggregates, although acid pH hinders this process. Powerful noncovalent forces maintain H chains as a dimer at all concentrations accessible to measurement. The L chain presents fewer sta-bility problems. Its molecular state may vary, depending on its source: usually a monomer-dimer equilibrium exists within some pH range, while outside this range, the monomer predominates [(149); F. Friedman, and S. Beychok, unpublished observations].

Owing to the insolubility of H chain under neutral conditions, most recombination experiments are performed in acid pH. The dynamics of subunit interactions have been studied utilizing intrinsic spectroscopic probes. The ultraviolet absorption difference spectrum between separated and recombined subunits was utilized to monitor association of autologous and heterologous chains (150, 151), and of L chain with the amino-terminal half (fragment) of H chain (Fd), which encompasses the L chain binding region (150). Association kinetics are second order in both cases, although a more rapid reaction observed with Fd' was presumably due to steric factors (150). Rates also vary among different myeloma proteins (150, 151),

Table 1 Interaction parameters of multisubunit proteins

Protein	Reaction[a]	Temperature (°C)	pH	k_1 (M^{-1} sec^{-1} × 10^{-4})	k_{-1} (sec^{-1})	K_{assoc} (M^{-1})	Reference
Aspartate transcarbamoylase[b] (c$_6$r$_6$)	c$_6$r$_4$ + r$_2$ → c$_6$r$_6$	22	7.0	10			34
	2c$_3$ + 3r$_2$ → c$_6$r$_6$	25	8.5	12–70			35
	c$_3$ + 3r$_2$ ⇌ c$_3$r$_6$	25	8.5			3.0 × 10^7	35
Hemoglobin ($\alpha_2\beta_2$)	2$\alpha\beta$ ⇌ ($\alpha\beta$)$_2$	24	7.0			3.0 × 10^{11}	211
deoxy[c]		20–22	7.0–7.4	43–73	5–83 × 10^{-6}		41–45, 51, 52
		20–24	7.0–7.4				42–44, 48, 49
	$\alpha + \beta \rightarrow \alpha\beta$	20	7.0				57
oxy	$\alpha \rightleftharpoons \alpha_2$	18	7.5	50		1.2 × 10^4	53
		21.5	7.4			8.4 × 10^3	54
	$\beta \rightleftharpoons \beta_4$	21.5	7.4			1.4 × 10^{16} (M^{-3})	54
RNA polymerase ($\alpha_2\beta\beta'\sigma$)	$\alpha_2\beta\beta' + \sigma \rightleftharpoons \alpha_2\beta\beta'\sigma$		8.0			1–100 × 10^6	77
		22	7.5	300	0.23	>3.1 × 10^9	78
Tryptophan synthetase ($\alpha_2\beta_2$)	$\alpha + \beta_2 \rightleftharpoons \alpha\beta_2 \overset{\alpha}{\rightleftharpoons} \alpha_2\beta_2$	37	7.8	2–60	2–48 × 10^{-4}	4–2,600 × 10^6	79
	$\beta \rightleftharpoons \beta_2$	20	7.3			4.7 × 10^5	82
	$\beta + \beta_2 \rightleftharpoons \beta_3$					6.3 × 10^4 (M^{-3})	82
cAMP-dependent protein kinase (C$_2$R$_2$)	CR → C + R	0	7.5		0.038		212
Lactate dehydrogenase muscle-M$_4$	4M → M$_4$	20	7.6	2.3			99, 100
heart-H$_4$	4H → H$_4$	20	7.6	0.5			100, 101

Protein	Reaction					Ref.
Aldolase (A_4)	$4A \rightarrow A_4$	0	7.5	0.14		105
		0	7.6	1.0		107
Glyceraldehyde 3-	$4Y \rightarrow Y_4$	15	7.6	4.5		115
Phosphate dehy-		25	8.5	7		117
drogenase	$Y_4 \rightarrow 4Y$	0	8.5		1.2×10^{-4}	117
yeast-Y_4	$Y_4 \rightarrow 2Y_2$	28	7.8		1.8×10^{-4}	109
rabbit-R_4	$2R_2 \rightleftharpoons R_4$	5	7.0		2×10^6	111
	$R_4 \rightarrow 2R_2$	28	7.8		5.9×10^{-5}	109
Phosphofructo-kinase (P_4)	$2P \rightarrow P_2$	23	8.0	0.79		126
Apotryptophan-ase (T_4)	$2T_2 \rightarrow T_4$	25	7.8	5	6.3×10^4	138
Enolase (E_2)	$2E \rightarrow E_2$	20	8.0	>20		139
Insulin (A_2)	$2A \rightleftharpoons A_2$	23	6.8	11,400	1.48×10^4	142
β-Lactoglobulin (A_2)	$2A \rightleftharpoons A_2$	35	3.7	4.7	2.1	142
α-Chymotrypsin (A_2)	$2A \rightleftharpoons A_2$	25	4.3	0.37	0.68	142
		25	3.9	0.95	1.9	143
Hemerythrin (A_8)	$8A \rightleftharpoons A_8$	25	7.0		3.4×10^{36} (M^{-7})	204b
Immunoglobulin G (H_2L_2)	$H_2 + L \rightarrow H_2L \xrightarrow{L} H_2L_2$	25	5.0–5.6	0.01–1		150–155
		20	7.5	600		156
	$2L \rightleftharpoons L_2$	25	5.5		$1-20 \times 10^5$	174
		25	7.5		10^4–10^5	149, 175
Immunoglobulin M $(H_2L_2)_5$	$2(HL) \rightleftharpoons (HL)_2$	22	8.0		4.3×10^5	196

a The overall reaction is listed. The actual mechanism may be a multistep process.
b Parameters refer to the formation of c : r bonding domains.
c Dimer-tetramer equilibrium constants for the liganded forms of hemoglobin under a variety of solvent conditions can be found in (19).

the observed rate constant in these studies ranging from 10^2–$10^4 M^{-1} sec^{-1}$. A microcalorimetry study indicated similar enthalpy changes accompanying the reactions of L chain with H chain or the Fd' fragment (150).

Similar kinetic data were obtained utilizing circular dichroism as a probe. The rate and equilibrium constants, as determined from ellipticity changes occurring upon association, increase with pH (152, 153). A pH-dependent equilibrium between differentially reactive H chain isomers was postulated. A decreased rate at higher KCl concentrations was attributed to direct interaction between the salt and L chains, since spectroscopic studies on the individual chains indicate a perturbation of L, but not H chain structure (154). Equilibrium studies revealed negative cooperativity in the reaction of H and L chains, but noninteracting, identical sites in the association of Fd with H chain (155). Since Fd and H react with L at the same rate in this study, their binding behavior was attributed to differences in dissociation rates. Separation of L from H chain apparently occurs more readily when the remaining H chain dimer binding site is occupied. These intersite reactions are presumably mediated by interactions between the carboxyl terminal half of the heavy chain.

Limitations in the spectroscopic methods confined the protein concentrations in the above work to the micromolar concentration range. Labeling the COOH-terminal cysteine of L chain with the sulfhydryl-specific fluorescent probe N-(iodoacetylaminoethyl)-8-naphthylamine-1-sulfonic acid provides a more sensitive method for detection of the subtle conformational changes accompanying association (156). This method offers the advantage that the location of the reported group is known. Also, at the low concentrations suitable for fluorescence measurements, the concentration-dependent aggregation of heavy chain at neutral pH is inhibited. The recombination reaction at pH 7.5 follows second order kinetics, with a rate constant of $6 \times 10^6 M^{-1} sec^{-1}$, which is far greater than previously measured for other immunoglobulins by alternative methods.

The noncovalent recombination experiments with H and L chains were all performed under conditions in which heavy chain is dimeric. The reactions thus involve sequential addition of light chain to H_2. Obviously this is not the only pathway of assembly. One can envisage mechanisms in which monomeric heavy chain participates, as in the covalent assembly schemes to be discussed. At present, instability of the monomer in the solvents that promote association precludes noncovalent assembly studies with this species; however, the Fd fragment is a good approximation for the monomer. The observed second order kinetics in the various studies demonstrates that in kinetic terms the two binding sites on H_2 are equivalent and noninteracting. The large variation in rate constants, from 10^2–$10^7 M^{-1} sec^{-1}$, for these structurally similar proteins, probably results from differences in primary structure of the variable region. The importance of amino acid sequence was demonstrated by competition experiments in which extensively denatured chains were renatured, and preferential association of autologous over heterologous chains was observed (157).

Studies with fragments of the chains help elucidate the role of the individual domains in the association process. Both constant and variable regions of L chain (COOH- and NH_2-terminal halves, respectively) combine with H chain (158) or its Fd' fragment (159), although the presence of variable region accelerates the association of constant region and Fd' (159), and enhances the association of constant region with H chain or Fd' (158, 159). The L chain variable region and its H chain complementary fragment, obtained from a mouse myeloma protein, can recombine into an active complex (160) and the kinetics of the renaturation process has been studied for the urea-denatured fragments diluted into neutral buffer (161).

Covalent assembly   Following the rapid noncovalent assembly, the proximities of the sulfhydryls permit formation of interchain disulfide bonds. Activation occurs through metal-catalyzed oxidation by oxygen in air (162–164), or through a thiol-disulfide interchange agent (162). Covalent assembly is monitored by transferring the protein to an oxidizing medium, withdrawing aliquots at appropriate times, quenching further oxidation by alkylating the remaining sulfhydryls, and analyzing the band patterns obtained from sodium dodecyl sulfate gel electrophoresis of the samples. Intermediates are identified by migration distance and quantified from the staining intensity. The six observed species correspond to those expected from consideration of the three possible assembly sequences:

1. $H + H \rightarrow H_2 \xrightarrow{L} H_2L \xrightarrow{L} H_2L_2$

2. $H + L \rightarrow HL \xrightarrow{HL} H_2L_2$

3. $H + L \rightarrow HL \xrightarrow{H} H_2L \xrightarrow{L} H_2L_2$

Analysis of the data can in principle assign specific assembly pathways for proteins.

The most direct approach is to perform reoxidation experiments on proteins in which disulfides are reduced and the chains unseparated. The IgG1 subclass, with two interheavy chain disulfides, reoxidizes with a half-time of about 80 min after the raising of the pH from 4.8–8.2 (162). The assembly pathways include HL and H_2L as principal intermediates. With an IgG4 molecule HL is predominant. Several environmental factors, such as temperature, pH, and metal ions influence the rate (162), as does pH of the medium prior to establishment of oxidizing conditions. For example, protein solutions transferred from a pH 3.2 to a pH 7.5 medium reoxidize more rapidly than protein initially exposed to a pH of 5.5 (163, 164). The low pH apparently alters the environment about the sulfhydryls, rendering them more reactive upon restoration of neutral pH. A conformational change is also supported by the observation of a UV absorption change which accompanies an increase in pH in a mouse protein assembly experiment (165). Even though environmental conditions influence the rate, the relative order of appearance of intermediates, and therefore of covalent bond formation, remains the same (163, 164). Factors intrinsic to the protein must therefore dictate its assembly pathway, while the medium affects the rate.

To examine covalent assembly and remove complications resulting from environmental variability, a method has been developed to eliminate time as a parameter in the analysis (163, 164). Instead, assembly is considered as a function of total free sulfhydryl during a particular stage of the reoxidation, with sulfhydryl titer measured directly using 5,5'-dithiobis-(2-nitrobenzoic acid). This mode of presenting the data proves to be more convenient for comparisons with various models of assembly (166).

Reoxidation studies have also been extended to include cases where the chains are separated prior to recombination and reoxidation (167–169). The patterns of assembly are similar to those in the unseparated chain experiments (168, 169). Since noncovalent association of chains occurs almost instantaneously, this result might have been anticipated. When H chain is reacted with excess L chain, the reaction rate decreases. Transient, nonproductive intermediates consisting of more than one L chain per H chain were postulated to account for the rate difference.

Resolution of the complex reoxidation process into its individual steps is a primary objective in these studies. Toward this end, experiments were performed in which L chains with irreversibly blocked sulfhydryls were mixed with H chains (168, 169). Interheavy chain bonds are then the only ones capable of forming, and reoxidation proceeds smoothly. Similar results are obtained when the variable region (NH_2-terminal half of L chain) rather than L chain is used. An interesting incidental finding is that the fragment enhances heavy chain solubility at neutral pH (169).

The number of alternative pathways of assembly complicates analysis of the data. Thus, although identification of particular intermediates indicates which pathways are operative, there are usually several. Figure 1 illustrates the possible pathways for IgGl, which is a relatively simple constrained system in that only correct (native) pairing of half-cysteines is considered. While quantitation of intermediates suggests the relative importance of different pathways, a rigorous kinetic analysis is required to establish the distribution of microscopic species at any time, and the necessary analytic

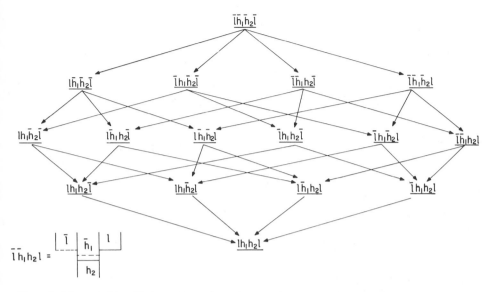

Figure 1   Theoretical reoxidation pathways for an immunoglobulin molecule with four interchain disulfide bonds. A four letter code is used to indicate the arrangement of closed and open disulfides in each species as illustrated by the schematic in the lower left. Specifically, each letter corresponds to one of the four disulfides in the molecule, and the absence or presence of a bar over the letter indicates that the bond is either oxidized or reduced, respectively. The horizontal solid and dashed lines of the immunoglobulin in the schematic indicate that a particular bond is either oxidized or reduced, respectively. Because the molecule is presumably symmetrical, no distinction is made between the two bonds involving L. The two inter-HH bonds are distinguishable, however, h_1 being proximal to and h_2 being distal to the N-terminal end of the H chain. The molecule in the inset, lh_1h_2l, is one in which (reading from left to right) one HL and one HH bond are reduced and the other two disulfides are oxidized. [Adapted from (166)].

solution to the problem is not available. The approach taken has been to compare the simulated data derived from theoretical models with experimental results, and the conclusions depend on the model under consideration. When no kinetic cooperativity in bond formation was assumed, data could be fitted to a model in which the ratio of susceptibility to oxidation of interheavy to heavy-light chain sulfhydryls was less than unity for human IgG1 and IgG4 (170), and greater than one for several mouse myeloma proteins (165). A model for in vivo assembly of mouse proteins allows comparisons with in vitro assembly (171). A more general model which incorporates kinetic cooperativity leads to the conclusion that bonds do not form independently of one another in human IgG1 (163, 164, 166). The heavy-light chain disulfide initially forms more readily than the interheavy chain bonds, but this difference in reactivity diminishes as reoxidation progresses (166). Formation of a second heavy-light chain disulfide is apparently hindered by a preexisting heavy-light or heavy-heavy bond.

Intracellularly, light chain is often present in excess over heavy chain. Several in vitro studies have focused on the covalent (169, 172, 173) and noncovalent (149, 174, 175) association of isolated light chain, and of its variable region (175, 176). In vitro, they react to form covalent dimers, and this occurs at least to some extent intracellularly. In vitro, these dimers rapidly react with heavy chains via disulfide interchange (169, 177), and assemble to H_2L_2 via the same intermediates produced when light chain monomer is the reactant.

To further explore the interactions between the thiol-disulfide reaction sites of IgG1, reduction of interchain disulfides in immunoglobulins was studied (178). All bonds are reduced with equal facility by dithiothreitol, a result that contrasts with the reoxidation behavior which indicates unequal facility for bond formation.

IMMUNOGLOBULIN M   The IgM molecule may be visualized as a cyclic pentamer of the basic four-chain immunoglobulin structure, in which the five subunits are disulfide bonded through their COOH-terminal regions. An additional component, J-chain (179), plays a role in assembly and is incorporated into the pentamer.

HL, H_2L_2, and polymers of the latter are the primary intermediates observed upon reduction (180–183) or subsequent reoxidation of a mouse myeloma IgM (182, 183). HL and H_2L_2 intermediates are also detected intracellularly (184, 185). Assembly thus appears to proceed by an HL pathway. The additional presence of H_2 and H_2L intermediates in an in vitro (186) and intracellular (187) study was attributed to the use of a cell line in which the intracellular excess production of L over H is not as great as that used in previous studies (187).

Both the covalent and noncovalent aspects of human IgM assembly have been examined. Several studies demonstrate that the intersubunit disulfides are more labile than the interchain bonds (188–190). Sedimentation patterns of the protein with different concentrations of reducing agent indicate several oligomeric intermediate species, which suggests that reduction of the first intersubunit bond in the pentamer facilitates subsequent cleavage of the remaining intersubunit bonds (189, 191). More extensive reduction cleaves the interchain disulfides (192, 193), and yields the same intermediates seen in assembly of IgG. The relative susceptibilities to reduction of the different disulfides were assessed with the aid of a theoretical model for IgM reduction (193). One interesting outcome of this study is that, based on the small difference in lability between the intersubunit and intrasubunit H–H bonds, two isomeric H_2L_2 reduction products are predicted: in one, two HL species within the same subunit are bonded through interheavy chain disulfides; in the second isomer, two HL units from neighboring subunits are bridged through the intersubunit disulfide.

Reduction and alkylation of IgM yields a noncovalently bound HL species (194–196) that is in equilibrium with H_2L_2 (196). The H–H interactions are apparently weaker than the H–L interactions, which suggests that noncovalent assembly proceeds via an HL intermediate. This observation is in contrast to that of IgG, which assembles by sequential addition of L to H_2. Furthermore, since HL is not strongly bound within a tetramer, but is relatively free, it may covalently assemble into one of the two tetrameric isomers postulated above (193). The progress of covalent assembly might then overlap that of noncovalent assembly, in sharp contrast to IgG assembly, where strong noncovalent interactions are rapidly formed and the tetramer assembles prior to reoxidation.

While an oxidation of extensively reduced IgM results in formation of disulfide bonds and active polymeric products (197, 198), heterogeneity in the size of reoxidation products is often observed (197–200). Several studies suggest the importance of J chain in the regulation of the assembly process (183, 200–202). The additional presence of a sulfhydryl-disulfide exchange system is most effective for efficient reassembly (183, 202). The reassembly data are consistent with a proposed model (191) in which J-chain polymerizes the subunits via a series of successive disulfide exchanges with their free sulfhydryls.

COMPARISONS OF PATHWAYS OF ASSEMBLY

Multisubunit proteins with at least four identical, or three nonidentical subunits may assemble by multiple pathways. The maximum number is readily determined from combinatorial considerations, although steric constraints may limit the possibilities. Thus, for example, reconstituted ATCase does not occur by direct association of two c_3r_6 intermediates. Highly specific interactions govern the affinity of the various species for one another, as demonstrated by the reconstitution of denatured oligomeric enzymes in the presence of foreign proteins (203). Hydrophobic interactions between subunits are a major stabilizing force, but complementarity between the amino acid side chains of the combining units probably play a key role in protein-protein recognition (204a).

The concepts of pathway and intermediate are applicable only when the initial and final states of the system are specified. Denaturant or ligand-induced dissociation of an oligomer often yields polymeric reactants. The observed assembly intermediates and inferred pathway in such cases may differ from those obtained with other initial states, such as one with monomeric reactants. Interpretation of reassembly data after denaturation of the protein is often complicated by refolding processes. With ligand-induced reassembly, the initial states are usually structured, thus obviating this difficulty.

Tetramers with identical subunits can associate by two alternative pathways:

$$1. \ P + P \rightarrow P_2 \overset{P}{\rightarrow} P_3 \overset{P}{\rightarrow} P_4$$

$$2. \ P + P \rightarrow P_2 \overset{P_2}{\rightarrow} P_4$$

Of the examples presented, the second scheme is predominant. When nonidentical subunits assemble, a greater variety of pathways is observed.

The various assembly pathways of multisubunit proteins are exemplified by the reaction mechanisms in Table 2. Only those processes involving intermediates directly on the assembly pathway are listed. Unimolecular isomerizations of a species, though potentially kinetically significant, are not included.

With L-asparaginase no intermediate species are observed (135). A similar all or none situation exists in the association-dissociation behavior of hemerythrin (204b), in which successive incorporation of subunits during assembly is apparently a highly cooperative process. In both cases, assembly must proceed through some intermediate state, but the instability of the intermediate species precludes detection by the methods used. These exam-

Table 2 Suggested pathways of noncovalent assembly of representative multisubunit proteins[a]

Protein	Subunit structure	Intermediates identified	Suggested assembly mechanism
Aspartate transcarbamoylase	$c_6 r_6$	$c_3 r_6, c_6 r_4$	1. $c_3 + n r_2 \rightarrow c_3 r_{2n}$ ($n = 0, 1, 2, 3$) $c_3 r_{2n} + c_3 r_{6-2n} \rightarrow c_6 r_6$ 2. $r_2 + $ excess $c_3 \rightarrow c_6 r_4$ $c_6 r_4 + r_2 \rightarrow c_6 r_6$
Hemoglobin	$\alpha_2 \beta_2$	$\alpha\beta$	$\alpha_2 \rightarrow 2\alpha, \beta_4 \rightarrow 4\beta$ $\alpha + \beta \rightarrow \alpha\beta$ $2\alpha\beta \rightarrow \alpha_2\beta_2$
E. coli RNA polymerase	$\alpha_2 \beta\beta'\sigma$	$\alpha_2\beta, \alpha_2\beta\beta'$	$\alpha_2 + \beta \rightarrow \alpha_2\beta$ $\alpha_2\beta + \beta' \rightarrow \alpha_2\beta\beta'$ $\alpha_2\beta\beta' + \sigma \rightarrow \alpha_2\beta\beta'\sigma$
Tryptophan synthetase	$\alpha_2 \beta_2$	$\alpha\beta_2$	$\alpha + \beta_2 \rightarrow \alpha\beta_2$ $\alpha + \alpha\beta_2 \rightarrow \alpha_2\beta_2$
cAMP-dependent protein kinase	$C_2 R_2$	CR_2	$C + R_2 \rightarrow CR_2$ $C + CR_2 \rightarrow C_2 R_2$
Lactate dehydrogenase	H_4	H_2	$2H \rightarrow H_2$ $2H_2 \rightarrow H_4$
Phosphofructokinase	P_4	P_2	$2P \rightarrow P_2$ $2P_2 \rightarrow P_4$

[a]See text for references.

ples illustrate an important point in analysis of assembly data in general, namely that failure to observe a species does not mean it does not exist. Rather, its existence may be inferred either indirectly, or by utilizing an alternative detection system.

RELATION TO FOLDING OF SINGLE-CHAIN PROTEINS

Structurally, a multisubunit protein is most obviously likened to a polypeptide chain with multiple domains, such as an IgG heavy chain. However, its study has a much broader application, extending to the simplest single-chain proteins. Recently, the concept of nucleation centers in folding has emerged (205, 206): rapidly formed structural regions within a chain are presumed to guide the course of subsequent folding by interacting to yield, eventually, a particular tertiary structure. Assembly of a multisubunit protein is the analogous process performed at a higher level of structure: separate monomers combine in the proper sequence to yield the intermediates that direct further assembly. Formation of tertiary and quarternary structure in this manner may be further correlated by research on recombination of fragments.

The measured rates of association of subunits allows an estimation of the rate of a similar association of the more constrained structures within a chain. Owing to the rapidity of formation of supersecondary structures, slow refolding or reactivation of a protein is probably due to other factors.

Knowledge of the role of specific amino acid residues would aid in understanding folding and assembly pathways. A single amino acid substitution can effect a drastic change if it plays a pivotal role in the overall energy balance (207). X-ray studies provide the most information on interactions in a stationary state; other approaches are necessary in investigating the dynamics. One such technique involves assessment of the interactions between various synthetic polypeptides and proteins, and inference of the reactive residues therefrom (208).

CONCLUSION

Comprehensive criteria for reassembly are required. Recovery of physical properties is a necessary, but not sufficient, criterion for recovery of structure. Perhaps the best single test of native-like structure is biological activity because it is dependent on preformation of a specific structure. One example from among the many possible is the fragment from the β_2 subunit of tryptophan synthetase: although several physical parameters are similar to that of the native subunit, it is inactive (85).

Knowledge of the relative amounts of intermediates present during assembly is likewise not sufficient for determining the relative importance of alternative assembly pathways. Kinetic analysis and/or models are necessary. Standard biochemical methods are continually being adapted to enable the study of rapid protein-protein associations. The recent development of a stopped-flow far UV circular dichroism instrument permits direct monitoring of changes in secondary structure resulting from recombination (61, 209). The advent of stopped-flow NMR offers another powerful probe of conformation (210). Measurement of light scattering by the stopped-flow or relaxation method is the most direct method for determining state of aggregation. Its disadvantage in requiring high concentrations or large molecular weight changes may be overcome by using more powerful laser light sources.

In vitro assembly mechanisms are covered in this review, and no attempt is made to extrapolate the results to the intracellular or in vivo situations. Aside from the different environments, the concentrations of polypeptide chains and the molecular state are better controlled in vitro. In vivo, a newly synthesized chain may assemble through a different intermediate because the availability of the various polypeptide chains is not the same as in vitro. Alternatively, if association with another chain occurs before folding or complete release from the ribosome is achieved, an altered protein recognition pattern may result in a different assembly pathway.

ACKNOWLEDGMENTS

Work reporting previously unpublished results was supported by grants from the National Science Foundation, #PCM 77–08537, and the National Cancer Institute, #CA 13014. The authors are indebted to Rhea McDonald for skilled typing assistance.

Literature Cited

1. Wetlaufer, D. B., Ristow, S. 1973. *Ann. Rev. Biochem.* 42:135–58
2. Anfinsen, C. B., Scheraga, H. A. 1975. *Adv. Protein Chem.* 29:205–300
3. Baldwin, R. L. 1975. *Ann. Rev. Biochem.* 44:453–75
4. Némethy, G., Scheraga, H. A. 1977. *Q. Rev. Biophys.* 10:239–352
5. Richards, F. M. 1977. *Ann. Rev. Biophys. Bioeng.* 6:151–76
6. Schulz, G. E. 1977. *Angew. Chem. Int. Ed. Engl.* 16:23–32
7. Chou, P. Y., Fasman, G. D. 1978. *Ann. Rev. Biochem.* 47:251–76
8. Liljas, A., Rossmann, M. G. 1974. *Ann. Rev. Biochem.* 43:475–507
9. Matthews, B. W. 1976. *Ann. Rev. Phys. Chem.* 27:493–524
10. Kleine, R. 1972. *Arzneim. Forsch.* 16:364–446
11. Klotz, I. M., Darnall, D. W., Langerman, N. R. 1975. In *The Proteins,* ed. H. Neurath, R. L. Hill, 1:243–411. New York: Academic. 547 pp. 3rd ed.
12. Williamson, A. R. 1969. *Essays Biochem.* 5:139–75
13. Klotz, I. M., Langerman, N. R., Darnall, D. W. 1970. *Ann. Rev. Biochem.* 39:25–62
14. Frieden, C. 1971. *Ann. Rev. Biochem.* 40:653–96
15. Matthews, B. W., Bernhard, S. A. 1973. *Ann. Rev. Biophys. Bioeng.* 2:257–317

16. Hammes, G. G., Wu, C. W. 1974. *Ann. Rev. Biophys. Bioeng.* 3:1–33
17. Lauffer, M. A. 1975. *Entropy Driven Processes in Biology.* Berlin: Springer. 264 pp.
18. Perham, R. N. 1975. *Philos. Trans. R. Soc. London Ser. B.* 272:123–36
19. Antonini, E., Chiancone, E. 1977. *Ann. Rev. Biophys. Bioeng.* 6:239–71
20. Thusius, D. 1977. In *Chemical Relaxation in Molecular Biology,* ed. I. Pecht, R. Rigler, pp. 339–70. New York: Springer. 418 pp.
21. Snyder, J. A., McIntosh, J. R. 1976. *Ann. Rev. Biochem.* 45:699–720
22. Clark, M., Spudich, J. A. 1977. *Ann. Rev. Biochem.* 46:797–822
23. Casjens, S., King, J. 1975. *Ann. Rev. Biochem.* 44:555–611
24. Butler, P. J. G., Durham, A. C. H. 1977. *Adv. Protein Chem.* 31:187–251
25. Ikai, A., Tanford, C. 1971. *Nature* 230:100–2
26. Ikai, A., Tanford, C. 1973. *J. Mol. Biol.* 73:145–63
27. Schachman, H. K. 1972. In *Protein-Protein Interactions,* ed. R. Jaenicke, E. Helmreich, pp. 17–56. New York: Springer. 464 pp.
28. Lipscomb, W. N., Evans, D. R., Edwards, B. F. P., Warren, S. G., Pastra-Landis, S., Wiley, D. C. 1974. *J. Supramol. Struct.* 2:82–98
29. Cohlberg, J. A., Pigiet, V. P. Jr., Schachman, H. K. 1972. *Biochemistry* 11:3396–3411
30. Nagel, G. M., Schachman, H. K. 1975. *Biochemistry* 14:3195–3203
31. Gerhart, J. C., Schachman, H. K. 1965. *Biochemistry* 4:1054–62
32. Chan, W. W.-C., Mort, J. S. 1973. *J. Biol. Chem.* 248:7614–16
33. Mort, J. S., Chan, W. W.-C. 1975. *J. Biol. Chem.* 250:653–60
34. Bothwell, M., Schachman, H. K. 1974. *Proc. Natl. Acad. Sci. USA* 71:3221–25
35. Chan, W. W.-C. 1978. *Eur. J. Biochem.* 90:271–81
36. Jacobson, G. R., Stark, G. R. 1973. *J. Biol. Chem.* 248:8003–14
37. Yang, Y. R., Syvanen, J. M., Nagel, G. M., Schachman, H. K. 1974. *Proc. Natl. Acad. Sci. USA* 71:918–22
38. Evans, D. R., Pastra-Landis, S. C., Lipscomb, W. N. 1974. *Proc. Natl. Acad. Sci. USA* 71:1351–55
39. Subrami, S., Bothwell, M. A., Gibbons, I., Yang, Y. R., Schachman, H. K. 1977. *Proc. Natl. Acad. Sci. USA* 74:3777–81
40. Baldwin, J. M. 1975. *Prog. Biophys. Mol. Biol.* 29:225–320
41. Kellett, G. L., Gutfreund, H. 1970. *Nature* 227:921–26
42. Ip, S. H. C., Johnson, M. L., Ackers, G. K. 1976. *Biochemistry* 15:654–60
43. Ackers, G. K., Johnson, M. L., Mills, F. C., Ip, S. H. C. 1976. *Biochem. Biophys. Res. Commun.* 69:135–42
44. Ip, S. H. C., Ackers, G. K. 1977. *J. Biol. Chem.* 252:82–87
45. Andersen, M. E., Moffat, J. K., Gibson, Q. H. 1971. *J. Biol. Chem.* 246:2796–2807
46. Nagel, R. L., Gibson, Q. H. 1971. *J. Biol. Chem.* 246:69–73
47. Nagel, R. L., Gibson, Q. H. 1972. *Biochem. Biophys. Res. Commun.* 48:959–66
48. Park, C. M. 1973. *Ann. NY Acad. Sci.* 204:237–49
49. Bunn, H. F., McDonough, M. 1974. *Biochemistry* 13:988–93
50. Benesch, R. E., Benesch, R. 1974. *Adv. Protein Chem.* 28:211–37
51. Gray, R. D. 1974. *J. Biol. Chem.* 249:2879–85
52. Wiedermann, B. L., Olson, J. S. 1975. *J. Biol. Chem.* 250:5273–75
53. Terpstra, F. A., Smith, D. B. 1976. *Can. J. Biochem.* 54:1002–10
54. Valdes, R. Jr., Ackers, G. K. 1977. *J. Biol. Chem.* 252:74–81
55. Antonini, E., Bucci, E., Fronticelli, C., Chiancone, E., Wyman, J., Rossi-Fanelli, A. 1966. *J. Mol. Biol.* 17:29–46
56. Antonini, E., Brunori, M. 1971. *Hemoglobin and Myoglobin in Their Reactions with Ligands,* p. 121. Amsterdam: North-Holland. 436 pp.
57. McGovern, P., Reisberg, P., Olson, J. S. 1976. *J. Biol. Chem.* 251:7871–79
58. Yip, Y. K., Waks, M., Beychok, S. 1972. *J. Biol. Chem.* 247:7237–44
59. Waks, M., Yip, Y. K., Beychok, S. 1973. *J. Biol. Chem.* 248:6462–70
60. Yip, Y. K., Waks, M., Beychok, S. 1977. *Proc. Natl. Acad. Sci. USA* 74:64–8
61. Luchins, J. I. 1977. *Stopped-flow circular dichroism.* PhD thesis. Columbia Univ., New York. 261 pp.
62. Zillig, W., Palm, P., Heil, A. 1976. In *RNA Polymerase,* ed. R. Losick, M. Chamberlin. pp. 101–125. Cold Spring Harbor, New York: Cold Spring Harbor Lab. 899 pp.
63. Ishihama, A., Taketo, M., Saitoh, T., Fukuda, R. 1976. See Ref. 62, pp. 485–502
64. Ishihama, A. 1972. *Biochemistry* 11:1250–58
65. Yarbrough, L. R., Hurwitz, J. 1974. *J. Biol. Chem.* 249:5394–99

66. Harding, J., Beychok, S. 1974. *Proc. Natl. Acad. Sci. USA* 71:3395–99
67. Harding, J., Beychok, S. 1976. See Ref. 62, pp. 355–67
68. Lowe, P. A., Malcolm, A. D. B. 1976. *Biochim. Biophys. Acta* 454:129–37
69. Lowe, P. A., Malcolm, A. D. B. 1975. *Biochem. Soc. Trans.* 3:652–53
70. Lowe, P. A., Malcolm, A. D. B. 1976. *Eur. J. Biochem.* 64:177–88
71. Ishihama, A., Ito, K. 1972. *J. Mol. Biol.* 72:111–23
72. Palm, P., Heil, A., Boyd, D., Grampp, B., Zillig, W. 1975. *Eur. J. Biochem.* 53:283–91
73. Ito, K., Ishihama, A. 1973. *J. Mol. Biol.* 79:115–25
74. Fischman, B. 1978. *Denaturation and reactivation of E. coli RNA polymerase.* PhD thesis. Columbia Univ., New York. 181 pp.
75. Saitoh, T., Ishihama, A. 1976. *J. Mol. Biol.* 104:621–35
76. Travers, A. 1975. *FEBS Lett.* 53:76–79
77. Campbell, A. M., Lowe, P. A. 1977. *Biochem. J.* 163:177–79
78. Wu, F. Y.-H., Yarbrough, L. R., Wu, C.-W. 1976. *Biochemistry* 13:3254–58
79. Creighton, T. E., Yanofsky, C. 1966. *J. Biol. Chem.* 241:980–90
80. Jackson, D. A., Yanofsky, C. 1969. *J. Biol. Chem.* 244:2426–38
81. Jackson, D. A., Yanofsky, C. 1969. *J. Biol. Chem.* 244:4539–46
82. Hathaway, G. M., Crawford, I. P. 1970. *Biochemistry* 9:1801–8
83. Hathaway, G. M., Kida, S., Crawford, I. P. 1969. *Biochemistry* 8:989–97
84. Goldberg, M. E., Creighton, T. E., Baldwin, R. L., Yanofsky, C. 1966. *J. Mol. Biol.* 21:71–82
85. Högberg-Raibaud, A., Goldberg, M. E. 1977. *Proc. Natl. Acad. Sci. USA* 74: 442–46
86. Hofmann, F., Beavo, J. A., Bechtel, P. J., Krebs, E. G. 1975. *J. Biol. Chem.* 250:7795–7801
87. Rangel-Aldao, R., Rosen, O. M. 1976. *J. Biol. Chem.* 251:3375–80
88. Rangel-Aldao, R., Rosen, O. M. 1977. *J. Biol. Chem.* 252:7140–45
89. Kamen, R. 1970. *Nature* 228:527–33
90a. Young, R. A., Blumenthal, T. 1975. *J. Biol. Chem.* 250:1829–32
90b. Blumenthal, T., Landers, T. A., Weber, K. 1972. *Proc. Natl. Acad. Sci. USA* 69:1313–17
91. Blumenthal, T., Landers, T. A. 1976. *Biochemistry* 15:422–25
92. Teipel, J. W., Koshland, D. E., Jr. 1971. *Biochemistry* 10:792–98
93. Teipel, J. W., Koshland, D. E., Jr. 1971. *Biochemistry* 10:798–805
94. Anderson, S., Weber, G. 1966. *Arch. Biochem. Biophys.* 116:207–23
95. Levitzki, A., Tenenbaum, H. 1974. *Isr. J. Chem.* 12:327–37
96. Vallee, R. B., Williams, R. C., Jr. 1975. *Biochemistry* 14:2574–80
97. Levitzki, A. 1972. *FEBS Lett.* 24:301–4
98. Tenenbaum-Bayer, H., Levitzki, A. 1976. *Biochim. Biophys. Acta* 445: 261–79
99. Rudolph, R., Jaenicke, R. 1976. *Eur. J. Biochem.* 63:409–17
100. Rudolph, R., Heider, I., Jaenicke, R. 1977. *Biochemistry* 16:5527–31
101. Rudolph, R., Heider, I., Westhof, E., Jaenicke, R. 1977. *Biochemistry* 16: 3384–90
102. Teipel, J. W. 1972. *Biochemistry* 11: 4100–7
103. Vimard, C., Orsini, G., Goldberg, M. E. 1975. *Eur. J. Biochem.* 51:521–27
104. Chan, W. W.-C., Mort, J. S., Chong, D. K. K., MacDonald, P. D. M. 1973. *J. Biol. Chem.* 248:2778–84
105. Rudolph, R., Westhof, E., Jaenicke, R. 1977. *FEBS Lett.* 73:204–6
106. Engelhard, M., Rudolph, R., Jaenicke, R. 1976. *Eur. J. Biochem.* 67:447–53
107. Rudolph, R., Englehard, M., Jaenicke, R. 1976. *Eur. J. Biochem.* 67:455–62
108. Spotorno, G. M. L., Hollaway, M. R. 1970. *Nature* 226:756–57
109. Osborne, H. H., Hollaway, M. R. 1974. *Biochem. J.* 143:651–62
110. Osborne, H. H., Hollaway, M. R. 1975. *Biochem. J.* 151:37–45
111. Hoagland, V. D. Jr., Teller, D. C. 1969. *Biochemistry* 8:594–602
112. Mockrin, S. C., Byers, L. D., Koshland, D. E. Jr. 1975. *Biochemistry* 14: 5428–37
113. Meighen, E. A., Schachman, H. K. 1970. *Biochemistry* 9:1177–84
114. Deal, W. C. Jr. 1969. *Biochemistry* 8:2795–2805
115. Rudolph, R., Heider, I., Jaenicke, R. 1977. *Eur. J. Biochem.* 81:563–70
116. Stancel, G. M., Deal, W. C. Jr. 1969. *Biochemistry* 8:4005–11
117. Bartholmes, P., Jaenicke, R. 1978. *Eur. J. Biochem.* 87:563–67
118a. Bartholmes, P., Jaenicke, R. 1975. *Biochem. Biophys. Res. Commun.* 64:485–92
118b. Constantinides, S. M., Deal, W. C., Jr. 1969. *J. Biol. Chem.* 244:5695–5702
119. Paetkau, V., Lardy, H. A. 1967. *J. Biol. Chem.* 247:2035–42
120. Pavelich, M. J., Hammes, G. G. 1973. *Biochemistry* 12:1408–14

121. Aaronson, R. P., Frieden, C. 1972. *J. Biol. Chem.* 247:7502–9
122. Bock, P. E., Frieden, C. 1974. *Biochemistry* 13:4191–96
123. Lad, P. M., Hill, D. E., Hammes, G. G. 1973. *Biochemistry* 12:4303–9
124. Lad, P. M., Hammes, G. G. 1974. *Biochemistry* 13:4530–37
125. Parr, G. R., Hammes, G. G. 1975. *Biochemistry* 14:1600–5
126. Parr, G. R., Hammes, G. G. 1976. *Biochemistry* 15:857–62
127. Siezen, R. J., van Driel, R. 1974. *J. Mol. Biol.* 90:91–102
128. Siezen, R. J., van Bruggen, E. F. J. 1974. *J. Mol. Biol.* 90:77–89
129. Engelborghs, Y., Lontie, R. 1973. *J. Mol. Biol.* 77:577–87
130. Kegeles, G. 1977. *Arch. Biochem. Biophys.* 180:530–36
131. Tai, M., Kegeles, G., Huang, C. 1977. *Arch. Biochem. Biophys.* 180:537–42
132. Siezen, R. J., van Driel, R. 1973. *Biochim. Biophys. Acta* 295:131–39
133. Siezen, R. J. 1974. *J. Mol. Biol.* 90:103–13
134. Shifrin, S., Luborsky, S. W., Grochowski, B. J. 1971. *J. Biol. Chem.* 246:7708–14
135. Shifrin, S., Parrott, C. L. 1974. *J. Biol. Chem.* 249:4175–80
136. Waley, S. G. 1973. *Biochem. J.* 135:165–72
137. London, J., Skrzynia, C., Goldberg, M. E. 1974. *Eur. J. Biochem.* 47:409–15
138. Raibaud, O., Goldberg, M. E. 1976. *J. Biol. Chem.* 251:2820–24
139. Brewer, J. M. 1972. *J. Biol. Chem.* 247:7941–47
140. MacKenzie, R. E., Rabinowitz, J. C. 1971. *J. Biol. Chem.* 246:3731–36
141. Harmony, J. A. K., Shaffer, P. J., Himes, R. H. 1974. *J. Biol. Chem.* 249:394–401
142. Koren, R., Hammes, G. G. 1976. *Biochemistry* 15:1165–71
143. Gilleland, M. J., Bender, M. L. 1976. *J. Biol. Chem.* 251:498–502
144. Bevan, M. J., Parkhouse, R. M. E., Williamson, A. R., Askonas, B. A. 1972. *Prog. Biophys. Mol. Biol.* 25:131–62
145. Baumal, R., Scharff, M. D. 1973. *Transplant. Rev.* 14:163–83
146. Scharff, M. D. 1974. *Harvey Lect.* 69:125–42
147. Cathou, R. E., Dorrington, K. J. 1975. In *Biological Macromolecules: Subunits in Biological Systems,* ed. S. N. Timasheff, G. D. Fasman, Part C, pp. 91–224, New York: Marcel Dekker
148. Nisonoff, A., Hopper, J. E., Spring, S.

B. 1975. *The Antibody Molecule.* New York: Academic. 542 pp.
149. Green, R. W. 1973. *Biochemistry* 12:3225–31
150. Bigelow, C. C., Smith, B. R., Dorrington, K. J. 1974. *Biochemistry* 13:4602–8
151. Bunting, P. S., Kells, D. I. C., Kortan, C., Dorrington, K. J. 1977. *Immunochemistry* 14:45–52
152. Azuma, T., Isobe, T., Hamaguchi, K. 1975. *J. Biochem.* 77:473–79
153. Azuma, T., Hamaguchi, K. 1975. *J. Biochem.* 78:341–47
154. Azuma, T., Isobe, T., Hamaguchi, K. 1975. *J. Biochem.* 78:335–40
155. Azuma, T., Hamaguchi, K. 1976. *J. Biochem.* 80:1023–38
156. Friedman, F. K., Chang, M. D. Y., Beychok, S. 1978. *J. Biol. Chem.* 253:2368–72
157. de Préval, C., Fougereau, M. 1976. *J. Mol. Biol.* 102:657–78
158. Smith, B. R., Dorrington, K. J. 1972. *Biochem. Biophys. Res. Commun.* 46:1061–65
159. Klein, M., Kortan, C., Kells, D. I. C., Dorrington, K. J. 1978. *Fed. Proc.* 37:1278
160. Hochman, J., Inbar, D., Givol, D. 1973. *Biochemistry* 12:1130–35
161. Hochman, J., Gavish, M., Inbar, D., Givol, D. 1976. *Biochemistry* 15:2706–10
162. Peterson, J. G. L., Dorrington, K. J. 1974. *J. Biol. Chem.* 249:5633–41
163. Sears, D. W. 1974. *In vitro reoxidation and reduction study of the interchain disulfides of a human, monoclonal IgGκ protein.* PhD thesis. Columbia Univ., New York. 154 pp.
164. Sears, D. W., Mohrer, J., Beychok, S. 1975. *Proc. Natl. Acad. Sci. USA* 72:353–57
165. Percy, M. E., Baumal, R., Dorrington, K. J., Percy, J. R. 1976. *Can. J. Biochem.* 54:675–87
166. Sears, D. W., Beychok, S. 1977. *Biochemistry* 16:2026–31
167. Chan, K. C., Smith, B. R., Latner, A. L. 1974. *Biochem. Soc. Trans.* 2:860–61
168. Sears, D. W., Kazin, A. R., Mohrer, J., Friedman, F., Beychok, S. 1977. *Biochemistry* 16:2016–25
169. Kazin, A. R. 1977. *In vitro covalent assembly of a human monoclonal immunoglobulin G1 protein.* PhD thesis. Columbia Univ., New York. 193 pp.
170. Percy, J. R., Percy, M. E., Dorrington, K. J. 1975. *J. Biol. Chem.* 250:2398–2400
171. Percy, J. R., Percy, M. E., Baumal, R. 1976. *Can. J. Biochem.* 54:688–98

172. Kishida, F., Azuma, T., Hamaguchi, K. 1976. *J. Biochem.* 79:91–105
173. Tanaka, Y., Azuma, T., Hamaguchi, K. 1978. *J. Biochem.* 83:1249–63
174. Azuma, T., Hamaguchi, K., Migita, S. 1974. *J. Biochem.* 76:685–93
175. Azuma, T., Kobayashi, O., Goto, Y., Hamaguchi, K. 1978. *J. Biochem.* 83:1485–92
176. Maeda, H., Engel, J., Schramm, H. J. 1976. *Eur. J. Biochem.* 69:133–39
177. Kazin, A. R., Beychok, S. 1978. *Science* 199:688–90
178. Sears, D. W., Mohrer, J., Beychok, S. 1977. *Biochemistry* 16:2031–35
179. Koshland, M. E. 1975. *Adv. Immunol.* 15:41–69
180. Parkhouse, R. M. E., Askonas, B. A., Dourmashkin, R. R. 1970. *Immunology* 18:575–84
181. Parkhouse, R. M. E. 1974. *Immunology* 27:1063–71
182. Askonas, B. A., Parkhouse, R. M. E. 1971. *Biochem. J.* 123:629–34
183. Della Corte, E., Parkhouse, R. M. E. 1973. *Biochem. J.* 136:597–606
184. Parkhouse, R. M. E. 1971. *Biochem. J.* 123:635–41
185. Buxbaum, J., Scharff, M. D. 1973. *J. Exp. Med.* 138:278–88
186. Percy, M. E., Percy, J. R., Baumal, R., Dorrington, K. J. 1974. *Fed. Proc.* 34:972
187. Percy, M. E., Feinstein, A., Baumal, R. 1978. *Can. J. Biochem.* 56:190–96
188. Morris, J. E., Inman, F. P. 1968. *Biochemistry* 7:2851–57
189. Beale, D., Feinstein, A. 1969. *Biochem. J.* 112:187–94
190. Mukkur, T. K. S., Inman, F. P. 1970. *Biochemistry* 9:1031–37
191. Chapuis, R. M., Koshland, M. E. 1974. *Proc. Natl. Acad. Sci. USA* 71:657–61
192. Kownatzki, E. 1973. *Scand. J. Immunol.* 2:433–37

193. Percy, M. E., Percy, J. R. 1975. *Can. J. Biochem.* 53:923–29
194. Schrohenloher, R. E., Hester, R. B. 1976. *Scand. J. Immunol.* 5:637–46
195. Egorov, A. M., Chernyak, V. Ya., Dunaevsky, Ya. E., Gavrilova, E. M., Moiseev, V. L. 1971. *Immunochemistry* 8:1157–63
196. Solheim, B. G., Harboe, M. 1972. *Immunochemistry* 9:623–34
197. Schrohenloher, R. E., Mestecky, J. 1973. *J. Immunol.* 111:1699–1711
198. Schrohenloher, R. E. 1974. *Immunol. Comm.* 3:553–64
199. Mukkur, T. K. S., Inman, F. P. 1971. *J. Immunol.* 107:705–14
200. Kownatzki, E. 1973. *Immunol. Commun.* 2:105–13
201. Wilde, C. E. III, Koshland, M. E. 1972. *Fed. Proc.* 31:755
202. Wilde, C. E. III, Koshland, M. E. 1978. *Biochemistry* 17:3209–14
203. Cook, R. A., Koshland, D. E. Jr. 1969. *Proc. Nat. Acad. Sci. USA* 64:247–54
204a. Chothia, C., Janin, J. 1975. *Nature* 256:705–8
204b. Langerman, N. R., Klotz, I. M. 1969. *Biochemistry* 8:4746–52
205. Levinthal, C. 1968. *J. Chim. Phys. Phys. Chim. Biol.* 65:44–45
206. Wetlaufer, D. B. 1973. *Proc. Natl. Acad. Sci. USA* 70:697–701
207. Yutani, K., Ogasahara, K., Sugino, Y., Matsushiro, A. 1977 *Nature* 267:274–5
208. Kubota, S., Chang, C. T., Samejima, T., Yang, J. T. 1976. *J. Am. Chem. Soc.* 98:2677–78
209. Luchins, J. I., Beychok, S. 1978. *Science* 199:425–26
210. Sykes, B. D., Grimaldi, J. J. 1978. *Methods Enzymol.* 29:295–321
211. Thomas, J. O., Edelstein, S. J. 1972. *J. Biol. Chem.* 247:7870–74
212. Tsuzuki, J., Kiger, J. A., Jr. 1978. *Biochemistry* 17:2961–70

Ann. Rev. Biochem. 1979. 48:251–74
Copyright © 1979 by Annual Reviews Inc. All rights reserved

CHEMISTRY AND BIOLOGY OF THE NEUROPHYSINS

❖12009

Esther Breslow[1]

Department of Biochemistry, Cornell University Medical College,
New York, New York 10021

CONTENTS

PERSPECTIVES AND SUMMARY

The neurophysins are small disulfide-rich proteins found in noncovalent association with the hormones oxytocin and vasopressin within the posterior pituitary; their principal function appears to be the stabilization of the hormones within neurosecretory granules. Evidence has been obtained in-

[1]This work was supported by NIH grant GM-17528.

251

dicating that neurophysin biosynthesis involves a precursor of higher molecular weight and it has been suggested, but not established, that there are common precursors for the neurophysins and the hormones. The common precursor concept is particularly attractive since it explains the existence of different structural classes of neurophysins within each species, each separately compartmentalized with a different hormone, but not all available data are consistent with this hypothesis. From a biological point of view, studies of neurophysin biosynthesis are therefore important to an understanding of the mechanism of hormone biosynthesis and packaging within neurosecretory granules.

Chemically the neurophysins are of particular interest as models for the study of protein-peptide interactions and of the mechanism and significance of the phenomenon of half-of-the-sites reactivity. Their use as models for the study of protein-peptide interactions derives from the fact that the sites on the hormones which bind to neurophysin have been largely defined and, by study of the binding of systematically related small peptides, the thermodynamic contributions of individual bonding interactions, and of conformational changes associated with binding, can be isolated. Their use as half-of-the-sites reactivity models derives from the fact that there are duplicated segments within each neurophysin polypeptide chain and there is also the potential for two binding sites on each chain, only one of which is most generally expressed. Additionally, preliminary studies place the putative neurophysin active site carboxyl within one of the duplicated segments, although a residue from a nonduplicated segment, Tyr-49, also appears to be near at least one of the binding sites. The present review places greatest emphasis on what is known about the chemistry of neurophysin and its interactions with hormones, but will do so in the context of what is known about neurophysin biosynthesis and function. Several models are proposed for the role of the duplicated segments in neurophysin half-of-the-sites reactivity. Because the review is necessarily selective, the reader is referred to other general reviews on neurophysin (1–3) for a more complete picture of the field.

BIOSYNTHESIS OF POSTERIOR PITUITARY HORMONES AND NEUROPHYSIN

Oxytocin and vasopressin are synthesized within nerve cell bodies of the supraoptic and paraventricular regions of the hypothalamus and transported within neurosecretory granules along the axons that extend from the cell bodies to the posterior pituitary; upon stimulation of the nerve fibers in the posterior pituitary the granule contents are released into the blood by exocytosis (1–8). Most evidence indicates that oxytocin and vasopressin are synthesized within separate neurons (5, 6, 9–12) although some studies suggest that they can coexist in the same cell (13, 14; for review see 15). Within the neurosecretory granules of the posterior pituitary the hormones are found in noncovalent combination with neurophysin (16–21). All mammalian species examined so far have at least two different neurophysins (1, 22–25). Guinea pig neurophysin, which appeared initially to be a single protein (26), has since been shown to be heterogeneous (27). Following the studies of Hope and co-workers (21) which indicated that the two principal neurophysins of the cow are differentially compartmentalized with the two hormones, considerable evidence has been obtained in support of thesis that, in each species, one class of neurophysin is associated in vivo with oxytocin and one with vasopressin although in vitro each neurophysin can bind either hormone (1, 28–30); this concept, the one hormone-one neurophysin hypothesis, is further discussed below. The concentration of the hormones and neurophysins within neurosecretory granules is very high, estimates for each ranging from 0.02 M (19) to 0.1 M (31); it is probable that the intragranular hormone-protein complex is in the precipitated or crystalline state (32, 33).

Sachs and co-workers (5, 6), in classic pulse-labelling studies, provided strong evidence for the concept that vasopressin was synthesized by a ribosomal mechanism via a higher molecular weight precursor which was subsequently cleaved to give active hormone. A variety of ancillary data supported this hypothesis (for review see 33) and a similar mechanism has been proposed for the biosynthesis of oxytocin (34). The biosynthesis of neurophysin also occurs in the cell bodies of the supraoptic and paraventricular regions of the hypothalamus (5, 6, 33, 35). Sachs and collaborators suggested that the biosynthesis of neurophysin and vasopressin might proceed via a common high molecular weight precursor (6) and subsequently demonstrated that the biosynthesis of neurophysin followed a lag period similar to that observed for vasopressin biosynthesis (36, 37). The concept that neurophysin and hormone synthesis are coupled was also supported by studies of the Brattleboro rat in which the synthesis of both vasopressin and one of the normal rat neurophysins were found to be absent (24, 38–40). Thus the common precursor theory offers an explanation of two classes of neurophysins each compartmentalized in vivo with a different hormone, suggesting that each neurophysin is the cleavage product of a different hormone precursor. Physical-chemical studies also argued for the existence of a biosynthetic precursor for neurophysin. Neurophysin disulfides were found to be highly susceptible to reduction and rearrangement with very little regeneration of the native structure accompanying reoxidation of the fully reduced protein; such behavior parallels that of insulin which is synthesized via a precursor in which disulfide pairing is achieved (1, 41, 42).

The existence of a neurophysin precursor is supported almost conclusively by recent studies (33, 43, 44–47). Gainer and co-workers (33, 44–47) injected pulses of [^{35}S]-cysteine into the supraoptic nuclei of rats and isolated different sections of the hypothalamo-neurohypophyseal tract—the supraoptic nucleus (cell body), the median eminence (axon), and the posterior pituitary (axon terminus)—at different time intervals after the pulse. Labelled protein appeared first in the supraoptic nucleus and last in the posterior pituitary. The major labelled protein isolated from the supraoptic nucleus one hour after label injection was assigned a molecular weight of 19,000–20,000 daltons by electrophoresis and Sephadex chromatography, compared to a molecular weight of 10,000–12,000 for native rat neurophysin, and was subsequently shown to react with antineurophysin antibodies. The amount of label in this protein (hereafter called the precursor) decreased with time after injection while the amount in native neurophysin increased. Both precursor and neurophysin were demonstrable in the median eminence 2 hr after injection of label, but only labelled neurophysin was isolable from this region or from the posterior pituitary after 24 hr. These labelling patterns are clearly in accord with synthesis of neurophysin via a higher molecular weight precursor which is converted to the native protein within the granules as these are transported from the cell body to the posterior pituitary. In related studies, precursors of vasopressin-associated rat neurophysin and oxytocin-associated rat neurophysin were distinguished by a comparison of the isolectric focusing behavior of neurophysin precursors in normal and Brattleboro rats (46); 17,000 molecular weight intermediates in precursor processing have also been detected (46, 47).

Other studies support the concept of processing of a neurophysin precursor and/or of the hormone precursor during intragranule transport (for review see 33). For example, a progessively decreased stability of the neurosecretory granule core to alkaline fixation during granule transport to the posterior pituitary has been observed (48). Dax & Johnston (49) have shown, in the pig, that the amount of immunoreactive neurophysin and posterior pituitary hormone increases between the hypothalamic cell nuclei and the posterior pituitary. It has been suggested (50, 51) that the enzyme which cleaves the precursor is the same as that responsible for the subsequent intragranular cleavage of carboxyl-terminal residues from neurophysin (see below).

While the evidence points clearly to the existence of precursors for the neurophysins and the hormones, the question remains as to whether there are *common* precursors for each of the hormones and their associated neurophysins. Preliminary evidence has been obtained that the precursor of vasopressin-associated rat neurophysin reacts with antivasopressin antibodies (47); it has also been noted (46) that the higher pI values of neurophysin precursors compared with those of neurophysins are compatible with removal of the basic hormones during precursor processing. On the other hand, Moore et al (52) chemically synthesized peptides in which sequences resembling the amino and carboxyl termini of bovine neurophysin-II were covalently attached to arginine vasopressin and found no intragranular enzyme capable of cleaving these to active hormone; they suggest that, if a common precursor is involved, the neurophysin and hormone segment are not directly linked. The existence of a common precursor would necessitate that the ratio of each hormone to its associated neurophysin within neurosecretory granules be constant, but this has yet to be conclusively demonstrated (see also below). For example, significant disparities between neurophysin and hormone ratios in sheep pituitary have been reported (53) and several human tumors have been shown to secrete vasopressin in the absence of any immunoreactive neurophysin (54, 55). While alternative explanations of these results are possible, they are clearly consistent with the concept that neurophysin and hormone synthesis need not be synchronous and that a common precursor may not be involved.

BIOLOGICAL ROLE OF NEUROPHYSINS

Neurophysin-hormone interactions are undoubtedly important in stabilizing the hormones within the granules during intra-axonal transport, either preventing hormone leakage across the granule membrane into the cytoplasm, preventing disulfide interchange to which the hormones are particularly susceptible, or protecting the hormones from proteolytic digestion. Additionally, the insolubility of the hormone-neurophysin complex within the granules (32, 33) should reduce the osmotic threat of the high hormone concentration to granule integrity; the suggestion that this insolubility also thermodynamically drives the packaging of the hormones at high concentrations is difficult to reconcile with the probability (33) that packaging occurs at the precursor stage—unless there are analogous insolubilizing interactions among precursor molecules. Within the blood, neurophysin is unlikely to aid in either the transport or stabilization of the hormones, since blood neurophysin levels are sufficiently low that the complexes are dissociated (1, 29, 56).

Does neurophysin play a role separate from that involving its interactions with posterior pituitary hormones? The posterior pituitary enzyme "protein carboxymethylase" methylates neurophysin in vitro (57), but the biological consequences of this are unknown. Several metabolic effects of neurophysin have been reported (58–60), but the most recent evidence suggests that these arise from contaminants in the neurophysin samples used (61–63). Nonetheless, protein which reacts with antineurophysin antibodies has been demonstrated in several peripheral tissues (1, 64–67), including the posterior pituitary hormone target tissues, kidney, uterus, and mammary gland. While its presence in kidney might be accounted for by the fact that neurophysin is cleared by the kidney (68), this is not the situation with uterine neurophysin which appears to be synthesized by the uterus rather than transported to it (69). The significance of relatively high neurophysin concentrations in the pituitary portal system, which feeds the anterior pituitary (14, 70), and in cerebrospinal fluid (71) is uncertain (69); in this context it is provocative but possibly fortuitous that the anterior pituitary hormone ACTH contains the key residues necessary for binding to neurophysin (63).

IMUNOLOGICAL INVESTIGATIONS USING NEUROPHYSINS AS BIOLOGICAL MARKERS

This active field has been reviewed elsewhere (15, 69, 72) and is discussed here only in outline. Most such studies fall within one of three categories. First, as indicated above, antibodies directed against neurophysin have been used to demonstrate the presence of neurophysin in tissues outside the hypothalamo-neurohypophyseal tract. These studies have recently also indicated the presence of neurophysin in the pineal gland where it is probably associated with arginine vasotocin (67, 73). Second, since antibodies have been raised that can distinguish between oxytocin-associated and vasopressin-associated neurophysins, immunohistochemical investigations of neurophysins have been used to trace axonal pathways within the hypotholamo-pituitary system (14, 15) and to distinguish between nerve fibers associated with the different hormones (e.g. 14), to confirm the association of a particular neurophysin with one of the hormones, and to determine the quantitative relationship between the neurophysin and hormone (e.g. 49, 74, 75). Finally, the fact that release into the blood of posterior pituitary hormones is accompanied by release of hormone-associated neurophysin has led to the quantitation of specific neurophysins in the blood by radioimmunoassay following different physiological stimuli; this quantitation is used both as a guide to the release of the different hormones (which are more difficult to measure directly) and also as a means of determining the hormone with which each neurophysin is associated. A particularly interesting result here has been the report in humans of estrogen-stimulated serum neurophysin (cf 76; see also 77, 78) which suggests a previously unknown effect of estrogen on posterior pituitary function.

PRIMARY STRUCTURE OF THE NEUROPHYSINS; EVOLUTION, CLASSIFICATION, AND INTERNAL DUPLICATION

Neurophysins have been isolated and sequenced from a number of mammalian species and from the cod (e.g. 1, 22, 25, 27, 79–85). All neurophysins studied to date are acidic proteins of approximately 10,000 molecular weight with a high content of disulfides, glycine, and proline. The aromatic amino acid content is characterized by a single tyrosine (Tyr-49) in the middle of the polypeptide chain, the absence of tryptophan, and, typically, 3 phenylalanines; a single histidine is occasionally present (Figure 1). Neurophysins as typically isolated do not contain significant quantities of nonprotein components, although some studies have suggested that they are associated in vivo with lipid (86–88) and carbohydrate (87). In vitro studies of the ability of purified neurophysins to bind lipid suggest the presence of limited probably weak interactions between lipid and protein (88, 89), binding apparently being stronger to the monomeric form of neurophysin (see below) than to the dimer (89).

Procedures for the isolation of neurophysin have been described in detail elsewhere (cf. 25, 90). The different neurophysins within a species have typically, but not uniformly, been named according to their relative electrophoretic mobility (25). Since, as will be evident below, this nomenclature does not accurately reflect the important structural or functional relationships among neurophysins of different species, alternative nomenclatures have been proposed, based either on the hormone with which the neurophysin is biologically associated (25) or on key residues of the sequence (79).

Figure 1 shows whole or partial sequences of several neurophysins. All sequences obtained to date (79–85) indicate a very high degree of evolutionary conservation particularly reflected in the central virtually invariant core containing residues 10–74. In this respect it is relevant that the original sequence proposed for bovine neurophysin-II (91) has been revised (79, 85) and that, among other changes, the provocative substitution (1) of Gln for Cys at position 34 in this protein does not occur. Also evident from Figure 1 is the fact that within a species, neurophysins are to be found (see porcine-I and -III) that are identical except for truncation at the carboxyl terminus. Truncation at the carboxyl or amino termini has been shown to account for differences among several neurophysins (e.g. 80, 82) and appears to occur in vivo (40, 50, 51) as well as during inappropriate isolation conditions (1, 92, 93). Based on studies in the rat, it has been suggested (50), that in all mammalian species where more than two neurophysins are found, the extra neurophysins are cleavage products of a single oxytocin-associated neuro-

Figure 1 (rotated, amino acid sequences of neurophysins)

	1	5	10	15	20
(MSEL) Bovine II [79]	Ala-Met-Ser-Asp-Leu-Glu-Leu-Arg-Gln-Cys-Leu-Pro-Cys-Gly-Pro-Gly-Gly-Lys-Gly-Arg-Cys-Phe-Gly-Pro				
(MSEL) Porcine-III [80]					
(MSEL) Porcine-I [80]					
(MSEL) Rat-II [27]	——Thr——Met-				
(VLDV) Bovine-I [83]	——Val-Leu——Asp-Val——Thr-				
(VLDV) Rat-I [27]	——Ala-Leu——Asp-Met——Lys-				

	25	30	35	40	45
Bovine-II	Ser-Ile-Cys-Cys-Gly-Asp-Glu-Leu-Gly-Cys-Phe-Val-Gly-Thr-Ala-Glu-Ala-Leu-Arg-Cys-Gln-Glu-Glu-Asn				
Porcine-III					
Porcine-I					
Rat-II	——Ala-				
Bovine-I					
Rat-I	——Ala-				

	50	55	60	65	70
Bovine II	Tyr-Leu-Pro-Ser-Pro-Cys-Gln-Ser-Gly-Gln-Lys-Pro-Cys-Gly-Ser-Gly-Gly-Arg-Cys-Ala-Ala-Ala-Gly-Ile				
Porcine-III					
Porcine-I					
Bovine-I					

	75	80	85	90	95
Bovine II	Cys-Cys-Asn-Asp-Glu-Ser-Cys-Val-Thr-Glu-Pro-Glu-Cys-Arg-Glu-Gly-Ile/Val-Gly-Phe-Pro-Arg-Arg-Val				
Porcine-III	Ala-Ser——Leu——Ala				
Porcine-I	——Ala-Ser——Leu				
Bovine-I	——Ser-Pro-Asp-Gly——His-Glu-Asp——Ala——Asp-Pro-Glu-Ala-Ala——Ser-Leu				

Figure 1 Amino acid sequences of neurophysins. The complete sequence of bovine neurophysin-II is shown; sequences of the other proteins are the same except for the substitutions listed. Rat neurophysins have been sequenced only through residue 35. (Note in proof: the Ala substitution shown at position 30 of rat neurophysins belongs at position 29).

physin and a single vasopressin-associated neurophysin. This conclusion has been supported by the immunological relationships among the different rat neurophysins (50). Similarly, among the three principal neurophysins (22) of the cow, an immunological identity between neurophysin-II and neurophysin-C has been demonstrated which contrasts with major immunological differences between neurophysins-I and -II (94) and which suggests, in agreement with other studies (94a) that neurophysin-C is derived from neurophysin-II.

The existence of different structural classes of the neurophysins was first shown by a comparison of the sequences of bovine neurophysins-I and -II and porcine neurophysin-I (95). As seen in Figure 1, the amino terminal sequence of bovine-I differs in 5 positions from that of the other two proteins. The fact that two neurophysins within the same species differed from each other more than two from different species led to the conclusion that the different structural classes of neurophysins arose by gene duplication early in mammalian evolution. As more neurophysins were sequenced, Chauvet et al (79) proposed that all neurophysins fall within two structural categories defined by the residues in positions 2, 3, 6, and 7; these were the MSEL class (analogous to bovine-II) and the VLDV class (analogous to bovine neurophysin-I). Since several members of the MSEL group (bovine-II, porcine-I, and porcine-III) are biologically compartmentalized with vasopressin (21, 96), and bovine-I is compartmentalized with oxytocin (21), it has also been suggested (97) that the two classes can be categorized as vasopressin-associated (MSEL) and oxytocin-associated (VLDV) respectively, such categorization also being compatible with the hormonal associations of human neurophysins (76, 84, 97). In Figure 1, the structures of the neurophysins are classified with respect to the MSEL-VLDV nomenclature. Schlesinger et al (27) have pointed out some of the limitations of this nomenclature, illustrated by the fact that the sequences of rat neurophysins-II and -I (Figure 1) fit imperfectly into the MSEL and VLDV classes, respectively. Additionally, rat neurophysin-II has been classified as oxytocin-associated and rat-I as vasopressin-associated (39, 40). Therefore, while the presence of different neurophysins within a species indicates that gene duplication occurred during evolution, the classification of neurophysins into two distinct structural groups, each biosynthetically related to a different hormone, and each maintaining its unique primary structural features during evolution, is not strictly supported by the present state of the field.

Figure 2 illustrates an important aspect of all neurophysin sequences, duplication of the same sequence within a single polypeptide chain (95). As shown for the bovine neurophysins, residues 12–31 show a high degree of homology ($\geqslant 50\%$) with residues 60–77 within the same chain; an evolutionary mechanism for the duplication has been suggested (95). It is inter-

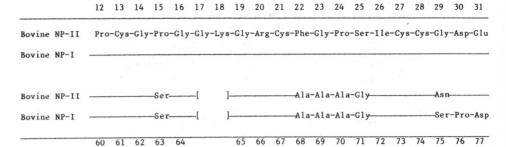

Figure 2   Internal duplication in bovine neurophysins (*NP*).

esting that the internal duplication in bovine-I is slightly less complete than in bovine neurophysin-II but, according to the criteria of Chou & Fasman (98) the substitutions at positions 75 and 76 should not generate important differences in preferred local conformation. The relative conformations of the duplicated segments within each chain are of particular significance when considering the interaction of neurophysin with hormones. A key determinant of these conformations will be disulfide pairing. However, proposed neurophysin disulfide-pairing schemes (84, 99) merit reassessment in view of the erroneous bovine neurophysin-II sequence upon which the original assignments were based (cf. 79, 85, 91).

NEUROPHYSIN CONFORMATION AND SELF-ASSOCIATION IN THE ABSENCE AND PRESENCE OF LIGANDS

Neurophysin conformation has been the subject of a number of physical-chemical investigations. A model has also been proposed solely from theoretical considerations (100) but utilizing the subsequently revised structure for bovine neurophysin-II. The high glycine and proline content of neurophysin is reflected in a relatively low α-helix content, circular dichroism (CD) studies suggesting less than 5% α-helix and approximately 40% β-structure in the absence of ligands (1, 93), and Raman studies indicating $\sim$ 20% α-helix and < 40% β-structure (101); disparate results from the two techniques may reflect contributions of the highly optically active disulfides (102) to far UV optical activity. Ligand-binding to neurophysin is associated with major circular dichroism changes but these have not been interpretable in terms of changes in secondary structure (103); Raman studies, however, suggest a ligand-induced increase in α-helical content (101).

With reference to tertiary structure, potentiometric titrations in the absence of ligands (93) suggest that none of the prototropic residues participates in strong intramolecular interactions. However, spectroscopic studies indicate weak intramolecular interactions involving several titratable residues. The slightly elevated pK_a of Tyr-49, when nitrated, suggests its proximity to a carboxylate (103). This proximity has been confirmed by fluorescence studies of the native protein which indicate that Tyr-49 is partially quenched at neutral pH by the ionized side chain of either Glu-46 or Glu-47 (94a); weak carboxylate-phenol hydrogen bonding is possible. NMR shifts of Tyr-49 ring protons suggest ring stacking of Tyr-49 with a phenylalanine (104, 105), an interpretation compatible with the incomplete exposure of Tyr-49 to solvent demonstrated by UV absorbance studies (30, 106, 107); it is unlikely that either the NMR shift or the apparent solvent-shielding result from the carboxylate effect since neither is affected by carboxyl protonation (S. Lord, S. Sur, and E. Breslow, unpublished observations). In the presence of bound hormones or peptides, Tyr-49 appears to become completely solvent-exposed (30, 106, 107) and its weak interactions with carboxylates and phenylalanines are lost as evidenced by the normalization of its pK and NMR behavior (103, 104). However, in the bound state, the fluorescence of Tyr-49 (94a) and the optical activity of nitrated Tyr-49 (103, 105) are so enhanced as to indicate an unusual and probably restricted environment. Other titratable residues participating in weak interactions are the α-amino of bovine neurophysin-II and one or more neurophysin carboxylates. Proton NMR studies suggest that the protonated α-amino of bovine neurophysin-II forms a weak intramolecular salt bridge with an unprotonated side-chain carboxyl; this salt bridge is absent in bovine neurophysin-I (108). Differences between neurophysins-I and -II in the pH-dependence of hormone binding (30, 109, 110) and optical activity (111) suggest conformationally imposed differences in the pK_a of one or more critical carboxyls (108), and it has been suggested that these conformational differences may be mediated by differences in the amino-terminal interactions (108). The pK_a of the single histidine of neurophysin-I is normal (93, 112), and this group can be modified with no observable effects on neurophysin properties (113, 114); only trivial effects of binding on the histidine are seen (112, 113).

Self-association of neurophysin has been demonstrated by gel-filtration (115) and analytical ultracentrifugation (24, 93, 116–118). Sedimentation equilibrium studies of bovine neurophysins-I and -II near neutral pH in the absence of ligands indicate the presence of a monomer ⇌ dimer equilibrium with a dimerization constant of 5–8 $\times$ $10^3 M^{-1}$ (93, 117, 118); however, it has not been strictly established that higher oligomers are not also present. The groups involved in dimerization are unknown. Tyr-49 does not appear to be involved (94a, 118) but both the kinetics of dimerization and an apparent lack of pH-dependence suggest that apolar interactions predominate (119). In the presence of bound oxytocin, vasopressin, or small peptide analogues of the hormones (see below), an increase in sedimentation velocity occurs (120) which has been shown to be due to preferential binding by the dimer (117, 118); a report (117) that high concentrations of vasopressin cause reversion to monomer has since been rescinded (118). Recent NMR studies (121, 122) suggest that in the liganded state aggregates of molecular weight higher than the dimer may be present. This conclusion is supported by preliminary X-ray crystallographic analysis of a peptide complex of bovine neurophysin-II; the assymmetric unit has a molecular weight of 40,000 (122a).

NUMBER AND RELATIVE AFFINITIES OF BINDING SITES ON NEUROPHYSIN; SITE-SITE INTERACTIONS

The presence of duplicated segments within the neurophysin polypeptide chain raises the question as to whether there is more than one binding site per chain. Under standard salt conditions, almost all studies are in agreement on a single binding site per chain for oxytocin with a maximum binding constant at 25°C between 10^5 and 10^6 M^{-1} (cf 29, 109, 117, 120, 123); one study however suggests the presence of a second very weak site for oxytocin under these conditions (124). Additionally, a single thermodynamically significant site for di- and tripeptide analogues of the amino-terminal region of the hormones has been demonstrated (1, 30, 41, 113); the specificity requirements for binding of these peptides and the identity of physical-chemical changes associated with peptide-binding and oxytocin-binding have been interpreted to indicate that the oxytocin- and peptide-binding sites are largely identical (93, 105, 120). Vasopressin has been shown to bind to neurophysin with an affinity similar to that of oxytocin and to displace bound oxytocin (28, 29); although there has been one report of noncompetitive interactions between oxytocin and vasopressin on binding to neurophysin (109) the validity of these competition studies has been questioned (1). Nonetheless, equilibrium dialysis studies by Cohen and co-workers (30, 107, 109, 118) have consistently demonstrated two binding sites per chain for vasopressin at standard salt concentrations. The initial claim by this group that these two sites were thermodynamically equivalent (30, 109) was disputed (105, 120) and more recent studies from Cohen's laboratory support the concept of a significant thermodynamic inequality between the two sites (117, 118), the second site on bovine neurophysin-II being particularly weak; the weakness of the second vasopressin-binding site on bovine neurophysin-II may account for the failure to detect it in earlier investigations (29). Equilibrium dialysis studies elsewhere (124) also support the concept of a second but weaker site for vasopressin, and indicate variable expression of this second site depending on conditions of pH and temperature. There is good reason to believe that, at standard ionic strength, the stronger vasopressin site is similar or identical to the single oxytocin site. With the exception of small CD differences, binding of either hormone in 1:1 ratio has so far led to the same physical-chemical changes in both the hormone and the protein (103, 105, 106, 118, 120, 125, 126) and both hormones show similar binding kinetics (127). Additionally, nitration of Tyr-49 (128) has no effect, at standard ionic strength, on the binding of oxytocin, small peptides, or the first mole of vasopressin, but does block binding of the second mole of vasopressin (107). The reality of a second hormone- or peptide-binding site and its relationship to Tyr-49 has been given further support by two recent findings. First, in the presence of 1.4 M LiCl, two sites for oxytocin are present (117, 118); the second oxytocin site in 1.4 M LiCl, like the second vasopressin site under normal salt conditions, is blocked by nitration of Tyr-49 (118). Second, spin labelled small peptide analogues of the amino-terminal sequence of the hormones have been shown by NMR studies to bind very weakly to a second site immediately adjacent to Tyr-49 (129). Therefore, the presence of a second site or the potential for a second site, proximal to Tyr-49, appears to have been demonstrated by two completely different experimental approaches and presumably in part reflects the internal duplication.

Related to the presence of the second site on each chain is the presence of interactions among neurophysin-binding sites. NMR spin label studies indicate that the second site on each chain is thermodynamically linked to occupancy of the stronger site (129), a view shared elsewhere (118); possible mechanisms of this linkage will be discussed below. In addition to the thermodynamic linkage between the strong and weak sites on each chain, there is a weak linkage between the two strong sites on each neurophysin dimer. Thus, under conditions where the weak site is unoccupied (i.e. oxytocin binding, peptide binding etc) Scatchard binding isotherms are nonlinear at neutral pH, reflecting weak positive interactions between the two strong sites on the dimer, and preferential binding by the dimer (111, 117, 123, 130). Preferential binding by the dimer is also reflected in the dependence of binding on protein concentration (111, 117, 130).

Does the existence of a second site have any physiological significance? As it turns out, the extremely high hormone concentrations in the posterior pituitary (see above) should allow occupancy of the second site, even if it were weaker than the first by a factor of 10^3, if hormone-protein ratios are greater than 1. If the two sites are not grossly disparate in affinity, the second site could be partially occupied in the presence of 1:1 hormone-protein ratios. The available data on intragranular hormone-protein ratios do not settle the question. Studies in bovine and rat pituitary point to a 1:1 ratio between each hormone and its associated neurophysin (131, 75).

However, in porcine pituitary, although the total hormone: neurophysin ratio is reported to be 1 : 1 (49), there are several reports (49, 68, 96) of a 2 : 1 excess of oxytocin over its putative neurophysin (porcine neurophysin-II). Poor correlation between neurophysin and hormone in sheep pituitary has also been observed (53), although in this instance neurophysin appears to be in excess. When considering the binding of hormone in vivo, potential physiological modifiers of neurophysin-hormone interaction are also relevant. There are conflicting reports (87, 88) on the affinity for hormones of lipid-associated neurophysin relative to that of lipid-free neurophysin. Earlier reports (e.g. 132) of an effect of calcium ion on hormone-neurophysin interaction have not been confirmed (1, 30, 93, 102).

MECHANISM OF HORMONE-NEUROPHYSIN INTERACTION: RESIDUES INVOLVED IN THE INTERACTION, THERMODYNAMICS, AND KINETICS

Based on data presented above, the binding sites on neurophysin can be most conservatively described as follows: (a) there is a principal site, capable of binding either oxytocin, vasopressin, or small peptide analogues of the amino terminus of the hormones; this site is unaffected by nitration of Tyr-49. (b) There is also a second site, generally markedly weaker than the principal site, which is proximal to Tyr-49 as judged by spin label studies (129) and the effects of nitration of Tyr-49 (118); the affinity of this second site depends on the neurophysin (117), the ligand (30), and environmental variables (117, 118, 124, 129). Relatively little is known about interactions at the weaker site. However, as described below, the mechanism of interaction at the stronger site has been extensively investigated.

Figure 3 shows the structure of oxytocin; vasopressin is the same but has phenylalanine in position 3 and arginine or lysine in position 8. As shown in Figure 3, the principal residues identified as binding to the strong site are the α-amino and the side chains of residues 1–3; direct interactions of elements of the hormone backbone are possible but have not been demonstrated. Interactions involving the side chains of residues 1–3 appear to be hydrophobic (105, 120) and have been observed by a variety of techniques. The importance of Tyr-2 was shown by the effects of substitutions for Tyr-2 on binding; only peptides containing a hydrophobic aromatic residue in position 2 bind, with phenylalanine equal in effectiveness to tyrosine (28, 105, 120, 133). The role of Tyr-2 has been confirmed by NMR (104, 112, 125, 134–137), CD (103), UV absorption (106), and fluorescence (94a); no evidence has been obtained supporting ring stacking of Tyr-2 with an aromatic residue on the protein, the data instead indicating apolar interactions of Tyr-2 in a sterically restricted environment (105, 112). Comparable studies have indicated participation of the side chains of residues 1 and 3 in apolar interactions, although the specificity requirements of the protein for these residues are far less restrictive than for residue 2 (104, 105, 120, 121, 134, 137, 138). Salt bridge (ion-pair) formation involving the protonated α-amino of the hormones and an ionized neurophysin side-chain carboxyl has been deduced from the pH-dependence of binding, changes in H^+ ion equilibria associated with binding, and from the loss of binding associated with replacement of the α-amino by a hydrogen or hydroxyl group (28, 56, 93, 109, 111, 139, 140).

The importance of residues 1–3 to binding is thermodynamically demonstrated by the fact that the free energy of binding of tripeptides resembling the amino-terminal hormone sequence is 2/3 that of the binding energy of the hormones (105, 120). The difference in binding affinity between the hormones and tripeptides suggests either that other residues on the hormone remain to be identified as binding participants or that the unbound hormone is preconstrained into a conformation more favorable to binding than the unbound peptide (105, 120). Strong interactions involving residues 7–9 of oxytocin have been discounted (29, 41, 137, 138, 141) although weak interactions at residue 9 (121), residue 4 (28), and perhaps elsewhere in the ring are possible. Preliminary thermodynamic studies (105) and possibly spectroscopic studies (134) suggest that interactions at residue 3 are

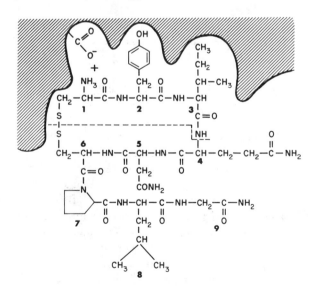

Figure 3 Oxytocin and its principal interactions with neurophysin at the strong hormone binding site. The shaded region represents the binding site on the protein.

stronger in the hormone than in the tripeptides, arguing for a contributory role of hormone conformation. However, Blumenstein & Hruby (121) have interpreted NMR changes in the hormone associated with binding as suggesting a conformational change in the hormone; this is also suggested by changes in hormone disulfide optical activity accompanying binding (120).

The thermodynamics of neurophysin-peptide binding reveal several provocative features of the interaction. First, binding of small peptides to the hormone site is an enthalpy-driven reaction accompanied by a negative entropy change (111); effects of temperature on hormone binding (124) indicate similar thermodynamics for the intact hormones. The origin of the large negative enthalpy change is unknown; it does not appear to be associated with salt bridge formation per se (111). Second, from the loss in binding at neutral pH associated with loss of either the hormone α-amino or the aromatic side chain in position 2, the combined free energy contribution to binding of both of these groups has been calculated as more negative than -9 Kcal/mole (105). Comparison of this minimum contribution with that of the weak binding ($\Delta G^0 = -2.3$ Kcal/mole) of Gly-Tyr NH_2 (which contains both residues), indicates that energetically uphill changes associated with changes in conformation and loss of translational, vibrational, and rotational degrees of freedom on binding, contribute a positive free energy to binding of at least 7 Kcal/mole (105). In this context, peptides lacking either an α-amino or aromatic residue in position 2 cannot bind with measurable affinity because the binding energy contribution of the residual group, even when combined with that of the side chains of residues 1 and 3 (105), is insufficient to counterbalance the positive free energy change associated with conformational change and loss of motional freedom.

A particularly surprising result from thermodynamic studies is the indication of a complex role for the hormone α-amino. Below pH 5, binding affinity initially diminishes with decreasing pH, as expected for the titration of the protein carboxyl with which the hormone α-amino forms a salt-bridge (109, 111). But binding persists and becomes independent of pH below pH 2 (111). Binding at very low pH has been attributed to binding to protein in which the active site carboxyl is protonated and in which the salt bridge of the neutral pH complex is absent (111). Based on this assumption, the difference in binding free energy between low and neutral pH has been used to calculate a free energy of -2.4 Kcal/mol for the contribution of the salt bridge to the interaction (111). This value is significantly less negative than the value of -4.6 Kcal/mole calculated for the minimum total contribution of the α-amino to binding at neutral pH (105). Moreover, at very low pH, where salt bridge formation is unimportant, no thermodynamically significant binding to the principal hormone site occurs with deaminooxytocin or hydroxymercaptopropionic oxytocin (111), both of which lack the α-amino, although NMR evidence suggestive of at least some deaminooxytocin binding at low pH has been reported (137). While these data may signify that the protonated α-amino alters the conformation of the hormone to one more favorable to binding (111), thermodynamically important conformational effects of the α-amino have yet to be firmly established. An alternate explanation is that the protonated α-amino participates in important but as yet unidentified bonding interactions with the protein additional to that of classic salt bridge formation. It is relevant in this context that different explanations of the pH-dependence of peptide-neurophysin interaction have been considered that would generate a more negative free energy for salt bridge formation, but are, nevertheless, considered unlikely (111). In either event, the critical role of the α-amino in binding has interesting implications with respect to the common precursor theory of neurophysin and hormone biosynthesis. Thus, the intermolecular interactions between hormone and neurophysin cannot be identical to the intramolecular interactions between hormonal and neurophysin segments of a common precursor unless the half-cystine at position 1 of the hormones also represents the first residue of the precursor. If the hormone does not represent the amino-terminal segment of the precursor, then the system has evolved such that new interactions are assured between the precursor porducts; in this context, these interactions have significance of their own and cannot be considered as vestigial.

The kinetics of neurophysin-hormone interaction are controversial and have been of particular interest because of their implications for the interpretation of NMR data. The rate of exchange of tripeptides between the free and bound state appears to be fast on an NMR time scale (104, 135). There is also general agreement on fast exchange for the hormones at low pH (104, 122, 137). However, the conclusion from early proton NMR studies that vasopressin-neurophysin exchange rates were slow at neutral pH (104) was subsequently challenged by proton NMR studies elsewhere (134, 136) which were interpreted as indicating fast hormone exchange at neutral pH. Relaxation kinetic studies of hormone-neurophysin interaction give strong support to the "slow exchange" school, indicating a dissociation rate for oxytocin- and vasopressin-complexes at neutral pH of 10–20/sec (127). Additionally, Blumenstein et al (125) have shown that broadening of oxytocin proton resonances by neurophysin at neutral pH, originally attributed to fast exchange (134), arises from a nonspecific viscosity increase, and they have provided compelling evidence from ^{13}C NMR studies that oxytocin-neurophysin exchange is slow at neutral pH (121). However, as observed by NMR, the kinetics of vasopressin binding at neutral pH may become complex at hormone:protein ratios in excess of unity (125, 126) and the "observed" exchange rates for oxytocin depend to some extent on the hormone residue being monitored (cf 121, 125, 126, 138). Possible explanations of these phenomena include subsets of binding sites (122), different microscopic exchange rates of different segments of the hormone in the bound state (121), and possibly, also, an effect of secondary site occupancy on the kinetics of binding to the primary site. A probable explanation of the increase in hormone exchange rate at low pH is the absence of the amino-carboxylate salt bridge in the low pH complex (see above).

Relatively little information is available on the identity of residues at the active site of the protein. Walter & Hoffman (142) in preliminary affinity labelling studies identified either Asp-30 or Glu-31 as a binding site on bovine neurophysin for the α-amino of arginine vasopressin. Since human neurophysin-I contains an Asp→Glu substitution in position 30 (84), it might be predicted that the exact residue is Glu-31 and this has been independently confirmed (R. Walter, personal communication). The demonstration of a negative Nuclear Overhauser effect between the ortho ring protons of Tyr-49 and the aromatic ring protons in position 2 of bound tripeptides has indicated proximity between these two residues (104, 135). Because only a single peptide site is thermodynamically significant, the original inference from these data was that Tyr-49 was close to residue 2 of peptides bound to the strong binding site (e.g. 105). However, the true picture appears more complex. Recently, two peptide spin labels were synthesized in which nitroxides were placed in positions analogous to position 3 of binding peptides and to an extended side chain at position 1 (113, 129). NMR studies of the effects of these on the relaxation rate of Tyr-49 ring protons indicate that the nitroxide of residue 3 is $\geqslant 15$ Å from the ring of Tyr-49, when the spin label is bound to the strong site (143). Since this nitroxide and the aromatic ring protons in position 2 of the peptide are approximately 10 Å apart, the results suggest that the distance between Tyr-49 and Tyr-2 of hormones bound to the strong site must be $\geqslant 5$ Å. This conclusion has been supported by recent fluorescence studies (94a) which indicate relatively little energy transfer between Tyr-49 and Tyr-2 when only the strong site is occupied. Moreover, the demonstration from spin label studies of a second very weak peptide site immediately adjacent to Tyr-49 (see above) suggests that the Nuclear Overhauser effect is, at least partially, mediated by binding of peptide to the weaker site (129). On the other hand, spin label studies allow proximity between Tyr-49 and an extended side chain of residue 1 of peptides bound to the strong site (129). Photochemical studies have indicated that hormone binding ability is reduced when Tyr-49 is photooxidized (114); this result is compatible with some role of Tyr-49 in binding to the principal site, but the effect of this oxidation on protein conformation was not determined.

POSSIBLE MECHANISMS FOR NEUROPHYSIN
HALF-OF-THE-SITES REACTIVITY

The identification of Glu-31, which lies in the first of the duplicated segments, as a participant in binding, together with the apparent potential for two binding sites per polypeptide chain, one only of which is most generally expressed, suggests that neurophysin exhibits half-of-the-sites reactivity (117). If this is correct, neurophysin is a particularly interesting model for the elucidation of the mechanism of this phenomenon, since the duplicated segments are nonidentical in sequence and are therefore distinguishable by chemical and physical-chemical means. There is no reason at present to exclude possible direct participation in binding of the nonduplicated segments, particularly those within the central core (144). Nonetheless, we suggest here (Figure 4) three possible mechanisms to explain the role of the duplicated segments, in the context of which existent and future data can be assessed. Each of the models can be modified to include participation of the nonduplicated core; models I and II have been previously cited (129). In Model I, the duplicated segments are viewed as nonidentical in conformation and binding properties in the unliganded state, this nonidentity arising either from the differences in primary structure and/or from differences in interactions with nonduplicated segments. Binding to the stronger site (here shown as the amino-terminal duplicated segment) leads to a conformational change which increases the binding affinity of the weaker site; i.e. in this model interactions between the sites are positively cooperative (118, 129). In Model II, the duplicated segments are shown as functionally alike in the unliganded state but are so oriented that binding to one segment reorients and weakens the affinity of the second segment; i.e. in this model the interactions between the sites are negative (129, 143). Model III is termed the "chelate" model. Here the duplicated segments are proposed to both participate in binding of the first ligand although each is capable of binding separately; on binding of the second ligand, the chelate is disrupted and the individual segments each bind a single hormone. The participation of both segments in binding of the first ligand strengthens the first binding constant relative to the second. As shown in Figure 4, all models are equally capable of explaining a greater proximity of Tyr-49 to the second site than to the first site. Also, as previously indicated (129), both models I and II explain interactions between the two sites, i.e. the appearance of a second very weak site for peptide spin labels immediately adjacent to Tyr-49 and of increased availability after the stronger site is occupied. The chelate model can also be reconciled with this observation. However, there are two implications to the chelate model that are significantly different from those of the other models. First, it suggests that the internal duplication arose not to allow binding of a second hormone molecule, but to facilitate binding the first hormone, a concept compatible with the lack of completeness of internal duplication. Second, the chelate model predicts that binding of the second ligand will alter the mode of binding of the first. Interestingly, Bothner-By (145) has presented NMR evidence that a chemical shift of the vasopressin tryosine protons in the 1:1 complex is absent when vasopressin is present in excess. The chelate model is also particularly compatible with the sensitivity of the apparent number of sites to experimental variables. The greater number of binding sites available to vasopressin than to oxytocin under most conditions can be accommodated particularly easily by this model by assuming that vasopressin, with its extra charged side chain, positions less favorably than oxytocin between the dup-

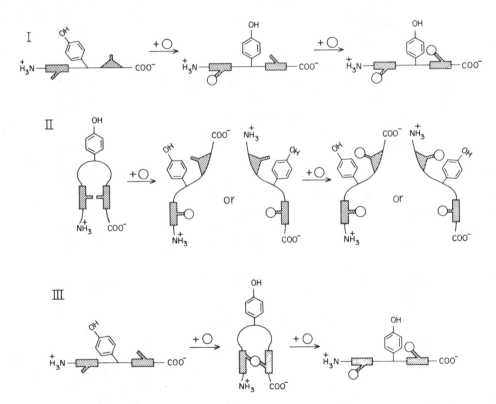

Figure 4 Three models for the role of the duplicated segments in half-of-the-sites reactivity. The duplicated segments are shown as shaded areas. A change in conformation from triangular to rectangular signifies an increase in binding affinity accompanying binding of ligand (*O*) while the reverse change signifies a decrease in affinity; the relative affinities of similarly depicted segments within a single chain are not necessarily identical but are indicated by the order in which ligands bind to each (see text for details).

licated segments, and binds more strongly than oxytocin to the individual segments. It is tempting to explain the effects of LiCl, using any of the models, by assuming that it causes a conformational change in the protein; however, no such LiCl-induced change has been detected either elsewhere (117) or in our own laboratory. Alternate possibilities are either that LiCl alters the conformation of oxytocin to resemble vasopressin or that, in the context of the chelate model, LiCl salt effects weaken the bidentate mode of binding, leading to increased expression of the second site relative to the first. The latter explanation is supported by the observation (117) that the affinity of the stronger site diminishes in 1.4 M LiCl.

CONCLUSIONS

An evaluation of the biological significance of different classes of neurophysins and of the internal duplication in neurophysin awaits the determination of whether there is a common neurophysin-hormone precursor and more precise quantitation of hormone:neurophysin ratios in vivo. Although neurophysin-lipid interactions have been cited here only briefly, further studies of these are needed to assess both their biological importance and the sites on neurophysin participating in lipid-binding. From the physical-chemical point of view, neurophysin remains a very useful model system for the study of the thermodynamics of protein-protein interactions and of potential mechanisms of half-of-the-sites reactivity. An understanding of the chemical significance of the internal duplication in neurophysin and of the half-of-the-sites phenomenon awaits determination of the three-dimensional conformation of the protein and further identification of the residues on neurophysin involved in interaction with hormone. In this respect, affinity labelling studies are in progress in several laboratories (e.g. 110, 142) and crystals of a neurophysin-peptide complex, suitable for X-ray diffraction, have recently been obtained (122a).

Literature Cited

1. Breslow, E. 1974. *Adv. Enzymol.* 40: 271–333
2. Walter, R. ed. 1975. *Ann. NY Acad. Sci.* 248:1–512
3. Pickering, B. T., Jones, C. W. 1978. In *Hormonal Proteins and Peptides,* ed. C. H. Li, 5:103–58. New York: Academic
4. Scharrer, E., Scharrer, B. 1954. *Recent Prog. Horm. Res.* 10:183–240
5. Sachs, H. 1969. *Adv. Enzymol.* 32: 327–72
6. Sachs, H., Fawcett, P., Takabatake, Y., Portonova, R. 1969. *Recent Prog. Horm. Res.* 25:447–91

7. Nagasawa, J., Douglas, W. W., Schulz, R. A. 1970. *Nature.* 227:407–9
8. Matthews, E. K., Legros, J. J., Grau, J. D., Nordmann, J. J., Dreifuss, J. J. 1973. *Nature New Biol.,* 241:86–88
9. Bisset, G. W., Clark, B. J., Errington, M. L. 1971. *J. Physiol.* 217:111–31
10. Schrier, R. W., Verroust, P. J., Jones, J. J., Fabian, M., Lee, J., de Wardener, H. E. 1968. *Clin. Sci.* 35:433–43
11. Vandesande, F., Dierickx, K. 1975. *Cell Tissue Res.* 164:153–62
12. Dierickx, K., Vandesande, F., DeMey, J. 1976. *Cell Tissue Res.* 168:141–51

13. Sokol, H. W., Zimmerman, E. A., Sawyer, W. H., Robinson, A. G. 1976. *Endocrinology* 98:1176–88
14. Zimmerman, E. A., Defendini, R., Sokol, H., Robinson, A. G. 1975. *Ann. NY Acad. Sci.* 248:92–111
15. Livitt, B. G. 1978. *Int. Rev. Cytol.* Suppl. 7:53–237
16. Acher, R., Chauvet, J., Olivry, G. 1956. *Biochim. Biophys. Acta* 22:421–27
17. Acher, R. G., Manoussos, G., Olivry, G. 1955. *Biochim. Biophys. Acta* 16:155–56
18. Sawyer, W. H. 1961. *Pharmacol. Rev.* 13:225–77
19. Ginsburg, M., Ireland, M. 1966. *J. Endocrinol.* 35:289–98
20. Dean, C. R., Hope, D. B. 1967. *Biochem. J.* 104:1082–88
21. Dean, C. R., Hope, D. B., Kazic, T. 1968 *Br. J. Pharmacol.* 34:192P–93P
22. Rauch, R., Hollenberg, M. D., Hope, D. B. 1969. *Biochem. J.* 115:473–79
23. Uttenthal, L. O., Hope, D. B. 1970. *Biochem. J.* 116:899–909
24. Burford, G. D., Jones, C. W., Pickering, B. T. 1971. *Biochem. J.,* 124:809–13
25. Walter, R., Breslow, E. 1974. *In Methods in Neurochemistry* ed. N. Marks, R. Rodnight, 2:247–79. New York: Plenum
26. Ellis, H. K., Watkins, W. B., Evans, J. J. 1972. *J. Endocrinol.* 55:565–75
27. Schlesinger, D. H., Pickering, B. T., Watkins, W. B., Peek, J. C., Moore, L. G., Audhya, T. K., Walter, R. 1977. *FEBS Lett.* 80:371–73
28. Breslow, E., Abrash, L. 1966. *Proc. Natl. Acad. Sci. USA* 56:640–46
29. Breslow, E., Walter, R. 1972. *Mol. Pharmacol.* 8:75–81
30. Cohen, P., Camier, M., Wolff, J., Alazard, R., Cohen, J., Griffin, J. H. 1975. *Ann. NY Acad. Sci.* 248:463–79
31. Dreifuss, J. J. 1975. *Ann. NY Acad. Sci.* 248:184–89
32. Livingstone, A., Lederis, K. 1971. *Mem. Soc. Endocrinol.* 19:233–61
33. Gainer, H., Peng Loh, Y., Sarne, Y. 1977. In *Peptides in Neurology* ed. H. Gainer, pp. 183–219. New York: Plenum
34. Pickering, B. T., Jones, C. W. 1971. In *Subcellular Organization and Function in Endocrine Tissues* ed. H. Heller, K. Lederis, pp. 337–51. New York: Cambridge Univ. Press
35. Sachs, H., Goodman, R., Osinchak, J., McKelvy, J. 1971. *Proc. Natl. Acad. Sci. USA* 68:2782–86
36. Sachs, H., Saito, S., Sunde, D. 1971. See Ref. 34, pp. 325–36
37. Walter, R., Audhya, T. K., Schlesinger, D. H., Shin, S., Saito, S., Sachs, H. 1977. *Endocrinology,* 100:162–74
38. Friesen, H. G., Astwood, E. B. 1967. *Endocrinology* 80:278–87
39. Sunde, D. A., Sokol, H. W. 1975. *Ann. NY Acad. Sci.* 248:345–61
40. Pickering, B. T., Jones, C. W., Burford, G. D., McPherson, M., Swann, R. W., Heap, P. F., Morris, J. F. 1975. *Ann. NY Acad. Sci.* 248:15–32
41. Chaiken, I. M., Randolph, R. E., Taylor, H. C. 1975. *Ann. NY Acad. Sci.* 248:442–50
42. Menendez-Botet, C. J., Breslow, E. 1975. *Biochemistry* 14:3825–35
43. Mendelson, I. S., Walter, R. 1978. In *Hypothalamic Hormones—Chemistry, Physiology, and Clinical Applications* ed. D. Gupta, W. Voelter, pp. 185–97. Germany: Verlag Chem.
44. Gainer, H., Sarne, Y., Brownstein, M. J. 1977 *Science* 195:1354–56
45. Brownstein, M. J., Robinson, A. G., Gainer, H. 1977. *Nature* 269:259–61
46. Brownstein, M. J., Gainer, H. 1977. *Proc. Natl. Acad. Sci. USA* 74:4046–49
47. Gainer, H., Brownstein, M. J. 1978. in *The Cell Biology of Hypothalamic Neurosecretory Processes,* ed. C. Kordon, J. D. Vincent. Paris, France: Centre National de la Recherche Scientifique (CNRS)
48. Morris, J. F., Cannata, M. A. 1973. *J. Endocrinol.* 57:517–29
49. Dax, E. M., Johnston, C. I. 1979. *J. Neurochem.* In press
50. North, W. G., Morris, J. F., LaRochelle, F. T., Valtin, H. 1976. In *Neurohypophysis,* ed. A. M. Moses, L. Share. pp. 43–52. Basel: S. Karger
51. North, W. G., Valtin, H., Morris, J. F., LaRochelle, F. T. Jr. 1977. *Endocrinology* 101:110–18
52. Moore, G. J., Swann, R. W., Lederis, K. 1977. *J. Endocrinol.* 75:341–42
53. Chauvet, M. T., Coffe, G., Chauvet, J., Acher, R. 1975. *FEBS Lett.* 53:331–33
54. Pettengill, O. S., Faulkner, C. S., Wurster-Hill, D. H., Maurer, L. H., Sorenson, G. D., Robinson, A. G., Zimmerman, E. A. 1977. *J. Natl. Cancer Inst.* 58:511–18
55. Hamilton, B. P., Upton, G. V., Amatruda, T. T. Jr. 1972. *J. Clin. Endocrinol. Metab.* 35:764–67
56. Ginsburg, M., Ireland, M. 1964. *J. Endocrinol,* 30:131–45
57. Diliberto, E. J. Jr., Axelrod, J., Chaiken, I. M. 1976. *Biochem. Biophys. Res. Commun.* 73:1063–1067

58. Rudman, D., Del Rio, A. E., Garcia, L. A., Barnett, J., Howard, C. H., Walker, W., Moore, G. 1970. *Biochemistry* 9:99–108
59. Robinson, A. G., Michelis, M. F., Warms, P. C., Davis, B. B. 1974. *J. Clin. Endocrinol. Metab.* 39:913–18
60. Trygstad, O., Foss, I., Sletten, K. 1975. *Ann. NY Acad. Sci.* 248:304–15
61. Robinson, A. G., Michelis, M. F., Warms, P. C., Davis, B. B. 1975. *Ann. NY Acad. Sci.* 248:317–21
62. Rudman, D., Chawla, R. K., Khatra, B. S., Yodaiken, R. E. 1975. *Ann. NY Acad. Sci.* 248:324–35
63. Cort, J. H., Sedlakova, E., Kluh, I. 1975. *Ann. NY Acad. Sci.* 248:336–44
64. Fawcett, C. P., Powell, A. E., Sachs, H. 1968. *Endocrinology* 83:1299–1310
65. Ginsburg, M., Jayasena, K. 1968. *J. Physiol.* 197:53–63
66. Robinson, A. G., Zimmerman, E. A., Engleman, E. G., Frantz, A. G. 1971. *Metabolism* 20:1138–47
67. Legros, J. J., Louis, F., Groeschel-Stewart, U., Franchimont, P. 1975. *Ann. NY. Acad. Sci.* 248:157–69
68. Johnston, C. I., Hutchinson, J. S., Morris, B. J., Dax, E. M. 1975. *Ann. NY Acad. Sci.* 248:272–80
69. Sinding, C., Robinson, A. G. 1977. *Metabolism* 26:1355–70
70. Zimmerman, E. A., Carmel, P. W., Husain, M. K., Ferin, M., Tannenbaum, M., Frantz, A. G., Robinson, A. G. 1973. *Science* 182:925–27
71. Robinson, A. G., Zimmerman, E. A. 1973. *J. Clin. Invest.* 52, 1260–67
72. Legros, J. J. 1976. *Les Neurophysines.* Paris, France: Masson. 265 pp.
73. Reinharz, A. C., Czernichow, P., Vallottan, M. B. 1975. *Ann. NY Acad. Sci.* 248:172–81
74. Vandesande, F., Dierickx, K., DeMey, J. 1975. *Cell Tissue Res.* 156:189–200
75. North, W. G., LaRochelle, F. T. Jr., Morris, J. F., Sokol, H. W., Valtin, H. 1978. In *Current Studies of Hypothalamic Function* Part I, ed. K. Lederis Basel: S. Karger In press
76. Robinson, A. G., 1975. *Ann NY Acad. Sci.* 248:246–54
77. Tissot-Berthet, M. C., Reinharz, A. C., Vallotton, M. B. 1975. *Ann NY Acad. Sci.* 248:257–71
78. Legros, J. J. 1975. *Ann. NY Acad. Sci.* 248:281–302
79. Chauvet, M. T., Chauvet, J., Acher, R. 1976. *Eur. J. Biochem.* 69:475–85
80. Wuu, T. C., Crumm, S. E. 1976. *J. Biol. Chem.* 251:2735–39

81. Chauvet, M. T., Codogno, P., Chauvet, J., Acher, R. 1978. *FEBS Lett.* 88: 91–93
82. Chauvet, M. T., Codogno, P., Chauvet, J., Acher, R. 1977. *FEBS Lett.* 80:374–76
83. Schlesinger, D. H., Audhya, T. K., Walter, R. 1978. *J. Biol. Chem.* 253: 5019–24
84. North, W. G., Walter, R., Schlesinger, D. H., Breslow, E., Capra, J. D. 1975. *Ann NY Acad. Sci.* 248:408–22
85. Wuu, T. C., Crumm, S. E. 1976. *Biochem. Biophys. Res. Commun.* 68: 634–39
86. Koenig, H., Mylroie, R. 1975. *Ann NY Acad. Sci.* 248:218–34
87. Pliska, V., Meyer-Grass, M. 1975. *Ann NY Acad. Sci.* 248:235–45
88. Audhya, T. K., Walter, R. 1979. *J. Biol. Chem.* In press
89. Camier, M., Nicolas, P., Cohen, P. 1976. *FEBS Lett.* 67:137–42
90. Audhya, T. K., Walter, R. 1977. *Arch. Biochem. Biophys.* 180:130–39
91. Walter, R., Schlesinger, D. H., Schwartz, I. L., Capra, J. D. 1971. *Biochem. Biophys. Res. Commun.* 44: 293–98
92. Dean, C. R., Hollenberg, M. D., Hope, D. B. 1967. *Biochem. J.* 104:8C–10C
93. Breslow, E., Aanning, H. L., Abrash, L., Schmir, M. 1971. *J. Biol. Chem.* 246:5179–88
94. DeMey, J., Vandesande, F. 1976. *Eur. J. Biochem.* 69:153–62
94a. Sur, S. S., Rabbani, L. D., Libman, L., Breslow, E. 1979. *Biochemistry.* In press
95. Capra, J. D., Kehoe, J. M., Kotelchuck, D., Walter, R., Breslow, E. 1972. *Proc. Natl. Acad. Sci. USA* 69:431–34
96. Pickup, J. C., Johnston, C. I., Nakamura, S., Uttenthal, L. O., Hope, D. B. 1973. *Biochem. J.* 132:361–71
97. Soloff, M. S., Pearlmutter, A. F. 1978. In *Biochemical Actions of Hormones,* Vol. 6, ed. G. Litwak. New York: Academic. In press
98. Chou, P. Y., Fasman, G. D. 1978. *Ann. Rev. Biochem.* 47:251–76
99. Schlesinger, D. H., Frangione, B., Walter, R. 1972. *Proc. Natl. Acad. Sci. USA* 69:3350–54
100. Smythies, J. R., Beaton, J. M., Benington, F., Bradley, R. J., Morin, R. F. 1976. *J. Theor. Biol.* 63:33–48
101. Liu, C. 1975. PhD thesis. *Raman spectroscopic studies of synthetic polypeptides and protein-peptide complexes.* Mass. Inst. Technol., Cambridge, Mass. 177 pp.

102. Breslow, E. 1970. *Proc. Natl. Acad. Sci. USA* 67:493–500
103. Breslow, E., Weis, J. 1972. *Biochemistry* 11:3474–82
104. Balaram, P., Bothner-By, A. A., Breslow, E. 1973. *Biochemistry* 12:4695–4704
105. Breslow, E. 1975. *Ann. NY Acad. Sci.* 248:423–41
106. Griffin, J. H., Alazard, R., Cohen, P. 1973. *J. Biol. Chem.* 248:7975–78
107. Wolff, J., Alazard, R., Camier, M., Griffin, J. H., Cohen, P. 1975. *J. Biol. Chem.* 250:5215–20
108. Lord, S. T., Breslow, E. 1978. *Int. J. Peptide Protein Res.* In press
109. Camier, M., Alazard, R., Cohen, P., Pradelles, P., Morgat, J. L., Fromageot, P. 1973. *Eur. J. Biochem.* 32:207–14
110. Klausner, Y. S., McCormick, W. M., Chaiken, I. M. 1978. *Int. J. Peptide Protein Res.* 11:82–90
111. Breslow, E., Gargiulo, P. 1977. *Biochemistry* 16:3397–3406.
112. Cohen, P., Griffin, J. H., Camier, M., Caizergues, M., Fromageot, P., Cohen, J. S. 1972. *FEBS Lett.* 25:282–85
113. Lundt, S. L., Breslow, E. 1976. *J. Phys. Chem.* 80:1123–26
114. Fukuda, H., Hayakawa, T., Kawamura, J., Aizawa, Y. 1976. *Chem. Pharm. Bull.* 24:2043–51
115. Burford, G. D., Ginsburg, M., Thomas, P. J. 1969. *J. Physiol.* 205:635–46
116. Burford, G. D., Ginsburg, M., Thomas, P. J. 1971. *Biochim. Biophys. Acta* 229:730–38
117. Nicolas, P., Camier, M., Dessen, P., Cohen, P. 1976. *J. Biol. Chem.* 251:3965–71
118. Nicolas, P., Wolff, J., Camier, M., DiBello, C., Cohen, P. 1978. *J. Biol. Chem.* 253:2633–39
119. Pearlmutter, A. F. 1979. *Biochemistry* In press
120. Breslow, E., Weis, J., Menendez-Botet, C. J. 1973. *Biochemistry* 12:4644–53
121. Blumenstein, M., Hruby, V. J. 1977. *Biochemistry* 16:5169–77
122. Deslauriers, R., Smith, I. C. P., Stahl, G. L., Walter, R. 1978. *Int. J. Peptide Protein Res.* In press
122a. Yoo, C. S., Wang, B. C., Sax, M., Breslow, E. 1978. *J. Mol. Biol.* In press
123. Hope, D. B., Wälti, M. Winzor, D. J. 1975. *Biochem. J.* 147:377–79
124. Glasel, J. A., McKelvy, J. F., Hruby, V.

J., Spatola, A. F. 1976. *J. Biol. Chem.* 251:2929–37
125. Blumenstein, M., Hruby, V. J., Yamamoto, D. M. 1978. *Biochemistry.* 17:497–77
126. Blumenstein, M., Hruby, V. J., Yamamoto, D., Yang, Y. 1977. *FEBS Lett.* 81:347–50
127. Pearlmutter, A. F., McMains, C. 1977. *Biochemistry* 16:628–33
128. Furth, A. J., Hope, D. B. 1970 *Biochem. J.* 116:545–53
129. Lord, S. T., Breslow, E. 1978. *Biochem. Biophys. Res. Commun.* 80:63–70
130. Nicolas, P., Dessen, P., Camier, M., Cohen, P. 1978. *FEBS Lett.* 86:188–92
131. Dean, C. R., Hope, D. B., Kazic, T. 1968. *Br. J. Pharmacol.* 34:193P–94P
132. Ginsburg, M., Jayasena, K., Thomas, P. J. 1966. *J. Physiol.* 184:387–401
133. Hope, D. B., Walti, M. 1973. *Biochem. J.* 135:241–43
134. Alazard, R., Cohen, P., Cohen, J. S., Griffin, J. H. 1974. *J. Biol. Chem.* 249:6895–6900
135. Balaram, P., Bothner-By, A. A., Breslow, E. 1972. *J. Am. Chem. Soc.* 94:4017–18
136. Griffin, J. H., Cohen, J. S., Cohen, P., Camier, M. 1975. *J. Pharm. Sci.* 64:507–11
137. Glasel, J. A., Hruby, V. J., McKelvy, J. F., Spatola, A. F. 1973. *J. Mol. Biol.* 79:555–75
138. Griffin, J. H., DiBello, C., Alazard, R., Nicolas, P., Cohen, P. 1977. *Biochemistry* 16:4194–98
139. Stouffer, J. E., Hope, D. B., duVigneaud, V. 1963. In *Perspectives in Biology,* ed. C. F. Cori, G. Foglia, L. F. Leloir, S. Ochoa, p. 75. Amsterdam: Elsevier
140. Hope, D. B., Walti, M. 1971. *Biochem. J.* 125:909–11
141. Convert, O., Griffin, J. H., DiBello, C., Nicolas, P., Cohen, P. 1977. *Biochemistry* 16:5061–65
142. Walter, R., Hoffman, P. L. 1973. *Fed. Proc.* 32:567 (Abstr.)
143. Lundt, S. L. 1977. *Magnetic resonance studies of neurophysin and its interactions with peptide hormone analogs.* PhD thesis. Cornell Univ. Med. Coll., New York. 187 pp.
144. Capra, J. D., Walter, R. 1975. *Ann. NY Acad. Sci.* 248:397–407
145. Bothner-By, A. A. 1975. *Ann. NY Acad. Sci.* 248:504–12

Ann. Rev. Biochem. 1979. 48:275–92
Copyright © 1979 by Annual Reviews Inc. All rights reserved

ENERGY INTERCONVERSION BY THE Ca^{2+}-DEPENDENT ATPase OF THE SARCOPLASMIC RETICULUM

♦12010

Leopoldo de Meis and Antonio Luiz Vianna

Departamento de Bioquímica, Instituto de Ciências Biomédicas,
Centro de Ciências da Saúde, Universidade Federal do Rio de Janeiro,
Ilha do Fundão, Rio de Janeiro—21.910, Brasil

CONTENTS

275

0066-4154/79/0701-0275$01.00

PERSPECTIVES AND SUMMARY

The sarcoplasmic reticulum is a network of tubules and cysternae surrounding the myofibrils of striated muscle. The membrane of the sarcoplasmic reticulum can be isolated in vesicular form from skeletal muscle homogenates by differential centrifugation. These vesicles retain a highly efficient Ca^{2+} transport system mediated by a membrane-bound ATPase. The ATPase represents 60–80% of the total protein content of the vesicles membrane (1–4). The enzyme can be solubilized (3, 5, 6) and purified as a lipoprotein complex (4, 7–9) with the use of mild detergents. Its molecular weight is in the range of 100,000–120,000 (4, 10). Indirect evidence suggests that the ATPase forms oligomers within the membrane, presumably tri- or tetrameric units (11, 12). In isolated vesicles the hydrolysis of ATP and the translocation of Ca^{2+} into the vesicles lumen involve a transfer of the γ-phosphate of ATP to the membrane-bound ATPase to form a phosphoprotein. This represents an intermediary product in the sequence of reactions leading to Ca^{2+} transport and P_i liberation (1, 3, 13, 14).

In 1971 it was shown that the entire process of Ca^{2+} transport can be reversed (15–17). When vesicles previously loaded with Ca^{2+} are incubated in a medium containing ADP, P_i, Mg^{2+}, and EGTA,[1] it is observed that P_i interacts with the enzyme and forms a phosphoprotein (18, 19), Ca^{2+} is released at a fast rate and ATP is formed (15, 17). These data have led to the proposal that the energy required for the synthesis of ATP is derived from the transmembrane Ca^{2+} gradient (15, 17–19). The simplicity of the sarcoplasmic reticulum membrane and the possibility of identifying a phosphoenzyme during ATP hydrolysis or synthesis has led to the use of this system as a tool for the study of interconversion of osmotic into chemical energy by membrane-bound ATPases. The outcome of recent studies indicates that a small but significant amount of ATP can be synthesized from ADP and P_i with the use of purified ATPase in the absence of Ca^{2+} gradient. The amount of ATP synthesized in these conditions is several orders of magnitude larger than that which could be derived from the chemical equilibrium between ATP, ADP, P_i, and water (20, 21). The synthesis of ATP is promoted by the asymmetrical binding of Ca^{2+} on the two sides of the vesicles membrane. In order to phosphorylate the enzyme by P_i, it is necessary that Ca^{2+} not be bound to a high affinity site located on the outer surface of the membrane. In order to transfer the phosphate from the phosphoenzyme to ADP Ca^{2+} must bind to a site of low affinity located in the inner surface of the vesicle membrane. Thus, for the synthesis of ATP, a large difference of Ca^{2+} concentration on the two sides of the membrane

---

[1]The abbreviations used are: EGTA, ethlylene glycolbis (β-aminoethyl ether) N,N'-tetraacetic acid; NTP, nucleoside triphosphate.

is only needed to meet the difference of affinities of the external and internal Ca^{2+} binding sites. The energy derived from the binding of Ca^{2+} and substrates to the enzyme seems to be involved in the process of ATP synthesis. Therefore the possibility has been raised that the sarcoplasmic reticulum ATPase might interconvert binding energy into chemical energy. The Ca^{2+} affinity of the two binding sites involved in the synthesis of ATP is altered by changing the pH of the medium. This recently led to experiments indicating that the sarcoplasmic reticulum ATPase can also drive the synthesis of ATP when there is a H$^+$ and not a Ca^{2+} concentration difference on the two sides of the membrane (21). If confirmed, this finding raises the possibility that when there is a transmembrane proton gradient, other transport ATPases can be mobilized for the synthesis of ATP, besides those that specifically transport H$^+$.

Finally, the Ca^{2+}-dependent ATPase is similar to the (Na$^+$ + K$^+$) ATPase provided that Ca^{2+} and Na$^+$ are assigned corresponding roles for the two enzymes. Both enzymes are phosphorylated from ATP or P$_i$, both catalyze the synthesis of ATP, and both have a similar primary structure in the neighborhood of the active site of phosphorylation (22–24).

Several excellent reviews have been written on the mechanism of Ca^{2+} transport, ATP hydrolysis, and properties of the sarcoplasmic reticulum ATPase (1, 3, 4, 13, 14, 25–28). In this review emphasis is on the mechanism of ATP synthesis. A recent review on the reversibility of the sarcoplasmic reticulum ATPase has been written by Hasselbach (14).

INTERCONVERSION OF Ca^{2+} BINDING SITES OF DIFFERENT AFFINITIES

The different reaction schemes proposed for the active transport of Ca^{2+} postulate that the Ca^{2+} affinity of the ATPase should be different at the external and internal surface of the membrane (29–37). The part of the ATPase facing the external surface of the membrane should have a high affinity site for Ca^{2+} in order to be able to bind this ion even when the Ca^{2+} concentration in the assay medium is lower than 10^{-6} M. After translocation through the membrane, the affinity of the binding site for Ca^{2+} should decrease significantly in order to permit the release of the bound Ca^{2+} into the vesicles lumen. The translocation of the Ca^{2+} binding site and the modification of its affinity should involve a conformational change of the ATPase.

The sarcoplasmic reticulum ATPase contains at least two different classes of Ca^{2+} binding sites which can be distinguished by their respectives affinities: K_S 0.3–2 X 10^{-6}M and 0.6–1 X 10^{-3}M at pH 7.0 (33, 38–42). Ikemoto

(33, 43), reports that during the process of ATP hydrolysis the Ca^{2+} binding site of high affinity is converted into a site of low affinity. A purified ATPase preparation that did not accumulate Ca^{2+} was incubated into a medium containing a Ca^{2+} concentration sufficient to allow binding of this cation only to the site of high affinity. Upon the addition of a small amount of ATP, part of the Ca^{2+} bound was released to the assay medium. After hydrolysis of the added ATP, the parcel of Ca^{2+} released did rebind back to the enzyme.

Evidence that the Ca^{2+} transport ATPase undergoes conformational changes has been reported by several authors (44–58). Coan & Inesi (50, 55) detected a modification of the electron spin resonance spectrum upon binding of Ca^{2+} and either acetyl phosphate or ATP to vesicles previously labeled with an iodoacetamide spin label. These authors discussed the conflicting results previously reported that were obtained using spin labels that react with sulfhydryl groups (44–49). Dupont (51) and Dupont & Leigh (58) have shown that the binding of Ca^{2+} to the site of high affinity of the ATPase results in an increase of the intrinsic fluorescence of the enzyme. The rate of the signal change was too slow to represent directly the binding of Ca^{2+}. Therefore, a two-step process was proposed, in which fast binding would be followed by a slow isomerization of the enzyme to a form that has a higher fluorescence strength. Measuring the rate of reaction of 5,5'-dithio-bis (2-nitrobenzoate) with the vesicles, Murphy (52) detected a modification of the reactivity of sulfhydryl groups upon binding of Ca^{2+} to the high affinity site. Thorley-Lawson & Green (56) as well as Anderson & Møller (54) detected a modification of reactivity of sulfhydryl groups only in the presence of both ATP and Ca^{2+}.

From these data it is concluded that a conformational change is promoted by either the simple binding of Ca^{2+} or of both Ca^{2+} and substrate to the enzyme. As far as we know, at present there is no direct evidence for a conformational change associated with the translocation of Ca^{2+} through the membrane.

A BRIEF ACCOUNT OF ATP HYDROLYSIS AND Ca^{2+} TRANSPORT

The substrate of the enzyme is the $Mg \cdot ATP$ complex (59, 60). The hydrolysis of ATP is initiated by the transfer of the terminal phosphate of ATP to an aspartyl residue of the enzyme, forming an acylphosphate. This reaction depends upon the binding of Ca^{2+} to a high affinity site of the enzyme located on the outer surface of the membrane (3, 25, 26, 57, 59–66). Ca^{2+} is translocated through the membrane and released in the vesicular lumen prior to the hydrolysis of the phosphoenzyme (31, 33, 43, 67, 68).

Mg^{2+} accelerates the phosphoenzyme hydrolysis (29, 64, 69–72). Two calcium ions are translocated for each molecule of ATP consumed (67, 73–75). The ATPase activity is inhibited by raising Ca^{2+} concentrations in the vesicular lumen (75–80) because of the binding of Ca^{2+} to a site of low affinity located in the inner surface of the vesicle (41, 80–83). The ATPase catalyzes a rapid ATP $\rightleftharpoons$ ADP exchange (31, 68, 73, 84–86) which is promoted by the reversal of the enzyme phosphorylation by ATP. Accordingly, the rate of ATP hydrolysis and the steady-state level of phosphoenzyme are decreased by increase in the ADP concentration (37, 63, 72, 87–89). Based on ADP sensitivity, two different forms of phosphoenzyme have been distinguished; only one of them is able to transfer its phosphoryl group back to ADP. The binding of Ca^{2+} to the low affinity site leads to accumulation of the ADP-sensitive phosphoenzyme (29, 37, 89).

With the use of a rapid quenching method, an initial burst of P_i liberation has been detected, the size of which varies with the ATP concentration in the assay medium. The interpretation of these findings is still controversial (29, 67, 90–93). The ATPase activity is modified by the addition of K^+ or Na^+ to the assay medium. When the Ca^{2+} concentration in the medium is insufficient to saturate the high affinity site, these monovalent cations inhibit both Ca^{2+} transport and ATP hydrolysis (88, 94–100). At saturating Ca^{2+} concentrations, K^+ and Na^+ activate the rate of hydrolysis (94, 99–103). This effect depends upon the Mg^{2+} and ATP concentrations in the medium (88, 96, 100). The ATP concentration dependence of ATPase activity is complex and can not be fitted by a single straight line on double-reciprocal plots (29, 59, 60, 100, 103–108). This has been attributed to an activating effect of ATP on an intermediate step of the reaction cycle. The mechanism of this regulatory role of ATP is still controversial (29, 32, 34, 88, 90, 92, 100, 103, 106–110). Besides ATP, other nucleoside triphosphates (NTP) (7, 60, 74, 88, 109–111), acetyl phosphate (87, 93, 95, 112, 113), p-nitrophenyl phosphate (114–116), and carbamyl phosphate (113) can support Ca^{2+} transport. The affinity of the enzyme for these substrates is one or more orders of magnitude lower than for ATP (88). The kinetic behavior of the enzyme is modified when ATP is replaced by other substrates (34, 60, 87, 88, 93, 94, 109–111, 116). For instance, when Ca^{2+} binds only to the high affinity site, the steady state level of phosphoenzyme obtained with the use of ITP is 2–3 times smaller than that formed with ATP. Upon saturation of the low-affinity Ca^{2+} binding site, the level of phosphoenzyme formed from ATP does not vary, while that obtained with ITP increases to a level similar to the one obtained with ATP (83, 109, 110). This finding contributed to the conclusion that the cycle of NTP hydrolysis includes a transient dephosphoenzyme form that is not capable of being phosphorylated by NTP. ATP but not ITP would speed the transformation of this enzymatic form into another form that is phosphorylated by NTP. Further data supporting this conclusion are discussed below.

A number of reaction sequences have been proposed to define the steps involved in the process of substrate hydrolysis and Ca^{2+} transport (29–37). The following minimal reaction sequence (Figure 1) is adopted by the authors (34).

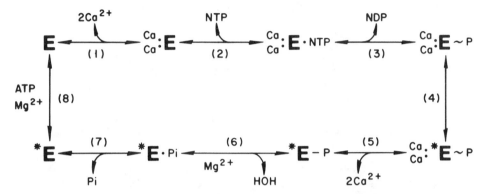

This sequence includes two distinct functional states of the enzyme, E and *E. The Ca^{2+} binding site in the E form faces the outer surface of the vesicle and has an apparent K_m for Ca^{2+} in the range of 0.2–2 μM at pH 7.0 (high affinity)[2]. In the *E form the Ca^{2+} binding site faces the inner surface of the vesicle and has an apparent K_m for Ca^{2+} in the range of 1–3 mM at pH 7.0 (low affinity). The E form is phosphorylated by NTP but not by P_i, while the form *E is phosphorylated by P_i but not by NTP. Reaction 8 can be the rate-limiting step depending on the NTP used. Conversion of *E into E occurs very slowly in the absence of NTP and increasing concentrations of NTP activates the rate of interconversion, ITP being much less effective than ATP.

THE REVERSAL OF THE Ca^{2+} PUMP

The reversal of the entire process of Ca^{2+} transport was first demonstrated by Makinose & Hasselbach (14–17) who have shown that the Ca^{2+}-dependent ATPase can catalyze the synthesis of ATP from ADP and P_i when a Ca^{2+} concentration gradient is formed across the vesicles membrane. This was demonstrated using two different experimental approaches: the net synthesis of ATP coupled with Ca^{2+} efflux (15, 17) and the ATP $\rightleftharpoons$ Pi exchange coupled with Ca^{2+} exchange between the pools of Ca^{2+} contained in the vesicles and in the assay medium (16, 31)

---

[2]The different values of the apparent K_m for Ca^{2+} reported are derived from the use of different values for the Ca^{2+}:EGTA dissociation constant.

ATP Synthesis

When vesicles previously loaded with Ca^{2+} are incubated in a medium containing EGTA, Ca^{2+} flows out of the vesicles at a slow rate due to the low Ca^{2+} permeability of the membrane. The Ca^{2+} efflux is sharply increased when ADP, P$_i$, and Mg^{2+} are added to the incubation medium. The increment of Ca^{2+} efflux is not observed when one of these reactants is omitted from the medium (15). The fast efflux of Ca^{2+} is coupled with the synthesis of ATP (17). For every two calcium ions released from the vesicles, one molecule of ATP is synthesized (17). The fast efflux of Ca^{2+} is inhibited by ionic calcium concentrations in the medium identical to those that activate ATP hydrolysis and Ca^{2+} uptake. This led Barlogie et al (15) to conclude that inhibition occurs on the outer surface of the vesicles at the same site where Ca^{2+} transport is activated, and that the membrane component involved in the active uptake of Ca^{2+} also takes part in Ca^{2+} efflux. The synthesis of ATP is initiated by the phosphorylation of the enzyme by P$_i$ forming a phosphoprotein. This was measured by incubating vesicles previously loaded with Ca^{2+} in a medium containing EGTA, P$_i$, and Mg^{2+}. In these conditions 2–4 μmole of phosphoenzyme per gram of protein were formed (18, 19, 117). The phosphorylation of the enzyme by P$_i$ is not accompanied by an increment of the Ca^{2+} efflux. The subsequent addition of ADP to the medium leads to a decrease of the steady state level of phosphoenzyme, synthesis of ATP, and enhancement of the rate of Ca^{2+} efflux (18, 19, 118). Therefore, the fast release of Ca^{2+} is triggered by the transfer of the phosphate from the phosphoenzyme to ADP. The part of the enzyme phosphorylated by P$_i$ faces the outer surface of the vesicle (119). Yamada et al (19) reported that the degree of enzyme phosphorylation by P$_i$ was dependent on the dimension of the Ca^{2+} gradient formed across the membrane. The steady state level of phosphoenzyme was increased by higher Ca^{2+} concentration in the vesicles and by lower Ca^{2+} concentration in the assay medium. The synthesis of ATP is not inhibited by either dinitrophenol or azide (120). Under appropriate conditions arsenate uncouples the fast efflux of Ca^{2+} and the synthesis of ATP. Although the fast release of Ca^{2+} is maintained, both the enzyme phosphorylation by P$_i$ and the synthesis of ATP are impaired. This effect of arsenate is reversible (120).

On the basis of information then available, in 1972 it was concluded that the energy required for the synthesis of ATP was derived from the Ca^{2+} concentration gradient formed across the membrane (15–19). The osmotic energy of the calcium ions was regarded as driving the first step of the process of ATP synthesis, the phosphorylation of the ATPase by P$_i$. The finding that the protein phosphate was an energy rich acyl phosphate (18, 19) contributed to this conclusion. The ΔG calculated from the ratio of

Ca^{2+} concentration in the vesicles and in the medium yield values compatible with those required for the synthesis of ATP or the formation of an acyl phosphate residue (18, 19). The experiments described above were performed at pH 7.0 and were reproduced in several laboratories (118, 119, 121–123). Recently, Hasselbach & Migala (124) reported the synthesis of pyrophosphate from inorganic phosphate. This is dependent on the formation of Ca^{2+} gradient and is inhibited by ATP.

$ATP \rightleftharpoons P_i$ Exchange

In 1971 Makinose (16) observed that when vesicles were incubated in a medium containing 0.2 mM $CaCl_2$, ATP, ADP, Mg^{2+}, and $^{32}P_i$, Ca^{2+} was accumulated by the vesicles and a Ca^{2+} concentration gradient was built up until a steady state was reached in which the Ca^{2+} efflux was balanced by the ATP-driven influx. When this condition was reached, a steady rate of exchange between $^{32}P_i$ and the γ-phosphate of ATP was observed. The formation of γ-[^{32}P] ATP started to occur as soon as the net uptake of Ca^{2+} ceased. During the initial phase of net Ca^{2+} accumulation, the vesicles did catalyze the hydrolysis of ATP but did not catalyze the incorporation of $^{32}P_i$ into the ATP pool. The ATP $\rightleftharpoons$ P_i exchange was abolished when the membrane of the vesicles was rendered leaky by different procedures that did not affect the transport ATPase. The ATP $\rightleftharpoons$ P_i exchange indicates that the ATPase operates simultaneously forward (ATP hydrolysis) and backwards (ATP synthesis from ADP and $^{32}P_i$). These data infer that when the steady state between Ca^{2+} influx and Ca^{2+} efflux was reached, the energy derived from the hydrolysis of ATP was used to maintain the Ca^{2+} gradient and the energy derived from the gradient was used for the synthesis of ATP (156, 125).

$P_i \rightleftharpoons$ HOH EXCHANGE AND PHOSPHORYLATION OF THE Ca^{2+}-DEPENDENT ATPase BY P_i IN THE ABSENCE OF Ca^{2+} CONCENTRATION GRADIENT

In 1973 Kanazawa & Boyer (30) reported that vesicles that were not previously loaded with Ca^{2+} were able to catalyze a rapid incorporation of water oxygen atoms into P_i when incubated in a medium containing EGTA, P_i, and Mg^{2+}. The $P_i \rightleftharpoons$ HOH exchange suggested that the ATPase was phosphorylated by P_i with elimination of water. The reversal of this reaction would lead to the incorporation of water oxygen into phosphate oxygen. Accordingly, it was found that a quite small but significant fraction of the enzyme was phosphorylated by P_i. The level of phosphoenzyme measured was 1/50 of the level of phosphoenzyme measured by Makinose (18) and Yamada et al (19) using vesicles previously loaded with Ca^{2+}. Both the

$P_i \rightleftharpoons HOH$ exchange and the phosphorylation by P_i were inhibited by the addition to the assay medium of Ca^{2+} concentrations similar to those required for the activation of ATP hydrolysis (1–10 μM). In the absence of free calcium in the medium, the $P_i \rightleftharpoons HOH$ exchange was inhibited by the addition of small concentrations of Triton X-100 which render the vesicles leaky without damaging the ATPase. It was already reported that vesicles contain a small amount of endogenous Ca^{2+} (38, 78). This, and the inhibition by Triton X-100, led the authors not to exclude definitively the requirement of intravesicular Ca^{2+} for the phosphorylation of the enzyme by P_i and activation of the $P_i \rightleftharpoons HOH$ exchange. The experiments of Kanazawa & Boyer were also performed at pH 7.0. Shortly afterward it was reported that the Ca^{2+}-dependent ATPase can be phosphorylated by P_i forming an acyl phosphate protein in the absence of a transmembrane Ca^{2+} concentration gradient (126). This was shown using vesicles rendered leaky with the use of diethyl ether. A small level of phosphoenzyme was measured when these vesicles were incubated in a medium at pH 7.0 containing Mg^{2+}, EGTA, and 2–4 mM P_i. The level of phosphoenzyme increased several fold when the pH of incubation medium was decreased to 6.0, reaching values in the same range as those measured in the presence of a Ca^{2+} gradient. The level of phosphoenzyme varies with the P_i concentration in the medium (126). The concentration of P_i required for half-maximal phosphorylation (K_{P_i}) varies significantly according to the pH of the medium and whether leaky vesicles or vesicles previously loaded with Ca^{2+} are used (118).

Table 1 shows that the presence of a Ca^{2+} gradient across the membrane promotes an increase of the apparent affinity of the enzyme for P_i. This is more pronounced at pH 7.0 than at pH 6.0. This peculiar pH dependence might account for earlier failure to measure high levels of phosphoenzyme in the absence of a transmembrane Ca^{2+} concentration gradient. The phosphorylation of leaky vesicles by P_i is inhibited by Ca^{2+} (K_i 5–10 μM),

Table 1 Ca^{2+} gradient and modification of the ATPase affinity for P_i[a]

Vesicles	K_{P_i} for E-P formation (mM)	
	pH 6.1	pH 7.0
Leaky	3.5 ± 0.5 (8)	25.3 ± 4.2 (3)
Loaded with calcium phosphate	0.7 ± 0.1 (7)	2.5 ± 0.4 (4)

[a] The assay medium composition was 15 mM EGTA, 25 mM $MgCl_2$, 20 mM Tris maleate buffer pH 6.1 or 7.0, and different $^{32}P_i$ concentrations. The reaction was performed at 30°, and was started by the addition of vesicles (0.7 mg protein/ml). It was arrested after 15 sec by the addition of ice-cold solution of perchloric acid. The values represent the average ± S.E. of the number of experiments shown in parentheses. For further details see (118).

Na^+, and K^+ (K_i 100 mM) and by both ATP and ADP. Acetyl phosphate, ITP, GTP, AMP, and IMP, up to a concentration of 4 mM, were ineffective as inhibitors. The inhibition promoted by ATP and ADP is overcome by raising the P_i concentration of the medium which indicates a competitive process (118, 126). These findings were reproduced in several laboratories (23, 42, 127, 128). In the absence of a Ca^{2+} gradient, essentially the same rate of $P_i \rightleftharpoons HOH$ exchange was measured with the use of intact or leaky vesicles (129). From the data reported, Masuda & de Meis (126) concluded that both ATP and P_i are substrates of the Ca^{2+}-dependent ATPase which undergoes a conformational change depending on the binding of Ca^{2+} to a site of high affinity. The binding of Ca^{2+} would therefore trigger the choice of P_i or ATP as substrate for the phosphorylating reaction. This Ca^{2+} binding site should be located on the outer surface of the vesicles since the inhibition profile of the phosphorylation by P_i observed by the addition of raising Ca^{2+} concentration in the medium is essentially the same whether leaky vesicles or vesicles previously loaded with Ca^{2+} are used, that is, it is independent of the Ca^{2+} concentration inside the vesicles. According to the reaction sequence proposed, the phosphorylation by P_i, the $P_i \rightleftharpoons HOH$ exchange, and the inhibition by Ca^{2+}, is accounted for by reactions 6, 7, 8, 1, and 2 in Figure 1. In the absence of Ca^{2+} and at pH 6.0, the equilibrium between the enzymatic forms *E-P, *E.P_i, *E, and E will be such as to favor the accummulation of the enzymatic form *E-P. Upon the addition of Ca^{2+}, the enzymatic form $^{Ca}_{Ca}$:E would be accumulated.

SIMULTANEOUS PHOSPHORYLATION OF THE Ca^{2+}-DEPENDENT ATPase BY NTP AND P_i

In both leaky and intact vesicles, the inhibition by Ca^{2+} of the enzyme phosphorylation by P_i is overcome by the addition of acetyl phosphate or an NTP to the assay medium (130). In these conditions, the enzyme is simultaneously phosphorylated by P_i and by the NTP. The portion of enzyme phosphorylated by P_i depends on the NTP used and its concentration in the assay medium. In all conditions tested, the sum of the enzyme phosphorylated by NTP and by P_i is constant (34, 89, 110, 130, 131). In the presence of 0.1 mM Ca^{2+} and 1 mM ATP, only 5–10% of the phosphorylation occurs with P_i. If ATP is replaced by ITP or GTP, then about 50% of the phosphorylation occurs with P_i and the remainder by ITP or GTP. Similar results are obtained with ATP provided that its concentration is decreased to the μmolar range (34, 110). The inhibition of $P_i \rightleftharpoons HOH$ exchange is also overcome by the addition of acetyl phosphate (30) or an NTP (110) to the medium. The rate of $P_i \rightleftharpoons HOH$ exchange varies with the NTP used and its concentration, in a fashion similar to that observed for the phosphorylation of the enzyme by P_i (110). These data indicate that the cycle of NTP hydrolysis includes a transient enzyme form capable of reacting with P_i but not with NTP. In the reaction sequence proposed, this enzymatic form was denoted as *E. During ITP or GTP hydrolysis *E would be accumulated because of the rate limitation of Reaction 8 in Figure 1. In the millimolar range ATP would speed the conversion of *E into E and consequently decrease the steady state level of both *E and *E-P.

PHOSPHATE TRANSFER FROM THE PHOSPHOENZYME TO ADP

When leaky vesicles are phosphorylated by P_i, the phosphoenzyme formed is not able to transfer its phosphate to ADP (42, 118). This could imply that the energy from the Ca^{2+} concentration gradient is not used for the phosphorylation of the enzyme by P_i, but for the transfer of the phosphate from the phosphoenzyme to ADP (118). Alternatively, the transfer of the phosphate to ADP could be related to the binding of Ca^{2+} to a site of low affinity of the enzyme located on the inner surface of the vesicles membrane. Makinose & Hasselbach (17), Makinose (18), and Yamada et al (19) reported that when a gradient is formed, the Ca^{2+} concentration inside the vesicle required to activate the synthesis of ATP is in the range of 1–20 mM. The first data indicating an involvement of the Ca^{2+} binding site of low affinity in the process of ATP synthesis were obtained from kinetics studies of the ATP $\rightleftharpoons$ P_i exchange reactions (34, 81, 131). Soluble ATPase or leaky vesicles were incubated in media containing millimolar concentrations of ATP, ADP, $^{32}P_i$, and different Ca^{2+} concentrations. At pH 6.8 the ATPase activity was progressively inhibited when the Ca^{2+} concentration was raised from 0.1 to 10 mM. No incorporation of $^{32}P_i$ into the ATP pool was detected in presence of 0.1 mM Ca^{2+}. Raising the Ca^{2+} concentration from 0.3 to 10 mM resulted in a progressive activation of γ-[^{32}P]ATP formation reaching a maximum in the Ca^{2+} concentration range of 4–5 mM (34, 81, 131). The Ca^{2+} concentration required for half maximum activation (K_{Ca}) of ATP $\rightleftharpoons$ P_i exchange was found to vary widely with the pH of the assay medium (83). At pH 6.0 saturation was not reached even in the presence of 20 mM Ca^{2+}. At pH 7.0 and 8.0 the values of K_{Ca} were found to be respectively 1.8 and 0.3 mM. At pH 6.8 the enzyme is phosphorylated by $^{32}P_i$ in the presence of either 0.1 or 4.0 mM Ca^{2+}. However, formation of γ-[^{32}P]ATP was measured only in the presence of 4 mM Ca^{2+} (34, 131). From these data it was concluded that the osmotic energy derived from the Ca^{2+} gradient is not required for the activation of the ATP $\rightleftharpoons$ P_i exchange reaction and that the binding of Ca^{2+} to a site of low affinity allows the transfer of phosphate from the phosphoenzyme to ADP. This Ca^{2+} binding site should be located in the inner surface of the vesicle, since the Ca^{2+} concentrations required in the medium for activation of the ATP $\rightleftharpoons$ P_i exchange in leaky vesicles or soluble ATPase are in the same range as those to be found inside the vesicles when a Ca^{2+} gradient sufficient to activate ATP $\rightleftharpoons$ P_i is formed. There is no net synthesis of ATP during the exchange reaction. In the different experimental conditions tested, the rate of NTP hydrolysis was found to be 2.5–50 times faster than the rate of NTP synthesis (34, 81, 131). Therefore, the data obtained with leaky vesicles and soluble enzyme indicate that the system must be able to conserve some of the energy released from NTP hydrolysis in a form that permits resynthesis of NTP.

P_i DEPENDENCE

The rate of ATP $\rightleftharpoons$ P_i exchange varies with the P_i concentration in the medium both in the presence and absence of a transmembrane Ca^{2+} concentration gradient. However the apparent K_m of P_i varies in these two conditions, being in the range of 3–4 mM when a gradient is formed, and in the range of 35–50 mM when measured in leaky vesicles (81, 82). Thus, the variation of the enzyme affinity for P_i seems to be a recurring theme (Table 1).

A possible link between osmotic energy derived from the Ca^{2+} gradient and change in the apparent K_m for P_i of the ATP $\rightleftharpoons$ P_i exchange was excluded by two different experimental approaches involving the use of silver (82, 131) and of ITP (34). In intact vesicles, when a gradient is formed, the rate of ATP $\rightleftharpoons$ P_i exchange is activated when the Ca^{2+} concentration in the medium decreases to a level insufficient to allow the saturation of the Ca^{2+} binding site of high affinity located on the outer surface of the vesicles (16, 34, 81, 82). When this condition is reached, a concomitant decrease of the degree of enzyme phosphorylation by ATP and enhancement of the phosphorylation by P_i is observed (34). This effect can be imitated with Ag^+ and leaky vesicles incubated in a medium containing 4 mM Ca^{2+} and 6 mM P_i (82, 131). Under appropriate conditions the binding of Ag^+ to the enzyme promotes a simultaneous inhibition of enzyme phosphorylation by ATP and of the rate of ATP hydrolysis. This is accompanied by an increase of the portion of enzyme phosphorylated by P_i and activation of the rate of γ-[^{32}P]ATP synthesis. Exploration of the requirement of ADP, P_i, and Ca^{2+} showed that the activation of ATP $\rightleftharpoons$ P_i exchange by Ag^+ was due to a decrease of the apparent K_m of P_i from 50 to 8 mM (82, 131). Replacement of ATP and ADP by ITP and IDP showed that when a Ca^{2+} concentration gradient is formed, the rate of ITP $\rightleftharpoons$ P_i exchange is no longer regulated by the binding of Ca^{2+} to the outer surface of the vesicles (34). The level of enzyme phosphorylated by P_i and the rate of ITP

$\rightleftharpoons$ P$_i$ exchange did not vary when the Ca^{2+} concentration in the medium varied from 100 to < 3 μM. ITP $\rightleftharpoons$ P$_i$ exchange was also observed when leaky vesicles were incubated in a medium containing 4 mM Ca^{2+}, but with the use of ITP the apparent K_m of P$_i$ for the exchange reaction was in the range of 2–4 mM (34).

From these data it was concluded that the rate of exchange is regulated by different mechanisms depending on the NTP used (34). According to the reaction sequence proposed, (see Figure 1), with the use of ITP Reactions 1 to 7 are faster than Reaction 8. This leads to the accumulation of the enzymatic form *E regardless of the Ca^{2+} concentration in the assay medium. In the presence of P$_i$ and high Ca^{2+} concentrations in the vesicular lumen Reactions 7 and 5 flow backwards in the direction of ITP synthesis. In the presence of millimolar concentrations of ATP, Reaction 8 is no longer the rate-limiting step. Thus, the enzymatic form *E is only accumulated when the Ca^{2+} concentration in the medium becomes insufficient to saturate the Ca^{2+} binding site of high affinity (Reaction 1). The binding of Ag$^+$ to the enzyme blocks the effect of ATP on Reaction 8.

NET SYNTHESIS OF ATP IN THE ABSENCE OF A Ca^{2+} GRADIENT

In 1975 Knowles & Racker (20) reported that leaky vesicles prepared from purified Ca^{2+}-dependent ATPase can synthesize a small amount of ATP in the absence of Ca^{2+} gradient. This was achieved by a two-step procedure where, initially, the enzyme was phosphorylated by P$_i$ at pH 6.3 in the presence of Mg^{2+} and EGTA. Subsequently, upon addition of ADP and 3.3–10 mM CaCl$_2$ (Ca^{2+} jump), it was found that about half of the phosphate of the phosphoenzyme was transferred to ADP, forming ATP. The success of this experimental approach could be predicted from the data of Taniguchi & Post (24) which showed that the Na$^+$ + K$^+$-dependent ATPase can drive the synthesis of ATP in the absence of a Na$^+$ or K$^+$ transmembrane gradient. The experiments of Knowles & Racker (20) were repeated in two different laboratories with different results. Beil et al (42) did not detect net synthesis of a significant amount of ATP at pH 6.0 upon the addition of 5 mM ADP and 5 mM CaCl$_2$ to leaky vesicles previously phosphorylated by P$_i$. In their report the authors did not exclude the possibility that the addition of 5 mM CaCl$_2$ might have produced a phosphoryl group acceptable to ADP. de Meis & Tume (21) were able to measure net synthesis of ATP by the Ca^{2+} jump procedure. The Ca^{2+} concentration required in the second step for the synthesis of ATP was found to be highly dependent upon the pH of the medium. This was shown by adding to the medium a small amount of KOH together with the mixture of ADP and $CaCl_2$. The Ca^{2+} concentration required for the transfer of 50% of the phosphoenzyme phosphate to ADP was found to be in the range of 20–40 mM at pH 6.0 and in the range of 1–2 mM at pH 7.4. This pH-dependence might account for the discrepancy of the data reported by Knowles & Racker (20) and Beil et al (42) since in both cases, suboptimal Ca^{2+} concentrations were used and the pH values reported varied between 6.0 and 6.3. Millisecond mixing and quenching experiments demonstrated that after the addition of ADP and Ca^{2+}, the disappearance of phosphoenzyme and the synthesis of ATP were synchronous (21). At pH 6.0 the time for half-maximal ATP synthesis and phosphoenzyme dephosphorylation was in the range of 20–40 msec. The amount of ATP synthesized was proportional and never exceeded the number of enzyme sites phosphorylated by P_i (20, 21).

From the data discussed above it is concluded that each of the individual reactions involved in the hydrolysis or synthesis of ATP is reversible and can flow forward and backwards without using the osmotic energy that might be derived from the concentration gradient of Ca^{2+} across the membrane (21). For the synthesis of ATP in sealed vesicles, the large difference of Ca^{2+} concentration on the two sides of the membrane is only needed to meet the differences of affinities of the external and internal Ca^{2+}-binding sites (21).

RATE OF PHOSPHOENZYME HYDROLYSIS

Boyer et al (129) measured the rate of interchange between P_i and the phosphoenzyme under steady-state conditions at pH 6.0 and in presence of EGTA. The enzyme was initially phosphorylated with nonradioactive P_i. Then the reaction medium was mixed with an equal volume of a solution of identical composition but without enzyme and with $^{32}P_i$. In order to be labeled, the nonradioactive phosphoenzyme must be hydrolyzed and then phosphorylated by the added $^{32}P_i$. The time for half-maximum labeling was found to be in the range of 30–40 msec. This shows that the rates of phosphoenzyme formation and hydrolysis, (Figure 1, Reactions 6 and 7), forward and backwards, are fast. A slow rate of phosphoenzyme disappearance (half time 2–4 sec) was measured when Ca^{2+} (0.1 or 20 mM) was added to a medium containing enzyme, $^{32}P_i$, and Mg^{2+} (24, 128). This indicates that when Ca^{2+} is added, the enzyme form capable of reacting with P_i (*E) is slowly converted into another form (Reaction 8) capable of binding Ca^{2+} on the outer surface of the membrane but no longer capable of reacting with P_i (E). Reactions 6 and 7 flow continuously forward and backwards until all the enzymatic form *E is converted into E. This conclusion can be related to the finding of Dupont & Leigh (58) that a slow increase of the enzyme fluorescence is promoted by the binding of Ca^{2+} to the site of high affinity.

The slow disappearance of phosphoenzyme induced by Ca^{2+} accounts for the ATP synthesis described above. The Ca^{2+} concentrations added in the second step of the Ca^{2+} jump procedure are sufficient to saturate both Ca^{2+} binding sites of high and low affinity. Synthesis of ATP is observed because the rate at which the phosphoenzyme transfers its phosphate to ADP is about two orders of magnitude faster than the rate of formation of the enzymatic complex $\substack{Ca \\ Ca}:E$ (24).

PROTON GRADIENT AND SYNTHESIS OF ATP BY THE Ca^{2+}-DEPENDENT ATPase

The binding of Ca^{2+} to the high- and low-affinity sites is altered by changing the pH of the medium. Both sites exhibit an increased affinity for Ca^{2+} in high pH, and a reduced affinity at low pH values (24, 40, 42, 83, 127), the relative differences in affinity of the two sites remaining unchanged at all pH values studied (24, 83). This was monitored either directly, by measuring the binding of Ca^{2+} to the enzyme (40, 42), or indirectly, by measuring the Ca^{2+} requirement for activation of the NTP $\rightleftharpoons$ P_i exchange, phosphorylation of leaky vesicles by ITP (83), and inhibition by Ca^{2+} of the enzyme phosphorylation by P_i (21, 42, 127). Based on this finding the rationale was developed that no ATP synthesis will be observed when the same Ca^{2+} concentration exists on both sides of the membrane and at a level sufficient to saturate the high affinity site but not sufficient to allow binding to the low affinity site. However, the system could synthesize ATP if the affinity of the two Ca^{2+} binding sites is modified by varying the pH on the two sides of the membrane, making it acidic in the assay medium and alkaline inside the vesicles. In such a situation the phosphorylation of the enzyme by P_i will no longer be inhibited, and the Ca^{2+} concentration inside the vesicles could become sufficient to allow the transfer of the phosphate to ADP. This hypothesis was tested using leaky vesicles reformed from purified ATPase and the technique described for the Ca^{2+} jump (21) except that the Ca^{2+} concentration was maintained constant and the affinity of the Ca^{2+} binding sites was varied by a sudden change of the pH of the medium. In the presence of 0.6 mM Ca^{2+}, the enzyme was not phosphorylated by P_i at pH 6.0. By decreasing the pH of the medium to 5.0, 1–2 μmole of phosphoenzyme per gram of protein were formed. Net synthesis of ATP was measured upon the addition to the medium of ADP and KOH in an amount sufficient to raise the pH of the mixture from 5.0 to 8.0. Synthesis of ATP was not detected if Ca^{2+} was omitted from the assay medium. When extended to sealed vesicles, these data indicate that the Ca^{2+}-dependent ATPase of the sarcoplasmic reticulum can catalyze the synthesis of ATP when there are equal Ca^{2+} concentrations and different pH readings on the two sides of the membrane. In such a situation there is a transmembrane H^+ gradient concentration and not a Ca^{2+} gradient; but the synthesis of ATP occurs without using the energy that is perhaps derived from the electrochemical potential of H^+. The difference of pH, only, is required to modify the affinity of the two Ca^{2+} binding sites of the enzyme.

At present, there is no evidence that under physiological conditions the two sides of the sarcoplasmic reticulum membrane are exposed to media with different pH. The aim of the above considerations is simply the understanding of the mechanism by which a concentration gradient of an ion can be utilized for the synthesis of ATP by biological membranes.

ACKNOWLEDGMENTS

Preparation of this review was aided by grants from the Conselho Nacional de Desenvolvimento Científico e Tecnológico (CNPq), the Conselho de Ensino para Graduados da UFRJ, and the Financiadora de Estudos e Projetos (FINEP).

Literature Cited

1. Inesi, G. 1972. *Ann. Rev. Biophys. Bioeng.* 1:191–210
2. Meissner, G., Conner, G. E., Fleischer, D. 1973. *Biochim. Biophys. Acta* 298:246–69
3. Hasselbach, W. 1974. *The Enzymes.* 10:431–67
4. MacLennan, D. H., Holland, P. C. 1975. *Ann. Rev. Biophys. Bioeng.* 4:377–404
5. Ikemoto, N., Bhatnagar, G. M., Gergely, J. 1971. *Biochem. Biophys. Res. Commun.* 44:1510–17
6. MacFarland, B. H., Inesi, G. 1971. *Arch. Biochem. Biophys.* 145:456–64
7. MacLennan, D. H. 1970. *J. Biol. Chem.* 245:4508–18
8. MacLennan, D. H., Seeman, P., Iles, G. H., Yip, C. C. 1971. *J. Biol. Chem.* 246:2702–10
9. Meissner, G., Fleischer, S. 1974. *J. Biol. Chem.* 249:302–9
10. Rizzolo, L. J., le Maire, M., Reynolds, J. A., Tanford, C. 1976. *Biochemistry* 15:3433–37
11. Scales, D., Inesi, G. 1976. *Biophys. J.* 16:735–51
12. le Maire, M., Møller, J. V., Tanford, C. 1976. *Biochemistry* 15:2336–42
13. Tada, M., Yamamoto, T., Tonomura, Y. 1978. *Physiol. Rev.* 58:1–79
14. Hasselbach, W. 1978. *Biochim. Biophys. Acta* 463:23–53
15. Barlogie, B., Hasselbach, W., Makinose, M. 1971. *FEBS Lett.* 12:267–68
16. Makinose, M. 1971. *FEBS Lett.* 12:269–70
17. Makinose, M., Hasselbach, W. 1971. *FEBS Lett.* 12:271–72
18. Makinose, M. 1972. *FEBS Lett.* 25:113–15
19. Yamada, S., Sumida, M., Tonomura, Y. 1972. *J. Biochem. (Tokyo)* 72:1537–48
20. Knowles, A. F., Racker, E. 1975. *J. Biol. Chem.* 250:1949–51
21. de Meis, L., Tume, R. K. 1977. *Biochemistry* 16:4455–63
22. Bastide, F., Meissner, G., Fleischer, S., Post, R. L. 1973. *J. Biol. Chem.* 248:8385–91
23. Post, R. L., Toda, G., Rogers, F. N. 1975. *J. Biol. Chem.* 250:691–701
24. Taniguchi, D., Post, R. L. 1975. *J. Biol. Chem.* 250:3010–18
25. Hasselbach, W. 1972. In *Molecular Bioenergetics and Macromolecular Biochemistry,* ed. H. H. Weber, pp. 149–71. Berlin: Springer
26. Tonomura, Y. 1972. *Muscle Proteins, Muscle Contraction and Cation Transport.* pp. 305–56. Tokyo: Univ. Tokyo Press

27. Martonosi, A. 1972. In *Curr. Top. Membr. Trans.* 3:83–183
28. MacLennan, D. H., Holland, P. C. 1976. In *The Enzymes of Biological Membranes*, ed. A. Martonosi, pp 221–59. New York: Plenum
29. Kanazawa, T., Yamada, S., Yamamoto, T., Tonomura, Y. 1971. *J. Biochem. (Tokyo)* 70:95–123
30. Kanazawa, T., Boyer, P. D. 1973. *J. Biol. Chem.* 248:3163–72
31. Makinose, M. 1973. *FEBS Lett.* 37:140–43
32. Martonosi, A., Lagwinska, E., Oliver, M. 1974. *Ann. NY Acad. Sci.* 227:549–67
33. Ikemoto, N. 1975. *J. Biol. Chem.* 250:7219–24
34. Carvalho, M. G. C., Souza, D. O., de Meis, L. 1976. *J. Biol. Chem.* 251:3629–36
35. Yamamoto, T., Tonomura, Y. 1976. *J. Biochem. (Tokyo)* 79:693–707
36. Hasselbach, W., Beil, F. U. 1977. In *Biochemistry of Membrane Transport FEBS-Symp. 42*, ed G. Semenza, E. Carafoli, pp. 416–28. Berlin: Springer
37. Shigekawa, M., Dougherty, J. P. 1978. *J. Biol. Chem.* 253:1458–64
38. Chevallier, J., Butow, R. A. 1971. *Biochemistry* 2733–37
39. Fiehn, W., Migala, A. 1971. *Eur. J. Biochem.* 20:245–48
40. Meissner, G. 1973. *Biochim. Biophys. Acta* 298:906–26
41. Ikemoto, N. 1974. *J. Biol. Chem.* 249:649–51
42. Beil, F. U., Chak, D., Hasselbach, W. 1977. *Eur. J. Biochem.* 81:151–64
43. Ikemoto, N. 1976. *J. Biol. Chem.* 251:7275–77
44. Landgraf, W. C., Inesi, G. 1969. *Arch. Biochem. Biophys.* 130:111–18
45. Inesi, G., Landgraf, W. C. 1970. *Bioenergetics* 1:355–65
46. Nakamura, H., Hori, H., Mitsui, T. 1972. *J. Biochem. (Tokyo)* 72:635–46
47. Pang, D. C., Briggs, F. N. 1974. *Arch. Biochem. Biophys.* 164:332–40
48. Tonomura, Y., Morales, M. F. 1974. *Proc. Natl. Acad. Sci. USA* 71:3687–91
49. Champeil, P., Bastide, F., Taupin, C., Gary-Bobo, C. M. 1976. *FEBS Lett.* 63:270–72
50. Coan, C. R., Inesi, G. 1976. *Biochem. Biophys. Res. Commun.* 71:1283–88
51. Dupont, Y. 1976. *Biochem. Biophys. Res. Commun.* 71:544–50
52. Murphy, A. J. 1976. *Biochemistry* 15:4492–96
53. Yamamoto, T., Tonomura, Y. 1976. *J. Biochem. (Tokyo)* 79:693–707
54. Andersen, J. P., Møller, J. V. 1977. *Biochim. Biophys. Acta* 485:188–202
55. Coan, C. R., Inesi, G. 1977. *J. Biol. Chem.* 252:3044–49
56. Thorley-Lawson, D. A., Green, N. M. 1977. *Biochem. J.* 167:739–48
57. Yamamoto, T., Tonomura, Y. 1977. *J. Biochem. (Tokyo)* 82:653–60
58. Dupont, Y., Leigh, J. B. 1978. *Nature* 273:396–98
59. Yamamoto, T., Tonomura, Y. 1967. *J. Biochem. (Tokyo)* 62:558–75
60. Vianna, A. L. 1975. *Biochim. Biophys. Acta* 410:389–406
61. Makinase, M. 1967. *Pflügers Arch.* 294:82
62. Yamamoto, T., Tonomura, Y. 1968. *J. Biochem. (Tokyo)* 64:137–45
63. Makinose, M. 1969. *Eur. J. Biochem.* 10:74–82
64. Inesi, G., Maring, E., Murphy, A. J., McFarland, B. H. 1970. *Arch. Biochem. Biophys.* 138:285–94
65. Bastide, F., Meissner, G., Fleischer, S., Post, R. L. 1973. *J. Biol. Chem.* 248:8385–91
66. Degani, C., Boyer, P. D. 1973. *J. Biol. Chem.* 248:8222–26
67. Kurzmack, M., Verjovski-Almeida, S., Inesi, G. 1977. *Biochem. Biophys. Res. Commun.* 78:772–76
68. Sumida, M., Tonomura, Y. 1974. *J. Biochem. (Tokyo)* 75:283–97
69. Panet, R., Pick, U., Selinger, Z. 1971. *J. Biol. Chem.* 246:7349–56
70. Yamada, S., Tonomura, Y. 1972. *J. Biochem. (Tokyo)* 72:417–25
71. Garrahan, P. J., Rega, A. F., Alonso, G. L. 1976. *Biochim. Biophys. Acta* 448:121–32
72. Shigekawa, M., Dougherty, J. P. 1978. *J. Biol. Chem.* 253:1451–57
73. Hasselbach, W., Makinose, M. 1963. *Biochem. Z.* 339:94–111
74. Makinose, M., The, R. 1965. *Biochem. Z.* 343:383–93
75. Weber, A., Herz, R., Reiss, I. 1966. *Biochem. Z.* 345:329–69
76. Makinose, M., Hasselbach, W. 1965. *Biochem. Z.* 343:360–82
77. Fiehn, W., Hasselbach, W. 1969. *Eur. J. Biochem.* 9:574–78
78. Duggan, P. F., Martonosi, A. 1970. *J. Gen. Physiol.* 56:147–67
79. Weber, A. 1971. *J. Gen. Physiol.* 57:50–63
80. Balzer, H. 1972. *Naunyn-Schmiedebergs Arch. Exp. Pathol-Pharmakol.* 274:256–63
81. de Meis, L., Carvalho, M. G. C. 1974. *Biochemistry* 13:5032–38

82. de Meis, L., Carvalho, M. G. C., Sorenson, M. M. 1975. In *Concepts of Membranes in Regulation and Excitation*, ed. M. R. Silva, G. S. Kurtz, pp. 7–19. New York: Raven
83. Verjovski-Almeida, S., de Meis, L. 1977. *Biochemistry* 16:329–34
84. Ebashi, S., Lipmann, F. 1962. *J. Cell. Biol.* 14:389–400
85. Hasselbach, W., Makinose, M. 1962. *Biochem. Biophys. Res. Commun.* 7: 132–36
86. Inesi, G., Almendares, J. 1968. *Arch. Biochem. Biophys.* 126:733–35
87. Friedman, Z., Makinose, M. 1970. *FEBS Lett.* 11:69–72
88. de Meis, L., de Mello, M. C. F. 1973. *J. Biol. Chem.* 248:3691–3701
89. Masuda, H., de Meis, L. 1977. *J. Biol. Chem.* 252:8567–71
90. Froehlich, J. P., Taylor, E. W. 1974. *J. Biol. Chem.* 250:2013–21
91. Froehlich, J. P., Taylor, E. W. 1975. *J. Biol. Chem.* 251:2307–15
92. Sumida, M., Kanazawa, T., Tonomura, Y. 1976. *J. Biochem. (Tokyo)* 79: 259–64
93. Kurzmack, M., Inesi, G. 1977. *FEBS Lett.* 74:35–37
94. de Meis, L. 1969. *J. Biol. Chem.* 244:3733–39
95. de Meis, L., Hasselbach, W. 1971. *J. Biol. Chem.* 246:4759–63
96. de Meis, L. 1971. *J. Biol. Chem.* 246:4764–73
97. de Meis, L. 1972. *Biochemistry* 11: 2460–65
98. Gattass, C. R., de Meis, L. 1975. *Biochim. Biophys. Acta.* 389:506–15
99. Duggan, P. F. 1977. *J. Biol. Chem.* 252:1620–27
100. Ribeiro, J. M. C., Vianna, A. L. 1978. *J. Biol. Chem.* 253:3153–57
101. Duggan, P. F. 1968. *Life Sci.* 7:913–19
102. Shigekawa, M., Pearl, L. S. 1976. *J. Biol. Chem.* 251:6947–52
103. Shigekawa, M., Dougherty, J. P., Katz, A. M. 1978. *J. Biol. Chem.* 253:1442–50
104. Inesi, G., Goodman, J. J., Watanabe, S. 1967. *J. Biol. Chem.* 242:4637–43
105. The, R., Hasselbach, W. 1972. *Eur. J. Biochem.* 28:357–63
106. Yates, D. W., Duance, V. C. 1976. *Biochem. J.* 159:719–28
107. Dupont, Y. 1977. *Eur. J. Biochem.* 72:357–63
108. Neet, K. E., Green, N. M. 1977. *Arch. Biochem. Biophys.* 178:588–97
109. Souza, D. O., de Meis, L. 1976. *J. Biol. Chem.* 251:6355–59
110. de Meis, L., Boyer, P. D. 1978. *J. Biol. Chem.* 253:1556–59
111. Makinose, M. 1966. *Biochem. Z.* 345: 80–86
112. de Meis, L. 1969. *Biochim. Biophys. Acta.* 172:343–4
113. Pucell, A., Martonosi, A. 1971. *J. Biol. Chem.* 246:3389–97
114. Inesi, G. 1971. *Science* 171:901–3
115. Hasselbach, W., Suko, J. 1974. *Biochem. Soc. Spec. Publ.* 4:159–73
116. Nakamura, Y., Tonomura, Y. 1978. *J. Biochem. (Tokyo)* 83:571–83
117. Yamada, S., Tonomura, Y. 1973. *J. Biochem. (Tokyo)* 74:1091–96
118. de Meis, L. 1976. *J. Biol. Chem.* 251:2055–62
119. de Meis, L., Carvalho, M. G. C. 1976. *J. Biol. Chem.* 251:1413–17
120. Hasselbach, W., Makinose, M., Migala, A. 1972. *FEBS Lett.* 20:311–15
121. Panet, R., Selinger, Z. 1972. *Biochim. Biophys. Acta* 255:34–42
122. Deamer, D. W., Baskin, R. J. 1972. *Arch. Biochem. Biophys.* 153:47–54
123. Masuda, H., de Meis, L. 1974. *Biochem. Biophys. Acta* 332:313–15
124. Hasselbach, W., Migala, A. 1977. *Z. Naturforsch.* 32c:992–96
125. de Meis, L., Carvalho, M. G. C. 1974. *Biochemistry* 13:5032–38
126. Masuda, H., de Meis, L. 1973. *Biochemistry* 12:4581–85
127. Kanazawa, T. 1975. *J. Biol. Chem.* 250:113–19
128. Rauch, B., Chak, D., Hasselbach, W. 1977. *Z. Naturforsch.* 32c:828–34
129. Boyer, P. D., de Meis, L., Carvalho, M. G. C., Hackney, D. D. 1977. *Biochemistry* 16:136–40
130. de Meis, L., Masuda, H. 1974. *Biochemistry* 13:2507–62
131. de Meis, L., Sorenson, M. M. 1975. *Biochemistry* 14:2739–44

Ann. Rev. Biochem. 1979. 48:293–325

PHOTOAFFINITY LABELING OF BIOLOGICAL SYSTEMS[1]

♦12011

Vinay Chowdhry[2]

Central Research & Development Department, E. I. du Pont de Nemours & Co., Experimental Station, Bldg. 328, Wilmington, Delaware 19898

F. H. Westheimer

Department of Chemistry, Harvard University, Cambridge, Massachusetts 02138

CONTENTS

[1]The following abbreviations are used: AChE, acetylcholine esterase; cAMP, cyclic 3',5' adenosine monophosphate; Chy, chymotrypsin; Chy–CH$_2$OH, chymotrypsin, showing hydroxymethyl group of "active site" serine; cGMP, cyclic 3',5' guanosine monophosphate; NAP, 4-azido-2-nitrophenyl; O–CM cysteine, O-carboxymethylcysteine; O–CM histidine, O-carboxymethylhistidine; O–CM serine, O-carboxymethylserine; O–CM tyrosine, O-carboxymethyltyrosine; RPRAd, adenosine diphosphoribosyl.

[2]Address correspondence to this author.

293

0066-4154/79/0701-0293$01.00

PERSPECTIVES AND SUMMARY

The elucidation of molecular interactions leading to biochemical processes is a major objective of chemical and biochemical research. Essential to an understanding of the interactions involved in biological processes at the molecular level is the identification and structural characterization of the components of the complex systems involved. Although a multitude of enzymes has been isolated, their biological substrates identified, and some understanding of their mechanism of action reached, the identification and isolation of hormone receptors, transport proteins, proteins regulating nucleic acid transcription and translation, and the topographical distribution and function of membrane proteins have not progressed as much. An understanding of the mechanism of action of enzyme-catalyzed reactions requires the identification of the amino acids in the active site involved in binding and those more directly involved in catalysis. The classical approach to acquiring such information has been the use of general and specific reagents for chemical modification of active site residues, commonly referred to as affinity labeling (1–3).

Other techniques are also widely used. Spectroscopy is not yet fully applicable to macromolecules. Determination of structures of crystalline proteins (4, 5) and nucleic acids (6, 7) by X-ray diffraction has provided, in many instances, molecular information in fascinating detail. Such structural information, although limited by the static picture it provides of these conformationally dynamic systems, complements that obtained by chemical studies conducted in solution. A promising new approach combining solution studies with X-ray crystallographic analysis has been recently introduced (8–10). NMR spectroscopy has provided valuable information on the dynamic structure of some macromolecules in solution but is as yet limited by macromolecular size and the need for relatively large quantities of material (11, 12). Other spectroscopic techniques, such as laser Raman spectroscopy (13), fluorescence spectroscopy (14, 15), ESR spectroscopy (16, 17), and neutron diffraction (18, 19), are being increasingly used.

More recently, another approach to chemical modification of active site residues has been introduced—suicide inhibition (20–24). This method makes use of the catalytic function of enzymes to generate a reactive molecule in the active site from an "inert" precursor. The reactive molecule, commonly an allenic moiety, then may function as a chemical trap for nucleophilic residues of the enzyme. Affinity labeling in general, and suicide inhibition in particular, however, require the presence of appropriately positioned reactive functional groups on the enzyme to bring about inactivation and labeling. Such a requirement, though generally satisfied in the active sites of enzymes (electrophilic and nucleophilic groups responsible for catalytic action are usually present in the active site of enzymes), does not necessarily obtain in proteins other than enzymes or indeed even in regions of enzymes, such as allosteric binding sites where the primary events of catalysis do not take place.

Ideal labeling reagents should react not only with nucleophiles but even with the significantly less reactive but ubiquitous hydrophobic regions (C–H bonds) of proteins. The high (indiscriminate) reactivity of such reagents should be "masked" but capable of activation in situ at the discretion of the experimentalist. Such reagents can provide valuable information on the constitution of binding and/or catalytic sites of isolated macromolecules in addition to tagging and therefore identifying different kinds of macromolecular receptors. Photogenerated reagents have the potential for satisfying both the requirement of high reactivity and that of activation in situ.

The use of photogenerated reagents to study biological macromolecules was introduced for the photolysis of diazoacetylchymotrypsin by Singh, Thornton & Westheimer (25) in 1962. Photoaffinity labeling has since evolved into a major technique for studying molecular interactions in biological systems (26–29). A photolabile reagent is anchored to the macromolecule, when possible by a covalent bond; photolysis of the complex then leads to the generation of a highly reactive species that, by reacting (by insertion) rapidly with the immediate environment, labels the macromolecule or specifically the active site. The technique has the potential for identifying complex biological receptors and providing information on their molecular structure. Since the initial experiment the method has been used to investigate receptors and transport proteins, membrane structure and function, protein-nucleic acid interaction, and antibodies, in addition to enzymes. These and other applications are discussed here to highlight the potential of the method, and its limitations; finally, the properties of various photolabile groups are evaluated. In the next section early experiments by different groups are presented to illustrate the basic approaches that have subsequently been amplified in the diverse uses presented in the section on recent applications.

HISTORICAL DEVELOPMENT

Diazoacetyl Chymotrypsin

Diazoacetylchymotrypsin was prepared by reacting *p*-nitrophenyl diazoacetate with chymotrypsin at pH 6.2, in analogy with the preparation of acetylchymotrypsin by Balls & Wood (30, 31). The inactivation of the enzyme by 10^{-3} M p-nitrophenyl diazoacetate is rapid, being 95% complete in 12 min at 25°C; the esterolytic activity was restored with 2 M hydroxylamine at pH 8. Reaction of chymotrypsin with p-nitrophenyl-[2-^{14}C]-diazoacetate yielded inactive enzyme having 0.98 mole of label per mole of enzyme, which suggests stoichiometric diazoacylation of the reactive serine residue in the active site (Equation 1).

$$\text{Chy} - \text{CH}_2\text{OH} + \overset{\text{N}_2}{\overset{\|}{\text{HC}}} - \overset{\text{O}}{\overset{\|}{\text{C}}} - \text{O} - \langle \bigcirc \rangle - \text{NO}_2 \quad \xrightarrow{\text{pH 6.2}}$$

1.

$$\text{Chy} - \text{CH}_2 - \text{O} - \overset{\text{O}}{\overset{\|}{\text{C}}} - \text{CHN}_2 + \text{HO} - \langle \bigcirc \rangle - \text{NO}_2$$

Although acetylchymotrypsin was reactivated by hydrolysis in 10 min at pH 7, diazoacetylchymotrypsin remained unchanged on standing in the dark at pH 6.2 for 48 hr. Diazoacetylchymotrypsin, in contrast to acetylchymotrypsin, could therefore conveniently be isolated and stored as the stable acyl enzyme. Photolysis of ^{14}C-labeled diazoacetylchymotrypsin with long wavelength UV light for 3–4 hr led to 20–25% irreversible incorporation of the radioactive label into the protein with regeneration of 70–75% of the enzymatic activity (25, 32). Regeneration of activity proceeds at least in part according to the reactions shown in Equation 2; the asterisk denotes the position of the radiolabel.

$$\text{Chy-CH}_2 - \text{O} - \overset{\text{O}}{\overset{\|}{\text{C}}} - \overset{*}{\text{CHN}_2} \longrightarrow \text{Chy-CH}_2 - \text{O} - \overset{\text{O}}{\overset{\|}{\text{C}}} - \overset{..}{\underset{*}{\text{CH}}} + \text{N}_2 \xrightarrow{\text{H}_2\text{O}}$$

2.

$$\text{Chy-CH}_2 - \text{O} - \overset{\text{O}}{\overset{\|}{\text{C}}} - \underset{*}{\text{CH}_2\text{OH}} \xrightarrow{\text{H}_2\text{O}} \text{Chy-CH}_2\text{OH} + \text{HO} - \text{CH}_2 - \overset{*}{\text{CO}_2\text{H}}$$

$$\text{active} \qquad\qquad \text{glycolic}$$
$$\text{enzyme} \qquad\qquad \text{acid}$$

Isolation of 2-^{14}C-labeled glycolic acid accounted for 55% of the label bound to enzyme prior to photolysis. Acid hydrolysis of the photolyzed radiolabeled enzyme followed by amino acid analysis revealed two major radioactive products, O-carboxymethylserine and O-carboxymethyltyrosine in addition to glycolic acid (32, 33). N-Carboxymethylhistidine was also obtained together with three other minor and as yet unidentified products. Hexter has subsequently shown (34) that the material identified as O-carboxymethylserine (32) may have been contaminated with 7–8% of

S-carboxymethylcysteine. A brief discussion of the mode of formation of these four carboxymethylated amino acids will help illustrate the potential of and problems associated with the use of photogenerated reagents to obtain information at the molecular level.

O-Carboxymethylserine is produced by a photoreaction analogous to the Wolff rearrangement of α-diazoketones, which may proceed either via the carbene or the excited state of the diazo group. The mechanism written below is intended only as a formal representation, and not a statement of the precise sequence of events.

$$Chy\text{-}CH_2\text{—}O\text{—}\underset{\underset{O}{\|}}{C}\text{—}CHN_2 \xrightarrow{h\nu} Chy\text{-}CH_2\text{—}O\text{—}\underset{\underset{O}{\|}}{C}\text{—}\ddot{C}H \xrightarrow{\text{Wolff rearrangement}}$$

$$Chy\text{-}CH_2\text{—}O\text{—}\underset{\underset{H}{|}}{C}{=}C{=}O \xrightarrow{H_2O} Chy\text{-}CH_2\text{—}O\text{—}CH_2\text{—}CO_2H$$

3.

$$\xrightarrow{6\text{NHCl}/110°C} O\text{-Carboxymethylserine}$$

O-Carboxymethyltyrosine is formed by "insertion" of the carbene, or of an equivalent species, into the hydroxyl group of tyrosine. The yield of O–CM tyrosine was found to be dependent on enzyme concentration (increasing with concentration of enzyme) and on pH (the yield was 33% at pH 4.0 but diminished to 21% at pH 6.0). Subsequent analysis of peptides (33) established that the labeled tyrosine was number 146 in the amino acid sequence, and was produced by insertion of the carbene, initially bound to the macromolecule, into the hydroxyl group of tyrosine 146 of another molecule of chymotrypsin. Proof that the chemical reaction occurs in a noncovalent dimer of trypsin is based not only on the dependence of the yield on concentration, but also on isolation of the "dimeric" product by gel chromatography. The dimeric product is believed to be linked from the glycolate ester of serine 195 of one molecule to tyrosine 146 of another:

$$Chy\text{—}CH_2\text{—}O\text{—}\underset{\underset{O}{\|}}{C}\text{—}CH_2\text{—}O\text{—}\bigcirc\text{—}CH_2\text{—}Chy$$

SERINE 195   TYROSINE 146

The mode of formation of N-carboxymethylhistidine is however not as clear. Although some of it may be formed by an insertion mechanism via the carbene, control experiments suggest that most, if not all, of it may be formed by an acid-catalyzed light-independent process. Finally, the S-carboxymethylcysteine obtained is likely formed by insertion of the carbene into the disulfide bond, between cysteines 42 and 58. X-ray crystallographic analysis of chymotrypsin places this disulfide bond in the proximity of serine 195 which is presumed to carry the diazoacetyl moiety. Significantly, diazomalonylchymotrypsin on photolysis yields S-carboxymethylcysteine as one of the primary products (35).

The formation of O-CM serine suggested that α-diazoesters could undergo the Wolff rearrangement on photolysis, as do α-diazoketones. The Wolff rearrangement of diazoacetylchymotrypsin, which accounts for 40% of the irreversibly incorporated ^{14}C label, "wastes" the carbene produced because it labels the serine which had been known to be in the active site prior to the experiment. The objective of photoaffinity labeling, to map the active site by reaction of a carbene with amino acid residues other than the one used to anchor the bifunctional reagent in the active site, is therefore not fulfilled by this reaction. Although the ketene generated by the rearrangement may in other instances react with active site nucleophiles, thereby salvaging some of the carbene that would otherwise be wasted by the Wolff rearrangement pathway, trapping of such a ketene may suffer from some of the same difficulties as those encountered in conventional affinity labeling studies. The formation of O–CM tyrosine and S–CM cysteine, on the contrary, illustrates the potential of photoaffinity labeling. O-Carboxymethyltyrosine in particular, because it is formed in an intermacromolecular reaction, demonstrates the novelty and potential of photoaffinity labeling in providing information on dynamic structures in solution and on the nature of intermacromolecular contact. The rapid reaction of a photogenerated carbene could therefore, in principle, be used to trap short-lived macromolecular complexes, in addition to mapping stable macromolecular ensembles. The high percentage yield of the insertion of carboxymethylcarbene into water, leading to glycolic acid, though undesirable, will depend on the accessibility of the carbene to water in the system being studied. Chymotrypsin, a hydrolytic enzyme, is clearly expected to hold a substantial amount of water in the active site. The formation of N–CM histidine identifies another potential difficulty. Acid-catalyzed and copper ion-catalyzed decomposition of diazo groups leading to labeling of enzymes has been demonstrated (36). Such labeling does not however enjoy the advantages, elucidated before, of photoaffinity labeling, and may lead to nonspecific labeling.

Other Diazoacyl Enzymes

Subsequent investigations have dealt with the photolysis of diazomalonylchymotrypsin (35), diazomalonyltrypsin (35, 37) and diazoacetylsubtilisin

(38). The photolysis of both diazomalonylchymotrypsin and diazoacetyl-chymotrypsin produced the same products, although their relative yields were different. Photolysis of both diazomalonyltrypsin and diazomalonylchymotrypsin yielded the same products, but in the former case, 1–3% of ^{14}C-radiolabeled glutamic acid was also obtained (37). This product is presumably formed by insertion of the carbene into the C–H bond of the methyl group of an alanine residue found in the active site (Figure 1). This result is an important demonstration of the potential of photoaffinity labeling to tag the ubiquitous but difficult to label C–H bonds found in biological macromolecules.

Photolysis of diazoacetyl glyceraldehyde-3-phosphate dehydrogenase identified another important problem (L. J. Crane, unpublished results). The diazoacetyl group may be attached as a thioester to a reactive cysteine in the active site. The only product of photolysis observed was S-carboxymethylcysteine, formed by the Wolff rearrangement. This rearrangement, which wastes ∼ 30% of the carbene on photolysis of the diazo moiety of an O-ester, thus leads to a complete waste of the reagent on photolysis of this thioester. Model studies suggest that diazomalonylthioesters also suffer exclusive Wolff rearrangement of the thioalkyl group on photolysis (39). By contrast, thioesters of the recently introduced 2-diazo-3,3,3-trifluoropropionyl chloride (40) undergo significant insertion on photolysis, and therefore hold promise for use in photoaffinity labeling as α-diazothioesters of biological molecules.

Labeling of Enzyme-Photolabile Reagent Complexes

The majority of small molecules that interact with proteins (coenzymes, inhibitors, allosteric modulators, neurotransmitters, antigens, hormones) function primarily by forming a specific complex with the protein which either alters its conformational state or, as in the case of coenzymes, in-

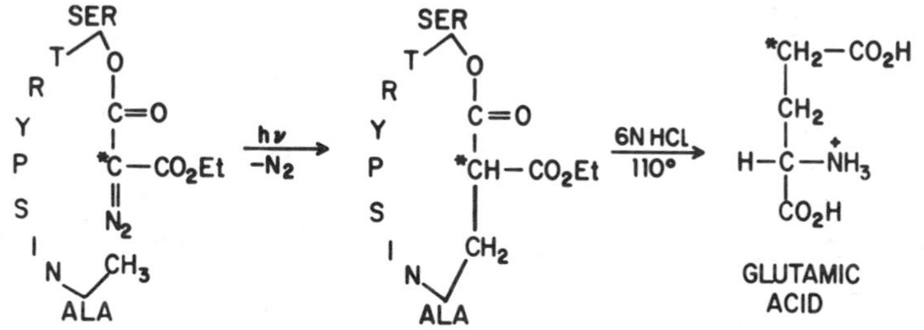

Figure 1   Insertion of diazomalonyl carbene into the C–H bond of the methyl group of alanine in the active site of trypsin. Acid hydrolysis followed by decarboxylation led to incorporation of a carboxymethyl moiety into the methyl group of alanine, thereby transforming it to glutamic acid.

creases greatly the effective concentration of the reactant, thereby leading to enhanced rate of chemical reaction. The potential for generation of a highly reactive reagent in situ by photoexcitation, and for labeling of sites not necessarily rich in reactive nucleophilic residues, in principle provides photoaffinity labeling with a special advantage in studying such binding sites. In a generally unsuccessful attempt to take advantage of this possibility, Browne, Hixon & Westheimer prepared a 3-diazoacetoxymethyl analogue of NAD (I) and investigated its use in photoaffinity labeling of yeast alcohol dehydrogenase (41).

The dissociation constant for the enzyme-NAD analogue complex was 5×10^{-4} M; the analogue, then, does not bind tightly. Although 20% incorporation of the radiolabel was observed on photolysis, the reduction in the amount incorporated in the presence of saturating concentrations of NAD was far less than expected. The labeling result is therefore ambiguous. Subsequently an azido analogue of NAD (II) was prepared and used to photolabel yeast alcohol dehydrogenase. There was only 7% incorporation of the radiolabel, of which about half may have been due to nonspecific labeling (42). These examples illustrate the problems inherent in photoaffinity labeling of "loose" receptor ligand complexes. In principle, if the dissociation constant is low (10^{-5}–10^{-6} M), and binding of the analogue is competitive with that of the natural ligand, specific labeling of the receptor would be expected provided the lifetime of the reactive species generated in situ is short (10^{-5}sec or less).

The ketenes generated by Wolff rearrangement of the diazoacyl moiety and the nitrenes produced on photolysis of aryl azides may have lifetimes long compared to the times for the dissociation of loosely bound complexes; if so, labeling may be inefficient or nonspecific. The problem of low yield of incorporated radiolabel on photolysis of loose complexes may sometimes be avoided by continuous replacement of the reagent being used for photoinactivation (43).

Labeling of Antibodies

The use of photoaffinity labeling to map the combining sites of antibodies was introduced by Knowles and co-workers (44), and simultaneously by

Richards et al (45), in 1969. Knowles and co-workers isolated antibodies to the antigenic determinant, 4-azido-2-nitrophenyl (NAP). The dissociation constant for purified anti-NAP antibody with NAP-^{3}H-lysine was 1.4 X 10^{-7} M at 4°C. This high association constant facilitated isolation of the complex by gel chromatography; two moles of NAP-^{3}H-lysine were complexed per mole of antibody. Photolysis of the isolated complex for 18 hr led to incorporation of about 1 mole of radiolabel per mole of antibody. This experiment not only illustrated the potential for successful specific labeling when "tight" complexes are photolyzed, but introduced the use of aromatic azides as reagents for photoaffinity labeling. They have since been widely used for this purpose (26–29).

Pseudophotoaffinity Labeling

Singer and co-workers, in an attempt to label the acetylcholine esterase of erythrocyte membranes (AChE), encountered problems due to nonspecific labeling that presumably arose from the relatively long lifetime of the reactive species generated on photolysis of an aromatic azide; they named such nonspecific incorporation pseudophotoaffinity labeling (46). Photolysis of AChE in the presence of 10^{-5} M [^{3}H]4-azido-2-nitrobenzyltrimethylammonium ion (III) caused 82% irreversible inactivation of the enzyme. Photolysis of the azide was needed for inactivation; the presence of good competitive inhibitors protected the enzyme from such photoinactivation. The amount of irreversibly incorporated radiolabel far exceeded the extent of inactivation both in the presence and in the absence of competitive inhibitor. Thus, photolysis is apparently accompanied by significant nonspecific incorporation. The latter was reduced tenfold upon photolysis in the presence of p-aminobenzoate, a scavenger that is believed to react with photoactivated reagent present in the solution rather than that at the binding site. More importantly, no photoinduced inactivation of AChE was observed in the presence of added scavenger. This result strongly suggests that inactivation did not occur by reaction of photogenerated nitrene before it diffused out of the active site, but that the reactive species diffused in and out of the active site several times before reacting. Such a labeling process is, of course, similar to conventional affinity labeling and does not enjoy the special advantages of photoaffinity labeling.

In an extension of their study cited earlier (45) Richards and co-workers (47–50) have conducted photoaffinity labeling studies on protein 460, a mouse IgA myeloma immunoglobulin (capable of binding the 2,4-dinitrophenyl group) with two different reagents 2,4-dinitrophenylalanyldiazoketone (IV), and 2,4-dinitrophenyl azide (V).

The dissociation constants of the myeloma protein complexes with compounds IV and V are shown with their structures. Photolysis with either reagent resulted in incorporation of radiolabel into protein and with IV there was an equivalent loss of binding sites for ε-2,4-dinitrophenyllysine. Although the diazoketone IV was found bonded primarily to the light chain, only 15% of the azido compound V that was incorporated was detected on this chain. Peptide analysis showed that the diazo compound reacted almost quantitatively with the ε-amino group of lysine 54 of the light chain to form an amide bond, but the azido reagent that bound to the light chain was found in three different peptides. However, the majority of the azide label (85%) was incorporated into tyrosines 33 and 88 in the variable region of the heavy chain. Assuming these residues to be homologous with residues in the human myeloma protein NEW [for which the crystal structure has been determined (51)], the two tyrosines are estimated to be too far apart (23 Å) to be contact residues at the same site.

Labeling by the azido reagent of residues 23 Å apart points to the possible ambiguities in interpretation of results when the labeling process and the diffusion of reactive species from the site of photogeneration proceed at comparable rates. Since lysine 54 was attached by an amide bond to the residue from photolysis of the diazo reagent, labeling must have proceeded via ketene obtained by Wolff rearrangement. As ketenes are expected to be less reactive than the corresponding carbenes, the exclusive reaction of the antibody with the ketene generated on photolysis of IV raises the possibility of nonspecific labeling. This result is not entirely unexpected as α-diazoketones such as IV are indeed more prone to undergo the Wolff rearrangement than α-diazoesters such as the α-diazoacyl enzymes discussed earlier.

Even when the dissociation constant is low, as in the case of the interaction of carbonic anhydrase and p-azidobenzene-sulfonamide ($K_i \sim 10^{-6}$M, with $\sim 99\%$ active site occupancy), photolysis leading to covalent labeling "occurs largely at a point removed from the active site" (52).

Photolabile Natural Ligands

Ideally, in studies of affinity labeling, the perturbation caused by the reactive group, or its precursor, to the molecular interaction being examined is kept to a minimum. If precise information about molecular interaction is desired, it is imperative to establish that the probe not only binds to the same site but also that it binds in a mode identical to that of the true substrate. If, however, the experiment is designed only to identify and tag binding sites or to identify specific receptors in a complex milieu, demonstration of competitive binding and true photoaffinity labeling will suffice even if the precise mode of binding is not known. The presence in a natural ligand of a functional group, such as an α,β-unsaturated system, might provide in some instances an advantage over reagents that incorporate synthetic photolabile groups, especially if the photochemical behavior is well understood. Martyr & Benisek have used photoexcited α,β-unsaturated keto-steroids to inactivate Δ^5-ketosteroid isomerase (53). Photolysis, although slow, leads to up to 90% inactivation. Reinvestigation, however, has shown that in contrast to the earlier report of covalent attachment of steroid to enzyme (53) the radiochemically labeled steroid, although tenaciously bound to the enzyme, is released on denaturing the enzyme (W. F. Benisek, private communication). Recently, amino acid analysis and sequence analysis (54) have shown that the photoinactivation is accompanied by the transformation of aspartic acid 38 to an alanine residue! No mechanism for this novel photochemical reductive decarboxylation has been proposed. Although the excited singlet states of α,β-unsaturated ketones are known to be short-lived, $\sim 10^{-5}$–10^{-9} sec, an efficient process of intersystem crossing to the longer lived triplet state (10^{-3} sec) is also known, posing the potential for problems, discussed earlier, arising from long lifetimes of reactive species. The excited state generally reacts with C–H bonds by hydrogen abstraction but a high degree of selectivity is observed. Galardy and co-workers have similarly used aromatic ketones for photoaffinity labeling (55).

Several groups have shown that cyclic nucleotides undergo photoincorporation into different receptors. Cooperman and co-workers observed incorporation of N^6-butyryl cAMP into erythrocyte ghost membranes on photolysis (56). Kallos has used the same derivative of cAMP to photolabel a cytoplasmic receptor for cAMP (57). Ferguson and co-workers have used both cAMP and cGMP in photoaffinity labeling studies with cyclic nucleotide receptors in extracts from testis, adrenal cortex (58, 59), and in messenger ribonucleoprotein-like particles (60).

Many different groups have studied the photochemical cross-linking of proteins to nucleic acids, generally utilizing the photoreactivity of nucleic

acids (61–64). The photochemical processes involved are not well understood and product analysis has rarely been achieved. A brief discussion of
such studies is presented later.

The experiments discussed in this section have been presented to focus
attention on the potential problems possible in photoaffinity labeling of
biological macromolecules. Since 1972 the use of photogenerated reagents
has proliferated in spite of these and other problems. The method has been
limited however by these difficulties and definitive interpretation of results
at a molecular level has often proven difficult. Use of the approach has been
extended to widely different systems, and many new reagents, primarily
incorporating the α-diazoacyl and p-azidonitrophenyl photolabile groups,
have been prepared (26–29). However, as has been repeatedly stressed in
recent years, the development of new photolabile groups free of the defects
discussed here is expected to greatly enhance the general utility of
photogenerated reagents (26–29). The experiments presented here fall into
two principal groups: one in which the photolabile reagent is covalently
bound to the macromolecule at a specific site prior to photolysis, and the
other in which the reagent is associated with the macromolecule as a
reversible complex, but where photoactivation leads to irreversible incorporation of the bound ligand. The binding of small molecules to proteins and
intermacromolecular interaction have been studied by the use both of synthetic photolabile groups and of photoreactive natural ligands. The diverse
uses of photoaffinity labeling presented in the next section extend the potential applicability of the method.

RECENT APPLICATIONS

Although the initial experiments presented above were conducted on purified proteins, the method has now been applied to complex ensembles such
as macromolecular receptors, ribosomes, and membranes, wherein the use
of photogenerated reagents has special and perhaps unique advantages
compared to other methods. Emphasis in this discussion is on photoaffinity
labeling studies of macromolecular assemblies, as these represent novel
areas of application of the technique. As a consequence new photoaffinity
labeling studies on purified proteins are discussed only briefly.

Receptors and Transport Proteins

The term receptors is used broadly to encompass both soluble and membrane-bound proteins that function as receptors for hormones, neurotransmitters, cyclic nucleotides, sugars, amino acids, peptides, and drugs. Most
hormone and drug receptors are membrane-bound and are not easily solubilized from the membrane, nor do they always retain physiological func-

tion or specific binding capacity upon such solubilizatión (65–67). The isolation of functional receptors has generally not been feasible. Significant advances in the experimental detection of receptors, usually present only at low concentrations, have been made (68–71), facilitated in part by the availability of analogues of natural ligands with radiolabel of high specific activity. Affinity labeling of receptors has been utilized by several groups, but successful isolation of an intact, functional receptor has not yet been achieved.

ACETYLCHOLINE RECEPTOR   The first attempts to utilize photoaffinity labeling to study receptors were directed by Schwyzer, Waser, and co-workers (72, 73) at the acetylcholine receptor in mouse phrenic nerve diaphragm, and simultaneously by Singer and co-workers (74) at the receptor in whole-frog sartorius muscle. Schwyzer and co-workers synthesized diazoacetylcholine bromide and studied its chemical and photochemical properties. They found it was hydrolyzed by acetylcholine esterase 10^4 times more slowly than was acetylcholine iodide, a finding reminiscent of the hydrolytic stability of diazoacetylchymotrypsin noted earlier. Neuromuscular transmission in isolated mouse phrenic nerve diaphragm was blocked by III (p. 302) (10^{-5}M) through prolonged depolarization of post-synaptic membranes; photolysis at long wavelength caused an irreversible depolarizing block of neuromuscular transmission. Acetylcholine esterase activity was, however, inhibited only 20% by a 5×10^{-4}M solution of III, which suggests that the active enzyme and the cholinergic receptor are present at different locations. These experiments suggested the potential use of photogenerated reagents to switch on or off the physiological function of specific receptors in complex biological systems.

The arylazide III was used to label the peptide chains of detergent-solubilized purified acetylcholine receptor from the electric tissue of *Torpedo californica* (75). Four polypeptide chains of different molecular weight are found in the isolated protein, and all are labeled on photolysis and are protected against labeling by agonists and antagonists containing a quarternary ammonium group; the neurotoxin from *Naja naja siamensis,* however, protected only the α-chain. The authors conclude that the α-chain is the binding site of the neurotoxin but explain that the labeling of all four chains is observed because a "photo-activated intermediate of the label exists long enough to react with polypepdide chains in the neighborhood of the initial binding site." This result then, is similar to that obtained earlier by Singer (46) with the same arylazide, and suggests that the problem of nonspecific labeling may be caused by the apparently long lifetime of the species obtained on photolysis of this photolabile group. Raftery & Witzemann have synthesized and used 1,10-bis(3-azidopyridinium)-decane diiodide as a photoaffinity label for the acetylcholine receptor of the electric organ of *Torpedo californica* (76). The receptor is composed of four different polypeptides of molecular weight 40, 50, 60, and 65 thousand. In addition to labeling the 40K subunit (the specific ligand binding site), photoaffinity labeling led to incorporation of the radiolabel into the 50K subunit in the membrane-bound receptor preparation, but into the 60K subunit when the purified receptor was utilized. This difference in the labeling pattern may reflect changes in the structural arrangement of the subunits upon solubilization and purification. The authors further suggest that the labeling of the 50K and 60K subunits may be due to nonspecific labeling caused by diffusion of the long-lived nitrene from the specific site into a neighboring subunit where reaction could have occurred with appropriately positioned nucleophiles.

Since proteins associated with changes in conductivity during synaptic transmission show a high affinity for positively charged ligands (77), studies on such proteins are included in this section. Hucho, Kiefer, and co-workers have used the N-triethylammonium analogue of III to photolabel the potassium transport channel in myelinated nerve fibers (78). Prior to irradiation, the arylazide, at 4 mM, caused 50% reversible inhibition of the potassium current. Photolysis, however, led to an 80% decrease in potassium current but less than a 10% decrease in sodium current. Even after washing the nerve fiber with Ringer's solution, 65% inhibition of potassium conductance was irreversible. When reagent was applied to the internal side of the nodal membrane by diffusion from the cut ends of the nerve fiber and then photolyzed, a surprising increase in potassium conductance was observed. This was interpreted to mean that "the binding site for the quaternary ammonium group is situated close to the internal opening of the potassium channel, because the 'reactive tail' of the photoaffinity label can react only with sites outside the channel—for example with proteins from the axoplasm, the latter thereby being removed from the channel." Such detailed conclusions, derived in the absence of radiolabeled studies and in light of the problems of nonspecific labeling by III, (46, 74, 75) though fascinating, must be regarded as speculative. Recently, the triethylammonium binding sites of unmyelinated crayfish nerve fibers have also been photolabeled with the same N-triethylammonium analogue of III (79).

Guillory and co-workers have used an arylazido-β-alanine tetrodotoxin derivative to photolabel the receptor sites associated with the sodium pores of excitable membranes in amphibian skeletal muscle (80). Tetrodotoxin is known to inhibit the sodium-dependent depolarization of crayfish (81) and amphibian (82) muscle. The arylazido analogue, at 8.5×10^{-7} M, produced an inhibition of depolarization similar to that observed with unmodified tetrodotoxin.

Photolysis is believed to have led to irreversible incorporation of the tetrodotoxin moiety into the receptor, since repeated washings in the Ca-free medium after photolysis led to no detectable decrease in inhibition of the sodium ion-dependent depolarization. This experiment provides another demonstration of the potential of photoaffinity labeling to attach small effector molecules permanently onto their macromolecular receptors thereby enabling either isolation of the receptor or perhaps opening the way to a study of the complex system in vivo.

NUCLEOTIDE RECEPTORS   Since Sutherland's discovery of the action of cAMP as a second messenger (83, 84) this compound has been implicated in a vast array of biological functions such as regulation of enzyme and membrane function (85), growth and contact inhibition of mammalian cells (86, 87), synaptic transmission (88, 89), and modulation of inflammation and immunity (90). Photolabile cAMP derivatives could help study these diverse functions. Cooperman & Brunswick were the first to synthesize and use a variety of cAMP derivatives (VI), incorporating the diazomalonyl photolabile group into both the adenine and ribose moieties (91, 92). Haley and co-workers synthesized 8-azido cAMP VII (93) and an 8-azido ATP analogue (94).

i) $R^1 = COCN_2CO_2C_2H_5$, $R^2 = H$

ii) $R^1 = R^2 = COCN_2CO_2C_2H_5$

iii) $R^1 = H$, $R^2 = COCN_2CO_2C_2H_5$

VI   VII

Both the α-diazoacyl and azido derivatives have been used to study cAMP receptors on erythrocyte membranes. Rubin, in extending the results of Cooperman and co-workers (56), showed that the single protein that was labeled was a regulatory subunit of a membrane-bound protein kinase (95). The 8-azido analogue used by Haley (93, 96, 97) labeled two proteins on the membrane in slightly higher yield than did the α-diazo analogue; one of the two proteins appears to be identical to that labeled by VI while the other may be a glycoprotein. A small amount of radiolabel was found in four other proteins, a finding that may have been the result of nonspecific labeling or the presence of secondary cAMP binding sites of low affinity. The mechanism of ATP regulation of the membrane-bound protein kinase has been investigated through the use of the 8-azido analogues of cAMP and ATP (98). The enzyme (type I protein kinase) has a membrane-bound regulatory subunit that contains binding sites for cAMP and ATP; the binding of ATP at this site appears to inhibit both cAMP binding and the cAMP-induced release of the catalytic unit from the membrane bound regulatory subunit. Recently, cAMP receptors on porcine venal cortical plasma membranes have been labeled by utilizing the intrinsic photolability of cAMP (99). Greengard et al have used VII to examine differences in the cAMP receptors in the cytosol of various tissues of rats (100). Primarily, only two receptors of different molecular weight were labeled; these are believed to be the regulatory subunits of Types I and II protein kinases. This example illustrates the potential of photogenerated reagents to identify, and perhaps even estimate the concentration of, different receptors in cytosol from different tissues or animal species.

Haley and co-workers have used 8-azido cAMP to label and thus to help distinguish cyclic AMP binding sites in whole cells of the sarcoma 37 line (101). In principle, such labeling studies on whole cells followed by cell fractionation may lead to information on the distribution of such receptors in nuclei, mitochondria, etc. The 8-azido analogue may enjoy a special advantage in that it is expected to be more stable to hydrolysis in whole cells than are the acyl derivatives, VI. Both acyl and alkyl derivatives of cAMP similar to VI have, however, been shown to be more potent than cAMP in provoking hormonal responses in intact cells (102, 103). An 8-azido analogue of GTP has been used to label the GTP-binding site on tubulin dimers (104). Synthesis of 8-azido GTP and other 8-azido purine analogues has been reported (105). Two newly introduced fluorescent photoaffinity labels for nucleotides, analogues of cAMP (106) and of guanosine phosphate (107), hold much promise for studies directed at identifying receptors by photolabeling of membranes, whole cells, and nerve fibers.

HORMONE RECEPTORS   Hormone-receptor interactions are being investigated for membrane-bound and cytoplasmic receptors, in studies largely centered on steroid hormones (108, 109) and on peptide hormones, e.g. insulin, gonadotropin, and melanotropin (110, 111).

Steroid receptors   The use of photogenerated reagents in the study of estrogen receptors was introduced by Katzenellenbogen and co-workers (112–118). They prepared diazo and azido derivatives of estradiol, estrone, and hexestrol to try to photolabel the cytoplasmic receptor that, despite several different approaches by other workers, has so far proved too difficult to isolate. The binding affinities of these various derivatives were obtained by a competition assay; photolysis showed incorporation of radiolabel into the receptor to the extent of 5–20%. Studies with partially purified receptor however yielded better results, with decreased nonspecific labeling. A problem encountered in early experiments using crude cytoplasm was the hydrolytic lability of the diazoacyl group when present as the ester of the steroid molecule. This work on steroid receptors has been recently reviewed by Katzenellenbogen (119–121).

Edelman and co-workers have used photogenerated reagents to study corticosteroid receptors (122, 123). They used 21-diazo corticosteroids (α-diazoketones) which, on photolysis in methanol, yielded exclusively the product of addition of methanol to the ketene generated by Wolff rearrangement. As this model study shows, exclusive generation of ketene could be a test of the potential for nonspecific labeling by photogenerated ketene. The apparent K_d for 21-diazo-21-deoxy-[6,7-^{3}H]corticosterone was found to be 3.2×10^{-8} M, only 1.5 times greater than that of corticosterone itself. Photolysis led to incorporation of label into the corticosteroid-binding globulin. Polyacrylamide gel electrophoresis revealed label in three proteins, but photolysis in presence of competitive inhibitors showed that only one of the three had been labeled specifically. Furthermore, photolysis in the presence of tris buffer instead of phosphate buffer resulted in incorporation of label in the specifically labeled protein but very little in the others. Tris might be expected to scavenge ketene present in solution; the specifically labeled protein had apparently reacted with the ketene generated at the active site and not that equilibrated with such species in solution. Independent evidence for site-specific labeling was provided by fluorescence quenching experiments. This study suggests that when the binding is tight ($\sim 10^{-8}$ M) and when, presumably, reactive nucleophiles are present in the binding site, photoaffinity labeling with diazoketones via ketenes may prove to be specific enough to be useful. However, in the absence of product analysis, the results of labeling by way of a ketene should be regarded as preliminary. These investigations with steroids also demonstrate the utility of initial partial purification of receptors to reduce nonspecific labeling, and emphasize the need for reagents of high specific radioactivity to facilitate isolation of receptors found in low abundance.

Peptide hormones   Photolabile derivatives of peptide hormones have been synthesized, and labeling of specific receptors can be expected. Schwyzer and co-workers have synthesized *p*-azidophenylalanine, and incorporated the 3,5-ditritiated analogues into different peptides and peptide hormones (124, 128). In photolysis experiments using various ring substituted azido and nitro derivatives of phenylalanine, they observed irreversible incorporation of peptides containing the p-nitrophenyl moiety (126) into chymotrypsin. This unexpected result prompted a model study of the photochemistry of p-nitrophenylalanine and the corresponding N-acetyl ethyl ester (129). Photolysis in aqueous media led only to tar-like residues, and in aqueous ethanol, the exclusive identifiable product, isolated in only 4% yield, was $p,p,'$-azoxy-di(N-acetyl-phenylalanine ethyl ester), a coupling product not useful from the point of view of photoaffinity labeling. These results cast doubt upon the utility of the aromatic nitro moiety for photo-affinity labeling.

A diazomalonyl and a 2-nitro-5-azidobenzoyl derivative of the neurohypophyseal hormone, oxytocin, have been prepared and used to photolabel the receptor believed to mediate oxytocin-stimulated water transport in toad urinary bladder (130). In a separate study a tripeptide containing p-azidophenylalanine has been used recently to study the interaction of neurophysin II with oxytocin (131). Photolysis led to irreversible inactivation of the ligand-binding capacity of neurophysin II. These studies, although preliminary, hold promise for the elucidation of ligand receptor interaction at a molecular level.

Glucagon (132) and insulin (133) receptors have also been studied by using photolabile derivatives of these hormones. Both studies used an aromatic azido photolabile group, and appear to have led to specific labeling of the corresponding receptor. A 2-nitro-5-azidobenzoyl pentapeptide that mimics the action of the potent secretagogue cerulein in stimulating secretion of proteins from acinar cells of guinea pig pancreas in vitro, has been used to investigate the cerulein receptor by Galardy & Jamieson (134). Photolysis of the peptide in the presence of pancreatic lobules led to irreversible stimulation of protein discharge, although no protection against irreversible labeling was provided by cerulein. The absence of protection by cerulein and other peptide secretagogues raises questions as to the interpretation of these experiments.

Galardy and co-workers had earlier introduced the use of aryl ketone derivatives, and had obtained up to 50 mole% photolabeling of bovine serum albumin (known to bind carboxyl terminal tetrapeptide of gastrin) with arylazido and aryl ketone derivatives of pentagastrin (55). As Martyr & Benisek had suggested earlier, the triplet state of the ketone is relatively inert to reaction with water (53). The rate constant for hydrogen abstraction from water by benzophenone on photoactivation is $10^2 M^{-1} s^{-1}$, whereas that for abstraction from alcohol is 10^5–$10^7 M^{-1} s^{-1}$ (135). This, in principle, may be a major advantage since, as noted before, both carbenes and nitrenes are "wasted" to a considerable degree by reaction with water during photoaffinity labeling studies of biological systems. In model studies, Galardy and co-workers photolyzed acetophenone or benzophenone together with Nacetylglycine methyl ester in water but obtained C–H insertion products in only 6–8% yield after prolonged photolysis ($t_{1/2} = 55$ hr for 0.03M acetophenone).

TRANSPORT PROTEINS   The identification and study of specific transport systems for inorganic ions, amino acids, and sugars is an active area of biochemical research (136–138). Like other specific receptors, these transport systems are found in low concentration and are usually imbedded in membranes; they are ideal targets for investigation by photoaffinity labeling. Photoaffinity labeling studies have been conducted on β-galactoside transport in *E. coli* (139, 140), anion transport in erythrocytes (141–144), glucose transport in erythrocytes (145) and adipocyte (146) plasma membranes, adenine nucleotide translocation (147–151), and peptide transport in *E. coli* (152). In initial studies, Kaback and co-workers showed inactivation of lactose transport on irradiation of *E. coli* membrane vesicles in the presence of 2-nitro-4-azidophenyl-1-thio-β-D-galactopyranoside and D-lactate (139). No inactivation was observed in the absence of D-lactate, which supports the hypothesis that the lac carrier system functions only when energized. The azidophenylgalactoside, although a competitive inhibitor of lactose transport ($K_i = 75$ μM), is itself accumulated in the vesicles; therefore labeling may have occurred from the inside surface. More recently, Kaback and co-workers (140) have repeated the study with 2'-N-(2-nitro-4-azidophenyl)-aminoethyl-1-thio-β-D-galactopyranoside. Although it is not transported, this compound is a competitive inhibitor of lactose transport ($K_i = 30$–40 μM). Photolysis leads again to inactivation of lac transport only in the presence of D-lactate or upon imposition of a membrane potential by valinomycin-induced potassium efflux, which supports the hypothesis concerning energized transport stated above.

Adenine nucleotide translocation in mitochondria has been investigated by using three different reagents: a photolabile derivative 8-azidoadenosine diphosphate (147), a 4-azido-2-nitrophenyl aminobutyrl derivative of the competitive inhibitor atractyloside (148), and a derivative of ADP that carries a photolabile group at the 2'-hydroxyl group (149). As the translocation system is readily damaged by ultraviolet light, the nitroarylazido derivative, which can be photoactivated by visible light, was utilized. The atractyloside derivative and the 2'-hydroxyl derivative of ADP appear to label the same protein, which supports the contention that the label was incorporated onto the specific translocation system.

Staros & Knowles have used glycyl-4-azido-2-nitro-L-phenylalanine to irreversibly inhibit the transport of glycylglycine by live *E. coli* W cells upon photolysis at 350 nm (152). The photolabile dipeptide is a reversible inhibitor of Gly-Gly transport in the dark and is itself transported. An analysis of the reversible inhibition suggests the presence of multiple transport systems; use of radiolabeled inhibitor is expected to lead to an identification of the dipeptide transport system of *E. coli.*

MISCELLANEOUS APPLICATIONS   Leonard and co-workers have synthesized 4-azido-2-chlorophenoxyacetic acid and shown that it has auxin activity (153). They hope to use the reagent to identify and isolate an auxin receptor. Recently, three different azido derivatives of adenine have been reported as possible photoaffinity labeling agents for cytokinin binding sites on plant cells (154, 155). The compounds have been shown to possess cytokinin-like activity.

In a fascinating new application of photoaffinity labeling, Menevse and co-workers have used aromatic azido compounds in an attempt to label frog olfactory receptors (156). Nakanishi and co-workers have synthesized the diazoacetyl analogue of (6E,11Z)-6,11-hexadecadienyl acetate, a sex pheromone for the moth *Antheraca polyphemus,* to study pheromone receptor interaction (157). By recording electrophysiological response from a single olfactory hair they have established that the photolabile analogue exhibits 10% of the activity of the natural pheromone and is specific in eliciting electrical response only from the acetate receptor cell dendrites and not from the aldehyde sensitive cells. These preliminary experiments are exciting in that they open a possible new area of application for photolabile reagents.

These studies on receptors and transport systems illustrate several important features of photogenerated reagents. For studies in vivo or with crude cytoplasm high specific radioactivity is necessary. The natural ligand, especially if it is small, should be perturbed as little as possible upon attachment of the photolabile group. Stringent controls are necessary to establish specific photoaffinity labeling; photolabile groups that generate reactive species of short lifetimes are particularly attractive. In systems where the dissociation constants are low, $\leqslant 10^{-8}$ M, and reactive nucleophiles are accessible in the binding site, longer-lived species may be acceptable, as seen in the labeling of the corticosteroid receptor.

Membrane Function and Structure

The structure, topographical distribution, and function of membrane-bound proteins is an area of current interest, as is the nature of the hydrophobic interaction of such proteins with the phospholipids of the bilayer (158–161). Photogenerated reagents of high indiscriminate reactivity are ideally suited to the study of the fluid mosaic of lipid and protein. Staros & Richards have used N-(4-azido-2-nitrophenyl)-2-aminoethyl sulfonate to label the surface proteins of human erythrocyte membrane (162). The permeability of the membrane to the reagent is temperature-dependent and in subsequent studies use was made of this fact to label the cytoplasmic surface of the membrane (163, 164). The results suggest that although some proteins are embedded either in the exterior or interior surface others may span the bilayer. Whole erythrocyte cells were incubated with the reagent at 37° (the membrane is permeable to the reagent at this temperature), then the cells were cooled to 0°, washed, and finally photolyzed. Electrophoresis of the photolyzed membranes revealed that oligomers ($n = 1$–4) of hemoglobin had been produced in significant yield. Polymerization is attributed to a "parallel side reaction accompanying the insertion process which is of principal interest for labeling." The authors suggest that radicals produced by the nitrene in the reaction with peptide chains could lead to polymerization by a radical recombination mechanism. The enhanced reactivity of the photogenerated species over conventional reagents was demonstrated by the labeling of proteins not previously tagged by other reagents.

Klip et al used 1-azidonaphthalene and 1-azido-4-iodobenzene in an attempt to photolabel membrane components from within the lipid core (165, 166). The idea in such studies and others subsequently published (167–171) is to diffuse into the lipid bilayer a hydrophobic photoaffinity labeling reagent that, on photolysis, will label protein components buried in the lipid core. Interpretation of such studies, however, may prove to be hazardous because such reagents are mobile in the bilayer and the lifetime of the nitrene or equivalent generated on photolysis will be an important variable.

Two careful studies by Bayley & Knowles highlight such concerns and cast doubt upon the usefulness of nitrenes in photolabeling hydrophobic regions of biological systems (172, 173). Phenylnitrene, generated by photolysis of phenylazide, was found to label fatty acid chains in phospholipid vesicles only in low yield (0.25–3.3%). In addition, the labeling was reduced to essentially zero when photolysis was conducted with glutathione present in the aqueous phase. Such scavenging of the photoactivated species in solution, even though the reagent may be generated in the bilayer, led the authors to conclude that "nitrenes are unsatisfactory reagents for the labeling not only of lipids but also of the hydrophobic amino acid chains of membrane proteins that are contiguous with lipid." Bayley & Knowles utilized 3-phenyl-3H-diazirine and spiro[adamantane-2,2'-diazirine] to generate carbenes in lipid bilayers and compared their utility as photolabeling reagents with that of phenylazide. These carbenes do undergo intramolecular rearrangement, but, nevertheless, label phospholipids in yields ranging from 3–10%; the yield, although significantly better than that obtained with nitrenes, is still quite low. Importantly, however, added glutathione had no effect on the yield. Carbenes, then, are clearly superior to nitrenes for photoaffinity labeling of hydrophobic regions of biological systems.

Khorana and co-workers synthesized fatty acids incorporating different photolabile groups (174) and used them to support the growth of auxotrophs of *E. coli* (175). Thus, viable biological membranes containing defined and characterized photolabile probes were obtained. Photolysis would lead to cross-linking with proteins in the immediate vicinity, thus providing a reliable picture of the membrane topography. This elegant approach has great potential, and these workers have outlined a comprehensive program directed at elucidating membrane structure, and lipid-lipid and lipid-protein interactions. Stoffel et al have also synthesized various fatty acids containing the azido group and demonstrated their incorporation into phospho- and spingholipids of eukaryotic cells by adding them to the growth medium in tissue culture (176). Preliminary results of experiments with high density apolipoprotein have been recently published (177).

A new α-diazo photolabile group, trifluorodiazopropionate, introduced by Chowdhry, Vaughan & Westheimer (40), has been used successfully to study membranes. Khorana and co-workers reported the synthesis of fatty acids of variable length, containing the trifluorodiazopropionyl and diazirinophenoxy groups; the photolabile fatty acids can be incorporated into mixed diacyl phospholipids, and into vesicles containing such phospholipids, either alone or together with ^{14}C-dipalmitoylphosphatidylcholine or ^{14}C-cholesterol (178). Photolysis of these vesicles led to insertion of the carbene into C–H bonds at the expected positions along the fatty acid chain. Further, fully reactivated Ca^{2+}/Mg^{2+} ATPase of sarcoplasmic reticulum and bacteriorhodopsin was obtained on treatment of delipidated proteins with the above-mentioned phospholipids and "photolysis of bacteriorhodopsin reconstituted with trifluorodiazopropionylphosphatidylcholine resulted in the expected level of phospholipid-protein crosslinking" (178). These results demonstrate the potential of appropriate photolabile groups to elucidate the lipid-lipid and phospholipid-protein interactions that are responsible for the stability, complexity, and function of biological membranes.

Erecinska and co-workers attached one to two 2,4-dinitrophenylazide moieties to cytochrome *c* (presumed to be at lysine residues on the protein surface) to investigate whether cytochrome *c* is active in the mitochondrial respiratory chain when bound to a mitochondrial site and, if so, to identify the binding site (179). The apparent dissociation constant of cytochrome *c* from the mitochondrial membrane is reported to be 5×10^{-8}M, and that of the derivatized cytochrome *c*, $\sim 10^{-7}$ M. Photolysis yielded cytochrome *c* covalently bound to the mitochondria; subsequent fractionation in the presence of detergents and salts showed that cytochrome *c* had been covalently attached to cytochrome *c* oxidase, although in poor yield. Recently Erecinska has reported a higher yield of the covalent complex in this same system when a different arylazide derivative of cytochrome *c* was used (180). The cytochrome *c* : cytochrome oxidase complex was shown to be capable of mediating electron transfer between N,N,N',N'-tetramethyl-*p*-phenylenediamine/ascorbate and the oxidase. A similar independent study has been published recently by Bisson et al (181). In another approach to the study of protein complexes embedded in membranes Ji et al (182) have used flash photolysis of cleavable heterobifunctional reagents incorporating an arylazide as the photolabile group. The reagent is first attached covalently to membrane proteins by an imidoester function, and then subjected to flash photolysis of millisecond duration leading to cross-links in proximate proteins. As the specificity of such cross-linking depends on the diffusion rate of proteins in the membrane and the lifetime of nitrenes, neither of which is well documented, rigorous control experiments and cautious interpretation of results are necessary.

The studies discussed in this section illustrate the potential for the use of photogenerated reagents in the study of membrane topography, the structure of multienzyme complexes, and the topographical distribution of proteins in organelles. Photoaffinity labeling is likely to emerge as a major technique for elucidation of structure and function of biological membranes.

Ribosomes

The structure and function of ribosomes from *E. coli* have been intensively studied in recent years (184, 185). Bispink & Matthaei used ethyl-2-diazomalonyl Phe-tRNA in the first study of the *E. coli* ribosome by photogenerated reagents (186). Several groups have used both α-diazoacyl and arylazide derivatives of different ribosomal ligands to probe the structure of ribosomes (186–188). Despite the importance of this research, and the impressive literature on the subject, limitations of space do not permit a discussion of these studies; they have, however, been recently reviewed (189).

Nucleic Acids

Photoaffinity labeling has been used to study protein-nucleic acid interactions (62–64, 190, 191) and drug-nucleic acid interactions (192, 193). Studies directed at elucidating protein-nucleic acid interactions have been undertaken with DNA (62) and RNA (63), polymerase complexes with DNA, lac repressor-DNA complex (190, 191), and complexes of tRNA with aminoacyl tRNA synthetases (64, 194, 195). Usually these studies have utilized the photochemical lability of nucleic acids at short wavelength to bring about photochemical cross-linking of nucleic acid and protein. The mechanism of photochemical labeling in such studies is often not well understood. This raises doubts about how well the cross-linked complex reflects the geometry of the natural complex, especially because in many of these studies significant photoinduced damage accompanies cross-linking.

Yielding et al have photolyzed an azido derivative of ethidium bromide to covalently attach this intercalating drug to DNA, both in vivo and in vitro (197–199). Photolysis enhanced production of petite mutants in *Saccharomyces* and produced frameshift mutations in *Salmonella*. Production of these mutants by irreversible attachment of the intercalating drug, ethidium bromide, to DNA has been cited as evidence for the need to produce covalent attachment in mutagenesis. Such studies could prove useful in elucidating not only the mechanism of action of drugs that interact with DNA, but also the mechanism of mutagenesis induced by covalent attachment of small molecules to DNA.

Antibodies

The results of initial studies on photoaffinity labeling of antibodies are presented above. Several new studies have since been published (200–204). Knowles and co-workers have reported on the photoaffinity labeling of antibodies produced against the nitroazidophenyl haptenic group (200, 201). Antibodies exhibiting low dissociation constants ($\sim 10^{-7}$ M) were used and photolysis of the isolated antibody antigen complex led to labeling of the combining sites. These studies, together with those reported by Fisher & Press (202) show that when the ligand receptor complex is tight (dissociation constant of 10^{-6} M or less), photoaffinity labeling with the arylazide photolabile group does not appear to suffer from the problems of pseudo-photoaffinity labeling discussed earlier. Applications of photogenerated reagents to the study of antibody combining sites have been recently reviewed (189c).

Enzymes

Enzymes catalyzing several different chemical reactions have been labeled with photogenerated reagents incorporating α-diazoacyl and arylazido groups. These applications have been listed in recent reviews (28, 29, 189c); only a few examples are presented here. Different azido derivatives of nicotinamide adenine dinucleotide have been prepared and their interactions with various dehydrogenases studied (205–207), although identification of the photochemically labeled amino acids has not as yet appeared. Several studies with photogenerated reagents have been directed at different adenosine triphosphatases; recent studies have utilized arylazido derivatives attached to the 3'-hydroxyl moiety of ATP (207–210). Such multiple investigations of specific enzymes by different reagents is expected to produce a more credible picture of the active sites. Bridges & Knowles photolyzed p-azido-^{14}C-cinnamoyl α-chymotrypsin to test the labeling characteristics of arylazides on a protein of known tertiary structure (211). Photolysis led to 60% incorporation of the radiolabel into the enzyme, and peptide analyses showed the radioactivity to be localized on the C chain in the tryptic fragment which forms the aromatic binding site of the enzyme. The high yield of label, localized in a hydrophobic pocket, further confirms the ability of photogenerated reagents to label biological receptor sites.

Henkin has recently introduced the diazoacyl mixed disulfide, VIII, to attach a photolabile reagent to thiol enzymes (212). This reagent was used successfully to label serine and threonine residues of creatine kinase, and should prove useful for the labeling of other sulfhydryl enzymes.

VIII   IX

Hixson & Hixson had earlier introduced p-azidophenacyl bromide (IX) as a reagent suitable for photoaffinity labeling of sulfhydryl enzymes, and reported a preliminary study of its inactivation of glyceraldehyde 3-phosphate dehydrogenase (213). The extent of photoincorporation or the identity of labeled amino acid residues has not yet been reported. Barden and co-workers have prepared p-azidobenzoyl CoA and used it to photolabel acyl CoA:glycine N-acyltransferase (214). The enzyme has also been photolabeled by an 8-azidoadenine derivative of the coenzyme (215). Finally, transcarboxylase has been photolabeled by the radioactive analogue of p-azidobenzoyl CoA (216). The labeling was specific for the CoA ester sites on the carboxylase and revealed the presence of two binding sites per polypeptide in the 26S-transcarboxylase. This result is believed to support the theory of gene duplication and fusion leading to structural homology in the polypeptides of $12S_H$ subunit.

New Photolabile Groups

The preceding discussion highlights the desirability of having several different photolabile groups of varying steric, electronic, and photochemical properties to facilitate a judicious choice, depending on the particular application being considered. Further, the reliability of results would be enhanced by studying the biological system of interest with different reagents and different photolabile groups, not only to ensure that the labeling patterns observed reflect true interactions but also because, in the absence of ideal reagents, negative results have little significance. The properties desirable in a photolabile group are chemical stability prior to photoactivation, rapid rate of photolysis at long wavelength, and very short half life for the photogenerated reactive species leading to labeling of the environment in which it is generated. The mechanism of labeling should not involve intramolecular rearrangement to a less reactive labeling species, such as ketene. None of the photolabile groups currently in use meet these requirements fully although some come closer than others. Arylazides are rapidly reduced to the corresponding amines by dithiothreitol at room temperature and physiological pH (105, 217). Such chemical instability is undesirable in a photoaffinity labeling reagent because dithiothreitol is commonly added to buffer solutions used in biochemical studies.

Smith & Knowles studied the photochemical behavior of aryl diazirines to evaluate their potential use in photoaffinity labeling studies (218, 219). Unlike the corresponding diazo compounds, these diazirines are stable for hours in 1 M acetic acid, and though of limited stability at room temperature in the pure state, are stable for several months as dilute (<0.1M) solutions in inert solvents at –20°. The diazirines show a characteristic absorption at 350–400 nm ($\epsilon \sim 200$–300). Photolysis leads to some isomerization to the corresponding acid-labile diazo compound which is also however photolabile. Thus photolysis of phenyl diazirine in hexane/acetic acid led to a mixture of benzyl acetate and products of insertion of the benzyl carbene into the C–H bonds of hexane. In spite of these complications, the photolabile diazirine group holds much promise, as illustrated by a recent study on photolabeling of vesicles (172, 173).

Despite the inherent advantages in the use of photogenerated carbenes for photoaffinity labeling (26–29, 172, 173) the α-diazoesters most used so far suffer from three principal defects which have restricted their general utility. These defects are 1. instability to heat and acid, 2. the 30–60% Wolff rearrangement observed on photolysis of α-diazoesters (32, 220) and the 100% Wolff rearrangement on photolysis of α-diazothioesters (39, 221), and 3. the long time required for photolysis at 350 nm. Chemical instability, in addition to being a problem during synthesis and handling of the photolabile reagents, may lead to nonspecific labeling. The Wolff rearrangement decreases the amount of insertion into the macromolecule under study. The new photolabile groups, 2-diazo-3,3,3-trifluoropropionate, [X, (40)] and p-toluenesulfonyldiazoacetate, [XI, (222, 223)], have been recently introduced. They are essentially free of these defects, and hold much promise as

precursors to carbenes for photoaffinity labeling of biological macromolecules (178, 224).

$$CF_3-\overset{\overset{\displaystyle N_2}{\|}}{C}-\overset{\overset{\displaystyle O}{\|}}{C}-R \qquad\qquad CH_3-\langle\bigcirc\rangle-\overset{\overset{\displaystyle O}{\|}}{\underset{\underset{\displaystyle O}{\|}}{S}}-\overset{\overset{\displaystyle N_2}{\|}}{C}-\overset{\overset{\displaystyle O}{\|}}{C}-R$$

$$\mathbf{X} \qquad\qquad\qquad\qquad \mathbf{XI}$$

$$R = CL, \quad -O-\langle\bigcirc\rangle-NO_2, \ OR', \ SR'$$

The compounds are stable in 1M hydrochloric acid! Photolysis of O-esters
of trifluorodiazopropionic acid yields only 15% of the Wolff rearrangement
product. In contrast to the photolysis of other α-diazothioesters (39, 221),
photolysis of N-acetyl-O-methyl-cysteinyl-2-diazo-3,3,3-trifluoropropion-
ate led to 40% insertion into the –OH bond of solvent methanol. α-Diazothioesters of biological thiols containing the photolabile group, X, may now
be used as precursors of carbenes for photoaffinity labeling studies. The
^{19}F nucleus provides a sensitive NMR probe of the macromolecule under
study (V. Chowdhry and R. A. Hudson, unpublished data). 2-Diazo-2-p-
toluenesulfonyl diazoacetates are stable crystalline solids. Ethyl-2-diazo-2-
p-toluenesulfonyl acetate yields 95% insertion into solvent on photolysis;
no Wolff rearrangement product was detected. The thioester, thioethyl-2-
diazo-2-p-toluenesulfonyl acetate, like the trifluorodiazopropionyl thioester, did not lead exclusively to the Wolff rearrangement product;
25% insertion into the –OH bond of solvent methanol was observed.
p-Toluenesulfonyl diazoacetates, like other α-diazoesters, exhibit broad
absorption maxima at long wavelength (~370 nm); the extinction coefficients and rates of photolysis are however significantly greater than those
of other diazoesters. This is a distinct advantage in photoaffinity labeling
studies of macromolecules, particularly those that are especially sensitive to
short wavelength UV radiation.

Recently, a series of photolabile α-diazophosphonate dianions has been
synthesized (225, 226). Their chemical stability depends on the substituent
attached to the diazo carbon. Breslow, Feiring & Herman have studied the
photochemistry of phosphoryl azides (227a). Photolysis of the thermally
stable alkoxy and phenoxy derivatives of these compounds in t-butanol
produced no intramolecular rearrangement to the metaphosphoroimidate,
and reasonable yields of the C–H insertion product on photolysis in hydrocarbon solvents were observed. The phosphoryl nitrenes also exhibited the
lowest selectivity between primary, secondary, and tertiary carbon-hydro-
gen bonds so far reported for nitrenes. The absence of intramolecular rearrangement and indiscriminate reactivity suggest nitrenes derived from these

compounds may be useful for photoaffinity labeling. Chládek and co-workers have synthesized 5'-phosphorylazide derivatives of guanosine and adenosine nucleotides (227b). Preliminary biochemical and photochemical studies suggest that these compounds may prove to be useful photoaffinity labeling reagents.

Nitrophenyl ethers undergo nucleophilic photosubstitution by aliphatic amines on photoactivation (228). Recently, new photoaffinity labeling reagents incorporating two nitrophenyl ethers, 4-nitro-methoxybenzene and 2-methoxy-4-nitro-methoxybenzene, have been used to cross-link proteins (229). Photoaffinity labeling reagents incorporating nitrophenyl ethers could prove useful in synthesizing covalent oligomers of proteins and studying topographical distribution of proteins in organelles.

CONCLUSION

Photoaffinity labeling has evolved into a widely used technique for the study of molecular interactions in complex biological systems. One of the more exciting areas of future application promises to be in vivo studies, including turning on specific receptors of specialized cells. Increased understanding of the mechanism of photoinactivation, through detailed studies on the photochemical behavior of reagents and photogenerated reactive species, should provide new capabilities, particularly for studying transient phenomena. Improvements in structural identification of small amounts of labeled products will greatly enhance the utility of photogenerated reagents. Their use in biochemical studies is now well established and is expected to grow both in volume and in level of sophistication.

ACKNOWLEDGMENTS

This review is drawn in part from the PhD thesis of V. Chowdhry (39); we thank the Institute of General Medical Sciences of the National Institutes of Health (grant 04712) and the donors of the Petroleum Research Fund, administered by the American Chemical Society, for support of the research in that thesis. We are grateful to Professors B. S. Cooperman, B. E. Haley, and J. R. Knowles for providing manuscripts reporting their work prior to publication, and to Dr. A. E. Feiring, Dr. T. Fukunaga, and Ms. S. Vladuchick for careful reading of the manuscript and helpful comments.

Literature Cited

1. Singer, S. J. 1967. *Adv. Protein Chem.* 22:1–54
2. Shaw, E. 1970. *The Enzymes* 1:91–146
3. Wold, F. 1977. *Methods Enzymol.* 46:3–14
4. Davies, D. R., Padlan, E. A., Segal, D. M. 1975. *Ann. Rev. Biochem.* 44: 639–67
5. Rossmann, M. G., Liljas, A., Branden, C., Banaszak, L. J. 1975. *The Enzymes* 11A:61–102
6. Kim, S. H., Suddath, F. L., Quigley, G. J., McPherson, A., Sussman, J. L., Wang, A. H. J., Seeman, N. C., Rich, A. 1974. *Science* 185:435–40
7. Robertus, J. D., Ladner, J. E., Finch, J. T., Rhodes, D., Brown, R. S., Clark, B. F. C., Klug, A. 1974. *Nature* 250: 546–51
8. Fink, A. L. 1977. *Acc. Chem. Res.* 10:233–39
9. Petsko, G. A. 1975. *J. Mol. Biol.* 96:381–92
10. Douzou, P., Hoa, G. H. B., Petsko, G. A. 1975. *J. Mol. Biol.* 96:367–80
11. Mildvan, A. S. 1977. *Acc. Chem. Res.* 10:246–52
12. Lilley, D. M. J., Pardon, J. F., Richards, B. M. 1977. *Biochemistry.* 16: 2853–60
13. Spiro, T. G., Gaber, B. P. 1977. *Ann. Rev. Biochem.* 46:553–72
14. Wu, C.-W., Yarbrough, L. R., Wu, F. Y.-H. 1976. *Biochemistry.* 15:2863–68
15. Brand, L., Gohlke, J. R. 1972. *Ann. Rev. Biochem.* 41:843–68
16. Dugas, H. 1977. *Acc. Chem. Res.* 10: 47–54
17. Butterfield, D. A. 1977. *Acc. Chem. Res.* 10:111–16
18. Engelman, D. M., Moore, P. B. 1976. *Sci. Am.* 235(4):44–54
19. Engelman, D. M., Moore, P. B. 1975. *Ann. Rev. Biophys. Bioeng.* 4:219–41
20. Helmkamp, G. M., Jr., Rando, R. R., Brock, D. J. H., Bloch, K. 1968. *J. Biol. Chem.* 243:3229–31
21. Abeles, R. H., Maycock, A. L. 1976. *Acc. Chem. Res.* 9:313–19
22. Rando, R. R. 1975. *Acc. Chem. Res.* 8:281–88
23. Walsh, C. T. 1977. In *Horizons in Biochemistry and Biophysics,* ed. E. Quagliariello, 3:36–81. Reading, Mass: Addison-Wesley. 340 pp.
24. Miesowicz, F. M., Bloch, K. E. 1975. *Biochem. Biophys. Res. Commun.* 65:331–35
25. Singh, A., Thornton, E. R., Westheimer, F. H. 1962. *J. Biol. Chem.* 237:PC3006–8
26. Knowles, J. R. 1972. *Acc. Chem. Res.* 5:155–60
27. Creed, D. 1974. *Photochem. Photobiol.* 19:459–62
28. Cooperman, B. S. 1976. In *Aging, Carcinogenesis and Radiation Biology,* ed. K. C. Smith, pp. 315–340. New York: Plenum 340 pp.
29. Bayley, H., Knowles, J. R. 1977. *Methods Enzymol.* 46:69–114
30. Balls, A. K., Aldrich, F. L. 1955. *Proc. Natl. Acad. Sci. USA* 41:190–96
31. Balls, A. K., Wood, H. N. 1956. *J. Biol. Chem.* 219:245–56
32. Shafer, J., Baronowsky, P., Laursen, R., Finn, F., Westheimer, F. H. 1966. *J. Biol. Chem.* 241:421–27
33. Hexter, C. S., Westheimer, F. H. 1971. *J. Biol. Chem.* 246:3928–33
34. Hexter, C. S. 1971. *Mapping of enzymic active sites with photochemically generated carbenes.* PhD thesis. Harvard Univ., Cambridge, Mass. 82 pp.
35. Hexter, C. S., Westheimer, F. H. 1971. *J. Biol. Chem.* 246:3934–38
36. Rajagopalan, T. G., Stein, W. H., Moore, S. 1966. *J. Biol. Chem.* 241:4295–97
37. Vaughan, R. J., Westheimer, F. H. 1969. *J. Am. Chem. Soc.* 21:217–18
38. Stefanovsky, Y., Westheimer, F. H. 1973. *Proc. Natl. Acad. Sci. USA* 70:1132–36
39. Chowdhry, V. 1977. *New reagents for photoaffinity labeling.* PhD thesis. Harvard Univ., Cambridge, Mass. 396 pp.
40. Chowdhry, V., Vaughan, R., Westheimer, F. H. 1976. *Proc. Natl. Acad. Sci. USA* 73:1406–8
41. Browne, D. T., Hixson, S. S., Westheimer, F. H. 1971. *J. Biol. Chem.* 246:4477–84
42. Hixson, S. S., Hixson, S. H. 1973. *Photochem. Photobiol.* 18:135–38
43. Cooperman, B. S., Brunswick, D. J. 1973. *Biochemistry* 12:4079–84
44. Fleet, G. W. J., Porter, R. R., Knowles, J. R. 1969. *Nature* 224:511–12
45. Converse, C. A., Richards, F. F. 1969. *Biochemistry* 8:4431–36
46. Ruoho, A. E., Kiefer, H., Roeder, P. E., Singer, S. J. 1973. *Proc. Natl. Acad. Sci. USA* 70:2567–71
47. Yoshioka, M., Lifter, J., Hew, C.-L., Converse, C. A., Armstrong, M. Y. K., Konigsberg, W. H., Richards, F. F. 1973. *Biochemistry* 12:4679–85
48. Hew, C.-L., Lifter, J., Yoshioka, M., Richards, F. F., Konigsberg, W. H. 1973. *Biochemistry* 12:4685–89

49. Lifter, J., Hew, C.-L., Yoshioka, M., Richards, F. F., Konigsberg, W. H. 1974. *Biochemistry* 13:3567–71
50. Richards, F. F., Lifter, J., Hew, C.-L., Yoshioka, M., Konigsberg, W. H. 1974. *Biochemistry* 13:3572–75
51. Amzel, L. M., Poljak, R. J., Varga, J. M., Richards, F. F. 1974. *Proc. Natl. Acad. Sci. USA* 71:1427–30
52. Hixson, S. H., Hixson, S. S. 1975. *Biochemistry* 14:4251–54
53. Martyr, R. J., Benisek, W. F. 1973. *Biochemistry* 12:2172–78
54. Ogez, J. R., Tivol, W. F., Benisek, W. F. 1977. *J. Biol. Chem.* 252:6151–55
55. Galardy, R. E., Craig, L. C., Jamieson, J. D., Printz, M. P. 1974. *J. Biol. Chem.* 249:3510–18
56. Guthrow, C. E., Rasmussen, H., Brunswick, D. J., Cooperman, B. S. 1973. *Proc. Natl. Acad. Sci. USA* 70:3344–46
57. Kallos, J. 1977. *Nature (London)* 265:705–10
58. Antonoff, R. S., Ferguson, J. J. Jr. 1974. *J. Biol. Chem.* 249:3319–21
59. Antonoff, R. S., Ferguson, J. J. Jr., Idelkope, G. 1976. *Photochem. Photobiol.* 23:327–29
60. Obrig, T. G., Antonoff, R. S., Kirwin, K. S., Ferguson, J. J. Jr. 1975. *Biochem. Biophys. Res. Commun.* 66:437–43
61. Smith, K. C. 1976. In *Photochemistry and Photobiology of Nucleic Acids,* ed. S. Y. Wang, Vols 1, 2. New York: Academic. 596 pp. 430 pp.
62. Markovitz, A. 1972. *Biochim. Biophys. Acta* 281:522–34
63. Strniste, G. F., Smith, D. A. 1974. *Biochemistry* 13:485–93
64. Schimmel, P. R., Budzik, G. P., Lam, S. S. M., Schoemaker, H. J. P. 1976. See Ref. 28, pp. 123–48
65. Singer, S. J. 1971. In *Structure and Function of Biological Membranes,* ed. L. I. Rothfield, pp. 145–222. New York: Academic. 486 pp.
66. Singer, S. J. Nicolson, G. L. 1972. *Science* 175:720–31
67. Lübke, K., Schillinger, E., Töpert, M. 1976. *Angew. Chem. Int. Ed. Engl.* 15:741–48
68. Karlin, A. 1974. *Life Sci.* 14:1385–1415
69. Cuatrecasas, P. 1974. *Ann. Rev. Biochem.* 43:169–214
70. Haber, E., Wrenn, S. 1976. *Physiol. Rev.* 56:317–38
71. Kahn, C. R. 1976. *J. Cell Biol.* 70:261–86
72. Frank, J., Schwyzer, R. 1970. *Experientia* 26:1207–9
73. Waser, P. G., Hofmann, A., Hopff, W. 1970. *Experientia* 26:1342–43
74. Kiefer, H., Lindstrom, J., Lennox, E. S., Singer, S. J. 1970. *Proc. Natl. Acad. Sci. USA* 67:1688–94
75. Hucho, F., Layer, P., Kiefer, H. R., Bandini, G. 1976. *Proc. Natl. Acad. Sci. USA* 73:2624–28
76. Witzemann, V., Raftery, M. A. 1977. *Biochemistry* 16:5862–68
77. Cohen, J. B., Changeux, J.-P. 1975. *Ann. Rev. Pharmacol.* 15:83–103
78. Hucho, F., Bergman, C., Dubois, J. M., Rojas, E., Kiefer, H. 1976. *Nature* 260:802–4
79. Hucho, F. 1977. *Nature* 267:719–20
80. Guillory, R. J., Rayner, M. D., D'Arrigo, J. S. 1977. *Science* 196:883–85
81. Reuben, J. P., Brandt, P. W., Girardier, L., Grundfest, H. 1967. *Science* 155:1263–66
82. Rayner, M. D. 1972. *Fed. Proc.* 31:1139–45
83. Robison, G. A., Butcher, R. W., Sutherland, E. W. 1968. *Ann. Rev. Biochem.* 37:149–74
84. Robison, G. A., Butcher, R. W., Sutherland, E. W. 1971. *Cyclic AMP,* New York: Academic. 531 pp.
85. Rubin, C. S., Rosen, O. M. 1975. *Ann. Rev. Biochem.* 44:831–87
86. Sheppard, J. R. 1971. *Proc. Natl. Acad. Sci. USA* 68:1316–20
87. Anderson, W. B., Russell, T. R., Carchman, R. A., Pastan, I. 1973. *Proc. Natl. Acad. Sci. USA* 70:3802–5
88. Daly, J. W. 1977. *Cyclic Nucleotides in the Nervous System,* New York: Plenum. 401 pp.
89. Nathanson, J. A., Greengard, P. 1977. *Sci. Am.* 237(2):108–19
90. Bourne, H. R., Lichtenstein, L. M., Melmon, K. L., Henney, C. S., Weinstein, Y., Shearer, G. M. 1974. *Science* 184:19–28
91. Brunswick, D. J., Cooperman, B. S. 1971. *Proc. Natl. Acad. Sci. USA* 68:1801–4
92. Brunswick, D. J., Cooperman, B. S. 1973. *Biochemistry* 12:4074–78
93. Haley, B. 1975. *Biochemistry* 14:3852–57
94. Haley, B. E., Hoffman, J. F. 1974. *Proc. Natl. Acad. Sci. USA* 71:3367–71
95. Rubin, C. S. 1975. *J. Biol. Chem.* 250:9044–52
96. Haley, B. E. 1977. *Methods Enzymol.* 46:339–46
97. Owens, J. R., Haley, B. E. 1976. *J. Supramol. Struct.* 5:91–102
98. Owens, J. R., Haley, B. E. 1978. *J. Supramol. Struct.* In press
99. Walkenbach, R. J., Forte, L. R. 1977. *Biochim. Biophys. Acta* 464:165–78

100. Walter, U., Uno, I., Liu, A. Y.-C., Greengard, P. 1977. *J. Biol. Chem.* 252:6494–500
101. Skare, K., Black, J. L., Pancoe, W. L., Haley, B. E. 1977. *Arch. Biochem. Biophys.* 180:409–15
102. Henion, W. F., Sutherland, E. W., Posternak, Th. 1967. *Biochim. Biophys. Acta* 148:106–13
103. Blecher, M., Ro'Ane, J. T., Flynn, P. D. 1970. *J. Biol. Chem.* 245:1867–77
104. Geahlen, R. L., Haley, B. E. 1977. *Proc. Natl. Acad. Sci. USA* 74:4375–77
105. Czarnecki, J., Geahlen, R., Haley, B. 1978. *Methods Enzymol.* In press
106. Keeler, E. K., Campbell, P. 1976. *Biochem. Biophys. Res. Commun.* 75: 575–80
107. Wiegand, G., Kaleja, R. 1976. *Eur. J. Biochem.* 65:473–79
108. Gorski, J., Gannon, F. 1976. *Ann. Rev. Physiol.* 38:425–50
109. Yamamoto, K. R., Alberts, B. M. 1976. *Ann. Rev. Biochem.* 45:721–46
110. Gospodarowicz, D., Moran, J. S. 1976. *Ann. Rev. Biochem.* 45:531–58
111. Reichlin, S., Saperstein, R., Jackson, I. M. D., Boyd, A. E., III, Patel, Y. 1976. *Ann. Rev. Physiol.* 38:389–424
112. Katzenellenbogen, J. A., Myers, H. N., Johnson, H. J. Jr. 1973. *J. Org. Chem.* 38:3525–33
113. Katzenellenbogen, J. A., Johnson, H. J. Jr., Myers, H. N. 1973. *Biochemistry* 12:4085–92
114. Katzenellenbogen, J. A., Johnson, H. J. Jr., Carlson, K. E., Myers, H. N. 1974. *Biochemistry* 13:2986–94
115. Katzenellenbogen, J. A., Hsiung, H. M. 1975. *Biochemistry* 14:1736–41
116. Katzenellenbogen, J. A., Ruh, T. S., Carlson, K. E., Iwamoto, H. S., Gorski, J. 1975. *Biochemistry* 14:2310–16
117. Katzenellenbogen, J. A., Myers, H. N., Johnson, H. J. Jr., Kempton, R. J., Carlson, K. E. 1977. *Biochemistry* 16:1964–69
118. Katzenellenbogen, J. A., Carlson, K. E., Johnson, H. J. Jr., Myers, H. N., 1977. *Biochemistry* 16:1970–76
119. Katzenellenbogen, J. A., Carlson, K. E., Johnson, H. J. Jr., Myers, H. N. 1976. *J. Toxicol. Environ. Health Suppl.* 1:205–30
120. Katzenellenbogen, J. A. 1978. *Fed. Proc.* 37:174–78
121. Katzenellenbogen, J. A., Johnson, H. J. Jr., Myers, H. N., Carlson, K. E., Kempton, R. J. 1978. In *Bioorganic Chemistry*, ed. E. E. van Tamelan, 4:207–37
122. Wolff, M. E., Feldman, D., Catsoulacos, P., Funder, J. W., Hancock, C., Amano, Y., Edelman, I. S. 1975. *Biochemistry* 14:1750–59
123. Marva, D., Chiu, W.-H., Wolff, M. E., Edelman, I. S. 1976. *Proc. Natl. Acad. Sci. USA* 73:4462–66
124. Schwyzer, R. Caviezel, M. 1971. *Helv. Chim. Acta* 54:1395–1400
125. Escher, E., Jost, R., Zuber, H., Schwyzer, R. 1974. *Isr. J. Chem.* 12:129–38
126. Escher, E., Schwyzer, R. 1974. *FEBS Lett.* 46:347–50
127. Escher, E., Schwyzer, R. 1975. *Helv. Chim. Acta* 58:1465–71
128. Fischli, W., Caviezel, M., Eberle, A., Escher, E., Schwyzer, R. 1976. *Helv. Chim. Acta* 59:878–79
129. Escher, E. 1977. *Helv. Chim. Acta* 60:339–41
130. Stadel, J. M. 1977. *Photoaffinity labeling of the antidiuretic hormone receptor in the toad urinary bladder.* PhD thesis. Univ. Penn., Phila., Penn. 110 pp.
131. Klausner, Y. S., McCormick, W. M., Chaiken, I. M. 1978. *Int. J. Pept. Protein Res.* 11:82–90
132. Bregman, M. D., Levy, D. 1977. *Biochem. Biophys. Res. Commun.* 78: 584–90
133. Yip, C. C., Yeung, C. W. T., Moule, M. L. 1978. *J. Biol. Chem.* 253:1743–45
134. Galardy, R. E., Jamieson, J. D. 1977. *Mol. Pharmacol.* 13:852–63
135. Helene, C. 1972. *Photochem. Photobiol.* 16:519–22
136. Oxender, D. L. 1972. *Ann. Rev. Biochem.* 41:777–814
137. Boos, W. 1974. *Ann. Rev. Biochem.* 43:123–46
138. Pressman, B. C. 1976. *Ann. Rev. Biochem.* 45:501–30
139. Rudnick, G., Kaback, H. R., Weil, R. 1975. *J. Biol. Chem.* 250:1371–75
140. Rudnick, G., Kaback, H. R., Weil, R. 1975. *J. Biol. Chem.* 250:6847–51
141. Cabantchik, Z. I., Knauf, P. A., Ostwald, T., Markus, H., Davidson, L., Breuer, W., Rothstein, A. 1976. *Biochim. Biophys. Acta* 455:526–32
142. Rothstein, A., Cabantchik, Z. I., Knauf, P. 1976. *Fed. Proc.* 35:3–10
143. Kaplan, J. H., Fasold, H. 1976. *Biochim. Biophys. Acta* 443:525–33
144. Knauf, P. A., Breuer, W., Davidson, L., Rothstein, A. 1976. *Biophys. J.* 16:107a
145. Farley, R. A., Collins, K. D., Konigsberg, W. H., 1976. *Biophys. J. 16:169a*
146. Trosper, T., Levy, D. 1977. *J. Biol. Chem.* 252:181–86

147. Schaefer, G., Schrader, E., Rowohl-Quisthoudt, G., Penades, S., Rimpler, M. 1976. *FEBS Lett.* 64:185–89
148. Lauquin, G., Brandolin, G., Vignais, P. 1976. *FEBS Lett.* 67:306–11
149. Lauquin, G. J. M., Brandolin, G., Lunardi, J., Vignais, P. V. 1978. *Biochim. Biophys. Acta* 501:10–19
150. Scala, A. M. 1977. *Modulation of the nucleoside transport system of mammalian cells by polyoma virus and photoaffinity probes.* PhD thesis. Univ. Rochester, Rochester, New York. 165 pp.
151. Rosenblit, P. D., Levy, D. 1977. *Biochem. Biophys. Res. Commun.* 77:95–103
152. Staros, J. V., Knowles, J. R. 1978. *Biochemistry* 17:3321–25
153. Leonard, N. J., Greenfield, J. C., Schmitz, R. Y., Skoog, F. 1975. *Plant Physiol.* 55:1057–61
154. Theiler, J. B., Leonard, N. J., Schmitz, R. Y., Skoog, F. 1976. *Plant Physiol.* 58:803–5
155. Sussman, M. R., Kende, H. 1977. *Planta* 137:91–96
156. Menevse, A., Dodd, G. H., Poynder, T. M., Squirrel, D. 1977. *Biochem. Soc. Trans.* 5:191–94
157. Ganjian, I., Pettei, M. J., Nakanishi, K., Kaissling, K.-E. 1978. *Nature (London)* 271:157–58
158. Singer, S. J., Nicolson, G. L. 1972. *Science* 175:720–31
159. Singer, S. J. 1974. *Ann. Rev. Biochem.* 43:805–33
160. DePierre, J. W., Ernster, L. 1977. *Ann. Rev. Biochem.* 46:201–62
161. Peters, K., Richards, F. M. 1977. *Ann. Rev. Biochem.* 46:523–51
162. Staros, J. V., Richards, F. M. 1974. *Biochemistry* 13:2720–26
163. Staros, J. V., Haley, B. E., Richards, F. M. 1974. *J. Biol. Chem.* 249:5004–7
164. Staros, J. V., Richards, F. M., Haley, B. E. 1975. *J. Biol. Chem.* 250:8174–78
165. Klip, A., Gitler, C. 1974. *Biochem. Biophys. Res. Commun.* 60:1155–62
166. Klip, A., Darszon, A., Montal, M. 1976. *Biochem. Biophys. Res. Commun.* 72:1350–58
167. Mikkelsen, R. B., Wallach, D. F. H. 1976. *J. Biol. Chem.* 251:7413–16
168. Mohinddin, G., Power, D. M., Thomas, E. M. 1976. *FEBS Lett.* 70:85–86
169. Nieva-Gomez, D., Gennis, R. B. 1977. *Proc. Natl. Acad. Sci. USA* 75:1811–15
170. Abu-Salah, K. M., Findlay, J. B. C. 1977. *Biochem. J.* 161:223–228
171. Bercovici, T., Gitler, C. 1978. *Biochemistry* 17:1484–89
172. Bayley, H., Knowles, J. R. 1978. *Biochemistry* 17:2414–19
173. Bayley, H., Knowles, J. R. 1978. *Biochemistry* 17:2420–23
174. Chakrabarti, P., Khorana, H. G., 1975. *Biochemistry* 14:5021–33
175. Greenberg, G. R., Chakrabarti, P., Khorana, H. G. 1976. *Proc. Natl. Acad. Sci. USA* 73:86–90
176. Stoffel, W., Salm, K., Körkemeir, U. 1976. *Hoppe-Seylers Z. Physiol. Chem.* 357:917–24
177. Stoffel, W., Därr, W., Salm, K.-P. 1977. *Hoppe-Seylers Z. Physiol. Chem.* 358:453–62
178. Gupta, C. M., Radhakrishnan, R., Gerber, G. E., Takagaki, Y., Khorana, H. G. 1977. *Natl. Meet. Am. Chem. Soc., 174th Chicago.* [Abstr. 179 (Biology)]
179. Erecinska, M., Vanderkooi, J. M., Wilson, D. F. 1975. *Arch. Biochem. Biophys.* 171:108–16
180. Erecinska, M. 1977. *Biochem. Biophys. Res. Commun.* 76:495–501
181. Bisson, R., Azzi, A., Gutweniger, H., Colonna, R., Montecucco, C., Zanotti, A. 1978. *J. Biol. Chem.* 253:1874–80
182. Ji, T. H. 1977. *J. Biol. Chem.* 252:1566–70
183. Kiehm, D. J., Ji, T. H. 1977. *J. Biol. Chem.* 252:8524–31
184. Nomura, M., Tissieres, A., Lengyel, P. 1974. *Ribosomes,* (Monograph Series) Cold Spring Harbor, NY: Cold Spring Harbor Lab. 930 pp.
185. Wittmann, H. G. 1976. *Eur. J. Biochem.* 61:1–13
186. Bispink, L., Matthaei, H. 1973. *FEBS Lett.* 37:291–94
187. Cooperman, B. S., Jaynes, E. N., Brunswick, D. J., Luddy, M. A. 1975. *Proc. Natl. Acad. Sci. USA* 72:2974–78
188a. Hsiung, N., Cantor, C. R. 1974. *Nucleic Acids. Res.* 1:1753–62
188b. Hsiung, N., Reines, S. A., Cantor, C. R. 1974. *J. Mol. Biol.* 88:841–55
188c. Sonenberg, N., Wilchek, M., Zamir, A. 1977. *Eur. J. Biochem.* 77:217–22
188d. Maassen, J. A., Möller, W. 1974. *Proc. Natl. Acad. Sci. USA* 71:1277–80
189a. Cooperman, B. S. 1978. In *Bioorganic Chemistry. A Treatise to Supplement Bioorganic Chemistry, an International Journal,* ed. E. van Tamelen, 4:81–115. New York: Academic. 478 pp.
189b. Cooperman, B. S., Grant, P. G., Goldman, R. A., Luddy, M. A., Minnella, A., Nicholson, A. W., Strycharz, W. A. 1979. *Methods Enzymol.* In press
189c. Jakoby, W. B., Wilchek, M. eds. 1977. *Methods Enzymol.* Vol. 46. 774 pp.

190. Lin, S.-Y., Riggs, A. D. 1974. *Proc. Natl. Acad. Sci. USA* 71:947–51
191. Anderson, E., Nakashima, Y., Konigsberg, W. 1975. *Nucleic Acids Res.* 2:361–71
192. Cantrell, C. W., Yielding, K. L. 1977. *Photochem. Photobiol.* 25:189–91
193. Bastos, R. N. 1975. *J. Biol. Chem.* 250:7739–46
194. Schimmel, P. R. 1977. *Acc. Chem. Res.* 10:411–18
195. Schimmel, P. R., Budzik, G. P. 1977. *Methods Enzymol.* 46:168–80
196. Yue, V. T., Schimmel, P. R. 1977. *Biochemistry* 16:4678–84
197. Hixon, S. C., White, W. E. Jr., Yielding, K. L. 1975. *J. Mol. Biol.* 92:319–29
198. Hixon, S. C., White, W. E. Jr., Yielding, K. L. 1975. *Biochem. Biophys. Res. Commun.* 66:31–35
199. Yielding, L. W., White, W. E. Jr., Yielding, K. L. 1976. *Mutat. Res.* 34:351–58
200. Fleet, G. W. J., Knowles, J. R., Porter, R. R. 1972. *Biochem. J.* 128:499–508
201. Smith, R. A. G., Knowles, J. R. 1974. *Biochem. J.* 141:51–56
202. Fisher, C. E., Press, E. M. 1974. *Biochem. J.* 139:135–49
203. Cannon, L. E., Woodard, D. K., Woehler, M. E., Lovins, R. E. 1974. *Immunology.* 26:1183–94
204. Lindeman, J. G., Woodard, D. K., Woehler, M. E., Chism, G. E., Lovins, R. E. 1975. *Immunochemistry.* 12:849–54
205. Koberstein, R. 1976. *Eur. J. Biochem.* 67:223–29
206. Chen, S., Guillory, R. J. 1977. *J. Biol. Chem.* 252:8990–9001
207. Guillory, R. J., Jeng, S. J. 1977. *Methods Enzymol.* 46:259–88
208. Jeng, S. J., Guillory, R. J. 1975. *J. Supramol. Struct.* 3:448–68
209. Russell, J., Jeng, S. J., Guillory, R. J. 1976. *Biochem. Biophys. Res. Commun.* 70:1225–34
210. Lunardi, J., Lauquin, G. J. M., Vignais, P. V. 1977. *FEBS Lett.* 80:317–23
211. Bridges, A. J., Knowles, J. R. 1974. *Biochem. J.* 143:663–68
212. Henkin, J. 1977. *J. Biol. Chem.* 252:4293–97
213. Hixson, S. H., Hixson, S. S. 1975. *Biochemistry* 14:4251–54
214. Lau, E. P., Haley, B. E., Barden, R. E. 1977. *Biochem.* 16:2581–85
215. Lau, E. P., Haley, B. E., Barden, R. E. 1977. *Biochem. Biophys. Res. Commun.* 76:843–49
216. Poto, E. M., Wood, H. G., Barden, R. E., Lau, E. P. 1978. *J. Biol. Chem.* 253:2979–83
217a. Staros, J. V., Bayley, H., Standring, D. N., Knowles, J. R. 1978. *Biochem. Biophys. Res. Commun.* 80:568–72
217b. Cartwright, I. L., Hutchinson, D. W., Armstrong, V. W. 1976. *Nucleic Acids Res.* 3:2331–39
218. Smith, R. A. G., Knowles, J. R. 1973. *J. Am. Chem. Soc.* 95:5072–73
219. Smith, R. A. G., Knowles, J. R. 1975. *J. Chem. Soc. Perkin Trans. 2* 686–94
220. Chaimovich, H., Vaughan, R. J., Westheimer, F. H. 1968. *J. Am. Chem. Soc.* 90:4088–93
221. Hixson, S. S., Hixson, S. H. 1972. *J. Org. Chem.* 37:1279–80
222. Chowdhry, V., Westheimer, F. H. 1978. *J. Am. Chem. Soc.* 100:309–10
223. Chowdhry, V., Westheimer, F. H. 1978. *Biorg. Chem.* 7:189–205
224. Westheimer, F. H. 1976. *21st Ann. Rep. Res. under Sponsorship of Pet. Res. Fund, Am. Chem. Soc.,* p. 2. (Research conducted by J. Stackhouse)
225. Goldstein, J. A., McKenna, C., Westheimer, F. H. 1976. *J. Am. Chem. Soc.* 98:7327–32
226. Bartlett, P. A., Long, K. P. 1977. *J. Am. Chem. Soc.* 99:1267–68
227a. Breslow, R., Feiring, A., Herman, F. 1974. *J. Am. Chem. Soc.* 96:5937–39
227b. Chládek, S., Quiggle, K., Chinali, G., Kohut, J. III., Ofengand, J. 1977. *Biochemistry* 16:4312–19
228a. Hajer, J. D., Shadid, O. B., Cornelisse, J., Havinga, E. 1977. *Tetrahedron* 33: 779–86
228b. Cornelisse, J., Havinga, E. 1975. *Chem. Rev.* 75:353–88
229a. Jelenc, P. C. 1977. *High yield photoreagents for protein cross-linking and affinity labeling.* PhD thesis. Columbia Univ., New York. 188 pp.
229b. Jelenc, P. C., Cantor, C. R., Simon, S. R. 1978. *Proc. Natl. Acad. Sci. USA* 75:3564–68

Ann. Rev. Biochem. 1979. 48:327–86

REGULATION OF OXYGEN AFFINITY OF HEMOGLOBIN:

◆12012

Influence of Structure of the
Globin on the Heme Iron

M. F. Perutz

Medical Research Council Laboratory of Molecular Biology,
Cambridge, CB2 2QH, England

CONTENTS

327

0066-4154/79/0701-0327$01.00

PERSPECTIVES AND SUMMARY

Vertebrate hemoglobins (Hb's) are in equilibrium between two alternative structures, the deoxy or T and the oxy or R structure, whose oxygen dissociation constants differ by the equivalent of over 3 kcal/mol heme. The dissociation constant of myoglobin (Mb) is slightly higher than that of the R structure. Many years ago David Keilin asked me how nature could use the same heme for so many different functions, simply by attaching it to different proteins. In this review I try to answer a small part of his question: how does the structure of the globin influence the oxygen affinity of the heme?

Ideally, the problem should be approached by accurate X-ray analyses of the heme geometries in these proteins, but even the best resolution available does not allow bond lengths to be determined with the required accuracy ($\pm$ 0.005 Å). The exact bond lengths must therefore be inferred from X-ray analyses of synthetic porphyrins designed to mimic the states of the heme in Hb. In deoxyHb the iron is five-coordinated and high-spin, while in HbO_2 it is six-coordinated and low-spin. Six-coordinated hemes are usually planar or nearly so, while in five-coordinated ones the iron lies at the apex of a square pyramid whose base is formed by the pyrrollic nitrogens. The influence of coordination number and spin on the height of the pyramid and on the Fe–N bond lengths can be gauged from Table 1.

The table shows that both bond lengths and height of the pyramid increase on going from six- to five-coordination and from lower to higher spin. In synthetic iron chelates differences in iron ligand bond lengths are

Table 1 Influence of the coordination number on the properties of Fe–N bonds

Valency of Fe	Coordination number	Spin (S)	Fe–N_{porph} (Å)	Fe–Ct (height of pyramid) (Å)
2	6	0 (low)	1.98–2.00	0–0.2
2	6	2 (high)	2.06	0
2	5	0 (low)	2.00	0.2
2	5	2 (high)	2.07–2.09	0.45 ($\pm$ 0.06)
3	6	1/2 (low)	1.99	0–0.1
3	6	5/2 (high)	2.04	1–0.3
3	5	5/2 (high)	2.07	0.45 ($\pm$ 0.06)

accompanied by spin changes, and vice versa, even when the coordination number and nature of the ligands remains unaltered.

As far as limited resolution allows us to judge, these general principles are borne out by the heme geometries of Mb and Hb. In their deoxy forms the iron atoms are displaced from the mean porphyrin plane by 0.55–0.63 Å; some of this displacement is probably due to doming of the porphyrin. In six-coordinated forms the iron atoms lie closer to the porphyrin plane. The hemes lie in isolated pockets into which they are wedged tightly by about 16 side chains of the globin; these allow no movement of the iron atom and the proximal histidine relative to the porphyrin plane without a change in tertiary structure of the globin. In the β-subunits of the T structure the oxygen site is obstructed by the distal valine, and this obstruction, again, cannot be cleared without a change of tertiary structure of the globin. This suggests that the oxygen affinity could be regulated by simple mechanical devices that impede the movement of the iron and of the distal valine relative to the heme center in different ways or by different degrees in the T and R structures. Comparison of the tertiary structures of deoxyHb and HbCO suggest that this is true, combination with oxygen being unhindered in the R, and hindered in the T structure. Conversely the displacement of the iron with its proximal histidine and, in the β-subunits only, of the distal valine from the heme center, appear to be the main factors regulating the allosteric equilibrium between the T and R structures.

Convincing chemical evidence for a restraint which impedes the movement of the iron towards the porphyrin plane on ligand binding came from HbNO, where a forced transition from the R to the T structure tore the heme iron away from its bond with the proximal histidine in the α-subunits. To measure the energy equivalent of the restraint, we turned to metHb derivatives in which there exists a thermal equilibrium between two alternative spin states. A switch in quaternary structure from R to T is shown to bias that equilibrium toward higher spin, i.e. toward longer Fe–N bonds. Conversely, the higher the spin, the lower the energy needed to effect the switch to the T structure. Accurate measurements of thermal spin equilibria of carp azide metHb over a wide temperature range showed that the change in spin equilibrium is equivalent to a free energy change of 1 kcal/mol heme averaged over the tetramer, but it seems more likely that the spin change occurs mainly in one pair of subunits (possibly the α) with a free energy change of 2 kcal/mol heme. This suggests that restraint of the heme iron's movement may be dominant in regulating the oxygen affinity in one pair of subunits and steric hindrance of the ligand site by the distal valine in the other. Azide metMb at room temperature has a higher spin than azide metHb in the R structure, which is consistent with the lower oxygen affinity of myoglobin.

Can steric restraint of the type envisaged here really lower the oxygen affinity? Steric restraint by the globin has been mimicked in synthetic iron porphyrins by attaching a methyl group to the proximal imidazole in position 2, i.e. at the carbon next to the $Fe-N_\epsilon$ bond. Close contact between this methyl and the porphyrin nitrogen hinders the approach of the histidine and of its attached iron atom to the porphyrin plane. This hindrance lowers the oxygen affinity by about the same factor as the transition from the R to the T structure in Hb, which shows that the proposed mechanism is chemically viable.

INTRODUCTION

The hemoglobins are essential to the life of all vertebrates; they also occur in some invertebrates and in the root nodules of leguminous plants. They all carry the same prosthetic heme group, iron (II) protoporphyrin IX, associated with a polypeptide chain of between 136 and 153 residues. In all of them the ferrous iron of the heme is linked to N_ϵ of a histidine, the porphyrin is wedged into its pocket by a phenylalanine, and about 35 other specific sites along the polypeptide chain are occupied by nonpolar residues. These requirements seem to suffice to determine the characteristic fold of the polypeptide chain that is common to the hemoglobins of all species (Figure 1). For the rest, the amino acid sequences of the hemoglobins are variable, the number of amino acid differences between any two species rising with their distance of separation on the evolutionary tree (1–3).

The carriage of oxygen by the hemoglobins is based on the ability of their ferrous iron to combine reversibly with molecular oxygen instead of being irreversibly oxidized; the heme and the amino acid to which the iron is linked are the same in all hemoglobins, but their oxygen affinities vary widely. The affinity is expressed as p_{50}, the partial pressure of oxygen, in mm Hg, at which the solution is half saturated with oxygen. The highest affinity (0.004 mm Hg) has been recorded in *Ascaris* hemoglobin and the lowest (~1000 mm Hg) in certain fish hemoglobins at acid pH. This 250,000-fold variation, equivalent to a difference in binding energy of 7 kcal/mol, shows that the structure of the globin has a strong influence on the chemical affinity of the heme iron.

The oxygen affinities of single-chain hemoglobins, such as those found in the body fluids of invertebrates and in the cytoplasm of vertebrate muscle (myoglobin), are generally independent of the ionic composition of the solution, and their oxygen equilibrium curves are rectangular hyperbolae. On the other hand, the hemoglobins found in the erythrocytes of vertebrates are tetrameric, consisting of four hemes plus two pairs of polypeptide chains of unequal length and different amino acid sequence. Their oxygen affinities rise with increasing oxygen saturation so that their oxygen equilibrium curves are sigmoid. Under physiological conditions the rise in oxygen affinity of human hemoglobin between zero and full oxygen saturation is about 500-fold, corresponding to a free energy difference of 3600 cal/mol heme. This is known as the free energy of heme-heme interaction. Its magnitude and the value of p_{50} vary with pH and with the concentrations of CO_2, Cl^-, HPO_4^{2-}, 2,3-diphosphoglycerate (DPG), and other phosphate esters. A change in the activity of any of these agents affects the affinity of hemoglobin for all the others. The interactions just described are known collectively as the cooperative effects of hemoglobin. They are a property of the hemoglobin tetramer and vanish upon splitting of the molecule into two dimers or into free subunits. This shows that they are somehow related to the bonds between the subunits.

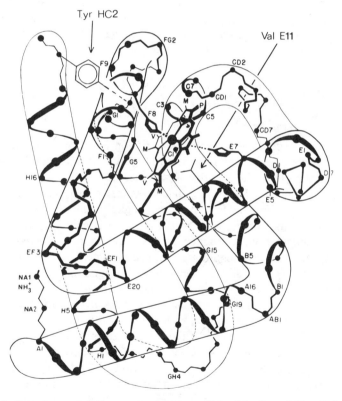

Figure 1 Secondary and tertiary structures characteristic of the hemoglobins. Helical segments are denoted *A* to *H;* nonhelical ones *NA, AB* to *GH,* and *HC.* Residues along segments are numbered in sequence. The diagram shows the proximal His F8, the distal His E7, and the distal Val E11, and Tyr HC2 which play an important part in the allosteric mechanism.

Structure analysis aims at interpreting these remarkable properties in stereochemical terms. We now know that the cooperative effects arise from a transition between two alternative structures, the deoxy or T and the oxy or R structure. These differ in the conformation of the individual subunits, the tertiary structure, and the relative arrangement of the four subunits, the quaternary structure. In the R structure the oxygen affinity is slightly higher than in free α- and β-subunits, but in the T structure it is several hundred times lower. By what kind of mechanism does the change in the structure of the globin influence the chemical reactivity of the heme iron? Or conversely, by what kind of mechanism does the combination of the heme iron with oxygen change the structure of the globin? There must be a stereochemical change at the heme accompanying the reaction with oxygen which sets in train the transition from the R to the T structure. Conversely the low oxygen affinity of the T structure must be due to constraints of the globin which oppose this change. I now review the stereochemical changes at the heme brought about by the reaction of oxygen with model compounds designed to mimic hemoglobin.

STEREOCHEMISTRY OF IRON PORPHYRINS AND RELATED COMPOUNDS

Coordination of the Iron Atom

The ferrous ion contains six and the ferric ion five valency (d) electrons which can populate the orbitals shown in Figure 2. In octahedral or square pyramidal complexes the T_g orbitals are oriented between the bond direction, so that repulsion by the iron ligands is minimized, and their energy remains low. The E_g orbitals point in the bond direction. Repulsion by the ligand atoms causes their energy to rise with the strength of the bonds to the iron atom. In five-coordinated iron porphyrins the energy of the E_g orbitals is usually low enough for them to be populated, so that the ferrous complexes contain four and the ferric ones five unpaired electrons, which make them strongly paramagnetic or "high-spin." In six-coordinated iron porphyrin complexes the distribution of electrons varies with the strength of the bonds to the distal ligands, with strong, short bonds favoring "low-spin", i.e. weakly paramagnetic or diamagnetic states, and weak, long bonds favoring high-spin states. There also exist states of intermediate spin with two unpaired electrons in the ferrous and three in the ferric state; in iron porphyrins these have been observed only in four-coordinated or very weakly five-coordinated complexes where the d_{z^2} orbital is occupied but the $d_{x^2-y^2}$ orbital remains empty.

In all Hb's the iron is coordinated to four nitrogens of the porphyrin and to N_ϵ of the proximal histidine. In deoxyHb the iron is five-coordinated and paramagnetic with a spin of $S = 2$. We now know the structures of

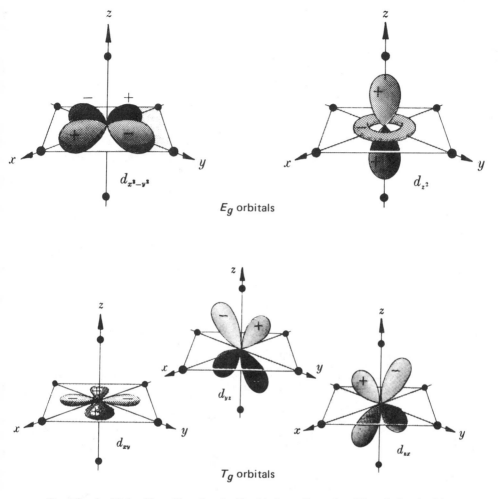

Figure 2 *d* orbitals of iron. Note that the T_g orbitals are directed at 45° to the bonds with the octahedrally coordinated ligands, while the E_g orbitals point in the bond directions. They normally have higher energies than the E_g's.

two model compounds of deoxyHb, both having as the axial base 2-methylimidazole, which is hindered from close approach to the porphyrin by the bulk of its methyl group (Figure 3 and Table 2), (4–6). In (2-methylimidazole)*meso*-tetraphenylporphinato Fe(II), (2-MeImTPPFeII for short) the porphyrin is markedly domed toward the imidazole, so that the plane of the four nitrogen atoms lies 0.13 Å above the plane of the 24 atom porphyrin skeleton. The iron lies (0.418 ± 0.005) Å above the plane of the nitrogens. The average length of the bonds from the iron to the

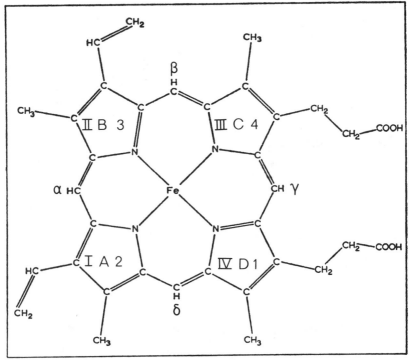

Figure 3a   Fe-protoporphyrin IX. The arabic numbers are those adopted in (7, 8), the roman ones are from (9–11), and the letters are from (12).

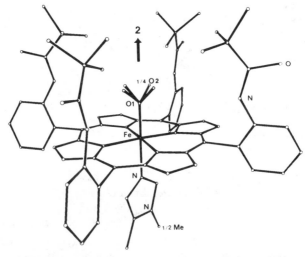

Figure 3b   The oxygenated picket fence complex of Collman and his collaborators. The terminal groups of the "pickets" are methyls. This has the sterically unhindered 1-methylimidazole as the proximal ligand. Reproduced, with permission, from (13).

Table 2 Geometry of iron porphyrins

Compound	Coordination number	Spin (S)	Fe–N$_{porph}$[b] (Å)	Fe–Ct[c] (Å)	Ref.
2-MeIm TPP Fe II	5	2	2.086 (4)	0.42	4, 6
2-MeIm piv PP Fe II	5	2	2.072 (5)	0.40	7
2-MeIm piv PP Fe II O_2	6	0	1.996 (4)	0.09	14
1-MeIm piv PP Fe II O_2	6	0	1.980 (10)	−0.03	15
bis THF TPP Fe II	6	2	2.057 (4)	0	114
1-MeIm TPP Fe II NO	6	1/2	2.008 (12)	−0.07	16
TPP Fe II NO	5	1/2	2.001 (3)	−0.211	17
bis Im TPP Fe III Cl⁻CH$_3$OH	6	1/2	1.989 (4)	0	4, 6
bis H_2O TPP Fe III	6	5/2	2.045 (8)	0	18
TPP Fe III N$_3^-$	5	5/2	2.067	−0.45	4
Pyr TPP Fe III N$_3^-$	6	1/2	1.990	−0.03	4

[a] TPP: tetraphenylporphyrin; piv: pivalamido; Me–Im: methylimidazole; THF: tetrahydrofuran.
[b] Numbers in brackets are RMS errors.
[c] Ct: center of porphyrin nitrogens; Fe–Ct: displacement of iron perpendicular to the plane of the four nitrogens, positive toward imidazole, negative toward O_2, NO, N$_3^-$, or SCN⁻.

porphyrin nitrogens is 2.086 (±0.004) Å and the length of its bond to the imidazole nitrogen is 2.161 (±0.005) Å.[1] The second model is (2-methylimidazole)*meso*-tetra($\alpha,\alpha,\alpha,\alpha$-$o$-pivalamidophenyl)porphinato-Fe(II), (2-MeImTpivPPFeII for short). Here the porphyrinato skeleton is less domed, so that the plane of the four nitrogens lies only 0.03 Å from the mean plane of the skeleton. The iron lies 0.399 Å above the plane of the four nitrogens. The average length of the bonds from the iron to the nitrogens is 2.072 (±0.005) Å and the length of its bond to the imidazole nitrogen is 2.095 (±0.006) Å. This compound is generally known as the "picket fence" complex because the pivalamidephenyl side chains form a fence around the iron ligand site. The crystal structures of two different oxygen adducts of this complex have now been determined, one having the sterically unhindered 1-methylimidazole as the fifth heme ligand, which might be regarded as a model of HbO_2 in the R structure (8), and the other with the sterically hindered 2-methylimidazole as the fifth ligand, which might be regarded as a model of HbO_2 in the T structure, because repulsion between the methyl group in the 2 position and the porphyrin nitrogens appears to mimic the restraints of the globins. In both structures the bonds from the iron to the porphyrin nitrogens are shorter by 0.09 Å (with an error of ±0.006 Å for the 2-MeIm compound) than in the oxygen-free complexes. In the unhindered complex the porphyrinato skeleton is slightly ruffled, so that neighboring methine carbons are displaced by 0.13 Å above

[1]See Fig. 8b, p. 354.

and below the mean porphyrinato plane, and the iron atom is displaced by 0.03 Å from the plane of the four nitrogens toward the oxygen ligand. In the sterically hindered complex the porphyrinato skeleton is buckled with a mean displacement from the least squares plane of 0.066 Å and the iron atom is displaced by 0.086 Å toward the imidazole ligand. In the unhindered complex the sum of $Fe-N_{Im}$ and Fe–O separations is 3.813 Å; in the hindered one the sum is 4.005 Å, the Fe–Im distance being 0.10 and the Fe–O distance 0.15 Å longer, consistent with its lower oxygen affinity (see section on implications for heme-heme interaction).

In summary, analysis of these structures has shown that the iron porphyrin bonds in models designed to mimic deoxyHb are longer by 0.09 Å than in those designed to mimic HbO_2. In the latter the iron and porphyrin atoms lie within 0.1 Å of one plane, while in the former the iron lies at the apex of a pyramid with the four porphyrin nitrogens as its base. The exact height of the pyramid and the doming of the four nitrogens relative to the rest of the porphyrin depend on the angle that the plane of the imidazole makes with the N–Fe–N bonds and also on crystal packing. Similar effects are seen in ferric porphyrins. Six-coordinated low-spin complexes are planar, with $Fe-N_{porph}$ distances of 1.99 Å; five-coordinated high spin ones are pyramidal, with $Fe-N_{porph}$ distances of 2.067 Å and displacements of the iron from the mean porphyrin plane of about 0.45 Å. These differences in geometry arise from two effects: van der Waals repulsion of the axial ligands by the porphyrin nitrogens, and repulsion between the E_g orbitals of the iron and the π orbitals of the porphyrin. The steric effect of a change in the number of axial ligands without change of spin is exemplified by the structures of two ferrous nitrosyl porphyrins: (1-methylimidazole)-a,β,γ,δ-tetraphenylporphinato Fe(II)NO (1-MeImTPPFeIINO) and the same complex without 1-methyl methylimidazole (TPPFeIINO). Both complexes have spins of $S = 1/2$. In the six-coordinated one the iron is displaced from the the porphyrin plane toward the NO by 0.07 Å, but in the five-coordinated one that displacement is increased to 0.21, even though the mean $Fe-N_{porph}$ bond lengths have increased by less than 0.01 Å. The iron is displaced further from the porphyrin plane because repulsion between the porphyrin nitrogens and the NO is no longer balanced by repulsion between the porphyrin nitrogens and the imidazole nitrogen. Note, however, that the iron in this five-coordinated *low-spin* complex is displaced by only 0.21 Å, while in the five-coordinated *high-spin* TPPFeIII N_3^- the displacement of the iron is 0.45 Å. It has been argued that repulsion between the fifth ligand and the porphyrin nitrogens is more important than spin in determining the displacement of the iron from the porphyrin plane. If that were true, then the displacement of the iron in TPPFeII NO should be greater than in TPPFeIII N_3^-, because it has the shorter Fe–N (ligand) distance (1.743 as against 1.91 Å), so that steric repulsion is greater in the nitrosyl complex. However, its Fe–N_{porph} distances are 2.001 Å, as against 2.065 Å in the azide complex, so that the iron is displaced more than twice as far in the latter.

The steric effects of filling the d_{z^2} and $d_{x^2-y^2}$ orbitals in turn can be gauged from several other comparisons. In TPP bisImFeIII both orbitals are empty, Fe–N_{porph} = 1.997 Å and Fe–N_{Im} = 2.014 Å. In 1-MeImTPPCoII the $d_{x^2-y^2}$ orbitals are empty and the d_{z^2} orbitals are filled; Fe–N_{porph} = 1.997 Å, the same as before, but Fe–N_{Im} = 2.157 Å, an increase of 0.143 Å.

It is also interesting to compare the position of the metal atom relative to the porphyrin ring in 2-MeImTPPFeII where the $d_{x^2-y^2}$ orbitals are filled, with that in 1-MeImTPPCoII where they are empty. In the former the iron lies 0.418 Å out of the porphyrin plane; in the latter only 0.13 Å, which again shows that five-coordination alone does not cause a substantial displacement of the metal if the $d_{x^2-y^2}$ is empty, but that repulsion between the $d_{x^2-y^2}$ orbitals of the metal and the π orbitals of the porphyrin pushes the five-coordinated metal out of the porphyrin plane. [For references to these structures see (6).]

The steric effect of filling the $d_{x^2-y^2}$ orbitals without change in the number of axial ligands is exemplified by the difference between the structures of $\alpha,\beta,\gamma,\delta$-tetraphenylporphinato-bis-imidazoleFeIII and $\alpha,\beta,\gamma,\delta$-tetraphenylporphinato-bis-H_2OFeIII (TPPbisImFeIII and TPPbisH$_2$OFeIII). Both have planar hemes. The former is low-spin ($S = 1/2$) and has average Fe–N distances of 1.989 Å; the latter is high-spin ($S = 5/2$) and has average Fe–N distances of 2.045 Å. The longer Fe–N distances are due to repulsion between the occupied $d_{x^2-y^2}$ orbitals of the iron and the π orbitals of the porphyrin nitrogens. The balancing effect of the two axial water molecules holds the iron in the porphyrin plane, but the strain which this planarity entails makes the Fe–N distances 0.026 Å shorter than in five-coordinated ferric high-spin complexes. The same planar geometry is found in the ferrous high-spin six-coordinated bis-tetrahydrofuran tetraphenylporphin Fe(II) (bis THF.TPPFeII for short). Again, the Fe–N_{porph} distances of 2.057 (4) Å are shorter by 0.02–0.03 Å than in five-coordinated ferrous high-spin compounds; on the other hand they are distinctly longer than in the ferric TPPbis H$_2$OFeIII, which shows that the valency of the iron does have a measurable effect on the Fe–N_{porph} distances in high-spin iron porphyrins, though it seems to be without effect in low-spin ones. These six-coordinated high-spin structures have an important bearing on heme proteins, because they show that the porphyrin is flexible, so that the position of the iron relative to the porphyrin plane and the lengths of the Fe–N bonds may be influenced by the constraints that the protein exercises on the distal ligands, and they may therefore differ from those observed in unconstrained model compounds.

Interaction between Structure and Spin State in Iron Chelates

Much can be learned about the influence of steric constraints on the state of the iron from X-ray and magnetic studies of iron chelates in which the iron is coordinated to sulphur, nitrogen, or oxygen. Especially well studied are the iron dithiocarbamates. In all these complexes the iron is coordinated to six potentially strong ligands, so that one would expect them all to be low-spin. However, steric effects which weaken and lengthen the iron ligand bonds may also change the spin state of the iron, so that some of these complexes are intermediate spin ($S = 3/2$), some are high-spin ($S = 5/2$), and others again are balanced in temperature-dependent equilibria between two alternative spin states. In such compounds changes of temperature may bring about changes in iron ligand bond length many times larger than those expected from thermal expansion or contraction alone (Table 3).

In Fe III (mcd)$_3$ the iron is coordinated to six sulphur atoms. (For the formula of this and other complexes quoted here see Table 3.) If this complex is crystallized from water or other hydrogen-bonding solvents, its ground state is $S = 3/2$ with $\langle Fe-S \rangle = 2.443$ Å. If the identical complex is crystallized from benzene its ground state is $S = \frac{1}{2}$ with $\langle Fe-S \rangle = 2.318$ Å. The difference is due to the occlusion of solvent molecules in the crystal lattice; hydrogen-bonding solvent molecules may withdraw electrons from the sulphur atoms, thus weakening the Fe–S bonds and favoring higher spin. The complex [Fe II (6-Me-Pyr)$_3$tren]$^{2+}$ in solution is high-spin, while the complex [Fe II (pyr)$_3$tren]$^{2+}$ in solution is low-spin; solutions of complexes in which either one or two of the three pyridines carry methyl groups in position 6 show thermal spin equilibria. The iron atoms are octahedrally coordinated to the three pyridine and the three imino nitrogens. On transition from the high- to the low-spin form, the Fe–N$_{pyr}$ bond lengths shorten by 0.31 Å, and the Fe–N$_{im}$ bond length shortens by 0.20 Å. Note that in the high-spin form these bonds are unusually long; this is believed to be due to overcrowding of the complex by the 6-methyl groups which prevent as close an approach of the iron atoms to the nitrogens as would be needed in a low-spin complex. A similar phenomenon had been found in [tris(2-methyl-1,10-phenanthrolin)Fe II].

In the pair of complexes [Fe III(acac)$_2$trien] PF$_6$ and [Fe III(sal)$_2$trien]-Cl$_2$H$_2$O the iron atoms are coordinated octahedrally to four atoms of nitrogen and two of oxygen, yet the former is pure high-spin and the latter pure low-spin. This difference is brought about, apparently, through the solvent water molecules in the latter which act as proton acceptors in hydrogen bonds from NH groups coordinated to the iron (H$_2$O$\cdots$HN$\cdot$Fe). These hydrogen bonds weaken the N–H bonds and thereby strengthen the N–Fe bonds, so that the complex becomes low-spin, with a reduction of 0.12 Å in the average iron-ligand bond distance. Steric effects of the water molecules may also play a part.

Especially striking are the crystallographic results on the dithiocarbamates, some of which show temperature-dependent spin equilibria, while the spin states of others are independent of temperature. For instance, in

Table 3 Effect of spin on bond distances in iron chelates

Compound	Spin(S)	$\langle$Fe–S$\rangle$ (Å)		Ref.
Fe III (mcd)$_3$ · H_2O^a	5/2	2.443		19
Fe III (mcd)$_3$ · $C_6H_6{}^a$	1/2	2.318		19
Fe III Tris (N,N-diethyldithiocarbamate)^b				
297°K; μ_e = 5.6 BM	high	2.357		20
79°K; μ_e = 2.8 BM	low	2.306		
		Fe–N (pyridine)	Fe–N (imine)	
Fe II (6-Me-Pyr)$_3$ tren $^{2+}$ (PF$_6$)$_2{}^c$	5/2	2.282	2.143	21
Fe II (Pyr)$_3$ tren $^{2+}$ (PF$_6^-$)$_2{}^d$	1/2	1.966	1.942	21
		$\langle$Fe–N$\rangle$	$\langle$Fe–O$\rangle$	
Fe III (acac)$_2$ trien PF$_6^-$ ^{e}	5/2	2.136	1.928	22
Fe III (sal)$_2$ trien Cl$^-$ · H_2O^e	1/2	1.968	1.884	22

^{a}mcd: morpholinocarbodithioato-SS'

^{b}Fe III (S$_2$C.N Ethyl$_2$)$_3$

^{e}Six-coordinated complexes of Fe III derived from triethylenetetramine (trien) and acetylacetone (acac) or salicyladehyde (sal). In both compounds the iron is coordinated octahedrally to four nitrogens and two oxygens.

$Fe[S_2CN(CH_3)_2]_3$ and $Fe[S_2CN(C_4H_4OH)_2]_3$ the effective magnetic moment μ_{eff} decreases from 4.2 to 2.2 EM on going from 295° to 150°K. This reduction in paramagnetic susceptibility is accompanied by a contraction of the Fe–S bond length by 0.058 Å and 0.062 Å respectively. On the other hand, in

$$Fe\left[\begin{array}{c} S_2CN \begin{array}{c} CH_2-CH_2 \\ | \\ CH_2-CH_2 \end{array} \end{array}\right]_3 \quad ,$$

where μ_{eff} remains constant over the same temperature range, the contraction in Fe–S bond length amounts to only 0.008 Å (23). In summary, Table 2 shows that a change of spin is always accompanied by a change in the iron ligand bond lengths and that in the fully low-spin compounds these bonds are shorter by at least 0.1 Å than in the fully high-spin compounds.

The Iron-Oxygen Bond

The nature of the iron-oxygen bond has been the subject of much discussion. The oxygen molecule has 12 valency electrons, and in the ground state these populate seven of the eight molecular orbitals shown in Figure 4. Five pairs go into the five lower lying orbitals, and one electron each goes into the two $1\pi_g^*$ antibonding orbitals. These give the oxygen molecule its spin of $S = 1$.

Pauling & Coryell found HbO_2 to be diamagnetic, whence they argued that the iron oxygen bond should have the resonating structure

$$Fe^- \underset{\cdot\cdot}{\overset{+}{\text{O}}}\!\!\diagup\!\!\overset{\cdot\cdot}{\underset{\cdot\cdot}{\text{O}}}: \quad , \quad Fe = \overset{+}{\text{O}}\!\!\diagup\!\!\overset{\cdot\cdot}{\underset{\cdot\cdot}{\text{O}}}:^-$$

which makes all electrons paired (25). In modern terms this means that the two $1\pi_g^*$ orbitals no longer have the same energy, so that their electrons pair in the single π_y^* which has a lower energy than π_x^*, because it lies at right angles to the Fe–O–O plane. On Pauling's model, the bond between the iron and oxygen would be made by hybridization between the π orbitals of the oxygen and the d_{xz} and d_{yz} orbitals of the iron, with some net transfer of charge from the oxygen to the iron. This model was challenged by J. J. Weiss who suggested that the bond might be ionic between a ferric ion and a superoxide ion, net charge being transferred from the iron to the oxygen ($Fe^{3+}O_2^-$) (26). Experimental support for Weiss' model was first advanced

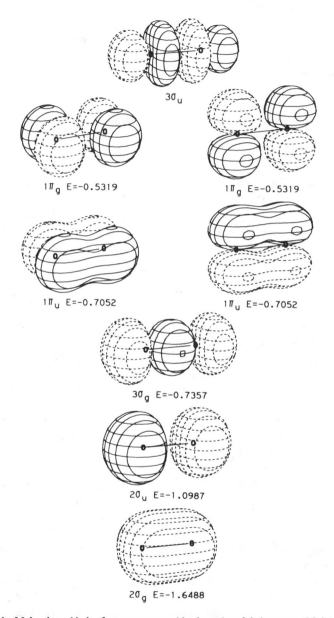

$3\sigma_u$

$1\pi_g$ E=-0.5319 $1\pi_g$ E=-0.5319

$1\pi_u$ E=-0.7052 $1\pi_u$ E=-0.7052

$3\sigma_g$ E=-0.7357

$2\sigma_u$ E=-1.0987

$2\sigma_g$ E=-1.6488

Figure 4   Molecular orbitals of oxygen arranged in the order of their energy. Of the six pairs of p electrons, five pairs fill the five lowest lying orbitals. The remaining pair is split between the two $1\pi_g$ orbitals. These are antibonding and are referred to in the text as π_x^* and π_y^*. Reproduced, with permission, from (24).

by Misra & Fridovich (27). They showed the autoxidation of HbO_2 to be a first order reaction depending only on $[HbO_2]$; when epinephrin was added to the solution it was oxidized to adrenochrome, but this oxidation was inhibited in the simultaneous presence of superoxide dismutase and catalase, which suggests that superoxide ion is liberated on autoxidation of HbO_2. Recently, Demma & Salhany (28) have shown that liberation of oxygen by flash photolysis of HbO_2 reduces oxidized cytochrome c, and this too is inhibited by superoxide dismutase and catalase. The $(Fe^{3+}O_2^-)$ structure is also supported by spectroscopic evidence. The infrared O–O stretching frequency in HbO_2 is 1107 cm^{-1} (29), which is in the superoxide ion range (1150–1100 cm^{-1}), much lower than that of the oxygen molecule (1556 cm^{-1}), and higher than that of a single O–O bond ($\sim$800 cm^{-1}). X-ray fluorescence also points to the presence of unpaired electron density on the iron atom (115). Cobalt porphyrins and Hb's combine reversibly with molecular oxygen and provide a probe for exploring the bond in the form of an unpaired electron on the cobaltous d^7 ion. This gives an ESR signal with nuclear hyperfine splitting from which the unpaired electron density can be located. In the deoxy derivatives hyperfine splitting from the Co is combined with that of a single nitrogen atom of the proximal histidine, which shows that the unpaired electron occupies the d^{z^2} orbital pointing toward the histidine. In the oxy derivatives, the nitrogen splitting disappears and the separation of the hyperfine lines due to Co is reduced to about a third. Since that separation is directly proportional to the unpaired electron density on the Co, it is inferred that a substantial fraction of the density has gone to the oxygen (30–32). This has been confirmed by a similar experiment in reverse, using $CoHb^{17}O_2$. ^{17}O has a nuclear spin of $I = 5/2$, so that it should give rise to hyperfine splitting if the unpaired electron density is transferred from the Co to O_2. This is indeed observed, and its magnitude suggests that about 60% of the unpaired electron density is transferred (33). It might be argued that the metal-oxygen bonds might be different in the Fe and Co derivatives, but the similarity of the O–O stretching frequencies, 1107 cm^{-1} for Fe and 1106 cm^{-1} for Co, suggests that they are similar (34). Sadly Weiss died before his prediction was confirmed experimentally. The $Fe^{3+}O_2^-$ model can be reconciled with diamagnetism or weak paramagnetism of the complex by postulating that the transferred d electron of the iron pairs with one of the two π^* electrons of the oxygen, and that the spin of the other π^* electron is paired with the odd d electron left behind on the iron by antiferromagnetic coupling. The diamagnetism of HbO_2 has recently been challenged by Cerdonio et al, who produced evidence of a low lying triplet state which makes it weakly paramagnetic at room temperature (35, 36). Their observations, though apparently flawless, do raise problems. For example, if oxyhemoglobin at room temperature had a molar susceptibility per heme of $+2460 \times 10^{-6}$ cgs/mol, as they report, one would expect its NMR spectrum to exhibit hyperfine shifted heme proton resonances, but these have not been observed.

The oxygen adducts of the picket fence complex are definitely diamagnetic (5). The oxygen molecules are bent to the heme axis and lie in four alternative orientations. Due to that disorder it has not been possible to determine the coordinates of the terminal oxygen as accurately as those of the other atoms. In the 1-MeIm complex O–O = 1.16 Å and Fe–O–O = 131°, and in the 2-MeIm complex O–O = $1.22 \pm (0.02)$ Å and Fe–O–O = 129($\pm$1)°. The authors state that they may have underestimated the O–O distance in the unhindered complex by as much as 0.15 Å. The geometry agrees with Pauling's prediction of a bent FeO_2 bond, and the O–O distance is close to that of 1.27 Å, predicted by him in a recent paper. It is slightly shorter than that of 1.34 Å in the superoxide anion, in agreement with the ESR results which show that no more than 2/3 of the density of one electron is transferred from the metal to the antibonding π^* orbitals of the oxygen.

STEREOCHEMISTRY OF HEMOGLOBINS

Monomeric and Dimeric Hemoglobins

COMMON TERTIARY STRUCTURE   These hemoglobins include mammalian myoglobin, and hemoglobins of the lamprey, of many invertebrates, and of the root nodules of leguminous plants.

The folding of the polypeptide chain (tertiary structure) of all these hemoglobins is similar to that of sperm whale myoglobin, the first protein structure to be solved (37). It is made up of eight helical and seven nonhelical segments forming a kind of basket for the hemes. Following Kendrew's notation, the helices are lettered A to H and the nonhelical segments NA, AB . . . GH, HC, from the amino to the carboxyl end. Residues along any of the segments are numbered from the amino end. In this way structurally equivalent residues carry the same notation in all hemoglobins regardless of additions or deletions.

Takano (38, 39) has recently refined the structures of sperm whale aquomet- and deoxymyoglobin by X-ray analysis at 2.0 Å resolution, using alternate cycles of real space and Fourier refinement; and Norvell et al (40) have refined the structure of carboxymyoglobin using real space refinement on neutron diffraction data to 1.8 Å resolution. Phillips has solved and refined the structure of oxymyoglobin at 2.0 Å resolution (41), which he is extending to 1.5 Å. The structures of several derivatives of chironomus erythrocruorin have been refined at 1.4 Å resolution (42, 43). The results obtained from these two proteins are quite different.

SURROUNDINGS AND STRUCTURE OF THE HEME IN MYOGLOBIN
The heme is bound to the globin by the mainly covalent bond between the iron atom and Nϵ of the proximal His F8; by hydrogen bonds from one of the propionate side chains to His FG2 and Gln H5, and from the other propionate to Arg CD3; and by about 80 mainly nonpolar interactions with 27 atoms of the globin which lie within less than 4 Å of the heme [see Table II of (44)].

There are three clusters of hydrophobic residues near the heme: one on the side of His E7, another on the side of His F8, and the third at the bottom of the heme. The first cluster accommodates the heme-linked oxygen, but contains enough room beneath His E7 for heme ligands as large as n-butyl isocyanide. Takano (38) suggests that the spare room may be needed to let the oxygen molecule in and out. The surface of the heme pocket does not have a hole large enough for oxygen to slip through, and the imidazole of His E7 is like a door that blocks the entrance by leaning against the heme; perhaps it opens by swinging into the free space below.

A xenon atom with a van der Waals radius of 2.2 Å can bind to the second cluster without disturbance to its hydrophobic side chains. The function of the "xenon hole" is unknown and so is that of the third cluster, which is the largest of the three. Takano suggests that they may make the molecule more flexible so that less activation energy is needed for the changes in tertiary structure that may occur on binding and dissociation of oxygen. In met- and deoxy-Mb three of the heme pyrroles seem to be tilted so as to bring their nitrogens closer to His F8, while the fourth seems tilted the other way. This buckling of the pyrroles is not yet very well resolved and needs checking at higher resolution which should be attainable because the X-ray diffraction pattern extends to about 1.4 Å^{-1}.

Table 4 shows the displacements of the iron atom and of N$_\epsilon$ of the proximal histidine (F8) from the mean plane of the porphyrin in various derivatives. The displacements in deoxyMb are exactly the same as in the model compound 2-MeImTPPFeII and, within error, also the same as in deoxyHb; but the displacements in met and oxyMb are larger than in any of the model compounds or in the corresponding Hb derivatives, which suggests that the tertiary structure of Mb is designed to hold the iron farther from the porphyrin plane. As would be expected from this feature, its oxygen affinity is lower than that of Hb in the R structure, but only by a factor of about three and much higher than that of Hb in the T structure. In MbO$_2$ the oxygen molecule occupies a single ordered position with Fe–O–O = 112°. The imidazole of His F8 and Fe–O–O are approximately coplanar and eclipse the Fe–N(Pyrrol II) bond. N$_\epsilon$ of His F8 lies 2.3 Å from the mean porphyrin plane compared to 2.07 Å in the picket fence complex. One explanation for this difference and for the displacement of the iron

Table 4 Heme stereochemistry in erthrocruorin, myoglobin, and hemoglobin

Compound	Derivative	Fe–Ct[a]	Fe–porph[b]	Fe–N$_{pyrrole}$	Fe–ligand	Fe–N$_\epsilon$	N$_\epsilon$–porph	Angle of imidazole plane — With N$_1$–Fe–N$_3$	Angle of imidazole plane — With heme normal	Ref.
Erythrocruorin	H$_2$O-met	0.16	0.08	2.02	3.0	2.25	2.3	7°	0°	43
	CN-met	0.13	0.06	2.01	2.2	2.1	2.1	7°	0°	43
	CO	0.11	0.01	2.01	2.4	2.1	2.2	7°	0°	43
	O$_2$	0.38	0.30	2.04	1.8	2.1	2.3	7°	0°	42, 43
	deoxy	0.23	0.17	2.02	3.1	2.2	2.3	7°	0°	43
Myoglobin	H$_2$O-met	—	0.40	2.04	2.0 (constrained)	2.1	2.5	19°	4°	38
	CO	—	0	—	—	—	2.3	19°	—	40
	O$_2$	0.16	0.26	2.05	2.0	2.1	2.3	6°	4°	41
	deoxy	—	0.55	2.06	—	2.1	2.6	19°	11°	39
Hemoglobin	H$_2$O-met α	—	0.07	2.0 (constrained)	2.0 (constrained)	2.1	2.2	21°	—	59
	H$_2$O-met β	—	0.21	2.0 (constrained)	2.0 (constrained)	2.2	2.4	15°	—	59
	CO α	—	0.04	—	—	—	—	—	—	61
	CO β	—	0.2	—	—	—	—	—	—	61
	deoxy α	—	0.60	2.1	—	2.0	2.6	20°	8°	10
	deoxy β	—	0.63	2.1	—	2.2	2.8	25°	7°	10

[a]Ct, center of porphyrin N's.
[b]porph, mean plane of porphyrin N's and C's, including the first C's of each side chain.

atom toward the proximal side in MbO_2 may be sought in the orientation of the imidazole with respect to the porphyrin. In MbO_2 this imidazole eclipses the Fe–N(pyrrole II & IV) bonds, while in the picket fence complex the imidazole plane lies at about 45° to them, which minimizes van der Waals repulsion and maximizes overlap of the Fe d orbitals with the π orbitals of the imidazole nitrogen. In MbO_2 repulsion is maximized and overlap minimized. The oxygen molecule is also constrained to the eclipsed orientation by steric hindrance of the distal residues Phe CD1, Val E11, and His E7.

In metMb the sixth coordination position at the iron atom is filled by a water molecule which lies at a distance of about 2 Å from the iron and is hydrogen-bonded to N_ϵ of the distal E7 His at a distance of 2.8 Å; it is also in van der Waals contact (3.2 Å) with methyl γ_2 of Val E11. In deoxyMb the sixth coordination position is empty, and the distal His side chain swings inwards and away from the position where the heme-linked water molecule has been. A positive peak lies at a distance of 3.2 Å from its N_ϵ, farther inside the heme pocket than the heme-linked water in metHb and at a distance of 3.0 Å from one of the pyrrolic carbons; this may be a weakly bound water molecule. In MbCO the heme is planar and the CO oxygen is displaced 0.8 Å from the heme axis into the heme pocket (40); the distal His is within van der Waals distance of the CO, but is not hydrogen-bonded to it. Similar bending was found in horse HbCO (45). This is surprising because the Fe–C–O line in iron carbonyls is normally straight; the bending in MbCO and HbCO was found to be due to steric hindrance by methyl γ_2 of Val E11, and N_ϵ of His E7. Bending of the CN group in cyanmetMb (46) and in lamprey and horse cyanmetHb probably arises from the same cause (47, 48).

The azide ion in metMb (49) and metHb (50) is also inclined to the heme axis, but more steeply than CN or CO (111° between N–N–N and the porphyrin plane) in a form characteristic for covalently linked azide (4). Clearly, steric hindrance by the distal residues in the heme pocket does not permit the azide ion to lie on the heme axis, as it might in an ionic complex.

SURROUNDINGS AND STRUCTURE OF THE HEME IN ERYTHRO-CRUORIN   Erythrocruorin (CTIII) is a monomeric Hb from the larva of the fly Chironomus tummi with a tertiary structure similar to that of myoglobin, but rather different coordination of the heme (43, 51, 52). The porphyrin is inverted, so that pyrroles 1 and 2 with their asymmetrically placed methyls and vinyls are interchanged and the convex side points toward the distal ligands rather than toward the proximal His as in myoglobin. (For numbering of pyrroles see Figure 3.) The β carbon of the vinyl on pyrrole 1 occupies a fixed position, but that at pyrrole 2 does not appear on the electron density map, which shows that it must be rotating freely about the pyrrole–Cα bond. Fermi (10) had found a similar situation in human deoxyHb. The distal Val E11 found in vertebrate Hb's and Mb's is replaced by Ile; the distal His E7 does not lie in the heme pocket, but is external in all except the cyanomet derivative and forms a salt bridge with the heme propionate of pyrrole 4. The other heme propionate forms a salt bridge with Arg FG2. In Mb the propionates form salt bridges with Arg CD3 and His FG2 instead. In Mb residue F4 is Leu, E14 Ala, E15 Leu, and G8 Ile; in erythrocruorin all these positions are taken up by Phe's.

DeoxyMb has a distal water molecule hydrogen bonded to His E7 but not coordinated to the iron; deoxyerythrocruorin has a water molecule coordinated to the iron at a distance of 3.1; the iron is displaced by only 0.23 Å from the plane of the porphyrin nitrogens, compared to 0.42 Å in myoglobin and in 2-MeImTPPFeII. The displacement from the mean porphyrin plane is 0.17 Å less than that from the nitrogens, due to the reversed doming of the ring just described. While in metHb and metMb the heme-linked water molecule lies at a distance of 2.1 Å from the iron atom, this distance is 3.0 in meterythrocruorin, the same as in the deoxy-derivative, the only difference being that in the met derivative this position is fully occupied, while in the deoxy derivative it is only half occupied. At first sight this suggested that the deoxy-crystals were half oxidized to met, but careful checks excluded such an artefact. The iron atom in the deoxy derivative is only 0.06 Å farther from the plane of the porphyrin nitrogens than in the met derivative, compared to 0.15 Å in Mb, presumably because the iron is six-coordinated in deoxyerythrocruorin, but five-coordinated in deoxyMb. In the CO derivative the CO axis is inclined to the heme normal by 19° and the angle between Fe–C and the heme normal is 8°. Comparison with MbCO and HbCO is not possible because their carbon positions have not yet been determined. The deviation from the heme normal is due to steric hindrance by Ile E11. The strangest geometry is that found in oxyerythrocruorin. According to unrefined coordinates the iron atom is 0.11 Å *farther* displaced from the porphyrin nitrogens than in the deoxy derivative. The iron oxygen complex is almost linear, with an Fe–O–O angle of 170° ($\pm$20°), compared to 112° in Mb and 129° and 133° in the picket fence complex. A water molecule is hydrogen-bonded to the second oxygen atom. The cyanide complex is nonlinear with Fe–C–N = 164° and an angle of 5° between Fe–C and the heme normal. The cyanide anion pulls the imidazole of the distal His E7 into the heme pocket so that equal fractions of its electron density are found in the internal and external positions. The differences in heme stereochemistry between erythrocruorin and Mb are undoubtedly real and not due to errors in chemistry or in crystallographic analysis. They illustrate the profound influence that lattice forces can have on the structure of the heme complex with different ligands, but the reasons for these differences are unclear, especially the extraordinary orientation of the oxygen molecule, which is at variance with theoretical predictions as well as with the picket fence and MbO_2 structures.

Tetrameric Hemoglobins

From coelacanth to man, the hemoglobins of bony vertebrates have the constitution $\alpha_2\beta_2$ and exhibit cooperative ligand binding. Crystals of the liganded (R) form of any one species are isomorphous, regardless of the valency of the iron and the nature of the ligand, and their structures have been found to differ only in details in and around the hemes. Crystals of the unliganded (T) form are unique. Table 5 shows the derivatives whose structures have been determined by X-ray analysis. The only ones to be refined and have their atomic coordinates determined are human deoxyHb and HbCO, and horse metHb. Table 4 shows their heme geometries. In human deoxyHb, the displacement of the iron from the mean porphyrin plane was calculated by a computer program which fitted a flat porphyrin to the observed electron density and allowed the iron atom to move. This resulted in a displacement of the iron from the porphyrin plane toward the proximal histidine of 0.6 Å in all four subunits. The displacement of N_ϵ of the proximal histidine (F8) from the porphyrin plane is 2.6 Å in the α- and 2.8 Å in the β-subunits. In human HbCO the displacements of the iron atoms are 0.04 Å in α and 0.21 Å in β, and the displacements of N_ϵ of His F8 are 2.2 Å in α and 2.4 Å in β. In horse metHb the displacements of the iron are 0.07 Å in α and 0.21 Å in β, and those of N_ϵ of His F8 are 2.2 Å in α and 2.4 Å in β, almost exactly the same as in human HbCO.

Table 5 Structure determinations of hemoglobins[a]

Species	Derivative	Resolution (Å)	Ref.
Human A	deoxy	2.5	10
	CO	2.8	45
	fluoromet + IHP	3.5	53
Human F	deoxy	2.5	54
Human S	deoxy	3.0	55, 56
Horse	deoxy	2.8	57
	CO	2.8	45
	NO	2.8	125
	fluoromet	2.8	58
	aquomet	2.0	59
	cyanomet	2.8	48
	azidomet	5.5, 2.8	49, 111
	metmangano	2.8	60

[a] (44) lists the abnormal Hb's solved up to 1976.

In view of the predominantly high-spin character of metHb, the displacements in the α-subunits seemed anomalously small at first but they are now seen to be consistent with the structure of the high spin $TPPbisH_2OFe(III)$ described in the section on coordination of the iron atom, where it is shown that the constraints of the axial ligands confine the iron to the porphyrin plane. The structure of HbO_2 has yet to be determined, but if the heme is planar with a distance $Fe-N_\epsilon = 2.1$ Å, as in the unhindered picket fence complex, then the proximal histidine would have to move towards the porphyrin plane by 0.6 ($\pm$0.16) Å on ligation of oxygen, coupled with transition of the entire molecule from the T to the R structure.

I have suggested that the low oxygen affinity of the T structure is due mainly to constraints of the globin that oppose that movement of the proximal histidines. Baldwin & Chothia have now compared the structures of human deoxyHb, human HbCO, and horse metHb in detail and have discovered the stereochemical nature of these constraints. The description which follows is based mainly on their paper (8).

CHANGES IN QUATERNARY STRUCTURE   Transition from the T to the R structure is brought about without significant changes at the contact which holds together the components of the $\alpha\beta$-dimer (i.e. the $\alpha_1\beta_1$-contact and the symmetry related $\alpha_2\beta_2$-contact). Significant changes are defined as greater than the RMS error of 0.43 Å of the individual sets of coordinates. If the interfaces of the $\alpha_1\beta_1$-dimers of human deoxyHb and HbCO are superimposed, the overall movement on ligation of the $\alpha_2\beta_2$-dimer relative to the $\alpha_1\beta_1$-dimer is an anticlockwise rotation of 14.9° and a translation of 0.8 Å. The rotation axis is normal to and intersects the molecular dyad at a point between the α-subunits 12 Å from the molecular centre, and makes an angle of 18° with the molecular x axis. The translation of 0.8 Å is along the same axis as the rotation, and toward the molecular centre (Figure 5).

The hemoglobin tetramer is formed by association of the $\alpha_1\beta_1$-dimer with the $\alpha_2\beta_2$-dimer. The C helix and FG corners of α_1 contact the FG corner and C helix of β_2 in both deoxy and liganded forms. The change of quaternary structure results from a large movement of about 6 Å at the contact between α_1C and β_2FG and a small movement at the contact between α_1FG and β_2C, [see Figure 7 of (44)].

CHANGES IN TERTIARY STRUCTURE   Having noted the absence of significant changes at the $\alpha_1\beta_1$-interface Baldwin & Chothia superimposed the electron density maps of human deoxy and HbCO at this interface and used it as a reference frame to search for differences in atomic coordinates due to changes in tertiary structure. They found that the following helical segments show no significant shifts: Bα and β, Cα, Dβ, Gα and β, and

Hα and β, except for the last few residues. Clear differences are seen for the hemes, the F and F' helices, (the F' helix is made up of residues usually referred to as EF5 to EF8), the FG segments of both subunits, and helix Eβ. The remaining regions show differences that are marginally significant. Figure 6 shows the important changes. On ligation the heme of the α-subunit moves 0.6 Å farther into its pocket in a direction parallel to heme nitrogens N_1 and N_3; the backbone of helix F moves 0.4 Å perpendicular to and toward the heme plane. It also moves the same distance as the heme parallel to CH_1–CH_3 and about 1 Å more than the heme parallel to CH_2–CH_4. The helix also tilts and turns relative to the heme with Ala F9

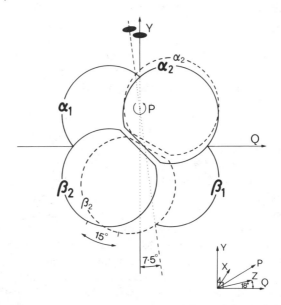

Figure 5   Schematic diagram illustrating the rotation of the $\alpha_2\beta_2$ dimer relative to $\alpha_1\beta_1$ that occurs in the quaternary structure change from deoxyHb (solid lines) to HbCO (broken lines). The molecule is viewed along the rotation axis (P) which intersects the molecular dyad of deoxyHb (Y) at the point shown. The relationship between this view and the standard molecular axes is shown in the bottom right-hand corner. Y is at right angles to X, Z, P, and Q. In addition to the rotation, the $\alpha_2\beta_2$ dimer moves into the paper by 0.8 Å.

The detailed geometry of the quaternary change is given below for the transitions from human deoxy to human HbCO and to horse metHb.

	deoxy/CO			deoxy/met		
Direction cosines of rotation axis	0.951	0.006	0.309	0.948	0.003	0.319
Rotation angle		−14.9°			−12.3°	
Shift along rotation axis		+0.8 Å			+1.19 Å	
A point on rotation axis	−0.03	11.91	−0.13	0.05	17.94	−0.32

Figures 5–9 and their captions are reproduced, with permission, from (8).

moving toward the heme plane and across toward pyrrole 3. A change of conformation at His FG1 and Lys FG2 allows the helix F and especially His F8 to tilt toward the heme without the side chains of Leu FG3 and Val FG5 altering their positions relative to the heme plane. The F' helix continues the motion of the F helix, but the turn parallel to the heme plane is lost and the motion is almost parallel to CH_2–CH_4. Helix E also turns and tilts slightly relative to the heme plane, but its center hardly moves perpendicular to that plane. Residues FG3 to G4 shift as illustrated in Figure 6, with little change in the conformation of the main chain. The side chain of Leu FG3 and Val FG5 show small changes of conformation to stay in contact with the heme. The movements of these residues are dissipated in the next two turns of helix G but Phe G5 and Leu G8 move in the same direction as the heme, toward the central cavity of the molecule. Residues H18 to HC1 adapt their main chain conformation slightly to the changes in the

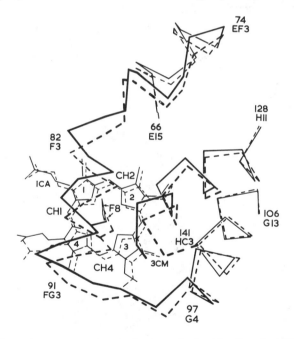

Figure 6   The tertiary structure changes that occur in the subunit on ligand binding. The figure shows Cα atoms, some of which are labeled, the hemes, and the side chain of His 87(F8) in deoxyHb (solid lines), and HbCO (broken lines). The relative positions of the two structures are derived by superimposing residues at the $\alpha_1\beta_1$ interface as described in the text. The view is perpendicular to the plane of the hemes. The 0.6 Å shift of the heme in the N1–N3 direction is clearly seen, as is the shift of the FG corner, and the F helix. The movements fade out at the beginning of the G helix and in the EF corner. 3CM denotes the methyl of pyrrole 3, 1CA the α-carbon of propionate 1.

adjacent helices, F and F'. In deoxyHb the C-terminal Arg HC3 has its α carboxyl salt bridged to Lys H10 and Val NA1 of the opposite α chain. Its side chain is salt-bridged to Asp H9 and also to an anion that lies between it and Val NA1 of the opposite α-chain. In the map of HbCO Arg HC3 has little density and must be rotating freely; the same is partly true of Tyr HC2 whose OH is hydrogen-bonded to CO of Val FG5 in deoxy, but spends only some of its time hydrogen-bonded to that group in HbCO.

The differences in tertiary structure in the β-subunits are similar to those in the α-subunits. The relative positions of the hemes and the E, F, G, and H helices of deoxyHb and HbCO are shown in Figure 7. On ligation the heme moves 1.5 Å farther into its pocket parallel to N_1–N_3 and rotates by 9.6° about an axis joining N_2 to N_4. This moves pyrrole 1 toward the side chain of Lys E10 and pyrrole 3 toward the side chain of Phe G5. The shifts of helices F and F' are similar to, but slightly smaller than, those in the α-subunit. The side chains of Phe F1, Leu F4, and His F8 move slightly

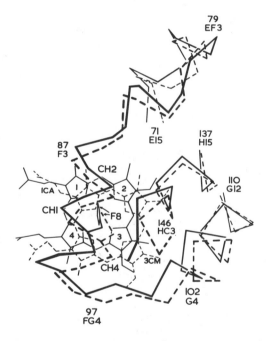

Figure 7 The tertiary structure changes that occur in the β-subunit on ligand binding. The figure shows Cα atoms, the hemes, and the side chain of His 92(F8), in deoxyHb (solid lines), and HbCO (broken lines). The relative positions are derived by superimposing the residues at the $α_1β_1$ interface as described in the text. The view is perpendicular to the plane of the heme in deoxyHb. The figure illustrates the 1.5 Å shift of the heme in the N1–N3 direction but its 9° rotation about an axis close to the N2–N4 direction is not apparent in this projection. For the sake of clarity only part of the E helix is shown.

to stay in contact with the heme. Cys F9 points out into the solution in deoxy and into the pocket between helices F and H in HbCO, the pocket having been vacated by Tyr HC2. Helix E tilts and turns relative to the heme plane. In deoxyHb the heme ligand site is blocked by $CH_3(\gamma_2)$ of Val E11 which lies only 3.5 Å from the heme centre. On ligand binding, the shift and rotation of the heme combine with the small shift of helix E to increase that distance to 4.9 Å, thus removing the obstruction to ligand binding. In deoxyHb, His HC3 has its α-carboxyl salt bridges to Lys C5α and its imidazole to Asp FG1 of the same β-chain; Tyr HC2 has its OH hydrogen bonded to CO of Val FG5, as in the α-subunit. In HbCO both these residues are freely rotating.

The similar motions of helix F relative to the heme in the α- and β-subunits have important effects on the proximal histidines. In deoxyHb, the imidazole rings are in asymmetric positions with respect to the hemes (Figures 8a and 8b). The imidazole plane is approximately normal to the heme and its projection on the heme plane is a line at about 20° to the line joining the nitrogens of pyrroles 1 and 3. The imidazole is tilted in its own plane so that $C_\epsilon-N_1$ is 3.2 Å in α and 3.5 Å in β, while $C_\delta-N_3$ is 3.8 Å in α and 4.2 Å in β. The distance of the iron atom from the mean plane is 0.6 Å in α and β. Note the similarity to 2-MeImTPPFeII. The changes on binding ligand, namely the translation of helix F across the hemes and the tilt of the end of helix F towards the heme, bring the histidines into positions where the imidazole rings can be made symmetrical relative to the heme by a small change of conformation of the side chains, and brought closer to the heme. Consequently in HbCO $C_\epsilon-N_1$ is 3.1 Å in α and 3.2 Å in β, and $C_\delta-N_3$ is 3.4 Å in α and 3.2 Å in β; the displacement of the iron from the heme plane has shrunk to 0.04 Å in α and 0.2 Å in β. Baldwin & Chothia show that the position of the heme relative to helix E in the β-subunits and the tilt of the proximal histidines relative to the hemes in both subunits are features intrinsic to the T structure which cannot be changed without a transition to the quaternary R structure, because tertiary and quaternary changes are intimately coupled. The hemes and helix F cannot move much without also moving the FG segment which forms part of the $\alpha_1\beta_2$-contact. The C helix of α_2 can make stable contacts with the FG segment of β_1 at the same time as the C helix of β_2 makes stable contacts with the FG segment of α_1 only when the two FG segments are the correct distance apart for either the T or the R structure. Due to the changes in tertiary structure this distance is 2.5 Å farther for the T than for the R structure.

SIGNIFICANCE OF STRUCTURAL CHANGES FOR LIGAND BINDING
To form a strong bond with oxygen, the iron must move from its position 0.6 Å above the plane of the porphyrin in deoxyHb into the plane in

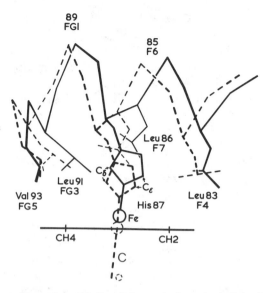

Figure 8a The positions of the F helix relative to the heme in the subunit of deoxy (solid lines) and HbCO (broken lines). The positions are shown relative to superimposed hemes. The Cα atoms of residues 81–93 are shown and also the side chains of those in contact with the heme; Leu 83, Leu 86, His 87, Leu 91 and Val 93. The heme direction CH1 – CH3 is perpendicular to the figure. The figure illustrates how, on ligand binding, the F helix moves in the CH2 – CH4 direction, and tilts toward the heme plane. The tilt is dissipated in residues 89 and 90 so that Leu 91 and Val 93 remain at the same distance from the heme plane. A figure drawn for the β subunits of deoxy and CO hemoglobins would be very similar. The angle between the plane of the imidazole and the line joining the heme center to the nitrogens of pyrroles 1 and 3 is 20°, compared to 7.4° with N_2–Fe–N_2' in 2–MeImTPPFeII in Figure 8b.

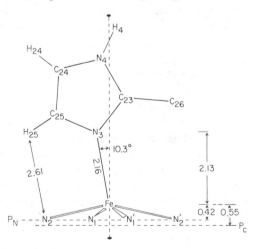

Figure 8b Structure of (2-methylimidazole)*meso*-tetraphenylporphinato FE(II) (2-MeIm-TPPFeII for short) viewed parallel to the mean plane of the porphyrin nitrogens and carbons, and nearly parallel (within 6.5°) of the imidazole plane. Fe – N_1 = 2.080 Å and Fe – N_2 = 2.091 Å. Note the similarity to deoxyHb. Reproduced by permission from (123).

HbO$_2$ and pull his F8 with it, which would reduce the shorter one of the two C–N distances in the α-subunits from 3.2 to 2.6 Å. However, these two atoms cannot approach closer than 3.0 Å before strong repulsion sets in. So unless the tilt can be righted, the iron atom has to stay out of the porphyrin plane, or the porphyrin N$_1$ must be pushed toward the distal side. In either case the iron oxygen bond would be much weakened. The spectroscopic observations described in the section on interactions between the heme and the globin indicate that in the β-subunits steric hindrance opposing the movement of the iron into the porphyrin plane is less severe, but the T structure shows the ligand position to be obstructed by Cγ_2 of Val E11; this obstruction is cleared by the relative motion of the heme and helix F on transition to the R structure (Figure 9). In the α-subunit Val E11 offers no obstruction to ligand binding in either quaternary structure. The results therefore suggest that in the α-subunits the oxygen affinity of the T structure is lowered mostly by the tilt of the proximal histidine relative to the heme which opposes its movement toward the prophyrin plane, and in the β-subunits mainly by the obstruction of Val E11. This picture is consistent with the results of experiments described below.

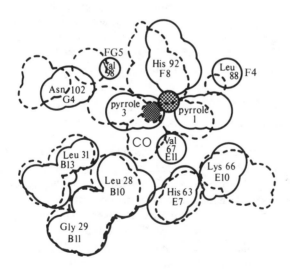

Figure 9 Space-filling diagram showing how the environment of the ligand binding site changes on going from deoxy (solid lines) to CO (broken lines) hemoglobin in the β-subunits. The figure is of a slice through a space-filling model cut so as to be perpendicular to the heme plane and to pass through N1 and N3 of the heme. In the β-subunit, room for the ligand is created by the slide (1.5 Å) and rotation (9°) of the heme. A small movement of Val 67(E11) in a direction perpendicular to the figure brings it out of the plane of the section by 0.8 Å. For the sake of clarity, the vinyl group of pyrrole 3 that occupies the space between pyrrole 3, Asn G4, and Val FG5 (see Figure 6) is omitted, as is the propionic group of pyrrole 1.

STRUCTURAL DIFFERENCES AT THE HEME IRONS BETWEEN HEMO-
GLOBIN AND MYOGLOBIN Table 4 compares the iron ligand geometries of the different hemoglobins, as far as they can be resolved, with those
of myoglobin, erythrocruorin, and the relevant model compounds. The
geometries were determined for hemoglobin by computer programs which
fitted a flat porphyrin to the electron density and allowed the distance of
the iron and proximal His from the porphyrin to vary so as to give the best
fit. The displacement of the iron in deoxyHb is similar to that in deoxyMb,
but the displacements of the iron in metHb, 0.07 Å in the α- and 0.21 Å
in the β-subunits, are much smaller than the displacement of 0.40 Å in
metMb, which suggests that in metHb, especially the α-subunits, repulsion
between the occupied E_g orbital of the iron and the π orbitals of the
porphyrin nitrogen causes an expansion of the porphyrin ring in its own
plane, as in TPPbisH$_2$OFeIII, while in metMb the constraints of the globin
favor a movement of the iron out of the porphyrin plane. Baldwin &
Chothia suggest that the slightly larger displacement of the iron atom in
metMb as compared to metHb may be due to the positions FG1 and FG3
being occupied by Leu and Val in the α- and β-subunits of Hb, but by the
slightly longer side chains of His and Ile in Mb. This may be the explanation
for the threefold lower oxygen affinity of Mb, compared to that of the R
structure of Hb. This effect also causes mixed spin derivatives of Mb to have
higher paramagnetic susceptibilities than the equivalent mixed spin derivatives of Hb (59): for instance at 20°C azide metMb is about 25% high-spin,
while azide metHb is only about 10% high-spin.

INTERACTIONS BETWEEN THE HEME
AND THE GLOBIN

For understanding heme-heme interaction the two basic questions are: how
does combination of ligands with the heme irons change the quaternary
structure of the globin from T to R?; and conversely, how does the change
from R to T lower the ligand affinity of the heme iron? I proposed that the
equilibrium between the two structures is governed by the displacement of
the iron atoms and the proximal histidines from the plane of the porphyrins
and by the steric effect of the ligand on the distal valines in the β-subunits
(62). By the laws of action and reaction, if movement of the iron and the
proximal histidine toward the porphyrin on ligand binding changes the
structure from T to R, then a transition from R to T must put the gears
into reverse and pull the iron and histidine away from the porphyrin. In that
case the T structure should exercise a tension on the heme that restrains the
iron from moving into the porphyrin plane (63). The existence of such a
restraint should be detectable by physical methods. In the remainder of this
review I describe experiments designed to test this hypothesis. I also de-

scribe an experiment designed to test the contribution that the obstruction by Val E11 β makes to the free energy of heme-heme interaction, even though its results were inconclusive (see the section on the role of distal residues).

Physical and Chemical Criteria for Quaternary Structure

In order to study the influence of the quaternary structure of the globin on the state of the heme, we needed a method of changing that structure without changing the sixth ligand at the heme. There are two ways of doing this: one is to use a valency hybrid in which the hemes in either the α- or the β-subunits are ferric; combination of the ferrous hemes with ligand is used to change the quaternary structure and the effect on the ferric hemes is studied spectroscopically. Alternatively, addition of the allosteric effector inositolhexaphosphate (IHP) may switch the quaternary structure from R to T. In certain fish hemoglobins the transition may be accomplished by merely lowering the pH.

The safest criterion for assessing the quaternary structure is X-ray analysis, but neither valency hybrids nor some of the complexes with IHP can be crystallized. Investigators have therefore tried to find spectroscopic and chemical criteria. These have been tested in derivatives whose quaternary structure had been determined crystallographically and then applied to those that could not be crystallized. We define the two alternative quaternary structures by the distances between the iron atoms and the structure of the $\alpha_1\beta_2$ contacts. In the R structure, $Fe(\alpha_1-\alpha_2) = 36$ Å, $Fe(\beta_1-\beta_2) = 33$ Å, Thr C3α_1 occupies the notch at Val FG5β_2 and Asp G1α_1 makes a hydrogen bond with Asn G4β_2. In the T structure, $Fe(\alpha_1-\alpha_2) = 35$ Å, $Fe(\beta_1-\beta_2) = 40$ Å, Thr C6α_1 fills the notch at Val FG5β_2, and Tyr C7α_1 makes a hydrogen bond with AspG1β_2 [see Figure 7 of (44)]. The physical and chemical criteria used to distinguish the two quaternary structures are the following.

1. *NMR*   The two alternative hydrogen bonds just described give rise to two alternative exchangeable proton resonances, one at −5.8 ppm from water characteristic for the R structure and another at −10.0 ppm from water characteristic for the T structure (64).

2. *UV circular dichroism*   DeoxyHb and fluorometHb + IHP, which have been proved to have the T structure by X-ray analysis, exhibit a prominent band of negative ellipticity with a single maximum at 287 nm. Derivatives known to have the R structure show weak positive ellipticity in this region with two slight dips near 285 and 290 nm. The negative band is due mainly to Trp C3β which lies at the $\alpha_1\beta_2$ contact (65, 66).

3. *UV spectra*   Known transitions from the R to the T structure produce a UV difference spectrum (T − R) with peaks in the aromatic region at 279, 287, 294, and 302 nm. The difference in optical density between the

minimum at 283 and the maximum at 287 nm should be $\epsilon_m \geqslant 0.035$ (Figure 10). The peaks at 294 and 302 nm are due to Trp C3β (65, 66).

4. *SH reactivity*   The rates of reaction of Cys F9 with p-mercuribenzoate or 2,2'-dithiobis(pyridine) are slowed down by the R → T transition, due to the formation of a salt bridge between His HC3β and Asp FG1β in the T structure which screens the SH group from these reagents (66). However, this effect is harder to interpret than the others because the reaction rates vary with changes of both tertiary and quaternary structure which have antagonistic effects. In the R structure the SH reactivity rises with rising spin of the β hemes, so that HbCO and Hb$^+$CN$^-$ have the lowest and Hb$^+$F$^-$ has the highest reactivity. This happens because low spin loosens the hydrogen bond between Tyr HC2β and Val FG5β, so that the tyrosine swings out of its pocket between helices H and F, and the SH group takes its place. High spin tightens that hydrogen bond so that the SH group is

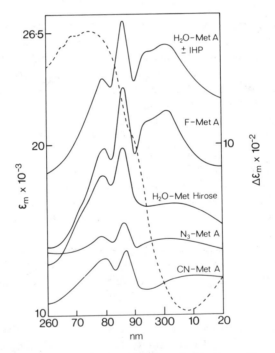

Figure 10   The broken line shows the ultraviolet absorption spectrum of human aquometHb; the solid lines show difference spectra, magnified ten times, of various human metHb derivatives ± IHP. Aquo- and fluorometHb A, and aquometHb Hirose all undergo the R → T transition, but Hb Hirose has Ser replacing Tyr at position C3β and therefore lacks the peaks 294 and 302 nm. Azide and cyanide metHb remain in the R structure which is modified by IHP in an as yet unknown manner. Reproduced by permission from (66).

excluded from the pocket and has to take up an external position instead (45, 59). In deoxyHb (T structure) the SH group is also external, but its reactivity is restricted by the surrounding salt bridge [see Figure 5 of (44)].

Methods of Changing Liganded Hb Derivatives to the T Structure

The allosteric equilibrium of valency hybrids of human Hb is governed by the spin state of the ferric irons, high spin biassing the equilibrium toward the T, low spin toward the R structure. Addition of IHP to solutions of human Hb below pH 7.0 can switch the predominantly high-spin ferric F^-, H_2O, OCN^-, SCN^-, $SeCN^-$, NO_2^-, $HCOO^-$, and CH_3COO^- derivatives to the T structure, but not the predominantly low-spin HbO_2, HbCO, Hb^+CN^-, and $Hb^+N_3^-$ (66, 67). One exception to that rule is HbNO which is low-spin, but can be switched to the T structure, because the unpaired NO electron is transferred to the d_{z^2} orbital of the iron where it weakens the $Fe-N_\epsilon$ bond (see below). There has been a suggestion that IHP can also switch HbO_2 to the T structure (68), but HbO_2 + IHP shows the exchangeable proton resonance at –6.4 ppm from water, and not the one at –9.4 ppm, which proves that it has the R structure (69).

Fish hemoglobins which exhibit the Root effect (very low oxygen affinity at pH <6.0) are more easily switched from R to T than mammalian ones. All derivatives, no matter whether high- or low-spin, can be switched to the T structure by lowering the pH and adding IHP, but the pH at which the transition occurs is lower the lower the spin state of the complex; in structural terms this implies that the number and energy of the salt bridges needed to stabilize the T structure is greater the lower the spin of the heme irons (Figure 11).

The validity of the criteria used by my co-workers and myself for the T structure has been questioned because aquometHb + IHP does not show the same SH reactivities and tetramer-dimer dissociation constants as deoxyHb + IHP. This argument typifies a widespread misunderstanding of the hemoglobin system for which I must perhaps blame myself, because I have failed to emphasize sufficiently the distinction between changes in tertiary and quaternary structure, and the associated thermodynamic states. In solution the two quaternary structures can be distinguished by spectroscopic criteria which reflect mainly the structure of the $\alpha_1\beta_2$ contact. On the other hand *thermodynamic properties* such as ligand affinities, SH reactivities, etc, must be interpreted with caution, because they depend on the *equilibrium* between R and T which varies with the spin state and ligation of the hemes and with the many other reciprocal interactions between tertiary and quaternary structures and their surrounding ions.

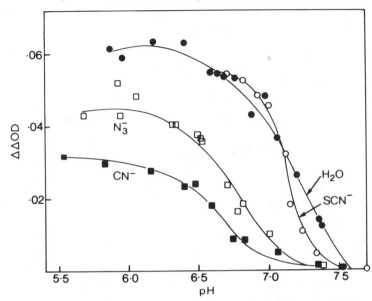

Figure 11   Magnitude of IHP-induced absorbance change between 293 and 287 nm plotted against pH for carp thiocyanide (O) metHb, aquo (●), azide (□), and cyano (■). Note the influence of the spin state of the heme on both the magnitude of the spectral change and the pH at which the R → T transition sets in. Reproduced by permission from (67).

Effects of Changes in Quaternary Structure on the Heme

In this section I consider the effect of changes in quaternary structure on the spectral and magnetic properties of the heme in a variety of derivatives of human and fish hemoglobins. Some of the changes cannot yet be interpreted; those that can, show that the predicted restraint exists in certain six-coordinated derivatives in the T structure. Its energy equivalent can be gauged from changes in thermal spin equilibria of mixed spin derivatives.

DEOXYHEMOGLOBIN   Gibson (70) first observed a change in the Soret (γ) band on transition from what are now known to have been deoxyHb dimers to tetramers in the T structure; fuller descriptions of the spectral changes have been given by Brunori et al (71) and by Perutz et al (65). They consist of blue shifts of the Soret, visible, and near infrared bands, together with the appearance of a shoulder at 590 nm flanking the peaks at 556 nm. The spectral changes in mammalian and fish Hb's are similar. Sugita (72) showed that they are due almost entirely to the hemes in the α-subunits. I tentatively interpreted them in terms of increased Fe–N_ϵ distances in the

T structure, but so far there is no support for my interpretation from other experiments, nor has any alternative interpretation been proposed.

The hyperfine shifted heme proton resonances of deoxyHb differ in the R and T structures. In the T structure they lie at −7.4, −12, and −17.5 ppm from HDO; in the R structure at −7.6, −11.4 and −15.3 ppm from HDO. The first two resonances are due to the α-, and the third to the β-subunits. Note that in the NMR spectrum the main change appears in β-hemes (73). The origin of these resonances and the reasons for their shifts are unknown. Significant changes between deoxyHb in the R and T structures have been observed in the resonance Raman spectra which are sensitive to changes in conformation of the porphyrin (121, 124) but not in X-ray absorption fine structure which is sensitive to changes in Fe–N bond length (77), or in the magnetic susceptibility (65).

LIGANDED FERROUS HEMOGLOBINS   Absorption spectra of mammalian HbO_2 and HbCO in the T structure have been observed only in the abnormal human Hb Kansas (Asn G4(102)β→Thr) (78). In fish Hb's showing the Root effect, the absorption spectra of HbCO in the T structure are observable, but not those of HbO_2, because the oxygen dissociates. As far as can be judged, the spectral changes in mammalian and fish Hb's are the same. In HbCO the Soret band is red-shifted by about 1 nm on transition from R to T; the difference spectrum is the exact reverse of that seen in deoxyHb. The visible α- and β-bands show slight red shifts and the β-band develops a shoulder at 580 nm (79–81). The visible changes in HbO_2 are similar but the red shifts are larger, and the shoulder appears at 583 nm (78).

HbNO proved the most revealing of the ferrous derivatives. The R → T transition gives rise to blue shifts and reduction in intensity of the α-, β-, and γ-bands, and the appearance of high-spin bands at 495, 518, and 603 nm, even though the complex remains low-spin [$S = 1/2$, Figure 12, (78, 82)]. Studies of hybrid Hb's show that these spectral changes are due mainly to the α-subunits, just as in deoxyHb. HbNO in the R structure shows a single infrared stretching frequency characteristic for six-coordinated nitrosyl hemes; transition to the T structure causes the appearance of a second infrared band, of intensity equal to the first, characteristic of five-coordinated hemes (83). Similarly, HbNO in the R structure shows an ESR spectrum similar to that of six-coordinated nitrosyl hemes, while HbNO in the T structure shows a composite of five- and six-coordinated nitrosyl hemes (84). Similar results have been obtained from comparisons of the resonance Raman spectra of five- and six-coordinated nitrosyl hemes with those of HbNO in the R and T structures (85, 86). Taken together, these results imply that in the R structure all four hemes are six-coordinated, but

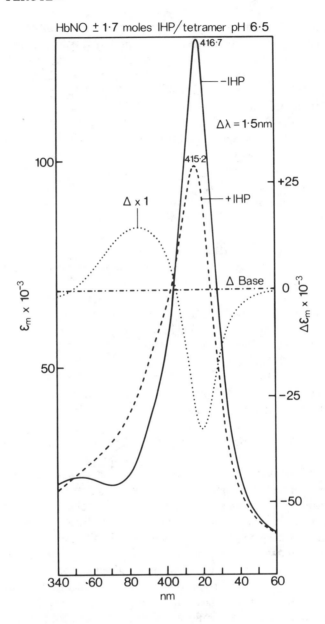

Figure 12 The Soret band of human HbNO ± IHP and difference spectrum. Most of the spectral change is due to the rupture of the Fe–N$_\epsilon$ bond in the α-subunits. Reproduced by permission from (78).

in the T structure only the hemes in the β-subunits remain six-coordinated; those in the α-subunits become five-coordinated, the iron histidine bond having been broken by the restraints that impede the movement of the proximal histidine toward the porphyrin. This experiment corroborates the restraint at the α-hemes in the T structure due to the tilt of the proximal histidines deduced by X-ray analysis, but it does not determine its energy equivalent, though kinetic experiments showed that an activation energy of 17 kcal/mol was needed to break the Fe–N_ϵ bond (78, 87, 88). The bond breaks because occupation of the antibonding Fe d_{z^2} orbital by the unpaired NO electron has weakened it.

LIGANDED FERRIC HEMOGLOBINS   FluorometHb is pure high-spin ($\mu_e = 5.9$ BM; $S = 5/2$) and aquometHb, at 20°C, a mixed-spin derivative ($\mu_e = 5.3$ BM) above the transition temperature, which obeys the Curie law and contains high- and low-spin components in proportion to their spin degeneracies (89). The same may be true of cyanate metHb. In aquometHb in the R structure a water molecule is coordinated to the iron at a distance of 2.1 Å and hydrogen-bonded to the distal histidine at a distance of 2.8 Å (59). In fluorometHb in the R structure the fluoride ion takes up the same position as the heme-linked water in aquomet, but the fluoride ion is hydrogen-bonded to both the distal histidine and, for part of the time, to another water molecule deeper inside the heme pocket (58).

In these derivatives the R → T transition shifts all heme absorption bands to the red or produces shoulders on their red edge, just as it does in the ferrous, low-spin HbO$_2$ and HbCO (90). There is a decrease in OD at the positions of the low-spin bands even though there is no change in magnetic susceptibility. I tentatively interpreted the red shifts of the near infrared charge transfer bands in terms of lengthened Fe–N bonds, but so far there is no firm evidence to support this. A difference Fourier synthesis of fluorometHb in the T structure minus deoxyHb A showed a large positive peak on the distal and a negative peak on the proximal side of the α-heme, as though the iron atom had moved from its position 0.6 Å on the proximal side of the porphyrin either into the porphyrin plane or beyond it to the distal side, but the resolution of 3.5 Å was insufficient to solve the structure (53). Asher et al (91) have tentatively identified resonance Raman peaks at 443 and 471 cm^{-1} as Fe–F stretching frequencies and found that these remain unchanged on addition of IHP. If their assignment is correct then the Fe–F constellation must be the same in the two structures.

The water proton relaxation of aquometHb shows a large paramagnetic effect which proves that the iron is accessible to water protons approaching it to within 3.4 Å. These protons exchange with the bulk water at a rate of 10^4 s^{-1} and with an activation energy of only 7.4 kcal mol^{-1}, which suggests that proton rather than water transfer is involved. Addition of IHP causes these water protons to dissociate three times more rapidly from their sites near the iron atoms, presumably because the water molecules are bound more loosely in the T than in the R structure. In fluorometHb proton exchange is much faster than in aquometHb. In this case the results allow the distance of the proton from the iron atom to be calculated as 4.11 Å in the R and 3.94 Å in the T structure (92). Anderson did an X-ray analysis of human metHb in the T structure, but unfortunately the resolution of 3.5 Å does not allow its Fe – H$_2$O distances to be compared with those of metHb in the R structure, so that we are not yet able to interpret those interesting magnetic effects (116).

CyanometHb is a pure low-spin compound [$\mu_e = 2.23$ BM, (93); $\mu_e = 2.59$ BM, (89)]. CyanometHb's of fish which show the Root effect can be switched to the T structure, but this has no significant effect on either the optical (67) or the resonance Raman spectra (76), presumably because all six iron ligand bonds are so strong that the steric strains of the T structure are dissipated elsewhere.

We now come to the class of ferric derivatives which have provided the most useful information concerning the effect of changes of quaternary structure on the state of the heme. These are the mixed-spin derivatives in which there exists a thermal equilibrium between the spin states of $S = 5/2$ and $S = 1/2$. The derivatives investigated include hydroxyl, azide, thiocyanate, and nitrite metHb (67, 89). The spectral changes induced by the R $\rightarrow$ T transition in these derivatives include blue shifts of the Soret band, increases in intensity of the visible high-spin bands and of the charge transfer bands in the near infrared, and decreases in intensity of the low-spin bands (Figure 13). The most striking spectral changes were seen in nitrite metHb of carp which has the red color characteristic of low spin ferric Hb's in the R structure and the brown color characteristic of high spin ferric Hb's in the T structure. The significance of this color change is discussed in the section on high pressure and magnetic measurements.

Magnetic measurements of human hemoglobins by NMR indicated that the R $\rightarrow$ T transition caused the paramagnetic susceptibility of hydroxymetHb to rise by 45% and that of thiocyanatemetHb by 11%. Human azide metHb cannot be converted to the T structure, but two abnormal Hb's form valency hybrids which allowed the effect of the R $\rightarrow$ T transition on the paramagnetic susceptibility of the α- and β-subunits to be measured separately by infrared absorption spectroscopy. This showed that in the R structure the α-hemes are less than 10% high spin, while in the T structure with IHP the high-spin fraction rises to 27%. The high-spin fraction of the β hemes in the R structure is probably about 10%; in the T structure with

IHP it rises to 35%. In trout IV and carp azide metHb measurements of the magnetic susceptibility and of the azide-stretching frequencies showed a rise of the high-spin fraction in the tetramer from about 10% in the R structure to 50% on the T structure, which is equivalent to a free energy change of ~ 1 kcal per mol heme. All these measurements were done at a

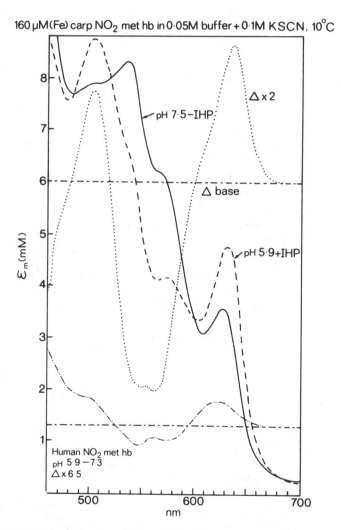

160 μM(Fe) carp NO₂ met hb in 0·05M buffer + 0·1M KSCN. 10°C

Figure 13 Visible spectra of carp NO₂metHb [– IHP at pH 7.5 (R structure), and + IHP at pH 5.9 (T structure)] and difference spectrum magnified two times. The ——··———··—— curve shows the pH-dependent difference spectrum of human NO₂ metHb which undergoes no change in quaternary structure, magnified 6.5 times. Reproduced by permission from (67).

single temperature near 20°C. Qualitatively these results showed that in mixed-spin derivatives the R → T transition causes a change to higher spin (67). We have seen, in the sections on coordination of the iron atom and the interaction between structure and spin state in iron chelates that this is equivalent to a stretching of the Fe–N bonds.

EFFECT OF QUATERNARY STRUCTURE ON THERMAL SPIN EQUI-LIBRIA (89) Most ferric iron complexes remain in the same spin state over a wide temperature range and thus obey the Curie-Weiss Law, $\chi = 1/(T-\Theta)$, where χ is the magnetic susceptibility, T is the absolute temperature, and Θ is a constant; but in some complexes the separation in energy between the low ($^2T_2$) and high ($^6A_1$) spin states is so small that it approaches the thermal energy (Figure 14). Such compounds exhibit a temperature-dependent spin equilibrium. If the low-spin state is the ground state, they show a region at low temperature where they are pure low spin ($S = 1/2$) and obey the Curie-Weiss Law; then a transitional region where their paramagnetic susceptibility rises with rising temperature until the high- ($S = 5/2$) and low-spin states are equally populated; and, finally, a region of mixed-spin where the Curie-Weiss Law is obeyed again (Figure 14). This state of mixed-spin may be contrasted with the true intermediate-spin of $S = 3/2$ which has been observed in only few synthetic compounds. One example is [Fe (III) octaethylporphin] (ClO$_4$) in which the coordination about the iron is square planar because the ClO$_4^-$ is bound to the iron only very weakly (94). The state of $S = 3/2$ is best distinguished from a spin equilibrium by its temperature-dependent magnetic susceptibility and Mössbauer spectrum.

Thermal spin equilibria in metHb derivatives have been studied extensively (for review see 95), but measurements have been confined to derivatives in the R structure. Messana et al (89) have determined the effect of the R → T transition on spin equilibria by measuring magnetic susceptibilities between 300° and 90°K with a high resolution superconducting magnetometer. The derivatives used were carp azide, nitrite, and thiocyanate metHb. The authors expected to find a dependence upon $1/T$ like the theoretical curves for mixed-spin derivatives shown in Figure 14, i.e. a low-spin ground state followed by a gradual transition to higher-spin above some critical temperature, but the actual results were rather different.

At the lowest temperatures all the plots of χ versus $1/T$ were linear, (Figure 15). From the linear parts of the curves the effective magnetic moments, μ_e, could be derived:

$$\mu_e = \sqrt{\frac{3\chi kT}{N\beta^2}} = 2.828 \ \sqrt{\chi T} , \qquad\qquad 1.$$

where N is Avogadro's number and β the Bohr magneton. The magnetic moments are listed in Table 6. For all but one of the hemoglobins in the R structure the moments have values characteristic of low-spin heme complexes, but for all hemoglobins in the T structure and for the thiocyanate derivative in the R structure their values are intermediate between low- and

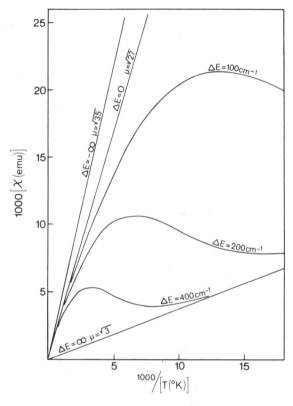

Figure 14 Temperature-dependence of paramagnetic susceptibility calculated from

$$\left(35 \exp \frac{-\Delta E}{kT} + 3\right) \bigg/ \left(3 \exp \frac{-\Delta E}{kT} + 1\right)$$

where k is the Boltzman constant, T the temperature in °K, and ΔE the difference in energy between the high-spin $^6A_1$ and the low-spin $^2T_2$ states. The effective magnetic moment $\mu_e = 2.828 \sqrt{\chi_A\,T}$ BM. The top line, $\Delta E = -\infty$ represents a pure high-spin compound; the next, $\Delta E = 0$, a mixed-spin compound, and the lowest line, $\Delta E = +\infty$ a pure low-spin compound. The curves represent thermal spin equilibria for various values of ΔE. Note that at high temperature these all converge on to the line of $\Delta E = 0$. The low-spin case represents $\mu_e = 2\sqrt{S(S+1)} = \sqrt{3}$ BM. Due to spin orbit coupling values of $\mu_e \leqslant \sqrt{5}$ BM are known to occur. (BM = Bohr magnetons).

Figures 14–17 are reproduced, with permission, from (89).

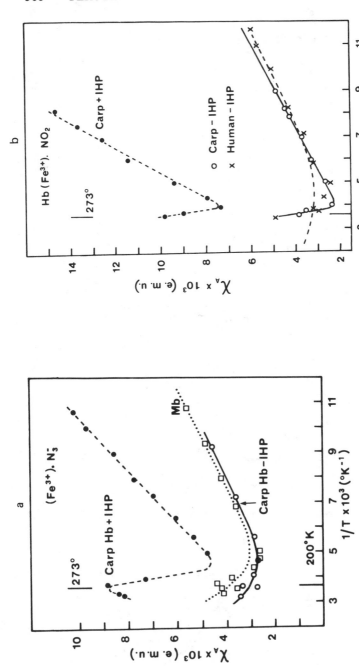

Figure 15 Magnetic susceptibilities of (*a*) azide and (*b*) nitrite metHb of carp ± IHP and of azide metMb. The dotted curve in *a* represents an attempt to fit the myoglobin data to a curve calculated from Equation 3, with the exponential term multiplied by a factor $\gamma = 10$ and with $\Delta E = 1000$ cm^{-1}. The broken line curve in *b* represents an attempt to fit the data to a curve calculated from Equation 3 with $\gamma = 1$. Note that the fit of both theoretical curves is poor.

Table 6 Effective magnetic moments of hemoglobin and myoglobin derivatives in the Curie range at low temperature

Derivative	Species	IHP	μ_e (BM)[a] ($\pm$ 0.05 BM)	α[b] (high-spin fraction)
$Hb^+ NO_2^-$	carp	+	3.80	0.34
$Hb^+ NO_2^-$	carp	−	2.07	0
$Hb^+ N_3^-$	carp	+	2.80	0.12
$Hb^+ N_3^-$	carp	−	1.98	0
$Hb^+ SCN^-$	carp	+	3.49	0.26
$Hb^+ SCN^-$	carp	−	3.09	0.18
$Hb^+ CN^-$	human	−	2.59[c]	0
			(2.23)[d]	
$Hb^+ H_2O$	human (0–30° C)	−	5.3 ($\pm$ 0.1 BM)	0.78
$Mb^+ N_3^-$	sperm whale		2.03	0
			(2.17)[d]	

[a] BM = Bohr magnetons
[b] α is calculated by assuming $\mu_{LS}^2 = 4.0$ and $\mu_{HS}^2 = 35$ BM^2; $\alpha = \mu_e^2 - 4/31$
[c] The higher value of μ_e for human $Hb^+ CN^-$ than for the derivatives of carp hemoglobin − IHP must be attributed to a larger orbital contribution, since all these derivatives are low spin, $S = 1/2$.
[d] The values in brackets are those found by Iizuka & Kotani (93).

high-spin. In all instances the effective magnetic moment for the frozen hemoglobins in the T structure is larger than in the R structure. In solution, several derivatives show reverse Curie behavior characteristic of compounds in which there exists a thermal equilibrium between two spin states. This behavior persists after freezing down to temperatures of between 250 and 200°K. The magnetic moments in solution in the T structure are also larger than in the R structure (Table 7); the rise just above the freezing point varies from 1.26-fold in thiocyanate to 1.84-fold in the nitrite derivative. The 1.52-fold rise in carp azide metHb at room temperature corresponds to a 2.4-fold rise in magnetic susceptibility which agrees with the 2.5-fold increase found in the closely related trout IV azide metHb by NMR and also with the increase calculated from the change in the relative intensities of the high- and low-spin infrared azide stretching frequencies of carp azide metHb mentioned in the preceding section (67).

If a transition metal compound exhibits two alternative spin states in thermal equilibrium, their populations should be proportional to

$$W_L \exp - \frac{E_L}{kT} \text{ and } W_H \exp - \frac{E_H}{kT} \qquad 2.$$

where W_L and W_H are proportional to the spin degeneracies $(2S_L + 1)$ and $(2S_H + 1)$, and E_L and E_H are the energies of the low- and high-spin states. If $\Delta E = E_H - E_L$, and μ_L and μ_H are the magnetic moments of the low-

Table 7 Rise in effective magnetic moment μ_e of carp metHbs on addition of IHP in solution

		−IHP		+IHP	
Derivative	T	μ_e	α[a]	μ_e	α
N_3^-	275°	2.6	0.09	4.4	0.50
	300°	2.9	0.14	4.4	0.50
SCN^-	275°	3.9	0.36	4.9	0.64
NO_2	275°	2.5	0.07	4.6	0.56

[a] α is the fraction high spin, calculated as in Table 5.

and high-spin states respectively, the paramagnetic susceptibility χ_{Fe} should follow the equation

$$\chi_{Fe} = \frac{1}{8T} \frac{\dfrac{2S_H + 1}{2S_L + 1} \exp -\dfrac{\Delta E}{kT} \quad \mu_H^2 + \mu_L^2}{\dfrac{2S_H + 1}{2S_L + 1} \exp -\dfrac{\Delta E}{kT} + 1}; \qquad 3.$$

$$\chi_{Fe} = \frac{1}{8T} \frac{K\mu_H^2 + \mu_L^2}{K + 1}, \text{ where } K = \frac{[HS]}{[LS]} \qquad 4.$$

i.e. the equilibrium constant between the high- and low-spin states.

However, Iizuka & Kotani (93, 96) found that this was not true of any of their metMb or metHb derivatives; the initial fall of χ with $1/T$ was always steeper than predicted by the simple theory. The results of Messana et al (89) confirmed their findings.

Iizuka & Kotani (93, 96) were able to fit their data by allowing the equilibrium constant K in Equation 4 to have the more general form

$$K = \exp\left(-\Delta G/RT\right) = \exp\left(-\frac{\Delta H - T\Delta S}{RT}\right), \qquad 5.$$

with constant, empirically chosen values for ΔH and ΔS.

A difficulty in applying the procedure of Iizuka & Kotani to fit the data of Messana et al by means of Equations 3–5 was that the empirical values of μ_L^2 (as determined from the slope of the χ versus $1/T$ plots at low temperature and listed in Table 6) for the T state Hb's were much higher than the maximum plausible value of about 6 BM^2 expected for the low spin $S = \frac{1}{2}$ state.

A plausible hypothesis to account for this anomaly is to suppose that below some temperature, T_0, the globin and surrounding ice become too rigid to accommodate the change in iron bond lengths that accompany spin changes, so that a random mixture of high- and low-spin hemes becomes frozen in and thermal equilibrium cannot be attained at temperatures below T_0; this hypothesis may be incorporated into Equation 5 by assuming that for temperatures below T_0 the equilibrium constant K has the fixed value $K = K(T_0)$. Such freezing-in of spin equilibria has been observed by Mössbauer spectroscopy in cyanate metMb (97) and in [Fe (II) 2-aminomethyl-pyridine)$_3$]$^{2+}$I$_2$ (98).

Messana et al determined the parameters of Equations 4 and 5 required to fit their data for the azide derivatives by means of a least squares procedure in which the values $\mu_H^2 = 35$ BM^2 and $\mu_L^2 = 4.0$ BM^2 were assumed. The data for azide metHb – IHP were fitted together with a single pair of values, ΔH, ΔS; in fitting the data for azide Hb + IHP the additional parameter T_0 was included to represent the temperature below which the spin equilibrium is assumed to be frozen-in. A reasonably good fit between theory and experiment was obtained when Equation 4 was plotted for the values $\Delta H = 2.6$ kcal mol^{-1} and $\Delta S = 5.7$ cal mol^{-1} deg^{-1} for Hb–IHP; and with $\Delta H = 2.4$ kcal mol^{-1}, $\Delta S = 8.3$ cal mol^{-1} deg^{-1}, and freeze-in temperature $T_0 = 193°$K for Hb + IHP. However, the values of ΔH and ΔS were subject to considerable uncertainty, especially in the case of Hb + IHP, because the two parameters are highly correlated ($C = -0.99$) in the fitting procedure. This situation can be illustrated graphically by replotting the data in terms of the apparent equilibrium constant K', between the high and low spin states, defined as

$$K' = \frac{8\chi T - \mu_L^2}{\mu_H^2 - 8\chi T}$$ 6.

The right-hand side of Equation 6 is the solution of Equation 4 for K; hence, if the data fit Equations 4 and 5, then $K' = K$ and

$$-\ln K' = \frac{\Delta G}{RT} = \frac{\Delta H}{R} \frac{1}{T} - \frac{\Delta S}{R},$$

so that ΔH and ΔS can be determined from the slope and intercept, respectively, of a plot of $-\ln K'$ against $1/T$. Such a plot is illustrated in Figure 16; the straight lines are drawn to correspond to the values of ΔH and ΔS corresponding to the curves in Figure 15.

The equilibrium at temperatures below $T_0 \approx 200°$K appears to be frozen-in, so that K' is constant below that temperature. Thus, it is only the data in the narrow range $1/T = 3-5 \times 10^{-3}$ deg^{-1} that can be used to determine the values of ΔH and ΔS, and these data are too meager to determine separately the slope and intercept of the fitting lines. However, if it is accepted that a simple spin equilibrium does obtain within this temperature range, then one can estimate the free energy difference $\Delta(\Delta G) = \Delta G(\text{-IHP}) - \Delta G(\text{+IHP})$ that stabilizes the high spin state in Hb + IHP relative to its stability in Hb − IHP. Since the vertical scale in Figure 16 is $\Delta G/RT$, the separation in height of the data points ± IHP gives $\Delta(\Delta G/RT)$. Thus at 300°K the difference in −lnK' is about 1.5–2.0; therefore $\Delta(\Delta G) = (1.5\text{–}2.0)$ × 1.98 cal mol^{-1} deg^{-1} × 300 deg = 0.9–1.2 kcal, in agreement with the value

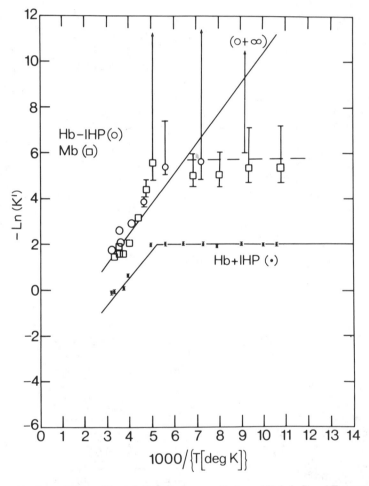

Figure 16 Data of Figure 15*a* and theoretical curve of spin equilibria in R and T structures plotted as the negative logarithm of the apparent equilibrium constant, K', calculated from Equation 4 with $\mu_H^2 = 35$ BM^2 and $\mu_L^2 = 4.0$ BM^2. Error bars are drawn for an arbitrarily assumed error of ±3% in χ; they become very large when the compound approaches pure low spin and K' tends to infinity. (BM = Bohr magnetons.)

derived from the infrared measurements mentioned in the preceding section. At 250°K the difference is 1.3–2.3, so that $\Delta(\Delta G) = 0.6$–1.1 kcal.

A survey of the literature on the magnetic properties of synthetic iron chelates shows that there is nothing unique about the properties of Hb. If the rise in χ with T in the non-Curie portions of the Hb curves is steeper than predicted from Boltzmann statistics with weighting based on spin degeneracy alone, this is also true of the majority of synthetic iron chelates. The only compound I could find that does follow Boltzmann statistics based on such weighting is crystalline [FE(II)(5R-phenanthroline)$_2$(SCN)$_2$]H$_2$O, where R is either Cl or NO$_2$ (98). Most magnetic work has been done on crystalline powders which often show very sharp transitions from low to high-spin at some critical temperature, probably because of cooperative effects in the crystal lattices; but even iron complexes in solutions show steeper transitions than predicted by the simple theory (21, 99–101). In crystalline samples, plots of the spin equilibrium constant against $1/T$ are often nonlinear, showing that there is not a simple intramolecular equilibrium; in solution such plots are mostly linear. If the spin equilibrium constant is defined as in Equation 4 and the low-spin state is the ground state, then ΔH and ΔS are always positive, but ΔS is always larger than the value of 2.2 kcal/mol expected from the change in spin degeneracy alone. The excess ΔS has been attributed to the stereochemical changes accompanying the spin changes.

Finally, one might envisage that an intermediate-spin state of $S = 3/2$ might play a part in some of the phenomena observed here, but this has been found only in porphyrins where axial bonds are either absent or extremely weak, such as the [FE (III) octaethylporphin] (ClO$_4$) structure already mentioned (94). It is unlikely, therefore, that this spin state exists in any of the Hb compounds, but the question can be answered with certainty only by Mössbauer spectroscopy. A pure $S = 3/2$ state is ruled out for azide metHb of carp in the T structure because its squared magnetic moment is far below 15 BM^2.

PRESSURE-DEPENDENCE OF PARAMAGNETIC SUSCEPTIBILITY IN MIXED-SPIN DERIVATIVES   A change to higher spin is expected to involve a lengthening of the Fe–N distances (see Tables 1 and 2) and should therefore be accompanied by an expansion in molecular volume. Conversely, a volume contraction produced by hydrostatic pressure should bring about a transition to lower spin. The pressure dependence of the spin equilibrium should allow determination of the volume change accompanying the spin transition.

The pressure-dependence of the visible absorption spectra of Hb and Mb derivatives has been studied extensively (102–105). High-spin metHb in which the sixth ligand is weakly bound are reversibly denatured, at pressures as low as 1 kbar at some pH's, to a low-spin form in which the distal imidazole probably becomes the sixth ligand. Low-spin derivatives such as cyanomethemoglobin, in which the iron ligand bonds are much stronger, can withstand pressures of up to 8 kbar without denaturation. There have also been studies of the effect of pressure on the electronic and Mössbauer spectra of ferric iron porphyrins that showed the appearance of mixed- or intermediate-spin states and a reduction of the iron to the ferrous state (106). However, the pressures used in these experiments ranged from 20–140 kbars, whereas those in the experiments reviewed here did not exceed 4 kbars.

Messana et al (89) used a pressure bomb that recorded optical rather than magnetic changes, and concentrated on the azide derivatives, because the relationship between their optical absorption and paramagnetic susceptibilities can be calibrated on the basis of both magnetic measurements and infrared spectra. Assuming values of $\chi = 0.014$ and 0.0022 emu for the high- and low-spin components respectively, the authors showed that the absorption coefficient ϵ_m (630 nm) is related to a, the fraction of high-spin component, by the equation $a = 0.032 + 0.348(\epsilon_{630}-2)$. Absorption at the high-spin band, at 630 nm, was found to rise, and at the low-spin bands, at 541 and 573 nm, to fall with rising pressure. These spectral changes were reversible and took place at lower pressures than the gross spectral changes that characterize denaturation. From the spectral changes the spin equilibrium $K = [^6A_1]/[^2T_2]$ was calculated as a function of pressure (Figure 17). Thermodynamics gives

$$[\partial (\ln K / \partial P)]_T = -\Delta V/RT \qquad\qquad 7.$$

where ΔV is the change in volume that occurs when protein containing one mole of heme is converted from the high- to the low-spin form. The volume changes derived from Equation 7 and Figure 17 are as follows:

	$-\Delta V$ (ml/mole heme)
Sperm whale azide metMb	12.5 ± 1.7
Carp azide metHb − IHP	6.7 ± 1.3
Carp azide metHb + IHP	13.3 ± 1.0

This shows that the transition from the high- to low-spin state is accompanied by a volume contraction, as in other iron chelates, and that in Hb in the T structure this contraction is twice as large as in the R structure. The contraction in Mb is near to that in the T structure of Hb.

DISCUSSION OF HIGH PRESSURE AND MAGNETIC MEASUREMENTS

The volume changes associated with $^6A_1 \rightarrow {}^2T_2$ transition in azide metMb and Hb are in the same direction as those observed in simple iron chelates,

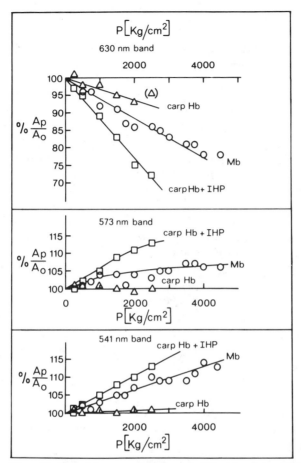

Figure 17a Effect of hydrostatic pressure P on the optical densities of the visible absorption bands of Mb and carp Hb ± IHP. A_p = optical density at pressure P; A_o = optical density at atomospheric pressure. All bands have been corrected for compression of water.

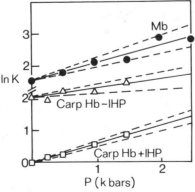

Figure 17b Spin equilibria (K) calculated from Figure 17a (top) as a function of hydrostatic pressure (P). The broken lines indicate the error limits.

but larger. For instance, the volume changes associated with the spin transitions of the octahedral FeS_6 complex in [Fe (III) tris(dithiocarbamate)] were between 3.7 and 4.1 ml/mol, depending on the value assumed for $\mu_e(^2T_2)$. A molecular model showed that the solvent would have relatively free access to the FeS_6 core, and the authors therefore felt justified in assuming the entire volume change to be due to a change in radius of that core. Taking Fe–S = 2.75 Å, this gave a contraction in Fe–S bond length of 0.07 Å (99). In azide metHb, the maximum likely contraction in average Fe–N bond length would be from 2.10 to 1.98 Å. If the solvent had access to the Fe–N core, this would produce a volume change of 3.7 ml/mol, close to the value found in the irondithiocarbamate. In fact, the volume changes observed in azide metHb are 6.7 ml/mol in the R and 13.3 ml/mol in the T structure. This means that there must be stereochemical changes in the protein associated with the spin transition and these must be larger in the T than in the R structure, presumably because the changes in Fe–N distances themselves are larger.

In going from 300° to 200–250°K the magnetic moments of all Hb derivatives drop much more steeply than can be accounted for by the difference in spin degeneracy of the $^2T_2$ and $^6A_1$ states. Again, this must be due to stereochemical changes in the surrounding protein. We do not know what these are, but the simplest proposition would be that they consist mainly in changes in the width of the heme pocket. A rough calculation shows that thermal contraction of the heme pocket over this temperature range may be of the same order as the movement of the iron atom relative to the porphyrin plane, and could therefore magnify the influence of temperature on the *spin equilibrium*. On the other hand, the freezing-in of the spin equilibrium at 250°–200°K implies that *spin transitions* cannot take place without structural changes in the protein which require a thermal activation energy greater than $\Delta E(^2T_2-^6A_1)$.

The large spectral and susceptibility changes observed in nitrite and thiocyanatemetHb may be connected with the fact that each of these ligands can combine with transition metals in two alternative ways—M–ONO or M–NO$_2$; M–SCN or M–NCS. In the first mode of ligation they lie nearer the high-spin and in the second nearer the low-spin end of the spectrochemical series. The first example of an M–ONO $\rightleftharpoons$ M–NO$_2$ isomerization was discovered by Jorgensen in [Co(NH$_3$)$_5$NO$_2$] Cl$_2$ in 1894; it is sterically possible because NO$_2$ is not linear: the angle O–N–O = 134°. There is no reason why such isomerization should not take place in these Hb derivatives on transition from the R to the T structure; inversion of the ligand on transition from low to high-spin would weaken the Fe–ligand bond so that the steric requirements of the T structure could be satisfied by stretching it.

IMPLICATIONS FOR HEME-HEME INTERACTION   All the measurements on carp and trout azide metHb point to the same answer, namely that in the T structure ΔE, the free energy gap between the six-coordinated high- and low-spin states is about 1 kcal/mol heme lower than in the R structure. This answers the question as to the energy equivalent of the restraint that opposes transition to the low spin state in the T structure, when the only chemical difference between the heme complexes in the two alternative quaternary structures consists of a change of spin. It amounts to about one third of the energy equivalent of the difference in oxygen affinity between the R and T structures of carp Hb. However, in the reaction with oxygen the change in spin is accompanied by a change in coordination number, so that the contribution made by the spin change cannot be separated experimentally from that due to the steric effects of the oxygen. Furthermore, the change of spin occurs in the ferrous form where the difference in energy between the high and low spin states is larger. Finally, it is important to realize that the ligand affinities of the a- and β-subunits in fish Hb's exhibiting the Root effect are very unequal. The free energies of heme-heme interaction of the two different subunits reacting with CO in menhaden Hb are 3.6 and 1.95 kcal/mol heme; it is not yet clear to which subunits these should be assigned (89). This heterogeneity suggests that the large changes in magnetic susceptibility are concentrated mainly in one pair of subunits, so that their free energy equivalent per heme would amount to 2 kcal/mol, compared to either 1.95 or 3.6 kcal/mol for the free energy of heme-heme interaction of either one or the other pair of subunits. The free energy of heme-heme interaction of the reaction with CO is generally equal to that with oxygen.

So much for the physical measurements to determine the energy equivalent of the restraint that opposes transition to the low-spin state in the T structure. Collman and his collaborators (107, 108) asked whether restraint of the kind I had envisaged could actually lead to a reduced oxygen affinity. They measured the thermodynamic constants of oxygen binding to some of the cobalt and iron picket-fence complexes, described in the section on coordination of the iron atom, with either N-methylimidazole or 1,2-methylimidazole as the fifth ligand. The former combines with the cobalt or iron atom without steric hindrance, while the latter is restrained by close contact of the 2-methyl group with the porphyrin, so that it opposes the movement of the metal atom into the plane of the porphyrin on ligation with oxygen, in the same way as I imagined the globin does in the T structure of Hb. The results show that the mean oxygen affinity (p_{50}) for the unhindered N-methyl cobalt complex is 150 torr, compared to 960 torr in the hindered 1,2-methylimidazole cobalt complex. In the iron complexes the corresponding values are 0.59 and 38 torr, corresponding to a difference in free energy of oxygen binding of 2.5 kcal/mol, comparable to the free energy of heme-heme interaction in Hb. The answer to the question asked at the beginning of this paragraph is therefore in the affirmative, restraint does lead to reduced oxygen affinity.

On the basis of this, or indeed of any theory of heme-heme interaction, one would have expected the Fe–O bond to be weaker in the T than in the R structure, weaker bonds being normally associated with lower affinity. Nagai et al (121) have tested this point by Resonance Raman Scattering, but found the Fe–O stretching frequency of HbO_2 to be within $5cm^{-1}$ the same in the two structures. Their result seems paradoxical, but they have proved it beyond any reasonable doubt. They point out that a frequency shift of 28 cm^{-1} should have been observed if the free energy of heme-heme interaction of 3 kcal/mol were stored in the iron oxygen bond, or a shift of 40 cm^{-1}, if the bond had been stretched by 0.02 Å in the T structure, as suggested by Warshel (110). Nagai et al have not measured the Fe–N_ϵ stretching frequency in HbO_2, and it remains to be seen if this bond is stretched in the T structure. More probably, their result implies that in HbO_2 in the T structure nearly all the free energy is stored in the protein rather than the heme, because the protein is "softer" and could take up strain energy in hydrogen bonds, van der Waals interactions, and a variety of small torsions. By contrast, in the mixed spin derivatives discussed above, the heme itself is "soft," in the sense that the spin equilibrium is readily shifted to the high-spin state with the longer Fe–N bonds.

THEORETICAL STUDIES OF HEME-HEME INTERACTION

Investigators have also tried to find out if my mechanism is reasonable on theoretical grounds. Gelin & Karplus (109) used empirical energy functions and structural data to calculate the most likely change in conformation of the heme complex on ligation and the pathway of its transmission to the subunit boundaries. They discovered the tilt of the proximal histidines, described in the section on changes in tertiary structure, as the source of the restraint to oxygenation in the T structure. They suggested that movement of the proximal histidines towards the porphyrin would force a change of tilt of the hemes which in its turn would trigger stereochemical changes at the subunit boundaries.

Warshel (110) used a different approach. He calculated the distribution of potential energy of the heme complex in Hb by molecular orbital theory, taking care to calibrate his potential functions from known structures and vibrational frequencies of metal porphyrins. He asked where the energy that determines the observed conformations of iron porphyrins comes from, where the free energy of heme-heme interaction arises, and where it is stored. He found that the pyramidal structure of five-coordinated high spin hemes is due both to the high-spin iron being too large for the size of the porphyrin hole, and to the steric repulsion between fifth ligand and the porphyrin, in agreement with the structural data reviewed in the section on the coordination of the iron atom. However, the potential well that keeps the five-coordinated high-spin iron out of the porphyrin plane has gentler slopes than that of the well that confines the six-coordinated low-spin iron to the porphyrin plane, because the iron nitrogen bonds and the van der Waals interactions between the porphyrin and the distal ligands are stronger. In other words, it costs less energy to push the iron atom of deoxyHb into the porphyrin plane than to pull the iron of HbO_2 out of the plane.

Warshel calculated the energies of the deoxygenated and oxygenated T and R structures. In agreement with Gelin & Karplus, his results indicated that the largest part of the free energy of heme-heme interaction is concentrated in only one of these four forms, namely the oxygenated T structure. He calculated that its low oxygen affinity is paid for by stretching the Fe–O bond by 0.02 Å and the Fe–N_ϵ bond by 0.004 Å, and by displacing the iron atom from the porphyrin plane by 0.03 Å; but mainly it is paid for by straining the surrounding protein. However, the structure of the sterically hindered picket fence complex shows that these estimates of bond stretching are too low; in fact the Fe–O bond is stretched by 0.15 Å and the Fe–N_ϵ bond by 0.10 Å (see the section on the coordination of the iron atom). Warshel's calculated iron displacement agrees with the observed one. Difference electron density maps of ligated Hb in the T structure do indeed show strain in the protein in the T structure, concentrated mainly in helix F and the FG segment, but they show much greater strain in deoxygenated Hb in the R structure. The constancy of the Fe–O stretching frequency (121) speaks against significant stretching of the Fe–O bond.

EXTENDED X-RAY ABSORPTION FINE STRUCTURE

Eisenberger et al (77) have challenged the decisive role played by the movement of the iron atom and the proximal histidine relative to the porphyrin plane in the allosteric mechanism of Hb on the basis of their extended X-ray absorption fine structure (EXSAF) determination of the iron-nitrogen bond distances in Hb. This new method measures the absorption of X-rays at wavelengths just below the absorption edge of the iron atom. At these wavelengths absorption is modulated by interference with low-energy photoelectrons scattered by neighboring atoms. From this interference pattern the distance between the iron atom and its neighbors can be calculated (117). The method has the great advantage of being applicable to amorphous materials and has recently led to a proposal for a tentative structure of the active Fe–Mo–S complex in nitrogenase.

Eisenberger et al used it to determine the Fe–N_{porph} distances in human deoxyHb and N-MeImTpivPPFe(II) compared to HbO_2 and N-MeImT-pivPPFe(II)O_2. Their absorption spectra for deoxyHb and the oxygen-free picket complex are identical, as are the spectra for HbO_2 and the oxygenated picket fence complex. For the oxygenated forms they calculate Fe–N_{porph} = 1.979($\pm$0.1) Å, in agreement with the X-ray crystallographic measurements. For the deoxygenated forms they find Fe–N_{porph} = 2.055 ($\pm$0.003) Å, substantially shorter than in any high-spin five-coordinated iron porphyrin, either ferrous or ferric (see Table 8). The reason for this anomalous result is unclear. Eisenberger et al now argue that if the distance from the center of the porphyrin ring to the nitrogens is 2.055 Å, as observed in 2-MeImTipvPPFe(II), the strain energy needed to expand the ring so as to accommodate the iron in its own plane would be smaller than the thermal energy, and therefore the iron position is being determined only by the steric requirements of the axial ligand, now no longer counter-balanced by a sixth ligand. However, we have seen in the section on the stereochemistry of iron porphyrins and related compounds that in fact these requirements and electronic factors play a roughly equal part in displacing the iron atom from the porphyrin plane, and theory shows that the geometry of a five-coordinated transition metal complex cannot be predicted from Morse potentials, but requires molecular orbital calculations. Such calculations indicate that only low-spin d^6 complexes prefer a square pyramidal conformation with an angle Θ = 90° between apical and basal bonds (112). High-spin complexes in which the d_{z^2} orbital is occupied favor a pyramid with Θ > 90°. A molecular orbital calculation carried out on a model Fe$(NH_3)_5^{2+}$ showed that the low-spin system has an energy minimum at Θ = 90°. A high-spin system in which both d_{z^2} and $d_{x^2-y^2}$ are singly occupied, as in deoxyHb, had its minimum at Θ = 112°, which corresponded to the iron atom being 0.77 Å out of the plane of the four nitrogens, even though there was no constraint on the center-to-nitrogen distance.

Eisenberger et al suggest that doming and buckling of the porphyrin may play a more important part in the Hb mechanism than the movement of the iron atom. In theory this could produce just as viable a mechanism, provided it leads to a movement of the proximal histidine relative to the mean porphyrin plane, but in fact there is no evidence that doming is any greater in deoxyHb than in the 2-MeImTPPFeII structure described in the section on the stereochemistry of the iron porphyrins, where the porphyrin nitrogens are out of the mean porphyrin plane by only 0.13 Å. Baldwin & Chothia (8) find that on transition from deoxy to HbCO the hemes shift and turn relative to the globin, but they find no change in heme conformation other than the movement of the iron atom.

The strongest evidence for the pyramidal geometry of the Fe–N complex in unstrained five-coordinated high-spin iron porphyrins comes from the many known structures. Table 8 contains no instance where the iron atom is displaced from the plane of the porphyrin nitrogens by less than 0.4 Å; the mean displacement is 0.46 Å. In addition, the porphyrin nitrogens are displaced from the mean plane of the carbons toward the iron by an average of 0.043 Å. In ferric iron porphyrins the Fe–N$_{porph}$ bond distance is never less than 2.060 Å; in ferrous ones it is at least 0.01 Å longer. The geometry of 2-MeImTPPFeII corresponds closely to that of the heme in deoxyMb which is sufficiently well resolved for us to be sure of its structure. Only where the high-spin ferrous iron has acquired a weak sixth ligand does the iron atom lie closer than 0.4 Å to the plane of the porphyrin nitrogens. One such example is the polymeric picket fence complex catena-{μ-[meso-Tetrakis($a,a,a,$ $a,-o$-pivalamidophenyl)- porphinato-N,N',N'',N''':O]-aquo-iron(II)-tetrahydrothiophene}, {Fe(TpivPP)(OH$_2$)(THT)} for short (122). Here a water molecule is coordinated to the iron at a distance of 2.9 Å and in consequence the distance of the iron from the plane of the porphyrin nitrogens is reduced to 0.32 Å. A similar situation is found in deoxyerythrocruorin where a water molecule is coordinated to the iron atom at a distance of 3.1 Å and the iron atom is displaced only by 0.23 Å from the plane of the porphyrin nitrogens (43).

Table 8 Parameters of five-coordinated high-spin iron porphyrin complexes from crystal structure data[a]

Compound	Fe–Ct (Å)[b]	Δ (Å)[c]	Fe–N$_{porph}$ (Å)
O [Fe III (Di MeOEP)]$_2$	0.512	0.018	2.065
O [Fe III (TPP)]$_2$	0.496	0.044	2.087
Fe III (TPP) Br	0.489	0.070	2.069
Fe III (TPP) (NCS)	0.485	0.068	2.065
Fe III (TPP) I	0.480	0.068	2.066
Fe III (Proto) Cℓ	0.474	0.080	2.062
Fe III (Meso-DME) (OCH$_3$)	0.456	0.034	2.073
Fe III (Proto-DME) (SR)	0.434	0.015	2.064
Fe II (TPP) (2-Me-Im)	0.418	0.134	2.086
Fe II (T piv PP) (2-Me-Im)	0.399	0.027	2.072
Fe III TPP Cℓ	0.389	0	2.060
Average	0.457	0.051	2.069

[a] I wish to thank Dr. Lynn Hoard for this table. For references to these structures see (4).

[b] Ct is at the centre of the porphyrin nitrogens.

[c] Δ = (Fe . . P$_C$) – (Fe – P$_N$) where P$_C$ is the mean plane of the porphyrin carbons and P$_N$ the mean plane of the porphyrin nitrogens.

ROLE OF DISTAL RESIDUES

The roles of the distal His E7 and Val E11 have been studied in two abnormal hemoglobins: Hb Zürich (His E7 → Arg) and Hb Sydney (Val E11 → Ala) (113). The replacement of His E7 by Arg leaves a large gap in the heme pocket, because the Arg side chain is external and forms a salt bridge with one of the heme propionates. The gap allows small molecules easy access to the heme iron and causes hemolytic inclusion body anemia on treatment of heterozygous carriers with sulfanilamides. Hb Zürich forms the normal quaternary T and R structures, but the association constant of the first oxygen to combine with it is 7.5 times larger than for Hb A. The second order rate constant for the combination of CO with the T structure of Hb Zürich is much larger for Hb A. The partition coefficient $M = (p\,O_2[HbCO]/p\,CO[HbO_2])$ for the abnormal β subunit of Hb Zürich is 500 compared to 250 for Hb A. The rate of autoxidation in vitro and the acid catalyzed reductive displacement of superoxide by azide are faster than in Hb A. The CO stretching frequency is raised from 1952 cm^{-1} in Hb A to 1958 cm^{-1} in Hb Zürich, but the O–O stretching frequency is not affected.

These properties show that the distal histidine plays a multiple role. Its C_δ and N_ϵ atoms are in contact with pyrrole 1 of the porphyrin and oppose the rotation of the heme on ligand binding which brings that pyrrole closer to Lys E10 (see the section on changes in tertiary structure). The histidine is therefore essential for maintaining the restraint responsible for the low oxygen affinity of the T structure. It also lowers the CO affinity and thus protects Hb from CO poisoning, which is vital because of the CO produced endogenously on porphyrin breakdown. The protective mechanism may be part steric and part electronic. Sterically the histidine helps to force the CO into the off axial orientation which probably weakens the Fe–C bond. The sp^2 orbital of N_ϵ overlaps the empty π^* antibonding orbital of the carbon. $sp^2 \rightarrow \pi^*$ donation would weaken the CO bond and reduce the CO stretching frequency, in accordance with observation. It would also weaken $d\pi \rightarrow \pi^*$ donation from the iron to the CO, thus weakening the Fe–C bond, but this is a second order effect. Finally, the distal His protects the iron from oxidation. It has a pK of only 5.4, so that at physiological pH its single proton must be on either N_δ or on N_ϵ. It has been argued that it is on N_ϵ and forms a hydrogen bond with the bound oxygen, but if N_ϵ were permanently protonated, the distal histidine should promote rather than inhibit oxidation of the heme iron. However, the point can be settled with certainty only by neutron diffraction.

We have seen that Val E11 β obstructs the ligand binding site in the T structure. Study of Hb Sydney should have allowed us to measure the contribution which that obstruction makes to the free energy of heme-heme interaction, but surprisingly its oxygen affinity was found to be normal. X-ray analysis provided the explanation. The space vacated by the γ_2 carbons of the valine was occupied in Hb Sydney by a water molecule attached by a hydrogen bond to His E7. Binding of oxygen requires this bond to be broken, which costs energy; on the other hand entropy would be gained by the return of the water molecule into free solution. There is no way of estimating the relative magnitudes of these two opposing effects so that the question we asked remained unanswered. Replacement of Val E11 by Ala does not alter the partition coefficient between CO and O_2 and has a smaller effect on the CO stretching frequency than the replacement of the histidine (1952–1955 cm^{-1}). [References to the literature on Hb's Zürich and Sydney are given in (113)].

CONCLUSIONS

The main theme of this article is the regulation of the oxygen affinity of the heme by the structure of the globin. In principle, this could be accomplished by inductive effects. For instance, there exists an abnormal hemoglobin in which an alanine in the heme pocket is replaced by an aspartate (Ala E14β→Asp). This Hb has an abnormally low oxygen affinity, apparently because the negative field generated by the carboxylate group opposes the formation of the iron-linked superoxide anion (118). However, analysis of the structures of normal Hb and Mb suggest that steric rather than inductive effects are dominant. They influence both the change in Fe–N bond lengths accompanying the transition from the high-spin five-coordinated deoxy to the low-spin six-coordinated oxy form and, by steric hindrance, the actual presence of the oxygen molecule at its binding site. Therefore, the physical basis of heme-heme interaction is mechanical rather than inductive.

I have not discussed the nature of the constraints that oppose the transition to the six-coordinated low-spin state of the heme in the T structure, or the mode of action of the allosteric effectors: [H$^+$], [Cl$^-$], [CO$_2$], and [2,3-diphosphoglycerate]. The constraints take the form of salt bridges between neighbouring subunits, and all the allosteric effectors act either by strengthening those salt bridges or by forming additional ones. I have also glossed over the possible mechanism by which ligand binding causes rupture of the salt bridges. These aspects have been dealt with in a recent review (44).

What remains to be done? We still do not know the exact structure of HbO$_2$, but can only infer it from the closely related structures of HbCO and MetHb. The low temperature techniques of X-ray analysis now available may allow this problem to be solved. More challenging is the dynamic mechanism of the Hb molecule, which we know so far only from static pictures of the end states. For instance, the heme pocket is inaccessible to oxygen unless some door opens. Frauenfelder and his colleagues have begun to study the barriers that oppose the uptake of ligands by low-temperature kinetics (119); the continuing great advances in NMR techniques may eventually allow us to reinterpret the crystallographers' static structures of proteins in dynamic terms (120).

Literature Cited

1. Dayhoff, M. O. 1972. *Atlas of Protein Sequence and Structure,* 5:51–85 Washington, DC: Natl. Biomed. Res. Found.
2. Romero-Herrera, A. E., Lehmann, H., Joysey, K. A., Friday, A. E. 1973. *Nature* 246:389–95; and 1978. *Philos. Trans. R. Soc. London* B 283:61–163
3. Goodman, M., Moore, G. W., Matsuda, G. 1975. *Nature* 253:603–8
4. Hoard, J. L. 1975. *Porphyrins and Metalloporphyrins,* ed. K. M. Smith, pp. 317–80 Amsterdam: Elsevier
5. Collman, J. P. 1977. *Acc. Chem. Res.* 10:265–72
6. Scheidt, W. R. 1977. *Acc. Chem. Res.* 10:339–45
7. Antonini, E., Brunori, M. 1971. *Hemoglobin and Myoglobin in their Reactions with Ligands.* Amsterdam: North-Holland 436 pp.
8. Baldwin, J., Chothia, C. 1979. *J. Mol. Biol.* In press
9. Perutz, M. F. 1969. *Proc. R. Soc. London* B 173:113–40
10. Fermi, G. 1975. *J. Mol. Biol.* 97:237–56
11. Dickerson, R. E., Geis, I. 1970. *Structure and Action of Proteins.* New York: Harper & Row
12. *Handbook of Biochemistry* 1968. p. J210. Cleveland, Ohio: Chem. Rubber Co.
13. Collman, J. P., Gagne, R. R., Reed, C. A., Robinson, W. T., Rodley, G. A. 1974. *Proc. Natl. Acad. Sci. USA* 71:1326–29
14. Jameson, G. B., Molinaro, F. S., Ibers, J. A., Collman, J. P., Brauman, J. I., Rose, E., Suslick, K. S. 1978. *J. Am. Chem. Soc.* 100:6769–70
15. Jameson, G. B., Rodley, G. A., Robinson, W. T., Gagne, R. R., Reed, C. A., Collman, J. P. 1978. *Inorg. Chem.* 17:850–57
16. Scheidt, R. W., Piciulo, P. L. 1976. *J. Am. Chem. Soc.* 98:1913–19
17. Scheidt, R. W., Frisse, M. E. 1975. *J. Am. Chem. Soc.* 97:17–21
18. Kastner, M. E., Scheidt, W. R., Mashiko, T., Reed, C. A. 1978. *J. Am. Chem. Soc.* 100:6354–62
19. Butcher, R. J., Ferraro, J. R., Sinn, E. 1976. *J. Chem. Soc. D:* 910–12
20. Leipold, G. J., Coppins, P. 1973. *Inorg. Chem.* 12:2269–74
21. Hoselton, M. A., Wilson, L. J., Drago, R. S. 1975. *J. Am. Chem. Soc.* 97: 1722–29
22. Sinn, E., Sim, G., Dose, E. V., Tweedle, M. F., Wilson, L. J. 1978. *J. Am. Chem. Soc.* 100:3375–90
23. Albertson, J., Oskarsson, Å. 1977. *Acta Crystallogr.* B 33:1871–77
24. Jorgensen, L., Salem, L. 1973. *An Organic Chemist's Book of Orbitals.* New York: Academic 302 pp.
25. Pauling, L., Coryell, C. 1936. *Proc. Natl. Acad. Sci. USA* 22:210–16
26. Weiss, J. J. 1964. *Nature* 202:83–4
27. Misra, H. P., Fridovich, I. 1972. *J. Biol. Chem.* 247:6960–62
28. Demma, L. S., Salhany, J. M. 1977. *J. Biol. Chem.* 252:1226–31
29. Barlow, C. H., Maxwell, J. C., Wallace, W. J., Caughey, W. S. 1973. *Biochem. Biophys. Res. Commun.* 55:91–95
30. Hoffmann, B. M., Petering, D. H. 1970. *Proc. Natl. Acad. Sci. USA* 67:637–43
31. Chien, J. C. W., Dickinson, L. C. 1972. *Proc. Natl. Acad. Sci. USA* 69:2783–87
32. Yonetani, T., Yamamoto, H., Iizuka, T. 1974. *J. Biol. Chem.* 249:2168–73
33. Gupta, R. K., Mildvan, A. S., Yonetani, T., Srivastava, T. S. 1975. *Biochem. Biophys. Res. Commun.* 67:1005–12
34. Maxwell, J. C., Caughey, W. S. 1974. *Biochem. Biophys. Res. Commun.* 60: 1309–14
35. Cerdonio, M., Congiu-Castellano, A., Mogno, F., Pispisa, B., Romani, G. L., Vitale, S. 1977. *Proc. Natl. Acad. Sci. USA* 74:398–400
36. Cerdonio, M., Congiu-Castellano, A., Calabresi, L., Morante, S., Pispisa, B., Vitale, S. 1978. *Proc. Natl. Acad. Sci. USA* 75:4916–19
37. Kendrew, J. C., Dickerson, R. E., Strandberg, B. E., Hart, R. G., Davies,

D. R., Phillips, D. C., Shore, V. C. 1960. *Nature* 185:422–27
38. Takano, T. 1977. *J. Mol. Biol.* 110: 537–68
39. Takano, T. 1977. *J. Mol. Biol.* 110: 569–84
40. Norvell, J. C., Nunes, A. C., Schoenborn, B. P. 1975 *Science* 190:568–70
41. Phillips, S. E. V. 1978. *Nature* 273: 247–48
42. Weber, E., Steigemann, W., Alwyn-Jones, T., Huber, R. 1978. *J. Mol. Biol.* 120:327–36
43. Steigemann, W., Weber, E. 1979. *J. Mol. Biol.* 127:309–38
44. Perutz, M. F. 1976. *Br. Med. Bull.* 32:195–208
45. Heidner, E. J., Ladner, R. C., Perutz, M. F. 1976. *J. Mol. Biol.* 104:707–22
46. Bretscher, M. S. 1968. *The X-ray analysis of cyanide and carboxymethylated metmyoglobin.* PhD thesis. Univ. Cambridge, Cambridge. 106 pp.
47. Hendrickson, W. A., Love, W. E. 1971. *Nature New Biol.* 232:197–203
48. Deatherage, J. F., Loe, R. S., Moffat, K. 1976. *J. Mol. Biol.* 104:723–28
49. Perutz, M. F., Mathews, F. S. 1966. *J. Mol. Biol.* 21:199–202
50. Stryer, L., Kendrew, J. C., Watson, H. C. 1964. *J. Mol. Biol.* 8:96–104
51. Huber, R., Epp, O., Formanek, H., 1970. *J. Mol. Biol.* 52:349–54
52. Huber, R., Epp, O., Steigemann, W., Formanek, H. 1971. *Eur. J. Biochem.* 19:42–50
53. Fermi, G., Perutz, M. F. 1977. *J. Mol. Biol.* 114:421–31
54. Frier, J. A., Perutz, M. F. 1977. *J. Mol. Biol.* 112:97–112
55. Wishner, B. C., Ward, K. B., Lattman, E. E., Love, W. E. 1975. *J. Mol. Biol.* 98:179–94
56. Love, W. E., Fitzgerald, P. M. D., Hanson, J. C., Roger, W. E. Jr. 1979. In *Development of Therapeutic Reagents for Sickle Cell Anemia,* ed. E. Beuzard, J. Hercules, J. Rosa. (INSERM Symp. No. 9.) Amsterdam: Elsevier; North-Holland
57. Bolton, W., Perutz, M. F. 1970. *Nature* 228:551–52
58. Deatherage, J. F., Loe, R. S., Moffat, K. 1976. *J. Mol. Biol.* 104:723–28
59. Ladner, R. C., Heidner, E. J., Perutz, M. F. 1977. *J. Mol. Biol.* 114:385–414
60. Moffat, K., Loe, R. S., Hoffmann, B. M. 1976. *J. Mol. Biol.* 104:669–85
61. Baldwin, J. M. 1979. *J. Mol. Biol.*
62. Perutz, M. F. 1970. *Nature* 228:726–39
63. Perutz, M. F. 1972. *Nature* 237:495–99
64. Fung, L. W.-M., Ho, C. 1975. *Biochemistry* 14:2526–35
65. Perutz, M. F., Ladner, J. E., Simon, S. R., Ho, C. 1974. *Biochemistry* 13: 2163–73
66. Perutz, M. F., Fersht, A. R., Simon, S. R., Roberts, J. C. K. 1974. *Biochemistry* 13:2174–86
67. Perutz, M. F., Sanders, J. K. M., Chenery, D. H., Noble, R. W., Pennelly, R. R., Fung, L. W.-M., Ho, C., Giannini, I., Pörschke, D., Winkler, H. 1978. *Biochemistry* 17:3640–52
68. Adams, M. R., Schuster, T. M. 1974. *Biochem. Biophys. Res. Commun.* 58. 525–31
69. Huang, T.-H. 1979. In preparation
70. Gibson, Q. H. 1959. *Biochem. J.* 71:293–303
71. Brunori, M., Antonini, E., Wyman, J., Anderson, S. R. 1968. *J. Mol. Biol.* 34:357–59
72. Sugita, Y. 1975. *J. Biol. Chem.* 250: 1251–56
73. Davis, D. G., Lindstrom, T. R., Mock, N. H., Baldassare, J. J., Charache, S., Jones, R. T., Ho, C. 1971. *J. Mol. Biol.* 60:101–11
74. Deleted in proof
75. Deleted in proof
76. Scholler, D. M., Hoffmann, B. M., Schrier, D. F. 1976. *J. Am. Chem. Soc.* 98:7866–68
77. Eisenberger, P., Shulman, R. G., Brown, G. S., Ogawa, S. 1976. *Proc. Natl. Acad. Sci. USA* 73:491–95
78. Perutz, M. F., Kilmartin, J. V., Nagai, K., Szabo, A., Simon, S. R. 1976. *Biochemistry* 15:378–87
79. Knowles, F. C., McDonald, M. J., Gibson, Q. H. 1975. *Biochem. Biophys. Res. Commun.* 66:556–63
80. Giardina, B., Ascoli, F., Brunori, M. 1975. *Nature* 256:761–62
81. Saffran, W. A., Gibson, Q. H. 1979. *J. Biol. Chem.* In press
82. Cassoly, R. 1974. *C. R. Acad. Sci. Ser. D:* 278:1417–19
83. Maxwell, J. C., Caughey, W. S. 1976. *Biochemistry* 15:388–96
84. Szabo, A., Perutz, M. F. 1976. *Biochemistry* 15:4427–28
85. Szabo, A., Barron, L. D. 1975. *J. Am. Chem. Soc.* 97:660–662
86. Burke, J. M., Daly, P., Wright, P., Spiro, T. G. 1979. In press
87. Salhany, J. M., Ogawa, S., Shulman, R. G. 1974. *Proc. Natl. Acad. Sci. USA* 71:3359–62
88. Salhany, J. M., Ogawa, S., Shulman, R. G. 1975. *Biochemistry* 14:2180–90

89. Messana, C., Cerdonio, M., Shenkin, P., Noble, R. W., Fermi, G., Perutz, R. N., Perutz, M. F. 1978. *Biochemistry* 17:3652–62

90. Perutz, M. F., Heidner, E. J., Ladner, J. E., Beetlestone, J. G., Ho, C., Slade, E. F. 1974. *Biochemistry* 13:2187–2200

91. Asher, S. A., Vickery, L. E., Schuster, T. M., Sauer, K. 1977. *Biochemistry* 16:5849–56

92. Gupta, R. K., Mildvan, A. S. 1975. *J. Biol. Chem.* 250:246–53

93. Iizuka, T., Kotani, M. 1969. *Biochim. Biophys. Acta* 194:351–63

94. Dolphin, D. H., Sams, J. R., Tsin, T. B. 1977. *Inorg. Chem.* 16:711–17

95. Iizuka, T., Yonetani, T. 1970. *Adv. Biophys.* 1:157–211

96. Iizuka, T., Kotani, M. 1969. *Biochim. Biophys. Acta* 181:275–86

97. Winter, M. R. C., Johnson, C. E., Lang, G., Williams, R. J. P. 1972. *Biochim. Biophys. Acta* 263:515–34

98. Renovitch, G. A., Baker, W. A. 1967. *J. Am. Chem. Soc.* 89:6377–78

99. Ewald, A. H., Martin, R. L., Ross, I. G., White, A. H. 1964. *Proc. R. Soc. London* A 280:235–57

100. Beattie, J. K., Sunin, N., Turner, D. H., Flynn, G. W. 1973. *J. Am. Chem. Soc.* 95:2052–54

101. Tweedle, M. F., Wilson, L. J. 1976. *J. Am. Chem. Soc.* 98:4824–34

102. Fabry, T. L., Hunt, J. W. 1968. *Arch. Biochem. Biophys.* 123:428–29

103. Gibson, Q. H., Carey, F. G. 1975. *Biochem. Biophys. Res. Commun.* 67:747–51

104. Zipp, A., Ogunmola, G. B., Newman, R. G., Kauzmann, W. 1972. *J. Am. Chem. Soc.* 94:2541–42

105. Ogunmola, G. B., Zipp, A., Chen, F., Kauzmann, W. 1977. *Proc. Natl. Acad. Sci. USA* 74:1–4

106. Grenoble, D. C., Frank, C. W., Bergeron, C. B., Drickamer, M. G. 1971. *J. Chem. Phys.* 55:1633–44

107. Collman, J. P., Brauman, J. I., Doxsee, K. M., Halbert, T. R., Hayes, S. E., Suslick, K. S. 1978. *J. Am. Chem. Soc.* 100:2761–66

108. Collman, J. P., Brauman, J. I., Doxsee, K. M., Halbert, T. R., Suslick, K. S. 1978. *Proc. Natl. Acad. Sci. USA* 75:564–68

109. Gelin, B. R., Karplus, M. 1977. *Proc. Natl. Acad. Sci. USA* 74:801–5

110. Warshel, A. 1977. *Proc. Natl. Acad. Sci. USA* 74:1789–93

111. Deatherage, J. F., Obendorf, S. K., Moffat, K. 1979. *J. Mol. Biol.* In press

112. Elian, M., Hoffmann, R. 1975. *Inorg. Chem.* 14:1058–1076

113. Tucker, P. W., Phillips, S. E. V., Perutz, M. F., Houtchens, R., Caughey, W. S. 1978. *Proc. Natl. Acad. Sci. USA* 75:1076–80

114. Scheidt, R. W., Reed, C. A. 1979. To be published

115. Koster, A. S. 1975. *J. Chem. Phys.* 63:3284–86

116. Anderson, L. 1973. *J. Mol. Biol.* 79:495–506

117. Cramer, S. P., Eccles, T. K., Kutzler, F., Hodgson, K. O., Doniach, S. 1976. *J. Am. Chem. Soc.* 98:8059–69

118. Anderson, N. L., Perutz, M. F., Stamatoyannopoulos, G. 1973. *Nature New Biol.* 243:274–75

119. Alberding, N., Chan, S. S., Eisenstein, L., Frauenfelder, H., Good, D., Gunsalus, I. C., Nordlund, T. M., Perutz, M. F., Reynolds, A. H., Sorensen, L. B. 1978. *Biochemistry* 17:43–51

120. Wagner, G., De Marco, A., Wüthrich, K. 1976. *Biophys. Struct. Mech.* 2:139–58

121. Nagai, K., Kitagawa, T., Morimoto, H. 1979. *Biochemistry.* In press

122. Jameson, G. B., Robinson, W. T., Collman, J. P., Sorrell, T. N. 1978. *Inorg. Chem.* 17:858–64

123. Buckingham, D. A., Collman, J. P., Hoard, J. L., Lang, G., Radonovich, L. J., Reed, C. A., Robinson, W. T. To be published

124. Shelnutt, J. A., Rousseau, D. L., Friedman, J. M., Simon, S. R. 1979. To be published

125. Deatherage, J. F., Moffat, K. 1979. *J. Mol. Biol.* In press

Ann. Rev. Biochem. 1979. 48:387–418

STRUCTURE AND FUNCTION ♦12013
OF NITROGENASE[1]

Leonard E. Mortenson

Department of Biological Sciences, Purdue University, West Lafayette,
Indiana 47907

Roger N. F. Thorneley

Agricultural Research Council Unit of Nitrogen Fixation, University of Sussex,
Brighton BN19QJ, England

CONTENTS

[1]The following abbreviations are used: The MoFe and Fe proteins of the nitrogenases of
Azotobacter vinelandii, Azotobacter chroococcum, Clostridium pasteurianum, Klebsiella pneumoniae, and *Bacillus polymyxa* are referred to as Av1 and Av2, Ac1 and Ac2, Cp1 and Cp2, Kp1 and Kp2, and Bp1 and Bp2, respectively. For oxidized proteins, "ox" is added after the designation, e.g. Av2ox. The reduced proteins have no designation, e.g. Av2. Super reduced MoFe protein is defined as the EPR silent form produced only on reduction of the MoFe protein by Fe protein·MgATP^{2-} (this has also been referred to as the fully reduced MoFe protein). S* is acid labile sulfur.

0066-4154/79/0701-0387$01.00

PERSPECTIVES AND SUMMARY

Biological nitrogen fixation is responsible for the major reduction of dinitrogen to the ammonia that is used by plants and thereby indirectly by animals. The process is very energy consuming since it requires in addition to $6e^-$ from a low potential reductant, such as reduced ferredoxin, a minimum of four ATP molecules per electron pair involved in the reduction of N_2 to $2NH_3$. The nitrogenase system (reduced carrier: N_2 oxidoreductase) is present only in procaryotes and is composed of two dissociating protein components. One, called the Fe protein, contains four iron and four acid labile sulfur atoms; the other, the MoFe protein, contains two molybdenum, 28–32 iron, and about 28 acid labile sulfur atoms. The Fe protein is a dimer of molecular weight from 57,674 (for the Fe protein of *C. pasteurianum*) to 73,000 (Fe protein of a *Corynebacterium*), while the MoFe protein is a tetramer of mol wt 220,000–245,000. In *C. pasteurianum* the subunits of the Fe protein are identical and at the present state of catalytic activity, the dimer contains one $Fe_4S_4^*$ cluster. No other Fe proteins have been characterized but all are isolated as dimers and have four iron atoms. It seems likely that two cysteine thiols from each subunit are liganded to the $Fe_4S_4^*$ center. The MoFe protein of *C. pasteurianum*, *K. pneumoniae*, and *A. vinelandii* contain two dissimilar subunits (of molecular weights 50,000 and about 60,000) complexed with (*a*) up to four $Fe_4S_4^*$ centers, (*b*) two uncharacterized centers that appear to contain $MoFe_8S_6^*$ (called the FeMo cofactor or FeMo-co), and (*c*) possibly also a $Fe_2S_2^*$ center. At low concentration the MoFe protein dissociates on ultracentrifugation and may be active as a dimer as well as a tetramer. Both nitrogenase proteins and the protein-free FeMo-co are sensitive to dioxygen with the sensitivity in the order FeMo-co>>>Fe protein>>MoFe protein. The cell must prevent this oxidation, and various mechanisms for this have been discussed extensively in recent reviews (1, 2, 3).

Both MoFe and Fe proteins are capable of donating and accepting electrons, and the sequence of the transfer of electrons to reducible substrates has been developed (Figure 1).

The structure of the MoFe protein is complex. It is certain that there are $Fe_4S_4^*$ centers but whether $Fe_2S_2^*$ centers also exist is equivocal. The MoFe protein contains two FeMo-co centers; part of this center may consist of a cubane structure like $Fe_4S_4^*$ with Mo substituted for one of the iron atoms, i.e. $MoFe_3S_4^*$. An additional three to five iron atoms and two to four S^* atoms apparently must be accounted for, perhaps as an associated $Fe_4S_4^*$ cluster or two $Fe_2S_2^*$ clusters.

The Fe protein, but not the MoFe protein, binds MgATP and MgADP and the binding constant for MgATP appears to be about 60 μM. Between two and four MgATP binding sites have been estimated but only two of

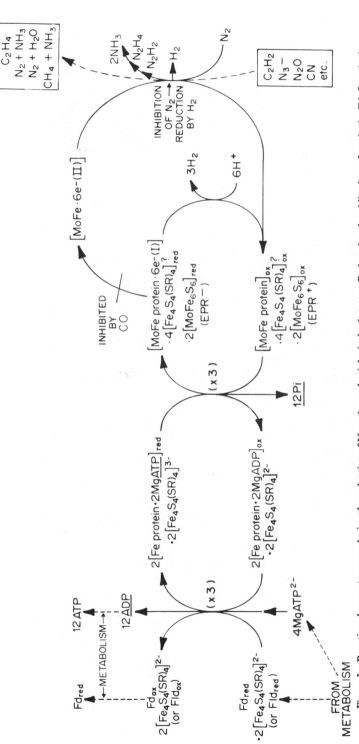

Figure 1 Reactions known to occur during the reduction of N_2 and nonphysiological substrates. Reduced, red; oxidized, ox; ferredoxin, Fd; flavodoxin, Fld; molybdenum-iron protein of nitrogenase, MoFe protein; iron protein of nitrogenase, Fe protein. The ATP and P_i (inorganic phosphate) are underlined to show that ATP hydrolysis occurs when electrons are transferred from the Fe protein to the MoFe protein; ADP is released in the generation of the second cycle of reduction of the Fe protein. There are 12 or more ATP hydrolyzed for each N_2 reduced, or 4 ATP per electron pair transferred to MoFe protein or to substrate. EPR (+ and −) show whether or not the state indicated has an EPR signal. The EPR silent and paramagnetic centers of both Fd and the Fe protein are represented as $[Fe_4S_4(SR)_4]^{2-}$ and $[Fe_4S_4(SR)_4]^{3-}$, respectively. The total oxidation states of the $[Fe_4S_4(SR)_4]$ centers of the MoFe protein are not established yet. The FeMo-co centers of the MoFe protein are listed as $MoFe_6S_6$ since the exact structure is not known, although a structure of the form $MoFe_6S_4^*$ seems to be a part of the structure (S^* is acid labile sulfur). There is also the possibility that the MoFe protein may have one or two $Fe_2S_2(SR)_4$ centers during its functioning.

these appear to be involved in nitrogenase catalysis. MgADP, a product of the utilization of ATP by nitrogenase, strongly inhibits catalysis: e.g., N_2 fixation is inhibited 90% when ADP and ATP are present at a 1:1 ratio of 10 mM. The binding of MgATP to the Fe protein results in a conformational change of the protein such that (a) it becomes more sensitive to O_2, (b) the Fe of its $Fe_4S_4^*$ center becomes easily removable by chelators, (c) its E_m becomes 120 mV more negative, (d) its EPR spectrum changes, and (e) its –SH groups are changed in reactivity with SH reagents. Changes c and d also are affected by MgADP but to a different extent than with MgATP.

Equilibria, reaction rates, and potentiometric titrations of the various partial steps in nitrogenase catalysis allow insight into several aspects of the catalytic mechanism. For example, Ac2ox is reduced rapidly by SO_2^- with $k > 10^8 M^{-1}$ sec^{-1}; only one electron is required for the reduction and MgADP inhibits it. The reduction of Ac2ox·MgADP had a limiting rate $K_{obs} = 25$ sec^{-1} which suggests that it undergoes a rate-limiting conformational change before being rapidly reduced by SO_2^-. The rate constant for the reduction of the Fe protein during nitrogenase turnover is 10^{-3} times that for the isolated protein. This suggests that during turnover MgADP is bound to the oxidized Fe protein, greatly affecting its reduction, and that ATP hydrolysis occurs while ATP is complexed to its Fe protein binding site.

Complex formation between the Fe protein and the MoFe protein is rapid ($k > 10^7 M^{-1}$ sec^{-1}) and the complex is fairly tight with a dissociation constant < 0.5 μM. Present data suggest a 1:1 or 2:1 ratio of Fe protein:MoFe protein for the nitrogenase complex during electron transport. Both complexes seem to be active, with the 2:1 complex having the greater rate. Free Fe protein not in the complex may transfer electrons to Fe protein in the complex.

N_2 appears to be reduced in steps that include enzyme-bound dinitrogen hydride intermediates. An intermediate at the diimide level has been suggested since D_2 reacts with nitrogenase during the reduction of N_2 to give some HD. An intermediate at the hydrazido(2–) level has been proposed to account for the hydrazine detected on quenching the enzyme during N_2 reduction. The concentration of this intermediate increases from zero when nitrogenase is activated with reductant and MgATP, and decreases as ATP is exhausted.

STRUCTURE OF THE NITROGENASE COMPONENTS

Fe Protein

One of the most intriguing proteins from a structural and functional point of view is the Fe protein of nitrogenase. This protein, from all organisms examined, has (a) a molecular weight of around 57,000 (exactly 57,674 daltons for *C. pasteurianum*) – 73,000, (b) at the present state of purity, four iron atoms and four labile sulfur atoms (designated S*), (c) two identical subunits (from *C. pasteurianum* (4, 5, 6) and probably also from *A. vinelandii* (7)), and (d) an extreme O_2 sensitivity (half-life in air of ~30 sec). Two different groups (3, 8) have shown that the four iron atoms of the Fe protein of *C. pasteurianum* are arranged in one Fe_4S_4* cluster. This was shown by the recently developed extrusion or displacement technique during which the four cysteine thiols of the Fe protein liganded to the four Fe atoms of the Fe_4S_4* cluster are replaced quantitatively by four thiol ligands of the added phenylthiol. The extruded product is easily identified since the properties of synthetic Fe_4S_4* (phenylthiol)$_4$ are well established and are the same as the extruded product.

The twelve cysteine groups of the Fe protein are not clustered in the sequence as might be expected from the structure of ferredoxins, but appear to be randomly distributed in the protein (6). Thus, the identical monomers of the Fe protein from *C. pasteurianum* which have been completely sequenced have 273 amino acid residues each, with cysteine residues appearing at positions (from the NH_2 terminal end) 37, 82, 94, 129, 181, and 231. Since there are two such peptides in the Fe protein, it would appear that the single Fe_4S_4* cluster is liganded to two cysteine thiols from each subunit. However, other arrangements are possible and the exact positioning may have to await X-ray studies. The function of the remaining eight cysteine residues is unknown. Four of them react with DTNB without any affect on complex formation or acetylene reduction (see later section).

From the amino acid sequence of the Fe protein of *C. pasteurianum* one can deduce the following: (a) there is no known sequence homology between it and other known iron-sulfur proteins, (b) it has a higher content of Gly-Gly sequences (one/39 residues) than all other proteins examined, with four Gly-Gly sequences predicted to form β-bends, (c) there are several dipeptide and tripeptide (xx and xxx) sequences, (d) the isoelectric point is about 4.4 with an overall negative charge at neutral pH, (e) the acid residues, unlike the basic residues, occur in clusters, (f) there are eight regions with five or more hydrophobic residues, and (g) when analyzed by the Chou-Fasman procedure (9) an α·helical content of about 35%, a β-structure content of 34%, and 18 β-bends were suggested. These characteristics reflect the known properties of the protein—that the reduced protein undergoes one type of conformational change when it binds MgATP (see later section) and different conformational changes when it binds MgADP or when it is oxidized. The protein obviously is designed to be flexible. There are four possible oxidation states (10, 11) in which the Fe_4S_4* cluster of the Fe protein could exist, i.e. $[x]^{-4}$, $[x]^{-3}$, $[x]^{-2}$, and $[x]^{-1}$ where $[x]$ = $[Fe_4S_4*(Cys)_4]$. The isolated Fe protein of all nitrogenases has an EPR signal of the "1.94 type" (see later sections) which is oxidized when the protein, complexed with $MgATP^{2-}$, transfers its electron to the MoFe protein (12–14). From this fact and from Mössbauer (15, 16) and potentiometric studies (17), it can be concluded that the $Fe_4S_4^*$ center of the Fe protein, like that of the four and eight Fe ferredoxins, operates between the $[Fe_4S_4(Cys)_4]^{-2}$ and $[Fe_4S_4(Cys)_4]^{-3}$ states and is a one electron donor (or acceptor).

Potentiometric titration, together with the measurement of either the loss of its EPR spectrum or changes in its visable absorption at 440 nm, shows that in the absence of $MgATP^{2-}$, the Fe protein has an E_m (midpoint potential) of about –250 to –295 mv (17). The same protein with 2 $MgATP^{2-}$-bound has an E_m of about –400 mv. Because the ATP/ADP ratio of growing N_2-fixing cells appears to be between 2:1 and 3:1 [(17a); R. G. Upchurch and L. E. Mortenson, unpublished] and because the binding of MgADP is greater than that of MgATP, one would expect a substantial amount of MgADP·Fe protein in the growing, N_2-fixing cell (a more detailed discussion of this appears later). Binding of MgADP also changed the potential of the Fe protein to more negative values but the titration curve was different from that in the presence of MgATP (18).

Binding of $MgATP^{2-}$ to the Fe protein also increased its O_2 sensitivity (19), exposed or concealed –SH groups when reacted with DTNB or iodoacetamide, respectively (18, 20), and exposed the $Fe_4S_4^*$ center, as shown (21) by following the rate of complex formation of Fe^{2+} with 2,2'-bipyridryl. With this technique it was demonstrated (a) that the conformation was changed on binding $MgATP^{2-}$, (b) that the effect of $MgATP^{2-}$-binding could be competitively inhibited by MgADP, like the nitrogenase reactions, (c) that the MgATP concentration required for half-maximum rate of Fe release was close to that for nitrogenase catalysis (about 0.4 mM), (d) that two MgATP molecules had to be complexed to the Fe protein in order for the conformational change resulting in $Fe_4S_4^*$ exposure to occur, (e) that the Fe^{2+} of the $Fe_4S_4^*$ center of oxidized Fe protein (not inactivated) became accessible to chelators, and (f) that a decrease in temperature led to an increase in exposure of the $Fe_4S_4^*$ center. Whether the latter effect correlates with the change in enthalpy at 17°C seen with the nitrogenase reaction is unknown (see later section). Using this technique, and replacing the 2,2'bipyridyl with the faster reacting bathophenanthrolinesulfonate, Ljones & Burris (22) have shown that from the Fe released one can determine the amount of active Fe protein present in a preparation. The preparation they used (1700 units/mg) was calculated to be only 57% active. Their data fit a model where the two MgATP binding sites are equivalent and interact with a third site, the $Fe_4S_4^*$ site. The latter possibility, however, can not be distinguished from the possibility that the

MgATP sites react not with the $Fe_4S_4^*$ site directly but rather with a remote site on the protein that in turn affects the $Fe_4S_4^*$ site. The two studies (21, 22) agree well, but do not suffice to distinguish between cooperative or noncooperative binding of the MgATP. The Fe proteins from some organisms are inactive even when isolated in the absence of O_2. Ludden & Burris (23) found that the Fe protein of the nitrogenase system of *Rhodospirillum rubrum* had to be activated by a peptide factor present in the extracts and that a divalent metal ion (preferably Mn^{2+}) and ATP were required for the activation. This activation was also seen by Nordlund et al (24) and several other Fe proteins, (e.g. those of *Azospirillum* and certain *Rhizobia*) were shown to require similar activation. Since the inactive form does not show O_2 resistance, and since the Fe protein is neither adenylylated nor phosphorylated by reaction with ATP, one has to assume that activation plays some role in regulation of activity.

MoFe Protein

The MoFe protein is much more complex than the Fe protein for it is isolated as a tetramer with a molecular weight in the range of 220,000 for Cp1 (25) and Kp1 (26) and 245,000 for Av1 (7, 27, 28). It contains two atoms of molybdenum, and although the precise number of Fe atoms is not known, the best values range from 24–32 (2, 3). Mössbauer analysis of Av1 and Cp1 suggest 30 ± 2 Fe/molecule (29). Acid-labile sulfur and Fe are present in about equal amounts, i.e. about 24–30 atoms of each per molecule. This strongly suggests that the Fe is present in classical iron-sulfur clusters. In fact Mössbauer studies indicate the presence of 16 Fe atoms in what appear to be four $Fe_4S_4^*$ clusters, and 12 Fe atoms in two clusters each with six Fe atoms. The latter clusters are responsible for the EPR spectrum of the MoFe proteins with g values of about 4.3, 3.7, and 2.01 (15, 29). In addition, Mössbauer analysis sees a third Fe environment that could be a $Fe_2S_2^*$ type cluster. If only one of the latter clusters is present per tetramer, the suggestion that the tetrameric MoFe protein is actually two dimers would be ruled out.

Extrusion of the iron-sulfur centers by the methods already described for the Fe protein shows the presence of at least three and probably four $Fe_4S_4^*$ centers (W. O. Gillum, L. E. Mortenson, and R. H. Holm, unpublished), but the absorption spectrum of the extruded cluster is influenced by the absorption of the other nonextrudable Mo, Fe, S center (discussed below) and the results are equivocal. Modified methods to identify the iron-sulfur clusters in the presence of interfering chromophores are under study. For example, G. Wong, D. Kurtz, R. H. Holm, and L. E. Mortenson (unpublished) use trifluormethylphenylthiol as the extruding ligand. The resulting highly fluorinated extrusion product is readily identified by fluorine NMR, i.e. the $Fe_2S_2^*$ and $Fe_4S_4^*$ clusters with the fluorinated ligands are separated by 2.6 ppm. The number of clusters extruded can be quantitated by adding, after extrusion and NMR analysis, a known fluorine standard and again analyzing under identical conditions. Bale, Averill, and Orme-Johnson (3) displaced the cluster to a new ligand, then identified the extruded cluster by again displacing it to the cysteine thiols of an apoprotein of a simple FeS protein with specificity for either the $Fe_2S_2^*$ or $Fe_4S_4^*$ clusters, i.e. one of the ferredoxin apoproteins. The ferredoxin produced was then identified and quantitated by comparing it with the known electron paramagnetic resonance (EPR) characteristics of the original holoprotein. Yields as high as 90% were obtained but no supporting data have yet appeared.

Both of the above methods demonstrate presence of $Fe_4S_4^*$ centers in the MoFe protein as predicted by Mössbauer, but suggest in addition the presence of $Fe_2S_2^*$.

Pertinent to this discussion of the iron-sulfur centers of the MoFe protein is the molybdenum-free MoFe protein from *C. pasteurianum* (30). This protein contains an average of eight Fe atoms/tetramer. Its 1.94-type EPR signal relaxed rapidly at temperatures above 40°C, indicating presence of two noninteracting $Fe_4S_4^*$ clusters (31). Based on current knowledge of the "MoFe Cofactor" (see following discussion) this protein undoubtedly is a precursor of the MoFe protein. This result again supports the presence of at least two $Fe_4S_4^*$ clusters in the MoFe protein.

Shah & Brill (32) discovered that appropriate extraction of the MoFe proteins from nitrogenases with N-methylformamide separated a component containing Fe, Mo, and S in the ratio 8:1:6 from the protein. This "cofactor" contained less than 0.1 mole of any amino acid per mole of Fe_8MoS_6 (J. B. Howard and B. B. Elliott, personal communications). It was prepared by (a) treating the MoFe protein anaerobically with citric acid to destroy and remove iron-sulfur centers other than cofactor, (b) bringing the cofactor-protein complex to pH 5 with Na_2HPO_4 and centrifuging the now precipitated protein, (c) washing the precipitated protein with dimethylformamide, and (d) releasing the cofactor from the protein with N-methylformamide. All steps had to be strictly anaerobic. Why only N-methylformamide extracts the cofactor is uncertain; its effectiveness may be concerned with its high dielectric constant and with the possibility of its acting as a ligand to replace the cysteine thiol ligands of the native protein. The latter explanation is reasonable since added phenylthiol considerably narrowed the lines of the EPR spectrum of the FeMo cofactor and made them more like those seen in the native protein (33).

Addition of the extracted FeMo complex to a molybdenum-free MoFe protein present in crude extracts of an *A. vinelandii* mutant, UW 45, results in production of fully active MoFe protein (32). This procedure has been repeated with minor modification by several groups.

The properties of this coenzyme (FeMo-co) show many similarities to those seen in the native protein. For example, from Mössbauer and EPR spectra, the $S = 3/2$ center of the cofactor was similar to each of the two $S = 3/2$ centers of the MoFe protein (33), i.e. the native protein has two FeMo-co centers. Such EPR spectra of FeMo-co in N-methylformamide plus thiophenol also are similar to those seen when the MoFe protein is placed in 80% N-methylformamide:20% buffer, provided the latter spectrum is corrected for the EPR of the $Fe_4S_4^*$ clusters that are seen in this solvent. The EPR spectrum of the $Fe_4S_4^*$ centers of the MoFe protein in 80% N-methylformamide is almost identical to that seen in the Mo-free MoFe protein of *C. pasteurianum* described earlier (30), which again suggests that this protein is a MoFe protein precursor that lacks FeMo-co and perhaps an additional iron-sulfur center. Finally, the X-ray absorption spectra of the Mo of FeMo-co in both the absorption edge and the fine structure region is almost identical to that found for the MoFe protein of *C. pasteurianum* (34) and of *A. vinelandii* (35).

A study to determine how and in what form Mo is incorporated into protein showed (36–38) that during growth of *C. pasteurianum* on ammonia, cells take up a low but constant amount of Mo. Although the rate of uptake increases only slightly when the ammonia is depleted and the cells are forced to fix dinitrogen, the accumulation of Mo dramatically increases. This increased accumulation is explained by the fact that the cells synthesize a new intracellular Mo-binding protein which, at the present state of purity, binds up to six Mo atoms per molecule of about 50,000 daltons. This type of accumulation system can be distinguished from those that involve binding proteins external to the membrane and from the phosphotransferase and membrane transport energy-linked systems. The Mo enters the cell via a membrane uptake system coupled to energy utilization then, by complexing to this binding protein, it accumulates within the cell to a concentration twenty times that found in NH_3-grown cells. Since (*a*) the membrane transport system for Mo is the same as that for sulfate, (*b*) sulfate is present in the culture medium and in nature in concentrations 100 times (or more) that of Mo, and (*c*) the apparent K_m for Mo and sulfate (also tungstate) are similar, this protein seems to be synthesized to allow the cell to obtain the needed concentrations of Mo, even in the presence of high sulfate. When the N-source for growth is switched from N_2 to ammonia, synthesis of the Mo binding protein stops abruptly and the protein is diluted out by further growth of cells. At high concentrations of Mo the steady state concentration of the binding protein decreases to about one third of normal; however, under N_2-fixing conditions, even with high Mo, there is always binding protein present. This suggests that a second role for the binding protein is to transfer Mo to the FeMo-co synthesizing system.

In 1974 a high intensity, broad-band X-ray source, the SPEAR electron storage ring, became available at Stanford University. With this source, a technique called X-ray absorption spectroscopy (XAS), provided a tool for studying the immediate environment of Mo in a protein, without the need for crystalline materials and without "interference" from other metal ions (39). XAS can be compared to absorption spectroscopy (39–41). For Mo the photon energy needed is in the range of 20,000eV ($\lambda = \sim 0.6$Å) with the inflection point (absorption maximum) in the range of 20,004–20,017 depending on the charge of Mo and its ligands (42). With increasing eV through the Mo K edge (where a $1S$ electron is promoted) one sees increasing absorption followed by decreased absorption when past the edge region. However, unlike simple spectroscopy where one would expect the absorption to smoothly decrease past the edge, one sees considerable fine structure that results from an interaction between the photoelectron wave emanating from the atom when it absorbs X rays and the wave backscattered by other atoms in the immediate vicinity (usually within 6A). What effect the backscattered wave has depends upon whether or not it returns to the Mo atom in or out of phase with the outgoing photoelectron wave [see references by Cramer & Hodgson (34, 35, 39–42) and references therein for more detailed information]. Thus the region between 20,050 and 20,300 eV contains a great deal of structural information concerning the environment around the Mo atom. The absorption edge data alone indicate that (a) the MoFe protein of nitrogenase has no Mo–O bonding and no Mo–Mo bonding less than 5Å, (b) there is primarily sulfur ligation to the Mo, and (c) the charge of the Mo is either +3 or +4 (34). From the fine structure region one can deduce further that there exists (a) a set of three or four bound sulfur atoms with an average Mo–S distance of 2.36 $\pm$ 0.02Å, (b) a set of two or three iron atoms at a distance of 2.72 $\pm$ 0.03Å from the Mo, and (c) one or two sulfur atoms at a longer Mo–S distance of about 2.49Å. Based on these data and chemical background from known structures two possible types of Mo, Fe, S clusters for the MoFe protein were proposed (34, 35). The most appealing suggests that Mo replaces one Fe atom in a typical $Fe_4S_4^*$ cluster. Such a compound has now been synthesized (43) and its primary Mo–S and Mo–Fe distances are similar to those of the MoFe protein and the MoFe cofactor. In this case it would appear that the structure of the MoFeS cluster may be determined by synthesis more easily than by crystallization and analysis.

Both isolated nitrogenase components have EPR signals. These signals change when the combined components function in N_2 fixation such that during steady state with no limited substrates, the EPR signal of the MoFe protein is about 90% diminished, whereas the EPR signal of the Fe protein is still prominent (see section on nitrogenase catalysis). However, when the reductant is depleted, the EPR signal of the MoFe protein returns and the EPR signal of the Fe protein becomes silent [for a discussion, see (2, 3)]. From these changes it was inferred that the oxidized form of the MoFe protein was the form with an EPR signal and the oxidized form of the Fe protein was the form without a signal. It would follow that the form of the MoFe protein present during steady state should be more reduced than the form with the EPR signal. Direct evidence for this was presented by Walker & Mortenson (44) who oxidatively titrated (with thionine) the MoFe protein in its 90% EPR silent form during steady state turnover and found it contained at least four more electrons than the paramagnetic (partially reduced) form. Titration with thionine monitored by absorption at 460 nm also showed (45) that four electrons were removed in going from the partially reduced form to the oxidized EPR silent form (E_m ; -70mV). However, because the E_m of the partially reduced Cp1 : oxidized Cp1 half cell at pH 7.5 was very positive (~-70mV, pH 7.5) relative to the electron donors for nitrogenase, reduced ferredoxin, and Cp2, it was concluded that this oxidation reduction system with $n = 1$, was not involved in the reduction of substrates by nitrogenase (45).

Rawlings et al (46) repeated this oxidative titration of the MoFe protein with thionine but monitored it by measuring the loss of the EPR signal. They found that four electrons were removed during the titration without any loss of the EPR signal; on further titration an additional two electrons were removed during which the EPR centers were oxidized. In the earlier titration Walker & Mortenson (45), monitoring only the absorption change at 460 nm, did not see this differential oxidation, and, since they removed only four electrons, the MoFe centers may not have been titrated at all. The two titrations also yield different midpoint potentials—about -70 mV by 460 nm absorption and -20 to 0 mV by EPR (17). These results suggest that the two methods are detecting different centers. Since there is little absorption change at 460 nm (44) when the MoFe center is reduced from the partially reduced paramagnetic form to the more reduced EPR silent form, it is possible that no absorption change at 460 nm occurs in going from the partially reduced paramagnetic form to the oxidized EPR silent form.

By analogy with these results one can suggest a simple explanation for the reduction of the MoFe protein to its EPR silent form when it is present in N_2-fixing reaction mixture in high concentration relative to the Fe protein (17). If one assumes that at least six reducible centers are present in the MoFe protein and if individual centers have different redox potentials when going from the partially reduced paramagnetic form to the oxidized form (E_m -20 to 0 mV), it is likely that they would be reduced differentially by the reduced iron protein·2MgATP. For example, with the reduced Fe protein at low concentration relative to the MoFe protein (1:20 vs the normal 2–4:1), the MoFe center might become reduced, whereas at least one of the remaining centers whose reduction is needed for substrate reduction would remain mostly in the partially reduced state. Thus even with its MoFe center fully reduced the MoFe protein would not be able to reduce substrates and would remain in the reduced EPR silent form. The fact that the "reduced" MoFe center remains under these conditions also argues that it probably is not the center for the reaction of H^+ or that the other centers must be reduced before it can function.

The presence in the MoFe protein of at least four centers in addition to the FeMo-co centers, makes the terms "super reduced" and "fully reduced" MoFe protein (based on whether or not the center is reduced and EPR silent) ambiguous. Such terms should not be used unless one can show that all centers are reduced or unless they are redefined to specify only the fact that the $g = 4.3, 3.7, 2.01$ center is reduced.

THE EFFECT OF MgATP AND MgADP ON THE Fe-PROTEIN

The binding of [14C]ATP to Cp2 (47, 48) and to Ac2 (19) in the presence of Mg^{2+} ions has been qualitatively shown by gel-filtration techniques. A more quantitative treatment involved a gel equilibration method that demonstrated two noncooperative binding sites for MgATP($K = 16.7$ μM) in Cp2 (49). However, cooperativity was observed when MgADP was bound to one site ($K = 5.2$ μM) since this caused MgATP to bind with approximately twofold greater affinity at the remaining site (49). Eady & Smith (50) criticized the latter experiments because Sephadex G-50 may not have completely excluded the Cp2 (mol. wt. 55,000). Orme-Johnson & Davis (3) pointed out that Tso & Burris's data (49) show significant binding of ATP in the absence of Mg^{2+} and that this may indicate a site capable of chelating metal ions but not hydrolyzing ATP. A more likely explanation is that ATP can compete with MgATP at the same site. This would be consistent with the steady state kinetic data for acetylene reduction by Kp nitrogenase which showed ATP to be a competitive inhibitor of MgATP (51). A reinvestigation of MgATP binding to Cp2 protein using Tso & Burris's technique (49) indicated 5 ± 2 binding sites with $K = 60 \pm 20$ μM (P. Wyeth and W. Orme-Johnson, unpublished observation). Proton relaxation rate NMR studies on Kp2 solutions (52) indicated four binding sites for Mg^{2+} ($K = 1.7 \pm 0.3$ mM) and Mn^{2+} ($K = 0.35 \pm 0.05$ mM). Although the formation of ternary protein-metal bonding was demonstrated, the number of MgATP binding sites could not be determined. A major difficulty was the instability of Kp2 at concentrations greater than 15 mg/ml.

The EPR spectra of all nitrogenase Fe-proteins change from rhombic to axial forms when MgATP binds (12–14, 53); the g values for these forms of Kp2 are given in Table 1. Stoichiometric binding of 2 moles of MgATP per mole of Cp2 protein occurs (13) consistent with the binding constants obtained by the gel-equilibration method (49). The unusual EPR spectrum of the Fe-protein and the change in signal symmetry on MgATP binding have been simulated by Lowe for Kp2 protein (54) by assuming a rapidly relaxing paramagnetic center that is unobservable by EPR but which causes anisotropic broadening of the EPR signal from the $Fe_4S_4^*$ cluster, (Figure 2). The low integrated intensity of the $g = 1.94$ signal, 0.2, 0.45, and 0.17 electron equivalents/mole for Cp2 (13), Kp2 (14), and Ac2 (53) proteins respectively, is also explained by the coupling between this center and the $Fe_4S_4^*$ cluster. The simulation requires the center to be $\sim$10Å from the $Fe_4S_4^*$ cluster and for the spin axes to rotate by about 70° relative to each other when MgATP binds. A smaller rotation simulates the EPR spectrum obtained when MgADP binds. This suggests that phosphate-protein interactions are important in inducing the conformation change. Cp2 protein has been reported to lose 30% integrated EPR signal intensity when MgATP binds (3). This was not observed with Kp2 protein (14) nor with an earlier study of the Cp2 protein (13). Apparent loss of signal intensity on MgATP binding is easily accommodated in the two center model of Lowe (54) and may be a species-dependent phenomenon. Although the second center has not been identified, the Mössbauer data for Kp2 (16) do not preclude the presence of a $Fe_4S_4^*$ cluster plus one additional magnetic Fe atom. Orme-Johnson et al (55) have discussed the quantitation of the EPR signals of Fe protein and the absorbance changes on oxidation-reduction and have also concluded that a second center may exist. Cp2 protein

Table 1   Parameters used in simulations of Kp2 protein EPR spectra of Figure 2

Figure	Spectrum type	g_x	g_y	g_z	Linewidth (mT)	$\theta_R{}^a$	$\phi_R{}^a$	$C(mt)^b$
2(a')	Na$_2$S$_2$O$_4$ reduced uncomplexed	1.85	1.942	2.05	1.0	25	63	3.5
2(b')	Na$_2$S$_2$O$_4$ reduced MgATP complexed	1.85 (g_m)	1.942	2.05	1.0	67	4	3.5
2(c)	No Interaction	1.85	1.942	2.05	1.0	—	—	0

[a] θ_R and ϕ_R are the azimuthal and equatorial angles (degrees) specifying the direction to the rapidly relaxing center relative to the principal g-value tensor of the observable spin; $\theta_R = 90°$, $\phi_R = 0°$ corresponds to the x-axis and $\theta_R = 0°$ corresponds to the z-axis.
[b] C is the interaction parameter between the two centers (54).

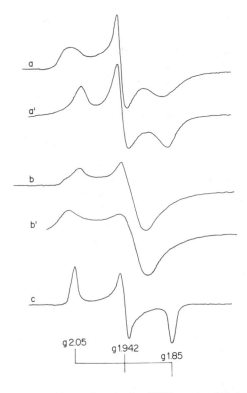

Figure 2 Experimental- and interacting-simulated EPR spectra of Kp2 protein: Kp2 protein, activity 700 nmol of C_2H_4 produced/min per mg, at 11 mg/ml was prepared in the presence of (*a*) 50 mM $MgCl_2$ alone and (*b*) plus 1.9 mM ATP. The buffer was 25 mM-Tris-HCl, pH 8.1 (at 0°C) with 1 mM $Na_2S_2O_4$. Spectra were recorded at about 18 K at a microwave frequency of 9.1 GHz and 20 mW microwave power. They are the spectra of Figures 3*a* and *b* respectively of Smith et al (14). *a'* and *b'* are simulations of *a* and *b* using the parameters given in Table 1. Curve *c* is a simulated spectrum with no interaction present.

binds MgATP at pH 7.0 (49) but at this pH the EPR signal does not change its symmetry (3). This is either a consequence of the inability of MgATP to induce a conformation change, or because Cp2 is in a protonated form that does not undergo this conformation change (3). Protolytic equilibria of this sort make EPR data for binding studies difficult to interpret since the pK_a values for buffers, substrates, and groups on the protein may all change on freezing and subsequent cooling (56). The rate of freezing may also be critical since different EPR spectra have been obtained with samples of Ac2 with MgATP or MgADP (frozen in 1–2 sec) and others frozen in a few milliseconds using a rapid-freezing apparatus (57).

A consequence of the MgATP-induced conformation change is that the redox potential of Cp2 protein changes from -294 to -400 mV at pH 7.5 (17). The small free energy change (2.5 Kcal/mol) associated with the lowering of redox potential could be obtained by a 70-fold tighter binding of MgATP to oxidized Cp2 than to reduced Cp2 and would not require cleavage of the γ-phosphate group of bound MgATP. This is consistent with the failure to detect any ATPase activity with the Fe protein alone. Binding constants for MgATP to oxidized Fe proteins have not been reported.

The MgATP-induced conformations of Ac2 (19) and Kp2 (58) proteins are more susceptible to inactivation by O_2 than the uncomplexed forms. For Ac2, the rate of inhibition is higher at 4°C than at 30°C even after correction for the increased solubility of O_2 at 4°C. These data indicate a preequilibrium with a negative enthalpy change prior to the rate limiting step for O_2 inactivation. The iron-sulfur protein II from *A. vinelandii*, first characterized by Shethna et al (59), forms a complex with Av nitrogenase, regulates nitrogenase activity, and protects nitrogenase from inactivation by O_2 (60, 61).

MECHANISM OF NITROGENASE CATALYSIS

A generally accepted scheme for the mechanism of nitrogenases has been deduced from EPR (12, 14, 17, 62, 63), Mössbauer (15, 16), and stopped-flow spectrophotometric studies (64–66). The following four equations indicate the direction of electron flow but do not define stoichiometries or the order of assembly and dissociation of the various complexes.

Reduction of Fe protein by an electron donor:

$$\text{Fe protein}_{ox} + e^- \longrightarrow \text{Fe protein} \qquad \qquad 1.$$

Assembly of a complex containing both Fe and MoFe proteins and MgATP:

$$\text{MoFe protein} + \text{Fe protein} + \text{MgATP} \rightleftharpoons$$
$$\text{Fe protein} \cdot 2\text{MgATP} \cdot \text{MoFe protein} \qquad \qquad 2.$$

Electron transfer from Fe protein to MoFe protein coupled to ATP hydrolysis:

$$\overset{e^-}{\overbrace{\text{Fe protein MgATP}}} \cdot \text{MoFe protein} \longrightarrow$$
$$\text{Fe protein}_{ox} \, (\text{MgADP}) \cdot \text{MoFe protein}_{reduced} \qquad \qquad 3.$$
$$\text{P}_i^+$$

Transfer of electrons to reducible substances:

$$\text{Fe protein}_{ox} \text{ (MgATP)} \cdot \text{MoFe Protein}_{reduced} \xrightarrow{\quad e^- \quad} N_2 \xrightarrow{\quad H^+ \quad}$$

Fe protein$_{ox}$ (MgATP) $\cdot$ MoFe Protein$_{reduced}$ N$_2$ $\xrightarrow{H^+}$

or   or

(MgADP + P$_i$)   H$^+$

or

C$_2$H$_2$ 4.

Fe protein$_{ox}$ (MgATP) $\cdot$ MoFe protein + NH$_3$

or   or

(MgADP + P$_i$)   H$_2$

or

C$_2$H$_4$

Each of the above steps is reviewed below.

Reduction of Fe Protein by Electron Donors

No detailed kinetic studies of this step have been reported that used reduced ferredoxins or flavodoxins, the in vivo electron donors. Ac2 protein has been oxidized with phenazine methosulphate without the loss of enzyme activity (53) and the kinetics of the reduction of this Ac2ox by sodium dithionite were studied by stopped-flow and rapid-freezing EPR spectroscopy (57, 65). As was found for the reduction of oxidized ferredoxins (66, 67), SO_2^-, obtained by the predissociation of $S_2O_4^{2-} \rightleftharpoons 2SO_2^-$, is the active reductant. The $Fe_4S_4^*$ cluster in Ac2ox is reduced by SO_2^-, ($k > 10^8 M^{-1}Sec^{-1}$) at least a hundred times faster than a $Fe_4S_4^*$ cluster in *C. pasteurianum* ferredoxin. This may indicate that the $Fe_4S_4^*$ cluster in Ac2 protein is more accessible to solvent and may also explain the rapid inactivation of Ac2 by O_2. A one electron reduction of Ac2ox ($k > 10^8 M^{-1}Sec^{-1}$) caused the EPR signal at $g = 1.94$ (integrated intensity 0.24 electron equivalents per mole Ac2) to become fully developed. However, slower reductions with no change in EPR signal intensity, occurring in three phases, accounted for an additional electron equivalent. These slow phases may be due to inactive protein or may be associated with the second center in the Fe protein proposed by Lowe (54). Cp2 protein that has been enzymically oxidized by a small amount of Cp1 protein in the presence of MgATP accepts one electron from dithionite (68).

MgADP, but not MgATP, inhibits the rate of reduction of Ac2ox by SO_2^- (57, 69). The observed first order rate constant for the reduction of Ac2ox·MgADP exhibits a hyperbolic dependence on $[S_2O_4^{2-}]^{1/2}$ with a limiting value of $K_{obs} = 25$ sec^{-1}. A rate-limiting conformation change in Ac2ox·MgADP prior to rapid reduction ($k > 10^8 M^{-1}sec^{-1}$) by SO_2^- accounts for these data (69). This mechanism can also explain the difference, first commented on by Watt (70), between the rate constant for the reduction of Fe protein by SO_2^- during turnover ($k = 3 \times 10^5 M^{-1}sec^{-1}$ for Av2ox) and that for the reduction of isolated Fe protein ($k > 10^8 M^{-1}sec^{-1}$ for Ac2ox). It suggests that during turnover (a) oxidized Fe protein is largely in a conformation with MgADP-bound which cannot be reduced rapidly, (b) that MgATP is hydrolyzed to MgADP at a site on the Fe protein created when it complexes with the MoFe protein during electron transfer, and (c) that ATP hydrolysis must occur before the reduction of Fe protein by SO_2^- in the catalytic cycle.

Assembly of a Complex Containing Both Fe and MoFe Proteins and MgATP

Ultracentrifuge studies with mixtures of Kp1 and Kp2 showed that they exist in a 1:1 complex (71, 72). In the absence of dithionite the sedimentation coefficient of Kp1 decreases at protein concentrations below 1mg/ml (F. Reithel and R. R. Eady, unpublished observation). This is consistent with the dissociation of Kp1 tetramer and indicates that Kp2 binds tighter to the tetrameric Kp1 than to lower n-mers. The dissociation of Cp1 tetramer to dimer was previously reported (25, 73). Modification of four accessible sulphydryl groups of Kp1 with DTNB does not diminish acetylene reduction activity or decrease the extent of complex formation with Kp2. This supports a conclusion from kinetic studies that complex formation is too rapid ($k > 10^7 M^{-1}sec^{-1}$) for disulfide bonds to form prior to electron transfer in the Kp1·Kp2·MgATP complex (74). In contrast to earlier studies (71, 72) limiting values of the sedimentation coefficient s, were not obtained at a 1:1 component protein ratio but increased to a value of 12.7s with a fivefold excess of Kp2 over Kp1 protein (F. Reithel and R. R. Eady, unpublished).

The observed first order rate constant for electron transfer from Kp2· MgATP to Kp1 (see next section) is independent of protein concentration. This requires a lower limit ($K < 0.5$ μM) for the dissociation constant of a 1:1 protein complex (64). A nonlinear dependence of acetylene reduction activity at low protein concentrations results from dissociation of the protein complex (dilution effect) (75–85). Analysis of such kinetic data for a 1:1 mole ratio of Kp1:Kp2 proteins gave a dissociation constant for the protein complex of 0.5 μM (72). An attempt to reinterpret these data in terms of higher order complexes gave equivocal results (55).

The formation of a catalytically inactive complex between Cp and Av component proteins has been exploited (using acetylene reduction as assay) to determine dissociation constants of ~1.0 μM for Cp1 with Cp2, ~0.1 μM for Av1 with Av2, and ~0.025 μM for Av1 with Cp2 (86). Substrate reduction kinetics indicate a ratio of Fe:MoFe proteins in the complex of 2:1 for Cp (87–89), Rj (90), and the heterologous Av1:Cp2 (86) nitrogenases. Although three mole equivalents of Kp2 were oxidized by Kp1 in a stopped-flow kinetic study (64), this does not necessarily indicate a 3:1 stoichiometry for the Kp1:Kp2 (64, 91). Orme-Johnson et al (55) compared the change in extinction coefficients at 420 nm on oxidation for Kp2 (64) with that for Cp2 (91) and concluded that only one mole of Kp2 was oxidized by Kp1 in the kinetic study of Thorneley (64). However, pulse labelling in vivo of Kp nitrogenase with ^{14}C-amino acids showed that Fe protein is synthesized at three times the rate of MoFe protein, and these rates of synthesis are reflected in a 3:1 ratio of the steady state concentrations of Fe protein:MoFe protein (92). The Kp1·Kp2 complex (72) has thermodynamic and kinetic parameters that are similar to those for the ferredoxin-NADP reductase complex (93). Both have dissociation constants of ∼0.5 μM, both are formed with a rate constant $k > 10^7$M^{-1}sec^{-1}, and both have $\Delta H = 0$ kJ mole^{-1} and $\Delta S = + 140$ kJ^{-1}mol^{-1} for their enthalpies and entropies of formation. These parameters have been interpreted in terms of electrostatic interactions between the component proteins (93).

At 23°C, a dissociation constant of 0.5 μM for the Kp nitrogenase complex and a lower limit of 10^7M^{-1} sec^{-1} for the association rate constant, require the first order dissociation rate constant to be > 0.5sec^{-1}. This lower limit is close to the turnover number of nitrogenase and may indicate that complex dissociation is the rate-limiting step (94) (Table 2). This calculation refers to the dissociation of reduced Fe protein(s) from MoFe protein (EPR state). If complex dissociation occurs after each electron transfer, it is the rate of dissociation of oxidized Fe protein from reduced MoFe protein EPR silent that is relevant. This dissociation reaction is rapid relative to the turnover time for H$_2$ evolution in the *A. vinelandii* nitrogenase, and indicates that the Fe protein and the MoFe protein forms only a transient complex with a lifetime long enough for electron transfer but not sufficient for the Fe protein to be directly involved in substrate (H$^+$) reduction (95). Whether a similar mechanism exists for C$_2$H$_2$ and N$_2$ reduction under physiological conditions of excess Fe over MoFe protein has yet to be demonstrated. Under these conditions, a high steady state concentration of the relatively stable complex of reduced Fe protein with MoFe protein could be maintained if the rate of reduction of bound oxidized Fe protein by free excess Fe protein is rapid compared to the rate of dissociation of the complex of oxidized Fe protein with reduced MoFe protein. This complex may be necessary in order to reduce substrates other than protons, since under conditions when its concentration would be expected to be low (limiting Fe protein), the system, although capable of reducing protons, reduces C$_2$H$_2$ and N$_2$ at relatively low rates.

MgATP and MgADP bind to either Kp2 or the Kp1·Kp2 protein complex with rate constants in the range of 10^6–$10^7 M^{-1} sec^{-1}$ which are typical of small molecule protein reactions (64, 96).

MgATP-Dependent Electron Transfer from the Fe Protein to the MoFe Protein Coupled to ATP Hydrolysis

This reaction was first observed for Kp (14, 62) and Cp (12, 17, 63) nitrogenases using EPR spectroscopy to monitor the decrease of the $g = 3.6$ signal of the MoFe protein and the $g = 1.94$ signal of the Fe protein. When the supply of electrons from dithionite ion was exhausted, the $g = 3.6$ but not the $g = 1.94$ signal returned. The conclusion that the Fe protein-MgATP complex was reducing the MoFe protein was further supported by Mössbauer studies with ^{57}Fe enriched MoFe protein (16). When MgATP was added to a mixture of MoFe and Fe proteins in the presence of sodium dithionite, the Mössbauer parameters of the MoFe protein changed, on reduction, in a similar manner to those of the high potential ion protein (HiPIP). EPR experiments using the rapid-freezing technique developed by Bray et al (97) showed that this reduction occurred with a time constant of ~10 msec at 23°C (14, 17). A kinetic analysis of absorbance changes at 420 nm associated with the oxidation of Kp2 protein showed that the observed first order rate constant for electron transfer ($k = 2 \times 10^2 sec^{-1}$) is independent of the component protein concentration or ratio but is a function of MgATP concentration (64). MgADP is a competitive inhibitor ($K_i = 20 \mu M$) of this MgATP-induced ($K_s = 0.4$ mM) electron transfer reaction (96). MgADP dissociates from the Kp nitrogenase complex at a rate comparable to that of electron transfer. However these processes are too fast to be rate-limiting during turnover. Rate and equilibrium constants for these reactions are given in Table 2. The ratio of the number of moles of MgATP hydrolyzed to electron pairs transferred to substrate (ATP:2e ratio) increases from about four when the Fe- and MoFe-proteins are equimolar to greater than twenty when the MoFe protein is in large excess (98, 99). At high MoFe:Fe protein ratios, the rate of ATP hydrolysis remains constant as the MoFe protein concentration increases, hence the reduced MoFe protein (EPR silent) is not itself acting as an ATPase. Orme-Johnson & Davis (3) have elaborated on a mechanism first proposed by Mortenson et al (18) in which electron transfer from the Fe to the MoFe protein (coupled to MgATP hydrolysis) is followed by electron transfer back from the MoFe to Fe protein with MgADP bound. This futile cycle would be favored when the Fe-protein is limiting and is proposed to explain the uncoupling of ATP hydrolysis from substrate reduction at high MoFe:Fe protein ratios.

A different futile cycle involving a more oxidized form of the MoFe protein could be the cause of the reductant-independent ATP hydrolysis

Table 2 Rate and equilibrium constants for ternary complex assembly and electron transfer with Kp and Ac nitrogenases proteins at 23°, pH 7.4

Reaction	Rate and equilibrium constants	Comments	References
Reduction of Fe protein by dithionite ion			
$S_2O_4^{2-} \overset{k_1}{\underset{k_{-1}}{\rightleftharpoons}} 2SO_2^{\cdot -}$ $SO_2^{\cdot -} + Ac2ox \overset{k_2}{\longrightarrow} Ac2$	$k_1 = 1.75s^{-1}$ $k_{-1} = 1.2 \times 10^9 M^{-1}s^{-1}$ $k_2 > 10^8 M^{-1}s^{-1}$	The half order dependence of k_{obs} on dithionite ion concentration is accounted for by this predissociation. At low dithionite ion and high Ac2ox concentrations k_1 becomes rate limiting.	65 66 67
MgADP-induced conformation change in the oxidized Fe protein			
$Ac2oxMgADP \overset{k_3}{\underset{k_{-3}}{\rightleftharpoons}} Ac2ox^aMgADP$	$k_{-3} = 25s^{-1}$	$Ac2ox^aMgADP$ reduced slowly by $SO_2^{\cdot -}$ therefore k_{-3} limits the rate of reduction at high dithionite concentrations.	68, 72

Protein complex formation

Reaction	Kinetic constants	Comments	Ref.
$Kp1 + Kp2(MgATP) \underset{k_{-4}}{\overset{k_4}{\rightleftharpoons}} Kp1Kp2MgATP$	$k_4 > 10^7 M^{-1}s^{-1}$ $0.5 < k_{-4} < 50 s^{-1}$ K complex $= 0.5\ \mu M$	MgATP may be necessary to induce complex formation. k_{-4} could be rate-limiting in turnover.	64, 71, 94

MgATP binding to reduced Fe protein or protein complex

Reaction	Kinetic constants	Comments	Ref.
$\begin{matrix} Kp2 \\ \text{or} \\ Kp1Kp2 \end{matrix} + MgATP \underset{k_{-5}}{\overset{k_5}{\rightleftharpoons}} \begin{matrix} Kp2MgATP \\ \text{or} \\ Kp1Kp2MgATP \end{matrix}$	$k_5 > 2.5 \times 10^6 M^{-1}s^{-1}$ $k_{-5} > 1 \times 10^3 s^{-1}$ $K = 0.4\ mM$	Not certain whether MgATP binds to Kp2 or to Kp1 in the Kp1Kp2 complex.	64

MgADP binding to reduced Fe protein or protein complex

Reaction	Kinetic constants	Comments	Ref.
$\begin{matrix} Kp2 \\ \text{or} \\ Kp1Kp2 \end{matrix} + MgADP \underset{k_{-6}}{\overset{k_6}{\rightleftharpoons}} \begin{matrix} Kp2MgADP \\ \text{or} \\ Kp1Kp2MgADP \end{matrix}$	$k_6 \sim 10^7 M^{-1}s^{-1}$ $k_{-6} \sim 2 \times 10^2 s^{-1}$ $K \sim 20\ \mu M$	Release of MgADP or a conformation change occurs at a similar rate to electron transfer from Kp2 to Kp1.	96

Electron transfer from Fe to MoFe protein within the protein complex

Reaction	Kinetic constants	Comments	Ref.
$Kp1Kp2MgATP \overset{k_2}{\longrightarrow} Kp1_{super\ reduced}Kp2_{ox}MgADP + P_i$	$k_2 = 2.0 \times 10^2 s^{-1}$	Electron transfer only occurs in the presence of MgATP. MgATP hydrolysis is coupled to electron transfer.	64, 65, 111

[a] Second species of $Ac2_{ox} \cdot MgADP$.

that occurs after reductant is consumed by nitrogenase (100). Since now the Fe protein is oxidized (EPR silent) and the MoFe protein is partially reduced (EPR active) electrons would have to be transfered from the EPR-active form of the MoFe protein back to the oxidized Fe protein and then from the Fe protein again to the oxidized form of the MoFe protein (EPR silent).

The coupling of ATP hydrolysis to electron transfer from the Fe protein to the MoFe protein has been demonstrated using stopped-flow spectrophotometry together with a rapid quenching technique (R. R. Eady, R. N. F. Thorneley and D. J. Lowe, unpublished). The correlation between the optical change at 420 nm associated with the oxidation of the Fe protein (Figure 3a) and the time course for a "burst" phase in the appearance of P_i (Figure 3b) indicates that ATP is hydrolyzed with a time constant ($\tau = 42$msec) indistinguishable from that of electron transfer between the component proteins. Hageman & Burris (95) observed a lag phase for H_2 evolution but not for ATP hydrolysis when Av2 protein was limiting, and concluded that ATP hydrolysis was coupled to electron transfer between the component proteins and not to substrate reduction.

Substrate Reduction

Nitrogenase reduces a number of substrates other than N_2, e.g. N_3^- to NH_3+N_2, CN^- to CH_4+NH_3, N_2O to N_2+H_2O, C_2H_2 to C_2H_4, and H^+ to H_2. All these reductions require both proteins and are coupled to MgATP hydrolysis. Steady state kinetic studies (101, 102), in which apparent K_m and K_i values for various substrates and inhibitors were determined, have been reviewed previously (2, 81). Dihydrogen is a competitive inhibitor of N_2 but does not inhibit N_3^-, C_2H_2, CN^-, or H^+ reduction (101). This is consistent with N_2 binding to Mo in the enzyme since certain molybdenum dinitrogen complexes react reversibly with H_2 to give molybdenum dihydrides and free N_2 (103). CO is a noncompetitive inhibitor of N_2, C_2H_2, and N_3^- reductions but does not inhibit H^+ reduction (101, 102). C_2H_2 is a noncompetitive inhibitor of N_2 reduction. These data led Hwang et al (101) to favor a model similar to that originally proposed by Hardy et al (104), which involved five substrate binding sites. An alternative model in which the ability of MoFe protein to bind and reduce various substrates is determined by the concentration of Fe protein and MgATP has been developed by Davis et al (105) following the initial observations of Silverstein & Bulen (106) that limiting ATP favored proton relative to N_2 reduction. Bergerson & Turner (90) showed that the apparent K_m and V_{max} values for N_2 and C_2H_2 depend on the ratio of Fe to MoFe protein and on ATP concentration. It is clear that one role for the Fe protein-MgATP complex is to transfer electrons to the MoFe protein. Hence, when either Fe protein or MgATP is limiting, electron supply to MoFe protein will

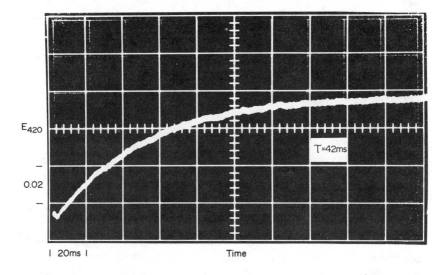

E$_{420}$

0.02

| 20ms | Time

T=42ms

a

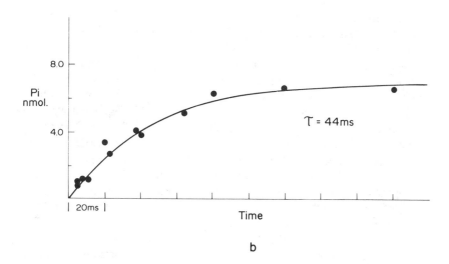

8.0

P$_i$
nmol.

4.0

| 20ms | Time

T = 44ms

b

Figure 3 Coupling of ATP hydrolysis to electron transfer from Fe protein to MoFe protein for *K. pneumoniae* nitrogenase. (*a*) Stopped-flow oscillograph of the absorbance change at 420 nm associated with the MgATP (5 mM)-induced oxidation of Kp2 (50 μM) by Kp1 (10 μM) at 10°C, pH 7.4, in the presence of Na$_2$S$_2$O$_4$ (10mM). $\tau = 42 \pm 3$ msec. (*b*) Time course for appearance of P$_i$ under identical conditions obtained using rapid quenching in trichloracetic acid (30% w/v) followed by colorimetric analysis for P$_i$. $\tau = 44 \pm 4$ msec.

be restricted and it is possible that two electron reductions of protons or C_2H_2 will be favored under these conditions relative to the reduction of N_2, which requires six electrons with electron donating sites, possibly of more negative redox potential (102, 105).

Reported rates of substrate reduction at high mole ratios of MoFe:Fe protein should be regarded with caution if only single time, stopped-assays have been used. Nonlinear rates of substrate reduction were observed by Thorneley & Eady (107) at 30°C with a Kp1:Kp2 ratio > 9:1 and at 10°C with Kp1:Kp2 > 1:1. A lag phase of 1–4 min duration, dependent on the Kp1:Kp2 ratio, occurred before linear rates of C_2H_2 reduction were achieved. A complementary burst phase for H_2 evolution in the presence of C_2H_2 was also observed. Slow lag and burst phases defined in terms of their rate relative to the overall catalytic reaction rate at longer times, are an increasingly common feature of enzymes involved in metabolic regulation (108). The term "hysteretic" has been suggested by Frieden (108) to describe these enzymes. Mechanisms involving ligand-induced isomerizations of the enzyme, displacement of tightly bound ligands by other ligands, and polymerization or depolymerization have been discussed to explain this behavior (108–110). The relevance of any of these mechanisms to nitrogenase is yet to be determined.

Lag phases of 10 min for C_2H_2 reduction (111) and 35 min for N_2 reduction (94) have been reported for a heterologous nitrogenase comprised of Cp2 with Kp1 protein at 1:1 mol ratio. The conditions that induce lags in C_2H_2 and N_2 reduction, i.e. limiting Fe protein, low temperature, and the use of heterologous nitrogenase, are also the conditions that uncouple ATP hydrolysis from electron transfer to reducible substrates and result in high ATP:2e ratios (98, 99, 111–113). At 10°C, MgATP-dependent electron transfer from Kp2 to Kp1 occurs rapidly $k = 24$ sec^{-1}, R. N. F. Thorneley, unpublished). At 23°C in the heterologous enzyme, Kp1·Cp2, it occurs with essentially the same rate constant (2.5×10^2 sec^{-1}) and MgATP dependence ($K_D = 0.4$ mM) as in the homologous Cp1·Cp2 and Kp1·Kp2 nitrogenases (111). MgADP is a simple competitive inhibitor ($K_i = 20$ μM) of the MgATP-induced electron transfer from Kp2 to Kp1 at 23°C (96). However, heterotropic interactions between MgATP and MgADP were observed for hydrogen evolution at 23°C (96). These have been interpreted in terms of both concerted and sequential models with MgATP or MgADP binding to two sites on separate subunits in the nitrogenase complex. Cooperative binding of 1,N^6-etheneoadenosine 5'-triphosphate to the Av1·Av2 protein complex (Hill coefficient $n = 2.3$) but noncooperative binding to Av2 alone ($n = 1.04$) has also been reported (114). These data support the hypothesis that there are two roles for MgATP; one in electron transfer from the Fe to MoFe protein and the other, involving heterotropic interactions between MgATP and MgADP, within the nitrogenase complex for the control of substrate reduction. Moustafa & Mortenson first suggested that the MgATP:MgADP ratio may control nitrogenase activity (115). This has been confirmed by kinetic and EPR studies in vitro in which the concentration of creatine kinase and the pH were used to control this ratio (116) and by studies in vivo with *A. vinelandii* (117) and intact bacteroids (118). The tight binding of MgADP relative to MgATP for Kp nitrogenase and the value of the "allosteric constant" $L = 3.7 \pm 1.1$, ensures rapid inhibition of substrate reduction for small decreases in the MgATP :MgADP ratio (96). The weak binding of MgATP presumably facilitates the selective switching off of the expensive (in terms of ATP requirement) nitrogenase while allowing other ATP-dependent enzymes to continue functioning under conditions of maintenance rather than growth of the organism.

Under conditions that result in lags in substrate reduction dinitrogen is a poor substrate relative to acetylene (107). This may explain some of the differences in estimates of nitrogen fixation rates when data from $^{15}N_2$ and acetylene reduction assays are used (119). The relative rates of N_2 reduction and concomitant H_2 evolution are significant with respect to the efficiency of nitrogen fixation. In vivo, the coupling of nitrogenase with an H_2 uptake hydrogenase can recycle H_2 to generate both reducing power and ATP thereby increasing the efficiency of nitrogen fixation in both symbiotic (120–123) and free living microorganisms (124–127). In vitro, the altered relative rates of H^+, C_2H_2, and N_2 reduction at 10°C, and at 30°C with limiting Fe protein, are reflected in a different distribution of electrons over the various iron-sulfur centers of Kp1 protein. This allowed Lowe et al (128) to detect a number of new EPR signals that are observed only under turnover conditions. These new signals are designated 1c, VI, VII, and VIII in Table 3 where they are listed with the EPR signals previously observed for Kp nitrogenase. Signals Ia and Ib are associated with different pH forms of Kp1 protein as isolated in the presence of dithionite ion. Shifts in this protolytic equilibrium (pK_a 8.7) have been used to monitor the binding of acetylene (14). Similar effects are observed for Bp1 protein but over a lower pH range ($pK_a = 5.5$) (B. E. Smith, unpublished). Signal 1c, which is observed only under turnover conditions, probably arises from an increased rhombic distortion of $Fe_4S_4^*$ center at the −1 oxidation level with $S = 3/2$ (128). Signals IIa, IIb, and IIc, which are associated with Fe protein, have been discussed earlier. Signal VI arises from a $Fe_4S_4^*$ cluster at the −1 oxidation level. It is seen only in the presence of ethylene under turnover conditions, and provides the first evidence for an enzyme-product complex for nitrogenase. ^{13}C-Ethylene did not broaden signal VI, hence there is no evidence for direct binding of the ethylene to the $Fe_4S_4^*$ cluster. However, a broadened signal VI was obtained when ^{57}Fe-substituted Kp1 was used indicating that this signal arises from a $Fe_4S_4^*$ cluster in the MoFe protein.

Table 3 EPR signals associated with the nitrogenase proteins of *K. pneumoniae* as isolated in the presence of sodium dithionite and in mixtures of component proteins in the steady state during turnover

Protein	Signal[a] designation	g values	g_{av}	Oxidation level of $Fe_4S_4^*$ center	Spin state	Comments	References
Kp1	I_a	4.32, 3.63, 2.009	—	-1	3/2	Protein as isolated, low pH form; decreased intensity during turnover	14
Kp1	I_b	4.27, 3.73, 2.018	—	-1	3/2	Protein as isolated, high pH form; decreased intensity during turnover	14
Kp1	I_c	4.67, 3.37, ~2.0		-1	3/2	Only observed during turnover under acetylene; intensity enhanced by ethylene	128
Kp2	II_a	2.053, 1.942, 1.865	1.953	-3	1/2	Protein as isolated	14
Kp2	II_b	2.036, 1.929 (g_m)	1.965	-3	1/2	MgATP bound form	14
Kp2	II_c	g values poorly defined; intermediate between II_a and II_b		-3	1/2	MgADP bound form	14
Kp1	III	2.073, 1.969, 1.927	1.990	-3	1/2	CO bound tightly; only observed during turnover	14
Kp1	IV	2.17, 2.06, 2.06	2.10	-1	1/2	CO bound weakly; only observed during turnover	128
Kp1	V	2.139, 2.001, 1.977	2.039	-1^b	1/2	Only observed during turnover; may be associated with H_2 evolution	128
Kp1	VI	2.125, 2.000, 2.000	2.042	-1	1/2	Ethylene bound form only observed under turnover; decreased by acetylene	128
Kp1	VII	5.7, 5.4			>3/2	Only observed during turnover	128
Kp1	VIII	2.092, 1.974, 1.933	2.000	-3^b	1/2	Only observed during turnover	128

[a] Signals arising from the same $Fe_4S_4^*$ center have been assigned the same number with subscripts to indicate different forms of the enzyme.
[b] Assignment of oxidation level difficult; most likely level indicated.

The amplitudes of signals 1c, IV, V, VI, VII, and VIII are a function of ethylene and acetylene concentration. These data indicate two types of binding site for acetylene ($K < 16$ μM and $K \sim 13$ mM) and a single site for ethylene, ($K = 1.5$ mM) (128). High and low affinity acetylene binding sites resulting in K_m values of 0.003 atm and 0.23 atm at 30° for Cp nitrogenase were detected (129).

Lowe et al (128) present three schemes involving two $Fe_4S_4^*$ clusters in the MoFe protein, in order to account for all the EPR signals given in Table 3 and the complex interconversions of these signals under various conditions. These schemes, which are intended as working hypotheses, require $Fe_4S_4^*$ clusters to operate at the –1, –2, –3, and –4 oxidation levels. The –4 oxidation level is required for the schemes to be consistent with the Mössbauer data of Smith & Lang (16) which showed that the EPR silent Fe atoms in Kp1 (isolated in the presence of dithionite ion) are all ferrous. There is no precedent for this oxidation level in FeS proteins but it has been detected in polarographic (130) and cyclic voltametric (131) studies of model $Fe_4S_4^*$ clusters. The schemes are also consistent with the suggestion of Orme-Johnson et al (129) that signal III of Cp nitrogenase, observed at low CO concentrations during turnover, arises from a reduced $Fe_4S_4^*$ cluster at the –3 oxidation level. At high concentration of CO this center becomes oxidized to the –1 oxidation level. Signal V was first observed by Yates & Lowe (132) with Ac nitrogenase, and later by Orme-Johnson et al (129) with Av nitrogenase; the latter workers suggested that it may be associated with a form of the enzyme with H_2 bound. The g values for signal V do not allow an unequivocal assignment of either the –1 or –3 oxidation level to the responsible $Fe_4S_4^*$ cluster. Lowe et al (128) present two schemes for an H_2 evolution cycle in order to accommodate this difficulty. Signal V increases in intensity in the presence of N_2 (129) and has been used to determine an apparent binding constant of 0.1 atm for N_2 with Kp nitrogenase (69). This value is close to the apparent K_m for N_2 determined from steady state kinetic data for the rate of NH_3 formation (26). All the EPR signals in Table 3 arise from iron-sulfur centers and no signal that could be assigned to a Mo center has been detected, although the center containing Mo, the FeMo-co, has been assigned (33).

Various mechanisms for the reduction of N_2 to NH_3 via partially reduced dinitrogen intermediates such as diimine (N_2H_2) or hydrazine (N_2H_4) have been proposed (104, 133–140) but early attempts to detect these species, either enzyme-bound or free in solution, failed (141, 142). However, Bulen (143) did show that N_2H_4 was a substrate with low affinity for nitrogenase (apparent $K_m \sim 14$ mM). Thorneley et al (144), by using high concentrations of purified Kp nitrogenase, have detected an enzyme-bound dinitrogen-hydride intermediate. Kp nitrogenase which was actively reducing

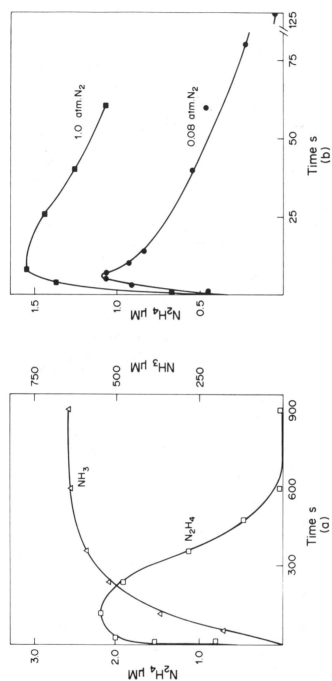

Figure 4 (*a*) Time course for the concentration of N_2H_4 ($\square$) and NH_3 produced ($\triangle$) during catalytic reduction of N_2 by *K. pneumoniae* nitrogenase at pH 7.4, 30°C. Assays (0.55 ml) contained MoFe protein, 20 μM; Fe protein, 18 μM; ATP, 18 mM; MgCl$_2$, 20 mM; creatine phosphate, 18 mM; Na$_2$S$_2$O$_4$, 27 mM; creatine kinase, 50 μg; and Hepes, 25 mM. N_2H_4 was determined in assays quenched with 2 ml of ethanol containing HCL (1M) and paradimethylaminobenzaldehyde (PDMAB) 0.07 M. After centrifuging to remove precipitated protein the N_2H_4 concentration was determined spectrophotometrically at 458 nm in a 4 cm path length cell using an experimentally determined calibration curve. The points at times less than 1 min were obtained with a rapid-quench apparatus when 0.4 ml of mixed solutions was shot into 2 ml of PDMAB solution. Reprinted from reference (144) with permission of *Nature*. (*b*) Time courses for the concentration of N_2H_4 obtained from nitrogenase under 100% ($\blacksquare$) and 61% saturation with N_2 ($\bullet$). Conditions and assay procedures were as in *a* except all the data points were obtained with the rapid-quench apparatus and MoFe protein (17 μM), and Fe protein (15 μM) were used. Under these conditions N_2 becomes limiting and H_2 inhibits to a greater extent than in *a*. Reprinted from reference (144) with permission of *Nature*.

N_2 to NH_3 was quenched with an ethanolic solution of p-dimethylamino-benzaldehyde (PMAB). The concentration of N_2H_4 derived from an enzyme-bound intermediate was determined spectrophotometrically as its PMAB derivative. The correlation between the concentrations of N_2H_4 detected and total NH_3 produced at various times is shown in Figure 4 a. The maximal rate of NH_3 production corresponds with the maximal concentration of N_2H_4. As MgATP becomes limiting and MgADP is inhibitory, the concentration of N_2H_4 falls as does the rate of NH_3 formation. The concentration of hydrazine depends on the degree of saturation of the enzyme with N_2 (Figure 4 b.) The intermediate need not be N_2H_4 because the chemistry of Mo and W dinitrogen complexes shows that other enzyme-bound dinitrogen hydride intermediates species could undergo nonenzymic hydrolysis on quenching with acid or alkali to give N_2H_4 (145, 146). The same yield of N_2H_4 was obtained when an alkali quench replaced the acid quench procedure. The structure of the intermediate is not likely to be $Mo-N\equiv N$ or $Mo-NH=NH$ since the analogous W and Mo complexes give different yields of N_2, N_2H_4, and NH_3 in acid and alkaline conditions (146). $Mo-NH_2-NH_2$ together with its protonated form $Mo-NH_2-NH_3^+$ are also unlikely structures since N_2H_4 has low affinity for enzyme, and intermediates of this type would be expected to dissociate leading to accumulation of N_2H_4 in solution. An intermediate in which the dinitrogen molecule is reduced to the –4 level with retention of some multiple bond character between the Mo and dinitrogen hydride is favored, i.e. $Mo=N-NH_2$ or $Mo=N-NH_3^+$ (147).

ACKNOWLEDGMENTS

The research of one of the authors (LEM) was supported by NIH grant AI 04865–16 and NSF grant PCM 77–24655. We wish to thank P. K. Mortenson for her substantial help in the production of this manuscript.

Literature Cited

1. Dalton, H., Mortenson, L. E. 1972. *Bacteriol. Rev.* 36:231–60
2. Zumft, W. G., Mortenson, L. E. 1975. *Biochim. Biophys. Acta* 416:1–52
3. Orme-Johnson, W. H., Davis, L. C. 1977. In *Iron Sulfur Proteins,* ed. W. Lovenberg, 3:15–60. New York: Academic
4. Tanaka, M., Hainu, M., Yasunobu, K. T., Mortenson, L. E. 1977. *J. Biol. Chem.* 252:7081–88
5. Tanaka, M., Hainu, M., Yasunobu, K. T., Mortenson, L. E. 1977. *J. Biol. Chem.* 252:7089–92
6. Tanaka, M., Hainu, M., Yasunobu, K. T., Mortenson, L. E. 1977. *J. Biol. Chem.* 252:7093–7100
7. Swisher, R. H., Landt, M., Reithel, F. J. 1975. *Biochem. Biophys. Res. Commun.* 66:1476–82
8. Gillum, W. O., Mortenson, L. E., Chen, J. S., Holm, R. H. 1977. *J. Am. Chem. Soc.* 99:584–95
9. Chow, P. Y., Fasman, G. D. 1974. *Biochemistry* 13:222–45
10. Carter, C. W., Kraut, J., Freer, S. T., Alden, R. A., Sieker, L. C., Adman, E., Jensen, L. H. 1972. *Proc. Natl. Acad. Sci. USA* 69:3526–29

11. Holm, R. H. 1977. In *Biol. Aspects Inorg. Chem. (Symp.)*, ed. A. W. Addison, W. R. Collen, D. Dolphin, pp. 71–111. New York: Wiley

12. Orme-Johnson, W. H., Hamilton, W. D., Ljones, T., Tso, M.-Y. W., Burris, R. H., Shak, V. K., Brill, W. J. 1972. *Proc. Natl. Acad. Sci. USA* 69:3142–45

13. Zumft, W. G., Palmer, G., Mortenson, L. E. 1973. *Biochim. Biophys. Acta* 292:413–21

14. Smith, B. E., Lowe, D. J., Bray, R. C. 1973. *Biochem. J.* 135:331–41

15. Münck, E., Rhodes, H., Orme-Johnson, W. H., Davis, L. C., Brill, W. J., Shah, V. K. 1975. *Biochim. Biophys. Acta* 400:32–53

16. Smith, B. E., Lang, G. 1974. *Biochem. J.* 137:169–80

17. Zumft, W. G., Mortenson, L. E., Palmer, G. 1974. *Eur. J. Biochem.* 46:525–35

17a. Haaker, H., De Kok, A., Veeger, C. 1974. *Biochim. Biophys. Acta* 357:344–57

18. Mortenson, L. E., Walker, M. N., Walker, G. A. 1976. *Proc. Int. Symp. Nitrogen Fixation, 1st* 1:117–49, ed. W. E. Newton, C. J. Nyman. Wash. State Univ. Press

19. Yates, M. G. 1972. *Eur. J. Biochem.* 29:386–92

20. Thorneley, R. N. F., Eady, R. R. 1973. *Biochem. J.* 133:405–8

21. Walker, G. A., Mortenson, L. E. 1974. *Biochemistry* 13:2382–88

22. Ljones, T., Burris, R. H. 1978. *Biochemistry* 17:1866–72

23. Ludden, P. W., Burris, R. H. 1976. *Science* 194:424–26

24. Nordlund, S., Eriksson, U., Baltscheffsky, H. 1977. *Biochim. Biophys. Acta* 462:187–95

25. Huang, J. C., Zumft, W. G., Mortenson, L. E. 1973. *J. Bacteriol.* 113:884–90

26. Eady, R. R., Smith, B. E., Cook, K. A., Postgate, J. R. 1972. *Biochem. J.* 128:655–75

27. Swisher, R. H., Landt, M. L., Reithel, F. J. 1977. *Biochem. J.* 163:427–32

28. Lundell, D. J., Howard, J. B. 1978. *J. Biol. Chem.* 253:3422–26

29. Huynh, B. H., Zimmerman, R., Zogg, J. A., Münck, E. 1978. *Proc. Int. Symp. Nitrogen Fixation, 3rd.* In press

30. Zumft, W. G., Mortenson, L. E. 1973. *Eur. J. Biochem.* 35:401–9

31. Cammack, R., Dickson, D. P. E., Johnson, C. E. 1977. See Ref. 3, 3:283–330

32. Shah, V. K., Brill, W. J. 1977. *Proc. Natl. Acad. Sci. USA* 74:3249–53

33. Rawlings, J., Shah, V. K., Chisnell, J. R., Brill, W. J., Zimmerman, R. J., Münck, E., Orme-Johnson, W. H. 1975. *J. Biol. Chem.* 4:1001–1104

34. Cramer, S. P., Hodgson, K. O., Gillum, W. O., Mortenson, L. E. 1978. *J. Am. Chem. Soc.* 100:3398–407

35. Cramer, S. P., Gillum, W. O., Hodgson, K. O., Mortenson, L. E., Stiefel, E. I., Chisnell, J. R., Brill, W. J., Shah, V. K. 1978. *J. Am. Chem. Soc.* 100:3814–19

36. Elliott, B. B., Mortenson, L. E. 1975. *J. Bacteriol.* 124:1295–301

37. Elliott, B. B., Mortenson, L. E. 1977. In *Proc. Int. Symp. Nitrogen Fixation*, ed. W. Newton, J. R. Postgate, C. Rodriguez-Barrueco, 2:205–17. London: Academic

38. Mortenson, L. E. 1978. *Curr. Top. Cell. Regul.* 13:179–232

39. Kincaid, B. M., Eisenberger, P., Hodgson, K. O., Doniach, S. 1975. *Proc. Natl. Acad. Sci. USA* 72:2340–42

40. Cramer, S. P., Eccles, T. K., Kutzler, F., Hodgson, K. O., Doniach, S. 1976. *J. Am. Chem. Soc.* 98:8059–69

41. Hodgson, K. O., Winick, H., Chu, G. 1976. In *Synchrotron Radiation Research.* Stanford Synchrotron Radiat. Res. Lab: Stanford, Calif.

42. Cramer, S. P., Hodgson, K. O., Stiefel, E. I., Newton, W. 1978. *J. Am. Chem. Soc.* 100:2748–61

43. Wolff, T. E., Berg, J. M., Warrick, C., Hodgson, K. O., Holm, R. H. 1978. *J. Am. Chem. Soc.* 100:4630–32

44. Walker, M., Mortenson, L. E. 1973. *J. Biol. Chem.* 249:6356–58

45. Walker, M., Mortenson, L. E. 1973. *Biochem. Biophys. Res. Commun.* 54:669–76

46. Rawlings, J., Henzl, M. T., Orme-Johnson, W. H. 1979. In *Proc. Int. Symp. Nitrogen Fixation.* 3:

47. Bui, P. T., Mortenson, L. E. 1968. *Proc. Natl. Acad. Sci. USA* 61:1021–27

48. Biggins, D. R., Kelly, M. 1970. *Biochim. Biophys. Acta* 205:288–99

49. Tso, M.-Y. W., Burris, R. H. 1973. *Biochim. Biophys. Acta* 309:263–70

50. Eady, R. R., Smith, B. E. 1978. In *Dinitrogen Fixation*, ed. R. W. F. Hardy, 3:399–490. New York: Wiley (Intersci.)

51. Thorneley, R. N. F. 1974. *Biochim. Biophys. Acta* 358:247–50

52. Bishop, E. O., Lambert, M. D., Orchard, D., Smith, B. E. 1977. *Biochim. Biophys. Acta* 482:286–300

53. Yates, M. G., Planque, K. 1975. *Eur. J. Biochem.* 60:467–76

54. Lowe, D. J. 1978. *Biochem. J.* 175:

55. Orme-Johnson, W. H., Davis, L. C., Henzl, M. T., Averill, B. A., Orme-Johnson, W. R., Münck, E., Zimmerman, R. 1977. See Ref. 37, pp. 131–78
56. Williams-Smith, D. L., Bray, R. C., Barber, M. J., Tsopanakis, A. D., Vincent, S. P. 1977. *Biochem. J.* 167:593–600
57. Yates, M. G., Thorneley, R. N. F., Lowe, D. J. 1975. *FEBS Lett.* 60:89–93
58. Eady, R. R., Smith, B. E., Thorneley, R. N. F., Yates, M. G., Postgate, J. R. 1975. In *Nitrogen Fixation by Free Living Microorganisms,* ed. W. D. P. Stewart pp. 377–88. London: Cambridge Univ. Press 471 pp.
59. Shethna, Y. I., Der Vartanian, D. V., Beinert, H. 1968. *Biochem. Biophys. Res. Commun.* 31:862–68
60. Haaker, H., Veeger, C. 1977. *Eur. J. Biochem.* 77:1–10
61. Scherings, G., Haaker, H., Veeger, C. 1977. *Eur. J. Biochem.* 77:621–30
62. Smith, B. E., Lowe, D. J., Bray, R. C. 1972. *Biochem. J.* 130:641–43
63. Mortenson, L. E., Zumft, W. G., Palmer, G. 1973. *Biochem. Biophys. Acta* 292:422–35
64. Thorneley, R. N. F. 1975. *Biochem. J.* 145:391–96
65. Thorneley, R. N. F., Yates, M. G., Lowe, D. J. 1976. *Biochem. J.* 155:137–44
66. Lambeth, D. O., Palmer, G. 1973. *J. Biol. Chem.* 248:6095–103
67. Creutz, C., Sutin, N. 1973. *Proc. Natl. Acad. Sci. USA* 70:1701–3
68. Ljones, T., Burris, R. H. 1978. *Biochem. Biophys. Res. Commun.* 80:22–25
69. Postgate, J. R., Eady, R. R., Lowe, D. J., Smith, B. E., Thorneley, R. N. F., Yates, M. G. 1978. In *Mechanisms of Oxidizing Enzymes,* ed. T. P. Singer, R. N. Ondarza, 1:173–80. New York: Elsevier/North Holland
70. Watt, G. D. 1977. See Ref. 37, pp. 179–90
71. Eady, R. R. 1973. *Biochem. J.* 135:531–35
72. Thorneley, R. N. F., Eady, R. R., Yates, M. G. 1975. *Biochim. Biophys. Acta* 403:269–84
73. Dalton, H., Morris, J. A., Ward, M. A., Mortenson, L. E. 1971. *Biochemistry* 10:2066–72
74. Thorneley, R. N. F., Eady, R. R., Smith, B. E., Lowe, D. J., Yates, M. G., O'Donnell, M. J., Postgate, J. R. 1978. In *Nitrogen Fixation of Plants,* ed. E. J. Hewitt, C. V. Cutting
75. Mortenson, L. E. 1964. *Proc. Natl. Acad. Sci. USA* 52:272–79
76. Bulen, W. A., Burns, R. C., LeCompte, J. R. 1965. *Proc. Natl. Acad. Sci. USA* 53:535–39
77. Burns, R. C., Bulen, W. A. 1965. *Biochim. Biophys. Acta* 105:437–45
78. Dilworth, M. J. 1966. *Biochim. Biophys. Acta* 127:285–94
79. Hardy, R. W. F., Knight, E. Jr. 1967. *Biochim. Biophys. Acta* 139:69–90
80. Hardy, R. W. F., Holsten, R. D., Jackson, E. K., Burns, R. C. 1968. *Plant Physiol.* 43:1185–207
81. Winter, H. C., Burris, R. H. 1968. *J. Biol. Chem.* 243:940–44
82. Yates, M. G. 1970. *FEBS Lett.* 8:281–85
83. Sorger, G. J. 1971. *Biochem. J.* 122:305–9
84. Shah, V. K., Davis, L. C., Brill, W. J. 1972. *Biochim. Biophys. Acta* 256:498–511
85. Burns, R. C., Hardy, R. W. F. 1975. In *Nitrogen Fixation by Free Living Micro-Organisms,* ed. W. D. P. Stewart, 1:437–66. London: Cambridge Univ. Press
86. Emerich, D. W., Burris, R. H. 1976. *Proc. Natl. Acad. Sci. USA* 73:4369–73
87. Tso, M.-Y. W., Ljones, T., Burris, R. H. 1972. *Biochim. Biophys. Acta* 267:600–4
88. Vandecasteele, J.-P., Burris, R. H. 1970. *J. Bacteriol.* 101:794–801
89. Mortenson, L. E., Zumft, W. G., Huang, T. C., Palmer, G. 1973. *Biochem. Soc. Trans.* 1:35–37
90. Bergersen, F. J., Turner, G. L. 1973. *Biochem. J.* 131:61–75
91. Ljones, T. 1973. *Biochim. Biophys. Acta* 321:103–13
92. Eady, R. R., Isaak, R., Kennedy, C., Postgate, J. R., Radcliffe, H. D. 1978. *J. Gen. Microbiol.* 104:277–85
93. Foust, G. P., Mayhew, S. G., Massey, V. 1969. *J. Biol. Chem.* 244:964–70
94. Smith, B. E., Eady, R. R., Thorneley, R. N. F., Yates, M. G., Postgate, J. R. 1977. See Ref. 37, pp. 191–203
95. Hageman, R. V., Burris, R. H. 1978. *Proc. Natl. Acad. Sci. USA* 75:2699–702
96. Thorneley, R. N. F., Cornish-Bowden, A. 1977. *Biochem. J.* 165:255–62
97. Bray, R. C., Lowe, D. J., Capeillère-Blandin, C., Fielden, E. M. 1973. *Trans. Biochem. Soc.* 1:1067–72
98. Ljones, T., Burris, R. H. 1972. *Biochim. Biophys. Acta* 275:93–101
99. Eady, R. R., Postgate, J. R. 1974. *Nature* 249:805–8
100. Jeng, D. Y., Morris, J. A., Mortenson, L. E. 1970. *J. Biol. Chem.* 245:2809–13
101. Hwang, J. C., Chen, C. H., Burris, R.

H. 1973. *Biochim. Biophys. Acta* 292:256–70
102. Rivera-Ortiz, J. M., Burris, R. H. 1975. *J. Bact.* 123:537–45
103. Archer, L. J., George, T. A. 1973. *J. Organometallic Chem.* 54:C25–C27
104. Hardy, R. W. F., Burns, R. C., Parshall, G. W. 1971. *Adv. Chem. Ser.* 100:219
105. Davis, L. C., Shah, V. K., Brill, W. J. 1975. *Biochim. Biophys. Acta* 384:353–59
106. Silverstein, R., Bulen, W. A. 1970. *Biochemistry* 9:3809–15
107. Thorneley, R. N. F., Eady, R. R. 1977. *Biochem. J.* 167:457–61
108. Frieden, C. 1970. *J. Biol. Chem.* 245:5788–99
109. Ainslie, G. R., Shill, J. P., Neet, K. E. 1972. *J. Biol. Chem.* 247:7088–96
110. Neet, K. E., Ainslie, G. R. 1976. *Trends in Biol. Sci.* 1(7):145–47
111. Smith, B. E., Thorneley, R. N. F., Eady, R. R., Mortenson, L. E. 1976. *Biochem. J.* 157:439–47
112. Watt, G. D., Bulen, W. A., Burns, A., Hadfield, K. L. 1975. *Biochemistry* 14:4266–72
113. Smith, B. E. 1977. In *The Potential Use of Isotopes in the Study of Biological Dinitrogen Fixation,* ed. R. Rennie, M. Fried. Vienna: Int. At. Energy Agency
114. Van Rossen, A. R., Peeters, J. F. M., Heremans, K. A. H. 1973. *Arch. Int. Physiol. Biochim.* 81:987
115. Moustafa, E., Mortenson, L. E. 1967. *Nature* 216:1241–42
116. Davis, L. C., Orme-Johnson, W. H. 1976. *Biochim. Biophys. Acta* 452:42–58
117. Haaker, H., De Kok, A., Veeger, C. 1974. *Biochim. Biophys. Acta* 357:344–57
118. Appleby, C. A., Turner, G. L., Macnicol, P. K. 1975. *Biochim. Biophys. Acta* 387:461–74
119. Hardy, R. W. F., Burns, R. C., Holsten, R. D. 1973. *Soil Biol. Biochem.* 5:47–81
120. Schubert, K. R., Evans, H. J. 1976. *Proc. Natl. Acad. Sci. USA* 73:1207–11
121. Schubert, K. R., Evans, H. J. 1977. See Ref. 37, pp. 469–85
122. Evans, H. J., Ruiz-Argueso, T., Jennings, N., Hanus, J. 1977. In *Genetic Engineering for Nitrogen Fixation,* ed. A. Hollaender, pp. 333–53. New York: Plenum
123. Ruiz-Argueso, T., Hanus, J., Evans, H. J. 1978. *Arch. Microbiol.* 116:113–18
124. Smith, L. A., Hill, S., Yates, M. G. 1976. *Nature* 262:209–10
125. Bothe, H., Tennigkeit, J., Eisbrenner, G., Yates, M. G. 1977. *Planta* 133:237–42
126. Bothe, H., Eisbrenner, G., Tennigkeit, J. 1977. *Arch. Microbiol.* 114:43–49
127. Peterson, R. B., Burris, R. H. 1978. *Arch. Microbiol.* 116:125–32
128. Lowe, D. J., Eady, R. R., Thorneley, R. N. F. 1978. *Biochem. J.* 173:277–90
129. Orme-Johnson, W. H., Davis, L. C., Henzl, M. T., Averill, B. A., Orme-Johnson, N. R., Münck, E., Zimmerman, R. 1977. See Ref. 37, pp. 131–78
130. DePamphilis, B. V., Averill, B. A., Herskovitz, T., Que, L., Holm, R. H. 1974. *J. Am. Chem. Soc.* 96:4159–67
131. Cambray, J., Lane, R. W., Wedd, A. G., Johnson, R. W., Holm, R. H. 1977. *Inorg. Chem.* 16:2565–71
132. Yates, M. G., Lowe, D. J. 1976. *FEBS Lett.* 72:121–26
133. Chatt, J. 1969. *Proc. R. Soc. London* 172:327–37
134. Burns, R. C., Hardy, R. W. F. 1975. In *Nitrogen Fixation in Bacteria and Higher Plants,* pp. 133–47. New York: Springer
135. Chatt, J. 1976. In *Biological Aspects of Inorganic Chemistry,* ed. A. W. Addison, W. R. Cullen, D. Dolphin, R. R. James, pp. 229–43. London: Wiley
136. Nikonova, L. A., Shilov, A. E. 1977. See Ref. 37, pp. 41–51
137. Sellman, D. 1977. See Ref. 37, pp. 53–67
138. Stiefel, E. I. 1977. See Ref. 37, pp. 69–108
139. Schrauzer, G. N. 1977. See Ref. 37, pp. 109–18
140. Newton, W. E., Bulen, W. A., Hadfield, K. L., Stiefel, E. I., Watt, G. D. 1977. See Ref. 37, pp. 119–30
141. Garcia-Rivera, J., Burris, R. H. 1967. *Arch. Biochem. Biophys.* 119:167–72
142. Burris, R. H., Winter, H. C., Munson, T. O., Garcia-Rivera, J. 1965. In *Non-Heme Iron Proteins,* ed. A. San Pietro, pp. 315–21. Yellow Springs, Ohio: Antioch Press
143. Bulen, W. A. 1976. See Ref. 18, pp. 177–86
144. Thorneley, R. N. F., Eady, R. R., Lowe, D. J. 1978. *Nature* 272:557–58
145. Chatt, J., Pearman, A. J., Richards, R. L. 1975. *Nature* 253:39–40
146. Chatt, J., Pearman, A. J., Richards, R. L. 1977. *J. Chem. Soc. Dalton Trans.* 1852–60
147. Thorneley, R. N. F., Chatt, J., Eady, R. R., Lowe, D. J., O'Donnell, M. J., Postgate, J. R., Richards, R. L., Smith, B. E. 1979. *Proc. Int. Symp. Nitrogen Fixation,* 4:

Ann. Rev. Biochem. 1979. 48:419–41
Copyright © 1979 by Annual Reviews Inc. All rights reserved

MITOCHONDRIAL GENES AND TRANSLATION PRODUCTS

♦12014

Alexander Tzagoloff and Giuseppe Macino

Department of Biological Sciences, Columbia University,
New York, New York 10027

Walter Sebald

Institute of Biotechnical Research, Braunschweig-Stöckheim, West Germany

CONTENTS

419

0066-4154/79/0701-0419$01.00

INTRODUCTION

As a sequel to the discovery that mitochondria contain an independent system of protein synthesis (1–3), many efforts have been devoted toward identifying the products formed and understanding their role in the morphogenesis of the organelle. Since mitochondria synthesize relatively few proteins, this work has progressed rapidly and there is presently a substantial body of knowledge about the synthesis, genetic origin, and, in some instances, the functions of this interesting class of proteins. This review summarizes some of the information that has emerged from studies of the past ten years and speculates on the future course of research in this area.

MITOCHONDRIAL TRANSLATION PRODUCTS

In Vivo Studies

Some of the earliest evidence, particularly from work done with yeast, indicated that the proteins synthesized in mitochondria were likely to be components of the terminal respiratory pathway. For example, it was known that in yeast, chloramphenicol and other selective inhibitors of mitochondrial ribosomes, blocked the development of a functional cytochrome system but had no pronounced effect on most of the other enzymatic machinery of the organelle (4, 5). A similar phenotype was found in cytoplasmic "petite" mutants of yeast (6) which are deficient in mitochondrial protein synthesis due to the loss of genetic information coding for mitochondrial ribosomal and transfer RNAs (7–9).

Based on these observations, a number of laboratories began to systematically study the relationship of mitochondrial translation products to the enzymes that were known to catalyze electron transport (10). These included the NADH- and succinate-coenzyme QH_2-reductase (10–12), coenzyme QH_2-cytochrome c reductase (13), and cytochrome oxidase (14). In addition to the four respiratory complexes, the oligomycin-sensitive ATPase which normally functions as an ATP synthetase (15, 16) was also suspected to contain mitochondrially derived polypeptides since only one part of this important inner membrane complex was synthesized in petite mutants or in yeast grown in the presence of chloramphenicol (17, 18).

The biosynthesis of the respiratory and ATPase complexes was first examined in the yeast *Saccharomyces cerevisiae* and in *Neurospora crassa*, each organism offering distinct experimental advantages for such studies. Several circumstances facilitated the translational sites of the subunit polypeptides to be established by rather straightforward in vivo labeling experiments.

1. Total cellular proteins could be labeled with high specific activity when cells were incubated in the presence of a radioactive precursor.
2. Antibiotics were available which selectively inhibited either mitochondrial or cytoplasmic protein synthesis.
3. The successful purification of the ATPase and some of the respiratory complexes of yeast and *Neurospora* made it possible to characterize their subunit compositions and at the same time provided the antibodies necessary for rapid isolation of the enzymes from small samples of cells.

In most studies, the protocol used to identify the mitochondrially translated subunits consisted of labeling cells with a radioactive amino acid in the presence of either an inhibitor of mitochondrial or cytoplasmic protein synthesis. Subsequent purification of the enzymes and analysis of the distribution of radioactivity in the various subunit polypeptides under each set of labeling conditions allowed their translation to be ascribed to either one of the two protein synthesizing systems.

When *S. cerevisiae*, *N. crassa*, or mammalian cells are labeled in the presence of cycloheximide, 8–10 prominent radioactive protein bands are seen in polyacrylamide gels of mitochondria solubilized with sodium dodecyl sulfate (19, 20). Minor bands are also occasionally observed but they most likely arise from incomplete inhibition of cytoplasmic protein synthesis. Similar results have been obtained in the absence of inhibitors by using a conditional mutant of yeast in which protein synthesis on cytoplasmic ribosomes is suppressed at the nonpermissive temperature (21) or by labeling wild type cells with fMet. Mitochondrial products can be selectively labeled with fMet since it is used as the initiator and the formyl group is retained in the completed polypeptide chain due to the absence of a deformylase (22). Most of the major mitochondrial translation products of yeast and *Neurospora* have been ascertained to be subunit polypeptides of the respiratory and ATPase complexes. The one exception is the protein designated as var 1. This mitochondrial product exhibits molecular weight variant forms (40,000–43,000 daltons) in different strains of yeast (23). The function of var 1 is still obscure but there are some indications that it might be a component of mitochondrial ribosomes (24).

Identity of Mitochondrial Products

The information currently available on the synthesis of the subunits of coenzyme QH_2-cytochrome *c* reductase, cytochrome oxidase, and ATPase is summarized below. Since both NADH- and succinate-coenzyme Q-reductases are synthesized exclusively on cytoplasmic ribosomes (25–28), relatively few studies have been devoted to these complexes and they are not further considered here.

CYTOCHROME OXIDASE: Cytochrome oxidase is composed of seven nonidentical subunit polypeptides with molecular weights ranging from 40,000–9,000 (29–32). The functional groups of the native enzyme are heme a (cytochromes a and a_3) and copper. Following depolymerization of cytochrome oxidase with denaturing agents, heme a has been found to copurify with subunit 1 (40,000 daltons) (33) and subunit 7 (9,000 daltons) (34) which suggests that they are the hemoprotein components of the complex. In general, however, the precise relationship of the redox carriers to the enzyme subunits is still not agreed upon.

The first demonstration that cytochrome oxidase is synthesized jointly on mitochondrial and cytoplasmic ribosomes came from studies on $N.$ $crassa$ (29). It was conclusively established from in vivo labeling experiments that the three largest polypeptides (subunits 1–3) of the enzyme are translated in mitochondria and that the low molecular weight polypeptides (subunits 4–7) are made externally on cytoplasmic ribosomes (35, 36). This was later confirmed in $S.$ $cerevisiae$ (37, 38) and $Xenopus$ $laevis$ (39) and appears to be an invariant feature of the mechanism of biosynthesis of cytochrome oxidase.

COENZYME QH$_2$-CYTOCHROME REDUCTASE   This intermediary respiratory complex was first purified and characterized from bovine heart mitochondria (13). Homogeneous preparations of coenzyme QH$_2$-cytochrome c reductase have recently also been obtained from yeast (40) and $N.$ $crassa$ (41). It is composed of seven or eight different subunits not all of whose functions are known. The best understood components of the enzyme are cytochromes b, c_1, and a nonheme iron protein. Each has been shown to be an obligatory electron carrier in the catalysis of cytochrome c reduction (13, 42). Weiss (43, 44) found that in $N.$ $crassa,$ cytochrome b is synthesized in mitochondria. Although it was originally thought that there are two different cytochrome b apoproteins (45), more recent genetic and biochemical evidence suggests the presence of only one chemical species of this cytochrome (46, 47). Cytochrome b has also been shown to be synthesized by yeast mitochondria (48) and in fact appears to be the only subunit of coenzyme QH$_2$-cytochrome c reductase that has a mitochondrial origin—all the other subunits, including cytochrome c_1 (49), are synthesized in the cytoplasm. There are no reports at present that deal with the biosynthesis of this complex in higher animal and plant cells.

OLIGOMYCIN-SENSITIVE ATPase   This is probably the most intricate enzyme of the mitochondrial inner membrane. It is concerned not only with the coupling of the energy of oxidation to ATP synthesis but also with the utilization of the energy of hydrolysis of ATP for various energy-dependent functions such as ion transport, pyridine nucleotide transhydrogenation, and reverse electron flow (15, 16, 50). This multiplicity of functions is reflected in the large number of polypeptides that have been attributed to the ATPase. The yeast complex consists of at least ten different protein subunits (51) none of whose functions can be clearly stated at present.

The oligomycin-sensitive ATPase complexes of bovine heart and yeast mitochondria have been dissected into three fractions that can spontaneously reassociate to form the native enzyme (52, 53). The three reconstitutively active fractions are F_1 (54), OSCP (55), and a set of hydrophobic proteins referred to as the membrane factor (52). F_1 is a water-soluble polymeric protein with a molecular weight of 340,000 (56). It is comprised of five distinct subunit polypeptides and catalyzes the hydrolysis of ATP— this hydrolytic reaction, however, is not inhibited by oligomycin and other potent inhibitors of the native complex (16, 54). In yeast (18), $N.$ $crassa$ (57, 58), and $X.$ $laevis$ (39), the five subunits of F_1 are synthesized in the cytoplasm. It is interesting that chloroplast F_1 which is structurally and functionally related to the mitochondrial F_1, contains some subunits that are made on chloroplast ribosomes (59).

OSCP is a water-soluble protein (17,000 daltons) that has been postulated to act as a link between F_1 and the membrane factor (60, 61). This protein has been shown to be synthesized on cytoplasmic ribosomes in yeast (61). A cytoplasmic origin may also be assumed for the OSCP component of other mitochondrial ATPases, although this has not been experimentally verified.

The membrane factor is composed of hydrophobic proteins that are highly insoluble in water. Unlike F_1 and OSCP which are peripherally bound to the inner membrane, the polypeptides of the membrane factor are intrinsic proteins that are lodged in the lipid bilayer (62). The biosynthesis of the membrane factor has been studied in yeast, $N.$ $crassa,$ and $X.$ $laevis$ and intriguing differences have been noted. The ATPase of $S.$ $cerevisiae$ contains four mitochondrially translated subunits which correspond to the membrane factor (63). In $N.$ $crassa,$ only two of the ATPase subunits have been found to originate in mitochondria (57, 58). The best evidence concerns the DCCD-binding or proteolipid subunit of the membrane factor. This is a low molecular weight protein (7,800 daltons) that has been clearly demonstrated to be a mitochondrial product in yeast (64, 65) and a cytoplasmic product in $N.$ $crassa$ (66). The ATPase of $X.$ $laevis$ has been reported to have three mitochondrially synthesized subunits (39). The relationship of the mitochondrial products of this enzyme to those of $S.$ $cerevisiae$ and $N.$ $crassa,$ however, has not been established (58).

Regulation of Synthesis of Mitochondrial Products

Normally mitochondrial and cytoplasmic products are continuously synthesized and used for the assembly of the larger enzyme entities. In order for this process to be efficient, the two sets of proteins must be produced in a stoichiometric fashion, which implies the existence of some sort of coordinate regulation of the two translational systems (for review see 20). In fact there is good evidence that the rate and extent of mitochondrial protein synthesis are influenced by cytoplasmic translation products. When yeast is inhibited with cycloheximide, mitochondrial protein synthesis proceeds at low rates and for only short duration (67). The rate of synthesis in the presence of cycloheximide is substantially increased if the cells are first incubated in chloramphenicol (67, 68). Presumably, the preincubation leads to an accumulation of cytoribosomal products which stimulate mitochondrial translation upon shifting to cycloheximide.

Cytoplasmic factors have also been observed to affect mitochondrial protein synthesis in vitro. Poyton & Kavanagh (69) have reported that the synthesis of cytochrome oxidase subunits in isolated mitochondria is stimulated by a crude postribosomal fraction of yeast and that the stimulation is proportional to the amount of supernatant added. More significantly, the effect was abolished when antibodies were used to adsorb cytochrome oxidase specific proteins from the cytosolic fraction (69). These experiments suggest that the synthesis of mitochondrial subunits of cytochrome oxidase depends on the presence of their cytoplasmically made counterparts. It is not unreasonable that the synthesis of mitochondrial translation products is tightly coupled to their integration with cytoplasmic subunits and that interruption of the assembly sequence due to the absence of one or more proteins may suppress further synthesis on mitochondrial ribosomes. The trivial explanation, namely that unassembled mitochondrial subunits are more susceptible to degradative enzymes, cannot be excluded however.

In addition to the biochemical evidence there are indications from genetic data of the involvement of cytoplasmic products in the regulation of mitochondrial protein synthesis. Ebner et al (70) have isolated nuclear mutants of S. cerevisiae in which the synthesis of just one of the three mitochondrial products of cytochrome oxidase is shut off. Similarly, other nuclear mutations have been reported to specifically block the synthesis of cytochrome b or of the ATPase proteolipid without affecting the synthesis of the other mitochondrial products (71).

Although there is abundant evidence that mitochondrial translation and/ or transcription are controlled by the nucleocytoplasmic system of the cell (20, 72), the reverse does not appear to hold true. Most of the cytoplasmic products, including those that are destined to be integrated into enzymes containing mitochondrially derived subunits, are synthesized at near normal levels even when mitochondrial synthesis is inhibited. This is best seen in petite mutants of yeast that are totally unable to elaborate any mitochondrial products. Such mutants, nonetheless, have respiratory-deficient mitochondrial organelles which, with the exception of the mitochondrial products, have a protein composition identical to those of wild-type yeast (73, 74).

Properties of Mitochondrial Products

Mitochondrial translation products represent some of the most hydrophobic proteins of the inner membrane. They are highly insoluble in water and tend to form large aggregates that can only be dispersed by means of powerful ionic detergents. These properties have been attributed to their high content of nonpolar amino acids (32, 33, 64, 75, 76). The mitochondrially synthesized subunits of cytochrome oxidase, for example, average 10–15% more nonpolar residues than do the other four subunits of the enzyme (32, 33, 75, 76). The ATPase proteolipid is an even more extreme case, having only 23% polar amino acids (64)—this is to be contrasted with values of 50–60% observed in most water soluble proteins (77).

It has been speculated that the hydrophobic character of certain mitochondrial components necessitated an intraorganellar site of synthesis, thereby avoiding the logistic problem of transporting highly insoluble proteins through the aqueous cytosolic phase. This interpretation, however, is not entirely satisfactory in view of the fact that many other hydrophobic proteins of mitochondria are known to be synthesized on cytoplasmic ribosomes and subsequently transported by still undefined mechanisms to the interior of the organelle. The ATPase proteolipid of N. crassa is a case in point. Even though the amino acid compositions and primary structures of the yeast and N. crassa proteolipids are very similar (78), they have been shown to have different synthetic origins (58). The yeast proteolipid is a well established mitochondrial product (64)—in N. crassa, the same protein has been shown to be synthesized in the cytoplasm (57, 65). The hydrophobic properties of this component, therefore, did not interfere with the evolutionary process of transfer of a mitochondrial gene to the nucleus and concomitantly of its transcription and translation in the cytoplasm.

The notion that mitochondria are concerned with the synthesis of proteins with structural functions has not been borne out by the existent evidence. Both cytochrome oxidase (33) and coenzyme QH_2-cytochrome c reductase (43, 44) contain at least one hemoprotein carrier synthesized in mitochondria. Our inability to assign functions to the other mitochondrial products arises from a general lack of knowledge about the functions of the different subunit polypeptides of the respiratory and ATPase complexes.

In Vitro Synthesis of Mitochondrial Products

Initial attempts to identify the proteins synthesized by isolated mitochondria were hampered by poor incorporation of radioactive precursors and technical difficulties in purifying the labeled products from small quantities of mitochondria. These problems have been largely overcome and it is now generally recognized that mitochondria synthesize completed proteins that are identical to the in vivo products. Poyton & Groot (79) have shown that isolated yeast mitochondria are capable of synthesizing the three large subunits of cytochrome oxidase. The in vitro products were integrated into a larger complex which could be precipitated with antisera to some of the smaller cytoplasmically made subunits. Presumably there is a sufficient endogenous pool of the cytoplasmic subunits to allow post-translational assembly of cytochrome oxidase. Although such experiments have not been extended to coenzyme QH_2-cytochrome c reductase and the ATPase complex, there is substantial evidence from studies with yeast and mammalian cells (80, 81) that the relative proportions and electrophoretic mobilities of the major mitochondrial products formed in vitro and in vivo are very similar.

Mitochondrial gene products have also been synthesized using poly(A)-RNA purified from yeast mitochondria and translated in an *E. coli* or wheat germ ribosomal system. Padmanaban et al (82) found that total mitochondrial poly(A) containing messenger RNA gave a fourfold stimulation of [³H] leucine incorporation in a cell-free ribosomal system. Some of the proteins made under these in vitro conditions cross-reacted with cytochrome oxidase antibodies and comigrated with the authentic mitochondrially synthesized subunits of the enzyme (82).

A number of laboratories have tried to develop coupled transcription-translation systems to study the gene products of mitochondrial DNA. While there have been some reports of the synthesis of immunochemically reactive cytochrome oxidase gene products when mitochondrial DNA was transcribed and translated in an *E. coli* system (83), in other studies the products formed were of low molecular weight and could not be related to any bona fide mitochondrial proteins (84). In view of recent advances in mitochondrial genetics and the tentative identification of many of the genes on mitochondrial DNA, the usefulness of cell free transcription-translation studies has been somewhat lessened.

MITOCHONDRIAL GENES

Mitochondria contain circular duplex DNA with a molecular weight ranging from 1×10^7– 5×10^7, depending on the organism. The larger genome size is more prevalent among fungi, protists, and higher plants. Mammalian mitochondria tend to have smaller DNAs that are generally 1×10^7 daltons (9).

Mit and Syn Genes

Most of our current knowledge of mitochondrial genes has come from studies of *S. cerevisiae* which has proven to be especially suitable for genetic analysis (85). The DNA of this yeast has a molecular weight of 5×10^7 which is equivalent to about 75 kilobases (9). The first genes to be recognized in the yeast genome coded for components of the mitochondrial protein synthetic machinery (9, 86). These have been designated as *syn* genes and they include the transfer and the ribosomal RNAs of mitochondria (9, 86, 87). At least 30 different mitochondrial tRNAs have been detected in *S. cerevisiae* (88). The major tRNA species of this yeast, representing the twenty common amino acids, all have been shown by hybridization to be gene products of mitochondrial DNA (88–91). The total number of isoaccepting species is compatible with the Wobble hypothesis which suggests that in yeast there is no need for importation of cytoplasmic tRNAs into mitochondria. This is probably also true of *N. crassa* (92, 93) but not of animal mitochondria (94–95). For example, most of the tRNAs present in *Tetrahymena* mitochondria are transcribed from nuclear DNA; the mitochondrially encoded species are confined to the four amino acids, leucine, phenylalanine, tryptophan, and tyrosine (96). In at least some organisms, therefore, mitochondria make use of cytoplasmic tRNAs.

In contrast to the RNA components, aminoacyl synthetases, protein synthesis initiation factors, and most of the ribosomal proteins have been shown to be nuclear gene products that are synthesized on cytoplasmic ribosomes (9). There is some evidence, however, that at least one ribosomal protein may be a mitochondrial product. The classical "poky" mutation in *N. crassa* has been correlated with the absence of a specific protein of the small ribosomal subunit (97). In *S. cerevisiae,* the mitochondrial gene product, var 1, has also been found to be a component of the small ribosomal subunit (24).

The second important class of mitochondrial genes are the mit genes which code for proteins that function in electron transport and oxidative phosphorylation. The existence of this class of genes has been suspected for a long time but was only recently established as the result of the isolation of new antibiotic-resistant mutants (98–100) and mit⁻ mutants (101–104) of *S. cerevisiae.* The mit⁻ mutants in particular have been instrumental in showing that many of the proteins synthesized by mitochondria are also gene products of mitochondrial DNA. To date, mit⁻ mutations have been found to affect the three inner membrane complexes, cytochrome oxidase (101–104), coenzyme QH_2-cytochrome *c* reductase (101–104), and the oligomycin-sensitive ATPase (105, 106). Based on genetic analyses of a large number of different mit⁻ strains, six complementation groups have been found (107) (Table 1). Three of the complementation groups (*oxi* 1, *oxi* 2, and *oxi* 3) code for products that have been tentatively identified to be the three mitochondrially synthesized subunits of cytochrome oxidases (108–109). Mutations that map in the *cob* complementation group are deficient in cytochrome *b* and have been convincingly shown to be in the structural of this respiratory carrier (46, 110). One group of ATPase-deficient mutants (*pho* 2) is now known to have lesions in the proteolipid component of the ATPase (106, 111). This gene also contains the *oil* 1 and *oil* 3 resistance loci (111). The gene products of the other ATPase mutants (*pho* 1) has not yet been identified. Another mitochondrial translation product known to be encoded in mitochondrial DNA is the var 1 protein (23, 112).

Although our information concerning mitochondrial genes is still fragmentary, it is nonetheless evident that of the nine proteins known to be synthesized in mitochondria, six have been documented to be specified by mitochondrial DNA.

Map of Mitochondrial DNA

All of the known mit⁻ and antibiotic resistance markers of the mitochondrial genome of *S. cerevisiae* have now been localized on the circular map by a combination of genetic and physical mapping methods. Two approaches have been used in genetic mapping. 1. Co-retention and co-deletion of the mutated alleles in the DNA of ρ^- mutants (102, 113, 114). This type of analysis allows mutations to be positioned relative to each other on the wild-type genome and at the same time places limits on the retained segments of DNA in the ρ^- mutants. 2. In some instance mutations have been mapped relative to each other and map distances obtained by recombinational analysis in two and three factorial crosses (46, 102).

Various methods have been devised to locate mitochondrial genes on the physical map. Ribosomal and tRNA genes have been mapped by hybridiza-

Table 1   Complementation groups and gene products of the yeast genome

Enzyme deficiency	Complementation group	Gene product
Cytochrome oxidase	*oxi* 1	subunit 2
Cytochrome oxidase	*oxi* 2	subunit 3
Cytochrome oxidase	*oxi* 3	subunit 1
Coenzyme QH$_2$-cytochrome *c* reductase	*cob*	cytochrome *b*
ATPase	*pho* 1	?
ATPase	*pho* 2	subunit 9 (proteolipid)

tion to restriction fragments generated from wild-type (115, 116) or ρ^- DNAs (90). A number of mit$^-$ and antibiotic resistance markers have been localized by restriction analysis of genetically marked mit$^-$ and ρ^- clones (117–119). In general, the results obtained by genetic and physical methods have been in good agreement and have provided a fairly unambiguous map of the mitochondrial genome of *S. cerevisiae*. The map presented in Figure 1 shows the positions assigned to the known syn and mit genes as well as the antibiotic resistance markers, most of which have now been related to specific gene products. The map of the yeast genome reveals a considerable dispersity of genes that code for related functions. The three structural genes of cytochrome oxidase are separated by intervening regions containing syn and mit genes. The ribosomal RNA, tRNA, and ATPase genes are also scattered in different regions of the genome. This organizational feature argues against a coordinately controlled expression of genes by means of polycistronic messengers. By analogy with eucaryotic genes, those of the mitochondrion appear to be transcribed individually.

Saturation of the Genome

How inclusive is our present information on the genetic content of the *S. cerevisiae* mitochondrial genome? The evidence that most of the mit type of genes are now known is quite compelling. This is attested to by the fact that the more than a thousand mit$^-$ strains isolated and studied in different laboratories all tend to fall into a limited number of genetic loci that compromise some six complementation groups. It must be kept in mind, however, that the mit$^-$ class of mutants is selected on the basis of its inability to grow on nonfermentable substrates (101, 103, 104). Since this phenotype is most likely to result from mutations in gene products directly involved in respiratory functions (namely electron transport carriers), the selection procedure may exclude mutations that alter mitochondrial functions in more subtle ways. Mutations in regulatory genes that might be expected to reduce or enhance the transcription of a structural gene without necessarily abolishing the capacity for respiration entirely, would not necessarily be recognized by the selection procedures used.

There are also gaps that need to be filled in regard to syn genes. While it is true that in *S. cerevisiae* the major tRNAs have been shown to be transcribed from mitochondrial DNA, there are many isoaccepting species present in mitochondria whose genetic origin and function have not been clarified. Nor is it known whether mitochondria code for protein factors that participate in or regulate mitochondrial protein synthesis. For example, the paramomycin resistance marker, formerly thought to be in the 16S ribosomal RNA gene, has recently been shown to map in a restriction fragment that is adjacent to, but distinct from, the RNA gene (116). This

raises the possibility that paramomycin acts on some yet unknown component of the translation machinery. These and related questions are most
likely to be answered when alternative selection methods are devised for the
isolation of new types of mitochondrial mutants.

Constancy of the Mitochondrial Genome

The map of the *S. cerevisiae* genome shown in Figure 1 is based on studies
of different laboratory strains of this yeast. All the strains examined appear
to have the same composition and relative position of genes. This is not to
say, however, that intra-species differences do not exist. Both genetic and
physical mapping data indicate considerable heterogeneity of the DNA
among yeast strains. Detailed physical maps of mitochondrial DNA have
now been obtained for at least four different strains. Based on the analysis
of HpaII and HaeIII digests, Prunell & Bernardi (120) have concluded that
there are frequent strain-specific insertions of deletions of DNA in different
parts of the genome. Insertions of up to 3,000 base pairs have been noted
in the segment of DNA included between the *oli* 1 and *par* 1 resistance loci
(121). This region of the genome seems to be the most variable in the strains
studied so far (121).

Genetic data also point to considerable differences in gene spacer regions
and even in the internal organization of genes. This is particularly evident
in the cytochrome *b* (or *cob* region). In some strains mutations in cytochrome *b* are genetically linked to the *oli* 1 resistance marker—in other

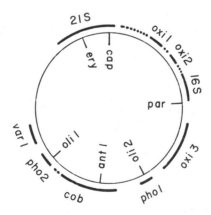

Figure 1 Map of the mitochondrial genome of *S. cerevisiae*. The following designations are
used for the antibiotic resistance loci: *cap*, chloramphenicol; *ery*, erythromycin; *par*, paromo-
mycin; *ant*, antimycin; *oli*, oligomycin. The *mit* complementation groups of Table 1 are
indicated by the heavy lines. The 16S and 21S refer to the two ribosomal RNA genes and var
1 to the 40,000–43,000 dalton mitochondrial products studied by Butow and co-workers (23,
112). The approximate positions of the tRNA genes are indicated by dots.

strains they are completely unlinked, which suggests that the spacer between the ATPase proteolipid and cytochrome *b* genes varies in length
depending on the strain (46). Even more intriguing discrepancies have been
found in the properties of cytochrome *b* mutants of different strains. Thus,
in some strains of *S. cerevisiae,* cytochrome *b* mutations fall into two
genetically unlinked clusters or loci (*cob* 1 and *cob* 2) which behave as a
single complementation group (46, 107). In other strains, the cytochrome
b complementation group has been found to consist of six unlinked loci
(*box* 1–6) (122)—many of the *cob* 1 and *cob* 2 mutations are nonetheless
allelic to mutations in the various *box* loci (100). These observations are
most simply explained by assuming the existence of insertions within the
structural gene of cytochrome *b*. The genetic data indicate that the number
and length of inserted sequences in the cytochrome *b* gene may be strain-
dependent.

When mitochondrial DNA of *S. cerevisiae* is compared to other organisms, the divergences are much more profound. Aside from differences in
physical size, buoyant density, and restriction maps, there is also considerable variation in the genetic content of mitochondrial DNA. At present
such information is limited almost entirely to the ribosomal and transfer
RNA genes. In *N. crassa* the ribosomal RNAs are transcribed as a single
32S precursor (123) and hybridization of the mature RNAs to restriction
fragments indicates that they have a proximal location on the genome (93).
This also appears to be true of most animal mitochondrial genomes (124,
125). There are equally significant differences in the number of tRNA genes.
In *S. cerevisiae* and *N. crassa,* tRNA species corresponding to the twenty
amino acids have been shown to be transcribed from mitochondrial DNA
(88–91, 93). Animal mitochondria, however, have fewer tRNA genes (86,
87, 95). It was already pointed out that in *Tetrahymena* isoacceptors for
only four amino acids are capable of hybridizing to mitochondrial DNA
(96). In mammalian mitochondria, the number of hybridizable species is
larger but again some of the organellar tRNAs appear to be imported from
the cytoplasm. Attardi and co-workers have reported that in HeLa cells the
tRNAs for twelve amino acids hybridize to the heavy strand of mitochondrial DNA, five other amino acids hybridize to the light strand, and four
fail to hybridize altogether. Both strands of this genome are therefore
transcribed (126). It is not known if this is also true for mitochondrial DNA
of *S. cerevisiae.*

ATPase PROTEOLIPID

The proteolipid of the yeast ATPase is presently the best understood mitochondrial gene and translation product. This protein is particularly interest-

ing since there is now very good evidence that in *N. crassa* it is coded by a nuclear gene (127) and is translated on cytoplasmic ribosomes (58). The transfer of a gene from the mitochondrion to the nucleus (or vice versa) represents an important change in the cell which might be experimentally exploited to test some of the current theories of mitochondrial evolution (128, 129).

DCCD-Binding Properties

Dicyclohexylcarbodiimide (DCCD) is a potent inhibitor of oxidative phosphorylation and of the mitochondrial ATPase (130). Cattell et al (131) showed that DCCD reacts covalently with a low molecular weight proteolipid of beef heart mitochondria which could be extracted with chloroform: methanol and purified on silicic acid. The proteolipid was named the DCCD-binding protein and was later shown to be a component of the oligomycin-sensitive ATPase (58, 132). The DCCD-binding protein has also been found to be present in the yeast ATPase (65). The yeast protein is one of the mitochondrially synthesized subunits (subunit 9) of the ATPase (63, 64). It is interesting that there are approximately six molecules of the protein per enzyme and these form a hexameric complex that resists depolymerization by SDS and other detergents (58, 63). The hexamer, however, is dissociated by organic solvents and by strong alkali or base (64). Although the DCCD-binding protein has also been demonstrated to be a component of the *N. crassa* ATPase, it is synthesized in this organism on cytoplasmic rather than mitochondrial ribosomes. Jackl & Sebald (57) have shown that the *N. crassa* proteolipid becomes labeled with a radioactive amino acid in the presence of chloramphenicol but not cycloheximide. Similar to the yeast enzyme, the *N. crassa* ATPase appears to have six proteolipid subunits but a stable hexameric complex has not been detected (57, 58).

Function

In addition to binding DCCD, the proteolipid probably also contains the oligomycin binding site of the ATPase. Criddle et al (133) have postulated that one of the keto groups of oligomycin may form a Schiff base with an amino group on the protein. When the yeast ATPase was reacted with oligomycin and reductively alkylated with [^{3}H] sodium borohydride, the radioactive label was associated exclusively with a protein that was soluble in chloroform: methanol and whose electrophoretic migration on SDS polyacrylamide gels was identical to the proteolipid. Since ATPase preparations obtained from oligomycin-resistant mutants failed to incorporate the radioactive label into the proteolipid, these authors concluded that the oligomycin-binding site of the ATPase resides in the proteolipid (133).

Although it is reasonable to conclude that the binding sites for DCCD, oligomycin, and perhaps other inhibitors of the ATPase are present in the proteolipid, the precise function of this protein is still obscure. There is some recent evidence, however, suggesting that the proteolipid may act as a protonophore. It has been reported that the yeast proteolipid increases the permeability of artificial lipid membranes to protons (134). These experiments were especially interesting in view of the fact that proton transport in this model system was inhibited by oligomycin when the source of proteolipid was a wild-type oligomycin-sensitive strain of yeast. When the protein was obtained from an oligomycin-resistant mutant, the sensitivity to the inhibitor was decreased. Whether the proteolipid acts as a proton carrier or channel needs to be studied further but the above results are consistent with the earlier suggestion that oligomycin inhibits oxidative phosphorylation by blocking the discharge of protons generated during electron transport across the inner membrane (135).

Amino Acid Sequence

It is possible to obtain high yields of pure ATPase proteolipid directly from mitochondria by a relatively simple procedure involving extraction with chloroform : methanol and thin layer chromatography. This fact, combined with its small size, have made it possible to sequence the ATPase proteolipids of yeast and other types of mitochondria (111, 136). The primary structure of the various proteolipids was established by solid phase sequencing of the total polypeptide and of cyanogen bromide fragments.

The primary structures of the proteolipids from *S. cerevisiae* and *N. crassa* are shown in Figure 2. Both proteins are extremely hydrophobic as was already evident from their amino acid compositions. Even though the yeast and *N. crassa* proteins are synthesized in two different compartments of the cell, they show a great deal of sequence homology—this is especially evident in the amino acid sequences spanning residues 17–41 and 51–68. It is also significant that DCCD has been found to bind covalently to the glutamic acid residues that occur at positions 59 and 64 in the yeast and *N. crassa* proteolipids, respectively. The sequence homologies and identity of the DCCD-binding residues argue strongly in favor of a common genetic derivation of the two proteins. It is also interesting to note that the *N. crassa* proteolipid contains five extra amino acids at the amino terminus and has a tyrosine instead of a formyl methionine as the amino terminal residue. The biosynthetic evidence on the sites of translation of the two proteins is thus nicely complemented by the sequence data. Since mitochondria utilize formyl methione as the initiator (22), the occurrence of this amino acid in the yeast but not the *N. crassa* protein is consistent with their proposed synthetic origins (58).

```
                    5                  10                 15
     f -Met-Gln-Leu-Val-Leu-Ala-Ala-Lys-Tyr-Ile-Gly-Ala-Gly-Ile-Ser-Thr-Ile-Gly-Leu-

         5                 10                 15                 20                 25
Tyr-Ser-Ser-Glu-Ile-Ala-Gln-Ala-Met-Val-Glu-Val-Ser-Lys-Asn-Leu-Gly-Met-Gly-Ser-Ala-Ala-Ile-Gly-Leu-


    20                 25                 30                 35                 40
Leu-Gly-Ala-Gly-Ile-Gly-Ile-Ala-Ile-Val-Phe-Ala-Ala-Leu-Ile-Asn-Gly-Val-Ser-Arg-Asn-Pro-Ser-Ile-Lys-

       30                 35                 40                 45                 50
Thr-Gly-Ala-Gly-Ile-Gly-Ile-Gly-Leu-Val-Phe-Ala-Ala-Leu-Leu-Asn-Gly-Val-Ala-Arg-Asn-Pro-Ala-Leu-Arg-


    45                 50                 55                 60                 65
Asp-Thr-Val-Phe-Pro-Met-Ala-Ile-Leu-Gly-Phe-Ala-Leu-Ser-Glu-Ala-Thr-Gly-Leu-Phe-Cys-Leu-Met-Val-Ser-

         55                 60                 65                 70                 75
Gly-Glu-Leu-Phe-Ser-Tyr-Ala-Ile-Leu-Gly-Phe-Ala-Phe-Val-Glu-Ala-Ile-Gly-Leu-Phe-Asp-Leu-Met-Val-Ala-


    70                 75
Phe-Leu-Leu-Leu-Phe-Gly-Val

                   80
Leu-Met-Ala-Lys-Phe-Thr
```

Figure 2 Primary structures of the ATPase proteolipids (DCCD-binding protein) of *S. cerevisiae* and *N. crassa*. The yeast protein (upper sequence) starts with an fMet at the amino terminus and the *Neurospora* protein (lower sequence) with a tyrosine.

Genetic Specification of the Proteolipid

There was some earlier evidence that the *oli* 1 resistance locus is in the structural gene of the proteolipid (137, 138). Mutations in *oli* 1 were found to modify certain properties of the yeast proteolipid (137). The localization of alleles conferring the resistance phenotype in the structural gene of the proteolipid has been fully confirmed by protein sequence data (111). At present two different amino acid substitutions are known to occur in *oli* 1 resistant mutants. In one group of mutants a serine replaces cysteine at position 65 (111); in another, a phenylalanine replaces leucine at position 53 (78, 111). *Oli* 3 (*oli* 3 and *oli* 1 resistant mutants are distinguished by the cross-resistance of the former but not the latter to venturicidin) and *pho* 2 mutants also have amino acid substitutions in the proteolipid (111). These data established conclusively that in yeast, the ATPase proteolipid is a product of a mitochondrial gene.

Sebald et al (127) have isolated oligomycin-resistant mutants of *N. crassa*. The growth of these strains is inhibited by oligomycin and as in yeast this property results from a decreased sensitivity of the mitochondrial ATPase to the antibiotic. In contrast to yeast, however, all the oligomycin-resistant strains of *N. crassa* were determined to have nuclear mutations that were linked to markers on chromosome VII (127). The amino acid sequence of the proteolipid from one such strain indicated a substitution of a serine for a phenylalanine at residue 61 (127). These results indicate that the structural gene of the proteolipid of *N. crassa* is in nuclear DNA and explains the extramitochondrial site of translation of the protein in this organism (47).

The biosynthesis and genetics of the proteolipid of animal cells have not been studied. There is some evidence that in *X. laevis* the proteolipid is synthesized in the cytoplasm and is probably a nuclear gene product (39). If this can be confirmed it would suggest that there may have been an early evolutionary transfer of this mitochondrial gene to the nucleus and that in more highly evolved eucaryotic cells it is a nuclear gene product whose messenger is translated in the cytoplasm.

Sequence of the Proteolipid Gene of Saccharomyces cerevisiae

Mitochondrial DNA is an especially attractive material for sequencing since it is present only in eucaryotic organisms, but in view of its postulated procaryotic origin, it may preserve some features of both types of genomes. In addition, it is of sufficiently small size that one may hope to obtain its complete nucleotide sequence. This information would be useful in understanding how mitochondrial genes are organized and perhaps how their expression may be regulated. Since DNA sequence data can be most meaningfully interpreted if the primary structure of the gene products is known, the ATPase proteolipid gene has been an obvious choice for sequencing.

As a result of recent studies, the complete nucleotide sequence of the yeast proteolipid gene has been obtained (Figure 3). These data were obtained from the analysis of mitochondrial DNA of a ρ^- mutant that contained the genetic markers *oli* 1 and *pho* 2 but had lost all the other currently known markers of the yeast genome (140). The retained segment of mitochondrial DNA in the ρ^- mutant was ascertained to be 1.8 kilobases long or approximately 2.5% of the original genome. Since *oli* 1 and *pho* 2 were known to be within the structural gene of the proteolipid, the 1.8-kilobase piece of DNA could safely be assumed to have the structural gene sequence.

The DNA sequence shows that there is an almost complete agreement between the primary structure of the proteolipid and the nucleotide sequence of its gene. The one exception is the amino acid residue at position 46 which, from the DNA data, should be a leucine rather than a threonine. The DNA sequence also confirms the utilization of the codons of the

```
          1                    5                    10                   15
          fMet-Gln-Leu-Val-Leu-Ala-Ala-Lys-Tyr-Ile-Gly-Ala-Gly-Ile-Ser-Thr-Ile-
TAATAAATAAATATT ATG CAA TTA GTA TTA GCA GCT AAA TAT ATT GGA GCA GGT ATC TCA ACA ATT


          20                   25                   30                   35
Gly-Leu-Leu-Gly-Ala-Gly-Ile-Gly-Ile-Ala-Ile-Val-Phe-Ala-Ala-Leu-Ile-Asn-Gly-Val-Ser-
GGT TTA TTA GGA GCA GGT ATT GGT ATT GCT ATC GTA TTC GCA GCT TTA ATT AAT GGT GTA TCA


          40                   45                   50                   55
Arg-Asn-Pro-Ser-Ile-Lys-Asp-Leu-Val-Phe-Pro-Met-Ala-Ile-Phe-Gly-Phe-Ala-Leu-Ser-Glu-
AGA AAC CCA TCA ATT AAA GAC CTA GTA TTC CCT ATG GCT ATT TTT GGT TTC GCC TTA TCA GAA


60                   65                   70                   75
Ala-Thr-Gly-Leu-Phe-Cys-Leu-Met-Val-Ser-Phe-Leu-Leu-Leu-Phe-Gly-Val-Ochre        Ochre
GCT ACA GGT TTA TTC TGT TTA ATG GTT TCA TTC TTA TTA TTA TTC GGT GTA TAA TATATA TAA
```

Figure 3 Nucleotide sequence of the ATPase proteolipid gene of *S. cerevisiae*. The DNA used for sequencing was obtained from a strain that was resistant to oligomycin (*oli* 1 locus) and was shown to have a substitution of a phenylalanine for a leucine at residue 53 of the proteolipid (142).

universal code by mitochondria and of one of the standard termination codons (ochre). Despite the fact that the universal codons are used, there appears to be little degeneracy in the code. This is seen in the preferential utilization of the UUA codon for leucine and UUC for phenylalanine. Whether this is true for other mitochondrial genes will have to await further sequence data. Another feature of the DNA that is not shown in Figure 3, is a long sequence rich in A+T that follows the gene. Some 1,000 nucleotides following the structural gene have been sequenced and, with the exception of two short regions that are rich in G+C, the DNA consists almost exclusively of A+T. The function of the A+T- and G+C- rich sequences which occur throughout the entire genome (141) is not yet known but may be important in the regulation of gene expression.

FUTURE PROBLEMS

Much of the recent work in mitochondrial biogenesis has revolved around the following three questions. What are the mechanisms of assembly of inner membrane complexes that contain both cytoplasmically and mitochondrially made subunits? What are the constituent genes of mitochondrial DNA? How are mitochondrial genes organized and what regulates their expression?

The progress made on all three fronts has been impressive and, as a consequence, many young investigators have been attracted to an already fast growing field. There are still outstanding problems that need to be resolved and it is possible to make some predictions as to what approaches will be taken during the next few years.

The assembly of multisubunit enzymes such as cytochrome oxidase and the ATPase is still a poorly understood process. There are several related questions that can best be studied with mutants that have lesions in either structural or regulatory genes that control the biosynthesis of these enzymes. Fortunately there are now a large number of mitochondrial as well as nuclear mutants of *S. cerevisiae* that can be put to use in exploring the sequence in which the different subunits are integrated during the course of enzyme assembly. Many steps in bacteriophage genesis have been deduced from the intermediates found in assembly-deficient mutants and this basic approach may be equally successful in probing the assembly of mitochondrial enzymes. In addition, mutants can be used to study possible regulatory proteins that modulate the expression of mitochondrial structural genes. It was pointed out that in *S. cerevisiae,* there are nuclear mutations that selectively shut off the synthesis of single mitochondrial translation products. Such strains are likely to have mutations in regulatory proteins that may repress or activate the transcription or translation of specific mitochondrial genes. This interesting class of mutants has received little attention up to now and may provide clues as to how cytoplasmic factors regulate mitochondrial gene expression.

Even though most of the structural genes of mitochondrial DNA are now known, it is not excluded that other genes may be present that have not yet been identified. It is also still questionable whether the minor isoaccepting species of tRNAs present in mitochondria are separate gene products of mitochondrial DNA or post-transcriptional modification products. The complete saturation of the mitochondrial genome requires more extensive searches for mutants employing new selection procedures. It is difficult to understand, for example, why only a few tRNA and ribosomal RNA mutants have been found up to now. Equally strange is that no single mutant in the var 1 protein has been isolated despite the fact that it is the largest gene product of mitochondrial DNA. These observations tend to indicate that mutations in certain mitochondrial genes may not be clearly expressed and may not necessarily lead to total absence of growth on nonfermentable substrates, the standard test currently used to obtain mitochondrial mutants.

The third and probably most promising line of research is the sequence analysis of mitochondrial DNA. The first efforts in this direction have yielded promising results in the sense that most of the techniques that have been developed in recent years for the sequencing of procaryotic and eucaryotic DNAs can be applied without major complications to mitochondrial DNA (139). Since the mitochondrial genome is many orders of magnitude smaller than the smallest nuclear DNA, it is safe to assume that this will be the first eucaryotic DNA to be completely sequenced. Knowledge of the DNA sequence should answer the already mentioned question of whether the organization of mitochondrial genes follows the general outlines of eucaryotic or procaryotic genomes. Furthermore, it will tell us something about the regulation of the genome. Finally, the DNA sequence may turn out to be the least cumbersone way of obtaining the primary structure of mitochondrial translation products that have proven to be difficult to sequence by conventional methods.

ACKNOWLEGMENTS

Some of the studies reported were supported by Grants HL 22174 and GM 25250 from the National Institutes of Health, United States Public Health Service

Literature Cited

1. McLean, J. R., Cohn, G. L., Brandt, I. K., Simpson, M. V. 1958. *J. Biol. Chem.* 233:657–63
2. Roodyn, D. B., Reis, P. J., Work, T. S. 1961. *Biochem. J.* 80:9–21
3. Beattie, D. S., Basford, R. E., Koritz, S. B. 1967. *J. Biol. Chem.* 242:3366–68
4. Lamb, A. J., Clark-Walker, G. D., Linnane, A. W. 1968. *Biochim. Biophys. Acta* 161:415–27
5. Mahler, H. R., Perlman, P. S. 1971. *Biochemistry.* 10:2977–90
6. Slonimski, P. P. 1953. *La Formation des Enzymes Respiratoire Chez la Levure.* Paris: Masson
7. Mounolou, J. C., Jakob, H., Slonimski, P. P. 1966. *Biochem. Biophys. Res. Commun.* 24:281–24
8. Nagley, P., Linnane, A. W. 1970. *Biochem. Biophys. Res. Commun.* 39:989–96
9. Borst, P. 1972. *Ann. Rev. Biochem.* 41:333–76
10. Hatefi, Y. 1963. *Enzymes* 7:495–515
11. Hatefi, Y., Haavik, A. G., Griffiths, D. E. 1962. *J. Biol. Chem.* 237:1676–80
12. Ziegler, D. M., Doeg, K. A. 1962. *Arch. Biochem. Biophys.* 97:41–50
13. Hatefi, Y., Haavik, A. G., Griffiths, D. E. 1962. *J. Biol. Chem.* 237:1681–85
14. Fowler, L. R., Richardson, S. H., Hatefi, Y. 1962. *Biochim. Biophys. Acta* 64:170–73
15. Racker, E. 1977. *Ann. Rev. Biochem.* 46:1006–14

16. Pedersen, P. L. 1975. *Bioenergetics* 6:243–75
17. Schatz, G. 1968. *J. Biol. Chem.* 243:2192–99
18. Tzagoloff, A. 1969. *J. Biol. Chem.* 244:5027–33
19. Ashwell, M., Work, T. S. 1970. *Ann. Rev. Biochem.* 39:251–90
20. Schatz, G., Mason, T. L. 1974. *Ann. Rev. Biochem.* 43:51–87
21. Bandlow, W. 1976. In *Genetics and Biogenesis of Chloroplasts and Mitochondria,* ed. Th. Bücher, W. Neupert, W. Sebald, S. Werner, pp. 819–26. Amsterdam: North-Holland
22. Feldman, F., Mahler, H. R. 1974. *J. Biol. Chem.* 249:3202–9
23. Douglas, M. G., Kendrick, E., Boulikas, P., Perlman, P., Butow, R. A. 1976. In *The Genetic Function of Mitochondrial DNA,* ed. C. Saccone, A. M. Kroon, pp. 199–207. Amsterdam: North-Holland
24. Groot, G. S. P., Grivell, L. A., van Harten-Loosbroek, N., Kreike, J., Moorman, A. F. M., van Ommen, G. J. B. 1979. In *Structure and Function of Energy Transducing Membranes,* ed. K. Van Dam, B. F. Van Gelder. Amsterdam: North-Holland. In press
25. Mahler, H. R., Perlman, P., Henson, C., Weber, C. 1968. *Biochem. Biophys. Res. Commun.* 31:474–80
26. Gorts, C. P. M., Hasilik, A. 1972. *Eur. J. Biochem.* 29:282–87

27. Rubin, M. S., Tzagoloff, A. 1973. *Fed. Proc.* 32:641 (Abstr.)
28. Lin, L. F. H., Kim, I. C., Beattie, D. S. 1974. *Arch. Biochem. Biophys.* 160: 458–64
29. Weiss, H., Sebald, W., Bücher, Th. 1971. *Eur. J. Biochem.* 22:19–26
30. Mason, T. L., Poyton, R. O., Wharton, D. C., Schatz, G. 1973. *J. Biol. Chem.* 248:1346–54
31. Rubin, M. S., Tzagoloff, A. 1973. *J. Biol. Chem.* 248:4269–74
32. Capaldi, R. A. 1978. In *Molecular Biology of Membranes,* ed. S. Fleischer, Y. Hatefi, D. H. MacLennan, A. Tzagoloff, pp. 103–19. New York: Plenum
33. Tzagoloff, A., Akai, A., Rubin, M. S. 1974. In *The Biogenesis of Mitochondria,* ed. A. M. Kroon, C. Saccone, pp. 405–21. New York: Academic
34. Schatz, G., Groot, G. S. P., Mason, T. L., Rouslin, W., Wharton, D. C., Saltzgaber, J. 1972. *Fed. Proc.* 31:21–29
35. Sebald, W., Weiss, H., Jackl, G. 1972. *Eur. J. Biochem.* 30:413–17
36. Schwab, A. J., Sebald, W., Weiss, H. 1972. *Eur. J. Biochem.* 30:511–16
37. Mason, T. L., Schatz, G. 1973. *J. Biol. Chem.* 248:1355–60
38. Rubin, M. S., Tzagoloff, A. 1973. *J. Biol. Chem.* 248:4275–79
39. Koch, G. 1976. *J. Biol. Chem.* 251: 6097–6107
40. Katan, M. B., Pool, L., Groot, G. S. P. 1967. *Eur. J. Biochem.* 65:95–105
41. Weiss, H., Juchs, B. 1978. *Eur. J. Biochem.* 88:17–28
42. Rieske, J. S. 1976. *Biochim. Biophys. Acta* 456:195–247
43. Weiss, H. 1972. *Eur. J. Biochem.* 30: 469–78
44. Weiss, H., Ziganke, B. 1974. *Eur. J. Biochem.* 41:63–71
45. Weiss, H., Ziganke, B. 1976. See Ref. 21, pp. 259–66
46. Tzagoloff, A., Foury, F., Akai, A. 1976. *Mol. Gen. Genet.* 149:33–42
47. Weiss, H., Ziganke, B. 1977. In *Mitochondria 1977, Genetics and Biogenesis of Mitochondria,* ed. W. Bandlow, R. J. Schweyen, K. Wolf, F. Kaudewitz, pp. 441–49. Berlin: Walter de Gruyter
48. Katan, M. B., Van Harten-Loosbroek, N., Groot, G. S. P. 1976. *Eur. J. Biochem.* 70:409–17
49. Ross, E., Schatz, G. 1976. *J. Biol. Chem.* 251:1997–2004
50. Tzagoloff, A. 1977. In *The Enzymes of Biological Membranes,* ed. A. Martonosi, 2:103–24. New York: Plenum
51. Tzagoloff, A., Meagher, P. 1971. *J. Biol. Chem.* 246:7328–36
52. Kagawa, Y., Racker, E. 1966. *J. Biol. Chem.* 241:2467–74
53. Tzagoloff, A., MacLennan, D. H., Byington, K. H. 1968. *Biochemistry* 7: 1596–602
54. Pullman, M. E., Penefsky, H. S., Datta, A., Racker, E. 1960. *J. Biol. Chem.* 235:3322–29
55. MacLennan, D. H., Tzagoloff, A. 1968. *Biochemistry.* 7:1603–10
56. Senior, A. E., Brooks, J. C. 1970. *Arch. Biochem. Biophys.* 140:257–66
57. Jackl, G., Sebald, W. 1975. *Eur. J. Biochem.* 54:97–106
58. Sebald, W. 1977. *Biochim. Biophys. Acta* 463:1–27
59. Mendiola-Morgenthaler, L. R., Morgenthaler, J. J., Price, C. A. 1976. *FEBS Lett.* 62:96–100
60. MacLennan, D. H., Akai, J. 1968. *Biochem. Biophys. Res. Commun.* 33: 441–47
61. Tzagoloff, A. 1970. *J. Biol. Chem.* 245:1545–51
62. Kagawa, Y., Racker, E. 1966. *J. Biol. Chem.* 247:2475–82
63. Tzagoloff, A., Meagher, P. 1971. *J. Biol. Chem.* 247:594–603
64. Sierra, M. F., Tzagoloff, A. 1973. *Proc. Natl. Acad. Sci. USA* 70:3155–59
65. Sebald, W. 1979. *Eur. J. Biochem.* In press
66. Sebald, W., Graf, T., Wild, G. 1976. See Ref. 21, pp. 167–74
67. Tzagoloff, A. 1971. *J. Biol. Chem.* 246:3050–56
68. Beattie, D. S., Lin, L.-F., Stuchell, R. N. 1974. See Ref. 33, pp. 465–75
69. Poyton, R. O., Kavanagh, J. 1976. *Proc. Natl. Acad. Sci. USA* 73:3947–51
70. Ebner, E., Mason, T. L., Schatz, G. 1973. *J. Biol. Chem.* 248:5369–78
71. Tzagoloff, A., Akai, A., Needleman, R. B. 1975. *J. Biol. Chem.* 250:8228–35
72. Mahler, H., Bastos, R. N., Feldman, F., Flury, U., Lin, C. C., Perlman, P. S., Pham, S. H. 1976. In *Membrane Biogenesis,* ed. A. Tzagoloff, pp. 15–61. New York: Plenum
73. Groot, G. S. P., Rouslin, W., Schatz, G. 1972. *J. Biol. Chem.* 247:1735–42
74. Weislogel, P. O., Butow, R. A. 1971. *J. Biol. Chem.* 246:5113–19
75. Sebald, W., Machleidt, W., Otto, J. 1973. *Eur. J. Biochem.* 38:311–24
76. Poyton, R. O., Schatz, G. 1975. *J. Biol. Chem.* 250:752–61
77. Capaldi, R. A., Vanderkooi, G. 1972. *Proc. Natl. Acad. Sci. USA* 69:930–32
78. Wachter, E., Sebald, W., Tzagoloff, A. 1977. See Ref. 47, pp. 441–49

79. Poyton, R. O., Groot, G. S. P. 1975. *Proc. Natl. Acad. Sci. USA* 72:172–76
80. Ibrahim, N. G., Burke, J. P., Beattie, D. S. 1973. *FEBS Lett.* 29:73–76
81. Lederman, M., Attardi, G. 1973. *J. Mol. Biol.* 78:275–83
82. Padmanaban, G., Hendler, F., Patzer, J., Ryan, R., Rabinowitz, M. 1975. *Proc. Natl. Acad. Sci. USA* 72:4293–97
83. Scragg, A. H., Thomas, D. Y. 1975. *Eur. J. Biochem.* 56:183–192
84. Moorman, A. F. M., Grivell, L. A. 1976. See Ref. 23, pp. 281–89
85. Coen, D., Deutsch, J., Netter, P., Petrochilo, E., Slonimski, P. P. 1970. In *Control of Organelle Development,* pp. 449–96. London: Cambridge Univ. Press
86. Nass, M. M. K., Buck, C. A. 1970. *J. Mol. Biol.* 54:187–98
87. Suyama, Y. 1976. In *Cell Biology,* ed. P. L. Altmann, D. D. Katz, pp. 228–30. Bethesda: FASEB
88. Martin, N. C., Rabinowitz, M. 1968. *Biochemistry.* 17:1628–34
89. Casey, J. W., Hsu, H. J., Getz, G. S., Rabinowitz, M., Fukuhara, H. 1974. *J. Mol. Biol.* 88:735–47
90. Martin, N. C., Rabinowitz, M., Fukuhara, H. 1977. *Biochemistry.* 21:4672–77
91. Martin, R., Schneller, J. M., Stahl, A. J. C., Dirheimer, G. 1976. See Ref. 21, pp. 755–58
92. Barnett, W. E., Brown, D. H. 1967. *Proc. Natl. Acad. Sci. USA* 57:452–58
93. Kroon, A. M., Terpstra, P., Holtrop, M., deVries, H., van der Bogert, C., de Jonge, J., Agsteribbe, E. 1976. See Ref. 21, pp. 685–96
94. Dawid, I. B. 1972. *J. Mol. Biol.* 63:201–16
95. Lynch, D. C., Attardi, G. 1976. *J. Mol. Biol.* 102:125–41
96. Suyama, Y., Hamada, J. 1976. See Ref. 21, pp. 763–70
97. Lambowitz, A. M., Chua, N.-H., Luck, D. J. L. 1976. *J. Mol. Biol.* 107:223–53
98. Avner, P. R., Coen, D., Dujon, B., Slonimski, P. P. 1973. *Mol. Gen. Genet.* 125:9–52
99. Pratje, E., Michaelis, G. 1977. *Mol. Gen. Genet.* 152:167–74
100. Colson, A. M., Slonimski, P. P. 1978. See Ref. 47, pp. 185–98
101. Tzagoloff, A., Akai, A., Needleman, R. B., Zulch, G. 1975. *J. Biol. Chem.* 250:8236–42
102. Slonimski, P. P., Tzagoloff, A. 1976. *Eur. J. Biochem.* 61:27–41
103. Rytka, J., English, K. J., Hall, R. M., Linnane, A. W., Lukins, H. B. 1976. See Ref. 21, pp. 427–34
104. Mahler, H. R., Bilinski, T., Miller, D., Hanson, D. 1976. See Ref. 21, pp. 857–63
105. Foury, F., Tzagoloff, A. 1976. *Eur. J. Biochem.* 68:113–19
106. Coruzzi, G., Trembath, M. K., Tzagoloff, A. 1979. *Eur. J. Biochem.* In press
107. Foury, F., Tzagoloff, A. 1978. *J. Biol. Chem.* 253:3792–97
108. Cabral, F., Rudin, Y., Solioz, M., Schatz, G., Clavilier, L., Slonimski, P. P. 1978. *J. Biol. Chem.* 243:297–304
109. Tzagoloff, A., Foury, F., Macino, G. 1978. In *Biochemistry and Genetics of Yeast,* ed. M. Bacila, B. L. Horecker, A. Stoppani, pp. 477–88. New York: Academic
110. Claisse, M. L., Spyridakis, A., Slonimski, P. P. 1977. See Ref. 47, pp. 337–44
111. Sebald, W., Wachter, E., Tzagoloff, A. 1979. *Eur. J. Biochem.* Submitted for publication
112. Perlman, P. S., Douglas, M. G., Strausberg, R. L., Butow, R. A. 1977. *J. Mol. Biol.* 115:675–94
113. Nagley, P., Spriprakash, K. S., Rytka, J., Choo, K. B., Trembath, M. K., Lukins, H. B., Linnane, A. W. 1976. See Ref. 23, pp. 231–42
114. Schweyen, R. J., Kaudewitz, F. 1976. *Mol. Gen. Genet.* 149:311–22
115. Van Ommen, G. J. B., Groot, G. S. P., Borst, P. 1977. *Mol. Gen. Genet.* 154:255–62
116. Borst, P., Bos, J. L., Grivell, L. A., Groot, G. S. P., Heyting, C., Moorman, A. F. M., Sanders, J. P. M., Talen, J. L., Van Kreiyl, C. F., Van Ommen, G. J. B. 1977. See Ref. 47, pp. 255–70
117. Lewin, A., Morimoto, R., Merten, S., Martin, N., Berg, P., Christianson, T., Levens, D., Rabinowitz, M. 1977. See Ref. 47, pp. 271–89
118. Grivell, L. A., Moorman, A. F. M. 1977. See Ref. 47, pp. 371–84
119. Morimoto, R., Merten, S., Lewin, A., Martin, N., Rabinowitz, M. 1978. *Mol. Gen. Genet.* 163:241–55
120. Prunell, A., Bernardi, G. 1977. *J. Mol. Biol.* 110:53–74
121. Sanders, J. P. M., Heyting, C., Borst, P. 1976. See Ref. 21, pp. 511–17
122. Pajot, P., Wambire-Kluppel, M. L., Kotylak, Z., Slonimski, P. P. 1976. See Ref. 21, pp. 443–51
123. Kuriyama, Y., Luck, D. J. L. 1973. *J. Mol. Biol.* 73:425–37
124. Wellauer, P. K., Dawid, I. B. 1973. *Proc. Natl. Acad. Sci. USA* 70:2827–31
125. Wu, M., Davidson, N., Attardi, G., Aloni, Y. 1972. *J. Mol. Biol.* 71:81–83
126. Attardi, G., Albring, M., Amalric, F.,

Gelfand, R., Griffith, J., Lynch, D., Merkel, C., Murphy, W., Ojala, D. 1977. See Ref. 21, pp. 573–85
127. Sebald, W., Sebald-Althaus, M., Wachter, E. 1977. See Ref. 47, pp. 433–40
128. Margulis, L. 1970. *Origin of Eukaryotic Cells.* New Haven: Yale Univ. Press
129. Bogorad, L. 1975. See Ref. 72, pp. 201–45
130. Beechey, R. B., Roberton, A. M., Holloway, T., Knight, I. G. 1967. *Biochemistry* 6:3867–79
131. Cattell, K. J., Lindop, C. R., Knight, I. G., Beechey, R. B. 1971. *Biochem. J.* 125:169–77
132. Stekhoven, F. S., Waitkus, R. F., Van Moerkerk, T. B. 1972. *Biochemistry* 11: 1144–50
133. Criddle, R. S., Arulanadan, C., Edwards, T., Johnston, R., Scharf, S., Enns, R. 1977. See Ref. 21, pp. 151–57

134. Criddle, R. S., Packer, L., Shieh, P. 1977. *Proc. Natl. Acad. Sci. USA* 74: 4306–10
135. Mitchell, P. 1973. *FEBS Lett.* 33: 267–74
136. Wachter, E., Sebald, W. 1979. *Eur. J. Biochem.* Submitted for publication
137. Tzagoloff, A., Akai, A., Foury, F. 1976. *FEBS Lett.* 65:391–96
138. Groot Obbink, D. J., Hall, R. M., Linnane, A. W., Lukins, H. B., Monk, B. C., Spithill, T. W., Trembath, M. K. 1976. See Ref. 23, pp. 163–73
139. Macino, G., Tzagoloff, A. 1979. *Proc. Natl. Acad. Sci. USA.* 76:131–35
140. Dujon, B., Colson, A. M., Slonimski, P. P. 1977. See Ref. 47, pp. 579–669
141. Bernardi, G. 1977. See Ref. 21, pp. 503–10
142. Macino, G., Tzagoloff, A. 1979. *J. Biol. Chem.* In press

Ann. Rev. Biochem. 1979. 48:443–70

DEVELOPMENTAL BIOCHEMISTRY OF PREIMPLANTATION MAMMALIAN EMBRYOS[1]

❖12015

Michael I. Sherman[2]

Roche Institute of Molecular Biology, Nutley, New Jersey 07110

CONTENTS

[1]The following abbreviations are used: ICM, inner cell mass of the blastocyst; conA, concanavalin A.

[2]I am grateful to Drs. P. Gage and M. Sellens for their comments on the manuscript.

0066-4154/79/0701-0443$01.00

PERSPECTIVES AND SUMMARY

Until recently, limitations on availability of material have precluded extensive investigations on the control at the molecular level of the early development of mammalian embryos. Consequently, most of the ideas concerning regulation of early embryogenesis were derived from experiments on embryos of more primitive species, such as amphibians and echinoderms, which are relatively large and can be obtained in great numbers. Now, with the use of microanalytic techniques and micromanipulative procedures, it has become possible to study, at the biochemical level, developmental patterns in preimplantation mammalian embryos. The results indicate that to some extent generalizations about the control of developmental processes from studies with other species do not apply to mammalian embryos. Consequently, the purpose of this review is to consider the degree to which investigations on early mammalian embryos, particularly mouse embryos (since they have been used most extensively), provide insights into their developmental processes. Specifically, the issue is to determine whether the cells in mammalian embryos undergo differentiation prior to the time of implantation; beyond implantation, organogenesis begins and the differentiation of developing cell types is overt.

The relatively small mammalian eggs do not have the extensive reserves of maternally inherited macromolecules that are present in ova from other species. Consequently, RNA and protein biosynthesis begin early in development and accelerate rapidly by the 16-cell stage. At the same time, transport systems become more sophisticated, which facilitates the uptake of precursors. The analysis of electrophetic patterns of total proteins synthesized at various stages indicates a shift, beginning shortly after fertilization, in the types of mRNAs translated. Enzyme and cell surface antigen studies make it clear that at least part of this shift is due to translation during cleavage of newly synthesized mRNAs. Nevertheless, the response of mammalian embryos to antimetabolites supports the view that their development through cleavage to the morula and then the blastocyst stage is at least partially dependent upon translation of relatively stable, maternally inherited, mRNAs.

The first obvious change in embryo morphology takes place at the morula stage when the extent of association between cells of the embryo becomes more intimate, a process known as compaction. Ultrastructural analyses reveal that a change in the nature of membrane interactions accompanies compaction. These ultrastructural modifications, together with the characteristics of amino acid transport systems, suggest that at and beyond the compact morula stage, the outer cells of the embryo resemble an organized epithelium. Also at the morula stage, cells that are in an internal position divide at a faster rate than those that are in the outer layer, which suggests an early difference in the two populations of blastomeres.

The cells at the blastocyst stage are segregated clearly into two groups, an enclosed cluster called the inner cell mass (ICM) and an outer, single-cell layer, the trophectoderm. Aside from their location and morphology, the two groups of cells can be distinguished by cleavage rate, protein synthetic patterns, enzyme production, cell surface properties, and susceptibility to antimetabolites and irradiation. It is very likely that to a large degree these differences result from activation of embryonic genes rather than post-transcriptional regulation. Therefore, by these criteria, the cells of the blastocyst have embarked upon differential pathways of gene expression. However, whether the cells at this stage differ in terms of production of *specialized* gene products remains to be determined.

OBJECTIVE

The objective of this review is to consider the biochemical evidence for, or against, the premise that cells of the mammalian embryo differentiate prior to implantation or, more precisely, by the mid-blastocyst stage. Accordingly, several biochemical studies on early mammalian embryos are not discussed here because present knowledge does not suggest a relationship between these investigations and the specific question at hand. The reader will find many of the omitted studies covered in other reviews, cited in subsequent sections, which deal with preimplantation mammalian development on a more general level or with a somewhat different emphasis.

PREIMPLANTATION STAGES AND CELL FATES

Within 24–48 hr of fertilization, the eggs of most mammals undergo cell division to give two blastomeres of approximately equal size. This cell division and the three subsequent ones comprise the cleavage stage of embryogenesis and, depending upon the mammal, the embryo progresses from the 2-cell stage to the 16-cell stage in 1–3 days from the first division (1, 2). The next stage reached is the morula stage in which a dramatic morphological alteration occurs. The blastomeres compact to give a smooth, solid ball in which individual cells are no longer discernible by gross observation. Cavitation characterizes the onset of the blastocyst stage, about 1 day after the morula stage is reached. The early blastocyst consists of an outer single layer of trophectoderm cells which surrounds a fluid-filled, blastocoelic cavity and a solid cluster of cells called the inner cell mass (ICM).

Whereas most mammalian embryos progress to the early blastocyst stage as described above, subsequent development can differ dramatically between species. For example, although all mammals undergo implantation, a process by which the outer cells of the blastocyst anchor the conceptus to the uterine wall, an implanting mouse blastocyst (on the fifth day of gestation), a spherical or slightly ovoid structure with a diameter of about 100 μm containing approximately 250 cells, hardly resembles an implanting pig blastocyst (on about the eighteenth day of gestation), an elongated structure greater than a meter in length and probably containing millions of cells! Clearly, it becomes difficult to make useful comparisons between such different organisms. Consequently, this review focuses on the biochemistry of mammalian embryonic development through those stages common to most species, i.e. from fertilization to the spherical blastocyst stage. The studies to be described generally involve mouse embryos because of the ease, economy, and convenience with which relatively large numbers of these embryos are obtainable. Although some parallel investigations have been carried out with embryos of other mammalian species, particularly rabbit, our knowledge of the developmental biochemistry of mammalian, nonmurine embryos is sparse. It may, therefore, be inadvisable to assume a priori that the developmental principles derived from studies on mouse embryos necessarily apply to embryos of other species.

Present knowledge allows us to trace the fate of trophectoderm and ICM cells of the mouse embryo through a number of generations and transitions. Several recent reviews on the biology of these developmental processes (3–9) and extensive descriptions of the anatomy of the early mouse embryo (including preimplantation stages) (10–12) are available. For the purposes of this review, it is pertinent to state that the trophectoderm cells of the mouse blastocyst appear to contribute only to cell types (trophoblast, ectoplacental cone, chorion) that eventually constitute the fetal moiety of the placenta. On the other hand, the cells of the ICM are more dynamic; derivatives of this group of cells [numbering only about 20 in a 64-cell mouse blastocyst (13)] proliferate extensively during and following implantation, and ultimately give rise to all the germ layer components of the embryo proper, as well as to most or all of the cell types in the extraembryonic membranes, i.e. parietal endoderm, yolk sac, amnion, and allantois.

DETERMINATION AND DIFFERENTIATION

The progeny of an embryonic cell may pass through a number of intermediate stages on the way to their ultimate fates. For example, the cells of the trophoblast layer of the placenta (an end state) are derived from blastomeres that occupied an external position at the morula stage and subsequently participated in the formation of the trophectoderm layer of the blastocyst (14). Conversely, the cells of the ICM of the blastocyst are derived from cells that occupied an internal position at the morula stage (14). The daughter cells in the ICM which are at the blastocoelic surface are likely to give rise to primitive endoderm cells (15, 16), which serve in turn as progenitors of the parietal endoderm layer and the visceral endoderm moiety of the yolk sac (5, 7). With this latter conversion, the cells have become fully differentiated. During these developmental progressions, the cells attain a stage at which they become determined, i.e. they become irrevocably programmed to develop along their particular pathway. Attendant upon this process is a progressive restriction in developmental potential. For example, at the mid-blastocyst stage, the fate of ICM cells becomes limited such that they can give rise only to primitive endoderm and primitive ectoderm cells, but not to trophoblast cells (15–19). One day later, the ICM cells at the junction with the blastocoel become determined to form primitive endoderm (15, 16, 18); presumably, these cells are unable to produce other types of derivatives associated with ICM development.

One must attempt to cope with certain conceptual difficulties when considering differentiation during early embryogenesis. In essence, these problems are born of our lack of insight into the nature and control of developmental processes. One source of confusion centers about the definition of the term differentiation, another on the relationship between differentation and determination. Davidson (20), for example, has defined differentiation operationally as "the active manifestation of a specialized function particular to each cell type." He indicates further that prior to the acquisition of specialized gene activity, a cell might differ from its neighbors by virtue of the maternal cytoplasm that it has inherited. Therefore, according to his definition, it is feasible that a cell might have, or be synthesizing (due to translation of maternal mRNAs), a population of proteins different from those of adjacent cells, without undergoing active differentiation. On the other hand, Johnson et al (18) define differentiation as "the acquisition of functional, morphological or molecular differences amongst cells of presumed homogeneous genotypic composition." This more flexible definition neither requires that the differentiated elements distinguishing one cell type from another be direct products of transcription of the embryonic genome, nor does it specify that the discriminatory factors need relate to *specialized* function.

As regards the relationship between determination and differentiation, one might ask whether a finite period always separates the two states; that is, is there always a hiatus between the time that a cell's fate is fixed and the earliest time at which it actually begins to differentiate? To some extent, the answer to this question depends, once again, on the definition of differentiation that is chosen, but no matter how flexible the definition, there is always the limitation of the biochemical information available: whereas biological techniques sometimes allow us to fix with reasonable precision the time of determination, e.g. in the case of cells in the preimplantation mouse embryo (see 7–9, 18), we can say only that differentiation takes place *no later than the time of expression of the earliest differentiated marker of which we are aware.*

After the appropriate evidence has been presented, the final section of this review considers whether there is justification for assuming that cells in the preimplantation mammalian embryo are differentiated, in terms of both the relatively rigid (20) and more flexible (18) definitions of the word. The relationship between differentiation and determination of cells of early mammalian embryos is also dealt with.

NUCLEIC ACID ANALYSES OF PREIMPLANTATION EMBRYOS

DNA Synthesis and Cell Division

From 2 to 4 hr following fertilization, the male and female pronuclei are detectable (and distinguishable from each other), and after a short hiatus (4–6 hr in the mouse) DNA synthesis begins (21–23). This initial S phase has a duration of $\sim$4 hr and for reasons that are unclear, replication proceeds more rapidly in the male than in the female pronucleus (21–23). In the second and subsequent cell cycles in the preimplantation mouse embryo, DNA replication requires 6–7 hr (23–25).

The first cell division in zygotes of mammals such as mouse, rat, and rabbit takes place within 24 hr of fertilization, although other mammalian eggs require a longer period (2). The time required for subsequent cell cycles generally diminishes as cleavage progresses (2); however, there is enough variation in cell cycle time among species so that cell numbers can differ dramatically within a few days of fertilization. For example, whereas a 96 hr rabbit embryo contains about 1000 cells (26), mouse and rat embryos of the same age possess only 64–128 cells (13, 27) and 16–32 cells (27), respectively. The shortening of the cell cycle time as preimplantation development proceeds is probably due in large part to a decrease in the duration of the G_2 phase. For instance, several investigators have observed that the 2-cell mouse embryo has a protracted (12–15 hr) G_2 phase (24, 25, 28), but the estimated time for G_2 in the 4- to 8-cell transition is only 0.5 hr (25).

Disagreement exists over the question of a G_1 phase (see 29). Initially, utilizing microspectrophotometry and/or thymidine incorporation into DNA, Dalcq & Pasteels (30), Oprescu & Thibault (31), and Gamow & Prescott (28) found no evidence of a G_1 period during early cell division in rat, rabbit, and mouse embryos. In more recent studies, Barlow et al (13), Luthardt & Donahue (24), and Sawicki et al (25) have claimed that a short G_1 period (1–2 hr) does exist during cleavage stages in mouse embryos. Although Mukherjee (23) still maintains that there is no G_1 phase in the mouse embryo prior to the blastocyst stage, he presents no data to substantiate this view.

In a thorough study involving autoradiographic analysis and reconstruction of serially sectioned embryos previously incubated with [^{3}H]thymidine, Barlow et al (13) reported that from the 16-cell to the blastocyst stage, the internal cells have a higher thymidine labeling index and presumably a shorter cell cycle than do the outer blastomeres that enclose them. This observation is contrary to an earlier report (32); Barlow et al (13) claimed that the latter studies were subject to error for technical reasons. Johnson et al (18) have pointed out that the difference in labeling index is one of the earliest properties by which internal and external cells can be distinguished.

In conclusion, all the apparatus necessary for replication and cell division is presumably present in the mammalian embryo within a short time of fertilization, although the macromolecules involved in DNA synthesis and control over the cell cycle during preimplantation stages have not been characterized. Inner cells appear to differ from external ones as early as the 16-cell stage by virtue of their faster cell cycle times. Speculations that cell division in cleavage stage embryonic cells is unregulated due to a lack of a G_1 phase (28), and that the presumptive differentiation of cells at the blastocyst stage is correlated with the earliest time of detection of a G_1 phase (23), do not merit serious attention at this time, first because there are no good reasons for believing that such relationships are valid, and second because a G_1 period might, in fact, be present from the earliest stages of embryogenesis.

RNA Analyses

There have been numerous reports on RNA content and synthesis in preimplantation mouse and rabbit embryos; the reader is directed to recent reviews (33–40) for thorough summaries of these studies. In both mouse (41) and rabbit (42) embryos, there is little increase in the net RNA content through cleavage, i.e. during the first 2–3 days of development. However, RNA synthesis might begin soon after fertilization occurs (38, 43–45), and in any event, the synthesis of all the commonly recognized forms of RNA is detectable in mouse and rabbit embryos before the 8-cell or 128-cell stage, respectively: 4s RNA by the 2- or 4-cell stage in the mouse (43, 46, 47) and by the 2-cell stage in the rabbit (48); 5s RNA by the 2- or 4-cell stage in the mouse (43, 44, 47) and rabbit (48); ribosomal RNA (rRNA) by the 4-cell stage (and possibly the late 2-cell stage) in the mouse (43, 44, 46, 47,

49) and by the 128-cell (early blastocyst) stage in the rabbit [one unpublished result, cited by Schultz & Church (36), suggests that small amounts of rRNA might be synthesized in 16-cell rabbit embryos]; and heterogeneous RNA throughout the preimplantation stages of both mouse (43, 44, 46, 50, 51) and rabbit (48) embryos. More recently, investigators have reported the synthesis of polyadenylated mRNAs during the early cleavage stages of mouse (39, 52) and rabbit (53, 54) embryos. Young (55) has claimed that 1-cell mouse embryos are capable of incorporating guanosine into Me^{7}G cap structures, but the evidence presented is not conclusive.

Quantitative estimates of RNA polymerase activities in preimplantation mouse embryos (56–60) must be viewed with some caution because assays were carried out on crude homogenates and the possible degradation of newly synthesized product by endogenous nucleases was not taken into account. Nevertheless, several of these studies have provided qualitative evidence for the presence during cleavage stages of an RNA polymerase activity that is sensitive to α-amanitin (57, 59, 60) and that operates optimally at relatively high concentrations of ammonium sulfate (58, 60), properties characteristic of RNA polymerase II, the enzyme that has been implicated in mRNA synthesis (see 39). In fact, there is evidence to suggest that types I, II, and III polymerases are all present in the mouse embryo by the morula state (60).

Taken together, these studies indicate that differentiative events *could* take place by the blastocyst stage, i.e. there is evidence for mRNA synthesis and it is likely that there is a sufficient supply of ribosomes and tRNA to allow for the production of differentiated gene products. More pertinent to the question of whether differentiation *does* take place during preimplantation stages would be studies on the diversity of the mRNAs synthesized; for example, one might ask whether new mRNA species are synthesized at a particular preimplantation stage but not at any earlier stage. Such studies involve analyses of unique sequence RNA transcripts that are very difficult due to limitations on the amount of tissue available (see below). In the one study that has been performed on late rabbit blastocysts, Schultz et al (61) reported that the RNA synthesized at that stage hybridizes with ∼ 1.8% of the unique sequence DNA in the genome. These authors estimated that if all these sequences were in fact in mRNA and if these mRNAs were, on average, 1000 nucleotides in length, then there would be as many as 60,000 different sequences. Schultz et al (61) did not compare the RNAs from sixth day blastocysts with earlier embryos, presumably for reasons of impracticality: for example, to obtain the same amount of RNA from early (fourth day) rabbit blastocysts that they obtained from 268 sixth day blastocysts, they would have required about 5400 of the earlier embryos (more than 500,000 mouse blastocysts would be needed to provide an equivalent amount of RNA).

Notwithstanding the difficulties in obtaining reasonable amounts of RNA from mouse embryos for sequence diversity analyses, one series of studies has been carried out (36, 62). Because of the paucity of details and data given, it is difficult to evaluate these experiments critically. Nevertheless, Church & Brown (62) presented a figure indicating that the RNA synthesized by 8- to 16-cell embryos, morulae, and blastocysts hybridizes to between .5% and 1% of the unique sequence DNA; there does not appear to be any substantial increase in the diversity of the transcripts of unique sequence DNA through these stages, although, of course, this does not necessitate that the mRNA population remains the same. There does, however, appear to be a doubling in the complexity of RNA species transcribed from repeated DNA between late cleavage stage embryos and blastocysts (36). The significance of the observed increase is unclear because some members of this class are mRNAs whereas others are not.

It is apparent that for a thorough analysis of mRNA diversity by these conventional hybridization procedures, more RNA is required than can be easily obtained from preimplantation embryos. Until increased amounts of RNA or RNA transcripts become available, perhaps through cloning, the question of whether embryos undergo differentiation prior to implantation can better be approached by protein analyses and antimetabolite studies. These subjects are considered in the following sections.

PROTEIN ANALYSES OF PREIMPLANTATION EMBRYOS

As is the case with RNA content, there appears to be no net increase in protein content in cleavage stage mammalian embryos. In fact, in mouse (27, 63) and rat (27) embryos, there seems to be a slight decrease in total protein content from fertilization through cleavage and even into the blastocyst stage. Despite this, it is apparent from a large number of studies that protein synthesis takes place prior to fertilization and at all stages of embryogenesis [see the review by Epstein (37) for a thorough account of pertinent publications]. After taking into consideration problems of amino acid uptake and endogenous pools, Epstein & Smith (64) concluded that there is a significant increase in the rate of protein synthesis in mouse embryos between the 8-cell and blastocyst stages. On a per cell basis, however, the protein synthetic rate remains fairly constant. Equally thorough analyses of protein synthetic rates have not been carried out in rabbit embryos; an early study by Manes & Daniel (65) suggests that a steep rise in protein synthetic activity takes place beyond the early blastocyst state, i.e. at a later time and stage than it occurs in mouse embryos. On a per cell basis, however, the protein synthetic rate appears to be less than that of 2-cell embryos until the late blastocyst stage.

Protein Electrophoretic Profiles

A number of investigators have analyzed the profiles of proteins synthesized by preimplantation mouse and rabbit embryos both by one-dimensional electrophoresis [mouse: (66, 67); rabbit: (68)] and by a combination of isoelectric focusing and electrophoresis [mouse: (18, 40, 69–71); rabbit: (38, 40, 72)]. It is virtually impossible to relate data from one study to the next, and even within a single study poor photoreproduction of the gels and inability to superimpose them often precludes a critical judgement of the data. Therefore, one can only assume that the investigators have analyzed their gels scrupulously and objectively, and accept the interpretations of their data. One can, perhaps, draw some reassurance from the fact that most of these investigators have come to the same conclusions from their studies.

In general, electrophoretic analyses of proteins synthesized by early rabbit and mouse embryos at progressive stages from fertilization through late blastocyst have revealed that a large population of proteins is "constitutive", that is, they are present throughout the preimplantation period. The most notable profile changes occur between fertilization and the 8-cell stage, prior to the time at which the rate of protein synthesis increases dramatically; by comparison, relatively few changes are apparent during the morula and blastocyst stages, when cells become geographically distinguishable as trophectoderm or ICM (40, 64, 67, 68, 71, 72). This pattern might be characteristic of most or all mammals since it appears also to apply to rat and hamster embryos [unpublished work, cited in (40)].

Most investigators have recognized the danger of attributing stage- or cell-specific status to any of the proteins under analysis: for example, the earliest time a protein is detected on the gel is not necessarily the earliest stage at which it is synthesized, nor is it easy to determine the time at which a protein band or spot "disappears;" also, post-translational modifications of proteins at certain stages of development may alter their electrophoretic mobilities, and thus create the impression that they are stage-specific. Notwithstanding these reservations, it appears that there are several classes of stage-specific proteins, among them proteins synthesized prior to fertilization and during the first few cleavages (38, 40, 67, 71, 72); proteins first synthesized in the fertilized egg or at the 2-cell stage that continue to be produced only during early cleavage (40, 67, 68, 71, 72); proteins absent during early cleavage, but produced from the 8-cell through the blastocyst stages (38, 67, 68, 71, 72); and proteins that appear first at the late morula stage, following compaction (18, 68, 71, 72). In mouse embryos, the latter group, which, as mentioned, is small in number compared to the former ones, can be subdivided further into proteins found predominantly in trophectoderm cells, proteins largely restricted to ICM cells, or proteins common to both (18, 40, 69, 70). The earliest stage at which individual proteins in these categories are detectable varies from the late morula to the mid-blastocyst stages, an interval of about 1.5 days (18, 70). Since there are limitations to the numbers of proteins that can both be resolved on these gels and labeled at high enough levels to be detected (about 100 on one-dimensional gels and about 600 on two-dimensional gels), it is likely that each of these classes of stage-specific proteins is substantially larger than indicated, and possible that other classes exist.

Enzymes

Because of limitations on the amount of material available, it is usually necessary to use microassay or histochemical procedures in the investigation of enzyme contents in preimplantation embryos. In large degree, the enzymes that have been chosen for study are those that might be related to shifts in the ability of developing preimplantation embryos to utilize various metabolites. There is no set pattern to the activity levels of these enzymes from fertilization to the blastocyst stage: of the 20 or so enzymes analyzed, some activities increase, whereas others decrease, and still others increase and then decrease, decrease and then increase, or remain constant (see 34, 37, 73–76). It is likely that this complex pattern reflects a combination of cessation of translation of maternal mRNAs, degradation of enzyme molecules synthesized prior to fertilization (77), and new enzyme synthesis stemming from the translation of mRNAs transcribed from the embryonic genome. Chapman et al (78) provided the first unequivocal evidence for the presence in early mammalian embryos of enzymes translated from embryonic, as opposed to maternal, mRNAs. These investigators detected the presence of the paternal isozymic form of glucose phosphate isomerase in late mouse blastocysts. Using larger numbers of embryos, Brinster (79) subsequently demonstrated that expression of the paternal gene for this enzyme takes place by the 8-cell stage. Analyses of β-glucuronidase activity in early mouse embryos also indicated that paternal gene expression takes place by the 8-cell stage and possibly as early as the 2-cell stage (80–82).

In view of recent evidence that inactivation of the X-chromosome accompanies differentiation of a particular embryonal carcinoma cell line (83) and because these cells resemble early embryonic cells in several respects (see 84), it is appropriate to consider biochemical studies concerning the time of X-chromosome inactivation in mammalian embryos. Biological and cytological studies have led to the view that both X-chromosomes are active in mammalian oocytes and early female embryos, but that one X-chromosome becomes inactive by the late blastocyst stage (see 85). In support of the first part of this proposal, several investigators have observed that the activities of three X-linked enzymes, glucose-6-phosphate dehydrogenase (86), hypoxanthine-guanine phosphoribosyl transferase (87–89), and phosphoglycerate kinase (90), in unfertilized eggs from XO mice are only half as high as those from (normal) XX females. On the other hand, in all of these studies, levels of selected autosomal enzymes in XO and XX eggs are similar. Efforts to determine the time of gene dosage compensation during embryogenesis, presumably due to X-inactivation, have been complicated by the presence of stores of enzyme, and probably mRNAs, synthesized prior to fertilization. Nevertheless, Kozak & Quinn (90) demonstrated that gene dosage compensation for phosphoglycerate kinase has occurred by the early egg-cylinder stage (seventh day of gestation) and more recent observations with hypoxanthine-guanine phosphoribosyl transferase (88, 89, 91) and α-galactosidase (92), another X-linked enzyme, are consistent with the view that X-chromosome inactivation occurs between the early and late blastocyst stage, in concurrence with proposals from biological and cytological analyses (85).

Although the aforementioned enzyme studies provide information on the time of activation of embryonic genes, they do not deal with the question of differential gene expression by cells in the preimplantation embryo. The enzymes investigated most commonly in this regard are the phosphatases, especially alkaline phosphatase. These studies have been fraught with confusion and disagreement. Initially, Mulnard (93; see also 94) carried out histochemical analyses on rat embryos in which he observed that alkaline phosphatase activity, using β-glycerophosphate as substrate, appears first at the morula stage. Enzyme activity was restricted to internal blastomeres at that stage and subsequently to the ICM cells of the blastocyst. Studies by Ortiz et al (95) on mouse embryos support this view. However, Solter et al (96) later claimed that all cells at the morula stage in the mouse possess approximately equal amounts of alkaline phosphatase activity. Izquierdo & Marticorena (97) also found this to be the case when they used naphthyl AS-BI phosphate as substrate. In more recent studies, Johnson et al (98) incubated frozen sectioned mouse embryos (as opposed to whole mounts which were used previously) with naphthyl AS-BI phosphate. They found no evidence of selective localization of alkaline phosphatase activity in cells at the morula stage but they did observe that ICM cells in the blastocyst are much more heavily stained than trophectoderm cells surrounding the blastocoelic cavity.

Two studies have appeared in which mouse embryos stained for alkaline phosphatase activity were inspected by electron microscopy (99, 100). In one (99), with either β-glycerophosphate or cytidine monophosphate as substrate, activity was observed in fertilized eggs but not at the 2- or 4-cell stages. In 8-cell embryos, activity was observed on plasma membranes, both at the exposed surface of the embryo and between blastomeres. At the morula stage, activity was more intense on outside than on inside blastomeres and by the blastocyst stage, activity was present on all trophectoderm surfaces but was weak or absent on membranes of ICM cells. The results of the second study are almost diametrically opposed to those of the first. Mulnard & Huygens (100), using β-glycerophosphate as substrate, found no activity in 1-cell mouse embryos, but activity was evident from the 2-cell stage. Only interblastomeric plasma membranes were stained during cleavage and at the morula stage. By the late blastocyst stage, trophectoderm cells were negative and activity was localized along the membranes of the inner (primitive ectoderm) cells of the ICM.

Clearly, it is impossible to draw a unified picture of alkaline phosphatase localization in preimplantation rodent embryos from these conflicting data. A number of variables can be cited to explain the discrepancies from one series of experiments to the next, including the fixative used and time of fixation, the choice of substrate and time of incubation, the pH of the incubation mix, and the means of inspection of the embryos, i.e. whole mounts vs sections, light vs electron microscopy. Until standardized and reproducible procedures are used, it seems premature to draw any conclusions about differential gene expression by cells in the preimplantation embryo from alkaline phosphatase analyses.

Cell Surface Properties

Ducibella (101) has reviewed data from ultrastructural studies that indicate that at late cleavage stages the surface membranes of the mammalian embryo begin to undergo major changes, with the end result that cellular junctions and associations become more intimate. Presumably as a direct consequence of these alterations, compaction of the blastomeres occurs, the outer cells pump fluid from the environment, and eventually the blastocoelic cavity forms, signalling the onset of the blastocyst stage. The acquisition of sophisticated junctional complexes is much more prevalent on outside than on inside cells. Indeed, based on the nature of these membrane alterations, Ducibella (101) has proposed that by the blastocyst stage, trophectoderm cells "differentiate into a well-developed epithelium." Biochemical studies dealing with alterations in surface membrane structure during preimplantation development have lagged far behind ultrastructural analyses. Generally, the former studies fall into three categories: direct biochemical analyses, investigations of membrane-associated transport systems and enzymes, and experiments utilizing probes such as lectins or antibodies.

BIOCHEMICAL ANALYSES   To date only two studies involving direct biochemical analysis of early embryonic membranes have appeared (102,

103). In the first, Pinsker & Mintz (102) labeled embryos from the 2- to 8-cell stage or from the 16-cell to early blastocyst stage with radioactive glucosamine, incubated the cells with trypsin to release surface glycopeptides, and then subjected the released material to exhaustive treatment with pronase. Analysis by Sephadex chromatography suggests a slight upward shift in mean molecular weight of the resultant glycopeptide fragments from the earlier to the later developmental stage embryos. As the authors point out, however, these experiments do not allow them to distinguish between the appearance during development of new glycoprotein components and the reorganization of existing glycoproteins, leading to susceptibility of a different population of glycoproteins to the trypsinization.

In the second study, Muramatsu et al (103) labeled cells continuously from the 2-cell to late morula/early blastocyst stages with fucose and analyzed, in a way similar to that of Pinsker & Mintz, the size distribution of the resultant pronase-resistant, fucose-containing material. They observed that a substantial proportion of the radioactivity was contained in a fraction excluded from the Sephadex G-50 column; a similar result was observed with the relatively undifferentiated embryonal carcinoma cells, but not with their differentiated derivatives. The authors were tempted to speculate that the presence of high molecular weight, fucose-containing material correlates with the early embryonic phenotype. It should be noted that Pinsker & Mintz (102) in their experiments found that only a small proportion of the pronase-treated, glucosamine-labeled material was excluded from a Sephadex G-50 column.

TRANSPORT SYSTEMS   Amino acid uptake studies have provided some information about the function of the cell membrane of preimplantation embryos. Initial studies indicated an increase in uptake of a number of amino acids between mouse 8-cell and morula stages (64, 104–106). Epstein & Smith (64) provided the first evidence that amino acid uptake in this system is an active process by demonstrating that it is temperature-dependent, subject to competition by related amino acids, and capable of concentration against a gradient. Borland & Tasca (107, 108) confirmed these results and showed further that the uptake of two amino acids, leucine and methionine, is, respectively, partially and completely sodium-dependent by the early blastocyst stage, but completely sodium-independent at the 4-cell stage. Subsequently, in an elegant study, DiZio & Tasca (109) demonstrated that sodium-dependent transport in mouse blastocysts, in this case of alanine and glycine, is blocked by ouabain, an inhibitor of sodium-potassium ATPases, but only when blastocoel is collapsed by administration of cytochalasin B. The effect is specific in that the transport of lysine, which is not sodium-dependent, is affected neither by ouabain, cytochalasin B, nor a combination of the two. On the basis of these observations, DiZio & Tasca concluded that the sodium-dependent transport of amino acids involves ATPases on the blastocoelic, as opposed to external surface, membranes of trophectoderm cells. This conclusion is supported by histochemical analyses of ATPase activities in blastocysts [(99); also see below]. The similarities between amino acid transport in this system and in differentiated epithelial systems (see 109) is, therefore, striking, and perhaps provides the best biochemical support for this relationship, suggested by Ducibella (101) on morphological grounds.

ENZYMES   Vorbrodt et al (99) studied the localization of several membrane-associated enzymes in preimplantation mouse embryos by histochemistry and electron microscopy. In some cases, as with cyclic AMP phosphodiesterase and adenylate cyclase, enzyme activity is detectable at early stages and there is no significant selectivity in distribution among cells. However, two other enzymes, 5'-nucleotidase and magnesium-activated ATPase, appear, by the blastocyst stage, to be restricted largely to the surface membrane of trophectoderm cells. Sodium-potassium ATPase activity is observed on both ICM and trophectoderm cells but, as mentioned above, not on the free external surfaces of the latter. As indicated previously, the observation that alkaline phosphatase activity is much more evident on trophectoderm than on ICM cells (99) is inconsistent with the results of a similar study (100).

LECTINS   Other attempts to detect changes in the cell surface membranes of preimplantation embryos have involved the use of lectins. Pienkowski (110) found little difference in the ability of concanavalin A (conA) to bind to, or agglutinate, mouse embryos from the 1- to 8-cell stage. Rowinski et al, however, reported that agglutination by conA requires higher concentrations of the lectin beyond the 8-cell stage; blastocysts do not agglutinate even in the presence of very large amounts (5 mg/ml) of conA (111). This resistance is specifically a property of trophectoderm surface membranes, since isolated (16) ICMs are agglutinated readily by low concentrations (10 μg/ml) of conA. Webb et al (112) obtained direct evidence of a change in the population of conA-binding proteins on the external surfaces of preimplantation embryos by radioiodinating surface proteins, treating the embryos with conA followed by anti-conA antiserum, and then analyzing the precipitated proteins by electrophoresis. Sobel & Nebel (113) claimed, from conA-induced hemadsorption studies, that there are differences between the plasma membranes of trophectoderm cells in contact either with the ICM or the blastocoel at the late mouse blastocyst. The observation of diminished binding of three different lectins to hamster blastocysts when compared with earlier stages (114) is also consistent with the view that the nature or organization of the outer surface membranes of the mammalian embryo changes by the blastocyst stage.

ANTIBODIES   The use of immunological probes to characterize surface membranes of early mammalian embryos has become widespread. Only those studies that bear directly upon the question of embryo differentiation are considered here; the reader is referred to recent reviews by Edidin (115) and Jenkinson & Billington (116) for a more general assessment of the subject.

A number of studies relate to the question of the time of expression of the embryonic genome in mouse embryos. In one such study (117) using antisera against the H-Y antigen, expressed only on male cells, and another (118) utilizing antisera against a group of so-called "non-H-2" antigens, there is the suggestion of transcription and translation of paternally inherited genes by the 8-cell stage. In a third study, Krco & Goldberg (119) proposed that paternal major histocompatibility (H-2) antigens are also expressed on the surface of 8-cell embryos. However, a good deal of controversy surrounds the question of expression of H-2 antigens by preimplantation mouse embryos. With the exception of the study by Krco & Goldberg (119), investigators have consistently failed to detect H-2 production prior to the blastocyst stage (112; see 116). Even at the blastocyst stage, by some techniques H-2 expression is detected on the trophectoderm surface (116, 120) whereas in others it appears to be limited to cells of the ICM (112). Furthermore, Håkansson and his colleagues (121, 122) and Searle et al (120) have proposed that the appearance of H-2 antigens on the trophectoderm surface is transiently restricted to a narrow window of developmental time prior to implantation and that this period can be prolonged by preventing blastocysts from implanting by ovariectomizing the mother.

A further series of experiments dealing with the time of expression of the embryonic genome involves the use of antisera against sperm from mice carrying t mutations [(123); for reviews of t mutations and their possible relationship to early embryonic differentiation, see (124, 125)]. These incompletely defined and undoubtedly complex (126) antisera are reactive with 8-cell mouse embryos obtained from crosses between wild-type females and $+/t^x$ heterozygous males (123). Since the mutant t alleles are paternally contributed, it appears likely that the presence of these antigens reflects embryonic gene expression.

Among the immunological probes used to characterize the surface constituents of preimplantation mouse embryos are antisera generated against embryonal carcinoma cells (127–131). The most widely used of these is an antiserum directed against one or more surface antigens of the F9 embryonal carcinoma cell line. Apparently, all embryonic cells at all stages of preimplantation development react with anti-F9 antiserum (132). Although intact rabbit anti-F9 antibodies do not interfere with preimplantation development in vitro, the possible developmental significance of "F9 antigens" is underscored by the observation that Fab fragments of anti-F9 antibodies prevent compaction of mouse morulae (133, 134). Notwithstanding several control experiments carried out by these investigators, the possibility of a nonspecific effect due to swamping of the cell surface by very large numbers of Fab fragments (as opposed to Fab fragments from other antibodies against antigens that are likely to be present on the cell surface at relatively low frequencies) merits consideration in view of the report that appropriate concentrations of conA will also prevent compaction (135).

Two antiteratocarcinoma antisera are of interest in that they react in a selective manner with cells of the blastocyst. One, generated against embryonal carcinoma line PCC4 and absorbed with F9 cells, does not react with any cells up to the late blastocyst stage, at which time it reacts with cells of the ICM, but not the trophectoderm (131). Another antiserum, prepared against 402AX cells, cultures of which contain embryonal carcinoma cells but also some endoderm-like cells, reacts, after appropriate absorption, at the early blastocyst stage, with trophectoderm cells bordering the blastocoel, but not with those overlying the ICM (129, 130). Although preimplantation ICMs were not tested for reactivity, when cultured to stages analogous to postimplantation stages, ICM cells were positive (129, 130).

In a single successful study in which antisera were generated against the preimplantation embryo per se, Wiley & Calarco (136) analyzed an antimouse blastocyst antiserum for its ability to bind to and affect in vitro development of preimplantation embryos. They observed minimal binding of the antiserum until the 4-cell stage; unexpectedly, the antiserum binds maximally at the 8- to 12-cell stages, and declines notably at the morula and blastocyst stages, even though blastocysts were used for immunization. The antiblastocyst serum has development-arresting properties as early as the 2-cell stage. Searle & Jenkinson (137) have recently obtained an antiserum against eighth gestation day mouse ectoplacental cones. This antiserum is first reactive with embryos at the 8-cell stage; reactivity segregates thereafter such that by the blastocyst stage trophectoderm cells are positive whereas ICM cells are negative.

Immunological probes have also provided some information about the production of extracellular matrix material by preimplantation embryos. Zetter & Martin (138) observed that the large, external, transformation-sensitive (LETS) protein is observed for the first time at the late blastocyst stage and that it is localized on ICM cells. Sherman et al [(139); M. I. Sherman, S. Gay, and R. Gay, manuscript in preparation] have observed that mouse embryos are capable of producing type III procollagen and collagen as early as the morula stage. Types I and II collagen were not detected immunologically in preimplantation embryos.

ANTIMETABOLITES

Inhibitors of DNA Synthesis and Cell Division

In view of the concept of a "quantal cell cycle" necessary for expression of differentiated properties of certain cell types (see 140), it is appropriate to consider the effects on early development of inhibitors of replication and cell division. Snow (141, 142) has reported that cleavage stage embryos cultured in the presence of concentrations of thymidine high enough to elicit a "thymidine block" effect do indeed contain a reduced cell number after 72 hr of culture but do not show any developmental abnormalities; the resultant embryos form undersized, but otherwise morphologically normal, blastocysts. Different results were obtained when embryos were cultured in thymidine of high specific activity but the abnormalities in this case were attributed to radiation effects [(141, 142); see section on radiation below]. The immediate development of blastocysts treated with cytosine arabinoside, an inhibitor of DNA synthesis, i.e. expansion of the blastocoel and hatching of the blastocyst, is apparently not interfered with (143). Thereafter, the drug, at high concentrations, is toxic to all cells, but at lower concentrations, toxicity is selective: trophectoderm cells give rise to trophoblast derivatives which, although they appear to have less DNA than their untreated counterparts, nevertheless seem to differentiate normally. On the other hand, ICM cells die within 72 hr of exposure to cytosine arabinoside (143).

Colcemid at high concentrations does appear to exert a developmental block on blastocysts in that it prevents them from hatching from their zona pellucidas (143). However, this is probably due to an effect on microtubule assembly unrelated to cell division. ICM cell derivatives are also more sensitive to Colcemid than are trophectoderm derivatives (143).

One type of experiment not involving antimetabolites but concerned with the issue of a critical cell cycle during differentiation concerns the development of cells disaggregated from 4-cell and 8-cell embryos. These studies indicate that neither cell number nor further cell division is important for compaction and cavitation (144, 145).

In summary, the bulk of the pertinent data suggests that insofar as early developmental steps in mammalian embryogenesis is concerned, a "quantal cell cycle" either does not exist or, if it does, occurs at very early stages in cleavage.

Inhibitors of RNA and Protein Synthesis

The literature contains at least fifteen papers on the effects of actinomycin D on preimplantation mouse embryos; the earliest report is by Silagi (146) and the most recent by Epstein & Smith (147). These studies are listed in the latter reference, and since for the most part they describe similar results and reach similar conclusions, they are not referred to here individually except for those containing points that are of particular pertinence to this review. Generally, the administration of actinomycin D at a concentration of 0.1 μg/ml blocks cell division and further development of mouse embryos at any preimplantation stage. At this level of antimetabolite, RNA synthesis is inhibited by more than 85% at all stages (49). Lesser concentrations of actinomycin D have progressively less dramatic effects upon early embryogenesis: 0.01 μg/ml actinomycin D, particularly during the first few cleavages, often allows one further cell cycle to occur followed by a subsequent block to development. In most instances, concentrations of 0.001 μg/ml and lower of actinomycin D are without effect. RNA synthesis is still effectively inhibited ($>$ 85%) during early cleavage stages by actinomycin D concentrations of 0.01 μg/ml (49), whereas at 0.001 μg/ml of the antimetabolite, interference with RNA synthesis is minimal (105). There are three interesting observations on the developmental effects of actinomycin D: Epstein & Smith (147) observed that compaction takes place even when 8-cell embryos are treated with as much as 10 μg/ml actinomycin D, which suggests that the RNA and/or proteins necessary for that event have already been synthesized; conversely, blastocysts demonstrate a marked sensitivity to actinomycin D with regard to viability. Although development of embryos at earlier stages is blocked by the antimetabolite, the cells do not die within a 24 hr incubation period; on the other hand, blastocysts tend to degenerate in concentrations of actinomycin D as low as 0.001 μg/ml (147–149). Epstein & Smith (147) investigated this phenomenon in detail but were unable to attribute the effects to any specific action of the drug. Finally, Rowinski et al (149) observed that whereas blastocysts can survive only for a few days with continuous exposure to 0.005 μg/ml actinomycin D, about half develop to some extent following a 24 hr pulse at that concentration. The result after 6 days of culture is similar to that described in the previous section for cytosine arabinoside: trophectoderm cells appear to differentiate normally into trophoblast, whereas ICM cells die. Similar effects are found with another RNA synthesis inhibitor, cordycepin (149).

The greatest problems in interpreting the results of treating embryos with actinomycin D are its side and secondary effects. Several investigators have in these studies cautioned the reader that actinomycin D can have detrimental effects on respiration; a single study by Golbus et al (150) which shows that ATP levels in 1-cell embryos do not change in response to actinomycin D does not substantially alleviate this concern. Observations that actinomycin D ultimately inhibits protein synthesis severely (49, 147) also complicate efforts to determine whether the cause of developmental failure is inability to transcribe adequately new mRNAs or to translate preformed ones. To some extent these difficulties in interpretation are alleviated by the use of another antimetabolite affecting RNA synthesis, α-amanitin.

α-Amanitin is presumed to be freer of side effects than actinomycin D, and is thought to act primarily at the level of inhibition of mRNA synthesis (see 39). However, even though this antimetobolite is supposed to be devoid of action on the nucleolar RNA polymerase, type I, at relatively high concentrations it has adverse effects, presumably secondary, on nucleolar morphology (150) and rRNA synthesis (151) in preimplantation embryos. Warner & Hearn (151) have also detected an inhibitory effect on DNA synthesis by α-amanitin at 50 μg/ml.

Notwithstanding these incidental effects of α-amanitin, some useful facts with implications concerning differentiation have emerged from studies with it. First, unlike actinomycin D, α-amanitin rarely causes immediate cessation of cell division in cleavage stage mouse (150, 152) or rabbit (153) embryos, even at concentrations as high as 450 μg/ml. Characteristically, two or three cell divisions take place prior to developmental arrest. This is in spite of the fact (153) that in rabbit embryos, α-amanitin causes almost complete cessation of RNA synthesis (there being no synthesis of rRNA at this time; see section on RNA analyses, above). Furthermore, as is the case with actinomycin D, compaction of blastomeres is relatively resistant to α-amanitin added at the 8-cell stage, whereas doses as low as 0.1 μg/ml can interfere with cavitation (18, 151, 152). These observations tend to suggest that preformed, relatively stable, mRNAs can support the initial stages of preimplantation development but that normal development from the morula stage requires the synthesis of new mRNA.

Golbus et al (150) have demonstrated that the characteristic developmental increase in hypoxanthine-guanine phosphoribosyl transferase activity and decrease in guanine deaminase activity do not occur by the morula stage in the presence of α-amanitin from the 2-cell stage, despite the fact that cell numbers are normal. On the other hand, these changes in enzyme activity are observed when the drug is not added until the 4- or 8-cell stage (150). These results could be interpreted to provide further support for the idea that stable mRNAs are synthesized shortly after fertilization. Finally, and perhaps most importantly in the context of this review, Johnson et al

(18) have investigated the effect of incubation of morulae with α-amanitin on the appearance of stage-specific and cell-specific (i.e. trophectoderm vs. ICM) proteins at the blastocyst stage (see section on protein electrophoretic profiles, above). Whereas in some cases the antimetabolite does not suppress the appearance of these stage-specific proteins, in most instances, particularly those involving cell-specific proteins, the drug blocks or reduces their synthesis (18). If the assumption that the α-amanatin effect is mediated in this case by interference with mRNA synthesis is correct, then the results are consistent with the view that many of the genes coding for cell- and stage-specific proteins are activated only shortly before the time of expression of these proteins.

Some investigators have examined the effects of protein synthesis inhibitors on early embryonic development and, not surprisingly, they observed dose-related developmental blocks. The only study of interest here is that of Rowinski et al (149): these investigators observed that culture of blastocysts in low levels of cycloheximide leads to selective killing of ICM cells.

Bromodeoxyuridine

In a number of systems in which cell types are capable of differentiating in culture, 5-bromodeoxyuridine treatment appears to prevent the appearance of the differentiated phenotype, i.e. the production of "luxury" macromolecules, without affecting cell viability or proliferation (reviewed in 154). Accordingly, a number of investigators have studied the effects of bromodeoxyuridine on the development of mouse embryos in vitro in order to determine whether any of the early steps are selectively affected (143, 155–159). There is general agreement that all but very high doses of bromodeoxyuridine do not interfere with early cleavage, and that the most notable effect of administering the drug is an eventual block to normal cavitation (155, 156, 158, 159). In large part, the explanation of this delayed effect is probably related to poor permeability, as is generally the case with nucleosides, until the morula and blastocyst stages (155). Most investigators observed inhibition of cell division such that by the late morula or blastocyst stage, surviving bromodeoxyuridine-treated embryos contain 20–70% fewer cells than controls (155, 158, 159). This is contrary to what is observed in most other systems wherein bromodeoxyuridine does not interfere with cell division, even at concentrations one or more orders of magnitude greater (154). Consequently, it is difficult to assess whether the block to cavitation truly represents an effect on differentiation or merely a nonspecific response reflecting the unhealthy state of the embryos.

The result of treating morulae or blastocysts with high concentrations of bromodeoxyuridine is failure of the blastocysts to hatch from their zona pellucidas (143, 156, 157, 159). At lower concentrations of the antimetabolite, hatching, outgrowth, and normal differentiation of trophoblast cells will occur, but ICM cells will not survive (143, 150, 157). At still lower concentrations, ICM cells will survive for several days but will not become organized into characteristic bilayered structures (157, 159). All these effects can be preprogrammed irreversibly by exposing the embryos to appropriate concentrations of bromodeoxyuridine at the morula or blastocyst stage for periods as short as 24 hr (143, 155, 156, 159). In all cases in which the symptoms appear, there is evidence of incorporation of the analog into DNA and the adverse effects of bromodeoxyuridine are largely avoided in a competitive manner by the addition of thymidine to the culture medium (143, 155, 157–159).

Tunicamycin

Tunicamycin is an antibiotic that interferes with the normal glycosylation of proteins, leading subsequently to the secretion and array on the cell surface of glycoproteins deficient in their carbohydrate moieties (see 160). The above section on protein analyses of preimplantation embryos considers the nature of alterations in cell surface structure of developing embryos. In an effort to determine the developmental effects of preventing normal membrane biogenesis, embryos were treated with tunicamycin. Preliminary experiments have revealed that preimplantation embryos are relatively sensitive to the antimetabolite. Whereas untransformed cells in culture are resistant to tunicamycin concentrations of 1 μg/ml (160), 50% of 2-cell mouse embryos are blocked from development to the compacted morula and hatched blastocyst stages in the continuous presence of only 0.03 and 0.009 μg/ml, respectively (S. B. Atienza-Samols and M. I. Sherman, unpublished observations). It remains to be determined whether the observed effects are related specifically to incomplete protein glycosylation or are a consequence of low levels of inhibition of protein synthesis observed with this drug (160).

Radiation

The effects of treating embryos with X-rays are similar to those caused by α-amanitin (see above), i.e. developmental arrest occurs, but only after a period of continued, apparently normal, development. Thus, embryos irradiated at the 2-cell or 4-cell stage begin to show developmental abnormalities only late in cleavage or at the morula stage; whereas many of the embryos do form morulae following doses of 200–400 R, about half fail to cavitate (161–164). Of those irradiated embryos forming blastocysts, only some will hatch from their zona pellucidas. Embryos treated beyond the 4-cell stage require high doses of irradiation for the blockage of cavitation and hatching. As is the case with several of the antimetabolites described

above, the effect of treatment of embryos with X-rays during preimplantation stages is the ultimate disappearance of the ICM at stages in vitro
analogous to those following implantation *in utero* (164). This effect is also
evidenced when cleavage stage embryos are incubated continuously with
[^{3}H]thymidine at 0.01–0.1 μCi/ml (141, 142). Although there is no evidence of adverse effects over the next three cell divisions, when control
embryos have reached the 32-cell stage, embryos incubated with [^{3}H]thymi-
dine contain about half the number of blastomeres. Cytological and biological studies indicate that when these treated embryos reach the blastocyst
stage, they lack ICM cells (141, 142). This effect is presumably due to
radiation damage because embryos exposed to equivalent concentrations of
unlabeled thymidine are unaffected and even those treated with high enough
doses to cause a thymidine block (see above) do not behave in the same way.
Furthermore, morulae and blastocysts developed from embryos treated
with [^{3}H]thymidine show evidence of increased frequency of chromosomal
abnormalities (141, 142). Snow (141, 142) has explained that the ICM
would be more susceptible to radiation effects because ICM progenitor cells
have a shorter cell cycle time than the other blastomeres in the embryo (13)
and might incorporate labeled thymidine at a faster rate. A larger endogenous pool of thymidine in the outside blastomeres which eventually give rise
to trophectoderm cells might also explain the differential effect.

Horner & McLaren (165) and Kelly & Rossant (166) observed some
interesting long-term effects of [^{3}H]thymidine labeling. These investigators
reported that cells of embryos exposed to the labeled nucleoside either at
very low concentrations or for very short periods participated in the formation of normal looking blastocysts, but showed evidence of an adverse effect
following implantation.

OVERVIEW

In his consideration of differential gene expression, Davidson (20) mentions
the possibility that cytoplasmic inheritance might play a role in determination of, and the acquisition of differential properties by, early embryonic
cells. A prime example is the difference in protein synthetic profiles between
the first two blastomeres, one of which contains a polar lobe, in the embryo
of the snail, *Ilyanassa* (167). A differential synthetic pattern between normal embryos and those in which the lobe has been removed persists through
cleavage (168). These experiments, taken together with the observation that
actinomycin D does not influence the extent of protein synthesis during
cleavage (168), provide evidence that cytoplasmic inheritance, and not early
activation of the embryonic genome, is responsible for these results. A more
direct demonstration of cytoplasmic inheritance at the molecular level is the

recent finding by Rodgers & Gross (169) that the population of maternally inherited RNAs transcribed from unique sequence DNA in the macromeres plus mesomeres portion of the 16-cell sea urchin embryo is more diverse than that in the micromeres portion. To date, technical difficulties have prevented analyses of this sort on cleavage stage mammalian embryos. However, even though antimetabolite studies, enzyme analyses, and total protein synthetic profiles together support the view that stable maternal mRNAs are translated in mammalian embryos throughout much of preimplantation development, there is no evidence of a direct effect of these maternal mRNAs on differential behavior of blastomeres at any developmental stage. In fact, biological studies suggest that in mammals, blastomeres do not inherit from the egg *any* cytoplasmic determinants that ultimately influence the direction of their differentiation, or, if such determinants are present, their effects can be overruled by manipulation of blastomere organization (4, 14, 144). It is, therefore, reasonable to assume that differentiation in the mammalian embryo depends upon activation and expression of embryonic genes.

Based on indirect evidence from RNA synthetic patterns and direct evidence from enzyme and cell surface analyses, it is clear that embryonic gene expression in the mouse embryo, and probably other mammalian embryos as well, takes place by the 8-cell stage and possibly as early as the 2-cell or even 1-cell stage. These embryonic mRNAs probably assume increasing developmental importance at the morula and blastocyst stages. This is quite different from embryos of lower animals such as amphibians and echinoderms in which there is a substantial hiatus between fertilization and the time of expression of many, although not all, embryonic genes (see 20). The nature of the stimulus for embryonic gene expression in mammalian embryos is unknown. It is possible that time or number of cell divisions from fertilization, duration of the phases of the cell cycle, or some combination of the above is involved.

The developmental pattern that emerges from an analysis of the experimental data described above is one in which a continuum of changes occurs in the population of macromolecules in, and on the surface of, preimplantation blastomeres. The result appears to be an increase in sophistication at each stage, at least insofar as metabolism, enzyme content, and surface characteristics of the cells are concerned. Not only do the cells acquire properties different from those at previous stages, but eventually cells at the same stage begin to differ from each other. Convincing evidence for this comes from cell cycle times, total protein analyses, enzyme and cell surface composition, and responses to antimetabolites and radiation. From these studies, the earliest estimate so far available for the stage at which cell populations differ in terms of gene expression in the mouse embryo is the morula (16- to 32-cell) stage. It should be noted that this is close to the time that trophectoderm cells in the mouse embryo are assumed to be determined but *prior* to the time at which the developmental potential of most ICM cells is fixed (9, 18). This is the rationale for the contention by Johnson et al (18) that determination represents nothing more than an early stage, and not necessarily the earliest one, of cell differentiation, and the argument is a compelling one. After all, some critical and differential biochemical change must have taken place in a cell that appears suddenly to have lost the option to develop in all but a particular direction.

Finally, the evidence presented above is consistent with the idea that differentiation is a preimplantation event in mammalian embryogenesis if one accepts the kind of flexible definition of "differentiation" offered by Johnson et al [(18); see section above on determination and differentiation]. Do preimplantation embryos perform "specialized functions" as required by Davidson's (20) more stringent definition of the term differentiation? The cells of the blastocyst have acquired several properties characteristic of "adult" differentiated cell types, but not observed, for instance, in relatively undifferentiated embryonal carcinoma cells, e.g. X-chromosome inactivation and, probably, expression of H-2 antigens. Most impressive, perhaps, is evidence from amino acid transport and ultrastructural studies that the trophectoderm layer of the blastocyst resembles organized, functional epithelia. However, one could easily argue that these properties as such do not necessarily imply specialized functions. A final decision on this issue must, therefore, await further study.

Literature Cited

1. Davies, J., Hesseldahl, H. 1971. In *The Biology of the Blastocyst*, ed. R. J. Blandau, pp. 27–48. Chicago, Ill: Univ. Chicago Press
2. McLaren, A. 1972. In *Reproduction in Mammals*, ed. C. R. Austin, R. V. Short, 2:1–42. London: Cambridge Univ. Press
3. Graham, C. F. 1971. In *Control Mechanisms of Growth and Differentiation*, ed. D. D. Davies, M. Balls, pp. 371–78. London: Cambridge Univ. Press
4. Herbert, M. C., Graham, C. F. 1974. *Curr. Top. Dev. Biol.* 8:151–78
5. Gardner, R. L., Papaioannou, V. E. 1975. In *The Early Development of Mammals*, ed. M. Balls, A. E. Wild, pp. 107–32. London: Cambridge Univ. Press
6. Gardner, R. L. 1975. In *The Developmental Biology of Reproduction*, ed. C. L. Markert, pp. 207–38. New York: Academic

7. Gardner, R. L., Rossant, J. 1976. In *Ciba Found. Symp.* 40:5–25
8. Rossant, J. 1977. In *Development in Mammals*, ed. M. H. Johnson, 2:119–50. Amsterdam: North-Holland
9. Rossant, J., Papaioannou, V. E. 1977. In *Concepts in Mammalian Embryogenesis*, ed. M. I. Sherman, pp. 1–36. Cambridge, Mass: MIT Press
10. Snell, G. D., Stevens, L. C. 1966. In *Biology of the Laboratory Mouse*, ed. E. L. Green, pp. 205–45. New York: Dover Publ. 2nd ed.
11. Rugh, R. 1968. *The Mouse. Its Reproduction and Development.* Minneapolis, Minn: Burgess Publ. Co. 430 pp.
12. Theiler, K. 1972. *The House Mouse.* Berlin: Springer. 168 pp.
13. Barlow, P. W., Owen, D. A. J., Graham, C. F. 1972. *J. Embryol. Exp. Morphol.* 27:431–45
14. Hillman, N., Sherman, M. I., Graham,

C. F. 1972. *J. Embryol. Exp. Morphol.* 28:263–78
15. Rossant, J. 1975. *J. Embryol. Exp. Morphol.* 33:991–1001
16. Solter, D., Knowles, B. B. 1975. *Proc. Natl. Acad. Sci. USA* 72:5099–5102
17. Rossant, J. 1975. *J. Embryol. Exp. Morphol.* 33:979–90
18. Johnson, M. H., Handyside, A. H., Braude, P. R. 1977. See Ref. 8, pp. 67–97
19. Pedersen, R. A., Spindle, A. I., Wiley, L. M. 1977. *Nature* 270:435–37
20. Davidson, E. H. 1976. *Gene Activity in Early Development.* New York: Academic. 452 pp. 2nd ed.
21. Luthardt, F. W., Donahue, R. P. 1973. *Exp. Cell Res.* 82:143–51
22. Abramczuk, J., Sawicki, W. 1975. *Exp. Cell Res.* 92:361–71
23. Mukherjee, A. B. 1976. *Proc. Natl. Acad. Sci. USA* 73:1608–11
24. Luthardt, F. W., Donahue, R. P. 1975. *Dev. Biol.* 44:210–16
25. Sawicki, W., Abramczuk, J., Blaton, O. 1978. *Exp. Cell Res.* 112:199–205
26. Daniel, J. C. Jr. 1964. *Am. Nat.* 98:85–87
27. Schiffner, J., Spielmann, H. 1976. *J. Reprod. Fertil.* 47:145–47
28. Gamow, E. I., Prescott, D. M. 1970. *Exp. Cell Res.* 59:117–23
29. Graham, C. F. 1973. In *The Cell Cycle in Development and Differentiation,* ed. M. Balls, F. S. Billett, pp. 293–310. London: Cambridge Univ. Press
30. Dalcq, A., Pasteels, J. 1955. *Exp. Cell Res.* 3:72–97 (Suppl.)
31. Oprescu, S. T., Thibault, C. 1965. *Ann. Biol. Anim. Biochem. Biophys.* 5:151–56
32. Zavarzin, A. A., Samoshkina, N. A., Dondua, A. K. 1976. *Zh. Obshch. Biol.* 27:697–709
33. Graham, C. F. 1973. In *The Regulation of Mammalian Reproduction,* ed. S. J. Segal, R. Crozier, P. A. Corfman, P. G. Condliffe, pp. 286–301. Springfield, Ill: Charles C Thomas
34. Biggers, J. D., Stern, S. 1973. *Adv. Reprod. Physiol.* 6:1–59
35. Manes, C. 1975. See Ref. 6, pp. 133–63
36. Schultz, G. A., Church, R. B. 1975. In *The Biochemistry of Animal Development,* ed. R. Weber, 3:47–90. New York: Academic
37. Epstein, C. J. 1975. *Biol. Reprod.* 12:82–105
38. Schultz, G. A., Tucker, E. B. 1977. See Ref. 8, Vol. 1, pp. 69–97
39. Warner, C. M. 1977. See Ref. 8, Vol. 1, pp. 99–136

40. Van Blerkom, J., Manes, C. 1977. See Ref. 9, pp. 37–94
41. Olds, P. J., Stern, S., Biggers, J. D. 1973. *J. Exp. Zool.* 186:39–46
42. Manes, C. 1969. *J. Exp. Zool.* 172:303–10
43. Woodland, H. R., Graham, C. F. 1969. *Nature* 221:327–32
44. Knowland, J., Graham, C. 1972. *J. Embryol. Exp. Morphol.* 27:167–76
45. Young, R. J., Sweeney, K., Bedford, J. M. 1978. *J. Embryol. Exp. Morphol.* 44:133–48
46. Church, R. B. 1970. In *Congenital Malformations,* ed. F. C. Fraser, V. A. McKusick, R. Robinson, pp. 19–28. Amsterdam: Excerpta Medica
47. Clegg, K. B., Pikó, L. 1975. *J. Cell Biol.* 67:72a (Suppl.)
48. Manes, C. 1971. *J. Exp. Zool.* 176:87–95
49. Monesi, V., Molinaro, M., Spalletta, E., Davoli, C. 1970. *Exp. Cell Res.* 59:197–206
50. Ellem, K. A. O., Gwatkin, R. B. L. 1968. *Dev. Biol.* 18:311–30
51. Pikó, L. 1970. *Dev. Biol.* 21:257–79
52. Levey, I. L., Stull, G. B., Brinster, R. L. 1978. *Dev. Biol.* 64:140–48
53. Schultz, G., Manes, C., Hahn, W. E. 1973. *Dev. Biol.* 30:418–26
54. Schultz, G. A. 1973. *Exp. Cell. Res.* 82:168–74
55. Young, R. J. 1977. *Biochem. Biophys. Res. Commun.* 76:32–39
56. Siracusa, G. 1972. *Exp. Cell Res.* 78:460–62
57. Warner, C. M., Versteegh, L. R. 1974. *Nature* 248:678–80
58. Siracusa, G., Vivarelli, E. 1975. *J. Reprod. Fertil.* 43:567–69
59. Moore, G. P. M. 1975. *J. Embryol. Exp. Morphol.* 34:291–98
60. Versteegh, L. R., Hearn, T. F., Warner, C. M. 1975. *Dev. Biol.* 46:430–35
61. Schultz, G. A., Manes, C., Hahn, W. E. 1973. *Biochem. Genet.* 9:247–59
62. Church, R. B., Brown, I. R. 1972. In *Nucleic Acid Hybridization in the Study of Cell Differentiation,* ed. H. Ursprung, pp. 11–24. New York: Springer
63. Brinster, R. L. 1967. *J. Reprod. Fertil.* 13:413–20
64. Epstein, C. J., Smith, S. A. 1973. *Dev. Biol.* 33:171–84
65. Manes, C., Daniel, J. C. Jr. 1969. *Exp. Cell Res.* 55:261–68
66. Epstein, C. J., Smith, S. A. 1974. *Dev. Biol.* 40:233–44
67. Van Blerkom, J., Brockway, G. O. 1975. *Dev. Biol.* 44:148–57

68. Van Blerkom, J., Manes, C. 1974. *Dev. Biol.* 40:40–51
69. Van Blerkom, J., Barton, S. C., Johnson, M. H. 1976. *Nature* 259:319–21
70. Handyside, A. H., Johnson, M. H. 1978. *J. Embryol. Exp. Morphol.* 44:191–9
71. Levinson, J., Goodfellow, P., Vadeboncoeur, M., McDevitt, H. 1978. *Proc. Natl. Acad. Sci. USA.* 75:3332–36
72. Van Blerkom, J., McGaughey, R. W. 1978. *Dev. Biol.* 63:151–64
73. Biggers, J. D. 1971. *J. Reprod. Fertil.* 14:41–55 (Suppl.)
74. Wolf, U., Engel, W. 1972. *Humangenetik* 15:99–118
75. Brinster, R. L. 1973. See Ref. 33, pp. 302–20
76. Brinster, R. L. 1974. *J. Anim. Sci.* 38:1003–12
77. Spielmann, H., Erickson, R. P., Epstein, C. J. 1974. *J. Reprod. Fertil.* 40:367–73
78. Chapman, V. M., Whitten, W. K., Ruddle, F. W. 1971. *Dev. Biol.* 26:153–58
79. Brinster, R. L. 1973. *Biochem. Genet.* 9:187–91
80. Wudl, L., Chapman, V. 1976. *Dev. Biol.* 48:104–9
81. Chapman, V. M., Adler, D. A., Labarca, C., Wudl, L. 1976. See Ref. 7, pp. 115–31
82. Chapman, V. M., West, J. D., Adler, D. A. 1977. See Ref. 9, pp. 95–135
83. Martin, G. R., Epstein, C. J., Travis, B., Tucker, G., Yatziv, S., Martin, D. W. Jr., Clift, S., Cohen, S. 1978. *Nature* 271:329–33
84. Graham, C. F. 1977. See Ref. 9, pp. 315–94
85. Lyon, M. F. 1974. *Proc. R. Soc. London Ser. B* 187:243–68
86. Epstein, C. J. 1969. *Science* 163:1078–79
87. Epstein, C. J. 1972. *Science* 175:1467–68
88. Monk, M., Kathuria, H. 1977. *Nature* 270:599–601
89. Monk, M. 1979. In *Genetic Mosaics and Chimeras in Mammals,* ed. L. B. Russel, pp. 239–46. New York: Plenum Tenn: ORNL. In press
90. Kozak, L. P., Quinn, P. J. 1975. *Dev. Biol.* 45:65–73
91. Epstein, C. J., Smith, S. A., Travis, B., Tucker, G. 1978. *Nature* 274:500–2
92. Chapman, V. M., West, J. D., Adler, D. A. 1979. See Ref. 89, pp. 227–37
93. Mulnard, J. 1955. *Arch. Biol. Paris* 66:528–688
94. Dalcq, A. M. 1957. *Introduction to General Embryology,* Chapter 11. Oxford: Oxford Univ. Press. 177 pp.
95. Ortiz, M. E., Carranza, C., Izquierdo, L. 1969. *Arch. Biol. Med. Exp.* 6:37–43
96. Solter, D., Damjanov, I., Škreb, N. 1973. *Z. Anat. Entwicklangsgesch.* 139:119–26
97. Izquierdo, L., Marticorena, P. 1975. *Exp. Cell Res.* 92:399–402
98. Johnson, L. V., Calarco, P. G., Siebert, M. L. 1977. *J. Embryol. Exp. Morphol.* 40:83–89
99. Vorbrodt, A., Konwinski, M., Solter, D., Koprowski, H. 1977. *Dev. Biol.* 55:117–34
100. Mulnard, J., Huygens, R. 1978. *J. Embryol. Exp. Morphol.* 44:121–31
101. Ducibella, T. 1977. See Ref. 8, Vol. 1, pp. 5–30
102. Pinsker, M. C., Mintz, B. 1973. *Proc. Natl. Acad. Sci. USA* 70:1645–48
103. Muramatsu, T., Gachelin, G., Nicolas, J. F., Condamine, H., Jakob, H., Jacob, F. 1978. *Proc. Natl. Acad. Sci. USA* 75:2315–19
104. Monesi, V., Salfi, V. 1967. *Exp. Cell Res.* 46:632–35
105. Tasca, R. J., Hillman, N. 1970. *Nature* 225:1022–25
106. Brinster, R. L. 1971. *J. Reprod. Fertil.* 27:329–38
107. Borland, R. M., Tasca, R. J. 1974. *Dev. Biol.* 30:169–82
108. Borland, R. M., Tasca, R. J. 1975. *Dev. Biol.* 46:192–201
109. DiZio, S. M., Tasca, R. J. 1977. *Dev. Biol.* 59:198–205
110. Pienkowski, M. 1974. *Proc. Soc. Exp. Biol. Med.* 145:464–69
111. Rowinski, J., Solter, D., Koprowski, H. 1976. *Exp. Cell Res.* 100:404–8
112. Webb, C. G., Gall, W. E., Edelman, G. M. 1977. *J. Exp. Med.* 146:923–32
113. Sobel, J. S., Nebel, L. 1976. *J. Reprod. Fertil.* 47:399–402
114. Yanagimachi, R., Nicolson, G. L. 1976. *Exp. Cell Res.* 100:249–57
115. Edidin, M. 1976. In *The Cell Surface in Animal Embryogenesis and Development,* ed. G. Poste, G. L. Nicolson, pp. 127–43 Amsterdam: Assoc. Sci. Publ.
116. Jenkinson, E. J., Billington, W. D. 1977. See Ref. 9, pp. 235–66
117. Krco, C. J., Goldberg, E. H. 1976. *Science* 193:1134–35
118. Muggleton-Harris, A. L., Johnson, M. H. 1976. *J. Embryol. Exp. Morphol.* 35:59–72
119. Krco, C. J., Goldberg, E. H. 1977. *Transplant. Proc.* 9:1367–70
120. Searle, R. F., Sellens, M. H., Elson, J.,

Jenkinson, E. J., Billington, W. D. 1976. *J. Exp. Med.* 143:348–59

121. Håkansson, S., Heyner, S., Sundqvist, K.-G., Bergström, S. 1975. *Int. J. Fertil.* 20:137–40

122. Håkansson, S., Sundqvist, K.-G. 1975. *Transplantation* 19:479–84

123. Kemler, R., Babinet, C., Condamine, H., Gachelin, G., Guenet, J. L., Jacob, F. 1976. *Proc. Natl. Acad. Sci. USA* 73:4080–84

124. Bennett, D. 1975. *Cell* 6:441–54

125. Sherman, M. I., Wudl, L. R. 1977. See Ref. 9, pp. 136–234

126. Artzt, K., Bennett, D. 1977. *Immunogenetics* 5:97–107

127. Artzt, K., Dubois, P., Bennett, D., Condamine, H., Babinet, C., Jacob, F. 1973. *Proc. Natl. Acad. Sci. USA* 70:2988–92

128. Stern, P. L., Martin, G. R., Evans, M. J. 1975. *Cell* 6:455–65

129. Edidin, M., Gooding, L. R. 1975. In *Teratomas and Differentiation*, ed. M. I. Sherman, D. Solter, pp. 109–21. New York: Academic

130. Gooding, L. R., Hsu, Y.-C., Edidin, M. 1976. *Dev. Biol.* 49:479–86

131. Gachelin, G., Kemler, R., Kelly, F., Jacob, F. 1977. *Dev. Biol.* 57:199–209

132. Jacob, F. 1977. *Immunol. Rev.* 33:3–32

133. Kemler, R., Babinet, C., Eisen, H., Jacob, F. 1977. *Proc.Natl. Acad. Sci. USA* 74:4449–52

134. Babinet, C., Kemler, R., Dubois, P., Jacob, F. 1977. *CR Acad. Sci. D* 284:1919–22

135. Ducibella, T., Anderson, E. 1975. *J. Cell Biol.* 67:101a (Suppl.)

136. Wiley, L. M., Calarco, P. G. 1975. *Dev. Biol.* 47:407–18

137. Searle, R. F., Jenkinson, E. J. 1978. *J. Embryol. Exp. Morphol.* 43:147–56

138. Zetter, B., Martin, G. R. 1978. *Proc. Natl. Acad. Sci. USA* 75:2324–28

139. Sherman, M. I., Shalgi, R., Rizzino, A., Sellens, M. H., Gay, S., Gay, R. 1979. In *Ciba Found. Symp.* 64:33–52

140. Bischoff, R., Holtzer, H. 1969. *J. Cell Biol.* 41:188–200

141. Snow, M. H. L. 1973. *J. Embryol. Exp. Morphol.* 29:601–15

142. Snow, M. H. L. 1973. See Ref. 29, pp. 311–24

143. Sherman, M. I., Atienza, S. B. 1975. *J. Embryol. Exp. Morphol.* 34:467–84

144. Tarkowski, A. K., Wroblewska, J. 1967. *J. Embryol. Exp. Morphol.* 18:155–80

145. Sherman, M. I. 1975. See Ref. 5, pp. 145–65

146. Silagi, S. 1963. *Exp. Cell Res.* 32: 149–52

147. Epstein, C. J., Smith, S. A. 1978. *Exp. Cell Res.* 111:117–26

148. Skalko, R. G., Morse, J. M. D. 1969. *Teratology* 2:47–54

149. Rowinski, J., Solter, D., Koprowski, H. 1975. *J. Exp. Zool.* 192:133–42

150. Golbus, M. S., Calarco, P. G., Epstein, C. J. 1973. *J. Exp. Zool.* 186:207–16

151. Warner, C. M., Hearn, T. F. 1977. *Differentiation* 7:89–97

152. Levey, I. L., Troike, D. E., Brinster, R. L. 1977. *J. Reprod. Fertil.* 50:147–50

153. Manes, C. 1973. *Dev. Biol.* 32:453–59

154. Rutter, W. J., Pictet, R. L., Morris, P. W. 1973. *Ann. Rev. Biochem.* 42: 601–46

155. Golbus, M. S., Epstein, C. J. 1974. *Differentiation* 2:143–49

156. Garner, W. 1974. *J. Embryol. Exp. Morphol.* 32:849–55

157. Pedersen, R., Spindle, A. I. 1976. See Ref. 7, pp. 133–54

158. Pollard, D. R., Baran, M. M., Bachvarova, R. 1976. *J. Embryol. Exp. Morphol.* 35:169–78

159. Spindle, A. I. 1977. *J. Exp. Zool.* 202:17–26

160. Duksin, D., Bornstein, P. 1977. *Proc. Natl. Acad. Sci. USA* 74:3433–37

161. Fisher, D. L., Smithberg, M. 1973. *Teratology* 7:57–64

162. Kirkpatrick, J. F. 1974. *Biol. Reprod.* 11:18–21

163. Alexandre, H. L. 1974. *J. Reprod. Fertil.* 36:417–20

164. Goldstein, L. S., Spindle, A. I., Pedersen, R. A. 1975. *Radiat. Res.* 62: 276–87

165. Horner, D., McLaren, A. 1974. *Biol. Reprod.* 11:553–57

166. Kelly, S. J., Rossant, J. 1976. *J. Embryol. Exp. Morphol.* 35:95–106

167. Donohoo, P., Kafatos, F. C. 1973. *Dev. Biol.* 32:224–29

168. Newrock, K. M., Raff, R. A. 1975. *Dev. Biol.* 42:242–61

169. Rodgers, W. H., Gross, P. R. 1978. *Cell* 14:279–88

Ann. Rev. Biochem. 1979. 48:471–89

SLOW TRANSITIONS AND HYSTERETIC BEHAVIOR IN ENZYMES

❖12016

Carl Frieden

Department of Biological Chemistry, Washington University
School of Medicine, St. Louis, Missouri 63110

CONTENTS

PERSPECTIVES AND SUMMARY

It has been clearly recognized for at least 15 years that the kinetic properties of certain enzymes serve an essential role in the regulation of competing metabolic pathways. The study of such regulatory enzymes has been extensive and has borne out the original concept that the unusual kinetic properties observed in vitro may be correlated with the regulation of metabolic pathways in vivo. For the most part, the kinetic properties that have been examined are the sigmoidal dependence of initial velocity on substrate concentration or the ability of specific metabolites to either activate or inhibit the reaction rate by binding to sites that are distinct from the active

471

0066-4154/79/0701-0471$01.00

site. It is becoming more evident, however, that as useful as initial velocity studies are, they may not display the full range of regulatory behavior of these enzymes. Thus, many enzyme systems show time-dependent behavior that is related to slow conformational changes which may arise in response to a change in ligand concentration. These types of conformational changes need to be more carefully analyzed. It is also becoming evident that these conformational changes may represent rate-limiting steps in the enzymatic reaction. Description of the kinetic properties of these enzymes demands further investigation not only for a clearer understanding of their role in metabolic pathways but also for the role of structural changes in the mechanism of enzyme action. This chapter discusses the analysis of this time-dependent behavior. It is noted that a proper analysis can be a difficult task, but represents yet another aspect of the relationship between enzyme structure and function.

INTRODUCTION

This review deals primarily with enzymes in which slow transitions or conformational changes may contribute to limiting the overall reaction rate. That conformational changes occur in enzymes is well established. An extensive review by Citri (1) discusses methods which have been used and lists experimental examples for a number of enzymes. Slow transitions exist in many nonenzyme proteins, but the ability to measure a distinctive property (i.e. catalytic activity) conveniently and under a wide variety of conditions which are nondisruptive to the normal function makes an enzyme the system of choice.

It is becoming evident that the actual chemical transformation, the step in which substrates are converted to products, is frequently not the rate-determining process in an enzymatic reaction (2, 3). One must examine, therefore, what step or steps do limit this rate. If we assume the simplest enzymatic mechanism of the type $E + S \rightleftharpoons ES \rightleftharpoons EP \rightleftharpoons E + P$ the above conclusion would place the rate-limiting step either at the point of substrate addition or product release. Substrate addition steps, however, when measured under conditions of reasonable substrate and enzyme concentrations, are quite rapid relative to the overall turnover rate of the enzyme (4). This would suggest that product release may be rate-limiting. While this appears to be true in many cases, no enzymatic reaction is as simple as that given above. Rather there are intermediate steps involving isomerization processes for either the enzyme, the substrate, the enzyme-substrate complex, or the enzyme-product complex. Thus conformational changes after substrate addition but preceding the chemical transformation, or after the chemical transformation but preceding product release may be rate-limiting.

Cleland has discussed possible mechanisms for several such enzyme systems (2). Frieden pointed out several years ago (5) that if conformational changes were slow relative to the measurement of the rate of the reaction, and if the conformational change led to an enzyme form with different kinetic properties, then the progress curve of the reaction could exhibit a lag or burst. He termed these time-dependent phenomena, which are considered distinct from those related to substrate depletion or product accumulation, hysteresis. At that time there were about a dozen enzymes that exhibited hysteretic properties (5), but the list has been considerably expanded since then (6). While another list is not compiled here, it is of interest that the majority of enzymes exhibiting this type of behavior can be classed as regulatory enzymes.

Enzymes that show hysteretic behavior are defined as those that respond slowly (in terms of their kinetic behavior) to a change in ligand concentration, e.g. the addition of substrate to initiate the reaction. Clearly the change in kinetic behavior is related to the time used to observe that behavior. If the velocity is measured over a period of one minute, changes that occur on the order of milliseconds or hours will not be observed. By the same token, the stopped-flow apparatus will be able to detect kinetic changes that are not observed in normal kinetic assays.

The availability of a continuous measurement of activity is almost essential for any detailed analysis of this type of kinetic behavior. It has been assumed in all the theoretical treatments of hysteretic behavior that the substrate level does not change over the time course of the measurement or, that if it does change, the change is not important to the observed kinetic properties. One of the most difficult problems to deal with, however, is that of substrate depletion (or product accumulation) over the time course used for measuring the kinetic behavior. This important limitation on the analysis of data is discussed later.

SOME TIME RANGES FOR DIFFERENT PROCESSES

Time events in enzymes such as the relaxation of tightly or loosely bound water, proton transfer steps, side-chain rotations and small (or local) movements, are believed to be quite rapid on the basis of experimental evidence (7). Evidence for local conformational motions in the range of 10^{-9} sec is provided by fluorescence decay and ^{13}C NMR measurements. The quenching of tryptophanyl fluorescence by oxygen depends on conformational changes that occur in this time range (8). The authors (8) concluded that proteins in general undergo structural fluctuations on the nanosecond time scale, and that to describe proteins as existing in a restricted number of equilibrium conformations (9) is probably incorrect; instead there are a multiplicity of equilibria.

In this context it is of interest to consider the proposal of McCammon et al (10) which is concerned with the dynamics of folded proteins. These authors, using bovine pancreatic trypsin inhibitor as a model, solved the equations of motion with an empirical potential energy function starting with the known crystallographic structure of the protein. They concluded that internal motion was fluid-like with different regions of the protein not equally flexible. Changes in shape, i.e. a concerted motion, appeared to oscillate as a whole with a period of about 6 psec, and smaller motions were even faster.

Obviously, changes in the three-dimensional structure of a protein that occur in the nano- or picosecond range are much too fast to be discerned in measurements of catalytic activity and are much faster than the conformational changes that are discussed in this chapter. However, it is possible that several small (or local) conformational changes must occur simultaneously or must occur in such a way that a stable (or metastable) configuration is formed between the side chains involved. For example, if a particular side chain or portion of the side chain was in a "correct" position only 10% of the time and if six such groups had to be properly positioned, the probability for all six events occurring at the same time would be 10^6 times longer than for the oscillation of one group. Such times would approach those observed in hysteretic behavior.

This, of course, is a greatly oversimplified picture of what may happen. It does not include, for example, solvent interactions or the making (or breaking) of noncovalent bonds such as the hydrogen bond, or electrostatic effects which undoubtedly occur on the surface of the protein. At least one additional mechanism may account for relatively slow conformational changes (i.e. ones that can be measured in seconds). Thus, in many enzymes the substrate fits into a pocket in the enzyme surface. While the sides of the pocket may be relatively constrained, the pocket itself may undergo opening and closing motions driven by solvent fluctuations or controlled by the diffusion of one large segment relative to the other, and such movements may also be relatively slow. Any mechanism should also be considered in the light of what may happen on addition of ligand. For example, the addition of the ligand may be a multistep process, as has been proposed for some nucleotides binding to bovine liver glutamate dehydrogenase (11, 12), where one portion of the ligand may bind to the surface and the remainder move to the surface more slowly (12). The process of ligand binding may induce conformational changes in the quaternary structure of the protein. The allosteric model of Koshland et al (13) assumes ligand-induced conformational changes that directly affect subunit-subunit interactions that in turn affect ligand binding on a different subunit. There is no reason to assume, as Koshland et al did (13), that such conformational changes will occur rapidly relative to the catalytic reaction. While slow conformational changes may occur after substrate addition but prior to the chemical transformation, or after the chemical transformation but prior to the release of product, it is not easy to distinguish such mechanisms, especially if one considers that most enzymatic reactions do not involve a single substrate and product, but rather two or more substrates and two or more products.

These conformational changes should be distinguished from ones that involve gross unfolding of the enzyme or disulfide bond formation. Unfolding transitions are usually considered to be much slower than enzyme catalysis. Presumably such unfolding is measured by loss of catalytic activity and may occur in minutes, hours, or longer, depending on the conditions. Similarly the formation of disulfide bonds may be a rather long-term process relative to catalytic activity. However, it is not possible to assign definite time ranges to either of these processes. For the purpose of this chapter, they are not considered even though it may be difficult to distinguish them from the processes that occur during the enzymatic reaction.

SOME PROPOSED MECHANISMS FOR HYSTERETIC BEHAVIOR

Monomeric Enzymes

Most mechanisms for hysteretic behavior assume at least two forms of the enzyme which differ in their kinetic behavior. However, we should make some distinction between those models that use slow transitions to explain kinetic cooperativity and those models that use slow transitions to describe time-dependent effects. This is not a clear-cut distinction, but some confusion in the literature could be clarified by noting that time-dependent effects always relate to kinetic aspects of the enzymatic reaction while cooperativity may arise from either kinetic or ligand binding characteristics. It has been recognized (14–18), that when two states of either free enzyme or enzyme-ligand complex are introduced into a mechanism, kinetic cooperativity with respect to the ligand may be observed when the kinetic equations are derived by steady state methods. Thus the simple mechanism given above will not under any conditions give rise to kinetic cooperativity with respect to substrate. Mechanism I can give such behavior because alternative pathways for substrate addition exist.

$$E + S \rightleftharpoons ES \rightleftharpoons E + P$$
$$\updownarrow \qquad\quad \updownarrow$$
$$E' + S \rightleftharpoons E'S \rightleftharpoons E' + P$$

Mechanism I

The kinetic equation derived by steady state methods yields terms in S^2, and under certain conditions of the rate constants such terms may become predominant in the velocity expression. Authors who have derived such equations using Mechanism I (or variations) have referred to the enzyme system as having memory (19, 20) or being mnemonical (17, 21, 22), presumably because the enzyme relaxes slowly from one state (i.e. E') to the other (i.e. E) after the product is released. While it was clear that the system may show cooperative behavior, the possible time dependence manifested by a lag or a burst in substrate utilization was not discussed (17, 19–22). Ainslie et al (16) were the first to point out, for the slightly expanded Mechanism II, that both hysteretic behavior and cooperative behavior can arise in this system depending on the values of rate constants chosen.

$$E + S \rightleftharpoons ES \rightleftharpoons EP \rightleftharpoons E + P$$
$$\updownarrow \qquad \updownarrow \quad \updownarrow$$
$$E' + S \rightleftharpoons E'S \rightleftharpoons E'P \rightleftharpoons E' + P$$

Mechanism II

These authors, using a computer program to calculate values of sets of rate constants, determined conditions under which either bursts or lags in substrate utilization were observed, together with either positive or negative cooperativity in plots of velocity vs substrate concentration. This type of mechanism has been used to describe the hysteretic kinetic data observed for hexokinase (21, 23) and for other enzymes containing only a single substrate binding site, such as ribonuclease (15) and glucokinase (22). More recently, Neet & Ainslie (24) have discussed in detail the conditions for obtaining hysteretic cooperativity in a monomeric enzyme. They have also discussed the conditions under which cooperativity may be lost. For example, for Mechanism II to give cooperativity, they have noted that three factors are involved: 1. the transition is on the same order of magnitude as the maximum velocity; 2. the substrate addition step(s) are not in rapid equilibrium; and 3. changing substrate concentration shifts the proportion of the reaction through alternate reaction pathways. These authors also give conditions for hysteretic behavior in a monomeric enzyme.

Hysteretic behavior and cooperative kinetics are not necessarily linked. In the original discussion of hysteresis, Frieden made the assumption that ligand binding steps were established rapidly (5). For the single substrate case (Mechanisms I and II) this assumption precluded the possibility of kinetic cooperativity and isolated the rate-determining steps in the conversion of one form of either free enzyme or enzyme-substrate complex to a kinetically different form. While Ricard et al (17) did not consider this possibility when they derived equations for the mnemonic enzyme, they later made this assumption to describe the kinetic behavior of wheat-germ hexokinase LI (19, 20). The assumption of rapid equilibration of ligand binding steps can now be examined. We know, from a good deal of experimental evidence (4) that the rate constant for the second order step $E + S \rightarrow ES$ is on the order of 10^7 sec^{-1} mol^{-1} for normal substrates (so-called poor substrates have on rates that may be smaller than this). We then assume that the step after ligand binding is involved in hysteresis and that the largest rate constant one can measure for this step (by stopped-flow methods) is on the order of 20 sec^{-1}. For a more or less rapid equilibrium system, we could then assign a value of about 100 sec^{-1} to the dissociation step $ES \rightarrow E + S$. The dissociation constant for the ligand is then calculated as $100/10^7 = 10^{-5}$ M. Thus, for a relatively fast hysteretic process ($\sim$20 sec^{-1}) and relatively tight ligand binding (10 μM), the rapid equilibrium assumption is a good approximation. If the hysteretic process is slower (as a good many are) or if the dissociation constant is larger, then the assumption becomes even better. By the same token, however, the assumption becomes less valid for tightly bound ligands. In fact, one mechanism that itself generates a hysteretic response is the slow off rate of a ligand (or a conformational change preceding that off rate).

A different interpretation has been presented by Solling & Esmann (25) for the glycogen synthase I from polymorphonuclear leukocytes. This enzyme is activated by glycogen and the time required for activation can be quite long. The substrate is of very high molecular weight and the authors conclude that the hysteretic behavior arises not from a rapid ligand binding followed by an isomerization, but from a slow binding of the glycogen itself. This result may be the consequence of the nature of the glycogen and it would be of interest to examine the rate of activation with different molecular weight substrates or substrate analogs.

The rapidity of ligand binding steps can be examined another way by comparing ligand binding with the enzymatic velocity as a function of ligand concentration. Hysteresis which gives rise to cooperativity in kinetic studies will not affect equilibrium binding studies. Thus the latter will be noncooperative. This distinction has been made by Ainslie et al (16) and by Neet & Ainslie (24).

Oligomeric Enzymes

The assumption of rapid equilibrium in ligand-binding steps has been made by investigators describing models for hysteretic behavior with allosteric models. This type of model differs from that generated by Ainslie et al (16), described above, in that it assumes that the allosteric behavior arises from multiple substrate binding sites (6, 26, 27), based on the assumption used by Monod et al (28) or Koshland et al (13) to describe cooperative ligand binding. Hysteretic behavior can occur if the transition between the two enzyme states (each of which contains several ligand-binding sites) is slow rather than instantaneous as assumed in the original models (13, 28). In applying the model of Monod et al as Janin (26) has done for the threonine inhibition of the aspartokinase I-homoserine dehydrogenase I, one makes a number of assumptions that are inherent in the allosteric model. The hysteresis, as indicated above, arises solely from the slow transition involved in conversion from one enzyme form to the other. Threonine inhibition of this enzyme was one of the first processes shown to be time-dependent (29) and has been investigated not only by Janin (26, see above), but more recently by Bearer & Neet (30–32). These authors show that while threonine inhibition of both the aspartokinase activity and the homoserine dehydrogenase activity is hysteretic, they are distinct processes. For the homoserine dehydrogenase, the cooperative nature of threonine inhibition is due to the cooperative nature of threonine binding and is not a kinetic argument. For the aspartokinase reaction, the authors conclude that the inhibition is the sum of kinetic and binding cooperativity (32).

It is useful to measure a conformational change by methods that are independent of activity measurements. Such a system has been described by Dubrow & Pizer (33) for the *E. coli* phosphoglycerate dehydrogenase which is inhibited by serine. The authors studied not only the rate constant for the inhibition process, but also the quenching of fluorescence of enzyme-bound NADH by serine, and an enhancement of protein fluorescence in the presence of NADH, substrate, and serine. They attempted to correlate these processes with particular isomerization steps in the overall mechanism. In the absence of the catalytic reaction, however, the process may have a different ligand or time dependence, since the substrate(s) may influence ligand binding or the rate of the conformational change. Apparent hysteretic effects also could arise from isomerization of the substrate rather than the enzyme, i.e. the conversion of one anomeric form to another. This rate should be independent of enzyme concentration. Such an effect is observed for the conversion of the inactive to the active form of glyceraldehyde 3-phosphate (34), and similar effects are seen with different forms of fructose-6-phosphate (35) or oxalacetate (36).

A more complex situation arises when the allosteric model of Koshland et al (13), in which ligand binding to one subunit influences binding to another, is used. In that case, the hysteretic behavior would be solely ligand-induced and would involve the rate of the change in subunit-subunit interaction induced by the binding. Some models of this type have been discussed by Mouttet et al (37) in attempts to fit the hysteretic behavior of L-phenylalanine ammonia lyase. In addition, Loudon & Koshland have considered the relationship between relaxation times and substrate concentrations for the sequential allosteric model (38). Such models are quite different from those in which cooperativity is a kinetic argument and for which the cooperativity could be observed for a monomeric enzyme containing a single substrate binding site (15, 16).

Polymerizing Enzymes

Hysteretic behavior may also result from reversible polymerization of an enzyme if the different polymeric states have different kinetic properties. This type of mechanism was originally discussed by Frieden (5) and expanded upon by Kurganov and co-workers (39–41).

Two examples of this type are the human erythrocyte phosphoribosylpyrophosphate synthetase (42) and the human glutamine phosphoribosylpyrophosphate amido transferase (43). The enzymes differ, however, with regard to which molecular weight form is active. For the synthetase, the authors conclude that lower molecular weight forms are inactive while higher molecular weight forms, which can be induced by ATP, Mg, and P_i are active (42). As expected, the rate of activation is dependent on enzyme concentration. For the transferase, the conversion of the high to low molecular weight form was responsible for the observed lag in the kinetic behavior (43). This model resembles that in which a slow isomerization occurs except that the resultant kinetic behavior will be dependent on enzyme concentration. As discussed later, analysis of this mechanism may be difficult even though tests to see whether it is operative are simple. A more familiar example, perhaps, is muscle phosphofructokinase which exists as an inactive dimer at pH values below about seven and an active tetramer at higher pH values (44–46). Raising or lowering the pH can give rise to hysteretic effects (47). The effect of ligands and temperature on the association-dissociation process has been fully investigated (46, 48). In a similar way, cold labile enzymes may show hysteretic behavior because of their tendency to undergo dissociation at colder temperatures. A list of some of these enzymes has recently been published (49).

Ligand Off Rates

Hysteretic behavior can also arise if the rate at which a ligand comes off the enzyme surface is sufficiently slow (5). Such a mechanism might not apply to substrates and products as much as to allosteric effectors, some of which may bind with dissociation constants less than 10^{-6} M. Displacement of one allosteric effector by another, either directly or indirectly via a conformational change, could lead to hysteretic behavior. This appears to be the case for muscle AMP deaminase, which shows normal kinetic behavior in the absence of any allosteric behavior, but hysteretic behavior in the presence of the inhibitor GTP (50). Analysis of this system (51) indicates three types of nucleotide binding sites: an active site (binding AMP); an inhibitory site (binding nucleoside triphosphates like GTP); and an activating site which is relatively non-specific. GTP inhibition is time-dependent. However, when the enzyme is preincubated with GTP, a time-dependent release of inhibition is observed, partly because the substrate AMP and perhaps the product IMP can bind to the activating site and weaken GTP binding to the inhibitory site. Whether the activation reflects the slow off rate of GTP or a conformational change preceding the GTP release is not clear (51). With respect to substrate or products, it was indicated earlier that the rate of the process ES → E + S (or EP → E + P) may be rapid relative to the hysteretic process and many enzymatic reactions appear to be limited by the rate of product release. This process alone will not give rise to hysteric behavior unless the free enzyme from which the product is released isomerizes slowly back to a form that binds substrate readily.

Hysteretic behavior also could arise from dissociation of tightly bound inhibitor whether the inhibitor is a small ligand or another protein as in the cAMP protein kinase (52). A recent kinetic study of cAMP binding and the resulting dissociation of the catalytic subunit has been presented for this enzyme from *Drosophila* (53).

ANALYSIS OF HYSTERETIC BEHAVIOR

By far the most convenient way to monitor hysteretic behavior is by continuous measurement of absorbancy or fluorescence changes. Sometimes it may be necessary to use coupling enzyme systems to do this and one should be clear that the hysteretic effect is not an artifact of the coupling system. A range of coupling enzyme concentrations should be tested and the observed effect should be independent of coupling enzyme concentration. This system can give artifacts if, for example, one of the products of the reaction activates the enzyme and is not kept at a low enough concentration by the coupling system. This is exactly the problem in the assay of phosphofructokinase where the product, fructose 1,6-bisphosphate, activates the enzyme. Its concentration builds up even in the presence of high levels of aldolase and glycerophosphate dehydrogenase and an apparent lag is observed (54).

Any burst or lag in product appearance (or substrate utilization) that is followed by a linear accumulation of product can be described by the expression

$$v_t = v_f + (v_i - v_f)e^{-kt} \qquad 1.$$

where v_t is the velocity as a function of time, v_i is the initial velocity, v_f the final (linear) velocity after the burst or lag, and k is the apparent rate constant for the transition. This equation is applicable to all of the mechanisms so far discussed. As discussed below, it assumes that there is no substrate depletion during the kinetic measurement. What may be different between different mechanisms is the dependence of k on ligand concentration or (in the case of a polymerizing system) on enzyme concentration. Before discussing this issue, we discuss the evaluation of the apparent rate constant, some of the assumptions involved, and problems that may arise in that evaluation.

Evaluation of k

The most obvious way to evaluate the apparent rate constants and other parameters of Equation 1 is shown in Figure 1. The values for v_i and v_f are obtained from the linear portions of the progress curve before and after the transition. The intersection of the lines representing the initial and final velocities is the relaxation time or the reciprocal of the apparent rate constant.

Assuming Equation 1 is an accurate representation of the real data, the progress of product accumulation is given by the expression

$$P_t = v_f t - (v_f - v_i)(1 - e^{-kt})/k \qquad\qquad 2.$$

Extrapolation of the linear portion of the curve representing v_f to $P = 0$ yields the intercept

$$t_{(P=0)} = (v_f - v_i)/(v_f k) \qquad\qquad 3.$$

Extrapolation of this portion of the curve to $t = 0$ yields the intercept

$$P_{(t=0)} = -(v_f - v_i)/k \qquad\qquad 4.$$

Either of these intercepts or the intersection of the two linear velocities v_i and v_f can yield values of the apparent rate constant.

Whether or not the value of k so obtained is correct rests upon whether the assumptions used in the derivation of Equations 1 and 2 are correct. An important assumption is that only a single exponential function is involved, i.e. that the transition can be represented as if it were a single step. This is best tested not by evaluating the apparent rate as shown by Figure 1 but by either plotting $\ln(v_t - v_f)$ vs time, which should be linear, or by fitting the real data with a computer simulation of a single exponential process according to Equation 2. Neet & Ainslie (24) have discussed treatment of multiple exponential decay. Deviation from a single exponential process could result from any of several reasons which include the possibility of more than one hysteretic step or of artifacts due to product accumulation or substrate depletion.

These latter problems can be quite serious. For example, one would like to examine the hysteretic behavior over a wide range of substrate concentra-

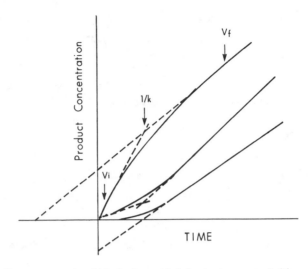

Figure 1 Progress curve in which the hysteretic behavior is expressed either as a lag or a burst. v_i refers to the initial velocity preceding the hysteretic transition while v_f is the velocity following the transition. From the intersection of the linear regions of the curves, one can obtain the apparent rate constant, k. The values of the extrapolated values to either the time or product concentration axes are given in the text.

tions. However, an assumption in the derivation of Equation 1 is that the substrate level does not change during the experiment. This assumption is satisfactory if the experiment utilizes perhaps 10% of the substrate, but such a restriction is not always possible at low substrate levels. The problem is compounded if the enzymatic reaction is reversible and one has to contend with the approach to equilibrium. This problem can be alleviated somewhat by using different enzyme concentrations, an approach that is useful in determining the parameters of hysteretic behavior (see below). However, it is important to maintain some relation between the rate constant for the hysteretic behavior and the rate of enzymatic reaction, the latter being defined by the enzyme concentration; this is best done by using the whole progress curve of the reaction (see below).

Product accumulation also gives rise to problems if the product is an allosteric effector, or if there is strong product inhibition by abortive complex formation. Addition of one of the products (in a multiproduct reaction) to the enzyme, prior to initiating the enzymatic reaction, might be useful in detecting this effect. In multisubstrate enzymes, hysteretic behavior may occur from abortive complex formation. Thus in a mechanism like

$$E + A \rightleftharpoons EA$$
$$EA + B \rightleftharpoons EAB \rightleftharpoons EPQ \rightleftharpoons EQ + P$$
$$EQ \rightleftharpoons E + Q$$

complexes such as EAP or EQB might form and inhibit the steady state rate as the substrate concentration is increased and also give rise to hysteretic behavior if a slow conformational change is associated with abortive complex formation. Such a mechanism has been suggested, for example, for the pigeon liver malic enzyme (55).

Another artifact arises if the enzyme inactivates during the time course of the measurement. This would appear as a burst in enzymatic activity and could be particularly confusing if the extent or time-dependence of inactivation is a function of substrate concentration. Changing the enzyme concentration or adding compounds that may stabilize the enzyme (i.e. albumin) may eliminate this problem. Performing experiments at high enzyme concentrations using stopped-flow techniques is probably the best way to avoid enzyme inactivation.

The Use of the Full Time Course of the Reaction

The above section discusses ways in which the apparent rate constant can be evaluated, and notes some problems that can arise particularly with substrate depletion. Such problems, if they exist, may be minimized by using the full progress curve of the enzymatic reaction. This is especially easy when using the stopped-flow apparatus where the full time course can be recorded over a relatively short time period. Along with this, a computer simulation program is needed to fit the real data. A review of computer applications has been given by Garfinkel et al (56), and an iterative computer program (such as that developed by Bates & Frieden (57) is particularly suitable for this purpose. In this type of program, a specific mechanism is chosen and the time course is computed by solving all the pertinent differential equations for each species. The real data are superimposed on the simulated data and parameters are varied to obtain the best fit. The program has been used successfully to describe the hysteretic behavior of glutamate dehydrogenase (58) and AMP deaminase (51).

Programs of this type require that one choose a specific mechanism and that one has available a relatively sophisticated computer with a program capable of simulating complex mechanisms. Finding the correct mechanism may not be too difficult since, if one chooses an inappropriate mechanism, a single set of parameters will not fit the experimental data over a wide set of initial substrate concentrations. Having available the computer simulation system is more of a problem, but some simplifications can be made which yield a situation for which a relatively simple computer program can be designed. Thus, one can write the velocity terms of Equation 1 in terms of the change in substrate concentration for a given time period (Δt) and solve the expression

$$\Delta S_t = \Delta S_f + (\Delta S_i - \Delta S_t)e^{-kt}$$
$$\text{and } S_t = S_{t=0} - \Delta S_t \text{ and } t = \Sigma \Delta t.$$

5.

Values for ΔS_f and ΔS_i are evaluated from the velocity expression for each of the two forms of the enzyme, E and E', as a function of substrate concentration and the assumption that $-\Delta S = v\Delta t$. For each Δt, a new velocity is calculated at the lower substrate level. This relatively simple calculation works well if product accumulation is not a problem and its applicability can be increased by using more complex functions for the velocity expressions for each form of the enzyme. Thus it can be applied to allosteric enzymes where the transition between the two forms is slow. The relevant velocity expressions are given by Kurganov et al (6).

An advantage of using computer simulation plots is that one need not solve the frequently complex expressions for the dependence of the apparent rate constant on substrate or ligand concentration.

In evaluating hysteretic behavior, it can be particularly useful to obtain full-time course curves for a reaction over as wide a range of enzyme concentrations as possible. The rationale for this is quite obvious if an association-dissociation reaction is involved and different molecular weight forms of the enzyme have different kinetic properties. Under such conditions, full-time course plots should not match when normalized to a given enzyme concentration (a nonmatch may also indicate changes in enzyme stability as a function of enzyme concentration). Such nonmatches are an important diagnostic test for this type of system. However, even for a unimolecular transition, i.e. a conformational change, one may also see different behavior at different enzyme concentrations even though the predicted behavior should be independent of enzyme concentration. Deviations are related to differences in the rate of substrate depletion simply because the rate constant is independent of enzyme concentration while the rate of substrate depletion is not. Thus, at high enzyme concentrations, the substrate may be depleted before the conversion of one enzyme form to another, while at low enzyme concentration the conversion may be completed before the activity measurement occurs. An example of this behavior has been observed for the hysteretic inhibition of glutamate dehydrogenase by NADH (58).

The Effect of Ligand or Substrate Concentration on k

Many investigators have observed either a lag or burst in kinetic behavior and have calculated the apparent rate constant under those particular conditions of substrate concentration, pH, temperature, etc. Where feasible, they may also preincubate the enzyme with the ligand that induces the transition to determine whether the hysteretic effect occurs in the absence of other ligands and whether the preincubation abolishes the hysteretic behavior. Preincubation with the ligand may not totally abolish the hysteretic effect. In such cases, either more than one ligand may be required or the conformational change occurs in a central complex (i.e. a complex involved in the interconversion of substrates to products).

As discussed earlier, however, many mechanisms obey Equation 1 when tested under a given set of conditions and thus cannot be distinguished on the basis of one experiment. Rather, the apparent rate constant needs to be measured as a function of substrate or ligand concentration. These measurements may also permit the evaluation of specific rate constants for a given mechanism. The hysteretic behavior of the Na-K ATPase from various sources has been measured by determining proton release with a pH-stat (59–61). Cantley & Josephson, in particular (60), examined the behavior of the dog kidney enzyme as a function of ligand concentration (Mg,K,Na,ATP) and concluded that ATP hydrolysis depended on a slow conformational change between enzyme forms with very different catalytic activities.

However, the variation of k with substrate or ligand concentration can be a very complex function and it may be considerably easier to check a specific mechanism by use of the full-time course and computer simulation, as discussed above. The equations for the simplest case (Mechanism I) as well as for this mechanism in the presence of an effector have been given by Frieden (5).

Mechanism I can be expanded to the case of an allosteric enzyme according to the model of Monod et al (28). Let us rewrite Mechanism I in this terminology as

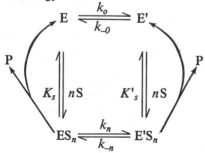

Mechanism Ia

It is assumed that the binding steps are in rapid equilibrium, that the reaction to give product is irreversible, and that the substrate binds with equal affinity to i sites within any given conformation. For convenience, the apparent rate constant k can be considered as the sum of two rate constants, one in the forward direction and one in the reverse direction (similar to any unimolecular reversible reaction):

$$k = k_f + k_r$$

For the case where $i = 1$ (one binding site/monomer)

$$k = (k_0 + k_1 S/K_s)/(1 + S/K_s) + (k_{-0} + k_{-1} S/K'_s)/(1 + S/K'_s)$$

As the substrate concentration changes from 0 to high levels, k varies from $k_0 + k_{-0}$ to $k_1 + k_{-1}$. As a function of S, k may increase to a maximum, decrease to a minimum, or go through a minimum. A similar dependence of k may also occur for sequential mechanisms for allosteric behavior (13). The ligand dependence of the apparent rate constant has been discussed in detail by Loudon & Koshland for these cases (38). The variation of k for a similar model has been also discussed by Mouttet et al (37).

For the allosteric case (n sites), the corresponding equation is (6, 27)

$$k = k_{-0}[(1 + L')/(1 + S/K'_s)^n] \sum_{i=0}^{i=n} [k_{-i} n!/k_{-0} i!(n-i)!] [(S/K'_s)^i]$$

where

$$L' = [(1 + S/K'_s)/(1 + S/K_s)]^n (k_0/k_{-0})$$

The expression becomes simpler if one assumes a geometric progression of k_{-1} and k_{+1} as did Janin in his analysis of the effect of threonine on the aspartokinase I-homoserine dehydrogenase I from *E. coli* (26). Unfortunately, the expressions for k become much more complex when an allosteric modifier is included in addition to the substrate. A complete study of the variation of k with substrate or modifier concentration is not an easy undertaking although it may yield information about the ligand dependence of individual rate constants of the conformational change.

Kurganov et al have discussed the variation of k with substrate and enzyme concentration for cases where the enzyme undergoes an association-dissociation reaction (39–41). Kurganov (41) discusses the application of the theory to several specific enzymes including data obtained for threonine deaminase (62), phosphorylase *a* (63, 64), and aspartic semialdehyde dehydrogenase (65). This approach may be particularly important since there is a variation with the enzyme concentration. Although enzyme levels that occur in vivo may give results that are very different from those studied in vitro, it seems likely that the association-dissociation system could provide a rather important metabolic control in vivo. The kinetic properties of associating-dissociating systems were discussed in a previous review (66).

Variation of v_i and v_f with Substrate Concentration: Lags or Bursts

Equation 1 does not define the nature of the initial velocity, v_i, (that before the transient) or v_f (that after the transient) as a function of substrate concentration. To observe hysteretic behavior in the systems we have discussed here, it is necessary that the enzyme be primarily in one form before ligand addition. For allosteric enzymes or for those systems described by Mechanism I and in which ligand binding steps are rapid (which is the allosteric case for $n = 1$), the enzyme must initially be primarily in the form that has a lower kinetic activity. Thus in these cases, only a lag will be observed. The substrate dependence of the initial velocity will be hyperbolic (or the sum of two hyperbolic dependencies) while that of the final velocity will be either hyperbolic or sigmoidal. The only condition where a burst in activity would be observed is if the form of the enzyme that binds substrate tightly has less (or no) catalytic activity. In this case, substrate inhibition would be observed at high substrate levels and along with such inhibition a burst in catalytic activity if the isomerization is slow enough. Only lags would be observed in the associating-dissociating systems as well, because in these systems, substrate binds more tightly to one form. If the enzyme is all in the tight binding form, no hysteretic behavior will be observed (since the isomerization has occurred prior to substrate addition). Models based on kinetic cooperativity can produce either a lag or a burst (16). In addition to the fact that cooperativity and hysteretic behavior are linked in this case, the observation of a burst in the velocity on substrate addition may suggest the kinetic model.

This rationale does not hold when considering an allosteric effector. One would expect, in this case, to see an initial burst if the allosteric effector is an inhibitor, or a lag if the effector is an activator. Burst kinetics also may be produced by artifacts, i.e. enzyme instability.

CONCLUSIONS

The observation of hysteretic behavior has now been clearly established in a number of enzyme systems. There is no single mechanism responsible, but several different ones, all of which give rise to similar time-dependent results. This diversity of mechanism makes it difficult to interpret a given enzyme by kinetic means alone. In addition, analysis can be complex when the substrate of the reaction is measurably depleted or when the accumulated product influences the kinetic behavior. Nevertheless, there are relatively simple experiments that can demonstrate the presence of hysteretic behavior. The details of the mechanism may depend upon analysis of the full-time course of the enzymatic reaction by computer simulation, plus independent measurements on the enzyme in the absence of the catalytic reaction. It is hoped that methods for analysis of these complex systems will become more generally available.

Enzymes showing hysteretic behavior are a diverse group, but the literature shows that a surprisingly large number can be classified as regulatory enzymes which show allosteric behavior. It seems likely that hysteretic behavior has a physiological function, although this may be difficult to prove since one would have to show the same response for a given enzyme under in vivo and in vitro conditions. There are, however, good reasons for such a slow kinetic response in terms controlling the flux through different metabolic pathways.

ACKNOWLEDGMENTS

This work was supported in part by grant AM 13332 from the United States Public Health Service. I would like to thank Dr. Kenneth Neet for a manuscript of his review prior to its publication.

Literature Cited

1. Citri, N. 1973. *Adv. Enzymol.* 37:397–648
2. Cleland, W. W. 1975. *Acc. Chem. Res.* 8:145–51
3. Albery, W. J., Knowles, J. R. 1976. *Biochemistry* 15:5631–40
4. Hammes, G. G., Schimmel, P. R. 1970. *Enzymes* 2:67–114
5. Frieden, C. 1970. *J. Biol. Chem.* 245:5788–99
6. Kurganov, B. I., Dorozhko, A. I., Kagan, Z. S., Yakovlev, V. A. 1976. *J. Theor. Biol.* 60:247–69
7. Careri, G., Fasella, P., Gratton, E. 1975. *Crit. Rev. Biochem.* 3:141–64
8. Lakowicz, J. R., Weber, G. 1973. *Biochemistry* 12:4171–79
9. Weber, G. 1972. *Biochemistry* 11:864–78
10. McCammon, J. A., Gelin, B. R., Karplus, M. 1977. *Nature* 267:585–90
11. Huang, C. Y., Frieden, C. 1972. *J. Biol. Chem.* 247:3638–46
12. Colen, A. H., Cross, D. G., Fisher, H. F. 1974. *Biochemistry* 13:2341–47
13. Koshland, D. E. Jr., Nemethy, G., Filmer, D. 1966. *Biochemistry* 5:365–85
14. Rabin, B. R. 1967. *Biochem. J.* 102:22c
15. Rubsamen, H., Khandker, R., Witzel, H. 1974. *Hoppe-Seylers Z. Physiol. Chem.* 355:687–708
16. Ainslie, G. R. Jr., Shill, J. P., Neet, K. E. 1972. *J. Biol. Chem.* 247:7088–96
17. Ricard, J., Meunier, J.-C., Buc, J. 1974. *Eur. J. Biochem.* 49:195–208
18. Whitehead, E. 1970. *Prog. Biophys.* 21:321–97
19. Ricard, J., Buc, J., Meunier, J.-C. 1977. *Eur. J. Biochem.* 80:581–92
20. Buc, J., Ricard, J., Meunier, J.-C. 1977. *Eur. J. Biochem.* 80:593–601
21. Meunier, J. C., Buc, J., Navarro, A., Richard, J. 1974. *Eur. J. Biochem.* 49:209–22
22. Storer, A. C., Cornish-Bowden, A. 1977. *Biochem. J.* 165:61–69
23. Shill, J. P., Neet, K. E. 1975. *J. Biol. Chem.* 250:2259–68
24. Neet, K. E., Ainslie, G. R. Jr. 1979. *Methods Enzymol.* In press
25. Solling, H., Esmann, V. 1977. *Eur. J. Biochem.* 81:129–39
26. Janin, J. 1971. *Cold Spring Harbor Symp. Quant. Biol.* 36:193–98
27. Janin, J. 1973. *Prog. Biophys. Molec. Biol.* 27:79–120
28. Monod, J., Wyman, J., Changeux, J.-P. 1965. *J. Mol. Biol.* 12:88–118
29. Barber, E. D., Bright, H. J. 1968. *Proc. Natl. Acad. Sci. USA* 60:1363–70
30. Bearer, C. F., Neet, K. E. 1978. *Biochemistry* 17:3512–16
31. Bearer, C. F., Neet, K. E. 1978. *Biochemistry* 17:3517–22
32. Bearer, C. F., Neet, K. E. 1978. *Biochemistry* 17:3523–30
33. Dubrow, R., Pizer, L. I. 1977. *J. Biol. Chem.* 252:1527–38
34. Trentham, D. R., McMurray, C. H., Pogson, C. I. 1969. *Biochem. J.* 114:19–24
35. Wurster, B., Hess, B. 1974. *FEBS Lett.* 38:257–60
36. Frieden, C., Fernandez-Sousa, J. 1975. *J. Biol. Chem.* 250:2106–11
37. Mouttet, C., Fouchier, F., Nari, J., Ricard, J. 1974. *Eur. J. Biochem.* 49:11–20
38. Loudon, G. M., Koshland, D. E. Jr. 1972. *Biochemistry* 11:229–41
39. Kurganov, B. I., Dorozhko, A. I., Kagan, Z. S., Yakovlev, V. A. 1976. *J. Theor. Biol.* 60:271–86
40. Kurganov, B. I., Dorozhko, A. I., Kagan, Z. S., Yakovlev, V. A. 1976. *J. Theor. Biol.* 60:287–99
41. Kurganov, B. I. 1977. *J. Theor. Biol.* 68:521–43

42. Meyer, L. J., Becker, M. A. 1977. *J. Biol. Chem.* 252:3919–25
43. Singer, S. C., Holmes, E. W. 1977. *J. Biol. Chem.* 252:7959–63
44. Pavelich, M. J., Hammes, G. G. 1973. *Biochemistry* 12:1408–14
45. Paetkau, V., Lardy, H. A. 1967. *J. Biol. Chem.* 242:2035–42
46. Bock, P. E., Frieden, C. 1976. *J. Biol. Chem.* 251:5630–36
47. Frieden, C. 1968. In *Regulation of Enzyme Activity and Allosteric Interactions,* ed. E. Kvamme, A. Pihl, pp. 59–71, New York: Academic
48. Bock, P. E., Frieden, C. 1976. *J. Biol. Chem.* 251:5637–43
49. Bock, P. E., Frieden, C. 1978. *TIBS* 3:100–3
50. Tomozawa, Y., Wolfenden, R. 1970. *Biochemistry* 9:3400–4
51. Ashby, B., Frieden, C. 1978. *J. Biol. Chem.* 253:8728–35
52. Krebs, E. G. 1972. *Curr. Top. Cell. Regul.* 5:99–133
53. Tsuzuki, J., Kiger, J. A. Jr. 1978. *Biochemistry* 17:2961–70
54. Emerk, K., Frieden, C. 1975. *Arch. Biochem. Biophys.* 168:210–18

55. Reynolds, C. H., Hsu, R. Y., Matthews, B., Pry, T. A., Dalziel, K. 1978. *Arch. Biochem. Biophys.* 189:309–16
56. Garfinkel, L., Kohn, M. C., Garfinkel, D. 1977. *Crit. Rev. Bioeng.* 2:329–61
57. Bates, D. J., Frieden, C. 1973. *Comp. Biomed. Res.* 6:474–86
58. Bates, D. J., Frieden, C. 1973. *J. Biol. Chem.* 248:7885–90
59. Grisar, T., Frere, J. M., Grisar-Charlier, J., Franck, G., Schoffeniels, E. 1978. *FEBS Lett.* 89:173–76
60. Cantley, L. J. Jr., Josephson, L. 1976. *Biochemistry* 15:5280–86
61. Recktenwald, D., Hess, B. 1977. *FEBS Lett.* 80:187–89
62. Feldberg, R. S., Datta, P. 1971. *Eur. J. Biochem.* 21:447–54
63. Metzger, B., Helmreich, E., Glaser, L. 1967. *Proc. Natl. Acad. Sci. USA* 57:994–100
64. Wang, J. H., Black, W. J. 1968. *J. Biol. Chem.* 243:4641–49
65. Holland, M. J., Westhead, E. W. 1973. *Biochemistry* 12:2270–75
66. Frieden, C. 1971. *Ann. Rev. Biochem.* 40:653–96

Ann. Rev. Biochem. 1979. 48:491–523

SURFACE COMPONENTS AND CELL RECOGNITION[1]

◆12017

William Frazier and Luis Glaser

Department of Biological Chemistry, Division of Biology and Biomedical Sciences, Washington University School of Medicine, St. Louis, Missouri 63110

CONTENTS

PERSPECTIVES AND SUMMARY

At one time or another we have all contemplated with a sense of wonder the finely detailed cellular architecture of a microscopic fragment of tissue. From the smallest multicellular creatures to the human brain, we are presented with thousands of examples of highly ordered arrangements of cells uniquely organized to accomplish precise functions. The ubiquity of su-

[1]Work in the authors' laboratories has been supported by grants from National Institutes of Health NS 13269 (WF), GM 18405 (LG), and the National Science Foundation PCM 78–04303 (WF), PCM 77–15972 (LG).

491

pracellular order convinces us of the important role of precise cellular interactions, while the diversity of organized forms, often within small regions of the same organ, serves fair warning that unraveling the subtleties of these interactions may prove a formidably complex task. The problem of understanding the biochemistry of specific recognition between cells is of course only one facet of the larger problem concerning the regulation of development of complex organisms. For example, temporal and spatial variations in the concentrations of growth factors and/or hormones during development undoubtedly play a critical role in determining the final tissue organization observed. A steadily increasing body of evidence, however, indicates that cells do contain on their surfaces the means to specifically recognize and bind to other cells, and that in some cases intercellular contacts are themselves an important regulatory event during development.

One of the most easily demonstrated examples of specific cellular interaction is that which initiates embryonic development, namely fertilization. Even though only two cell types, sperm and egg, participate, other aspects of the fertilization process have made this an extremely difficult system to study in higher organisms. Recently, however, rapid progress has been made in elucidating the mechanism of sperm-egg recognition in the sea urchin. For the study of specific cell-cell interactions that occur in embryonic development after fertilization, the chick embryo has proven to be the most widely studied system. Its advantages are the short developmental time course and the ready availability of material at all stages of development. More recently the use of cultured cells for the study of certain types of cell contact phenomena has provided a valuable system that circumvents the problems of cell type heterogeneity inherent in the study of embryonic tissues. These systems have the disadvantage, however, that the biological relevance of the types of interactions displayed is somewhat difficult to assess. As is seen below one of the most severe limitations in dealing with all of these systems, in terms of the biochemistry of the components responsible for cellular adhesive specificity, is the minute quantities of the cell surface components that can be obtained. Two organisms in which this problem of quantity can be overcome to some degree are the marine sponges and the cellular slime molds. These systems, as well as being of interest in their own right, are also viewed as potential model systems for defining some of the basic biochemical aspects of intercellular adhesive specificity. The marine sponges, in fact, provided the first direct demonstration of differential cellular affinities. In 1907, H. V. Wilson (1) published an account of the sorting out of mixtures of mechanically dissociated marine sponges to produce discrete organisms containing cells of only one species. This process was easily observed, since the species of marine sponges used contained brightly pigmented cells of different colors. Many of the conclusions that form the basis of our understanding of the specificity of supracellular architecture and cellular adhesion are based on elegant descriptive work of this type. Only within the past decade or two have the first inroads been made into quantitating cell adhesive phenomena. Even more recently, a few experimental systems have begun to yield to biochemical approaches to an understanding of those processes termed cell recognition, intercellular adhesion, or cell-cell interaction. In this review we describe several biological systems in which the biochemistry of cellular adhesive specificity is becoming elucidated. These include sea urchin fertilization, reaggregation of dissociated chick embryo cells, and contact phenomena in cultured 3T3 fibroblasts and Schwann cells. In addition, two "simpler" systems, sponge reaggregation and aggregation of cellular slime molds are discussed.

SPECIFICITY OF CELLULAR ADHESION

Before proceeding with a description of methods used to assay intercellular adhesion, it is important to introduce the concept of specificity as applied to this phenomenon. Specificity becomes more difficult to define the more complex the organism from which the cells are derived. In many cases, it is not obvious just which cells should preferentially adhere to which others in complex tissues such as brain. Should the pattern of intercellular affinities directly reflect the cytoarchitecture that existed at the time of tissue disaggregation, or should the cellular affinities reflect an earlier embryonic hierarchy? In addition, dissociation procedures for embryonic tissues that use trypsin or other proteases destroy many cell surface proteins. Thus, another question concerns the identity of the proteins that "grow back" with those removed during tissue dissociation, and the possible concomitant changes in cellular adhesive specificity. Even in simpler systems like the cellular slime molds, the definition of specificity often reduces to an arbitrarily mechanical one. Adhesion of cells is, in general, not an all or none phenomenon, and the best that can usually be achieved is the differential adhesion of cell A for A over A for B. Thus it is often found that only a very narrow range of frequency of cell collisions and opposing shear force allows the demonstration of specificity or "differential intercellular affinities." In fact, it is a general observation in the chick embryo system that most dissociated embryonic cells will stick to nearly all other cell types from the embryo. In long term (24–48 hr) reaggregation experiments, many different cell types associate rapidly into clumps of cells (1 hr). With time, the cells tend to sort out within these aggregates to form islands of rather homogeneous cell type (2). In short term assays, evidence has been obtained for a multistep ordered model for cell adhesion. It has been proposed that cells initially recognize each other in a readily reversible step, and that subsequent events lead to cellular adhesions that are much less reversible (3). Thus, the terms cell recognition and adhesion which are often used interchangeably, may in some instances refer to reversible and irreversible associations respectively and in others to a combination of these events. It is apparent that even defining what we mean by specificity is difficult and must vary somewhat from one biological system to another, and from one type of assay to another.

Table 1 and Figure 1 illustrate the different types of assays that have been used for the measurement of cell adhesion. A detailed description of these methods can be found in the original references and in several recent reviews (5, 25, 26). Each of these methods has advantages as well as limitations. In addition, since cell adhesion appears to be a multistep process, these assays may not reflect the same rate-limiting step and may therefore

Table 1 Methods for the study of cell adhesion

Method	Limitation	References	Advantages
I Disappearance of single cells into aggregates	Can only be used to measure specificity of blocking agents or aggregation promoting agents, including lectins, antibodies, and membranes. Artifactual loss of single cells often encountered.	(4–7)	Often most rapid and convenient. Requires only counting of cells.
II Binding of single cells or small aggregates to immobilized cells or large aggregates	Relation of surface of aggregates or immobilized cells to original cells is not known. This is method most sensitive to mechanical parameters.	(8–18)	Easily quantitated by counting of radioactive label in single cells.
III Formation of mixed or segregated aggregates from differentially labeled cells	Relatively clumsy and requires manual separation of individual aggregates; adhesion between homogeneous aggregates, may be identical to random adhesion. Potentially very useful but has not been widely used.	(19)	Allows dynamic picture of *both* cell populations.
IV Binding of plasma membranes to cells	Membranes may not retain the adhesive components of the original cell surface, or the components may be altered. Purity of membranes is difficult to assess.	(5, 20, 21)	Easily quantitated with labeled membranes. Not as dependent on shear force as whole cells.
V Aggregate size after prolonged incubation *in vitro*	Measures complex differentiation and survival processes, and is not a direct reflection of adhesion. May reflect the effects of tissue-specific hormones and nutrients on cells.	(2, 22–24)	Cell differentiation and organization within the aggregate are often physiologically relevant.

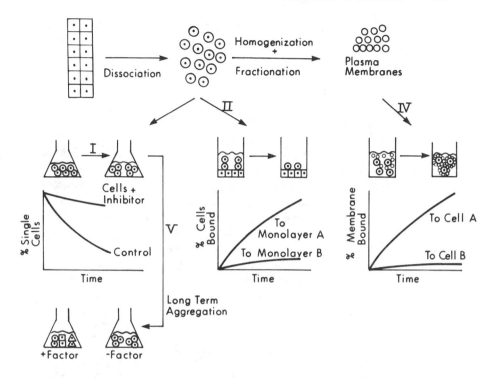

Figure 1 Methods for the measurements of cell adhesion. A tissue is dissociated into single cells either mechanically or by the use of proteolytic enzymes. Cell to cell adhesion can be measured by the methods described in Table 1 and indicated by numerals I to V. Specificity is shown either by the use of inhibitors or by the ability of cells or membranes to discriminate between two different cell types. In method V differentiation and aggregation induced by tissue-specific factors is schematically indicated by showing cells of different shape.

be a measurement of the "specificity" of a different portion of the adhesive sequence.

A simple and useful model for the consideration of cell to cell adhesion is illustrated in Figure 2. This model assumes that specific cell to cell recognition is brought about by the interaction of at least two mutually complementary ligands, much in the same manner as an antibody, an enzyme, or a lectin binds a ligand.[2] This initial step, which is reversible, is

[2]Several in vitro adhesion phenomena cannot be simply explained by the interaction of two complementary adhesive molecules. Most notably, the segregation of cells from an initial random aggregate such that cell A is always exterior to cell B, requires different interactions which may relate to the physicochemistry of the cell surface, as has been carefully discussed by M. Steinberg (27, 28), or possibly chemotactic effects of components released by cells into the surrounding medium. Such interaction is not discussed in this review.

followed by one or more irreversible reactions (on the time scale of a laboratory assay), pictured in Figure 2 as the formation of multiple adhesion sites and changes in the cell surface. In order to be biologically meaningful this "irreversible" step of adhesion must be reversible, on the time scale of development, to allow for cell migration and other adjustments of cellular contact. This reversal may be due to the selective destruction (turnover) of the adhesive sites.

Any model that assumes that specific intercellular adhesion is the result of the interaction of complementary structures, has implicit in it the assumption that at least one of the interacting molecules is a protein. These are the only molecules that have the required specificity. The model makes no assumption as to the nature of the protein or the second ligand which could be another protein or a specific carbohydrate, as has been assumed in more specific versions of this model (29, 30). It is clear that nonspecific surface forces can alter cell to cell adhesion, but cannot by themselves account for the specificity observed in many cell to cell adhesion assays.

It is generally assumed that homotypic adhesion is preferred, and that cells from a defined anatomical region prefer to adhere to each other as compared to cells from a different anatomical region. This is based on the

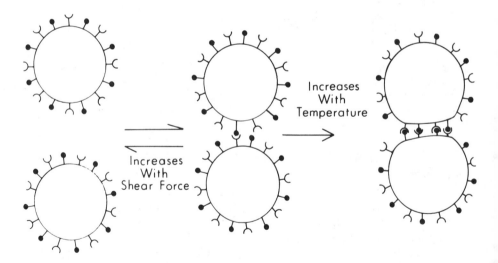

Figure 2 A simple model for cell to cell adhesion. The model assumes that specific cell adhesion is determined by complementary cell surface ligands. Under most assay conditions an initial reversible event (3) is followed by one or more apparently irreversible events, denoted schematically by the formation of multiple cell to cell attachments and a change in cell shape. These latter events are responsible for the striking temperature-dependence of cell adhesion. A detailed analysis of the effect of the physical properties of the membrane on this event has been presented (26a).

obvious fact that cells in a given organ remain together at the end of the developmental process. The in vitro assay is seldom an equilibrium measurement [see however (27, 28, 31)], and is most frequently a rate whose precise relation to a possible equilibrium parameter is not known. Indeed, if adhesion involves a chemically irreversible step that is reversed only by destruction of the ligands then equilibrium as such may not have any meaning in these systems. Rate measurements in cell to cell adhesion assays provide information that allows us to distinguish in an operational sense one cell surface from another, but it is not clear that the relative values of these rate measurements can be predicted from anatomical considerations. This can be illustrated by a consideration of the adhesive properties of neural retinal cells obtained from the chick embryo.

Dissociated cells obtained from either neural retina or optic tectum will rapidly adhere to each other under a variety of assay conditions. The neural retina clearly consists of several types of cells, and no methods are available that allow the separation of these different cell types so that their properties can be investigated separately. It is possible however to ask whether different regions of the retina show identical adhesive behavior. This question has been explored in a number of different assay systems. All of these experiments have been motivated by the Sperry model of retino-tectal connectivity (32) and variations of this model involving specific carbohydrate ligands (33, 34).

The ganglion cells of the neural retina project to the optic tectum in such a way that the dorsal-ventral and medial-lateral axes are inverted. It is postulated that this is due to the presence of surface ligands on both the retinal axons and the tectal cells such that an axon from the extreme dorsal region of the retina would preferentially connect to a cell at the extreme ventral region of the tectum, and the converse would be true for an axon derived from an extreme ventral region of the retina (32–34). With this in mind, Barbera et al (35, 36) tested the rate of adhesion of single cells derived from the dorsal and ventral halves of the retina to tectal halves and found that cells from the dorsal half of the retina adhered faster to ventral tectal halves than to dorsal tectal halves and that the converse was true for cells from the ventral half of the neural retina. This data would agree with the Sperry model for retino-tectal connectivity, provided the following assumptions hold true: 1. all of the retinal cells contain the same adhesive markers that are important for retino-tectal connectivity, i.e. these markers are not restricted to the ganglion cells or their axons which represent less than 10% of the cell population, and 2. the rate of cell adhesion in this system reflects the equilibrium binding affinity between retinal axons and tectal cells. The validity of these assumptions is not known at this time and no tests for them are available at present.

The simplest version of the retino-tectal adhesion model would assume that the same two ligands are present in complementary gradients on the surface of the retina cells as on tectal cells. With this model in mind, Gottlieb et al (37) tested retinal cells obtained from the extreme dorsal or extreme ventral areas of the retina for their ability to adhere to monolayers prepared from cells from different regions along the dorsal ventral axis of the retina. The results show that there is a dorsal-ventral adhesive gradient in the retina such that dorsal cells adhere most rapidly to monolayers prepared from ventral cells, and ventral cells adhere most rapidly to monolayers prepared from dorsal cells. No similar medio-lateral gradient could be found. In addition, Gottlieb & Arington (38) find that dorsal tectal cells adhere preferentially to ventral tectal monolayer as compared to dorsal tectal monolayers. These three sets of data are at least in partial agreement with the postulate that retino-tectal adhesion reflects the presence of gradients of adhesive molecules on the surface of the retina or tectum. It is, however, not clear that the observed gradients actually have a function in retino-tectal connectivity.

Gottlieb & Arington (38) have also examined the binding of chick embryo neural cells to heterologous areas of the nervous system. In many cases they find that heterotypic binding is preferred to homotypic binding; for example single cells prepared from telencephalon prefer to adhere to mesencephalon, and cells from mesencephalon prefer to adhere to telencephalon. Dorsal neural retina cells prefer mesencephalon to telencephalon, but the converse it true for ventral retinal cells.[3]

The conclusions from this series of experiments are the following:

(*a*) When examined under in vitro conditions, homotypic adhesion is not necessarily the preferred mode of cell interaction.

(*b*) Within any given defined anatomical region of the nervous system there occur clear regional differences in adhesive rate (probably reflecting quantitative but perhaps also qualitative changes). In the cases examined these differences appear as gradients along one of the major axes of the embryo. These gradients may be of importance in providing cells with positional information, but their relevance to synaptogenesis is not clear. Adhesive preference may be exhibited by cell pairs that normally do not form synapses.

(*c*) Within any given tissue or region, cells with the highest rate of adhesion do not necessarily occur next to each other (for example, dorsal

---

[3]It is possible (26) that type II assays measure only those ligands on the cell surface that have not bound to adjacent cells in the monolayer or large preformed aggregate; the conclusions about homologous vs heterologous specificity are not altered by this assumption.

and ventral retinal cells), thus if a cell's position is in part determined by the affinity of surface ligands for neighboring cells, then the initial rates of cell to cell adhesion measured in an in vitro system and the equilibrium position are not necessarily related.

A separate and very important question is whether cell to cell adhesion is a permanent or a transient event in vivo. Consider an embryonal cell that is migrating in a direction specified by an adhesive gradient, or an axon which, having found the optic tectum, now tries to find the region of the optic tectum for which it has highest affinity. In both of these situations one must assume that contacts are made and are then reversed. How this is accomplished is not understood. Are all adhesive processes in vivo readily reversible so that a thermodynamically preferred conformation is reached, or does a cell simultaneously sample several adjacent cells to which it can adhere, and then release all contacts except the ones for which it has the strongest affinity? If so, how is release accomplished?

In this context it is worthwhile to consider the binding of isolated flagella to chlamydomonas. This binding is transitory in that flagella bind and appear to be inactivated during the binding process and then released by the cells. The cell remains competent to bind new flagella but the released flagella have been inactivated (39). This transitory binding is a reflection of the steps involved in mating in chlamydomonas, which initially involves the attachment of flagellar tips between cells of opposite mating types followed by cell fusion and separation of the adhering flagella (see e.g. 40, 41).

In more complex systems transient synapse formation followed by selective stabilization has been suggested as a mechanism for the generation of specific neuronal connections (42). The rapid reversal of inappropriate (retina to muscle) synapses has been demonstrated in tissue culture and compared to formation of stable synapses between appropriate cells (spinal cord to muscle) (43). Thus the specificity for reversal may be as important as the specificity of binding in accounting for the ultimate position of a cell within an organized structure.

CELL TO CELL ADHESION STUDIES WITH DISSOCIATED EMBRYONAL CELLS

There is an extensive literature on the adhesion properties of dissociated embryonal cells, studied by a number of variations of the methods shown in Table 1. We will restrict our discussion to recent work which is directed at the elucidation of the chemical basis of cell to cell adhesion. Additional information may be found in two very comprehensive reviews by Roth and co-workers (25, 30) and in the proceedings of a recent symposium (44).

Studies With Plasma Membrane Fractions in the Nervous System

If cell to cell adhesion is described by the simple model shown in Figure 2, then isolated plasma membrane vesicles should be able to bind to cells and show some of the same specificity as is shown in cell to cell adhesive assays. The membranes may also be expected to modify cell to cell adhesion, either by promoting the formation of aggregates if membranes can act as cell to cell cross-linking reagents, or by inhibiting cell to cell adhesion if cells coated with vesicles cannot adhere to each other. This later situation would arise for example if membrane vesicles contained only one of the two complementary components that are a minimum requirement for specific cell adhesion.

Plasma membranes have been prepared from neural retina and cerebellum of 8 day chick embryos. These membranes bind preferentially to homologous cells. The retinal membranes, when added in excess, inhibited the aggregation of retinal cells in a type I assay (Table 1) (45).

When plasma membranes were obtained from retinal and tectal cells of different age embryos, they preferentially blocked the aggregation of homologous cells of the same developmental stage (46). Retinal membranes also blocked the aggregation of tectal cells but the converse was not true, i.e. tectal membranes did not prevent the aggregation of retinal cells. Thus, these data show that there occur significant and rapid changes in cell adhesion during development. These observations are based on type I assays in which membranes compete with cells for adhesive sites. For technical reasons it is only possible to carry out these assays over limited concentrations of membranes (5) and the temporal differences may be quantitative rather than qualitative. Thus, the fact that membrane prepared from 8 day neural retina does not inhibit the aggregation of 7 day neural retina cells does not necessarily mean that the ligands present on 8 day cells are absent from 7 day cells.[4]

The molecules responsible for the ability of retinal membranes to inhibit cell aggregation were extracted with lithium diiodosalicylate from an acetone powder of the membranes. They appeared to be low molecular weight proteins that could not be purified (49).

Recently, a low molecular weight protein (ligatin) has been purified to homogeniety from plasma membranes of chick neural retina. The molecular

[4]Several other parameters in neural retina have been shown to change dramatically between day 7 and 10 of development; for example, there are major changes in cell surface glycoproteins during this time period (47) as well as changes in the ability of neural retina to form synapses in culture (42). The ability of retinal and tectal cells to coaggregate and induce acetyl cholinesterase also changes during this period of development (48).

weight is 10,000, and in the presence of Ca^{2+} the protein polymerizes to form filaments. This protein blocks the aggregation of retinal cells and prevents the binding of labelled molecules released from retinal cells in culture to freshly dissociated retinal cells (50). Ligatin represents about 1% of the retinal plasma membrane protein and may be identical to the aggregation inhibitory components of Merrell et al (49), although additional factors have to be postulated to account for the temporal specificity observed by these authors. Ligatin or a protein very similar to it had previously been isolated from rat ileum, and shown to be required for the attachment of β-D-N-acetylhexosaminidase to the plasma membrane (51).

Studies With Hepatocyte Plasma Membranes

In a number of instances, dissociated cells from the same organ or tissue but from different species will coaggregate (9, 52, 53). A notable exception to this is liver cells, where it has been shown that species specificity predominates. In particular, chick and rat hepatocytes will not bind to each other (19, 54). These observations have been extended to liver plasma membranes. Plasma membranes bind to and stimulate the rate of aggregation of homologous but not heterologous liver cells (21, 55). In order to observe this effect, chicken liver plasma membranes had to be dialyzed against pH 9.6 buffer, and rat liver plasma membranes required dialysis against EGTA. These observations suggest that the binding sites in the membrane may be occupied as the membranes are isolated and that dissociation of this complex is required before the membranes can act to cross-link cells. There is a similarity between these activating conditions and those required to dissociate the glycoprotein binding proteins, previously known to be present in the plasma membrane of liver cells, from their ligands (56–58). These proteins are known to be present in the hepatocyte plasma membrane. Whether these proteins are involved in the cross-linking of liver cells by plasma membranes is not known. It is suggestive, however, that rat hepatocytes will bind to acrylamide sheets containing galactose residues, and chicken hepatocytes to gels containing GlcNAc residues (59, 59a). This binding would be in agreement with the specificity of glycoprotein binding components of these cells (56–58). If this speculation is correct, then the hepatocyte aggregation system is one in which carbohydrate is one of the specific ligands involved in cell to cell adhesion. The liver system is particularly attractive for the study of cell adhesion because livers from young animals rather than embryos can be used and hepatocytes can be prepared as a relatively homogenous cell population in large quantities.

A number of carbohydrate binding proteins have been identified on the surface of hepatocytes (56–58, 60), Kupfer cells (61–63), fibroblasts (64, 65), macrophages (66), muscle (67, 71b), and embryonic rat and chick brain

(72–74). In chick brain (73, 74a) these lectins have been shown to be developmentally regulated. Some of these proteins are involved in the specific uptake of glycoproteins (including lysozomal enzymes) by these tissues (57, 58, 60–66) while others have no known function (67–74a). It is attractive to consider the possibility that they are also important for cell recognition, but no evidence is available in this regard. The evidence that a galactose binding lectin present on the surface of myoblasts is involved in cell recognition and cell fusion is contradictory (68–71).

Use of Antibodies for the Study of Cell Adhesion

If the cell surface contains specific molecules involved in cell adhesion, and if these molecules are antigenic then it should be possible to prepare antibodies against such molecules. The monovalent Fab fragments prepared from such antibodies should inhibit cell aggregation if the antigenic site(s) and the adhesive site are in close enough proximity. It is necessary in such experiments to show that only antibodies against specific cell surface molecules have this effect and that the inhibition is not due to an unspecific effect of any antibody that can bind to the cell surface.

This approach, originally used by Gerisch and colleagues to study slime mold aggregation (see below), has been used to study the aggregation of neural retinal cells by Rutishauser, Edelman, and co-workers (7, 75, 76). They have shown that antibody against a polypeptide of molecular weight 140,000 specifically inhibits retinal cell aggregation. This molecule, designated as CAM (cell adhesion molecule), appears in the culture supernatant when neural retina cells are incubated in tissue culture medium in the absence of serum. No evidence is yet available to show that the molecule (CAM) itself can block adhesion, which would be expected if it were a component of the cell adhesive machinery, but this may reflect alterations in the molecule during isolation. Since cell to cell adhesion is a complex multistage phenomenon, it is not yet clear which step in the adhesive process is blocked by these antibodies. The immunological approach, however, is extremely important since it allows the identification and isolation of adhesive components by assay methods that do not depend directly on their biological activity. This is often, however, a drawback in determining the precise functional relevance of the antigen(s).

Similar studies have been carried out by McClay et al (77) who have immunized rabbits with plasma membranes from chick neural retinal cells. Although the specificity of these antibodies is less well characterized, they are directed against trypsin-sensitive (i.e. protein) components at the retinal cell surface (77). The relation of the antibodies isolated by McClay to those isolated against CAM (76) remains to be established.

It will be of interest in the future to characterize the antigens detected by these antibodies and try to determine their role in the adhesion process as well as their tissue specificity. One can imagine for example that adhesion requires both tissue specific components as well as a series of common molecules involved in cell to cell adhesion. Thus in the simple model in Figure 2, antibodies could be directed against either of the two complementary ligands assumed to be tissue specific or against other components on the cell surface responsible for the formation of a stable cell to cell adhesion that could be common to many different cell types in an organism. Alternatively, one can assume that the cell-specific ligands are attached to a common membrane component, such as the postulated base plate in sponge cell adhesion (44, 78) and antibodies against this component will block cell adhesion in many different cells.

Enzymatic Alterations of the Cell Surface as a Probe for Adhesive Molecules

In principle, any group-specific reagent that will not penetrate the cell can be used to alter cell surface molecules and provide information regarding the chemical nature of the adhesive components present on the cell surface. In practice, the reagents used most effectively in this work are enzymes that modify the cell surface. The most extensive study of this type has been carried out by Marchase et al (33, 79).

The system used was the adhesion of retinal cells to tectal halves. Cells from the dorsal half of the retina adhere preferentially to the ventral half of the tectum, and cells from the ventral half of the retina adhere preferentially to the dorsal half of the tectum. This preferential adhesion of trypsin dissociated cells is only expressed after recovery from trypsinization (79).

The *preferential* adhesion of the ventral retinal cells to the dorsal half of the tectum is prevented by pretreatment of the cells with proteases (trypsin or chymotrypsin), the presence of cycloheximide or puromycin during the recovery period, or by treatment of the tectal half with β-N-acetylhexosaminidase (79).

Conversely, the *preferential* adhesion of dorsal retinal cells to the ventral half of the tectum was prevented by treatment of the retinal cells by β-N-acetylhexosaminidase, or by treatment of the tectum with proteases. These and other observations suggest that one of the ligands involved in retina tectal adhesion is a ganglioside (GM-2) that has a terminal GalNAc, and that the other is a protein molecule that can bind to GM-2, possibly a glycosyl transferase. This model was supported by the observation that liposomes containing GM-2 bound preferentially to dorsal tectal halves (by a factor of about 1.5) (79).

The potential identification of the receptor for cell adhesion by this approach is ingenious. It is somewhat surprising, however, that treatment with proteases or glycosidases only abolishes the preferential binding and leaves the basal level of binding intact. This is at least 50% of the total binding that is observed in this system. These findings suggest that at least two different systems of ligands must be responsible for the adhesion of retinal cells to tectal halves. One of these does not distinguish between dorsal and ventral cells and is neither protease nor glycosidase sensitive, thus making this system extremely complex.

Isolation of Aggregation-Promoting Factors

Work initiated primarily in the laboratory of A. A. Moscona has been directed at the isolation of components that stimulate the formation of large aggregates and that show tissue specificity, or specificity for different regions of the nervous system. The assay used is Method V of Table 1, which requires prolonged incubation periods. It is not clear whether factors that stimulate the formation of large aggregates are components of the cell to cell adhesion apparatus per se or are components that influence cell metabolism, such as hormones, and only indirectly affect cell adhesion. Factors that specifically promote growth of aggregate size under these conditions have been described from a number of sources including various areas of the embryonal nervous system (23, 24, 80–82), teratocarcinomas (83–85), and embryonal liver (86). The purification of two of these factors and their chemical characterization have been reported from chick neural retina (22, 23) and from embryonal liver (86).

Hausman & Moscona (22, 23) have purified a protein from chick neural retina that specifically promotes retinal aggregation. The protein has a molecular weight of 50,000 and contains 15% carbohydrate. The carbohydrate appears not to be essential for biological activity. The material was originally prepared from supernatants of retinal cultures maintained in the absence of serum; it can, however, also be prepared by butanol extraction of retinal plasma membranes, which is suggestive evidence for a cell surface location of this protein.

An aggregation-promoting factor from liver has been obtained in pure form, and has been identified as taurine (86). Since taurine is unlikely to act directly as a cell to cell ligand, it most likely acts indirectly by affecting cellular metabolism. There are at least three recorded instances in which the adhesive properties of cells are affected by hormones or trophic factors. These are 1. the induction over a 24 hr period of a change in the adhesive properties of isolated tectal cells by high concentrations of nerve growth factor (87), a change that mimics a normal developmental change of these cells; 2. the induction of adhesive changes in Pc12 (a cloned pheochromocytoma) by nerve growth factor, which over a 30 min period becomes more adhesive to plastic as well as to other cells (88); and 3. changes in the adhesive characteristics of fibroblasts to Concanavalin A–coated nylon fibers induced by epidermal growth factor (89).

The aggregation-promoting factors are of interest because they represent tissue or organ specific effectors. The molecular basis of their action however remains unknown, at least in the embryonal systems. Carbohydrate has been implicated as a functional component of the teratoma aggregation factor, but whether this aggregation factor (which consists of several components) acts to cross-link cells in the manner of a lectin or an antibody is not known (83–85).

Balsamo & Lilien (90, 91) have obtained proteins from culture supernatants of dissociated retinal or cerebral cells labelled with radioactive GluNAc which bound specifically to retinal and cerebral cells. The binding to retinal cells appeared to be blocked by GalNAc and the retinal protein contained GalNAc. It is interesting that GalNAc is present in this protein and that it has also been postulated to be involved in retina-tectal adhesion (79). The relation of these binding proteins to the cell adhesion process is not known.

CELL TO CELL ADHESION STUDIES WITH CULTURED CELLS

The adhesive properties of a number of permanent cultured cell lines have been examined. Since the precise anatomical origin of such cell lines is not always known, it is not always obvious which adhesive events measured with these cells are biologically meaningful and which are laboratory curiosities. Cell to cell adhesion has been measured with normal and transformed fibroblasts such as 3T3 cells (e.g. see 92, 93) and with neuronal cell lines (20, 94, 95). The attractive feature of such cells is that they are available as a cloned uniform cell population in almost unlimited quantities.

We primarily discuss some of the results obtained with neuronal cells, because of their implications for cell recognition in general. Cell adhesion has been examined in two separate laboratories, using neuronal cell lines most of which have been developed in the laboratory of D. Schubert (96). In one instance, the specificity was examined primarily by measuring adhesion of plasma membranes to cells (Method IV in Table 1) (20, 95) and in another by a modified monolayer adhesion assay (Method II in Table 1) (94). Although the data do not agree in detail, each of the two laboratories concluded that cell adhesion between various neuronal cell lines is mediated by a number of different pairs of complementary ligands. Any given cell may express one or more of these cell ligands and a cell need not express simultaneously the two components of a complementary ligand pair. In addition, with some cells, different ligands were expressed at low cell density and at high cell density (95).

These observations lead to the interesting speculation that a large number of cell surfaces, differing in their adhesive properties, may be generated by controlling the expression of a limited number of ligands. Thus for n types of ligands where a given cell can express anywhere from 0 to n types of ligand molecules, it is possible to generate 2^n different cell surfaces. The implications of such a model for the generation of the extensive cell surface diversity required for synaptogenesis is obvious (20).

FUNCTIONAL CONSEQUENCES OF CELL TO CELL ADHESION

When two cells can be shown to bind to each other in the laboratory, is the binding meaningful in terms of cell function or is it an artifactual consequence of the assay conditions? With embryonal cells, binding is usually considered meaningful when it appears to correspond to known developmental patterns. A better identification of physiologically meaningful binding would be possible if one were to study situations in which cell to cell binding is followed by metabolic changes in the cells, such as induction of specific proteins (97). The systems where this approach is possible are limited. The most extensive observations so far have been made with the study of contact inhibition of growth with Swiss 3T3 cells and the mitogenic action of neurites on Schwann cells.

When grown in 10% calf serum, Swiss 3T3 cells cease to grow at a cell density of $5 \times 10^4/cm^2$ even if the growth medium is changed frequently. Such confluent cells can be made to undergo one or two additional rounds of cell division by increasing the serum concentration (98). Transformed 3T3 cells do not show this phenomenon and will grow to much higher densities when in low serum. Two general theories have been proposed to explain contact inhibition. The first assumes that the cessation of growth at contact is due to a negative signal for growth which arises as a result of cell to cell contact (99). The second assumes that growth at confluency ceases because cells are exhausting some component from the growth medium or that the rate of delivery of mitogens from the bulk medium to the cell surface becomes rate-limiting at confluency (98, 100, 101). These two theories are not mutually exclusive and each phenomenon may contribute to a different extent to growth control with different cell lines. It is beyond the scope of this review to discuss in detail growth control of cells in culture. It should be clear, however, that if there is a contact component to growth control, and if plasma membranes retain this component in functional form, then addition of plasma membranes to sparse cells should inhibit the growth of these cells, in a manner resembling that observed at confluency. It should then be possible to purify these membrane components and study their mode of action in detail. This implies that just as mitogens can initiate growth by binding to the cell surface, the binding of other ligands to the cell surface can inhibit growth. A model system for such an inhibitory phenomenon is the inhibition of growth of 3T3 cells and SV40-transformed 3T3 cells by succinylated Concanavalin A (102, 102a).

When a purified plasma membrane fraction prepared from confluent 3T3 cells is added to a sparse culture of 3T3 cells, the growth of these cells is inhibited (103). The following data suggest that inhibition resembles that observed at contact. The inhibition by plasma membranes is reversible (by trypsinization) and is not due to a toxic effect of the membranes on cells. SV3T3 cells are not affected by membranes. The cells are blocked in the G1 portion of the cell cycle, and the membranes do not deplete the growth medium of mitogenic factors. The activity in the membranes is heat labile, and active membranes, but not heat-inactivated membranes, bind to cells (102, 104). The membrane-inhibited cells show decreased rates of transport of uridine and α-aminoisobutyric acid, but contrary to what is observed at confluency show normal rates of transport of glucose and phosphate (105).[5] The active component(s) have been extracted from the membranes with the nonionic detergent, octylglucoside, and remain active after removal of the detergent by dialysis (107). The extracted material is stable and it should be possible to purify it by standard techniques.

In contrast to this system, where membranes and, by inference, cell contact act as an "antimitogenic" signal, is the interaction of Schwann cells and neurites. Schwann cells isolated from dorsal root ganglia by methods developed in the laboratory of R. Bunge (108, 109), only grow in the vicinity of dorsal root ganglia neurites. It has now been shown that a membrane fraction prepared from neurites acts as a very specific mitogen for Schwann cells (110). The mitogenic activity appears to be localized to the surface of the neurites, and is trypsin- and heat-sensitive. The mitogenic activity cannot be detected in a variety of other cells, including fibroblasts and permanent neuronal cell lines. Thus in this system there occurs a heterologous cell-cell contact in which Schwann cells interact with the neurite cell surface to generate a mitogenic signal for the Schwann cells.

It seems reasonable to consider cell to cell interactions of the type observed with 3T3 cells and with Schwann cells in the same way as the interaction of polypeptide hormones with the cell surface. In each case the

[5]Recent independent evidence suggests that glucose and phosphate transport are not obligatorily coupled to growth control (106).

binding of a ligand to the cell surface has a transmembrane effect with profound metabolic consequences. The two examples listed are complex in the sense that the metabolic response (growth) is extremely complicated and at present poorly understood. It seems likely, however, that simpler systems, where better understood metabolic pathways than those involved in growth control are affected by cell contact, will be found and will be easier to understand at the molecular level.

Several systems show specific cell interactions that will be difficult to investigate at the molecular level. Aggregates of heart cells (7 day chick embryos) will adhere to monolayers prepared from the same cells. Both aggregates and layers will beat and after 6–12 hr of attachment beat synchronously. The synchrony is associated with the formation of gap junctions between the aggregates and the monolayers. The development of synchrony is prevented by cycloheximide, and presumably some proteins required to form functional junctions are induced as a consequence of cell to cell adhesion (111, 112). The system is extremely interesting but does not appear to readily allow the development of an assay for the isolation of adhesive components. Cell to cell contact is often followed by the formation of junctions through which small molecules can pass from cell to cell. The formation of such junctions is an indication of a physiologically meaningful cell contact, and one which shows specificity (114, 114a). However, like the electric coupling between heart cells, this is a system that cannot readily be used for the isolation of cell surface adhesive components.

Similarly, aggregation of retinal cells, or mixtures of retinal and tectal cells of corresponding ages, induces the production of enzymes involved in neurotransmitter synthesis (48, 113). The time course of this induction is, however, extremely slow and this system also, if in fact the cell surface components are responsible for this induction, cannot be readily studied.

SEA URCHIN FERTILIZATION

Fertilization appears to involve a relatively simple case of specific cell-cell recognition, adhesion, and, ultimately, fusion. In higher organisms where fertilization occurs within the uterus or oviduct, many physiologically complex steps precede the events of fertilization per se (115, 116). Thus simple marine organisms, such as sea urchins, for which fertilization takes place in sea water, are much easier to study in terms of the primary interactions of sperm with egg (117). Even in sea urchins, some confusion existed until recently about which of the three layers surrounding the egg—the jelly coat (JC), the vitelline layer, (VL) or the plasma membrane (PM)—was the real site of species-specific sperm binding (118–120). It is now clear that while sperm adhere to the JC, this process is not species-specific (120). The fertilization process is quite species-specific and is so easy to monitor that direct observation of the fertilization reaction and its inhibition by putative sperm receptor preparations is the best assay available (120, 121, 122). Within 5 min of mixing sperm and eggs of the same species, in sea water, the fertilization membrane (vitelline layer) becomes highly refractive under phase optics and lifts away from the underlying plasma membrane (121).

Using this assay, developed originally by Vacquier & Payne (121), Lennarz and co-workers (122) have shown that eggs of *Arbacia punctulata* contain on their cell surface (not the JC) a glycoprotein that inhibits the fertilization of eggs of this species, but not of *Strongylocentrotis purpuratus* eggs. The inhibitory activity of a membrane fraction was destroyed by treatment with trypsin. A soluble factor was derived from egg membranes, simply by incubation in sea water, which also inhibits fertilization in a species-specific way. The most likely reason for its release from the membranes under such mild conditions is that a protease from the cortical granules associated with the egg PM (123, 124) cleave an active fragment from the membranes. It is thought that this process is normally important as a block to polyspermy (116); such blockage occurs very rapidly (121) after a single sperm fertilizes the egg (120). The cortical granules (125) immediately fuse with the PM (126–128) and empty their contents, including trypsin-like "fertilization protease" (127–129), into the perivitelline space. These enzymes then strip the egg surface of sperm receptors (130) which may also become soluble inhibitors of further sperm binding (122). Previously bound sperm are also released (121). The "autosolubilized" factor binds to sperm, thus inhibiting fertilization (122). The soluble fraction contained seven major bands on SDS gel electrophoresis which stained for both protein and carbohydrate. The soluble fertilization inhibitor bound to con A-Sepharose (122), and Con A was also found to inhibit fertilization in this and other species (131–133). The factor has not yet been purified.

Aketa and co-workers (134) have solubilized a factor from *Hemicentrotus pulcherrinus* eggs with 1 M urea which appears to inhibit fertilization in a species-specific manner. They used an inhibition-of-fertilization assay (135) (described above) and also an assay involving the ability of soluble fractions to neutralize antibody to the crude soluble fraction (136). This antibody inhibited fertilization, so that the assay relies upon reversal of inhibition of fertilization (134, 136). The soluble extract was fractionated stepwise by acid, calcium acetate, and ammonium sulfate precipitation, followed by gel filtration on Sepharose 4B at pH 7.2 and CM cellulose chromatography (134). The purified material gave a single band on cellulose acetate strip electrophoresis and did not enter a 7% polyacrylamide SDS gel. Gel filtration on Sepharose CL-4B in 6 M guanidine HCl also indicated a very large size; the chemical composition and subunit structure of the factor remain undetermined (134).

Vacquier and associates have obtained a protein from sea urchin sperm acrosomal vesicles which is believed to be responsible for the species-specific binding of sperm to egg receptors (137–140). This protein, named bindin (137), occurs in pure form as the sole internal component of acrosomal vesicles (137), the structure implicated by Summers & Hylander as being responsible for species-specific binding of sperm to eggs (120). Bindin from both *S. purpuratus* and *S. franciscanus* sperm has a molecular weight in SDS of 30,500 and is insoluble in the absence of detergents or guanidine HCl. Both bindins are very similar in amino acid composition and tryptic peptide maps (138). Partial sequence data from the N terminus indicate extensive homology between the two bindins and an interesting internal repetition of a 10 residue segment (140). Fresh or glutaraldehyde-fixed unfertilized eggs or their isolated vitelline layers (139, 140) were agglutinated by bindin. A glycopeptide fraction of VL inhibited the agglutination as did prior trypsin treatment or periodate oxidation of the eggs. This agglutination reaction is species specific (139, 140). Bindin also agglutinates trypsinized glutaraldehyde-fixed rabbit red cells and the agglutination is inhibited by low concentrations of galactose, lactose, and fucose. (140).

To investigate the bindin receptor, intact eggs were labeled with ^{125}I (141). Labeled eggs were then activated by the Ca^{2+} ionophore A23187 in the presence of low concentrations of soybean trypsin inhibitor (128, 130) to allow partial proteolysis of receptors (122) by released cortical granule enzymes (fertilization protease) to occur (141). This treatment resulted in solubilization of radioactively labeled material, possibly the same factor identified by Lennarz's group (122), which bound specifically to bindin from the same species. The crude receptor preparation was fractionated by gel filtration on Biogel A5M and the activity was found in the void volume. It also did not enter a 5% polyacrylamide SDS gel (141). Treatment of this fraction with fertilization protease abolished its binding activity and reduced the apparent molecular size of the radioactively labeled components in the fraction. Isoelectric focusing of the active fraction from gel filtration yielded an active peak at pI 4.02 and another peak of radioactivity at pH 2.5, believed to be egg jelly (141).

The current view of the recognitive event of fertilization in sea urchins which emerges from these data is that glycoprotein receptors (122, 134, 141) exist on the surface (under the jelly coat) of unfertilized eggs which can recognize in a species-specific way an acrosomal carbohydrate binding protein (bindin) on activated sperm (137–140). These sperm, which have undergone the acrosomal reaction (120), have their acrosomal vesicles exposed

(137). The insoluble nature of bindin (137, 138) may be important for maintaining a locally high concentration of the protein in sea water. After fusion of one sperm at the egg PM, the cortical reaction occurs (126–128) in which the egg releases fertilization protease (127, 128) to cleave off all remaining sperm receptors (121, 122, 130, 141), which provides a block to polyspermy. While extensive proteolysis appears to completely inactivate the sperm receptor (141), more limited proteolysis appears to produce a soluble inhibitor of sperm binding (122, 141). The combination of bindin with receptor is inferred to be a protein-carbohydrate interaction (140) the details of which await further characterization.

SPONGE REAGGREGATION

In 1907, H. V. Wilson (1) described the "coalescence and regeneration" of dissociated marine sponges. He used species containing different colored cells and concluded that different species tended not to reaggregate with one another. Since then this observation has been confirmed in several laboratories with a variety of techniques and with a great number of different sponge species (142–144). Sponges dissociated by pressing through cheesecloth reaggregate rapidly when gyrated at either 24° or 4° (145, 146). If the sponge is dissociated in the presence of Ca^{2+}- and Mg^{2+}-free (CMF) sea water, the cells do not reaggregate at 5°, but, with time, aggregate above 18°. However if the CMF supernatant is added back, the cells rapidly reaggregate even at 5°, and Ca^{2+} enhanced this aggregation. For aggregates to develop the morphological appearance of sponges the temperature must be above 18–20° (142). It appears that an aggregation factor was released in CMF sea water which along with Ca^{2+} (or Mg^{2+}) is required for cell adhesion (142). Such aggregation factors have been solubilized from many species and the general (but not universal) finding is that they are species specific, enhancing reaggregation of only their homotypic cells (142, 143, 147–149).

These aggregation factors (AFs) are some of the most interesting cell adhesion ligands thus far described. They appear to be extremely large proteoglycan structures that display polydispersity of charge and size (148–152). Because of these features, chemical homogeneity of AF preparations has been difficult to establish. Humphreys and co-workers (150–152) have developed purification procedures for AFs from *Microciona parthena* and Burger's group adopted these methods for the *M. prolifera* AF (148). The procedure (150) employs differential centrifugation with the AF sedimenting at 150,000 X *g*. This pellet is redissolved in Ca^{2+}-supplemented sea water and chromatographed on Sepharose 2B. Aggregation-promoting activity is recovered primarily in the void volume. This material was characterized by zonal density gradient sedimentation (70S), zonal electrophoresis (high negative charge), and CsCl density gradient equilibrium centrifugation (density = 1.46 g/cm^3). A constant ratio of protein:carbohydrate: activity across peaks generated in all three procedures is the evidence presented for purity of the AF (150, 151). The overall purification is about 10,000-fold from sponge tissue or 10-fold from the CMF supernatant (crude factor) (150). In negatively stained or shadowed electron micrographs (151, 152), the AF from *M. prolifera, M. parthena,* and *Geodia cydonium* appear as "sunbursts" with a circular center and about 10–16 radiating arms. However, AF from *Halichondria bowerbanbii, Terpios zeketi,* and *Haliclona occulata* prepared by identical methods, appear as linear backbone structures with arms. EDTA rapidly (<1 min) inactivates AF of all species (151, 152); however, several weeks of EDTA treatment are required before dissociation of the sunbursts is seen (151, 152). Thus it is not clear what the structure-function relationships of the AF might be. EDTA, as well as prolonged incubation with sodium dodecyl sulfate (SDS) or guanidine HCl, breaks down the arms of the sunbursts to heterodisperse subunits, but leaves the circles intact (152). The ratio of protein to carbohydrate appears relatively constant in all fractions (152).

Muller and co-workers isolated AF from *G. cydonium* (153, 154) and *Suberites domuncula* (155) and also found subunits or fibrous polymers even larger than those of *M. Protifera* (2×10^9 vs 2×10^7 daltons) (149, 153). They dissociated a low molecular weight AF from the large particle by treatment with Nonidet P-40 (148, 155). Two low molecular weight AFs, purified by G-100 gel filtration on Sephadex G100 and DEAE Sephadex chromatography of the detergent-treated high molecular weight polymers, have molecular weights of about 16,000 (*G. cydonium*) and 55,000 (*S. domuncula*) and consist of >90% protein. They also promote reaggregation of sponge cells in a species-specific way even at 0° and their activity is stimulated by Ca^{2+} (148, 155). These workers suggest that the large fibrous structures are carriers or large aggregates that have AF bound to them (148). A sialyl transferase activity has also been dissociated from the large particles and purified (156). This enzyme can transfer sialic acid to an "aggregation receptor" from the homologous species (*G. cydonium*) but not to a receptor from another species. The enzyme has a mol wt of 52,000 and is apparently distinct from AF which has a mol wt of 16,000–20,000 (148). A galactosyl- and a glucuronosyltransferase activity have also been detected in *G. cydonium* AF preparations (157). As in other work with sponge AF, purity of the preparations is somewhat dubious, as is the relationship of the small AFs to the large AF particles.

Burger's group has obtained evidence that carbohydrate recognition may be involved in *M. prolifera* aggregation (158). Cells treated with CMF sea water are aggregated by solubilized AF in the presence of Ca^{2+}. If glucuronic acid is added to cells before AF, no aggregation is seen. While high concentrations of glucuronic acid ($\sim$25–100 mM) are required for this inhibition, the effect is a specific one in that no other sugars, including galacturonic acid (a better Ca^{2+}-binding sugar than glucuronic acid), inhibit (158). Digestion of AF preparations with crude *Helix pomatia* glycosidase also inhibited its aggregation-promoting activity and only glucuronic acid would protect AF from enzymatic attack (158). In addition, glucuronic acid would not inhibit reaggregation of *Cliona celata* cells by AF from this species. The high carbohydrate content (148, 150–152) (including uronic acids) and glycosyl transferase activity (156, 157) of isolated AF from several species makes the involvement of carbohydrate recognition in this system an attractive hypothesis.

Burger's group has shown that a receptor for AF can be removed from cells of *M. parthena* (stripped of AF with CMF sea water) by hypotonic shock (159). This material, which apparently forms a bridge between AF and the cell membrane is called "baseplate." Baseplate is assayed by its ability to restore AF-mediated aggregability to shocked cells, its inhibition of AF-mediated aggregation of CMF-treated cells, and its ability to adsorb AF when covalently attached to Sepharose beads (148, 154). Using the baseplate beads and purified AF, specific adhesion has been demonstrated in this artificial system which mimics the process observed in *M. prolifera* cells (148, 154). While this constitutes an interesting system for study of the components involved, the facts that the baseplate preparations are not pure (148) and that some doubts exist about the nature (and purity) of the AF (148, 150–152) point to the need for further characterization of AF and baseplate before a detailed understanding of the mechanism of sponge reaggregation is possible. As presently characterized, the baseplate has a molecular weight of about 50,000, is stable to 5 mM EDTA, freezing, lyophilization, and pHs of 3–12, but is inactivated by heat (60° for 10 min). It has been partially purified by gel filtration on Sephadex G-200 and G-100 and by DEAE cellulose chromatography (148). By sodium dodecyl sulfate extraction of *S. domuncula* Muller et al (160) have obtained a soluble AF receptor that contains carbohydrate (including uronic acid) and has a molecular weight of about 43,000. At the present state of purification, this factor consists primarily of carbohydrate (160).

Significant differences in the properties of the AFs and baseplates obtained by various investigators may reflect, in part, species differences, differences in assay conditions, and differences in the methods for preparation of cells depleted of these factors. Their clarification will require that the same methodology be applied successfully to different species under comparable conditions.

Thus species-specific cell adhesion in sponges involves a multicomponent system consisting of an AF, the "bridging" molecule, a baseplate or AF receptor, and the anchoring of this receptor to the membrane. That enzymes (glycosyl transferases) which may be important for later steps in cell adhesion (and for modification of the adhesion apparatus itself) are part of the large AF particles is a very interesting suggestion that may have parallels in some of the embryonic systems discussed in an earlier section. In addition the influence of cell adhesion on subsequent sponge development is an area of important future work (149).

THE CELLULAR SLIME MOLDS

For part of their life cycle, the cellular slime molds such as *Dictyostelium discoideum,* exist as single amoeboid cells as long as a bacterial food source is available. Upon starvation, the cells rapidly develop the ability to aggregate and adhere to one another in a species-specific way to form a coherent multicellular slug or pseudoplasmodium. This readily induced transition from a single cell to a multicellular phase has long attracted the attention of developmental biologists who have seen it as a model for differentiation in eukaryotic organisms [for reviews of differentiation of the cellular slime molds see (161–167)]. This same feature also makes these organisms attractive objects in which to study intercellular adhesion, since the problems of cell heterogeneity and the dissociation and subsequent repair of cells are minimized (164, 165, 168–170). In addition, the growth and differentiation of cells in culture (163, 165) is quite easy, even in very large quantity (10^{11} cells), and culture conditions closely approximate the normal habitat of these organisms (162, 163, 165, 168–171).

A clear distinction must be made between "aggregation" and cell adhesion in these organisms. Aggregation is the result of a chemotactic signal relay system that directs cells to migrate into aggregation centers; [for reviews of the chemotactic system see (163, 165, 169, 171–173)]. While they are migrating, the cells also become cohesive, as reflected by their ability to adhere to one another in vivo and in several types of in vitro cell adhesion assays (171, 174, 177, 178). In addition, cells can be differentiated to a cohesive state in suspension (179), thus bypassing the chemotactic phase which occurs only on a moist surface.

Immunologically Defined Contact Sites

Gerish and co-workers have used antibodies (Fab fragments) prepared against differentiated *D. discoideum* cells as a probe for cell surface sites (termed contact sites) involved in cell adhesion (see 170–173, 180). The potential difficulties in using antibodies directed against cell surface components, which may have secondary effects on cell adhesion, motility, chemotaxis, or membrane topography appear to have been avoided (176, 181). Two types of contact sites have been defined in *D. discoideum*. Both vegetative and differentiated cells aggregate in gyrated suspension. EDTA prevents the cohesion of vegetative cells but not differentiated ones (175, 176). Fab from antibody prepared against whole lysates of differentiated cells and then absorbed extensively with vegetative cells inhibits the EDTA-resistant cohesion of differentiated cells (176). The cell surface (182) antigens to which the Fab binds have been termed contact sites A (cs-A). Fab from antiserum against vegetative cells blocks the EDTA-sensitive cohesion of these cells but not the cohesion of differentiated cells in the presence of EDTA. These antigens are called contact sites B (cs-B) (176). The number of cs-A sites for Fab binding have been estimated at less than 3×10^5 per cell (182), a relatively small number, which suggests that the effect on cell adhesion is not due to a general covering of the cell surface (182). This inference is supported by the observation that Fab directed against other cell surface antigens (oligosaccharides) binds to a greater extent (2.5×10^6 Fab fragments per cell) but has no effect on EDTA-resistant cohesiveness (182). Contact sites A are species-specific since Fab against *D discoideum* cs-A has only marginal effects on adhesion of *Polysphondylium pallidum* cells (176). Both detergent and n-butanol extracts of differentiated *D. discoideum* membranes contain material that neutralizes the effect of cs-A specific Fab on cell adhesion (183). This activity has been purified by DEAE-cellulose ion exchange chromatography and sucrose gradient centrifugation. The only detectable species is a single glycoprotein which binds Con A (184) and has a molecular weight of 80,000–120,000 (185). This glycoprotein may thus represent purified cs-A. The possibility that the Fab-neutralizing activity of the preparation resides in a group of polydisperse contaminants of the glycoprotein remains to be discounted. Using a similar immunological approach, Bozzaro & Gerisch (186) have prepared adhesion-blocking Fab against *P. pallidum* cells. This Fab does not inhibit cohesion of *D. discoideum*.

Lectins as Cell Adhesion Ligands

The concept that endogenous carbohydrate-binding proteins (lectins) may be cell adhesion ligands in cohesive cellular slime molds began with the observation by Rosen (187) that extracts of differentiated but not vegetative *D. discoideum* cells agglutinate sheep erythrocytes. The specific agglutination activity increased 400-fold from the beginning of differentiation to 12 hr of development (188). The soluble agglutination activity is inhibited by sugars of the D-galactose configuration (177) and is easily purified in a one-step procedure using Sepharose 4B (a polymer containing D-galactose) as an affinity matrix (188). Two carbohydrate binding proteins are obtained in the D-galactose eluate of the affinity column (189). The major component (70–90%) has a subunit molecular weight of 26,000 on SDS gel electrophoresis, a native molecular weight of about 100,000, and is designated discoidin I. The minor component, called discoidin II, has a subunit molecular weight of 24,500 in SDS and is also a tetramer (188, 190). The two proteins are coded by different genes as indicated by differences in their amino acid compositions, tryptic fingerprints, different time course of appearance (189, 191), and the presence of distinct mRNAs for each (192).

Polysphondylium pallidum also produces multiple lectins (193) (pallidins), identified as three distinct isoelectric forms after purification on acid-treated Sepharose [(68) S. D. Rosen, unpublished]. These three forms are made up of at least two distinct peptide chains. One of the three pallidins was first purified by affinity adsorption on fixed rabbit or human erythrocytes and elution with galactose (193, 194). This pallidin has a native weight average molecule weight of 250,000 and a subunit weight of 25,000 in SDS (194). Its carbohydrate specificity is clearly distinguishable from that of discoidin I and II by sugar hapten inhibition of agglutination activity (195). Its amino acid composition is similar to that of the discoidins, but the tryptic peptide maps are clearly different (193).

All six species of cellular slime mold examined were found to produce agglutinins when differentiated (196). Rabbit red cells are agglutinated by extracts of all species, and, using this cell type, the sugar specificity of each of the six species could be distinguished using hapten inhibition of agglutination (196). The lectins of five species have been purified (193) and found to have similar subunit molecular weights (24,000–26,000), amino acid compositions and, in the case of the major lectin of D. mucoroides (mucoroidin I) and discoidin I, very similar tryptic peptide maps (193). Thus, the slime mold lectins appear to constitute a class of related carbohydrate-binding proteins of similar structure and similar, but distinguishable, carbohydrate-binding specificity (169).

The role of slime mold lectins in cell adhesion has been explored in D. discoideum and P. pallidum (168, 171). In both species, antibody prepared against the purified lectins binds to the surface of differentiated cells as detected by a fluorescent or ferritin double antibody technique (197, 198). Surface iodination of differentiated D. discoideum cells followed by detergent extraction of the particulate fraction and immunoprecipitation also supports a cell surface localization for discoidin. Quantitative immunoprecipitation (191) studies indicate that about 15% of the discoidin is on the cell surface while the remainder is intracellular (199). Of three noncohesive mutants examined two had no detectable cell surface discoidin and the third had a reduced amount compared to the wild type.

Receptors for discoidin and pallidin appear on the cells' surface with a time course similar to that of lectin appearance (190). When cohesive cells are fixed with 1% glutaraldehyde (pH 7.2) their cohesiveness is lost, but the cells are readily agglutinated by exogenous slime mold and plant lectins. With a binding assay in which free lectin was determined in a quantitative agglutination assay, the number of sites and their affinities were estimated for the binding of discoidin I and II and pallidin to *D. discoideum* and *P. pallidum* cells (190). The apparent number of sites for all three lectins is 3–5 X 10^5 per cell. The lectins bind to vegetative cells with no apparent species specificity and with a K_a of 5–10 X 10^7 liters/mole. After differentiation the number of sites remains about the same, but the discoidins and pallidin bind to their homospecific cell types with 20-fold greater affinity. This high affinity binding is species-specific (190). Current work includes more direct binding studies with ^{125}I-labeled discoidins and pallidins, and solubilization and isolation of their receptors. Also of interest is the isolation of the cell surface discoidin and pallidin for comparison to the intracellular, soluble forms of the proteins. While the mode of attachment of the lectins at the cell membrane is not known (171), both the lectins and high affinity receptors are present on the cell surface of differentiated, cohesive cells.

If cell adhesion is mediated by the interaction of cell surface lectin and receptor, then compounds that antagonize this interaction should block cell adhesion. Two types of inhibitors have been used in this approach—antibodies (Fab fragments) to purified lectin (200) and carbohydrates that bind to the lectin (201). Initially it was shown that simple sugars that inhibited pallidin's agglutination activity could inhibit the adhesion of differentiated *P. pallidin* cells in a gyratory assay (195). Very high concentrations (> 0.1 M) were required; thus inhibitors of higher affinity were tested. Both asialofetuin (201) and the Fab fragments prepared from antipallidin IgG (200) inhibit cohesion of *P. pallidum* cells under hypertonic conditions or in the presence of azide or 2,4,dinitrophenol. No inhibition of cell adhesion could be demonstrated under isotonic conditions. The adhesion observed in hypertonic or antimetabolic solutions appears to reflect a functionally relevant process since vegetative cells showed less cohesiveness than differentiated cells and Fab prepared from nonimmune serum did not reduce cell cohesion (200, 201). In addition, modifications of the carbohydrate moities of fetuin alter its potency as an inhibitor of both the palladin-mediated agglutination of erythrocytes and cohesion of *P. pallidum* cells in a parallel manner (201).

While experiments of this type with *D. discoideum* cells and discoidin have not yet been reported, a mutant of *D. discoideum* has now been found that lacks functional discoidin I (but has cross-reactive material to antidiscoidin) and fails to aggregate. Mutant cells will, however, coaggregate with wild-type cells, which indicates a functional receptor for discoidin. Revertants of the mutant regain the ability to aggregate along with functional discoidin I (R. Lerner, personal communication).

The model for cell adhesion in differentiated slime mold cells suggested by these results is as follows:

1. Cell surface lectins are bound to the membrane through a carbohydrate binding site or by integral associations with the lipid bilayer.
2. High affinity receptors containing oligosaccharide moieties (glycoprotein or glycolipid) are anchored in the membrane.
3. The interaction of, at first, one lectin binding site with one receptor oligosaccharide and then binding at multiple sites gives rise to very rapid and extremely stable cell cohesion.
4. The inhibitory effect of metabolic poisons on cell adhesion implies that energy-dependent processes may be involved in the redistribution of surface lectin and/or receptor.

Clearly more work is needed in defining the structure of the soluble and surface-bound lectins and their receptors as well as their path of synthesis and insertion into the membrane. Since the event of cell contact during aggregation of slime mold cells is an important developmental signal (164, 166, 169), it is possible that the lectin-receptor interaction is also a developmental regulatory system. At present it is not clear what relationship, if any, exists between the lectin-receptor system and the contact sites A system described by Gerisch and co-workers (171). Anti cs-A Fab does not react with lectin (183, 186) and solubilized cs-A does not inhibit lectin-mediated agglutination (185). The simplest explanation is that two or more separate systems have been developed in the cellular slime molds for cell adhesion. Alternatively, cs-A and the lectin-receptor systems may represent different components of a single cell adhesion system that is more complex than presently appreciated (171, 185).

OTHER CELLULAR ADHESION SYSTEMS

Space does not permit the discussion of many topics related to cell adhesion. The following references to recent papers or reviews are presented to serve as useful guides to the literature in these areas. Cell adhesion to substrata has been carefully reviewed by Grinnell (202). Surface components involved in yeast mating (203) and bacterial conjugation (204, 205) have been described, as well as the binding of flagella to chlamydomonas, an initial step in the mating process (40, 41).

ACKNOWLEDGMENT

We are grateful to all of our colleagues who provided unpublished manuscripts and to Dr. D. I. Gottlieb for careful and thoughtful reading of this review.

Literature Cited

1. Wilson, H. V. 1907. *J. Exp. Zool.* 5:245–58
2. Moscona, A. A. 1965. In *Cells and Tissues in Culture,* ed. E. N. Willmer, pp. 489–529. New York: Academic
3. Umbreit, J., Roseman, S. 1975. *J. Biol. Chem.* 250:9360–68
4. Orr, C. W., Roseman, S. 1969. *J. Membr. Biol.* 1:110–24
5. Merrell, R., Gottlieb, D. I., Glaser, L. 1976. In *Neuronal Recognition,* ed. S. Barondes, pp. 249–73. New York: Plenum
6. Huesgen, A., Gerisch, G. 1975. *FEBS Lett.* 56:46–49
7. Brackenbury, R., Thiery, J. P., Rutishauser, U., Edelman, G. M. 1977. *J. Biol. Chem.* 252:6835–40
8. Roth, S., Weston, J. A. 1967. *Proc. Natl. Acad. Sci. USA* 58:974–78
9. Roth, S. 1968. *Dev. Biol.* 18:602–13
10. Roth, S., McGuire, E. J., Roseman, S. 1971. *J. Cell. Biol.* 51:525–35
11. McClay, D. R., Baker, S. R. 1975. *Dev. Biol.* 43:109–27
12. Vosbeck, K., Roth, S. 1976. *J. Cell Sci.* 22:657–70
13. McGuire, E. J., Burdick, C. L. 1976. *J. Cell Biol.* 68:80–89
14. Walther, B. T., Ohman, R., Roseman, S. 1973. *Proc. Natl. Acad. Sci. USA* 70:1569–73
15. Gottlieb, D. I., Glaser, L. 1975. *Biochem. Biophys. Res. Commun.* 63: 815–21
16. Edelman, G. M., Rutishauser, U. 1974. *Methods Enzymol.* 34:195–228
17. Cassiman, J. J., Bernfield, M. R. 1976. *Dev. Biol.* 52:231–45
18. Springer, W. R., Barondes, S. H. 1978. *J. Cell Biol.* In press
19. Grady, S. R., McGuire, E. J. 1976. *J. Cell Biol.* 71:96–106
20. Santala, R., Gottlieb, D. I., Littman, D., Glaser, L. 1977. *J. Biol. Chem.* 252:7625–34
21. Obrink, B., Wärmegard, B., Pertoft, H. 1977. *Biochem. Biophys. Res. Commun.* 77:665–70
22. Moscona, A. A. 1961. *Exp. Cell Res.* 22:455–75
23. Hausman, R. E., Moscona, A. A. 1975. *Proc. Natl. Acad. Sci. USA* 72:916–20
24. Hausman, R. E., Moscona, A. A. 1976. *Proc. Natl. Acad. Sci. USA* 73:3594–98
25. Marchase, R. B., Vosbeck, K., Roth, S. 1976. *Biochim. Biophys. Acta* 457:385–416
26. Glaser, L. 1978. *Rev. Physiol. Biochem. Pharmacol.* 83:89–122
26a. Moya, F., Silbert, D., Glaser, L. 1978. *Biochim. Biophys. Acta.* In press
27. Steinberg, M. S. 1978. In *Cell-Cell Recognition.* Presented at SEB Symp. 32 ed. A. S. G. Curtis, pp. 25–49. London/New York: Cambridge Univ. Press
28. Steinberg, M. S. 1978. In *Specificity of Embryonal Interaction,* ed. D. Garrod, London: Chapman & Hall. In press
29. Roseman, S. 1970. *Chem. Phys. Lipids* 5:270–97
30. Shur, B. D., Roth, S. 1975. *Biochim. Biophys. Acta* 415:473–512
31. Beug, H., Gerisch, G. 1972. *J. Immunol. Methods* 2:49–57
32. Sperry, R. W. 1963. *Proc. Natl. Acad. Sci. USA* 50:703–10
33. Marchase, R. B., Barbera, A. J., Roth, S. 1975. *Ciba Found. Symp.* 29:315–27
34. Barondes, S. H. 1970. In *Neurosciences Second Study Program,* ed. F. O. Schmitt, pp. 747–60. New York: Rockefeller Univ. Press
35. Barbera, A. J., Marchase, R. B., Roth, S. 1973. *Proc. Natl. Acad. Sci. USA* 70:2482–86
36. Barbera, A. J. 1975. *Dev. Biol.* 46: 167–91
37. Gottlieb, D. I., Rock, K., Glaser, L. 1976. *Proc. Natl. Acad. Sci. USA* 73:410–14
38. Gottlieb, D. I., Arington, C. 1978. *Dev. Biol.* In press
39. Snell, W. J., Roseman, S. 1977. *Fed. Proc.* 36:811
40. Goodenough, U. W., Weiss, R. L. 1975. *J. Cell Biol.* 67:623–37
41. Snell, W. J. 1976. *J. Cell. Biol.* 68:70–79
42. Changeux, J. P., Danchin, A. 1976. *Nature* 264:705–12
43. Ruffolo, R. R. Jr., Eisenbarth, G. S., Thompson, J. M., Nirenberg, M. 1978. *Proc. Natl. Acad. Sci. USA* 75:2281–85

520   FRAZIER & GLASER

44. Lash, J., Burger, M. M. 1977. *Cell and Tissue Interaction*. New York: Raven
45. Merrell, R., Glaser, L. 1973. *Proc. Natl. Acad. Sci. USA* 70:2794–98
46. Gottlieb, D. I., Merrell, R., Glaser, L. 1974. *Proc. Natl. Acad. Sci. USA* 71:1800–2
47. Mintz, G., Glaser, L. 1978. *J. Cell Biol.* In press
48. Ramirez, G., Seeds, N. W. 1977. *Dev. Biol.* 60:153–62
49. Merrell, R., Gottlieb, D. I., Glaser, L. 1975. *J. Biol. Chem.* 250:5655–59
50. Marchase, R. B., Jakoi, E. R. 1978. *J. Cell Biol.* In press
51. Jakoi, E. R., Zampighi, G., Robertson, J. D. 1976. *J. Cell Biol.* 20:97–111
52. Garber, B. B., Moscona, A. A. 1972. *Dev. Biol.* 27:217–34
53. Yaffe, D., Feldman, M. 1965. *Dev. Biol.* 11:300–17
54. Burdick, M. L. 1972. *J. Exp. Zool.* 180:117–26
55. Obrink, B., Kuhlenschmidt, M. S., Roseman, S. 1977. *Proc. Natl. Acad. Sci. USA* 74:1077–81
56. Pricer, W. E. Jr., Ashwell, G. 1971. *J. Biol. Chem.* 246:4825–33
57. Hudgin, R. L., Pricer, W. E. Jr., Ashwell, G., Stockert, R. J., Morell, A. G. 1974. *J. Biol. Chem.* 249:5536–43
58. Kawasaki, T., Ashwell, G. 1977. *J. Biol. Chem.* 252:6536–43
59. Weigel, P. H., Schmell, E., Lee, Y. C., Roseman, S. 1978. *J. Biol. Chem.* 253:330–33
59a. Schnaar, R. L., Wiegel, P. H., Kuhlenschmidt, M. S., Lee, Y. C., Roseman, S. 1978. *J. Biol. Chem.* 253:7940–51
60. Prieels, J. P., Pizzo, S. V., Glasgow, L. R., Paulson, J. C., Hill, R. L. 1978. *Proc. Natl. Acad. Sci. USA* 75:2215–19
61. Stahl, P., Schlesinger, P. H., Rodman, J. S., Doebber, T. 1976. *Nature* 264:86–88
62. Stockert, R. J., Morell, A. G., Scheinberg, I. H. 1976. *Biochem. Biophys. Res. Commun.* 68:988–93
63. Achord, D. T., Brot, F. E., Sly, W. S. 1977. *Biochem. Biophys. Res. Commun.* 77:409–15
64. Kaplan, A., Achord, D. T., Sly, W. S. 1977. *Proc. Natl. Acad. Sci. USA* 74:2026–30
65. Sando, G. N., Neufeld, E. F. 1977. *Cell* 12:619–27
66. Stahl, P. R., Rodman, J. S., Miller, M. J., Schlesinger, P. H. 1978. *Proc. Natl. Acad. Sci. USA* 75:1399–1403
67. Teichberg, V. I., Silman, I., Beitsch, D. D., Resheff, G. 1975. *Proc. Natl. Acad. Sci. USA* 72:1383–87

68. Gartner, T. K., Podleski, T. R. 1975. *Biochem. Biophys. Res. Commun.* 67:972–78
69. DeWaard, A., Hickman, S., Kornfeld, S. 1976. *J. Biol. Chem.* 251:7581–87
70. Den, H., Malinzak, D. A., Rosenberg, A. 1976. *Biochem. Biophys. Res. Commun.* 69:621–27
71. Den, H., Malinzak, D. A. 1977. *J. Biol. Chem.* 252:5444–48
71a. Nowak, T. P., Kobiler, D., Roel, L. E., Barondes, S. H. 1977. *J. Biol. Chem.* 252:6026–30
71b. Mir-Lechaire, F. J., Barondes, S. H. 1978. *Nature* 272:256–58
72. Simpson, D. L., Thorne, P. R., Loh, H. H. 1977. *Nature* 266:367–69
73. Kobiler, D., Barondes, S. H. 1977. *Dev. Biol.* 60:326–30
74. Kobiler, D., Beyer, E. C., Barondes, S. H. 1978. *Dev. Biol.* 66:265–72
74a. Gremo, F., Kobiler, D., Barondes, S. H. 1978. *J. Cell Biol.* 79:491–99
75. Rutishauser, U., Thiery, J. P., Brackenbury, R., Sela, B. A., Edelman, G. M. 1976. *Proc. Natl. Acad. Sci. USA* 73:577–81
76. Thiery, J. P., Brackenbury, R., Rutishauser, U., Edelman, G. M. 1977. *J. Biol. Chem.* 252:6841–45
77. McClay, D. R., Gooding, L. R., Fransen, M. E. 1977. *J. Cell. Biol.* 75:56–66
78. Burger, M. M., Jumblatt, J. 1977. In *Cell and Tissue Interactions*, ed. J. Lash, M. M. Burger, pp. 155–72. New York: Raven
79. Marchase, R. B. 1977. *J. Cell Biol.* 75:237–57
80. Lilien, J. E. 1968. *Dev. Biol.* 17:657–78
81. McClay, D. R., Moscona, A. A. 1974. *Exp. Cell Res.* 87:438–43
82. Hausman, R. E., Knapp, L. W., Moscona, A. A. 1976. *J. Exp. Zool.* 198:417–22
83. Oppenheimer, S. B., Humphreys, T. 1971. *Nature* 232:125–27
84. Oppenheimer, S. B. 1975. *Exp. Cell Res.* 92:122–26
85. Meyer, J. T., Oppenheimer, S. B. 1976. *Exp. Cell Res.* 102:359–64
86. Sankaran, L., Proffitt, R. T., Petersen, J. R., Pogell, B. M. 1977. *Proc. Natl. Acad. Sci. USA* 74:4486–90
87. Merrell, R., Pulliam, M. W., Randono, L., Boyd, L. F., Bradshaw, R. A., Glaser, L. 1975. *Proc. Natl. Acad. Sci. USA* 72:4270–74
88. Schubert, D., Whitlock, C. 1977. *Proc. Natl. Acad. Sci. USA* 74:4055–58
89. Aharonov, A., Vlodavsky, I., Pruss, R.

M., Fox, C. F., Herschman, H. R. 1978. *J. Cell Physiol.* 95:195–202

90. Balsamo, J., Lilien, J. 1974. *Proc. Natl. Acad. Sci. USA* 71:727–31
91. Balsamo, J., Lilien, J. 1975. *Biochemistry* 14:167–71
92. Cassiman, J. J., Bernfield, M. R. 1975. *Exp. Cell. Res.* 91:31–35
93. Dorsey, J. K., Roth, S. 1973. *Dev. Biol.* 33:249–56
94. Stallcup, W. B. 1977. *Brain Res.* 126:475–89
95. Santala, R., Glaser, L. 1977. *Biochem. Biophys. Res. Commun.* 79:285–91
96. Schubert, D., Heinemann, S., Carlisle, W., Tarikas, H., Kimes, B., Patrick, J., Steinback, J. H., Culp, W., Brandt, B. L. 1974. *Nature* 249:224–27
97. McMahon, D. 1973. *Proc. Natl. Acad. Sci. USA* 70:2396–400
98. Holley, R. W. 1975. *Nature* 258:487–90
99. Dulbecco, R. 1970. *Nature* 227:802–6
100. Stoker, M., Piggott, D. 1974. *Cell* 3:207–15
101. Whittenberger, B., Glaser, L. 1978. *Nature* 272:821–23
102. Mannino, R. J., Burger, M. M. 1975. *Nature* 256:19–22
102a. Mannino, R. J., Ballmer, K., Burger, M. M. 1978. *Science* 201:824–26
103. Whittenberger, B., Glaser, L. 1977. *Proc. Natl. Acad. Sci. USA* 74:2251–55
104. Whittenberger, B., Raben, D., Glaser, L. 1979. *J. Supramol. Struct.* In press
105. Liberman, M. A., Raben, D., Whittenberger, B., Glaser, L. 1978. Manuscript in preparation
106. Barsh, G. S., Cunningham, D. D. 1978. *J. Supramol. Struct.* 7:425–41
107. Whittenberger, B., Raben, D., Lieberman, M. A. 1978. *Proc. Natl. Acad. Sci. USA* 75:5457–61
108. Wood, P. M., Bunge, R. P. 1975. *Nature* 256:662–64
109. Wood, P. M. 1976. *Brain Res.* 15: 361–75
110. Salzer, J. L., Glaser, L., Bunge, R. P. 1977. *J. Cell Biol.* 75:118
111. Griepp, E. B., Bernfield, M. R. 1978. *Exp. Cell Res.* 113:263–72
112. Griepp, E. B., Peacock, J. H., Bernfield, M. R., Revel, J. P. 1978. *Exp. Cell Res.* 113:273–82
113. Ramirez, G. 1977. *Neurochem. Res.* 2:417–25
114. Fentiman, I., Taylor-Papadimitriou, J., Stoker, M. 1976. *Nature* 264:760–62
114a. Pitts, J. D., Bark, R. R. 1976. *Nature* 264:762–64
115. McRoric, R. A., Williams, W. L. 1974. *Ann. Rev. Biochem.* 43:777–803

116. Gwatkin, R. B. L. 1976. *The Cell Surface in Animal Embryogenesis and Development,* ed. G. Poste, G. L. Nicolson, pp. 1–54. Amsterdam: North-Holland
117. Lillie, F. E. 1914. *J. Exp. Zool.* 16:523–90
118. Tyler, A. 1948. *Am. Nat.* 83:195–219
119. Dan, J. L. 1956. *Int. Rev. Cytol.* 5:365–440
120. Summers, R. G., Hylander, B. L. 1975. *Exp. Cell Res.* 96:63–68
121. Vacquier, V. D., Payne, J. E. 1973. *Exp. Cell Res.* 82:227–35
122. Schmell, E., Earles, B. J., Breaux, C., Lennarz, W. J. 1977. *J. Cell Biol.* 72:35–46
123. Anderson, E. 1968. *J. Cell Biol.* 37:514–39
124. Millonig, G. 1969. *J. Submicrosc. Cytol.* 1:69–84
125. Detering, N. K., Decker, G. L., Schmell, E. D., Lennarz, W. J. 1977. *J. Cell Biol.* 75:899–914
126. Epel, D., Weaver, A. M., Muchmore, A. V., Schimke, R. T. 1964. *Science* 163:294–96
127. Grossman, A., Levy, M., Troll, W., Weissmann, G. 1973. *Nature* 243: 277–78
128. Vacquier, V. D., Tegner, M. J., Epel, D. 1973. *Exp. Cell Res.* 80:111–19
129. Fodor, E. J. B., Ako, H., Walsh, K. A. 1975. *Biochemistry* 14:4923–27
130. Carroll, E. J., Epel, D. 1975. *Dev. Biol.* 44:22–32
131. Aketa, K. 1975. *Exp. Cell Res.* 90: 56–62
132. Howe, C. W. S., Metz, C. B. 1972. *Biol. Bull. (Woods Hole, Mass.)* 143:465
133. Lallier, R. 1972. *Exp. Cell Res.* 72: 157–63
134. Tsuzuki, H., Hoshida, M., Onitake, K., Aketa, K. 1977. *Biochem. Biophys. Res. Commun.* 76:502–11
135. Aketa, K. 1973. *Exp. Cell Res.* 80: 439–41
136. Aketa, K., Onitake, K. 1969. *Exp. Cell Res.* 56:84–86
137. Vacquier, V. D., Moy, G. W. 1977. *Proc. Natl. Acad. Sci. USA* 74:2456–60
138. Bellet, N. F., Vacquier, J. P., Vacquier, V. D. 1977. *Biochem. Biophys. Res. Commun.* 79:159–65
139. Glabe, C. G., Vacquier, V. D. 1977. *Nature* 267:836–38
140. Vacquier, V. D., Moy, G. W. 1978. *Cell Reproduction ICN-UCLA Symp. Mol. Cell. Biol.,* Vol. 12 ed. E. R. Dirksen, D. Prescott, C. F. Fox. New York: Academic. In press

141. Glabe, C. G., Vacquier, V. D. 1978. *Proc. Natl. Acad. Sci. USA* 75:881–85
142. Humphreys, T. 1963. *Devel. Biol.* 8:27–47
143. Moscona, A. A. 1968. *Devel. Biol.* 18:250–77
144. McClay, D. R. 1971. *Biol. Bull.* 141:319–30
145. Humphreys, T., Humphreys, S., Moscona, A. A. 1960. *Biol. Bull.* 119:294
146. Humphreys, T., Humphreys, S., Moscona, A. A. 1960. *Biol. Bull.* 119:295
147. Margoliash, E., Schenck, J. R., Hargie, M. P., Burokas, S., Richter, W. R., Barlow, G. H., Moscona, A. A. 1965. *Biochem. Biophys. Res. Commun.* 20:383–88
148. Jumblatt, J. E., Weinbaum, G., Turner, R., Ballmer, K., Burger, M. M. 1976. In *Surface Membrane Receptors*, ed. R. A. Bradshaw, W. A. Frazier, R. C. Merrell, D. I. Gottlieb, R. A. Hogue-Angeletti, pp. 73–86. New York: Plenum
149. Muller, W. E. G., Muller, I., Zahn, R. K. 1978. *Res. Mol. Biol. (Mainz)* 8:7–87
150. Henkart, P., Humphreys, S., Humphreys, T. 1973. *Biochemistry* 12:3045–50
151. Cauldwell, C. B., Henkart, P., Humphreys, T. 1973. *Biochemistry* 12:3051–55
152. Humphreys, S., Humphreys, T., Sano, J. 1977. *J. Supramol. Struct.* 7:339–51
153. Muller, W. E. G., Zahn, R. K. 1973. *Exp. Cell Res.* 80:95–104
154. Zahn, R. K., Muller, W. E. G., Geisert, M., Reinmuller, J., Michaelis, M., Pondeljak, V., Beyer, R. 1976. *Cell Differ.* 5:129–37
155. Muller, W. E. G., Muller, I., Pondeljak, V., Kurelec, B., Zahn, R. K. 1978. *Cell Differ.* 10:45–53
156. Muller, W. E. G., Arendes, J., Kurelec, B., Zahn, R. K., Muller, I. 1977. *J. Biol. Chem.* 252:3836–42
157. Muller, W. E. G., Zahn, R. K., Kurelec, B., Uhlenbruck, G., Vaith, P., Muller, I. 1978. *Hoppe-Seyler's Z. Physiol. Chem.* 359:529–37
158. Turner, R. S., Burger, M. M. 1973. *Nature* 244:509–10
159. Weinbaum, G., Burger, M. M. 1973. *Nature* 244:510–12
160. Muller, W. E. G., Zahn, R. K., Kurelec, B., Muller, I. 1978. *Differentiation* 10:55–60
161. Raper, K. B. 1941. *Growth 3rd Symp.* 5:41–76
162. Raper, K. B. 1973. *The Fungi*, ed. G. C.

Ainsworth, F. K. Sparrow, A. S. Sussman, pp. 9–36. New York: Academic
163. Bonner, J. T. 1967. *The Cellular Slime Molds*, New Jersey: Princeton Univ. Press. 205 pp.
164. Newell, P. C. 1971. *Essays Biochem.* 7:87–126
165. Loomis, W. F. 1975. *Dictystelium discoideum: A Developmental System*, New York: Academic. 214 pp.
166. Jacobson, A., Lodish, H. F. 1975. *Ann. Rev. Genet.* 9:145–85
167. Sussman, M., Brackenbury, R. 1976. *Ann. Rev. Plant Physiol.* 27:229–65
168. Barondes, S. H., Rosen, S. D. 1976. See Ref. 148, pp. 39–55
169. Frazier, W. A. 1976. *Trends Biochem. Sci.* 1:130–33
170. Gerisch, G. 1976. See Ref. 148, pp. 67–72
171. Rosen, S. D., ed. Barondes, S. H. 1977. In *Specificity of Embryological Interactions*, ed. D. R. Garrod, pp. 235–64. London: Chapman & Hall
172. Gerisch, G., Malchow, D. 1976. *Adv. Cyclic Nucleotide Res.* 7:49–68
173. Newell, P. C. 1977. In *Microbial Interactions* Vol. 3, ed. J. L. Reissig, pp. 3–57. London: Chapman & Hall
174. Shaffer, B. M. 1957. *Q. J. Microsc. Sci.* 98:383–405
175. Gerisch, G. 1961. *Exp. Cell Res.* 25:535–54
176. Beug, H., Katz, F. E., Gerisch, G. 1973. *J. Cell Biol.* 56:647–58
177. Rosen, S. D., Kafka, J. A., Simpson, D. L., Barondes, S. H. 1973. *Proc. Natl. Acad. Sci. USA* 70:2554–57
178. Alexander, S., Brackenbury, R., Sussman, M. 1975. *Nature* 254:698–99
179. Gerisch, G. 1960. *Wilhelm Roux Arch. Entwicklungsmech. Org.* 152:632–54
180. Gerisch, G., Beug, H., Malchow, D., Schwartz, H., Stein, A. V. 1974. *Miami Winter Symp.* 7:49–66
181. Beug, H., Gerisch, G., Kempff, S., Reidel, V., Cremer, G. 1970. *Exp. Cell Res.* 63:147–58
182. Beug, H., Katz, F. E., Stein, A., Gerisch, G. 1973. *Proc. Natl. Acad. Sci. USA* 70:3150–54
183. Huesgen, A., Gerisch, G. 1975. *FEBS Lett.* 56:46–49
184. Eitle, E., Gerisch, G. 1977. *Cell Differ.* 6:339–46
185. Muller, K., Gerisch, G. 1978. *Nature* 274:445–49
186. Bozzaro, S., Gerisch, G. 1978. *J. Mol. Biol.* 120:265–79
187. Rosen, S. D. 1972. *A possible assay for intercellular adhesion molecules* PhD thesis, Cornell Univ., New York

188. Simpson, D. L., Rosen, S. D., Barondes, S. H. 1974. *Biochemistry* 13:3487–93
189. Frazier, W. A., Rosen, S. D., Reitherman, R. W., Barondes, S. H. 1975. *J. Biol. Chem.* 250:7714–21
190. Reitherman, R. W., Rosen, S. D., Frazier, W. A., Barondes, S. H. 1975. *Proc. Natl. Acad. Sci. USA* 72:3541–45
191. Siu, C. H., Lerner, R. A., Ma, G., Firtel, R. A., Loomis, W. F. Jr. 1976. *J. Mol. Biol.* 100:157–78
192. Ma, G. C. L., Firtel, R. A. 1978. *J. Biol. Chem.* 253:3924–32
193. Frazier, W. A., Rosen, S. D., Reitherman, R. W., Barondes, S. H. 1976. See Ref. 148, pp. 57–66
194. Simpson, D. L., Rosen, S. D., Barondes, S. H. 1975. *Biochim. Biophys. Acta* 412:109–19
195. Rosen, S. D., Simpson, D. L., Rose, J. E., Barondes, S. H. 1974. *Nature* 252:128; 149–51
196. Rosen, S. D., Reitherman, R. W., Barondes, S. H. 1975. *Exp. Cell Res.* 95:159–66
197. Chang, C.-M., Reitherman, R. W., Rosen, S. D., Barondes, S. H. 1975. *Exp. Cell Res.* 95:136–42
198. Chang, C.-M., Rosen, S. D., Barondes, S. H. 1977. *Exp. Cell Res.* 104:101–9
199. Lu, C. H., Loomis, W. F., Lerner, R. A. 1977. *Proc. Int. Symp. Mol. Basis Cell-Gel Interaction 1st,* pp. 439–58. New York: Alan R. Liss
200. Rosen, S. D., Haywood, P. L., Barondes, S. H. 1976. *Nature* 263:425–27
201. Rosen, S. D., Chang, C.-M., Barondes, S. H. 1977. *Devel. Biol.* 61:202–13
202. Grinnell, F. 1978. *Int. Rev. Cytol.* 53:65–144
203. Sing, V., Yeh, Y. F., Ballou, C. E. 1976. See Ref. 148, pp. 87–98
204. Achtman, M., Kennedy, N., Skurray, R. 1977. *Proc. Natl. Acad. Sci. USA* 74:5104–08
205. Kennedy, N., Beutin, L., Achtman, M., Skurray, R. 1977. *Nature* 270:580–85

Ann. Rev. Biochem. 1979. 48:525–48
Copyright © 1979 by Annual Reviews Inc. All rights reserved

RNA REPLICATION: FUNCTION AND STRUCTURE OF Qβ-REPLICASE[1]

Thomas Blumenthal

Program in Molecular and Cellular Biology, and Department of Biology,
Indiana University, Bloomington, Indiana 47401

Gordon G. Carmichael

Department of Pathology, Harvard Medical School,
Boston, Massachusetts 02115

CONTENTS

[1]The following abbreviations are used: EF-Tu, EF-Ts, and EF-Tu·Ts indicate the protein synthesis elongation factors Tu, Ts, and the 1:1 complex of Tu and Ts, respectively; aatRNA, aminoacyl-tRNA.

525

0066-4154/79/0701-0525$01.00

PERSPECTIVES AND SUMMARY

In spite of the fact that the small single-stranded RNA-containing bacterio-phages (e.g. Qβ, R17, f2, MS2) contain only enough genetic material to code for three or four polypeptides, they must perform a complex biosynthetic process—RNA replication in a host organism, *Escherichia coli,* in which this process does not normally occur. Furthermore, template recognition by the replication machinery must be of sufficient specificity to ensure that among all of the various RNA species in the cell, only the phage RNA is recognized. In order to accomplish this task the phage assembles a complex enzyme composed of one phage-coded polypeptide and three pre-existing host polypeptides borrowed from a different biosynthetic process, protein synthesis. In addition, a single-stranded RNA-binding polypeptide known as host factor is required for replication.

In 1963 Haruna et al (1) and Weissmann et al (2) showed that cells infected with Type I RNA phages contained an enzymatic activity, not present in uninfected cells, capable of RNA-dependent RNA synthesis. A similar activity was subsequently detected in Qβ-infected cells and found to be highly specific for added Qβ RNA (3, 4). The enzyme was named Qβ replicase. This crude enzyme preparation was capable of catalyzing the net synthesis of infectious viral RNA when primed with Qβ RNA in vitro (5, 6). It was also shown to synthesize RNA only in the 5'–3' direction and to initiate synthesis only with GTP (7–13). After extensive purification, Kamen (14, 15) and Kondo et al (16) demonstrated that the enzyme was composed of one phage-coded polypeptide of molecular weight 65,000 (called subunit II) and three host-coded polypeptides of molecular weights 70,000, 45,000, and 35,000. These latter three polypeptides were shown later to be, respectively, 30S ribosomal protein S1 (17, 18) and protein synthesis elongation factors EF-Tu and EF-Ts (19).

The major issues discussed in this review include: 1. What functions do the various polypeptides perform in the RNA replication process? Recent experiments have suggested that ribosomal protein S1 performs a function in Qβ RNA replication related to its protein synthetic function—RNA site recognition. On the other hand, the RNA replication functions of the elongation factors apparently do not derive directly from their known protein synthesis activities. 2. How is the enzymatic machinery organized structurally? Qβ-replicase is composed of two subcomplexes, S1-II and EF-Tu·Ts, which are most tightly associated with one another at high ionic strength. 3. What features of the template are recognized by Qβ replicase and host factor? Weismann's group has suggested that the enzyme is bound to one or more internal RNA sites prior to initiation of synthesis at the 3' end. The host factor may be required for enzymatic interaction with the 3' end of Qβ RNA. 4. Can Qβ-replicase serve as an enzyme capable of making accurate RNA copies of RNA templates? While Qβ-replicase normally exhibits a high degree of template specificity and thus would seem unsuitable for this purpose, a number of techniques to overcome this specificity have recently been described. The sequence of events in RNA phage replication has received little attention in recent years and so is not considered in depth here. This topic along with many of those listed above have been discussed in previous reviews. (20–29).

THE ENZYMATIC MACHINERY: Qβ-REPLICASE AND HOST FACTOR

Identification of the Host Components and their Functions in Uninfected Escherichia coli

In order to facilitate discussion of possible RNA synthetic functions performed by the various polypeptides involved in Qβ RNA replication, we first briefly describe their physical properties and known functions in uninfected cells.

SUBUNIT 1   The largest polypeptide chain involved in Qβ RNA replication (mol wt 70,000) was first identified as protein synthesis interference factor i (30–34) and subsequently as 30S ribosomal protein S1 (17, 18, 35), the largest ribosomal protein and the only 30S protein that has a clearly identifiable protein synthetic function. S1 is used for translation of synthetic and natural mRNAs (36–44). In protein biosynthesis it functions in the binding of the mRNA to the 30S ribosomal subunit (36, 38, 41, 45). In polysomes there is approximately one molecule of S1 per ribosome (43, 45). Although S1 can bind to ribosomes by binding to the 3' end of the 16S RNA (46–48), it is bound only at a different site in active ribosomes (49). It is not known whether the binding of S1 to the 3' end of 16S RNA is required for S1 function. The "interference factor" activity of S1 apparently results from an interaction between mRNA and excess free S1, such that the ribosome-bound S1 is no longer able to bind the mRNA (40). There is no evidence that this activity has any physiological significance.

Although the precise mode of action of S1 in protein synthesis is not yet known, recent experiments have shed some light on this problem. S1 has a low S value (3·3S) for a polypeptide of 70,000 molecular weight, which suggests that it has an elongated conformation (50, 51). Each S1 molecule may contain two polynucleotide binding sites (52); one site that binds either single-stranded DNA or RNA noncooperatively and apparently recognizes the polynucleotide backbone (53), and another that binds cooperatively to polypyrimidines and noncooperatively to polypurines (54). This latter site binds only RNA and thus recognizes both the ribose and bases (54). The existence of only a single site has also been argued (55, 56). S1 has been found to unfold stacked or helical single-stranded polynucleotides (57–59), and two results suggest that these activities are physiologically significant. First, if S1 is treated with N-ethylmaleimide, it loses its unstacking activity (56, 60). This chemically altered S1 still binds to ribosomes and to mRNA but fails to catalyze the binding of mRNA to the 30S subunits (56, 60). Second, if mRNA is treated with formaldehyde to reduce secondary structure, the requirement for S1 for translation is abolished (41). Thus, the RNA unfolding activity of S1 seems to be required in order to alter the conformation of the mRNA so that it can interact with ribosomes.

SUBUNITS III AND IV   Subunits III and IV are the protein synthesis elongation factors EF-Tu and EF-Ts (19). These factors function in protein synthesis by catalyzing the transfer of aminoacylated transfer RNAs to the elongating ribosome-mRNA complex (61):

$$\text{EF-Tu·GTP} + \text{aatRNA} \leftrightarrows \text{EF-Tu·aatRNA·GTP}$$

$$\text{EF-Tu·aatRNA·GTP} + \text{ribosome} \rightarrow \text{ribosome·aatRNA} + \text{EF-Tu·GDP} + P_i$$

$$\text{EF-Tu·GDP} + \text{EF-Ts} \leftrightarrows \text{EF-Tu·Ts} + \text{GDP}$$

$$\text{EF-Tu·Ts} + \text{GTP} \leftrightarrows \text{EF-Tu·GTP} + \text{EF-Ts}$$

EF-Tu has been found to be abundant in *E. coli,* with between eight to fourteen copies per ribosome (62, 63). There are between four and fourteen copies of EF-Tu per copy of EF-Ts and S1 (62–64). This excess EF-Tu may be needed for the binding of all of the aatRNA in the cell (62) or it may be involved in other cellular functions. EF-Tu has been reported to be associated with the inner membrane (63), and furthermore, to have properties similar to those of eucaryotic actin (65, 66). *E. coli* has two widely separated EF-Tu genes (67) which may code for two slightly different EF-Tu's (68, 69). The significance of these observations has yet to be determined. Other possible roles for EF-Tu and EF-Ts in the cell have not yet been demonstrated conclusively, although there have been reports that EF-Ts and/or EF-Tu may be involved in host RNA synthesis (70–72). These reports have been disputed (73–75).

An EF-Tu-specific antibiotic, kirromycin (76), causes a conformational change in EF-Tu (77–79). It also allows EF-Tu to catalyze the hydrolysis of GTP in the absence of ribosomes (76, 80) and apparently inhibits protein synthesis by preventing the release of EF-Tu from the ribosome (81). It will prevent the interaction between EF-Tu and EF-Ts but will not dissociate an already formed EF-Tu·Ts complex (80, 82). Kirromycin-resistant mutants of *E. coli* have recently been isolated and are currently under study (83–85).

THE HOST FACTOR   The host factor is the only host component of the Qβ RNA replication machinery whose function in uninfected cells has not been determined. It consists of a hexamer of identical 12,000 molecular weight polypeptide chains (86), and binds tightly to single-stranded RNA with a higher affinity for A-rich regions than for other sequences (86–89). This RNA-binding specificity has been the basis for the recent purification schemes for host factor, involving affinity chromatography on poly(A)-cellulose (87, 88). Host factor does not bind to single- or double-stranded DNA, or to double-stranded RNA (86).

Although host factor is found bound to ribosomes in uninfected cells, it is not identical to any of the known ribosomal proteins and is not as abundant as ribosomes (88). Host factor associates specifically with the 30S ribosomal subunit but it is not known whether this association is physiologically significant (90). That host factor may be a translation control protein is suggested by its ribosomal location as well as by the observation that host factor inhibits efficiently, in vitro, poly(A)-primed polylysine synthesis, while not affecting poly(U)-dependent polyphenylalanine synthesis (88). Host factor has proven to be a very effective translational inhibitor of natural mRNAs (88), but as with S1, this may be a nonspecific phenomenon related to the ability of host factor to bind tightly to single-stranded RNA (89). Antibody to host factor does not affect the amount or pattern of translation in vitro (M. S. Dubow, personal communication; A. Wahba, personal communication). Even though a role for host factor in protein biosynthesis has not been discovered, it has been found to be present in similar form in a wide variety of bacterial species, which suggests that it performs an important function in the cell (90, 91).

Structure of the Replicase

ENZYME SHAPE   Qβ-replicase contains one each of the four subunits that have an aggregate molecular weight of 215,000 (14, 16, 92). Yet the enzyme has a sedimentation coefficient of only 6.7S (16). This low S value might be caused by either a rapidly reversible subunit dissociation or by an oblate shape. There is evidence for both: First, subunit exchange can take place during purification of the replicase (16); the enzyme can donate EF-Tu to an in vitro poly(U)-dependent polyphenylalanine synthesizing system (93); and a plot of RNA synthesis activity versus enzyme concentration is sigmoidal (T. Blumenthal, unpublished). The oblate shape of the enzyme is suggested by the higher molecular weight calculated by gel filtration (at least 230,000) compared to that calculated from the S value ($\sim$130,000) (35). As previously pointed out by Kamen (23), the enzyme's oblate shape seems to be conferred by S1, since a form of the enzyme lacking this subunit has an apparent molecular weight of $\sim$130,000 calculated by either gel filtration or sucrose gradient sedimentation (23).

SUBUNIT SUBCOMPLEXES   Several lines of evidence indicate that the enzyme consists of a relatively easily dissociated complex of two tight subcomplexes (S1-II and EF-Tu·Ts). The tightness of association of these two complexes in whole enzyme is increased by increasing ionic strength. When the enzyme is sedimented in glycerol gradients either in a low ionic strength buffer (14) or in the presence of Qβ RNA (16) or poly(A,C) (93) the enzyme partially dissociates into subcomplexes. When the enzyme is treated with bifunctional protein cross-linking reagents, three cross-linked products are seen—whole enzyme and the two subcomplexes (92). Again, either decreasing ionic strength or the presence of Qβ RNA increases the amount of subcomplexes at the expense of whole enzyme (92). These experiments raised the possibility that the EF-Tu·Ts is released from the rest of the enzyme when RNA is present, but this is apparently not the case. Several characteristics of EF-Tu·Ts are different when it forms a part of Qβ replicase than when it is free (e.g. sensitivity of GDP binding to ionic strength and dissociability by GDP) and these properties of EF-Tu·Ts in Qβ replicase are not influenced by the presence of Qβ RNA (94). Furthermore, antibodies to EF-Tu and EF-Ts are still able to inhibit RNA synthesis activity when added after initiation (95).

As mentioned above, a form of Qβ-replicase lacking S1 has been isolated (35). Since this enzyme is active in the transcription of some templates [poly(c), Qβ minus strand, etc] and since normal enzyme can be reconstituted simply by adding back S1, this subunit is not essential for maintaining the enzyme in a conformation in which RNA synthesis can occur (35). On the other hand, the EF-Tu·Ts complex appears to be required to maintain the phage-coded subunit in its proper conformation. In order to recover RNA synthesis activity from the subunit II-S1 complex denatured in 8M urea, EF-Tu·Ts complex must be present during renaturation (93).

Although the EF-Tu·Ts in Qβ-replicase can be donated to an in vitro protein synthesis system (93), it is clearly changed in some ways by being part of the enzyme. The EF-Tu GDP binding site is less accessible (93, 94) and thus the complex is resistant to dissociation by GDP, particularly under conditions (e.g. high salt) in which the enzyme forms a tight complex (93). The EF-Tu·Ts in replicase is also far more resistant to cleavage by trypsin, and the ratio of the cleavage products to one another is altered (T. Blumenthal, unpublished). Qβ replicase is stabilized by high salt and glycerol and is destabilized by the presence of guanine nucleotides (by virtue of their interaction at the EF-Tu binding site) (96). Since EF-Tu·Ts covalently linked by dimethyl suberimidate is fully functional in Qβ-replicase (97), it is clear that these two polypeptides act as a complex during RNA synthesis (see below). Enzyme made from the covalently linked EF-Tu·Ts is more stable than untreated enzyme and is not destabilized by guanine nucleotides, since the EF-Tu nucleotide binding site is blocked by the EF-Ts (96, 97). The rate of renaturation of denatured Qβ replicase containing covalently linked EF-Tu·Ts is much greater than normal enzyme, and this rate is decreased by GDP with the latter but not with the former (96, 97). Thus a picture of the enzyme emerges in which both the rate of assembly and stability of Qβ-replicase are dependent upon the presence of an EF-Tu·Ts complex.

Proof that the host-coded polypeptides form part of the Qβ RNA replication machinery in vivo is difficult to obtain. When the phage-coded polypeptide is made in an in vitro protein synthesizing system, only a small percentage of it is assembled into complete four subunit Qβ-replicase, and the quarternary structure is apparently achieved only after $(NH_4)_2SO_4$ precipitation of the in vitro product (98). Convincing evidence for the involvement of the host proteins in vivo must come from the analysis of mutants. Although evidence has been presented that a mutant bacterial strain (HAK88) has an altered EF-Ts and that this strain makes an altered Qβ-replicase (99), subsequent work with it has failed to confirm the presence of an altered EF-Ts (100, 101; T. Blumenthal, unpublished). HAK88 does have an electrophoretically changed EF-Tu (102), but Qβ-replicase made from this strain has normal RNA synthesis activity (T. Blumenthal, unpublished).

Qβ-REPLICASE-HOST FACTOR ASSOCIATION   Host factor may be weakly associated with Qβ-replicase. This is suggested by the fact that it purifies together with the enzyme up to the final sizing step, in which it is completely separated from Qβ-replicase [(86); T. Blumenthal, T. A. Landers, G. G. Carmichael, unpublished)]. Furthermore, host factor can partially protect replicase from inactivation by sulfhydryl-blocking reagents (103). Accordingly, the host factor hexamer could be a part of Qβ-replicase in vivo, but it is removed during purification. Further experiments are required to resolve this question.

THE GENERAL SCHEME OF IN VITRO Qβ RNA REPLICATION

Plus Strands and Minus Strands

Each infectious Qβ virion contains one molecule of single-stranded RNA of molecular weight $\sim 1.5 \times 10^6$ (104). This is termed the viral "plus strand" RNA and is the strand utilized as mRNA to direct viral protein synthesis (105). RNA molecules complementary to the plus strand are "minus strands." Although both strands behave as fully single-stranded entities by several different criteria, including sensitivity to pancreatic RNase and to RNase III, and chromatographic elution from cellulose columns (106–111), the nucleotide sequence of Qβ RNA suggests the presence of a high degree of secondary and tertiary structure, with perhaps as many as 70–80% of the bases being involved in hydrogen-bonded structures (28, 104). The plus and minus strands must have quite different three-dimensional structures: their RNA sequences do not allow identical hydrogen-bonded "loop" structures (because of the involvement of GU base pairing); plus strands are recognized by ribosomes whereas minus strands are not (112); and the mechanisms of enzymatic recognition of these strands for replication are different (see below).

Replicating RNA

When the Qβ-replicase and the host factor are incubated in vitro with Qβ plus strand RNA, Mg^{2+}, and the four ribonucleoside triphosphates, RNA synthesis may proceed linearly for hours (13), and results in the extensive net synthesis of infectious, single-stranded progeny plus strands (13, 110) as well as free, single-stranded minus strands (110, 113, 114). The product consists of greater than 90% completed, full-length molecules (6, 110, 115). The ratio of plus to minus strands synthesized in vitro is determined by the concentration of the host factor: when host factor is present in saturating amounts, equal quantities of plus and minus strands are synthesized, while when host factor is limiting, excess plus strands are made (23, 25). It is not known how many replicase molecules can polymerize on a single template at one time. The early replicating complex has been designated the "replicative intermediate." It is a labile structure sedimenting as a broad peak at 40–50S (nonreplicating viral RNA sediments at 30S) in sucrose gradients (106, 115). The replicative intermediate is largely single-stranded in character as judged by chromatography on cellulose columns, sensitivity to pancreatic RNase and to RNase III, and electron microscopy (107, 109, 111, 116). Nevertheless, it is converted readily by protein denaturants into a form in which the nascent minus strands are fully annealed to the plus strand template (13, 106, 110). This must mean that, although the strands are prevented from duplex formation during replication, either by the enzyme or by RNA secondary structure or by both, they are, nevertheless, positioned in such a way as to facilitate the stable interaction between strands when the enzyme is denatured. This "single-stranded" replicating structure is necessary for the reuse of the plus strand as template, since the replicase cannot initiate synthesis from any duplex RNA, including the Qβ plus strand/minus strand hybrid (110).

When the replicase has finished copying the plus strand into a full-length minus strand, a termination event occurs, followed either by enzymatic strand-switching, or by enzyme release, and binding to a free, released minus strand. The termination event involves the addition of an A residue to the 3' end of the minus strand (117, 118); this residue cannot be added by a classical base-pairing scheme, since the 5' terminal residue of the plus strand is pppG. The hypothesis that the terminal addition of A is a termination event is supported by the finding that replicase will not add an A to free completed plus or minus strands lacking this residue, but only to nascent chains (118). The "6S" RNA molecules which can be replicated by Qβ-replicase (see below) also have 3' terminal A residues which must be added post-transcriptionally (119, 120).

Several possible replication schemes which account for the observed intermediates have been discussed (27, 110). One is the "zipper" model, in which the strands are separated at the growing fork by the enzyme and the template and product strands are stabilized by their secondary structure. Since the internal T_m of viral RNA is quite high, this model seems plausible; the stable structure of the RNA is a kinetic barrier to annealing (110). A second possible scheme would have the enzyme physically holding strands apart at more than one place (110). Alternatively, the enzyme may impose a topological constraint against duplex formation, such as by binding both the 3' end of one strand and the 5' end of the nascent strand, thus forming a "loop" structure (110). One testable prediction of this model is that only one replicase molecule can transcribe a given RNA strand at one time. This model has been applied to a hypothesis for the in vivo scheme of Qβ RNA replication (108). There is no compelling evidence to support or reject any of the above hypothetical replication schemes.

THE ENZYMATIC MECHANISM OF IN VITRO Qβ RNA REPLICATION

Template Specificity

Qβ-replicase is remarkable both in its discrimination between the use of natural RNA molecules as templates and in its lack of discrimination between many artificial or nonphysiological templates. On the one hand, the enzyme normally replicates, in vitro, $Q\beta$ plus and minus strands (3, 121) and the RNAs of very closely related viruses (122), while ignoring all other natural cellular or viral RNAs tested (3); on the other hand, the enzyme readily transcribes poly(C) [or poly(dC)] (123) and other synthetic polymers rich in cytidylate (124, 125). It also replicates "variant" $Q\beta$ RNA molecules (formed by forcing the enzyme to copy its template faster and faster in vitro) (126, 127) and "6S" RNA (128) [found only late in infected cells and possibly the product of template-independent polymerization by the enzyme (129)]. The discrimination between natural RNAs must involve the recognition of a specific secondary and/or tertiary structure of the template: other RNA phage RNAs having 3' terminal nucleotide sequences similar to $Q\beta$ RNA are not used (28). Template recognition results in tight and specific binding of replicase to $Q\beta$ RNA, as evidenced by the fact that the enzyme binds to $Q\beta$ RNA about ten times more tightly than to heterologous RNAs (7, 130). Subunit I and the host factor are not required for utilization of any template other than $Q\beta$ plus strands: replicase lacking subunit I can itself transcribe poly(C), $Q\beta$ minus strands, and 6S RNA almost as well as normal enzyme containing a full complement of S1 (35). Enzyme specificity must reside in subunit II: another RNA phage replicase, from the related phage f2, shares the host subunits of $Q\beta$-replicase, yet has a different template specificity (21, 131).

Overcoming Template Specificity

At present there are three ways to circumvent the extreme template specificity of the replicase using natural RNAs. The first is by adding manganese ions to the reaction mixture, which allows the enzyme to initiate and to transcribe RNA from templates not generally used (132, 133). This synthesis begins with pppG (as do all chains initiated by $Q\beta$-replicase) and is limited to the production of the complementary strand, fully annealed to its template (132, 133). There is considerable variability in the efficiency of transcription of different templates. R17 RNA is a relatively good template in the presence of Mn^{2+}, while tRNA and 16S rRNA are poor templates. Other RNA species, including 23S rRNA and the RNAs of eucaryotic viruses, are transcribed with intermediate efficiencies (132). This technique has also been used to make apparently full length copies of eucaryotic rRNA species and of histone mRNA (133). Mn^{2+} also overcomes the requirement for host factor in the plus strand–directed reaction (132).

The second way to relax replicase specificity is by adding an oligonucleotide primer to the reaction (134). If the primer perfectly matches a complementary sequence on the template, then the replicase will elongate RNA complementary to the template, beginning at the 3' end of the primer

(134–136). It should be emphasized that this primer technique overcomes template specificity by eliminating the initiation step of RNA synthesis. The technical advantage of the primer method, as opposed to the Mn^{2+} method, is that synthesis is initiated at precisely controllable locations rather than at the 3' end of the template as is the case with pppG (134). This technique has been used to prepare nearly full-length cRNA from 9S globin RNA, using oligo(U) or oligo(dT) as primers complementary to the 3' poly(A) (135, 136). In addition, discrete-length partial transcripts were seen (135, 136).

The third way to relax replicase specificity is by adding a stretch of C (or dC) residues to the 3' end of the RNA to be copied (123, 137, 138). This has allowed copying of synthetic polymers and globin RNA (but not fd DNA). However, the nature of the product has not been reported. The fact that Qβ-replicase is capable of RNA synthesis in the absence of added template, after a long incubation period, decreases somewhat the utility of all of the above methods (28, 129, 133).

Recognition of Synthetic RNAs

The recognition of cytidilate-containing synthetic templates seems to differ in a number of respects from the recognition of RNA species replicated by Qβ-replicase (Qβ RNA and the "6S" RNAs, which are bound at one or more internal sites, see below). In order to be recognized by the enzyme an oligo(C) molecule must be at least 15 nucleotides long (138), and must be present at high concentration (137). In experiments with defined sequence heteropolymers it has been found that synthesis always initiates with pppG, no matter what the 3' terminal nucleotide is (137). Apparently any synthetic RNA can be copied if an oligo(C) at least 5 bases long is present at the 3' end (137). Transcription of oligo(C) is not prevented by a single G residue at the 3' end but is prevented by poly(A) attached to the 3' end (137).

While poly(U) is a strong competitive inhibitor of initiation of transcription of poly(C), poly(A) is not (139). Qβ RNA (but not f2 RNA) also inhibits the poly(G) synthesis (apparently noncompetitively) and this inhibition is reversed by host factor (139).

Recognition of Qβ RNA

The first stage of in vitro Qβ RNA replication is enzyme recognition of, and binding to, the Qβ template. Although replicase alone binds preferentially to Qβ RNA (7, 130), the addition of HF and GTP to the reaction (in the absence of Mg^{2+}) results in a complex in which the replicase is bound much more tightly, and will not exchange with free RNA molecules (140). This irreversible complex cannot initiate synthesis in the absence of Mg^{2+} (140).

BINDING TO INTERNAL REGIONS   The idea that template recognition might involve binding of Qβ-replicase to internal regions of Qβ RNA was first suggested by the finding that fragmented Qβ RNAs fail to serve as template unless they contain the 3' terminal half of the molecule (4, 141). More recently it has been found that ribonuclease T1 oligonucleotides derived from two widely separated internal sites on Qβ RNA are bound to nitrocellulose filters by Qβ-replicase (141–145). The first of these, termed the S site, overlaps the ribosome binding site for the coat protein gene (approximately 70% from the 3' end of the RNA) and is apparently involved in repression of translation by Qβ-replicase (144, 146, 147). The second site, the M site, is in the proximal portion of the gene which codes for the phage-specified subunit of Qβ-replicase, approximately 50% from the 3' end of the RNA. This second site is bound by the enzyme only in the presence of Mg^{2+} (145). An initiation inhibitor, polyethylene sulfonate (148), has been found to specifically inhibit enzyme binding to this site (149). Based on these observations, Weissmann has proposed a model in which template recognition and initiation involves the interaction of the enzyme with one or both of the internal sites on the RNA molecule, as well as with the 3' end where transcription initiates (27). The internal binding may help to position the enzyme in such a way that the 3' end of the RNA is near the initiation site of the enzyme.

The question of whether enzyme binding to one or both internal sites is a prerequisite for initiation of replication has yet to be resolved. Indeed, since not all of the Qβ-replicase used for these experiments was active in RNA binding (149), it has not been possible to rigorously show that the enzymes bound to the internal binding sites represent intermediates in RNA synthesis. The best evidence that internal binding is necessary for replication is provided by the observation that both M-site and S-site RNA fragments inhibit in vitro RNA replication catalyzed by Qβ-replicase, but do not inhibit the poly(G) polymerase reaction (142). This does not necessarily imply that S-site binding is required since M-site and S-site fragments compete with each other for enzyme binding (149). In fact, three experiments demonstrate that S-site binding is not required for transcription of Qβ RNA under all conditions. First, Qβ RNA with a ribosome bound at the coat protein ribosome binding site, which overlaps the S site, is able to serve as a template for transcription (146, 150). Second, the 3' half of Qβ RNA which contains the M site but lacks the S site can be recognized and transcribed by Qβ-replicase (141). And third, a deletion mutant of Qβ lacking the S site is efficiently replicated by Qβ-replicase (151). Nevertheless, it is possible that under normal circumstances an S-site replicase interaction is required in order to alter the RNA structure so that the M site becomes available for enzyme binding.

Some evidence that both M and S sites are bound simultaneously by Qβ replicase has been provided by an electron microscopic investigation of Qβ RNA-enzyme complexes. Loop structures were seen, with the enzyme connecting two linearly distant sites on the RNA (152). The two sites were a constant distance from the ends of the RNA: one was consistent with the known location of the S site, while the other was apparently just outside the location of the M site. A single RNA with two enzyme molecules bound at the two sites was never seen, even when enzyme was present at a 30-fold molar excess. Thus, assuming that the two sites visualized are the same as the two identified biochemically, this experiment suggests that the binding of Qβ replicase to the M and S sites is a single event, in spite of the linear distance between the two sites. The alternative hypothesis, suggested by the competition between S- and M-site fragments for enzyme binding, is that Qβ replicase can bind to either RNA site, the former for inhibition of coat protein synthesis or the latter for initiation of replication.

The observation that excess Qβ RNA inhibits the in vitro reaction (independently of host factor) (153) could be explained if there are two sites on the enzyme, each capable of binding two different sites on the RNA independently and if both RNA sites must be bound by the same enzyme molecule for initiation of replication. Excess RNA, then, would inhibit by increasing the chance of one enzyme binding two different RNA molecules and thereby forming an inactive complex. This hypothesized mode of inhibition by excess Qβ RNA is supported by the observation that, with host factor–independent, Mn^{2+}-dependent RNA synthesis, excess Qβ RNA is inhibitory, while great excesses of other transcribed RNAs are not (132). Any two of the three sites, M, S, or the 3' end, could be involved in this specific inhibition.

FUNCTION OF S1   S1 appears to be involved in recognition of the specific sites on Qβ RNA bound by Qβ replicase. S1 alone will protect a portion of the S site (154), as well as another site close to the 3' end of the RNA (89), from RNase T1 digestion. Replicase lacking S1 is able to transcribe all RNAs transcribed by complete enzyme except Qβ RNA which is transcribed poorly and only at low ionic strength (35). If S1 is simply added back to this enzyme, full activity with Qβ RNA is restored (35). Antibody to S1 is a powerful inhibitor of the initiation of transcription of Qβ RNA, but not of poly(C) (95). This antibody is far more effective when added before initiation than it is when added afterwards (95). Furthermore, enzyme lacking S1 is no longer capable of protecting either internal site (M or S) against RNase T1 digestion (23).

The roles played by S1 in protein synthesis and Qβ replicase, while related, are apparently not identical. S1 isolated from *Caulobacter crescen-*

tus, which can replace *E. coli* S1 in in vitro protein synthesis (60), does not form a complex with the other Qβ-replicase subunits (P. Cole, personal communication). Also, S1 which has been treated with N-ethylmaleimide and, therefore, has lost its ability to unstack bases and support protein synthesis (60) shows nearly full activity in Qβ-replicase (P. Cole, personal communication). This suggests that the base unstacking activity of S1 is not crucial to its function in Qβ-replicase.

BINDING OF THE 3' TERMINAL REGION   The only sequence common to all of the "natural" templates transcribed by Qβ-replicase is the $CpCpC(pA)_{OH}$ at the 3' end (119). This sequence alone, however, is clearly only one component of template recognition: The 3' ends of the RNAs of closely related RNA phages also share this sequence but are not transcribed by Qβ-replicase (3). Furthermore, the enzyme does not itself bind tightly to the 3' end of the RNA (27, 144). Only when Qβ-replicase is incubated with Qβ RNA under initiating conditions in the presence of GTP is the 3' terminal oligonucleotide (along with those from the M and S sites) bound to nitrocellulose filters (149). The yield of the 3' terminal fragment is greatly enhanced when Mn^{2+} is present (149).

Support for the idea that the structure or sequence of the 3' ends is important in replication is supplied by two mutant phages isolated after site-directed mutagenesis (151, 155) with mutations in the 3' extracistronic region. One of these mutations, at position −16 from the 3' end, increases the in vitro rate of replication (156), while the other, at position −40, decreases the rate (157). The fact that this latter mutation also reduces the tightness of binding of S1 to the 3' end region of Qβ RNA (J. A. Steitz, personal communication), suggests that the S1-3' end binding may be an important aspect of RNA recognition. It is not known whether the mutation at position −16 alters the S1-3' end binding.

FUNCTION OF HOST FACTOR   Host factor is absolutely required for the initiation of synthesis with Qβ RNA template (86, 158). Antibody to host factor prevents all initiation of Qβ RNA–dependent transcription, but has only a small effect when added after initiation (95). Host factor acts by interacting with the template since the ratio of host factor to RNA, rather than to replicase is of critical importance. Maximal activity is achieved at a ratio of one or two host factor hexamers per Qβ RNA (158). The replicase of another RNA phage, SP, is able to copy SP RNA without host factor but requires host factor in order to copy Qβ RNA (159). Host factor has been shown to bind specifically to two A-rich sites on Qβ RNA: One A-rich site has been located at approximately 670 nucleotides from the 3' end (160) and the other is at the site spanning residues -38 to -63 from the 3' terminus which also binds S1 (89, 154). The fact that $Q\beta$-replicase alone will form a nondissociable complex with $Q\beta$ RNA and GTP but will not initiate without host factor (140), circumstantially implicates this protein in 3' end binding. This idea is further supported by the observation that Mn^{2+}, which increases 3' end binding (149) can replace host factor to allow initiation on $Q\beta$ RNA (132).

A second protein, called host factor II, previously thought to be required for in vitro $Q\beta$ RNA replication (25), has been found to be necessary only to remove an inhibitor (probably Bentonite) present in some RNA preparations (161).

Comparison of the Reactions Catalyzed by Qβ-Replicase

There are important differences between the two well-studied reactions catalyzed by $Q\beta$-replicase: 1. $Q\beta$ RNA–directed synthesis requires S1 and host factor, while poly(C)-directed synthesis does not (35); 2. the product of the former reaction is single-stranded (110), but the product of the latter is double-stranded (162); 3. the poly(G) polymerase reaction is extremely salt-sensitive (94), while $Q\beta$ RNA replication is not (35; see also 163); 4. M- and S-site RNA fragments inhibit $Q\beta$ RNA replication but not poly(C)-directed RNA synthesis (142); 5. the poly(C)-directed reaction requires more than 100-fold higher template concentration than does the $Q\beta$ RNA reaction (153, 162); 6. the $Q\beta$ RNA–dependent reaction is much more sensitive to inhibition by antibodies to EF-Tu and EF-Ts than is the poly(G) polymerase (95). [On the other hand, both reactions are strongly inhibited by poly(U) (139, 153, 164, 165).]

It seems likely that transcription of poly(C) requires interaction only of the 3' end site on $Q\beta$-replicase. When the enzyme is in its high salt (tightly-associated) conformation it cannot initiate synthesis at this site. The specific reaction with $Q\beta$ RNA appears more complex. It may involve recognition of internal RNA sites (M and S) for which ribosomal protein S1 is required, as well as recognition of the 3' end, for which host factor is apparently necessary. These binding reactions can occur at higher ionic strength and they result in altered relationships among the $Q\beta$-replicase subunits (92). The differential sensitivity of the two reactions to antibodies (95) could be explained by the apparently much larger conformational change induced in $Q\beta$-replicase by $Q\beta$ RNA than by poly(C) (92), which could make the elongation factors more available to the antibodies.

The Mn^{2+}-dependent reaction is inhibited by salt and does not require host factor (132) (Replicase lacking S1 has not been tested.) Also, since the RNA molecules transcribed in this reaction do not have the M and S sites, it is presumably a poly(C)-like reaction. The Mn^{2+} seems to facilitate interaction between the 3' end site on the enzyme, and the 3' ends of RNA molecules (149).

Recognition of "6S" RNAs

Qβ-replicase replicates a variety of small (6S) RNA molecules, called midi- (166), micro- (167), and nano-variant (119) RNAs. Many of these small RNAs, ranging from 91–220 nucleotides in length, have now been sequenced (119, 167, 168). These molecules do not show significant nucleotide sequence homology either with each other or with the internal sites that Qβ replicase protects on Qβ RNA (119). All of these small RNA molecules have a number of stems and loops including one near the 3' end (119). Furthermore, the enzyme binds to internal site(s) on the these molecules (119, 169) which suggests that Qβ-replicase recognizes structure rather than sequence in binding to internal regions of RNA molecules. RNA synthesis with "6S" RNAs is salt-sensitive and does not require S1 or host factor (169). This reaction is also "poly(C)-like" with respect to sensitivity to anti-EF-Tu and anti-EF-Ts (P. Cole, personal communication). Thus, this reaction shares some properties with both the Qβ RNA–directed and poly(C)-directed reactions. Assuming that the replicase without S1 is binding to internal regions of the small 6S molecules in order to replicate them, it seems likely that S1 functions in *specific* site selection with Qβ RNA.

Function of EF-Tu and EF-Ts

While the fact that protein synthesis elongation factors are present in an RNA-dependent RNA-synthesizing enzyme was originally perceived as bizarre, two possible functions in RNA synthesis derived from their known protein synthetic functions have been hypothesized (19, 93). First, since EF-Tu binds aatRNA, and since the 3' end of RNA phage RNA can be drawn in a structure resembling tRNA, it has been suggested that EF-Tu could be responsible for binding the enzyme to the 3' end of Qβ RNA, prior to the initiation of RNA synthesis. Second, because EF-Tu binds GTP tightly and specifically, and because all Qβ-replicase-catalyze RNA synthesis initiates with this nucleotide, it is possible that EF-Tu could supply the GTP to initiate replication. The fact that Qβ-replicase from which the elongation factors have been removed following initiation is able to elongate RNA chains at normal rates (93), lends credence to the idea that EF-Tu and EF-Ts are indeed functioning in initiation. Studies with Qβ-replicase containing a temperature-sensitive EF-Ts also suggest that the elongation factors are involved in initiation (99). Recent studies, however, have indicated that both of the hypothesized functions (aatRNA and GTP binding) may be incorrect.

The experiments that have provided insight into the role of EF-Tu and EF-Ts are based on the observation that the rate of renaturation of Qβ-replicase denatured in 8M urea is controlled by the rate of refolding of EF-Tu, and that Qβ-replicase containing exogenously added EF-Tu is formed when native EF-Tu or EF-Tu·Ts is added to the renaturation mixture at a temperature too low to permit refolding of the endogenous denatured EF-Tu (170). This allows the replacement of endogenous EF-Tu with EF-Tu which has been chemically or enzymologically altered to eliminate one or more of the hypothesized functions (77, 82, 97).

When the exogenous EF-Tu is treated with either N-ethylmaleimide or L-1-tosylamido-2-phenylethyl chloromethyl ketone, or cleaved with trypsin to eliminate its ability to bind aatRNA, no reduction in its ability to function in Qβ-replicase is observed (77, 82). In fact, EF-Tu can be covalently linked to EF-Ts with the protein cross-linking reagent dimethyl suberimidate, without loss of ability to work in Qβ replicase (96, 97). This treatment abolishes both aatRNA binding activity and the ability of the EF-Tu to bind GTP to nitrocellulose filters (96, 97). Thus, the functions performed by the elongation factors in Qβ RNA replication do not appear to derive directly from the partial reactions they catalyze in protein biosynthesis. Nevertheless, the possibility that the treated EF-Tu could retain the ability to bind the enzyme to the 3' end of Qβ RNA even though it has lost the ability to bind aatRNA cannot be excluded. Numerous attempts to demonstrate an interaction between EF-Tu and Qβ RNA have met with failure (P. Kaesberg, personal communication; T. Blumenthal, unpublished). It has also been shown that the S1-II complex alone is able to protect the M and S sites on Qβ RNA against RNase T1 digestion (144). Thus, all available data do not implicate the EF-Tu in Qβ replicase in RNA binding.

Purified Qβ-replicase made from the cross-linked EF-Tu·Ts is capable of initiating 0.8 polynucleotide chain per enzyme present, even though it can bind at most only 0.07 GTP molecule to nitrocellulose filters per enzyme (97). Still, the possibility that the cross-linked EF-Tu·Ts can bind GTP to initiate replication but cannot bind GTP to nitrocellulose filters in the standard assay has not been excluded. Hence while it has not been unequivocally demonstrated that the EF-Tu in Qβ-replicase does not function by binding GTP or the 3' end of the RNA, the experiments performed so far indicate that the EF-Tu probably performs some other function in the EF-Tu·Ts complex form (see below).

Kirromycin, which prevents the association between EF-Tu and EF-Ts, but does not dissociate a preformed EF-Tu·Ts complex [(82); J. Douglass and T. Blumenthal, unpublished], has been found to be an effective inhibitor in the EF-Tu-dependent Qβ replicase renaturation assay, but does not inhibit when native EF-Tu·Ts is added (82). These data suggest that the elongation factors must be bound together in order to participate in Qβ-replicase formation. In support of this idea, GDP, which can dissociate the EF-Tu·Ts complex, is an effective inhibitor of the EF-Tu·Ts-dependent replicase renaturation assay, but has no effect when cross-linked EF-Tu·Ts is used (82, 96, 97). Furthermore, the cross-linked EF-Tu·Ts is far more effective than even the untreated EF-Tu·Ts (97). These results strongly suggest that the elongation factors act as a complex in Qβ-replicase. Although they are known to exist in this form in uninfected cells, they have not been shown to participate directly in protein synthesis as EF-Tu·Ts (61).

More recent studies using this renaturation assay have shown that EF-Tu·Ts complexes from *Caulobacter crescentus* and from *Bacillus subtilis* can replace the *E. coli* EF-Tu·Ts in Qβ-replicase without loss of RNA synthetic function (L. Stringfellow and T. Blumenthal, unpublished data). These results, taken together with the fact that elongation factors treated with chemicals or cleaved with trypsin still function in Qβ-replicase, demonstrate that EF-Tu·Ts complex, with a wide variety of alterations, can function in RNA synthesis.

Inhibitors of RNA Synthesis

Qβ RNA replication is inhibited by polyethylene sulfonate (PES) (148). This polyanion has been shown to prevent initiation of RNA synthesis without preventing the binding of Qβ-replicase to Qβ RNA. Prebound enzyme remains sensitive to the inhibitor unless it has been initiated by the addition of GTP, in which case the elongation reaction is completely resistant (148). As already mentioned, PES interferes with the interaction between Qβ-replicase and the M site on Qβ RNA (149).

Another inhibitor of Qβ-replicase specific to the initiation reaction is a polymeric contaminant of aurintricarboxylic acid (ATA) a preparation that inhibits many proteins (e.g. *E. coli* RNA polymerase, T7 RNA polymerase, lac repressor) by preventing the binding of nucleic acids (171, 172). Again, elongation of preinitiated chains is completely resistant to the inhibitor (171).

Kirromycin prevents the assembly of Qβ-replicase, but enzyme that is already formed is unaffected (82). Similarly, L-1-tosylamido-2-phenylethyl chloromethyl ketone which inhibits EF-Tu, is not an inhibitor of Qβ-replicase (82). The interaction of guanine nucleotides with the enzyme is complicated because of the multiple binding sites for these nucleotides on the enzyme. GDP and ppGpp are competitive inhibitors of initiation of

Qβ-replicase, presumably by interaction at the initiation site specific for GTP (93, 96). The subunit location of this site has not been determined. On the other hand, elongation of initiated chains is not inhibited even by very high concentrations of nucleoside diphosphates (96). GDP, ppGpp, and GTP, can also increase the rate of inactivation of enzyme and can prevent the assembly of Qβ-replicase by binding to the EF-Tu site, thereby dissociating or preventing formation of the EF-Tu·Ts complex (96).

Accuracy of Base Selection

Both natural populations of Qβ RNA and Qβ RNA synthesized in vitro have been found to be heterogeneous (173). It has been estimated that Qβ-replicase is responsible for approximately 1.6 transition mutations per doubling, or 10^{-3}–10^{-4} mutations/nucleotide/doubling (173). Thus the RNA replication reaction is not nearly as accurate as DNA replication, presumably because DNA polymerases have proofreading activities (174), while Qβ-replicase does not.

REPLICATION OF OTHER RNA PHAGE RNAs

The other RNA replicating enzyme to be purified and extensively studied in vitro is from the Type I coliphage, f2 (21). The f2 enzyme has been found to be very similar to Qβ-replicase: It contains one phage-coded polypeptide of molecular weight 63,000 and the same three host-coded polypeptides present in the Qβ enzyme (131). It shows a salt-sensitive poly(C)-dependent poly(G) polymerase activity (163), a complicated interaction with GTP (163), and a high degree of specificity for its homologous template (175). Like Qβ-replicase, the f2 enzyme requires an additional host protein in order to transcribe its plus strand, but unlike Qβ-replicase, the host factor is also required for minus strand transcription (176). The f2 host factor also binds tightly to RNA and has a sedimentation coefficient similar to the Qβ host factor, but it is not known whether it is the same protein (176). However, there are reports that replication of R17, another Type 1 RNA phage, involves the host DNA-dependent RNA polymerase (177, 178). Considering that all the other closely related RNA phages that have been studied do not use RNA polymerase, and since RNA polymerase can utilize RNA templates (179–181), it remains to be shown that this in vitro activity is actually involved in the replication of R17 RNA.

CONCLUSIONS AND FUTURE DIRECTIONS

Qβ-replicase is a complex enzyme capable of making RNA copies of a variety of RNA templates under the proper conditions. In the cell it must differentiate between Qβ RNA and all other RNA species. This is apparently accomplished by recognition of internal RNA sequences. It is perhaps unique among well-studied replication enzymes in that it recognizes only single-stranded nucleic acids and its product is also single-stranded. The mechanism by which the fully complementary strands are prevented from annealing remains a mystery. There are many other issues remaining to be resolved, as well. While we have now identified the host functions of the host-coded enzyme subunits, we have not yet determined their precise roles in RNA replication. It is particularly important to discover how polypeptides which were evolved to function in one biosynthetic process (protein synthesis) are utilized in a different biosynthetic process (RNA replication). The answers to these questions should provide new insights both into the processes of evolution and the mechanisms of protein action. The functions of the RNA binding protein, host factor, both in Qβ RNA replication and in the host, also have yet to be determined. With the exception of the initial interaction between Qβ-replicase and Qβ RNA, the entire sequence of events in RNA replication is still unknown. An understanding of this relatively easily studied system would be desirable to provide a model for the study of replication of many eucaryotic viral RNAs.

ACKNOWLEDGEMENTS

We thank Drs. B. Polisky, J. Richardson, and H. Weber for their useful comments on this manuscript, and our many colleagues who provided access to unpublished material. The work of T. B. Blumenthal is supported by research grant GM-21024 and research career development award GM-00016 from the National Institute of General Medical Sciences.

Literature Cited

1. Haruna, I., Nozu, K., Ohtaka, Y., Spiegelman, S. 1963. *Proc. Natl. Acad. Sci. USA* 50:905–11
2. Weissmann, C., Simon, L., Ochoa, S. 1963. *Proc. Natl. Acad. Sci. USA* 49:407–14
3. Haruna, I., Spiegelman, S. 1965. *Proc. Natl. Acad. Sci. USA* 54:579–87
4. Haruna, I., Spiegelman, S. 1965. *Proc. Natl. Acad. Sci. USA* 54:1189–93
5. Pace, N. R., Spiegelman, S. 1966. *Science* 153:64–67
6. Spiegelman, S., Haruna, I., Holland, I. B., Beaudreau, G., Mills, D. R. 1965. *Proc. Natl. Acad. Sci. USA* 54:919–27
7. August, J. T., Banerjee, A. K., Eoyang, L., Franze de Fernandez, M. T., Hori, K., Kuo, C. H., Rensing, U., Shapiro, L. 1968. *Cold Spring Harbor Symp. Quant. Biol.* 33:73–81
8. Banerjee, A. K., Eoyang, L., Hori, K., August, J. T. 1967. *Proc. Natl. Acad. Sci. USA* 57:986–93
9. Banerjee, A. K., Kuo, C. H., August, J. T. 1969. *J. Mol. Biol.* 40:445–55
10. Billeter, M. A., Dahlberg, J. E., Goodman, H. M., Hindley, J., Weissmann, C. 1969. *Cold Spring Harbor Symp. Quant. Biol.* 34:635–46
11. Bishop, D. H. L., Pace, N. R., Spiegelman, S. 1967. *Proc. Natl. Acad. Sci. USA* 58:1790–97
12. Goodman, H. M., Billeter, M. A., Hindley, J., Weissmann, C. 1970. *Proc. Natl. Acad. Sci. USA* 67:921–28
13. Spiegelman, S., Pace, N. R., Mills, D. R., Levisohn, R., Eikhom, T. S., Taylor, M. M., Peterson, R. L., Bishop, D. H. L. 1968. *Cold Spring Harbor Symp. Quant. Biol.* 33:101–24

14. Kamen, R. I. 1970. *Nature* 228:527-33
15. Kamen, R. I. 1972. *Biochim. Biophys. Acta* 262:88-100
16. Kondo, M., Gallerani, R., Weissmann, C. 1970. *Nature* 228:525-27
17. Inouye, H., Pollack, Y., Petre, J. 1974. *Eur. J. Biochem.* 45:109-17
18. Wahba, A. J., Miller, M. J., Niveleau, A., Landers, T. A., Carmichael, G. G., Weber, K., Hawley, D. A., Slobin, L. I. 1974. *J. Biol. Chem.* 249:3314-16
19. Blumenthal, T., Landers, T. A., Weber, K. 1972. *Proc. Natl. Acad. Sci. USA* 69:1313-17
20. August, J. T., Eoyang, L., Franze de Fernandez, M. T., Hasegawa, S., Kuo, C. H., Rensing, U., Shapiro, L. 1969. *J. Cell Physiol.* 74: Suppl. 1, pp. 187-95
21. Fedoroff, N. 1975. In *RNA Phages,* ed. N. Zinder, 235-58. Cold Spring Harbor, NY: Cold Spring Harbor Lab. 428 pp.
22. Hindley, J. 1973. *Progr. Biophys. Mol. Biol.* 26:269-320
23. Kamen, R. I. 1975. See Ref. 21, pp. 203-34
24. Kozak, M., Nathans, D. 1972. *Bacteriol. Rev.* 36:109-34
25. Kuo, C.-H., Eoyang, L., August, J. T. 1975. See Ref. 21, pp. 259-77
26. Stavis, R. L., August, J. T. 1970. *Ann. Rev. Biochem.* 39:527-60
27. Weissmann, C. 1974. *FEBS Lett.* 43: S10-18
28. Weissmann, C., Billeter, M. A., Goodman, H. M., Hindley, J., Weber, H. 1973. *Ann. Rev. Biochem.* 42:303-28
29. Weissmann, C., Ochoa, S. 1968. *Prog. Nucleic Acid Res. Mol. Biol.* 6:353-99
30. Groner, Y., Pollack, Y., Berissi, H., Revel, M. 1972. *Nature New Biol.* 239:16-19
31. Groner, Y., Scheps, R., Kamen, R., Kolakofsky, D., Revel, M. 1972. *Nature New Biol.* 239:19-20
32. Jay, G., Kaempfer, R. 1974. *J. Mol. Biol.* 82:193-212
33. Miller, M. J., Niveleau, A., Wahba, A. J. 1974. *J. Biol. Chem.* 249:3803-7
34. Miller, M. J., Wahba, A. J. 1974. *J. Biol. Chem.* 249:3808-13
35. Kamen, R., Kondo, M., Romer, W., Weissmann, C. 1972. *Eur. J. Biochem.* 31:44-51
36. Hermoso, J. M., Szer, W. 1974. *Proc. Natl. Acad. Sci. USA* 71:4708-12
37. Sobura, J. E., Chowdhury, M. R., Hawley, D. A., Wahba, A. J. 1977. *Nucleic Acids Res.* 4:17-29
38. Szer, W., Leffler, S. 1974. *Proc. Natl. Acad. Sci. USA* 71:3611-15
39. Tal, M., Aviram, M., Kanarek, A., Weiss, A. 1971. *Biochim. Biophys. Acta* 222:381-92
40. Van Dieijen, G., Van der Laken, C. J., Van Knippenberg, P. H., Van Duin, J. 1975. *J. Mol. Biol.* 93:351-66
41. Van Dieijen, G., Van Knippenberg, P. H., Van Duin, J. 1976. *Eur. J. Biochem.* 64:511-18
42. Van Dieijen, G., Van Knippenberg, P. H., Van Duin, J., Koekman, B., Pouwels, P. H. 1977. *Mol. Gen. Genet.* 153:75-80
43. Van Duin, J., Van Knippenberg, P. H. 1974. *J. Mol. Biol.* 84:185-95
44. Szer, W., Hermoso, J. M., Leffler, S. 1975. *Proc. Natl. Acad. Sci. USA* 72:2325-29
45. Van Knippenberg, P. H., Hooykaas, P. J. J., Van Duin, J. 1974. *FEBS Lett.* 41:323-26
46. Dahlberg, A. E. 1974. *J. Biol. Chem.* 249:7673-78
47. Dahlberg, A. E., Dahlberg, J. E. 1975. *Proc. Natl. Acad. Sci. USA* 72:2940-44
48. Kenner, R. A. 1973. *Biochem. Biophys. Res. Commun.* 51:932-38
49. Laughrea, M., Moore, P. B. 1978. *J. Mol. Biol.* 121:411-30
50. Laughrea, M., Moore, P. B. 1977. *J. Mol. Biol.* 112:399-421
51. Yokota, T., Arai, K.-I., Kaziro, Y. 1977. *J. Biochem.* 82:1485-89
52. Draper, D. E., Pratt, C. W., von Hippel, P. H. 1977. *Proc. Natl. Acad. Sci. USA* 74:4786-90
53. Draper, D. E., von Hippel, P. H. 1978. *J. Mol. Biol.* 122:321-38
54. Draper, D. E., von Hippel, P. H. 1978. *J. Mol. Biol.* 122:339-60
55. Lipecky, R., Kohlschein, J., Gassen, H. G. 1977. *Nucleic Acids Res.* 4:3627-42
56. Thomas, J. O., Kolb, A., Szer, W. 1978. *J. Mol. Biol.* 123:163-76
57. Bear, D. G., Ng, R., Van der Veer, D., Johnson, N. P., Thomas, G., Schleich, T., Noller, H. F. 1976. *Proc. Natl. Acad. Sci. USA* 73:1824-28
58. Szer, W., Thomas, J. O., Kolb, A., Hermoso, J. M., Boublik, M. 1977. In *Nucleic Acid-Protein Recognition,* ed. H. J. Vogel, pp. 519-29. New York: Academic. 587 pp.
59. Szer, W., Hermoso, J. M., Boublik, M. 1976. *Biochem. Biophys. Res. Commun.* 70:957-64
60. Kolb, A., Hermoso, J. M., Thomas, J. O., Szer, W. 1977. *Proc. Natl. Acad. Sci. USA* 74:2379-83
61. Lucas-Lenard, J., Lipmann, F. 1971. *Ann. Rev. Biochem.* 40:409-48

62. Furano, A. V. 1976. *Eur. J. Biochem.* 64:597–606
63. Jacobson, G. R., Rosenbusch, J. P. 1976. *Nature* 261:23–26
64. Miyajima, A., Kaziro, Y. 1978. *J. Biochem.* 83:456–62
65. Beck, B. D., Arscott, P. G., Jacobson, A. 1978. *Proc. Natl. Acad. Sci. USA* 75:1250–54
66. Rosenbusch, J. P., Jacobson, G. R., Jaton, J.-C. 1976. *J. Supramol. Struct.* 5:391–96
67. Jaskunas, S. R., Lindahl, L., Nomura, M., Burgess, R. R. 1975. *Nature* 257:458–62
68. Furano, A. V. 1977. *J. Biol. Chem.* 252:2154–57
69. Miller, D. L., Nagarka, S., Laursen, R. A., Parker, J., Friesen, J. D. 1978. *Mol. Gen. Genet.* 159:57–62
70. Travers, A. 1973. *Nature New Biol.* 244:15–18
71. Travers, A., Buckland, R. 1973. *Nature New Biol.* 243:257–60
72. Travers, A., Kamen, R., Schleif, R. 1970. *Nature* 228:748–51
73. Haseltine, W. A. 1972. *Nature* 235:329–33
74. Hussey, C., Pero, J., Shorenstein, R. G., Losick, R. 1972. *Proc. Natl. Acad. Sci. USA* 69:407–11
75. Pettijohn, D. E. 1972. *Nature New Biol.* 235:204–6
76. Wolf, H., Chinali, G., Parmeggiani, A. 1974. *Proc. Natl. Acad. Sci. USA* 71:4910–14
77. Blumenthal, T., Douglass, J., Smith, D. 1977. *Proc. Natl. Acad. Sci. USA* 74:3264–67
78. Wilson, G. E., Cohn, M. 1977. *J. Biol. Chem.* 252:2004–9
79. Wilson, G. E., Cohn, M., Miller, D. 1978. *J. Biol. Chem.* 253:5764–68
80. Chinali, G., Wolf, H., Parmeggiani, A. 1977. *Eur. J. Biochem.* 74:55–65
81. Wolf, H., Chinali, G., Parmeggiani, A. 1977. *Eur. J. Biochem.* 75:67–75
82. Brown, S., Blumenthal, T. 1976. *J. Biol. Chem.* 251:2749–53
83. Fischer, E., Wolf, H., Hantke, K., Parmeggiani, A. 1977. *Proc. Natl. Acad. Sci. USA* 74:4341–45
84. Van de Klundert, J. A. M., Den Turk, E., Borman, A. H., Van der Meide, P. H., Bosch, L. 1977. *FEBS Lett.* 81:303–7
85. Van de Klundert, J. A. M., Van der Meide, P. H., Van de Putte, P., Bosch, L. 1978. *Proc. Natl. Acad. Sci. USA* 75:4470–73
86. Franze de Fernandez, M. T., Eoyang,
L., August, J. T. 1968. *Nature* 219:588–90
87. Carmichael, G. G. 1975. *J. Biol. Chem.* 250:6160–67
88. Carmichael, G. G., Weber, K., Niveleau, A., Wahba, A. 1975. *J. Biol. Chem.* 250:3607–12
89. Senear, A., Steitz, J. A. 1975. *J. Biol. Chem.* 251:1902–12
90. DuBow, M. S., Ryan, T., Young, R. A., Blumenthal, T. 1977. *Mol. Gen. Genet.* 153:39–43
91. DuBow, M. S., Blumenthal, T. 1975. *Mol. Gen. Genet.* 141:113–19
92. Young, R. A., Blumenthal, T. 1975. *J. Biol. Chem.* 250:1829–32
93. Landers, T. A., Blumenthal, T., Weber, K. 1974. *J. Biol. Chem.* 249:5801–8
94. Blumenthal, T., Young, R. A., Brown, S. 1975. *J. Biol. Chem.* 251:2740–43
95. Carmichael, G. G., Landers, T. A., Weber, K. 1976. *J. Biol. Chem.* 251:2744–48
96. Blumenthal, T. 1977. *Biochim. Biophys. Acta* 478:201–8
97. Brown, S., Blumenthal, T. 1976. *Proc. Natl. Acad. Sci. USA* 73:1131–35
98. Happe, M., Jockusch, H. 1975. *Eur. J. Biochem.* 58:359–66
99. Hori, K., Harada, K., Kuwano, M. 1974. *J. Mol. Biol.* 86:699–708
100. Lemaux, P. G., Miller, D. L. 1978. *Mol. Gen. Genet.* 159:47–56
101. Yamamoto, M., Strycharz, W. A., Nomura, M. 1976. *Cell* 8:129–38
102. Pedersen, S., Blumenthal, R. M., Reeh, S., Parker, J., Lemaux, P., Laursen, R. A., Nagarkatti, S., Friesen, J. D. 1976. *Proc. Natl. Acad. Sci. USA* 73:1698–701
103. Ohki, K., Hori, K. 1972. *Biochim. Biophys. Acta* 281:233–43
104. Boedtker, H., Gesteland, R. F. 1975. See Ref. 21, pp. 1–28
105. Zinder, N., ed. 1975. See Ref. 21. 428 pp.
106. Feix, G., Slor, H., Weissmann, C. 1967. *Proc. Natl. Acad. Sci. USA* 57:1401–8
107. Libonati, M., Viñuela, E., Weissmann, C. 1967. *FEBS Abstr.* p. 144
108. Robertson, H. D. 1975. See Ref. 21, pp. 113–46
109. Robertson, H. D., Webster, R. E., Zinder, N. D. 1968. *J. Biol. Chem.* 243:82–91
110. Weissmann, C., Feix, G., Slor, H. 1968. *Cold Spring Harbor Symp. Quant. Biol.* 33:83–100
111. Weissmann, C., Feix, G., Slor, H., Pollet, R. 1967. *Proc. Natl. Acad. Sci. USA* 57:1870–77
112. Steitz, J. A. 1975. See Ref. 21, pp. 319–54

113. Mills, D. R., Pace, N. R., Spiegelman, S. 1966. *Proc. Natl. Acad. Sci. USA* 56:1778–85
114. Weissmann, C., Feix, G. 1966. *Proc. Natl. Acad. Sci. USA* 55:1264–68
115. Hori, K. 1970. *Biochim. Biophys. Acta* 217:394–407
116. Thach, S. S., Thach, R. E. 1973. *J. Mol. Biol.* 81:367–80
117. Rensing, U., August, J. T. 1969. *Nature* 224:853–56
118. Weber, H., Weissmann, C. 1970. *J. Mol. Biol.* 51:215–24
119. Schaffner, W., Ruegg, K. J., Weissmann, C. 1977. *J. Mol. Biol.* 117:877–907
120. Trown, P. W., Meyer, P. L. 1973. *Arch. Biochem. Biophys.* 154:250–62
121. Feix, G., Pollet, R., Weissmann, C. 1968. *Proc. Natl. Acad. Sci. USA* 59:145–52
122. Miyake, T., Haruna, I., Shiba, T., Itoh, Y. H., Yamane, K., Watanabe, I. 1971. *Proc. Natl. Acad. Sci. USA* 68:2022–24
123. Feix, G., Sano, H. 1976. *FEBS Lett.* 63:201–4
124. Eikhom, T. S., Spiegelman, S. 1967. *Proc. Natl. Acad. Sci. USA* 57:1833–40
125. Hori, K., Eoyang, L., Banerjee, A. K., August, J. T. 1967. *Proc. Natl. Acad. Sci. USA* 57:1790–97
126. Levisohn, R., Spiegelman, S. 1969. *Proc. Natl. Acad. Sci. USA* 60:866–72
127. Mills, D., Peterson, R. L., Spiegelman, S. 1967. *Proc. Natl. Acad. Sci. USA* 58:217–24
128. Banerjee, A. K., Rensing, U., August, J. T. 1969. *J. Mol. Biol.* 45:181–93
129. Sumper, M., Luce, R. 1975. *Proc. Natl. Acad. Sci. USA* 72:162–66
130. Silverman, P. M. 1973. *Arch. Biochem. Biophys.* 157:234–42
131. Fedoroff, N. V., Zinder, N. 1971. *Proc. Natl. Acad. Sci. USA* 68:1838–43
132. Palmenberg, A., Kaesberg, P. 1974. *Proc. Natl. Acad. Sci. USA* 71:1371–75
133. Obinata, M., Nasser, D. S., McCarthy, B. J. 1975. *Biochem. Biophys. Res. Commun.* 64:640–47
134. Feix, G., Hake, H. 1975. *Biochem. Biophys. Res. Commun.* 65:503–9
135. Feix, G. 1976. *Nature* 259:593–94
136. Vournakis, J. N., Carmichael, G. G., Efstratiadis, A. 1976. *Biochem. Biophys. Res. Commun.* 70:774–82
137. Feix, G., Sano, H. 1975. *Eur. J. Biochem.* 58:59–64
138. Kuppers, B., Sumper, M. 1975. *Proc. Natl. Acad. Sci. USA* 72:2640–43
139. Hori, K. 1973. *J. Biochem.* 74:273–78
140. Silverman, P. M. 1973. *Arch. Biochem. Biophys.* 157:222–33
141. Schwyzer, M., Billeter, M. A., Weissmann, C. 1972. *Experientia* 28:750
142. Meyer, F., Weber, H., Weissmann, C. 1976. *Experientia* 32:804
143. Meyer, F., Weber, H., Vollenweider, H. J., Weissmann, C. 1975. *Experientia* 31:743
144. Weber, H., Billeter, M. A., Kahane, S., Weissmann, C., Hindley, J., Porter, A. 1972. *Nature New Biol.* 237:166–70
145. Weber, H., Kamen, R., Meyer, F., Weissmann, C. 1974. *Experientia* 30:711
146. Kolakofsky, D., Weissmann, C. 1971. *Nature New Biol.* 231:42–46
147. Kolakofsky, D., Weissmann, C. 1971. *Biochim. Biophys. Acta* 246:596–99
148. Kondo, M., Weissmann, C. 1972. *Biochim. Biophys. Acta* 259:41–49
149. Meyer, F. 1978. Structure and function of QB replicase binding sites on QB RNA. PhD thesis. Univ. Zurich, Zurich, Switzerland. 120 pp.
150. Kolakofsky, D., Billeter, M. A., Weber, H., Weissmann, C. 1973. *J. Mol. Biol.* 76:271–84
151. Sabo, D. L., Domingo, E., Bandle, E. F., Flavell, R. A., Weissmann, C. 1977. *J. Mol. Biol.* 111:235–52
152. Vollenweider, H. J., Koller, T., Weber, H., Weissmann, C. 1976. *J. Mol. Biol.* 101:367–77
153. Kondo, M., Weissmann, C. 1972. *Eur. J. Biochem.* 24:530–37
154. Goelz, S., Steitz, J. A. 1977. *J. Biol. Chem.* 252:5177–79
155. Flavell, R. A., Sabo, D. L., Bandle, E. F., Weissmann, C. 1974. *J. Mol. Biol.* 89:255–72
156. Flavell, R. A., Sabo, D. L., Bandle, E. F., Weissmann, C. 1975. *Proc. Natl. Acad. Sci. USA* 72:367–71
157. Domingo, E., Flavell, R. A., Weissmann, C. 1976. *Gene* 1:3–25
158. Franze de Fernandez, M. T., Hayward, W. S., August, J. T. 1972. *J. Biol. Chem.* 247:824–31
159. Nishihara, T., Haruna, I., Yamaguchi, N., Watanabe, I. 1972. *Nature New Biol.* 238:141–42
160. Billeter, M. 1978. *J. Biol. Chem.* 253:8381–89
161. Kamen, R. I., Monstein, H.-J., Weissmann, C. 1974. *Biochim. Biophys. Acta* 366:292–99
162. Mitsunari, Y., Hori, K. 1973. *J. Biochem.* 74:263–71
163. Fedoroff, N. V., Zinder, N. D. 1972. *J. Biol. Chem.* 247:4586–92
164. Haruna, I., Spiegelman, S. 1966. *Proc. Natl. Acad. Sci. USA* 55:1256–63

165. Kondo, M. 1976. *Biochem. J.* 155: 461–64
166. Kacian, D. L., Mills, D. R., Kramer, F. R., Spiegelman, S. 1972. *Proc. Natl. Acad. Sci. USA* 69:3038–42
167. Mills, D. R., Kramer, F. R., Dobkin, C., Nishihara, T., Spiegelman, S. 1975. *Proc. Natl. Acad. Sci. USA* 72:4252–56
168. Mills, D. R., Kramer, F. R., Spiegelman, S. 1973. *Science* 180:916–27
169. Mills, D. R., Nishihara, T., Dobkin, C., Kramer, F. R., Cole, P. E., Spiegelman, S. 1977. See Ref. 58, pp. 533–47
170. Blumenthal, T., Landers, T. A. 1975. *Biochemistry* 15:422–25
171. Blumenthal, T., Landers, T. A. 1973. *Biochem. Biophys. Res. Commun.* 55: 680–88
172. Gonzalez, R. G., Blackburn, B. J., Schleich, T. 1979. *Biochim. Biophys.* *Acta.* In press
173. Domingo, E., Sabo, D., Taniguchi, T., Weissmann, C. 1978. *Cell* 13:735–44
174. Kornberg, A. 1974. *DNA Synthesis* pp. 88–90. San Francisco: Freeman. 399 pp.
175. Fedoroff, N. V., Zinder, N. D. 1972. *J. Biol. Chem.* 247:4577–85
176. Fedoroff, N. V., Zinder, N. D. 1973. *Nature New Biol.* 241:105–8
177. Igarashi, S. J., Bissonnette, R. P. 1971. *J. Biochem.* 70:835–44
178. Igarashi, S. J., Bissonnette, R. P. 1971. *J. Biochem.* 70:845–54
179. Fox, C. F., Robinson, W. S., Haselkorn, R., Weiss, S. B. 1964. *J. Biol. Chem.* 239:186–95
180. Melli, M., Pemberton, R. E. 1972. *Nature New Biol.* 236:172–74
181. Robertson, H. D. 1971. *Nature New Biol.* 229:169–72

Ann. Rev. Biochem. 1979. 48:549–80

REGULATION OF PROTEIN SYNTHESIS IN EUKARYOTES[1]

♦12019

Severo Ochoa and Cesar de Haro

Roche Institute of Molecular Biology, Nutley, New Jersey 07110

CONTENTS

PERSPECTIVES AND SUMMARY

Gene expression in prokaryotes is controlled mainly at the level of transcription. In eukaryotes, with more stable messenger RNAs (mRNAs), there is control also at the translational level. In reticulocytes, synthesis of globin occurs only in the presence of heme, the prosthetic group of hemo-

[1]The following abbreviations are used: HCI, hemin-controlled inhibitor (an eIF-2 kinase); eIF-2, eukaryotic initiation factor 2; eIF-2(P), phosphorylated eIF-2; ESP, eIF-2-stimulating protein; Met-tRNA$_i$, initiator methionyl transfer RNA; dsRNA, double-stranded RNA; BHK, bovine heart protein kinase.

549

0066-4154/79/0701-0549$01.00

globin. In the absence of hemin, an inhibitor of polypeptide chain initiation is activated. The available evidence is consistent with the view that the hemin-controlled inhibitor (HCI) is a cAMP-independent protein kinase (eIF-2 kinase) that inhibits chain initiation through phosphorylation of the smallest (38,000 daltons) of the three subunits of which the initiation factor eIF-2 is composed. In the presence of the recently discovered eIF-2-stimulating protein (ESP), eIF-2 forms a ternary complex with GTP and initiator methionyl transfer RNA which binds to a 40S ribosomal subunit giving rise to a 40S initiation complex. Inhibition of translation is a consequence of the phosphorylation of the small subunit of eIF-2 because this blocks the interaction of eIF-2 with ESP and prevents formation of the ternary complex. ESP probably functions by displacing the equilibrium to favor formation of the ternary complex. There is evidence that HCI, like phosphorylase kinase, is activated by phosphorylation catalyzed by cAMP-dependent protein kinase. Cyclic AMP-dependent protein kinases consist of two regulatory (R) and two catalytic (C) subunits (R_2C_2); cAMP activates the enzyme by binding to R and releasing active C. Recent studies are consistent with the view that hemin blocks HCI activation by interfering with the activation of cAMP-dependent protein kinase by cAMP. Hemin binds to a site of the regulatory subunit of cAMP-dependent protein kinase that is apparently different from the site to which cAMP binds, and blocks the binding of cAMP to this subunit in an allosteric fashion. A similar system of translational control exists in cells other than reticulocytes, although the dependence of translation on the presence of hemin may be less pronounced or nil. Further study of the structure and nature of HCI should throw light on the mechanism of activation of this kinase which, as in the case of phosphorylase kinase, might involve phosphorylation of a regulatory subunit. The phosphorylation of eIF-2 and eIF-2 kinase is probably reversed by dephosphorylation catalyzed by specific protein phosphatases, as in the case of phosphorylase and phosphorylase kinase. Further studies of the mechanism of regulation of translation should throw light on this aspect of the problem.

There are other mechanisms of translational regulation in reticulocytes and other cells that are not subject to control by hemin. Addition of small amounts of dsRNA or oxidized gluthathione to hemin-containing reticulocyte lysates activates HCI-like inhibitors of chain initiation that, like HCI, are associated with a cAMP-independent protein kinase that phosphorylates the 38,000 dalton subunit of eIF-2. The mode of action of oxidized gluthathione is unknown but dsRNA may activate a protein kinase that, in the presence of ATP, activates an HCI-like inhibitor. Incubation of hemin-containing reticulocyte lysates with dsRNA and ATP also gives rise to the formation of a potent translational inhibitor, a novel oligonucleotide (pppA2'p5'A2'p5'A) that inhibits protein synthesis at exceedingly low concentrations (1–10 nM) by activating an endonuclease that degrades both cell and viral RNAs.

Treatment of animal cells (e.g. mouse L cells, ascites tumor cells) with the antiviral agent interferon induces enzymes that, in the presence of low levels of dsRNA, activate an HCI-like translational inhibitor and give rise to the synthesis of pppA2'p5'A2'p5'A from ATP; these are the same enzymes that are present, apparently in a constitutive manner, in rabbit reticulocytes. The chief enzyme induced by interferon appears to be a protein kinase that is activated by dsRNA. In the presence of ATP the active kinase probably activates (a) an HCI-like translational inhibitor and (b) a pppA2'p5'A2'p5'A-synthesizing enzyme. Interferon is induced by viruses in animal cells. In the presence of interferon, dsRNA produced during viral replication possibly activates the interferon-induced kinase and triggers the events leading to the production of translational inhibitors thus inhibiting the growth of the virus and the virus-infected cells.

Although one might expect phosphorylation of ribosomal proteins and of initiation factors other than eIF-2 to play a part in translational regulation, there are thus far no indications that this may be the case. Incubation of eukaryotic ribosomes with protein kinases (both cAMP-dependent and independent), in the presence of [^{32}P]ATP, phosphorylates several proteins of the 60S and 40S subunits, whereas in vivo mainly one protein of the 40S subunit (S6) of rat liver is phosphorylated. This phosphorylation is markedly enhanced by cAMP and by glucagon which suggests the involvement of a cAMP-dependent protein kinase. No functional alteration of the ribosomes has been detected. The middle subunit (~50,000 daltons) of eIF-2, several subunits of the initiation factor eIF-3 (composed of ten or more subunits), and the initiation factors eIF-5 (the so-called ribosome joining factor) and eIF-4B are phosphorylated by cAMP-independent protein kinases. Here again, these phosphorylations appear to lead to no functional modification of the factors. Whereas there seem to be several ways of shutting off translation (inhibition of chain initiation, degradation of mRNA) more specific mechanisms for control of translation of individual or group mRNAs are still unknown.

Finally, certain small oligonucleotides found in eukaryotic cells, apparently formed as a result of endonuclease digestion of cell RNA, can affect translation at the chain elongation level. They may be involved in the transition from a dormant or quiescent state to an active state, as in emergence from cryptobiosis (e.g. *Artemia salina* embryos) or egg fertilization, but there is no evidence for this.

INTRODUCTION

In prokaryotic cells, with short lived RNAs, regulation of gene expression occurs mainly at the transcriptional level. In eukaryotic cells, on the other hand, in which mRNAs have a longer life span, gene expression is controlled not only at the transcriptional but also at the translational level. Regulation of translation in eukaryotic cells takes place, in part, during the initiation of polypeptide chains and this may explain why eukaryotic initiation factors are more numerous and structurally more complex than their prokaryotic counterparts (1, 2). Whereas the initiation apparatus of prokaryotes consists of two or three initiation factors, the number of initiation factors in eukaryotes is at least seven or eight. Moreover, in prokaryotes the initiator methionyl tRNA binding factor IF-2 consists of a single polypeptide chain of molecular weight 80,000 or 90,000, and the mRNA recognizing factor IF-3 of one polypeptide chain of molecular weight 23,000. Eukaryotic eIF-2, however, is composed of three subunits with a combined molecular weight of about 150,000; eIF-3 has no less than 10 subunits with an aggregate molecular weight of over 500,000.

It has become apparent that control of translation in eukaryotes involves protein kinases which, when activated, inhibit translation. This review is largely concerned with the role of protein kinases in translational control.

REGULATION OF PROTEIN SYNTHESIS BY HEMIN

The main function of reticulocytes, the immediate precursors of the erythrocytes, is to synthesize globin, the protein component of hemoglobin. In reticulocytes, the synthesis of globin, together with that of other less abundant proteins, depends on the presence of heme, the prosthetic group of hemoglobin (3). Protein synthesis in lysates of rabbit reticulocyte starts at a high rate but drops to very low values in a few minutes unless the system is supplemented with hemin (3–6). The iron-free compound (protoporphyrin IX) is inactive (4). Optimal concentrations of hemin (20–40 μM) maintain protein synthesis for prolonged periods, but higher concentrations are inhibitory. The fact that heme synthesis is stimulated by iron explains the early observation that protein synthesis in reticulocytes and bone marrow erythroid cells is greatly enhanced by addition of iron to the incubation medium. Addition of hemin to a suspension of reticulocytes causes a decrease in the synthesis of protoporphyrin IX and a simultaneous increase in globin synthesis. Since reticulocytes contain no nucleus, the enhancing effect of hemin on globin synthesis must be effected at the level of translation. Reticulocytes can synthesize heme; they also synthesize globin, although at a reduced rate, in the absence of added hemin. However, reticulocyte lysates, lacking mitochondria, do not synthesize heme and are markedly dependent on the addition of hemin for protein synthesis. In the absence of hemin, an inhibitor of polypeptide chain initiation is formed from a proinhibitor of similar molecular weight (7–9). This inhibitor, when added to a lysate that is actively synthesizing protein in the presence of hemin, rapidly inhibits protein synthesis. The time course of protein synthesis in the presence of both hemin and inhibitor is essentially the same as in the absence of hemin. Under the conditions of protein synthesis absence of hemin leads to a rapid conversion of proinhibitor to inhibitor. When hemin is not present from the start of incubation but is added at various times thereafter, the rate of protein synthesis is restored to a greater or lesser extent depending on the time lapse between the start of incubation and the addition of hemin (3). Addition of hemin 1 min after beginning the incubation restores the translation rate to almost the rate observed when hemin is present from the start. However, after 5 min or more the restoration is incomplete. This suggests that with short incubation times the inhibitor formed can be fully converted back to proinhibitor but that, with increasing incubation time, this is no longer the case. This means that the irreversible reaction b becomes increasingly prominent as the incubation becomes longer.

$$\text{Proinhibitor} \underset{}{\overset{(a)}{\rightleftharpoons}} \text{inhibitor} \xrightarrow{(b)} \text{irreversible inhibitor}$$

The proinhibitor-inhibitor conversion has been studied with crude proinhibitor preparations obtained from the postribosomal supernatant of rabbit reticulocyte lysates by chromatography on carboxymethyl-Sephadex (9). These preparations are largely free of ATP. We see later that they are also virtually free of cAMP but contain a very active cAMP-dependent protein kinase. Addition of proinhibitor to hemin-containing reticulocyte lysates has no effect on translation but, if it has previously been converted to inhibitor, translation is inhibited. In this way, it seems that irreversible inhibitor is formed on incubation for several hours at 34–37°. The same conversion can occur in a few minutes at 30° upon addition of small concentrations of N-ethylmaleimide or o-iodosobenzoate (9) and the conversion is quantitative. To reach the same level of inhibitor by incubation at 34° requires 16 or more hours. This conversion is prevented by hemin but the concentration required is higher than the optimal concentration needed to maintain protein synthesis in lysates (10–12). Within the first 30 min of incubation of proinhibitor at 34°, the inhibitor formed can disappear following a few more minutes of incubation in the presence of hemin (11, 13). This form has been referred to as reversible inhibitor. After longer incubation, however, the addition of hemin has little or no effect (11, 13). This form has been referred to as irreversible inhibitor. The slow conversion of proinhibitor to inhibitor that takes place on incubation at 34° can also be largely prevented by the addition of dithiothreitol (11). However, dithiothreitol has no effect upon the rapid conversion occurring in lysates incubated at 30° in the absence of hemin, under the conditions of protein synthesis, (O. Vicente, J. M. Sierra, and S. Ochoa, unpublished results). This suggests that the slow conversion obtained on incubation of partially purified proinhibitor preparations at 34° is not related to the rapid one occurring at the same or lower temperature in lysates in the absence of hemin, and probably has no bearing on the physiological mechanism of activation of the proinhibitor. Since formation of the inhibitor is prevented by hemin, it has been referred to as the hemin-controlled repressor or the hemin-controlled inhibitor. Antibody against irreversible HCI also neutralized reversible HCI formed within 20 min of incubation of proinhibitor preparations at 34° but, as judged by competition experiments, has no effect on the HCI present before incubation (14). This finding indicates the occurrence of a significant structural change even in formation of the so-called reversible inhibitor by the warming procedure and suggests that this species of HCI, although sensitive to hemin, may not be a truly reversible form of HCI, i.e. one that can be converted back to proinhibitor (see Equation 1 below). HCI has also been found in postribosomal supernatants from human erythrocytes (15).

Nature of Hemin-Controlled Translational Inhibitor

Inhibition of translation by HCI is accompanied by polysome disaggregation and by a marked decrease in formation of the 40S initiation complex and this points to chain initiation as the site of inhibition (16–19). However, the actual mechanism was unknown. Translational inhibition in reticulocyte lysates, resulting from either the absence of hemin or the addition of HCI in the presence of hemin, is relieved by the addition of the chain initiation factor eIF-2 (18, 20–22), which suggests that HCI might interfere with the function of this factor. Previously it had been reported that inhibition due to the absence of hemin is prevented by addition of a ribosomal salt wash rich in initiation factors (16, 17). eIF-2, the initiation factor that recognizes and promotes the ribosomal binding of the initiator aminoacyl tRNA, forms a ternary complex with GTP and Met-tRNA$_i$ which, in the presence of a 40S ribosomal subunit, gives rise to a 40S initiation complex (23, 24). Formation of this complex is followed by binding to the 60S ribosomal subunit whereupon GTP is hydrolyzed and eIF-2 is released. Formation of the 80S initiation complex sets the stage for polypeptide chain elongation. In reticulocyte lysates, or in crude ribosome preparations, HCI decreases the formation of 40S ribosomal complexes (17–19, 22) and this effect is dependent on the presence of ATP (25). These studies implicated a phosphorylation reaction and, together with the relief of inhibition by eIF-2, suggested phosphorylation of this factor as the cause of inhibition of chain initiation. Consistent with this suggestion, highly purified preparations of HCI are found to be associated with a cAMP-independent protein kinase activity that catalyzes the phosphorylation of the small subunit of eIF-2 by ATP (26–29). Thus, HCI is an eIF-2 kinase. eIF-2 consists of three polypeptide chains with molecular weights of $\sim$55,000, 50,000, and 38,000 (30, 31). Proof for phosphorylation of the 38,000 dalton subunit has been obtained by incubation of eIF-2 and HCI with [γ-^{32}P]ATP followed by sodium dodecyl sulfate-gel electrophoresis and autoradiography. In all cases, translational inhibitory activity purified together with the activity phosphorylating the 38,000 dalton subunit of eIF-2. Furthermore, preparations of proinhibitor that have no inhibitory activity in hemin-containing reticulocyte lysates have no phosphorylating activity for the small subunit of eIF-2 (29). eIF-2 is isolated from reticulocyte lysates in a nonphosphorylated form (32). Irreversible HCI has been purified in several laboratories from the postribosomal supernatant of rabbit reticulocyte lysates following activation by warming for several hours at 34–37° or brief exposure to N-ethylmaleimide (26, 28, 33, 34). Sodium dodecyl sulfate-gel electrophoresis of highly purified HCI has revealed the presence of two or three main protein bands with molecular weights of $\sim$86,000, 91,000, and 96,000 (26). The molecular weight of highly purified irreversible HCI determined by gel filtration is $\sim$140,000 (34). On the other hand, crude proinhibitor and the (presumably) irreversible inhibitor derived from it by incubation for several hours at 34° have a molecular weight by gel filtration of $\sim$300,000 (9). Recently, two species of reversible inhibitor have been extensively purified, apparently making use of the stabilizing effect of dithiothreitol (12) or β-mercaptoethanol (35). One of these preparations, of $\sim$2,000-fold purification, shows a single band on polyacrylamide gel electrophoresis under nondenaturing conditions, and a single band (mol wt $\sim$95,000) on sodium dodecyl sulfate-gel electrophoresis (35). The apparent molecular weight from glycerol gradient centrifugation is $\sim$140,000 (35), similar to that obtained by gel filtration for irreversible HCI (34).

Phosphorylation of the 38,000 dalton subunit of eIF2, on incubation with HCI and [γ-^{32}P]ATP, is accompanied by phosphorylation of the 96,000 dalton protein band of HCI (26). Incubation of highly purified reversible HCI with [γ-^{32}P]ATP, in the absence of eIF-2, also leads to phosphorylation of this band (12), [given in (12) as 100,000, in (35) as 95,000 daltons]. The activation of HCI, by warming for a few hours at 34° or by brief incubation with N-ethylmaleimide does not require ATP (9). However, if conducted in the presence of [γ-^{32}P]ATP, activation is followed by rapid phosphorylation of the 96,000 dalton polypeptide (complete in 50 sec at

34°) that has the features of a self-phosphorylation reaction (12). Blocking the activation of HCI with hemin blocks the phosphorylation of the 96,000 dalton band of HCI as well as that of the 38,000 dalton subunit of eIF-2 (12). Hemin can also block the activity of irreversible HCI (36, 37), but the activity of the enzyme appears to be less sensitive to hemin than its activation. Whether HCI activity is inhibited or not seems to depend on the hemin concentration and, since hemin binds nonspecifically to a number of proteins (see below), its effective concentration will depend on the protein content of the samples (9). Thus, 30-μM hemin has been reported to inhibit phosphorylation of the 38,000 dalton subunit of eIF-2, as catalyzed by small amounts of partially purified irreversible HCI (37); but, as judged by the inhibition of protein synthesis, the same concentration of hemin does not prevent HCI from phosphorylating eIF-2 in reticylocyte lysates (18) which contain rather large amounts of protein.

Translational inhibition in reticulocyte lysates can also be prevented or relieved by high levels of GTP (25, 38) as well as cAMP, 2-aminopurine, and other purine derivatives (39). GTP appears to block HCI activation (25), possibly by competing with ATP for a site on cAMP-dependent protein kinase (see next section); high concentrations of cAMP (40) and 2-aminopurine (26) inhibit protein kinases.

The fact that HCI preparations, capable of inhibiting translation in hemin-containing reticulocyte lysates, contain a cAMP-independent protein kinase (eIF-2 kinase) that phosphorylates the 38,000 dalton subunit of eIF-2 strongly suggests that translational inhibition by HCI is a consequence of this modification of the initiation factor. However, although addition of HCI and ATP to hemin-containing reticulocyte lysates leads to a rapid inhibition of translation, similar treatment of purified eIF-2 does not appear to affect its ability to form ternary or 40S initiation complexes (26, 41–43). On the other hand, using partially purified preparations of eIF-2, addition of HCI and ATP leads to inhibition of both ternary and 40S complex formation (41, 42), which suggests that an additional factor is required for inhibition of translation.

Table 1 lists the known groups of protein kinases [(44) see also the chapter by E. G. Krebs in this volume] although it does not include some kinases that are activated by cGMP and others that may by activated by dsRNA. It may be noted that eIF-2 kinase (HCI) belongs with phosphosphorylase kinase to a group of protein kinases that are highly substrate-specific.

Mechanism of HCI Activation

As already mentioned, HCI can be activated or, in other words, proinhibitor can be converted to inhibitor: (a) slowly by incubation at 34–37° for

Table 1 Protein kinases; ATP (GTP) + protein → ADP (GDP) + phosphoprotein)

cAMP-independent	
Type A	Type B
phosphate donor: ATP or GTP	phosphate donor: ATP
substrates: acidic proteins (casein, phosvitin), certain subunits of some initiation factors	substrates: phosphorylase (phosphorylase kinase), 38K subunit of eIF-2 (eIF-2 kinase)

cAMP-dependent (R_2C_2[a] + cAMP $\rightleftharpoons$ R_2cAMP$_2$ + 2 C)

phosphate donor: ATP
substrates: basic proteins (protamines, histones, ribosomal proteins), enzymes (glycogen synthetase, hormone-sensitive lipase, phosphorylase kinase, eIF-2 kinase)[b]

[a] R = protein kinase regulatory subunit; C = catalytic subunit.
[b] Some other enzymes, including phosphofructokinase, pyruvate kinase, hydroxymethylglutaryl CoA reductase, acetyl CoA carboxylase, glycerophosphate acyltransferase, and cholesterol ester hydrolase, may have to be added to this list.

several hours, or (*b*) rapidly by brief incubation (5 min at 34°) with 5 mM N-ethylmaleimide. The former, but not the latter, can be prevented by hemin. After short incubation without N-ethylmaleimide the inhibitor formed can be inactivated by the addition of hemin (reversible inhibitor). Longer incubation without or short incubation with N-ethylmaleimide leads to the formation of inhibitor that is less susceptible to inactivation by hemin (irreversible inhibitor). The mechanism of these activations is unknown. In any case, the slow activation observed on incubation of proinhibitor, in the absence of hemin, is not likely to be related to the physiological mechanism of activation which, under the conditions of protein synthesis in reticulocyte lysates, occurs in a few minutes if hemin is not added.

Protein synthesis in hemin-containing reticulocyte lysates is inhibited not only by HCI but also by cAMP-dependent protein kinases, such as bovine heart protein kinase, and even better by their catalytic subunit (42). Using a ternary complex formation assay with partially purified eIF-2, HCI inhibits complex formation and this inhibition has a stringent requirement for ATP. On the other hand, cAMP-dependent kinases, such as BHK or partially purified cAMP-dependent protein kinase from rabbit reticulocytes, or BHK catalytic subunit, have no effect on this reaction, which suggests that inhibition of protein synthesis in lysates by cAMP-dependent kinases is indirect (42). In the control of glycogen metabolism in animal tissues, inactive phosphorylase is activated by phosphorylation catalyzed by phosphorylase kinase (44). This is a cAMP-independent protein kinase that exists in an inactive, nonphosphorylated form and an active phosphorylated form. Phosphorylation of the inactive enzyme is catalyzed by cAMP-dependent protein kinase in the presence of cAMP. The possibility that a similar mechanism activates eIF-2 kinase is consistent with the observation that ternary complex formation with partially purified preparations of eIF-2, in the presence of crude reticulocyte proinhibitor, is inhibited by incubation for a few minutes at 30° with ATP and BHK catalytic subunit, but not by either alone (42). Significant inhibition of ternary complex formation has been obtained by addition of cAMP to samples containing proinhibitor, partially purified eIF-2, and ATP (45). As previously mentioned, crude reticulocyte proinhibitor preparations contain a very active cAMP-dependent protein kinase but virtually no cAMP (45). cAMP is active at physiological concentrations (46). In the presence of the cAMP phosphodiesterase inhibitor, 3-isobutyl-l-methylxanthine, 0.05 μM cAMP causes 65% conversion of proinhibitor to HCI; 0.02 μM cAMP produces 50% conversion (45). These results are consistent with the view that cAMP-dependent protein kinase is involved in the activation of HCI, in analogy to the activation of phosphorylase kinase, and suggest that cAMP may control protein synthesis.

Inhibition of protein synthesis in rabbit liver slices by glucagon or cAMP was reported in 1960 (47). More recently cAMP has been shown to inhibit protein synthesis (50–60%) in rat liver preparations including slices, hepatocytes, and cell-free systems (48–51), and the data are consistent with the involvement of cAMP-dependent protein kinase and HCI-like inhibitor. Incubation of rat liver postmitochondrial supernatant with cAMP and ATP, but not with either alone, results in a time-dependent loss of protein synthesis activity in the isolated microsomal fraction (49). Moreover, the microsomal fraction itself (which can synthesize protein without the addition of postribosomal supernatant) is not susceptible to cAMP unless supplemented with the supernatant (48, 49). These observations are consistent with the involvement of a phosphorylation reaction (requirement of ATP for the cAMP effect) and with the fact that, in rat liver, HCI and cAMP-dependent protein kinase are largely supernatant enzymes. Protein synthesis is inhibited in cell-free systems from livers of rats injected with dibutyryl cAMP (52). The occurrence of polysome disaggregation suggests inhibition of chain initiation. A production of fatty livers by glucagon has been attributed to decreased synthesis of the protein moiety of the lipoproteins involved in lipid transport (53).

Some workers failed to observe an effect of cAMP on either protein synthesis in intact cells or HCI activation. In one instance, a marked elevation of intracellular cAMP, caused by a catecholamine (isoproterenol) in cultured rat glioma cells or by prostaglandin E_1 in mouse lymphocytes, had no effect on the rate of protein synthesis (54). Results of this kind should be interpreted with caution for it is conceivable that some cells may bring into play compensatory mechanisms, e.g. increased protein phosphatase activity, to offset the effects of an increased cAMP concentration on protein synthesis. In another instance, the slow rate of activation of HCI upon incubation at 34° in the presence of ATP was not changed by cAMP (12). However, cAMP would not have any effect if the HCI preparation contained no cAMP-dependent protein kinase or consisted largely of "reversible" inhibitor rather than proinhibitor (see Equation 1). The so-called reversible HCI is rendered inactive by hemin and does not inhibit translation in a hemin-containing reticulocyte lysate.

$$\text{inactive HCI (proinhibitor)} \underset{-P}{\overset{+P}{\rightleftharpoons}} \text{active HCI (inhibitor)}$$
$$\searrow \qquad \swarrow$$
$$\text{"reversible" HCI} \tag{1.}$$
$$\downarrow$$
$$\text{irreversible HCI}$$

A model for translational regulation by hemin, based on studies with reversible inhibitor, assumes that the proinhibitor is simply an inactive hemin-inhibitor complex from which active inhibitor is formed by removal of hemin as in Equation 2 (35):

$$\text{hemin-inhibitor} \xrightarrow[\text{plus hemin}]{\text{minus hemin}} \text{inhibitor} \xrightarrow{\substack{\text{prolonged incubation} \\ \text{or N ethylmaleimide}}}$$
$$\text{irreversible inhibitor} \tag{2.}$$

but this is open to question.

Since HCI is probably very similar to phosphorylase kinase as regards its activation mechanism, one would expect some structural similarity between the two proteins. For this reason it may be helpful to summarize here the salient features of the latter enzyme (55, 56). Phosphorylase kinase from rabbit skeletal muscle contains three types of subunits, *A, B,* and *C.* When activated by incubation with cAMP-dependent protein kinase and ATP, subunit *B* is phosphorylated faster than subunit *A* and its phosphorylation results in full activation of the enzyme; subunit *C* is not phosphorylated. Phosphorylase kinase can also be activated by proteolysis (56, 57), e.g. by trypsin or by proteases that contaminate the purified enzyme. Activation by proteolysis results from limited cleavage of subunit *A* with little change in molecular weight but the enzyme dissociates to lower molecular weight species in the presence of ATP. More extensive proteolysis leads to the disappearance of subunits *A* and *B.* These two subunits may be considered to be the equivalent of regulatory subunits, whereas subunit *C,* somewhat resistant to proteolysis, would correspond to a catalytic subunit. The normal mechanism of activation of phosphorylase kinase is by phosphorylation by cAMP-dependent protein kinase (55–56). However, limited activation of the enzyme may occur as a result of self-phosphorylation (55). This reaction requires higher ATP and Mg^{2+} concentrations than does the cAMP-dependent protein kinase-catalyzed reaction, and the pattern of phosphorylation of the *A* and *B* subunits is different from that obtained with cAMP-dependent kinase; the incorporation of phosphate into subunit *B* is very low in the self-activation reaction. The nonphysiological activation of HCI by warming at 34° might conceivably result from a conformational change leading to alteration or release of the catalytic subunit. This change would be prevented by hemin; in addition hemin would bind to and inactivate the catalytic subunit (reversible inhibitor). Prolonged warming or treatment with N-ethylmaleimide may decrease the affinity of the catalytic subunit for hemin, thus giving rise to irreversible inhibitor.

Mode of Action of Hemin

Under the conditions of protein synthesis in lysates HCI is rapidly activated in the absence but not in the presence of hemin. Thus hemin appears to block the conversion of proinhibitor to inhibitor. The involvement of a cAMP-dependent protein kinase in HCI activation (42, 45) and the finding that hemin inhibits the phosphorylation of histone and ribosomal proteins by cAMP-dependent protein kinase from rabbit reticulocytes (58), suggests that hemin blocks the activation of cAMP-dependent protein kinase by cAMP, i.e. it interferes with the cAMP-mediated dissociation of the enzyme (Equation 3) thus preventing release of the active catalytic subunit.

$$R_2C_2 + 2 \text{ cAMP} \underset{\text{hemin}}{\rightleftharpoons} 2 \text{ R-cAMP} + 2 \text{ C} \qquad\qquad 3.$$

This hypothesis is supported by the observation that hemin inhibits the phosphorylation of histone by the cAMP-dependent BHK but has much less effect on the phosphorylation by the catalytic subunit (45). Consistent with these observations, the formation of inhibitor elicited by addition of cAMP to proinhibitor preparations from reticulocytes is blocked by hemin at concentrations optimal for maintenance of protein synthesis in lysates, whereas the formation of inhibitor promoted by BHK catalytic subunit is not affected (45). Moreover, hemin inhibits the binding of [³H]cAMP to cAMP-dependent protein kinase holoenzyme (e.g. BHK or cAMP-dependent protein kinase from rabbit reticulocytes) or to BHK regulatory subunit (45). [³H]hemin binds to BHK or its regulatory subunit with high affinity, but not to its catalytic subunit (59). The dissociation constant of the reaction, protein kinase + hemin ⇌ protein kinase-hemin complex, is about

10^{-6} M. Labeled hemin binds nonspecifically to a number of proteins, but its binding to BHK is specific. At saturation, approximately 2 moles of hemin or of cAMP are bound per mole of BHK (R_2C_2) or 1 mole per regulatory subunit. Labeled cAMP bound to BHK can be displaced by an excess of either hemin or nonlabeled cAMP. On the other hand, bound labeled hemin is displaced by nonlabeled hemin but not by cAMP. This result has been interpreted to mean that hemin binds to a site on the regulatory subunit other than that to which cAMP binds and affects the binding of cAMP through an allosterically induced conformational change (59).

Figure 1 is a schematic representation of the activation of HCI (eIF-2 kinase) by catalytic subunit (C), released through the cAMP-promoted, hemin-sensitive dissociation of the protein kinase holoenzyme (top left) and the ensuing phosphorylation of eIF-2. Also shown in Figure 1 is the hypothetical reversal of the phosphorylation of eIF-2 kinase and eIF-2 by specific protein phosphatases. This kind of reversal is well established for phosphorylase kinase and phosphorylase (60–62). A protein phosphatase of this kind could be a factor in the postribosomal supernatant of reticulocyte lysates that neutralizes or reverses the effect of HCI (63, 64). Phosphoprotein phosphatases that remove phosphoryl groups from ribosomal proteins have been partially purified from rabbit reticulocyte lysates (65, 66). The presence of a phosphatase activity that dephosphorylates a 67,000 dalton polypeptide as well as eIF-2(P) in interferon-treated mouse L cells has been reported (67, see also 26).

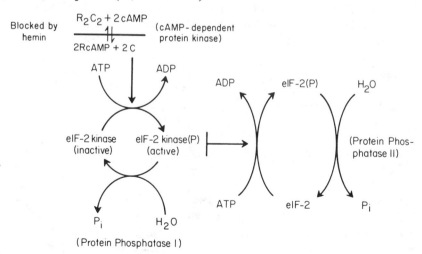

Figure 1   A model for the regulation of eIF-2 phosphorylation in reticulocytes.

Mode of Action of HCI.

Although the sharp decline in the rate of protein synthesis, occurring a few minutes after the start of incubation of hemin-containing reticulocyte lysates with HCI and ATP, would seem to be the result of the phosphorylation of eIF-2, treatment of purified eIF-2 with HCI and ATP does not interfere with its ability to form ternary or 40S initiation complexes (41–43). On the other hand, formation of the complex is impaired if partially purified eIF-2 is used (26, 41, 42). This suggests that inhibition of translation involves another factor(s) present in lysates and in partially purified preparations of eIF-2 (68). Salt washes of *Artemia salina* embryos and rabbit reticulocyte ribosomes contain, along with eIF-2, a factor required for inhibition of ternary complex formation when purified eIF-2 is first incubated with HCI and ATP (68, 69). This factor, called eIF-2-stimulating protein (ESP), enhances the ability of eIF-2 to form ternary or 40S initiation complexes. ESP stimulates the activity of unmodified eIF-2 but it has no effect on phosphorylated eIF-2 (68, 69). The enhancement of eIF-2 activity by ESP is stronger the lower the concentration of eIF-2, and at the low concentrations prevailing in reticulocyte lysates (68) eIF-2 is virtually inactive in the absence of ESP (68, 69). Thus, HCI inhibits chain initiation by abolishing the effect of ESP on eIF-2. Since preincubation of ESP with HCI and ATP, followed by removal of ATP with charcoal and ternary complex formation assay, does not prevent ESP from stimulating eIF-2 activity, whereas similar treatment of eIF-2 does (68), it must be concluded that inhibition of ternary complex formation, when samples containing both eIF-2 and ESP are preincubated with HCI and ATP, is the consequence of a modification of eIF-2 and not of ESP. ESP probably complexes with eIF-2 and displaces the equilibrium towards ternary complex and, therefore, 40S complex formation. At high concentrations of eIF-2, such as are used for assay of the factor, ESP is not required for ternary complex formation and, since this formation is not affected by phosphorylation, the addition of large amounts of eIF-2 to lysates inhibited by the presence of HCI and ATP relieves the inhibition. This explains the reversal by eIF-2 (whether phosphorylated or not) of inhibition by hemin deprivation, reported from several laboratorie (18, 20–22).

ESP has been partially purified (68, 69) and its molecular weight, as determined by gel filtration, is in the neighborhood of 200,000. Like eIF-2, ESP is sensitive to SH-binding reagents such as N-ethylmaleimide or *p*-chloromercuribenzene sulfonate. ESP increases both the extent and the rate of ternary complex formation. This effect is analogous to that of a factor isolated earlier from rabbit reticulocyte ribosomal wash (70). The presence of a similar factor in Ehrlich ascites cells has been reported. As with reticulocyte ESP, the stimulation of complex formation by the ascites cell factor is inhibited by reticulocyte or ascites cell HCI (71). ESP from *Artemia salina* embryos has the same properties and the same molecular weight as reticulocyte ESP. *A. salina* ESP interacts with reticulocyte eIF-2 and the same is true of reticulocyte ESP and *A. salina* eIF-2 (69, 72).

As already stated, phosphorylation of eIF-2 does not alter its capacity to form either a ternary or a 40S initiation complex. This contrasts with reports that, whereas ternary complex formation is not affected, the formation of a 40S complex is decreased (22, 73). These results can be explained by the fact that ternary complex formation was assayed (73) with Met-$tRNA_i$, GTP, and purified eIF-2 (or eIF-2(P)) as the only components, but 40S complex formation was assayed in the presence of an additional ribosomal salt wash fraction that probably contained ESP.

The fact that in reticulocyte lysates, upon addition of HCI and ATP, translation proceeds at normal rates for a few minutes and then declines abruptly, suggests that under these conditions the eIF-2 present is used only once, i.e. stoichiometrically rather than catalytically. This has been interpreted to mean that phosphorylation of eIF-2 prevents recycling of the factor (74). The finding that ESP is essential for eIF-2 function at physiological concentrations of eIF-2 and that it has no activity on eIF-2(P) (68, 69), makes the recycling hypothesis unnecessary. Furthermore, eIF-2(P) can recycle, for, on formation of the 80S initiation complex, it leaves the ribosome as readily as does the nonphosphorylated factor (75). However, since initiation complex formation requires higher concentrations of eIF-2 when the factor is phosphorylated, phosphorylation converts eIF-2 into a less efficient catalyst. There is evidence that formation of the ternary complex is preceded by that of the binary complex eIF-2·GTP and that ESP acts at the level of binary complex formation (C. de Haro, and S. Ochoa, unpublished results). This suggests that ESP may act by binding to the binary complex displacing the equilibrium toward increased binding of GTP by e IF-2 (Equation 4, *a* and *b*). ESP would then be released by Met-$tRNA_i$ to form the ternary complex proper (Equation 4, *c*).

(*a*) eIF-2 + GTP$\rightleftharpoons$eIF-2·GTP

(*b*) eIF-2·GTP + ESP$\rightleftharpoons$eIF-2·GTP·ESP   4·

(*c*) eIF-2·GTP·ESP + Met-$tRNA_i$$\rightleftharpoons$eIF-2·GTP·Met-$tRNA_i$ + ESP

Other Translational Inhibitors

Translational inhibitors are produced by treatment of hemin-containing reticulocyte lysates with small amounts of dsRNA (26, 37, 76–81) or oxidized glutathione (82–84). The inhibition of translation by HCI, dsRNA, and oxidized glutathione has several features in common (3): (*a*) the initial rate of protein synthesis remains normal for a few minutes and then declines sharply, (b) at the time of shutoff, polysomes disaggregate as nascent polypeptides are completed and released from the ribosomes, (c) the number of native 40S subunits carrying Met-tRNA$_i$ is markedly reduced, and (d) inhibition is relieved or prevented by high concentrations of eIF-2 and also by high levels of cAMP and 2-aminopurine. These observations suggest that the accumulation of a translational inhibitor that interferes with eIF-2 function is a common feature of all three inhibitions. Warming of hemin-supplemented reticulocyte lysates to 42–45° activates an HCI-like inhibitor of chain initiation (85–88).

There have been reports that hemin enhances or maintains protein synthesis in nonerythroid cells such as Krebs II ascites cells (89), HeLa cells (90), platelets (91), and hepatoma cells (92), and HCI-like inhibitors have been isolated from Ehrlich ascites tumor cells (93), Friend leukemia cells (94), and rat liver (95). Like HCI these inhibitors are present in the postribosomal supernatant from which they have been partially purified. They all inhibit protein synthesis in hemin-containing reticulocyte lysates with kinetics similar to HCI, and the inhibition is prevented or reversed by eIF-2. The Friend leukemia cell inhibitor (FLI) and the rat liver inhibitor (RLI) are associated with a cAMP-independent protein kinase activity catalyzing the phosphorylation of the 38,000 dalton subunit of eIF-2. The molecular weight of FLI is about 200,000 by gel filtration. Judging from the fact that the inhibitor activity of Friend leukemia cell supernatants is not increased by incubation for 16 hr at 37° (the effect of N-ethylmaleimide was not tested) and from the small stimulation of protein synthesis by hemin in Friend leukemia cell lysates (94), FLI seems to be present in these lysates in an active form rather than in the form of a proinhibitor. It would be interesting to see if, as in the case of A. salina and wheat germ extracts (see below), the synthesis of protein in Friend leukemia cell extracts is inhibited by the catalytic subunit of cAMP-dependent protein kinase. The fact that protein synthesis in rat liver extracts is inhibited by cAMP (48–51) suggests, on the other hand, that RLI is present as a proinhibitor. It may be predicted that translation in rat liver extracts would be strongly inhibited by addition of the catalytic subunit of cAMP-dependent protein kinase. A proinhibitor-inhibitor system similar to the one in reticulocytes appears to be present in organisms evolutionarily as far removed from mammals as A. salina and wheat germ (96). Protein synthesis in cell-free A. salina systems is strongly inhibited by BHK catalytic subunit, and the same is true of wheat germ extracts, which suggests that a proinhibitor is present in these preparations analogous to the HCI precursor. In A. salina systems, the inhibition of translation by protein kinase catalytic subunit is due to inhibition of initiation, for the subunit has no effect on chain elongation or on the translation of poly(U), which requires higher Mg^{2+} concentrations than translation of natural messengers, and is independent of initiation factors. *A. salina* extracts contain both cAMP-independent and cAMP-dependent protein kinase activities, and proinhibitor preparations made from *A. salina* extracts by chromatography on CM-Sephadex show inhibition of ternary complex formation upon addition of either cAMP or BHK catalytic subunit. As with the reticulocyte proinhibitor the effect of cAMP, but not that of the catalytic subunit, is blocked by hemin. Wheat germ extracts have only cyclic nucleotide–independent protein kinase activity, and proinhibitor preparations obtained by the CM-Sephadex procedure are activated (giving rise to inhibition of ternary complex formation) by BHK catalytic subunit and ATP, but cAMP has no effect.

Physiological Significance

A significant decrease of protein synthesis has been observed following the injection of BHK catalytic subunit into *Xenopus laevis* oocytes (F. Cavalieri, S. Pestka, A. Datta, and S. Ochoa, unpublished results). This observation may be related to the inhibition of progesterone-induced stimulation of meiotic cell division in *Xenopus* oocytes following the microinjection of cAMP-dependent protein kinase catalytic subunit and the direct induction of meiotic division, in the absence of progesterone, after microinjection of regulatory subunit (97). Induction of meiotic cell division is also produced by microinjection of the heat-stable protein inhibitor of cAMP-dependent protein kinase, which binds, and thereby inactivates the catalytic subunit (97). These responses are probably mediated by eIF-2 kinase. Consistent with these results, the maturation of *Xenopus* oocytes induced by various agents (human corionic gonadotropin, progesterone, testosterone, lanthanum ions) is delayed by inhibitors of cAMP-phosphodiesterase (theophylline, papaverine) and by cAMP (98). Phosphodiesterase inhibitors decrease the activity of oocyte phosphodiesterase and they, as well as cAMP, cause significant inhibition of oocyte protein synthesis in vivo. It has been postulated that decreased protein synthesis limits the availability of a protein factor or factors that promote oocyte maturation (98). The observed inverse relationship between cell growth and cAMP levels (99–101) may be explained in terms of the identical relationship between protein synthesis and cAMP levels. cAMP and prostaglandin E_1 (which raises the intracellular level of cAMP) inhibit the growth of cultured mouse lymphoma S49 cells but are much less effective on a variant that has lower levels of cAMP-dependent protein kinase (102). Thus the inhibition by cAMP of both protein synthesis and cell growth is mediated by cAMP-dependent protein kinase. Inhibition of cell growth is probably a consequence of protein synthesis inhibition.

The control of HCI activation by hemin provides a purposeful mechanism of regulation of globin synthesis that ensures that globin is synthesized only when heme, its prosthetic group, is available. However, hemin also appears to inhibit eIF-2 kinase activation in systems that, unlike reticulocyte lysates, can synthesize protein in the absence of added hemin. This is the case in the *A. salina* (96) and Friend leukemia (94) cell-free systems. Hemoglobin is the respiratory pigment in *A. salina* but it is not known whether hemin blocks eIF-2 kinase activation in organisms that are devoid of hemolobin. Why hemin is required for protein synthesis in reticulocytes but not in other cell-free systems is not clear. Through the development of a cAMP-dependent kinase that is highly sensitive to cAMP and hemin, the reticylocyte may have evolved as a cell adjusted to synthesize globin only as long as hemoglobin can be assembled, i.e. when heme is present. Since cAMP cannot displace protein kinase–bound hemin, it cannot be responsible for reversal of the translational effect of the porphyrin. However, globin can completely reverse this effect (59). Globin has a high affinity for hemin and effectively competes with the regulatory subunit for hemin bindings. Thus the reticulocyte synthesizes globin as long as hemin is available and synthesis stops when enough globin has been produced to bind all of the hemin present.

TRANSLATIONAL INHIBITION IN INTERFERON-TREATED CELLS

Interferons (103) are glycoproteins synthesized by various animal cells as a result of viral infection or of treatment with interferon inducers. The best inducer is dsRNA, e.g. reovirus RNA and poly(I)-poly(C). Although dsRNA is formed during the replication of some viruses it is not known whether it mediates the viral induction of interferon. Interferons are released from the cells in which they are produced and act on other cells, presumably after binding to specific receptors on the cell surface that recognize only those interferons that are produced by cells of the same species. Cells exposed to interferon fail to support the multiplication of a wide range of DNA and RNA viruses (103).

Induction of Translational Inhibitors

Protein synthesis in extracts of cells, e.g. mouse L cells and ascites cells, that are treated with species-specific interferon, is markedly depressed by the addition of small amounts of dsRNA (104–107). The translation of host and viral messengers is inhibited to a similar extent (105, 108, 109). Translation is also inhibited when dsRNA is added to a cell-free system from control cells (that contains added ATP) supplemented with small amounts of a dialyzed 30S supernatant (sap) from interferon-treated cells (110). The addition of dsRNA and ATP to interferon cell sap, but not control cell sap, activates a protein kinase(s) that phosphorylates a 64,000–67,000 dalton polypeptide, and to a lesser extent a 35,000–38,000 dalton polypeptide, present in the sap (106, 107, 111). This coincides with the appearance of a protein kinase activity that phosphorylates histone which indicates that the phosphorylation of the sap proteins promoted by dsRNA is due to activation of a protein kinase (111–113). These findings suggest that interferon induces the synthesis of a protein kinase(s) that, in the presence of ATP and upon activation by dsRNA, phosphorylates various proteins which are thereby activated. One of these proteins is a protein kinase (eIF-2 kinase II in Figure 2) which, like HCI (eIF-2 kinase I in Figure 2), phosphorylates the 38,000 dalton subunit of eIF-2 (107, 111–116). Since the effects

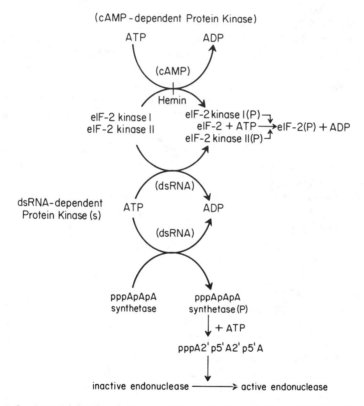

Figure 2 A model for the mechanism of translational control by cAMP-dependent and dsRNA-dependent protein kinases. In reticulocytes the enzymes are constitutive but in other cells some of them are inducible by interferon. Possibly the induced enzyme(s) is (are) the dsRNA-dependent protein kinase(s).

of interferon are abolished when cellular mRNA synthesis is inhibited by actinomycin D, or cellular protein synthesis by cycloheximide (115), it is assumed that interferon does not activate a preexisting enzyme but induces the synthesis of a protein kinase(s) that is (are) activated by dsRNA. Moreover, addition of dsRNA and ATP to extracts of interferon-treated cells stimulates an endonuclease that degrades single-stranded RNA, e.g. reovirus mRNA and R17 phage RNA (117–120). Once the nuclease is activated the dsRNA and ATP can be removed without impairing its activity (120).

Interferon treatment has also been found to decrease methylation of the reovirus RNA cap (121, 122). Other reports suggest that decreased translational activity in interferon-treated cells may be due to deficiency of one or more tRNAs whose addition can restore translation to normal levels (123–130).

A further effect of incubating interferon, but not control cell sap with dsRNA and ATP, is the formation of a heat-stable low-molecular-weight compound that at subnanomolar concentrations inhibits translation in hemin-containing reticulocyte lysates or in L cell systems (131, 132). The enzyme responsible for the synthesis of the low-molecular-weight inhibitor and the induced protein kinase can be bound to a dsRNA-Sepharose column and low-molecular-weight inhibitor can be isolated by elution with ATP-containing buffer. The low-molecular-weight inhibitor synthetase and the kinase can also be eluted from the column with considerable purification. Inhibitor has been prepared with ATP labeled with ^{3}H or with ^{32}P in the α or in the γ positions. Low-molecular-weight inhibitor is sensitive to treatment with bacterial alkaline phosphatase, snake venom phosphodiesterase, and alkali, but insensitive to a number of nucleases, including P1, T1, T2, U2, and pancreatic, spleen, and micrococcal nucleases (132). Using a combination of enzyme treatments, chromatography, and periodate treatment followed by β elimination, the inhibitor has been identified as pppA2'p5'A2'p5'A (132, 133). This structure is further supported by NMR data and chemical synthesis [cited in (134)]. The trinucleotide is the product formed in largest amounts, but smaller amounts of tetramers and pentamers, which are also inhibitory, are obtained. The dinucleotide obtained by periodate treatment and β elimination is inactive (133). Since, as seen above, dsRNA activates a protein kinase, it is possible that pppApApA synthetase is activated by this kinase as shown in Figure 2. Alternatively, the synthetase itself might be activated by dsRNA.

Mode of Action of Inhibitors.

Protein synthesis in reticulocyte lysates is inhibited by small concentrations of dsRNA or oxidized glutathione under conditions (e.g. presence of hemin) that exclude HCI activation. Translational inhibitors are activated when dsRNA and ATP are added to reticulocyte lysates in the presence of hemin. One of these inhibitors, dsRNA-activated inhibitor (DAI) is, like HCI, a cAMP-independent protein kinase that phosphorylates the 38,000 dalton subunit of eIF-2 (26, 37). DAI is activated in the presence of ATP and low levels of dsRNA (10–50 ng/ml), but high concentrations of dsRNA (20–50 μg/ml) are inactive. Unlike HCI, which is present in the postribosomal supernatant, DAI is ribosome-associated. Crude ribosomes, prepared by filtration of rabbit reticulocyte lysates through Sepharose 6B (6B ribosomes), form active DAI on incubation with low levels of dsRNA and ATP, but not with either alone or if AMPPCP is substituted for ATP. DAI is assayed through its inhibitory effect on translation in hemin-containing reticulocyte lysates, in the presence of high concentrations of dsRNA, which prevent further activation (26). Inhibition of translation by DAI, like that due to HCI, is relieved by eIF-2 but by no other initiation factors. DAI activated on 6B ribosomes is associated with the ribosome pellet upon centrifugation of the reaction mixture but it can be extracted from the pellet by washing with 0.5 M KCl. However, the dsRNA used for activation appears to remain ribosome-bound because the salt-treated ribosomes inhibit protein synthesis in the absence, but not in the presence, of high concentrations of dsRNA (37) On incubation with low levels of dsRNA and [γ-^{32}P]ATP, 6B ribosomes phosphorylate a 67,000 dalton protein in addition to the 38,000 dalton component of eIF-2 (26, 37). However, in contrast to the 38,000 dalton polypeptide, no phosphorylation of a 67,000 dalton protein is apparent in incubations with preactivated (with unlabeled ATP) DAI in the presence of high levels of dsRNA (26). These findings suggest that phosphorylation of the 67,000 dalton protein is associated with DAI activation rather than with DAI activity (26). The above observations are compatible with a model for activation of DAI (eIF-2 kinase II, Figure 2) similar to that proposed for HCI activation. It would involve phosphorylation of inactive DAI (a ribosome-associated, cAMP-independent protein kinase) by a dsRNA rather than a cAMP-dependent protein kinase. Another possibility is that there is only one kinase involved (eIF-2 kinase II) that is activated by dsRNA and ATP, to form eIF-2 kinase II (P), through a self-phosphorylation reaction. The interferon-induced, dsRNA-ATP-activated protein kinases from Ehrlich ascites tumor cells and mouse fibroblasts (L929 cells) have been partially purified (112, 113). For activation dsRNA cannot be replaced by single-stranded RNA, DNA, or RNA-DNA hybrids. Moreover, dsRNA broken down to small fragments by digestion with RNase III is inactive. In the crude state both kinases phosphorylate histone as well as a 65,000–67,000 dalton polypeptide and the 38,000 dalton subunit of eIF-2. The ribosomal salt wash, used as the starting material for purification of the Ehrlich ascites cell kinase, has considerable phosphohistone phosphatase activity, but the approximately

1,000-fold purified enzyme (112) is virtually free of phosphatase; it phosphorylates histone and the 38,000 dalton subunit of eIF-2 but shows very low endogenous phosphorylation. Once activated by dsRNA and ATP, the dsRNA can be degraded with RNase III without affecting the activity of the enzyme.

Low-molecular-weight inhibitor identical to that synthesized by interferon-treated cells is also formed in reticulocyte lysates (135). It may be that reticulocytes contain enzymes in a constitutive form which in other cells are induced by interferon. Interferon appears to induce the synthesis of one or more enzymes which, when activated by dsRNA and ATP, inhibit translation in various ways.

Inhibition of translation by the low-molecular-weight inhibitor begins after a lag of a few minutes. This, together with the low concentrations at which the compound is active, suggest that the low-molecular-weight inhibitor does not inhibit translation directly. Examination of polypeptide patterns obtained upon translation of globin or encephalomyocarditis virus RNA messenger, in the absence or presence of low-molecular-weight inhibitor or HCI, indicates that the effect of the oligonucleotide is different from that of HCI (135). Recently it has been proposed that the low-molecular-weight inhibitor activates a nuclease that prevents mRNA from being utilized for protein synthesis (136). This is based on the observation that, in the presence of inhibitors of chain elongation, the oligonucleotide promotes partial degradation of the polysomes while it does not affect the association of [^{35}S]Met-tRNA$_i$ with initiation complexes. Moreover, RNA extracted from control lysates and from lysates inhibited with low-molecular-weight inhibitor, and assayed for active mRNA content by retranslation in a fresh mRNA-dependent system, show extensive loss of template activity in the latter case. The conclusion that low-molecular-weight inhibitor activates an endonuclease that degrades mRNA is supported by work from other laboratories (116, 120, 137–140). Two partially purified enzyme fractions from interferon-treated Ehrlich ascites tumor cells can degrade R17 RNA in the presence of dsRNA [poly(I)·poly(C)] and ATP (120). Incubation of one of these fractions (DE1) with dsRNA and ATP yields heat-stable low-molecular-weight inhibitor (pppApApA) which, when incubated with the other fraction (DE2), promotes the hydrolysis of [^{32}P]R17 RNA as assayed by polyacrylamide-gel electrophoresis. Hence, DE2 contains inactive nuclease. The pppApApA-synthesizing activity is virtually absent in DE1 from control cells (i.e. those not treated with interferon) but the DE2 fraction has substantial nuclease content equal to about 50% of that of interferon-DE2. These results could mean that interferon-DE1 contains both inactive pppApApA synthetase and dsRNA-dependent protein kinase(s) whereas control (or interferon) DE2 contains inactive endonuclease. In the presence of dsRNA and ATP, the interferon-induced protein kinase activates the pppApApA synthetase and dsRNA-dependent protein kinase(s) whereas control (or interferon) DE2 contains inactive endonuclease. In the presence of dsRNA and ATP, the interferon-induced protein kinase activates the pppApApA synthetase as assumed in Figure 3. The low-molecular-weight inhibitor appeared to have similar effects on the translation of globin or encephalomyocarditis virus RNA (135), but mengo virus RNA has been reported to be degraded to a greater extent than globin mRNA by the pppApApApA-activated endonuclease (139). Interestingly, the activated endonuclease can degrade pppApApA, its own activator (138). In the presence of interferon, dsRNA produced during viral infection activates the interferon-induced kinase(s) and triggers the events leading to the production of translational inhibitors. The translational effects of interferon are multiple, involving (*a*) effects on initiation by interference with the activity of the initiation factor eIF-2, (*b*) effects on translation through degradation of mRNA, and (*c*) effects concerned with deficient methylation of mRNA caps and and/or tRNA imbalance.

Activation of a protein kinase and an endonuclease by dsRNA and ATP has also been observed in extracts of human HeLa cells treated with human interferon (141).

PHOSPHORYLATION OF RIBOSOMAL PROTEINS AND VARIOUS INITIATION FACTORS

Incubation of eukaryotic ribosomes or ribosomal subunits (e.g. from liver, reticulocytes, and yeast) with a variety of protein kinase preparations, leads to phosphorylation of a number of ribosomal proteins, about 10 in the 60S and 2–4 in the 40S subunit (44, 142–148). On the other hand, administration of ^{32}P-labeled phosphate to animals leads to the phosphorylation of only one, at most two, proteins of the 40S subunit (149–152); this phosphorylation is strongly enhanced by the prior administration of glucagon or cAMP (153). The molecular weight of the phosphorylated protein(s) is between 37,500 and 28,000 (see 154). One-dimensional gel electrophoresis shows the presence of radioactivity in only one protein band (153). Two-dimensional gel electrophoresis separates two bands, the labeling of only one of which is increased by glucagon or cAMP (154). The protein whose labeling is increased by cAMP or glucagon has been identified as protein S6 in Wool's nomenclature (153). S6 (given as S13 in some publications) is the only protein that is phosphorylated in vitro on incubation of rabbit reticulocyte ribosomal subunits with cAMP-dependent protein kinase (155, 156). Protein S6, phosphorylated in chicken liver in vivo, turns over rapidly (157).

An acidic protein, apparently analogous to L7/L12 of *E. coli,* is the major 60S protein phosphorylated in mouse ascites cells and hamster fibroblasts (158, 159). Cell suspensions of a transplantable tumor of the islet cells of the hamster pancreas respond to addition of glucagon or cAMP to the medium with an increased secretion of insulin, concomitant with cellular accumulation of cAMP, and this effect is enhanced by cAMP-phosphodiesterase inhibitors (154). One of the many protein bands labeled when the incubations are carried out in the presence of ^{32}P-labeled ATP displays a significant increase in ^{32}P-incorporation in animals first treated with glucagon or cAMP. This corresponds to a 40S ribosomal protein that may be identical to S6 (154). Attempts to determine whether phosphorylation of ribosomal proteins is accompanied by changes in the translational activity of the ribosomes have mostly yielded negative results (160), although perhaps a more exhaustive search is required. However, the glucagon-stimulated phosphorylation of the 40S ribosomal protein in islet tumor cells coincides with increased secretion of insulin and may be related to this physiological effect. An effect on protein secretion by modification of a ribosomal protein could be mediated through attachment of the modified ribosomes to membranes. A marked increase in the incorporation of ^{32}P into protein S6 following induction of cell division has been reported (161). Phosphorylation of ribosomal proteins could conceivably affect the ribosomal binding of specific mRNAs. It has been reported that phosphorylation of a 30,000 dalton polypeptide of the 40S ribosome by a cAMP-independent protein kinase inhibits 40S complex formation (162). Since the 40S kinase activity was a contaminant of HCI, incubations were conducted in stages as follows:

1. Phosphorylation of the 40S subunit by HCI-containing 40S kinase, in the presence of [^{32}P] ATP.
2. Removal of excess ATP with hexokinase and glucose.
3. Assay for 40S complex formation after supplementing the samples with eIF-2, GMPPCP, and other additions.

The conclusion that inhibition of 40S complex formation was due to phosphorylation of the 40S subunit rather than of eIF-2 would be more convincing if the 40S ribosome kinase activity had been free of HCI. It is conceivable, for example, that the HCI-40S kinase fraction may have contained adenylate kinase. If so, ATP would be formed from unlabeled ADP at stage *2* and inhibition would be the result of phosphorylation of eIF-2 by unlabeled ATP at stage *3*.

cAMP-independent protein kinases from rabbit reticulocytes or erythrocytes phosphorylate the middle subunit (49,000–53,000 daltons) of the initiation factor eIF-2 (36, 37, 163–165) and two subunits (110,000–

120,000, 65,000–70,000 daltons) of eIF-3 (163–165). Two other initiation factors are phosphorylated; eIF-5 (the so-called ribosomal joining factor) and eIF-4B (80,000 daltons) (163, 165). The cAMP-independent kinases phosphorylate casein and phosvitin. The 130,000 dalton subunit of eIF-3 is phosphorylated by a cAMP-dependent protein kinase (166). These preparations can also catalyze the phosphorylation of histone. Most of the cAMP-independent kinase activity can use ATP or GTP as phosphate donors. A similar phosphorylation pattern is obtained by incubation of reticulocytes in media containing ^{32}P (165). As in the case of ribosomal protein phosphorylation, the phosphorylation of initiation factors by kinases other than eIF-2 kinase has not been found to be associated with changes in activity of the factors (163). It has been pointed out, however, that phosphorylation of factor eIF-4B could have physiological significance because of recent results suggesting a role for this factor in mRNA selection (165). Evidently further investigation of the possible role of these phosphorylation reactions is needed for it seems unlikely that this modification is without effect and it is attractive to think that phosphorylation of some of the many subunits of the initiation factor eIF-3 may modulate messenger selection.

OTHER FACTORS THAT MAY AFFECT TRANSLATION

The translational inhibition caused in reticulocyte lysates by hemin or glucose deficiency, or by oxidized glutathione, can be reversed or prevented, to a greater or lesser extent, by phosphorylated sugars (167–170), including glucose-6-phosphate, 2-deoxyglucose-6-phosphate, fructose-6-phosphate, and sedoheptulose-7-phosphate (0.05–0.5 mM). Glucose, ribose-5-phosphate, galactose-6-phosphate, and fructose-1,6-diphosphate (0.5–1 mM) are less active. Glycolytic intermediates have no effect. Since some of the active compounds are members of the hexosemonophosphate shunt, it would appear that the effect is mediated through the generation of NADPH which might act by reducing glutathione. As seen in a previous section, oxidized glutathione can promote the formation of an HCl-like translational inhibitor. However, a study of NADPH levels, in the presence and absence of phosphorylated sugars, and the inability to maintain protein synthesis by the addition of NADPH, do not favor such an explanation (169, 170). Phosphorylated sugars also restore translation if added to extracts of cells (e.g. ascites, L, HeLa, and myeloma) when the rate of protein synthesis declines (169). In this case, fructose-1,6-diphosphate is the most active sugar. The phosphorylated sugars appear to affect chain initiation. Although they do not seem to act in the same way as hemin, the increased rate of protein synthesis caused by phosphorylated sugars when the translation rate declines is reminiscent of that produced by hemin under similar conditions. The exact mode of action of the sugar phosphates remains to be elucidated.

Poliovirus mRNA is translated in extracts of both infected and uninfected HeLa cells, but vesicular stomatitis virus (VSV) mRNA, like host cell mRNA, is translated only in extracts from uninfected cells. Translation of VSV mRNA in extracts of infected cells is restored by addition of the chain initiation factor eIF-4B which is apparently inactivated by poliovirus infection (171). Since VSV mRNA, like cell mRNA, bears an m^7 GPPPGN cap at the 5' end whereas poliovirus mRNA does not, it appears that inactivation of eIF-4B provides a selective mechanism for inhibition of translation of capped messengers.

Certain oligoribonucleotides from eukaryotic cell extracts have been implicated in translational control. For example, the activity of a ribosomal salt wash (containing initiation factors) from rabbit reticulocytes, that is lost upon dialysis, is largely restored by the addition of an oligonucleotide fraction from the dialysate (172). The base composition of this oligonucleotide (mol wt ~11,000) is 33% U, 46% A, 14% C, and 7% G. Oligonucleotides isolated from crude initiation factor fractions of chick erythroblasts and embryonic chick muscle have translational effects (173, 174). One of the better characterized compounds (mol wt ~6,500) is rich in pyrimidines, particularly uridylic acid residues (48% U, 12% A, 28% C, and 12% G), and has been reported to inhibit the translation of heterologous, but not homologous, mRNAs (173). This compound is referred to as translational control RNA (tcRNA).

Postribosomal supernatants of eukaryotic cells have long been known to contain low molecular weight inhibitors of translation that are routinely removed by Sephadex G-25 gel filtration. An inhibitor from *A. salina* embryos (175) is very similar to tcRNA (mol wt ~6,000; 47% U, 11% A, 26% C, and 16% G); it is sensitive to alkali and RNase A, but not to RNase T1. This oligonucleotide inhibits the translation of heterologous and homologous mRNAs and is active on both prokaryotic and eukaryotic systems. Its effect is not very specific. The reaction that is most sensitive (50% inhibition with about 2 molecules of oligonucleotide per ribosome) is the EF-1 and GTP-dependent binding of aminoacyl tRNA to 80S ribosomes, but chain initiation reactions are also inhibited. The oligonucleotide probably binds to the ribosome and blocks the binding of aminoacyl tRNA. The level of oligonucleotide inhibitor is about the same in dormant or developing *A. salina* embryos but the latter contain an activator oligonucleotide (mol wt ~9,000; 33% U, 10% A, 6% C, and 51% G) that complexes with the inhibitor and counteracts its effect. This oligonucleotide is sensitive to alkali and RNase T1 but insensitive to RNase A. The inhibitor and activator appear to arise by hydrolysis of embryo RNA by RNase T1 and RNase A, respectively. There is a significant increase of RNase A activity when development of *A. salina* embryos is resumed upon hydration. The *A. salina* activator is similar to the oligonucleotide activator from reticulocytes, in the sense that both are purine-rich, although the reticulocyte compound is rich in A whereas its *A. salina* counterpart is rich in G (175).

A ribosome-associated inhibitor that blocks ribosomal binding of aminoacyl tRNA is present in unfertilized sea urchin eggs (176, 177) and dormant *A. salina* embryos (178); it is released or inactivated after fertilization or hydration (176, 178). In another study, the inhibitor has been found in decreased amounts in sea urchin embryos, particularly in the monoribosome fraction (179). This inhibitor, throught to be a protein, has not been thoroughly characterized. In the case of *A. salina* embryos it is not clear whether there are two different inhibitors of the same reaction or whether the inhibitor believed to be a protein actually contained U-rich oligonucleotide. A translational inhibitor, possibly a low molecular weight oligonucleotide, is present in association with ribosomes during certain phases of development of the aquatic fungus *Blastocladiella emersonii* (180). Postribosomal supernatants of unfertilized sea urchin eggs contain a translational oligonucleotide inhibitor that is similar in molecular weight and base composition to the *A. salina* oligonucleotide inhibitor and tcRNA (S. Lee-Huang and S. Ochoa, unpublished results). However, although this inhibitor may decrease somewhat after fertilization, no counterpart to the *A. salina* activator oligonucleotide has been yet found in fertilized sea urchin eggs. The possibility that inhibitor-activator oligonucleotide pairs may control the transition of cells from a state of dormancy or quiescence to one of development (e.g. emergence from cryptobiosis, fertilization) is interesting but there is no evidence to support it. Since the *A. salina* oligonucleotides appear to be the end products of digestion of RNA by intracelluar endonucleases, their formation could be an artifact of the manipulations involved in the preparation of cell-free extracts.

CONCLUDING REMARKS

We know at this time of two major mechanisms of translational control in eukaryotes. The first controls polypeptide chain initiation by blocking the function of the initiation factor eIF-2 through phosphorylation of one of its three subunits. This phosphorylation is catalyzed by a cAMP-independent protein kinase (eIF-2 kinase). There seem to be at least three kinds of eIF-2 kinases that may be activated in different ways: (*a*) by phosphorylation catalyzed by a cAMP-dependent protein kinase, (*b*) by phosphorylation catalyzed by a dsRNA-activated protein kinase, and (c) by oxidized gla-tathione in an as yet unknown manner. The activation of kinase as in a is blocked by heme. The second mechanism shuts off translation by degrading mRNA. This is operated via activation of an endonuclease by an oligonu-cleotide of unusual structure, pppA2'p5'A2'p5'A. The oligonucleotide is synthesized from ATP by an enzyme that may be activated by a dsRNA-dependent protein kinase.

In most cells the dsRNA-dependent protein kinase(s), and possibly the oligonucleotide synthetase, are induced by interferon but the other enzymes are constitutive. In reticulocytes, all the translational control enzymes ap-pear to be constitutive.

As far as is known, the above mechanism (blocking of eIF-2 function, mRNA degradation) inhibit translation in a nonselective fashion. However, partial inhibition of initiation could affect the translation of different mRNAs to differing extents depending on their ribosomal affinity (181). There are indications that selective translational control may occur in special cases. Thus, poliovirus infection appears to cause inactivation of the initiation factor eIF-4B. This factor is believed to be concerned with recog-nition of mRNAs bearing an $m^7GpppGN$ cap at the 5' end and may be required for translation of capped host mRNAs but not for that of un-capped poliovirus RNA. The next few years should bring not only more detailed knowledge of the mechanisms of regulation of protein synthesis but might also disclose the existence of selective ones for modulation of transla-tion of different messengers. It is also to be hoped that future research may throw more light on the role of phosphorylation of ribosomal proteins and factors other than eIF-2 in translational control.

ACKNOWLEDGMENTS

We wish to thank Drs. B. L. Horecker and H. Weissbach for helpful comments. We are also indebted to the many colleagues who kindly pro-vided us with copies of their manuscripts before publication.

Literature Cited

1. Weissbach, H., Ochoa, S. 1976. *Ann. Rev. Biochem.* 45:191–216
2. Ochoa, S. 1977. *J. Biochem.* 81:1–14
3. London, I. M., Clemens, M. J., Ranu, R. S., Levin, D. H., Cherbas, L. F., Ernst, V. 1976. *Fed. Proc.* 35:2218–22
4. Zucker, W. V., Schulman, H. M. 1968. *Proc. Natl. Acad. Sci. USA* 59:582–89
5. Hunt, T., Vanderhoff, G., London, I. M. 1972. *J. Mol. Biol.* 66:471–81
6. Mathews, M. B., Hunt, T., Brayley, A. 1973. *Nature New Biol.* 243:230–33

7. Maxwell, C. R., Kamper, C. S., Rabino-vitz, M. 1971. *J. Mol. Biol.* 58:317–27
8. Gross, M., Rabinovitz, M. 1972. *Proc. Natl. Acad. Sci. USA* 69:1565–68
9. Gross, M., Rabinovitz, M. 1972. *Bio-chim. Biophys. Acta* 287:340–52
10. Gross, M. 1974. *Biochim. Biophys. Acta* 340:484–97
11. Gross, M. 1978. *Biochim. Biophys. Acta* 520:642–49
12. Gross, M., Mendelewski, J. 1978. *Bio-chim Biophys. Acta.* 520:650–63

13. Gross, M. 1974. *Biochim. Biophys. Acta* 366:319–32
14. Gross, M. 1974. *Biochem. Biophys. Res. Commun.* 57:611–19
15. Freedman, M. L., Geraghty, M., Rosman, J. 1974. *J. Biol. Chem.* 249: 7290–94
16. Mizuno, S., Fisher, J. M., Rabinovitz, M. 1972. *Biochim. Biophys. Acta* 272: 638–50
17. Balkow, K., Mizuno, S., Fisher, J. M., Rabinovitz, M. 1973. *Biochim. Biophys. Acta* 324:397–409
18. Clemens, M. J., Henshaw, E. C., Rahmimoff, H., London, I. M. 1974. *Proc. Natl. Acad. Sci. USA* 71:2946–50
19. Lenz, J. R., Baglioni, C. 1977. *Nature* 266:191–93
20. Kaempfer, R. 1974. *Biochem. Biophys. Res. Commun.* 61:541–47
21. Ranu, R. S., Levin, D. H., Delaunay, J., Ernst, V., London, I. M. 1976. *Proc. Natl. Acad. Sci. USA* 73:2720–24
22. Clemens, M. J. 1976. *Eur. J. Biochem.* 66:413–22
23. Trachsel, H., Erni, B., Schreier, M. H., Staehelin, T. 1977. *J. Mol. Biol.* 116: 755–67
24. Benne, R., Hershey, J. W. B. 1978. *J. Biol. Chem.* 253:3078–87
25. Balkow, K., Hunt, T., Jackson, R. J. 1975. *Biochem. Biophys. Res. Commun.* 67:366–75
26. Farrell, P. J., Balkow, K., Hunt, T., Jackson, R. J., Trachsel, H. 1977. *Cell* 11:187–200
27. Levin, D. H., Ranu, R. S., Ernst, V., London, I. M. 1976. *Proc. Natl. Acad. Sci. USA* 73:3112–16
28. Kramer, G., Cimadevilla, J. M., Hardesty, B. 1976. *Proc. Natl. Acad. Sci. USA* 73:3078–82
29. Gross, M., Mendelewski, J. 1977. *Biochem. Biophys. Res. Commun.* 74: 559–69
30. Benne, R., Wong, C., Luedi, M., Hershey, J. W. B. 1976. *J. Biol. Chem.* 251:7675–81
31. Schreier, M. H., Erni, B., Staehelin, T. 1977. *J. Mol. Biol.* 116:727–53
32. Barrieux, A., Rosenfeld, M. G. 1977. *J. Biol. Chem.* 252:3843–47
33. Gross, M., Rabinovitz, M. 1973. *Biochem. Biophys. Res. Commun.* 50: 832–38
34. Ranu, R. S., London, I. M. 1976. *Proc. Natl. Acad. Sci. USA* 73:4349–53
35. Trachsel, H., Ranu, R. S., London, I. M. 1978. *Proc. Natl. Acad. Sci. USA* 75:3654–58
36. Tahara, S. M., Traugh, J. A., Sharp, S. B., Lundak, T. S., Safer, B., Merrick, W. C. 1978. *Proc. Natl. Acad. Sci. USA* 75:789–93
37. Levin, D., London, I. M. 1978. *Proc. Natl. Acad. Sci. USA* 75:1121–25
38. Ernst, V., Levin, D. H., Ranu, R. S., London, I. M. 1976. *Proc. Natl. Acad. Sci. USA* 73:1112–16
39. Legon, S., Brayley, A., Hunt, T., Jackson, R. J. 1974. *Biochem. Biophys. Res. Commun.* 56:745–52
40. Iwai, H., Inamasu, M., Takeyama, S. 1972. *Biochem. Biophys. Res. Commun.* 46:824–30
41. Clemens, M. J., Pain, V. M., Henshaw, E. C. 1976. *EMBO Workshop Cytoplasmic Control Eukaryotic Protein Synth.* Cambridge, England. (Abstr.)
42. Datta, A., de Haro, C., Sierra, J. M., Ochoa, S. 1977. *Proc. Natl. Acad. Sci. USA* 74:1463–67
43. Merrick, W. C., Peterson, D. T., Safer, B., Lloyd, M. 1978. *Fed. Proc.* 37:1659
44. Rubin, C. S., Rosen, O. M. 1975. *Ann. Rev. Biochem.* 44:831–87
45. Datta, A., de Haro, C., Sierra, J. M., Ochoa, S. 1977. *Proc. Natl. Acad. Sci. USA* 74:3326–29
46. Beavo, J. A., Bechtel, P. J., Krebs, E. G. 1974. *Proc. Natl. Acad. Sci. USA* 71: 3580–83
47. Pryor, J., Berthet, J. 1960. *Biochim. Biophys. Acta* 43:556–57
48. Bloxham, D. P., Akhtar, M. 1972. *Int. J. Biochem.* 3:294–308
49. Sellers, A., Bloxham, D. P., Munday, K. A., Akhtar, M. 1974. *Biochem. J.* 138:335–40
50. Klaipongpan, A., Bloxham, D. P., Akhtar, M. 1975. *FEBS Lett.* 58:81–84
51. Klaipongpan, A., Bloxham, D. P., Akhtar, M. 1977. *Biochem. J.* 168:271–75
52. Hoshino, J., Kühne, U., Studinger, G., Kröger, H. 1977. *Biochem. Biophys. Res. Commun.* 74, 663–69
53. De Oya, M., Prigge, W. F., Swenson, D. E., Grande, F. 1971. *Am. J. Phys.* 221:25–30
54. Horak, I., Koschel, K. 1977. *FEBS Lett.* 83:68–70
55. Hayakawa, T., Perkins, J. P., Krebs, E. G. 1973. *Biochemistry* 12:574–80
56. Cohen, P. 1973. *Eur. J. Biochem.* 34:1–14
57. Graves, D. J., Hayakawa, T., Horvitz, R. A., Beckman, E., Krebs, E. G. 1973. *Biochemistry* 12:580–85
58. Hirsch, J. D., Martelo, O. J. 1976. *Biochem. Biophys. Res. Commun.* 71: 926–32
59. Datta, A., de Haro, C., Ochoa, S. 1978. *Proc. Natl. Acad. Sci. USA* 75:1148–52

60. Riley, W. D., DeLange, R. J., Bratvold, G. E., Krebs, E. G. 1968. *J. Biol. Chem.* 243:2209–15
61. Antoniw, J. F., Cohen, P. 1976. *Eur. J. Biochem.* 68:45–54
62. Cohen, P., Antoniw, J. F. 1973. *FEBS Lett.* 34:43–47
63. Gross, M. 1975. *Biochem. Biophys. Res. Commun.* 67:1507–15
64. Gross, M. 1976. *Biochim. Biophys. Acta* 447:445–59
65. Lightfoot, H. N., Mumby, M., Traugh, J. A. 1975. *Biochem. Biophys. Res. Commun.* 66:1141–46
66. Traugh, J. A., Sharp, S. B. 1977. *J. Biol. Chem.* 252:3738–44
67. Revel, M., Schmidt, A., Shulman, L., Zilberstein, A., Kimchi, A. 1979. *FEBS Meet. 12th, Dresden.* In press
68. de Haro, C., Datta, A., Ochoa, S. 1978. *Proc. Natl. Acad. Sci. USA* 75:243–47
69. de Haro, C., Ochoa, S. 1978. *Proc. Natl. Acad. Sci. USA* 75:2713–16
70. Dasgupta, A., Majumdar, A., George, A. D., Gupta, N. K. 1976. *Biochem. Biophys. Res. Commun.* 71:1234–41
71. Mastropaolo, W., Smith, K., Ickowicz, R., Henshaw, E. 1978. *Fed. Proc.* 37:1623
72. Malathi, V. G., Mazumder, R. 1978. *FEBS Lett.* 86:155–59
73. Pinphanichakarn, P., Kramer, G., Hardesty, B. 1976. *Biochem. Biophys. Res. Commun.* 73:625–31
74. Cherbas, L., London, I. M. 1976. *Proc. Natl. Acad. Sci. USA* 73:3506–10
75. Trachsel, H., Staehelin, T. 1978. *Proc. Natl. Acad. Sci. USA* 75:204–8
76. Ehrenfeld, E., Hunt, T. 1971. *Proc. Natl. Acad. Sci. USA* 68:1075–78
77. Graziadei, W. D. III, Lengyel, P. 1972. *Biochem. Biophys. Res. Commun.* 46:1816–23
78. Darnbrough, C., Hunt, T., Jackson, R. J. 1972. *Biochem. Biophys. Res. Commun.* 48:1556–64
79. Hunter, T., Hunt, T., Jackson, R. J., Robertson, H. D. 1975. *J. Biol. Chem.* 250:409–17
80. Lenz, J. R., Baglioni, C. 1978. *J. Biol. Chem.* 253:4219–23
81. Beuzard, Y., London, I. M. 1974. *Proc. Natl. Acad. Sci. USA* 71:2863–66
82. Kosower, N. S., Vanderhoff, G. A., Benerofe, B., Hunt, T., Kosower, E. M. 1971. *Biochem. Biophys. Res. Commun.* 45:816–21
83. Clemens, M. J., Safer, B., Merrick, W. C., Anderson, W. F., London, I. M. 1975. *Proc. Natl. Acad. Sci. USA* 72:1286–90

84. Ernst, V., Levin, D. H. 1977. *Fed. Proc.* 36:868
85. Mizuno, S. 1977. *Arch. Biochem. Biophys.* 179:289–301
86. Bonanou-Tzedaki, S. A., Smith, K. E., Sheeran, B. A., Arnstein, H. R. V. 1978. *Eur. J. Biochem.* 84:591–600
87. Bonanou-Tzedaki, S. A., Smith, K. E., Sheeran, B. A., Arnstein, H. R. V. 1978. *Eur. J. Biochem.* 84:601–10
88. Spieler, P. J., Ibrahim, N. G., Freedman, M. L. 1978. *Biochim. Biophys. Acta* 518:366–79
89. Beuzard, Y., Rodvien, R., London, I. M. 1973. *Proc. Natl. Acad. Sci. USA* 70:1022–26
90. Weber, L. A., Feman, E. R., Baglioni, C. 1975. *Biochemistry* 14:5315–21
91. Freedman, M. L., Karpatkin, S. 1973. *Biochem. Biophys. Res. Commun.* 54:475–81
92. Jennings, C., Ranu, R. S., London, I. M. 1978. *Fed. Proc.* 37:1623
93. Clemens, M. J., Pain, V. M., Henshaw, E. C., London, I. M. 1976. *Biochem. Biophys. Res. Commun.* 72:768–75
94. Pinphanichakarn, P., Kramer, G., Hardesty, B. 1977. *J. Biol. Chem.* 252:2106–12
95. Delaunay, J., Ranu, R. S., Levin, D. H., Ernst, V., London, I. M. 1977. *Proc. Natl. Acad. Sci. USA* 74:2264–68
96. Sierra, J. M., de Haro, C., Datta, A., Ochoa, S. 1977. *Proc. Natl. Acad. Sci. USA* 74:4356–59
97. Maller, J. L., Krebs, E. G. 1977. *J. Biol. Chem.* 252:1712–18
98. Bravo, R., Otero, C., Allende, C. C., Allende, J. E. 1978. *Proc. Natl. Acad. Sci. USA* 75:1242–46
99. Otten, J., Johnson, G. S., Pastan, I. 1971. *Biochem. Biophys. Res. Commun.* 44:1192–98
100. Otten, J., Johnson, G. S., Pastan, I. 1972. *J. Biol. Chem.* 247:7082–87
101. Pastan, I. H., Johnson, G. S., Anderson, W. B. 1975. *Ann. Rev. Biochem.* 44:491–522
102. Daniel, V., Litwack, G., Tomkins, G. M. 1973. *Proc. Natl. Acad. Sci. USA* 70:76–79
103. Friedman, R. M. 1977. *Bacteriol. Rev.* 41:543–67
104. Kerr, I. M., Brown, R. E., Ball, L. A. 1974. *Nature* 250:57–59
105. Kerr, I. M., Brown, R. E., Clemens, M. J., Gilbert, C. S. 1976. *Eur. J. Biochem.* 69:551–61
106. Lebleu, B., Sen, G. C., Shaila, S., Cabrer, B., Lengyel, P. 1976. *Proc. Natl. Acad. Sci. USA* 73:3107–11

107. Zilberstein, A., Federman, P., Shulman, L., Revel, M. 1976. *FEBS Lett.* 68:119–24
108. Gupta, S. L., Sopori, M. L., Lengyel, P. 1973. *Biochem. Biophys. Res. Commun.* 54:777–83
109. Gupta, S. L., Graziadei, W. D. III, Weideli, H., Sopori, M. L., Lengyel, P. 1974. *Virology* 57:49–63
110. Roberts, W. K., Clemens, M. J., Kerr, I. M. 1976. *Proc. Natl. Acad. Sci. USA* 73:3136–40
111. Roberts, W. K., Hovanessian, A., Brown, R. E., Clemens, M. J., Kerr, I. M. 1976. *Nature* 264:477–80
112. Sen, G. C., Taira, H., Lengyel, P. 1978. *J. Biol. Chem.* 253:5915–21
113. Zilberstein, A., Kimchi, A., Schmidt, A., Revel, M. 1978. *Proc. Natl. Acad. Sci. USA* 75:4734–38
114. Cooper, J. A., Farrell, P. J. 1977. *Biochem. Biophys. Res. Commun.* 77:124–31
115. Ohtsuki, K., Dianzani, F., Baron, S. 1977. *Nature* 269:536–38
116. Lewis, J., Falcoff, E., Falcoff, R. 1978. *Eur. J. Biochem.* 86:497–509
117. Brown, G. E., Lebleu, B., Kawakita, M., Shaila, S., Sen, G. C., Lengyel, P. 1976. *Biochem. Biophys. Res. Commun.* 69:114–22
118. Sen, G. C., Lebleu, B., Brown, G. E., Kawakita, M., Slattery, E., Lengyel, P. 1976. *Nature* 264:370–72
119. Ratner, L., Sen, G. C., Brown, G. E., Lebleu, B., Kawakita, M., Cabrer, B., Slattery, E., Lengyel, P. 1977. *Eur. J. Biochem.* 79:565–77
120. Ratner, L., Wiegand, R. C., Farrell, P. J., Sen, G. C., Cabrer, B., Lengyel, P. 1978. *Biochem. Biophys. Res. Commun.* 81:947–54
121. Sen, G. C., Lebleu, B., Brown, G. E., Rebello, M. A., Furuichi, Y., Morgan, M., Shatkin, A. J., Lengyel, P. 1975. *Biochem. Biophys. Res. Commun.* 65:427–34
122. Sen, G. C., Shaila, S., Lebleu, B., Brown, G. E., Desrosiers, R. C., Lengyel, P. 1977. *J. Virol.* 21:69–83
123. Gupta, S. L., Sopori, M. L., Lengyel, P. 1974. *Biochem. Biophys. Res. Commun.* 57:763–70
124. Content, J., Lebleu, B., Nudel, U., Zilberstein, A., Berissi, H., Revel, M. 1975. *Eur. J. Biochem.* 54:1–10
125. Falcoff, R., Lebleu, B., Sanceau, J., Weissenbach, J., Dirheimer, G., Ebel, J. P., Falcoff, E. 1976. *Biochem. Biophys. Res. Commun.* 68:1323–31
126. Samuel, C. E. 1976. *Virology* 75:166–76
127. Sen, G. C., Gupta, S. L., Brown, G. E., Lebleu, B., Rebello, M. A., Lengyel, P. 1976. *J. Virol.* 17:191–203
128. Zilberstein, A., Dudock, B., Berissi, H., Revel, M. 1976. *J. Mol. Biol.* 108:43–54
129. Hiller, G., Winkler, I., Viehhauser, G., Jungwirth, C., Bodo, G., Dube, S., Ostertag, W. 1976. *Virology* 69:360–63
130. Mayr, U., Bermayer, H. P., Weidinger, G., Jungwirth, C., Gross, H. J., Bodo, G. 1977. *Eur. J. Biochem.* 76:541–51
131. Hovanessian, A. G., Brown, R. E., Kerr, I. M. 1977. *Nature* 268:537–40
132. Kerr, I. M., Brown, R. E., Hovanessian, A. G. 1977. *Nature* 268:540–42
133. Kerr, I. M., Brown, R. E. 1978. *Proc. Natl. Acad. Sci. USA* 75:256–60
134. Hunt, T. 1978. *Nature* 273:97–98
135. Hovanessian, A. G., Kerr, I. M. 1978. *Eur. J. Biochem.* 84:149–59
136. Clemens, M. J., Williams, B. R. G. 1978. *Cell* 13:565–72
137. Baglioni, C., Minks, M. A., Maroney, P. A. 1978. *Nature* 273:684–87
138. Schmidt, A., Zilberstein, A., Shulman, L., Federman, P., Berissi, H., Revel, M. 1978. *FEBS Lett.* 95:257–64
139. Farrell, P. J., Sen, G. C., Dubois, M. F., Ratner, L., Slattery, E., Lengyel, P. 1978. *Proc. Natl. Acad. Sci. USA* 75:5893–97
140. Clemens, M. J., Vaquero, C. M. 1978. *Biochem. Biophys. Res. Commun.* 83:59–68
141. Shaila, S., Lebleu, B., Brown, G. E., Sen, G. C., Lengyel, P. 1977. *J. Gen. Virol.* 37:535–46
142. Eil, C., Wool, I. G. 1973. *J. Biol. Chem.* 248:5122–29
143. Traugh, J. A., Mumby, M., Traut, R. R. 1973. *Proc. Natl. Acad. Sci. USA* 70:373–76
144. Traugh, J. A., Traut, R. R. 1974. *J. Biol. Chem.* 249:1207–12
145. Becker-Ursic, D., Davies, J. 1976. *Biochemistry* 15:2289–96
146. Floyd, G. A., Merrick, W. C., Traugh, J. A. 1979. *Eur. J. Biochem.*
147. Hebert, J., Pierre, M., Loeb, J. E. 1977. *Eur. J. Biochem.* 72:167–74
148. Kudlicki, W., Grankowski, N., Gasior, E. 1978. *Eur. J. Biochem.* 84:493–98
149. Gressner, A. M., Wool, I. G. 1974. *J. Biol. Chem.* 249:6917–25
150. Stahl, J., Bohm, H., Bielka, H. 1974. *Acta Biol. Med. Germ.* 33:667–76
151. Hoffman, W. L., Ilan, J. 1975. *Mol. Biol. Rep.* 2:219–24
152. Leader, D. P., Coia, A. A. 1978. *FEBS Lett.* 90:270–74
153. Gressner, A. M., Wool, I. G. 1976. *J. Biol. Chem.* 251:1500–4

154. Schubart, U. K., Shapiro, S., Fleischer, N., Rosen, O. M. 1977. *J. Biol. Chem.* 252:92–101
155. Traugh, J. A., Porter, G. G. 1976. *Biochemistry* 15:610–16
156. DuVernay, V. H. Jr., Traugh, J. A. 1978. *Biochemistry* 17:2045–49
157. Hochkeppel, H. K. 1977. *Ann. Rep.* p. 21. Basel: Friedrich Miescher Inst.
158. Leader, D. P., Coia, A. A. 1978. *Biochim. Biophys. Acta* 519:213–23
159. Leader, D. P., Coia, A. A. 1978. *Biochim. Biophys. Acta* 519:224–32
160. Eil, C., Wool, I. G. 1973. *J. Biol. Chem.* 248:5130–36
161. Thomas, G., Haselbacher, G., Humbel, R. E. 1977. See Ref. 157, p. 21
162. Kramer, G., Henderson, A. B., Pinphanichakarn, P., Wallis, M. H., Hardesty, B. 1977. *Proc. Natl. Acad. Sci. USA* 74:1445–49
163. Traugh, J. A., Tahara, S. M., Sharp, S. B., Safer, B., Merrick, W. C. 1976. *Nature* 263:163–65
164. Issinger, O. G., Benne, R., Hershey, J. W. B., Traut, R. R. 1976. *J. Biol. Chem.* 252:6471–74
165. Benne, R., Edman, J., Traut, R. R., Hershey, J. W. B. 1978. *Proc. Natl. Acad. Sci. USA* 75:108–12
166. Traugh, J. A., Lundak, T. S. 1978. *Biochem. Biophys. Res. Commun.* 83:379–84
167. Giloh, H., Mager, J. 1975. *Biochim. Biophys. Acta* 414:293–308
168. Levin, D. H., Ernst, V., London, I. M. 1977. *Fed. Proc.* 36:868
169. Lenz, J. R., Chatterjee, G. E., Maroney, P. A., Baglioni, C. 1978. *Biochemistry* 17:80–87
170. Ernst, V., Levin, D. H., London, I. M. 1978. *J. Biol. Chem.* 253:7163–72
171. Rose, J. K., Trachsel, H., Leong, K., Baltimore, D. 1978. *Proc. Natl. Acad. Sci. USA* 75:2732–36
172. Bogdanovsky, D., Hermann, W., Schapira, G. 1973. *Biochem. Biophys. Res. Commun.* 54:25–32
173. Heywood, S. M., Kennedy, D. S., Bester, A. J. 1974. *Proc. Natl. Acad. Sci. USA* 71:2428–31
174. Bester, A. J., Kennedy, D. S., Heywood, S. M. 1975. *Proc. Natl. Acad. Sci. USA* 72:1523–27
175. Lee-Huang, S., Sierra, J. M., Naranjo, R., Filipowicz, W., Ochoa, S. 1977. *Arch. Biochem. Biophys.* 180:276–87
176. Metafora, S., Felicetti, L., Gambino, R. 1971. *Proc. Natl. Acad. Sci. USA* 68:600–4
177. Gambino, R., Metafora, S., Felicetti, L., Raisman, J. 1973. *Biochim. Biophys. Acta* 312:377–91
178. Huang, F. L., Warner, A. H. 1974. *Arch. Biochem. Biophys.* 163:716–27
179. Hille, M. B. 1974. *Nature* 249:556–8
180. Adelman, T. G., Lovett, J. S. 1974. *Biochim. Biophys. Acta* 335:236–45
181. Lodish, H. F. 1976. *Ann. Rev. Biochem.* 45:39–72

Ann. Rev. Biochem. 1979. 48:581–600

ACTIVATION OF ADENYLATE CYCLASE BY CHOLERAGEN[1,2]

Joel Moss and Martha Vaughan

Laboratory of Cellular Metabolism, National Heart, Lung, and Blood Institute, National Institutes of Health, Bethesda, Maryland 20014

CONTENTS

[2]The following abbreviations are used: G_{M1}, galactosyl-N-acetylgalactosaminyl-[N-acetylneuraminyl]-galactosylglucosylceramide; G_{M2}, N-acetylgalactosaminyl-[N-acetylneuraminyl]-galactosylglucosylceramide; G_{D1a}, N-acetylneuraminyl-galactosyl-N-acetylgalactosaminyl-[N-acetylneuraminyl]-galactosylglucosylceramide; G_{T1}, N-acetylneuraminylgalactosyl-N-acetylgalactosaminyl-[N-acetylneuraminyl-N-acetylneuraminyl]-galactosylglucosylceramide; Gpp(NH)p, guanylyl imidodiphosphate.

PERSPECTIVES AND SUMMARY

Choleragen (cholera toxin) is an enterotoxin of *Vibrio cholerae* that is responsible for the watery diarrhea characteristic of clinical cholera. The toxin is an oligomeric protein of ~84,000 daltons composed of three dissimilar peptides, A_1, A_2, and B. A_1 (23,500) is linked through a single disulfide bond to A_2, a peptide of 5500 daltons. Five B subunits (11,600 daltons) are present in the oligomer. [For reviews on the clinical, biochemical, and historical aspects of choleragen, see (1–28)]. Choleragen is believed to exert its effects on cells through activation of adenylate cyclase. The initial event appears to be the binding of the B subunits of the toxin to ganglioside G_{M1} on the cell surface. Although the mechanism by which choleragen gains access to adenylate cyclase or its regulatory component is unclear, lateral mobility of the toxin on the cell surface or within the membrane may be necessary. Multivalent binding of the B subunits to the cell surface, followed by patching and/or capping of the toxin may facilitate entry.

It has been suggested that the toxin activates adenylate cyclase by catalyzing the ADP-ribosylation of either the cyclase itself or of a regulatory component of the cyclase system. This hypothesis is supported by the demonstration that: 1. NAD is required for choleragen activation of adenylate cyclase in cell-free homogenates or membrane preparations. 2. Choleragen, in the absence of cellular components, catalyzes the hydrolysis of NAD to ADP-ribose and nicotinamide. 3. Choleragen catalyzes the NAD-dependent ADP-ribosylation of arginine and several proteins. Although the specific cellular substrate(s) for ADP-ribosylation by choleragen has not been purified to homogeneity, there is evidence that it may be a GTP-binding protein. It has been postulated that choleragen, by catalyzing the covalent modification of its protein substrate, inhibits the GTP hydrolysis that is associated with reversion of the cyclase from an active to an inactive form and thereby maintains the activated state of the enzyme. Consistent with this hypothesis are the findings that: 1. Choleragen inhibits a specific GTPase in turkey erythrocytes, and 2. the protein in turkey erythrocytes that is ADP-ribosylated by choleragen binds to GTP-Sepharose, is eluted with GTP, is necessary for demonstration of adenylate cyclase activation by GTP and, when purified from choleragen-treated membranes, supports equivalent activation of adenylate cyclase by GTP and Gpp(NH)p.

E. coli heat-labile enterotoxin is similar in many ways to choleragen. The toxin utilizes a ganglioside, probably G_{M1}, as a cell surface receptor and requires NAD for in vitro activation of adenylate cyclase. In addition, it catalyzes the hydrolysis of NAD, the ADP-ribosylation of proteins, and the stereospecific β-NAD-dependent ADP-ribosylation of arginine to yield α-ADP-ribose-L-arginine. Recently, an enzyme has been isolated from avian erythrocytes that also possesses NAD glycohydrolase and ADP-ribosyltransferase activities. It will activate brain adenylate cyclase in the presence of NAD. Although a physiological role for this enzyme remains to be established, its presence is consistent with the possibility that vertebrate cells may use mechanisms for physiological activation of adenylate cyclase that are similar to those through which the choleragen and *Escherichia coli* toxins produce disease.

ROLE OF GANGLIOSIDE G_{M1} AS THE CELL SURFACE RECEPTOR FOR CHOLERAGEN

The first step in choleragen action is the binding of the toxin to the cell surface as a result of its interaction with ganglioside G_{M1} in the plasma membrane. Gangliosides are glycolipids that consist of a ceramide moiety and a carbohydrate moiety. It is the composition of the oligosaccharide portion of the molecule that distinguishes the major types of gangliosides (Figure 1) and is responsible for the specificity of the interaction of G_{M1} with choleragen. The determinants of this specificity have been defined using choleragen and its A and B subunits in studies with G_{M1} in solution (29–51), incorporated into artificial lipid membranes (52–55), or present in cell membranes (31, 34, 43, 44, 47, 51, 56–72). Evidence from several types of studies supports the conclusions that G_{M1} in the plasma membrane serves as the cell surface receptor for choleragen by interacting with the B subunits of the toxin (32, 33, 35, 40, 43, 44, 73–82), and that it is the presence of G_{M1} that makes cells sensitive to the toxin (31, 43, 47, 51, 60–62, 65–67, 69, 72). Despite the fact that G_{M1} seems to be required for activation of

G_{M2} GalNAc → Gal → Glc → Cer
                        ↑
                      NANA

G_{M1} Gal → GalNAc → Gal → Glc → Cer
                               ↑
                             NANA

G_{D1a} Gal → GalNAc → Gal → Glc → Cer
               ↑   ↑
             NANA   NANA

G_{T1} Gal → GalNAc → Gal → Glc → Cer
               ↑   ↑
             NANA   NANA → NANA

Figure 1 Ganglioside structure. Abbreviations used are: Gal, galactose; GalNAc, N-acetyl-galactosamine; Glc, glucose; NANA, N-acetylneuraminic acid: Cer, ceramide.

adenylate cyclase by choleragen in intact cells, in broken-cell systems, where the toxin can gain access to the cyclase without the necessity of traversing the plasma membrane, the A subunit is active alone and neither the B (or binding) subunit nor cellular G_{M1} is required (43, 67, 83–90).

Interaction of Gangliosides and Their Oligosaccharides with Choleragen and Its Subunits

W. E. Van Heyningen and co-workers initially demonstrated that mixed brain gangliosides inhibited the action of choleragen on the gut (29). Inhibition in those experiments was presumed to result from the binding of choleragen to ganglioside in the medium, thus making the toxin unavailable for binding to the cell surface (29). Subsequent studies further documented the inhibitory effects of mixed gangliosides on choleragen action in various systems (37, 44, 45). The ganglioside primarily responsible for these effects is G_{M1}. G_{M1} was the most potent ganglioside in inhibiting ^{125}I-choleragen binding (30, 47, 50), uptake (91), and action (31–36, 39–44, 46, 48, 49, 51); gangliosides with other oligosaccharides were less active (30, 92). G_{M1} appears to interact directly with choleragen, since it will precipitate the toxin from solution (33, 38, 39, 48), and the ganglioside, when attached to Agarose beads (93), or present in a ganglioside-cerebroside mixture (94), will bind toxin. The oligosaccharide derived from G_{M1} can also interact with choleragen (38, 39, 48, 95); it can inhibit the binding of ^{125}I-choleragen to fibroblasts, although it is less potent than the complete ganglioside (95).

Gangliosides incorporated into liposomal model membranes can react with antibodies specific for the oligosaccharide portions of the glycolipids (96). Liposomes containing G_{M1} bind ^{125}I-choleragen much more effectively than do liposomes containing other gangliosides (52). Incubation of liposomes containing G_{M1} with choleragen causes alterations in membrane permeability that are not observed with liposomes containing G_{M2} or G_{D1a} (52, 53). Similar permeability changes occur when liposomes containing G_{M1} interact with the B protomer, but not the A protomer, of choleragen (53, 54). Further evidence that the liposomes containing G_{M1} do bind the toxin and its B protomer is the observation that the addition of antitoxin (which reacts with choleragen or its A or B protomers) plus a source of complement leads to lysis of such liposomes (52, 53). The A protomer of choleragen can bind to liposomes independent of their ganglioside content (53). It is unclear whether this interaction mimics that which occurs during activation of membrane adenylate cyclase by A protomer. Choleragen also causes aggregation of liposomes containing G_{M1} (but not those containing other gangliosides) (55). This probably reflects the multivalent nature of choleragen binding, i.e. it would appear that choleragen has the properties of a lectin (55).

The presence of G_{M1} causes a "blue shift" in the tryptophanyl fluorescence spectrum of choleragen (68, 97). Other gangliosides do not produce this effect (68, 97). The interaction of the B protomer of choleragen with G_{M1} (but not with other gangliosides) produces changes in its fluorescence spectrum like those resulting from the G_{M1}-choleragen interaction (97). Although the ganglioside binding site in the B subunit has not been identified, it is postulated that a lysyl residue may be involved (98). Several gangliosides interact in a relatively nonspecific manner with the A protomer of choleragen and increase the intensity of its tryptophanyl fluorescence (97).

Studies of the interaction of choleragen and ganglioside in aqueous solution are complicated by the tendency of gangliosides to self-associate (66, 99, 100). Perhaps for this reason the molar ratios of G_{M1} to choleragen apparently required to inactivate the toxin have varied from $1:1-3:1$ (33, 38, 39, 49, 101). By using the oligosaccharide of G_{M1} (generated from the ganglioside by ozonolysis and alkaline hydrolysis) which is free of the lipid backbone (102), problems of self-association can be avoided. Choleragen is capable of binding 4–6 moles of ^{3}H-labeled G_{M1} oligosaccharide, as demonstrated both by gel permeation chromatography and equilibrium dialysis (95, 103). These findings are consistent with data from other types of studies, all of which support the conclusion that the holotoxin contains 4–6 B subunits (81, 88, 95, 98, 103–115). The G_{M1} oligosaccharide has effects on the fluorescence and circular dichroic spectra of choleragen and the B protomer similar to those produced by the intact ganglioside (95). Oligosaccharides derived from other gangliosides do not (95). None of the oligosaccharides induced changes in the fluorescence spectrum of the A protomer (95).

The results of all of the studies with G_{M1} present in solution or incorporated into artificial membranes point to the conclusion that it is the B protomer of choleragen that allows the toxin to recognize and interact specifically with the oligosaccharide moiety of the ganglioside.

Activation of Adenylate Cyclase by Choleragen in Intact Cells

Based initially on the observation that G_{M1} can bind choleragen and prevent its effects on cells, it was suggested that G_{M1} might be very similar to or even identical with the membrane receptor for the toxin. The studies in which choleragen binding to and effects on cells were used to evaluate the amounts of functionally available toxin in the presence of G_{M1} or other gangliosides were useful, because, as is now clear, the interaction of choleragen with G_{M1} in solution probably reflects quite closely the biologically significant interaction of the toxin with G_{M1} on the cell surface.

The capacity of cells to bind and respond to choleragen appears to be related to their G_{M1} content (31, 34, 43, 44, 47, 51, 56–62, 65–67, 69, 72). Cells and tissues possessing a high G_{M1} content generally exhibit more choleragen binding (58–60). It was found that treatment of cells with neuraminidase, which catalyzes the removal of sialic acid residues from di- and trisialogangliosides and thus converts them to G_{M1}, increased the G_{M1} content (72) as well as the binding of (30, 34, 56, 69, 72) and the response to (44, 57) choleragen. Similarly, when cellular or membrane G_{M1} was increased by incorporation of exogenous G_{M1}, choleragen binding to cells or membranes (31, 47, 60, 62, 69, 72) and the responsiveness of cells (31, 43, 51, 60, 61, 65, 66) were increased. The relationship between G_{M1} content and choleragen responsiveness was quantified using ganglioside-deficient fibroblasts which, when grown in chemically defined medium, lack G_{M1} and do not respond to choleragen (67). The fibroblasts can take up exogenous gangliosides from the medium. Using ^{3}H-labeled G_{M1} to determine amounts taken up by the cells, it was shown that responsiveness to choleragen and binding of ^{125}I-choleragen were a function of cellular G_{M1} content. Other gangliosides were much less effective (63).

Relatively small amounts of G_{M1} may be sufficient to make cells sensitive to choleragen (67, 71). The cultured fibroblasts mentioned above were responsive to choleragen after taking up, on the average, ~20,000 molecules of $[^3H]G_{M1}$ per cell (67), an amount of the ganglioside that would not be detectable by the usual procedures for analysis. Fat cells contain very low levels of G_{M1} that are demonstrable unequivocally only with special methodology (71). The responsiveness of these cells is increased following incubation with exogenous G_{M1} (31). Only a portion of cell-associated G_{M1}, whether of endogenous or exogenous origin, is apparently available for choleragen binding (66, 80). The oligosaccharide moiety of G_{M1} on the surface of intact cells can be oxidized with galactose oxidase or periodate; its accessibility to oxidation is diminished in fibroblasts that have bound choleragen (68, 70).

Modifications in choleragen structure that reduce its capacity to bind G_{M1} also reduce its activity toward cells (64). Choleragenoid, a biologically inactive molecule (73, 78, 81, 116–120) immunologically related to choleragen but composed solely of the B subunits (81, 94, 108–111), is capable of interacting with G_{M1} and its oligosaccharide (33, 40, 73, 79, 82). It will inhibit choleragen binding to cells and membranes (73, 76) and block the action of choleragen in intact cells (32, 35, 43, 44, 73–75, 77, 78, 80, 81), but not in lysed cells (43). Like the interaction that occurs in defined systems, the interaction of choleragen with G_{M1} in the plasma membrane, which is required for toxin action on intact cells, appears to involve the B subunit of the toxin and the oligosaccharide portion of the ganglioside.

Activation of Adenylate Cyclase by Choleragen in Cell-Free Systems

In most intact cells, the G_{M1} requirement for activation of adenylate cyclase by choleragen is paralleled by a dependence on the B or binding subunit of the toxin (88). Usually the A subunit or A_1 peptide exhibits only a fraction of the activity of the holotoxin (43, 84, 87, 88, 90, 121, 122).[2] In contrast, in homogenates or membrane preparations, activation of adenylate cyclase is independent of G_{M1} and the B subunits of choleragen (43, 79, 84). Intact fibroblasts lacking G_{M1} did not respond to choleragen unless previously incubated with this ganglioside (67). When the cells were disrupted and then incubated with choleragen, however, adenylate cyclase activity was enhanced approximately ten-fold, whether or not they had been incubated with G_{M1} prior to homogenization (67). Choleragen activation of adenylate cyclase can be observed in isolated membranes and in detergent-solubilized or solubilized and partially purified preparations (43, 83, 84, 86, 87, 89, 90, 123–126). In such systems, activation of adenylate cyclase usually occurs without the delay characteristically observed following the addition of choleragen to intact cells (85, 87, 124). Choleragenoid, which consists entirely of B subunits (81, 94, 108–111), inhibits the action of the toxin on intact cells (32, 35, 43, 44, 73–75, 77, 78, 80, 81). In disrupted cells it does not interfere with adenylate cyclase activation (43) and G_{M1} does not block the action of choleragen (43) or of the A subunit (85, 87, 88).

Activation of adenylate cyclase in cell-free systems has been demonstrated with the A subunit or the A_1 peptide (43, 84, 87, 90, 127, 128) and with a choleragen breakdown product believed to be derived from the A subunit that was generated by incubation of the toxin with cell membranes (83, 86, 89). The single cysteine –SH in A_1 is not critical, since blocking it with iodoacetate does not abolish activity [(90) J. Moss, and M. C. Lin, in preparation]. The A_1 peptide must be released from A_2 by reducing the disulfide bond that links them in order to demonstrate its enzymatic activity (129). It has recently been reported that a cross-linked complex consisting of A_1, A_2, and B_{4-6} can activate adenylate cyclase, but it is not clear that the release of an active A_1 peptide through proteolysis was ruled out in these studies (51). At present, the bulk of evidence supports the view that activation of adenylate cyclase is a function of the A_1 peptide and does not involve the B subunit of the toxin or ganglioside G_{M1}. Thus, although G_{M1} is responsible (and necessary) for the specific binding of choleragen to the cell surface, it apparently does not function as a receptor in the sense that the term is used to refer to those receptors that are an obligatory part of the mechanisms through which β-adrenergic agonists and certain peptide hormones activate adenylate cyclase.

---

[2]In pigeon erythrocytes the A subunit is reported to be as active as intact choleragen (85).

REQUIREMENTS FOR DEMONSTRATION OF ADENYLATE CYCLASE ACTIVATION BY CHOLERAGEN IN CELL-FREE SYSTEMS

Activation of adenylate cyclase in broken-cell systems is catalyzed by the A subunit or A_1 peptide of choleragen (43, 84, 87, 90). In the intact toxin A_1 is linked to A_2 and probably must be released by reduction of a single disulfide bond in order to be active (83, 124, 129). NAD and GTP appear to be necessary in most systems for demonstration of maximal choleragen effects on adenylate cyclase (67, 68, 83, 86, 87, 89, 90, 123–126, 130–133). Activation of brain adenylate cyclase requires the addition of a low molecular weight calcium-dependent regulatory protein (126). In some systems, unidentified cytosolic factors are also necessary (124, 126). It is unclear, in general, whether individual factors are required for activation per se or for expression of the catalytic activity of adenylate cyclase.

NAD

Gill first demonstrated that NAD was necessary for the activation of pigeon erythrocyte adenylate cyclase by choleragen (123, 124). An NAD requirement has subsequently been established in almost all systems studied (67, 68, 83, 86, 87, 89, 90, 125, 126, 128, 130, 132, 133); optimal activation requires NAD concentrations in the millimolar range (89, 123–126). In some instances in which an NAD effect was not immediately evident, incubation of the preparation with an NAD glycohydrolase resulted in NAD-dependence (89, 123, 124). Presumably, in those systems, endogenous NAD was present in concentrations sufficient to permit optimal choleragen activation, and the further addition of exogenous NAD was not necessary.

The NAD requirement is quite specific (123, 125). Although thionicotinamide adenine dinucleotide was more active than NAD in pigeon erythrocytes, NADP, deamino-NAD, NMN, nicotinyl hydroxamic acid adenine dinucleotide, and ethyl nicotinate adenine dinucleotide were less active; α-NAD, nicotinamide, ADP-ribose, 3-acetyl- 4-methyl- or 3-carbonylpyridine adenine dinucleotide, FAD, ATP, and GTP were inactive (123). In the rat liver system, NADH, NADP, and NADPH were inactive (125). In view of the crude adenylate cyclase systems used for studying choleragen activation, it is possible that some of the compounds that could apparently substitute for NAD, like NADH, were converted to NAD during incubation.

GTP

Whether choleragen activation is carried out with intact cells or in a broken-cell system, the presence of GTP in the assay appears to be required for demonstration of enhanced adenylate cyclase activity (7, 77, 118, 126, 131–134). Following activation of the cyclase by choleragen in intact cells, GTP was as effective as the GTPase-resistant analogue, Gpp(NH)p, for increasing enzyme activity (132, 133). With a solubilized preparation of brain adenylate cyclase, no activation by choleragen was observed when Gpp(NH)p replaced GTP (126). The effect of GTP added only during the assay period was very small relative to that seen when it was present along with choleragen (and the calcium-dependent regulatory protein mentioned below) during the incubation period that preceded assay.

Calcium-Dependent Regulatory Protein

A low molecular weight, heat-stable protein which in the presence of Ca^{2+} stimulates the activity of a cyclic nucleotide phosphodiesterase has been purified from many tissues (135–144). This protein also (in the presence of Ca^{2+}) increases the activity of ATPase, protein kinase, and adenylate cyclase (145–154; for review, see 155). Separation of a solubilized brain adenylate cyclase from the calcium-dependent protein by various chromatographic procedures resulted in substantial reduction of its activity. With the addition of pure calcium-dependent regulatory protein or GTP alone, very little activation by choleragen was observed; when both were present, the effect of choleragen on adenylate cyclase activity was readily demonstrable (126).

ROLE OF NAD AS SUBSTRATE IN REACTIONS CATALYZED BY CHOLERAGEN

A model for choleragen action was developed from the comparison of its NAD requirement and enzymatic activities to those of Pseudomonas exotoxin A and diphtheria toxin (156–170). These toxins use NAD as an ADP-ribose donor and inhibit protein synthesis in susceptible cells by catalyzing the ADP-ribosylation of elongation factor II (156, 161, 163, 164, 166–169, 171–180). In the absence of the protein acceptor, the toxins catalyze the slow hydrolysis of NAD to ADP-ribose and nicotinamide (162, 165). Choleragen also possesses NAD glycohydrolase activity (97, 129) and catalyzes the transfer of ADP-ribose from NAD to arginine, related guanidino compounds, and several polypeptides or proteins that contain arginine (181–183). It was proposed that activation of adenylate cyclase occurs as a result of the ADP-ribosylation of either the cyclase itself or of a regulatory component of the cyclase system (123, 124, 129, 181, 184, 185). Preliminary evidence is consistent with the suggestion that the natural acceptor for ADP-ribose in turkey erythrocytes is a GTP-binding protein (184, 185). Cassel & Pfeuffer (184) have proposed that the ADP-ribosylation of this protein causes inhibition of a specific GTPase associated with it and thereby activates adenylate cyclase.

Mechanism of Action of NAD-Dependent Pseudomonas Exotoxin A and Diphtheria Toxin

The mechanism of action of diphtheria toxin and Pseudomonas exotoxin A, two NAD-dependent bacterial toxins, has been extensively investigated (156–170). Both toxins cause cell death, apparently by inhibiting protein synthesis (161, 169, 171–180). Cell susceptibility differs and may be determined by the presence of specific, as yet unidentified, cell surface receptors. Both toxins inhibit protein synthesis by catalyzing the transfer of the ADP-ribose moiety of NAD to a single amino acid in elongation factor II (EF II, Reaction 1) which is then inactive in protein synthesis (156, 161, 163, 164, 166–169). This reaction apparently requires the formation of a ternary complex between toxin, NAD, and EF II (162).

$$\text{NAD} + \text{elongation factor II} \rightleftharpoons \text{ADP-ribose-EF II} + \text{nicotinamide} + \text{H}^+ \qquad 1.$$

In the absence of EF II, the toxins catalyze the hydrolysis of NAD to ADP-ribose and nicotinamide (Reaction 2) (162, 165).

$$\text{NAD} + \text{H}_2\text{O} \rightarrow \text{ADP-ribose} + \text{nicotinamide} + \text{H}^+ \qquad 2.$$

This reaction is much slower than the ADP-ribosylation of EF II and is believed not to be relevant to toxin action (166). The fact that both toxins catalyze this reaction indicates that both can activate the ribosyl-nicotinamide bond of NAD in the absence of the appropriate acceptor protein. Presumably, the catalytic sites are accessible to water, and the enzymes can utilize water as an alternative acceptor.

Diphtheria toxin and Pseudomonas exotoxin A are single polypeptide chains and their native conformations are critical to their action in intact cells. The native toxins exhibit relatively little enzymatic activity (157–160, 170). Limited tryptic digestion of diphtheria toxin plus thiol treatment which gives rise to fragment A (157–160), or denaturation of Pseudomonas toxin with urea and thiol (170), results in enhancement of enzymatic activity, presumably by unmasking the catalytic site, but causes marked diminution in activity against intact cells. The catalytically active fragment of diphtheria toxin has a molecular weight of about 25,000 compared to a molecular weight of 60,000 for the native toxin (157–160). Although denatured Pseudomonas toxin is active as a single polypeptide chain of 71,000 daltons (170), a 26,000 dalton species that is enzymatically active has been isolated from cultures (165). The catalytic sites of both toxins probably represent only small segments of the complete proteins, as evidenced by the activity of proteolytic fragments (157, 160, 162, 165). The remainders of the peptide chains are thought to contain determinants that facilitate binding and entry of the toxin or its active fragment into cells (158).

NAD-Glycohydrolase and ADP-Ribosyltransferase Activities of Choleragen

NAD is required for activation of adenylate cyclase by choleragen, and choleragen, like the Pseudomonas and diphtheria toxins which require NAD for activity, catalyzes the hydrolysis of NAD to ADP-ribose and nicotinamide (Reaction 2) (97, 129). The fact that choleragen catalyzes the hydrolysis of NAD indicates that, like the toxins that catalyze the NAD-dependent ADP-ribosylation of EF II, it can activate and disrupt the ribosyl-nicotinamide bond, which results in the formation of ADP-ribose and nicotinamide. In contrast to some other NAD glycohydrolases, choleragen cannot utilize methanol as a substitute for water (186). It does recognize the guanidino moiety in arginine and related compounds as a preferred ADP-ribose acceptor (Reaction 3), and such compounds accelerate nicotinamide release (181, 183).

NAD + arginine → ADP-ribose-L-arginine +
nicotinamide + H$^+$ 3.

NAD + acceptor protein → ADP-ribose-acceptor protein +
nicotinamide + H$^+$ 4.

In addition, choleragen catalyzes the transfer of the ADP-ribose moiety of NAD to several proteins, such as histones, lysozyme, polyarginine, and the A subunit of choleragen (Reaction 4) (182, 187, 188).

Both NAD glycohydrolase and ADP-ribosyltransferase activities co-chromatograph with the A subunit of choleragen and are demonstrable in purified iodoacetylated preparations of A$_1$ [(129); J. Moss, and M. C. Lin, in preparation]. They are independent of the A$_2$ and B peptides, cellular components, or gangliosides [(97, 129); J. Moss, and M. C. Lin, in preparation]. Thus, the A$_1$ peptide responsible for the NAD-dependent activation of adenylate cyclase also catalyzes the model reactions (Reactions 3 and 4). When holotoxin or A subunit was used to initiate the reaction, a thiol was necessary for activity (129); no such requirement was observed when the iodoacetylated A$_1$ peptide was used (J. Moss, and M. C. Lin, in preparation). The glycohydrolase and transferase activities of the A subunit and the A$_1$ peptide were enhanced by high salt concentrations (129, 181) and at low concentrations, enzymatic activity was stabilized by nonionic detergents and proteins (189).

D- and L-Arginine were equipotent in stimulating toxin-catalyzed nicotinamide release from NAD, which suggests that the stereochemistry at the α-carbon was not critical for acceptor activity (181). Citrulline, in which the guanidino moiety is replaced by a ureido function, was inactive (181). Since guanidine itself was active, it is apparent that the remainder of the arginine serves as a carrier for the functional guanidino residue (181). Structural modification of arginine, however, altered its ability to stimulate nicotinamide release and presumably its capacity to serve as an ADP-ribose acceptor (183). Arginine methyl ester and agmatine, in which the carboxyl is methylated or absent, respectively, were more active than arginine, which suggests that the presence of a negatively charged residue on the α-carbon is inhibitory. In an acceptor protein, where the carboxyl and α-amino moieties of arginine are probably involved in peptide linkages, the environment of the guanidine group is perhaps more similar to that in arginine methyl ester than to that in free arginine. The inhibitory potential of a neighboring carboxyl on the activity of the guanidine moiety is also evidenced by the fact that guanidinopropionate was less active than guanidinobutyrate in accelerating nicotinamide release.

Choleragen can catalyze the ADP-ribosylation of several purified proteins including histones, lysozyme, polyarginine, and glycopeptide hormones (182, 188). The toxin also catalyzed the ADP-ribosylation of its A subunit (187); the relevance of this reaction to choleragen action is unclear. Although it has not been directly established that it is an arginine moiety in these proteins that is ADP-ribosylated, the fact that arginine inhibited ADP-ribosylation, whereas amino acids that are not alternative ADP-ribose acceptors did not, suggests that a substrate site on choleragen recognizes the guanidino moiety (182).

Oppenheimer (186) has established the stereochemistry of the choleragen-catalyzed synthesis of ADP-ribosyl arginine. As is the case with the pigeon liver poly(ADP-ribose) synthetases (190), choleragen cleavage of the nicotinamide-glycosyl bond in NAD is α-specific. The α-ADP-ribosyl arginine product, however, anomerizes relatively rapidly. If anomerization can also occur when the acceptor arginine is located in a peptide chain, this would obviously complicate any physiological regulation involving ADP-ribosylation (186). The fact that choleragen catalyzes NAD glycohydrolase and ADP-ribosyltransferase reactions does not, of course, prove that these are involved in the activation of adenylate cyclase. The preliminary reports summarized in the following section, however, are consistent with the hypothesis that choleragen catalyzes the ADP-ribosylation of a GTP-binding protein. Thus, the reactions described in this section may indeed be models for that involved in the activation of adenylate cyclase.

Effects of Choleragen on GTP-Binding Proteins and GTPase Activity

Following incubation of cells or membranes with choleragen, GTP is as effective as Gpp(NH)p for activation of adenylate cyclase (126, 132, 133). Cassel & Selinger proposed, therefore, that choleragen exerts its effects by inhibiting a specific GTPase that is associated with the GTP site of adenylate cyclase (191). In this view, adenylate cyclase with bound GTP is the active form and GTP bound to the cyclase is the substrate for GTPase. Thus, the observed activity of the specific GTPase when substrate-limited would be proportional to adenylate cyclase activity. Using conditions designed to reveal a high affinity GTPase, Cassel & Selinger (192) found in turkey erythrocytes, which contain an isoproterenol-responsive cyclase, a so-called isoproterenol-stimulated GTPase. The increased hydrolysis of GTP induced by isoproterenol was thought to reflect the increased amount of adenylate cyclase-GTP complex, i.e. the increased concentration of substrate for the GTPase. Choleragen reduced the activity of the specific isoproterenol-stimulated GTPase (191). It was proposed that by inhibiting GTPase, choleragen would increase the number of adenylate cyclase molecules maintained in the activated state.

The evidence presented in the previous section is consistent with choleragen exerting its effects through the ADP-ribosylation of a cellular protein. This protein(s) has not been purified to homogeneity nor has its role in activating adenylate cyclase been established definitively. There is some evidence, however, that it is a GTP-binding protein (184). Several workers (184, 185, 193, 194) have noted that membranes and cells incubated with choleragen show increased incorporation of ^{32}P from ^{32}P-NAD into several proteins. Cassel & Pfeuffer (184) found that on incubation of pigeon erythrocyte membranes with choleragen ^{32}P from ^{32}P-NAD was incorporated into proteins of 42,000, 80,000, and 200,000 daltons. The detergent-solubilized 42,000 dalton protein labeled with ^{32}P-NAD was bound to GTP-Sepharose and could be eluted with GTP or a hydrolysis-resistant analogue of GTP. Addition of this fraction was required in order to demonstrate GTP activation of the adenylate cyclase that did not bind to the affinity column. Acid treatment of the fraction containing the ^{32}P-labeled 42,000 dalton protein released products with mobilities identical to those of ADP-ribose and 5'-AMP. Although this protein has not been completely purified and characterized, the available information supports the view that choleragen acts by catalyzing the NAD-dependent ADP-ribosylation of the protein through which GTP influences adenylate cyclase activity.

SIMILARITIES BETWEEN CHOLERAGEN AND *ESCHERICHIA COLI* HEAT-LABILE ENTEROTOXIN

Certain strains of *E. coli* are believed to be responsible for "traveler's diarrhea" (195–204). In some instances, the symptoms may result from both heat-labile and heat-stable enterotoxins (197, 199–202, 204–206). The heat-labile enterotoxin (41, 197, 207–210), known as LT, can activate adenylate cyclase and cause accumulation of cyclic AMP in intact and broken-cell systems (41, 197, 209–216). Thus, it appears that the effects of LT are similar to those of choleragen (7, 20, 22, 41, 209, 210, 213, 217–220). LT, however, has not been as extensively purified or as well-characterized as choleragen. Preparations with molecular weights varying from 20,000 to 2×10^6 have been isolated by different procedures (210, 211, 215, 221–226). Structural similarities between the two toxins are suggested by the fact that antibodies to choleragen will react with LT (41, 208, 210, 211, 214, 215, 222, 227–230). The action of LT, like that of choleragen, was inhibited by G_{M1}, which was the most potent of all gangliosides tested (32, 41). In addition, the ganglioside-deficient fibroblasts which lack chemically detectable G_{M1} and did not respond to choleragen also did not respond to LT. After the cells were incubated with G_{M1} (but not with G_{M2} or G_{M3}), both choleragen (67) and LT (Moss et al in preparation) increased intracellular cAMP. These findings are consistent with the conclusion that both toxins use G_{M1} as a cell surface receptor. If this is the case, it might be expected that choleragenoid, which consists only of B or ganglioside binding subunits of choleragen, might block the action of LT. This has been observed in some studies but not in others (32, 35, 231). It has been suggested that the failure to demonstrate inhibition could be due to the use of insufficient choleragenoid or possibly to contamination of LT with the heat-stable enterotoxin (231).

LT and choleragen appear to activate adenylate cyclase through similar mechanisms. Gill et al (215) found that activation of the pigeon erythrocyte cyclase by LT required NAD, cytosolic factors, and ATP as was previously demonstrated with choleragen. LT exhibits NAD glycohydrolase and ADP-ribosyltransferase activities (232). Arginine and several proteins can serve as ADP-ribose acceptors in the transferase reaction as they do when choleragen is the catalyst (183, 232). The stereospecificity of the transferase reaction is identical for LT and choleragen (183, 186). The ADP-ribosylation of the proteins by LT was inhibited by arginine methyl ester, which itself stimulated nicotinamide release from NAD and was ADP-ribosylated (183, 232). These results are consistent with the possibility that arginine residues in the protein acceptors are the sites of ADP-ribosylation catalyzed by LT. Both LT and choleragen require dithiothreitol for enzymatic activity and exhibit some delay before maximal reaction rates are achieved (129, 232). Optimal assay conditions, affinities for NAD and arginine, and relative preferences for protein acceptors differ, however, with the two enterotoxins. These observations are consonant with other evidence that the toxins differ in structure. The presence of similar catalytic activities in two structurally distinct toxins strengthens the view that NAD-dependent ADP-ribosylation is the mechanism by which both activate adenylate cyclase.

AN ADP-RIBOSYLTRANSFERASE FROM TURKEY ERYTHROCYTES WITH CHOLERAGEN-LIKE ACTIVITY

The isolation of a low molecular weight, heat-labile protein from avian erythrocytes that catalyzes 1. the NAD-dependent activation of adenylate cyclase, 2. the hydrolysis of NAD to ADP-ribose and nicotinamide, and 3. the transfer of the ADP-ribose moiety of NAD to arginine and proteins, such as histones, lysozyme, and polyarginine, has recently been reported (182). The K_m for NAD in the ADP-ribosyltransferase reaction was 30μM, significantly less than those observed with choleragen (4 mM) and *E. coli* heat-labile enterotoxin (LT) (8 mM). Two K_ms of 5 and 50 mM were observed for arginine methyl ester as ADP-ribose acceptor; the lower value is significantly less than those found with choleragen (50 mM) and LT (240 mM). As observed with both toxins, arginine methyl ester was a considerably better substrate for the turkey enzyme than was D- or L-arginine or guanidine (which were approximately equivalent), and citrulline which was inactive. Although in the case of protein acceptors it has not been proven that arginine is ADP-ribosylated, free arginine almost completely inhibited ADP-ribosylation of the protein while itself acting as an acceptor. If arginine and acceptor protein compete for the same site on the transferase, it is likely that the locus of ADP-ribosylation in the protein is an arginine or similar amino acid. The enzymatic activities of the cytosolic protein from turkey erythrocytes thus appear to be similar to those previously demonstrated for choleragen and LT, although the optimal assay conditions as well as the kinetic constants for the reactants differ from those found for the two toxins.

The observations that these three structurally distinct proteins catalyze apparently identical enzymatic reactions involving activation of the ribosyl-nicotinamide bond of NAD and bring about the NAD-dependent activation of adenylate cyclase lends support to the hypothesis that the adenylate cyclase activation results from ADP-ribosylation of a protein that influences its activity. Whether the avian enzyme plays a role in the activation of adenylate cyclase in intact turkey erythrocytes remains to be determined. Nevertheless, the demonstration of this enzyme provides support for the view that mechanisms similar to those through which choleragen and *E. coli* toxin produce pathology may be employed by vertebrate cells for the physiological activation of adenylate cyclase.

Literature Cited

1. Finkelstein, R. A. 1972. *Toxicon* 10: 441–50
2. Finkelstein, R. A. 1973. *CRC Crit. Rev. Microbiol.* 2:553–623
3. van Heyningen, W. E. 1974. *Bull. Inst. Pasteur* 72:433–64
4. Finkelstein, R. A. 1975. In *Microbiology*, ed. D. Schlessinger, 236–41. Washington, DC: Am. Soc. Microbiol.
5. Finkelstein, R. A. 1975. *Curr. Top. Microbiol. Immunol.* 69:137–96
6. Finkelstein, R. A. 1975. *Dev. Ind. Microbiol.* 16:406–16
7. Bennett, V., Craig, S., Hollenberg, M. D., O'Keefe, E., Sahyoun, N., Cuatrecasas, P. 1976. *J. Supramol. Struct.* 4:99–120
8. Bennett, V., Cuatrecasas, P. 1976. In *Methods in Receptor Research*, ed. M. Blecher, Part 1, 73–98. New York: Dekker
9. Bennett, V., Cuatrecasas, P. 1976. In *The Specificity and Action of Animal, Bacterial and Plant Toxins. Receptors and Recognition*, ed. P. Cuatrecasas, Series B, Vol. 1, 1–66. London: Chapman & Hall
10. Cuatrecasas, P., Bennett, V., Craig, S., O'Keefe, E., Sahyoun, N. 1976. In *The Structural Basis of Membrane Function*, 275–91. New York: Academic
11. Finkelstein, R. A. 1976. *Dev. Biol. Stand.* 33:102–7
12. Finkelstein, R. A. 1976. In *Mechanisms in Bacterial Toxinology*, ed. A. W. Bernheimer, 54–84. New York: Wiley
13. Finkelstein, R. A. 1976. *Zentralbl. Bakteriol. Parasitenkd. Infektionskre. Hyg. Abt. 1: Orig. Reihe A* 235:13–19
14. Fishman, P. H., Brady, R. O. 1976. *Science* 194:906–15
15. Holmgren, J., Lindholm, L. 1976. *Immunol. Commun.* 5:737–56
16. van Heyningen, W. E., King, C. A. 1976. In *Ganglioside Function: Biochemical and Pharmacological Implications*, ed. G. Porcellati, B. Ceccarelli, G. Tettamanti, 205–14. New York: Plenum
17. van Heyningen, W. E., van Heyningen, S., King, C. A. 1976. *Ciba Found. Sym.* 42:73–88
18. Vaughan, M. 1976. In *Eukaryotic Cell Function and Growth. Regulation by Intracellular Cyclic Nucleotides*, ed. J. E. Dumont, B. L. Brown, N. J. Marshall, 113–21. New York: Plenum
19. Finkelstein, R. A. 1977. *Zbl. Bakt. Hyg., I. Abt. Orig. A* 239:283–93
20. Gill, D. M. 1977. In *Adv. Cyclic Nucleotide Res.* 8:85–118

21. Kohn, L. D. 1977. In *Annual Reports in Medicinal Chemistry*, ed. F. H. Clarke, 211–22. New York: Academic
22. van Heyningen, S. 1977. *Biol. Rev.* 52:509–49
23. Vaughan, M., Moss, J. 1978. *J. Supramol. Struct.* 18:473–88
24. Lönnroth, I. 1976. *Cholera toxin. Structure-function relationship.* PhD thesis, Inst. Med. Microbiol. Univ. Göteborg, Göteborg, Sweden, 41 pp.
25. Holmgren, J. 1978. In *Bacterial Toxins and Cell Membranes*, ed. J. Jeljaszewicz, T. Wadström, 333–66. New York: Academic
26. Holmgren, J. 1976. *Proc. Int. Congr. Endocrinol., 5th, Hamburg, 1976*, 1: 497–501
27. Holmgren, J. 1973. *Proc. US-Japan Cholera Conf., 9th*, pp. 196–213
28. Holmgren, J., Svennerholm, A.-M. 1977. *J. Infect. Dis.* 136:S105–12 (Suppl.)
29. van Heyningen, W. E., Carpenter, C. C. J., Pierce, N. F., Greenough, W. B. III. 1971. *J. Infect. Dis.* 124:415–18
30. Cuatrecasas, P. 1973. *Biochemistry* 12: 3547–58
31. Cuatrecasas, P. 1973. *Biochemistry* 12: 3558–66
32. Holmgren, J. 1973. *Infect. Immun.* 8: 851–59
33. Holmgren, J., Lönnroth, I., Svennerholm, L. 1973. *Infect. Immun.* 8: 208–14
34. King, C. A., van Heyningen, W. E. 1973. *J. Infect. Dis.* 127:639–47
35. Pierce, N. F. 1973. *J. Exp. Med.* 137:1009–23
36. van Heyningen, W. E., Mellanby, J. 1973. *Naunyn-Schmiedebergs Arch. Pharmacol.* 276:297–302
37. Wolff, J., Temple, R., Cook, G. H. 1973. *Proc. Natl. Acad. Sci. USA* 70:2741–44
38. Holmgren, J., Månsson, J.-E., Svennerholm, L. 1974. *Med. Biol.* 52:229–33
39. Staerk, J., Ronneberger, H. J., Wiegandt, H., Ziegler, W. 1974. *Eur. J. Biochem.* 48:103–10
40. van Heyningen, W. E. 1974. *Nature* 249:415–17
41. Zenser, T. V., Metzger, J. F. 1974. *Infect. Immun.* 10:503–9
42. Donta, S. T., Viner, J. P. 1975. *Infect. Immun.* 11:982–85
43. Gill, D. M., King, C. A. 1975. *J. Biol. Chem.* 250:6424–32
44. Haksar, A., Maudsley, D. V., Perön, F. G. 1975. *Biochim. Biophys. Acta* 381:308–23

45. Wolff, J., Cook, G. H. 1975. *Biochim. Biophys. Acta* 413:283–90
46. Donta, S. T. 1976. *J. Infect. Dis.* 133:S115–19 (Suppl.)
47. Manuelidis, L., Manuelidis, E. E. 1976. *Science* 193:588–90
48. Wiegandt, H., Ziegler, W., Staerk, J., Kranz, T., Ronneberger, H. J., Zilg, H., Karlsson, K.-A., Samuelsson, B. E. 1976. *Hoppe-Seylers Z. Physiol. Chem.* 357:1637–46
49. Wishnow, R. M., Lifrak, E., Chen, C.-C. 1976. *J. Infect. Dis.* 133:S108–14 (Suppl.)
50. Stoeckel, K., Schwab, M., Thoenen, H. 1977. *Brain Res.* 132:273–85
51. van Heyningen, S. 1977. *Biochem. J.* 168:457–63
52. Moss, J., Fishman, P. H., Richards, R. L., Alving, C. R., Vaughan, M., Brady, R. O. 1976. *Proc. Natl. Acad. Sci. USA.* 73:3480–83
53. Moss, J., Richards, R. L., Alving, C. R., Fishman, P. H. 1977. *J. Biol. Chem.* 252:797–98
54. Ohsawa, T., Nagai, Y., Wiegandt, H. 1977. *Japan. J. Exp. Med.* 47:221–22
55. Richards, R. L., Moss, J., Alving, C. R., Fishman, P. H., Brady, R. O. 1978. *Fed. Proc.* 37(6):1677(Abstr.)
56. van Heyningen, W. E. 1973. *Naunyn-Schmiedebergs Arch. Pharmacol.* 276: 289–95
57. Haksar, A., Maudsley, D. V., Peron, F. G. 1974. *Nature* 251:514–15
58. Hollenberg, M. D., Fishman, P. H., Bennett, V., Cuatrecasas, P. 1974. *Proc. Natl. Acad. Sci. USA* 71:4224–28
59. Gascoyne, N., van Heyningen, W. E. 1975. *Infect. Immun.* 12:466–69
60. Holmgren, J., Lönnroth, I., Månsson, J.-E., Svennerholm, L. 1975. *Proc. Natl. Acad. Sci. USA* 72:2520–24
61. Révesz, T., Greaves, M. 1975. *Nature* 257:103–6
62. Basu, M., Basu, S., Shanabruch, W. G., Moskal, J. R., Evans, C. H. 1976. *Biochem. Biophys. Res. Commun.* 71:385–92
63. Fishman, P. H., Moss, J., Vaughan, M. 1976. *J. Biol. Chem.* 251:4490–94
64. Holmgren, J., Lönnroth, I. 1976. *J. Infect. Dis.* 133:S64–74. (Suppl.)
65. Kanfer, J. N., Carter, T. P., Katzen, H. M. 1976. *J. Biol. Chem.* 251:7610–19
66. King, C. A., van Heyningen, W. E., Gascoyne, N. 1976. *J. Infect. Dis.* 133:S75–81 (Suppl.)
67. Moss, J., Fishman, P. H., Manganiello, V. C., Vaughan, M., Brady, R. O. 1976. *Proc. Natl. Acad. Sci. USA* 73:1034–37

68. Mullin, B. R., Aloj, S. M., Fishman, P. H., Lee, G., Kohn, L. D., Brady, R. O. 1976. *Proc. Natl. Acad. Sci. USA* 73:1679–83
69. Revesz, T., Greaves, M. F., Capellaro, D., Murray, R. K. 1976. *Br. J. Haematol.* 34:623–30
70. Moss, J., Manganiello, V. C., Fishman, P. H. 1977. *Biochemistry* 16:1876–81
71. Pacuszka, T., Moss, J., Fishman, P. H. 1978. *J. Biol. Chem.* 253:5103–8
72. Hansson, H.-A., Holmgren, J., Svennerholm, L. 1977. *Proc. Natl. Acad. Sci. USA* 74:3782–86
73. Cuatrecasas, P. 1973. *Biochemistry* 12:3577–81
74. Lichtenstein, L. M., Henney, C. S., Bourne, H. R., Greenough, W. B. III. 1973. *J. Clin. Invest.* 52:691–97
75. Field, M. 1974. *Proc. Natl. Acad. Sci. USA* 71:3299–3303
76. Walker, W. A., Field, M., Isselbacher, K. J. 1974. *Proc. Natl. Acad. Sci. USA* 71:320–24
77. Bennett, V., Cuatrecasas, P. 1975. *J. Membr. Biol.* 22:1–28
78. Bennett, V., O'Keefe, E., Cuatrecasas, P. 1975. *Proc. Natl. Acad. Sci. USA* 72:33–37
79. King, C. A., van Heyningen, W. E. 1975. *J. Infect. Dis.* 131:643–48
80. Lönnroth, I., Lönnroth, C. 1977. *Exp. Cell Res.* 104:15–24
81. Mekalanos, J. J., Collier, R. J., Romig, W. R. 1977. *Infect. Immun.* 16:789–95
82. Sattler, J., Schwarzmann, G., Staerk, J., Ziegler, W., Wiegandt, H. 1977. *Hoppe-Seylers Z. Physiol. Chem.* 358:159–63
83. Bitensky, M. W., Wheeler, M. A., Mehta, H., Miki, N. 1975. *Proc. Natl. Acad. Sci. USA* 72:2572–76
84. Sahyoun, N., Cuatrecasas, P. 1975. *Proc. Natl. Acad. Sci. USA* 72:3438–42
85. van Heyningen, S., King, C. A. 1975. *Biochem. J.* 146:269–71
86. Matuo, Y., Wheeler, M. A., Bitensky, M. W. 1976. *Proc. Natl. Acad. Sci. USA* 73:2654–58
87. van Heyningen, S. 1976. *Biochem. J.* 157:785–87
88. van Heyningen, S. 1976. *J. Infect. Dis.* 133:S5–13. (Suppl.)
89. Wheeler, M. A., Solomon, R. A., Cooper, C., Hertzberg, L., Mehta, H., Miki, N., Bitensky, M. W. 1976. *J. Infect. Dis.* 133:S89–96. (Suppl.)
90. Wodnar-Filipowicz, A., Lai, C. Y. 1976. *Arch. Biochem. Biophys.* 176:465–71
91. Joseph, K. C., Kim, S. U., Stieber, A., Gonatas, N. K. 1978. *Proc. Natl. Acad. Sci. USA* 75:2815–19

92. Holmgren, J., Lindholm, L., Lönnroth, I. 1974. *J. Exp. Med.* 139:801–19
93. Cuatrecasas, P., Parikh, I., Hollenberg, M. D. 1973. *Biochemistry* 12:4253–64
94. van Heyningen, S. 1974. *Science* 183:656–57
95. Fishman, P. H., Moss, J., Osborne, J. C. Jr. 1978. *Biochemistry* 17:711–16
96. Alving, C. R., Fowble, J. W., Joseph, K. C. 1974. *Immunochemistry* 11:475–81
97. Moss, J., Osborne, J. C. Jr., Fishman, P. H., Brewer, H. B. Jr., Vaughan, M., Brady, R. O. 1977. *Proc. Natl. Acad. Sci. USA* 74:74–78
98. Lönnroth, I., Holmgren, J. 1975. *J. Gen. Microbiol.* 91:263–77
99. Gammack, D. B. 1963. *Biochem. J.* 88:373–83
100. Yohe, H. C., Rosenberg, A. 1972. *Chem. Phys. Lipids* 9:279–94
101. Holmgren, J., Lönnroth, I., Svennerholm, L. 1973. *Scand. J. Infect. Dis.* 5:77–78
102. Wiegandt, H., Bücking, H. W. 1970. *Eur. J. Biochem.* 15:287–92
103. Sattler, J., Schwarzmann, G., Knack, I., Röhm, K.-H., Wiegandt, H. 1978. *Hoppe-Seylers Z. Physiol. Chem.* 359:719–23
104. Jacobs, J. W., Niall, H. D., Sharp, G. W. G. 1974. *Biochem. Biophys. Res. Commun.* 61:391–95
105. Holmgren, J., Lönnroth, I. 1975. *J. Gen. Microbiol.* 86:49–65
106. Sattler, J., Wiegandt, H., Staerk, J., Kranz, T., Ronneberger, H. J., Schmidtberger, R., Zilg, H. 1975. *Eur. J. Biochem.* 57:309–16
107. Gill, D. M. 1976. *Biochemistry* 15:1242–48
108. Klapper, D. G., Finkelstein, R. A., Capra, J. D. 1976. *Immunochemistry* 13:605–11
109. Kurosky, A., Markel, D. E., Touchstone, B., Peterson, J. W. 1976. *J. Infect. Dis.* 133:S14–22 (Suppl.)
110. Lai, C. Y., Mendez, E., Chang, D. 1976. *J. Infect. Dis.* 133:S23–30 (Suppl.)
111. Ohtomo, N., Muraoka, T., Tashiro, A., Zinnaka, Y., Amako, K. 1976. *J. Infect. Dis.* 133:S31–40 (Suppl.)
112. Kurosky, A., Markel, D. E., Peterson, J. W. 1977. *J. Biol. Chem.* 252:7257–64
113. Lai, C.-Y. 1977. *J. Biol. Chem.* 252:7249–56
114. Lai, C.-Y., Mendez, E., Chang, D., Wang, M. 1977. *Biochem. Biophys. Res. Commun.* 74:215–22
115. Sigler, P. B., Druyan, M. E., Kiefer, H. C., Finkelstein, R. A. 1977. *Science* 197:1277–79
116. Finkelstein, R. A., LoSpalluto, J. J. 1969. *J. Exp. Med.* 130:185–202
117. Vaughan, M., Pierce, N. F., Greenough, W. B. III. 1970. *Nature* 226:658–59
118. Bennett, V., Mong, L., Cuatrecasas, P. 1975. *J. Membr. Biol.* 24:107–29
119. Hart, D. A., Finkelstein, R. A. 1975. *J. Immunol.* 114:476–80
120. Holmgren, J., Lange, S., Lindholm, L., Lönnroth, C., Lönnroth, I. 1977. *Exp. Cell Res.* 108:31–39
121. Berkenbile, F., Delaney, R. 1976. *J. Infect. Dis.* 133:S82–88 (Suppl.)
122. Donta, S. T., Kreiter, S. R., Wendelschafer-Crabb, G. 1976. *Infect. Immun.* 13:1479–82
123. Gill, D. M. 1975. *Proc. Natl. Acad. Sci. USA* 72:2064–68
124. Gill, D. M. 1976. *J. Infect. Dis.* 133:S55–63 (Suppl.)
125. Martin, B. R., Houslay, M. D., Kennedy, E. L. 1977. *Biochem. J.* 161:639–42
126. Moss, J., Vaughan, M. 1977. *Proc. Natl. Acad. Sci. USA* 74:4396–4400
127. Gill, D. M., Rappaport, R. 1976. *Proc. Joint Conf. US-Japan Cooperative Med. Program, 13th, Sapporo, 1976,* pp. 159–65
128. Fischer, J., Kohler, T. R., Lipson, L. G., Flores, J., Witkum, P. A., Sharp, G. W. G. 1978. *Biochem. J.* 173:59–64
129. Moss, J., Manganiello, V. C., Vaughan, M. 1976. *Proc. Natl. Acad. Sci. USA* 73:4424–27
130. Flores, J., Witkum, P., Sharp, G. W. G. 1976. *J. Clin. Invest.* 57:450–58
131. Enomoto, K., Gill, D. M. 1978. *J. Supramol. Struct.* In press
132. Johnson, G. L., Bourne, H. R. 1977. *Biochem. Biophys. Res. Commun.* 78:792–98
133. Levinson, S. L., Blume, A. J. 1977. *J. Biol. Chem.* 252:3766–74
134. Glossmann, H., Struck, C. J. 1977. *Naunyn-Schmiedebergs Arch. Pharmacol.* 299:175–85
135. Cheung, W. Y. 1971. *J. Biol. Chem.* 246:2859–69
136. Wolff, D. J., Siegel, F. L. 1972. *J. Biol. Chem.* 247:4180–85
137. Teo, T. S., Wang, T. H., Wang, J. H. 1973. *J. Biol. Chem.* 248:588–95
138. Stevens, F. C., Walsh, M., Ho, H. C., Teo, T. S., Wang, J. H. 1976. *J. Biol. Chem.* 251:4495–4500
139. Watterson, D. M., Harrelson, W. G. Jr., Keller, P. M., Sharief, F., Vanaman, T. C. 1976. *J. Biol. Chem.* 251:4501–13
140. Lin, Y. M., Liu, Y. P., Cheung, W. Y. 1974. *J. Biol. Chem.* 249:4943–51

141. Cheung, W. Y. 1970. *Biochem. Biophys. Res. Commun.* 38:533–38
142. Dedman, J. R., Potter, J. D., Jackson, R. L., Johnson, J. D., Means, A. R. 1977. *J. Biol. Chem.* 252:8415–22
143. Klee, C. B. 1977. *Biochemistry* 16:1017–24
144. Wolff, D. J., Poirier, P. G., Brostrom, C. O., Brostrom, M. A. 1977. *J. Biol. Chem.* 252:4108–17
145. Brostrom, C. O., Huang, Y.-C., Breckenridge, B. M., Wolff, D. J. 1975. *Proc. Natl. Acad. Sci. USA* 72:64–68
146. Brostrom, M. A., Brostrom, C. O., Breckenridge, B. M., Wolff, D. J. 1976. *J. Biol. Chem.* 251:4744–50
147. Lynch, T. J., Tallant, E. A., Cheung, W. Y. 1976. *Biochem. Biophys. Res. Commun.* 68:616–25
148. Lynch, T. J., Tallant, E. A., Cheung, W. Y. 1977. *Arch. Biochem. Biophys.* 182:124–33
149. Gopinath, R. M., Vincenzi, F. F. 1977. *Biochem. Biophys. Res. Commun.* 77:1203–9
150. Jarrett, H. W., Penniston, J. T. 1977. *Biochem. Biophys. Res. Commun.* 77:1210–16
151. Cheung, W. Y., Bradham, L. S., Lynch, T. J., Lin, Y. M., Tallant, E. A. 1978. *Biochem. Biophys. Res. Commun.* 66:1055–62
152. Dabrowska, R., Sherry, J. M. F., Aromatorio, D. K., Hartshorne, D. J. 1978. *Biochemistry* 17:253–58
153. Waisman, D. M., Singh, T. J., Wang, J. H. 1978. *J. Biol. Chem.* 253:3387–90
154. Jarrett, H. W., Penniston, J. T. 1978. *J. Biol. Chem.* 253:4676–82
155. Wang, J. H., Teo, T. S., Ho, H. C., Stevens, F. C. 1975. In *Adv. Cyclic Nucleotide Res.* 5:179–94
156. Goor, R. S., Maxwell, E. S. 1970. *J. Biol. Chem.* 245:616–23
157. Collier, R. J., Kandel, J. 1971. *J. Biol. Chem.* 246:1496–503
158. Drazin, R., Kandel, J., Collier, R. J. 1971. *J. Biol. Chem.* 246:1504–10
159. Gill, D. M., Dinius, L. L. 1971. *J. Biol. Chem.* 246:1485–91
160. Gill, D. M., Pappenheimer, A. M. Jr. 1971. *J. Biol. Chem.* 246:1492–95
161. Honjo, T., Nishizuka, Y., Kato, I., Hayaishi, O. 1971. *J. Biol. Chem.* 246:4251–60
162. Kandel, J., Collier, R. J., Chung, D. W. 1974. *J. Biol. Chem.* 249:2088–97
163. Robinson, E. A., Henriksen, O., Maxwell, E. S. 1974. *J. Biol. Chem.* 249:5088–93
164. Collier, R. J. 1975. *Bacteriol. Rev.* 39:54–85

165. Chung, D. W., Collier, R. J. 1977. *Infect. Immun.* 16:832–41
166. Pappenheimer, A. M. Jr. 1977. *Ann. Rev. Biochem.* 46:69–94
167. Collier, R. J., Cole, H. A. 1969. *Science* 164:1179
168. Honjo, T., Nishizuka, Y., Hayaishi, O., Kato, I. 1968. *J. Biol. Chem.* 243:3553–55
169. Leppla, S. H., 1976. *Infect. Immun.* 14:1077–86
170. Vasil, M. L., Kabat, D., Iglewski, B. H. 1977. *Infect. Immun.* 16:353–61
171. Collier, R. J., Pappenheimer, A. M. Jr. 1964. *J. Exp. Med.* 120:1019–39
172. Collier, R. J. 1967. *J. Mol. Biol.* 25:83–98
173. Iglewski, B. H., Kabat, D. 1975. *Proc. Natl. Acad. Sci. USA* 72:2284–88
174. Strauss, N., Hendee, E. D. 1959. *J. Exp. Med.* 109:145–63
175. Kato, I., Pappenheimer, A. M. Jr. 1960. *J. Exp. Med.* 112:329–49
176. Strauss, N. 1960. *J. Exp. Med.* 112:351–59
177. Kato, I. 1962. *Jpn. J. Exp. Med.* 32:335–43
178. Goor, R. S., Pappenheimer, A. M. Jr. 1967. *J. Exp. Med.* 126:899–912
179. Pavlovskis, O. R., Gordon, F. B. 1972. *J. Infect. Dis.* 125:631–36
180. Pavlovskis, O. R., Shackelford, A. H. 1974. *Infect. Immun.* 9:540–46
181. Moss, J., Vaughan, M. 1977. *J. Biol. Chem.* 252:2455–57
182. Moss, J., Vaughan, M. 1978. *Proc. Natl. Acad. Sci. USA* 75:3621–24
183. Moss, J., Garrison, S., Oppenheimer, N. J., Richardson, S. H. 1978. *Proc. US-Jpn. Joint Cholera Conf., Karatsu, Japan, 1978.* In press
184. Cassel, D., Pfeuffer, T. 1978. *Proc. Natl. Acad. Sci. USA* 75:2669–73
185. Gill, D. M., Meren, R. 1978. *Proc. Natl. Acad. Sci. USA* 75:3050–54
186. Oppenheimer, N. J. 1978. *J. Biol. Chem.* 253:4907–10
187. Trepel, J. B., Chuang, D.-M., Neff, N. H. 1977. *Proc. Natl. Acad. Sci. USA* 74:5440–42
188. Trepel, J. B., Chuang, D.-M., Neff, N. H. 1978. *Fed. Proc.* 37(3):798 (Abstr.)
189. Moss, J., Ross, P. S., Vaughan, M. 1978. *Proc. Joint Conf. on Cholera, 13th, Atlanta, 1977,* pp. 382–95
190. Ferro, A. M., Oppenheimer, N. J. 1978. *Proc. Natl. Acad. Sci. USA* 75:809–813
191. Cassel, D., Selinger, Z. 1977. *Proc. Natl. Acad. Sci. USA* 74:3307–11
192. Cassel, D., Selinger, Z. 1976. *Biochim. Biophys. Acta* 452:538–51

193. Gill, D. M. 1978. *J. Supramol. Struct.* In press
194. Gill, D. M. 1978. See Ref. 189, pp. 229–48
195. Rowe, B., Taylor, J., Bettelheim, K. A. 1970. *Lancet* 1:1–5
196. Jacks, T. M., Wu, B. J. 1974. *Infect. Immun.* 9:342–47
197. Kantor, H. S., Tao, P., Gorbach, S. L. 1974. *J. Infect. Dis.* 129:1–9
198. Shore, E. G., Dean, A. G., Holik, E. J., Davis, B. R. 1974. *J. Infect. Dist.* 129:577–82
199. Gorbach, S. L., Kean, B. H., Evans, D. G., Evans D. J. Jr., Bessudo, D. 1975. *N. Engl. J. Med.* 292:933–36
200. Finkelstein, R. A., Vasil, M. L., Jones, J. R., Anderson, R. A., Barnard, T. 1976. *J. Clin. Microbiol.* 3:382–84
201. Merson, M. H., Morris, G. K., Sack, D. A., Wells, J. G., Feeley, J. C., Sack, R. B., Creech, W. B., Kapikian, A. Z., Gangarosa, E. J. 1976. *N. Engl. J. Med.* 294:1299–305
202. Ryder, R. W., Sack, D. A., Kapikian, A. Z., McLaughlin, J. C., Chakraborty, J., Mizanur Rahman, A. S. M., Merson, M. H., Wells, J. G. 1976. *Lancet* 1:659–62
203. Sack, D. A., Kaminsky, D. C., Sack, R. B., Itotia, J. N., Arthur, R. R., Kapikian, A. Z., Ørskov, F., Ørskov, I. 1978. *N. Engl. J. Med.* 298:758–63
204. Sack, R. B. 1975. *Ann. Rev. Microbiol.* 29:333–53
205. Field, M., Laird, W. J., Graf, L. H. Jr., Smith, P. L., Gill, D. M. 1978. See Ref. 189, pp. 127–36
206. Field, M., Graf, L. H. Jr., Laird, W. J., Smith, P. L. 1978. *Proc. Natl. Acad. Sci. USA* 75:2800–4
207. Gyles, C. L., Barnum, D. A. 1969. *J. Infect. Dis.* 120:419–26
208. Holmgren, J., Söderlind, O., Wadström, T. 1973. *Acta Pathol. Microbiol. Scand.* Sect. B 81:757–62
209. Hewlett, E. L., Guerrant, R. L., Evans, D. J. Jr., Greenough, W. B. III. 1974. *Nature* 249:371–73
210. Dorner, F., Jaksche, H., Stöckl, W. 1976. *J. Infect. Dis.* 133:S142–56 (Suppl.)
211. Evans, D. J. Jr., Chen, L. C., Curlin, G. T., Evans, D. G. 1972. *Nature New Biol.* 236:137–38
212. Mashiter, K., Mashiter, G. D., Hauger, R. L., Field, J. B. 1973. *Endocrinology* 92:541–49
213. Hynie, S., Rasková, H., Sechser, T., Vaneček, J., Matejovská, D., Matejovská, V., Treu, M., Polák, L. 1974. *Toxicon* 12:173–79
214. Kwan, C. N., Wishnow, R. M. 1974. *Infect. Immun.* 10:146–51
215. Gill, D. M., Evans, D. J. Jr., Evans, D. G. 1976. *J. Infect. Dis.* 133:S103–7 (Suppl.)
216. Bergman, M. J., Guerrant, R. L., Murad, F., Richardson, S. H., Weaver, D., Mandell, G. L. 1978. *J. Clin. Invest.* 61:227–34
217. Donta, S. T., Moon, H. W., Whipp, S. C. 1974. *Science* 183:334–36
218. Donta, S. T., Smith, D. M. 1974. *Infect. Immun.* 9:500–5
219. Keusch, G. T., Donta, S. T. 1975. *J. Infect. Dis.* 131:58–63
220. Evans, D. J. Jr., Evans, D. G., Richardson, S. H., Gorbach, S. L. 1976. *J. Infect. Dis.* 133:S97–102 (Suppl.)
221. Larivière, S., Gyles, C. L., Barnum, D. A. 1973. *J. Infect. Dis.* 128:312–20
222. Evans, D. J. Jr., Evans, D. G., Gorbach, S. L. 1974. *Infect. Immunol.* 10:1010–17
223. Söderlind, O., Möllby, R., Wadström, T. 1974. *Zentralbl. Bakteriol. Parasitenkd. Infektionskre. Hyg. Abt. 1: Orig.* 229:190–204
224. Finkelstein, R. A., LaRue, M. K., Johnston, D. W., Vasil, M. L., Cho, G. J., Jones, J. R. 1976. *J. Infect. Dis.* 133:S120–37 (Suppl.)
225. Miwatani, T., Takeda, Y., Takeda, T., Honda, T., Yano, T., Taga, S., Shimizu, M. 1978. See Ref. 189, pp. 369–81
226. Rappaport, R. S. 1978. See Ref. 189, pp. 443–74
227. Smith, N. W., Sack, R. B. 1973. *J. Infect. Dis.* 127:164–70
228. Gyles, C. L. 1974. *J. Infect. Dis.* 129:277–83
229. Dafni, Z., Robbins, J. B. 1976. *J. Infect. Dis.* 133:S138–41 (Suppl.)
230. Klipstein, F. A., Engbert, R. F. 1977. *Infect. Immun.* 18:110–17
231. Nalin, D. R., McLaughlin, J. C. 1978. *J. Med. Microbiol.* 11:177–86
232. Moss, J., Richardson, S. H. 1978. *J. Clin. Invest.* 62:281–85

Ann. Rev. Biochem. 1979. 48:601–48

AMINOACYL-tRNA SYNTHETASES: GENERAL FEATURES AND RECOGNITION OF TRANSFER RNAS

♦12021

Paul R. Schimmel

Department of Biology, Massachusetts Institute of Technology, Cambridge, Massachusetts 02139

Dieter Söll

Department of Molecular Biophysics and Biochemistry, Yale University, New Haven, Connecticut 06520

CONTENTS

601

0066-4154/79/0701-0601$01.00

PERSPECTIVES AND SUMMARY

The aminoacyl-tRNA synthetases are a class of enzymes that catalyze the first step in protein biosynthesis (1–15). In this reaction, each amino acid is attached to the 3' end of its cognate tRNA chain. An ester linkage is formed between the amino acid and one of the hydroxyl groups of the terminal adenosine. The energy required for formation of the ester bond is supplied by the hydrolysis of ATP according to the reaction

$$AA + ATP + tRNA \leftrightarrows AA\text{-}tRNA + AMP + PP_i \qquad 1.$$

where AA denotes amino acid. Although there are often several specific tRNA species corresponding to each amino acid, as far as is known at present, in bacteria there is only one aminoacyl-tRNA synthetase for each amino acid.

The aminoacylation reaction catalyzed by these enzymes must be controlled with extraordinary precision. If the wrong amino acid is stably attached to a tRNA, that amino acid can then be inserted into the growing polypeptide chain at a position determined by the interaction between the anticodon of the tRNA and the codon of the messenger RNA (16). Thus misacylation, like mutations, can give rise to mistakes in proteins.

During the past five years the amount of research on aminoacyl-tRNA synthetases has rapidly increased. The area has attracted investigators with widely different interests. For example, from the standpoint of enzyme mechanisms, the enzymes are of interest because their reactions are complex and involve three substrates, one of which is a macromolecule. In the field of protein-nucleic acid interactions, the synthetases and tRNAs are unusually attractive for study. This is because (*a*) a large number of synthetase-tRNA systems can be investigated and compared; (*b*) the nucleic acid (tRNA) is relatively small; (*c*) for approximately 100 of them the primary structure is known (17, 18). In one case even the three-dimensional structure has been elucidated (19, 20). In addition, the synthetases are interesting to study from an evolutionary standpoint, since they represent a large group of enzymes with similar functions.

But research on synthetases has not been confined simply to the study of their reactions, structural relationships, and role in protein synthesis. It has become increasingly clear that tRNAs are involved in the regulation of gene expression (21–24). In particular, it appears that the level of aminoacyl-tRNA serves as a signal to turn on or switch off specific genes. In some cases it is not clear whether the synthetase, or aminoacyl-tRNA, or both are the critical components. One possible mechanism for their action is suggested by the attenuation phenomenon in which tRNA, or aminoacyl-tRNA, or the synthetase influence transcription termination at a site between the promotor and the first structural gene of an operon (see e.g. 25–27). Even without knowledge of all the details, it is apparent that because synthetases catalyze the formation of aminoacyl-tRNA, they are directly or indirectly involved in regulating the expression of certain genes. For this reason, considerable effort has been made to study the control of aminoacyl-tRNA synthetase formation (28, 29).

In this article we summarize some of the general aspects of the recent research on aminoacyl-tRNA synthetases. In view of the explosive development of the field in recent years, it is not possible to cover adequately all topics in a short article. For example, it has not been practical to provide a comprehensive review of the enormous amount of work done on the kinetic and mechanistic features of these enzymes. Instead, we concentrate on some general features, including the still puzzling question of the recognition of tRNAs. To assist in summarizing this work, we have employed a number of figures and tables. These limitations notwithstanding, we believe we have touched upon most of the exciting achievements of the past few years.

GENERAL CHARACTERISTICS

Physical Characteristics

SIZES AND SUBUNIT STRUCTURES   Table 1 summarizes data on the molecular weights and subunit structures of aminoacyl-tRNA synthetases

Table 1 Molecular weight, quaternary structure, and substrate binding sites of highly purified aminoacyl-transfer RNA synthetases[a]

Synthetase	Source	Structural parameters				Binding sites	
		Mol. wt.	Subunits	Type	AA comp[b]	Type	Number
Alanyl-	yeast	128,000 (30)	none (30)	α	unknown	—	—
Arginyl-	Bacillus stearothermophilus	78,000 (31)	none (31)	α	unknown	amino acid ATP aminoacyl adenylate	1 (31) 1 (31) none (31)
	Escherichia coli	63,000 (32) 74,000 (33, 34)	none (32, 33) —	α —	known (34) —	— —	— —
	Neurospora crassa	85,000 (35)	none (35)	α	unknown	—	—
Aspartyl-	yeast	106,000 (36)	none (36)	α	unknown	—	—
Cysteinyl-	yeast	160,000 (37)	—	—	unknown	—	—
	rat liver	240,000 (38)	2 (38)	α_2	unknown	aminoacyl adenylate	2 (38)
Glutaminyl-	E. coli	69,000 (39)	none (39)	α	known (39)	—	—
Glutamyl-	E. coli	102,000 (40)	2: α = 56,000 β = 46,000 (40)	$\alpha\beta$	known (40)	tRNA	1 (41)
		59,000 (187)	none (187)	α	—	tRNA	1 (187)
Glycyl-	E. coli	227,000 (42, 43)	4: α = 33,000 β = 80,000 (42, 43)	$\alpha_2\beta_2$	known (43)	glycyl-AMP	2 (43)
Histidyl-	E. coli	84,000 (44)	2 (44)	α_2	known (44)	amino acid ATP	2 (44) 2 (44)
	Salmonella typhimurium	80,000 (45)	2 (45)	α_2	known (45)	—	—

	Source	Molecular weight	Number of subunits	Sequence	Subunit structure	Binding site	Number (ref)
Isoleucyl-	B. stearothermophilis	115,000 (46)	—	unknown	—	aminoacyl adenylate	1 (46)
						tRNA	1 (46)
	E. coli	112,000 (47, 48)	none (47)	known (48)	α	amino acid	1 (49)
						ATP	1 (50)
						aminoacyl adenylate	1 (51)
						tRNA	1 (52, 53)
Leucyl-	yeast	124,000 (30)	none (30)	unknown	α		—
	B. stearothermophilus	110,000 (54)	none (54)	known (54)	α		—
	E. coli	105,000 (55, 56)	none (55, 56)	known (55)	α	ATP	1 (56)
						aminoacyl adenylate	1 (56)
						tRNA	1 (56)
Lysyl-	Saccharomyces cerevisiae	120,000 (57)	2 (57)	unknown	α2		—
	Candida utilis	128,000 (58)	none (58)	unknown	α		—
	E. coli	104,000 (59)	2 (59)	known (59)	α2	amino acid	2 (59)
						ATP	2 (59)
	yeast	138,000 (59, 60)	2 (59, 60)	known (59)	α2	amino acid	2 (59)
						ATP	2 (59)
						aminoacyl adenylate	1 (61)
						tRNA	1 (61)
Methionyl-	rabbit reticulocytes	122,000 (45a)	2 (45a)	unknown	α2		—
	B. stearothermophilus	135,000 (54)	2 (54)	known (54)	α2	aminoacyl adenylate	2 (62)
						amino acid	2 (66)
	E. coli	170,000 (63, 64)	2 (63)	known (65)	α2	ATP	4 (66)
						aminoacyl adenylate	2 (62)
						methioninol-AMP	2 (66)
						tRNA	2 (67)
	wheat germ A	105,000 (68)	none (68)	known (68)	α		—
	wheat germ B	70,000 (68)	none (68)	known (68)	α		—

Table 1 *(continued)*

Synthetase	Source	Structural parameters				Binding sites	
		Mol. wt.	Subunits	Type	AA comp[b]	Type	Number
Phenylalanyl-	rat liver	287,000 (69)	4: α = 69,000 β = 75,000 (69)	—	unknown	—	—
	Drosophila melanogaster	180,000 (70)	—	—	unknown	—	—
	E. coli	267,000 (71, 72)	4: α = 39,000 β = 94,000 (71, 72)	α₂β₂	known (73)	aminoacyl adenylate tRNA	2 (74) 2 (74)
	yeast	220,000 (75)	4: α = 50,000 β = 60,000 (75)	α₂β₂	known (75)	amino acid aminoacyl adenylate tRNA	2 (76) 2 (77) 2 (76)
		262,000 (78)	4: α = 61,000 β = 70,000 (78)	α₂β₂	—	—	—
		286,000 (79)	4: α = 66,000 β = 77,000 (79)	α₂β₂	—	—	—
Prolyl-	*E. coli*	94,000 (80)	2 (80)	α₂	unknown	—	—
Seryl-	hen liver	120,000 (81)	2 (81)	α₂	unknown	—	—
	E. coli B	103,000 (82)	2 (82)	α₂	unknown	ATP aminoacyl adenylate	2 (83) 2 (83)
	E. coli K₁₂	95,000 (84)	2 (84)	α₂	known (84)	amino acid ATP tRNA	2 (85) 2 (85) 1 (86); 2 (85)

	Source	Molecular weight	Subunit number	Structure	Mechanism	Intermediate	Ref.
	yeast	95,000 (87)	2 (87)	α2	unknown	tRNA	1 (87, 88)
		120,000 (89)	2 (89)	α2	unknown	tRNA	2 (90–93)
						amino acid	2 (94)
						ATP	2 (94)
Threonyl-	*Lupinus luteus* seeds	110,000 (95)	2 (95)	α2	unknown	—	—
	E. coli	152,000 (96)	none (96)	α2	unknown	—	—
Tryptophanyl-	human placenta	116,000 (97)	2 (97)	α2	unknown		
	bovine pancreas	108,000 (98)	2 (98)	α2	known (99)	amino acid	2 (98)
						aminoacyl adenylate	2 (100)
						tRNA	2 (101)
	buffalo brain	155,000 (102)	3 (102)	α3	unknown	—	—
	B. stearothermophilus	70,000 (54)	2 (54)	α2	known (103)	aminoacyl adenylate	2 (62)
						amino acid	2 (106)
						tryptophanyl-ATP	2 (105)
						tRNA	2 (106)
	E. coli	74,000 (104)	2 (104)	α2	known (105)	—	—
	yeast	110,000 (107)	2 (107)	—	unknown	—	—
	L. luteus seeds	200,000 (95)	4 (95)	α4	unknown		
Tyrosyl-	*B. stearothermophilus*	88,000 (54)	2 (54)	α2	known (54)	amino acid	2 (108)
						ATP	2 (108)
						aminoacyl adenylate	2 (108)
	B. subtilis	88,000 (109)	—	—	known (109)	—	—
	E. coli	95,000 (109, 110)	2 (110)	α2	known (109, 110)	amino acid	2 (111)
						aminoacyl adenylate	2 (111)
						tRNA	1 (112, 113)
	yeast	80,000 (114)	2 (114)	α2	unknown	—	2 (110)
		116,000 (30)	4 (30)	α4	unknown	—	—

Table 1 (continued)

Synthetase	Source	Structural parameters				Binding sites	
		Mol. wt.	Subunits	Type	AA comb[b]	Type	Number
Tyrosyl- (continued)	rat liver	124,000 (115)	2 (115) α = 62,000 β = 61,000	αβ	unknown	—	—
	soybean cytoplasm	122,000 (116)	2 (116)	α₂	unknown	—	—
	soybean chloroplast	86,000 (116)	2 (116)	α₂	unknown	—	—
Valyl-	B. stearothermophilus	110,000 (54)	none (54)	α	known (54)	amino acid ATP aminoacyl adenylate	2 (111) 2 (111) 2 (111)
	E. coli	110,000 (117)	none (49)	α	known (117)	amino acid aminoacyl adenylate ttRNA	1 (49) 1 (49) 1 (118, 119)
	yeast	122,000 (59)	none (59)	α	known (59)	amino acid ATP tRNA	1 (59) 1 (59) 1 (120)
	L. luteus seeds	125,000 (95)	none (95)	α	unknown	—	—

[a] Criteria for including data for an enzyme were (a) a check of the purity of the enzyme preparation by several methods, and (b) a determination of the molecular weight, preferably by polyacrylamide gel electrophoresis in the presence of sodium dodecyl sulfate. Since the accuracy of these methods is limited, and since proteolytic cleavage is sometimes a problem during enzyme purification (60), values for the molecular weights must be regarded as approximate. Figures in brackets are reference numbers.
[b] Amino acid composition.

from various sources. Most of these data were obtained on bacterial enzymes. The molecular weights of synthetases are surprisingly variable. For example, considering just *Escherichia coli* enzymes, they range in molecular weight from 59,000 for Glu-tRNA synthetase to 227,000 for Gly-tRNA synthetase. There is even larger variation when enzymes from different organisms are considered.

A second striking feature is that synthetases have four different types of subunit structures: α, α_2, α_4, and $\alpha_2\beta_2$. Of course, this breakdown into four classes is somewhat tenuous. For example, one could imagine that an enzyme isolated as a single chain in vitro, and thus designated α, could in vivo occur as an α_2 or an α_4 species. But this qualification notwithstanding, it is apparent that the enzymes show substantial variations in subunit structure as determined in vitro.

At present, there is no apparent reason why such a large diversity in molecular weights and subunit structures exists. However, as discussed in a subsequent section, it is now known that the protomers of some synthetases contain amino acid sequences that are repeated twice. For example, there appears to be a twofold repeat of certain sequences in *E. coli* Ile-tRNA synthetase, a single-chain (α) enzyme of molecular weight 110,000 (121). Obviously, this finding adds a new dimension to our way of viewing the subunit structures of synthetases.

NUMBER OF SYNTHETASES   As mentioned earlier, only one aminoacyl-tRNA synthetase for each amino acid has been found in bacteria. A possible exception may be in *Bacillus subtilis* where no Gln-tRNA synthetase is found. Instead Glu-tRNA synthetase acylates both tRNA^{Glu} and tRNA^{Gln} with Glu (122). The Glu-tRNA^{Gln} is then converted to Gln-tRNA^{Gln} by a transamination reaction (123).

In eukaryotic cells different aminoacyl-tRNA synthetases are present in the cytoplasm, the mitochondria, and the chloroplasts. And even in the cytoplasm multiple synthetases may exist for some amino acids (see Table 1).

SPECIAL FEATURES OF SYNTHETASES FROM HIGHER EUKARYOTES
In isolating aminoacyl-tRNA synthetases from bacterial sources, it is routine to obtain the enzymes as free species, unbound to other material. In contrast, when synthetases are isolated from higher organisms, many investigators have found that they occur as high molecular weight (HMW) complexes. Typically, synthetase activities have been found in complexes whose molecular weights range from several hundred thousand to well over a million.

The HMW forms of synthetases have been isolated from sources as diverse as rat liver, rabbit reticulocytes, chick embryos, Chinese hamster ovary cells, HeLa cells, and human placenta (124–134). In one study, synthetases from six different eukaryote cell types were simultaneously examined (134). In all cases, HMW forms were found. The HMW complexes may also contain proteins other than synthetases. For example, tRNA modification enzymes and elongation factors may be associated with large complexes that also contain aminoacyl-tRNA synthetases (132, 135, 136).

Studies of synthetase complexes are still at a primitive stage. One study suggests that multiple copies of only two synthetases can give rise to HMW complexes that may also contain lesser amounts of other enzymes (137). Perhaps because of the fragile nature of these complexes there is no strong evidence to suggest that all enzymes are usually bound together in a single complex. But it is just as possible that there are heterogeneous complexes, each containing different groups of enzymes.

The physiological significance of these complexes is not apparent. There is no question that they occur, and they undoubtedly are present in vivo. But such complexes have not been detected in bacteria such as *E. coli*, although they may simply be more labile so that they escape undetected. Possibly the activity or stability of the enzymes may be different in the complex and in the "free state." In any event, the question of physiological significance is likely to be a long standing issue.

Mutants of Aminoacyl-tRNA Synthetases

RATIONALE FOR STUDY   An extraordinary effort has been directed at isolating and characterizing mutants of aminoacyl-tRNA synthetases. Many of these studies were done in the hope that the use of mutationally altered enzymes will shed light on the various roles in cellular processes of the synthetases themselves or of their product, aminoacyl-tRNA. Of course, such mutants should also be useful for learning more about the mechanism of catalysis and the organization of ligand binding sites on these enzymes.

Experiments with temperature-sensitive strains containing a thermolabile synthetase have provided much general information (for reviews see 22–24). If growth of the strain is conducted under semipermissive conditions, then the concentration of the aminoacyl-tRNA should be much reduced compared to that found in wild-type cells. The steady-state levels of specific cellular components can then be determined. Their amounts reflect the influence of the in vivo concentration of a particular aminoacyl-tRNA on the regulation of their levels. Such studies, for instance, showed the involvement of aminoacyl-tRNA or of the aminoacyl-tRNA synthetase in the regulation of certain amino acid operons (for example see 138), in the transport of branched-chain amino acids (139, 140), in the synthesis of the magic spot compounds (p)ppGpp (141), and in cell division (142).

Recently, synthetase mutants proved useful in defining the ligand binding sites of an enzyme. Using two mutants, a hybrid enzyme of *E. coli* Phe-tRNA synthetase that is analog resistant and temperature-sensitive was formed and used to demonstrate the presence of two ligand binding sites on Phe-tRNA synthetase (143).

So far a large variety of problems have been studied with the help of aminoacyl-tRNA synthetase mutants. Although the details of these studies are not considered further here, a summary of the available mutants follows.

SUMMARY OF MUTANTS   Table 2 summarizes the known aminoacyl-tRNA synthetase mutants from bacteria, yeast, *N. crassa,* and mammalian cells. Altogether mutants for 17 of the enzymes have been obtained. In some cases mutants for the same amino acid specific enzyme have been isolated from several different organisms.

The temperature-sensitive mutants are the most common. Next most common are the auxotrophic and analog-resistant mutants. Auxotrophic mutants are characterized by a requirement for high concentrations of the cognate amino acid; this can be due to an elevated amino acid K_m. Analog-resistant enzymes are mutants that can tolerate a substrate analog under conditions that are toxic to the wild-type protein. Resistance to an amino acid analog can result from a change in the amino acid binding site, so that the analog no longer inactivates or inhibits the protein.

Of the mutations that have been characterized, the most common ones affect the amino acid binding site. However, mutations affecting the ATP and tRNA sites have also been reported.

Thus far, the main use of these matters has been the study of many regulatory processes in the cell. It may be expected that, in the future, use will be made of mutants of synthetases for the study of their mechanism.

CLONING OF SYNTHETASE GENES   The availability of aminoacyl-tRNA synthetase mutants makes possible selection methods to clone specific synthetase genes into plasmid or phage vectors. For example, the *E. coli* strain KL 380, containing a temperature-sensitive Ala-tRNA synthetase, has been used as a host to clone the wild-type Ala-tRNA synthetase on the *pBR322* plasmid (D. LaDage and P. R. Schimmel, in preparation). Transformations were done with a pool of recombinant plasmids carrying a variety of *E. coli* genes. Transformed KL380 cells that were temperature-resistant had picked up the *pBR322* plasmid bearing the wild-type Ala-tRNA synthetase gene. By a somewhat different procedure, the

Table 2 Aminoacyl-transfer RNA synthetase mutants

Synthetase	Organism	Type of mutant	Enzyme defect[a]	Genetic location	Reference
Alanyl-	Chinese hamster	auxotrophic	—	—	144
	Escherichia coli	temperature-sensitive	tRNA	near *recA*	145–147
	mouse cells	temperature-sensitive	—	—	148
Arginyl-	Chinese hamster	temperature-sensitive	—	—	149
	E. coli	analog-resistant	ATP, tRNA	between *pheS* and *his*	33, 150, 151
Asparaginyl-	Chinese hamster	temperature-sensitive	amino acid		152, 153
	E. coli	temperature-sensitive	—	at min 21	154
	hamster	temperature-sensitive	—	—	155
Glutaminyl-	Chinese hamster	temperature-sensitive	—	—	149
	E. coli	temperature-sensitive	—	near *lip*	156
Glutamyl-	*E. coli*	temperature-sensitive	—	near *dsdA*	157
		streptomycin-dependent	amino acid	near *xyl*, near *his*	158, 159
Glycyl-	*E. coli*	auxotrophic	amino acid	near *xyl*	160, 161
		temperature-sensitive	—	near *xyl*	162
Histidyl-	Chinese hamster	temperature-sensitive	—	—	149
	E. coli	bradytrophic	amino acid	—	163
	Salmonella typhimurium	analog-resistant	amino acid	near *strB*	164
Isoleucyl-	*E. coli*	auxotrophic	ATP, amino acid	between *thr* and *pyrA*	165, 166
		analog-resistant	—	—	167
		temperature-sensitive	—	—	168
	S. typhimurium	auxotrophic	—	near *pyrA*	169
	Saccharomyces cerevisiae	temperature-sensitive	—	on chromosome 2	170, 171
Leucyl-	Chinese hamster	temperature-sensitive	amino acid	—	172–174
	E. coli	temperature-sensitive	—	near *lip*	175
	S. typhimurium	analog-resistant	amino acid	near *gal*	176, 177
	Neurospora crassa	temperature-sensitive, auxotrophic	amino acid	—	178, 179
Lysyl-	*Bacillus subtilis*	temperature-sensitive	amino acid, ATP	between *purA* and *sul*	180, 181
	E. coli	analog-resistant	—	—	182
Methionyl-	Chinese hamster	temperature-sensitive	—	—	149
	E. coli	auxotrophic	amino acid	at min 47	183
		analog-resistant	tRNA	—	184
	S. typhimurium	auxotrophic	amino acid	at min 67	185, 186
	S. cervisiae	temperature-sensitive	amino acid	—	171
Phenylalanyl-	*E. coli*	analog-resistant	amino acid	near *pps*	188, 189
		temperature-sensitive	—	near *pps*	157
Seryl-	*E. coli*	temperature-sensitive	—	near *serC*	175, 190
		serine hydroxamate-resistant	—	—	191
Threonyl-	*E. coli*	borrelidin-resistant	amino acid, ATP[b]	at min 37.7	96, 192
		auxotrophic	amino acid	near *trp*	193
Tryptophanyl-	*B. subtilis*	temperature-sensitive, analog-resistant	—	between *argG* and *metA*	194
	E. coli	auxotrophic	amino acid	—	195, 196
		temperature-sensitive	—	between *purA* and *aroB*	157
	N. crassa	auxotrophic	—	—	197
Tyrosyl-	*E. coli*	analog-resistant	amino acid	at min 35	157, 198, 199
	S. typhimurium	auxotrophic	—	—	200
Valyl-	*E. coli*	temperature-sensitive	ATP, amino acid	near *pyrB*	201, 202, 203

[a] The affinity of the enzyme for the specified substance is lowered unless otherwise indicated.
[b] Affinity increased.

gene for *E. coli* Gly-tRNA synthetase has also been cloned on a *colEl* plasmid (G. Nagel, private communication). The gene for *E. coli* Phe-tRNA synthetase has also recently been cloned in a plaque-forming λ-transducing phage (206). Similar in principle to the Ala-tRNA synthetase selection described above, the selection for this phage was based on complementation of a temperature-sensitive strain containing a thermolabile Phe-tRNA synthetase. An additional bonus of this λ phage was found later: it also carries the gene for *E. coli* Thr-tRNA synthetase (207). In an analogous fashion the gene for *E. coli* Trp-tRNA synthetase was cloned using a strain with a defective Trp-tRNA synthetase which caused Trp auxotrophy (208). A *colEl* plasmid, into which the gene for wild-type *E. coli* Trp-tRNA synthetase had been inserted by recombinant DNA techniques, was selected for its ability to restore prototrophy to this strain (209). A specialized λ-transducing phage carrying the gene for *E. coli* Ile-tRNA synthetase has been obtained by conventional genetic methods (210). The DNA of this phage has recently been used to synthesize Ile-tRNA synthetase in a coupled transcription-translation system and to determine the effect in vitro of magic spot compounds on its formation (211).

Cells carrying a synthetase gene on a recombinant plasmid have enzyme levels that are elevated several fold (204, 205). In the case of *trpS* plasmid over 30-fold amplification of Trp-tRNA synthetase was found (209). Thus, cloning is a useful procedure for preparing large amounts of these enzymes. In addition, the cloned genes may serve as useful vehicles for studying appropriate cellular regulatory processes. With so many synthetase mutants now available, it is anticipated that most of the aminoacyl-tRNA synthetase genes will be cloned in the future.

SEQUENCES OF AMINOACYL-TRANSFER RNA SYNTHETASES

Introduction

As a first step in understanding the mechanism and function of any enzyme, it is necessary to define the primary structure in terms of amino acid sequence. This is a common motivation for sequence studies. However, in the case of aminoacyl-tRNA synthetases, additional questions are particularly interesting and are best approached through sequence analysis.

For example, the different subunit structures and sizes of the synthetases are somewhat puzzling. One might expect that enzymes with such closely similar functions would show roughly similar structures, at least at the level of overall size as measured by molecular weight. It is easy to imagine that the evolution of synthetases proceeded from a common species that made suitable adjustments in its sequence to accommodate the different amino acids and transfer RNAs. There is no evidence for this from molecular weight data; sequence data may support this theory. Also, because the enzymes have closely similar catalytic functions, we might expect them to share certain structural homologies. Notwithstanding the differences in molecular weights and subunit structures, we can readily imagine that certain critical areas, such as those around the ATP and tRNA binding sites, might show structural homologies.

As discussed below, these kinds of questions can and have been analyzed by amino acid sequence determinations. Although only one enzyme has been sequenced in its entirety (103), partial sequences are now available for a number of others. Collectively, these data have given a surprising amount of interesting new information and insights into the questions posed above. A dramatic expansion of this work in the next few years is certain.

Internal Sequence Repeats

ENZYMES FROM BACILLUS STEAROTHERMOPHILUS Koch, Boulanger & Hartley undertook a study of sequence relationships in a number of enzymes from *B. stearothermophilus* (54). Investigations were carried out on the methionine (a dimer of 66,000 molecular weight subunits), valine (a single chain of 110,000 molecular weight), and leucine (a single chain of 110,000 molecular weight) enzymes. Using high resolution two-dimensional mapping of tryptic peptides, it was shown in each case that the number of peptides was approximately one half or less of the expected number. This determination was made by using more than one specific stain for the peptides. In addition, a number of specific tryptic peptides from each enzyme was purified. Amino acid compositions and yields were carefully determined. In the case of the leucine and valine enzymes, some peptides are clearly present in amounts of 1.7–1.9 moles per mole of single chain monomeric enzyme. The methionine enzyme also has some peptides that are present in amounts exceeding 1 per subunit. Thus, both the peptide mapping and the independent determination of the amounts of specific peptides per enzyme molecule or subunit, clearly prove that the three enzymes have internal sequence repeats.

The question of the positional location of the sequence repeats in the primary structure was also approached. One can imagine that, in an individual subunit or enzyme molecule, the repeated sequences are clustered together in tandem, or nearly so, or that they are widely separated. When the valine enzyme was split into two pieces consisting of 67,000 and 40,000 molecular weight segments, tryptic maps of these two fragments and of the native enzyme showed that with all three proteins the same peptides stand out on their chromatograms. This suggests that at least some of the repeated sequences must occur in widely separated regions. A similar conclusion was reached when the methionine enzyme was analyzed.

In the case of these enzymes, it seems likely that they arose by gene duplication and fusion (54). Conceivably this process was part of that used to evolve different enzymes from a common ancestor, although this is by no means certain.

Similar studies were also done with Tyr-tRNA synthetase from *B. stearothermophilus.* This enzyme is a dimer of 47,000 molecular weight subunits. In this instance, no evidence could be found for internal sequence repeats (54). This suggests, as have subsequent studies (see below), that only the large subunits, or single chain enzymes, carry the repeats.

SEQUENCE REPEATS IN E. COLI SYNTHETASES   The question of sequence repeats has been investigated in a number of other enzymes. These studies tend to confirm the impression that large, single chain enzymes, or enzymes with large subunits, contain repeats, while those with small subunits do not. One of the first sequence investigations was done by Kula on Ile-tRNA synthetase from *E. coli* MRE 600 (121). This enzyme is a single polypeptide chain with a molecular weight of 112,000. Attention was directed at the peptides containing cysteine. Altogether there are 14 cysteine residues in the enzyme. But after careful analysis, only 8 cysteines can be accounted for in unique sequences. The clear implication is that the cysteine residues occur within sequences that are internally repeated. This study gives rise to the idea that Ile-tRNA synthetase, although a large monomer, may in fact be equivalent to a dimeric enzyme in which the subunits have been covalently fused together.

Waterson & Konigsberg studied both the Ser-tRNA synthetase and Leu-tRNA synthetase from *E. coli* (212). The former is a dimer of identical 50,000 molecular weight subunits, while the latter is a single chain 105,000 molecular weight protein. Their data showed that the latter, but not the former, contains sequence repeats. Similarly, *E. coli* Met-tRNA synthetase, a dimer of 85,000 molecular weight subunits, has repeated sequences (213). In this case, as in the case of the *B. stearothermophilus* enzymes mentioned above, the repeats appear to occur in regions spaced widely apart in the sequence.

In the case of the Trp synthetase, sequence questions have been explored with enzyme from three different sources—*E. coli* (214, 215), *B. stearothermophilus* (103), and human placenta (214, 215). In the former two the enzyme is a dimer of identical 37,000 molecular weight subunits, while in the latter it is a dimer of 58,000 molecular weight subunits. In each instance, there is no evidence of sequence repeats.

Finally, the $\alpha_2\beta_2$ yeast Phe-tRNA synthetase has recently been examined by Robbe-Saul et al (215a). The subunits have molecular weights of about 60,000 (β) and 70,000 (α) (see Table 1). Repeated sequences appear to occur within each subunit.

Sequence Homologies Between Enzymes

Do sequence homologies exist between the various aminoacyl-tRNA synthetases? In view of the common catalytic functions of these enzymes this is certainly plausible. Although data at this point are quite limited, there is no evidence that synthetases specific for different amino acids contain sequence homologies. However, it may be expected that this question will be investigated more rigorously in the future.

However, some interesting sequence homologies exist between the same enzyme isolated from different organisms. For example, Muench and co-workers studied the Trp-tRNA synthetase from *E. coli* and human placenta (214). As mentioned above, these enzymes are both dimers, but differ considerably in molecular weight (see Table 1). The amino acid sequences of some of the cysteine-containing peptides from these two enzymes were determined. Interestingly enough, it was found that remarkable homologies exist between a nonapeptide isolated from the *E. coli* enzyme and a decapeptide isolated from the human enzyme. The region of homology extends over about 7 residues.

The complete amino acid sequence of Trp-tRNA synthetase from *B. stearothermophilus* has been reported by Winter & Hartley (103). The enzyme has no internal sequence repeats, but does have a significant number of homologies with the partially sequenced enzyme from *E. coli* (216). The conservation of sequence is substantially higher in the N-terminal third of the molecule, than in the C-terminal two thirds (216). In addition, the *B. stearothermophilus* sequence appears to have no homologies with the partial sequences of the *B. stearothermophilus* Tyr-tRNA synthetase or *E. coli* Met-tRNA synthetase (103). Finally, it should be mentioned that it has not yet been possible to convert the sequence information into useful concepts about the biological mechanism of action of aminoacyl-tRNA synthetases. Information on this three-dimensional structure is obviously needed.

Active Site Peptides

Affinity labeling procedures have been used to locate active site regions in aminoacyl-tRNA synthetases (217–224a). These studies have employed both chemical and photochemical methods. For example, one approach relied on the attachment of a chemically reactive group to the aminoacyl moiety on aminoacyl-tRNA. In this way, Met-tRNA synthetase from *E. coli* has been acylated with *p*-nitrophenyl-carbamyl-Met-tRNA (217) and Ile-tRNA synthetase has been affinity labeled with N-bromoacetyl-Ile-tRNA (218). This approach presumably labels that part of the synthetase that is close to the aminoacyl-moiety of the tRNA.

Using direct ultraviolet irradiation, it has been possible to photo-cross-link a number of synthetases to bound tRNAs (219–221). In this way, several enzyme-tRNA contact sites are labeled, not just those associated with the aminoacyl-moiety.

In spite of these efforts and possibly because of the difficulty of obtaining large amounts of these enzymes, very little has been done to isolate and sequence active site peptides. Only two cases are known to date. First, Bruton & Hartley isolated and sequenced an octapeptide from Met-tRNA synthetase that is labeled with p-nitrophenyl-carbamyl-Met-tRNA (217). Presumably labeling occurred at a cysteine thiol group. In another example, Yue & Schimmel photo-cross-linked ATP to its specific receptor site on Ile-tRNA synthetase. A hexapeptide was isolated and sequenced (223). Similar cross-linking of ATP to other synthetases is now being done to determine whether the labeled peptides have homologies with each other.

CRYSTALLOGRAPHIC STUDIES

The X-ray crystallographic studies of aminoacyl-tRNA synthetases have already yielded some results (225, 226). Most progress has been made with Tyr-tRNA synthetase from *B. stearothermophilus.* In this case, the enzyme readily forms crystals that are ordered and relatively stable to X-rays. The enzyme is a dimer with a molecular weight of 90,000. In the crystal, the dimer occupies two crystallographic asymmetric units. It has been possible to make three isomorphous heavy atom derivatives. These have been used to obtain diffraction data to a resolution of 2.7 Å (226). From the electron density map, it is known that the monomer has dimensions of 60 Å × 60 Å × 40 Å. In the dimer, the maximum dimension is about 130 Å. From the tracing of the polypeptide chain, it appears that the subunits each have 5 α-helices that are longer than 12 Å and a 6-stranded pleated-sheet structure in the center of each subunit.

Further progress on the interpretation of the X-ray data, and the localization of binding and catalytic sites, is awaiting completion of the primary structure determination. With a detailed structural interpretation based on the knowledge of the amino acid sequence, it may be possible to construct rather accurate models of enzyme-ligand complexes, such as the synthetase-tRNA complex. This will be facilitated by affinity labeling studies in which substrates are cross-linked at their active sites to the enzyme. For example, it is possible to photo-cross-link tRNA[Tyr] to Tyr-tRNA synthetase (219). Parts of the tRNA involved in photo-cross-linking can be identified (219), and if the counterparts from the protein are also determined (by isolating cross-linked peptides and determining their sequences), it should be possible to locate precisely in the X-ray structure the position of the tRNA contact sites. Together with the molecular scale model available for tRNA, the detailed structure of the enzyme, and the affinity labeling information, it should be possible to construct a molecular scale model of the enzyme-tRNA complex.

At present, crystallographic work is also being done on the methionine enzyme from *E. coli* (225). A large, active tryptic fragment has been isolated and crystallized. Although the native enzyme is a dimer of molecular weight 172,000, the active fragment is a single chain of molecular weight 64,000. A 4 Å map of the fragment indicates an ellipsoidal structure of dimensions 90 Å X 43 Å X 43 Å.

It may be expected that considerable efforts in the crystallographic studies will continue and develop over the next several years. At this time all attempts to obtain crystals of enzyme-tRNA complexes suitable for X-ray diffraction analysis have failed. Given the different solubility properties of enzyme and tRNA, the obstacles of obtaining such stable crystals are formidable, although in due course some will no doubt be obtained. However, with molecular structures of both a free enzyme and free nucleic acid and information from affinity labeling experiments, it should be possible to build a reasonably accurate model of the complex if both macromolecules do not undergo gross conformational changes upon complex formation.

RECOGNITION OF TRANSFER RNAs BY AMINOACYL-TRANSFER RNA SYNTHETASES

The attachment of an animo acid to a specific tRNA is the event during which information in a nucleic acid is first given correspondence with an amino acid. Because there is a separate synthetase and one or more tRNAs for each amino acid, the synthetase-tRNA systems have provided a rich area for the investigation of protein-nucleic acid interactions. In particular, because of the wide variety of proteins and nucleic acids available, it is possible to study not only cognate complexes, but also the many permutations involving noncognate synthetase-tRNA complexes.

In spite of a large number of ingenious investigations, unraveling the mechanism of specific synthetase-tRNA interactions has been and remains an extraordinary challenge. In part, the difficulty is raised by the many structural similarities among the various tRNA species. This is illustrated by Figure 1, which gives the sequence and cloverleaf structure of yeast tRNA[Phe]. Almost all tRNAs sequenced to date (18) can fit into this general structure, having common features such as the dihydrouridine (D) loop, anticodon loop, variable loop, TΨC loop, and acceptor stem. In addition, there are constant nucleotides or constant purines or pyrimidines that are shared by most tRNA species. As indicated in Figure 1, many of these common bases are used to stabilize tertiary interactions that give rise to a three-dimensional structure that is believed to be similar for all tRNA species (227). For these reasons the synthetases differentiate between molecules that are very similar. Although there are segments, such as the anticodon, which are clearly unique to each tRNA species, neither the anticodon nor any other region has proven to be the sole recognition site for synthetases. With these considerations in mind, it is not hard to see why the molecular basis of recognition is a difficult challenge.

In discussing the synthetase-tRNA recognition problem, we consider first those studies that have concentrated on characterizing the enzyme-tRNA complexes from a physical standpoint. These studies deal with thermodynamic and kinetic characterizations of complex formation. Following this, we take up the many investigations of the molecular basis for recognition.

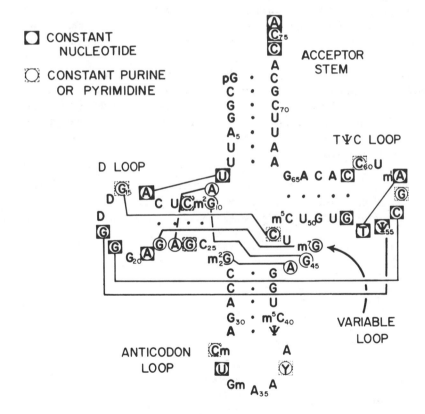

Figure 1 Cloverleaf model of yeast tRNA^Phe. Nucleotides that occur in constant positions for tRNAs involved in protein synthesis are indicated. Solid lines join residues that interact in the tertiary structure in the crystalline state. Adapted from (267).

Physical Studies of Synthetase-tRNA Interactions

STABILITY CONSTANTS MEASURED AT EQUILIBRIUM SHOW THAT
COGNATE COMPLEXES ARE STRONGER THAN NONCOGNATE ONES
Synthetase-tRNA complexes have been studied by nitrocellulose filter binding (52, 53, 119, 228), column chromatography (86, 229–231), sedimentation (41, 86, 87, 106, 119, 228, 230–234), fluorescence quenching (41, 87,
88, 118, 228, 231, 234–241), equilibrium partition (241a), and other methods. For the purpose of calculating stability constants, interpretation of data
seems least ambiguous with the fluorescence quenching measurements.
These studies, which are generally consistent with other investigations, have
shown that complex formation is stronger at lower pH values (e.g. association at pH 5.5 is stronger than that at pH 7.5) and that high concentrations
of a monovalent salt such as sodium chloride weakens complex formation
(240). Also, Mg^{2+} affects complex stability; a study of one system shows
that optimal stabilization is achieved with a Mg^{2+} concentration of between
1 and 5 mM (240).

Table 3 summarizes some of the data obtained. Results were obtained at
pH 5.5, where binding is stronger and easier to study. Data are given for
both cognate and noncognate systems. In addition, for some of these systems, both enzyme and nucleic acid are derived from the same organism
(homologous system) while in other cases they are derived from different
organisms (heterologous system). The data show that for the cases studied
the cognate, homologous complexes are strongest. For example, at pH 5.5
E. coli Ile-tRNA synthetase has an association constant of 10^8 M^{-1} with
E. coli tRNA[Ile]. Association constants of that synthetase with other E. coli
tRNAs tested are considerably less. On the other hand, noncognate associa-

Table 3 Association constants for cognate and noncognate enzyme-tRNA complexes[a]

Enzyme	tRNA	K (M^{-1})
Yeast ValRS	Val (yeast)	$\simeq 10^8$
	Val (E. coli)	3.2×10^7
	Ile (E. coli)	9.1×10^6
	Phe (E. coli)	2.8×10^6
	Glu (E. coli)	$< 10^4$
E. coli IleRS	Ile (E. coli)	$\simeq 10^8$
	Val (yeast)	2.0×10^7
	Phe (yeast, −Y)	2.5×10^6
	Phe (E. coli)	1.1×10^5
	Tyr (E. coli)	$< 10^4$
	Glu (E. coli)	$< 10^4$

[a] All measurements done in 67 mM sodium phosphate, pH 5.5, at 17°. [Data are from
(240)].

tions with the yeast tRNAs tested are considerably stronger. In particular, yeast tRNA^{Phe} binds at least an order of magnitude stronger than the homologous *E. coli* tRNA^{Phe}. Thus, it appears that binding specificity is greater in the homologous system (within the same organism) than in the heterologous cases.

THERMODYNAMIC PARAMETERS. COMPLEX FORMATION IS EN-TROPICALLY DRIVEN   Association constants for both homologous cognate and heterologous noncognate complexes have been studied as a function of temperature (234, 240). It is apparent from the data in Table 4 that in the case of the homologous cognate complexes, a large positive entropy change provides the bulk of the thermodynamic stabilization of the synthetase-tRNA complex. In fact, under the conditions used, for the *E. coli* Ile-tRNA synthetase-tRNA^{Ile} complex $\Delta H°$ is zero. In the case of the two heterologous, noncognate complexes studied, the enthalpic change actually discourages complex formation, while the positive entropy change is even greater than that observed for the homologous cognate complexes. These results suggest that electrostatic interactions play a major role in stabilizing the complexes. The large $\Delta S°$ values can arise from the liberation of solvating water molecules from charged residues upon complex formation (242). The results are reminiscent of those obtained by Riggs et al on the temperature dependence of the *lac* repressor-*lac* operator interaction, where a positive enthalpy and large positive entropy change occur (243). Thus, solvation changes associated with bringing together charged sites may in general dominate the thermodynamic features of protein-nucleic acid interactions.

Table 4   Thermodynamic parameters for formation of synthetase-tRNA complexes

Enzyme	tRNA	Conditions	$\Delta H°$ (kcal. mol^{-1})	$\Delta S°$ (cal. deg. mol^{-1})	Ref.
E. coli Ile-tRNA synthetase	*E. coli* tRNA^{Ile}	20 mM Na pipes, 5 mM MgCl$_2$, pH 6.5, 17°	0.0	34	240
E. coli Ile-tRNA synthetase	yeast tRNA^{Val}	20 mM Na pipes, 5 mM MgCl$_2$, pH 6.5, 17°	8.0	54	240
Yeast Phe-tRNA synthetase	yeast tRNA^{Phe}	30 mM K phosphate, 0.05 M KCl, 10 mM Mg^{2+}, pH 7.2, 24°	−3.5	20	234
Yeast Phe-tRNA synthetase	*E. coli* tRNA^{Tyr}	30 mM K phosphate, 0.05 M KCl, 10 mM Mg^{2+}, pH 7.2, 24°	15	75	234

DYNAMICS OF SYNTHETASE-tRNA INTERACTIONS. A TWO-STEP MECHANISM FOR FORMATION OF COGNATE COMPLEXES   The dynamics of synthetase-tRNA interactions have been studied using stopped-flow and temperature-jump methods (87, 234, 241, 244, 245). In both the yeast Ser-tRNA synthetase-yeast tRNA^{Ser} and the yeast-tRNA synthetase-yeast tRNA^{Phe} systems, two relaxation processes are associated with complex formation (234, 241, 245). These are interpreted according to a simple two-step mechanism given as

$$E + tRNA \underset{k_{-1}}{\overset{k_1}{\rightleftharpoons}} E \cdot tRNA \underset{k_{-2}}{\overset{k_2}{\rightleftharpoons}} E' \cdot tRNA' \qquad\qquad 2.$$

where E represents the enzyme. Thus, initial complex formation is followed by a unimolecular conformational change, indicated by the primed species. This kind of mechanism has also been found in studies of protein-small molecule complexes (246, 247).

Rate constants characterizing the two steps in the mechanism given by Equation 2 are given in Table 5. The bimolecular rate constant $k_1 \geqslant 10^8$ M^{-1} s^{-1}, which is near the diffusion-controlled limit. The disassociation rate constant k_{-1} and the rate constants associated with the unimolecular conformational change are in the range of $10^2 - 10^3$ s^{-1}.

The data have been interpreted to mean that initial complex formation is rapid and has a broad specificity, while the conformational change represents a more precise reading, or discrimination step (234, 241, 245). Support for this idea comes from data obtained in the heterologous noncognate pair yeast Phe-tRNA synthetase-E. coli tRNA^{Tyr}. In this case, only one relaxation process is detected, and this corresponds to bimolecular complex formation (Table 5). The bimolecular rate constant k_1 is near the diffusion controlled limit, which shows that initial complex formation proceeds with no difficulty. The absence of the second step in the mechanism indeed

Table 5   Rate constants for formation of synthetase-tRNA complexes

Enzyme	tRNA	Conditions	k_1 (M^{-1}s^{-1})	k_{-1} (s^{-1})	k_2 (s^{-1})	k_{-2} (s^{-1})	Ref.
Yeast Ser-tRNA synthetase	yeast tRNA^{Ser}	0.03 M K phosphate, 0.1 M KCl, 5 mM MgCl$_2$, pH 7.2, 24°	2.7×10^8	220	720	330	245[a]
Yeast Phe-tRNA synthetase	yeast tRNA^{Phe}	0.03 M K phosphate, 10 mM MgCl$_2$, 0.2 M KCl, pH 7.2, 24°	2×10^8	250	420	750	234
Yeast Phe-tRNA synthetase	E. coli tRNA^{Tyr}	0.03 M K phosphate, 10 mM MgCl$_2$, 0.05 M KCl, pH 7.2, 24°	8×10^8	1,600	—	—	234

[a] See also (241).

suggests that it is important for discrimination. In this regard, it should be noted that yeast Phe-tRNA synthetase cannot attach Phe to *E. coli* tRNA^{Tyr} (234).

STEADY-STATE KINETIC PARAMETERS INDICATE THAT THE MAXIMUM VELOCITY PLAYS A MAJOR ROLE IN RECOGNITION   As mentioned earlier (see Table 3), in a homologous system a synthetase binds most strongly to its cognate tRNA. However, studies of Ebel and co-workers have shown that discrimination depends not only on binding specificity, but in some cases an even bigger role is played by the maximum velocity (248). This is illustrated by data in Table 6A which give results of aminoacylations catalyzed by yeast Val-tRNA synthetase and Arg-tRNA synthetase, with homologous yeast tRNAs. In these cases, some misacylations can be detected. But as the table shows, while the K_m values for the misacylations are roughly 10^2 higher than those for the correct aminoacylations, the V_{max} values are $10^3 - 10^4$-fold smaller for the wrong aminoacylations. Thus, there probably are two distinguishable levels of recognition. One is at the level of substrate binding, while the other is at the level of catalysis.

This dual discrimination is also evident from data obtained by Roe, Sirover & Dudock (249). These investigators studied misacylations catalyzed by yeast Phe-tRNA synthetase using *E. coli* tRNAs. As discussed earlier in connection with Table 3, noncognate binding between heterologous enzymes and tRNAs is generally much stronger than the corresponding interactions in homologous systems. Furthermore, misacylations occur more readily in heterologous systems. In fact, yeast Phe-tRNA synthetase carries out a wide variety of misacylations with *E. coli* tRNAs. Some data are tabulated in Table 6B which show that the K_m values for the heterologous misacylations are similar to those for the cognate homologous aminoacylation. However, the misacylations proceed with considerably lower

Table 6A   Kinetic parameters for correct and incorrect homologous aminoacylations[a]

Enzyme	tRNA	K_m (μM)	V_{max} (arbitrary units)
Yeast Val-tRNA synthetase	yeast tRNA$_2^{Val}$	0.04	1
	yeast tRNA^{Phe}	14	0.0001
	yeast tRNA^{Ala}	2.5	0.0002
Yeast Arg-tRNA synthetase	yeast tRNA$_3^{Arg}$	0.1	1
	yeast tRNA^{Asp}	10	0.0004

[a] Experiments were done in 55 mM Tris-HCl (pH 7.5) at 30–37°. Data are taken from (248).

Table 6B Kinetic parameters for correct and incorrect heterologous aminoacylations of yeast Phe-tRNA synthetase

tRNA	K_m (μM)	V_{max} (arbitrary units)
Yeast tRNA^{Phe}	0.83	1.0
Wheat tRNA^{Phe}	0.56	1.2
E. coli tRNA^{Phe}	5.8	0.2
E. coli tRNA$_1^{Val}$	3.7	0.1
E. coli tRNA$_1^{Ala}$	4.2	0.05
E. coli tRNA^{Lys}	5.2	0.02
E. coli tRNA$_2^{Ala}$	3.4	0.02
E. coli tRNA$_{2A, 2B}^{Val}$	1.5	0.01
E. coli tRNA^{Ile}	1.3	0.007
E. coli tRNA^{Met}	1.0	0.007

Aminoacylations were carried out at 30° in 50 mM Tris, 40 mM MgCl$_2$, pH 8.2. [Taken from (249)].

maximum velocities. Here again, it is clear that substantial discrimination correlates with changes in V_{max}. Taken together, the data in Tables 6A and 6B establish a strong case for the concept of dual discrimination.

It should be noted that misacylations, even in the homologous system, can be induced by altering the solution conditions. For example, addition of organic solvents to the reaction mixture has been commonly used to induce a wide variety of misacylations (248, 250–255). This suggests that recognition is extremely subtle, and that through suitable manipulation the specificity can be varied. Such studies underscore the delicate nature of the recognition process.

Molecular Basis for Recognition

DIVERSE APPROACHES HAVE BEEN USED In attempting to elucidate the structural determinants of tRNA responsible for accurate enzyme recognition, a rich diversity of methods have been applied. These include nuclease digestion of synthetase-tRNA complexes (90, 256, 257), aminoacylation of dissected molecules (e.g. 258, 259), analysis of mutant tRNAs (260–264), chemical modifications (265), photochemical cross-linking (219–221), isotope labeling (266), and heterologous mischarging (249). These studies have given a wealth of information; and while not solving the recognition problem, they have succeeded in delineating some of its major features.

The investigations can be divided into two main areas. In one the tRNA is modified in some way and the effect of the modification on the interaction with synthetase is observed. Studies with chemically modified tRNAs, mutant tRNAs, and tRNA fragments (or dissected molecules) fall into this category. If the modification of a particular residue in the tRNA structure weakens or eliminates the interaction with synthetase, it is possible that this area is intimately involved in the synthetase-tRNA interaction. However, the modification or alteration may somehow disturb the tRNA structure and produce a new conformation that cannot interact with the synthetase. In this case, one cannot conclude that the site of modification is a synthetase interaction point. On the other hand, if the destruction or modification of a particular area of the tRNA has no effect on the synthetase-tRNA interaction, one may reasonably assume that it is not a critical locus for recognition by the enzyme.

The second class of studies aims to map directly the synthetase-tRNA contact points. Of this type are studies in which enzyme and tRNA after complex formation are photo-cross-linked. Determination of the cross-linking sites gives some idea of the synthetase-tRNA contact points. Isotope labeling studies, and, to a certain extent, nuclease digestion studies of enzyme-tRNA complexes, also fall into this category.

Another class of studies rely on comparisons of the primary structures of tRNA species that are acylated by the same homologous or heterologous aminoacyl-tRNA synthetase (14). In the different tRNAs, regions of similar sequence were found. Such sequences may be related to the enzyme recognition site. In the absence of detailed knowledge of the tertiary structure of these molecules, this approach is unfortunately limited to comparisons of primary and secondary structure features. As such this may not be sufficient, since enzyme recognition may depend critically on a certain three-dimensional conformation of the tRNA.

Many of these studies have been summarized elsewhere (13–15, 267–268). Here we discuss a few of the major investigations, and attempt to draw some general conclusions. Although our coverage is of necessity incomplete, the ideas behind the experiments and the general conclusions drawn are of broad applicability.

CHEMICAL MODIFICATION OF tRNAs   A large number of chemical modifications have been introduced into tRNAs. These include bisulfite-induced conversion of cytidine to uridine, cyanoethylation, kethoxalation, chloroacetaldehyde treatment, N-oxidation, site specific attachment of dyes or spin labels, photo-oxidation, and many more (269).

From the standpoint of chemical modification, the most heavily studied tRNA is *E. coli* tRNA^{fMet}. A useful summary of these studies is given by Schulman & Pelka (265). Many sites in this molecule have been modified and the altered tRNAs tested in their interaction with Met-tRNA synthetase. The general conclusions obtained are presented in Figure 2, which gives the nucleotide sequence and cloverleaf structure of tRNA^{fMet}. Areas that have been probed by chemical modifications are enclosed in boxes. Altogether 25 of the 77 bases have been probed. In most cases, the modification leaves the nucleic acid in an active form. However, modification of either of the first two bases of the anticodon, or of a base in the extra loop, or either of the two bases in the acceptor stem, gives rise to inactive tRNA species.

A major point can be drawn from Figure 2, as well as from other studies: sites that are apparently important for synthetase-tRNA interactions occur in diverse regions of the structure. For example, inactive tRNA^{fMet} molecules are produced by modifications in the anticodon or at the 3' end of the molecule (Figure 2). In the three-dimensional tRNA structure, these regions are separated by about 75 Å. This suggests that the synthetase may span a large part of the tRNA structure, and select critical residues in widely different areas.

A major criticism of chemical modification studies is that a modification in a particular area may disturb the conformation of the molecule as a

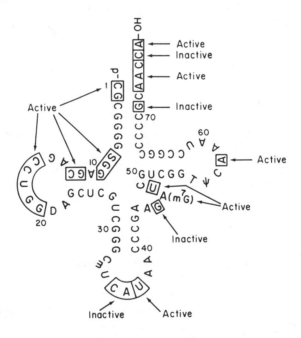

Figure 2 Modification sites in *E. coli* tRNA^{fMet} studied by Schulman and co-workers. Adapted from (265).

whole, thus introducing serious ambiguities into the interpretation of effects of the modification in synthetase interactions. Although this cannot be ruled out, the large number of studies done on tRNA^{fMet} lend credence to the conclusions. Since 20 out of 25 of the modifications still yield active molecules, nonspecific structural alterations are probably not common. In addition, the inactivation targets are close to sites that can be modified without inactivation; this makes it unlikely that inactivation targets are simply sites in a region that is generally more susceptible to producing a conformational change of the tRNA structure upon modification.

"DISSECTED MOLECULE" APPROACH   The goal of the dissected molecule approach is to cut away those parts of the tRNA structure that are not required for synthetase interaction, and to leave behind only the essential regions. This is done by aminoacylating fragments of a tRNA or combinations of fragments (7, 258, 259, 270). In this way partial molecules containing different parts of the structure are fabricated.

One of the most exhaustive studies of this type has been carried out by Zachau and co-workers on yeast tRNA^{Phe} (259). The molecule was split and degraded in various ways and artificial recombinations were generated. In a summary of the data, Figure 3 shows some of the structures tested and gives the phenylalanine acceptance as a percentage of that of intact tRNA^{Phe}. From the results it is clear that large portions of the anticodon loop and the dihydrouridine loop can be removed without destroying amino acid acceptance. On the other hand, a species in which a few nucleotides from each of these regions have been simultaneously removed is almost completely inactive. This apparent paradox could be due to the difficulty in forming the hydrogen bonded cloverleaf structure as it is depicted in Figure 3. That is, in constructing fragment recombinations, the different conformational states of the fragments may make it difficult to regenerate recombined molecules that contain all the normal cloverleaf hydrogen bonds that are possible.

This objection notwithstanding, it is clear that where removal of nucleotides results in a molecule that can still be aminoacylated, the bases in question must not be essential for recognition. In particular, the removal of anticodon nucleotides of tRNA^{Phe} results in a molecule that can still be aminoacylated with Phe-tRNA synthetase (Figure 3). Therefore, the anticodon is not crucial for the recognition. On the other hand, Figure 2 shows that chemical modification of certain anticodon bases in tRNA^{fMet} resulted in inactivation. Likewise, the removal of one anticodon nucleotide in yeast tRNA^{Val} leads to complete loss of the capacity to be aminoacylated by the Val-tRNA synthetase (258). These results could be harmonized by simply postulating that in some cases the enzyme (e.g. Met- or Val-tRNA synthe-

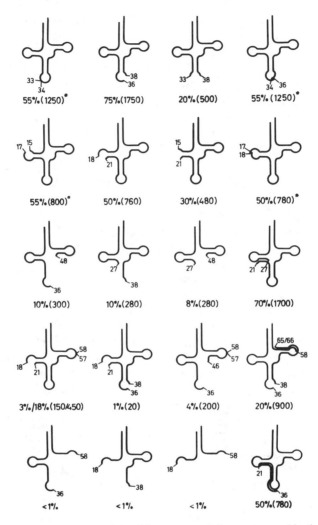

Figure 3 Maximal aminoacylation of fragments and fragment recombinations of yeast tRNA[Phe] relative to intact yeast tRNA[Phe]. Adapted from (259).

tase) interacts with the anticodon region, while in the other cases (e.g. Phe-tRNA synthetase) it does not. This suggests that different synthetases read different parts of the structure in their recognition process.

RECOGNITION AS STUDIED BY MUTATIONALLY ALTERED tRNAs
An elegant approach to the recognition problem takes advantage of the ability to produce by genetic means tRNA species differing from each other

by single base changes. [For a review see (260); (see also 264, 271)]. Since these are intact tRNA species, this approach avoids the objection of the dissected molecule method, in which the removal of nucleotides may give rise to a substantially altered tRNA conformation. In addition, the genetic approach has an advantage over chemical modification studies in which extraneous reagents and sometimes bulky groups are introduced into the tRNA.

A number of mutant E. coli tRNA species have been isolated that have altered synthetase recognition compared to wild-type tRNA (Table 7). In one series of investigations, genetic manipulation of the E. coli amber suppressor (su$_3^+$) tRNA^{Tyr} provided a number of mutant tRNAs (261, 262). Nucleotide analysis revealed that in each case the tRNA^{Tyr} species was

Table 7 Mutations in *E. coli* transfer RNAs that affect synthetase interactions

tRNA	Mutation	Location in structure	Effect of mutation	Reference
Gly$_1$		altered anticodon	greatly decreased rate of aminoacylation with Gly-tRNA synthetase	260
Gly$_2$	C36 → U36	third base in anti-codon	greatly decreased rate of aminoacylation with Gly-tRNA synthetase	260
Gly$_3$	C35 → U35	second base in anticodon	greatly decreased rate of aminoacylation with Gly-tRNA synthetase	260
Su$_3^+$Tyr	A82 → G82	acceptor stem	misacylation with glutamine	261–263
Su$_3^+$Tyr	C81 → A81	acceptor stem	misacylation with glutamine	262, 263
Su$_3^+$Tyr	C81 → U81	acceptor stem	misacylation with glutamine	262, 263
Su$_3^+$Tyr	G2 → A2	acceptor stem	misacylation with glutamine	262, 263
Su$_3^+$Tyr	G1 → A1	acceptor stem	misacylation with glutamine	263
Su$_7^+$Trp	C35 → U35	second base in anticodon	misacylation with glutamine	264
Su$_2^+$Gln$_2$	ψ37 → A37	anticodon loop	misacylation with unknown amino acid	a
Su$_2^+$Gln$_2$	ψ37 → C37	anticodon loop	misacylation with unknown amino acid	a
Su$_2^+$Gln$_2$	ψ38 → C38	anticodon loop	misacylation with unknown amino acid	a
Su$_2^+$Gln$_2$	ψ37 → A37/ G29 → A29		misacylation with Trp	a

[a] H. Ozeki, personal communication.

altered by a different single base change near the acceptor end and was then able to be aminoacylated in vivo and in vitro with glutamate. In addition, these tRNAs were still recognized by Tyr-tRNA synthetase and could be aminoacylated in vitro with tyrosine (263). By the use of a similar genetic scheme, starting from the amber suppressor (su$_2^+$) tRNA^{Gln}, mutant tRNAs were isolated in which single base changes in the anticodon region led to misacylation by an as yet unknown amino acid (271). However, a mutant tRNA, which differed in two positions from tRNA^{Gln}, could accept tryptophan in vitro (271).

Likewise, a mutation in the anticodon of the *E. coli* tRNA^{Trp} resulted in the formation of a new amber suppressor tRNA (su$_7^+$) which was aminoacylated with glutamine and responded to the amber codon by inserting glutamine into polypeptides (264). A nice confirmation of this genetic work came from the chemical modification of tRNA^{Trp} with sodium bisulfite. Upon a C → U transition in the anticodon the modified tRNA recognized the amber codon and could be charged with glutamine (272). In vivo the su$_7^+$ suppressor inserts low amounts of tryptophan in addition to glutamine. This is evidence that Trp-tRNA synthetase can still recognize the tRNA (273).

Using similar genetic schemes one should be able to isolate conditional mutants that contain aminoacyl-tRNA synthetases with altered amino acid specificity.

Studies of missense suppressor tRNAs also bear on the question of tRNA-synthetase interaction. Carbon and his colleagues have shown that mutations in the anticodon of some *E. coli* tRNA^{Gly} species result in a sharp reduction in the rate of aminoacylation by the cognate aminoacyl-tRNA synthetase [see (260) for a summary].

Thus, through genetic manipulation, it is possible to affect synthetase-tRNA recognition to the point where a tRNA is actually aminoacylated by a noncognate synthetase. Mutations of this type occur both in the anticodon region and in the acceptor terminus, regions that are widely separated in the three-dimensional structure. Although these studies do not define the molecular basis of recognition, they clearly show that it is extremely precise, and that recognition can be affected by only a single base change.

PHOTOCHEMICAL CROSS-LINKING AND ISOTOPE LABELING OF SYNTHETASE-tRNA COMPLEXES   Some studies have aimed at delineating topological features and contact points, without introducing significant structural perturbations into either the tRNA or the protein. For this purpose, both photochemical cross-linking of enzyme to tRNA and isotope labeling of tRNA have been useful.

Photochemical cross-linking is achieved by direct irradiation at 254 nm (219–221). Since a variety of nucleotide bases and amino acid side chains can covalently couple together under the action of UV light (274–276), this is an ideal procedure for introducing cross-links. When applied to synthetase-tRNA complexes, it has been shown that cross-linking is specific under appropriate conditions (219–221). An abundance of data indicate that the cross-linked species are representative of the native complexes. By determining the regions on the tRNA that are involved in cross-linking, it has been possible to estimate contact points in six different synthetase-tRNA complexes.

A different strategy is used in the isotope labeling experiments (266). The aim is the introduction of a reagent that will react with a wide number of bases in the tRNA molecule without disturbing the conformation. The reaction is carried out in the presence and absence of bound synthetase; those regions in tRNA that are covered by synthetase become inaccessible to the reagent. They can be identified through appropriate analysis and are probably contact or close proximity points.

Clearly, one of the main obstacles is finding a reagent that can react with a wide number of bases without perturbing the structure. For this purpose, advantage has been taken of the slow exchange of the C-8 hydrogen atoms of purine nucleotides (277–283). This hydrogen may be exchanged for solvent tritium. Thus, the tritium labeling pattern of free and bound tRNA is determined. Through appropriate nuclease digestions and chromatographic analysis, it is possible to determine the amount of tritium incorporated into each purine nucleotide throughout the structure. In this way, synthetase-tRNA contact points can be mapped (266).

In one case, an isotope labeling study was carried out on a complex that had also been studied by photochemical cross-linking (266). It was found that most of the sites perturbed in the isotope labeling experiment were also identified as contact regions in a photochemical cross-linking study (220). This confirmation not only supports the validity of both approaches, but also gives strong credence to the picture of synthetase-tRNA complexes that emerges from such studies.

The results of a number of investigations, particularly those involving photochemical cross-linking on six different complexes, suggests that there are some common features to the structural organization of synthetase-tRNA complexes. This is illustrated in Figure 4 which shows that in all photo-cross-linked complexes examined, cross-linking occurs in and around the 5'-side of the dihydrouridine stem. Since amino acids are attached to the 3'-termini of tRNAs, the enzymes must make contact at this point. Indeed, in some studies it has been possible to demonstrate directly that this

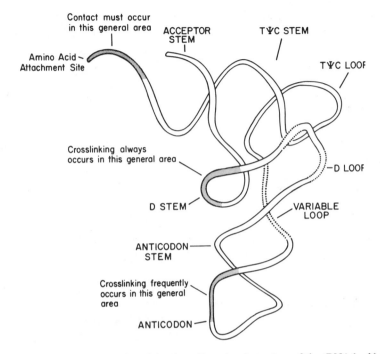

Figure 4   Schematic illustration of the three-dimensional structure of the tRNA backbone. Shaded areas identify regions that are close to the surface of the aminoacyl-tRNA synthetase, according to results on several systems. Adapted from (268).

is a contact area (220, 257, 284). Finally, the anticodon region is often implicated in photochemical cross-linking studies.

It is important to recognize that Figure 4 draws attention only to those regions that are generally, or frequently, observed as contact sites in synthetase-tRNA complexes. Certainly, in specific systems other regions may also be implicated, but these are not necessarily contact areas in a number of different synthetase-tRNA complexes. Thus, in addition to sharing some common features in the way in which they bind to tRNAs, individual variations among enzymes may also occur.

HETEROLOGOUS AMINOACYLATIONS   As discussed above, pure aminoacyl-tRNA synthetases can aminoacylate noncognate tRNA species from different cell origin (6). A systematic study (249) was undertaken with pure yeast Phe-tRNA synthetase which was shown to aminoacylate 11 pure tRNA species to high extents albeit at low reaction rates (Table 6B). A comparison of their nucleotide sequences revealed some common structural features (the same base found in the same position in the different tRNAs).

Based on this analysis, Dudock and his colleagues (249) proposed that 9 nucleotides in the D stem and the fourth nucleotide from the 3' end comprise the specific recognition sequence for this enzyme. Some doubts were cast on this appealing hypothesis when sequence analysis of yeast tRNA^{Met} showed this tRNA to contain all the nucleotides of the Phe-tRNA synthetase "recognition site" (285). Under standard aminoacylation conditions only a very low degree of mischarging of yeast tRNA^{Met} could be obtained; this level could be considerably raised when special aminoacylation conditions were used (286). Recently it was demonstrated that *Saccharomyces cerevisiae* Phe-tRNA synthetase efficiently charges tRNA^{Phe} from *Schizosaccharomyces pombe,* a different yeast species, although this tRNA lacks part of the nucleotide sequence of the proposed recognition site (287). These results make it clear that such a formalistic concept, based on the consideration of similarities in the two-dimensional cloverleaf structure, cannot be correct. Evidence has accumulated [e.g. from dissected molecule studies, (288)] that the correct tertiary structure plays an important role in the aminoacylation process. Thus, consideration of cloverleaf structures alone in deriving models of tRNA recognition is not enough.

PROPOSAL FOR THE STRUCTURAL ORGANIZATION OF SYNTHETASE-tRNA COMPLEXES   From considering a wide variety of data it has been suggested that tRNA synthetases in general bind along and around the inside of the L-shaped structure (267). This is illustrated in Figure 5, which depicts the three-dimensional structure of tRNA. The binding of synthetases is indicated by dashed lines. In some cases, the binding may extend all the way from the 3' terminus to the anticodon, while in other cases it may extend only part of this distance, as indicated by the different dashed lines. Since the diagram is only two-dimensional, and since no details of the synthetase binding cleft can be given, it is clear that there is some latitude as to the precise angle at which the tRNA molecule is inserted into the synthetase binding cleft. In particular, we can envision a plane passing through the tRNA molecule that intersects the 3'-terminal adenosine, the anticodon, and the corner near base 56. The angle at which this plane is inserted into the surface of the synthetase could be perpendicular to the protein surface, but could also be tilted in such a way that the dashed lines A-C in Figure 5 would not represent the central part of the contact area.

This proposal, which takes into account a large variety of diverse data, can also be harmonized with the different sizes of synthetases (see Table 1). A small enzyme may not be able to span the entire distance from the 3' terminus to the anticodon, while a larger enzyme can. Even with this rough working model for the structural features of the complexes we are still left with the question of the specificity of recognition. For this purpose, studies

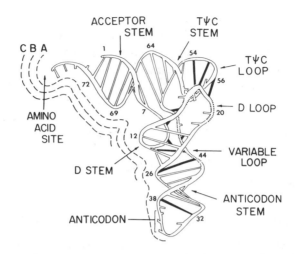

Figure 5 Schematic illustration of the three-dimensional structure of yeast tRNA[Phe]. The backbone is represented as a continuous tube and crossbars designate hydrogen bonds. Dashed lines A-C indicate different synthetase surfaces, which may extend part or all of the way from the 3' terminus to the anticodon. Interactions between enzyme and tRNA are not intended to be viewed as simply occurring along an edge, but also may involve regions on either side of the dotted lines. Adapted from (267).

that concentrate on specific kinds of bonding between amino acid side chains and bases within the part of the structure believed to be important for the synthetase-tRNA interactions, are most desirable.

IS THERE A UNIFORM RECOGNITION MECHANISM? The proposal outlined above suggests a general overall structure for all synthetase-tRNA pairs. Within such a complex there must be a few precise recognition points. They may involve specific recognition by the enzyme of several bases situated at distant points in the tRNA structure. The various approaches used give good evidence that certain regions in a tRNA are responsible for recognition by its cognate synthetase. However, these regions are not located at similar sites in the cloverleaf model for all the systems studied. Therefore, at this time we are not aware of regularities that may underlie the different results.

The question of synthetase-tRNA recognition has been, and will continue to be, a rich area for future investigations. Many physical approaches are being applied to the study of complexes, such as low angle X-ray scattering in solution (289) and neutron diffraction (290). New information should come from proton (291) or fluorine (292) NMR studies on tRNA or complexes (293). Also, important ideas concerning structural features will be forthcoming as investigators continue to piece together a variety of evidence. One interesting proposal is Kim's symmetry recognition hypothesis, which attempts to incorporate ideas of symmetry into the organization of the complexes (294). The merit of this proposal must await further investigations.

OTHER ASPECTS

Kinetics and Mechanism

Studies of the kinetics and mechanism of action of aminoacyl-tRNA synthetases have mushroomed during the past seven years. In the space of this article it is not possible to do justice to this interesting and important area. Instead, we attempt to summarize some of the most important conclusions.

AMINOACYL ADENYLATE MECHANISM   As indicated in Equation 1, the overall reaction involves the esterification of amino acid to tRNA with the accompanying hydrolysis of ATP to AMP and PP_i. Some controversy has centered on the question of whether the reaction proceeds in a concerted fashion (10), or whether an enzyme-bound aminoacyl adenylate intermediate is formed. Although the latter can be isolated and shown to transfer amino acid onto tRNA, this in itself is not considered proof of the obligatory nature of this intermediate, according to one viewpoint (10). However, in recent years a large number of studies have been done to prove, in various ways, the kinetic and mechanistic significance of the aminoacyl adenylate intermediate for the typical synthetase that carries out an ATP-PP_i exchange in the absence of tRNA (295–298). Thus, with the exception of those unusual cases where exchange does not occur in the absence of tRNA (cf 14), the issue of the adenylate intermediate appears to have been pretty much laid to rest.

Consequently, Equation 1 can be written as a sum of two overall reactions, indicated in Equation 3:

$$E + AA + ATP \rightleftharpoons E \cdot AA \sim AMP + PP_i$$
$$E \cdot AA \sim AMP + tRNA \rightleftharpoons AA\text{-}tRNA + AMP + E$$

3.

where AA~AMP denotes aminoacyl adenylate, which is generally bound firmly to the enzyme. Both of these individual reactions, as well as the overall reaction, have been studied in considerable depth.

RATE-DETERMINING STEP   The first step in Equation 3, known as amino acid activation, is typically studied by ATP-PP_i isotope exchange. In some systems, this reaction is found to be 1–2 orders of magnitude faster than the overall rate of aminoacylation (10). For these cases, the rate-limiting steps must occur after formation of the enzyme-bound aminoacyl adenylate.

A number of studies indicate that, at least with the Ile and Val enzymes under some conditions, release of newly formed aminoacyl-tRNA from the surface of the enzyme is rate-determining (53, 118, 295, 297). Although this conclusion has been called into question (299), the similar results of numerous investigations using different techniques (53, 118, 295, 297, 300), and confirmatory experiments (S. S. Lam and P. R. Schimmel, unpublished), argue strongly that release of the end product is rate-limiting in some situations. However, this is not the case under all circumstances. For example, data of Fersht & Kaethner indicate that for *E. coli* Ile-tRNA synthetase, the rate-determining step (under the conditions studied) occurs somewhere between formation of the adenylate and release of the aminoacyl-tRNA from the surface of the enzyme (298).

It cannot be concluded that the elementary mechanistic details are the same in each case. In the case of Tyr-tRNA synthetase from *E. coli*, it appears that under some conditions rate constants for adenylate formation and transfer are so close that both processes may contribute to the rate-determining step (298).

Finally, it should be appreciated that the scheme written in Equation 3 is much over simplified. A number of elementary steps have been omitted, such as the binding and dissociation of various ligands. In addition, the scheme takes no note of various ternary or other complexes that can form. For example, Fersht & Kaethner indicate that the dominant species at saturating reagent concentrations for Ile-tRNA synthetase is the E·tRNA·Ile ~ AMP complex (298).

Site of Initial Attachment of Amino Acid

At the 3' terminus of tRNA there are 2 hydroxyl groups (2'- and 3'-OH groups of the terminal ribose) that can be esterified to an amino acid. Once attached to either group, the amino acid can rapidly migrate back and forth between both of them (302). It can easily be imagined that the availability of two positions, and the ability to migrate back and forth between them, could play an important role in the mechanism of protein synthesis.

One question that has been extensively studied is the site of initial attachment, as catalyzed by aminoacyl-tRNA synthetases. To accomplish this, three different approaches have been used. All of these take advantage of tRNAs that have been modified at the 3' terminus. In one case, adenosine at the 3' terminus has been substituted with 2'-deoxy adenosine or 3'-deoxy adenosine (303, 304). Clearly, in tRNA molecules terminating in these deoxy analogs, no acyl migration is possible. Thus, if aminoacylation occurs with one analog and not the other, one can be reasonably confident that the initial site of attachment has been identified.

A second approach has been to make use of tRNA molecules that terminate in an adenosine whose ribose unit has been oxidized and subsequently reduced. In this way, the bond between the 2' and 3' carbon is broken and the resulting molecule, after reduction, has 2 vicinal hydroxyl groups that are no longer rigidly oriented with respect to each other; this means that acyl migration is not possible, and analysis of the product of aminoacylation can reveal whether the 2' or 3' OH is the attachment site (305).

Finally, in another group of studies, the adenosine at the terminus has been substituted with either 2'-amino-2'-deoxy adenosine or 3'-amino-3'-deoxy adenosine (306). Upon aminoacylation of these analogs, the amino acid ends up stably attached in an amide bond to the amino group. This happens because acyl migration, from the initially formed ester linkage with the hydroxyl group to the amino group, gives rise to a stable amide bond; back migration of the acyl moiety to give the original ester is not possible. Thus, if amino acid is stably attached to a 2'-amino-2'-deoxy analog, and not to the 3'-amino-3'-deoxy analog, it is concluded that the initial site of attachment is the 3'-OH group.

Taking advantage of these analogs, studies have been done to determine the initial site of aminoacylation of *E. coli,* yeast, calf liver, and wheat germ tRNAs (304, 307–310). A summary of many of these data is given in Table 8. The conclusions obtained with the various analogs are in reasonable agreement, although some discrepancies can be noted. Notwithstanding these discrepancies two major conclusions can be drawn. First, the initial site of attachment is not the same for all tRNAs. Some use the 3' position, some the 2' position, and some both. Although there have been attempts to rationalize this distribution of initial attachment sites among the different amino acids, at present there is no obvious biological rationale for these results. Possibly this distribution is a consequence of the evolutionary relationship between the different aminoacyl-tRNA synthetases (311). Second, there is a striking conservation in the position of the initial attachment site in tRNAs derived from very different sources, i.e. *E. coli,* yeast, and calf liver. This surprising phenomenon suggests that there may be some significance to the initial attachment site.

After aminoacylation of tRNA, the next step in protein synthesis is the interaction of the charged tRNA with elongation factors. For example, in *E. coli* the aminoacyl-tRNA is bound to EF-Tu. However, it appears that both positional isomers of aminoacyl-tRNA bind to EF-Tu (310, 312). This raises some questions as to the significance of the positional specificity of the initial aminoacylation, at least as far as its role in subsequent steps of protein synthesis.

Table 8 Comparison of *E. coli*, wheat germ, yeast, and calf liver tRNAs with regard to the site(s) at which a given amino acid is found to reside after aminoacylation of a 2′– or 3′–tRNA analogue[a]

Amino acid	E. coli amino tRNA (308)	E. coli deoxy tRNA (307)	E. coli deoxy tRNA (304)	Yeast deoxy tRNA (304)	Wheat germ amino tRNA (310)	Calf liver deoxy tRNA (309)
Ala	3′	U[b]	3′	3′	3′	3′
Arg	2′, 3′	2′	2′	2′	2′, 3′	2′
Asn	3′	2′	2′, 3′	2′, 3′	3′	2′, 3′
Asp	3′	U	U	2′, 3′	3′	3′
Cys	U	2′, 3′	2′, 3′	2′, 3′	3′	3′
Gln	2′	U	U	3′	2′	U
Glu	2′	U	2′	U	2′, 3′	U
Gly	3′	3′	3′	3′	3′	3′
His	3′	3′	3′	3′	3′	3′
Ile	2′, 3′	2′	2′	2′	2′	2′
Leu	2′	2′	2′	2′	2′	2′
Lys	3′	3′	3′	3′	2′, 3′	3′
Met	2′, 3′	2′	2′	2′	2′, 3′	U
Phe	2′	2′	2′	2′	2′, 3′	2′
Pro	2′, 3′	U	3′	3′	3′	U
Ser	2′, 3′	3′	3′	3′	3′	3′
Thr	3′	3′	3′	3′	3′	3′
Trp	2′, 3′	2′	2′	3′	2′, 3′	3′
Tyr	2′	2′, 3′	2′, 3′	2′, 3′	2′	2′, 3′
Val	2′	2′	2′	2′	2′	2′

[a] Adapted from (310).
[b] U designates uncertainty.

Deacylation Activity

Since aminoacyl-tRNA synthetases catalyze the aminoacylation of tRNA, microscopic reversibility requires that they also catalyze the deacylation of aminoacyl-tRNA, using the reverse pathway together with the products AMP and PP_i. However, several years ago, extensive investigations showed that synthetases have a deacylase activity that is quite distinct from the catalytic steps involved in the aminoacylation reaction (313). In particular, the deacylase activity operates in the absence of AMP and PP_i. This observation raises the possibility that the special deacylase activity may be of physiological significance.

Shortly thereafter it was reported that the deacylation activity can be greatly accelerated if the wrong amino acid is attached to a tRNA (314). For example, if the mischarged species Val-tRNA^{Ile} is mixed with Ile-tRNA synthetase, it is rapidly deacylated; in contrast, the deacylation activity is much slower when the same enzyme is presented with Ile-tRNA^{Ile} (turnover number of about 1 min^{-1} at 37°). This result sheds light on an earlier finding. It had been observed that Ile-tRNA synthetase can activate valine to give the valyl adenylate. However, when the enzyme-bound adenylate is mixed with tRNA^{Ile}, the complex breaks down to liberate free valine and AMP (48). Thus, in view of the deacylase activity, it was suggested that this activity quickly destroys any transiently formed Val-tRNA^{Ile} (314). The deacylation of other mischarged tRNAs has also been studied. For example, Ile-tRNA^{Phe} is rapidly deacylated by Phe-tRNA synthetase (315). Also, *B. stearothermophilus* Val-tRNA synthetase, which misactivates threonine to give threonyl adenylate, rapidly deacylates Thr-tRNA^{Val} (316). On the other hand, there are other mischarged tRNAs that are apparently not deacylated. However, these may correspond to species that are unlikely to arise naturally through enzyme errors.

When a misactivated adenylate, such as the valyl adenylate bound to Ile-tRNA synthetase, is rapidly broken down in the presence of tRNA^{Ile}, the question arises as to whether transfer of valine to tRNA^{Ile} actually occurs, or whether breakdown is induced before transfer. This question has been approached in studies with the elongation factor Tu, which binds strongly to aminoacyl-tRNAs. Using the elongation factor in a coupled assay, it has been possible to trap Val-tRNA^{Ile} in a mischarging reaction catalyzed by Ile-tRNA synthetase [V. T. Yue and P. R. Schimmel, unpublished, (318)]. In the absence of EF-Tu, the mischarged tRNA is rapidly destroyed so as to go undetected. Also, it has been possible to detect a transient Thr-tRNA^{Val} species in a mischarging reaction catalyzed by Val-tRNA synthetase (316).

Kinetic studies have investigated more deeply the mechanism of tRNA-induced hydrolysis of misactivated enzyme-bound valyl adenylate in the case of *E. coli* Ile-tRNA synthetase. The simplest explanation is that transfer occurs to the tRNA, and then the mischarged amino acid is rapidly removed by the enzyme's deacylase activity. However, some data suggest, although do not establish, that another hydrolysis reaction occurs even before transfer actually takes place (319).

As mentioned above, both the 2'- and 3'-OH groups of the 3' terminus are available for esterification by amino acids. Since synthetases show considerable specificity in the site of initial attachment (Table 8), the question arises as to whether one of the hydroxyl groups plays a role in the special deacylation reaction. This issue has been investigated by von der Haar & Cramer using yeast tRNAs (320). In the case of tRNA^{Ile}, the amino acid is initially attached to the 2'-OH. In tRNA^{Ile}, which terminates in a 3'-deoxyadenosine, it is found that Ile-tRNA synthetase stably attaches valine to tRNA^{Ile} without subsequent rapid hydrolysis. Thus, it would appear that the adjacent hydroxyl is necessary for the special deacylase activity. The idea of a role for the vicinal hydroxyl in this reaction is discussed elsewhere (319, 320).

CONCLUDING REMARKS

It is apparent that research on aminoacyl-tRNA synthetases is moving at a vigorous pace and continues to reach into new areas. For example, tRNAs and aminoacyl-tRNA synthetases in plant cytoplasm, chloroplasts, and mitochondria are being examined in detail. It appears that there may be a similarity between organelle-specific aminoacyl-tRNA synthetases and the corresponding bacterial enzymes (321). These studies bear on interesting structural, genetic, and evolutionary questions. The inhibition by amphetamines of protein synthesis in mammalian systems may be due to an effect of the drug on aminoacylation (321a). This question is being explored further. Two observations that show aminoacyl-tRNA synthetase in unusual roles are particularly noteworthy. First, it has been reported that Ser-tRNA synthetase is the agent responsible for the activation of the carcinogen 4-nitroquinoline 1-oxide (322). Second, it has been found that the compound diadenosine $5',5'$-P^1,P^4-tetra-phosphate (AppppA) is made as a by-product of the amino acid activation step, and that this compound may be an important pleiotropic regulator molecule (see 323 and references therein). Undoubtedly further work will show how general these observations are.

We comment briefly on some areas that will see considerable activity during the coming years. Studies on the mechanism of these enzymes are proceeding at a fast pace (101, 324–329). Both kinetic and thermodynamic analyses are being done. Undoubtedly these studies will continue to refine our understanding of the mechanistic aspects of the synthetases and define better the differences and similarities in the mechanism of various enzymes.

Efforts are also being directed at improving the purification methods of the synthetases (330–332). Taking advantage of agarose-hexyl-adenosine-$5'$-phosphate and blue dextran-sepharose (330, 331), affinity chromatography has been used with good success. Some of these promising procedures will certainly be adopted by many workers and further improved.

Studies on the primary structures of the synthetases will also continue and lead to many more sequences in the near future. It will be of great interest to determine whether there are homologies between some of the enzymes, and also to clarify the meaning of the sequence repeats. With the sequence information in hand, the molecular basis for the kinetic properties and the results of the affinity labeling studies may begin to be interpreted.

In the near future we may expect to have a high resolution structure of at least one aminoacyl-tRNA synthetase. Currently the best prospects are *B. stearothermophilus* Tyr-tRNA synthetase and *E. coli* Met-tRNA synthetase (225, 226). Once high resolution structures are known, it should be possible to construct fairly good models of the synthetase-tRNA complex using the approaches discussed in this review. Ideally, the complex itself will be crystallized and analyzed, although this goal will not be reached as quickly.

As another approach to structural issues, attempts are being made to dissect aminoacyl-tRNA synthetases with appropriate proteolytic enzymes (333–335). Here the goal is to obtain fragments that retain either catalytic activity or perhaps only tRNA binding activity. Analysis of such fragments may contribute to our knowledge of the substrate binding sites of the enzymes.

The occurrence of synthetases in large complexes in mammalian cells will continue to perplex and challenge investigators. Crucially needed are additional studies that accurately determine the composition and stoichiometry of enzymes in the complexes. Once this basic information is known, rational studies of the relationship between structural organization and function might be planned and carried out. However, one should keep in mind that the structural organization may not be solely for achieving some goal related to aminoacylation in protein synthesis, but rather for functions (possibly not yet known) of synthetases in cellular metabolism.

Novel approaches are needed to shed light on the basis for specificity of tRNA recognition. The genetic and biochemical analysis of mischarging mutants may be extended to the *E. coli* serine (su_1^+) or leucine (su_6^+) tRNAs. In addition, "mutant" tRNAs could be created by joining enzymatically derived or chemically synthesized tRNA fragments with RNA ligase (336) or by specific alterations of cloned tRNA genes by recombinant DNA techniques. The properties of these altered tRNAs could then be checked by established procedures.

An obviously important, but experimentally difficult, task is the further delineation of the role of synthetases in other cellular processes. The availability of cloning techniques and of good in vitro transcription-translation systems may make possible appropriate assays for the detailed examination of the way in which synthetases affect gene expression. The challenges of this work may also necessitate the collection of more mutant aminoacyl-tRNA synthetases. Because of the diverse involvement of synthetases and aminoacyl-tRNAs in cellular processes, elucidating the regulation of the biosynthesis or of the activity of these enzymes is of great importance. Much information about their metabolic control has been available (28, 29), but this has not provided a unified picture. More recently, work on the genetic control of synthetase biosynthesis has begun (29, 337–339) and has already given additional insight on synthetase involvement in some metabolic processes. Continuation of such studies may also uncover whether the mechanisms controlling tRNA and synthetase levels in the cell are linked.

It is impressive to see how much work on aminoacyl-tRNA synthetases was done in recent years. Much of this work has raised new questions which in turn has prompted novel further investigations. We foresee that this field will continue to grow at a fast pace; some of the most exciting problems are yet to be worked out.

Literature Cited

1. Berg, P. 1961. *Ann. Rev. Biochem.* 30: 292–324
2. Novelli, G. D. 1967. *Ann. Rev. Biochem.* 36:449–84
3. Lengyel, P., Söll, D. 1969. *Bacteriol. Rev.* 33:264–301
4. Mehler, A. H. 1970. *Prog. Nucleic Acid Res. Mol. Biol.* 10:1–22
5. Muench, K. H. 1971. *Methods Mol. Biol.* 1:213–33
6. Jacobson, K. B. 1971. *Prog. Nucleic Acid Res. Mol. Biol.* 11:461–81
7. Chambers, R. W. 1971. *Prog. Nucleic Acid Res. Mol. Biol.* 11:489–523
8. Loftfield, R. B. 1971. In *Protein Synthesis,* ed. E. McConkey, 1:1–28. New York: Dekker
9. Mehler, A. H., Chakraburtty, K. 1971. *Adv. Enzymol.* 35:443–501
10. Loftfield, R. B. 1972. *Prog. Nucleic Acid Res. Mol. Biol.* 12:87–128
11. Chapeville, F. 1972. In *The Mechanism of Protein Synthesis and its Regulation,* ed. L. Bosch, A. Neuberger, E. L. Tatum, p. 5–32. Amsterdam: North-Holland
12. Favorova, O. O., Parin, A. B., Lavrik, O. I. 1972. *Biofizika* 2:6–143
13. Kisselev, L. L., Favorova, O. O. 1974. *Adv. Enzymol.* 40:141–238
14. Söll, D., Schimmel, P. R. 1974. *Enzymes* 10:489–538
15. Ofengand, J. 1977. In *Molecular Mechanisms of Protein Biosynthesis,* ed. H. Weissbach, S. Pestka, p. 7–69. New York: Academic
16. Chapeville, F., Lipmann, F., von Ehrenstein, G., Weisblum, B., Ray, W. J. Jr., Benzer, S. 1962. *Proc. Natl. Acad. Sci. USA* 48:1086–92
17. Barrell, B. G., Clark, B. F. C. 1974. *Handbook of Nucleic Acid Sequences.* Oxford: Joynson-Bruvvers
18. Sprinzl, M., Grüter, F., Gauss, D. H. 1978. *Nucleic Acids Res.* 5:r15–27
19. Kim, S. H., Suddath, F. L., Quigley, G. J., McPherson, A., Sussman, J. L., Wang, A. H.-J., Seeman, N. C., Rich, A. 1974. *Science* 185:435–40
20. Robertus, J. D., Ladner, J. E., Finch, J. T., Rhodes, D., Brown, R. S., Clark, B.

F. C., Klug, A. 1974. *Nature* 250: 546–51
21. Littauer, U. Z., Inouye, H. 1973. *Ann. Rev. Biochem.* 42:439–70
22. Brenchley, J. E., Williams, L. S. 1975. *Ann. Rev. Microbiol.* 29:251–74
23. LaRossa, R., Söll, D. 1978. In *Transfer RNA,* ed. S. Altman, pp. 136–67. Cambridge, Mass: MIT Press
24. Cortese, R. 1979. In *Biological Regulation and Control,* ed. R. F. Goldberger, New York: Plenum. In press
25. Kasai, T. 1974. *Nature* 249:523–27
26. Artz, S. W., Broach, J. R. 1975. *Proc. Natl. Acad. Sci. USA* 72:3453–57
27. Yanofsky, C., Soll, L. 1977. *J. Mol. Biol.* 113:663–77
28. Neidhardt, F. C., Parker, J., McKeever, W. G. 1975. *Ann. Rev. Microbiol.* 29: 215–50
29. Morgan, S. D., Söll, D. 1978. *Prog. Nucleic Acid Res. Mol. Biol.* 21:181–207
30. Bhanot, O. S., Kucan, Z., Aoyagi, S., Lee, F. C., Chambers, R. W. 1974. *Methods Enzymol.* 29:547–76
31. Parfait, R., Grosjean, H. 1972. *Eur. J. Biochem.* 30:242–49
32. Craine, J., Peterkofsky, A. 1975. *Arch. Biochem. Biophys.* 168:343–50
33. Hirshfield, I. N., Bloemers, H. P. J. 1969. *J. Biol. Chem.* 244:2911–16
34. Marshall, R. D., Zamecnik, P. C. 1969. *Biochim. Biophys. Acta* 181:454–64
35. Nazario, M., Evans, J. A. 1974. *J. Biol. Chem.* 249:4934–42
36. Gangloff, J., Dirheimer, G. 1973. *Biochim. Biophys. Acta* 294:263–72
37. James, H. L., Bucovaz, E. T. 1969. *J. Biol. Chem.* 244:3210–16
38. Pan, F., Lee, H. H., Pai, S. H., Yu, T. C., Guoo, J. Y., Duh, G. M. 1976. *Biochim. Biophys. Acta* 452:271–83
39. Folk, W. R. 1971. *Biochemistry* 10: 1728–32
40. Lapointe, J., Söll, D. 1972. *J. Biol. Chem.* 247:4966–74
41. Lapointe, J., Söll, D. 1972. *J. Biol. Chem.* 247:4975–81
42. Ostrem, D. L., Berg, P. 1970. *Proc. Natl. Acad. Sci. USA* 67:1967–74
43. Ostrem, D. L., Berg, P. 1974. *Biochemistry* 13:1338–48

44. Kalousek, F., Konigsberg, W. H. 1974. *Biochemistry* 13:999–1006
45. De Lorenzo, F., Di Natale, P., Schechter, A. N. 1974. *J. Biol. Chem.* 249:908–13
45a. Kane, S. M., Vugrincic, C., Finbloom, D. S., Smith, D. W. E. 1978. *Biochemistry* 17:1509–14
46. Charlier, J., Grosjean, H. 1972. *Eur. J. Biochem.* 25:163–74
47. Arndt, D. J., Berg, P. 1970. *J. Biol. Chem.* 245:665–67
48. Baldwin, A. N., Berg, P. 1966. *J. Biol. Chem.* 241:831–38
49. Berthelot, F., Yaniv, M. 1970. *Eur. J. Biochem.* 16:123–25
50. Holler, E., Bennett, E. L., Calvin, M. 1971. *Biochem. Biophys. Res. Commun.* 45:409–15
51. Norris, A. T., Berg, P. 1964. *Proc. Natl. Acad. Sci. USA* 52:330–37
52. Yarus, M., Berg, P. 1967. *J. Mol. Biol.* 28:479–90
53. Yarus, M., Berg, P. 1969. *J. Mol. Biol.* 42:171–89
54. Koch, G. L. E., Boulanger, Y., Hartley, B. S. 1974. *Nature* 249:316–20
55. Hayashi, H., Knowles, J. R., Katze, J. R., Lapointe, J., Söll, D. 1970. *J. Biol. Chem.* 245:1401–6
56. Rouget, P., Chapeville, F. 1970. *Eur. J. Biochem.* 14:498–508
57. Chirikjian, J. G., Wright, H. T., Fresco, J. R. 1972. *Proc. Natl. Acad. Sci. USA* 69:1638–41
58. Murasugi, A., Hayashi, H. 1975. *Eur. J. Biochem.* 57:169–75
59. Rymo, L., Lundvik, L., Lagerkvist, U. 1972. *J. Biol. Chem.* 247:3888–99
60. Dimitrijevic, L. 1972. *FEBS Lett.* 25: 170–74
61. Rymo, L., Lagerkvist, U., Wonacott, A. 1970. *J. Biol. Chem.* 245:4308–16
62. Fersht, A. R., Ashford, J. S., Bruton, C. J., Jakes, R., Koch, G. L. E., Hartley, B. S. 1975. *Biochemistry* 14:1–4
63. Koch, G. L. E., Bruton, C. J. 1974. *FEBS Lett.* 40:180–82
64. Lemoine, F., Waller, J. P., van Rapenbusch, R. 1968. *Eur. J. Biochem.* 4: 213–21
65. Cassio, D., Waller, J. P. 1971. *Eur. J. Biochem.* 20:283–300
66. Blanquet, S., Fayat, G., Waller, J. P., Iwatsubo, M. 1972. *Eur. J. Biochem.* 24:461–69
67. Blanquet, S., Iwatsubo, M., Waller, J. P. 1973. *Eur. J. Biochem.* 36:213–26
68. Rosa, M. D., Sigler, P. B. 1977. *Eur. J. Biochem.* 78:141–51
69. Tscherne, J. E., Lanks, K. W., Salim, P. D., Grunberger, D., Cantor, C. R.,

Weinstein, I. B. 1973. *J. Biol. Chem.* 248:4052–59
70. Christopher, C. W., Sittman, D. B., Stafford, D. W. 1975. *Arch. Biochem. Biophys.* 166:94–101
71. Fayat, G., Blanquet, S., Dessen, P., Batelier, G., Waller, J. P. 1974. *Biochimie* 56:35–41
72. Hanke, T., Bartmann, P., Hennecke, H., Kosakowski, H. M., Jaenicke, R., Holler, E., Böck, A. 1974. *Eur. J. Biochem.* 43:601–7
73. Kosakowski, M. H. J. E., Böck, A. 1970. *Eur. J. Biochem.* 12:67–73
74. Bartmann, P., Hanke, T., Holler, E. 1975. *J. Biol. Chem.* 250:7668–74
75. Fasiolo, F., Befort, N., Boulanger, Y., Ebel, J.-P. 1970. *Biochim. Biophys. Acta* 217:305–18
76. Fasiolo, F., Remy, P., Pouyet, J., Ebel, J.-P. 1974. *Eur. J. Biochem.* 50:227–36
77. Fasiolo, F., Ebel, J.-P. 1974. *Eur. J. Biochem.* 49:257–63
78. Schmidt, J., Wang, R., Stanfield, S., Reid, B. R. 1971. *Biochemistry* 10: 3264–68
79. Lanks, K. W., Eng, R. K., Baltusis, L. 1973. *Fed. Proc.* 32:460 (Abstr. 1336)
80. Lee, M.-L., Muench, K. H. 1969. *J. Biol. Chem.* 244:223–30
81. Lemeur, M. A., Gerlinger, P., Clavert, J., Ebel, J.-P. 1972. *Biochimie* 34: 1391–97
82. Boeker, E. A., Hays, A. P., Cantoni, G. L. 1973. *Biochemistry* 12:2379–83
83. Boeker, E. A., Cantoni, G. L. 1973. *Biochemistry* 12:2384–89
84. Katze, J. R., Konigsberg, W. 1970. *J. Biol. Chem.* 245:923–30
85. Waterson, R. M., Clarke, S. J., Kalousek, F., Konigsberg, W. H. 1973. *J. Biol. Chem.* 248:4181–88
86. Knowles, J. R., Katze, J. R., Konigsberg, W., Söll, D. 1970. *J. Biol. Chem.* 245:1407–15
87. Pingoud, A., Riesner, D., Boehme, D., Maass, G. 1973. *FEBS Lett.* 30:1–5
88. Engel, G., Heider, H., Maelicke, A., von der Haar, F., Cramer, F. 1972. *Eur. J. Biochem.* 29:257–62
89. Heider, H., Gottschalk, E., Cramer, F. 1971. *Eur. J. Biochem.* 20:144–52
90. Hörz, W., Zachau, H. G. 1973. *Eur. J. Biochem.* 32:1–14
91. Pachmann, U., Cronvall, E., Rigler, R., Hirsch, R., Wintermeyer, W., Zachau, H. G. 1973. *Eur. J. Biochem.* 39:265–73
92. Rigler, R., Cronvall, E., Hirsch, R., Pachmann, U., Zachau, H. G. 1970. *FEBS Lett.* 11:320–23
93. Rigler, R., Cronvall, E., Ehrenberg, M.,

Pachmann, U., Hirsch, R., Zachau, H. G. 1971. *FEBS Lett.* 18:193–98
94. Pachmann, U., Zachau, H. G. 1978. *Nucleic Acids Res.* 5:961–73
95. Jakubowski, H., Pawelkiewicz, J. 1975. *Eur. J. Biochem.* 52:301–10
96. Hennecke, H., Böck, A., Thomale, J., Nass, G. 1977. *J. Bacteriol.* 131:943–50
97. Penneys, N. S., Muench, K. H. 1974. *Biochemistry* 13:560–65
98. Gros, C., Lemaire, G., van Rapenbusch, R., Labouesse, B. 1972. *J. Biol. Chem.* 247:2931–43
99. Lemaire, G., van Rapenbusch, R., Gros, C., Labouesse, B. 1969. *Eur. J. Biochem.* 10:336–44
100. Dorizzi, M., Labouesse, B., Labouesse, J. 1971. *Eur. J. Biochem.* 19:563–72
101. Dorizzi, M., Merault, G., Fournier, M., Labouesse, J., Keith, G., Dirheimer, G., Buckingham, R. H. 1977. *Nucleic Acids Res.* 4:31–42
102. Liu, C.-C., Chung, C.-H., Lee, M.-L. 1973. *Biochem. J.* 135:367–73
103. Winter, G. P., Hartley, B. S. 1977. *FEBS Lett.* 80:340–42
104. Joseph, D. R., Muench, K. H. 1971. *J. Biol. Chem.* 246:7610–15
105. Joseph, D. R., Muench, K. H. 1971. *J. Biol. Chem.* 246:7602–9
106. Muench, K. H. 1976. *J. Biol. Chem.* 251:5195–99
107. Hossain, A., Kallenbach, N. R. 1974. *FEBS Letters* 45:202–5
108. Fersht, A. R., Mulvey, R. S., Koch, G. L. E. 1975. *Biochemistry* 14:13–18
109. Calendar, R., Berg, P. 1966. *Biochemistry* 5:1681–90
110. Chousterman, S., Chapeville, F. 1973. *Eur. J. Biochem.* 35:51–56
111. Fersht, A. R. 1975. *Biochemistry* 14:5–12
112. Buonocore, V., Schlesinger, S. 1972. *J. Biol. Chem.* 247:1343–48
113. Jakes, R., Fersht, A. R. 1975. *Biochemistry* 14:3344–50
114. Faulhammer, H. G., Cramer, F. 1977. *Eur. J. Biochem.* 75:561–70
115. Rao, Y. S. P., Srinivasan, P. R. 1978. *Nucleic Acids Res.* 4:3887–900
116. Locy, R. D., Cherry, J. H. 1978. *Phytochemistry* 17:19–27
117. Yaniv, M., Gros, F. 1969. *J. Mol. Biol.* 44:1–15
118. Hélène, C., Brun, F., Yaniv, M. 1971. *J. Mol. Biol.* 58:349–65
119. Yaniv, M., Gros, F. 1969. *J. Mol. Biol.* 44:17–30
120. Lagerkvist, U., Rymo, L. 1969. *J. Biol. Chem.* 244:2476–83
121. Kula, M. R. 1973. *FEBS Lett.* 35:299–302

122. Duplain, L., Lapointe, J. 1978. Manuscript submitted
123. Wilcox, M. 1969. *Eur. J. Biochem.* 11:405–12
124. Deutscher, M. P. 1967. *J. Biol. Chem.* 242:1123–31
125. Bandyopadhyay, A. K., Deutscher, M. P. 1971. *J. Mol. Biol.* 60:113–22
126. Vennegoor, C., Bloemendal, H. 1972. *Eur. J. Biochem.* 26:462–73
127. Goto, T., Schweiger, A. 1973. *Hoppe-Seyler's Z. Physiol. Chem.* 354:1027–33
128. Som, K., Hardesty, B. 1975. *Arch. Biochem. Biophys.* 166:507–17
129. Norton, S. J., Key, M. D., Scholes, S. W. 1965. *Arch. Biochem. Biophys.* 109:7–12
130. Hampel, A., Enger, M. D. 1973. *J. Mol. Biol.* 79:285–93
131. Moline, G., Hampel, A., Enger, M. D. 1974. *Biochem. J.* 143:191–95
132. Smulson, M., Lin, C. S., Chirikjian, J. G. 1975. *Arch. Biochem. Biophys.* 167:458–68
133. Denny, R. M. 1977. *Arch. Biochem. Biophys.* 183:156–67
134. Ussery, M. A., Tanaka, W. K., Hardesty, B. 1977. *Eur. J. Biochem.* 72:491–500
135. Agris, P. F., Woolverton, D. K., Setzer, D. 1976. *Proc. Natl. Acad. Sci. USA* 73:3857–61
136. Agris, P. F., Setzer, D., Gehrke, C. W. 1977. *Nucleic Acids Res.* 4:3803–18
137. Ehrlich, P. H. 1977. *Studies on high molecular weight aminoacyl tRNA synthetases* PhD thesis, Mass. Inst. Technol., Cambridge, Mass. 223 pp.
138. Neidhardt, F. C. 1966. *Bacteriol. Rev.* 30:701–19
139. Quay, S. C., Kline, E. L., Oxender, D. L. 1975. *Proc. Natl. Acad. Sci. USA* 72:3921–24
140. Moore, P. A., Jayme, D. W., Oxender, D. L. 1977. *J. Biol. Chem.* 252:7427–30
141. Cashel, M., Gallant, J. 1974. In *Ribosomes*, ed. M. Nomura, A. Tissieres, P. Lengyel, pp. 733–45. Cold Spring Harbor, New York: Cold Spring Harbor Lab.
142. Unger, M. W. 1977. *J. Bacteriol.* 130:11
143. Hennecke, H. 1976. *FEBS Lett.* 72:182–86
144. Hankinson, O. 1976. *Somatic Cell Genet.* 2:497–507
145. Buckel, P., Lubitz, W., Böck, A. 1971. *J. Bacteriol.* 108:1008
146. Yaniv, M., Gros, F. 1966. *Genet. Elem. Prop. Funct. Symp., 3rd, FEBS Meet.* p. 157
147. Theall, G., Low, K. B., Söll, D. 1977. *Mol. Gen. Genet.* 156:221

148. Sato, K. 1975. *Nature* 257:813
149. Thompson, L. H., Lofgren, D. J., Adair, G. M. 1977. *Cell* 11:157–68
150. Cooper, P. H., Hirshfield, I. N., Maas, W. K. 1969. *Mol. Gen. Genet.* 104: 383–90
151. Williams, L. S. 1973. *J. Bacteriol.* 113:1419–32
152. Thompson, L. H., Stanners, C. P., Siminovitch, L. 1975. *Somatic Cell Genet.* 1:187–208
153. Andrulis, I. L., Chiang, C. S., Arkin, S. M., Miner, T. A., Hatfield, G. W. 1978. *J. Biol. Chem.* 253:58–62
154. Yamamoto, M., Nomura, M., Ohsawa, H., Maruo, B. 1977. *J. Bacteriol.* 132: 127–31
155. Wasmuth, J. J., Caskey, C. T. 1976. *Cell* 9:655–62
156. Körner, A., Magee, B. B., Liska, B., Low, K. B., Adelberg, E. A., Söll, D. 1974. *J. Bacteriol.* 120:154–58
157. Russell, R. R. B., Pittard, A. J. 1971. *J. Bacteriol.* 108:790–98
158. Murgola, E. J., Adelberg, E. A. 1970. *J. Bacteriol.* 103:20–26
159. Murgola, E. J., Adelberg, E. A. 1970. *J. Bacteriol.* 103:178–83
160. Böck, A., Neidhardt, F. C., 1966. *Z. Vererbungsl.* 98:187–92
161. Folk, W. R., Berg, P. 1970. *J. Bacteriol.* 102:204–12
162. Roback, E. R., Friesen, J. D., Fiil, N. P. 1973. *Can. J. Microbiol.* 19:421–26
163. Nass, G. 1967. *Mol. Gen. Genet.* 100: 216–24
164. Roth, J. R., Ames, B. N. 1966. *J. Mol. Biol.* 22:325–34
165. Iaccarino, M., Berg, P. 1971. *J. Bacteriol.* 105:527–37
166. Treiber, G., Iaccarino, M. 1971. *J. Bacteriol.* 107:828–32
167. Coker, M., Umbarger, H. E. 1970. *Bacteriol. Proc.* p. 135. (Abstr.)
168. McGinnis, E., Williams, L. S. 1974. *Ann. Meet. Am. Soc. Microbiol.* p. 172. (Abstr.)
169. Blatt, J. M., Umbarger, H. E. 1972. *Biochem. Gen.* 6:99–118
170. Hartwell, L. H., McLaughlin, C. S. 1968. *Proc. Natl. Acad. Sci. USA* 59: 422–28
171. McLaughlin, C. S., Hartwell, L. H. 1969. *Genetics* 61:557–66
172. Thompson, L. H., Harkins, J. L., Stanners, C. P. 1973. *Proc. Natl. Acad. Sci. USA* 70:3094–98
173. Haars, L., Hampel, A., Thompson, L. 1976. *Biochim. Biophys. Acta.* 454:493–503
174. Farber, R. A., Deutscher, M. P. 1976. *Somatic Cell Genet.* 2:509–520
175. Low, B., Gates, F., Goldstein, T., Söll, D. 1971. *J. Bacteriol.* 108:742–50
176. Alexander, R. R., Calvo, J. M., Freundlich, M. 1971. *J. Bacteriol.* 106:213–20
177. Mikulka, T. W., Stieglitz, B. I., Calvo, J. M. 1972. *J. Bacteriol.* 109:584–93
178. Beauchamp, P. M., Horn, E. W., Gross, S. R. 1977. *Proc. Natl. Acad. Sci. USA* 74:1172–76
179. Gross, S. R., McCoy, M. T., Gilmore, E. B. 1968. *Proc. Natl. Acad. Sci. USA* 61:253–60
180. Racine, F. M., Steinberg, W. 1974. *J. Bacteriol.* 120:372–83
181. Racine, F. M., Steinberg, W. 1974. *J. Bacteriol.* 120:384–89
182. Hirshfield, I. N., Tomford, J. W., Zamecnik, P. C. 1972. *Biochim. Biophys. Acta* 259:344–56
183. Somerville, C. R., Ahmed, A. 1977. *J. Mol. Biol.* 111:77–81
184. Archibald, E. R., Williams, L. S. 1973. *J. Bact.* 114:1007–13
185. Sanderson, K. E., Demerec, M. 1965. *Genetics* 51:897–913
186. Gross, T. S., Rowbury, R. J. 1969. *Biochim. Biophys. Acta* 184:233–36
187. Willick, G. E., Kay, C. M. 1976. *Biochemistry* 15:4347–52
188. Fangman, W. L., Neidhardt, F. C. 1964. *J. Biol. Chem.* 239:1839–43
189. Comer, M. M., Böck, A. 1976. *J. Bacteriol.* 127:923–33
190. Clarke, S. J., Low, B., Konigsberg, W. 1973. *J. Bacteriol.* 113:1096–1103
191. Tosa, T., Pizer, L. I. 1971. *J. Bacteriol.* 106:972–82
192. Nass, G., Thomale, J. 1974. *FEBS Lett.* 39:182–86
193. Johnson, E. J., Cohen, G. N., Saint-Girons, I. 1977. *J. Bacteriol.* 129:66–70
194. Steinberg, W., Anagnostapoulos, C. 1971. *J. Bacteriol.* 105:6–19
195. Doolittle, W. F., Yanofsky, C. 1968. *J. Bacteriol.* 95:1283–94
196. Kano, Y., Matsushiro, A., Shimura, Y. 1968. *Mol. Gen. Genet.* 102:15–26
197. Nazario, M., Kinsey, J. A., Ahmad, M. 1971. *J. Bacteriol.* 105:121–26
198. Schlesinger, S., Nester, E. W. 1969. *J. Bacteriol.* 100:167–75
199. Buonocore, V., Harris, M. H., Schlesinger, S. 1973. *J. Biol. Chem.* 247: 4843–44
200. Heinonen, J., Artz, S. W., Zalkin, H. 1972. *J. Bacteriol.* 112:1254–63
201. Eidlic, L., Neidhardt, F. C. 1965. *J. Bacteriol.* 89:706–711
202. Tingle, M. A., Neidhardt, F. C. 1969. *J. Bacteriol.* 98:837–39
203. Yaniv, M., Gros, F. 1969. *J. Mol. Biol.* 44:31–45

204. Deleted in proof
205. Deleted in proof
206. Hennecke, H., Springer, M., Böck, A. 1977. *Mol. Gen. Genet.* 152:205–10
207. Hennecke, H., Böck, A., Thomale, J., Nass, G. 1976. *J. Bacteriol.* 131:943–50
208. Doolittle, W. F., Yanofsky, C. 1968. *J. Bacteriol.* 95:1283–94
209. Chang, P., Omnaas, J., Safille, A., Muench, K. H. 1978. Manuscript in preparation
210. Buckel, P. 1976. *Mol. Gen. Genet.* 149:291–96
211. Wirth, R., Buckel, P., Böck, A. 1977. *FEBS Lett.* 83:103–6
212. Waterson, R. M., Konigsberg, W. H. 1974. *Proc. Natl. Acad. Sci. USA* 71:376–80
213. Bruton, C. J., Jakes, R., Koch, G. L. E. 1974. *FEBS Lett.* 45:26–28
214. Muench, K. H., Lipscomb, M. S., Lee, M.-L., Kuehl, G. V. 1975. *Science* 187:1089–91
215. Kuehl, G. V., Lee, M., Muench, K. H. 1976. *J. Biol. Chem.* 251:3254–60
215a. Robbe-Saul, S., Fasiolo, F., Boulanger, Y. 1977. *FEBS Lett.* 84:57–62
216. Winter, G. P., Hartley, B. S., McLachlan, A. D., Lee, M., Muench, K. H. 1977. *FEBS Lett.* 83:348–50
217. Bruton, C. J., Hartley, B. S. 1970. *J. Mol. Biol.* 52:165–78
218. Santi, D. V., Marchant, W., Yarus, M. 1973. *Biochem. Biophys. Res. Commun.* 51:370–75
219. Schoemaker, H. J. P., Schimmel, P. R. 1974. *J. Mol. Biol.* 84:503–13
220. Budzik, G. P., Lam, S. S. M., Schoemaker, H. J. P., Schimmel, P. R. 1975. *J. Biol. Chem.* 250:4433–39
221. Schoemaker, H. J. P., Budzik, G. P., Giegé, R., Schimmel, P. R. 1975. *J. Biol. Chem.* 250:4440–44
222. Piszkiewicz, D., Duval, J., Rostas, S. 1977. *Biochemistry* 16:3538–43
223. Yue, V. T., Schimmel, P. R. 1977. *Biochemistry* 16:4678–84
224. Akhverdyan, V. Z., Kisselev, L. L., Knorre, D. G., Lavrik, O. I., Nevinsky, G. A. 1977. *J. Mol. Biol.* 113:475–501
224a. Fayat, G., Fromant, M., Blanquet, S. 1979. *Proc. Natl. Acad. Sci. USA* In press
225. Zelwer, C., Risler, J. L., Monteilhet, C. 1976. *J. Mol. Biol.* 102:93
226. Irwin, M. J., Nyborg, J., Reid, B. R., Blow, D. M. 1976. *J. Mol. Biol.* 105:577–86
227. Kim, S. H., Sussman, J. L., Suddath, F. L., Quigley, G. J., McPherson, A., Wang, A. H.-J., Seeman, N. C., Rich,

A. 1974. *Proc. Natl. Acad. Sci. USA* 71:4970–74
228. Parfait, R. 1973. *Eur. J. Biochem.* 38:572–80
229. Lagerkvist, U., Rymo, L., Waldenström, J. 1966. *J. Biol. Chem.* 241:5391–5400
230. Lagerkvist, U., Rymo, L. 1969. *J. Biol. Chem.* 244:2476–83
231. Fasiolo, F., Remy, P., Pouyet, J., Ebel, J.-P. 1974. *Eur. J. Biochem.* 50:227–36
232. Krauss, G., Römer, R., Riesner, D., Maass, G. 1973. *FEBS Lett.* 30:6–10
233. Krauss, G., Pingoud, A., Boehme, D., Riesner, D., Peters, F., Maass, G. 1975. *Eur. J. Biochem.* 55:519–29
234. Krauss, G., Riesner, D., Maass, G. 1976. *Eur. J. Biochem.* 68:81–93
235. Hélène, C., Brun, F., Yaniv, M. 1969. *Biochem. Biophys. Res. Commun.* 37:393–98
236. Bruton, C. J., Hartley, B. S. 1970. *J. Mol. Biol.* 52:165–78
237. Blanquet, S., Petrissant, G., Waller, J. P. 1973. *Eur. J. Biochem.* 36:227–33
238. Pachmann, U., Cronvall, E., Rigler, R., Hirsch, R., Wintermeyer, W., Zachau, H. G. 1973. *Eur. J. Biochem.* 39:265–73
239. Maelicke, A., Engle, G., Cramer, F., Staehelin, M. 1974. *Eur. J. Biochem.* 42:311–14
240. Lam, S. S. M., Schimmel, P. R. 1975. *Biochemistry* 14:2775–80
241. Rigler, R., Pachmann, U., Hirsch, R., Zachau, H. G. 1976. *Eur. J. Biochem.* 65:307–15
241a. Hustedt, H., Kula, M.-R. 1977. *Eur. J. Biochem.* 74:191–98
242. Kauzmann, W. 1959. *Adv. Protein Chem.* 14:1–63
243. Riggs, A. D., Bourgeois, S., Cohn, M. 1970. *J. Mol. Biol.* 53:401–17
244. Rigler, R., Cronvall, E., Hirsch, R., Pachmann, U., Zachau, H. G. 1970. *FEBS Lett.* 11:320–23
245. Riesner, D., Pingoud, A., Boehme, D., Peters, F., Maass, G. 1976. *Eur. J. Biochem.* 68:71–80
246. Hammes, G. G., Schimmel, P. R. 1970. *Enzymes* 2:67
247. Hammes, G. G. 1978. *Principles of Chemical Kinetics,* pp. 217–46, New York: Academic
248. Ebel, J.-P., Giegé, R., Bonnet, J., Kern, D., Befort, N., Bollack, C., Fasiolo, F., Gangloff, J., Dirheimer, G. 1973. *Biochimie* 55:547–57
249. Roe, B., Sirover, M., Dudock, B. 1973. *Biochemistry* 12:4146–54
250. Giegé, R., Kern, D., Ebel, J.-P., Taglang, R. 1971. *FEBS Lett.* 15:281–85

251. Yarus, M. 1972. *Biochemistry* 11:2352–61
252. Kern, D., Giegé, R., Ebel, J.-P. 1972. *Eur. J. Biochem.* 31:148–55
253. Giegé, R., Kern, D., Ebel, J.-P. 1972. *Biochimie* 54:1245–55
254. Yarus, M., Mertes, M. 1973. *J. Biol. Chem.* 248:6744–49
255. Giegé, R., Kern, D., Ebel, J.-P., Grosjean, H., De Henau, S., Chantrenne, H. 1974. *Eur. J. Biochem.* 45:351–62
256. Dube, S. K. 1973. *Nature New Biol.* 243:103–5
257. Dickson, L. A., Schimmel, P. R. 1975. *Arch. Biochem. Biophys.* 67:638–45
258. Mirzabekov, A. D., Lastity, D., Levina, E. S., Bayev, A. A. 1971. *Nature New Biol.* 229:21–22
259. Thiebe, R., Harbers, K., Zachau, H. G. 1972. *Eur. J. Biochem.* 26:144
260. Carbon, J., Squires, C. 1971. *Cancer Res.* 31:663–66
261. Shimura, Y., Aono, H., Ozeki, H., Sarabhai, A., Lamfrom, H., Abelson, J. 1972. *FEBS Lett.* 22:144–48
262. Hooper, M. L., Russell, R. L., Smith, J. D. 1972. *FEBS Lett.* 22:149–56
263. Smith, J. D., Celis, J. E. 1973. *Nature* 243:66–71
264. Yaniv, M., Folk, W. R., Berg, P., Soll, L. 1974. *J. Mol. Biol.* 86:245–60
265. Schulman, L. H., Pelka, H. 1977. *Biochemistry* 16:4256–65
266. Schoemaker, H. J. P., Schimmel, P. R. 1976. *J. Biol. Chem.* 251:6823–30
267. Rich, A., Schimmel, P. R. 1977. *Nucleic Acids Res.* 4:1649–65
268. Schimmel, P. R. 1977. *Acc. Chem. Res.* 10:411–18
269. Goodard, T. P. 1977. *Prog. Biophys. Molec. Biol.* 32:233–308
270. Bayev, A. A., Fodor, I., Mirzabekov, A. D., Axelrod, V. D., Kazarinova, L. Y. 1967. *Mol. Biol.* 1:714–23
271. Deleted in proof
272. Seno, T. 1975. *FEBS Lett.* 51:325–29
273. Celis, J. E., Coulondre, C., Miller, J. H. 1976. *J. Mol. Biol.* 104:729–34
274. Smith, K. C. 1969. *Biochem. Biophys. Res. Commun.* 34:354–57
275. Smith, K. C. 1975. In *Photochemistry and Photobiology of Nucleic Acids,* ed. S. Y. Wang, pp. 187–215. New York: Academic
276. Smith, K. C. ed. 1976. *Aging, Carcinogenesis, and Radiation Biology,* New York: Plenum
277. Maeda, M., Saneyoshi, M., Kawazoe, Y. 1971. *Chem. Pharm. Bull.* 19:1641–49
278. Elvidge, J. A., Jones, J. R., O'Brien, C. 1971. *Chem. Commun.* 394–95
279. Tomasz, M., Olson, J., Mercado, C. C. 1972. *Biochemistry* 11:1235–41
280. Livramento, J., Thomas, G. J. Jr. 1974. *J. Am. Chem. Soc.* 96:6529–31
281. Gamble, R. C., Schimmel, P. R. 1974. *Proc. Natl. Acad. Sci. USA* 71:1356–60
282. Gamble, R. C., Schoemaker, H. J. P., Jekowsky, E., Schimmel, P. R. 1976. *Biochemistry* 15:2791–99
283. Schoemaker, H. J. P., Gamble, R. C., Budzik, G. P., Schimmel, P. R. 1976. *Biochemistry* 15:2800–3
284. Schimmel, P. R., Uhlenbeck, O. C., Lewis, J. B., Dickson, L. A., Eldred, E. W., Schreier, A. A. 1972. *Biochemistry* 11:642–46
285. Gruehl, H., Feldmann, H. 1976. *Eur. J. Biochem.* 68:209–17
286. Feldmann, H., Zachau, H. G. 1977. *Hoppe-Seylers Z. Physiol. Chem.* 358:891–96
287. McCutchan, T., Silverman, S., Kohli, J., Söll, D. 1978. *Biochemistry* 17:1622–28
288. Wübbeler, W., Lossow, C., Fittler, F., Zachau, H. G. 1975. *Eur. J. Biochem.* 59:547–57
289. Osterberg, R., Sjöberg, B., Rymo, L., Lagerkvist, U. 1975. *J. Mol. Biol.* 99:383–400
290. Giegé, R., Jacrot, B., Moras, D., Thierry, J.-C., Zaccai, G. 1977. *Nucleic Acids Res.* 4:2421–27
291. Reid, B. R., Hurd, R. E. 1977. *Acc. Chem. Res.* 10:396–402
292. Horowitz, J., Ofengand, J., Daniel, W. E., Cohn, M. 1977. *J. Biol. Chem.* 252:4418–20
293. Shulman, R. G., Hilbers, C. W., Söll, D., Yang, S. K. 1974. *J. Mol. Biol.* 90:609–11
294. Kim, S.-H. 1975. *Nature* 256:679–81
295. Eldred, E. W., Schimmel, P. R. 1972. *Biochemistry* 11:17–23
296. Midelfort, C. F., Chakraburtty, K., Steinschneider, A., Mehler, A. H. 1975. *J. Biol. Chem.* 250:3866–73
297. Carr, A. C., Igloi, G. L., Penzer, G. R., Plumbridge, J. A. 1975. *Eur. J. Biochem.* 54:169–73
298. Fersht, A. R., Kaethner, M. M. 1976. *Biochemistry* 15:818–23
299. Lovgren, T. N. E., Pastuszyn, A., Loftfield, R. B. 1976. *Biochemistry* 15:2533–40
300. Eldred, E. W., Schimmel, P. R. 1973. *Anal. Biochem.* 51:229–39
301. Deleted in proof
302. Griffin, B. E., Jarman, M., Reese, C. B., Sulston, J. E., Trentham, D. R. 1966. *Biochemistry* 5:3638–49

303. Sprinzl, M., Cramer, F. 1973. *Nature New Biol.* 245:3–5
304. Hecht, S. M., Chinault, A. C. 1976. *Proc. Natl. Acad. Sci. USA* 73:405–9
305. Ofengand, J., Chládek, S., Robilard, G., Bierbaum, J. 1974. *Biochemistry* 13:5425–32
306. Fraser, T. H., Rich, A. 1973. *Proc. Natl. Acad. Sci. USA* 70:2671–75
307. Sprinzl, M., Cramer, F. 1975. *Proc. Natl. Acad. Sci. USA* 72:3049–53
308. Fraser, T. H., Rich, A. 1975. *Proc. Natl. Acad. Sci. USA* 72:3044–48
309. Chinault, A. C., Tan, K. H., Hassur, S. M., Hecht, S. M. 1977. *Biochemistry* 16:766–76
310. Julius, D. J., Fraser, T. H., Rich, A. 1979. *Biochemistry.* 18:604–9
311. Wetzel, R. 1979. *Origin of Life.* In press
312. Hecht, S. M., Tan, K. H., Chinault, A. C., Arcari, P. 1977. *Proc. Natl. Acad. Sci. USA* 74:437–41
313. Shreier, A. A., Schimmel, P. R. 1972. *Biochemistry* 11:1582–89
314. Eldred, E. W., Schimmel, P. R. 1972. *J. Biol. Chem.* 247:2961–64
315. Yarus, M. 1972. *Proc. Natl. Acad. Sci. USA* 69:1915–19
316. Fersht, A. R., Kaethner, M. M. 1975. *Biochemistry* 15:3342–46
317. Deleted in proof
318. Hopfield, J. J., Yamane, T., Yue, V., Coutts, S. M. 1976. *Proc. Natl. Acad. Sci. USA* 73:1164–68
319. Fersht, A. R. 1977. *Biochemistry* 16:1025–30
320. von der Haar, F., Cramer, F. 1976. *Biochemistry* 15:4131–38
321. Weil, J. H., Burkard, G., Guillemaut, P., Jeannin, G., Martin, R., Steinmetz, A. 1977. In *Biological Reactive Intermediates,* ed. O. J. Jollow, J. J. Kocsis, R. Snyder, H. Vainio. pp. 97–120. New York: Plenum

321a. Nowak, T. S., Munro, N. H. 1977. *Biochem. Biophys. Res. Commun.* 77:1280
322. Tada, M. 1975. *Nature* 255:510–12
323. Rapaport, E., Zamecnik, P. C. 1976. *Proc. Natl. Acad. Sci. USA* 73:3984–88
324. Hinz, H.-J., Weber, K., Flossdorf, J., Kula, M.-R. 1976. *Eur. J. Biochem.* 71:437–42
325. Santi, D. V., Webster, R. W. Jr. 1976. *J. Med. Chem.* 19:1276–79
326. Jacques, Y., Blanquet, S. 1977. *Eur. J. Biochem.* 79:433–41
327. Rainey, P., Hammer-Raber, B., Kula, M.-R., Holler, E. 1977. *Eur. J. Biochem.* 78:239–49
328. Merault, G., Graves, P.-V., Labouesse, B., Labouesse, J. 1979. *Eur. J. Biochem.* In press
329. Fersht, A., Gangloff, J., Dirheimer, G. 1978. *Biochemistry* 17:3740–46
330. Moe, J. G., Piszkiewicz, D. 1976. *FEBS Lett.* 72:147–50
331. Fayat, G., Fromant, M., Kahn, D., Blanquet, S. 1977. *Eur. J. Biochem.* 78:333–36
332. Kern, D., Dietrich, A., Fasiolo, F., Renaud, M., Giegé, R., Ebel, J.-P. 1977. *Biochimie* 59:453–62
333. Lee, M.-L. 1974. *Biochemistry* 13:4747–52
334. Piszkiewicz, D., Goitein, R. K., 1974. *Biochemistry* 13:2505–11
335. Epely, S., Gros, C., Labouesse, J., Lemaire, G. 1976. *Eur. J. Biochem.* 61:139–46
336. Kaufman, G., Littauer, U. Z. 1974. *Proc. Natl. Acad. Sci. USA* 71:3741–45
337. Morgan, S., Körner, A., Low, B., Söll, D. 1977. *J. Mol. Biol.* 117:1013–31
338. LaRossa, R., Vögeli, G., Low, B., Söll, D. 1977. *J. Mol. Biol.* 117:1033–48
339. LaRossa, R., Mao, J., Low, B., Söll, D. 1977. *J. Mol. Biol.* 117:1049–59

Ann. Rev. Biochem. 1979. 48:649–79

ELECTRON MICROSCOPIC VISUALIZATION OF NUCLEIC ACIDS AND OF THEIR COMPLEXES WITH PROTEINS

♦12022

Harold W. Fisher

Department of Biochemistry and Biophysics, University of Rhode Island, Kingston, Rhode Island 02881

Robley C. Williams

The Virus Laboratory and Department of Molecular Biology, University of California, Berkeley, California 94720

CONTENTS

649

0066-4154/79/0701-0649$01.00

PERSPECTIVES AND SUMMARY

Electron microscopy of nucleic acids and nucleoproteins has been marked, during the period under review, by improvements in methods of specimen preparation and a notable broadening of the investigative areas. A significant advance in methodology has been the development of ways to spread pure, "naked" nucleic acids for electron microscopic observation. Prior to such developments the only feasible way to visualize pure nucleic acids was to embed them in a film of denatured cytochrome c (the "Kleinschmidt" technique), and thereby surround each strand with a thick collar of bound molecules. However, purified nucleic acids free of cytochrome c, or other basic protein, can be spread satisfactorily if they are prevented from lateral aggregation upon drying. This has been accomplished by (a) spreading them in a monofilm of small molecules such as benzyldimethylalkylammonium chloride (BAC), (b) adsorbing them to immobile, positively charged residues on the specimen film surface, or (c) stiffening them through intercalation by some small molecule. It is now fairly routine to photograph duplex strands of nucleic acid whose width, even as augmented by materials used to enhance contrast, is no greater than 4.0 nm, as contrasted with the 15–20 nm width found in cytochrome c preparations. Single-stranded nucleic acids, however, appear severely kinked even after treatment with formamide or glyoxal unless they are spread in a monofilm such as BAC.

Unfortunately, the visualization of pure nucleic acids tells little about their structure whether they are embedded in cytochrome c or not. The nucleotide repeat distance is too small to be discerned, and earlier attempts to make chemical identifications by specific, heavy-atom tagging were overtaken by the great success of base-sequence analysis. A fiber of pure nucleic acid seen in the electron microscope is not rich in visual potential; the most that can be directly discerned is its contour length, its degree of convolution (as dehydrated), and the multiplicity of strandedness along its length. The inferences drawn from these primitive kinds of observations have been enormously valuable, of course, but their reliability and reach have not been appreciably affected by the presence or absence of a few dozen molecules of cytochrome c. The value of the advances made in the preparation of electron microscope specimens lies in the improved possibilities of seeing protein molecules that may be complexed with the nucleic acid. Only in the last three or four years, for example, has it been routinely feasible to visualize molecules of RNA polymerase that have been bound to DNA, while even now complexes with the smaller repressor molecules are not easy to photograph. The future will show increased exploitation of the capability to visualize the consequences of in vitro complexing of protein molecules and nucleic acids.

The recent years have seen a broadening of the use of electron microscopy in investigations of the structural aspects of "native" nucleic acids and nucleoproteins—those free from their intact environment with a minimum of manipulation. A notable landmark was the demonstration by Miller & Beatty in 1969 (78) that something as "living" as the signs of ongoing transcription and translation could be brought under the scrutiny of electron microscopy. Once the barrier to looking at something far removed from "pure" was so spectacularly breached the way was open in several directions, particularly for reexamining the structural framework of eukaryotic chromatin. It is the history of a methodology like electron microscopy that its early findings are rich in research significance, and are followed by their use as routine aspects of methodology. It seems a safe prediction that investigation of chromatin by electron microscopy will have a long, exciting life before it becomes routine.

INTRODUCTION

Electron microscopy of nucleic acids is now about 30 years old. In 1948 Scott (1) reported observations on calf thymus DNA that had been dried in air from a water solution and shadowed to enhance contrast. His distinctly unspectacular results demonstrated the primary difficulty encountered in forming a satisfactory preparation of elongated strands of DNA for electron microscopic visualization: their strong tendency to be laterally aggregated by forces of surface tension during the removal of the liquid in which they are suspended. In 1952 Williams (2) demonstrated that calf thymus DNA, suspended in water, could be microfreeze-dried for electron microscopy, a technique that greatly reduced the side-to-side aggregation of the DNA fibers. From measured lengths of shadows Williams estimated that the diameter of a fibril is about 1.5 nm, a value not out of accord with other estimates of DNA diameter that were being suggested in the early 1950s. During the next five years or so, several papers appeared on the electron microscopy of nucleic acids, notably one by Hart (3) showing that the single-stranded RNA of tobacco mosaic virus could be visualized, after shadowing, if the material were stretched out on a Parlodion substrate film, and one by Hall & Litt (4) showing micrographs of preshadowed replicas of well-distributed, double-stranded DNA that had been sprayed in a volatile solvent upon the surface of freshly cleaved mica. The absence of lateral clumping was correctly surmised to be due to the extremely hydrophilic nature of the mica surface, on which the spray droplets spread to form thin liquid sheets that dried very rapidly.

In 1959 Kleinschmidt & Zahn (5) showed that nucleic acids could be prepared for electron microscopy by incorporating them in a surface film of denatured, basic protein, such as cytochrome c. Their preparations showed unaggregated DNA strands that were loosely coiled, rather than kinked, and that apparently possessed intact secondary structure. Their measured lengths, in relation to their molecular weights, were such as to suggest the absence of breakage. This distinctly successful method of preparation has been modified and improved over the years and is by far the one most commonly employed for observation of nucleic acids. For many kinds of experiments, however, the protein monofilm method is not useful, inasmuch as the material seen in the electron microscope is not nucleic acid alone ("naked" nucleic acid), but rather a complex with molecules of cytochrome c. The resulting strand, nucleic acid *plus* cytochrome c, is so large in diameter that it is difficult to discern other molecules, such as those of RNA polymerase, bound to the DNA. Thus, while the protein monofilm method remains a convenient one for observation of the secondary structure of nucleic acids, it must be supplemented by ones that will allow visualization of naked nucleic acids.

METHODS FOR VISUALIZING PURIFIED NUCLEIC ACIDS

Specimen Support Films

A good specimen support film should be easy to prepare, immovable under the impact of the electron beam, and the electron image "noise" arising from its structure should be small compared to the "signal" from the specimen. Its surface should be free of objects large enough to be microscopically visible; i.e. "clean." It is usually preferable for the surface to be hydrophilic if nucleic acid is to be deposited directly upon it, but if the specimen has been preattached to another film (like denatured cytochrome c), or to another surface (like mica) from which a replica is made, this characteristic may be irrelevant.

Support films are usually made [see (6) for a recent review] of plastic, carbon, or carbon over plastic. Plastic films, of Parlodion or Formvar, are seldom used alone because of their dimensional instability in the electron beam. Films of pure carbon, although somewhat fragile, may be made by the deposition upon a mica surface of carbon evaporated from an arc in vacuo, followed by stripping in water and transfer to a specimen grid. Most frequently, probably, the open-face sandwich film is made: a plastic film upon which a film of carbon is vacuum-condensed to confer rigidity. Surfaces of unmodified carbon are always strongly hydrophobic; although films of Formvar and Parlodion are less so, they are never strongly hydrophilic (contact angle $\rightarrow$ 0°). Films made of materials other than the above have been used and two of them, aluminum (7) and silicon monoxide (R. C. Williams, personal observation), are hydrophilic when freshly made. A carbon surface can be made hydrophilic for several minutes to a few hours by subjecting it for a few seconds to a "glow" discharge (8)—a moderately high-voltage plasma created at a partial pressure of residual air (and, usually, pump-oil vapor).

Deposition of Specimen Material

The transfer of strands of nucleic acid from solution to a dry, two-dimensional array ready for electron microscopy can be accomplished in several ways. The ideal technique would produce no aggregation nor entanglement, nor breakage arising from shearing forces, and would effect no distortion of the conformation and interbase spacing. It should permit observation of the nucleic acid strands at high resolution, and, when desired, in their naked state.

SPREADING BY USE OF MONOFILMS   The now classical method of spreading nucleic acids with the aid of a protein monofilm was described by Kleinschmidt & Zahn in 1959 (5). The rationale underlying this method was to find a way to reduce the aggregation of nucleic acid strands during drying from solution by first immobilizing them on or within a semirigid monofilm of denatured protein. If the protein has basic side groups, like cytochrome *c,* it will form complexes with nucleic acid when in solution with it; the complexes will become immobilized as the bulk of the cytochrome *c* unfolds into a monofilm. In the original "spreading method" (5) a small amount of nucleic acid (usually at about 1 μg/ml) mixed with cytochrome *c,* in an appropriate buffer, is dribbled slowly upon the surface of a "hypophase"—a dilute buffer solution, preferably wholly volatile. The cytochrome *c,* with its enmeshed nucleic acid, spreads rapidly into a loose monofilm whose surface density can be increased with the aid of a movable bar touching the hypophase surface. If a filmed electron microscope grid is then touched lightly to the monofilm surface, film side down, it will, upon withdrawal, pick up a portion of the monofilm. Dehydration, staining, and shadowing (see below) follow.

Single-stranded nucleic acid will not be well spread out in a protein monofilm because of intra- and interchain base-pair bonding. A bond-disrupting agent like formamide (9), if present in sufficient concentration in both hypophase and hyperphase, will disrupt the bonds sufficiently to allow the cytochrome *c* to complex with the single strands and stiffen them (for review see 10). Single-stranded nucleic acid can usually be distinguished from double-stranded when formamide or glyoxal-formamide (11) is used, because they appear somewhat thinner and definitively less smooth in contour (more kinky). If such denaturing agents are not used the single-stranded fibrils will coil tightly to form a brush-like tangle, a "bush."

In 1964 Lang et al (12) developed a modification of the original method, called the "diffusion method," in order to reduce the chances of sheer deformation of DNA during the rapid spreading and denaturation of the cytochrome c. The cytochrome c monofilm was first formed, followed by introduction of the nucleic acid (without cytochrome c) into the body of the hypophase. Subsequent sampling of the monofilm showed attached strands of nucleic acid that had presumably reached the monofilm by diffusion. The methods of "spreading" and of "diffusion" are not identical in the routes by which the nucleic acid becomes attached to the monofilm. In the former case the nucleic acid strands doubtless bind with cytochrome c prior to incorporation within the monofilm. In the latter case the attachment mechanism would seem to be solely one of adsorption of naked nucleic acid strands to basic groups on the under surface of the monofilm. This viewpoint is evidently too simple, however, inasmuch as diameters of double-stranded DNA are about the same (~15 nm) when it is spread either way (13).

Spreading nucleic acids within a protein monofilm can readily be miniaturized (14–16) simply by decreasing, appropriately, the amounts of materials. The diffusion method has also been changed to a quasi-diffusion one employing only small volumes. But instead of forming a monofilm initially, beneath which the nucleic acid is insinuated into a hypophase, one large drop containing buffer, nucleic acid, cytochrome c, formaldehyde, and, optionally, formamide, is formed. The nucleic acid becomes complexed with some of the cytochrome c, the remaining protein forms a surface monofilm, and the complexed DNA diffuses to and attaches to it.

Inasmuch as strands of nucleic acid, when spread in a monofilm of cytochrome c, have a thick collar of that material around them, it should be advantageous to form the monofilm method with molecules smaller than those of cytochrome c. Vollenweider et al (17) have demonstrated that the quaternary ammonium salt, benzyldimethylalkylammonium chloride (BAC), with a molecular weight of about 350, can form a monofilm to which nucleic acids will attach. The method is effective in spreading both double- and single-stranded nucleic acid when performed on either a macro- or a micro-scale (18). Double-stranded DNA, after rotary shadowing, has been reported to be only 6.0 nm in diameter (17), which indicates very little increase in diameter arising from bound BAC. Single-stranded material appears kinky and in places difficult to distinguish from the background film (17). The most recently reported work with this technique (19) shows binding between S4 and S8 ribosomal proteins and 16S RNA.

Quite recently Zollinger et al (20) have reported that strands of DNA, when spread on an aqueous hypophase from a hyperphase containing both cytochrome c and sodium lauryl sarcosinate, appear less than half as thick as they do when spread from cytochrome c alone. Why addition of the detergent should have this effect is not clear, although it may be that it simply prevents formation of complexes of DNA and cytochrome c while still allowing formation of the protein monofilm. Alternatively, it could be that only the detergent forms the monofilm and that the DNA binds to it, as it does to BAC monofilms. There was no report of the effects of using sodium lauryl sarcosinate alone. Although the method seems to require stringent conditions (20) to achieve a good spread, it may well be a satisfactory alternative to the use of BAC (17).

DEPOSITION WITHOUT USE OF MONOFILMS   If strands of naked nucleic acid are to be satisfactorily deposited directly upon an electron microscope specimen film they must attach firmly to the film while still in solution. Otherwise the forces of surface tension, upon drying, will produce clumping. The support film almost universally used is carbon, either alone or deposited over a plastic film. Little is known about forces that determine attachment of nucleic acids to solid surfaces such as carbon, but a safe generalization seems to be that the attractiveness of a surface is related to the strength and disposition of any positive charges it may have. Information about the electrostatic character of films used for electron microscopy is almost wholly lacking, however. In one study, an unmodified carbon film, in air, was found to maintain a negative potential on its upper surface (7), but the relevance of this fact to adsorption of nucleic acid from solutions is not clear.

Carbon films that are untreated have frequently been used for the deposition of naked nucleic acids (1, 21–23). When a liquid drop containing nucleic acid is rapidly removed from such a film by, say, suction, most of the nucleic acid strands are swept away. Those remaining are almost always found to be "streaked" into linear forms, a phenomenon that is understandable if the nucleic acid strands are not entirely pure and contain a few specks of material that can preferentially adsorb to the hydrophobic carbon (23). There is some evidence that whatever bonding exists between an untreated carbon film and nucleic acid is of hydrophobic nature, since an increase in Na^+ increases the amount of nucleic acid attached (21). The use of untreated carbon films is no longer common, but, interestingly enough, such films have been quite helpful in delineating single-stranded nucleic acid (23) whose kinky contours, unless stretched, make visualization and length measurement uncertain. If a carbon film is made hydrophilic by treatment to ionic bombardment (8) the attachment of nucleic acid to it is quantitatively improved (21) and the proportion of strands showing "streaking" is decreased. Such carbon films are sometimes referred to as "charged" (24) but there is no evidence to support the use of this term except ex post facto inference. Charges need not be invoked to explain the improvement in nucleic acid deposition. If the carbon surface is completely hydrophilic, and if a drop of applied liquid is wholly volatile and is allowed to dry, it would seem that the deposition of suspended nucleic acid strands would be ideal. The last traces of liquid would form a sheet of uniform thickness, but eventually so thin that lateral motions of the particles would cease, and their eventual "beaching" would be in a well-spread array. The practical fact is that subtle variables, probably in glow discharge conditions, introduce considerable inconsistency in the hydrophilic character of glowed films. Their use for deposition of naked nucleic acids has largely been superseded by ones that involve modifications of more specific nature. For the deposition of *protein* molecules, however, the use of glowed carbon films is consistently effective (25).

In 1969 a first attempt was made (18) to improve the adsorptive properties of a carbon surface for nucleic acids by treating the surface with a material known to possess positively charged groups, in this case an ionic detergent, benzyldimethylalkylammonium chloride (BAC). Concentrations as low as 10^{-5} g/ml, when applied to an otherwise untreated carbon film for a few hours and dried without rinsing, permitted attachment of both double- and single-stranded DNAs. The former showed generally smooth contours, but with some sharp kinks, while the latter, applied in the presence of formaldehyde and at 60°C, was severely kinked. As mentioned above, BAC is now much more commonly used to form nucleic acid-adsorbing monofilms (17), rather than to pretreat carbon films.

A carbon film can be partially covered with charged groups if it is ionically bombarded in a reducing atmosphere of amylamine, as shown by Dubochet et al (26). Amylamine is water-soluble, but when it is polymerized in a plasma discharge it evidently becomes insoluble. The effect of the amylamine polymerization persists for at least several hours. The nucleic acid sample, at about 1 μg/ml, is applied to the treated film for about 2 min, and the specimen then rinsed and stained. Dubochet et al (26) found that the efficiency with which DNA was found adsorbed to the films decreased to zero as the salt concentration rose to 100 mM (a distinct limitation to the technique) and decreased with repeated rinsing of the applied specimen. They concluded from these facts that the carbon surface was by no means covered with basic poly amylamine sites. Single-stranded nucleic acids, after adsorption, appeared severely aggregated and clumped, but double-stranded DNA appeared well spread. It was smoothly curved, with a diameter of only about 2.5 nm after light uranyl acetate staining.

A recently described method (27) for depositing nucleic acids upon carbon films makes use of polylysine. DNA collapsed by polylysine had been previously examined by electron microscopy (28) and used as a titration procedure to study condensates of DNA. The method developed by Williams (27) specifically for mounting individual strands of nucleic acid for electron microscopy uses a water solution of polylysine of mol wt ~2000. Carbon-coated films are first ionically bombarded and then covered with a drop of a polylysine solution (~1 μg/ml) for a few seconds. The drop is removed by suction, with or without a rinse in water. Double-stranded DNA, at about 1 μg/ml, applied to the film for several seconds, adsorbs efficiently and without entanglement, collapse, or signs of streaking. Solution conditions seem not to be critical; adsorption proceeds well from solutions ranging from 5 mM–400 mM in Na$^+$, containing glycerol up to 20%, or containing, if desired, formamide up to 90%. After adsorption of the DNA, the specimen is rinsed thoroughly in water or, preferably, dilute ammonium acetate, followed by uranyl acetate staining, further rinsing, and final drying. Rotary shadowing greatly enhances the visibility of the nucleic acid strands. An example of DNA prepared by this method is shown in Figure 1. When single-stranded DNA is adsorbed, even when prepared with 80% formamide or 0.5 M glyoxal, it appears to be full of what might be called "microkinks," making it useless for contour-length measurements. This configuration is to be expected, since single-stranded nucleic acid in solution is an extremely flexible random coil with very short persistence

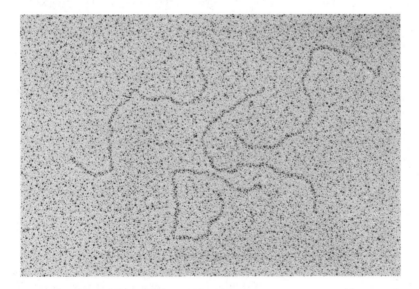

Figure 1   The (*Hinf* T7)$_{1100}$ DNA fragment absorbed at 0.4 μg/ml upon a glowed carbon film treated with polylysine. Stained with 5% aqueous uranyl acetate, and rotary shadowed with tungsten. Note that the duplex strands, including shadowing material, are less than 4.0 nm wide. X200,000. [From Williams (27)]

lengths. When it collapses by projection upon a plane surface its contour must contain numerous "microkinks." As mentioned earlier, when it is bound to cytochrome c, or BAC, and stretched during the formation of a monofilm, its kinks are removed.

The DNA-intercalating dyes, ethidium bromide and propidium diiodide, have been used to increase the stiffness of DNA strands and decrease the amount of entanglement during spreading for electron microscopy (29). The complex is reported to attach well to untreated carbon films and to the surface of freshly cleaved mica, but only if the concentration of ethidium bromide is between 50 and 250 μg/ml, with the DNA at about 0.5 μg/ml. It is not clear why the dye concentration must be so much greater than that required to saturate the DNA. Indeed, the attachment mechanism is in itself a mystery, since a carbon film remains hydrophobic when exposed to a water solution of ethidium bromide and dried. A hint of an answer is found in the results of Sogo et al (7) who found a negative surface potential for a carbon film in air, and in the finding of Paoletti & Le Pecq (30) that in a buffer of low ionic strength the DNA-ethidium complex behaves as a cation. As might be expected the diameter of double-stranded DNA is not increased appreciably by complexing with ethidium bromide; Koller et al (29) reported it to be about 2.5 nm. They found that the contour lengths of T3 and T7 DNA bound to ethidium bromide were closely comparable to those reported for DNA prepared in protein monofilms. Freifelder (31) had earlier found that DNA, spread in a cytochrome c film, increased about 25% in length after complexing with ethidium bromide.

Dehydration and Enhancement of Contrast

Common to all the methods outlined above is a stage in which strands of nucleic acid, either alone or in a monofilm, are attached to the surface of the support film on a specimen grid but are still submerged under a drop of liquid that may contain excess nucleic acid and nonvolatile salts. Dehydration by one means or other must now take place. The simplest procedure is, after rinsing with pure water, to blot the drop or remove it by vacuum aspiration. An alternative procedure is a rinse step in which the replacement fluid prior to final blotting is an organic solvent—ethanol, methanol, acetone, or isopentane, for example. There seems to be no universally accepted procedure for dehydration, although organic solvents appear to be most commonly used with monofilm spreading (10), but are frequently omitted in favor of water when monofilm methods are not used (22, 24, 26, 32). A recent report (33) indicates that conditions of dehydration may be important in determining contour lengths of DNA. It was found that phage λ DNA, spread in a BAC monofilm (17) and dehydrated in methanol, was 95% as long as it was after freeze-drying from a water rinse; after ethanol dehydration, however, three length classes were found, the shortest only 75% of the freeze-dried length. Complexing with ethidium bromide prior to freeze-drying increased its length by almost 40%. These results not only bear on the conformational effects of dehydrating solvents, but raise some concern as to the degree to which DNA is anchored to a BAC film.

Although a strand of DNA encased in a collar of cytochrome c can be seen by dark-field electron microscopy (34), a strand of naked DNA will be unambiguously visible only if it has been in some manner coated with atoms of a heavy metal. A nucleic acid strand has so little electron scattering power as to be relatively invisible, but what is worse, what mass it has will be largely volatilized by the impact of the electron beam before a photograph can be taken (35). After is has been stained or shadowed with heavy metal it will appear visible, but what is seen is not the organic structure but the heavy atoms that have been applied to it (36, 37). Staining may be either "positive" or "negative" depending upon whether the specimen objects appear darker or lighter than the surrounding background. Preparations containing only nucleic acids, either naked or in a cytochrome c film, are most usefully stained positively, but when nucleoproteins such as complexes of DNA with RNA polymerase are observed, they may be stained negatively (38). The stains almost universally used are solutions of uranyl acetate or, less frequently, uranyl formate in either an aqueous solvent or one in which the principal component is ethanol, methanol, or acetone. There are suggestions that DNA binds more uranium per residue when the solvent is organic rather than aqueous (39, 40). It is probably true that the effectiveness of nucleic acid staining depends not only upon the staining vehicle, but also upon the condition of the nucleic acid (naked or in a monofilm) and its solvent, and the character of the support film (plastic, untreated carbon, treated carbon). Two very different but commonly used positive-staining procedures are: (a) A water rinse of the specimen followed by dehydration in ethanol, then staining with uranyl acetate in ethanol, followed by rinsing in, and drying from, ethanol; (b) A water rinsing and aspiration of the residual drop, then a quick transfer of the grid to uranyl acetate in aqueous solution, followed by another water rinse and aspiration of the residual water drop.

Strands of nucleic acid complexed with cytochrome c show sufficient contrast after prolonged staining for visualization by bright-field electron microscopy. Naked nucleic acids, however, are only marginally visible in bright field after staining alone (26, 32). Electron microscopy by dark-field imagery greatly increases contrast and allows stained DNA to be photographed satisfactorily (26) but at the cost of more elaborate optical manipulation. Following staining it is common practice to increase contrast by some form of shadowing. In the original shadowing process (41) both the single source of vaporizing metal atoms and the specimen upon which they condensed remained stationary in order that a protruding object, like a virus particle, would cast a metal free "shadow" on its rearward side. Strands of nucleic acid so shadowed are visible, but only marginally so in those stretches that are parallel to the shadowing direction. Shadowing at two or more different angles partially remedies this situation, but the fullest degree of angular uniformity is achieved by allowing the specimen plane to rotate about a vertical axis during shadowing. Several different metals have been used for such rotary shadowing, their common characteristics being high density and high melting point. The latter characteristic is associated with the production of a shadowing film of desirably fine grain. The metal that is perhaps the most convenient to evaporate and the finest grained is tungsten (24, 42) which sublimes readily from a wire heated nearly to its melting point for several seconds to a few minutes. It has been proposed (24) that the condensation of tungsten vapors upon a specimen effects "selective nucleation," a process by which a "reactive" metal (such as tungsten) would selectively bind to proteins and nucleic acids, as distinct from the carbon support film, until a monolayer of tungsten metal is deposited. Such "vapor phase positive staining" would be interesting indeed, but so far there has appeared no experimental evidence supporting its existence.

OBSERVATIONS OF PURIFIED NUCLEIC ACIDS

Contour Lengths and Molecular Weights

The simplest and most direct way of obtaining information about such physical characteristics of nucleic acids as strandedness, degree of flexibility and extent of supercoiling, and contour length is by electron microscopy. The information, of course, strictly refers to two-dimensional representations of three-dimensional objects and must therefore lead to only approximate, and possibly misleading, inferences about their conformation. Measurements of contour length, however, can be made to any degree of statistical precision even in cases where nucleic acid strands lie in complex patterns such as the alternating single- and double-stranded regions of a heteroduplex. Whether or not contour lengths of strands of thoroughly dried nucleic acid are the same as their contour lengths in solution is not known, but for most purposes of electron microscopy the question is immaterial.

In measuring relative contour lengths in a sample of double-stranded DNA it is convenient to add a reference DNA, such as that of ϕX174 RF, whose length is assumed to be uniform and constant so long as preparative conditions remain unchanged. If contour lengths are to be expressed in absolute units, such as μm, the contour length of the reference material must obviously be determined. Once this determination is made there is no further need to know micrographic magnifications. Electron microscopy has been used extensively to determine not only contour lengths but also the molecular weights of various DNAs. If a reference DNA is available whose molecular weight is precisely known, and if the reasonable assumption is made that its mass per unit of dry length is the same (on a given specimen grid) as that of the target DNA, the determination of molecular weight is reduced to simple measurement with a contour measurer. Until quite recently, however, no DNA of precisely known molecular weight was available, and it was common practice to use an alternate approach wherein a value was taken for the base-pair repeat distance along the specimen strands. With this parameter assumed to be known, along with the molecular weight of a base-pair (662 daltons), a molecular weight determination was a simple matter. In the early years of nucleic acid electron microscopy the value taken for the base-pair repeat distance was that obtained by X-ray analysis for the B form of paracrystals of lithium salts of DNA—0.338 nm/base pair (43, 44)—a well-documented number but one unlikely to be closely applicable to electron microscope specimens.

Many attempts have been made to determine experimentally a valid number for the base-pair repeat distance of DNA spread in an electron microscope specimen, all based on the elementary procedure of measuring contour lengths of samples of reference DNAs whose molecular weights were determined by various physicochemical methods. Fairly large deviations developed (45), most of them arising from (a) differences in the molecular weights determined for the standard DNAs; (b) failure to use for electron microscopy a DNA sample that was from the stock used for molecular weight measurement; and (c) variations in contour length which resulted from differing conditions in preparation for electron microscopy. Recent advances in nucleic acid chemistry have eliminated the first source of variation, and improved genetic analysis has brought the second source of error sharply to attention. Several nucleic acids, or fragments, have been sequenced within a very small uncertainty, and one in particular—the ϕX174 RF DNA which has 5375 ± 20 base pairs totaling 3.558 X 10^6 daltons (46)—has recently been used as a reference standard for electron microscopy. Stüber & Bujard (45) have measured the contour length of this nucleic acid, using an aliquot of the stock that was used for the sequencing studies, in order to eliminate uncertainty about possible genetic and physical variations among ϕX174 isolates. They obtained a contour length of 1.71 ± 0.02 μm from material spread in a cytochrome c monofilm, without formaldehyde, which led to a linear mass density of 2.08 ± 0.03 X 10^6 daltons/μm, equivalent to a base-repeat distance of 0.318 ± 0.005 nm. This absolute determination was used to arrive at molecular weights of several phage DNAs; among them were those of T7 and λ, with molecular weights of 26.4 X 10^6 and 30.0 X 10^6, respectively. As a note of caution it should be mentioned that Stüber & Bujard (45) found that a sample of T7 DNA, previously used by them as a standard, had a 5% deletion compared to samples from other laboratories.

Two brief reports have recently appeared on the base-pair repeat distance of short, restriction enzyme fragments of T7 DNA prepared for electron microscopy by the poly amylamine (26) and the polylysine (27) techniques. In the former (32), a 970-nucleotide fragment of λ phage DNA was used, for which a surprisingly small base-pair repeat distance of 0.270 nm was reported. In the latter study a 263 ± 5 base-pair piece from PM2 DNA was examined and a repeat distance of 0.326 ± 0.018 nm was found. It is worth noting that this value and that found by Stüber & Bujard (45) for ϕX174 RF agree within the sum of their standard deviations despite considerable differences in the methods of spreading the samples for electron microscopy.

Although a double-stranded DNA with a precisely known number of bases is now available, that of ϕX174 RF, it does not follow that it will exhibit the same contour length every time it is measured electron microscopically. All experience indicates that variations in sample preparation can affect contour length by as much as 10%. Examples from recent work show the T7 genome to have a length ranging from (a) 11.7 μm (29) to 13.33 μm (47) when spread with ethidium bromide, (b) 11.8 μm when spread in BAC without formamide and 12.54 μm when formamide was used (17), (c) 12.25 μm when spread with cytochrome c and Sarkosyl (20), and (d) 12.37 μm with cytochrome c with no formaldehyde (45). Even the conditions of dehydration alone seem to be able to influence contour length significantly, as shown by Vollenweider et al (33), who obtained base-pair separations for the same stock of λ DNA ranging from 0.343 nm with freeze-dry dehydration to 0.317 nm with ethanol dehydration.

There has been no report on the determination of base-pair spacing for double-stranded RNA since the work of Granboulan & Franklin (48) with RF RNA from R17 bacteriophage. Their procedure was like that of Stüber & Bujard (45), but the reliability of the determination of the number of bases in the R17 genome was doubtless far below that of the recent work with ϕX174 (46). Granboulan & Franklin obtained 0.314 nm as the base-pair repeat distance for the RF form and, inasmuch as they found the same value for the single-stranded genome, they concluded that it had a stacked-base structure (48).

The electron microscope is clearly a useful instrument for measuring the relative lengths of nucleic acid fragments produced by the action of restriction enzymes. Quite recently it was used to map, by direct visualization, the order of occurrence of such fragments in ϕX174 RF DNA (49). Fragments produced by a restriction enzyme from *Arthrobacter luteus* and subsequently separated on a gel were hybridized in pairs and triples to whole ϕX174 DNA. The technique used a cytochrome c spread, with formamide, in which double-stranded regions could be distinguished from single-stranded ones and the lengths of both measured. Some 11 fragments were ordered, ranging from about 100–1000 nucleotides in length, and good agreement with the results of gel electrophoresis was found.

Physical Mapping

Physically mapping a one-dimensional object, like a highway or a strand of nucleic acid, consists of establishing a set of correspondences between distance measured along the object from one or more reference points and some distinguishable property. It is usually a two-step procedure: (*a*) a physical or chemical manipulation whose purpose is to create in the nucleic acid observable manifestations of the property being considered, and (*b*) their observation and localization. The electron microscope would seem to be a uniquely powerful aid to physical mapping, and, indeed, in 1961 Nomura & Benzer (50) suggested that the instrument be used for determining the length of a deletion in the DNA of a mutant bacterial virus. They noted that the manipulation step would be denaturation and mixed annealing of wild-type and mutant genomes, and the observation phase would be to look for a "loop" of unpaired, single-stranded DNA in an otherwise duplex structure. This innovative notion lay dormant for several years. In 1971, Davis et al (10) observed hybridized DNAs of wild-type and mutant λ phages and found "bush"-like structures which they interpreted as regions of collapsed, unpaired DNA. At about the same time Westmoreland et al (9) made the far-reaching discovery that addition of formamide to a cytochrome c spreading solution would prevent regions of single-stranded DNA from collapsing into the unmeasurable bush structures. Simplified and reliable techniques for creating and observing hybridization events, particularly those involving heteroduplex analysis, are reviewed by Inman & Schnös (51) and, quite recently, by Evenson (6).

It should be noted that in the type of physical mapping discussed above the only property investigated along the length of a nucleic acid single strand is its ability to form a duplex with another strand; the "resolution" of the method lies in the precision with which a point of structural discontinuity (single strand $\leftrightarrows$ double strand) can be located. The optical resolving power of almost any electron microscope is more than adequate for the observations, inasmuch as precision of localization is strongly limited by the graininess resulting from contrast-enhancing steps (staining and shadowing), and by the size of the cytochrome c collar around the nucleic acid. So far, there have been no reports of visualizing nucleic acid hybridization in the absence of a protein surface film. Not all discontinuities observed in heteroduplexes are significant, of course, since single-strand scissions and imperfect annealing produce artifacts. The popular methods introduced by Davis et al (10) reduce such disturbances by minimizing handling of the DNA and reducing the time spent in hybridization.

Even under the best observational conditions the minimal length of a segment of nucleic acid for which the strandedness can be ascertained is 30–40 nm, a length comparable with that of tRNAs. Wu & Davidson (52), however, succeeded in making regions of hybridized tRNA-DNA visible by linking ferritin to the former by a process allowing retention of its ability to hybridize with complementary sequences. In an accompanying paper Wu et al (53) mapped the genes for the tRNAs of three amino acids on a segment of bacterial DNA within a transducing phage. Gene mapping by use of the ferritin-tagging technique has also been reported by Angerer et al (54) who found 19 regions of homology between mitochondrial DNA and 4S RNA (tRNA) from HeLa cells, and by Ohi et al (55) who found 21 such regions in hybridization between mitochondrial DNA and 4S RNA from *Xenopus laevis.*

Physical mapping of nontrivial regions in strands of nucleic acid obviously requires not only pinpointing the start and end of the regions but also measuring their distances from some recognizable zero point. If the strands are linear and the sample is monodisperse either end can be taken as a reference point, although the statistical reliability of establishing the location of a map point decreases with the square root of its contour distance from the end chosen as reference. If the nucleic acid is circular the selection of reference points is not so simple. One solution has been to find regions where there is preferential denaturation (regions rich in A-T) brought about by elevated temperature, high pH, or high concentrations of formamide [see Inman & Schnös (51) and Inman (56) for reviews]. If a species of DNA has large fluctuations in base content it may show a highly reproducible pattern of denatured regions. Even if it is circular (and hence has no natural markers at ends) the denaturation map allows reference markers to be chosen with confidence, so long as the denatured regions are numerous and consistently formed. Denaturation mapping has furnished reference points for a study of replication forks (57) which show replication to be bidirectional in some DNAs and unidirectional in others, for identification of recombination junctions (58), and for the electron microscopic demonstration of repeated sequences in rDNA of *X. laevis* (59). Nowadays, restriction endonucleases, followed by gel separative methods, produce precisely cut, reproducible, short segments of double-stranded DNAs (60), thus providing ample zero-point markers for both linear and circular genomes. As one example, the *Eco*RI sites of mitochondrial DNA from *X. laevis* were recently used as reference markers in the study of hybridization with 4S RNA (55).

In recent years there have been several reports on the use of electron microscopy in mapping secondary structure of purified RNAs. In those few instances where the nucleotide sequence is known, as with tRNA and, recently, MS2 phage RNA (61), some aspects of secondary structure can be inferred from direct inspection of the sequence. In RNA spread for electron microscopy the regions of interest are hairpin loops and consistently located collapsed forms where secondary pairing is likely to have occurred. Wellauer & Dawid (62) prepared secondary structure maps of 45S precursor rRNA from HeLa cells, and found loops of fairly distinctive morphology at consistent locations. By use of a 3'-exonuclease they deduced that the 45S precursor molecule started with the 28S rRNA at the 5' end, was followed by a spacer, the 18S rRNA, and ended with a spacer at the 3' end. The processing pathway inferred from electron microscopy agreed well with earlier, biochemical studies. An elegant study by electron microscopy of the secondary structure of MS2 phage RNA was done by Jacobson (63). It was found that a gradation of hairpin loop formation, from simple to complex, could be obtained by increasing the Mg^{2+} concentration. One basic pattern could account for all the loops observed. Although the 3' and the 5' ends of the molecules seen in the electron microscope were not identified, a comparison of the structures seen by Jacobson, with the implications of the sequence data, indicated at least one region of satisfactory agreement: a large, central loop that was located in the region known to code for the coat protein cistron. The existence of a large, central loop has also been reported for the RNA of phages f2, Qβ, and PP7 (64), with the map of f2 having, in addition, hairpin loops in a pattern similar to that of MS2.

Replication and Recombination

An electron microscopic study of the replication events of phage ϕX174 DNA was recently performed by Koths & Dressler (65). Separation of the phage and bacterial DNA, and partial purification of the former, were accomplished in one step with the aid of differential density labeling of virus and host. An important feature was the inclusion, in the aliquots of phage DNA spread for electron microscopy, of a morphologically distinguishable plasmid DNA in a known number concentration. Inasmuch as the number of infected cells represented in each spread was also known, the various forms taken by the viral DNA could be expressed in numbers per cell. It was found that the net DNA synthesis, early in the viral cycle, accumulated into two structures: fully duplexed circles with single-strand tails of varying length; and circles having various degrees of duplicity from zero to full. Late in the cycle only the former, as well as single-stranded circles, were seen. These results were taken to confirm the rolling circle model of DNA replication, the duplex circles with tails being one of the replicative intermediates and the partially duplexed circles the other. The latter were not found in the late stage because the newly synthesized single-stranded circles were packaged into virions before their replication could start.

The phenomenon of DNA recombination, particularly as described by the model of Holliday (66) and its modifications (67), has been the recent subject of electron microscopic observations. An elementary inspection of a recombinational intermediate in the electron microscope is not exciting or unambiguous—simply a place where two DNA strands cross. Three kinds of observation can, in principle, unequivocally distinguish a recombination from an accidental juxtaposition of two duplex strands. One is a high-resolution micrograph that clearly identifies the four single strands involved in recombination; another is a demonstration that the junction occurs at homologous places on the two strands; and a third is a demonstration of branch migration. So far, the first test has not been a convincing one, although the preparation of DNA spreads without a protein monofilm may render it useful. Early intimations of the existence of visible recombination junctions came from examination of DNA from ϕX174-infected cells in which figure-8 forms were seen that crossed exactly in the middle (68); these forms were unlikely to represent the crossing of double-length DNA circles, but they could be two monomer rings interlocked like rings in a chain. Benbow et al (69) coinfected cells with wild-type ϕX174 and a deletion mutant and demonstrated the existence of figure 8s made up of monomers of both input sizes, an observation strongly indicating the existence of recombinational intermediates.

Valenzuela & Inman (58) were the first to attempt a demonstration of homology at putative recombination junctions by employing partial-denaturation mapping of DNA from bacteriophage λ. They concluded from measurements of the positions of these points that they occurred at homologous positions on the two strands, consistent with the popular models of recombination. Potter & Dressler (70) looked for recombination intermediates in the DNA of a colicine plasmid related to E1. By use of chloramphenicol the quantity of the plasmid in an infected cell could be increased to 1000 copies, making recombination an event more likely to be seen. As had earlier been seen with the DNA from ϕX174, the plasmid DNA exhibited some figure 8s. Proof of homology in the regions of strand crossing, as required by the Holliday model, was obtained by treating the figure 8s with *Eco*RI enzyme (which cleaves monomeric colicine DNA rings at one, unique site) and observing the production of chi-shaped forms whose arm lengths showed complete bilateral symmetry. Potter & Dressler (70) also saw several cases in which the four nucleic acid strands involved in a recombination could be distinguished. Simultaneous with the above-described work was a study by Thompson et al (71) which demonstrated the existence of recombination intermediates in G4 RF DNA, using as proof not only the creation of bilaterally symmetrical chi-forms by EcoRI endonuclease but also the disappearance of these forms through a random-walk generated branch migration. They found that at 37°C the time interval between random jumps was 170 μsec per nucleotide pair; in more familiar terms, there is a 0.32 chance that the junction will be more than 850 nucleotide pairs from its starting point in eight minutes.

Extrachromosomal Genetic Elements

Much of the recent literature on the electron microscopy of purified nucleic acids has dealt with the integration of one segment of genetic information into another. The extra chromosomal genetic elements worked with include a sex factor F, episomes, plasmids, mitochondria, insertion elements, and transposons. Research with genetic materials that can exist independently and also integrate with the chromosome of the cell has been very active owing to the great potential for gene manipulation, cloning, and genetic engineering. Most of these genetic operations are routinely confirmed by DNA spreading for electron microscopy. A recent book reviewing much of this work was based on the DNA Insertions Meeting at Cold Spring Harbor Laboratory (72).

OBSERVATIONS OF NATIVE NUCLEIC ACIDS AND NUCLEOPROTEINS

The distinction between "native" and "purified" nucleic acid cannot be a sharp one, depending as it does on the relative severity of the methods used to separate the nucleic acid from the biological structures with which it is associated in vivo. But if no particular effort has been made to separate nucleic acid from bound molecules, such as proteins and polyamines, by use of strong reagents like hot phenol and detergents, the nucleic acid may be called native. The methods first used to prepare native nucleic acids for electron microscopy made use of osmotic shock, a phenomenon readily adapted to the spreading of viral nucleic acids in a protein monofilm formed on a hypophase of pure water (73). Later, reagents such as perchlorate (74) and dilute detergents, and large pH changes, were used to facilitate nucleic acid release [see (75) for review of methods]. When specimens more struc-turally complex than viruses were examined some degree of subcellular fractionation was employed, such as production of protoplasts (76), nuclei (77), and nucleoli (78), sometimes followed by controlled sedimentation of the fractions upon the electron microscope specimen film.

Viral Nucleic Acids

Some of the more interesting of recent electron microscopic studies of viral nucleic acids have centered around the old puzzles of bacteriophage growth and maturation. One of these is the problem of how the DNA is packed into a preformed, empty phage head, a question specifically addressed as long ago as 1953 (79). Thomas (80) and Chattoraj & Inman (81) investigated, with phages λ, P2, 186, and P4, whether there was specificity as to the end of the DNA that was attached to the phage tail. Partial lysis was accomplished in a solution containing a cross-linking reagent, in which case one end of the partially ejected DNA was found attached near the head-tail junction. That it was always the same end (the A-T rich one in the case of phage λ) was shown by denaturation mapping. There has been recent electron microscopic evidence from Yamagishi & Okamoto (82) that supports in part the model for phage morphogenesis in which, (a) as one of the terminal steps in assembly, one end of a DNA strand is joined to a structural protein, either within the head or at the head-tail junction, (b) the DNA matures from concatemers by restriction enzyme cleavage into monomers, and (c) maturation is by the "headful" mode originally proposed as a model by Streisinger et al (83). Cells infected with phage λ were lysed at various times with the assistance of lysozyme and centrifuged upon electron microscope grids through a sucrose and formalin cushion. Counts were made of identifiable, intermediate structures seen near the ruptured cells. Empty heads were initially seen, followed by ones having protruding DNA strands. Next in time came partially filled heads that were round or angular, but both had attached DNA. The final structure seen was a filled head with hexagonal outline and no DNA strand showing.

The problem of DNA folding within phage heads was examined by Richards et al (84) who observed the form taken by the DNA within, or just outside, phage particles that had been partially ruptured during preparation for electron microscopy. The semicondensed material frequently appeared like concentric circles, or tightly wound spirals (not differentially distinguishable), from which observation it was suggested that the DNA in a mature phage head is packed like a ball of string or wound coaxially like a spool (84). Laemmli (28) investigated whether the condensation of DNA within a phage head could arise through repulsive interactions with a highly acidic internal peptide, by observing the morphological effects of treating T4 DNA with the acidic polymer, poly(ethylene oxide). Polylysine treatment was also used, to effect a sort of control in which a basic peptide produced the condensation. The latter treatment produced circular donuts and short stem structures like those reported by Olins & Olins (85), while treatment with the poly(ethylene oxide) yielded solid, ellipsoidal forms about the size of the heads of T4 phage. Digestion with a single-strand-specific endonuclease indicated that these forms have numerous enzyme-vulnerable regions, presumably collapsed hairpin loops. Chattoraj et al (86) investigated the effects of treatment with polyamines on the forms taken by several phage DNAs. They found structures not unlike those produced by polylysine (28); i.e. donuts of collapsed DNA with sizes comparable to the heads of the corresponding phages. Inasmuch as these two studies showed that either charge neutralization or charge repulsion can condense DNA into forms somewhat similar to those seen by Richards et al (84), the origination of these forms remains obscure.

Electron microscopy has recently been elegantly used in studying the mode of in vitro assembly of tobacco mosaic virus. The initiation site on the RNA was known to be about 1000 nucleotides from its 3'-hydroxy end, which requires that elongation proceed in two directions. Butler et al (87) compared the electron microscopic appearance, in negative stain, of tobacco mosaic virus that had been partially assembled, and partially stripped with alkali. The two ends of the rods could be distinguished by having either a concave or convex appearance. In the alkali-stripped rods a "puff" of un-coated RNA was seen at the concave end, and at the convex end in partially assembled rods. It is known that only the 5'-OH end of the RNA is uncov-ered during partial stripping, and in partially assembled rods both the 3' and 5' ends are free. It was concluded (87) that during the early times of assembly, at least, both ends of the RNA protrude from the end that is the 3'- one when the rod is fully assembled. A neater way of arriving at the same answer was devised by Lebeurier et al (88) who partially stripped rods of tobacco mosaic virus with dimethyl sulfoxide and spread the resulting material in a cytochrome c film. Short rods were seen with two RNA strands, generally of unequal length, protruding from one end. One strand remained constant in length (about 70 nm) during most of the assembly; the other had a length that decreased as the lengths of assembled rods increased. Both experiments agreed in showing that the assembly of tobacco mosaic virus proceeds rapidly from an internal initiation site to the 5'-OH end of the RNA, followed by slower encapsidation from the initiation site to the 3'-OH end.

Eukaryotic Chromatin

Electron microscopy has increasingly been used in the study of chromatin since the recognition of its fundamental, repeating unit, designated the "ν body" by Olins & Olins (89) and, later, called the "nucleosome" by Oudet et al (90). The nucleosome, it will be recalled, comprises a length of DNA of about 200 base pairs that is complexed with an octamer of histones H2A, H2B, H3, and H4, taken in two copies each, and is loosely associated with histone H1 (91). Recent work involving electron microscopy has centered on (*a*) the fine structure of the nucleosome and on (*b*) its manner of assembly to form intact chromatin [see (92) for a review of research on chromatin structure]. Fine structure analysis is best done with the relatively stable nucleosome "core," created by controlled nuclease digestion, that contains 140 base pairs and the histone octamer. The DNA that is cleaved from the nucleosome when the core is created, which varies in amount from 0–100 base pairs depending on species and tissue of origin, has been termed the "spacer" or "linker" DNA. Finch et al (93) and Finch & Klug (94) have recently examined core structure by X-ray analysis and electron microscopy. Crystals sufficiently large for X-ray analysis could be grown if the core material had undergone some proteolysis. They found the nucleosome core to be a flat bipartite disc 11.0 nm in diameter and 5.7 nm in height. The path taken by the DNA could not be discerned, but the X-ray data were consistent with its being helically wrapped into about 1-¾ turns on the outside. There is some direct electron microscopic evidence to support the idea, based on extensive studies of nuclease cleavage sites (95, 96) and histone interactions (97), that the nucleosome has a dyad axis and might be able to unfold into two half-nucleosomes (98). The minichromosome of SV40, normally containing 20–24 nucleosomes (99), was found to have 40–50 similar, but smaller, structures arranged occasionally in pairs, when it was prepared for electron microscopy in a medium of low ionic strength (100).

Nucleosomes have been reconstituted from SV40 (and polyoma virus) DNA mixed with the four core histones, to yield beaded-ring structures identical in appearance with the SV40 minichromosome (101). Indeed, the two arginine-rich histones, H3 and H4, alone were found to complex with viral DNA into beaded-ring structures, although the beads were distinctly smaller than normal nucleosomes and were complexed with less DNA (102–104).

The intriguing question of of the type of folding involved in the enormous DNA compaction achieved by native chromatin has been approached by electron microscopy. Primary attention has been given to the fundamental chromatin filament, about 10 nm in diameter, and to the next higher structural order, the thick fiber of 20–30 nm diameter. The former is most likely a linear array of nucleosome cores in near contact. The latter was proposed by Finch & Klug (105) to result from packing the thin filament into a "solenoid", or supercoil. They observed structures about 30 nm in diameter, and with 4–10 subunits per turn, if Mg^{2+} was present and histone H1 had not been depleted. A similar structure was reported by A. L. Olins (106), with the difference that the individual nucleosomes seem more distinctly visible and were frequently arranged in discrete clusters. The crucial role of histone H1 in higher-order packing of linear chains of nucleosomes has been verified by electron microscopic observations. If the H1 is depleted the minichromosome of SV40 changes from a compact form to an open one exhibiting the familiar "beads" (107, 108), a structural change that reverses upon in vitro addition of histone H1 (109). Similar results have been found for chromatin from nucleic of rat liver (110). The effect of histone H1 on the folding of the thin chromatin fiber is a sensitive function of ionic strength, as is the degree of binding of the H1, within a narrow range of salt concentration (111). Laemmli et al (112), and Paulson & Laemmli (113), removed all histones from metaphase HeLa cell chromosomes by extraction with polyanions and found that, in addition to multitudinous loops of DNA, the isolated, treated material exhibited protein "scaffolds" with morphological similarity to intact chromosomes.

Transcriptionally active nonribosomal genes have been examined by electron microscopy with use of the elegant spreading methods of Miller & Beatty (114) developed for exploring transcription of rRNA genes. The outcome of the experiments is not yet clear, inasmuch as the classical tree-like structures indicative of nascent ribonucleoprotein strands, and hence of transcriptional activity, are not easily identified on nonribosomal chromosomal elements. The direct question asked is whether segments of chromatin fibers showing nucleosomal beads also show the tree-like structures. A recent investigation (115) reached the conclusion that those regions exhibiting transcriptional activity are devoid of nucleosomes, but other papers have reported their presence on nonribosomal genes of *Oncopeltus fasciatus* (116), and of *Drosophila melanogaster* (117), particularly if the production of transcripts is moderate to sparse (117).

OBSERVATIONS OF NUCLEIC ACID-PROTEIN COMPLEXES

RNA Polymerase

Although the binding interactions between DNA and RNA polymerase that lead to transcription have been studied by biochemical methods for some time, they have been investigated by electron microscopy only in recent years. Such studies have been accelerated by the development of reliable methods of visualizing DNA in the absence of an enveloping coat of cytochrome *c* which thus allow easy visualization of molecules of bound RNA polymerase. In all electron microscopic studies of RNA polymerase binding so far reported the enzyme examined has been that from *E. coli,* while the most commonly used DNA is that of T7 bacteriophage. Emphasis has been on the localization of the strong binding site(s) at the left end of the T7 genome, sites known for several years to have promoter activity in RNA transcription (for review see 118). The earliest work by Davis & Hyman (119) made use of cytochrome c for spreading the DNA and, of necessity, the RNA polymerase molecules were not sought nor seen. Rather, transcription was allowed to proceed after the binding took place, and the elongating, single-stranded RNA transcripts were indirectly visualized, and distinguished from template DNA, by their "bushy" appearance in formamide-free preparations. From time-lapse measurements of the places along the DNA at which the bushes were situated it was possible to infer the location of the starting point(s); i.e. the binding site(s). Subsequently, a more definitive way of distinguishing transcripts from template was developed, in which gene 32 protein of T4 phage was bound to the single-stranded RNA in order to make it appear wider and more prominent than duplex T7 DNA (120). Both of these investigations located a promoter region (defined as the origin of transcripts) at about 1.3 map units from the left end of the T7 genome.

Since about 1974 it has become increasingly certain from electron microscopy, and from biochemical experiments, that the left-most 2% of the T7 genome contains four sites of strong polymerase binding, all of them probably being regions of initiation for RNA transcription. Bordier & Dubochet (121), and Darlix & Dausse (122), found the three strongest of these after spreading DNA-RNA polymerase complexes on carbon films made adsorptive by polymerization of amylamine (26). Portmann et al (123) and, quite recently, Koller et al (124) used ethidium bromide as a spreading agent (29) and identified the four anticipated binding sites; the latter investigators found three others located further to the right along the genome. It is interesting to note that these binding complexes existed in mixtures containing high concentrations of ethidium bromide. Bordier & Dubochet (121) had found that if T7 DNA is treated with low concentrations of ethidium bromide its binding capacity for RNA polymerase is lost. But Koller et al (124) formed the complexes and treated them with 0.1% glutaraldehyde before adding the dye, and, in so doing may have nullified its inhibitory activity. Zollinger et al (20) made use of cytochrome c mixed with Sarkosyl (a technique mentioned above) to spread monofilms containing T7 DNA complexed with RNA polymerase, and found the four left-end promoter sites, as did Williams (27), who used carbon support films pretreated with polylysine for spreading the specimen material (Figure 2). There is now ample evidence for the existence of strong binding sites on the T7 genome at positions approximately 0.55, 1.15, 1.45, and 1.70 map units from the left end. Three more are probably sufficiently strong to be identifiable by electron microscopy, at map positions ~4, 8, and 92 (124).

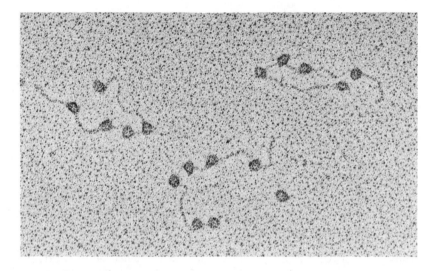

Figure 2   *E. coli* RNA polymerase bound at four early promoter sites on (*Hinf* T7)$_{1100}$ DNA. The three uniformly spaced particles are the A$_1$, A$_2$, and A$_3$ sites, while the fourth is at the D site. Some binding is also seen at the ends of the DNA fragments. X200,000. [From Williams (27)]

Some electron microscopic results have been reported by Williams & Chamberlin (125) concerning nonspecific (random) binding of RNA polymerase to a restriction fragment of T7 DNA that contains no promoters or, at most, one. Enzyme and DNA were mixed in various concentrations, in appropriate binding buffers, and after incubation the material was spread on polylysine-treated grids. Counts were made of the average number of polymerase particles found in contact with ("bound"—by morphological definition) the DNA fragment, *Mbo*I-C. The binding constant, K_a, under low-salt conditions was found to be 3 X 10^4 M^{-1}. This value is smaller by at least a factor of 10^8 than that estimated for strong promoter binding on T7 DNA. Williams & Chamberlin (125) also found that rapid dilution of the enzyme, or prolonged incubation with the DNA, increased the binding constant by an estimated factor of ~10^3–10^4, a result they provisionally ascribed to enzyme "damage."

Repressors

The binding of repressors to DNA has been reported in three recent electron microscopic investigations. The molecular weights of the monomeric units of the two repressors studied, λ repressor and *lac* repressor, are probably too small (25,000–35,000) to permit their visualization, but their common occurrence as tetramers allows them to be seen when bound to DNA in this form. Brack & Pirrotta (126) studied the morphology of λ repressor and of its binding complex with the operator. The poly amylamine procedure (26) was used for spreading the DNA, while the staining with uranyl formate was adjusted to show the repressor particles in negative stain and the λ DNA in positive stain. Free repressor was found as trimers as well as tetramers, but only the latter were seen attached to DNA. With increasing molar ratio of repressor to operator the repressor molecules were seen to be bound in multiples of one to four arranged in a row along the DNA with no visible gaps. Hirsh & Schleif (32), using similar techniques of electron microscopy, also observed the binding of λ repressor, and were able to demonstrate that its site of binding to the operator overlapped that of RNA polymerase. They also found *lac* repressor bound at a specific site on the *lac* operator. The specimens were observed after uranyl formate staining, but without shadowing. The micrographs show particles of RNA polymerase to be vaguely three-sided [similarly reported by Williams (27)], while those of the λ repressor appear distinctly smaller and more tightly associated with the DNA. Binding of *Escherichia coli lac*-repressor to nonoperator DNA was recently reported (127). Under conditions of excess repressor bound to calf thymus DNA, the closely packed repressor molecules were seen to line both "sides" of the DNA, an appearance interpreted to mean that any small segment of DNA, about 25 base pairs in length, could bind two repressor molecules. Experiments in which SV40 DNA was used as a length standard showed that multiple binding of repressor did not change its length—that it did not grossly unwind or kink as had been suggested (128). Hirsh & Schleif (32) had also found no gross conformational change upon repressor binding.

When complexes of DNA and protein, such as RNA polymerase, are observed by electron microscopy a degree of uncertainty exists as to whether what is seen reliably represents the reaction mixture. Failure of representation is most likely to occur because of dissociation of complexes during the steps of specimen preparation. If appreciable dissociation does take place it will have little effect on locating specific binding sites; it may eliminate those that are the least strong, but is unlikely to affect the positions of those that persist. With random binding, however, wherein the attachment of enzyme to DNA is far weaker than it is at specific sites, dissociation is much more likely. Furthermore, the effect of dissociation will be first-order inasmuch as the relevant determination to be made is the numbers of bound enzymes, not their positions. One group of investigators (29, 123, 124) have hoped to decrease dissociation of complexes by use of glutaraldehyde immediately following incubation of DNA and enzyme. Most investigators, however, have not used it (20, 27, 32, 121, 122, 125, 127). Perhaps surprisingly, no precise measurements have been reported on the effects of glutaraldehyde on the DNA-RNA polymerase interaction. It is universally found, however, that pretreatment of the enzyme with 0.1% glutaraldehyde prevents formation of complexes, and one report (125) indicates that glutaraldehyde treatment of random complexes extends their half-time of attachment by a few minutes.

Unwinding Proteins

The DNA-unwinding proteins so far isolated are too small to be discerned as individual molecules in the electron microscope, but they bind so strongly to single-stranded nucleic acids that their presence can be easily visualized and the conformational consequences of their binding studied. The first DNA-unwinding protein to be found and characterized was the gene 32 product of T4 bacteriophage (129). In an extensive and elegant study Delius et al (130) determined by electron microscopy the changes introduced in the appearance of nucleic acids when complexed with gene 32 protein. In confirmation of earlier conclusions (129) it was found that the protein bound extensively to, and only to, single-stranded nucleic acids (from fd and ϕX174 bacteriophages) and caused them to appear distinctly wider and much less flexible, although they preserved their open, circular form. If the protein was in excess over the DNA, it bound to it with uniform density throughout its length. Strongly cooperative binding was demonstrated by the fact that if the protein was not sufficient to coat all the DNA it was found in patches of contiguously packed molecules, leaving interspersed segments of bare DNA. Treatment of the DNA-protein complexes with glutaraldehyde at the optimal concentration (0.01 M) enhanced the stability of binding. The denaturing effect of binding of gene 32 protein on the double-stranded DNA of λ bacteriophage was demonstrated, but only if glutaraldehyde was used, by micrographs showing, dramatically, a conformational consequence of irreversible denaturation—two gene-32 protein-covered strands of length about equal to that of λ DNA, lying side by side with some intertwining.

Another DNA-unwinding protein, isolated from *E. coli* cells that were uninfected, was characterized with the aid of electron microscopy (131) with results very much like those found for the gene 32 protein (130). A distinct difference in the morphological consequences of DNA binding, however, was found for the gene 5 DNA-unwinding protein isolated from *E. coli* cells infected with M13 or fd bacteriophage (132). Whereas gene 32 protein when bound to fd DNA produced open circles with smooth contours, the gene 5 protein produced thick, rod-like forms that frequently contained branches. It was concluded that the protein causes coalescence of the single-stranded genome, with branches occurring if coalescence took place in two or more separated regions.

The ω-Protein

Several proteins are known that are capable of breaking and rejoining the backbone bonds of DNA. The first one found, designated the *E. coli* ω-protein, was reported by Wang (133) to have the property of reducing the amount of twist in a negatively superhelical DNA and leaving it covalently closed. Recently, the interaction of the ω-protein with single-stranded DNA (from fd bacteriophage) was studied (134) and the topological consequences of the interaction visualized by electron microscopy. Unlike the polymerases, repressors, and DNA-unwinding proteins, the ω-protein does not bind to DNA in a directly measurable way; hence, any anticipated consequences would have to be on the conformation of the DNA alone. What was found was that the fd genome (and φX174 as well), spread in a basic protein film with the aid of formamide after mixing with the ω-protein, was in the form of a knotted ring with about the same contour length as it had before mixing—a novel topological isomer indeed! Formation and "untying" of the knots were reversible, and resulted in an equilibrium mixture of the two topological forms under appropriate conditions. The phenomenon was believed to be a consequence of the dynamic state of pairing and unpairing of bases in complementary regions of the single-stranded DNA, combined with occasional backbone cleavage and ligation in one of the helical regions.

Ribosomal Proteins

Electron microscopy was used by Cole et al (19) to visualize directly the binding of ribosomal proteins to rRNA and hence locate the binding sites. The procedure was analogous to one of those used for studying RNA-polymerase binding: ribosomal proteins S8 and S4 were purified, bound to 16S rRNA and treated to formaldehyde, and spread for electron microscopy in a film formed with BAC. At two places on the rRNA strands the effects of protein binding were seen (although the protein molecules were not). Near the middle there was a small hairpin, which suggested that the protein, S8, bound at two adjacent sites; with ribosomal protein S4 there were two large loops at one end, which is indicative of protein binding at three sites fairly close together. If there had been any monovalent binding without production of loops, it would probably have escaped notice.

Literature Cited

1. Scott, J. F. 1948. *Biochim. Biophys. Acta* 2:1–5
2. Williams, R. C. 1952. *Biochim. Biophys. Acta* 9:237–39
3. Hart, R. G. 1955. *Proc. Natl. Acad. Sci. USA* 41:261–64
4. Hall, C. E., Litt, M. 1958. *J. Biophys. Biochem. Cytol.* 4:1–4
5. Kleinschmidt, A. K., Zahn, R. K. 1959. *Z. Naturforsch Teil B* 14:770–79
6. Evenson, D. P. 1977. *Methods Virol.* 6:219–64
7. Sogo, J. M., Portmann, R., Kaufmann, P., Koller, T. 1975. *J. Microscopy* 104:187–98
8. Reissig, M., Orrell, S. A. 1970. *J. Ultrastruct. Res.* 32:107–17
9. Westmoreland, B. C., Szybalski, W., Ris, H. 1969. *Science* 163:1343–48
10. Davis, R. W., Simon, M., Davidson, N. 1971. *Methods Enzymol.* 21:413–28
11. Hsu, M.-T., Kung, H.-J., Davidson, N. 1973. *Cold Spring Harbor Symp. Quant. Biol.* 38:943–50
12. Lang, D., Kleinschmidt, A. K., Zahn, R. K. 1964. *Biochim. Biophys. Acta* 88:142–54
13. Abermann, R., Salpeter, M. M. 1974. *J. Histochem. Cytochem.* 22:845–55
14. Inman, R. B., Schnös, M. 1970. *J. Mol. Biol.* 49:93–98
15. Mayor, H. D., Jordan, L. E. 1968. *Science* 161:1246–47
16. Lang, D., Mitani, M. 1970. *Biopolymers* 9:373–79
17. Vollenweider, H. J., Sogo, J. M., Koller, T. 1975. *Proc. Natl. Acad. Sci. USA* 72:83–87
18. Koller, T., Harford, A. G., Lee, Y. K., Beer, M. 1969. *Micron* 1:110–18
19. Cole, M. D., Beer, M., Koller, T., Strycharz, W. A., Nomura, M. 1978. *Proc. Natl. Acad. Sci. USA* 75:270–74
20. Zollinger, M., Guertin, M., Mamet-Bratley, M. D. 1977. *Anal. Biochem.* 82:196–203
21. Highton, P. J., Whitfield, M. 1974. *J. Microscopy* 100:299–306
22. Highton, P.J., Beer, M. 1963. *J. Mol. Biol.* 7:70–77
23. Nanninga, N., Meyer, M., Sloof, P., Reijnders, L. 1972. *J. Mol. Biol.* 72:807–10
24. Griffith, J. D. 1973. *Methods Cell Biol.* 7:129–46
25. Dubochet, J., Kellenberger, E. 1972. *Microscopica Acta* 72:119–30
26. Dubochet, J., Ducommun, M., Zollinger, M., Kellenberger, E. 1971. *J. Ultrastruct. Res.* 35:147–67
27. Williams, R. C. 1977. *Proc. Natl. Acad. Sci. USA* 74:2311–15
28. Laemmli, U. K. 1975. *Proc. Natl. Acad. Sci. USA* 72:4288–92
29. Koller, T., Sogo, J. M., Bujard, H. 1974. *Biopolymers* 13:995–1009
30. Paoletti, J., Le Pecq, J.-B. 1971. *J. Mol. Biol.* 59:43–62
31. Freifelder, D. 1971. *J. Mol. Biol.* 60:401–3
32. Hirsh, J., Schleif, R. 1976. *J. Mol. Biol.* 108:471–90
33. Vollenweider, H. J., James, A., Szybalski, W. 1978. *Proc. Natl. Acad. Sci. USA* 75:710–14
34. Haydon, G. B., Lemons, R. A., Clayton, D. A. 1971. *J. Microscopy* 94:Part 1, pp. 69–72
35. Williams, R. C., Fisher, H. W. 1970. *J. Mol. Biol.* 52:121–23
36. Breedlove, J. R. Jr., Trammell, G. T. 1970. *Science* 170:1310–13
37. Crewe, A. V., Wall, J., Langmore, J. 1970. *Science* 168:1338–40
38. Brack, C., Delain, E. 1975. *J. Cell Sci.* 17:287–306
39. Gordon, C. N., Kleinschmidt, A. K. 1968. *Biochim. Biophys. Acta* 155:305–7
40. Wetmur, J. G., Davidson, N., Scaletti, J. V. 1966. *Biochem. Biophys. Res. Commun.* 25:684–88
41. Williams, R. C., Wyckoff, R. W. G. 1944. *J. Appl. Phys.* 15:712–16
42. Hart, R. G. 1963. *J. Appl. Phys.* 34:434–36
43. Langridge, R., Marvin, D. A., Seeds, W. E., Wilson, H. R., Hooper, C. W., Wilkins, M. H. F., Hamilton, L. D. 1960. *J. Mol. Biol.* 2:38–64
44. Arnott, S., Hukins, D. W. L. 1973. *J. Mol. Biol.* 81:93–105
45. Stüber, D., Bujard, H. 1977. *Molec. Gen. Genet.* 154:299–303
46. Sanger, F., Air, G. M., Barrell, B. G., Brown, N. L., Coulson, A. R., Fiddes, J. C., Hutchison, C. A. III, Slocombe, P. M., Smith, M. 1977. *Nature* 265:687–95
47. Koller, T., Kübler, O., Portmann, R., Sogo, J. M. 1978. *J. Mol. Biol.* 120:121–31
48. Granboulan, N., Franklin, R. M. 1968. *J. Virol.* 2:129–48
49. Keegstra, W., Vereijken, J. M., Jansz, H. S. 1977. *Biochim. Biophys. Acta* 475:176–83
50. Nomura, M., Benzer, S. 1961. *J. Mol. Biol.* 3:684–92
51. Inman, R. B., Schnös, M. 1974. In *Principles and Techniques of Electron Mi-*

croscopy, ed. M. A. Hyatt, 4:64–80. New York: Van Nostrand Reinhold
52. Wu, M., Davidson, N. 1973. *J. Mol. Biol.* 78:1–21
53. Wu, M., Davidson, N., Carbon, J. 1973. *J. Mol. Biol.* 78:23–34
54. Angerer, L., Davidson, N., Murphy, W., Lynch, D., Attardi, G. 1976. *Cell* 9:81–90
55. Ohi, S., Ramirez, J. L., Upholt, W. B., Dawid, I. B. 1978. *J. Mol. Biol.* 121:299–310
56. Inman, R. B. 1974. *Methods Enzymol.* 29:451–58
57. Schnös, M., Inman, R. B. 1970. *J. Mol. Biol.* 51:61–73
58. Valenzuela, M. S., Inman, R. B. 1975. *Proc. Natl. Acad. Sci. USA* 72:3024–28
59. Brown, D. D., Wensink, P. C., Jordan, E. 1972. *J. Mol. Biol.* 63:57–73
60. Roberts, R. J. 1976. *Crit. Rev. Biochem.* 4:123–64
61. Fiers, W., Contreras, R., Duerinck, F., Haegeman, G., Iserentant, D., Merregaert, J., Min Jou, W., Molemans, F., Raeymaekers, A., Van den Berghe, A., Volckaert, G., Ysebaert, M. 1976. *Nature* 260:500–7
62. Wellauer, P. K., Dawid, I. B. 1973. *Proc. Natl. Acad. Sci. USA* 70:2827–31
63. Jacobson, A. B. 1976. *Proc. Natl. Acad. Sci. USA* 73:307–11
64. Edlind, T. D., Bassel, A. R. 1977. *J. Virol.* 24:135–41
65. Koths, K., Dressler, D. 1978. *Proc. Natl. Acad. Sci. USA* 75:605–9
66. Holliday, R. 1964. *Genet. Res.* 5:282–304
67. Holliday, R. 1974. *Genetics* 78:273–87
68. Thompson, B. J., Escarmis, C., Parker, B., Slater, W. C., Doniger, J., Tessman, I., Warner, R. C. 1975. *J. Mol. Biol.* 91:409–19
69. Benbow, R., Zuccarelli, A., Sinsheimer, R. 1974. In *Mechanisms in Recombination,* ed. R. Grell, pp. 3–18. New York: Plenum
70. Potter, H., Dressler, D. 1976. *Proc. Natl. Acad. Sci. USA* 73:3000–4
71. Thompson, B. J., Camien, M. N., Warner, R. C. 1976. *Proc. Natl. Acad. Sci. USA* 73:2299–2303
72. Bukhari, A. I., Shapiro, J. A., Adhya, S. L., eds. 1977. *DNA Insertion Elements, Plasmids, and Episomes.* Cold Spring Harbor, New York: Cold Spring Harbor Lab. 782 pp.
73. Kleinschmidt, A. K., Lang, D., Jacherts, D., Zahn, R. K. 1962. *Biochim. Biophys. Acta* 61:857–64
74. Dunnebacke, T. H., Kleinschmidt, A. K. 1967. *Z. Naturforsch.* 22b:159–64
75. Kleinschmidt, A. K. 1968. *Methods Enzymol.* 12B:361–77
76. Worcel, A., Burgi, E. 1974. *J. Mol. Biol.* 82:91–105
77. Solari, A. J. 1967. *J. Ultrastruct. Res.* 17:421–38
78. Miller, O. L. Jr., Beatty, B. R. 1969. *Science* 164:955–57
79. Levinthal, C., Fisher, H. W. 1953. *Cold Spring Harbor Symp. Quant. Biol.* 18:29–33
80. Thomas, J. O. 1974. *J. Mol. Biol.* 87:1–9
81. Chattoraj, D. K., Inman, R. B. 1974. *J. Mol. Biol.* 87:11–22
82. Yamagishi, H., Okamoto, M. 1978. *Proc. Natl. Acad. Sci. USA* 75:3206–10
83. Streisinger, G., Emrich, J., Stahl, M. M. 1967. *Proc. Natl. Acad. Sci. USA* 57:292–95
84. Richards, K. E., Williams, R. C., Calendar, R. 1973. *J. Mol. Biol.* 78:255–59
85. Olins, D. E., Olins, A. L. 1971. *J. Mol. Biol.* 57:437–55
86. Chattoraj, D. K., Gosule, L. C., Schellman, J. A. 1978. *J. Mol. Biol.* 121:327–37
87. Butler, P. J. G., Finch, J. T., Zimmern, D. 1977. *Nature* 265:217–19
88. Lebeurier, G., Nicolaieff, A., Richards, K. E. 1977. *Proc. Natl. Acad. Sci. USA* 74:149–53
89. Olins, A. L., Olins, D. E. 1974. *Science* 183:330–32
90. Oudet, P., Gross-Bellard, M., Chambon, P. 1975. *Cell* 4:281–300
91. Kornberg, R. D. 1974. *Science* 184:868–71
92. Felsenfeld, G. 1978. *Nature* 271:115–22
93. Finch, J. T., Lutter, L. C., Rhodes, D., Brown, R. S., Rushton, B., Levitt, M., Klug, A. 1977. *Nature* 269:29–36
94. Finch, J. T., Klug, A. 1977. *Cold Spring Harbor Symp. Quant. Biol.* 42:1–9
95. Lutter, L. C. 1977. *Cold Spring Harbor Symp. Quant. Biol.* 42:137–47
96. Noll, M. 1977. *Cold Spring Harbor Symp. Quant. Biol.* 42:77–85
97. D'Anna, J. A. Jr., Isenberg, I. 1974. *Biochemistry* 13:4992–97
98. Weintraub, H., Worcel, A., Alberts, B. 1976. *Cell* 9:409–17
99. Griffith, J. D. 1975. *Science* 187:1202–3
100. Oudet, P., Spadafora, C., Chambon, P. 1977. *Cold Spring Harbor Symp. Quant. Biol.* 42:301–12
101. Germond, J. E., Hirt, B., Oudet, P., Gross-Bellard, M., Chambon, P. 1975. *Proc. Natl. Acad. Sci. USA* 72:1843–47
102. Oudet, P., Germond, J. E., Sures, M., Gallwitz, D., Bellard, M., Chambon, P.

1977. *Cold Spring Harbor Symp. Quant. Biol.* 42:287–300
103. Woodcock, C. L. F., Frado, L.-L. Y. 1977. *Cold Spring Harbor Symp. Quant. Biol.* 42:43–55
104. Bina-Stein, M., Simpson, R. T. 1977. *Cell* 11:609–18
105. Finch, J. T., Klug, A. 1976. *Proc. Natl. Acad. Sci. USA* 73:1897–1901
106. Olins, A. L. 1977. *Cold Spring Harbor Symp. Quant. Biol.* 42:325–29
107. Griffith, J. D., Christiansen, G. 1977. *Cold Spring Harbor Symp. Quant. Biol.* 42:215–26
108. Keller, W., Müller, U., Eicken, I., Wendel, I., Zentgraf, H. 1977. *Cold Spring Harbor Symp. Quant. Biol.* 42:227–43
109. Müller, U., Zentgraf, H., Eicken, I., Keller, W. 1978. *Science* 201:406–15
110. Thoma, F., Koller, T. 1977. *Cell* 12:101–7
111. Renz, M., Nehls, P., Hozier, J. 1977. *Cold Spring Harbor Symp. Quant. Biol.* 42:245–52
112. Laemmli, U. K., Cheng, S. M., Adolph, K. W., Paulson, J. R., Brown, J. A., Baumbach, W. R. 1977. *Cold Spring Harbor Symp. Quant. Biol.* 42:351–60
113. Paulson, J. R., Laemmli, U. K. 1977. *Cell* 12:817–28
114. Miller, O. L. Jr., Beatty, B. R. 1969. *J. Cell. Physiol.* 74:Suppl. 1, pp. 225–32
115. Franke, W. W., Scheer, U., Trendelenburg, M., Zentgraf, H., Spring, H. 1977. *Cold Spring Harbor Symp. Quant. Biol.* 42:755–72
116. Foe, V. E., Wilkinson, L. E., Laird, C. D. 1976. *Cell* 9:131–46
117. McKnight, S. L., Bustin, M., Miller, O. L. Jr. 1977. *Cold Spring Harbor Symp. Quant. Biol.* 42:741–54
118. Chamberlin, M. J. 1976. In *RNA Polymerase*, ed. R. Losick, M. Chamberlin, pp. 159–91. Cold Spring Harbor, New York: Cold Spring Harbor Lab.
119. Davis, R. W., Hyman, R. W. 1970. *Cold Spring Harbor Symp. Quant. Biol.* 35:269–81
120. Delius, H., Westphal, H., Axelrod, N. 1973. *J. Mol. Biol.* 74:677–87
121. Bordier, C., Dubochet, J. 1974. *Eur. J. Biochem.* 44:617–24
122. Darlix, J.-L., Dausse, J.-P. 1975. *FEBS Lett.* 50:214–18
123. Portmann, R., Sogo, J. M., Koller, T., Zillig, W. 1974. *FEBS Lett.* 45:64–67
124. Koller, T., Kübler, O., Portmann, R., Sogo, J. M. 1978. *J. Mol. Biol.* 120:121–31
125. Williams, R. C., Chamberlin, M. J. 1977. *Proc. Natl. Acad. Sci. USA* 74:3740–44
126. Brack, C., Pirrotta, V. 1975. *J. Mol. Biol.* 96:139–52
127. Zingsheim, H. P., Geisler, N., Weber, K., Mayer, F. 1977. *J. Mol. Biol.* 115:565–70
128. Crick, F. H. C., Klug, A. 1975. *Nature* 255:530–33
129. Alberts, B. M., Frey, L. 1970. *Nature* 227:1313–18
130. Delius, H., Mantell, N. J., Alberts, B. 1972. *J. Mol. Biol.* 67:341–50
131. Sigal, N., Delius, H., Kornberg, T., Gefter, M. L., Alberts, B. 1972. *Proc. Natl. Acad. Sci. USA* 69:3537–41
132. Alberts, B., Frey, L., Delius, H. 1972. *J. Mol. Biol.* 68:139–52
133. Wang, J. C. 1969. *J. Mol. Biol.* 43:263–72
134. Liu, L. F., Depew, R. E., Wang, J. C. 1976. *J. Mol. Biol.* 106:439–52

Ann. Rev. Biochem. 1979. 48:681–717
Copyright © 1979 by Annual Reviews Inc. All rights reserved

THE ISOLATION OF EUKARYOTIC MESSENGER RNA

♦12023

John M. Taylor

Department of Microbiology and the Specialized Cancer Research Center,
The Milton S. Hershey Medical Center, The Pennsylvania State University,
College of Medicine, Hershey, Pennsylvania 17033

CONTENTS

681

0066-4154/79/0701-0681$01.00

PERSPECTIVES AND SUMMARY

Messenger RNA molecules are central components in the expression of eukaryotic structural genes. Thus, an important objective for many studies on gene regulation is to isolate specific mRNAs for structure-function analysis, and to prepare specific radioactively labeled complementary DNA (cDNA) to be used as a hybridization probe for the quantitation of mRNA. The cDNA can also be used as an affinity probe for the isolation of specific gene segments and RNA transcription products, and for the determination of nucleotide sequences.

Messenger RNA isolation depends upon the exploitation of unique properties of the molecule such as its chemical and physical properties, its nucleotide sequence, and the protein for which it codes. Several methods of isolation, usually involving combinations of different techniques, have been developed during the last decade. Among these techniques are size fractionation by means of gradient sedimentation and gel electrophoresis, polysome immunoprecipitation with specific antibodies, sequence enrichment through mRNA:cDNA hybridization, and specific nucleotide sequence-dependent identification. The advent of bacterial cloning of plasmid-carried cDNA sequences promises to become a powerful preparative tool to augment the already existing methods of mRNA purification.

The isolation of a particular mRNA species requires a specific assay. The most common and readily accessible method is mRNA translation in a cell-free protein-synthesizing system with identification of the specific protein product. Messenger RNA–dependent systems derived from rabbit reticulocytes and wheat germ extracts provide the basis for convenient and sensitive assays. Alternative identification techniques depend upon specific nucleotide sequence detection, such as the use of specific oligonucleotide probes, partial sequencing of fractionated cDNAs, or the identification of unique restriction endonuclease cleavage fragments of cDNAs.

The development of mRNA isolation methods has depended upon overcoming several common technical problems. The control of ribonuclease activity has usually been a major difficulty. The low levels of individual mRNAs present in most cells have required large amounts of starting material and efficient RNA extraction procedures. Close similarities in chemical and physical properties of most mRNAs, and the approximately equivalent concentrations of large numbers of mRNA species in most tissues, have made the exploitation of unique molecular properties difficult. Thus, isolation of a specific mRNA has often depended upon identifying particular tissues or physiological circumstances in which it shows an unusual degree of enrichment.

Determination of the purity of the isolated mRNA is best achieved by the combined use of several procedures, which include an examination of its physical properties, characterization of its cell-free translation product, hybridization of the mRNA to its cDNA (to examine related kinetic properties), and sequence-related methods such as oligonucleotide fingerprinting or partial nucleotide sequencing. Restriction endonuclease mapping for the cDNA can also be important for this purpose, as well as for subsequent bacterial cloning efforts and gene mapping studies.

The focus of this review is to examine the major methods of eukaryotic mRNA isolation and characterization, with appropriate references to experimental systems in which these techniques have been applied. First, general methods and common problems are examined, then specific approaches and selected experimental systems are reviewed.

GENERAL TECHNIQUES

Several general techniques are commonly employed in the isolation of specific mRNAs. The development and refinement of these techniques has been of central importance in the ability to isolate individual mRNAs. However, each of these methods may require modification in order to be applicable to a particular system.

Ribonuclease Inhibition

A major technical problem in the isolation of mRNA is the control of ribonuclease activity. A variety of ribonucleases are present in all tissues and occur as contaminants in many reagents derived from biological sources, such as sucrose, and are found on human fingers. Inhibiting or inactivating these enzymes is essential for success in mRNA isolation. Because the content, type, and specificity of ribonuclease varies widely in different tissues, procedures that work in one system may be quite ineffective in other systems. Combinations of methods are usually necessary, although a few general means of controlling ribonuclease activity have been established.

Diethyl pyrocarbonate is an effective inhibitor of ribonuclease, and its chemistry and biological applications have been extensively reviewed (1). At intermediate pH values, diethyl pyrocarbonate attacks ribonuclease as well as other proteins, and reacts especially with the imidazole nitrogens of histidine residues and free amino groups, with resultant loss of enzyme activity (1, 2). Thus, treatment of materials and reagents that cannot be autoclaved with diethyl pyrocarbonate is an effective way to inactivate ribonuclease (1). Diethyl pyrocarbonate has also been included in tissue homogenates in the preparation of undegraded RNA (3, 4). However, it also attacks single-stranded nucleic acids, and adenine bases are particularly sensitive (5, 6). The reaction of this reagent with nucleic acids is apparently slower than its reaction with proteins, so that appropriate conditions for mRNA isolation may be established (1, 7, 8). Nevertheless, diethyl pyrocarbonate can destroy mRNA activity (7, 8, 9), and its use in preparative procedures is unpredictable.

The preparation of translationally active mRNA from most tissues usually requires the inclusion of ribonuclease inhibitors in tissue homogenates, especially when subcellular fractionation has been involved. Heparin, a sulfated polysaccharide, is a particularly effective competitive inhibitor of ribonuclease (10, 11), which greatly improves the recovery of undegraded polysomes and RNA (9, 12, 13). Several other native and synthetic sulfated polymeric inhibitors have also been described, with a high charge density required for maximum activity (14, 15). These various polymers may act by adsorbing nucleases (15). Direct removal of ribonuclease by adsorption to bentonite is inefficient, and results in low yields of RNA (16).

Some tissues, such as rat liver, contain an endogenous protein that functions as a ribonuclease inhibitor (17). It is a sulfhydryl-containing protein that can be stabilized by the addition of low concentrations of reducing reagents, such as dithiothreitol (18). The rat liver ribonuclease inhibitor has been purified (19, 20) and has often been added to tissue homogenates and reagents for RNA isolation. Liver cytosol has also been added to reagents as a crude source of ribonuclease inhibitor (21). However, there may be some tissue specificity with regards to its inhibitory activity, and no particular advantage over the use of heparin has been demonstrated (22).

Many detergents, particularly dodecyl sulfate (23, 24), act as strong denaturants of proteins, and can thereby function as ribonuclease inhibitors. Partly for this reason, detergents are usually added to tissue homogenates or subcellular fractions in bulk RNA extraction procedures. Including dodecyl sulfate in solutions of isolated RNA, and in reagents for procedures that do not involve protein reactants, provides protection against ribonuclease contamination.

Extraction of Total Cellular RNA

Another problem in isolating cellular RNA is the removal of protein in such a way that the nucleic acids are not degraded. Since most cellular RNAs, including mRNAs, are tightly associated with specific proteins, the purification of nucleic acids requires reagents that will completely disrupt the ribonucleoprotein complex and allow extraction of undegraded RNA from the mixture. There are many different methods, but it is difficult to identify the one best approach because no systematic comparison has been made and individual requirements vary greatly.

Many commonly employed methods of nucleic acid deproteinization are based on the use of phenol as a protein denaturant (25, 26). The general procedure consists of first emulsifying a tissue homogenate with buffer-saturated liquified phenol. Then the mixture is subjected to centrifugation to separate the aqueous and organic phases, with compaction of the denatured protein at the interface. Total nucleic acid partitions into the aqueous phase and can be collected by ethanol precipitation in the presence of dilute salt. Detergents such as dodecyl sulfate (27), sarkosyl (28), or triisopropylnaphthalene sulfonate (29) are usually added to the tissue homogenates to dissolve nucleoprotein complexes and facilitate protein denaturation, improve RNA yields, and inhibit ribonuclease activity. Prior treatment of tissue homogenates with nonspecific proteases, such as self-digested pronase or proteinase K, reduces the protein interface and may facilitate RNA extraction (30–32).

Phenol extraction is sensitive to experimental conditions such as temperature, salt concentration, and pH. Some of the variables have been investigated by Brawerman et al (33, 34) and Perry et al (35). When phenol extractions were carried out in the cold, it was necessary to employ alkaline buffers (about pH 9) in order to allow most of the mRNA to partition into the aqueous phase. At neutral pH, the mRNA was mainly found in the denatured protein interface and the interaction was promoted by monovalent cations (33). The behavior of mRNA under these conditions was felt to be due to the capacity of the 3'-terminal poly(A) segment of mRNA to bind to denatured proteins (34, 36). However, when the extraction was carried out at room temperature with the addition of an equal volume of chloroform, the mRNA partitioned into the aqueous phase (35). Other additives, found in some procedural variations, include isoamyl alcohol as an antifoaming agent (37), chelating agents such as EDTA to destabilize RNA-protein interactions (38, 39), 8-hydroxyquinoline as an antioxidant and chelating agent (40), and m-cresol to potentiate protein denaturation and act as an antifreeze (41).

DNA can be removed from the nucleic acid extract before further RNA fractionation. Washing the salt-ethanol nucleic acid precipitate with 3 M sodium acetate (by resuspending and resedimenting) solubilizes the DNA, as well as tRNA, 5S rRNA, glycogen, and heparin (41–43). An additional benefit of the sodium acetate washing is that the remaining mRNA, with the above translational inhibitory material removed, becomes a more efficient template in protein synthesis assays (43). A potential disadvantage of sodium acetate washing is that small mRNA molecules might be solubilized, although this problem has not been reported. A variation of the phenol extraction technique, which is carried out at elevated temperatures, results in the separation of only RNA in the aqueous phase, and conditions for this procedure have been examined (39, 44).

Phenol extraction results in the aggregation of RNA, and rather specific aggregates can form between mRNA and other RNA species such as rRNA

(29, 45–47). Aggregates appear to form in virtually any modification of the phenol extraction procedures, and are apparently not due to ethanol precipitation (45). Treatment of the extracted RNA with strong denaturants such as formamide or dimethylsulfoxide can be effective in dissociating aggregates (30, 45, 48–50). Alternatively, heating the RNA to about 65° in a dilute buffer at neutral pH for 10–15 min, followed by rapid cooling, appears to dissociate these phenol-induced aggregates (41, 46, 47, 51).

Alternative procedures for RNA extraction that avoid phenol have been developed which employ concentrated salt solutions to dissociate ribonucleoprotein complexes. Cox (52) has described the use of 6 M guanidine· HCl for the isolation of RNA from ribosomes in which ribonuclease is rapidly denatured. Deeley et al (53) described a modification of this procedure, where tissues were homogenized in 8 M guanidine·HCl containing 1 mM dithiothreitol, cellular debris removed by brief centrifugation, and total RNA precipitated by the addition of a half volume of cold ethanol. The guanidinium salt was then exchanged for the sodium salt by dissolving the RNA precipitate in an EDTA solution and extracting with chloroform-butanol. In other related methods, RNA has been dissociated from protein and precipitated by 2 M LiCl at cold temperatures (51, 54, 55). Sedimentation through CsCl gradients to separate RNA from DNA and protein has also been described for small scale RNA preparations (56).

Poly(A) Adsorption Methods

Eukaryotic mRNA contains a covalently attached 3'-terminal poly(A) segment of heterogeneous length, ranging in size from 20–250 bases in length (reviewed in 36, 57, 58), and this provides the basis for a major technical advance in the isolation of mRNA. Poly(A)-containing RNAs can be separated from other cellular RNAs by affinity chromatography on oligo(dT)-cellulose (59–62) or poly(U)-agarose (63, 64), or by adsorption to nitrocellulose filters (65, 66) or unmodified cellulose (67–70). These methods are based on the selective ability of the poly(A) segment to bind to the affinity matrix in the presence of a moderately high salt concentration.

Affinity chromatography with small columns of oligo(dT)-cellulose has become widely used. The matrix is available commercially, and usually consists of oligo(dT) strands of about 12–18 nucleotides in length covalently attached to microgranular cellulose. Typically, a solution of RNA containing 0.1–0.5 M NaCl and 0.1–0.5 % sodium dodecyl sulfate at neutral pH is passed over the affinity support at a slow flow rate to allow the poly(A)-containing RNA to bind. Poly(A)-minus RNAs, including rRNA, pass through with little adsorption. The bound RNA is then eluted with a very low ionic strength buffer, and recovered by NaCl-ethanol precipitation. An intermediate low-salt washing step is often employed to help remove nonspecifically adsorbed RNA (62). After a single column step, the poly(A)-containing RNA preparation usually contains contaminating rRNA, which can account for up to 50% or more of the eluate. However, repassing the RNA sample in a second column step is usually sufficient to yield essentially pure poly(A)-containing RNA. By employing buffers containing moderate salt concentrations, thermal elution chromatography has been used for the fractionation of poly(A)-containing RNA (51, 71). Bantle et al (72) have further examined conditions for the general use of oligo(dT)-cellulose.

Poly(U)-agarose is another commonly used affinity chromatography support that requires mRNA binding conditions similar to those of oligo(dT)-cellulose (63, 64, 73). The poly(U) ligand is usually somewhat longer than the oligo(dT) ligand and therefore requires stronger elution conditions, such as 70–90% formamide or elevated temperature (63, 64, 73). Nonspecific adsorption of rRNA to poly(U)-agarose is minimal. Poly(U) bound to glass fiber filters through ultraviolet light activation has also been employed (74).

Adsorption techniques employing nitrocellulose filters were often used before the advent of readily available affinity chromatography supports. This approach was developed by Brawerman, and the details of its use have been described (66, 75). Although nitrocellulose filters are effective in the isolation of poly(A)-containing RNA, they also nonspecifically adsorb rRNA and small amounts of single-stranded DNA, even after a readsorption step (76). Lower recoveries of mRNAs from nitrocellulose filtration techniques are also observed, apparently due to a requirement for longer poly(A) segments (77). Lingrel and his associates have shown that nitrocellulose filters require a poly(A) length of about 50 nucleotides for binding whereas oligo(dT)-cellulose effectively binds mRNA with a poly(A) length as short as 20 bases (77). They were able to fractionate mouse globin mRNAs into discrete classes on the basis of the length of the poly(A) segments by differential adsorption to oligo(dT)-cellulose and nitrocellulose filters (77).

Unmodified cellulose has been shown to bind poly(A) homopolymers (78), and it has been employed for the partial purification of mRNA (67–70). This binding property may be due to an interaction between poly(A) and the aromatic lignins that usually occur as impurities in cellulose preparations (69, 70). However, mRNA prepared by this technique is usually contaminated with substantial amounts of rRNA (46, 79). Benzoylated DEAE-cellulose has been employed for the partial purification of HeLa cell mRNA (80), but it was not shown to be particularly effective for the purification of ovalbumin mRNA (79).

Even using the most efficient affinity chromatography techniques employing either oligo(dT)-cellulose or poly(U)-agarose, as much as 20% or more of the translatable mRNA activity from rat liver, chicken oviduct, and other tissues fails to bind to the affinity ligand with repeated readsorption steps (79, 81). This lack of binding may be the result of very short poly(A) segments (i.e. less than 10–20 bases in length), lack of a poly(A) segment, or a poly(A) segment that is blocked in some manner.

Translation Assays

The establishment of cell-free protein-synthesizing systems capable of translating exogenous heterologous mRNA was especially important for the isolation of individual mRNA species. Although many different biological sources have been developed for translation assays, a few systems have emerged that are of general importance for monitoring mRNA protein synthetic activity during purification procedures. They generally require the use of specific antibodies for the detection of radioactively labeled individual protein products by immunoprecipitation assays.

Supplemented rabbit reticulocyte lysates have proven to be highly efficient protein-synthesizing systems for the translation of exogenous heterologous mRNA. Procedures for the preparation and use of unfractionated reticulocyte lysates have been reviewed in detail (82–86). The principal disadvantage of the original system was a function of the high endogenous globin mRNA background, which limited the exogenous mRNA detection level and required sensitive antibody assays. However, prior treatment of the lysate with a readily inactivated nuclease caused the system to become mRNA-dependent, thus improving the sensitivity and versatility of the assay (87). Fractionated reconstituted reticulocyte lysate systems have also been described, but it is usually not necessary to resort to them for routine translation assays to monitor mRNA purification.

A cell-free protein-synthesizing system derived from untoasted wheat germ was developed by Marcus et al (88, 89), and a convenient modification has been described by Roberts & Paterson (90). Among the advantages of this system are that it is easy to prepare and use, it is economical, and it has a relatively low endogenous background of mRNA activity. Although there is considerable variability in translation activity with different batches of wheat germ, extracts are relatively easy to prepare and characterize. A disadvantage of this system is that all mRNA species in a particular tissue may not have the same assay requirements, and minor changes in conditions can have dramatic effects on translation efficiencies (91). Depending on individual needs, it may be necessary to evaluate translation requirements for individual mRNA species in each wheat germ extract employed, particularly with respect to potassium and magnesium ion levels (91, 92). Suboptimal ion concentrations cause the production of incomplete peptide fragments as a result of premature termination of protein synthesis with the release of peptidyl-tRNA (91). Longer RNA species would appear to be particularly susceptible to this problem. Thus the use of the wheat germ protein-synthesizing system to quantitate individual mRNA levels may not always yield reliable results. This problem is further complicated if translation yields incomplete peptide products. Another difficulty relates to the possibility that various mRNAs may exhibit concentration-dependent differences in their peptide initiation rates, as suggested by Lodish (93).

Mathews & Korner (94) have described a cell-free protein-synthesizing system derived from Krebs II mouse ascites cells that has become widely used. The system shows a moderately low endogenous translation background, and it is frequently employed as an mRNA-dependent assay method.

A whole cell in vitro translation system employing *Xenopus laevis* oocytes has been described by Gurdon and his associates (95, 96). Although somewhat inconvenient to use, the oocyte system offers the advantage of being quite sensitive to low concentrations of exogeneous microinjected mRNA. Translation of exogenous mRNAs can continue for several days in these cells (97). The oocyte system also permits metabolic studies of mRNAs and protein products not possible with the cell-free protein-synthesizing systems.

cDNA Synthesis and Hybridization

In 1972 Ross et al (98), Verma et al (99), and Kacian et al (100) demonstrated that the RNA-dependent DNA polymerase (reverse transcriptase) from avian myeloblastosis virus could synthesize complementary DNA copies of globin mRNA. The enzyme required the use of a primer, oligo(dT), which bound to the 3'-terminal poly(A) segment of the mRNA template. Deoxyribonucleoside triphosphates were employed as substrates, one of which was radioactively labeled. Following the reaction, the template mRNA was degraded by alkaline hydrolysis. The DNA product could easily be recovered and it showed hybridization specificity for globin mRNA. These findings had a dramatic effect on the study of eukaryotic gene expression. Specific radioactive cDNAs were then prepared and employed as hybridization probes in other experimental systems. Complementary DNAs have also been used preparatively in the isolation of specific mRNAs. The ability to produce large quantities of cDNAs by cloning with bacterial plasmids will facilitate the use of cDNAs for many other applications.

Reaction parameters for the synthesis of full-length cDNAs have been investigated by several laboratories (101–107). Although the nucleotide sequence and secondary structure of a particular mRNA species may require a unique set of reaction conditions for the production of a long cDNA, useful general guidelines for cDNA synthesis have emerged. The reaction is usually carried out in 50 mM Tris-HCL buffer at pH 8.3 with 2–20 mM dithiothreitol and 6–10 mM Mg^{2+}. A concentration range of 0–140 mM for KCL or NaCl has been reported. According to Monahan et al (103), omission of KCl is required for the synthesis of a full length cDNA, whereas Buell et al (107) found that the KCl concentration affected only the yield of cDNA and not its size distribution. Although it had been reported that relatively high concentrations of substrate deoxynucleotides were required for the synthesis of full length cDNA (102), other reports suggest that a minimum concentration of about 35–50 μM of deoxynucleotides is sufficient for the production of a long cDNA, with higher substrate levels only increasing the yield (103, 106, 107). However, Rothenberg & Baltimore (106) found that the highest proportion of large leukemia virus cDNA was produced at relatively high deoxynucleotide concentrations with the Mg^{2+} concentration slightly below that of the total substrates. A concentration range of 5–50 μg per ml of mRNA has usually been employed. For the primer, a mass ratio of about 0.1 μg oligo(dT) per μg of mRNA should be sufficient to saturate the poly(A) segments of most mRNAs; however, a mass ratio of primer to template of 1 or greater is often employed.

The most commonly used enzyme for cDNA synthesis is the reverse transcriptase from avian myeloblastosis virus (AMV). Properties of the enzyme and its reactions have been reviewed (108). About 2–10 units of enzyme per μg of mRNA are sufficient for cDNA synthesis, but up to 80 units per μg of RNA have been used. Since the reverse transcriptase also has a DNA-dependent DNA polymerase activity, actinomycin D is usually added (20–200 μg per ml) to prevent the synthesis of the second cDNA strand. Pyrophosphate has also been reported to block second-strand synthesis (109). Trace amounts of contaminating ribonuclease activity may affect the apparent efficiency of the AMV enzyme, as well as the proportion of long cDNA product (107). Variations in the reverse transcriptase preparations probably account for some of the differences in reaction conditions reported by various laboratories. The recommended reaction temperatures range from 25°–46°, and incubation times of 5–360 min have been used. Following the reaction, unused substrates as well as other salts are separated from the cDNA by gel filtration, and the mRNA is removed by alkaline hydrolysis.

The size of the cDNA can be estimated by centrifugation on alkaline sucrose gradients, which also allows fractionation of the cDNA if desired. McCarty et al (110) have described conditions for the preparation of isokinetic sucrose gradients for several different centrifuge rotors. Sedimentation through these gradients permits the calculation of a sedimentation coefficient. This value can then be used to estimate the nucleotide length of the cDNA, as described by Studier (111). More accurate size information can be obtained by gel electrophoresis in the presence of viral DNA restriction endonuclease cleavage fragments of known length.

The conditions for the hybridization of nucleic acids have been studied in several laboratories. The rate of hybridization is dependent upon several factors, including incubation temperature, the cation concentration, the concentration of the reactants, the length of the reactants, solution viscosity, and pH. A detailed examination of these effects has been presented elsewhere (112–119). The hybridization of cDNA with mRNA is usually carried out at neutral pH and at about 20–25° below the hybrid melting temperature. Sodium chloride can be added to the hybridization reaction mixture to increase the reaction rate, and a table for calculating the effect of salt concentration on the reaction rate is available (115). Including a small amount of tRNA as well as trace amounts of sodium dodecyl sulfate and EDTA in the reaction buffer protects the mRNA from degradation. Addition of formamide to the buffer allows a concomitant reduction in the temperature of incubation (120, 121). The formation of hybrids can be conveniently assayed according to their resistance to digestion by a single-strand specific nuclease (S_1-nuclease) (122). The reaction properties of the S_1-nuclease from *Aspergillus oryzae* have been described (122, 123). Hybrids can also be assayed by hydroxylapatite chromatography in the presence of a sodium phosphate buffer. Nucleic acids bind to hydroxylapatite at low phosphate concentrations, with single-stranded nucleic acids eluted at 0.12–0.14 M phosphate buffer and double-stranded nucleic acids at 0.4–0.5 M phosphate buffer. A detailed examination of the use of hydroxylapatite has been presented by Britten et al (115).

ISOLATION OF mRNA

Early approaches to the isolation of individual mRNA species depended upon the exploitation of tissues in which a specific mRNA constituted a relatively large portion of the total mRNA population. This condition permitted the use of size fractionation techniques for the isolation of certain mRNAs. As the limitations of these techniques became increasingly apparent, additional methods for the preparation of pure mRNAs were developed.

Size Fractionation

Several methods have been developed for the isolation of specific mRNAs on the basis of length. Many variations of size fractionation techniques have been used, and the usefulness of this approach can be illustrated by examining several experimental systems.

GLOBIN mRNA   Mammalian globin mRNA was the first eukaryotic mRNA to be isolated and studied extensively. In 1964, Marbaix & Burny (124) reported the identification in rabbit reticulocytes of a 9S RNA species that had the properties expected of globin mRNA. The 9S RNA was subsequently isolated by repeated fractionation of this material from successive sucrose gradient centrifugations (125–127). The subsequent use of zonal rotors facilitated the preparation of large amounts of reticulocyte 9S RNA (128). Clear evidence of its identity as globin mRNA was first presented by Lockard & Lingrel (129). They demonstrated that 9S reticulocyte polysomal RNA from mice was capable of directing the synthesis of mouse hemoglobin β-chains in a heterologous rabbit reticulocyte cell-free protein-synthesizing system. This RNA was also capable of directing the synthesis of globin when injected into frog oocytes (95, 97, 130).

A further advance in RNA purification was reported by Aviv & Leder, who isolated globin mRNA from phenol-extracted rabbit reticulocyte polysomal RNA by affinity chromatography on oligo(dT)-cellulose (62). Krystocek et al (131) subsequently showed that reticulocyte polysomes could be dissolved in sodium dodecyl sulfate and applied directely to an oligo(dT)-cellulose column without prior extraction of protein. Examination of the eluted mRNA fraction showed that 9S globin mRNA also occurred in specific larger aggregates with 18S and 28S rRNA (131, 132). These translationally active larger forms could be readily dissociated by heating at 65° for 10 min to yield 9S globin mRNA and rRNA.

The reticulocyte 9S RNA fraction contained a mixture of the α- and β-globin mRNAs. Sedimentation through sucrose gradients or electrophoresis in polyacrylamide gels under nondenaturing conditions did not separate them (133–135). However, Morrison et al (136) and Orkin et al (137) resolved the two mRNA species by electrophoresis in polyacrylamide gels containing 98% formamide. The gels were sliced, homogenized, and the RNA extracted by either buffer elution (136) or phenol-chloroform-isoamyl alchohol extraction (137). Nudel et al (138) separated human α- and β-globin mRNAs by electrophoresis in linear gradient polyacrylamide gels containing 90% formamide, with a preparative gel apparatus that allowed electrophoretic elution of RNA. The two mRNAs were separated to greater than 90% purity, and they were translationally active and capable of directing the synthesis of cDNAs. Partial purification of chicken α- and β-globin mRNAs has also been achieved by electrophoresis in polyacrylamide gels containing 98% formamide followed by RNA elution from gel slices (47).

Messenger ribonucleoprotein (mRNP) complexes have been used to facilitate the isolation of globin mRNA (134, 135). These naturally occurring complexes consist of mRNA in a tight association with several distinct protein species (139, 140–144). Reticulocyte polysomes can be dissociated by EDTA, and the globin mRNA liberated as a 15S mRNP particle that can be separated from ribosome subunits and tRNA by sedimentation in sucrose gradients (139, 145, 146). Incubation of polysomes with puromycin in the presence of 0.5 M KCl has also been employed to release mRNP particles (147). In other tissues, mRNP particles have been isolated from EDTA-dissociated polysomes by affinity chromatography on oligo(dT)-cellulose (148–150).

CRYSTALLIN mRNA   The mRNAs for lens crystallins, which are the major structural proteins of the lens (151–153), have been isolated by affinity chromatography on oligo(dT)-cellulose and repeated size fractionation of RNA sedimented through successive sucrose gradients (154–160). From both calf lens and rat lens material, 10S and 14S RNA fractions were isolated that directed the cell-free synthesis of β-crystallin and α-crystallin respectively (159, 161). Calf lens 10S and 14S mRNAs have also been prepared from 13S and 19S mRNP particles respectively (156, 157). Molecular hybridization and oligonucleotide fingerprint studies indicated considerable nucleotide sequence homology between calf lens 10S and 14S mRNAs, leading Chen & Spector (162) to suggest that the 14S species is a bicistronic molecule. However, Cohen et al (159) suggested that the rat lens 14S RNA fraction is composed of two discrete 14S mRNAs, each coding for closely related α-crystallin peptides.

PROTAMINE mRNA   One of the smallest mRNAs that has been isolated is protamine mRNA from trout testes. Gedamu & Dixon (163) employed phenol-chloroform-isoamyl alcohol extraction of the microsomal tissue fraction, affinity chromotography on oligo(dT)-cellulose, and sucrose gradient fractionation to isolate the 6S poly(A)-containing protamine mRNA species. Examination of this material by electrophoresis in polyacrylamide gels containing 8 M urea revealed the presence of four distinct mRNA subclasses (164). An alternative large scale isolation protocol was also established (165), which involved the preparation of 12S protamine mRNP particles. The mRNP particles were dissolved in sodium dodecyl sulfate, and the RNA was adsorbed to a column of DEAE cellulose. Ion-exchange chromatography was then employed to elute and separate the low-molecular-weight RNAs. A 4S poly(A)-minus, translationally-active protamine mRNA species was also identified in polysomal mRNP particles (165–167).

HISTONE mRNAs   The isolation of histone mRNAs was based upon the observation that histone synthesis and histone mRNA availability are associated with increased levels of DNA synthesis (168). Histone mRNA levels have been detected in relatively high abundance during the S phase of cultured HeLa cells (169) and in rapidly cleaving sea urchin embryos (170, 171). Histone mRNA purification has involved phenol extraction of polysomal RNA, sedimentation of the RNA through sucrose gradients, and preparative gel electrophoresis (172, 173). Translation in a cell-free protein-synthesizing system derived from mouse ascites cells suggested that about 90% of the 9S RNA fraction coded for histones (174). Ribosomal RNA fragments have been detected as significant contaminants of this material (175, 176). Preparative disc electrophoresis techniques employing polyacrylamide gels were further developed for the isolation and fractionation of the 5 major histone mRNA species (177–180). The finding that histone mRNA may not contain a 3'-terminal poly(A) segment (181, 182) has been exploited. Contaminating poly(A)-containing RNA species were adsorbed to oligo(dT)-cellulose, which resulted in an enrichment for histone mRNAs in the unbound material (179, 183). Alternatively, RNA extraction at neutral pH with cold phenol, which appears to leave most poly(A)-containing RNA trapped at the denatured protein interphase, could avoid the use of affinity chromatography (179). However, recent evidence indicates that histone mRNAs may contain poly(A) tracts and bind to oligo(dT)-cellulose (180, 184).

OVALBUMIN mRNA   Ovalbumin is an egg white protein that accounts for about 60% of the protein synthetic activity of the chicken oviduct (13, 185), where it has been shown to be under the hormonal control of estrogen and progesterone (186, 187). Two different approaches have been used to isolate ovalbumin mRNA. Polysome immunoprecipitation was developed by Schimke and his associates (discussed below), whereas size fraction techniques were employed by O'Malley and his colleagues.

Although the length of ovalbumin mRNA (1890 nucleotides) is similar to the average mRNA length of the oviduct (188, 189), its relative abundance greatly facilitated the use of size fractionation techniques in its isolation. Rosen et al (76) collected phenol-extracted total oviduct RNA, and the poly(A)-containing RNA was partially purified by adsorption to nitrocellulose filters. RNA was eluted from the filters and further purified by reverse-flow gel filtration through an agarose column, which allowed partial purification of the mRNA on the basis of size. Contaminating DNA was excluded from the gel and the elution of 28S rRNA was retarded due to adsorption. Electrophoresis in agarose gels containing 6 M urea at pH 3.5 was employed as a final purification step. A preparative gel apparatus was used to allow for electrophoretic elution of the ovalbumin mRNA band, which yielded an improved recovery compared to the extraction of homogenized gel slices (190). These same techniques were also employed in a similar approach for the isolation of casein mRNAs from mouse mammary glands (191, 192).

Sedimentation through sucrose gradients has been employed by others as the primary size fractionation technique for the separation of ovalbumin mRNA from total oviduct poly(A)-containing RNA (46, 193). Buell et al (193) employed 5–20% isokinetic sucrose gradients, prepared accoring to McCarty et al (110), to achieve a significantly improved resolution of ovalbumin mRNA from other mRNAs. Repeated size fractionation could yield an ovalbumin mRNA of ~ 90%, or greater, homogeneity. A partial purification of conalbumin, ovomucoid, and lysozyme mRNAs could also be achieved on these gradients (193).

SILK FIBROIN mRNA   Among the first large mRNA species to be isolated was that coding for silk fibroin from the posterior gland of the silk worm *Bombyx mori* (30, 194). Phenol-extracted nucleic acid from pronase-treated tissue was digested with DNase and the RNA was sedimented through linear sucrose gradients. A 40–60S fraction of aggregated RNA was collected, denatured in 70% formamide, and resedimented through sucrose gradients containing 70% formamide to yield purified 32S fibroin mRNA.

MYOSIN mRNA   Sucrose gradient fractionation was also employed to isolate 26S myosin heavy-chain mRNA from mRNP particles, as well as the large polysome fraction of embryonic chick skeletal muscle (195–197). However, subsequent examination by RNA-cDNA hybridization kinetics suggested that the 26S fraction was composed of two different mRNA sequences (197).

PROCOLLAGEN mRNA   For the isolation of procollagen mRNA from embryonic chick calvaria, poly(A)-containing RNA was sedimented through linear sucrose gradients, and the 27S fraction was isolated (198, 199). This material consisted of ~ 50% procollagen sequences. Further purification was carried out by Boedtker et al (199) by means of agarose adsorption chromatography (200). In a buffer containing a relatively high salt concentration, the guanine-rich procollagen mRNA preferentially bound to the agarose, with subsequent elution carried out at a low ionic strength. Under these conditions, most mRNAs were not adsorbed to the gel (201). This procedure was based on observations by Petrovic et al (202, 203) who showed that 28S rRNA was quantitatively retained on agarose in 0.5 M NaCl because of its high guanine content. Further characterization of the resolving properties of standard agarose and cross-linked agarose for mRNA purification have been examined by Zeichner & Stern (204).

VITELLOGENIN mRNA   Vitellogenin is a 240,000 molecular weight protein that can be induced, by estrogen treatment of egg-laying animals, to become the major protein synthesized by the male liver (205, 206). The isolation of the 29S vitellogenin mRNA has been particularly difficult due to the high endogenous ribonuclease content of amphibian and avian livers. Three quite different approaches to this problem have been followed. Shapiro & Baker (207) characterized the ribonuclease activity of *Xenopus laevis* liver and employed conditions for RNA extraction in which nuclease activity was minimal. Livers were homogenized in a buffer at pH 6.0 which contained 0.3 M NaCl, 0.1 M MgCl$_2$, 1 mM dithiothreitol, and 1 mg/ml of heparin. Total RNA was prepared by phenol-chloroform extraction in the presence of 1% sodium dodecyl sulfate. Sodium acetate–washed RNA was further purified on oligo(dT)-cellulose and sedimented through isokinetic sucrose gradients to isolate the 29S fraction. Goldberger and his associates (53) homogenized estrogen-treated rooster livers in 8 M guanidine·HCl containing 1 mM dithiothreitol at pH 5.0 to achieve rapid and effective denaturation of ribonuclease. Poly(U)-Sephadex G10 was found to yield improved recoveries of the large poly(A)-containing mRNA, compared to oligo(dT)-cellulose or poly(U)-Sepharose 4B. Jost et al (7) employed 1% diethyl pyrocarbonate to isolate vitellogenin mRNA from chicken liver polysomes. The liver-homogenizing buffer also included 6 mg/ml yeast RNA and 4 mg/ml heparin to further minimize ribonuclease activity. Isolated polysomes were then dissolved in a buffer containing 2% sodium dodecyl sulfate and phenol-chloroform extracted to obtain RNA.

Polysome Immunoprecipitation

In most tissues the relatively low amount of even the most abundant mRNA species, as well as their similar chemical and physical properties, prevents the use of size fractionation procedures as the primary method of mRNA isolation. In these situations it may be possible to employ immunochemical methods to isolate the specific polysomes synthesizing the protein of interest. The immunoprecipitation of specific polysomes depends upon the recognition of nascent peptide chains by antibodies that have been prepared against the native protein. The major technical limitation is a function of the amount of nonspecific adsorption and trapping in the immunoprecipitate. Most modifications of the method have usually been for the purpose of minimizing this problem. Many of the technical requirements have been examined by Schimke and his associates in their isolation of chicken ovalbumin-synthesizing polysomes (79, 185, 208–211). The method of indirect immunoprecipitation of polysomes, described by Shapiro et al (211), is probably the most general procedure, and it has been easily modified in different systems. This technique has been used to isolate or enrich for mRNAs constituting as little as 1% of the total mRNA population of a tissue.

The general technique of polysome immunoprecipitation is illustrated in Figure 1, which shows the isolation of rat liver albumin-synthesizing polysomes (81). The isolation of albumin-synthesizing polysomes involves the incubation of liver polysomes with antibody prepared against native albumin, which binds to the nascent albumin peptide chains on the ribosome. The single antibody binding step does not result in polysome precipitation. This first binding reaction is followed by the incubation of the polysome-antibody complex with a second antibody, which has been prepared against the first antibody (anti-antibody). The polysome-antibody-anti-antibody complex is then sedimented through a discontinuous sucrose gradient to remove unreacted polysomes and unreacted antibody. The immunoprecipitated material can be dissolved in detergent and the mRNA purified by affinity chromatography on poly(U)-agarose or oligo(dT)-cellulose.

Methods of polysome preparation for immunoprecipitation vary widely, but they can be grouped into three general procedures. Collecting polysomes from tissue homogenates by sedimenting them through layers of 0.5–2 M sucrose is a commonly used technique (212, 213). The polysome pellet must then be resuspended and any aggregated material removed. Sedimentation of polysomes through sucrose layers onto a 2.5 M sucrose

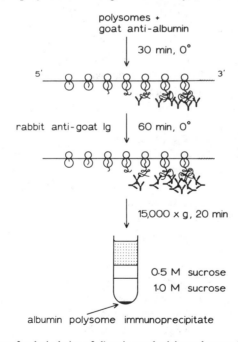

Figure 1   Procedure for the isolation of albumin-synthesizing polysomes by indirect immunoprecipitation.

cushion is an alternative procedure (185, 214). Polysomes are collected by means of a syringe, and the sucrose is subsequently removed by dialysis. Polysome aggregation does not occur and minimal ribonuclease damage results, but lower yields are obtained. Magnesium precipitation of polysomes has also been successfully employed. Incubation of tissue homogenates with $MgCl_2$ at concentrations of 60 mM or greater for about 1 hr causes a maximum aggregation of polysomes, which can then be collected by sedimentation (43). High yields of polysomes are obtained, but the relatively long incubation period can result in partial degradation of the polysomes in those tissues with high ribonuclease content, even though high magnesium concentrations have been reported to inhibit ribonuclease activity (215).

Isolation of mRNAs coding for secretory proteins has been facilitated by the separation of membrane-bound polysomes from free (unbound) polysomes. This approach is based on the observation that free polysomes will sediment through 2 M sucrose whereas rough endoplasmic reticulum does not penetrate this layer (212, 213, 215–219). However, it is quite difficult to prevent partial degradation of membrane-bound polysomes in tissues with high ribonuclease content. Before immunoprecipitation, polysomes must be removed from membranes by detergents.

Antibody preparation and characterization require particular attention. Antibodies can be made ribonuclease-free by ion-exchange chromatography on sterile columns of DEAE-cellulose and CM-cellulose (185, 211). To minimize the size of polysome immunoprecipitates and thereby decrease the extent of nonspecific adsorption and nonspecific trapping, the first antibody can be immunopurified by affinity chromatography with immobilized antigens (210, 211). Immunopurification has also been specifically employed to remove ribonuclease from antibodies (220).

The specificity of the first antibody for the immunoprecipitation of specific polysomes has been investigated by employing ^{125}I-labeled immunoglobulins in a variety of polysome binding experiments (185, 208, 214, 221). Polysomes were briefly incubated with radioactively labeled antibody, sedimented through linear sucrose gradients, and fractions were collected for the determination of radioactivity. These studies showed that antibody binding is highly specific, and occurs through an immunological recognition of the nascent peptide chains on individual ribosome monomers. Nonspecific binding to ribosomal structural components was not observed. Bound antibody did not cause significant changes in the rate of polysome sedimentation through sucrose gradients, so that specific polysome sizes could be estimated (185, 208, 214). The antibody binding to nascent peptide chains appears to be a saturable reaction, and has been used to quantitate the relative amounts of specific polysomes (208, 214).

Immunoprecipitation of specific polysomes has been found to require the addition of a second binding component to the polysome-antibody complex. A number of alternatives for the second component are possible. Shapiro et al (211) and Schechter (222, 223) developed procedures that employed a second antibody (antiimmunoglobulin) which was prepared against the first antibody to precipitate a polysome-antibody-anti-antibody complex. Shapiro et al (211) established conditions where the nonspecific adsorption and trapping in the immunoprecipitate was < 0.5% of the total polysomal starting material. The primary consideration was to minimize the size of the final immunoprecipitate, since nonspecific adsorption and trapping were approximately proportional to the bulk size of the immunoprecipitate and not the amount of polysomal material isolated. The reaction conditions required for polysome immunoprecipitation were determined by employing polysomes with radioactively labeled nascent peptide chains, and varying concentrations of both antibodies in a series of reactions (211, 214). Optimum conditions were found that permitted quantitative immunoprecipitation of specific polysomes, which provided an estimate of their relative concentration in the total polysome population (211, 214, 224). Examination of the nascent peptide chains of immunoprecipitated polysomes again demonstrated the specificity of the immunoprecipitation reaction (208, 211).

The double antibody or indirect immunoprecipitation technique for the isolation of specific polysomes has been applied to several different systems. Schechter has isolated κ- and λ- immunoglobulin light-chain mRNAs from mouse myelomas (222, 223, 225). Legler & Cohen have purified a different κ- light-chain mRNA as well as a γ- heavy-chain mRNA from a mouse myeloma (226). Ewe-α-casein and β-casein mRNAs from lactating mammary glands (227), rabbit reticulocyte α- and β-globin-synthesizing polysomes (220), and *Xenopus* liver vitellogenin-synthesizing polysomes (228) have been isolated by similar techniques. In all of these systems, the immunoprecipitated polysomes constituted a relatively large portion of the total polysome population. The indirect immunoprecipitation of chick embryo collagen-synthesizing polysomes (229) and rat liver fatty acid synthetase-synthesizing polysomes (230) has been more difficult because of their relatively low concentrations.

Direct immunoprecipitation of specific polysomes has also been employed. In this procedure, Palmiter et al (208) incubated chicken oviduct polysomes with anti-ovalbumin in a first incubation, then added ovalbumin in a second incubation, and finally added excess first antibody in a third incubation to immunoprecipitate ovalbumin-synthesizing polysomes. Ono et al used direct immunoprecipitation to isolate κ-light-chain mRNA and γ-heavy-chain mRNAs from a mouse myeloma (231, 232). The disadvantages of this particular approach to polysome immunoprecipitation are the relatively large amounts of antigen required for preparative purposes, and that the final immunoprecipitate can be somewhat bulky.

As an alternative to polysome immunoprecipitation in solution, Palacios et al used a solid matrix affinity support to isolate ovalbumin-synthesizing polysomes (210). Ovalbumin was cross-linked with glutaraldehyde then homogenized to form a particulate resin. The polysome-antibody complex was incubated with the cross-linked matrix, which was then washed to remove nonspecifically adsorbed material, and the mRNA and ribosomal subunits were eluted with EDTA. A cross-linked antigen matrix was also used by Scott & Wells for the isolation of chicken reticulocyte histone V-synthesizing polysomes (233). The requirement for a large quantity of antigen to form a cross-linked matrix for preparative purposes is a disadvantage. However, the solid matrix approach has been modified for indirect immunoprecipitation of polysomes. Innis & Miller (234) used an insoluble cross-linked second antibody (anti-antibody) to bind the first polysome-antibody complex in the isolation of rat α-fetoprotein mRNA from a Morris hepatoma. This same procedure was also used for the isolation of immuno-globulin light- and heavy-chain mRNAs from mouse plasmacytomas (235). Use of the insoluble, finely divided matrix facilitated the resuspension and washing of the immunoabsorbate to lower nonspecifically adsorbed and trapped background material. Further modification of the solid matrix support has been described by Schutz et al, who coupled the second antibody to p-aminobenzyl-cellulose via a diazo linkage (236). This system was used to isolate ovalbumin, ovomucoid, and lysozyme mRNAs from chicken oviduct (237).

A solid matrix for use in the indirect immunoprecipitation method was derived from *Staphylococcus aureus* (238). The protein A component of the *S. aureus* cell wall binds rabbit immunoglobulin molecules, and these bacteria, which have been inactivated by heat and formaldehyde treatment, can be used to precipitate antigen-antibody complexes (239). For polysome immunoprecipitation, the polysome-antibody complex was incubated with inactivated *S. aureus,* then washed by centrifugation through layers of 0.5 M and 1 M sucrose (238). The bound material was released by treatment with 0.5% sodium dodecyl sulfate. This procedure was used for the isolation of leukemia virus-specific polysomes from infected cells (238). The technique was found to be more efficient and to result in lower nonspecific backgrounds than the double antibody indirect immunoprecipitation method.

Some problems have been reported with the polysome immunoprecipitation technique. In some systems, the nonspecific binding of antigen to polysomes during their isolation could result in considerable nonspecific adsorption of antibody (240). Delovitch et al developed direct and indirect techniques for the immunoprecipitation of immunoglobulin light-chain polysomes from plasmacytomas (241, 242), and found considerable non-specific adsorption of antibody molecules to ribosomes. A similar non-specific binding problem has been reported for rat liver polysomes (243). Therefore, immunoglobulin molecules were treated with pepsin to produce F(ab')$_2$ fragments (241–243). The purified F(ab')$_2$ fragments resulted in a decrease in nonspecific polysome precipitation, which suggests that the nonspecific adsorption problem was due to the Fc region of the immuno-globulin molecules. This difficulty has not been a problem for most systems.

Molecular Hybridization Purification Techniques

Selective hybridization of partially purified mRNAs to their cDNAs can be employed to further enrich individual sequences, since size fractionation or polysome immunoprecipitation methods in most cases only results in a partially purified mRNA. A cDNA can be prepared against the partially purified mRNA and further enriched by hybridization to low R_0t with homologous tissue mRNAs or hybridization to high R_0t with heterologous tissue mRNAs, where R_0t is defined as the product of the RNA concentration in moles of nucleotide per liter and the time in seconds. When large amounts of cDNA can be prepared, such as from large amounts of mRNA or from cloned recombinant DNA, cDNA-cellulose can be made and used for both analytical and preparative procedures.

Purification of cDNA by hybridization to low R_0t was initially developed for the isolation of immunoglobulin mRNAs; however, the early prepara-tion of immunoglobulin mRNAs involved only size fractionation. Stavnezer & Huang first purified a 13S RNA fraction by sucrose gradient sedimenta-tion of microsomal RNA from a mouse plasmacytoma, which coded for mouse immunoglobulin light chain protein in a rabbit reticulocyte cell-free protein-synthesizing system (244). The mRNA was subsequently purified further by oligo(dT)-cellulose affinity chromatography and sedimentation through sucrose gradients containing dimethylsulfoxide (245). Similar size fractionation techniques were employed by several other laboratories for the purification of immunoglobulin light-chain mRNAs from different plasma-cytomas (32, 246, 247). Some preparations of size-fractionated mRNA were radioactively labeled with ^{125}I and hybridized to cellular DNA (248, 249). The hybridization kinetics indicated that the mRNA was contaminated by about 20% with material that hybridized with "reiterated" kinetics, whereas the main mRNA component hybridized with "unique" kinetics. However, further purification of the light-chain mRNA by more stringent size fractionation appeared to eliminate the "reiterated gene product" con-taminants (249, 250).

In order to obtain a highly purified radioactive hybridization probe, Honjo et al (251) further purified a mouse light-chain cDNA by hybridization. The cDNA preparation was enriched for long molecules by fractionation on alkaline sucrose gradients, then hybridized to excess mRNA template to a low R_0t value. The hybrid was collected by elution from hydroxylapatite and treated with alkali to destroy the RNA. The final cDNA preparation showed only "unique sequence" hybridization kinetics with cellular DNA. The low R_0t hybridization approach was subsequently used by others to purify the cDNAs to different immunoglobulin light-chain mRNAs (252–254).

Complementary DNAs to several other mRNAs have also been enriched by low R_0t hybridization, including procollagen mRNA (255), albumin mRNA (256), and three mouse heavy-chain immunoglobulin mRNAs (257). In the latter case, previous attempts to isolate heavy-chain mRNAs had occasionally been hampered by the apparently low translation efficiency of these species in cell-free protein-synthesizing systems (258). Size fractionation alone produced a heavy chain mRNA of only about 60% purity (259). Several other studies which indicated that heavy chain mRNA binds specifically to whole immunoglobulin molecules, thereby permitting purification, have proved to be irreproducible (260).

Modifications of the low R_0t hybridization technique have been useful for the enrichment of mRNAs that are present in relatively low cellular concentrations. The mRNA coding for α_{2u}-globulin, a male rat liver protein that represents ~1% of hepatic protein synthesis, was enriched to ~35% purity by matrix-antibody polysome immunoprecipitation (261). A cDNA was prepared and hybridized with female rat liver mRNA (which does not appear to contain α_{2u}-globulin mRNA) to a high R_0t value. The single-stranded cDNA was recovered by hydroxylapatite chromatography, then further enriched by hybridization with homologous male liver mRNA to a low R_0t value. The final cDNA preparation could hybridize to a limit of ~75% with male liver mRNA, and appeared to be specific for α_{2u}-globulin sequences. Alt et al (262) employed indirect polysome immunoprecipitation with a methotrexate-resistant variant of cultured murine sarcoma 180 cells to enrich for dihydrofolate reductase mRNA. This enzyme accounted for ~2% of the total protein synthesis in these cells. Polysome immunoprecipitation resulted in about a tenfold purification of dihydrofolate reductase mRNA. A cDNA was prepared and further enriched by hybridization to high R_0t with heterologous mRNA from cells containing very low levels of reductase mRNA. Nonhybridized cDNA was recovered by hydroxylapatite chromatography and further enriched by hybridization to low R_0t with homologous cell mRNA. This final cDNA preparation showed high specificity for dihydrofolate reductase sequences.

Sequence-Specific Isolation

Alternative methods have been developed that employ specific cDNAs, restriction endonuclease cDNA cleavage fragments, or synthetic oligonucleotides of defined sequence as the principal means of mRNA detection and isolation. These variations of mRNA isolation are likely to become of general importance for many eukaryotic systems as detection methods are developed and refined.

cDNA-CELLULOSE   Solid phase-bound cDNA sequences have been described by several laboratories for the purification of specific mRNAs. Venetianer & Leder (263) first reported the synthesis of cellulose-bound DNA that was complementary to globin mRNA, and demonstrated its potential for globin mRNA purification. The reaction mixture contained the components required for cDNA synthesis, with oligo(dT)-cellulose substituted as the primer for reverse transcriptase. Following the reaction, the matrix was washed with NaOH to remove the mRNA template, while the cDNA remained covalently attached to the cellulose matrix through the oligo(dT) primer. RNA samples were passed through the column under high salt binding conditions at the usual hybridization incubation temperature. The column was washed with binding buffer, and the mRNA eluted with low salt buffer. Levy & Aviv (264) used globin cDNA-cellulose as an analytical hybridization probe for the detection of low levels of globin mRNA in dimethyl sulfoxide-treated Friend leukemic cells. Wood & Lingrel (265) examined conditions required for optimal synthesis of globin cDNA, and explored the use of this affinity chromatographic medium for both analytical and preparative applications. An important application of globin cDNA-cellulose has been for the isolation of nuclear precursor molecules to globin mRNA (266, 267).

Optimum conditions for the synthesis and preparative applications of ovalbumin cDNA-cellulose were characterized by Rhoads & Hellmann (268). By optimizing conditions for cDNA synthesis, up to 35% by mass of the input mRNA was copied into cDNA. Batch-wise procedures for mRNA hybridization and elution were established for the preparation of ovalbumin mRNA. Anderson & Schimke (269) have prepared ovalbumin cDNA-cellulose and used it to enrich and partially purify ovalbumin gene sequences from chicken DNA fragments. In other applications, Hirsch et al (270) have employed cDNA-cellulose, which was synthesized from total unfractionated liver or hepatoma mRNAs, for the subsequent preparation of tissue-specific mRNAs.

Chemical coupling methods have also been established for linking DNA sequences to cellulose. Ovalbumin cDNA as well as SV40 DNA fragments have been coupled to phosphocellulose (271) and neutral cellulose powder

(272) by means of a water-soluble carbodiimide. Woo et al (271) employed the bound ovalbumin cDNA to isolate long fragments of chicken DNA enriched for ovalbumin gene segments. Noyes & Stark (273) coupled full-length linear SV40 DNA to m-aminobenzyloxymethyl-cellulose by diazotization. This affinity support was capable of functioning effectively for analytical hybridization experiments with low background adsorption, and it was capable of functioning preparatively for RNA isolation.

RESTRICTION ENDONUCLEASE FRAGMENTS   An important technique for the identification and purification of specific cDNAs which depends upon the use of restriction endonucleases has been described by Seeburg et al (274). This approach does not require extensive purification of RNA beyond the preparation of poly(A)-containing RNA by affinity chromatography. However, the limit of specific sequence detection requires that the mRNA of interest be not less than about 2% of the poly(A)-containing RNA fraction under study. This technique has been used for the preparation of cDNA fragments complementary to the mRNAs for human chorionic somatomammotropin from placenta (274), rat preproinsulin from islets of Langerhans (275), and rat somatotropin from cultured pituitary cells (276).

In this approach, radioactive single-stranded or double-stranded cDNAs were digested by appropriate restriction endonucleases, and the cleavage fragments were separated by gel electrophoresis (274–276). The more abundant cDNAs yielded prominent cleavage fragments, which could be visualized by autoradiography, and subsequently isolated. Only internal fragments containing restriction endonuclease cuts at both ends migrated as sharply defined bands during gel electrophoresis, which reduced the effect of size heterogeneity in cDNA preparations. The prominent fragments were characterized by DNA sequence analysis (277) and related to the known amino acid sequence of the protein under study. An alternative method of identifying cleavage fragments might be provided by hybrid-arrested cell-free translation (278), which is based on the observation that a mRNA hybridized with its cDNA is unable to direct the cell-free synthesis of its protein. In addition to yielding important nucleotide sequence information, the isolated cDNA fragments provided relatively pure hybridization probes which could subsequently be inserted into plasmids for cloning purposes (275, 276, 279).

Characteristic restriction endonuclease fragment patterns of cDNAs have also provided the basis for mRNA detection. Howard et al (280) isolated the procollagen-rich 28S mRNA fraction from chick embryo calvaria, prepared radioactive double-stranded cDNA, and determined the restriction endonuclease fragment pattern. Characteristic procollagen cDNA cleavage fragments were then identified in similarly prepared material from normal and virus-transformed chick embryo fibroblasts. Results from this technique might be difficult to interpret if the reverse transcriptase did not copy all templates with equal efficiency, if full-length cDNAs were not made, and if prominent fragments of unique length were not obtained.

Characteristic cDNA cleavage fragments might also be employed for the isolation of specific eukaryotic mRNAs. A useful technique was developed for the preparation of translatable virus mRNAs (281). Cytoplasmic RNA, isolated from cells late after infection by adenovirus type 2, was hybridized to Eco R_I endonuclease restriction fragments of virus DNA. Then, the hybridized mRNA was separated from single-stranded mRNA by hydroxylapatite chromatography. Hybrids were denatured by heating, and the mRNA was separated from the DNA fragments by sucrose gradient sedimentation or by adsorption to oligo(dT)-cellulose. Related methods have also been described for the isolation of virus-specific RNA by hybridization to filter-bound virus DNA (282).

SYNTHETIC OLIGODEOXYRIBONUCLEOTIDE PRIMERS Synthetic oligodeoxyribonucleotides have been employed as cDNA primers and as limited hybridization probes for the identification of specific gene sequences. A pentadecanucleotide complementary to a particular sequence in yeast cyctochrome C mRNA, has been shown by Szostak et al (283) to bind to this mRNA and serve as a primer for the synthesis of a cDNA by $E.$ $coli$ DNA polymerase I. Montgomery et al (284) employed a ^{32}P-tridecamer as a hybridization probe to identify cytochrome C gene sequences in fractionated yeast DNA fragments prepared by Eco R_I restriction endonuclease digestion. However, the small size of this probe required the use of hybridization conditions of low stringency such that binding of the probe occurred to several other DNA fragments. Rabbits (285) used a specific hexanucleotide that was complementary to a unique base sequence near the 3'-terminal portion of the light-chain immunoglobulin constant region, to prime for a cDNA for use as a defined hybridization probe. Specific binding of synthetic oligonucleotides to RNAs (containing the complementary sequence) has been demonstrated for other purposes. A specific tridecamer has been found to bind to Rous sarcoma virus RNA (286) and inhibit virus replication and translation. Purified globin mRNA has been shown to bind a specific hexanucleotide or octanucleotide at the initiation codon region for the synthesis of a cDNA to the 5'-terminal noncoding region to be used for sequencing purposes (287, 288).

Specific oligodeoxyribonucleotides may be useful for the identification and isolation of other mRNA species for which at least some amino acid sequence information is known. In this case, a hypothetical partial nucleotide sequence of the mRNA can be inferred. If possible, a region would be identified that shows minimum ambiguity with respect to the degeneracy of the genetic code. A short oligonucleotide complementary to this region would then be synthesized to serve as a primer for the synthesis of a cDNA. Wu (289) has suggested some guidelines for the synthesis of oligonucleotide primers. A technical limitation of this approach relates to the ability to synthesize an oligonucleotide of sufficient length to form a stable hybrid with an individual mRNA species when it is a minor constituent of a complex RNA mixture. The minimum necessary oligonucleotide length would be difficult to predict for the general case since helix stability for oligonucleotides is a function of both length and nucleotide sequence (290).

CHARACTERIZATION OF mRNA

Several criteria of purity in various combinations have been applied to isolated mRNAs. These criteria include examination by gel electrophoresis, translation in a mRNA-dependent cell-free protein-synthesizing system with analysis of the translation product, hybridization kinetics with its cDNA, and oligonucleotide fingerprinting.

Gel Electrophoresis

Gel electrophoresis has become a major analytical technique for the measurement of macromolecular weights because of its relatively simple technology and high resolution. Electrophoretic mobility can be considered as one criterion of mRNA purity when size fractionation techniques have not been employed in the isolation procedures. Maximum information can be gained by carrying out gel electrophoresis under denaturing conditions. Several denaturing systems have been described, including reaction with formaldehyde (291), electrophoresis in 99% formamide (292), electrophoresis in 6 M urea at pH 3.5 (76), and reaction with methylmercury (293). Lehrach et al (294) have examined and compared these four denaturing conditions for the determination of RNA molecular weights by gel electrophoresis. Reliable molecular weight measurements of large RNA molecules were obtained only after reaction with either 2.2 M formaldehyde or 10 mM methylmercury hydroxide. Incomplete denaturation of GC-rich RNAs occurred in formamide which led to anomalous mobilities that resulted in erroneous molecular weight estimations. In urea-containing gels at acid pH, not all RNA species were found to have a constant charge : mass ratio which resulted in anomalous mobilities for some RNA molecules. Agarose gels in which either formaldehyde or methylmercury served as the denaturant provided for complete RNA denaturation, stable high porosity gels, and reliable molecular weight estimates. Buell et al (193) found that 20 mM methylmercury hydroxide provided for fully denaturing conditions in electrophoresis. They also reported problems with the formamide system as a result of ionic discontinuities in the gels.

Direct measurement of RNA length can be made by electron microscopy. Techniques have been described for measuring the length of RNA molecules which remove the effects of secondary structure and nucleotide composition (295). Electron microscopy has been employed to advantage with larger mRNA molecules such as ovalbumin (296) and vitellogenin (297, 298). However, mRNA fragmentation during sample preparation can lead to an underestimate of the length.

The length of the poly(A) segment in an mRNA can be estimated by gel electrophoresis (299–301). Digestion of mRNA preparations with RNase A and RNase T_1 leaves only the poly(A) segment intact (299–301). An accurate estimation of length is then obtained by comparing its electrophoretic mobility to that of known poly(A) standards (301). When it is not possible to use endogenously labeled material, the poly(A) can be terminally labeled with [^{3}H]NaBH$_4$ before electrophoresis (189). Alternatively, the poly(A) can be eluted from gel slices and measured with a radioactive hybridization probe (302, 303). The amount of poly(A) present in any RNA sample can be measured indirectly by hybridization with ^{3}H-labeled poly(U) (302) or poly(dT) (304). This determination has been employed to estimate the amount of contaminating rRNA in preparations of poly(A)-containing RNA for sequence complexity studies (118, 189, 305). However, this latter approach requires that general assumptions be made with respect to the poly(A) content of mRNAs. Further characterization of poly(A) segments has been reviewed (36, 58, 306).

Translation

Translation of an isolated mRNA in a mRNA-dependent cell-free protein-synthesizing system is a useful criterion of mRNA purity. Examination of the total postribosomal translation product should reveal only the single peptide product of the exogenous mRNA. Proteins with relatively short precursor regions will essentially comigrate with standard native proteins when examined by gel electrophoresis. The translation product would be precipitable by a monospecific antibody prepared against the mature authentic protein. Comparison of the tryptic peptide map of the translation product with that of an authentic standard is a particularly good characterization of the translation product. The easiest comparisons can be made when the translation product is labeled with a ^{3}H-labeled amino acid and the protein standard with the homologous ^{14}C-amino acid. When the tryptic peptides (307) are resolved by ion exchange chromatography (308), all of the ^{14}C-labeled peptides should comigrate with a ^{3}H-labeled peptide. Differences would be observed when a precursor peptide from the translation product is detected, and possibly when the mature native protein is glycosylated, which would give rise to glycopeptides with mobilities different from those of the cell-free translation product peptides.

Hybridization

The RNA-excess hybridization of an isolated mRNA to its cDNA can provide useful information with respect to mRNA purity. Hybridization kinetics and data analysis have been described elsewhere (112–116, 309), and only a few aspects will be considered here. An RNA-driven hybridization is a pseudo first-order reaction. For a pure mRNA species, the reaction will be complete (100% hybridization) within a 100-fold range of R_0t values. The kinetic complexity of the isolated mRNA can be determined by comparing it to the hybridization reaction of a pure standard RNA of known length to its cDNA. Complexity can be defined as the length in nucleotides of the longest composite nonrepeating sequence in a nucleic acid preparation. The $R_0t_{1/2}$ values are calculated (where $R_0t_{1/2}$ is the R_0t value at which 50% hybridization has occurred), and comparison with the known standard can give an estimate of kinetic complexity for the isolated mRNA. For a pure mRNA molecule lacking a reiterated component, the kinetic complexity should be in reasonable agreement with the molecular weight estimation from gel electrophoresis under rigorously denaturing conditions. In this regard, reiterated sequence components have not yet been identified in highly purified eukaryotic mRNA molecules.

This determination assumes that the kinetics of mRNA-cDNA hybridization are independent of the RNA species being studied. It also assumes that a pure kinetic standard is being used. The most commonly employed kinetic standard is globin mRNA, which is usually a mixture of the α and β species. Chicken ovalbumin mRNA is an excellent kinetic standard because it is relatively easy to prepare, it is a singular molecular species, and its size is comparable to the average length of many mRNAs in most tissues. The hybridization assay also requires that the cDNAs be of equivalent length, or that an appropriate correction for the effect of length on the rate of hybridization be taken into account. Wetmur (310) has reported that in reactions between equal quantities of separated T4 DNA strands of different lengths, the rate of renaturation depends upon the square root of the length of the shorter strand. It has not yet been rigorously shown that this relationship holds for RNA driven cDNA hybridization reactions. However, the hybridization kinetics of purified ovalbumin mRNA with short and long cDNA are consistent with this relationship (103).

Hybridization of the isolated mRNA to a cDNA preparation prepared against the total poly(A)-containing RNA in the starting material is a particularly sensitive way of detecting contaminating sequences. The contaminants would be detected as a second component in a biphasic curve when the reaction is driven to high R_0t values. This approach is more sensitive than the hybridization of the purified mRNA to its cDNA copy. It is also more sensitive than hybridization of the cDNA copy of the purified mRNA to the poly(A)-containing RNA of the starting material.

Oligonucleotide Fingerprinting

The oligonucleotide "fingerprint" of an RNA molecule is a function of its nucleotide sequence. The technique involves complete digestion of a radioactively labeled RNA to oligonucleotides by a ribonuclease that, preferably, shows specificity for a single nucleotide. Separation of the digestion products is accomplished by any of several two-dimensional electrophoretic or chromatographic fractionation procedures. Oligonucleotides are then visualized by autoradiography and each individual RNA molecule will display its own characteristic pattern or fingerprint. The technique has found application in assessing the purity of isolated mRNA molecules (i.e. in searching for characteristic oligonucleotides of contaminating rRNA), and for gaining nucleotide sequence information (247, 259, 296). Fingerprinting has also been employed for comparing related RNA molecules, since species that have some sequences in common would be expected to have some of the same characteristic oligonucleotides. Beemon has reviewed the use of oligonucleotide fingerprinting in characterizing the structure of RNA tumor viruses (311).

Oligonucleotide fingerprinting provides the most information if uniformly ^{32}P-labeled RNA is employed. It is usually not possible to obtain such preparations of individual mRNA species from higher eukaryotes because of the great difficulty in labeling to high specific radioactivity in vivo. Methods for labeling nucleic acids in vitro with ^{125}I have been described which yield molecules of very high specific radioactivity (312, 313). Iodination is essentially specific for cytosine residues, and the specificity of several nucleases is not altered and hybridization can still occur (312–314). Since the iodination of cytosine residues appears to be a random event, fingerprints generated from ^{125}I-labeled RNA are reproducible, and therefore, have been useful for comparative studies (314). However, the use of iodination for sequence determinations is limited. For oligonucleotide mapping of unlabeled mRNAs, oligonucleotides can also be terminally labeled in vitro with $[\gamma\text{-}^{32}\text{P}]$ATP by polynucleotide kinase, resolved by electrophoresis, and employed for sequence analysis (315).

RNase T_1, which is specific for guanine residues, is usually used for RNA digestion. Oligonucleotides of various sizes are produced, if there is a random distribution of guanine residues throughout the RNA molecule. The largest oligonucleotides are the most characteristic of a particular RNA molecule since they are most likely to contain unique sequences. Oligonucleotides are usually resolved by electrophoresis in one dimension and by homochromatography (displacement by unlabeled heterogeneous oligonucleotides) in a second direction (316).

CONCLUSION

The ability to isolate, examine, and quantitate individual mRNAs will lead to a greater understanding of the regulatory mechanisms concerned with the expression of eukaryotic genes. For most experimental systems, the initial preparative goal will be the purification of a specific cDNA. The availability of recombinant DNA and bacterial cloning technology will then permit the preparation of large amounts of the cDNA, which will facilitate many studies of eukaryotic gene expression. In particular, the use of cDNA would simplify the isolation of additional mRNA, and would permit the identification and isolation of nuclear precursors to the mRNA (317–321). The cDNA could also be employed as a hybridization probe to identify cloned recombinant DNA fragments containing specific eukaryotic gene sequences (274–276, 279). Several specific gene sequences have already been identified by this approach, but a discussion of these advances is beyond the scope of this review.

ACKNOWLEDGEMENTS

I would like to thank Dr. Stanley Cohen of Vanderbilt University and Dr. Robert Schimke of Stanford University for their training and support. I would also like to thank Dr. David Shapiro for many stimulating discussions, and my colleagues Robert Hamilton, Warren Liao, George Ricca, and Stanley Hreniuk for their suggestions. I am grateful for the research support from the National Institutes of Health provided by grants CA 16746, CA 18450, and AM 22013, and for a Research Career Development Award, CA 00393.

Literature Cited

1. Ehrenberg, L., Fedorcsak, I., Solymosy, F. 1976. *Prog. Nucl. Acid Res. Mol. Biol.* 16:189–262
2. Melchior, W. B., Fahrney, D. 1970. *Biochemistry* 9:251–57
3. Fedorcsak, I., Ehrenberg, L. 1966. *Acta Chem. Scand.* 20:107–15
4. Solymosy, F., Fedorcsak, I., Gulyas, A., Farkas, G. L., Ehrenberg, L. 1968. *Eur. J. Biochem.* 5:520–27
5. Oberg, B. 1970. *Biochim. Biophys. Acta* 204:430–37
6. Leonard, N. J., McDonald, J. J., Reichman, M. E. 1970. *Proc. Natl. Acad. Sci. USA* 67:93–99
7. Jost, J. P., Pehling, G., Panyim, S., Ohno, T. 1978. *Biochim. Biophys. Acta* 517:338–48
8. Mendelsohn, S. L., Young, D. A. 1978. *Biochim. Biophys. Acta* 519:461–73
9. Rhoads, R. E., McKnight, G. S., Schimke, R. T. 1973. *J. Biol. Chem.* 238:2031–39
10. Zollner, N., Fellig, J. 1953. *Am. J. Physiol.* 173:223–28
11. Roth, J. S. 1953. *Arch. Biochem. Biophys.* 44:265–70
12. Rowley, P. T., Morris, J. 1967. *Exp. Cell Res.* 45:494–97
13. Palmiter, R. D., Christensen, A. K., Schimke, R. T. 1970. *J. Biol. Chem.* 245:833–45
14. Fellig, J., Wiley, C. E. 1959. *Arch. Biochem. Biophys.* 85:313–16
15. Mora, P. T., Young, B. G. 1959. *Arch. Biochem. Biophys.* 82:6–20
16. Payne, P. I., Loening, U. E. 1970. *Biochim. Biophys. Acta* 224:128–35
17. Roth, J. S. 1956. *Biochim. Biophys. Acta* 21:34–43
18. Gribnau, A. A. M., Schoenmakers, J. G. G., Van Kraaikamp, M., Hilak, M., Bloemendal, H. 1970. *Biochim. Biophys. Acta* 224:55–62
19. Gribnau, A. A. M., Schoenmakers, J. G. G., Bloemendal, H. 1969. *Arch. Biochem. Biophys.* 130:48–52
20. Gagnon, C., De Lamirande, G. 1973. *Biochem. Biophys. Res. Commun.* 51:580–86
21. Blobel, G., Potter, V. R. 1967. *J. Mol. Biol.* 28:539–48
22. Gerlinger, P., Le Meur, M. A., Clavert, J., Ebel, J. P. 1973. *Biochimie* 55:297–307
23. Schmidt, G. 1957. *Meth. Enzymol.* 3:687–91
24. Noll, H., Stutz, E. 1968. *Meth. Enzymol.* 12B:129–55
25. Kirby, K. S. 1956. *Biochem. J.* 64:405–12
26. Kirby, K. S. 1964. *Prog. Nucl. Acid Res. Mol. Biol.* 3:1–31
27. Hiatt, H. H. 1962. *J. Mol. Biol.* 5:217–28
28. Muramatsu, M. 1973. *Meth. Cell Biol.* 7:23–51
29. Parish, J. H., Kirby, K. S. 1966. *Biochim. Biophys. Acta* 129:554–62
30. Suzuki, Y., Brown, D. D. 1972. *J. Mol. Biol.* 63:409–18
31. Wiegers, U., Hilz, H. 1971. *Biochem. Biophys. Res. Commun.* 44:513–19
32. Mach, B., Faust, C., Vassalli, P. 1973. *Proc. Natl. Acad. Sci. USA* 70:451–55
33. Brawerman, G., Mendecki, J., Lee, S. Y. 1972. *Biochemistry* 11:637–44
34. Brawerman, G. 1973. *Meth. Cell Biol.* 7:1–22
35. Perry, R. P., La Torre, J., Kelley, D. E., Greenberg, J. R. 1972. *Biochim. Biophys. Acta* 262:220–26
36. Brawerman, G. 1976. *Prog. Nucl. Acid Res. Mol. Biol.* 17:117–48
37. Marmur, J. 1963. *Meth. Enzymol.* 6:726–38
38. Wagner, E. K., Katz, L., Penman, S. 1967. *Biochem. Biophys. Res. Commun.* 28:152–59
39. Girard, M. 1967. *Meth. Enzymol.* 12A:581–88
40. Kirby, K. S. 1962. *Biochim. Biophys. Acta* 55:545–51
41. Kirby, K. S. 1965. *Biochem. J.* 96:266–69
42. Kern, H. 1975. *Anal. Biochem.* 67:147–56
43. Palmiter, R. D. 1974. *Biochemistry* 13:3606–15
44. Markov, G. G., Arion, V. J. 1973. *Eur. J. Biochem.* 35:186–200
45. MacNaughton, M. C., Freeman, K. B., Bishop, J. O. 1974. *Cell* 1:117–25
46. Haines, M. E., Carey, N. H., Palmiter, R. D. 1974. *Eur. J. Biochem.* 43:549–60
47. Longacre, S. S., Rutter, W. J. 1977. *J. Biol. Chem.* 252:2742–52
48. Helmkamp, G. K., Ts'o, P. O. P. 1961. *J. Am. Chem. Soc.* 83:138–42
49. Holmes, D. S., Bonner, J. 1973. *Biochemistry* 12:2330–38
50. Fedoroff, N., Wellauer, P. K., Wall, R. 1977. *Cell* 10:597–610
51. Rhoads, R. E. 1975. *J. Biol. Chem.* 250:8088–97
52. Cox, R. A. 1968. *Meth. Enzymol.* 12B:120–29
53. Deeley, R. G., Gordon, J. I., Burns, A. T. H., Mullinix, K. P., Binastein, M., Goldberger, R. F. 1977. *J. Biol. Chem.* 252:8310–19

54. Kruh, J. 1967. *Meth. Enzymol.* 12A: 609–13
55. Schimke, R. T., Palacios, R., Sullivan, D., Kiely, M. L., Gonzalez, C., Taylor, J. M. 1974. *Meth. Enzymol.* 30:631–48
56. Glisin, V., Crkvenjakov, R., Byus, C. 1974. *Biochemistry* 13:2633–37
57. Darnell, J. E., Jelinek, W. R., Molloy, G. R. 1973. *Science* 181:1215–23
58. Molloy, G., Puckett, L. 1976. *Prog. Biophys. Mol. Biol.* 31:1–38
59. Edmonds, M. 1971. In *Procedures in Nucleic Acid Research,* ed. G. L. Cantoni, D. R. Davies, 2:629–40. New York: Harper & Row. 924 pp.
60. Astell, C., Smith, M. 1971. *J. Biol. Chem.* 246:1944–46
61. Nakazato, H., Edmonds, M. 1972. *J. Biol. Chem.* 247:3365–67
62. Aviv, H., Leder, P. 1972. *Proc. Natl. Acad. Sci. USA* 69:1408–12
63. Lindberg, U., Persson, T. 1972. *Eur. J. Biochem.* 31:246–54
64. Adesnik, M., Salditt, M., Thomas, W., Darnell, J. E. 1972. *J. Mol. Biol.* 71:21–30
65. Lee, S. Y., Mendecki, J., Brawerman, G. 1971. *Proc. Natl. Acad. Sci. USA* 68:1331–36
66. Brawerman, G., Mendecki, J., Lee, S. Y. 1972. *Biochemistry* 11:637–41
67. Kitos, P. A., Saxon, G., Amos, H. 1972. *Biochem. Biophys. Res. Commun.* 47: 1426–33
68. Schutz, G., Beato, M., Feigelson, P. 1972. *Biochem. Biophys. Res. Commun.* 49:680–89
69. DeLarco, J., Guroff, G. 1973. *Biochem. Biophys. Res. Commun.* 50:486–93
70. Sullivan, N., Roberts, W. K. 1973. *Biochemistry* 12:2395–403
71. Astell, C. R., Doel, M. T., Jahnke, P. A., Smith, M. 1973. *Biochemistry* 12: 5068–74
72. Bantle, J. A., Maxwell, I. H., Hahn, W. E. 1976. *Anal. Biochem.* 72:413–27
73. Firtel, R. A., Lodish, H. F. 1973. *J. Mol. Biol.* 79:295–314
74. Sheldon, R., Jurale, C., Kates, J. 1972. *Proc. Natl. Acad. Sci. USA* 69:417–21
75. Brawerman, G. 1974. *Meth. Enzymol.* 30:605–12
76. Rosen, J. M., Woo, S. L. C., Holder, J. W., Means, A. R., O'Malley, B. W. 1975. *Biochemistry* 14:69–78
77. Gorski, J., Morrison, M. R., Merkel, C. G., Lingrel, J. B. 1974. *J. Mol. Biol.* 86:363–71
78. Kitos, P. A., Amos, H. 1973. *Biochemistry* 12:5086–91
79. Shapiro, D. J., Schimke, R. T. 1975. *J. Biol. Chem.* 250:1759–64

80. Murphy, W., Attardi, G. 1973. *Proc. Natl. Acad. Sci. USA* 70:115–19
81. Taylor, J. M., Tse, T. P. H. 1976. *J. Biol. Chem.* 251:7461–67
82. Lingrel, J. B. 1972. In *Methods in Molecular Biology,* ed. J. Last, A. Laskin, 2:231–63. New York: 336 pp.
83. Schimke, R. T., Rhoads, R. E., McKnight, G. S. 1974. *Meth. Enzymol.* 30:694–701
84. Villa-Komaroff, L., McDowell, M. J., Baltimore, D., Lodish, H. F. 1974. *Meth. Enzymol.* 30:709–23
85. Woodward, W. R., Ivey, J. L., Herbert, E. 1974. *Meth. Enzymol.* 30:724–31
86. Palmiter, R. D. 1973. *J. Biol. Chem.* 248:2095–2106
87. Pelham, H. R. B., Jackson, R. J. 1976. *Eur. J. Biochem.* 67:247–56
88. Marcus, A. 1970. *J. Biol. Chem.* 245:955–61
89. Marcus, A., Efron, D., Weeks, D. P. 1974. *Meth. Enzymol.* 30:749–54
90. Roberts, B. E., Paterson, B. M. 1973. *Proc. Natl. Acad. Sci. USA* 70:2330–34
91. Tse, T. P. H., Taylor, J. M. 1977. *J. Biol. Chem.* 252:1272–78
92. Benveniste, K., Wilczek, J., Ruggieri, A., Stern, R. 1976. *Biochemistry* 15: 830–35
93. Lodish, H. F. 1976. *Ann. Rev. Biochem.* 45:39–72
94. Mathews, M. B., Korner, A. 1970. *Eur. J. Biochem.* 17:328–38
95. Gurdon, J. B., Lane, C. D., Woodland, H. R., Marbaix, G. 1971. *Nature* 233:177–82
96. Laskey, R. A., Mills, A. D., Gurdon, J. B., Partington, G. A. 1977. *Cell* 11:345–51
97. Gurdon, J. B., Lingrel, J. B., Marbaix, G. 1973. *J. Mol. Biol.* 80:539–51
98. Ross, J., Aviv, H., Scolnick, E., Leder, P. 1972. *Proc. Natl. Acad. Sci. USA* 69:264–68
99. Verma, I. M., Temple, G. F., Fan, H., Baltimore, D. 1972. *Nature New Biol.* 235:163–67
100. Kacian, D. L., Spiegelman, S., Bank, A., Terada, M., Metafora, S., Dow, L., Marks, P. A. 1972. *Nature New Biol.* 235:167–69
101. Faust, C. H., Diggelman, H., Mach, B. 1973. *Biochemistry* 12:925–31
102. Efstratiadis, A., Maniatis, T., Kafatos, F. C., Jeffrey, A., Vournakis, J. N. 1975. *Cell* 4:367–76
103. Monahan, J. J., Harris, S. E., Woo, S. L. C., Robberson, D. L., O'Malley, B. W. 1976. *Biochemistry* 15:223–33
104. Weiss, G. B., Wilson, G. N., Steggles,

A. W., Anderson, W. F. 1976. *J. Biol. Chem.* 251:3425–31
105. Friedman, E. Y., Rosbash, M. 1977. *Nucleic Acids Res.* 4:3455–71
106. Rothenberg, E., Baltimore, D. 1977. *J. Virol.* 21:168–78
107. Buell, G. N., Wickens, M. P., Payvar, F., Schimke, R. T. 1978. *J. Biol. Chem.* 253:2471–82
108. Temin, H. M., Baltimore, D. 1974. *Adv. Virus Res.* 17:129–87
109. Kacian, D. L., Myers, J. C. 1976. *Proc. Natl. Acad. Sci. USA* 73:2191–95
110. McCarty, K. S. Jr., Vollmer, R. T., McCarty, K. S. 1974. *Anal. Biochem.* 61:165–83
111. Studier, F. W. 1965. *J. Mol. Biol.* 11:373–90
112. Wetmur, J. G., Davidson, N. 1968. *J. Mol. Biol.* 31:349–70
113. Bishop, J. O. 1972. *Acta Endocrinol. Suppl.* 161:247–76
114. Hutton, J. R., Wetmur, J. G. 1973. *J. Mol. Biol.* 77:495–500
115. Britten, R. J., Graham, D. E., Neufeld, B. R. 1974. *Meth. Enzymol.* 29:363–418
116. Bishop, J. O., Beckmann, J. S., Campo, M. S., Hastie, N. D., Izquierdo, M., Perlman, S. 1975. *Phil. Trans. R. Soc. London B* 272:147–57
117. Weiss, G. B., Wilson, G. N., Steggles, A. W., Anderson, W. F. 1976. *J. Biol. Chem.* 251:3425–31
118. Hastie, N. D., Bishop, J. O. 1976. *Cell* 9:761–74
119. Chamberlin, M. E., Galan, G. A., Britten, R. J., Davidson, E. H. 1978. *Nucleic Acids Res.* 5:2073–94
120. McConaughy, B. L., Laird, C. D., McCarthy, B. J. 1969. *Biochemistry* 8:3289–95
121. Hutton, J. R. 1977. *Nucleic Acids Res.* 10:3537–55
122. Vogt, V. M. 1973. *Eur. J. Biochem.* 33:192–200
123. Wiegand, R. C., Godson, G. N., Radding, C. M. 1975. *J. Biol. Chem.* 250:8848–55
124. Marbaix, G., Burny, A. 1964. *Biochem. Biophys. Res. Commun.* 4:522–28
125. Burny, A., Marbaix, G. 1965. *Biochem. Biophys. Acta* 103:409–16
126. Marbaix, G., Burny, A., Huez, G., Chantrenne, H. 1966. *Biochim. Biophys. Acta* 114:404–6
127. Chantrenne, H., Burny, A., Marbaix, G. 1967. *Progr. Nucleic Acid Res. Mol. Biol.* 7:173–94
128. Lingrel, J. B., Lockard, R. E., Jones, R. F., Burr, H. E., Holder, J. W. 1971. *Ser. Haematol.* 4:37–69

129. Lockard, R. E., Lingrel, J. B. 1969. *Biochem. Biophys. Res. Commun.* 37:204–12
130. Lane, C. D., Marbaix, G., Gurdon, J. B. 1971. *J. Mol. Biol.* 61:73–91
131. Krystosek, A., Cawthon, M. L., Kabat, D. 1975. *J. Biol. Chem.* 250:6077–84
132. Kabat, D. 1975. *J. Biol. Chem.* 250:6085–92
133. Gaskill, P., Kabat, D. 1971. *Proc. Natl. Acad. Sci. USA* 68:72–75
134. Lanyon, W. G., Paul, J., Williamson, R. 1972. *Eur. J. Biochem.* 31:38–43
135. Kazazian, H. H. Jr., Moore, P. A., Snyder, P. G. 1973. *Biochem. Biophys. Res. Commun.* 51:564–71
136. Morrison, M. R., Brinkley, S. A., Gorski, J., Lingrel, J. B. 1974. *J. Biol. Chem.* 249:5290–95
137. Orkin, S. A., Swan, D., Leder, P. 1975. *J. Biol. Chem.* 250:8753–60
138. Nudel, U., Ramirez, F., Marks, P. A., Bank, A. 1977. *J. Biol. Chem.* 252:2182–86
139. Morel, C., Kayibanda, B., Scherrer, K. 1971. *FEBS Lett.* 18:84–88
140. Spohr, G., Kayibanda, B., Scherrer, K. 1972. *Eur. J. Biochem.* 31:194–208
141. Gander, E. S., Stewart, A. G., Morel, C. M., Scherrer, K. 1973. *Eur. J. Biochem.* 38:443–52
142. Nudel, U., Lebeu, B., Zehavi-Willner, T., Revel, M. 1973. *Eur. J. Biochem.* 33:314–22
143. Huynh-Van-Tan, Schapira, G. 1978. *Eur. J. Biochem.* 85:271–81
144. Preobrazhensky, A. A., Spirin, A. S. 1978. *Prog. Nucleic Acid Res. Mol. Biol.* 21:1–38
145. Huez, G., Burny, A., Marbaix, G., Lebleu, B. 1967. *Biochim. Biophys. Acta* 145:629–38
146. Lebleu, B. 1974. *Meth. Enzymol.* 30:613–21
147. Blobel, G. 1971. *Proc. Natl. Acad. Sci. USA* 68:832–35
148. Lindberg, U., Sundquist, B. 1974. *J. Mol. Biol.* 86:451–68
149. Irwin, D., Kumar, A., Malt, R. A. 1975. *Cell* 4:157–65
150. Kumar, A., Pederson, T. 1975. *J. Mol. Biol.* 96:353–65
151. Spector, A., Li, L. K., Augusteyn, R. C., Schneider, A., Freund, T. 1971. *Biochem. J.* 124:337–46
152. Bloemendal, H., Berns, T. J. M., Zweers, A., Hoenders, H., Benedetti, E. L. 1972. *Eur. J. Biochem.* 24:401–10
153. Spector, A., Kinoshita, J. H. 1964. *Invest. Ophthalmol.* 3:517–22
154. Berns, A. J. M., De Abreu, R. A., Van Kraaikamp, M., Benedetti, E. L., Bloe-

mendal, H. 1971. *FEBS Lett.* 18:159–63
155. Chen, J. H., Lavers, G. C., Spector, A. 1973. *Biochem. Biophys. Res. Commun.* 52:767–73
156. Chen, J. H., Lavers, G. C., Spector, A. 1976. *Biochem. Biophys. Acta* 418:39–51
157. Berns, A. J. M., Bloemendal, H. 1974. *Meth. Enzymol.* 30:675–94
158. Chen, J. H., Spector, A. 1977. *Biochemistry* 16:499–505
159. Cohen, L. H., Westerhuis, L. W., Smits, D. P., Bloemendal, H. 1978. *Eur. J. Biochem.* 89:251–58
160. Zelenka, P., Piatigorsky, J. 1976. *Exp. Eye Res.* 22:115–24
161. Mathews, M. B., Osborn, M., Berns, A. J. M., Bloemendal, H. 1972. *Nature New Biol.* 236:5–7
162. Chen, J. H., Spector, A. 1977. *Proc. Natl. Acad. Sci. USA* 74:5448–52
163. Gedamu, L., Dixon, G. H. 1976. *J. Biol. Chem.* 251:1455–63
164. Iatrou, K., Dixon, G. H. 1977. *Cell* 10:433–41
165. Gedamu, L., Iatrou, K., Dixon, G. H. 1978. *Biochem. J.* 171:589–99
166. Gedamu, L., Dixon, G. H., Davies, P. L. 1977. *Biochemistry* 16:1383–91
167. Gedamu, L., Iatrou, K., Dixon, G. H. 1977. *Cell* 10:443–51
168. Kedes, L. H. 1976. *Cell* 8:321–31
169. Borun, T. W., Scharff, M. D., Robbins, E. 1967. *Proc. Natl. Acad. Sci. USA* 58:1977–83
170. Nemer, M., Lindsay, D. T. 1969. *Biochem. Biophys. Res. Commun.* 35:156–60
171. Kedes, L. H., Gross, P. R. 1969. *J. Mol. Biol.* 42:559–75
172. Kedes, L. H., Gross, P. R., Cognetti, G., Hunter, A. L. 1969. *J. Mol. Biol.* 45:337–51
173. Breindl, M., Gallwitz, D. 1973. *Eur. J. Biochem.* 32:381–91
174. Jacobs-Lorena, M., Baglioni, C., Borun, T. W. 1972. *Proc. Natl. Acad. Sci. USA* 69:2095–99
175. Bos, E., Roskam, W., Gallwitz, D. 1976. *Biochem. Biophys. Res. Commun.* 73:404–10
176. Woods, D., Fitschen, W. 1977. *Nucleic Acids Res.* 4:3187–98
177. Grunstein, M., Levy, S., Schedl, P., Kedes, L. H. 1973. *Cold Spring Harbor Symp. Quant. Biol.* 38:717–24
178. Gross, K., Probst, E., Schaffner, W., Birnstiel, M. 1976. *Cell* 8:455–69
179. Stephens, R. E., Pan, C. J., Ajiro, K., Dolby, T. W., Borun, T. W. 1977. *J. Biol. Chem.* 252:166–72

180. Borun, T. W., Ajiro, K., Zweidler, A., Dolby, T. W., Stephens, R. E. 1977. *J. Biol. Chem.* 252:173–80
181. Schochetman, G., Perry, R. P. 1972. *J. Mol. Biol.* 63:591–96
182. Adesnik, M., Darnell, J. E. 1972. *J. Mol. Biol.* 67:397–406
183. Thrall, C. L., Park, W. D., Rashba, H. W., Stein, J. L., Maw, R. J., Stein, G. S. 1974. *Biochem. Biophys. Res. Commun.* 61:1443–49
184. Ruderman, J. V., Pardue, M. L. 1978. *J. Biol. Chem.* 253:2018–25
185. Palacios, R., Palmiter, R. D., Schimke, R. T. 1972. *J. Biol. Chem.* 247:2316–21
186. Schimke, R. T., McKnight, G. S., Shapiro, D. J., Sullivan, D., Palacios, R. 1975. *Recent Prog. Horm. Res.* 31:175–211
187. O'Malley, B. W., Means, A. R. 1974. *Science* 183:610–20
188. Woo, S. L. C., Rosen, J. M., Liarakos, C. D., Choi, Y. C., Busch, H., Means, A. R., O'Malley, B. W., Robberson, D. L. 1975. *J. Biol. Chem.* 250:7027–39
189. Monahan, J. J., Harris, S. E., O'Malley, B. W. 1976. *J. Biol. Chem.* 251:3738–48
190. Rosen, J. M., Woo, S. L. C., Means, A. R., O'Malley, B. W. 1976. *Meth. Mol. Biol.* 8:369–434
191. Rosen, J. M. 1976. *Biochemistry* 15:5263–71
192. Rosen, J. M., Woo, S. L. C., Comstock, J. P. 1975. *Biochemistry* 14:2895–3001
193. Buell, G. N., Wickens, M. P., Payvar, F., Schimke, R. T. 1978. *J. Biol. Chem.* 253:2471–82
194. Brown, D. D., Suzuki, Y. 1974. *Meth. Enzymol.* 30:648–54
195. Heywood, S. M., Kennedy, D., Bester, A. J. 1974. *FEBS Lett.* 53:69–72
196. Robbins, J., Heywood, S. M. 1978. *Eur. J. Biochem.* 82:601–8
197. Patrinou-Georgoulas, M., John, H. A. 1977. *Cell* 12:491–99
198. Rowe, D. W., Moen, R. C., Davidson, J. M., Byers, P. H., Bornstein, P., Palmiter, R. D. 1978. *Biochemistry* 17:1581–90
199. Boedtker, H., Frischauf, A. M., Lehrach, H. 1976. *Biochemistry* 15:4765–70
200. Frischauf, A. M., Lehrach, H., Rosner, C., Boedtker, H. 1978. *Biochemistry* 17:3243–49
201. Morris, G. E., Buzash, E. A., Rourke, A. W., Tepperman, K., Thompson, W. C., Heywood, S. M. 1972. *Cold Spring Harbor Symp. Quant. Biol.* 37:535–41
202. Petrovic, S., Novakovic, M., Petrovic, J. J. 1971. *Biochim. Biophys. Acta* 254:493–95

203. Petrovic, S. L., Novakovic, M. B., Petrovic, J. J. 1975. *Biopolymers* 14: 1905–13
204. Zeichner, M., Stern, R. 1977. *Biochemistry* 16:1378–82
205. Clemens, M. J. 1974. *Progr. Biophys. Mol. Biol.* 28:69–108
206. Follet, B. K., Redshaw, M. R. 1974. In *Physiology of the Amphibia,* ed. B. Lofts, 2:219–38. New York: Academic. 458 pp.
207. Shapiro, D. J., Baker, H. J. 1977. *J. Biol. Chem.* 252:5244–50
208. Palmiter, R. D., Palacios, R., Schimke, R. T. 1972. *J. Biol. Chem.* 247:3296–304
209. Palacios, R., Schimke, R. T. 1973. *J. Biol. Chem.* 248:1424–30
210. Palacios, R., Sullivan, D., Summers, N. M., Kiely, M. L., Schimke, R. T. 1973. *J. Biol. Chem.* 248:540–48
211. Shapiro, D. J., Taylor, J. M., McKnight, G. S., Palacios, R., Gonzalez, C., Kiely, M. L., Schimke, R. T. 1974. *J. Biol. Chem.* 249:3665–71
212. Wettstein, F. O., Staehelin, T., Noll, H. 1963. *Nature* 197:430–35
213. Blobel, G., Potter, V. R. 1967. *J. Mol. Biol.* 26:279–92
214. Taylor, J. M., Schimke, R. T. 1974. *J. Biol. Chem.* 248:7661–68
215. Dissous, C., Lempereur, C., Verwaerde, C., Krembel, J. 1978. *Eur. J. Biochem.* 83:17–27
216. Blobel, G., Potter, V. R. 1967. *J. Mol. Biol.* 28:539–44
217. Shore, G., Tata, J. R. 1977. *J. Cell Biol.* 72:726–43
218. Mechler, B., Vassalli, P. 1975. *J. Cell Biol.* 67:1–15
219. Ramsey, J. C., Steele, W. J. 1976. *Biochemistry* 15:1704–12
220. Boyer, S. H., Smith, K. D., Noyes, A. N. 1974. *Ann. NY Acad. Sci.* 241: 204–22
221. Bouma, H. III, Kwan, S. W., Fuller, G. M. 1975. *Biochemistry* 14:4787–92
222. Schechter, I. 1973. *Proc. Natl. Acad. Sci. USA* 70:2256–60
223. Schechter, I. 1974. *Biochemistry* 13:1875–85
224. Keller, G. H., Taylor, J. M. 1976. *J. Biol. Chem.* 251:3768–73
225. Burstein, Y., Kantor, F., Schechter, I. 1976. *Proc. Natl. Acad. Sci. USA* 73:2604–8
226. Legler, M. K., Cohen, E. P. 1976. *Biochemistry* 15:4390–99
227. Houdebine, L. M., Gaye, P. 1976. *Eur. J. Biochem.* 63:9–14
228. Jost, J. P., Pehling, G. 1976. *Eur. J. Biochem.* 66:339–46

229. Pawlowski, P. J., Gillette, M. T., Martinell, J., Lukens, L. N., Furthmayr, H. 1975. *J. Biol. Chem.* 250:2135–42
230. Flick, P. K., Chen, J., Alberts, A. W., Vagelos, P. R. 1978. *Proc. Natl. Acad. Sci. USA* 75:730–34
231. Ono, M., Kondo, T., Kawakami, M. 1977. *J. Biochem.* 81:941–47
232. Ono, M., Kondo, T., Kawakami, M., Honjo, T. 1977. *J. Biochem.* 81:949–54
233. Scott, A. C., Wells, J. R. E. 1975. *Biochem. Biophys. Res. Commun.* 64:448–55
234. Innis, M. A., Miller, D. L. 1977. *J. Biol. Chem.* 252:8469–9475
235. Okuyama, A., McInnes, J., Green, M., Pestka, S. 1978. *Arch. Biochem. Biophys.* 188:98–104
236. Schutz, G., Kieval, S., Groner, B., Sippel, A. E., Kurtz, D. T., Feigelson, P. 1977. *Nucleic Acids Res.* 4:71–84
237. Groner, B., Hynes, N. E., Sippel, A. E., Jeep, S., Huu, M. C. N., Schutz, G. 1977. *J. Biol. Chem.* 252:6666–74
238. Mueller-Lantzsch, N., Fan, H. 1976. *Cell* 9:579–88
239. Jonsson, S., Kronvall, G. 1974. *Eur. J. Immunol.* 4:29–35
240. Eschenfeldt, W. H., Patterson, R. J. 1975. *Biochem. Biophys. Res. Commun.* 67:935–45
241. Delovitch, T. L., Davis, B. K., Holme, G., Sehon, A. H. 1972. *J. Mol. Biol.* 69:373–86
242. Delovitch, T. L., Boyd, S. L., Tsay, H. M., Holme, G., Sehon, A. H. 1973. *Biochim. Biophys. Acta* 299:621–33
243. Ikehara, Y., Pitot, H. C. 1973. *J. Cell. Biol.* 59:28–39
244. Stavnezer, J., Huang, R. C. C. 1971. *Nature New Biol.* 230:172–76
245. Stavnezer, J., Huang, R. C. C., Stavnezer, E., Bishop, J. M. 1974. *J. Mol. Biol.* 88:43–63
246. Swan, D., Aviv, H., Leder, P. 1972. *Proc. Natl. Acad. Sci. USA* 69:1967–71
247. Brownlee, G. G., Cartwright, E. M., Cowan, N. J., Jarvis, J. M., Milstein, C. 1973. *Nature New Biol.* 244:236–40
248. Tonegawa, S., Steinberg, C., Dube, S., Bernardini, A. 1974. *Proc. Natl. Acad. Sci. USA* 71:4027–31
249. Farace, M. G., Allen, M. F., Briand, P. A., Faust, C. H., Vassalli, P., Mach, B. 1976. *Proc. Natl. Acad. Sci. USA* 73:727–31
250. Tonegawa, S. 1976. *Proc. Natl. Acad. Sci. USA* 73:203–7
251. Honjo, T., Packman, S., Swan, D., Nan, M., Leder, P. 1974. *Proc. Natl. Acad. Sci. USA* 71:3659–63

252. Honjo, T., Packman, S., Swan, D., Leder, P. 1976. *Biochemistry* 15:2780–85

253. Smith, M. M., Huang, R. C. C. 1976. *Proc. Natl. Acad. Sci. USA* 73:775–79

254. Storb, U., Hager, L., Wilson, R., Putnam, D. 1977. *Biochemistry* 16:5432–38

255. Rowe, D. W., Moen, R. C., Davidson, J. M., Byers, P. H., Bornstein, P., Palmiter, R. D. 1978. *Biochemistry* 17:1581–90

256. Strair, R. K., Yap, S. H., Shafritz, D. A. 1977. *Proc. Natl. Acad. Sci. USA* 74:4346–50

257. Marcu, K. B., Valbuena, O., Perry, R. P. 1978. *Biochemistry* 17:1723–33

258. Green, M., Zehavi-Willner, T., Graves, P. N., McInnes, J., Pestka, S. 1976. *Arch. Biochem. Biophys.* 172:74–89

259. Cowan, N. J., Secher, D. S., Milstein, C. 1976. *Eur. J. Biochem.* 61:355–68

260. Stevens, R. A., Williamson, A. R. 1975. *Proc. Natl. Acad. Sci. USA* 72:4679

261. Kurtz, D. T. Feigelson, P. 1977. *Proc. Natl. Acad. Sci. USA* 74:4791–95

262. Alt, F. W., Kellems, R. E., Bertino, J. R., Schimke, R. T. 1978. *J. Biol. Chem.* 253:1357–70

263. Venetianer, P., Leder, P. 1974. *Proc. Natl. Acad. Sci. USA* 71:3892–95

264. Levy, S., Aviv, H. 1976. *Biochemistry* 15:1844–47

265. Wood, T. G., Lingrel, J. B. 1977. *J. Biol. Chem.* 252:457–63

266. Smith, K., Rosteck, P. Jr., Lingrel, J. B. 1978. *Nucleic Acids Res.* 5:105–15

267. Ross, J. 1978. *J. Mol. Biol.* 119:21–35

268. Rhoads, R. E., Hellmann, G. M. 1978. *J. Biol. Chem.* 253:1687–93

269. Anderson, J. N., Schimke, R. T. 1976. *Cell* 7:331–38

270. Hirsch, F. W., Nall, K. N., Spohn, W. H., Busch, H. 1978. *Proc. Natl. Acad. Sci. USA* 75:1736–39

271. Woo, S. L. C., Monahan, J. J., O'Malley, B. W. 1977. *J. Biol. Chem.* 252:5789–97

272. Shih, T. Y., Martin, M. A. 1974. *Biochemistry* 13:3411–18

273. Noyes, B. E., Stark, G. R. 1975. *Cell* 5:301–10

274. Seeburg, P. H., Shine, J., Martial, J. A., Ullrich, A., Baxter, J. D., Goodman, H. M. 1977. *Cell* 12:157–65

275. Ullrich, A., Shine, J., Chirgwin, J., Pictet, R., Tischer, E., Rutter, W. J., Goodman, H. M. 1977. *Science* 186:1313–19

276. Seeburg, P. H., Shine, J. S., Martial, J. A., Baxter, J. D., Goodman, H. M. 1977. *Nature* 270:486–94

277. Maxam, A. M., Gilbert, W. 1977. *Proc. Natl. Acad. Sci. USA* 74:560–64

278. Paterson, B. M., Roberts, B. E., Kuff, E. L. 1977. *Proc. Natl. Acad. Sci. USA* 74:4370–74

279. Shine, J., Seeburg, P. H., Martial, J. A., Baxter, J. D., Goodman, H. M. 1977. *Nature* 270:494–99

280. Howard, P. H., Adams, S. L., Sobel, M. E., Pastan, I., deCrombrugghe, B. 1978. *J. Biol. Chem.* 253:5869–74

281. Lewis, J. B., Atkins, J. F., Anderson, C. W., Baum, P. R., Gesteland, R. F. 1975. *Proc. Natl. Acad. Sci. USA* 72:1344–48

282. Marotta, C. A., Lebowitz, P., Dhar, R., Zain, B. S., Weissman, S. M. 1974. *Meth. Enzymol.* 29:254–72

283. Szostak, J. W., Stiles, J. J., Bahl, C. P., Wu, R. 1977. *Nature* 265:61–63

284. Montgomery, D. L., Hall, B. D., Gillam, S., Smith, M. 1978. *Cell* 14:673–80

285. Rabbitts, T. H. 1977. *Immunol. Rev.* 36:29–50

286. Zamecnik, P. C., Stephenson, M. L. 1978. *Proc. Natl. Acad. Sci. USA* 75:280–84

287. Baralle, F. E. 1977. *Cell* 10:549–58

288. Baralle, F. E. 1977. *Cell* 12:1085–95

289. Wu, R. 1972. *Nature* 236:198–200

290. Borer, P. N., Dengler, B., Tinoco, I. Jr., Uhlenbeck, O. C. 1974. *J. Mol. Biol.* 86:843–53

291. Boedtker, H. 1971. *Biochim. Biophys. Acta* 240:448–53

292. Pinder, J. C., Staynov, D. Z., Gratzer, W. B. 1974. *Biochemistry* 13:5373–78

293. Bailey, J. M., Davidson, N. 1976. *Anal. Biochem.* 70:75–85

294. Lehrach, H., Diamond, D., Wozney, J. M., Boedtker, H. 1977. *Biochemistry* 16:4743–51

295. Robberson, D., Aloni, Y., Attardi, G., Davidson, N. 1971. *J. Mol. Biol.* 60:473–84

296. Woo, S. L. C., Rosen, J. M., Liarakos, C. D., Choi, Y. C., Busch, H., Means, A. R., O'Malley, B. W., Robberson, D. L. 1975. *J. Biol. Chem.* 250:7027–39

297. Wahli, W., Wyler, T., Weber, R., Ryffel, G. U. 1976. *Eur. J. Biochem.* 66:457–65

298. Wahli, W., Wyler, T., Weber, R., Ryffel, G. U. 1978. *Eur. J. Biochem.* 86:225–34

299. Edmonds, M., Vaughan, M. H. Jr., Nakazato, H. 1971. *Proc. Natl. Acad. Sci. USA* 68:1336–40

300. Darnell, J. E., Philipson, L., Wall, R., Adesnik, M. 1971. *Science* 174:507–10

301. Morrison, M. R., Merkel, C. G., Lingrel, J. B. 1973. *Mol. Biol. Rep.* 1:55–60

302. Bishop, J. O., Rosbash, M., Evans, D. 1974. *J. Mol. Biol.* 85:75–86

303. Rosbash, M., Ford, P. J. 1974. *J. Mol. Biol.* 85:87–101
304. Kaufman, S. J., Gross, K. W. 1975. *Biochim. Biophys. Acta* 353:133–45
305. Bishop, J. O., Morton, J. G., Rosbash, M., Richardson, M. 1974. *Nature* 250:199–204
306. Edmonds, M., Winters, M. A. 1976. *Prog. Nucleic Acid Res. Mol. Biol.* 17:149–79
307. Smyth, D. G. 1967. *Meth. Enzymol.* 11:214–18
308. Schroeder, W. A. 1967. *Meth. Enzymol.* 11:351–59
309. Monahan, J. J., Harris, S. E., O'Malley, B. W. 1977. In *Receptors and Hormone Action,* ed. B. W. O'Malley, L. Birnbaumer, 1:298–329. New York: Academic. 580 pp.
310. Wetmur, J. G. 1971. *Biopolymers* 10:601–13
311. Beemon, K. L. 1978. *Curr. Top. Microbiol. Immunol.* 79:73–110
312. Commerford, S. L. 1971. *Biochemistry*
10:1993–99
313. Prensky, W., Steffensen, D. M., Hughes, W. L. 1973. *Proc. Natl. Acad. Sci. USA* 70:1860–64
314. Robertson, H. D., Dickson, E., Model, P., Prensky, W. 1973. *Proc. Natl. Acad. Sci. USA* 70:3260–64
315. Frisby, D. 1977. *Nucleic Acids Res.* 4:2975–96
316. Brownlee, G. G., Sanger, F. 1969. *Eur. J. Biochem.* 11:395–99
317. Curtis, P. J., Weissmann, C. 1976. *J. Mol. Biol.* 106:1061–75
318. Curtis, P. J., Mantei, N., Weissmann, C. 1978. *Cold Spring Harbor Symp. Quant. Biol.* 42:971–84
319. Kwan, S. P., Wood, T. G., Lingrel, J. B. 1977. *Proc. Natl. Acad. Sci. USA* 74:178–82
320. Ross, J., Knecht, D. A. 1978. *J. Mol. Biol.* 119:1–20
321. Tilghman, S. M., Curtis, P. J., Tiemeier, D. C., Leder, P., Weissmann, C. 1978. *Proc. Natl. Acad. Sci. USA* 75:1309–13

Ann. Rev. Biochem. 1979. 48:719–54

THE STRUCTURE AND FUNCTION OF EUKARYOTIC RIBOSOMES

♦12024

Ira G. Wool

Department of Biochemistry, University of Chicago, Chicago, Illinois 60637

CONTENTS

PERSPECTIVES AND SUMMARY

The grand design in research on ribosomes is to know the structure and the activity of the organelle: to identify the precise location, and to define the exact function, of each of the molecules in the particle. A great deal is

719

0066-4154/79/0701-0719$01.00

known of prokaryotic ribosomes, especially *Escherichia coli* (1–9); far less is known of eukaryotic particles. In the circumstances it is not irrelevant (nor irreverent) to ask the importance of research on the latter. In fact, good reasons exist for attending to them. There are significant differences in the structure of eukaryotic and prokaryotic ribosomes that form the basis of what one might call the central dilemma. Eukaryotic ribosomes are appreciably larger than prokaryotic ones: they contain a greater number of proteins, about 80 rather than 53, and they have an extra molecule of RNA. In addition, the proteins and nucleic acids are, on the average, larger. The difference in size is a paradox since eukaryotic ribosomes perform the same general function, namely to catalyze the synthesis of protein. Moreover, they employ appreciably the same partial reactions. What then was the evolutionary pressure for the accretion of the extra proteins and for the increase in size? The question implies that there are functions of eukaryotic ribosomes to be discovered.

Some of the extra ribosomal proteins may be specialized for interaction with receptors in the endoplasmic reticulum (10–11) called ribophorins I and II (12–12b), others may be involved in regulation, especially the regulation of the translation of messenger RNA, although these functions are not necessarily unique to eukaryotic ribosomes. The particles may have cement proteins needed only to create or maintain their structure, proteins that do not participate in peptide synthesis. The need for them may derive from the peculiarities of the biogenesis of the organelle (13, 14). Ribosomal proteins are synthesized in the cytoplasm and transported to the nucleolus where they are assembled into subunits, probably by attachment to a large nascent precursor rRNA while the latter is still being transcribed from ribosomal DNA. The precursor has to be processed and 5S rRNA, which is synthesized at another site in the nucleus, has to be incorporated into the large subparticle. The two subunits have to be transported from the nucleus to the cytoplasm. The extra proteins may participate in this extraordinarily complicated process, with its need for two-way traffic between the cytoplasm and nucleus: they could take part in assembly or in the transport of assembled subunits.

Mitochondrial and chloroplast ribosomal proteins are almost certainly distinct subsets (15, 16). Moreover, most, if not all, of the organelle ribosomal proteins are encoded in nuclear DNA and synthesized on cytoplasmic ribosomes. Thus in eukaryotic cells having mitochondria and chloroplasts, as many as 200 ribosomal proteins need to be synthesized in the cytoplasm. The proteins then have to be partitioned between the nucleolus, mitochondria, and chlorpolast. Exactly how that is accomplished is a mystery. It may be that all the proteins are filtered through the three organelles, and that the correct group is retained by attachment to its ribosomal RNA: each organelle does synthesize its own unique rRNA. It is possible, albeit difficult to concoct a mechanism, that the supernumerary proteins abet the partition.

Finally, although it cuts against the grain to suggest it, it is conceivable that ribosomes have luxury proteins, proteins that serve no function. There are thiostrepton-resistant mutants of *Bacillus megaterium* (16a) and *Bacillus subtilis* (16b) that lack ribosomal protein L11. While the former is sick the latter is not (16c). Thus, it follows that either *B. subtilis* L11 has no function in protein synthesis or its role can be assumed by one or more of the remaining proteins.

One can look at the problem the other way around and ask why are prokaryotic ribosomes so small? Perhaps, because they had to be streamlined. During rapid growth in rich medium as much as 30% of all cellular protein is ribosomal (17, 18) and the bacterium might have difficulty supporting a particle with the protein content of eukaryotic ribosomes. The assumptions are that ribosomes evolved only once—which seems reasonable; that prokaryotic ribosomes initially had a greater number of proteins, a number comparable to that in eukaryotic particles; and that prokaryotic ribosomes having had more opportunity to evolve responded to selective pressure by discarding proteins size of those that were retained—the advantage being a greater number of smaller ribosomes performing the same function as fewer larger particles. All that would have to have happened after divergence of primitive eukaryotes. The argument, however, does not change the fundamental nature of the problem: What prokaryotic ribosomal proteins could be dispensed with without apparent loss of function? If the extra proteins could be discarded with impunity why have they been retained by eukaryotes? It is possible that in the putative process the function of several prokaryotic ribosomal proteins were combined, as seems to have happened with the initiation factors. Eukaryotes use a greater number of initiation factors than bacteria to catalyze what appear to be similar partial reactions (18a, 18b).

The purification and characterization of the components is necessary for analysis of the structure and function of eukaryotic ribosomes. It is easy to isolate eukaryotic ribosomal nucleic acids and there is a good deal of information on their chemistry (19–23). However, until recently, pure proteins had not been prepared in appreciable amounts, although, a start had been made on their characterization [see (24) for a review and for the earlier references]. Appreciably all the proteins of rat liver ribosomes have now been isolated (25–30), and in amounts sufficient not only to characterize them, but to begin to carry out sequence studies, to raise antibodies, and to determine binding to ribosomal RNA. This review describes, first, the general characteristics of eukaryotic ribosomes and then the isolation and characterization of the proteins. It continues with an account of the identification of the proteins that bind to ribosomal RNA, an overview of the phosphorylation of ribosomal proteins, and a narration of some immunochemical studies of structure and function. A good deal, in particular the chemistry of ribosomal RNA and ribosome biogenesis and its regulation, is left out. The emphasis is on mammalian ribosomes.

GENERAL CHARACTERISTICS OF EUKARYOTIC RIBOSOMES

Eukaryotic ribosomes are composed of two subunits, which together contain four molecules of RNA and 70–80 proteins. The sedimentation coefficient of the monomer and subparticle varies between species (as do other of the characteristics) but are close to 80 for the ribosome and 60 and 40 for the subunits (31).

The small subunit of rat liver ribosomes has a sedimentation coefficient of 36.9 (Table 1; D. Blair, W. E. Hill, and I. G. Wool, unpublished data). The particle contains one molecule of 18S RNA of molecular weight 0.7×10^6 (32, 33) and about 30 proteins whose combined mass is also near 0.7×10^6 daltons (see below). Thus the molecular weight of the particle should be 1.4×10^6 which is exactly what has been measured (Table 1). The particle might be expected, then, to have equal amounts of protein and RNA, whereas the chemical determination (34) reveals it to contain only 45% RNA (Table 2). The buoyant density of the small subparticle is 1.515 g/ml, from which the percent RNA can also be calculated to be 45 (34). The values obtained by others (35, 36) in a different way are similar. The extinction coefficient for the 40S subparticles is 94.5, and the diffusion coefficient 2.0×10^{-7} (Table 1).

The large subunit of rat liver ribosomes has a sedimentation coefficient of 56.3 (Table 1) and is composed of one molecule each of 5S, 5.8S, and 28S

Table 1   Physical properties of eukaryotic ribosomes[a]

Physical properties	40S subunit	60S subunit
Molecular weight[b], M_r	1.4×10^6	2.9×10^6
Sedimentation coefficient, $S^{\circ}_{20, w}$	36.9	56.3
Extinction coefficient, $E^{1\%}_{260}$	94.5	99.5
Density increment, $(\delta\rho/\delta c)_\mu$	0.330	0.394
Diffusion coefficient, $D^{\circ}_{20, w}$ ($\times 10^7$ cm^2/sec)	2.0	1.21

[a] The unpublished preliminary data are from D. Blair, W. E. Hill, and I. G. Wool.
[b] Calculated from the Svedberg equation: $M_r = S^{\circ}_{20, w} \, RT/D^{\circ}_{20, w} \, (\delta\rho/\delta c)_\mu$

RNA, and 45–50 proteins. The molecular weight of 28S RNA is given as 1.7×10^6 (32, 33), and the molecular weights of 5S and 5.8S RNA, calculated from the sequence (37, 38), are approximately 39,000 and 51,000, respectively. The total mass of RNA in the particle is about 1.79×10^6 daltons. The particle has 45–50 proteins whose masses are approximately 1×10^6 (see below). The expected molecular weight of the particle, then, is 2.79×10^6, and the actual measurement is 2.9×10^6 (Table 2) which, given the uncertainties, is in reasonable agreement; a similar value had been obtained before (35). The 60S subunit has a buoyant density of 1.600 g/ml (Table 2); the RNA content from the buoyant density was 59.4%, and the chemical determination gave the same value (34). Thus the large subparticle has an appreciably greater proportion of RNA than the 40S subunit.

Eukaryotic ribosomes are by no means a uniform group; "80S ribosomes" actually range in aggregate mass from about 3.9 (for plants) to 4.55×10^6 daltons (for mammals) (39–41). The change in ribosome mass is the result of an increase in the size of the large subunit from 2.4 to 3.05×10^6 daltons; the increase is in the mass of both RNA and protein. The 40S particle, on the other hand, has not changed appreciably in size during eukaryotic evolution.

There are detailed electron microscopic studies of eukaryotic ribosomes (42–45). Observers (44, 45) have been struck by the general similarity in the morphology of eukaryotic and prokaryotic ribosomal subunits, despite the obviously larger size of the former. The most prominent features of rat liver ribosomes are: the division of the small subunit into two unequal parts; the assymetric apposition of the small subunit to the large in ribosome couples which confers a handedness to the structure; and a notch in the profile of the large subparticle (43). The 40S subunit approximates a curved and flattened prolate elipsoid with dimensions of $230 \times 140 \times 115$ Å (43); it is divided into "head" (one third) and "body" (two thirds) regions by a transverse partition seen in electron micrographs as a dense line 80 Å from one end of the subunit. The 60S subunit is seen either as a rounded object, or as asymetric and skiff-shaped with a pointed end opposite to a blunt end (43).

Table 2   Protein and RNA content and buoyant density of eukaryotic ribosomes

Ribosomal particle	Protein ($\mu g/A_{260}$)	RNA ($\mu g/A_{260}$)	RNA (%)	Buoyant density (gm/cm^{-3})	RNA (%)
40S	57.3	45.5	44.3	1.515	44.9
60S	36.2	52.1	59.0	1.600	59.4
80S	51.7	52.3	50.3	1.577	55.6

A particularly promising approach to the solution of the structure of eukaryotic ribosomes is through three-dimensional reconstruction from electron micrographs of naturally occurring crystalline arrays which develop in the oocytes of lizards during hibernation (46, 47), or which can be induced in chicken embryos by cooling (43, 48–50).

ISOLATION AND CHARACTERIZATION OF EUKARYOTIC RIBOSOMAL PROTEINS

Ribosomal proteins of *E. coli* have been isolated and characterized (51–54), and the primary sequence of most has been determined (4, 7, 9). The preparation of eukaryotic ribosomal proteins has been less tractable since the ribosomes contain a greater number of proteins (24). Eukaryotic ribosomal proteins like those from *E. coli* are monotonously similar; although each protein has a unique covalent structure, they are almost all small, extremely basic (many have isoelectric points above 11), and terribly insoluble in ordinary buffers. Thus, it followed that an efficient procedure for the fractionation of the proteins into groups would facilitate purification (55).

There is also a special problem in obtaining an adequate quantity of material to begin the purification. The liver of an average size rat (250 gm) will yield about 13.5 mg of 80S ribosomes, but, because of the inefficiency of the preparation of subunits, only 1 mg of 40S, and 2 mg of 60S protein. The successful separation of appreciable amounts of all *E. coli* 30S ribosomal subunit proteins was only achieved after dozens of attempts, and the preliminary trials were made with as much as 2 gm of protein (H. G. Wittmann, personal communication). The tuition implicit in the experience did not go unheeded. It was obvious that since purification would require grams of ribosomal protein one would require rats in the order of the numbers led out of the city of Hamelin by the Pied-Piper.

Finally, the effort was faciliated by the development of a microprocedure for two-dimensional polyacrylamide gel electrophoresis (56). Since ribosomal proteins have no function when separated from the particle, the only practical method for following their purification is gel electrophoresis. The method of choice, developed by Kaltschmidt & Wittmann (57), is electrophoresis in two dimensions. By judicious selection of conditions one can display most of the proteins of ribosomal subunits as single spots. The coordinates of the separate spots provide the basis for the nomenclature (58–60): the proteins are numbered from left to right in successive tiers and the prefix S for small subunit, and L for large appended. It was apparent that the value of the Kaltschmidt-Wittmann procedure would be increased if less time and less protein were required. Both purposes were achieved by miniaturization (56). Substantial savings in material and time were effected because the purification could be rapidly assessed without expending significant amounts of valuable material.

Latterly, something of the same purpose was achieved by one-dimensional electrophoresis in gels containing sodium dodecyl sulfate (27, 28). Analysis of the proteins of the small subunit yields 20 distinct bands, of the large subparticle, 21. Most of the proteins of the subunits have been assigned to one or another of the bands; however, the assignment could only be made retrospectively by analysis of single proteins after their purification.

The separation of rat liver ribosomal proteins was assisted, then, by the development of an efficient means for fractionating ribosomal subunits into groups; by the development of suitable means of analyzing the proteins by electrophoresis; and by the realization that large amounts of material would be required.

Purification

Eukaryotic (rat liver) ribosomal proteins were resolved largely with the procedures that had served for the isolation of prokaryotic ribosomal proteins (51–54). Isolation begins with the preparation of 40S and 60S subunits (61). That is actually an initial fractionation into two groups. Protein is extracted from the subunits with 67% acetic acid in 100 mM $MgCl_2$ (34, 51), or, on occasion, with 2 M LiCl and 4 M urea (34, 62–64). The extracts are referred to as TP40 and TP60, meaning the total proteins of 40S or 60S subunits.

Ribosomal subunit proteins are separated into groups by stepwise elution from carboxymethylcellulose with LiCl at pH 6.5 (55). A number of procedures had been tried before with prokaryotic (52, 54, 65, 66) and eukaryotic (67) ribosomes. In general, they suffered the same failings: a great deal of overlap in the several fractions, or an insufficient number of groups, or both. The present method, however, proved efficient. Proteins of the 40S subunit were separated into five groups and those of the large subunit into seven; the groups contained two to twelve proteins and there was little overlap between them (55). A limitation of the procedure is that the same property (separation by charge) is exploited in the preliminary separation as is used later for purification of the proteins.

The proteins in the group fractions in general have been resolved by ion-exchange chromatography on carboxymethylcellulose, or on phosphocellulose, or, for the small subset of acidic proteins, on DEAE-cellulose (25–30). Some of the proteins in the groups were purified in that way in a single step, others were resolved after chromatography by gel filtration.

In this manner 33 proteins were purified from extracts of 40S subunits (25, 28, 30) and 49 from 60S subparticle (26, 27, 29, 30), making a total of 82 (Table 3). Earlier Terao & Ogata (68) had isolated 12 proteins from the

Table 3   Rat liver ribosomal proteins[a]

40S subunit		60S subunit	
Protein[b]	M_r ($\times 10^{-3}$)	Protein	M_r ($\times 10^{-3}$)
(Sa)	41.5	(La)	37.9
(Sc)	33.0	(Lb)	29.8
S2	33.1	(Lf)	14.6
S3	30.4	P1	16.1
S3a	32.0	P2	15.2
(S3b)	30.4	L3	37.8
S4	29.5	L4	41.8
S5	22.8	L5	32.5
(S5')	21.5	L6	33.0
S6	31.0	L7	29.2
S7	22.2	L7'	28.7
S8	26.8	L8	28.4
S9	24.3	L9	24.7
S10	20.1	L10	24.2
S11	20.7	L11	21.3
S12	14.9	L12	18.7
S13	18.6	L13	26.3
S14	17.3	L13'	24.6
(S15)	19.6	L14	25.8
S15'	15.7	L15	24.5
(S16)	17.1	L16	18.7
S17	18.0	L17	22.1
S18	18.5	L18	24.5
S19	17.1	L18'	21.3
S20	16.5	L19	25.3
S21	12.3	L20	16.2
S23/S24	18.8	L21	20.3
S25	17.0	L22	16.1
S26	16.5	L23	15.6
S27	14.5	(L23')	18.0
S27'	12.8	(L25)	17.5
S28	11.3	L26	18.6
S29	11.2	L27	17.8
—	—	L27'	18.0
—	—	L28	17.8
—	—	L29	20.5
—	—	L30	14.5
—	—	L31	15.6
—	—	L32	17.2
—	—	L33	15.6
—	—	L34	15.8
—	—	L35	17.5
—	—	L35'	13.7
—	—	L36	14.3
—	—	L36'	16.2
—	—	L37	15.4
—	—	L37'	12.8
—	—	L38	11.5
—	—	L39	11.6

[a] In 40S subunit ΣM_r = 707,000; number of proteins = 33; aver. M_r = 21,400. In 60S subunit ΣM_r = 1,035,000; number of proteins = 49; aver. M_r = 21,200.

[b] The proteins in parenthesis may not be components of the ribosome or may not be unique peptides (see text for details).

40S subparticle of rat liver ribosomes but the amounts were only sufficient for a determination of the molecular weight and not for further characterization. Westermann & Bielka (67) had reported the isolation and characterization (molecular weight, amino acid composition, and tryptic peptides) of 24 of the proteins of the small subunit of rat liver ribosomes; however the purification of the proteins was not documented. Some rat liver ribosomal proteins have been prepared by elution from plugs cut out of two-dimensional gels (69). A start has been made on the isolation of the proteins from the ribosomes of other eukaryotic species (70).

Characterization

Many of the isolated proteins had no detectable contamination (determined by gel electrophoresis); in no case were impurities greater than 9% tolerated. The character of the impurity is important: 9% of contamination with a single second protein is more significant than a total of 9% contamination with several proteins. The latter may be acceptable, the former generally is not.

The yield of individual ribosomal proteins has depended on a number of factors, but most critically, the quality of the chromatography, in particular the number of steps required for the purification of the protein, and the losses tolerated for the sake of purity. The actual amounts of individual proteins isolated (from several grams of ribosomal subunit protein) varied from 0.3–40 mg and generally was in the range of 10–15 mg.

The molecular weight of the purified proteins was determined by electrophoresis in gels containing sodium dodecyl sulfate (Table 3). The number average molecular weight for the 33 40S ribosomal subunit proteins is 21,400, (the range is 11,200–31,100). The number average molecular weight for the 49 proteins isolated from 60S subparticles is 21,200 (the range is 11,500–41,800). Therefore the average molecular weight of a eukaryotic (rat liver) ribosomal protein is about 21,000; the average for *E. coli* ribosomal proteins is approximately 18,000. Although there are some discrepancies the results, in general, do not differ greatly from values reported earlier (67, 68, 71–76), especially when one gives consideration to the variety of methods that have been used.

It is important to realize that the determination of the molecular weight of ribosomal proteins (Table 3) is an approximation, since the values are influenced by the method used for the analysis as well as the physical properties of the protein (26). The molecular weight of histone proteins, estimated from polyacrylamide gel electrophoresis in sodium dodecyl sulfate, is greater than the actual value determined from the amino acid sequence (77, 78). Since ribosomal proteins, like the histones, are, in general, relatively small and basic, the molecular weights determined by sodium dodecyl sulfate gel electrophoresis may also exceed their true values. Indeed, that has been shown to be the case for several *E. coli* ribosomal proteins (79–82).

The amino acid composition of the isolated rat liver ribosomal proteins are similar, but in most cases they are definitely unique (25–30). There are, however, important exceptions (see next section).

The Number of Ribosomal Proteins

Genetic analysis and reconstitution of subunits are two powerful procedures for testing whether a protein is part of a prokaryotic ribosome. Neither can be used in a determination of the status of putative eukaryotic ribosomal proteins. How formidable the problem can be is illustrated by the amendments that have had to be made in the number of *E. coli* ribosomal proteins. First, it was discovered that S20 and L26, originally thought to be two proteins, were in fact one (2). Apparently, the protein is at the interface and a portion partitions with each ribosome subunit when couples are dissociated. Then it was established that L8, once thought to be unique protein, was actually an aggregate of L7, L12, and L10 (83).

It is vexing that one still can not say how many proteins there are in eukaryotic (mammalian) ribosomes. The original estimate was from inspection of two-dimensional polyacrylamide gel plates (59, 72–76, 84–88). It was calculated that there were about 72; 31 in the 40S, and 41 in the 60S subunit (59). Eighty-two proteins have been isolated from rat liver ribosomes: 33 from the 40S, and 49 from the 60S subunit (25–30). That is 10 more than was estimated at first. Moreover, several proteins designated initially have not been isolated, indeed, were not even encountered during the purification; whereas, other proteins which were not resolved by electrophoresis were isolated. The proteins in the first subset were, in general, only rarely seen on two-dimensional gels or stained lightly; the proteins in the second group were, as a rule, difficult to resolve by electrophoresis. Proteins enumerated but not isolated include: S1, S22, S23 or S24, S30, and S31 from the small subunit; and L1, L2, L24, L40, and L41 from the large subparticle.

Proteins S1, S22, and S30 are only rarely seen on two-dimensional electrophoretograms and were not identified during the purification of the other proteins. It is possible they are not structural proteins of the small subunit. It is not certain whether S23 and S24 are a single protein or two distinct polypeptides. They generally are not resolved by two-dimensional gel electrophoresis; however, the configuration of the spot suggests two proteins (59). The protein that was isolated forms a single spot on two-dimensional gel electrophoresis, and a single diffuse band after electrophoresis in one dimension in a number of different concentrations of polyacrylamide either with sodium dodecyl sulfate, or in urea at pH 4.5 (25). For that reason, it is designated for the time S23/S24.

The proteins that were isolated include several (Sa, Sb, S3a, S3b, S5', S15', and S27') that were not in the original classification and whose exact status is still uncertain. Sa and Sb are relatively acidic proteins found in small amounts in extracts of the 40S subunit; there is no assurance they are ribosomal proteins. Comparison of the amino acid composition, molecular weight, and tryptic peptide fingerprints suggests that S3b may be a chemically modified form of S3a, and that S5' may be a derivative of S5. S15' occupies the same position on two-dimensional gels as L25, but they differ in molecular weight and amino acid composition; S15' also differs from S15. S27' is clearly different from S27: they form distinct spots on two-dimensional gel electrophoresis, they have different molecular weights (S27, 14,500; S27', 12,800), and their amino acid compositions are dissimilar (25, 28).

It was originally thought that S8 and L13 might be the same protein since they occupied the same position on two-dimensional gels (59, 89). However, even though S8 and L13 have similar molecular weights (S8, 26,800; L13, 26,300) their amino acid compositions are quite distinct. Hence, it is likely that they are unique proteins. For a like reason S31 and L20 were thought to be the same protein (89). While L20 has been isolated, S31 was not identified during the purification of the proteins of the 40S subunit (25, 28).

The large subunit proteins L1, L12, L24, L40, and L41 have not been isolated. L1, L12, and L24 were not found in any of the group fractions (36). They had occurred only occasionally on two-dimensional gel plates and then stained lightly.

On the other hand the 49 proteins isolated from the 60S subunit (26, 27, 30) include 13 (La, Lb, Lf, Pl, P2, L7', L13', L18', L23', L27', L35', and L37') that were not in the original classification. La, Lb, and Lf are acidic proteins found in small amounts and there is doubt they are structural proteins of the ribosome. Inspection of the amino acid composition, molecular weights, and tryptic peptide fingerprints raises the question of whether L23' may not be the same protein as S18. The other proteins in the group appear unique. My colleagues and I were from the first sensitive to the possibility that some of the isolated proteins might be chemically modified forms of another polypeptide. The chemical modification could either be physiological, as, for example, phosphorylation, acetylation, or methylation; or it could be adventitious, having occurred during purification. Since the proteins are exposed to urea during isolation, carbamylation of amino groups can occur even though the purification is done at acidic pH: carbamylation would change the net charge and later the migration of the protein during electrophoresis. A second change we were concerned about was proteolytic cleavage at a sensitive peptide bond. Of course, the enzyme would have to be specific since we assume most of the proteins are not affected.

The decision whether an isolated eukaryotic ribosomal protein is unique (which must be tentative until the sequence is available) has rested on evaluation of the molecular weight, the amino acid composition, and the behavior on electrophoresis. Most, but not all, of the proteins are distinct by all three criteria.

S3a and S3b form distinct spots on two-dimensional gel plates; moreover, they differ in mass (the apparent molecular weights are 32,000 and 30,400). Nonetheless, the amino acid composition of the proteins is very close (28). The relationship of S5 and S5' is very much the same: the proteins form distinct spots on two-dimensional electrophoretograms and they differ in molecular weight by 1,300 (S5, 22,800; S5', 21,500); again, the amino acid compositions are similar (25, 28). In the circumstances, it is possible that S3b is a chemically altered form of S3a, and S5' is a derivative of S5. The alteration that suggests itself is proteolysis. The possibility was tested by doing peptide fingerprints (K. Todokoro, and I. G. Wool, unpublished data). Tryptic peptide maps of S3a and S3b are so similar as to make it likely one is derived from the other. The comparison for S5 and S5' has not yet been made for lack of sufficient amounts of the latter.

S16 and S19 form distinct spots on two-dimensional gel plates but have the same molecular weight (17,100) and like amino acid compositions (25, 28). The tryptic peptide maps for the two proteins are similar, although there are a few peptides unique to each. Once again they may differ only by some chemical modification.

The small subunit protein S15 and the large subparticle protein L25 occupy distinct positions on two-dimensional gels and differ in molecular weight (S15, 19,600; L25, 17,500), yet their amino acid composition is similar (27, 28). Tryptic peptide fingerprints of the proteins have not yet been done.

S18 and L23' are found in the same zone on two-dimensional electro-pherograms, their molecular weights are similar (18,500 and 18,000), and their amino acid compositions hardly differ (27, 28). Since they have the same tryptic peptide maps it is likely they are one protein.

The amino acid composition of each rat liver ribosomal protein has been compared with all the others (A. H. Reisner, personal communication). The computer analysis is based on percentilely transformed data and yields correlation coefficients. That analysis suggests that other pairs of proteins may be related; for example, L6 and L18, L21 and L26, L3 and L36', and Sc and S28. Further chemical analyses are needed to test the prediction.

A striking feature of the computer comparison of the amino acid composition is the number of rat liver ribosomal proteins that are significantly similar statistically. A correlation coefficient of 0.623 is significant at the 1% level, and 0.742 at 0.1% (14° of freedom). For S6 the value is greater than 0.623 for the comparison with 12 proteins (S7, S9, S12, S13, S19, Lf, L13', L18, L19, L34, L36, and L38); and greater than 0.742 for two (S19 and L34). Many of the other proteins show a similar degree of relatedness. Ribosomal proteins are a unique subset of relatively small, basic polypeptides and similarity of amino acid composition and of short sequences is to be expected. Indeed, a computer analysis of sequence repeats found in *E. coli* ribosomal proteins has indicated that even hexa- and heptapeptides may occur by chance (90). The comparison, to have validity, must be with randomly generated proteins having the properties of ribosomal proteins; if the comparison (91) is with unrelated proteins (taken from the literature) the likelihood that the same hexapeptide would occur in two separate proteins is only 9.1×10^{-10}.

The similarity in structure suggested by the amino acid composition has potential significance with respect to the origin of the "extra" eukaryotic ribosomal proteins. They could be the result of the accretion of entirely new proteins during evolution from prokaryotes; or they could have arisen by duplication of the genes for preexisting ribosomal proteins. Gene duplication should lead to families of eukaryotic ribosomal proteins sharing similarities in amino acid sequence, which is what the comparison of amino acid composition indicates. There is, it should be noted, no evidence that the genes for *E. coli* ribosomal proteins arose by duplication (90).

Stoichiometry

The sum of the molecular weights of 28 of the proteins isolated from the 40S ribosomal subunit is approximately 0.6×10^6 (Table 3). For the time we do not include Sa and Sc whose status is not certain; S1, S22, S30, and S31 which were not isolated; and S3b, S5', and S16, which are likely to be related to S3a, S5 and S19. The molecular weight of 18S rRNA is reported (32) to be 0.7×10^6. Thus the total mass of the molecular constituents of the small ribosomal subunit is about 1.3×10^6 daltons, which is close to the molecular weight (1.4×10^6) determined for the particle (Table 1).

Forty-nine proteins have been isolated from the 60S subunit of rat liver ribosomes: the sum of the molecular weights is approximately 1×10^6 (Table 3). If one sets aside La, Lb, and Lf, whose status is not certain; and L23' and L25 which are probably the same as S18 and S15, the total mass of protein is 0.95×10^6 daltons. If we take the molecular weight of 28S rRNA to be 1.75×10^6 (32) and 5S and 5.8S to be about 0.09×10^6 (37, 38), then the mass of the rRNA in the 60S subparticle of mammalian ribosomes is approximately 1.84×10^6 daltons. The mass of the components of the large subunit then is about 2.79×10^6 daltons, near the estimate of 2.9×10^6 for the molecular weight of the particle (Table 1). However, the values for the molecular weight of eukaryotic ribosomes and subunits are not certain, nor for that matter are the determinations of the molecular weight of ribosomal proteins and ribosomal RNA precise, and hence one must be cautious in making judgments about the stoichiometry from the available data. It is, nonetheless, unlikely that there is more than one mole of most proteins, although it is still possible that there is more than one copy of a few.

Attempts have been made to determine the stoichiometry of eukaryotic ribosomal proteins from analysis of spots on two-dimensional gel plates (92) and by other means (93). Even if there were no technical problems (and there are many) the calculation yields only the number of copies of the protein in isolated ribosomes. The sad, painful experience with _E. coli_ was that the stoichiometry of the proteins was different for the population of isolated ribosomes (94–96) and for ribosomes in the cell (97). Fractional ribosomal proteins and ribosome heterogeneity (98–100) proved to be artifacts created during the preparation of the particles (97). Indeed, it was to be expected that ribosomal proteins would have a range of association constants and, hence, that some would be more easily lost than others. The same may be true of eukaryotic ribosomes.

Sequence Analyses

Eukaryotic ribosomes contain ~16,000 amino acids. To determine the sequence of all would be a monumental task, at least by conventional methods, and its achievement clearly is not imminent. It is possible that more rapid automated methods for sequencing small amounts of protein will be developed and make the task feasible. It is also conceivable that it will become a routine matter to determine the structure of proteins by sequencing their structural genes. One would still need the means to prepare large amounts of eukaryotic ribosomal protein DNA; it would be necessary to know at least the amino-terminal and the carboxy-terminal sequences of the protein, and it would be necessary to be able to recognize potential intervening DNA sequences. Nonetheless, one has the impression that if the sequence of all the proteins in eukaryotic ribosomes is ever to be had it will come from a combination of the analysis of DNA and of protein. What is being done for the time (B. Wittmann-Liebold, H. G. Wittmann, A. Lin, and I. G. Wool, unpublished data) is to determine the amino-terminal sequence of proteins (Table 4) since that gives the maximum information for the expenditure of practical amounts of time and effort. One purpose in

Table 4 Amino-terminal sequences of rat liver ribosomal proteins[a]

S4

1	5	10	15	19

Ala-Arg-Gly-Pro-Lys-Lys-His-Leu-Lys-(Arg)-Val-Ala-Ala-Pro-Lys-(His)-Trp-Met-Leu-

20	25	30	35

Asp-Lys-Leu-Leu-Gly-Phe-Val-Ala-(Pro)-(Thr)-Pro-(Lys)-Leu-Gly-(Pro)-(Asp)-Lys-Leu-

S6

1	5	10	15	19

Met-Lys-Leu-Asn-Ile-(Ser)-Phe-Pro-Ala-Thr-Gly-(Ser)-Gln-Lys-Leu-Leu-Glu-Val-Asp-

20	25	30

Asp-Glu-(Arg)-Lys-Leu-(Arg)-X-Phe-Tyr-(Glu)-Lys-X-Met-Ala-

S8

1	5	10	15	18

Gly-Ile-(Ser)-Arg-Asp-Asn-Trp-His-Lys-Arg-(Arg)-Lys-(Thr)-Gly-Gly-Lys-Arg-Lys-

19

Pro-Tyr-

L6

1	5	10	15	19

Ala-Gly-Glu-Lys-Ala-Glu-Lys-Pro-Asp-Lys-Lys-Glu-(Gln)-Lys-Pro-Ala-Ala-Lys-Lys-

20

Ala-Gly-(Gly)-Asp-Ala-

L7'

1	5	10	15	20

Pro-Lys-Gly-Lys-Lys-Ala-Lys-Gly-Lys-Lys-Val-Ala-Pro-Ala-Pro-Ala-Val-Val-Lys-Lys-

21	25	30	35

Gln-Glu-Ala-Lys-Lys-Val-(Val)-Asn-(Pro)-Leu-Phe-Gln-Lys-Ser-(Pro)-Lys-Asn-Phe-Gly-
                    Thr   Ser   Arg

L18

1	5	10	15

Gly-Val-Asp-Ile-(Arg)-His-Asn-Lys-Glu-(Ser)-Lys-Val-(Ser)-(Arg)-Lys-Leu-Pro-Lys-Leu-

L27

1	5	10	15	20 21

Gly-Lys-Phe-Met-Lys-Leu-Gly-Lys-Val-Val-Leu-Val-Leu-Lys-Gly-X-Tyr-X-Gly-X-Lys-
            Val

22	25

Ala-Val-Ile-Val-Lys-Asn-

L30

1	5	10	15	18

Val-Ala-Ala-Lys-Lys-Leu-Lys-Lys-(Ser)-Leu-Glu-(Ser)-Ile-(Asn)-(Ser)-Arg-Leu-Gln-
                          Thr   Thr   Thr

19	25	30

Leu-Val-Met-Lys-(Ser)-(Gly)-(Lys)-Tyr-Val-Leu-Gly-Tyr-Lys-Gln-
Thr   Thr

Table 4 *(continued)*

L37

1	5	10	15	18

X-Lys-(Gly)-Thr-(Ser)-(Ser)-Phe-Gly-Lys-Arg-Arg-Asn-Lys-(Thr)-(His)-Thr-Leu-(Gly)-

19	25	30

Arg-Arg-Lys-X-Lys-Gly-Lys-Gly-Ala-Leu-Gln-Lys-

L37'

1	5	10	15	20

Ala-Lys-Arg-Thr-Lys-Lys-Val-Gly-Ile-Val-Gly-Lys-Tyr-Gly-Thr-Arg-Tyr-Gly-Ala-(Ser)-

21	25

Leu-Arg-Lys-Met-(Val)-Lys-Lys-Ile-

L39

1	5	10	15

Ser-Ser-His-Lys-Thr-Phe-Arg-Ile-Lys-Arg-Phe-Leu-Ala-Lys-Lys-Gln-Lys-Gln-

P2

1	5	10	15	19

(Met)-X-Tyr-Val-Ala-(Gly)-Tyr-Leu-Leu-Ala-Ala-Leu-Gly-Gly-Asn-(Trp)-Asn-Pro-(Trp)-

20	25	30

Ala-(Ala)-Asp-Ile-(Gly)-(Met)-Ile-Leu-Leu-X-Val-
                                   Thr Thr

[a] Preliminary unpublished data from B. Wittmann-Liebold, H. G. Wittmann, A. Lin, and I. G. Wool. The sequences were determined on small samples of protein: the purpose was to characterize them and compare the sequence with other ribosomal proteins.

collecting the data is to correlate rat liver and *E. coli* sequences, in order to search for homologous ribosomal proteins; a second purpose is to determine the relatedness of individual eukaryotic ribosomal proteins. There surely will be proteins of special importance, S6 and P2 are examples (see below), that will justify the effort necessary for the determination of the complete sequence.

THE ASSOCIATION OF RIBOSOMAL PROTEINS AND RIBOSOMAL RNA

Information on the interaction of proteins and nucleic acids is important for a solution of the structure of ribosomes. An attempt is being made to identify, by affinity chromatography, the proteins that bind to 5S and 5.8S rRNA (101). The strategy is simple (102, 103): mixtures of ribosomal proteins are passed through a Sepharose column containing the immobilized nucleic acid and those that bind are eluted and identified by gel electrophoresis. The 5S rRNA-binding proteins revealed by this procedure are L6 and L19 (101); the former accounts for 42% of the bound protein, the latter

23%. The small amounts of several other proteins bound probably do not associate directly with 5S rRNA.

No 40S ribosomal protein associated with immobilized 5S rRNA, nor did the proteins of either the small or large subparticle of *E. coli* ribosomes; the latter includes L5, L18, and L25, the proteins that bind to *E. coli* 5S rRNA (104–107). The results are not entirely unexpected since Wrede & Erdmann have shown (107) that *E. coli* L18 and L25 bind to yeast 5.8S rRNA and not to 5S rRNA. That, and other evidence, indicate that the functionally related ribosomal nucleic acids are eukaryotic 5.8S and prokaryotic 5S (107, 108).

A 5S rRNA-protein complex (5S rRNP) can be extracted from the large subunit of eukaryotic ribosomes with EDTA (109–113). The single protein in the complex is L5 (113), and it was assumed that it was one of, or the only, protein associated with the nucleic acid in the ribosome. No L5, however, was bound to immobilized 5S rRNA. How then to reconcile the results? It is possible that EDTA unfolded the ribosomal subparticle and caused the proteins to lose their specific locations on ribosomal RNA. EDTA will randomize the distribution of proteins in *E. coli* ribosomal subparticles (114). A second possibility is that the particle prepared with EDTA preserves the physiological binding, whereas affinity chromatography on 5S rRNA-Sepharose gives spurious results. The circumstances in which the affinity chromatography was done—the ionic conditions were selected to favor detection of even weak binding of proteins to 5S rRNA —and the control experiments, mitigate against that interpretation. There is another explanation, that the binding site for L5 is at or near the 3' end of 5S rRNA, and that site is perturbed by periodate oxidation or coupling to the Sepharose column.

Affinity chromatography has also been used to identify the large ribosomal subunit proteins that bind to 5.8S rRNA (N. Ulbrich, A. Lin, and I. G. Wool, unpublished results). They are predominantly L19, L8, and L6; L19 accounted for 58% of the bound protein, L8 was 16%, and L6 about 13%. Two of those proteins, L6 and L19, also bind to 5S rRNA. However, the protein with the strongest affinity for 5S rRNA was L6; with 5.8S RNA the order is L19 > L8 or L6. The significance of the finding that L6 and L19 bind to both species of rRNA is not certain; it is possible that 5S and 5.8S RNA are in the same neighborhood, and that L6 and L19 bind to both to form a domain.

Appreciable amounts of the 40S ribosomal subunit proteins S13 and S9 also bind to 5.8S rRNA; perhaps 5.8S rRNA participates in the formation of ribosome couples by stimultaneously binding small and large subunit proteins.

No *E. coli* ribosomal proteins bind to rat liver 5.8S rRNA, although *E. coli* proteins L18 and L25 associate with yeast 5.8S rRNA (107). It is difficult to account for the discrepancy.

HeLa cell polysomes have been oxidized with periodate to cross-link nearby proteins to the 3' terminus of rRNA (115). The proteins covalently bound to 5.8S rRNA were tentatively identified as L7' and L23'; neither protein bound to immobilized 5.8S rRNA. Once again the discrepancy may be due to modification of the 3' terminus of the nucleic acid during construction of the affinity absorbant.

Little is known of the association of proteins with the large ribosomal ribonucleic acids, except that S3a is cross-linked by periodate oxidation to the 3'-OH of 18S rRNA, and L3 is cross-linked to 28S rRNA (115).

PHOSPHORYLATION OF EUKARYOTIC RIBOSOMAL PROTEINS

When analyzing two related objects it is difficult to know whether to emphasize similarities or differences, to decide which are more important for understanding their fundamental nature. The temptation is to stress the essential identity of eukaryotic and prokaryotic ribosomes. It is perhaps equally important not to neglect two striking differences, both related to phosphorus metabolism. Eukaryotic ribosomes do not participate with stringent factor in the synthesis of guanosine tetra- and pentaphosphate (116–118). Indeed, eukaryotic cells do not make ppGpp and ppGppp and, hence, must have some other means to regulate the synthesis of ribosomal components. In fact yeast cells have a fully coordinated stringent response to the deprivation of an amino acid which encompasses a parallel inhibition in the synthesis of rRNA and mRNA for ribosomal proteins, but spares synthesis of tRNA and mRNA for the bulk of the cells' proteins (118a). Secondly, eukaryotic ribosomal proteins can be phosphorylated (119) but except in the most unusual of circumstances prokaryotic ribosomes are not. What is most important, and quite unique, is that the phosphorylation of eukaryotic ribosomal proteins is affected by a variety of physiological stimuli.

That ribosomal proteins could be phosphorylated was discovered by Kabat (120) and by Loeb (121). The weakness, in what were generally exemplary experiments, was a failure to identify the phosphoproteins. The impression was of a substantial number but that is almost certainly not the case.

Shortly afterwards a number of investigators established that eukaryotic ribosomal proteins could be phosphorylated in vitro by cyclic AMP-activated and cyclic AMP-independent protein kinases (122–133). Those experiments are of value now mainly for the cautionary lessons learned. Many more proteins are phosphorylated in vitro than in vivo; moreover, phosphorylation of ribosomes in vitro by protein kinases did not affect any of the partial reactions of protein synthesis that were assayed (124), nor did dephosphorylation of reticulocyte ribosomes alter their ability to translate globin mRNA (119). Elaborate scholastic arguments were propounded to rationalize the failure (119). It is possible that phosphorylation alters a ribosome function that is not related to protein synthesis; or affects a function that was not, or could not be assayed in vitro (as, for example, fidelity of translation). Still, the fact is that no one has shown an effect of phosphorylation, and not for indolence in trying.

A result of the in vitro experiments was to raise doubts about the authenticity of the earlier in vivo findings. It was decided to test whether rat liver ribosomal proteins were actually phosphorylated in situ (134, 135), and, if they were, to identify the phosphoprotein(s). [^{32}P]orthophosphate was administered to rats, and ribosomal subunit proteins were separated by two-dimensional gel electrophoresis. Only S6 was phosphorylated (135).

One can, in some circumstances, recognize on two-dimensional gels as many as six forms of S6 (135), presumably containing increasing numbers of phosphoserine residues. Indeed, just that was shown in an experiment with two isotopes (135). Rats were injected with ^{32}P, then the extracted protein was labeled again in vitro by reductive alkylation with [^{14}C]formaldehyde. The ^{32}P is a measure of the relative number of phosphoserine residues, the ^{14}C of the amount of S6. The ratio of the two radioactivities in S6 indicated that the derivatives contained increasing amounts of phosphoserine (the only phosphorylated amino acid in S6). Since there is no reason to believe that any 40S ribosomal subunit contains more than one copy of S6 there may well be a kind of ribosome heterogeneity, i.e. ribosomes containing more or less extensively phosphorylated S6.

The results obtained first with rat liver ribosomes have been confirmed now for a number of other eukaryotes. S6 is the most prominent, if not the only, phosphoprotein in rabbit reticulocytes (136), in ascites cells (137), in HeLa cells (138, 139), and in others (128, 140, 141), including such primative eukaryotes as the plant cell *Lemna minor* (142), and the protozoan *Tetrahymena pyriformis* (143). In yeast, there are several other proteins besides S6 that are phosphorylated (144, 145), including an acidic large subunit protein [(144); also see below]. It is likely, then, that all eukaryotic ribosomes can be phosphorylated and that S6 (or a closely similar protein) is the major phosphoprotein of the particle.

A number of other ribosomal proteins have been reported to be phosphorylated in one condition or another; the proteins, which generally have less radioactive phosphate than S6, include S2 (138, 146–149), S3 (and/or S3a) (148–150), S16 (146), L6 (148), and L14 (138, 148, 149). The phosphorylation of acidic proteins of the large ribosomal subunit is discussed below.

Structure of S6

S6 has been isolated and characterized (25): it has a molecular weight of approximately 31,000 (Table 3) and contains about 15 serine residues. The sequence of the amino-terminal 33 residues has been determined (Table 4): there are serines at positions 6 and 12, but whether they are phosphorylated is not known. The identity and sequence of the phosphorylation sites have not yet been determined.

Conservation of S6

It is likely that some eukaryotic and prokaryotic ribosomal proteins are homologous[1] (see below). Antibodies against eukaryotic ribosomal proteins were tested for their effect on poly(U)-directed synthesis of polyphenylalanine by *E. coli* ribosomes (G. Stöffler, N. Fischer, R. Hasenbank, K. H. Rak, and I. G. Wool, unpublished data). Synthesis was inhibited by an antiserum against rat liver TP40, which indicated that one or more proteins in the small subunit was homologous with *E. coli* ribosomal protein(s). One protein appears to be S6, since an antiserum raised to that protein also inhibits the synthesis of polyphenylalanine by *E. coli* ribosomes. The weakness in the experiment is that neither the antigen, S6, which was prepared from spots cut out of two-dimensional gel plates, nor the antiserum has been characterized fully yet. Nonetheless, this may be an example of the appearance of the substrate before the evolution of its enzyme; the assumption is that the *E. coli* homologue of rat liver S6, which has not yet been identified, is not phosphorylated because the bacteria lack protein kinase.

In ordinary circumstances the proteins of prokaryotic ribosomes are not phosphorylated (151). However, if *E. coli* are infected with T7 (but not T4) a number of ribosomal proteins are phosphorylated (152, 153). *E. coli* probably has no endogenous protein kinase (153); indeed, the levels of phosphoserine and phosphothreonine in protein are an order of magnitude lower than in eukaryotes (154). A report to the contrary (155), probably was the result of the confusion of a polyphosphate kinase that is stimulated by histone (156) with a true protein kinase. The physiological consequence of the phosphorylation of ribosomal proteins after infection is not known; there is no change in the function of the particles when they are assayed in vitro (152), and a protein kinase mutant of T7 seems to develop normally (153).

---

[1]Homology, which implies a common ancestor, is an inference from the actual observations that are described.

Conditions Altering the Phosphorylation of S6

A wide variety of alterations in the physiological milieu affect the phosphorylation of S6 (128, 134, 135, 138–141, 157–168). The stimuli include hormones, cyclic AMP, viral infection, changes in growth conditions of cells in culture, and a number of others (Table 5).

The increase in the phosphorylation of S6 induced by the different stimuli is not associated with any obvious common change in cell function. There is, for example, no obligatory link to a net alteration in protein synthesis. There is an increase in phosphorylation of S6 when protein synthesis is increased, as during liver regeneration (135), when it is decreased, as in diabetes (157), or when there is no appreciable change, as after glucagon administration (159); indeed there is even an increase in phosphorylation of S6 in the absence of any protein synthesis, as after the administration of puromycin or cycloheximide (134). Phosphorylation of S6 is not correlated with, nor does it require, the synthesis of new ribosomes, at least not in rabbit reticulocytes or yeast. Reticulocytes do not synthesize ribosomes but do phosphorylate proteins in the particles (136); phosphorylation of yeast ribosomal proteins occurs at the restrictive temperature in a mutant in ribosome biogenesis (144). It should be noted that there is turnover in the phosphate groups i.e. there is both phosphorylation and dephosphorylation (160, 169), and this turnover is also independent of ribosome formation.

Table 5 Conditions altering the phosphorylation of ribosomal protein S6

Increased phosphorylation of S6	References
Hormones	
glucagon	141, 158, 159
thyroxin	162
diabetes	157
epinephrine	163
Cyclic AMP	128, 141, 159–161
Viral infection	
vaccina virus	138, 146
vesicular stomatitis virus	149
Changes in growth conditions of cells in culture	139, 140, 148, 164, 165
Liver regeneration	135
Cell cycle	166
Hepatic injury	167, 168
Inhibitors of protein synthesis (puromycin and cycloheximide)	134
Decreased phosphorylation of S6	References
Insulin administration to diabetic animals	157
Ethionine	163

There is a general correlation between the phosphorylation of S6 and an increase in the intracellular concentration of cyclic AMP (128, 141, 159–161), although there are exceptions. In some cells, usually malignant or transformed ones, cyclic AMP may not stimulate phosphorylation of S6 (139, 140).

There is, of course, no assurance that each of the stimuli lead to the phosphorylation of the same site on S6 or the same combination of sites. There are 15 serine residues in S6 and as many as six may be phosphorylated; the same subset of six serines may not be phosphorylated in all conditions. Phosphorylation of the many potential sites (or the many possible combinations of them) may not produce equivalent changes in ribosome function, indeed, it would be surprising if they did. If one treats each phosphorylation site separately there is a vast number of possible forms of S6 and hence a number of potential alterations in ribosome function.

Certain lessons can be adduced from the experimental results in hand. The impression is that little of fundamental importance is likely to come from enlarging the catalogue of stimuli that affect phosphorylation. The real problem is to learn what contribution S6 makes to the function of the ribosome, and how phosphorylation affects that function. A possibility that must be considered *ab initio* is that the phosphorylation of S6 has no effect at all. There is a precedent for this apparently subversive suggestion. A portion of *E. coli* ribosomal protein L12 is acetylated (170, 171), and yet it has never been demonstrated that the acetylation of the N-terminal serine of L12, to form L7, in any way affects ribosome function, although the degree of acetylation (i.e. the ratio of L7 to L12) is altered by changes in growth conditions (172, 173).

It was suggested early on that ribosomal protein Su or II (probably S6) was phosphorylated in single ribosomes but not in polysomes (120), and that phosphorylation was a means to exclude ribosomes from participation in protein synthesis. The proposal has been resurrected from time to time (174) but there is evidence against it (135, 169).

The conjecture is that S6 forms part of the mRNA-binding domain on the ribosome, and that phosphorylation of S6 provides a more favorable site for the attachment of certain classes of mRNA. Until recently all that was lacking was evidence. There are now at least preliminary indications from experiments in which poly(U) was cross-linked to S6 in ultra-violet irradiated 40S ribosomal subunits (174a), and from experiments with an inhibitor of initiation (174b), that S6 is located at the mRNA binding site.

Acidic Ribosomal Phosphoproteins

There was at first some reluctance to credit the sporadic reports of acidic ribosomal phosphoproteins. One reason is that most contaminants of the ribosome, including phosphoproteins, are acidic; a second that Gressner & Wool (135), and Traugh & Porter (136) had not found phosphorylated acidic proteins in liver or reticulocyte ribosomes, even though an assiduous search was made.

It was the resolution of the acidic proteins of the large subunit (29) that changed the perspective. Group A60 contains about fourteen 60S ribosomal proteins that do not bind to carboxymethylcellulose at pH 6.5. When the proteins are displayed on two-dimensional gels there are two striking arrays of three and four spots. The seven proteins form only two bands after electrophoresis in gels containing sodium dodecyl sulfate. The possibility that the proteins are phosphorylated derivatives of one or more polypeptides was tested. The Group A60 proteins were treated with alkaline phosphatase—two of the seven proteins remained after enzyme treatment; they were designated P1 and P2. Their derivatives are referred to as P1a and P1b, and P2a, P2b, and P2c. Subsequently, it was shown that the P1 and P2 proteins incorporate ^{32}P from orthophosphate in vivo and are phosphorylated in vitro by a cyclic AMP-independent protein kinase that transfers the γ-phosphoryl group of GTP (A. Lin, and I. G. Wool, unpublished observations).

P1, P2, and their derivatives were isolated by chromatography of Group A60 proteins on carboxymethylcellulose at pH 4.2, and on DEAE-cellulose at pH 4.6 (29). P1 has a molecular weight of 16,000; P2 of 15,200. The molecular weight and the amino acid composition of P1a, and P1b is the same as that of pure P1; that of P2a, b, and c are the same as P2. Thus it seems certain that P1a and P1b are phosphorylated derivates of P1; and that P2a, b, and c are derived from P2. From the positions of the proteins on two-dimensional gels we assume P1b has more phosphate residues than P1a, and that P2c, has more than P2b, which in turn has more than P2a. Whether P1 and P2 are phosphorylated is not certain: the suspicion is that they are not. Phosphatase converts the derivatives to P1 and to P2 but does not lead to the formation of new spots as would be expected if they were phosphorylated (29). The derivatives are phosphorylated in vivo, i.e. are radioactive after administration of [^{32}P] orthophosphate to animals, whereas P1 and P2 are not. Finally, the derivatives are phosphorylated when 60S subunits are incubated with protein kinase and [^{32}P]GTP; P1 and P2 are not.

A distinctive feature of the amino acid composition of P1 and P2 is the large amount of alanine (P1 has 20.4 mole %; P2 has 17.5); it also needs to be remarked that P1 has no arginine and P2 has very little (3.2 mole%) whereas most ribosomal proteins have large amounts of arginine.

Tryptic peptide maps of P1 and P2 are similar but have distinctive features (A. Lin, and I. G. Wool, unpublished observation). There is, nonetheless, a lingering suspicion that P1 and P2 may be related; that P2 may be formed from P1 by a proteolytic cleavage. The matter has been difficult to settle because it is hard to prepare appreciable amounts of P1. In general, P1 amounts to only 10% of P2 which is difficult enough to come by. When the tryptic peptide maps are of ^{32}P-labeled P1 proteins there are two radioactive spots; with the P2 proteins there are three. The radioactive peptides could account, then, for the phosphorylation sites leading to formation of the derivatives P1a and b, and P2a, b, and c.

Relation of Rat Liver P1 and P2 to E. coli L7/L12

E. coli ribosomes have a pair of acidic proteins, L7 and L12, that resemble acidic proteins in the large subunit of eukaryotic ribosomes. Moreover, if *E. coli* 50S ribosomal subunits are incubated with rabbit skeletal muscle protein kinase and [γ-^{32}P]GTP, L7/L12 are selectively and exclusively phosphorylated (175). The observations suggested that the structure or the function, or both, of *E. coli* L7/L12 and rat liver P1 and P2 might be related. However, despite the similarity in their characteristics, there is no evidence for structural identity. The amino acid composition of L7/L12 and of P1 and P2 are quite different (29). Moreover, it has not been possible to demonstrate homology between *E. coli* L7/L12 and rat liver P1/P2 by immunological procedures: there is no cross-reaction, either in radioimmunoassay or in immunodiffusion by the Ouchterlony procedure, when anti L7/L12 is tested with P1 and P2; nor does anti-P2 react with L7/L12 (E. Collatz, T. Tanaka, and I. G. Wool, unpublished data). Inasmuch as only positive results carry weight in immunological experiments of this kind, the findings must of necessity be considered tentative. Rat liver ribosomes have another pair of large subunit acidic proteins, which were designated L40 and L41 (89), and which were shown by immunological procedures to be structurally and functionally related to *E. coli* L7/L12 (see below). However, L40 and L41 have not been isolated so their relation to rat liver P1/P2 is not known.

Artemia Salina Acidic Ribosomal Phosphoprotein

Two acidic ribosomal proteins from the large subunit of the brine shrimp *Artemia salina* have been prepared and characterized (176). The proteins were at first called EL7 and EL12 (176) (E, for eukaryotic); the numbers clearly implied they were related to *E. coli* L7/L12. When it was found that EL12 is a phosphorylated form of EL7, the proteins were rechristened EL12 and EL12p (177).

EL12 is required by *Artemia* ribosomes for EF-1-dependent binding of aminoacyl-tRNA, and for EF-2-catalyzed GTP hydrolysis, just as L7 and L12 are required by *E. coli* ribosomes for the same reactions. The dependence of the two partial reactions of protein synthesis on EL12 was deduced from the observation that addition of the proteins to core particles (PI cores) lacking EL12 reconstituted activity (176). This was supported by the finding that serum raised against EL12 inhibited the reactions (177).

While anti-EL12 inhibits the function of *Artemia* ribosomes it does not cross-react with *E. coli* L7/L12 (177); nor does the reciprocal experiment (anti-*E. coli* L7/L12 against *Artemia* EL12) give evidence of immunological relatedness. Curiously, even though there is no cross-reaction (177), and even though anti-*E. coli* L7/L12 does not inhibit EF-2-catalyzed GTP hydrolysis by *Artemia* ribosomes (177), *E. coli* L7/L12 restores the function of *Artemia* PI cores as well as does EL12 (176). The clear implication is that no matter the extent of the structural identity, *Artemia* EL12 and *E. coli* L7/L12 serve the same ribosome function. The importance of deciding whether the proteins that serve the same function in prokaryotic and eukaryotic ribosomes are structurally similar (i.e. evolutionarily related) or entirely different can not be exaggerated.

The amino-terminal sequence of EL12 has been determined (178, 179) (in fact, the entire sequence is all but completed). The suggestion has been made that *Artemia* EL12 is structurally homologous with *E. coli* L7/L12, with *Halobacterium cutirubrim* L20 [a protein of the halophilic bacteria that is structurally related to *E. coli* L7/L12 (180)], with the yeast (*Saccharomyces cerevisiae*) acidic ribosomal proteins, and with an alanine-rich peptide from a rat liver ribosomal protein (referred to as rat liver L7/L12); and (not to omit anything) with the alanine-rich alkali light chain A_1 from rabbit myosin. The decision that the sequences are homologous is arbitrary, having been made by inspection without applying rigorous criteria. Moreover the conclusion was prejudiced by a comparison, in each case, of an alanine-rich region.

However, one sequence homology does seem likely, and that is between *Artemia* EL12 and rat liver P2 [compare Table 4 and (178, 179)]; of the thirty amino-terminal residues of P2, twenty correspond to the N-terminal sequence of EL12 (B. Wittmann-Liebold, H. G. Wittmann, A. Lin, and I. G. Wool, unpublished data). Thus one protein has been conserved during evolution from invertebrates (brine shrimp) to mammals (rat).

IMMUNOCHEMICAL STUDIES OF THE STRUCTURE AND FUNCTION OF EUKARYOTIC RIBOSOMES

Antibodies to eukaryotic ribosomes, to ribosomal subunits, and to mixtures of proteins from the particles have been used to compare the ribosomes and ribosomal proteins from different eukaryotic species (181–183), from different tissues of the same species (183), and from normal and malignant cells

(184–188); to establish that there are homologous proteins in prokaryotic and eukaryotic ribosomes (24, 189–191); and to demonstrate that there is little, if any, free ribosomal protein in the cytosol (192). However, a concerted effort to prepare monospecific antibodies to all the proteins of eukaryotic ribosomes has not been made.

Homology of Eukaryotic and Prokaryotic Ribosomal Proteins

If the proteins of eukaryotic and prokaryotic ribosomes are all chemically unique it would mean that the particles arose separately on two occasions, or that the transition species and their ribosomes are not known or have disappeared. The origin of ribosomes on two separate occasions seems, intuitively, unlikely. Since proteins in the ribosomes of a primitive eukaryote like yeast, in the brine shrimp *Artemia salina,* and in a mammal (rat) have sequence homologies, it is unlikely all the intervening species are gone.

The *E. coli* large ribosomal subunit acidic proteins L7 and L12, which differ only in that the N-terminal serine of the former is acetylated (193), have remarkable chemical and physical properties (170, 171, 193): they are rich in alanine, but lack arginine, cystine and the aromatic amino acids, although they contain a rare amino acid, methyllysine; and they have about 60% α-helical content and a large nonpolar region in the interior of the molecule. L7/L12 are involved in translocation—the movement of mRNA one codon with respect to the ribosome after each of the reiterative cycles of peptide bond formation (for review see 2, 194, and references therein). The proteins provide, in addition, part or all of the binding site for initiation, elongation, and termination factors; and they participate in the hydrolysis of GTP. Certainly, L7/L12 play an important role in ribosome function. The large subunit of rat liver ribosomes also contains a pair of acidic proteins (designated L40/L41) that behave during electrophoresis like *E. coli* L7/L12 (89). That resemblance suggested they might be related.

To test whether L7/L12 had been conserved during evolution from prokaryotes to eukaryotes, antisera were prepared (24, 189). Rabbits were immunized with a mixture of L40 and L41 obtained by extracting 60S ribosomal subunits with ethanol, and with L40 and L41 obtained by cutting the proteins out of two-dimensional gel plates. The rat liver ribosomal protein antisera (A-L40/L41, A-L40, and A-L41) cross-reacted in Ouchterlony double immunodiffusion with *E. coli* L7/L12 (189). In the reciprocal experiment a serum raised against *E. coli* L12 reacted with several antigen preparations that contained rat liver L40/L41, with TP80, and TP60 (but not TP40), and with L40/L41 (189). The formation of a spur on the Ouchterlony plate between the precipitin bands for *E. coli* L7 and rat liver L40/L41 indicates that the structure of the eukaryotic and prokaryotic proteins are similar but not identical. Yeast ribosomes also contain a protein that is related to *E. coli* L7/L12 and to rat liver L40/L41 (24).

An attempt was made to determine if the part of the structure of L7/L12 that had been conserved was required for function (24). The assay that was used was the EF-G-dependent binding of [^{3}H]GDP to *E. coli* 50S ribosomal subunits in the presence of fusidic acid (195, 196). The reaction, referred to as Complex 1 formation, is dependent on the integrity of L7/L12 (197). A number of sera raised against eukaryotic ribosomal proteins—A-TP60 (RL), A-L40/L41 (RL), A-L40 (RL), A-TP80 (yeast), and A-Acidic Protein (yeast)—inhibited Complex I formation (24). However, A-TP40 (RL) did not, since Complex I formation is a large ribosomal subunit function. The inhibitory activity of the sera against eukaryotic ribosomal proteins was lost if they were absorbed with *E. coli* L7/L12. It would seem that prokaryotic L7/L12 and eukaryotic L40/L41 not only share portions of their structure but have a common function.

Howard et al (190) have confirmed that antibodies to *E. coli* ribosomal proteins L7/L12 cross-react with proteins of eukaryotic 60S subparticles. Moreover, the antiserum inhibited a number of EF-1-dependent functions of chicken liver ribosomes. Several antisera specific for *E. coli* L7/L12 were tested by Grasmuk et al (198) for their effect on elongation reactions by mouse ascites cell ribosomes. The antisera were selected because they cross-reacted with rat liver L40/L41. One anti-L7/L12 serum inhibited EF-1-dependent phe-tRNA binding and GTP hydrolysis catalyzed by either EF-1 or EF-2. Thus antisera against eukaryotic L40/L41 inhibit partial reactions of protein synthesis by prokaryotic ribosomes, and antibodies against prokaryotic L7/L12 interfere with protein synthesis by eukaryotic ribosomes. The findings reinforce the conclusion that prokaryotic and eukaryotic acidic ribosomal proteins are structurally and functionally homologous.

If the proteins are related it seemed it might be possible to construct chimera, that is ribosomes with eukaryotic and prokaryotic proteins. Core particles, lacking only proteins L7/L12, were prepared from *E. coli* 50S subunits, by extraction with ethanol and ammonium chloride (199). Fully active 50S particles are reconstituted if L7/L12 are added to the cores (24, 199). If the same core particles were supplemented with yeast acidic protein (which is immunologically related to *E. coli* L7/L12) they were active in EF-G-dependent binding of [^{3}H]GDP (24). The results were qualitatively similar with rat liver acidic proteins, but the latter were far less active (G. Stöffler, R. Hasenbank, and I. G. Wool, unpublished data). It is remarkable that activity was observed with EF-G which is generally species (prokaryotic)-specific (200). The obverse experiment was done by Richter & Möller (201). They prepared core particles from yeast 60S subunits by extraction with ethanol and salt. The yeast cores, when reconstituted with *E. coli* L7/L12, were active in EF-2-dependent binding of [^{3}H]GDP, in EF-1-directed binding of phe-tRNA, and in polyphenylalanine synthesis.

The most salient unsolved problem is the chemical identity of the eukaryotic ribosomal proteins homologous with *E. coli* L7/L12. The difficulty is that L40/L41, which cross-react, have not been purified and characterized; whereas, *Artemia* EL12 which may serve the same function as *E. coli* L7/L12, and which have been isolated and characterized, do not cross-react.

There are other homologous eukaryotic and prokaryotic ribosomal proteins. Sera against rat liver TP40 and S6 inhibit polyphenylalanine synthesis by *E. coli* ribosomes (G. Stöffler, N. Fischer, R. Hasenbank, K. H. Rak, and I. G. Wool, unpublished data), which indicates that one (S6) or more small subunit proteins are homologous. The homologous *E. coli* protein(s) has not yet been identified.

The homologous ribosomal proteins are a rare, if not unique, instance of conservation during evolution from prokaryotes to eukaryotes.

Immunological Comparison of Ribosomal Proteins from Different Eukaryotic Species

There are, amongst eukaryotes, proteins peculiar to at least distant species; there are, as an instance, marked differences in the ribosomal proteins of rat liver and protozoa (202) and also distinct differences in the number and size of ribosomal proteins in animals and plants (39–41, 203, 204). It seems unlikely that ribosomal proteins have been so assiduously conserved as histones. However, some prokaryotic ribosomal proteins have been conserved during evolution to eukaryotes (even to mammals), which is presumably a greater span than that covered by the histones (24, 179, 189–191, 198, 201). Whether the ribosomal proteins of more closely related eukaryotic species differ is moot. Delaunay et al (205) found no definite differences in the pattern of ribosomal proteins on two-dimensional electropherograms between man, rat, birds, and lizards: the authors argue that at least the size and charge of the proteins have been conserved. On the other hand, when Noll & Bielka (183) compared ribosomal proteins from even more similar species (rat, mouse, cow, and chicken), by immunochemical techniques, they found definite differences.

Immunological techniques have been extremely valuable in establishing homologies between distant species [cf (2, 24) for reviews and detailed references]; they are equally useful in establishing differences in the structure of proteins of closely related species (206–208).

A detailed immunochemical comparison has been made of the proteins of chicken and rat liver ribosomes (191). Antisera specific for chicken or rat liver ribosomes recognize only about 20% of common determinants. While there are important reservations (see below), the results suggest differences in the proteins. Despite these differences, rat and chicken liver share homologous determinants with bacterial ribosomal proteins (191). An enriched antibody preparation against chicken 80S ribosomes inhibited the poly(U)-directed synthesis of polyphenylalanine, and the EF-G-catalyzed binding of [³H]GDP to *E. coli* ribosomes. Thus chicken liver ribosomes, like ribosomes from rat liver and yeast, must have proteins homologous with those of *E. coli* ribosomes (190, 191). Hence there is both conservation of the structure of ribosomal proteins of evolutionarily distant species and variation in the structure amongst closer species.

In immunochemical experiments of this kind there is an important parameter other than the ribosomal proteins of the species being compared (i.e. chicken and rat); that is the ribosomes in the animals (rabbit or sheep) that have been immunized. The ribosomal proteins of the latter may have extensive sequence homologies with the immunogens, and may only make antibodies to unlike sequences. Thus rat and chicken liver ribosomes may share most of the sequence of the majority of their proteins with each other and with rabbit and sheep ribosomal proteins. However, all of the antibodies may be against the small fraction of the total sequences that are different. In that case chicken and rat liver ribosomal proteins may be more similar than the immunochemical experiments indicate.

Ribosome Assembly

Ribosomal proteins are synthesized in the cytoplasm (208a–211) on free polysomes (211a), transported to the nucleolus (13), and assembled there, probably on a large precursor to ribosomal RNA (13, 19). The precursor has a sedimentation coefficient of about 45 and contains approximately 100% excess of nucleotides (13, 19). Many of the details of the process remain to be worked out. For example, it is not known how much of the assembly occurs on the nascent precursor rRNA still attached to ribosomal DNA (212); which ribosomal proteins are added in the nucleolus (213); which, if any, are added at some other site in the nucleus or even in the cytoplasm as a terminal step in ribosome biogenesis; and whether there are nucleolar specific proteins present in the particle during assembly but removed before the subunits are transported to the cytoplasm (213).

Active transcription units of ribosomal DNA can be prepared from some eukaryotic cells, spread on a grid, and observed with the electron microscope (212, 214, 215). The technique has been used to prepare ribosomal genes from *Drosophila melanogaster* oocyte nurse cells (Y. W. Chooi, unpublished data). The transcription unit contains rDNA and the nascent precursor rRNA in the process of being transcribed by RNA polymerase.

The nascent rRNA is decorated with what are presumed to be ribosomal proteins since they are resistant to ribonuclease and sensitive to protease. It is easy to identify the 5' end of the nascent rRNA merely by noting the orientation with respect to the rDNA that is being transcribed. One striking feature is that the nascent rRNA is shorter than the rDNA that has been transcribed, presumably because the binding of proteins initiates the folding of the rRNA and the formation of the ribosomal particles. It was recognized [Y. Chooi, G. Stöffler, H. Swift, and I. G. Wool, unpublished data, and (14)] that antibodies might be valuable in answering some of the questions concerning assembly, especially the sequence of addition of proteins. The idea was to localize with the electron microscope ferritin-labeled antibodies directed against ribosomal proteins.

If one decorates transcription units with ferritin-labeled Fab's prepared from IgG's against rat liver TP60 they bind only to the longest nascent rRNA molecules, which indicates the antibodies associate with proteins located toward the 3' rather than the 5' end. From this single experiment two conclusions emerge: the proteins associated with the transcription unit are, in fact, ribosomal; and *Drosophila* and mammalian ribosomal proteins have some homology.

A similar experiment was done with an IgG preparation against rat liver TP40. Antibodies associated with proteins on nascent rRNA of all sizes, even the smallest transcripts, which indicates that the IGG's were bound to proteins near the 5' end. Taken with the previous experiment, the results indicate that the sequence in the precursor is 5'-18S–28S–3'. There is a great deal of other evidence for that order (216–222).

An experiment was also done with ferritin-labeled Fab's prepared from IgG's against *E. coli* L7/L12. Those Fab's did not bind to the transcription unit, although they did bind to *Drosophila* 80S ribosomes. While one must be cautious in the interpretation of negative results, especially negative results of immunochemical experiments, the observations suggest that the eukaryotic homologues of *E. coli* L7/L12 are either added as a terminal step in the assembly process in the nucleolus, or even later in the cytoplasm. Addition of the proteins in the cytoplasm would be of physiological importance since it would provide a means of activating ribosomes. The preliminary experiments seem to establish the feasibility of the enterprise, although it is recognized that it would be best to use antibodies to homologous (i.e. *Drosophila*) ribosomal proteins.

Immune-Electron Microscopy of Eukaryotic Ribosomes

Perhaps the most spectacular use that has been made of antibodies in the study of the structure of bacterial ribosomes is the location of the proteins on the surface of the particle by immune-electron microscopy (7, 223–226).

The technique has now been applied to rat liver ribosomes (227). The location in the small ribosomal subunit of the protein S3 has been determined; indeed, the IgG prepared from the serum against S3 binds at three separate sites which indicates that the protein is elongated. While the prospects raised by these experiments are exciting it has to be noted that neither the characterization of the antigen (rat liver ribosomal protein S3) nor of the antisera has been reported. Thus there is no assurance the experiments were done with a mono-specific antibody, which is essential to the evaluation of the results.

ACKNOWLEDGMENTS

I am indebted to the people with whom I have had the pleasure of working: Ekkehard Collatz, Alan Lin, Georg Stöffler, Tatsuo Tanaka, Kazuo Todokoro, Kunio Tsurugi, Norbert Ulbrich, Heinz–Günther Wittmann, and Brigitte Wittmann-Liebold. I am especially grateful to Don Blair, Yean Chooi, Walter Hill, Alan Lin, A. Reisner, Georg Stöffler, Kazuo Todokoro, Heinz-Günther Wittmann, and Brigitte Wittmann-Liebold for allowing me to describe the results of unpublished experiments.

The expenses of the research were met by grants from the National Institutes of Health (GM-21769, CA-19265, and AM-04842). The review is dedicated to Dieter Roth.

Literature Cited

1. Nomura, M. 1970. *Bacteriol. Rev.* 34:228–77
2. Stöffler, G. 1974. In *Ribosomes,* ed. M. Nomura, A. Tissiéres, P. Lengyel, pp. 615–67. Cold Spring Harbor, NY: Cold Spring Harbor Lab.
3. Wittmann, H. G. 1974. See Ref. 2, pp. 93–114
4. Wittmann, H. G., Wittmann-Liebold, B. 1974. See Ref. 2, pp. 115–40
5. Nomura, M., Morgan, E. A., Jaskunas, S. R. 1977. *Ann. Rev. Genet.* 11:297–347
6. Kurland, C. G. 1977. In *Molecular Mechanism of Protein Synthesis,* ed. H. Weissbach, S. Pestka, pp. 81–116. New York: Academic
7. Stöffler, G., Wittmann, H. G. 1977. See Ref. 6, pp. 117–202
8. Kurland, C. G. 1977. *Ann. Rev. Biochem.* 46:173–200
9. Brimacombe, R., Stöffler, G., Wittmann, H. G. 1978. *Ann. Rev. Biochem.* 47:217–49
10. Blobel, G., Sabatini, D. D. 1971. In *Biomembranes,* ed. L. A. Manson, 2:193–95. New York: Plenum

11. Sabatini, D. D., Kreibich, G. 1976. In *The Enzymes of Biological Membranes,* ed. A. Martinosi, 2:531–79. New York: Plenum
12. Kreibich, G., Ulrich, B. L., Sabatini, D. D. 1978. *J. Cell Biol.* 77:464–87
12a. Kreibich, G., Freienstein, C. M., Pereyra, B. N., Ulrich, B. L., Sabatini, D. D. 1978. *J. Cell Biol.* 77:488–506
12b. Kreibich, G., Czako-Graham, M., Grebenau, R., Mok, W., Rodriguez-Boulan, E., Sabatini, D. D. 1978. *J. Supramol. Struct.* 8:279–302
13. Warner, J. R. 1974. See Ref. 2, pp. 461–88
14. Chooi, W. Y. 1976. In *Handbook of Genetics,* ed. R. C. King, 5:219–65. New York/London: Plenum
15. Chua, N. H., Luck, D. J. L. 1974. See Ref. 2, pp. 519–39
16. O'Brian, T. W., Matthews, D. E. See Ref. 14, 5:535–80
16a. Cundliffe, E., Dixon, B., Stark, M., Stöffler, G., Ehrlich, R., Geisser, M. Submitted for publication
16b. Wienen, B., Ehrlich, R., Geisser, M., Stöffler, G., Smith, I., Weiss, D., Vince,

R., Pestka, S. Submitted for publication
16c. Pestka, S., Weiss, D., Vince, R., Wienen, B., Stöffler, G., Smith, I. 1976. *Mol. Gen. Genet.* 144:235–41

17. Maaløe, O. 1969. *Dev. Biol.* 3:33–58 (Suppl.)

18. Kjeldgaard, N. O., Gausing, K. 1974. See Ref. 2, pp. 369–92

18a. Trachsel, H., Erni, B., Schreier, M. H., Staehelin, T. 1977. *J. Mol. Biol.* 116: 755–67

18b. Merrick, W. C., Peterson, D. P., Safer, B., Lloyd, M., Kemper, W. M. 1978. *Proc. FEBS Meeting, Copenhagen, 11th, July 1977,* 43:17–26. Amsterdam: Elsevier

19. Maden, B. E. H., Salim, M., Robertson, J. S. 1974. See Ref. 2, pp. 829–39

20. Dalgarno, L., Shine, J. 1973. *Nature New Biol.* 245:261–62

21. Maden, B. E. H. 1976. *Trends Biochem. Sci.* 1:196–99

22. Khan, M. S. N., Salim, M., Maden, B. E. H. 1978. *Biochem. J.* 169:531–42

23. Hagenbüchle, O., Santer, M., Steitz, J. A., Mans, R. J. 1978. *Cell* 13:551–63

24. Wool, I. G., Stöffler, G. 1974. See Ref. 2, pp. 417–60

25. Collatz, E., Wool, I. G., Lin, A., Stöffler, G. 1976. *J. Biol. Chem.* 251: 4666–72

26. Tsurugi, K., Collatz, E., Wool, I. G., Lin, A. 1976. *J. Biol. Chem.* 251: 7940–46

27. Tsurugi, K., Collatz, E., Todokoro, K., Wool, I. G. 1977. *J. Biol. Chem.* 252: 3961–69

28. Collatz, E., Ulbrich, N., Tsurugi, K., Lightfoot, H. N., MacKinlay, W., Lin, A., Wool, I. G. 1977. *J. Biol. Chem.* 252:9071–80

29. Tsurugi, K., Collatz, E., Todokoro, K., Ulbrich, N., Lightfoot, H. N., Wool, I. G. 1978. *J. Biol. Chem.* 253:946–55

30. Lin, A., Tanaka, T., Wool, I. G. 1979. *Biochemistry* In press

31. Petermann, M. L. 1964. *The Physical and Chemical Properties of Ribosomes,* Amsterdam/London/New York: Elsevier

32. Loening, U. E. 1968. *J. Mol. Biol.* 38:355–65

33. Weinberg, R. A., Penman, S. 1970. *J. Mol. Biol.* 47:169–78

34. Sherton, C. C., Wool, I. G. 1974. *Mol. Gen. Genet.* 135:97–112

35. Hamilton, M. G., Pavlovec, A., Petermann, M. L. 1971. *Biochemistry* 10: 3424–27

36. Bielka, H., Welfle, H., Böttger, M., Förster, W. 1968. *Eur. J. Biochem.* 5:183–90

37. Brownlee, G. G., Sanger, F., Barrell, B. G. 1967. *Nature* 215:735–36

38. Nazar, R. N., Sitz, T. O., Busch, H. 1975. *J. Biol. Chem.* 250:8591–97

39. Cammarano, P., Pons, S., Romeo, A., Galdieri, M., Gualerzi, C. 1972. *Biochim. Biophys. Acta* 281:571–96

40. Cammarano, P., Romeo, A., Gentile, M., Felsani, A., Gualerzi, C. 1972. *Biochim. Biophys. Acta* 281:597–624

41. Cammarano, P., Felsani, A., Gentile, M., Gualerzi, C., Romeo, A., Wolf, G. 1972. *Biochim. Biophys. Acta* 281: 625–42

42. Nonomura, Y., Blobel, G., Sabatini, D. D. 1971. *J. Mol. Biol.* 60:303–23

43. Lake, J. A., Sabatini, D. D., Nonomura, Y. 1974. See Ref. 2, pp. 543–57

44. Emanuilov, I., Sabatini, D. D., Lake, J. A., Freienstein, C. 1978. *Proc. Natl. Acad. Sci. USA* 75:1389–93

45. Boublik, M., Hellmann, W. 1978. *Proc. Natl. Acad. Sci. USA* 75:2829–33

46. Unwin, P. N. T., Taddei, C. 1977. *J. Mol. Biol.* 114:491–506

47. Unwin, P. N. T. 1977. *Nature* 269: 118–22

48. Morimoto, T., Blobel, G., Sabatini, D. D. 1972 *J. Cell. Biol.* 52:338–54

49. Morimoto, T., Blobel, G., Sabatini, D. D. 1972. *J. Cell. Biol.* 52:355–66

50. Byers, B. 1967. *J. Mol. Biol.* 26:155–67

51. Hardy, S. J. S., Kurland, C. G., Voynow, P., Mora, G. 1969. *Biochemistry* 8:2897–905

52. Mora, G., Donner, D., Thammana, P., Lutter, L., Kurland, C. G. 1971. *Mol. Gen. Genet.* 112:229–42

53. Hindennach, I., Stöffler, G., Wittmann, H. G. 1971. *Eur. J. Biochem.* 23:7–11

54. Hindennach, I., Kaltschmidt, E., Wittmann, H. G. 1971. *Eur. J. Biochem.* 23:12–16

55. Collatz, E., Lin, A., Stöffler, G., Tsurugi, K., Wool, I. G. 1976. *J. Biol. Chem.* 251:1808–16

56. Lin, A., Collatz, E., Wool, I. G. 1976. *Mol. Gen. Genet.* 144:1–9

57. Kaltschmidt, E., Wittmann, H. G. 1970. *Anal. Biochem.* 36:401–12

58. Wittmann, H. G., Stöffler, G., Hindennach, I., Kurland, C. G., Randall-Hazelbauer, L., Birge, E. A., Nomura, M., Kaltschmidt, E., Mizushima, S., Traut, R. R., Bickle, T. A. 1971. *Mol. Gen. Genet.* 111:327–33

59. Sherton, C. C., Wool, I. G. 1972. *J. Biol. Chem.* 247:4460–67

60. McConkey, E. H. et al. 1979. *Mol. Gen. Genet.* 169:1–6

61. Sherton, C. C., DiCamelli, R. F., Wool,

I. G. 1974. *Methods Enzymol.* 30(F): 354–67

62. Leboy, P. S., Cox, E. C., Flaks, J. G. 1964. *Proc. Natl. Acad. Sci. USA* 52:1367–74

63. Spitnik-Elson, P. 1965. *Biochem. Biophys. Res. Commun.* 18:557–62

64. Kaltschmidt, E., Wittmann, H. G. 1972. *Biochemie* 54:167–75

65. Waller, J. P. 1964. *J. Mol. Biol.* 10: 319–36

66. Spitnik-Elson, P. 1970. *FEBS Lett.* 7:214–16

67. Westermann, P., Bielka, H. 1973. *Mol. Gen. Genet.* 126:349–56

68. Terao, K., Ogata, K. 1972. *Biochim. Biophys. Acta* 285:473–82

69. Goerl, M., Welfle, H., Bielka, H. 1978. *Biochim. Biophys. Acta* 519:418–27

70. Howard, G. A., Ramjoué, H. P. R., Gordon, J. 1977. In *Translation of Natural and Synthetic Polynucleotides*, ed. A. B. Legocki, pp. 305–11. Poznan, Poland: Univ. Agric.

71. Lin, A., Wool, I. G. 1974. *Mol. Gen. Genet.* 134:1–6

72. Howard, G. A., Traugh, J. A., Croser, E. A., Traut, R. R. 1975. *J. Mol. Biol.* 93:391–404

73. Terao, K., Ogata, K. 1975. *Biochim. Biophys. Acta* 402:214–29

74. Martini, O. H. W., Gould, H. J. 1975. *Mol. Gen. Genet.* 142:317–31

75. Issinger, O. G., Beier, H. 1978. *Mol. Gen. Genet.* 160:297–309

76. Welfle, H., Goerl, M., Bielka, H. 1978. *Mol. Gen. Genet.* 163:101–12

77. Panyim, S., Chalkley, R. 1971. *J. Biol. Chem.* 246:7557–60

78. Hayashi, K., Matsutera, E., Ohba, Y. 1974. *Biochim. Biophys. Acta* 342: 185–94

79. Ritter, E., Wittmann-Liebold, B. 1975. *FEBS Lett.* 60:153–55

80. Vandekerckhove, J., Rombauts, W., Peeters, B., Wittmann-Liebold, B. 1976. *Hoppe-Seylers Z. Physiol. Chem.* 356: 1955–76

81. Wittmann-Liebold, B., Marzinzig, E., Lehmann, A. 1976. *FEBS Lett.* 68: 110–14

82. Wittmann-Liebold, B., Pannenbecker, R. 1976. *FEBS Lett.* 68:115–18

83. Pettersson, I., Hardy, S. J. S., Liljas, A. 1976. *FEBS Lett.* 64:135–38

84. Welfle, H., Stahl, J., Bielka, H. 1971. *Biochim. Biophys. Acta* 243:416–19

85. Welfle, H., Stahl, J., Bielka, H. 1972. *FEBS Lett.* 26:228–32

86. Delaunay, J., Creusot, F., Schapira, G. 1973. *Eur. J. Biochem.* 39:305–12

87. Chatterjee, S. K., Kazemie, M., Matthaei, H. 1973. *Hoppe-Seylers Z. Physiol. Chem.* 354:481–86

88. Lastick, S. M., McConkey, E. H. 1976. *J. Biol. Chem.* 251:2867–75

89. Sherton, C. C., Wool, I. G. 1974. *J. Biol. Chem.* 249:2258–67

90. Wittmann-Liebold, B., Dzionara, M. 1976. *FEBS Lett.* 65:281–83

91. Wittmann-Liebold, B., Dzionara, M. 1976. *FEBS Lett.* 61:14–19

92. Martini, O. H. W., Temkin, R., Jones, A., Riley, K., Gould, H. J. 1975. *FEBS Lett.* 56:205–11

93. Westermann, P., Heumann, W., Bielka, H. 1976. *FEBS Lett.* 62:132–35

94. Voynow, P., Kurland, C. G. 1971. *Biochemistry* 10:517–24

95. Weber, H. J. 1972. *Mol. Gen. Genet.* 119:233–48

96. Deusser, E. 1972. *Mol. Gen. Genet.* 119:249–58

97. Hardy, S. J. S. 1975. *Mol. Gen. Genet.* 140:253–74

98. Kurland, C. G., Voynow, P., Hardy, S. J. S., Randall, L., Lutter, L. 1969. *Cold Spring Harbor Symp. Quant. Biol.* 34:17–24

99. Kurland, C. G. 1970. *Science* 169: 1171–77

100. Kurland, C. G. 1972. *Ann. Rev. Biochem.* 41:377–408

101. Ulbrich, N., Wool, I. G. 1978. *J. Biol. Chem.* 253:9049–52

102. Burrell, H. R., Horowitz, J. 1975. *FEBS Lett.* 49:306–9

103. Burrell, H. R., Horowitz, J. 1977. *Eur. J. Biochem.* 75:533–44

104. Horne, J. R., Erdmann, V. A. 1972. *Mol. Gen. Genet.* 119:337–44

105. Gray, P. N., Bellemare, G., Monier, R., Garrett, R. A., Stöffler, G. 1973. *J. Mol. Biol.* 77:133–52

106. Yu, R. S. T., Wittmann, H. G. 1973. *Biochim. Biophys. Acta* 324:375–85

107. Wrede, P., Erdmann, V. A. 1977. *Proc. Natl. Acad. Sci. USA* 74:2706–9

108. Erdmann, V. A. 1976. *Progr. Nucl. Acid Res. Mol. Biol.* 18:45–90

109. Blobel, G. 1971. *Proc. Natl. Acad. Sci. USA* 68:1881–85

110. Lebleu, B., Marbaix, G., Huez, G., Temmerman, J., Burny, A., Chantrenne, H. 1971. *Eur. J. Biochem.* 19:264–69

111. Petermann, M. L., Pavlovec, A. 1971. *Biochemistry* 10:2770–75

112. Grummt, F., Grummt, I., Erdmann, V. A. 1974. *Eur. J. Biochem.* 43:343–48

113. Terao, K., Takahashi, Y., Ogata, K. 1975. *Biochim. Biophys. Acta* 402: 230–37

114. Newton, I., Rinke, J., Brimacombe, R. 1975. *FEBS Lett.* 51:215–18
115. Svoboda, A. J., McConkey, E. H. 1978. *Biochem. Biophys. Res. Commun.* 81:1145–52
116. Richter, D., Isono, K. 1977. In *Curr. Top. Microbiol.* 76:81–123
117. Martini, O., Irr, J., Richter, D. 1977. *Cell* 12:1127–31
118. Silverman, R. H., Atherly, A. G. 1977. *Develop. Biol.* 56:200–5
118a. Warner, J. R., Gorenstein, C. 1978. *Nature* 275:338–9
119. Krystosek, A., Bitte, L. F., Cawthon, M. L., Kabat, D. 1974. See Ref. 2, pp. 855–70
120. Kabat, D. 1970. *Biochemistry* 9:4160–75
121. Loeb, J. E., Blat, C. 1970. *FEBS Lett.* 10:105–8
122. Eil, C., Wool, I. G. 1971. *Biochem. Biophys. Res. Commun.* 43:1001–9
123. Eil, C., Wool, I. G. 1973. *J. Biol. Chem.* 248:5122–29
124. Eil, C., Wool, I. G. 1973. *J. Biol. Chem.* 248:5130–36
125. Kabat, D. 1971. *Biochemistry* 10:197–203
126. Stahl, J., Welfle, H., Bielka, H. 1972. *FEBS Lett.* 26:233–36
127. Traugh, J. A., Mumby, M., Traut, R. R. 1973. *Proc. Natl. Acad. Sci. USA* 70:373–76
128. Barden, N., Labrie, F. 1973. *Biochemistry* 12:3096–3102
129. Delaunay, J., Loeb, J. E., Pierre, M., Schapira, G. 1973. *Biochim. Biophys. Acta* 312:147–51
130. Martini, O. H. W., Gould, H. J. 1973. *Biochim. Biophys. Acta* 295:621–29
131. Walton, G. M., Gill, G. N. 1973. *Biochemistry* 12:2604–11
132. Ventimiglia, F. A., Wool, I. G. 1974. *Proc. Natl. Acad. Sci. USA* 71:350–54
133. Ashby, C. D., Roberts, S. 1975. *J. Biol. Chem.* 250:2546–55
134. Gressner, A. M., Wool, I. G. 1974. *Biochem. Biophys. Res. Commun.* 60:1482–90
135. Gressner, A. M., Wool, I. G. 1974. *J. Biol. Chem.* 249:6917–25
136. Traugh, J. A., Porter, G. G. 1976. *Biochemistry* 15:610–16
137. Rankine, A. D., Leader, D. P. 1975. *FEBS Lett.* 52:284–87
138. Kaerlein, M., Horak, I. 1976. *Nature* 259:150–51
139. Lastick, S. M., Nielsen, P. J., McConkey, E. H. 1977. *Mol. Gen. Genet.* 152:223–30
140. Leader, D. P., Rankine, A. D., Coia, A. A. 1976. *Biochem. Biophys. Res. Commun.* 71:966–74
141. Schubart, U. K., Shapiro, S., Fleischer, N., Rosen, O. M. 1977. *J. Biol. Chem.* 252:92–101
142. Trewavas, A. 1973. *Plant Physiol.* 51:760–67
143. Kristiansen, K., Plesner, P., Kruger, A. 1978. *Eur. J. Biochem.* 83:395–403
144. Zinker, S., Warner, J. R. 1976. *J. Biol. Chem.* 251:1799–1807
145. Becker-Ursic, D., Davies, J. 1976. *Biochemistry* 15:2289–96
146. Kaerlein, M., Horak, I. 1978. *Eur. J. Biochem.* 90:463–69
147. Rankine, A. D., Leader, D. P., Coia, A. A. 1977. *Biochim. Biophys. Acta* 474:293–307
148. Leader, D. P., Coia, A. A. 1975: *Biochim. Biophys. Acta* 519:224–32
149. Marvaldi, J., Lucas-Lenard, J. 1977. *Biochemistry* 16:4320–27
150. Anderson, W. M., Grundholm, A., Sells, B. H. 1975. *Biochem. Biophys. Res. Commun.* 62:669–76
151. Gordon, J. 1971. *Biochem. Biophys. Res. Commun.* 44:579–86
152. Rahmsdorf, H. J., Herrlich, P., Pai, S. H., Schweiger, M., Wittmann, H. G. 1973. *Mol. Gen. Genet.* 127:259–71
153. Rahmsdorf, H. J., Pai, S. H., Ponta, H., Herrlich, P., Roskoski, R., Schweiger, M., Studier, F. W. 1974. *Proc. Natl. Acad. Sci. USA* 71:586–89
154. Rask, L., Waolinder, O., Zetterqvist, Ö., Engström, L. 1970. *Biochim. Biophys. Acta* 221:107–113
155. Kuo, J. F., Greengard, P. 1969. *J. Biol. Chem.* 244:3417–19
156. Li, H. C., Brown, G. G. 1973. *Biochem. Biophys. Res. Commun.* 53:875–81
157. Gressner, A. M., Wool, I. G. 1976. *Nature* 259:148–50
158. Blat, C., Loeb, J. E. 1971. *FEBS Lett.* 18:124–26
159. Gressner, A. M., Wool, I. G. 1976. *J. Biol. Chem.* 251:1500–4
160. Cawthon, M. L., Bitte, L. F., Krystosek, A., Kabat, D. 1974. *J. Biol. Chem.* 249:275–78
161. Roberts, S., Ashby, C. D. 1978. *J. Biol. Chem.* 253:288–96
162. Correze, C., Pinell, P., Nunez, J. 1972. *FEBS Lett.* 23:87–91
163. Treloar, M. A., Treloar, M. E., Kisilevsky, R. 1977. *J. Biol. Chem.* 252:6217–21
164. Lastick, S. M., McConkey, E. H. 1978. In *ICN-UCLA Symposia on Molecular and Cellular Biology: Cell Reproduction* ed. E. R. Dirksen, D. M. Prescott, C. F. Fox, 12:61–69

165. Leader, D. P., Coia, A. A., Fahmy, L. H. 1978. *Biochem. Biophys. Res. Commun.* 83:50–58
166. Rupp, R. G., Humphrey, R. M., Shaeffer, J. R. 1976. *Biochim. Biophys. Acta* 418:81–92
167. Gressner, A. M., Greiling, H. 1977. *FEBS Lett.* 74:77–81
168. Gressner, A. M., Greiling, H. 1978. *Exp. Mol. Pathol.* 28:39–47
169. Kabat, D. 1972. *J. Biol. Chem.* 247: 5338–44
170. Möller, W., Groene, A., Terhorst, C., Amons, R. 1972. *Eur. J. Biochem.* 25:5–12
171. Terhorst, C., Wittmann-Liebold, B., Möller, W. 1972. *Eur. J. Biochem.* 25:13–19
172. Deusser, E., Wittmann, H. G. 1972. *Nature* 238:269–70
173. Thammana, P., Kurland, C. G., Deusser, E., Weber, J., Maschler, R., Stöffler, G., Wittmann, H. G. 1973. *Nature New Biol.* 242:47–49
174. Leader, D. P., Coia, A. A. 1978. *FEBS Lett.* 90:270–74
174a. Terao, K., Ogata, K. 1979. *J. Biochem.* In press
174b. Terao, K., Ogata, K. 1979. *J. Biochem.* In press
175. Issinger, O. G., Traut, R. R. 1974. *Biochem. Biophys. Res. Commun.* 59: 829–36
176. Möller, W., Slobin, L. I., Amons, R., Richter, D. 1975. *Proc. Natl. Acad. Sci. USA* 72:4744–48
177. van Agthoven, A. J., Maassen, J. A., Möller, W. 1977. *Biochem. Biophys. Res. Commun.* 77:989–98
178. Amons, R., van Agthoven, A., Pluijms, W., Möller, W., Higo, K., Itoh, T., Osawa, S. 1977. *FEBS Lett.* 81:308–10
179. Amons, R., van Agthoven, A., Pluijms, W., Möller, W. 1977. *FEBS Lett.* 86:282–84
180. Oda, G., Strøm, A. R., Visentin, L. P., Yaguchi, M. 1974. *FEBS Lett.* 43: 127–30
181. Alberghina, F. A. M., Suskind, S. R. 1967. *J. Bacteriol.* 94:630–49
182. Lamon, E. W., Bennett, J. C. 1970. *Proc. Soc. Exptl. Biol. Med.* 134:968–70
183. Noll, F., Bielka, H. 1970. *Mol. Gen. Genet.* 106:106–13
184. Meyer-Bertenrath, J. G. 1967. *Hoppe-Seylers Z. Physiol. Chem.* 348:645–50
185. Noll, F., Bielka, H. 1970. *Arch. Geschwaltsforsch.* 35:338–46
186. Wikman-Coffelt, J., Howard, G. A., Traut, R. R. 1972. *Biochim. Biophys. Acta* 277:671–76
187. Busch, H., Busch, R. K., Spohn, W. H., Wikman, J., Daskal, Y. 1971. *Proc. Soc. Exptl. Biol. Med.* 137:1470–78
188. Busch, R. K., Spohn, W. H., Daskal, Y., Busch, H. 1972. *Proc. Soc. Exptl. Biol. Med.* 140:1030–33
189. Stöffler, G., Wool, I. G., Lin, A., Rak, K. H. 1974. *Proc. Natl. Acad. Sci. USA* 71:4723–26
190. Howard, G. A., Smith, R. L., Gordon, J. 1976. *J. Mol. Biol.* 106:623–37
191. Fischer, N., Stöffler, G., Wool, I. G. 1978. *J. Biol. Chem.* 253:7355–60
192. Wool, I. G., Stöffler, G. 1976. *J. Mol. Biol.* 108:201–18
193. Terhorst, C., Möller, W., Laursen, R., Wittmann-Liebold, B. 1973. *Eur. J. Biochem.* 34:138–52
194. Möller, W. 1974. See Ref. 2, pp. 711–31
195. Bodley, J. W., Lin, L. 1970. *Nature* 227:60–61
196. Lelong, J. C., Gros, D., Gros, F., Bollen, A., Maschler, R., Stöffler, G. 1974. *Proc. Natl. Acad. Sci. USA* 71:248–52
197. Highland, J. H., Bodley, J. W., Gordon, J., Hasenbank, R., Stöffler, G. 1973. *Proc. Natl. Acad. Sci. USA* 70:142–50
198. Grasmuk, H., Nolan, R. D., Drews, J. 1977. *Eur. J. Biochem.* 79:93–102
199. Hamel, E., Koka, M., Nakamoto, T. 1972. *J. Biol. Chem.* 247:805–14
200. Lucas-Lenard, J., Lipmann, F. 1971. *Ann. Rev. Biochem.* 40:409–48
201. Richter, D., Möller, W. 1974. In *Lipmann Symposium: Energy, Biosynthesis and Regulation in Molecular Biology,* ed. D. Richter, pp. 524–33. New York, Berlin: Walter de Gruyter
202. Martin, T. E., Wool, I. G. 1969. *J. Mol. Biol.* 43:151–61
203. Jones, B. L., Nagabhushan, N., Gulyas, A., Zalik, S. 1972. *FEBS Lett.* 23: 167–70
204. Gualerzi, C., Janda, H. G., Passow, H., Stöffler, G. 1974. *J. Biol. Chem.* 249:3347–55
205. Delaunay, J., Creusot, F., Schapira, G. 1973. *Eur. J. Biochem.* 39:305–12
206. Geisser, M., Tischendorf, G. W., Stöffler, G., Wittmann, H. G. 1973. *Mol. Gen. Genet.* 127:111–28
207. Geisser, M., Tischendorf, G. W., Stöffler, G. 1973. *Mol. Gen. Genet.* 127:129–45
208. Hasenbank, R., Guthrie, C., Stöffler, G., Wittmann, H. G., Rosen, L., Apirion, D. 1973. *Mol. Gen. Genet.* 127:1–8
208a. Ogata, K., Terao, K., Morita, T., Sugano, H. 1968. *GANN. Monog.* 6:109–23
209. Heady, J. E., McConkey, E. H. 1970.

Biochem. Biophys. Res. Commun. 40:30–36

210. Craig, N., Perry, R. P. 1971. *Nature New Biol.* 229:75–80

211. Wu, R. S., Warner, J. R. 1971. *J. Cell Biol.* 51:643–52

211a. Nabeshima, Y. I., Tsurugi, K., Ogata, K. 1975. *Biochim. Biophys. Acta* 414: 30–43

212. Miller, O. L. Jr., Beatty, B. R., Hamkalo, B. A., Thomas, C. A. Jr. 1970. *Cold Spring Harbor Symp. Quant. Biol.* 35:505–12

213. Kumar, A., Warner, J. R. 1972. *J. Mol. Biol.* 63:233–46

214. Miller, O. L. Jr., Beatty, B. R. 1969. *Science* 164:955–57

215. Miller, O. L. Jr., Bakken, A. H. 1972. *Acta Endocrinol.* 168:155–77

216. Reeder, R. H., Brown, D. D. 1970. *J. Mol. Biol.* 51:361–77

217. Hackett, P. B., Sauerbier, W. 1975. *J. Mol. Biol.* 91:235–56

218. Liau, M. C., Hurlbert, R. B. 1975. *J. Mol. Biol.* 98:321–32

219. Trapman, J., Planta, R. J. 1975. *Biochim. Biophys. Acta* 414:115–25

220. Schibler, U., Hagenbüchle, O., Wyler, T., Weber, R., Boseley, P., Telford, J., Birnstiel, M. L. 1976. *Eur. J. Biochem.* 68:471–80

221. Dawid, I. B., Wellauer, P. K. 1976. *Cell* 8:443–48

222. Reeder, R. H., Higashinakagawa, T., Miller, O. Jr. 1976. *Cell* 8:449–54

223. Tischendorf, G. W., Zeichhardt, H., Stöffler, G. 1974. *Mol. Gen. Genet.* 134:187–208

224. Tischendorf, G. W., Zeichhardt, H., Stöffler, G. 1974. *Mol. Gen. Genet.* 134:209–23

225. Lake, J. A., Pendergast, M., Kahan, L., Nomura, M. 1974. *Proc. Natl. Acad. Sci. USA* 71:4688–92

226. Lake, J. A., Kahan, L. 1975. *J. Mol. Biol.* 99:631–44

227. Noll, F., Bommer, U. A., Lutsch, G., Theise, H., Bielka, H. 1978. *FEBS Lett.* 87:129–31

Ann. Rev. Biochem. 1979. 48:755–82

PEPTIDE NEUROTRANSMITTERS

◆12025

Solomon H. Snyder and Robert B. Innis

Departments of Pharmacology and Experimental Therapeutics, and Psychiatry and Behavioral Sciences, Johns Hopkins University School of Medicine, Baltimore, Maryland 21205

CONTENTS

PERSPECTIVES AND SUMMARY

In examining the role of peptides or any other substances as neurotransmitters, we must first review what is meant by a neurotransmitter and what evidence is required before one can conclude that a given substance fulfills a neurotransmitter role. Neurotransmitters are usually thought of as chemicals released by nerve endings which alter the firing activity of adjacent neurons or which, in the periphery, affect the physiological activity of various organs. A neurotransmitter must exist in neurons with high concentrations of the transmitter in nerve endings, be released upon depolarization of the neuron, and its actions should mimic the effects of nerve stimulation.

755

0066-4154/79/0701-0755$01.00

This last crucial criterion is often impossible to ascertain in the brain where one cannot readily examine in isolation a given neuron acting on another neuron. At the neuromuscular junction acetylcholine elicits a rapid excitation, while in the central nervous system amino acids such as GABA (γ-aminobutyric acid) and glycine produce potent inhibition by hyperpolarizing neurons. Some peptide neurotransmitter candidates do not display such clear-cut effects and have been proposed instead as neuromodulators, which modify responses of neurons to other neurotransmitters. Differentially labeling substances as neuromodulators or neurotransmitters may only be a semantic artifice to mask our ignorance of synaptic mechanisms. The numerous ways that a substance, localized in nerve endings and released upon depolarization, can affect other neurons may all "transmit" important information. For the purposes of simplicity we consider substances that are highly localized to specific neuronal systems in the brain, released on depolarization, and produce changes in neuronal activity as "neurotransmitters." Some of the more recently described peptides may not yet have been tested for all these criteria.

For many years the biogenic amines, acetylcholine, norepinephrine, serotonin, and dopamine were the neurotransmitters examined by investigators. Amino acids such as glycine, glutamate, and γ-aminobutyric acid, have been recognized as quantitatively major neurotransmitters in the past decade. Peptides are the most recent addition to the group of neurotransmitter candidates. During the past four years the number of recognized peptide neurotransmitter possibilities has increased to about a dozen. Most of these are small, comprising three to fifteen amino acids, though peptides up to a hundred amino acids in length have been proposed as possible transmitters. Some are excitatory, others inhibitory, while for many peptides the exact synaptic effects are unclear. Besides their localization in the brain, many peptides are concentrated in neuronal and other types of cells in the intestine, but hardly in any other place in the body, perhaps because their embryological precursor cells migrated only to the intestine and the central nervous system (1). No known peptides seem to account for more than 1% of synapses in most brain regions, which is similar to the density of biogenic amine synapses, but much less than amino acid transmitters such as γ-aminobutyric acid which, in some brain regions, accounts for 35–40% of synapses. However, several peptide transmitters are highly concentrated in these areas of the central nervous system and subserve important functions such as pain perception and emotional regulation.

The peptide transmitters reviewed here include the enkephalins, substance P, neurotensin, vasoactive intestinal polypeptide, cholecystokinin, bradykinin, thyrotropin-releasing hormone, and somatostatin. Of all the peptide transmitters, the greatest amount of research has dealt with the opioid peptides, enkephalins and endorphins. These were discovered as naturally occurring substrates for the opiate receptor. A thorough understanding of their functions requires some discussion of the opiate receptor, which is the best characterized peptide transmitter receptor.

OPIATE RECEPTORS AND ISOLATION OF OPIOID PEPTIDES

The potent, stereospecific effects of opiates that are blocked by selective antagonists suggested that these drugs must act at highly specific receptor sites. Successful labeling of these sites by the reversible binding of ^{3}H-opiates required ^{3}H-ligands of high specific activity, and careful washing to remove nonspecific binding which would greatly exceed the number of opiate receptors, about 30 pmoles/g in the brain (2–5). Evidence that these binding sites mediate the pharmacological effects of opiates involved demonstrations of a close correlation between biological and binding potencies in the same strips of guinea pig intestine in which opiates inhibit electrically induced contractions (6, 7), and general correlation between binding and analgesic potency. Studies labeling receptors in intact animals indicate that pharmacologic effects of opiates are determined by the proportion of receptors occupied by the drugs (73).

Opiate receptors were shown, in elegant studies by Kuhar and collaborators with autoradiographic techniques, to be highly concentrated in areas regulating functions affected by opiates, especially the integration of pain perception and euphoria (8–11). The clear functional relevance of opiate receptors in the brain suggested that they were not merely evolutionary vestiges but might serve to interact with some normally occurring opiate-like substance. Other investigations fitted with such a hypothesis. Electrical stimulation of some brain areas produces analgesia which can be blocked by the opiate antagonist naloxone which suggests that such analgesia involves release of a normally occurring morphine-like substance (12–14). Moreover, naloxone, while producing virtually no overt behavioral effects itself, does appear to worsen pain, an effect that can be reproduced by other chemically different opiate antagonists in a stereospecific manner (15, 16).

Some properties of opiate receptor binding also suggested the existence of an endogenous ligand for the opiate receptor. Injections of opiate agonists and antagonists in vivo elicit an increase in receptor binding measured in isolated membranes (17). Antagonists, which in the body's normal sodium environment have greater potency at opiate receptors than corresponding agonists, were more potent in enhancing receptor binding by unmasking new binding sites. Such findings suggested that the drugs displaced an endogenous ligand from opiate receptors. Similarly, incubation of receptor containing brain membranes released into the medium a substance with considerable affinity for the opiate receptor (18).

Two approaches were taken in searching for endogenous opioid substances in the brain. Hughes (19) showed that brain extracts would mimic morphine's effects upon electrically induced contractions of the mouse vas deferens and the guinea pig ileum. These effects on intestinal smooth muscle are exerted by opiates in proportion to their analgesic properties (7). The morphine-like action of brain extracts was blocked by naloxone. Pasternak et al (20) and Terenius & Wahlstrom (21) showed that brain extracts contain a material that competes for opiate receptor binding. The specificity of these effects is evident in that regional variations of the morphine-like substance within the brain closely parallel variations in opiate receptor density (20).

Using a bioassay of its effects on smooth muscle, Hughes et al (22) isolated the morphine-like activity from pig brain and observed that it consists of two pentapeptides, methionine enkephalin (met-enkephalin—Tyr-Gly-Gly-Phe-Met-OH) and leucine enkephalin (leu-enkephalin—Tyr-Gly-Gly-Phe-Leu-OH). Using as an assay the competition for receptor binding, Simantov & Snyder (23) isolated the same two pentapeptides from calf brain, and confirmed the findings of Hughes et al (22). Once the amino acid sequences of the enkephalins were established it became evident that the five amino acids of met-enkephalin were contained within the 91-amino acid sequence of the peptide β-lipotropin isolated a decade earlier from the pituitary by Li (24). β-Lipotropin had been identified on the basis of its relatively weak and nonselective influences upon lipolysis. It was known to contain within it the sequence of β-melanocyte-stimulating hormone (β-MSH) and the amino acids 4–10 of adrenocorticotropin (ACTH). Several groups showed that various fragments of β-lipotropin, all containing the met-enkephalin sequence, did possess opiate activity (25–29). The most potent of these, which appears to account for essentially all the opiate-like activity of pituitary gland extracts, is β-endorphin, the carboxyl-terminal 31 amino acids of β-lipotropin. Other fragments of β-lipotropin with opioid activity include α-endorphin (β-LPH$_{61-76}$), γ-endorphin (β-LPH$_{61-77}$), and C' fragment (β-LPH$_{61-87}$). Even prior to identification of the enkephalin amino acid sequence, Goldstein and collaborators (30) had observed an opiate-like effect of pituitary extracts upon the contractions of guinea pig ileum.

INTERACTIONS OF OPIATES AND OPIOID PEPTIDES WITH OPIATE RECEPTORS

Both met-enkephalin and leu-enkephalin as well as β-endorphin have potencies fairly similar to morphine in competing for opiate receptor binding (22, 31–33) and all have similar potencies at opiate receptor containing smooth muscle preparations (34). The apparent lesser potencies of enkephalins than of opiates and β-endorphin in some studies (25, 26) reflects the extremely rapid degradation of enkephalins by tissue membranes, which can be prevented by proteolysis inhibitors such as bacitracin (35, 36) and o-phenanthroline (37). Since the enkephalin receptor is apparently identical with the opiate receptor, which has been characterized in considerable detail, properties of the opiate receptor provide a paradigm in exploring receptors for other neurotransmitter peptides.

One of the most striking aspects of opiate pharmacology is the continuum of agonist to antagonist activity. Rather small chemical changes converting an N-methyl moiety to an N-allyl, N-cyclopropylmethyl, or N-cyclobutylmethyl, transform opiate agonists such as morphine into antagonists. Some pure antagonists, such as naloxone, lack any analgesic or euphoric "agonist" effects, while many antagonists are "contaminated" with some agonist properties. Some mixed agonist-antagonists are quite potent analgesics but are much less addictive than pure opiate agonists. One goal of opiate receptor research was to understand the difference between agonist and antagonist activities, which are fundamental properties of many drug classes. In initial studies no differences were detected in the relative affinities or slopes of displacement curves for matching agonists and antagonists (2, 3), because ^{3}H-naloxone-binding to brain membranes was routinely assayed with buffers that lacked metal ions. While binding of the antagonist ^{3}H-naloxone was unaffected by sodium (2, 3), binding of the agonist ^{3}H-etorphine was reduced by sodium (4). Adding low concentrations of sodium in parallel experiments dramatically differentiated agonists and antagonists (17, 38). Sodium enhanced binding of ^{3}H-antagonists while markedly reducing the binding of ^{3}H-agonists. The sodium effect was quite potent, being elicited by as little as 0.5 mM sodium chloride, and specific, since lithium, whose atomic radius and biological activity are similar to sodium, was the only other ion that mimicked the effects of sodium though it did so to a lesser degree (38). By contrast, potassium, rubidium, and cesium did not discriminate agonists and antagonists but merely decreased the binding of both at higher concentrations of the ion (38). Potencies of drugs in inhibiting the binding of ^{3}H-naloxone in the absence or presence of sodium showed variations along the continuum of pure agonist-mixed agonist and antagonist-pure antagonist. Pure antagonists were just as potent in the presence as in the absence of sodium. Antagonists with limited agonist activity were reduced in potency about twofold by sodium while pure agonists became 12–60 times weaker in the presence of sodium. Mixed agonist-antagonists, with potential as relatively nonaddicting analgesics, displayed a reduction in potency of about three- sixfold in the presence of sodium (38). This "sodium index" of opiates has proved useful in screening large numbers of drugs for therapeutic potential as analgesics. The potent effects of sodium upon opiate receptor binding suggested that it might be an ion whose permeability changes relate to the pharmacological effects of opiates. Indeed, Enero (39) showed that influences of morphine on contractions of the cat nictitating membrane were facilitated by low concentrations of sodium, while the antagonist effects of naloxone were more apparent at high sodium concentrations. Neurophysiological studies of opiate effects on neuronal firing showed that their inhibitory actions involve blockade of the excitatory effects of acetylcholine and glutamate by interfering at the sodium ion conductance modulator (40).

Other investigations suggest that besides sodium, cyclic AMP may play a second messenger role in the effects of opiates. In neuroblastoma-glioma cells in culture, opiates depress cyclic AMP formation with potencies that correspond to their affinities for opiate receptors (41, 47). A role for cyclic AMP is suggested by the observation that in vivo (48) and in vitro (49) morphine increases brain levels of cyclic GMP.

Adenylate cyclase activity in many cases is modulated by guanine nucleotides (42–46). Guanine nucleotides also affect receptor binding for substances whose effects may be mediated by adenylate cyclase. Guanosine triphosphate (GTP) selectively reduces the affinity of agonists but not antagonists to β-noradrenergic (45, 46) receptors and elicits similar effects at glucagon receptors (42, 44). This effect is thought to be related in some way to the coupling of hormone receptor to adenylate cyclase and is not manifest with antagonists. Antagonists presumably do not elicit physiological responses because they do not stimulate coupling of receptor to the cyclase. Recently, Blume observed a selective effect of GTP and its nonmetabolized analogue Gpp(NH)p on opiate receptor binding in brain membranes (50) and neuroblastoma-glioma cells (51) in which GTP and Gpp(NH)p but not GMP, ATP, ADP, or AMP reduced ^{3}H-opiate binding. While these studies showed an effect with the antagonist ^{3}H-naloxone as well as with ^{3}H-agonists, we found GTP to selectively lower only agonist binding to opiate receptors (52), a finding subsequently confirmed by A. J. Blume (personal communication). GTP also selectively decreases the affinity of agonists for the dopamine receptors, which is where dopamine selectively enhances adenylate cyclase activity (53). However, similar GTP effects are observed at α-noradrenergic receptors whose association with adenylate cyclase is unclear (54), and at angiotensin II receptors (55) which have no known relationship to cyclic AMP.

Adenylate cyclases are also regulated by divalent cations and there is evidence that regulatory sites of guanine nucleotides and divalent cations are related. At α-noradrenergic receptors, divalent cations specifically antagonize influences of guanine nucleotides (56). Divalent cations decrease the inhibitory potency of GTP on ^{3}H-agonist binding. These effects are most marked with manganese, less so with magnesium, and are minimally elicited by calcium. Similar interactions have also been observed at the β-receptor (57, 58).

Divalent cation-guanine nucleotide interactions also regulate opiate receptor binding. Earlier, divalent cations had been shown to influence opiate receptor binding (59, 60). In micromolar concentrations, manganese enhances the binding of ^{3}H-agonists but not ^{3}H-antagonists. The effect is apparent with higher concentrations of magnesium but not elicited by calcium at all. Chelation of all endogenous divalent cations by EDTA reduces ^{3}H-agonist binding, an effect opposite to that of the divalent cations themselves, which suggests that endogenous divalent cations regulate the opiate receptor. EGTA, which chelates calcium, but not magnesium or manganese, has no influence on receptor binding. The physiological relevance of these findings was established by experiments of Enero (39) which showed that effects of morphine on cat nictitating membrane contractions were enhanced selectively by low concentrations of manganese. As with α- and β-noradrenergic receptors the divalent cations interact at the opiate receptor selectively with guanine nucleotides. In the absence of guanine nucleotides, only modest elevations of ^{3}H-agonist binding to the opiate receptor are elicited with manganese while guanine nucleotides amplify the stimulatory effect of manganese and, to a lesser extent, of magnesium and calcium.

Thus, opiate receptors, like other neurotransmitter and hormone receptors, are regulated by guanine nucleotides and divalent cations. This regulation may provide a clue to mechanisms whereby recognition of opiates or opioid peptides at receptor sites is transformed into physiological and pharmacological alterations. It is unclear whether these effects indicate a direct relationship to adenylate cyclase. Microtubule function is also influenced by guanine nucleotides and divalent cations and might conceivably interact with the opiate receptor.

LOCALIZATION AND BIOSYNTHESIS OF OPIOID PEPTIDES

Immunohistochemical techniques have provided powerful tools for localizing peptides in the brain and providing evidence for neurotransmitter roles. In these procedures specific antisera to the peptides are applied to slides of thin sections of tissue and the antisera-peptide complexes visualized by use of a second antibody labeled with a fluorescent dye or horseradish peroxidase. Utilizing these techniques specific systems of enkephalin neurons have been localized to various brain areas (61–67). Enkephalin neurons are localized to areas that are highly enriched in opiate receptors. In the dorsal spinal cord, in layers I and II, opiate receptors are apparently concentrated on nerve endings of the sensory neurons which are relatively specific for pain perception (68–70). In this area, enkephalin is contained in small interneurons which appear to synapse upon the sensory terminals and may act by inhibiting the release of substance P, the peptide transmitter of thin unmyelinated pain fibers (71). Enkephalin and opiate receptors are also highly concentrated in parts of the limbic system, the emotion-regulating portion of the brain.

Though the highest concentrations of β-endorphin occur in the pituitary gland, the brain also contains it in levels about 10% those of enkephalin (72). β-Endorphin-containing neurons have been localized to specific areas of the brain, with their cell bodies being most enriched in the hypothalamus (74–76). The localization of β-endorphin does not correspond to that of opiate receptors which indicates that the physiological opiate receptors are for the most part "enkephalin receptors." Antisera used to localize β-endorphin also cross-react with β-lipotropin and some extraction procedures convert β-lipotropin to β-endorphin so that some immunoreactive β-endorphin represents β-lipotropin (77–79). Similarly, immunoreactive β-melanocyte-stimulating hormone (β-MSH) identified in tissues, human plasma, and spinal fluid may also represent β-lipotropin and/or β-endorphin (80).

Besides localization in a specific population of neurons, selective release of a substance upon depolarization of nerves provides valuable information that it may be a neurotransmitter. In brain slices (81), synaptosomes (82), and guinea pig ileum strips (83, 84) depolarization of nervous tissue is associated with enkephalin release.

Much understanding of biogenic amine function has been derived from studies of their biosynthesis and turnover regulation. Some limited studies have indicated an incorporation of radioactive amino acids into enkephalins (85, 86). However, the greatest success has come in studies of the biosynthesis of β-endorphin in the pituitary gland where its immediate precursor, β-lipotropin is well known. Conversion of β-lipotropin to β-endorphin by a trypsin-like activity is readily demonstrable (87–89). A major advance has come in the recognition that β-lipotropin and ACTH share a common precursor, a 31,000 molecular weight peptide referred to as big ACTH or 31-K ACTH (90–93) (Figure 1). Some of these studies, utilizing cell-free systems (92, 94), have been extended to characterize the detailed mechanisms of synthesis. Sequence determination of the 31-K precursor indicates that the N-terminal region comprising at least 40 amino acids does not contain either ACTH or β-lipotropin sequences. The messenger RNA for the 31-K precursor has been purified and used as a template to make a complementary copy of DNA with reverse transcriptase. This complementary DNA has been inserted into a bacterial plasmid and cloned in large quantities. They can then be used as a hybridization probe to measure levels of the corresponding messenger RNA in various tissues (E. Herbert, personal communication).

Several approaches have been used to evaluate functional roles of opioid peptides. Some workers have developed enkephalin analogues with increased potency and stability that may prove useful therapeutic agents. Substituting the glycine$_2$ with D-alanine or D-proline (95–98) enhances the stability of enkephalins while other modifications (99) result in analogues

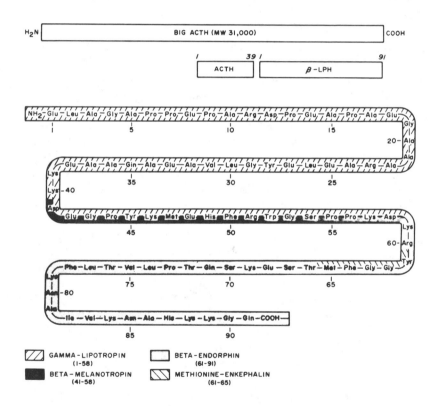

Figure 1 The precursor relationships of several endogenous opioid peptides. Adrenocorticotrophic hormone (ACTH) and β-lipotropin (β-LPH) are contained in the C-terminal portion of big ACTH or 31-K ACTH. The amino acid structure of β-LPH is given below in the snake-like diagram. Smaller peptides which are contained in β-LPH are shown by different shadings. For example, amino acids 61–65 of β-LPH are the same as those of methionine enkephalin.

up to 30,000 times more potent than enkephalin in vivo and active when given orally. Changes in opioid peptides levels have been examined under various conditions. Thus, chronic treatment with neuroleptic antischizophrenic drugs has been shown to produce substantial increases in total brain levels of enkephalin (100). No consistent changes in enkephalin levels have been detected in opiate-addicted animals (101–103). In the pituitary gland, β-endorphin, being part of the same precursor molecule as ACTH, is released together with ACTH when animals are stressed (104). Beneficial psychiatric effects of β-endorphin (105) and des-tyr-α-endorphin (β-LPH$_{62-77}$) (106) have been reported.

SUBSTANCE P

Substance P (SP) is the oldest putative peptide neurotransmitter. In 1931 von Euler & Gaddum (107) discovered SP (P for powder) from a lyophilyzed intestinal extract that caused contraction of the isolated rabbit jejunum. Studying extracts from various organs, these researchers found that SP is also present in the brain and later presented evidence that it is peptidergic in nature (108). Pernow (109) found that SP is at least ten times more concentrated in the dorsal or sensory root of the spinal cord than in the ventral or motor root. This discovery led Lembeck in 1953 (110) to postulate that SP is involved in the transmission of primary sensory neurons. Thus, SP joined the ranks of putative neurotransmitters with the distinction of being the first to be a peptide.

Nearly 40 years elapsed between the initial discovery of SP and its purification from bovine hypothalami by Chang & Leeman (111). These workers had been trying to isolate the still elusive corticotropin-releasing factor and serendipitously found fractions from bovine hypothalami that caused copious salivation in rats after intravenous administration (112). This sialogogic activity was not affected by cholinergic or adrenergic blocking agents but was destroyed by incubation with pepsin or pronase. Leeman and co-workers isolated this peptide and found it to be an undecapeptide (Table 1). The identification of this sialogogic peptide as SP was suspected because of their common biological actions, namely, contraction of isolated intestinal smooth muscle, lowering of blood pressure, and production of saliva. In addition, the sialogogic peptide shared many chemical and physical characteristics reported for the partially purified preparations of SP. The identity was definitively proven when Studer et al (113) reported the amino acid sequence of SP isolated from equine intestine. The structure was identical to that of the sialogogic peptide from hypothalamus.

The first criterion that a putative neurotransmitter must fulfill is that it be present in nervous tissue. The initial report of the discovery of SP noted

Table 1   Peptide structures

Peptide	Structure
Substance P	Arg-Pro-Lys-Pro-Gln-Gln-Phe-Phe-Gly-Leu-Met-NH$_2$
Bradykinin	Arg-Pro-Pro-Gly-Phe-Ser-Pro-Phe-Arg-COOH
CCK-33	Lys-Ala-Pro-Ser-Gly-Arg-Val-Ser-Met-Ile-Lys-Aln-Leu-Gln-Ser-Leu-Asp-Pro-Ser-His-Arg-Ile-Ser-Asp-Arg-Asp-Tyr$^{(SO3)}$-Met-Gly-Trp-Met-Asp-Phe-NH$_2$
CCK-8	Asp-Tyr$^{(SO3)}$-Met-Gly-Trp-Met-Asp-Phe-NH$_2$
Neurotensin	pyroGlu-Leu-Tyr-Glu-Asn-Lys-Pro-Arg-Arg-Pro-Tyr-Ile-Leu-OH
VIP	His-Ser-Asp-Ala-Val-Phe-Thr-Asp-Asn-Tyr-Thr-Arg-Leu-Arg-Lys-Gln-Met-Ala-Val-Lys-Lys-Tyr-Leu-Asn-Ser-Ile-Leu-Asn-NH$_2$
Angiotensin II	Asp-Arg-Val-Tyr-Val-His-Pro-Phe
TRH	pyroGlu-His-ProNH$_2$
Somatostatin	Ala-Gly-Cys-Lys-Asn-Phe-Phe-Trp-Lys-Thr-Phe-Thr-Ser-Cys-OH

its presence by bioassay in extracts of equine brain (107). To obtain a more precise idea of the location of a putative neurotransmitter, one can submit nervous tissue to subcellular fractionation. Many classical neurotransmitters are located in vesicles within the presynaptic bouton. By homogenizing brain in isotonic sucrose, the bouton breaks off from its axon, and the membrane edges fuse to form a spherical structure called a synaptosome. The synaptosomes may then be partially purified by differential centrifugation, and several investigators (114–116) have found SP in the synaptosome-rich fraction.

Another way to determine the presence and distribution of a putative neurotransmitter in the brain is to assay dissected regions of the brain. The limits of sensitivity of this technique are determined by the sensitivity of the assay [about 1 fmole by radioimmunoassay (117)] and the ability to dissect small regions. The limit of dissection has probably been attained by "punching out" with a needle small areas from a thin section of brain. Brownstein and co-workers (118) have used this technique and found especially high levels of SP in the substantia nigra, interpeduncular nucleus, and several hypothalamic nuclei. The punch technique does not allow one to say, for instance, whether a substance is in neurons or glia. The technique of indirect immunofluorescence staining shows the cellular localization of an antigen. Elegant immunohistochemical work by Hokfelt and co-workers (119) has shown that SP is contained within neuronal cell bodies (where it is presumably synthesized) and in fibers with varicosities (which presumably correspond to synaptic boutons strung along the axon). The most detailed anatomical distribution of SP has resulted from electron microscopic studies using the peroxidase antiperoxidase technique developed by Sternberger et al (120). Pickel and co-workers (121) have shown in electron micrographs that SP is present in presynaptic boutons and especially in large (60–80 nm diameter) vesicles.

At a higher level of fulfilling neurotransmitter criteria, one should be able to delineate neuronal pathways in the brain that utilize the transmitter. For SP three such tracts have been established by lesioning techniques. One extends from the striatum to the substantia nigra (122), another from the habenula to the interpeduncular nucleus (123), and the third from the dorsal spinal ganglion to the dorsal horn of the spinal cord. This last tract has received the greatest attention, and it appears likely that SP is a major neurotransmitter of primary sensory afferents. Immunohistochemical studies by Hokfelt et al (124) have shown SP in perhaps 20% of dorsal sensory neurons and in fibers of the substantia gelatinosa zone of the dorsal horn of the spinal cord. Dorsal rhizotomy causes a marked reduction of these fluorescent SP-positive fibers in the dorsal horn, which supports their direct connection with the dorsal ganglion SP-positive cell bodies. In addition, Takahashi & Otsuka (125) have shown that SP accumulates on the ganglion side of a dorsal root ligation, which suggests that SP is transported somatofugally (i.e. away from the cell body).

Neurotransmitters should be released upon neuronal depolarization. Otsuka & Konishi (126) have shown calcium-dependent release of radioimmunoassayable SP from rat spinal cord following stimulation by an electric field or a high potassium medium. A calcium-dependent release of SP has also been demonstrated in the brain using a synaptosomal preparation stimulated by a high potassium bathing medium (114). Following release from the presynaptic bouton, SP presumably binds to a postsynaptic receptor, and this interaction leads to a cascade of events that culminate in a change of ionic permeability of the postsynaptic membrane. Nakata et al (127) have reported a saturable, high affinity, specific binding of ^{3}H-SP to brain synaptic membranes. The regional distribution of ^{3}H-SP-specific binding correlated with the levels of endogenous SP, and the relative affinities of SP fragments and related peptides generally paralleled their biological activities. After binding to the postsynaptic receptor, the action of SP may be mediated by cAMP. Duffy & Powell (128) have shown stimulation of adenyl cyclase in human brain by SP and Narumi & Maki (129) have shown that SP increases cAMP levels and stimulates neurite extension in cultured neuroblastoma cells.

Exogenously administered neurotransmitters should mimic the actions of the endogenous compound. This has been partially demonstrated for SP as a primary sensory neurotransmitter. Primary afferent fibers form excitatory synapses with interneurons in the substantia gelatinosa and with motoneurons in the ventral horn of the spinal cord. SP is 1,000–10,000 times as potent as glutamate, another possible primary sensory neurotransmitter, in depolarizing spinal motoneurons of the newborn rat (130). The on-set and off-set of SP-induced depolarization is slower than that of glutamate, which suggests a longer diffusion time for the larger molecule SP. The depolarizing action appears to be a direct effect on motoneurons, because the response still occurs in a low calcium, high magnesium medium which blocks synaptic transmission. Furthermore, Wright & Roberts (131) have shown neurophysiologically that unilateral dorsal rhizotomy induces supersensitivity of interneurons in the deafferented side of the spinal cord to the iontophoretic application of an SP analogue, eledosin-related peptide.

Mechanisms for the synthesis and the termination of SP action are not clear. The neuronal cell body is certainly equipped to synthesize peptides, but experiments have not been reported on the specific synthesis of SP. Thus, it is unknown, for example, whether SP is synthesized as part of a larger precursor molecule. After release from the axon terminal, the mechanism of inactivation of SP is unknown. Although many neurotransmitters are inactivated by reuptake into the presynaptic neuron, Jessell et al (132) were unable to find evidence for an active uptake mechanism for SP by brain slices in vitro. However, SP can be degraded by peptidases in brain extracts, and Quik & Jessell (133) have shown disappearance of SP immunoreactivity on incubation with rat brain slices. Furthermore, they showed that the rate of disappearance was higher with slices from an SP-rich area (substantia nigra) than with slices from an SP-poor area (cerebellum).

Drug effects often provide convincing evidence for neurotransmitter identification. Except for the endogenous opiate peptides, very little pharmacology is known for the putative peptide neurotransmitters. β-(4-chlorophenyl)-γ-aminobutyric acid (Baclofen, Lioresal) blocks the actions of exogenously administered SP on spinal motorneurons (134), although in other areas of the central nervous system it appears to be a less specific antagonist of SP (135).

An interesting interaction between opiates and SP has been investigated by Jessell & Iversen (71, 136). They have shown potassium-stimulated release of SP from primary afferent terminals of rat trigeminal nerve, which corresponds to a dorsal root for the facial area. Morphine (10 μM), levorphanol (5 μM), and [D-Ala^2-Met^5]-enkephalin (1 μM) were able to suppress the release of SP. This action of the opiates was stereospecifically sensitive to naloxone, a pure opiate antagonist, and could not be elicited by the inactive enantiomer of levorphanol or dextrorphan (50 μM). Jessell & Iversen (136) hypothesize, therefore, that part of the mechanism of opiate-induced analgesia involves presynaptic inhibition of SP release. It is noteworthy that areas of the brainstem and spinal cord found by Atweh & Kuhar (9) to have high densities of opiate receptors correspond to areas containing SP terminals. Furthermore, LaMotte et al (68) have shown a marked reduction of opiate receptors in the substantia gelatinosa after lesion of the dorsal roots in monkeys, which suggests that opiate receptors are indeed located presynaptically on primary afferent terminals.

If SP is a primary afferent neurotransmitter, perhaps it is also the mediator of the axon reflex. The axon reflex refers to local vasodilatation that occurs around an injured area and is due to activity in sensory nerves. Dale (137) suggested that the vasodilatation was caused by release of the same neurotransmitter that the primary afferent terminal releases in the spinal cord. That is, the afferent neuron would contain and release only one neurotransmitter. Subsequently, this suggestion has been generalized into "Dale's Law"—"one neuron, one neurotransmitter." SP has been detected immunohistochemically in nerve endings of the skin and often in association with blood vessels (138). No one has been able to detect the release of SP following antidromic stimulation, but SP is one of the most potent vasodilator substances known (139), and it could mediate the axon reflex. However, Chan-Palay et al (140) have demonstrated both serotonin (a putative neurotransmitter) and SP in the same neurons of the rat's brainstem. Thus, SP may both prove Dale's original hypothesis regarding the axon reflex and disprove "Dale's Law" within the central nervous system.

NEUROTENSIN

Just as SP was serendipitously isolated by Chang & Leeman (111) while searching for corticotropin-releasing factor, so neurotensin (NT) was discovered by Carraway & Leeman (141) while trying to isolate SP. While testing fractions of hypothalamic extracts from a sulfoethyl Sephadex column for sialogogic activity, another distinct biological activity was detected. Certain fractions injected intravenously caused vasodilatation in the exposed cutaneous areas of anesthetized rats, and even larger doses caused gross cyanosis. The isolated peptide was named neurotensin because of its presence in neural tissue and its ability to affect blood pressure.

Although many biological actions of the tridecapeptide NT have been discovered, its normal physiologic role is unclear. NT is a potent vasodilatory and hypotensive agent, induces hyperglycemia, and may play a role in the inflammatory response (142). Peripherally administered NT induces release of adrenocorticotrophic hormone, luteinising hormone, and follicle-stimulating hormone. Nemeroff et al (143) have shown that intracerebroventricular administration of NT produces a marked dose-related decrease in body temperature of rats. Other effects of centrally administered NT include decreased locomotor activity and increased susceptibility to pentobarbital-induced sedation and mortality. Perhaps the most startling

CNS effect is that described by Clineschmidt & McGuffin (144)—as little as 0.2 fmol NT injected intracisternally in mice elicited analgesia, which makes neurotensin at least one thousand times more potent an analgesic than enkephalin.

The very first criterion of a neurotransmitter is that it be present in neural tissue. This was initially demonstrated, of course, by Carraway & Leeman (141), who isolated NT from bovine hypothalami. Radioimmunoassay revealed highest levels of NT immunoreactivity in the hypothalamus and basal ganglia (145, 146). Subcellular fractionation of rat hypothalamus showed a strong association of NT immunoreactivity with the synaptosomal fraction, consistent with a neurotransmitter role for NT in the brain. Uhl, Kuhar & Snyder (147) reported an immunohistochemical mapping of NT in the rat CNS. NT immunofluorescence is associated with neuronal cell bodies and fibers, with notable densities in the substantia gelatinosa zone of the spinal cord and the basal hypothalamus. The immunohistochemical distribution of NT resembles that of enkephalin (148), which suggests that they may share common functions.

Uhl et al (149, 150) have used ^{125}I-NT to bind to apparent NT receptors in membrane preparations from rat brain. Under conditions that minimize degradation of the polypeptide, ^{125}I-NT binds saturably, reversibly, and with high affinity (a dissociation constant of 3 nM). The binding is displaced by NT sequence fragments with potencies generally paralleling their activity in peripheral systems. Furthermore, the regional distribution of NT receptor binding roughly parallels the distribution of endogenous NT determined by radioimmunoassay. Lazarus et al (151) using ^{125}I-NT and Kitabgi et al (152) using ^{3}H-NT have found similar results in their binding studies.

Iversen et al (153) have demonstrated release of NT from hypothalamic slices of rat brain. The resting rate of release equalled 0.25% of the total tissue stores per minute and exposure to a high potassium medium caused a threefold increase. Potassium-stimulated release could not be measured in a calcium-free medium, a finding consistent with the role of NT as a neurotransmitter. There have been few investigations of the neurophysiologic actions of NT following its presumptive release from presynaptic terminals. Although Guyenet & Aghajanian (154) found no effect of iontophoretically applied neurotensin, Young, Uhl & Kuhar (155) found that iontophoresis in the area of the locus coeruleus produced a rapid and reversible inhibition of firing in 13 of 35 cells tested.

There have been no reports on the synthesis of NT or the termination of its actions. There are no pharmacologic agonists or antagonists of NT. Clearly, therefore, more work needs to be done to establish NT as a neurotransmitter.

VASOACTIVE INTESTINAL POLYPEPTIDE

Vasoactive intestinal polypeptide (VIP) was isolated in 1970 from porcine small intestine by Said & Mutt (156). Its amino acid sequence and biological actions are similar to those of the well known hormones, secretin and glucagon (157). VIP has a broad spectrum of biological activities, including vasodilatation, stimulation of glycogenolysis, lipolysis and insulin secretion, inhibition of gastric acid production, and stimulation of secretion by the pancreas and small intestine (158). In addition, VIP is likely to play a pathophysiologic role as the mediator of the human watery diarrhea syndrome called "pancreatic cholera" (159).

Bryant et al (160), Larsson et al (161), and Said & Rosenberg (162) all reported, in 1976, the existence of VIP in the central nervous system. By radioimmunoassay the highest levels of VIP are in the cortex. Fuxe et al (163) used immunohistochemistry to show that VIP is contained in cortical neurons. Thus, VIP became the first putative peptide neurotransmitter to be contained in cortical neurons, whereas most other peptide neurotransmitters had been found mainly in the hypothalamus. In addition, VIP has been found in peripheral nerves innervating the intestinal wall (160, 161). These fine nerves, demonstrated immunohistochemically, do not disappear following vagotomy or sympathectomy, which suggests that these fibers may be intrinsic to the intestine. Hokfelt et al (164) have studied sympathetic ganglia of the guinea pig and shown dense networks of VIP-positive varicose nerve fibers around principal ganglion cells. Larsson et al (165) have studied the distribution of VIP immunoreactivity in intrapancreatic ganglia of several other species. Some species (e.g. pig and cat) show dense VIP-positive nerves around nonimmunoreactive nerve cell bodies, while in the canine pancreas, ganglia contain strongly reactive nerve cell bodies that give off axons that appear to innervate vessels and endocrine and exocrine cells. Using the peroxidase antiperoxidase technique, Larsson (166) has shown, with the electron microscope, the localization of VIP immunoreactivity to vesicles of neuronal terminals, which gives further support to VIP's role as a neurotransmitter.

Giachetti et al (167) have found that VIP is present in highest concentration in the synaptosomal fraction from rat brain. In addition, they showed calcium-dependent, potassium-stimulated release of VIP from synaptosomal pellets. Emson et al (168) extended these results by showing the presence of VIP in the vesicles released after a brief lysing of the synaptosomal pellet. Furthermore, they showed calcium-dependent, potassium-stimulated release of immunoreactive VIP from superfused rat hypothalamus in vitro.

Robberecht et al (169) have studied the binding of [125] I-VIP to guinea pig brain membranes and found specific, saturable, and reversible binding to receptors that are regulated by guanine nucleotides. The biological effects of VIP may be mediated by cAMP, since VIP stimulates adenylate cyclase in plasma membrane preparations from fat cells, liver, and pancreas (170). Presumably this receptor and its associated adenylate cyclase mediate the effects of VIP released from peripheral nerves.

CHOLECYSTOKININ

Cholecystokinin (CCK) is an intestinal hormone that causes contraction of the gallbladder and secretion by the pancreas. It was originally isolated by Ivy et al (171) from duodenal mucosa as a substance that caused contraction of the gallbladder, and was called cholecystokinin. Harper & Raper (172) independently discovered a substance from duodenal mucosa that caused secretion by the pancreas and called it pancreozymin. From analysis of their common biological actions and comigration in all purification steps, Mutt & Jorpes (173) correctly suggested that CCK and pancreozymin were, in fact, the same hormone. CCK is the most commonly used name. It has been isolated from porcine intestine as a triacontatriapeptide, CCK-33 (174), which has been sequenced (175) and synthesized (176).

Vanderhaeghen et al (177) discovered a peptide from boiling water extracts of vertebrate brains which reacted with antigastrin antibodies. Dockray (178) was the first to recognize that this immunoreactivity is actually due to CCK and that the cross-reactivity derived from the fact that gastrin and CCK share the same C-terminal pentapeptide sequence. Muller, Straus & Yalow (179) found two components in porcine cerebral cortex: one resembling intact CCK-33 and the other resembling the C-terminal octapeptide of CCK (CCK-8). Using several different antisera Rehfeld (180) has studied the distribution of the molecular forms of CCK in the central nervous system and small intestine of man. In both tissues the predominant molecular form is CCK-8; (80% of cerebral immunoreactivity and 60% of intestinal immunoreactivity is due to CCK-8). The second most abundant form corresponds to the C-terminal tetrapeptide, with only a small percent of immunoreactivity due to CCK-33. Only the posterior pituitary contains any immunoreactive gastrin (181). Among different brain regions, the cerebral cortex contains by far the highest concentration of CCK-8—200–2300 pmol/g tissue. Considering the large size of human cerebral cortex, one can make the startling calculation that the human brain contains 1–2 mg of CCK-8, whereas other hormonal peptides are present in microgram amounts.

Straus et al (182) have demonstrated by peroxidase antiperoxidase immunohistochemistry the presence of CCK-8-like immunoreactivity in rabbit cortical neurons and white matter. Innis et al (183) have used the indirect immunohistofluorescence method to study the distribution of

CCK-8 in the rat brain. They have mapped CCK-8-positive cells in the cerebral cortex, the hypothalamus, and periaqueductal grey and shown a diffuse distribution of varicose fibers.

Gibbs, Young & Smith (184) discovered that intraperitoneal injections of CCK-8 elicit satiety in rats. One might have assumed that this effect was mediated by interaction with the hypothalamic satiety center and the endogenous CCK-8. However, Nemeroff et al (185) have shown that intravenous administration of CCK-8 inhibits eating in rats, whereas centrally administered CCK-8 is effective only at much higher doses. This would suggest that CCK-8-induced satiety is mediated by peripheral receptors and that centrally administered CCK-8 is only effective after passing the blood brain barrier.

In summary, CCK-8 has been demonstrated to be present in large quantities in vertebrate brain, especially in the cortex, only by radioimmunoassay, and, by immunohistochemistry, in neuronal cell bodies and varicose fibers. The functions of these CCK-8 systems are unclear, and much more evidence needs to be gathered before CCK-8 can be considered a neurotransmitter.

BRADYKININ

Bradykinin (BK) was discovered by Rocha e Silva et al (186) as a biologically active factor released from the α_2-globulin fraction of blood by incubation with trypsin or snake venom (from *Bothrops jararaca*). This factor was called bradykinin because it caused slow contraction of the guinea pig ileum. This nonapeptide may be involved in many pathophysiologic reactions, including inflammation, cardiovascular shock, hypertension, pain generation, and rheumatoid arthritis (187). Centrally administered BK causes stereotyped behavior and cardiovascular responses (188). By injecting into different regions of the brain, Correa & Graeff (189) have shown that the lateral septal area mediates the hypertensive response of centrally administered BK. In addition, Ribeiro, Corrado & Graeff (190) have shown that intraventricular BK is antinociceptive and that the site of this action may be along the cerebral aqueduct. Thus, BK is a centrally active peptide, although it is unclear whether these are physiologic or pharmacologic actions.

Hori (191) has shown by bioassay the presence of a BK-like substance in brain extracts. Using a sensitive and specific radioimmunoassay, Correa & Snyder (in preparation) have found immunoreactivity from rat brain extracts that coelutes on gel filtration columns with synthetic BK. Furthermore, Correa, Innis & Snyder (192) have used the indirect immunofluorescence technique to map the distribution of BK-like immunoreactivity in the rat brain. BK-positive neuronal cell bodies are only found in the hypothalamus. Fibers with large varicosities are in many regions of the brain, including the lateral septal area and periaqueductal grey, areas that may mediate the hypertensive and analgesic actions, respectively, of centrally administered BK.

At this point, therefore, there is growing evidence of a role for BK as a neurotransmitter.

ANGIOTENSIN II

Study of the renin-angiotensin system has centered predominantly around the kidney, but evidence has recently accumulated supporting the role of a central renin-angiotensin system in the physiologic and pathophysiologic control of water balance and blood pressure. In the periphery (193), the decapeptide angiotensin I is formed from the plasma precursor angiotensinogen by the proteolytic action of renin, originally discovered in the kidney by Tigerstedt & Bergman in 1898 (194). The physiologically potent octapeptide angiotensin II (Ang II) is formed by the removal of the C-terminal histidyl-leucine from angiotensin I by angiotensin-converting enzyme, found in particularly high concentration in lung. Ang II is a potent vasoconstrictor and stimulant of secretion by the adrenal cortex of aldosterone, which causes sodium retention by the kidney. Angiotensins I and II are degraded by a number of hydrolytic enzymes collectively referred to as angiotensinases. All components of the renin-angiotensin system have been found in the brain and have been named the isorenin-angiotensin system (195). Isorenin was discovered by Fischer-Ferraro et al (196) and Ganten et al (197) in dog and rat brain. By physical, biochemical, and immunological criteria, brain isorenin is similar to kidney renin. However, cathepsin D copurifies with isorenin activity (198) and it may be that these two activities derive from just one enzyme. Brain angiotensinogen was isolated by Ganten et al (197) and found to have physicochemical properties similar to plasma angiotensinogen. Its presence in the brain has been confirmed by the generation of angiotensin I following injection of renin into the brain (199). The concentration of angiotensin-converting enzyme activity in brain is lower than in lung but higher than that in plasma and kidney (199). And finally, in order to terminate the action of Ang II, angiotensinase activity is quite high in brain (200).

Angiotensin I-like immunoreactivity has been demonstrated in extracts of dog brain, with the highest concentration (85 ng/g tissue) found in the hypothalamus (196). Ang II has been visualized in the central nervous system with the indirect immunofluorescence technique by Fuxe et al (201) and with the peroxidase antiperoxidase technique by Changaris et al (202). These two groups agree that Ang II immunoreactivity exists in neuronal cell bodies and fibers, but there are many discrepancies as to its regional distribution. It is unclear whether these discrepancies are due to different techniques or to different specificities of the antisera.

Using ^{125}I-labeled Ang II Bennett & Snyder (203) and Sirett et al (204) have described specific binding to bovine and rat brain membranes. In addition, Bennett & Snyder (in preparation) have studied binding with the radioiodinated angiotensin antagonist, ^{125}I-sarcosine$_1$ leucine$_8$-angiotensin II. Unlike the opiate receptor, the concentration of sodium ion cannot differentiate agonists from antagonists but does have a marked effect on receptor binding, which was studied in detail using ^{125}I-Ang II. In the presence of 150 mM sodium, ^{125}I-Ang II-binding to calf cerebellar membranes can be resolved into distinct high affinity and low affinity sites with respective dissociation constants (K_D) of 0.08 nM and 0.5 nM. In the presence of 10 mM sodium ion, the high affinity ($K_D = 0.08$ nM) binding sites are lost and 90% of the lower affinity (now $K_D = 0.36$ nM) sites are also lost. Considering the close association of the renin-angiotensin system with fluid and electrolyte balance, the effect of sodium may play a physiologic role in the action of Ang II. These findings resemble the classic studies of angiotensin receptors in the adrenal by Catt and associates (205).

Ang II induces drinking following either intravenous or intracerebral administration (206). Mapping studies suggest that areas around the third ventricle mediate this dipsogenic response (207). Moreover, neurons in this area have been shown to be stimulated by iontophoretically applied Ang II, and this response is antagonized by the inhibitor saralasin (208). Intraventricularly administered Ang II stimulates release of antidiuretic hormone and thus increases intravascular volume (209). In addition to this indirect pressor response, Ang II has direct central vasopressor action. When administered intraventricularly, Ang II is believed to increase peripheral vascular resistance by stimulating the subnucleus medials of the midbrain; when administered intravenously it does so by stimulating the area postrema, which lacks a blood-brain barrier (210). The central vasopressor effects of Ang II may have pathophysiologic significance, because elevated levels of an Ang II-like peptide have been reported in the cerebrospinal fluid of patients with essential hypertension (211) and of spontaneously hypertensive rats (212). Furthermore, following perfusion of the brain's ventricular system with the antagonist saralasin, there is a significant decrease in blood pressure in these hypertensive rats (212).

THYROTROPIN-RELEASING HORMONE

Two hypothalamic hormones are putative peptide neurotransmitters: thyrotropin-releasing hormone (TRH) and growth hormone release-inhibiting hormone (somatostatin). These trophic hormones are synthesized in the hypothalamus and released from nerves into portal vessels. They are transported to the pituitary and either stimulate or inhibit the release of the target hormone by cells of the anterior and intermediate lobes of the pituitary. TRH induces release of thyroid-stimulating hormone. TRH was the first hypothalamic hormone to be isolated, and it has been identified as pyroGlu-His-Pro-NH_2 by Boler et al (213) and Burgus et al (214). In the rat brain only 20% of TRH-like immunoreactivity is in the hypothalamus (215). Significant amounts are also found in the thalamus (34%), cerebrum (27%), and brainstem (19%). Using the micropunch technique, Brownstein et al (216) reported the detailed distribution of radioimmunoassayable TRH in the rat hypothalamus. By far the highest concentration was found in the median eminence. After surgical isolation of the medial basal hypothalamus, Brownstein et al (217) showed a 75% decrease of TRH in this area but no change of TRH in other brain areas. This finding suggests that the majority of TRH in the basal hypothalamus is synthesized elsewhere and transported to this area. However, an alternative interpretation is that the hypothalamic cells respond to denervation by making and storing less TRH. Hokfelt et al (218) have visualized TRH with immunohistochemical methods and found TRH-positive fluorescence in several discrete regions of the rat brain and spinal cord. Winokur et al (219) have localized TRH-like immunoreactivity to subcellular fractions of rat brain consisting largely of synaptosomes and synaptic vesicles. Warberg et al (220) have demonstrated calcium-dependent, potassium-stimulated release of radioimmunoassayable TRH from hypothalamic synaptosome-rich fractions.

Very little is known of the synthesis of TRH. McKelvy & Grimm-Jorgensen (221) have shown that guinea pig hypothalamic cultures and whole newt brain incorporate ^{3}H-proline into TRH, but it is unknown, for instance, whether TRH is synthesized via an mRNA template. Parker et al (222) attempted to measure the uptake of ^{3}H-TRH into synaptosomes, but actually only measured the uptake of ^{3}H-proline. Thus, TRH may be inactived by proteolysis and subsequent uptake of its constituent amino acids.

Binding of ^{3}H-TRH to pituitary plasma membranes appears to be associated with a specific TRH receptor that mediates the known physiologic action of TRH at the pituitary (223). Burt & Snyder (224) have demonstrated binding of ^{3}H-TRH to extrapituitary brain membranes. Two binding sites with dissociation constants of 50 nM and 5 μM were found. The high affinity site resembles that in pituitary membranes. There is electrophysiological evidence of a TRH receptor in that iontophoretically applied TRH increases the firing of spinal motoneurons (225) and decreases the firing rate in several supraspinal regions (226).

Centrally administered TRH causes a variety of biological and behavioral effects, including antagonism of sedation and hypothermia induced by barbiturates and ethanol, increased locomotor activity, and anorexia (227). Some of the first evidence suggesting that TRH is a centrally active peptide actually came from human studies showing an improvement in depressive symptoms following intravenous administration (228). Many subsequent trials were performed to see whether TRH would be an effective treatment for depression, but the results have been generally disappointing (227).

SOMATOSTATIN

Somatostatin (SS) inhibits release of growth hormone by the pituitary. It was isolated from bovine hypothalami by Brazeau et al (229) and found to be a cyclic polypeptide containing 14 amino acids. In addition to inhibiting the secretion of growth hormone, SS inhibits the release of thyrotropin and prolactin (230). In the periphery SS is found in the stomach, intestine, and pancreas (231), and it inhibits the secretion of glucagon, insulin, and gastrin (232). Pancreatic tumors containing SS have been described in humans (233).

Only about 25% of the total SS in the brain is in the hypothalamus, although the median eminence has by far the highest concentration (234). Complete deafferentation of the medial basal hypothalamus or a knife cut just anterior to this region causes a 70–80% decrease of SS in this area (235). These results are consistent with the idea that SS is synthesized in cell bodies of the preoptic and periventricular nuclei and then transported posteriorly. In fact, SS-containing cell bodies have been demonstrated in these areas (231). Immunohistochemistry has also shown SS-like fluorescence in small neurons of dorsal ganglia and in fibers of the substantia gelatinosa region of the spinal cord (236).

Subcellular preparations from the hypothalamus and amygdala show that over 70% of SS immunoreactivity is localized in the synaptosomal fraction (237). A calcium-dependent, potassium-stimulated release of SS has been demonstrated in rat pituitary (238) and hypothalamus (153). To our knowledge, no one has described specific binding of radiolabeled SS to plasma membranes. Electrophysiologic evidence of a receptor has been reported as both excitatory (239) and inhibitory (240) following iontophoretic application. Neurotrophic actions of centrally administered SS include a decrease in motor activity, an increase in the anesthesia time of barbiturates, and an increase in appetite (241).

Other interesting and promising putative peptide neurotransmitters that can not be discussed because of space limitations include carnosine (242) and other hypothalamic peptides such as luteinising hormone-releasing hormone, oxytocin, and vasopressin (243).

ACKNOWLEDGMENTS

R. B. Innis is a recipient of a grant from the Insurance Medical Scientist Scholarship Training Fund sponsored by the Mutual of Omaha Insurance Company.

Literature Cited

1. Pearse, A. G. E. 1971. *Nature* 262:92–94
2. Pert, C. B., Snyder, S. H. 1973. *Science* 179:1011–14
3. Pert, C. B., Snyder, S. H. 1973. *Proc. Natl. Acad. Sci. USA* 70:2243–47
4. Simon, E. J., Hiller, J. M., Edelman, I. 1973. *Proc. Natl. Acad. Sci. USA* 70:1947–49
5. Terenius, L. 1973. *Acta Pharmacol. Toxicol.* 33:377–84
6. Creese, I., Snyder, S. H. 1975. *J. Pharm. Exp. Ther.* 194:205–19
7. Kosterlitz, H. W., Waterfield, A. A. 1975. *Ann. Rev. Pharmacol.* 15:29–47
8. Pert, C. B., Kuhar, M. J., Snyder, S. H. 1976. *Proc. Natl. Acad. Sci. USA* 73:3729–33
9. Atweh, S., Kuhar, M. J. 1977. *Brain Res.* 124:53–67
10. Atweh, S., Kuhar, M. J. 1977. *Brain Res.* 129:1–12
11. Atweh, S., Kuhar, M. J. 1977. *Brain Res.* 134:393–405
12. Mayer, D. J., Liebeskind, J. C. 1974. *Brain Res.* 68:73–93
13. Akil, H., Mayer, D. J., Liebeskind, J. C. 1976. *Science* 191:961–62
14. Oliveras, J. L., Hosobuchi, Y., Redjemi, F., Guilbaud, G., Besson, J. M. 1977. *Brain Res.* 120:221–29
15. Jacob, J. J., Tremblay, E. C., Colombel, M. C. 1974. *Psychopharmacologica* 37:217–23
16. Jacob, J. J. C., Ramabadran, K. 1978. *Br. J. Pharmacol.* 64:91–98
17. Pert, C. B., Pasternak, G. W., Snyder, S. H. 1973. *Science* 182:1359–61
18. Pasternak, G. W., Wilson, H. A., Snyder, S. H. 1975. *Mol. Pharmacol.* 11:478–84
19. Hughes, J. T. 1975. *Brain Res.* 88:295–308
20. Pasternak, G. W., Goodman, R., Snyder, S. H. 1975. *Life Sci.* 16:1765–69
21. Terenius, L., Wahlstrom, A. 1974. *Acta Pharmacol.* 35: Suppl. 1, p. 55
22. Hughes, J., Smith, T. W., Kosterlitz, H. W., Fothergill, L. A., Morgan, B. A., Morris, H. R. 1975. *Nature* 258:577–79
23. Simantov, R., Snyder, S. H. 1976. *Proc. Natl. Acad. Sci. USA* 73:2515–19
24. Li, C. H. 1964. *Nature* 201:924–25
25. Bradbury, A. F., Smyth, D. G., Snell, C. R., Birdsall, N. J., Hulme, E. C. 1976. *Nature* 260:793–95
26. Chretien, M., Benjannet, S., Dragon, N., Seidah, N. G., Lis, M. 1976. *Biochem. Biophys. Res. Commun.* 72:472–78
27. Cox, B. M., Goldstein, A., Li, C. H. 1976. *Proc. Natl. Acad. Sci. USA* 73:1821–23
28. Guillemin, R., Ling, N., Burgus, R. 1976. *CR Ac. Sci. D* 282:783–85
29. Ling, N., Burgus, R., Guillemin, R. 1976. *Proc. Natl. Acad. Sci. USA* 73:3942–46
30. Cox, B. M., Opheim, K. E., Teschemacher, H., Goldstein, A. 1975. *Life Sci.* 16:1777–87
31. Simantov, R., Snyder, S. H. 1976. *Endogenous Opioid Peptides,* ed. H. W. Kosterlitz, pp. 41–48. Amsterdam: North-Holland
32. Simantov, R., Snyder, S. H. 1976. *Mol. Pharmacol.* 12:987–98
33. Simantov, R., Childers, S. R., Snyder, S. H. 1978. *Eur. J. Pharmacol.* 47:319–31
34. Smith, T. W., Hughes, J., Kosterlitz, H. W., Sosa, R. P. 1976. *Opiates and Endogenous Opioid Peptides,* ed. H. W. Kosterlitz, pp. 57–62. Amsterdam: North-Holland
35. Miller, R. J., Chang, K.-J., Cuatrecasas, P., Wilkinson, S. 1977. *Biochem. Biophys. Res. Commun.* 74:1311–18
36. Miller, R. J., Chang, K.-J., Cooper, B., Cuatrecasas, P. 1978. *J. Biol. Chem.* 253:531–38
37. Meek, J. L., Yang, H.-Y. T., Costa, E. 1977. *Neuropharmacology* 16:151–54
38. Pert, C. B., Snyder, S. H. 1974. *Mol. Pharmacol.* 10:868–79
39. Enero, M. A. 1977. *Eur. J. Pharmacol.* 45:349–56
40. Zieglgansberger, W., Bayerl, H. 1976. *Brain Res.* 115:111–28
41. Sharma, S. K., Nirenberg, M., Klee, W. A. 1975. *Proc. Natl. Acad. Sci. USA* 72:590–94
42. Lad, P. M., Welton, A. F., Rodbell, M. 1977. *J. Biol. Chem.* 252:5942–46

778 SNYDER & INNIS

43. Rodbell, M. 1978. *Molecular Biology and Pharmacology of Cyclic Nucleotides*, ed. G. Folco, R. Paoletti, pp. 1–12. Amsterdam: Elsevier North-Holland

44. Lin, M. C., Nicosia, S., Lad, P. M., Rodbell, M. 1977. *J. Biol. Chem.* 252:2790–92

45. Lefkowitz, R. J., Williams, L. T. 1977. *Proc. Natl. Acad. Sci. USA* 74:515–19

46. Mukherjee, C., Lefkowitz, R. J. 1976. *Proc. Natl. Acad. Sci. USA* 73:494–98

47. Traber, J., Fischer, K., Latzin, S., Hamprecht, B. 1975. *Nature* 253:120–22

48. Racagni, G., Zsilla, G., Guidotti, A., Costa, E. 1976. *J. Pharm. Pharmacol.* 28:258–60

49. Minneman, K. P., Iversen, L. L. 1976. *Nature* 262:313–14

50. Blume, A. J. 1978. *Proc. Natl. Acad. Sci. USA* 75:1713–17

51. Blume, A. J. 1978. *Life Sci.* 22:1843–52

52. Childers, S. R., Snyder, S. H. 1978. *Life Sci.* 23:759–62

53. Creese, I., Snyder, S. H. 1978. *Eur. J. Pharmacol.* 50:459–61

54. U'Prichard, D. C., Snyder, S. H. 1978. *J. Biol. Chem.* 253:3444–52

55. Glossmann, H., Baukal, A., Catt, K. J. 1974. *J. Biol. Chem.* 249:664–66

56. U'Prichard, D. C., Snyder, S. H. 1979. *J. Neurochem.* In press

57. Bird, S. J., Maguire, M. E. 1978. *Fed. Proc.* 37:1788

58. Williams, L. T., Mullikin, D., Lefkowitz, R. J. 1978. *J. Biol. Chem.* 243:2984–89

59. Pasternak, G. W., Snowman, A. M., Snyder, S. H. 1975. *Mol. Pharmacol.* 11:735–44

60. Simantov, R., Snowman, A. M., Snyder, S. H. 1976. *Mol. Pharmacol.* 12:977–86

61. Elde, R., Hokfelt, T., Johansson, O., Terenius, L. 1976. *Neuroscience* 1:349–51

62. Hokfelt, T., Elde, R., Johannson, O., Terenius, L., Stein, L. 1977. *Neurosci. Lett.* 5:25–31

63. Hokfelt, T., Ljungdahl, D., Terenius, L., Elde, R., Nilsson, G. 1977. *Proc. Natl. Acad. Sci. USA* 74:3081–85

64. Simantov, R., Kuhar, M. J., Uhl, G. R., Snyder, S. H. 1977. *Proc. Natl. Acad. Sci. USA* 74:2167–71

65. Uhl, G. R., Goodman, R. R., Kuhar, M. J., Childers, S. R., Snyder, S. H. 1979. *Brain Res.* In press

66. Watson, S. J., Akil, H., Sullivan, S., Barchas, J. D. 1977. *Life Sci.* 21:733–38

67. Sar, M., Stumpf, W. E., Miller, R. J.,

Chang, K.-J., Cuatrecasas, P. 1978. *J. Comp. Neurol.* 182:17–38

68. LaMotte, C., Pert, C. B., Snyder, S. H. 1976. *Brain Res.* 112:407–12

69. Crain, S. M., Peterson, E. R., Crain, B., Simon, E. J. 1977. *Brain Res.* 133:162–66

70. Hiller, J. M., Simon, E. J., Crain, S. M., Peterson, E. R. 1978. *Brain Res.* 145:396–400

71. Jessell, T. M., Iversen, L. L. 1977. *Nature* 268:549–61

72. Rossier, J., Bayon, A., Vargo, T. M., Minick, S., Ling, N., Bloom, F. E., Guillemin, R. 1977. *Proc. Natl. Acad. Sci. USA* 74:5162–65

73. Hollt, V., Herz, A. 1978. *Fed. Proc.* 37:158–61

74. Watson, S. J., Barchas, J. D., Li, C. H. 1977. *Proc. Natl. Acad. Sci. USA* 74:5155–58

75. Bloom, F., Battenberg, E., Rossier, J., Ling, N., Guillemin, R. 1978. *Proc. Natl. Acad. Sci. USA* 75:1591–95

76. Pelletier, G., Desy, L., Lissitszy, J.-C., Labrie, F., Li, C. H. 1978. *Life Sci.* 22:1799–804

77. Rubinstein, M., Stein, M., Gerber, L. D., Udenfriend, S. 1977. *Proc. Natl. Acad. Sci. USA* 74:3052–55

78. Rubinstein, M., Stein, M., Udenfriend, S. 1977. *Proc. Natl. Acad. Sci. USA* 74:4969–72

79. Liotta, A. S., Suda, T., Krieger, D. T. 1978. *Proc. Natl. Acad. Sci. USA* 75:2950–54

80. Shuster, S., Smith, A., Plummer, N., Thody, A., Clark, F. 1977. *Br. Med. J.* 1:1318–19

81. Iversen, L. L., Iversen, S. D., Bloom, F. E., Vargo, T., Guillemin, R. 1978. *Nature* 271:679–81

82. Smith, T. W., Hughes, J., Kosterlitz, H. W., Sosa, R. P. 1976. *Opiates and Endogenous Opioid Peptides*, ed. H. W. Kosterlitz, pp. 57–62. Amsterdam: North-Holland

83. Schulz, R., Wuster, M., Simantov, R., Snyder, S. H., Herz, A. 1977. *Eur. J. Pharmacol.* 41:347–48

84. Puig, M. M., Gascon, P., Craviso, G. L., Musacchio, J. M. 1977. *Science* 195:419–20

85. Yang, H.-Y. T., Hong, J. S., Fratta, W., Costa, E. 1978. *The Endorphins*, ed. E. Costa, M. Trabucchi, pp. 149–60. New York: Raven

86. Hughes, J., Kosterlitz, H. W., McKnight, A. T. 1978. *Br. J. Pharm.* 63:396P

87. Lazarus, L. H., Ling, N., Guillemin, R.

1976. *Proc. Natl. Acad. Sci. USA* 73:2156–59

88. Crine, P., Benjannet, S., Seidah, N. G., Lis, M., Chretien, M. 1977. *Proc. Natl. Acad. Sci. USA* 74:4276–80

89. Bradbury, A. F., Smyth, D. G., Snell, C. R. 1976. *Biochem. Biophys. Res. Commun.* 69:950–56

90. Mains, R. E., Eipper, B. A., Ling, N. 1977. *Proc. Natl. Acad. Sci. USA* 74:3014–18

91. Mains, R. E., Eipper, B. A. 1978. *J. Biol. Chem.* 253:651–55

92. Roberts, J. L., Herbert, E. 1977. *Proc. Natl. Acad. Sci. USA* 74:5300–4

93. Eipper, B. A., Mains, R. E. 1978. *J. Supramol. Struct.* 8:247–62

94. Roberts, J. L., Phillips, M., Rosa, P. A., Herbert, E. 1978. *Biochemistry.* 17:3609–18

95. Pert, C. B., Pert, A., Chang, J.-K., Fong, B. T. W. 1976. *Science* 194:330–32

96. Szekely, J. I., Ronai, A. Z., Dunai-Kovacs, Z., Miglecz, E., Berzetri, I., Bajusz, S., Graf, L. 1977. *Eur. J. Pharmacol.* 43:293–94

97. Beddell, C. R., Clark, R. B., Hardy, G. W., Lowe, L. A., Ubatuba, F. B., Vane, J. R., Wilkinson, S., Chang, K.-J., Cuatrecasas, P., Miller, R. J. 1977. *Proc. R. Soc. B.* 198:249–65

98. Beddell, C. R., Clark, R. B., Lowe, L. A., Wilkinson, S., Chang, K.-J., Cuatrecasas, P., Miller, R. J. 1977. *Br. J. Pharmacol.* 61:351–56

99. Roemer, D., Buescher, H. H., Hill, R. C., Pless, J., Bauer, W., Cardinaux, F., Closse, A., Hauser, D., Huguenin, R. 1977. *Nature* 268:547–49

100. Hong, J. S., Yang, H.-Y. T., Fratta, W., Costa, E. 1978. *J. Pharmacol. Exp. Ther.* 205:141–47

101. Fratta, W., Yang, H.-Y. T., Hong, J., Costa, E. 1977. *Nature* 268:452–53

102. Childers, S. R., Simantov, R., Snyder, S. H. 1977. *Eur. J. Pharmacol.* 46:289–93

103. Simantov, R., Snyder, S. H. 1976. *Nature* 262:505–7

104. Guillemin, R., Vargo, T., Rossier, J., Minick, S., Ling, N., Rivier, C., Vale, W., Bloom, F. E. 1977. *Science* 197:1367–69

105. Kline, N. S., Li, C. H., Lehmann, H. E., Lajtha, A., Laski, E., Cooper, T. 1977. *Arch. Gen. Psychiatry* 34:1111–13

106. Verhoeven, W. M. A., van Praag, H. M., Botter, P. A., Sunier, A., van Ree, J. M., deWied, D. 1978. *Lancet* 1:1046–47

107. von Euler, U. S., Gaddum, J. H. 1931. *J. Physiol. London* 72:74–87

108. von Euler, U. S. 1936. *Scand. Arch. Physiol.* 73:142–44

109. Pernow, B. 1953. *Acta Physiol. Scand.* 291: Suppl. 105, pp. 1–90

110. Lembeck, F. 1953. *Arch. Exp. Pathol. Pharmakol.* 219:197–213

111. Chang, M. M., Leeman, S. E. 1970. *J. Biol. Chem.* 745:4784–90

112. Leeman, S. E., Hammerschlag, R. 1967. *Endocrinology* 81:803–10

113. Studer, R. O., Trzeciak, A., Lergier, W. 1973. *Helv. Chim. Acta* 56:860–66

114. Schenker, C., Mroz, E. A., Leeman, S. E. 1976. *Nature* 264:790–92

115. Duffy, M. J., Mulhall, D., Powell, D. 1975. *J. Neurochem.* 25:305–7

116. Powell, D., Leeman, S. E., Tregear, G. W., Niall, H. D., Potts, J. T. 1973. *Nature New Biol.* 241:252–54

117. Mroz, E. A., Brownstein, M. J., Leeman, S. E. 1977. *Substance P,* ed. U.S. von Euler, B. Pernow, pp. 147–54. New York: Raven

118. Brownstein, M. J., Mroz, E. A., Kizer, J. S., Palkovits, M., Leeman, S. E. 1976. *Brain Res.* 116:299–305

119. Nilsson, G., Hokfelt, T., Pernow, B. 1974. *Med. Biol.* 52:424–27

120. Sternberger, L. A., Hardy, P. H., Cuculis, J. J., Meyer, H. G. 1970. *J. Histochem. Cytochem.* 18:315–33

121. Pickel, V. M., Reis, D. J., Leeman, S. E. 1977. *Brain Res.* 122:534–40

122. Mroz, E. A., Brownstein, M. J., Leeman, S. E. 1977. *Brain Res.* 125:305–11

123. Mroz, E. A., Brownstein, M. J., Leeman, S. E. 1976. *Brain Res.* 113:597–99

124. Hokfelt, T., Kellerth, J. O., Nilsson, G., Pernow, B. 1975. *Brain Res.* 100:235–52

125. Takahashi, T., Otsuka, M. 1975. *Brain Res.* 87:1–11

126. Otsuka, M., Konishi, S. 1976. *Nature* 264:83–84

127. Nakata, Y., Kusaka, Y., Segawa, T., Yajima, H., Kitagawa, K. 1978. *Life Sci.* 22:259–68

128. Duffy, M. J., Powell, D. 1975. *Biochim. Biophys. Acta.* 385:275–80

129. Narumi, S., Maki, Y. 1978. *J. Neurochem.* 30:1321–26

130. Konishi, S., Otsuka, M. 1974. *Nature* 252:734–35

131. Wright, D. M., Roberts, M. H. 1978. *Life Sci.* 22:19–24

132. Jessell, T., Iversen, L. L., Kanazawa, I. 1977. *Nature* 268:549–51

133. Quik, M., Jessell, T. 1979. *J. Neurochem.* In press

134. Saito, K., Konishi, S., Otsuka, M. 1975. *Brain Res.* 97:177–80

135. Fotherby, K. J., Morrish, N. J., Ryall, R. W. 1976. *Brain Res.* 113:210–13
136. Jessell, T. 1977. *J. Physiol. Lond.* 270:56–57
137. Dale, H. H. 1935. *Proc. Roy. Soc. Med.* 28:319–32
138. Hokfelt, T., Kellerth, J. O., Nilsson, G., Pernow, B. 1975. *Science* 190:889–90
139. Pernow, B., Rosell, S. 1975. *Acta Physiol. Scand.* 93:139–41
140. Chan-Palay, V., Jonsson, G., Palay, S. 1978. *Proc. Natl. Acad. Sci. USA* 75:1582–86
141. Carraway, R. E., Leeman, S. E. 1973. *J. Biol. Chem.* 248:6854–61
142. Leeman, S. E., Mroz, E. A., Carraway, R. E. 1977. *Peptides in Neurobiology,* ed. H. Gainer, pp. 99–144. New York: Plenum
143. Nemeroff, C. B., Bissette, G., Prange, A. J., Loosen, P. T., Barlow, S., Lipton, M. A. 1977. *Brain Res.* 128:485–96
144. Clineschmidt, B. V., McGuffin, J. C. 1977. *Eur. J. Pharmacol.* 46:395–96
145. Carraway, R., Leeman, S. E. 1976. *J. Biol. Chem.* 251:7045–52
146. Uhl, G. R., Snyder, S. H. 1976. *Life Sci.* 19:1827–32
147. Uhl, G. R., Kuhar, M. J., Snyder, S. H. 1977. *Proc. Natl. Acad. Sci. USA* 74:4059–63
148. Snyder, S. H., Uhl, G. R., Kuhar, M. J. 1978. *Centrally Acting Peptides,* ed. J. Hughes, pp. 85–97. London: MacMillan
149. Uhl, G. R., Bennett, J. P. Jr., Snyder, S. H. 1977. *Brain Res.* 130:299–313
150. Uhl, G. R., Snyder, S. H. 1977. *Eur. J. Pharmacol.* 41:89–91
151. Lazarus, L. H., Brown, M. R., Perrin, M. H. 1977. *Neuropharmacol.* 16:625–29
152. Kitabgi, P., Carraway, R., van Rietschoten, J., Granier, C., Morgat, J. L., Menez, A., Leeman, S., Freychet, P. 1977. *Proc. Natl. Acad. Sci. USA* 74:1846–50
153. Iversen, L. L., Iversen, S. D., Bloom, F., Douglas, C., Brown, M., Vale, W. 1978. *Nature* 273:161–63
154. Guyenet, G., Aghajanian, K. 1977. *Brain Res.* 136:178–84
155. Young, W. S., Uhl, G. R., Kuhar, M. J. 1978. *Brain Res.* 150:431–35
156. Said, S. I., Mutt, V. 1970. *Science* 169:1271–72
157. Mutt, V., Said, S. I. 1974. *Eur. J. Biochem.* 42:581–89
158. Said, S. I. 1975. *Gastrointestinal Hormones,* ed. J. C. Thompson, pp. 591–97. Austin, Texas: Univ. Texas Press

159. Verner, J. V., Morrison, A. B. 1974. *Clin. Gastroenterol.* 3:595–608
160. Bryant, M. G., Polak, J. M., Modlin, I., Bloom, S. R., Albuquerque, R. H., Pearse, A. G. E. 1976. *Lancet* 1:991–93
161. Larsson, L. I., Fahrenkrug, J., Muckadell, O. S., Sundler, F., Hakanson, R., Rehfeld, J. F. 1976. *Proc. Natl. Acad. Sci. USA* 73:3197–200
162. Said, S. I., Rosenberg, R. N. 1976. *Science* 192:907–8
163. Fuxe, K., Hokfelt, T., Said, S. I., Mutt, V. 1977. *Neurosci. Lett.* 5:241–46
164. Hokfelt, T., Elfvin, L. G., Schultzberg, M., Fuxe, K., Said, S. I., Mutt, V., Goldstein, M. 1977. *Neurosci.* 2:885–96
165. Larsson, L. I., Fahrenkrug, J., Holst, J. J., Muckadell, O. S. 1978. *Life Sci.* 22:773–80
166. Larsson, L. I. 1977. *Histochem.* 54:173–76
167. Giachetti, A., Said, S. I., Reynolds, R. C., Koniges, F. C., 1977. *Proc. Natl. Acad. Sci. USA* 74:3424–28
168. Emson, P. C., Fahrenkrug, J., Schaffalitzkyde, O. B., Muckadell, O. S., Jessell, T. M., Iversen, L. L. 1978. *Brain Res.* 143:174–78
169. Robberecht, P., DeNeef, P., Lammens, M., Deschodt-Lanckman, M., Christophe, J.-P. 1978. *Eur. J. Biochem.* 90:147–54
170. Robberecht. P., Conlon, T. P., Gardner, J. D. 1976. *J. Biol. Chem.* 251:4635–39
171. Ivy, A. C., Kloster, H. M., Leuth, H. C., Drewyer, G. E. 1929. *Am. J. Physiol.* 91:336–44
172. Harper, A. A., Raper, H. S. 1943. *J. Physiol.* 102:115–25
173. Mutt, V., Jorpes, J. E. 1968. *Eur. J. Biochem.* 6:156–62
174. Mutt, V., Jorpes, J. E. 1967. *Biochem. Biophys. Res. Commun.* 26:392–97
175. Mutt, V., Jorpes, E. 1971. *Biochem. J.* 125:57P–58P
176. Ondetti, M. A., Pluscec, J., Sabo, E. F., Sheehan, J. T., Williams, N. 1970. *J. Am. Chem. Soc.* 92:195–99
177. Vanderhaeghen, J. J., Signeau, J. C., Gepts, W. 1975. *Nature* 257:604–5
178. Dockray, G. J. 1976. *Nature* 264:568–70
179. Muller, J. E., Straus, E., Yalow, R. S. 1977. *Proc. Natl. Acad. Sci. USA* 74:3035–37
180. Rehfeld, J. F. 1978. *J. Biol. Chem.* 253:4016–21
181. Rehfeld, J. F. 1978. *J. Biol. Chem.* 253:4022–30
182. Straus, E., Muller, J. E., Choi, H., Paronetto, F., Yalow, R. S. 1977. *Proc. Natl. Acad. Sci. USA* 74:3033–34

183. Innis, R. B., Correa, Uhl, G. R., Schneider, B., F. M., Snyder, S. H. 1979. *Proc. Natl. Acad. Sci. USA* 76:521–25
184. Gibbs, J., Young, R. C., Smith, G. P. 1973. *Nature* 245:323–25
185. Nemeroff, C. B., Osbahr, A. J., Bissette, G., Jahnke, G., Lipton, M. A., Prange, A. J. 1978. *Science* 200:793–94
186. Rocha e Silva, M., Beraldo, W. T., Rosenfeld, G. 1949. *Am. J. Physiol.* 156:261–73
187. Eisen, V. 1970. *Rheumatology* 3:103–68
188. Graeff, F. G., Pela, I. R., Rocha e Silva, M. 1969. *Br. J. Pharmacol.* 37:723–32
189. Correa, F. M. A., Graeff, F. G. 1975. *J. Pharmacol. Exp. Ther.* 192:670–76
190. Ribeiro, S. A., Corrado, A. P., Graeff, F. G. 1971. *Neuropharmacol.* 10: 725–31
191. Hori, S. 1968. *Jap. J. Physiol.* 18: 772–87
192. Correa, F. M. A., Innis, R. B., Uhl, G. R., Snyder, S. H. 1979. *Proc. Natl. Acad. Sci. USA* In press
193. Khairallah, P. A. 1971. *Kidney Hormones,* ed. J. W. Fisher, pp. 129–71. New York: Academic
194. Tigerstedt, R., Bergman, P. O. 1898. *Skand. Arch. Physiol.* 8:223
195. Ganten, D., Duxe, K., Phillips, M. I., Mann, F. J., Ganten, U. 1978. *Frontiers In Neuroendocrinology,* ed. W. F. Ganong, L. Martini, 5:66–100. New York: Raven
196. Fischer-Ferraro, C., Nahmod, V. E., Goldstein, D. J., Finkielman, S. 1971. *J. Exp. Med.* 133:353–61
197. Ganten, D., Marquez, J. A., Granger, P., Hayduk, K., Karsunky, K. P., Boucher, R., Genest, J. 1971. *Am. J. Physiol.* 221:1733–37
198. Day, R. P., Reid, I. A. 1976. *Endocrinology* 99:93–100
199. Reid, I. A., Day, R. P. 1977. *Central Actions of Angiotensin and Related Hormones* ed. J. C. Buckley, C. Ferrario, pp. 267–82. New York: Pergamon
200. Goldstein, D. J., Diaz, A., Finkielman, S., Nahmod, V. E., Fischer-Ferraro, C. 1972. *J. Neurochem.* 19:2451–52
201. Fuxe, K., Ganten, D., Hokfelt, T., Bolme, P. 1976. *Neurosci. Lett.* 2: 229–34
202. Changaris, D. G., Severs, W. B., Keil, L. C. 1978. *J. Histochem. Cytochem.* 26:593–607
203. Bennett, J. P. Jr., Snyder, S. H. 1976. *J. Biol. Chem.* 251:7423–30
204. Sirett, N. E., McLean, A. S., Bray, J. J., Hubbard, J. I. 1977. *Brain Res.* 122: 299–312
205. Douglas, J., Saltman, S., Fredlund, P., Kondo, T., Catt, K. J. 1976. *Circ. Res.* 38:108–12
206. Epstein, A. N., Fitzsimons, J. T., Rolls, B. J. 1970. *J. Physiol. London* 210: 457–74
207. Johnson, A. K., Epstein, A. N. 1976. *Brain Res.* 86:399–418
208. Phillips, M. I., Felix, D. 1976. *Brain Res.* 109:531–40
209. Keil, L. C., Summy-Long, J., Severs, W. B. 1975. *Endocrinology* 96:1063–65
210. Buckley, J. P. 1977. *Biochem. Pharmacol.* 26:1–3
211. Finkielman, S., Fischer-Ferraro, C., Diaz, A., Goldstein, D. J., Nahmod, V. E. 1972. *Proc. Natl. Acad. Sci. USA* 69:3341–44
212. Ganten, D., Hutchinson, J. S., Schelling, P., Ganten, U., Fischer, H. 1976. *Clin. Exp. Pharmacol. Physiol.* 3: 103–26
213. Boler, J., Enzmann, F., Folkers, K., Bowers, C. Y., Schally, A. V. 1969. *Biochem. Biophys. Res. Commun.* 37:705–10
214. Burgus, R., Ling, N., Butcher, M., Guillemin, R. 1973. *Proc. Natl. Acad. Sci. USA* 70:684–88
215. Oliver, C., Eskay, R. L., Ben-Jonathan, N., Porter, J. C. 1974. *Endocrinology* 95:540–53
216. Brownstein, M. J., Palkovits, M., Saavedra, J. M., Bassiri, R. M., Utiger, R. D. 1974. *Science* 185:267–69
217. Brownstein, M. J., Utiger, R. D., Palkovits, M., Kizer, J. S. 1975. *Proc. Natl. Acad. Sci. USA* 72:4177–79
218. Hokfelt, T., Fuxe, K., Johansson, O., Jeffcoate, S., White, N. 1975. *Eur. J. Pharmacol.* 34:389–92
219. Winokur, A., Davis, R., Utiger, R. D. 1977. *Brain Res.* 120:423–34
220. Warberg, J., Eskay, R. L., Barnea, A., Reynolds, R. C., Porter, J. C. 1977. *Endocrinology* 100:814–25
221. McKelvy, J. F., Grimm-Jorgensen, Y. 1975. *Hypothalamic Hormones: Chemistry, Physiology, Pharmacology and Clinical Uses,* ed. M. Motta, P. T. Crosignani, L. Martini, pp. 13–26. New York: Academic
222. Parker, C. R., Neaves, W. B., Barnea, A., Porter, J. C. 1977. *Endocrinology* 101:66–75
223. Grant, G., Vale, W., Guillemin, R. 1973. *Endocrinology* 92:1629–33
224. Burt, D. R., Snyder, S. H. 1975. *Brain Res.* 93:309–328
225. Nicoll, R. A. 1977. *Nature* 265:242–43
226. Winokur, A., Beckman, A. 1978. *Brain Res.* 150:205–9

227. Prange, A. J., Nemeroff, C. B., Lipton, M. A., Breese, G. R., Wilson, I. C. 1978. *Handbook of Psychopharmacology*, ed. L. L. Iversen, S. D. Iversen, S. H. Snyder, 13:1–107. New York: Plenum

228. Prange, A. J., Wilson, I. C., Lara, P. P., Alltop, L. B., Breese, G. R. 1972. *Lancet* 2:999–1002

229. Brazeau, P., Vale, W., Burgus, R., Ling, N., Butcher, M., Rivier, J., Guillemin, R. 1973. *Science* 179:77–79

230. Vale, W., Rivier, C., Brazeau, P., Guillemin, R. 1974. *Endocrinology* 95:968–77

231. Hokfelt, T., Efendic, S., Hellerstrom, C., Johansson, O., Luft, R., Arimura, A. 1975. *Acta Endocr.* 80: Suppl. 200: pp. 5–41

232. Bloom, S. R., Mortimer, C. H., Thorner, M. O., Besser, G. M., Hall, R., Gomez-Pan, A., Roy, V. M., Russell, R. C. G., Coy, D. H., Kastin, A. J., Schally, A. V. 1974. *Lancet* 2:1106–9

233. Ganda, O. P., Weir, G. C., Soeldner, J. S., Legg, M. A., Chick, W. L., Patel, Y. C., Ebeid, A. M., Gabbay, K. H., Reichlin, S. 1977. *New Engl. J. Med.* 296:963–67

234. Brownstein, M., Arimura, A., Sato, H., Schally, A. V., Kizer, J. S. 1975. *Endocrinology* 96:1456–61

235. Brownstein, M. J., Arimura, A., Fernandez-Durango, R., Schally, A. V., Palkovits, M., Kizer, J. S. 1977. *Endocrinology* 100:246–49

236. Hokfelt, T., Elde, R., Johansson, O., Luft, R., Nilsson, G., Arimura, A. 1976. *Neuroscience* 1:131–36

237. Epelbaum, J., Brazeau, P., Tsang, D., Brawer, J., Martin, J. B. 1977. *Brain Res.* 126:309–23

238. Patel, Y. C., Zingg, H. H., Dreifuss, J. J. 1977. *Nature* 267:852–53

239. Dodd, J., Kelly, J. S. 1978. *Nature* 223:674–75

240. Renaud, L. P., Martin, J. B., Brazeau, P. 1975. *Nature* 255:233–35

241. Vale, W., Rivier, C., Brown, M. 1977. *Ann. Rev. Physiol.* 39:473–527

242. Margolis, F. L. 1978. *Trends Neurosci.* 1:42–44

243. Brownstein, M. J. 1978. *Centrally Acting Peptides,* ed. J. Hughes, pp. 37–47. London: MacMillan

Ann. Rev. Biochem. 1979. 48:783–836
Copyright © 1979 by Annual Reviews Inc. All rights reserved

DNA REPAIR IN BACTERIA ◆12026
AND MAMMALIAN CELLS[1]

Philip C. Hanawalt, Priscilla K. Cooper, Ann K. Ganesan, and Charles Allen Smith

Department of Biological Sciences, Stanford University,
Stanford, California 94305

CONTENTS

[1]The following abbreviations are used: AP site, apurinic or apyrimidinic site; T4 endo V, T4 endonuclease V; BrUra, 5-bromouracil; XP, xeroderma pigmentosum; NAAAF, N-acetoxy-2-acetylaminofluorene; UV, ultraviolet light.

0066-4154/79/0701-0783$01.00

PERSPECTIVES AND SUMMARY

The maintenance of genetic continuity was once thought to be solely a consequence of the inherent stability of the DNA molecule but we now know that repair processes are essential for this continuity. The DNA in living cells is continually subject to deleterious alterations, some spontaneous, some induced by chemicals or radiation. Damage to DNA, even though attenuated by repair, is responsible for most mutagenesis and some lethality. The field of DNA repair has received a surge of interest recently with the evidence that most types of repairable damage in DNA are both mutagenic and carcinogenic. This correspondence has been supported by the discovery that a deficiency in the repair of DNA photoproducts is correlated with the high incidence of sunlight-induced cancer in the human hereditary disease known as xeroderma pigmentosum (XP). A number of other hereditary diseases conferring predisposition to cancer also appear to involve deficiencies in DNA repair processes.

The classic test lesion for the study of DNA repair has been the cyclobutane dimer, formed by the fusing of adjacent pyrimidine bases in a DNA strand upon the absorption of ultraviolet (UV). This intrastrand dimer poses a block to replication; in the absence of "tolerance" mechanisms one unrepaired dimer in the entire genome of an *Escherichia coli* bacterium may be lethal. The simplest mechanism for repair of this lesion is direct reversal by photoreactivation. An enzyme binds to the dimer-containing region of DNA thus generating a new enzyme-DNA chromophore that absorbs visible light to catalyze the cleavage of the joined bases without breaking any phosphodiester bonds. Photoreactivation is a special case because it operates only upon pyrimidine dimers and it requires light. Photoreactivation has been demonstrated in many living systems, including man, and it can provide a useful test for the role of dimers in measurable biological effects. However, we restrict our attention to the repair schemes that operate in the dark, with particular emphasis on excision repair and on mechanisms for bypassing damage in DNA.

The principle of excision repair is very simple: the damage is recognized; a phosphodiester strand scission is made; the lesion, along with adjacent nucleotides, is excised from the strand containing it; and the deleted stretch is reconstituted utilizing the intact complementary strand as template (Figure 1). In some instances a glycosylase may start the whole process by removing an altered or incorrect base (e.g. uracil) from DNA. The resulting apurinic or apyrimidinic (AP) site may then be recognized by a specific endonuclease or it may become the substrate for direct replacement of the missing base by an "insertase" activity.

Excision repair is possible only in duplex regions of DNA. It obviously cannot operate on single-stranded DNA, such as in the region of a replication fork, where the parental strands have separated. There are a number of hypothetical possibilities when a growing fork encounters a lesion: replication may be totally blocked, it may resume "downstream" from the lesion as the parental strands continue to unwind thereby leaving a gap opposite the lesion, or it may eventually replicate through the region containing the lesion. The available options may depend upon the nature of the lesion and the polarity of the strand containing it. Some damage may not interfere with replication at all while other lesions such as interstrand crosslinks should block replication completely by preventing parental strand separation. Any replication past a lesion must be error-prone unless the polymerase can somehow recognize the correct sequence in spite of the alteration. However, a polymerase-associated 3'-5' "proofreading" exonuclease in bacterial systems normally precludes the addition of most incorrectly paired nucleotides to a growing strand, a constraint that presumably must be relaxed for replication to proceed past damage. There is evidence in bacteria that a system induced by the damage itself permits replication to bypass the damage site, with the expected high frequency of errors. Alternatively, a gap

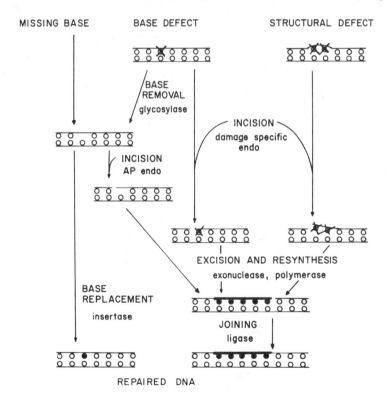

Figure 1 Schematic overview of excision repair.

left opposite the lesion in a daughter strand may be filled by strand exchange with intact homologous DNA. It should be noted that although gap-filling and bypass processes may promote recovery they do not *repair* DNA, since the lesions remain. Nevertheless, they may allow another opportunity for excision repair and may result in a gradual dilution of defects as the cells continue to synthesize DNA and divide.

We now understand in some detail the enzymology of excision repair in bacteria, aided by the isolation of many mutants and the characterization of the relevant enzymes. In mammalian systems the genetics is far less advanced and the analysis of repair is further complicated by the structure of chromatin and the nature of eukaryotic replicons. Nevertheless, the models, methodology, and interpretations from bacterial studies have been enthusiastically carried over to these complex eukaryotic systems. It is the purpose of this review to examine critically the apparent parallels and to try to identify a few of the pitfalls in a comparative analysis of DNA repair in bacteria and mammalian cells.

DNA repair has been treated comprehensively in the proceedings of a recent conference (1) and an earlier one (2). We choose illustrative examples and references for the points we wish to make and refer to more specialized reviews for further documentation. There are a number of recent reviews on selected topics in this field including: photoreactivation (3), base excision repair (4), lesion-recognizing enzymes as probes (5), chemically induced lesions (6), bacterial excision repair (7), mismatch repair and genetic recombination (8), inducible repair and mutagenesis (9), mammalian excision repair (10–13), repair deficiency in human genetic disease (14–18), and DNA repair and cancer (19, 20).

REPAIRABLE DAMAGE

We define damage as any modification of DNA that alters its coding properties or its normal function in replication or transcription. Some minor base modifications may result in an altered DNA sequence while other types of damage may distort the DNA structure and interfere with replication. To be repairable, damage must be recognized by a protein that can initiate a sequence of biochemical reactions leading to its elimination and the restoration of the intact DNA structure.

Various classification schemes for DNA damage and repair enzymes have been proposed. One scheme classifies lesions according to the repair pathways that act on them (4). Classification by this scheme should become increasingly precise as our knowledge of repair develops. Classification based upon the degree of DNA distortion caused by various lesions (21) has not proved especially useful in guiding exploration of DNA repair. The attempted classification of nucleases involved in repair has also encountered several problems (7). The proposed terms "correndonuclease" and "correxonuclease" are misleading because nuclease activity in itself does not constitute repair, and furthermore, the actions of these enzymes may not be restricted to repair processes. Classifying the endonucleases according to the number of bases involved in the lesion (e.g. "correndonuclease II" for "diadducts") (7) implies recognition mechanisms that may not be valid. For example, the T4 endonuclease V (T4 endo V) recognizes pyrimidine dimers but not other "diadducts", while the uvrABC system in *E. coli* recognizes both "monoadducts" and "diadducts". Classification of DNA-damaging agents has not proved useful because in general each produces more than one type of lesion (22). For example, UV-irradiated DNA contains 5,6-dihydroxydihydrothymine as well as pyrimidine dimers in ratios dependent upon wavelength (23). Moreover, different agents may form the same lesion; ionizing radiation and some chemicals also produce 5,6-dihydroxydihydrothymine. Some damaging agents may act indirectly by producing intermediate reactive species, e.g. superoxide radicals (22). For example, the amino acid tryptophan upon absorption of UV generates peroxide radicals that damage DNA (24). Some agents require metabolic activation (25). Many environmental carcinogens, such as benzo(a)pyrene (26) and aflatoxin (27), fall into this category.

Missing, Incorrect, or Altered Bases

Damage to DNA bases threatens the integrity of genetic information, and a variety of repair mechanisms have evolved to counter this threat. Base loss may occur spontaneously; it is also promoted by UV, ionizing radiation, and alkylation (4). In principle a missing base can be restored directly by reinsertion, using the complementary base as template. An enzyme activity from human fibroblasts has been described that inserts the appropriate base into apurinic sites (28), but the role in vivo of such an insertase has yet to be established. A more general repair mechanism for AP sites is the excision repair pathway initiated by an AP endonuclease (Figure 1). Several AP endonucleases have now been characterized from a variety of bacterial and eukaryotic sources (4). Any damaged or incorrect base that can be removed, leaving an AP site, would provide this common substrate for excision repair.

Incorrect bases incorporated during replication and not corrected by a proofreading function may be recognized by a mismatch repair system (8) which is yet to be characterized biochemically.

While not a normal constituent of DNA, except in certain bacteriophages, uracil is occasionally incorporated in place of thymine (29). Although they should not affect DNA function, these uracils are removed by uracil glycosylases found in both bacterial and eukaryotic cells, thus generating AP sites (4). An alternate mechanism for their removal in *E. coli* may by excision repair initiated by endonuclease V, which incises uracil-containing DNA (30), although this does not appear to be a major pathway in vivo (31).

Another source of incorrect bases in DNA is the deamination of cytosine, adenine, and guanine, which produces uracil, hypoxanthine, and xanthine respectively. Unique glycosylases for uracil and for hypoxanthine have been identified (4). In addition, all three products are incorrectly paired and therefore may be subject to mismatch repair (8).

The bases in DNA are subject to a wide variety of additions, some of which constitute damage while others represent normal processing of the DNA and may have important biological functions. Methylations at the 5 position of cytosine and 6 position of adenine do not alter their coding properties. Enzymatic methylation at these positions in newly replicated DNA produces characteristic patterns, and the asymmetry of methylation immediately after synthesis may allow identification of the template strand for purposes of mismatch repair (32–34). In bacteria, methylation of bases in specific sequences protects the DNA from endogenous restriction enzymes. Methylation patterns may also carry information related to gene expression (35). In eukaryotes, methylation occurs principally at the 5 position of cytosine in the statistically rare dinucleotide, CpG. Studies with ribosomal DNA of *Xenopus laevis* have suggested that methylation of this dinucleotide in a newly made daughter strand occurs only if the cytosine in the complementary sequence in the template is already methylated (36). This would maintain a unique methylation pattern from one generation to the next; changes in this pattern might then constitute a serious form of damage to the DNA. It is not known whether there are mechanisms to repair such changes.

Alkylations are produced by chemicals and perhaps also by inappropriate enzymatic action. Whether or not such alterations are innocuous depends upon the alkyl group and its position. Although 7-methyl and 7-ethyl guanine are apparently not specifically excised, they may be preferentially lost by spontaneous depurination; however, a glycosylase recognizing 3-methyl adenine and 3-ethyl adenine has been isolated from *E. coli* (4). Glycosylases for O^6-methyl guanine have been found both in bacteria and in human cells; two complementation groups of XP have been reported to be defective in this activity (37). O^6-methyl guanine in *E. coli* may also be subject to repair by the adaptive response described later (see the section on conditioned recovery responses). It has recently been shown that N-nitrosoalkylating agents also react significantly with the pyrimidine oxygens (38); however, little is yet known about removal of the products.

Major base alterations include pyrimidine dimers, some alkylations, and a variety of bulky adducts. These alterations, sometimes termed structural defects, distort the helical structure of the DNA and may interfere with replication through steric hindrance or by destruction of the pairing ability of the bases. Models based on the study of the repair of pyrimidine dimers have been presumed to be applicable to repair of other structural defects.

Interstrand Crosslinks

Covalent cross-linking of complementary strands of DNA may pose an absolute block to replication and transcription. In addition, since the damage occurs in both strands at the same or neighboring base pairs, neither strand can function directly as a template for error-free resynthesis even if some mechanism permits the strands to separate. Nevertheless, both bacteria and mammalian cells appear to repair this type of damage.

Chemical cross-linking agents include the nitrogen and sulfur mustards (39), mitomycin C (40), nitrous acid (41), various platinum derivatives (42), and certain intercalating furocoumerins (e.g. 8-methoxypsoralen) in a photocatalytic reaction (43). Cole et al (44, 45) have proposed a model that invokes both excision repair and strand exchange in the repair of psoralen crosslinks in *E. coli*. The *uvrA* and *uvrB* gene products are required for the initial incision. The 5'–3' exonuclease of DNA polymerase I is implicated in a second cut in the same strand on the other side of the bound psoralen to break the crosslink. The requirement for *recA* and possibly *recF* in a subsequent phase may reflect either recombination with an intact region of a homologous DNA molecule (45) or some other *recA*-dependent process.

Cells from patients with Fanconi's anemia are more sensitive to cross-linking agents than are normal human cells. This suggests the existence of a repair process in human cells for such damage. Biochemical evidence for crosslink removal in normal but not Fanconi's anemia cells has been obtained in studies with mitomycin C (46, 47). Interestingly, XP cells are only slightly sensitive to mitomycin C and remove the crosslinks as well as normal cells do; but, unlike Fanconi's anemia cells, they are sensitive to a monofunctional analog (46). This implies that recognition of crosslinks occurs by a different mechanism than that of monoadducts. In bacteria the initial incision for both types of damage appears to depend on the uvrABC system. Repair synthesis has been demonstrated in normal, but not XP, cells exposed to 8-methoxypsoralen plus long wave UV [(48); J. Kaye, unpublished]. However, it is difficult to assess the contribution, if any, of cross-link repair to this synthesis, since the treatment produces a high proportion of monoadducts.

A complex tolerance mechanism for replication of cross-linked DNA has been suggested in which a sister-chromatid exchange transfers the crosslinks to one of the two daughter DNA molecules (49). According to this model Fanconi's anemia involves a defect in the sister chromatid exchange process. However, cells from patients with dyskeratosis congenita, another disease thought to involve defective cross-link repair, display an enhanced frequency of sister-chromatid exchanges after psoralen plus UV treatment as compared to that in normal cells (50). The possible relationship of sister-chromatid exchanges to repair has been recently reviewed (51).

Strand Breaks

The single strand scissions that occur during replication (52) and possibly during transcription (53) need not be considered damage. Damage resulting from ionizing radiation includes several different kinds of single strand breaks, the repair of which has been extensively studied and recently reviewed (54, 55). Breaks that are more complicated than simple phosphodiester bond scissions and therefore not subject to direct ligation may require excision repair. Ionizing radiation also produces double-strand breaks directly as single events. Since such double-strand breaks are detected by sedimentation of DNA in neutral sucrose gradients they can include two single-strand breaks staggered by as much as a few bases as well as those that are directly opposed. If one or both breaks is a substrate for ligase, the only special requirement for repair would be keeping the ends together until they have been joined. However, if both breaks involve base or sugar damage, only faulty templates would be available for excision repair. Nevertheless, repair of double-strand breaks has been documented in *Micrococcus radiodurans* (56, 57), *Bacillus subtilis* (58), *E. coli* (59), and yeast (60). The current hypothesis for such repair involves recombination-like events since multiple copies of the genome and functional *recA* and *rad-52* genes are required by *E. coli* and *Saccharomyces cerevisiae* respectively (60–62). Repair of double-strand breaks in mammalian cells has also been reported but not conclusively documented (54).

EXCISION REPAIR

If we rigorously define "repair" as removal of lesions from DNA, then with the exception of photoreactivation all repair schemes so far investigated are forms of excision repair. The generality of such a process is immediately apparent: as long as the damage is confined to one strand, recognition of the damage need only initiate a common set of steps to effect repair. The well-defined genetic system of *E. coli* (Table 1) has promoted understanding of simple sequences of enzymatic steps for excision repair, particularly in response to UV-induced damage. Schemes based upon the results with *E. coli* have shaped ideas about the excision repair process in mammalian

Table 1 *E. coli* K-12 genes affecting responses to DNA damage[a]

Gene	Map location	References	Function and/or comment
uvr genes[b]			
uvrA	91	65, 66	Required for endonuclease activity which initiates pyrimidine dimer excision
uvrB	17	65, 66	Required for the same endonuclease activity determined by *uvrA*
uvrC	42	65–67	May act as part of a complex with *uvrA* and *uvrB* gene products
uvrD	84	65, 68	See *uvrE*
uvrE	84	65, 68	Probably same gene as *uvrD, mutU,* and *recL.* Mutation results in increased UV sensitivity and increased spontaneous mutation rate for frameshifts and transversions.
uvrF	(82)	65	Mutation inhibits UV induction of λ prophage. May be identical to *recF.*
rec genes[c]			
recA	58	65, 69–73	Structural gene for *recA* protein (protein X). May be same gene as *tif, zab,* and *lexB.* Mutations cause pleiotropic effects.
recB	60	65, 74	Structural gene for subunit of exonuclease V
recC	60	65, 74	Structural gene for another subunit of exonuclease V
recD	—	75	Unassigned; originally reserved for the unmapped mutation, *rec-34,* now tentatively identified as *recA*
recE	(27–30?)	76, 77	The structural gene for EcoExoVIII (LamExoVIII)
recF	82	65, 78, 79	Mutation causes UV sensitivity; blocks UV induction of λ prophage; causes recombination deficiency in *recB recC sbcB* mutants. May be identical to *uvrF*
recG	(82)	65	Mutation causes recombination deficiency
recH	(58)	65	Mutation causes recombination deficiency. May be the same as *recA*
recJ	(60–62)	80	Mutation causes recombination deficiency in *recB recC sbcB* strain
recK	(25–45)	80	Mutation causes recombination deficiency in *recB recC sbcB* strain
recL	84	68	Probably same as *uvrD.* Mutation causes same phenotype except it does not increase spontaneous mutation frequencies
Other significant genes in alphabetical order			
alk	43–44	81	Mutation causes sensitivity to MMS but not to UV or X-radiation
capR	10	65	See *lon*
dam	73	34, 65, 82, 83	Mutation causes: reduced amounts of an enzyme which methylates the N^6 position of adenine in DNA; increased UV sensitivity; spontaneous mutagenesis and spontaneous induction of prophage; and "hyper-rec" phenotype. May be involved in "mismatch" repair

Table 1 *(continued)*

Gene	Map location	References	Function and/or comment
dcm	43	65, 82	Mutation causes reduced amounts of an enzyme that methylates the 5 position of cytosine in DNA
dnaB	(91)	84–87	Structural gene for a protein required for DNA replication. At least one mutation causes deficient host cell reactivation of λ and enhanced UV mutability
dnaQ	(5)	88	Mutation causes conditional lethal mutator effect
dut	81	29	Mutation causes reduced dUTP-ase; "hyper-rec" phenotype
lexA	90	65, 71–73	Controlling gene for *recA*. Probably same gene as *tsl*. Also same as *exrA* in B/r
lexB	58	69	See *recA*, *umuB* (also referred to as *unmB*)
lexC	91	89	Mutant phenotype similar to *lexA*, except that it potentiates UV-induced filamentation and confers thermal sensitivity
lig	51	65, 90	Structural gene for DNA ligase
lon	10	65	UV sensitivity; filament formation (same as *capR*)
lop	51	65	Regulates production of DNA ligase
mfd	(17–27)	91	Mutation in B/r causing deficiency in "mutation frequency decline"
mul	81	65	Mutation increases mutagenesis in UV-irradiated phage λ. Suppressed by *recA* mutations
mutD	5	92	Mutation causes generalized hypermutability; thymidine stimulated. Suppressed by *nalA*, *sumD*
mutL	93	92	Mutation causes high rates of transitions and frameshifts
mutR	61	92	Mutation causes increased rates of frameshifts and transitions. Same as *mutH* of B/r?
mutS	58	92	Mutation causes high rates of transitions and certain frameshifts
mutT	2	92	Mutation causes high rate of AT → GC transversions
mutU	84	92	Same as *uvrE*
ntwA	38	93–95	Structural gene for the AP endonuclease activity of exoIII (originally designated endo II, now possibly endo VI). See *xthA*
phr	16	96, 97	Determines photoreactivating enzyme
polA	85	65, 98	Structural gene for DNA polymerase I and associated exonucleases
polB	2	65, 99	Structural gene for DNA polymerase II
polC	4	65	Structural gene for DNA polymerase III (same as *dnaE*)
rac	31	65	Suppresses *recB* and *recC* mutant phenotype in conjugational merozygotes
ras	9	65	Mutation increases sensitivity to UV and X rays
rnm	(90)	100	Mutation in B/r; may be same as *tsl*
rorA	60	65	Mutation causes sensitivity to X rays but not to UV
sbcA	30	65	Controlling gene for exo VIII (see *recE*)
sbcB	44	65	Structural gene for exo I; suppressor of *recB* and *recC*

Table 1 *(continued)*

Gene	Map location	References	Function and/or comment
sfiA	21	101	Suppressor of filamentation
sfiB	2	101	Suppressor of filamentation
sof	81	29	See *dut*
spr	(90)	102, 103	May be same gene as *lexA*
sul	22	65	Suppressor of *lon* mutation
tag	48	4	Structural gene for 3-methyl-adenine glycosylase
ter	—	104	Mutations cause resistance to T-even phage, UV-sensitivity, and defective excision of pyrimidine dimers
tif	58	69–73	See *recA*
tsl	90	73, 105, 106	May be same gene as *lexA*
ung	55	107	Gene determining uracil-DNA glycosylase
umuA	(90)	108	May be same gene as *lexA*
umuB	58	108, 109	Same as *lexB*
umuC	25	108	Mutation reduces UV mutagenesis
xonA	44	65, 110	Probably identical to *sbcB*, but *recB recC xonA* strains are recombination-deficient
xseA	54	111	Structural gene for exo VII
xthA	38	93–95	Structural gene for exo III. May be same gene as *ntwA*
zab	58	70, 112	See *recA*

[a] This table is an updating of one prepared by Clark in 1973 (63, 64). The location of each gene is given to the nearest whole unit on the genetic map of *E. coli* shown in (65). Numbers in parentheses indicate approximate map locations.

[b] All mutations confer sensitivity to UV-irradiation, and deficiency in host cell reactivation.

[c] All mutations reduce the yield of genetic recombinants in appropriate genetic backgrounds but have little or no effect on plasmid inheritance. All cause sensitivity to UV light and ionizing radiation.

cells. Unfortunately, the identification of gene products involved in repair in mammalian cells has been limited by the number and variety of mutations available for study; thus far these include only naturally occurring mutations causing repair-related human disorders (Table 2). These mutations are necessarily restricted to those that permit development and survival of the individual long enough for characterization of the disorder.

Interpretation of results with mammalian cells is complicated by the organization of DNA into chromatin. The effects of chromatin structure are discussed in a later section, but the fact that the substrate for damage and repair is a highly structured DNA-protein complex must be kept in mind when considering models for eukaryotic excision repair. Furthermore, studies of the molecular events involved in excision repair are best performed with cells in culture, which, because they have been selected for properties

Table 2 Human hereditary disorders that may be related to DNA repair deficiency[a]

Disorder	Syndrome	Inheritance	References
Ataxia telangiectasia (Louis-Bar syndrome)	Cerebellar ataxia; telangiectasia; neurological deterioration; often immune deficiency; hypersensitivity to ionizing radiation (may be blocked at early stage of excision of radioproducts); high frequency of spontaneous chromosome aberrations	autosomal recessive	16, 113, 114
Bloom's syndrome	Growth retardation; telangiectatic erythema (exposed areas); hypersensitivity to sun and to UV; high frequency of spontaneous chromosome aberrations, especially quadriradials, and sister-chromatid exchanges	autosomal recessive	115
Cockayne's syndrome	Cachetic dwarfism; mental retardation; microcephaly; premature aging; hypersensitivity to sun and to UV	autosomal recessive	116
Dyskeratosis congenita (Zinsser-Cole-Engman syndrome)	Cutaneous hyperpigmentation; dystrophy of the nails; mucosal leukokeratosis; continuous lacrimation; often thrombocytopenia; anemia; testicular atrophy; cultured cells show enhanced frequency of sister chromatid exchanges after treatment with psoralen and light	X-linked recessive	50
Fanconi's anemia or pancytopenia	Bone marrow deficiency; anatomical defects; growth retardation; high frequency of spontaneous chromosome aberrations; cultured cells show hypersensitivity to DNA cross-linking agents	autosomal recessive	46, 47
Progeria (Hutchinson-Gilford Syndrome)	Precocious senility; high incidence of coronary artery disease	autosomal recessive (?)	18, 117
Retinoblastoma (familial)	Malignant neoplasm of the eye. Cultured cells show hypersensitivity to X-radiation	autosomal dominant (chromosome 13)	118
Xeroderma pigmentosum (classical XP)	Hypersensitivity to sun and to UV; sun induced keratoses, skin carcinomata and melanomata; occasionally neurological abnormalities; probably defective in an early stage of excision repair	autosomal recessive	16, 119
Xeroderma pigmentosum (XP variant)	Similar to XP, but not defective in excision; may be defective in daughter strand closure	autosomal recessive (?)	119, 120

[a] For reviews and more complete description of these hereditary syndromes see (1, 14–18, and 20).

not necessarily related to their role in vivo, may have concomitantly altered repair responses.

A variety of methods have been used to study excision repair. Although each appears to focus upon one of the several enzymatic steps in the pathway, no one step can in fact be measured totally independently of the others because all the steps are coordinated in vivo.

Direct measurement of excision can be accomplished by monitoring the loss from DNA of radioactivity in adducts or by measuring the disappearance of identifiable lesions chromatographically. Indeed, the first biochemical evidence for excision repair was the observation of selective release of pyrimidine dimers from DNA in UV-irradiated bacteria (121–123). However, these methods have several limitations. Biologically important lesions may be undetectable if they are rare, and measurement of radioactive chemical adducts ignores lesions produced through nonradioactive intermediates. In addition, loss of radioactivity may result not only from excision but from chemical degradation of the adduct (124).

An indirect method for monitoring disappearance of damage from DNA relies upon enzymes that produce single-strand scissions at or near sites of damage (5). Only a few suitable enzymes, specific for certain lesions such as pyrimidine dimers, are currently available. Because the method compares the single-strand continuity of the DNA before and after treatment with the enzyme it requires intact DNA and its sensitivity is diminished if excision repair has been initiated but not completed.

The initiation and completion of excision repair can be followed by the appearance and disappearance of strand breaks in DNA, e.g. by sedimentation in alkaline sucrose gradients. This method does not require identification of the lesion(s) being repaired.

The determination of repair resynthesis monitors the total excision-repair activity in the cell and also does not depend upon knowledge of the specific lesions being repaired. The most general method involves the physical resolution of repair replication (synthesis in short stretches) from semiconservative replication of DNA. This is accomplished by combined density and radioactive labeling of the newly made DNA, then separating species according to density in isopycnic gradients (125–127). The radioactivity incorporated into unreplicated parental density DNA is then a measure of repair resynthesis. Although this method appears quantitative, additional details about the biochemistry of the process are necessary for meaningful interpretation of the results. Repair synthesis has also been resolved from semiconservative replication by the affinity of replicating regions for benzoylated-naphthoylated DEAE-cellulose, but this procedure relies upon hydroxyurea to slow the replication fork (128). Other methods have been devised that circumvent the necessity for separation of DNA species. Perhaps the most useful, but least quantitative, is unscheduled DNA synthesis, in which the incorporation of radioactive DNA precursors into individual eukaryotic cells is detected by autoradiography. Since chromosomal DNA replication is confined to the S phase of the cell cycle, it is assumed that cells in S phase will be heavily labeled and thus can be distinguished from the lightly labeled cells undergoing repair. Another technique measures DNA precursor incorporation in cells in which semiconservative synthesis is suppressed, either because cells are contact-inhibited or synchronized and not in S phase, or because they are treated with an inhibitor such as hydroxyurea. Because these methods dispense with the density-labeled precursor, much higher specific activities of radioactive precursors may be used, but they suffer from the fact that the suppression is never complete and from the uncertainties encountered when using hydroxyurea (129), which has been shown to affect the repair process (130, 131). The suppression of normal replication by the damaging agent itself has been used in some bacterial studies in which it was assumed that the precursor incorporation stimulated by the damage represented repair (132), an assumption that may be warranted only if control experiments utilizing density labeling show it to be valid.

Three methods have been employed to estimate the size of the resynthesized region. The amount of radioactive label incorporated during repair, with suitable assumptions, can be used to calculate the total repair synthesis; and using an independent estimate of the number of repair events, an average patch size can be calculated. A second method is an extension of the density labeling technique for measuring repair replication. It involves shearing by sonication of the isolated light (parental) density DNA to known size before analysis in an alkaline CsCl gradient; the observed increment in the density of fragments containing repair patches, together with the average size of the fragments, yields an estimate of the average patch size (133). A recent refinement of this method has increased its sensitivity and improved the accuracy of the patch-size determinations (134). Unlike other methods, this technique does not require estimating the number of sites repaired. Moreover, it allows visualization of heterogeneity in patch size. It is limited by the ability to obtain DNA fragments small enough to observe the contribution of the repair patches to the density of the fragment and by the broad distribution of fragment sizes produced by sonication.

A third method for estimating patch size utilizes photolysis of 5-bromouracil (BrUra) incorporated by repair synthesis in the presence of hydroxyurea (135). The cells are then subjected to increasing fluences of 313 nm light, which selectively produces alkali labile sites at positions where BrUra was incorporated. Alkaline sucrose gradient sedimentation is used to measure the extent of DNA fragmentation that results. If enough 313 nm light can be delivered to achieve a plateau level of fragmentation (i.e.

produce at least one labile site in each patch), then the average patch size can be derived directly from the known efficiency of BrUra photolysis. However, if this condition is not fulfilled, an independent estimate of the number of repair patches is required.

Bacterial Systems

INCISION Models for excision repair usually begin with the endonucleolytic cleavage of a phosphodiester bond (termed incision) at or near the site of DNA damage, assuming that recognition of the DNA alteration is provided by the endonuclease protein itself. A number of endonucleases have been isolated which are each specific for a particular lesion. These include the various AP endonucleases described earlier and the pyrimidine dimer specific endonucleases of bacteriophage T4, *E. coli,* and *Micrococcus luteus* (136, 137). [Recent evidence suggests that the latter may cleave a glycosylic bond at the site of a dimer as well as a phosphodiester bond (138)]. It is unlikely that specific endonucleases exist for each of the many possible kinds of damage known to stimulate excision repair. In some cases, complex systems may recognize distortion of the DNA structure rather than specific lesions (139). An example may be the uvrABC system of *E. coli.*

Mutations in the *uvrA, uvrB,* or *uvrC* genes render *E. coli* sensitive to UV and unable to excise dimers (140). Neither *uvrA* nor *uvrB* mutants perform incision (141, 142). The *uvrC* mutants do perform incision, but the breaks are introduced more slowly than in uvr^+ cells, and the steady-state number of strand breaks in the damaged cells is higher in *uvrC* than in uvr^+ (143). A defect in the ability to restore biological activity of transfecting UV-irradiated ϕX174 DNA in any of these mutants is largely corrected by pretreatment of the DNA with the dimer-specific T4 endo V (144). These results suggest that the defect in each mutant is in some aspect of incision. Incision appears to be an energy-requiring process in vivo as demonstrated by its marked stimulation by ATP in irradiated toluene–permeabilized (145) or sucrose-plasmolyzed (146) uvr^+ cells. However, the low-molecular-weight, dimer-specific endonucleases purified from *M. luteus* (147, 148), phage T4–infected cells (149), and *E. coli* itself (150) are not stimulated by ATP and are active in the presence of EDTA. One of the two *E. coli* enzymes described by Braun & Grossman (150) is present in *uvrA, uvrB,* and *uvrC* mutants and therefore probably does not contribute significantly to incision in vivo. The other, a protein with a molecular weight of about 12,000, although present in normal amounts in *uvrC* mutants, appears to be missing in *uvrA* and *uvrB* mutants and has been referred to as the *uvrA,B* endonuclease (150, 151). More recently, all three *uvr* gene products were shown by Seeberg et al (67) to be necessary for an ATP and Mg^{2+}-dependent endonuclease activity which attacks UV-irradiated DNA. Using a complementation assay based upon this activity, Seeberg (66) has partially purified the three gene products and separated the *uvrA* product from the other two. While neither the *uvrA*⁺ nor the *uvrB*⁺/*C*⁺ product has endonuclease activity individually, they complement each other to form an ATP-dependent activity that introduces approximately one nick per dimer into lightly irradiated DNA. Further purification of the *uvrA* product yielded a protein with a molecular weight of 100,000 which exhibited ATP- or GTP-stimulated binding to superhelical DNA and showed some preference for a UV-irradiated substrate (152). Functional *uvrB* and *uvrC* products may form a complex, since those two activities co-chromatograph on DEAE cellulose when purified from *uvr*⁺ cells and the chromatographic behavior of uvrC activity from *uvrB* mutants is altered (66). Thus the molecular weight of 70,000 determined for the uvrB activity may actually represent a uvrB/uvrC complex.

The ATP-dependence of the endonuclease activity of the *uvr*⁺ complex and the absence of this activity in the three types of excision-deficient mutants suggests that it is the major activity in vivo. At present the relationship between the complex and the small ATP-independent endonuclease is unknown. Since *uvrA* and *uvrB* mutants are also sensitive to a variety of other agents including nitrous acid (153), 4-nitroquinoline 1-oxide (4NQO) (154), mitomycin C (155), and psoralen plus light (45, 156), a critical test of the importance of the *uvr*⁺ complex in vivo will be its activity against DNA treated with these agents. The binding properties of the *uvrA* gene product to such DNA may help elucidate the mechanism by which this system recognizes damage. It is possible that the recognition element in the incision complex is not identical to the endonuclease element.

EXCISION AND RESYNTHESIS   The most striking feature of the excision and resynthesis steps of excision repair in *E. coli* is the multiplicity of enzymes with suitable properties in vitro and the even larger number of mutations that have been shown to alter these processes in vivo (64). No single mutation has been shown to eliminate either excision or resynthesis once the incision has occurred, and no single mutation affecting excision or resynthesis confers the degree of UV-sensitivity characteristic of *uvr*⁻ strains. It is therefore difficult to identify an enzymatic pathway that can confidently be designated the principal mechanism for excision repair in vivo. While much is now known from genetic studies about the effects of altering an enzyme activity or combination of activities on the excision-repair process, there is still relatively little information concerning coordination of the possible pathways, and it is in this direction that we can expect significant future advances.

Three enzymes in *E. coli* are known to remove dimers selectively from incised UV-irradiated DNA in vitro. These are the 5'-exonuclease activity of DNA polymerase I (157), the 5'-exonuclease activity of DNA polymerase III (158), and the single-strand specific exonuclease VII (159). In addition, exonuclease V, the *recB,C* gene product, is capable of degrading dimer-containing DNA but the selective release of dimers has only been demonstrated with mildly heated DNA (160). The excision capabilities of strains carrying mutations affecting each of these enzymes except the polymerase III 5'-exonuclease have been characterized. Mutants deficient only in exonuclease VII (*xseA*) or in exonuclease V (*recB,C*) excise dimers as efficiently as do wild-type cells (161, 162). Mutants deficient in polymerase I 5' exonuclease (*polAex*) also do not show appreciable alteration in dimer excision at UV doses below 40 J/m^2. There is an apparent decrease in excision ability at higher doses, but interpretation of this result is complicated by extensive DNA degradation under these conditions (161, 163). However, when degradation is reduced by the introduction of temperature-sensitive *recB* and *recC* mutations (164) into the *polAex* mutant, the measured excision ability is indistinguishable from that of the *polAex recB$^+$ recC$^+$* parent at all doses (165). If this result is interpreted to mean that most of the degradation does not discriminate against dimer-containing DNA and therefore does not affect quantitation of excision, then the reduced ability of *polAex* mutants to remove dimers after high doses is evidence for participation of the polymerase I 5'-exonuclease in excision. Evidence that exonuclease VII can function as a minor excision enzyme in vivo comes from the finding that a strain deficient in all three 5'-exonuclease activities (*polAex xseA tsrecB,C*) has a more pronounced excision deficiency than does the *polAex tsrecB,C* strain (166). The former does have a measurable level of excision, possibly due to leakiness of the mutations involved but more probably reflecting participation of yet another enzyme, perhaps the polymerase III 5' exonuclease. Since an additive effect of the *xseA* and *polAex* mutations on excision ability is not observable except in the presence of the *recB,C* mutations (161, 166), the possibility remains that exonuclease V may also function in dimer excision in vivo.

Dimer excision may occur more efficiently when it is "driven" by concomitant polymerization. Although the small fragment of polymerase I generated by proteolysis which contains the 5'-exonuclease activity is capable of dimer excision at a low rate, the addition of the large fragment containing the polymerizing activity results in greatly enhanced excision under conditions allowing DNA synthesis (167). Moreover, *polA1* mutants, deficient in polymerizing activity, are much less proficient in dimer excision than are *polAex* mutants, deficient in the 5'-exonuclease activity (165). One implication of this result is that the polymerizing activity can act in concert

with unassociated 5'-exonucleases to effect excision, a possibility worth
testing in vitro. It is likely that the intact polymerase I performs most
excision in vivo. The residual excision in *polA* mutants may reflect an
activity similar to the low rate of excision by the small fragment in vitro
or, alternatively, it might be due to polymerase III activity. The latter
possibility is favored by the finding that a mutant (*polA polC*) deficient in
both polymerizing activities has much less excision capability than the *polA*
strain even at low UV doses (165).

Several mutations in *E. coli* alter excision capability without affecting any
of the enzymes discussed above. One such is the *mfd* mutant isolated as
deficient in "mutation frequency decline", a poorly understood uvr^+-
dependent loss of potential suppressor mutations that occurs when protein
synthesis is inhibited following UV irradiation. Despite the fact that this
mutant is not UV-sensitive, it exhibits slower (though eventually as extensive) removal of dimers and enhanced UV mutagenesis (91). A particular
dnaB mutant (86) also shows normal UV resistance but hypermutability
and a reduced rate of dimer removal. In neither case is the molecular basis
for the slow excision rate known. Strains carrying *recL* (168) and *uvrE*
(169) mutations, which may be in the same gene (68), are UV-sensitive and
after high doses have a pronounced defect in excision but not in incision.
It has been proposed (68, 168) that the *recL* and *uvrE* genes are somehow
involved in regulation of either DNA polymerase I or III. Should this be
the case, this excision defect also would be consistent with the notion that
efficient excision is coupled to resynthesis.

Each of the three known DNA polymerases in *E. coli* is potentially able
to perform resynthesis; and in toluene-permeabilized cells of mutants deficient in all three, no UV-stimulated repair synthesis was detected (170).
Two of these, polymerases I and III, have associated 5'-exonuclease activities and hence are attractive candidates for coupled excision and resynthesis. As with excision enzymes, however, it is difficult to assess the relative
importance of the respective polymerases to excision repair in vivo. Conceptually, DNA polymerase I is uniquely suitable because unlike polymerases
II and III it is able to bind in vitro at nicks generated by dimer-specific
endonuclease. Also, there are about 400 polymerase I molecules per cell but
only 40 of polymerase II and 10 of polymerase III (171). In fact, *polA*
mutants, deficient in DNA polymerase I, are UV-sensitive, although less so
than *uvr⁻* mutants, whereas *polB* mutants, deficient in polymerase II, are
not (172). Since *polC* mutants, deficient in polymerase III, have only
been obtained as temperature-sensitive lethals, it is not possible to assess
their UV sensitivity under conditions in which the enzyme is not functional.

Studies of repair synthesis in vivo by means of the density-labeling procedure and alkaline CsCl gradient analysis of repaired DNA after shearing

have shown that repair patches in wild-type *E. coli* are heterogeneous in size (133). The majority of lesions result in short repair patches of approximately 20–30 nucleotides while a small proportion lead to long patches, recently estimated (P. K. Cooper, unpublished) to be several hundred nucleotides in extent. Other methods for estimating average patch size, which are incapable of resolving size classes, have yielded results in general agreement with the estimate for short patches obtained by the shearing technique, as expected if short patches are much more frequent. Estimates ranging from 13–30 nucleotides per patch have been obtained in vivo by the BrUra photolysis technique (168), by measurement of UV-stimulated DNA synthesis and dimer excision in toluene-treated cells (173), and by measurement in an in vitro system of repair synthesis and loss of sites sensitive to *M. luteus* UV endonuclease (174).

The two size classes of repair patches appear to result from the operation of different enzymatic pathways for repair synthesis, as suggested by the observation that polymerase I–deficient (*polA*) mutants actually perform *more* repair synthesis than *pol*$^+$ strains (175). Since the number of sites undergoing repair, as determined by measurements of dimer excision, is the same (after low doses) or less (after doses greater than 20 J/m^2) in *polA* compared to *pol*$^+$ strains, the increased repair synthesis presumably reflects an increased frequency of long patches. This result was interpreted to mean that polymerase I is primarily responsible for short-patch repair and that other polymerases can produce longer patches (175). Although studies of mutants additionally deficient in polymerases II and/or III are impracticable because of their extreme DNA degradation after UV irradiation, the analysis of toluene-treated cells of such mutants has shown that both of these polymerases can perform repair synthesis (170, 176). In addition, studies of the rejoining of incision breaks in vivo have implicated both polymerases II and III in excision repair in the absence of polymerase I (177, 178).

The frequency of long patches is also increased by the *recL* mutation (168) and by mutations (*polAex*) affecting the 5'-exonuclease activity of polymerase I, although to a lesser extent than by the *polA* mutation (163). It is not obvious why *polAex* mutations should lead to increased patch size, but the known specificities of polymerases II and III do provide some theoretical basis for increased patch size in *polA* mutants. Since neither appears to bind at nicks, it is possible that the incision must be expanded to a gap by some 3' exonuclease prior to resynthesis by polymerase II or III (7), although there is no direct evidence in support of this idea.

Another line of evidence for at least two qualitatively different pathways for repair synthesis comes from studies of rejoining of incision breaks by various mutants during incubation in buffer, in growth medium, or in growth medium containing chloramphenicol (179). By this criterion, repair was divided into two categories, growth medium–dependent and growth medium–independent. It was presumed that the growth medium dependence was due to a requirement for protein synthesis, since results obtained with cells incubated in medium containing chloramphenicol were similar to those in buffer. Repair in the absence of protein synthesis was primarily dependent on polymerase I and therefore probably related to short patch repair, while growth medium–dependent repair required $recA^+$, $recB^+$, and $exrA^+$ ($lexA^+$). Since repair synthesis studies have implicated the $recA$ gene in the long patch pathway (175), and since the presence of chloramphenicol during the repair period has been shown to reduce long-patch repair synthesis preferentially (165), growth medium–dependent repair of incision breaks and long patch repair synthesis may be related phenomena. Their common requirement for protein synthesis and the $recA^+$ genotype suggests that both phenomena might be attributable to the inducible SOS repair system (see the section on conditioned recovery responses). In support of this idea, it has recently been shown that the sensitivity of long-patch repair to chloramphenicol can be eliminated by exposing the cells to a prior inducing dose of UV followed by a period of growth (165). This result implies that long-patch repair is dependent on a protein induced in response to UV damage.

The involvement of an inducible function in long-patch repair could explain the failure to observe long patches in toluene-treated cells, in which protein synthesis does not occur (180). Examination of repair synthesis in toluene-treated cells that have been given an inducing treatment before permeabilization should be of interest in this regard (181). Some features of repair in toluene-treated $polA$ mutants are difficult to understand in these simple terms, however. Repair synthesis does occur under these conditions, so apparently the constitutive polymerases II and/or III must function in this capacity, but it is not clear how this relates to their roles in vivo after induction. Repair synthesis by polymerases II and III in toluene-treated $polA$ cells differs significantly from repair synthesis in toluene-treated pol^+ cells in that it exhibits a postincision requirement for ATP (182) and it does not lead to closure of incision breaks (182, 183). More information concerning the induced product is required before the roles of polymerases II and III in excision repair can be understood.

Completion of excision repair occurs when repair patches are covalently linked to parental DNA to restore strand continuity. This function is performed in $E.\ coli$ by polynucleotide ligase (184).

Mammalian Systems

INCISION   Most of our information on the incision step in mammalian excision repair derives from studies on cells from patients with XP. Fibroblasts from XP patients are UV-sensitive and, with the exception of the class known as XP-variants, are deficient in pyrimidine dimer excision and repair replication. Like the *uvr* mutants of *E. coli,* they are also sensitive to a variety of chemical agents that produce bulky adducts. Initially, it seemed likely that the defect was in the structural gene for a putative repair endonuclease, because XP cells are not sensitive to certain agents that produce strand breaks. However, when unscheduled DNA synthesis was examined in the nuclei of heterokaryons formed by fusing cells from different XP patients, certain combinations appeared normal in excision repair capacity. Seven complementation groups have been defined on this basis, designated A–G (119). There is considerable evidence that a defect in most, if not all, of these groups involves the incision activity. This may be studied directly by monitoring the frequency of strand breaks following UV irradiation. Experiments utilizing alkaline sucrose gradients for this purpose have been limited by the sensitivity of the method, since in normal cells the frequency of incisions during repair is near the level of reliable detection. Results with this technique have been contradictory (18). Using the more sensitive alkaline DNA unwinding technique, Fornace et al (185) detected a low frequency of strand breaks in normal human cells which rose to a maximum shortly after UV irradiation and declined thereafter. A much reduced frequency of strand breakage was detected for XP cells of groups A–D. The few breaks that were observed in these cells were not resealed and might have arisen by some degradative process. Erixon & Ahnström (131) obtained similar results with normal cells from measurements of the "rate of alkaline-induced strand separation" although they found no incision breaks in XP-A cells. Cook et al (186) examined the sedimentation properties of "nucleoids" of human lymphocytes after UV irradiation. Because nucleoids contain domains of superhelical DNA, their sedimentation properties are very sensitive to strand breakage. This study demonstrated incision breaks in normal cells but not in XP cells from groups A, C, and D.

Evidence for a defect in incision activity in XP has also been provided by studies utilizing T4 endo V. Tanaka et al (187) showed that when the enzyme was introduced into cells made permeable with hemagglutinating virus of Japan, normal levels of unscheduled DNA synthesis were observed in UV-irradiated XP cells of complementation groups A–E, and they presented evidence suggesting that this led to increased cell survival (188). Ciarrocchi & Linn (189) showed that T4 endo V restores repair replication activity in a cell-free system obtained from XP-A cells. Smith & Hanawalt (134) characterized repair replication activity in isolated human cell nuclei, and demonstrated that the activity restored to XP-A cells by addition of T4 endo V closely resembled repair replication in normal human cells. The observation that an incision endonuclease can allow cells from several different complementation groups to perform the subsequent steps in excision repair suggests that all are deficient in incision but not in excision-resynthesis.

It thus appears that mutation in any one of a number of genes leads to loss of incision activity in human cells. Furthermore, mutants within a single complementation group may be deficient in more than one activity. For example, XP-D cells appear to be deficient in at least one AP endonuclease activity (190), and XP-A and XP-C are both deficient in removal of O^6-methyl guanine (37). However, the existence of seven complementation groups does not necessarily indicate that seven different proteins are required for incision. Some of the groups may represent genes-determining polypeptides that interact to form a single functional complex, as in *E. coli.* Others may reflect intragenic complementation, mutations in regulatory genes, or mutations in genes whose products facilitate the endonucleolytic event.

Some support for the last possibility comes from the experiments of Mortelmans et al (191). They reported that normal human cells disrupted by sonication were capable of specifically excising pyrimidine dimers both from purified DNA that had been heavily irradiated, and from their endogenous cellular DNA. Surprisingly, sonicated XP cells from groups A, C, and D also excised dimers from purified DNA. However, the sonicated XP-A cells were unable to excise dimers from their endogenous DNA. This was shown to be a deficiency in enzyme activity rather than a property of the endogenous DNA, since DNA-free sonicates of normal cells were able to promote dimer excision from the XP-A DNA. It was thus argued that these XP cells are not deficient in the endonuclease per se but in some other activity necessary for incision in vivo. Alternatively, the excision of dimers from purified DNA may represent the activity of enzymes not normally associated with excision repair in vivo. Although the inability of sonicated XP-A cells to excise dimers from their endogenous DNA mimics their excision properties in vivo, this is not true for XP-D cells, in which excision is observed in vitro but not in vivo. Conversely, XP variant cells, which excise dimers efficiently in vivo, are unable to do so in this system (192). Further study with defined substrates is warranted, as the sonication conditions are apparently critical for excision (192), and the state of the endogenous DNA, although referred to as chromatin, has not been well-characterized. Thus, while the concept of an activity that facilitates recognition of, or access to, damage in DNA organized into chromatin, is an appealing one, it must still be regarded as hypothetical.

Although the notion of a single repair endonuclease that initiates excision repair has been replaced by the idea that a complex set of gene products is required, it is possible that some components of this system might individually be able to incise purified DNA containing certain lesions. Several endonuclease activities purified from mammalian sources as active against highly UV-irradiated DNA have been found to recognize photoproducts other than dimers (13, 193). Recently Waldstein et al (194) have reported isolation of a labile endonuclease activity from bovine thymus that does recognize dimers, and determination of its lesion specificity will be of interest.

At least four of the complementation groups of XP are also deficient in repair synthesis following treatment with N-acetoxy-2-acetylaminofluorine (NAAAF) (195, 196). Comparisons of the repair response to UV and NAAAF would seem to offer a good test of whether the identical factor is required for recognition of the damage produced by each of these agents. Ahmed & Setlow (197) reported that in normal human cells treated with both agents, the amount of repair as determined by BrUra photolysis, by removal of sites sensitive to dimer specific endonuclease, or by unscheduled DNA synthesis, was additive even when the doses of each damaging agent were sufficient to saturate the excision-repair system when given alone. This suggested that repair of at least the major lesions produced by these agents does not proceed by identical pathways. However, Brown et al (198), measuring repair replication in normal human cells, obtained the opposite result. Ahmed & Setlow (199) also reported that in partially deficient XP cells and in Chinese hamster cells (which do not remove lesions produced by the two agents as well as do normal human cells) repair was not additive even at low levels of damage. This result is not easily explained by a model in which a single factor recognizes both types of lesions, but since we do not know the actual defect(s) in these cells, the meaning of the result remains obscure. The NAAAF lesion whose repair appears to mimic that of pyrimidine dimers is probably not the only lesion produced by this agent (200). This complicates the interpretation of observed differences among rates of removal of pyrimidine dimers and of various radiolabeled adducts (201, 202) by human cells.

The limited excision of pyrimidine dimers in established cell lines from rodents and in adult animals in situ may represent partial loss of an ability present in embryos (203, 204). Data on the degree to which such loss affects excision of other products may lead to better understanding of the incision mechanisms.

EXCISION AND RESYNTHESIS   No single enzyme has been isolated from mammalian cells with the combined excision and polymerizing activities of *E. coli* DNA polymerase I (205, 206). Heterogeneity of mammalian polymerases has plagued biochemists, and it is possible that closely associated proteins are lost during purification (207, 208), perhaps by specific proteolytic cleavage (209). However, the following evidence suggests that mammalian excision and resynthesis may be coupled processes, as they appear to be in *E. coli.* The dimer excision activity in sonicated human cells is stimulated by the four dNTPs and requires Mg^{2+} (192). A 5'-exonuclease activity from human lymphoblasts which was purified together with an AP endonuclease was stimulated by polymerase α (208), and the combined activity of these enzymes resulted in repair resynthesis of depurinated DNA in vitro (210).

Several other exonuclease activities that might function in excision repair have been described (13). DNase IV from rabbit tissue (211) shows a preference for double-stranded DNA and can excise pyrimidine dimers by 5'–3' degradation of UV-irradiated DNA treated with an endonuclease from *M. luteus* specific for UV-irradiated DNA. Cook & Friedberg (212, 213) used selective loss of pyrimidine dimers from DNA incised with T4 endo V as a basis for purification of three different exonucleases from human KB cells.

Another exonuclease activity that seems suited to a role in excision repair in human cells has been purified from both placenta and tissue culture cells (214). It is inactive on circular DNA but hydrolyses single-stranded DNA from both 3' and 5' ends. This enzyme releases 5'-phosphorylated oligonucleotides averaging four residues from double-stranded DNA randomly nicked or specifically incised with *M. luteus* UV endonuclease, leaving gaps approximately forty nucleotides in length. Highly efficient release of pyrimidine dimers occurred with the latter substrate. No deficiency in this enzyme has been found in xeroderma pigmentosum cells of complementation groups A–E (J. Doniger, personal communication).

Four mammalian polymerases have been described (205, 206, 215). The γ-polymerase is probably the mitochondrial DNA polymerase (216–219); it is not clear whether this polymerase functions in the nucleus. The δ-polymerase, reported to have an associated 3'-exonuclease activity (215), has not been well-characterized and may prove to be a form of the α-polymerase.

Polymerase α undoubtedly participates in semiconservative synthesis. Its abundance is correlated with DNA synthetic activity, both during the cell cycle and in differentiating systems (205). Inhibitors that can be used to differentiate purified α from β in vitro implicate α in semiconservative replication in permeabilized cells (220–222) and isolated nuclei (223), and in a system for replication of SV40 DNA in vitro (224, 225). Repair synthesis in vitro is also sensitive to agents that inhibit α (134, 189); this result might implicate α-polymerase in repair in vivo. However, γ-polymerase is also inhibited by these agents, and the synthesis observed in vitro may not reflect the activity of the same polymerases that function in vivo. Inhibition of steps in repair prior to synthesis may also account for this result.

It has been suggested that the β-polymerase functions in repair synthesis because it is small and not regulated with the cell cycle. Evidence that β is capable of repair synthesis has been obtained in studies of brain or muscle cells which differentiate in culture, cease to divide, and lose most or all α-polymerase activity. Following UV irradiation, unscheduled DNA synthesis has been detected in such cells (226–229). Interpretation of these results is complicated by the difficulty in making quantitative comparisons between the "differentiated" and parent cells, and by the possibility that α-polymerase may persist in these cells in small quantity or in an altered form. If the endonucleolytic incision event is rate-limiting in repair of UV damage, different levels of polymerase would not be expected to have a great effect upon initial repair rates. Indeed, in HeLa cells, repair levels are not affected by cycloheximide for at least 24 hr after irradiation (230), while under these conditions polymerase levels drop dramatically (231). A report that muscle cells are unable to perform unscheduled DNA synthesis following treatment with methyl methane sulfonate (232) suggests a role for α in repair of this damage, which may not be rate-limited by incision (208). Unfortunately, this study did not include comparable experiments with UV-induced damage.

In human lymphocytes stimulated by phytohemagglutinin Bertazzoni et al (233) reported that α-polymerase activity rose and fell in conjunction with normal DNA synthesis, while β activity rose more slowly and reached a maximum at a time when α activity was declining. Repair, measured by hydroxyurea-insensitive incorporation of thymidine following large UV doses, increased by about a factor of two at a time when the α activity was declining but was still at a much higher level than β activity. It is not clear whether this slight effect justifies conclusions about the role of β in repair synthesis.

Properties of the highly purified polymerases α and β in vitro may help to clarify their roles, although in vivo their activities may be altered by modifying factors (207, 234) or by chromatin structure. Highly purified α apparently is unable to initiate synthesis at a nick but requires a single-strand gap of roughly 20–70 nucleotides (235). This contrasts with the properties of the less purified preparation described above which, in combination with an exonuclease, promoted insertion of a few nucleotides at nicks. The D-loop initiation sequence of mitochondrial DNA in human KB cells can be extended by highly purified β but not by α (235). The activities of these highly purified polymerases have not yet been studied with damaged DNA.

The isolation of mammalian mutants temperature-sensitive for DNA synthesis may assist in the assignment of functional roles to the polymerases (236, 237).

Two pathways for repair synthesis appear to operate in mammalian cells. Because they result in different patch sizes, they have been termed "long patch" and "short patch," by analogy with the designations for bacterial systems. However, these terms have completely different meanings in the two systems. In fact, the size of the short patch in bacteria resembles that of the long patch in mammalian cells. The lesion repaired appears to determine the pathway utilized in mammalian cells but the extent to which these pathways are precisely lesion-specific is difficult to determine without means for producing only lesions of a single type. Nevertheless the following framework is useful in considering experimental results.

The short-patch pathway operates on the major damage products of agents that have been called "X-ray like" (238), e.g. ionizing radiation and alkylating agents. Whether assayed by the disappearance of strand breaks, by loss of sites sensitive to nucleases such as γ endonuclease of $M.$ $luteus$ and S_1 nuclease, or by repair synthesis, repair by this pathway is largely completed within an hour or two (5, 6, 20, 114, 238, 239). For cells irradiated with X or γ rays, the size of the patch inserted has been estimated at 3–4 nucleotides by measurements of radioactivity incorporated (240) and by photolysis of BrUra (238, 241). In both cases, the number of repair events was assumed to be equal to the number of single-strand breaks sealed so that these may be overestimates. Qualitative aspects of the method of BrUra photolysis have been used to identify, but not quantitate, short-patch repair following damage by several other agents (238). It is possible, but not established, that repair initiated by AP endonuclease proceeds via this pathway.

The long-patch pathway operates on damage caused by agents that have been described as "UV-like" (238), e.g. UV, NAAAF, aflatoxin B_1, and psoralen plus light. These lesions are thought to distort the DNA helix, and they are generally repaired less efficiently in XP cells than in normal cells. The frequency of incision breaks remains low throughout the repair period, which continues for many hours for this class of lesions. The patch size for repair of UV-damaged DNA in human cells has been estimated by the density labeling technique to be about 35 nucleotides (242, 243). An early determination using photolysis of BrUra relied upon a rough estimate for the number of repaired sites, and was reported as 14–25 BrUra residues per patch, which corresponds to a patch size of 50–85 nucleotides (135). Later reports have cited a value of about 100 nucleotides, in conflict with the estimate from the density-labeling method. It is possible that the different kinds of errors involved in the alternative technique explain the discrepancy or that the average size of the repair patch synthesized in cells during prolonged incubation in hydroxyurea (photolysis of BrUra) is larger than that synthesized during a short interval in the absence of the drug (density gradient method). Indeed, there is some indirect evidence for this in Chinese hamster cells (244). The size range for the long patches may overlap that for the short patches. The difference between the average patch size of the two classes is not great enough to allow their resolution by the density gradient method.

Photolysis of BrUra has been used to characterize the patch size involved in repair of lesions produced by a number of other agents, and a range of long-patch sizes has been reported—from 30 for the mutagen ICR170 to 120 for NAAAF (238). The biological significance of these differences is unknown, but may be related to the spectrum of lesions produced by a given agent. UV also produces some damage other than pyrimidine dimers, the repair of which may be the barely detectable unscheduled DNA synthesis observed in XP-A cells (245) and the repair replication in chick embryo cells that was observed even after photoreactivation of most of the dimers (246). Similarly, γ rays have been reported to produce a minor species of damage which is repaired by the long-patch pathway (241). XP cells apparently can repair a component of this damage produced under aerobic conditions, but not that produced under anoxic conditions. This is an exception to the general finding that XP cells are defective in the long-patch pathway. Some lesions produced by 4-nitroquinoline-1-oxide are apparently repaired with long patches, while others result in short patches (238).

The details of the time course of long-patch excision repair continue to be the subject of controversy. This may be due in part to the use of different methods, which have optimal reliabilities at different extents of damage and which measure different aspects of the repair process. Discrepancies among different investigations using the same methods may also be due to genuine differences between cell types.

Direct determination of pyrimidine dimers is most reliable at dimer frequencies produced by doses above 20 J/m^2, which saturate excision repair, and which few cells survive. At biologically significant doses, the method is less reliable and discrepant results have been reported using human fibroblasts: some investigators (247, 248) have reported little or no excision during the first few hours after irradiation, while others (201) have observed 75% excision within an hour, or intermediate results [(249); C. A. Smith, unpublished results]. Failure to observe significant dimer loss at early times after irradiation by direct determination could be due to excision of dimers in acid-insoluble fragments which are then slowly degraded (248). However, experiments to test this, in which DNA was fractionated according to size before analysis, have not supported this hypothesis (250, 251).

Assay of sites sensitive to endonucleases specific for pyrimidine dimers is better suited to low doses, especially when high-molecular-weight DNA is analyzed with this method. Results with human cells have generally been intermediate between the extremes noted above. Significant loss is generally detected during the first few hours after irradiation and continues for 12–24 hours, at which time 70–90% of the sites have been removed (5, 243, 248). The observation has been consistently made that human cells fail to remove all endonuclease-sensitive sites. At high UV doses the effects of the damage on cell metabolism may inhibit repair. At low UV doses, the sensitivity of the technique is limited by the size of the DNA under study when purified DNA is used, and by the nonideal sedimentation properties of the large DNA from cells lysed directly in alkaline sucrose gradients. In addition, the time necessary for complete removal may be longer than the 12–24 hr incubation times typically used in these experiments, since cells held in a confluent state for several days show increased survival (252).

The use of repair synthesis measurements to determine the time course of excision repair depends upon the assumption that both the patch size and the specific activity of the radioactive precursors remain constant over the period of measurement. Recent studies with confluent WI-38 cells (243) indicated that the patches made in the later stages of excision repair were the same size as those made immediately after UV irradiation. Specific activity changes are minimal in cells treated with inhibitors of de novo precursor synthesis and high concentrations of low-specific-activity radio-active tracers, since in these cells the sum of repair incorporation observed during several consecutive intervals is similar to that observed during the entire period comprising the intervals (27, 253). Experiments using small concentrations of high-specific-activity tracers may be subject to greater changes of specific activity.

Measurements of repair synthesis are most sensitive and reliable during early stages of excision repair. Although the variety of experimental conditions used makes detailed comparisons of the results obtained by different investigators difficult, repair synthesis has generally been observed to begin at a rapid rate immediately following irradiation. This rate continually declines with time but synthesis can be detected for 24–48 hr (243, 247, 248, 253, 254).

For low doses the time course of repair synthesis agrees well with measurements of loss of endonuclease-sensitive sites (243). However, data of Paterson et al (5) suggest that above 20 J/m^2 the number of endonuclease-sensitive sites removed by human fibroblasts as a function of time is a constant for at least 12 hr, and not related to dose. On the other hand, Edenberg & Hanawalt (253) observed a decline in the rate of repair synthesis in HeLa cells at these high doses, beginning as soon as 1.5 hr after irradiation. A systematic study in a single cell type is clearly needed to interpret the significance of this discrepancy.

DISTRIBUTION OF DAMAGE AND REPAIR IN THE GENOME   Hypotheses and experiments based upon properties of purified DNA and enzymes in vitro have a limited ability to explain the processes occurring in the highly organized eukaryotic chromosome. Until recently, attempts to determine whether DNA damage or repair occurs preferentially in some regions of the DNA have been confined largely to those cases in which DNA can be fractionated according to base sequence or composition (255, 256). Now, however, the rapid advances in our understanding of chromatin organization (257) are beginning to allow the investigation of its influence on repair processes.

For the bulk of the DNA in the cell, the first level of organization involves a highly specific interaction in which segments of 140 base pairs form a "nucleosome core" by winding around an octamer of two each of the four histones that have been highly conserved during evolution. Alternating along the DNA with these cores are regions of "spacer" or "linker" DNA, probably associated with histone H_1. The lengths of these spacers are not as rigidly controlled as the size of the cores; the average length is characteristic of the species and varies from about 20–100 nucleotides. Spacers in most higher eukaryotes are about 50 nucleotides long.

Digestion of chromatin by staphylococcal nuclease and DNase I has been instrumental in the construction of this model. Initial cleavage by these enzymes takes place in the spacer regions, and liberates "nucleosomes," each containing a core and some spacer DNA, and multiples of nucleosomes. Further digestion degrades spacer DNA, and also produces cleavages at a limited number of sites within the cores. A limit of about 50% of the DNA has often been reported for staphylococcal nuclease. However, in many studies of repair, even more extensive digestion has occurred; thus "nuclease-sensitive" DNA includes but is not restricted to spacer DNA. At present, we know little about the dynamics of this structure in the cell.

An important question is whether nucleosome organization is affected by DNA damage, because an alteration in DNA organization by damage might be an important facet of the recognition process, and furthermore it might also cause misinterpretation of the results of digestion studies. For instance, if a lesion formed in core DNA disrupted the structure of the nucleosome sufficiently, that stretch of DNA might become accessible to repair enzymes or to nucleases used to determine the distribution of damage. Pyrimidine dimers apparently do not cause such disruption in the chromatin of irradiated human cells in culture: staphylococcal nuclease-resistant DNA contains the same dimer frequency as undigested DNA, both after low UV doses and limited digestion (258; C. A. Smith, unpublished) and after high doses and extensive digestion (J. Williams, personal communication). Although changes in nucleosome structure that are not revealed by this technique may occur in nucleosome cores containing dimers, these results do suggest that digestion studies are valid for determining the distribution of other types of DNA damage, such as those produced by chemical agents whose access to DNA may be limited by protein-DNA interactions.

In contrast to pyrimidine dimers, the distribution of which is apparently not affected by nucleosome structure, covalent adducts of the psoralens appear to be restricted to spacer DNA, possibly because of the requirement for intercalation of the molecules prior to bond formation. Digestion of DNA in chromatin of cells treated with radioactive psoralens has indicated that the adducts are located in nuclease-sensitive regions (259, 260), a result consistent with electron microscopic observations of the cross-link distribution along the DNA (261, 262).

The extent to which the distribution of other chemical lesions is affected by chromatin structure has been found to vary greatly. A study of the lesions produced by benz(a) pyrene showed that while the diol epoxide-II-dG was randomly distributed, the diol epoxide-I-dG was slightly more frequent in core DNA and the 4-5 diol-epoxide products were much more frequent in the linker DNA (22). Chromatin from rats given the alkylating agent dimethylnitrosamine contained a higher frequency of adducts in the nuclease-sensitive regions (263). The overall digestion properties of chromatin prepared from duck erythrocytes were not altered by reactions with NAAAF in vitro, but when about 40% of the DNA was digested with staphylococcal nuclease the sensitive fraction contained roughly twice the frequency of reaction products as the resistant fraction, over a wide range of extents of reaction with the NAAAF (264). At present it is not clear whether the distribution of the different NAAAF adducts are affected to different extents by nucleosome structure.

The operation of the excision-repair system also should be affected by nucleosome structure. The staphylococcal nuclease sensitivity of DNA synthesized by repair replication has been studied in a number of cases. In the experiments reported to date cells were irradiated or treated with chemical agents, then incubated with [^{3}H]thymidine in the presence of hydroxyurea. Nuclei from these cells were then incubated with nucleases and the digestion of the [^{3}H]DNA and bulk DNA were compared. In mouse mammary cells treated for 3 hr with methyl methane sulfonate, the [^{3}H]DNA made during a subsequent 2 hr period was markedly more sensitive than bulk

DNA: about 65% of the radioactivity was digested when only 30% of the DNA was made acid soluble (265). Similar results were reported for methyl nitrosourea (266). Most of the repair synthesis elicited by these agents, which is thought to be in short patches (much smaller than linkers), is completed in this period. Alkylation damage in other cells has been reported to be more frequent in the nuclease-sensitive DNA. The methyl methane sulfonate damage that elicits the repair synthesis in these cells is likely to be similarly distributed. The marked staphylococcal nuclease sensitivity of the [³H]DNA is therefore probably accounted for by damage occurring predominantly in linker DNA and repaired with short patches without disturbing nucleosome cores. The fraction of [³H]DNA resistant to staphylococcal nuclease may represent synthesis initiated in core DNA or movement of nucleosomes on to DNA synthesized by repair replication in spacer regions. Similar experiments with γ-irradiated human lymphocytes have not shown increased staphylococcal nuclease sensitivity of DNA synthesized by repair replication (258). Perhaps the strand breaks responsible for this synthesis are accessible to the repair system regardless of their location.

Similar experiments have been performed with UV-irradiated human cells in culture. In this case, the damage appears to be uniformly distributed, repair proceeds for many hours, and the patch size is as long as or longer than spacer size. The radioactivity incorporated during the first few hours after irradiation has been reported to be much more staphylococcal nuclease sensitive than bulk DNA (267–271). With increased labeling time, the difference in staphylococcal nuclease sensitivity becomes less apparent. Different explanations for this have been suggested. Cleaver (267) reported that incubation in nonradioactive medium following a short labeling period did not alter the sensitivity of the [³H]DNA to digestion and suggested that the repair immediately after irradiation occurs primarily in spacer DNA. However, Smerdon & Lieberman (269) reported that after reincubation the digestion properties of the labeled DNA gradually came to resemble those of bulk DNA. They suggested that just after synthesis, all repair patches are in nuclease-sensitive regions but then acquire the nuclease resistance of bulk DNA as nucleosome cores rearrange with time. This has been supported by the observation that DNA made in short periods at later times after irradiation is also more sensitive to staphylococcal nuclease than bulk DNA (M. J. Smerdon and M. W. Lieberman, personal communication). The reason for this discrepancy in results is not clear at present.

There are several possible explanations of the staphylococcal nuclease sensitivity of repair patches just after their synthesis. Repair may be initiated preferentially on damage in staphylococcal nuclease–sensitive regions, or the repair process itself may rearrange the nucleosome structure thus rendering the newly synthesized DNA more staphylococcal nuclease sensitive. Alternatively, the staphylococcal nuclease sensitivity may be due in part to the procedure used: in the presence of hydroxyurea, the repair process is slowed (272), and therefore a large fraction of repair events may be in progress at the time of digestion; such incomplete repair patches might be particularly sensitive to staphylococcal nuclease attack. In similar experiments using density label to distinguish repair replication from semiconservative synthesis in the absence of hydroxyurea, R. Wilder and C. A. Smith (unpublished) observed a slightly increased staphylococcal nuclease sensitivity of the DNA made by repair synthesis during 1 and 4 hr incubations after UV irradiation, a result consistent with published data for these time intervals (268–270). Experiments of Tlsty & Lieberman (273) using NAAAF instead of UV gave results similar to those reported by Smerdon et al (268). The postulated nucleosome rearrangement with increasing times (269) may be a constitutive process in normal cells, or may be induced by the repair process itself.

Wilkins & Hart (274) observed that dimer-specific endonuclease from *M. luteus* incised fewer sites in DNA in UV-irradiated cells made permeable by osmotic shock, than in purified DNA, and that additional sites were made accessible to the endonuclease by exposing the cells to high concentrations of NaCl. Furthermore, they found that the additional sites were repaired in the cells much more slowly than the sites originally accessible to the endonuclease. These results have been interpreted to mean that chromatin structure prevents access of endogenous repair enzymes to the protected sites. If the access of staphylococcal nuclease is similarly limited, then these results are not consistent with a nucleosome rearrangement that results in the randomization of endonuclease-sensitive sites and regions of repair synthesis. Sites protected from T4 endo V in irradiated cells made permeable by freezing and thawing are also made accessible by treatment with NaCl. However, in this case, the additional sites seemed to be repaired about as rapidly as the originally accessible sites (C. A. Smith and A. A. van Zeeland, unpublished). In agreement with this latter observation the frequency of pyrimidine dimers measured chromatographically in staphylococcal nuclease–resistant DNA was observed to be the same as that in undigested DNA from normal human fibroblasts several hours after irradiation (J. Williams, personal communication).

Nucleosome formation provides only a part of the packaging of DNA in eukaryotic cells; interactions among nucleosomes are likely to result in higher orders of organization. It is not clear how many different levels of interactions may be involved in constructing the chromosome, and there are at present no biochemical techniques for direct examination of the effects of this further organization on the repair process. We may, however, draw some inferences.

Higher order organization does not appear to partition the DNA into accessible and inaccessible regions at the chromosome level as do the histones at the nucleosome level. This is demonstrated by the unimodal molecular weight distributions observed on alkaline sucrose gradients of DNA from irradiated permeable cells treated with dimer-specific endonucleases.

Like the bacterial nucleoid, the mammalian chromosome contains domains of DNA, each of which loses superhelicity if a single nick occurs within it (186). It has been suggested that replication initiation anywhere within one of these domains is inhibited by a nick, such as that produced by ionizing radiation (275). It is possible that these domains may interact with repair systems in some coordinated manner.

In some cells the DNA associated with certain structural features of the chromosome may have properties that allow its isolation. For example, mouse satellite DNA seems to be associated with condensed heterochromatin. This DNA was reported to be repaired less efficiently than the main band DNA in cells treated with alkylating agents (276), but repaired equally well in cells exposed to UV or NAAAF (277).

It is possible that organization of DNA into a less accessible form in unstimulated lymphocytes and some highly differentiated cells in culture is responsible for decreased repair capability in these cells. However, even the highly condensed chromosomes in mitotic HeLa cells appear to be subject to normal repair activity following UV irradiation (130); so condensation per se does not seem to prevent enzymatic systems from monitoring the DNA for damage.

DNA undergoing transcription is an important class which may be more sensitive to certain damaging agents and more accessible to repair enzymes than inactive DNA. It is possible that ribosomal DNA is organized in a fashion different from other DNA, but nucleosome structure appears to be retained in other transcribing genes since they display the staphylococcal nuclease digestion properties characteristic of bulk DNA (257). Subtle changes in the structure of transcribing DNA have been detected by digestion studies with other nucleases, but no simple technique to distinguish transcribing DNA is available at present. Evidence for extra sensitivity of transcribing DNA to methyl cholanthrene in rats has been presented (278) and it has been suggested that preferential repair of transcribing DNA occurs in mouse cells treated with alkylating agents (265). The DNA in G_1 cells that has been reported to be in a single-stranded configuration (279) may also be more sensitive to damaging agents.

Because damage in replicating DNA may block fork progression or lead to misincorporation, it would seem advantageous for the excision-repair system to monitor the DNA for damage just ahead of replication forks. However, a recent study of the removal of endonuclease-sensitive sites in replicated DNA suggests that this does not occur in human fibroblasts (280). A similar conclusion was reached for UV-irradiated *E. coli* from earlier studies in which similar rates of repair replication were measured in the regions ahead of and behind the replication forks (281).

RESPONSES TO DAMAGE IN REPLICATING REGIONS

The structure of DNA in replicating regions poses a problem for excision repair because the complementary template strands unwind and separate. Clark & Volkert (282) have suggested that reactivation mechanisms that act on damage in these regions be called "intrareplication repair" even though most current hypotheses do not entail the removal of the primary lesion but rather its circumvention.

The phenomenon termed "postreplication repair" was originally described in excision-deficient mutants of *E. coli* K-12 (142, 283). When grown with [^{3}H]thymidine for 10 min after a UV dose of 6 J/m^2 these cells incorporated radioactivity into DNA that sedimented more slowly in alkaline sucrose gradients than DNA synthesized in unirradiated control cells under the same conditions. After reincubation for 70 min in nonradioactive medium the slowly sedimenting DNA was converted into a form that cosedimented with DNA from control cells. This experimental design and these observations have provided the basis for most subsequent studies of "postreplication repair", both in bacteria and in mammalian cells.

Bacterial Systems

MODEL FOR DAUGHTER-STRAND GAP REPAIR   According to the hypothesis proposed to account for the observations in bacteria (142), DNA synthesized after UV irradiation contains a gap opposite each pyrimidine dimer present in the template strand. The gaps are generated when strand elongation, blocked at a dimer, resumes beyond it, presumably at the next site for the initiation of an Okazaki fragment. These gaps, rather than pyrimidine dimers per se, are the substrates for "postreplication repair," and we therefore refer to this process as *daughter-strand gap repair.* The repair occurs by sister-strand exchanges which fill each gap with undamaged DNA from the isopolar parental strand. Resulting discontinuities in parental strands can then be eliminated by repair synthesis, using the undamaged regions of the complementary daughter strands as templates. The products of this process are intact strands of daughter and parental DNA still containing the primary lesions (142, 283). According to this hypothesis the response can be distinguished from excision repair in that it occurs in excision-deficient (*uvrA* or *uvrB*) mutants and permits the cell to tolerate the primary lesions by repairing the secondary lesions produced by replication of damaged templates.

Experiments with *E. coli* designed to test various predictions of the model have generally yielded results consistent with it. In *uvrA recA* and *uvrB recA* cells, deficient in both ER and daughter-strand gap repair, the size of DNA synthesized after UV irradiation approximated the estimated distance between pyrimidine dimers in the irradiated strands (284, 285). Removal of dimers by photoreactivation or excision prior to DNA synthesis resulted in larger daughter DNA (286). Sister-strand exchanges appear to accompany repair of daughter-strand gaps. Density transfer experiments indicated that parental DNA in stretches of about 1000 nucleotides became covalently attached to daughter DNA under conditions permitting daughter-strand gap repair (287). Results obtained with autoradiography and electron microscopy revealed inserts of radioactive parental DNA varying from 3×10^3–2.4×10^4 nucleotides in length in nonradioactive daughter strands (288). Photolysis of BrUra incorporated into DNA by *uvrA* cells after UV irradiation provided evidence that repair synthesis in parental strands is correlated with daughter-strand gap repair (289). After a dose of 1 J/m^2, about one BrUra-containing region averaging approximately 15,000 nucleotides in length occurred per dimer originally induced in the template strands. Estimates of the size of the gaps in daughter DNA, which range between 1000 nucleotides (290) and 1500–40,000 nucleotides (291) are probably compatible with these values for the length of exchanges associated with their repair, particularly since it is not clear how much of each gap is filled with parental DNA and how much might be filled by repair synthesis. In principle, only a small stretch of DNA need be transferred (292). However, if branch migration (293) occurred during the exchange, a parental insert larger than the original gap might result.

One provocative aspect of daughter-strand gap repair in UV-irradiated *E. coli* is that it does not seem to generate a dimer-free copy of the genome directly. Studies in which repair was monitored with T4 endo V indicated that the dimers were randomly transferred into newly synthesized DNA as gaps were repaired, thus becoming equally distributed among parental and progeny strands (294, 295). This observation supports the idea that sister-strand exchanges accompany repair and is in accord with current models which include DNA strand isomerization during genetic recombination (8). However, it shows that the production of dimer-free DNA by this mechanism in the absence of excision repair requires several rounds of replication (295).

GENETIC CONTROL OF DAUGHTER-STRAND GAP REPAIR Inactivation of several genes in *E. coli,* either singly or in suitable combinations, inhibits the conversion of DNA synthesized after UV irradiation to a form that cosediments with unirradiated control DNA in alkaline sucrose gradients. The genes implicated include *recA* (286), *lexA* (177, 285, 295, 296), *recB* (285) (and probably *recC* which codes for the same enzyme, Eco-ExoV), *recF* (79, 295, 297), *uvrD* (285), and possibly *dnaG* (298). Although *polA* or *polC* single mutants repair gaps efficiently, *polA polC* double mutants do not, which suggests that the activity of DNA polymerase I is required in the absence of polymerase III and vice versa (178, 299, 300). Recently the *uvrA* and *uvrB* genes have been implicated by observations that *uvrA* or *uvrB* derivatives of strains carrying *recA* (106), *recB* (285), or *recF* mutations (79) show less complete repair of daughter-strand gaps than corresponding *uvr*+ strains.

In general, daughter-strand gap repair has been measured by velocity sedimentation, which detects only the final step in repair, ligation. Intermediate reactions and reaction products have not been extensively characterized. Results of experiments using benzoylated naphthoylated DEAE-cellulose chromatography (290) to study gap filling independent of ligation indicated that *lexA* and *recA* mutants may be blocked at a stage after gaps have been at least partially filled (301). Extensive work will be needed to amplify this observation and to relate it to the role of *recA* in forming intermediates in genetic recombination (8, 302, 303).

Because daughter-strand gap repair requires a functional *recA* gene, also needed for genetic recombination, and because it involves exchanges of DNA, this response has been referred to as recombinational repair. However, it requires some biochemical functions not needed for genetic recombination, since *recF* and *lexA* mutations inhibit daughter-strand gap repair (177, 295–297) but not genetic recombination (80, 304). The *lexB* and *zab* mutations, which are located in the *recA* gene and cause UV sensitivity but not recombination deficiency (70), may also prove to inhibit daughter-strand gap repair.

Mammalian Systems

REPLICONS AND DAUGHTER-STRAND CLOSURE   An experimental design analogous to that devised for bacteria has revealed a response in mammalian cells resembling daughter-strand gap repair (305–308). For example, when mouse lymphoma cells were incubated with [^{3}H]thymidine for 25–60 min after UV irradiation, the labeled DNA sedimented more slowly in alkaline sucrose gradients than did DNA from unirradiated control cells. When the irradiated cells were reincubated in nonradioactive medium for 3 hr after labeling, the radioactive DNA cosedimented with the control DNA (307). This response has been reported to be impaired in cells from XP variants (120), and in some cases to be inhibited by caffeine (120,

309–311). Although these results resemble those obtained with bacteria, their interpretation is more difficult. An experimental problem is that the large DNA from mammalian cells sediments anomalously and appears to entangle smaller DNA, a problem usually resolved by intentionally fragmenting the DNA to approximately 2×10^8 daltons by alkaline hydrolysis or ionizing radiation immediately before sedimentation (55, 135, 307, 312–314). A more fundamental difficulty arises from the fact that, unlike bacteria, eukaryotic chromosomes contain multiple tandem replicons (315–317). These replicating units, which appear to be arranged in clusters, have unique internal initiation sites from which replication is bidirectional, although unidirectional replication has also been reported (316). Mammalian replicons range in size from 8–560 million daltons, with the average being 200 million daltons (cf *E. coli* in which each replicon is approximately 2800 million daltons). The overall pattern of replication in a mammalian chromosome thus involves joining of daughter strands at several different levels. Low-molecular-weight intermediates (Okazaki fragments) must be joined within individual replicons. Then daughter strands in adjacent replicons must also be joined. This latter process causes particular problems in interpreting results of sedimentation experiments designed to study repair because: (*a*) depending upon the amount of damage incurred by the DNA, the distance between lesions may approximate or exceed the size of many of the replicons; (*b*) the rate at which daughter strands in normal, adjacent replicons join may be similar to the rate at which daughter-strand discontinuities caused by damage are eliminated (307); (*c*) discontinuities in daughter strands may result not only from gaps left opposite lesions in template DNA, but also from the inhibition of replicon initiation or strand elongation or both. When strand elongation is inhibited, the discontinuities occur between adjacent active replicons. When initiation is inhibited, the discontinuities may also occur between nonadjacent active replicons separated by one or more inactive replicons. The alkaline sucrose gradient technique usually employed cannot distinguish among the three types of discontinuity. These limitations should be kept in mind when considering the various models for producing intact daughter strands. The extreme views are: 1. gaps are left in DNA opposite damaged sites, to be filled in later by (*a*) strand exchanges or (*b*) synthesis opposite the lesions; 2. replication does not resume beyond the lesion (*a*) at all, or (*b*) until some event occurs to permit synthesis opposite the lesion. There is controversy about each of these possibilities (318).

PRESENCE AND LOCATION OF GAPS IN DAUGHTER STRANDS
Pyrimidine dimers do not completely block DNA replication in vivo. For example, hamster cells, which do not excise dimers efficiently, can replicate

their entire genome after UV doses as high as 15 J/m² (319), a dose which should produce approximately 10⁷ dimers per genome or one dimer per 5 million daltons. However, DNA synthesis is inhibited by UV irradiation as shown by a dose-dependent reduction in the overall rate of [³H]thymidine-incorporation (320, 321) and of semiconservative replication (317). This might reflect an inhibition of replicon initiation or strand elongation or both. Results of experiments using fiber autoradiography to distinguish among these possibilities have indicated that inhibition of elongation is primarily responsible for the reduction in DNA synthesis (317, 322). However, in one case, in which HeLa cells were labeled immediately after irradiation, the majority of replicons appeared to terminate synthesis in a manner consistent with their being blocked by pyrimidine dimers (317). In another case, when the labeling of hamster V79 cells was delayed until 25 min after irradiation, active replicons were able to synthesize stretches of DNA 4–5 times longer than the expected distance between dimers, which indicates that in these replicons dimers did not block replication (322). Unfortunately, the resolution of these experiments did not permit direct visualization of gaps that might have been formed opposite the dimers. Evidence for gaps of approximately 800 nucleotides in DNA synthesized by UV-irradiated mammalian cells was obtained by photolysis of BrUra incorporated into mouse lymphoma cells (307). Supporting evidence for discontinuities in DNA synthesized after UV irradiation was reported by Meneghini (323) who found that treatment with *Neurospora* endonuclease specific for single-stranded DNA reduced the rate of sedimentation in neutral sucrose gradients of recently replicated DNA from irradiated, but not from unirradiated, human cells. These results suggested that single-stranded regions of template DNA separated the stretches paired with daughter strands but gave no indication of the size of the regions. In addition, no evidence was obtained for the presence of pyrimidine dimers in the single-stranded regions.

Treating the DNA with T4 endo V, instead of with the single-strand specific nuclease, did not reduce the sedimentation rate of newly synthesized DNA in neutral sucrose gradients. A reduction would have been expected if the enzyme had cleaved template strands at pyrimidine dimers located in single-stranded regions (324, 325). Similar results have also been reported for hamster (CHO) cells treated with a dimer-specific endonuclease activity from *Micrococcus luteus* (326). However, it has not been demonstrated that either of these enzymes is able to cleave DNA at a dimer opposite a terminus in the complementary strand. Thus, these experiments do not rule out the possibility that gaps are formed opposite pyrimidine dimers.

POSSIBLE MECHANISMS FOR TOLERATING DIMERS   In mammalian cells the amount of sister-strand exchange detectable in UV-irradiated cells after discontinuities in daughter strands have been closed is much less than in bacteria (307, 327, 328). Experiments using photolysis of BrUra indicated that gaps in the DNA are filled by de novo synthesis (307). However, pyrimidine dimers have been detected in DNA synthesized after irradiation by using specific endonucleases (318, 324, 325, 329–334). In some cases this may result from extension of daughter strands being synthesized at the time of irradiation (333, 335). However, the response has been reported to occur in human cells treated to minimize the proportion of cells in S phase and may therefore, at least in some cases, reflect sister chromatid exchanges (334).

Mammalian cells incubated for several hours after UV irradiation acquire the capacity to synthesize DNA of the same size as that made by unirradiated cells (308, 336, 337), a response similar to that observed in excision-deficient cells of *E. coli* (287, 338). In *E. coli* the apparently intact daughter molecules contained pyrimidine dimers as judged by their sensitivity to T4 endo V, which suggests that their high molecular weight resulted primarily from sister-strand exchanges (294, 295). However, in mammalian cells sister-strand exchanges appear to be infrequent; it has therefore been proposed that the production of high-molecular-weight daughter strands upon prolonged incubation of irradiated mammalian cells occurs either through a mechanism permitting synthesis of a new strand to continue past an unrepaired lesion in the template strand ("transdimer synthesis") (308, 339) or by synthesis using the complementary daughter strand as a template in the damaged region (292, 309, 340). Studies using density labels have provided evidence for limited pairing of daughter strands in UV-irradiated human cells (309, 340). However, it is not known what proportion of the paired regions is formed in vivo, and what is produced as an artifact during cell lysis (341).

Further investigations are needed not only to clarify the mechanisms underlying responses to damage in replicative regions but also to establish their importance for normal cell function. One might expect that tolerance mechanisms would be more important for the survival of cells that do not show efficient excision repair, e.g. rodent cells. However, the severity of the symptoms exhibited by XP-variant patients, if due to a deficiency in these responses, indicates a critical role for them even in cells proficient in excision repair.

CONDITIONED RECOVERY RESPONSES

Bacterial Systems

THE SOS FUNCTIONS   The designation "SOS functions" has been applied to a complex group of inducible responses in *E. coli* that appear to be coordinately regulated. Included among these responses are inhibition of cell division, inhibition of postirradiation DNA degradation, induced bacterial mutagenesis, Weigle reactivation, and Weigle mutagenesis [see review by Witkin (9)]. The last two responses were originally described by J. Weigle (342), who observed that UV-irradiated bacteriophage λ yielded more plaques and a higher proportion of mutants when plated on lightly UV-irradiated *E. coli* than when plated on unirradiated cells. The observations that the genetic and physiological requirements for these phenomena also applied to λ prophage induction (343) and that the mutagenic response appeared to operate on host DNA as well (344, 345) led to the proposal that such coordinately regulated functions represent a complex inducible response of bacteria to unrepaired damage in their DNA, i.e. an "SOS" (346, 347). The inducible responses are all dependent upon *recA*+ and *lexA*+ and are constitutively expressed at high temperature in a *tif* mutant (345, 348).

Induction of the SOS responses is accompanied by the appearance of a prominent 40,000 dalton protein, originally designated protein X (349), and now known to be the *recA* gene product (71–73, 350). Synthesis of the recA protein appears to be regulated by the unlinked gene *lexA*, thought to determine a repressor of *recA*, and by the recA protein itself. A current model for this regulation invokes a protease activity of the recA protein (103, 351, 352). Briefly, it is proposed that there is a constitutive low-level production of recA protein which, under inducing conditions, is converted to a form that proteolytically inactivates a variety of repressors including the lexA protein and other repressors controlling the coordinately expressed SOS functions, thereby derepressing not only its own synthesis but that of a number of other proteins. In addition, the recA protein has recently been shown to enhance the rate of renaturation of DNA in the presence of ATP (353, 353a).

Several different treatments known to inhibit DNA replication have been shown to induce the recA protein. These include UV irradiation, exposure to nalidixic acid, and thymine starvation (354). Products of the degradation of newly synthesized DNA near the replicating fork by exonuclease V appear to cause induction (355–357). However, since recA protein can be induced by UV in *recB* mutants (354), apparently exonuclease V is not essential. Further support for the idea that degradation products are the inducing signal comes from the finding that induction occurs after treatment with nalidixic acid (358) but not after treatment with novobiocin (356). The former is a specific inhibitor of replication that acts on gyrase subunit A, leading to selective degradation of newly replicated DNA, while the latter inhibits replication by acting on gyrase subunit B without causing DNA degradation (359). Unlike nalidixic acid, UV does not require active replication forks to cause induction, since recA protein can be induced by UV in a *dnaA* mutant in which replication has been terminated at the restrictive temperature (354, 360). Thus degradation occurring at incision sites, as well as at stalled replication forks, may result in induction. Whatever the mechanism responsible for the requisite DNA degradation, the induction signal may be an oligonucleotide (M. Oishi, personal communication; 361).

The principal model for Weigle reactivation and mutagenesis proposes that modification of one or more of the normal DNA polymerases by an inducible protein facilitates replication past lesions, termed "transdimer synthesis" (282), and increases the probability of error. Several lines of evidence support this idea. UV irradiation of the host cell leads to enhanced synthesis on UV-irradiated single-stranded ϕX174 DNA, whereas a pyrimidine dimer is evidently a block to replication of ϕX174 in unirradiated cells (362). Since blockage of replication at pyrimidine dimers in unirradiated cells is ascribed to the $3' \rightarrow 5'$ editing exonuclease activity associated with bacterial DNA polymerases, it is this editing activity that is presumed to be suppressed after induction. Consistent with the hypothesis that a stalled polymerase "idles," incorporating and removing nucleotides in place, is the observation of an increase in the conversion of dNTPs to free monophosphates during DNA synthesis in vitro by polymerase I on UV-irradiated ϕX174 DNA (339).

Indirect evidence supporting the involvement of DNA polymerase III in error-prone transdimer synthesis in vivo has been obtained in a study of UV mutagenesis in an E. coli uvrA polC mutant (363). Mutation fixation was measured by loss of the ability to remove potential mutations by photoreactivation (loss of photoreversibility). While loss of photoreversibility occurred normally in UV-irradiated cells of this strain during incubation at the permissive temperature, it ceased immediately after transfer to the restrictive temperature. This observation indicated that the mutational event (presumably error-prone transdimer synthesis) required active polymerase III. In contrast, in spite of the polC mutation, DNA replication and daughter-strand gap repair continued for 30 min at the nonpermissive temperature in the particular strain employed in this study. A possible explanation for this apparent discrepancy is that polymerase III-transdimer synthesis is involved in the repair of only a few daughter-strand gaps, a fraction too small to detect biochemically under these conditions but large enough to account for mutagenesis.

The proportionate contribution of transdimer synthesis to daughter-strand gap repair is difficult to assess. A detectable level of inducible error-prone repair of daughter-strand gaps has been inferred from the observation that in an excision-deficient strain complete inhibition of mutation fixation by chloramphenicol occurred concomitantly with partial inhibition of daughter-strand gap repair (364). Transdimer synthesis should produce gap-free daughter strands without requiring sister-strand exchanges and the accompanying transfer of pyrimidine dimers into daughter strands (see the section on responses to damage in replicative regions). However, under conditions conducive to induction of SOS functions, dimer-free daughter strands were not detected in excision-deficient (*uvrA*) cells by the use of T4 endo V (106, 295). Since more than 50% of the irradiated population survived in these experiments, dimer-free molecules should have been detected if they represented a large fraction of the synthetic activity of the surviving cells. Thus, transdimer synthesis may occur less frequently under these conditions than daughter-strand gap repair by sister-strand exchange.

Presumably the ability of polymerase III to perform error-prone transdimer synthesis is due to modification by some hypothetical inducible factor. A facet of the hypothesis that awaits clarification is whether or not the putative modified polymerase III *only* functions in a repair capacity or whether it can also still function in replication, presumably with reduced fidelity resulting in a mutator effect. In the former case, mutation should be "targeted" (restricted to sites of damage); in the latter case it should be "untargeted" (365).

Although all UV-induced mutations in *E. coli* are dependent upon the *recA*⁺ and *lexA*⁺ genotype, UV mutagenesis is probably mediated by more than one pathway. The simple interpretation that all induced mutations arise as a result of error-prone replication of unexcised lesions is probably invalid for wild-type strains (366). In *uvr*⁺ bacteria, fixation of most mutations as measured by loss of photoreversibility occurs in the absence of DNA replication and with a time course similar to that for excision repair; only suppressor mutations, which alter tRNA, are dependent on DNA replication (367–369). Indeed, it has been estimated that at least 75% of the UV-induced mutations in *uvr*⁺ strains arise during excision repair (366). However, because excision-proficient bacteria are much less mutable by UV than excision-deficient strains, the error frequency is presumed to be considerably lower for excision repair than for daughter-strand closure mechanisms. Moreover, it has been postulated that an excision-dependent mode of daughter-strand closure (282, 370) makes postreplicative events less error prone in *uvr*⁺ than in *uvr*⁻ strains. The postreplicative mechanism leading to mutation in *uvr*⁺ and *uvr*⁻ cells is thought to be transdimer synthesis by a modified polymerase; daughter-strand gap filling by sister-strand exchange is thought not to be involved, since induced mutagenesis can occur in *uvr recB recF* mutants deficient in both pathways of recombination (371).

The production of mutations during excision repair has been attributed to the long-patch pathway, since it depends on *recA*⁺ and probably *lexA*⁺ (9, 372–374). The recent demonstration that long-patch repair requires protein synthesis and is induced by exposure to UV (165) lends further support to the idea that long patches are a manifestation of the SOS error-prone system operating in excision repair. Closely spaced lesions on opposite strands are the postulated substrate for mutagenic excision repair. According to this idea, excision and resynthesis initiated at one lesion would stop at the second lesion and would require transdimer synthesis for completion, which would result in a long repair patch. In agreement with predictions of this hypothesis, the proportion of available sites repaired by long patches increases with dose in a nonlinear fashion (P. K. Cooper, unpublished). However, mutation fixation at nonsuppressor loci as measured by loss of photoreversibility has recently been found to occur in the presence of chloramphenicol (375), while long-patch repair is completely inhibited under similar conditions (P. K. Cooper, unpublished). It was concluded that mutagenic excision repair is constitutive (375), although requiring $recA^+$, $lexA^+$, and functional polymerase III (363). It is difficult to reconcile this conclusion with the known properties of excision-repair pathways, since the constitutive short-patch pathway is $recA^+$- and $lexA^+$-independent and mediated by polymerase I, while the long-patch pathway requires $recA^+$ and protein synthesis and is mediated by polymerases II and III.

Analysis of the repair characteristics of mutants specifically deficient in induced mutagenesis should help to elucidate the relationship of misrepair mutagenesis to known repair pathways. Although recA and lexA mutants exhibit no UV mutagenesis, they are unsuitable for this purpose because of their pleiotropic effects. Recently, however, a gene designated umuC has been identified, mutants of which show very little UV-induced mutagenesis or Weigle reactivation (at least in a uvr^- background) but are normal for other SOS functions (108). Although daughter-strand joining appeared to be normal in the umuC mutant (376), this result does not rule out the possibility that error-prone transdimer synthesis is responsible for a small fraction of daughter-strand closure.

The relationship of error-prone synthesis to survival of damaged cells is difficult to assess. The $umuC^-$ strains that are additionally $uvrA^-$ are slightly more UV-sensitive than a uvr^- single mutant (108), but the UV survival of $umuC^-$ in a uvr^+ background has not been reported. A provocative recent study (377) suggests that the inducible SOS pathway acts primarily in conjunction with uvr^+-dependent mechanisms to increase survival. In particular, UV irradiation dramatically increases the resistance of bacteria to a second dose of UV followed by chloramphenicol treatment, an effect dependent on uvr^+ as well as on $recA^+$ and $lexA^+$. Of course, a $recA^+$-, $lexA^+$-dependent process is not necessarily error prone, so this result may not reflect the survival value of the error-prone pathway. It will be of interest to know whether the umuC mutations confer greater relative UV sensitivity to uvr^+ cells than they do to uvrA cells.

ADAPTIVE RESPONSE TO ALKYLATING AGENTS   Exposure of *E. coli* to sublethal concentrations of N-methyl-N'-nitro-N-nitrosoguanidine causes "adaptation." Adapted cells show higher survival and lower mutation frequency when exposed to lethal concentrations of this compound than do nonadapted cells (378). This response clearly differs from any of the SOS functions, since all *recA* and *lexA* mutants tested (379) show adaptation. Higher concentrations of N-methyl-N'-nitro-N-nitrosoguanidine are required to elicit the SOS functions than the adaptive response. When the SOS functions are induced by growing *tif-1* cells at 42°, resistance to N-methyl-N'-nitro-N-nitrosoguanidine does not increase but mutation frequency does (379). UV light or 4-nitroquinoline-1-oxide can induce the SOS functions but not the adaptive response, and N-methyl-N'-nitro-N-nitrosoguanidine adapted cells are not more resistant to these two agents. In general, SOS functions are induced by treatments that block DNA replication, while adaptation usually occurs under conditions that do not detectably alter cell growth and DNA synthesis.

Other alkylating agents that elicit the adaptation response include methyl methane sulfonate, ethyl methane sulfonate, N-methyl-N-nitrosourea, and N-ethyl-N'-nitro-N-nitrosoguanidine (379). Cells adapted to any one of these compounds are also adapted to the others.

Separate biochemical pathways may control the response to lethal and to mutagenic damage, since *polA* mutants show decreased mutagenesis upon adaptation, but not enhanced survival (380, 381). *PolB* mutants show both enhanced survival and decreased mutagenesis (380).

Adaptation appears to reduce the accumulation of O^6-methyl guanine, but not of 3-methyl adenine in DNA. The level of a third purine adduct produced by N-methyl-N'-nitro-N-nitrosoguanidine, 7-methyl guanine, is not affected by adaptation (381, 382). As yet the molecular mechanism of the adaptive response is unknown. It might involve an inducible glycosylase, an endonuclease, or perhaps a specific demethylase for O^6-methyl guanine. Although such demethylase activities have not been reported in bacteria, there is evidence for an O^6-methyl guanine demethylase in mammalian cell extracts (383).

Mammalian Systems

Phenomena analogous to Weigle reactivation and Weigle mutagenesis have been reported in mammalian cells. UV irradiation of host cells prior to viral infection results in enhanced survival of UV-irradiated Herpes virus (384, 385), Kilham rat virus (386), adenovirus (387), or SV40 (387, 388), but not of vaccinia or poliovirus (387). Enhanced survival of UV-irradiated SV40 (388) or Herpes virus (389) has also been reported in cells exposed to low levels of chemical carcinogens such as aflatoxin B_1 metabolites, NAAAF, methyl methane sulfonate, or ethyl methane sulfonate prior to infection. Reactivation of Herpes in XP cells indicates that the excision-repair pathway is not involved (390). The enhancement of survival of Herpes is inhibited by cycloheximide (386), which suggests that protein synthesis is required for the effect in mammalian cells as it is in bacteria.

Weigle reactivation of irradiated phage λ in *E. coli* is accompanied by increased mutagenesis of the phage; mutagenesis of unirradiated λ is also enhanced by irradiation of the bacteria (391). Recent studies have indicated that UV irradiation of the mammalian host cells prior to infection is mutagenic for both normal (392) and irradiated (392, 393) Herpes virus. Attempts to demonstrate enhanced mutagenesis of UV-irradiated adenovirus in UV-irradiated normal human fibroblasts have not been successful, however (394). An analysis of the replication intermediates of UV-irradiated viral molecules in UV-irradiated host cells (395, 396) may in the future provide biochemical evidence for an inducible SOS response in mammalian cells.

Results of experiments analyzing daughter-strand closure have also been cited as evidence for an inducible "repair" system in mammalian cells (397, 398). In these experiments a low "inducing" dose of UV or chemical carcinogen was administered several hours before a large "challenge" dose, after which newly synthesized DNA was labeled by incubation of the cells in [³H]thymidine. The change in the molecular weight distribution of this [³H]DNA upon further incubation was determined by sedimentation in alkaline sucrose gradients. The rate at which the distribution of the molecular weight of the [³H]DNA returned to that obtained for unirradiated control cells was reported to be enhanced by an inducing dose given from 2–12 hr prior to the challenge dose. This result has been confirmed in other laboratories (399, 400). The presence of cycloheximide in the interval between the two doses eliminated the effect (397) but also further reduced the rate of DNA synthesis after the challenge dose (400).

The interpretation of the apparent effect of an inducing dose upon the rate of daughter-strand closure has recently been questioned by Painter (400), who suggested that this response can be accounted for by the damaging effects of the inducing exposure. In particular these effects include a reduced rate of DNA synthesis and an anomalous bimodal distribution of DNA sizes at the time of administration of the challenge dose. The nascent DNA available for elongation several hours after the inducing dose includes a prominent species of *higher* molecular weight not seen in the unirradiated control: this might well give the appearance of enhanced daughter-strand closure following the challenge dose (400).

Phenomena possibly analogous to the adaptive response to alkylating agents in bacteria have been described recently in a mammalian system. The removal of O^6-methyl guanine from DNA in rat livers may be due to a demethylase (383). Its removal is more rapid in animals given low levels of dimethylnitrosamine for several weeks prior to a large dose of the agent (401). Larger, possibly saturating doses, decrease removal (402).

CONCLUDING REMARKS

Since bacterial excision repair was last reviewed in this series (7), not only has our understanding of previously recognized mechanisms increased, but new mechanisms have been described which have broadened our concept of excision repair to include possible roles of glycosylases, insertases, and demethylases. In addition, the importance of interrelations between the inducible and constitutive responses to lesions in DNA has become evident. However, the relationships of biological end points, such as survival and mutagenesis, to the biochemical criteria of responses to DNA damage, remain obscure. This is a particularly serious problem in the case of mammalian cells, since we may not even know the relevant biological end point. Survival and mutagenesis of cells in culture are not necessarily directly related to the function of cells in the organism. However, the colony-forming ability of irradiated XP cells does appear to correlate with the severity of the patients' neurological symptoms (403), and the excision-repair capability of cultured human cells may correlate with mutation and survival after exposure to UV and chemical carcinogens (404). Further analysis of responses to DNA damage may clarify their role in abnormal differentiation and malignant transformation.

ACKNOWLEDGMENTS

We appreciate the careful reading and editorial assistance by Evelyn Parker and we express thanks to Jyl Simpson for expert rendition of the seemingly endless revisions.

Work by the authors is supported by the American Cancer Society, the Department of Energy, and the National Institute of General Medical Sciences.

Literature Cited

1. Hanawalt, P. C., Friedberg, E. C., Fox, C. F., eds. 1978. *DNA Repair Mechanisms.* New York: Academic. 813 pp.
2. Hanawalt, P. C., Setlow, R. B. eds. 1975. *Molecular Mechanisms for Repair of DNA,* Parts A, B. New York: Plenum. 843 pp.
3. Sutherland, B. M. 1978. *Int. Rev. Cytol.* Suppl. 8, pp. 301–34
4. Lindahl, T. 1979. *Prog. Nucleic Acid Res. Mol. Biol.* 22: In press
5. Paterson, M. C. 1978. *Adv. Radiat. Biol.* 7:1–53
6. Roberts, J. J. 1978. *Adv. Radiat. Biol.* 7:211–436
7. Grossman, L., Braun, A., Feldberg, R., Mahler, I. 1975. *Ann. Rev. Biochem.* 44:19–43
8. Radding, C. M. 1978. *Ann. Rev. Biochem.* 47:847–80
9. Witkin, E. M. 1976. *Bacteriol. Rev.* 40:869–907
10. Cleaver, J. E. 1974. *Adv. Radiat. Biol.* 4:1–75
11. Hewitt, R. R., Meyn, R. E. 1978. *Adv. Radiat. Biol.* 7:153–79
12. Strauss, B., Tatsumi, K., Karran, P., Higgins, N. P., Ben-Asher, E., Altamirano-Dimas, M., Rosenblatt, L., Bose, K. 1978. In *Polycyclic Hydrocarbons and Cancer,* ed. H. Gelboin, P. Ts'o, 2:177–201. New York: Academic
13. Friedberg, E. C., Cook, K. H., Duncan, J., Mortelmans, K. 1977. *Photochem. Photobiol. Revs.* 2:263–322
14. Robbins, J. H., Kraemer, K. H., Lutzner, M. A., Festoff, B. W., Coon, H. G. 1974. *Ann. Intern. Med.* 80: 221–48
15. Cleaver, J. E., Bootsma, D. 1975. *Ann. Rev. Genet.* 9:19–38
16. Kraemer, K. 1977. In *DNA Repair Processes,* ed. W. W. Nichols, D. G. Murphy, pp. 37–71. Miami: Symposia Specialists. 286 pp.
17. Arlett, C. F., Lehmann, A. R. 1978. *Ann. Rev. Genet.* 12:95–115
18. Friedberg, E. C., Ehmann, U. K., Williams, J. I. 1979. *Adv. Radiat. Biol.* In press
19. Setlow, R. B. 1978. *Nature* 271:713–17
20. Paterson, M. C. 1979. In *Carcinogens: Identification and Mechanisms of Action,* ed. C. R. Shaw, A. C. Griffin, New York: Academic. In press
21. Cerutti, P. A. 1975. See Ref. 2, pp. 3–12
22. Cerutti, P. A. 1978. See Ref. 1, pp. 1–14
23. Hariharan, P. V., Cerutti, P. A. 1977. *Biochemistry* 16:2791–95
24. McCormick, J. P., Fischer, J. R., Pachlatko, J. P., Eisenstark, A. 1976. *Science* 191:468–69
25. Miller, E. C., Miller, J. A. 1969. *Ann. NY Acad. Sci.* 163:731–50
26. Heflich, R. H., Dorney, D. J., Maher, V. M., McCormick, J. J. 1977. *Biochem. Biophys. Res. Commun.* 77:634–41
27. Sarasin, A. R., Smith, C. A., Hanawalt, P. C. 1977. *Cancer Res.* 37:1786–93
28. Deutsch, W. A., Linn, S. 1979. *Proc. Natl. Acad. Sci. USA.* In press
29. Tye, B.-K., Nyman, P.-O., Lehman, I. R., Hochhauser, S., Weiss, B. 1977. *Proc. Natl. Acad. Sci. USA* 74:154–57
30. Gates, F. T. III, Linn, S. 1977. *J. Biol. Chem.* 252:1647–53
31. Tye, B.-K., Chien, J., Lehman, I. R., Duncan, B. K., Warner, H. R. 1978. *Proc. Natl. Acad. Sci. USA* 75:233–37
32. Wagner, R. Jr., Meselson, M. 1976. *Proc. Natl. Acad. Sci. USA* 73:4135–39
33. Boiteux, S., Villani, G., Spadari, S., Zambrano, F., Radman, M. 1978. See Ref. 1, pp. 73–84
34. Glickman, B., van den Elsen, P., Radman, M. 1978. *Mol. Gen. Genet.* 163:307–12
35. Bird, A. P., Southern, E. M. 1978. *J. Mol. Biol.* 118:27–47
36. Bird, A. P. 1978. *J. Mol. Biol.* 118: 49–60
37. Goth-Goldstein, R. 1977. *Nature.* 267:81–82
38. Singer, B., Bodell, W. J., Cleaver, J. E., Thomas, G. H., Rajewsky, M. F., Thon, W. 1978. *Nature* 276:85–88
39. Brookes, P., Lawley, P. D. 1961. *Biochem. J.* 80:496–503
40. Iyer, V. N., Szybalski, W. 1963. *Proc. Natl. Acad. Sci. USA* 50:355–62
41. Geiduschek, E. P. 1961. *Proc. Natl. Acad. Sci. USA* 47:950–55
42. Roberts, J. J., Pascoe, J. M. 1972. *Nature* 235:282–84
43. Cole, R. S. 1970. *Biochim. Biophys. Acta* 217:30–39
44. Cole, R. S., Sinden, R. R. 1975. See Ref. 2, pp. 487–95
45. Cole, R. S., Sinden, R. R., Yoakum, G. H., Broyles, S. 1978. See Ref. 1, pp. 287–90
46. Fujiwara, Y., Tatsumi, M., Sasaki, M. S. 1977. *J. Mol. Biol.* 113:635–49
47. Sasaki, M. S. 1978. See Ref. 1, pp. 675–84
48. Baden, H. P., Parrington, J. M., Delhanty, J. D. A., Pathak, M. A. 1972. *Biochim. Biophys. Acta* 262:247–55
49. Shafer, D. A. 1977. *Hum. Genet.* 39:177–90

50. Carter, D. M., Gaynor, A., McGuire, J. 1978. See Ref. 1, pp. 671–74
51. Wolff, S. 1978. See Ref. 1, pp. 751–60
52. Strayer, D. R., Boyer, P. D. 1978. *J. Mol. Biol.* 120:281–95
53. Hanawalt, P., Grivell, A., Nakayama, H. 1975. See Ref. 2, pp. 47–50
54. Hutchinson, F. 1978. See Ref. 1, pp. 457–63
55. Ormerod, M. G. 1976. In *Biology of Radiation Carcinogenesis,* ed. J. M. Yuhas, R. W. Tennant, J. D. Regan, pp. 67–92. New York: Raven. 347 pp.
56. Burrell, A. D., Feldschreiber, P., Dean, C. J. 1971. *Biochim. Biophys. Acta* 247:38–53
57. Kitayama, S., Matsuyama, A. 1968. *Biochem. Biophys. Res. Commun.* 33: 418–22
58. Hariharan, P. V., Hutchinson, F. 1973. *J. Mol. Biol.* 75:479–94
59. Krisch, R. E., Krasin, F., Sauri, C. J. 1976. *Int. J. Radiat. Biol.* 29:37–50
60. Resnick, M. A. 1978. See Ref. 1, pp. 417–20
61. Resnick, M. A., Martin, P. 1976. *Mol. Gen. Genet.* 143:119–29
62. Krasin, F., Hutchinson, F. 1977. *J. Mol. Biol.* 116:81–98
63. Clark, A. J. 1973. *Ann. Rev. Genet.* 7:67–86
64. Clark, A. J., Ganesan, A. 1975. See Ref. 2, pp. 431–37
65. Bachmann, B. J., Low, K. B., Taylor, A. L. 1976. *Bacteriol. Rev.* 40:116–67
66. Seeberg, E. 1978. *Proc. Natl. Acad. Sci. USA* 75:2569–73
67. Seeberg, E., Nissen-Meyer, J., Strike, P. 1976. *Nature* 263:524–26
68. Kushner, S. R., Shepherd, J., Edwards, G., Maples, V. F. 1978. See Ref. 1, pp. 251–54
69. Morand, P., Goze, A., Devoret, R. 1977. *Mol. Gen. Genet.* 157:69–82
70. Castellazzi, M., Morand, P., George, J., Buttin, G. 1977. *Mol. Gen. Genet.* 153:297–310
71. Emmerson, P. T., West, S. C. 1977. *Mol. Gen. Genet.* 155:77–85
72. Gudas, L. J., Mount, D. W. 1977. *Proc. Natl. Acad. Sci. USA* 74:5280–84
73. McEntee, K. 1977. *Proc. Natl. Acad. Sci. USA* 74:5275–79
74. Tomizawa, J., Ogawa, H. 1972. *Nature New Biol.* 239:14–16
75. Hoekstra, W. P. M., Storm, P. K., Zuidweg, E. M. 1974. *Mutat. Res.* 23:319–26
76. Gottesman, M. M., Gottesman, M. E., Gottesman, S., Gellert, M. 1974. *J. Mol. Biol.* 88:471–87
77. Gillen, J. R., Karu, A. E., Nagaishi, H.,

Clark, A. J. 1977. *J. Mol. Biol.* 113: 27–41
78. Clark, A. J., Volkert, M. R., Margossian, L. J. 1978. *Cold Spring Harbor Symp. Quant. Biol.* 43: In press
79. Rothman, R. H., Clark, A. J. 1977. *Mol. Gen. Genet.* 155:279–86
80. Horii, Z.-I., Clark, A. J. 1973. *J. Mol. Biol.* 80:327–44
81. Yamamoto, Y., Katsuki, M., Sekiguchi, M., Otsuji, N. 1978. *J. Bacteriol.* 135: 144–52
82. Marinus, M. G., Morris, N. R. 1973. *J. Bacteriol.* 114:1143–50
83. Marinus, M. G., Konrad, E. B. 1976. *Mol. Gen. Genet.* 149:273–77
84. Schendel, P. F. 1977. *Mol. Gen. Genet.* 156:281–87
85. Bridges, B. A. 1975. *Mutat. Res.* 29: 489–91
86. Bridges, B. A., Mottershead, R. P., Lehmann, A. R. 1976. *Biol. Zentralbl.* 95:393–403
87. Wickner, S. H. 1978. *Ann. Rev. Biochem.* 47:1163–91
88. Horiuchi, T., Maki, H., Sekiguchi, M. 1978. *Mol. Gen. Genet.* 163:277–83
89. Johnson, B. F. 1977. *Mol. Gen. Genet.* 157:91–97
90. Konrad, E. B., Modrich, P., Lehman, I. R. 1973. *J. Mol. Biol.* 77:519–29
91. George, D. L., Witkin, E. M. 1975. *Mutat. Res.* 28:347–54
92. Cox, E. C. 1976. *Ann. Rev. Genet.* 10:135–56
93. White, B. J., Hochhauser, S. J., Cintrón, N. M., Weiss, B. 1976. *J. Bacteriol.* 126:1082–88
94. Verly, W. G. 1978. See Ref. 1, pp. 187–90
95. Weiss, B., Rogers, S. G., Taylor, A. F. 1978. See Ref. 1, pp. 191–94
96. Youngs, D. A., Smith, K. C. 1978. *Mutat. Res.* 51:133–37
97. Sutherland, B. M., Chamberlin, M. J., Sutherland, J. C. 1973. *J. Biol. Chem.* 248:4200–5
98. Uyemura, D., Eichler, D. C., Lehman, I. R. 1976. *J. Biol. Chem.* 251:4085–89
99. Kornberg, T., Gefter, M. L. 1971. *Proc. Natl. Acad. Sci. USA* 68:761–64
100. Volkert, M. R., George, D. L., Witkin, E. M. 1976. *Mutat. Res.* 36:17–28
101. George, J., Castellazzi, M., Buttin, G. 1975. *Mol. Gen. Genet.* 140:309–32
102. Mount, D. W. 1977. *Proc. Natl. Acad. Sci. USA* 74:300–4
103. Roberts, J. W., Roberts, C. W., Mount, D. W. 1977. *Proc. Natl. Acad. Sci. USA* 74:2283–87
104. Ohkawa, T. 1977. *Biochim. Biophys. Acta* 476:190–202

105. Mount, D. W., Walker, A. C., Kosel, C. 1973. *J. Bacteriol.* 116:950–56
106. Ganesan, A. K., Seawell, P. C., Mount, D. W. 1978. *J. Bacteriol.* 135:935–42
107. Duncan, B. K., Rockstroh, P. A., Warner, H. R. 1978. *J. Bacteriol.* 134:1039–45
108. Kato, T., Shinoura, Y. 1977. *Mol. Gen. Genet.* 156:121–31
109. Glickman, B. W., Guijt, N., Morand, P. 1977. *Mol. Gen. Genet.* 157:83–9
110. Kushner, S. R., Nagaishi, H., Clark, A. J. 1972. *Proc. Natl. Acad. Sci. USA* 69:1366–70
111. Chase, J. W., Richardson, C. C. 1977. *J. Bacteriol.* 129:934–47
112. Castellazzi, M., George, J., Buttin, G. 1972. *Mol. Gen. Genet.* 119:153–74
113. Paterson, M. C., Smith, B. P., Knight, P. A., Anderson, A. K. 1977. In *Research in Photobiology*, ed. A. Castellani, pp. 207–18. New York: Plenum. 776 pp.
114. Paterson, M. C. 1978. See Ref. 1, pp. 637–50
115. German, J. 1978. See Ref. 1, pp. 625–31
116. Wade, M. H., Chu, E. H. Y. 1978. See Ref. 1, pp. 667–70
117. DeBusk, F. L. 1972. *J. Pediatr.* 80:697–724
118. Little, J. B., Weichselbaum, R. R., Nove, J., Albert, D. M. 1978. See Ref. 1, pp. 685–90
119. Bootsma, D. 1978. See Ref. 1, pp. 589–601
120. Lehmann, A. R., Kirk-Bell, S., Arlett, C. F., Paterson, M. C., Lohman, P. H. M., de Weerd-Kastelein, E. A., Bootsma, D. 1975. *Proc. Natl. Acad. Sci. USA* 72:219–23
121. Setlow, R. B., Carrier, W. L. 1964. *Proc. Natl. Acad. Sci. USA* 51:226–31
122. Boyce, R. P., Howard-Flanders, P. 1964. *Proc. Natl. Acad. Sci. USA* 51:293–300
123. Riklis, E. 1965. *Can. J. Biochem.* 43:1207–19
124. Regan, J. D., Setlow, R. B., Francis, A. A., Lijinsky, W. 1976. *Mutat. Res.* 38:293–302
125. Pettijohn, D. E., Hanawalt, P. C. 1963. *Biochim. Biophys. Acta* 72:127–29
126. Pettijohn, D. E., Hanawalt, P. 1964. *J. Mol. Biol.* 9:395–410
127. Hanawalt, P. C., Cooper, P. K. 1971. *Methods Enzymol.* 21:221–30
128. Scudiero, D., Henderson, E., Norin, A., Strauss, B. 1975. *Mutat. Res.* 29:473–88
129. Smith, C. A., Hanawalt, P. C. 1976. *Biochim. Biophys. Acta* 432:336–47
130. Collins, A. R. S. 1977. *Biochim. Biophys. Acta* 478:461–73

131. Erixon, K., Ahnström, G. 1979. *Mutat. Res.* In press
132. Dorson, J. W., Moses, R. E. 1975. In *DNA Synthesis and Its Regulation,* ed. M. Goulian, P. Hanawalt, C. F. Fox, pp. 815–21. Menlo Park, Calif: Benjamin. 880 pp.
133. Cooper, P. K., Hanawalt, P. C. 1972. *J. Mol. Biol.* 67:1–10
134. Smith, C. A., Hanawalt, P. C. 1978. *Proc. Natl. Acad. Sci. USA* 75:2598–602
135. Regan, J. D., Setlow, R. B., Ley, R. D. 1971. *Proc. Natl. Acad. Sci. USA* 68:708–12
136. Waldstein, E. 1978. See Ref. 1, pp. 219–24
137. Grossman, L., Riazuddin, S. 1978. See Ref. 1, pp. 205–17
138. Grossman, L., Riazuddin, S., Haseltine, W., Lindan, K. 1978. *Cold Spring Harbor Symp. Quant. Biol.* 43: In press
139. Hanawalt, P. C., Haynes, R. H. 1965. *Biochem. Biophys. Res. Commun.* 19: 462–67
140. Howard-Flanders, P., Boyce, R. P., Theriot, L. 1966. *Genetics* 53:1119–36
141. Shimada, K., Ogawa, H., Tomizawa, J. 1968. *Mol. Gen. Genet.* 101:245–56
142. Rupp, W. D., Howard-Flanders, P. 1968. *J. Mol. Biol.* 31:291–304
143. Seeberg, E., Johansen, I. 1973. *Mol. Gen. Genet.* 123:173–84
144. Taketo, A., Yasuda, S., Sekiguchi, M. 1972. *J. Mol. Biol.* 70:1–14
145. Waldstein, E. A., Sharon, R., Ben-Ishai, R. 1974. *Proc. Natl. Acad. Sci. USA* 71:2651–54
146. Seeberg, E., Strike, P. 1976. *J. Bacteriol.* 125:787–95
147. Riazuddin, S., Grossman, L. 1977. *J. Biol. Chem.* 252:6280–86
148. Riazuddin, S., Grossman, L. 1977. *J. Biol. Chem.* 252:6287–93
149. Minton, K., Durphy, M., Taylor, R., Friedberg, E. C. 1975. *J. Biol. Chem.* 250:2823–29
150. Braun, A., Grossman, L. 1974. *Proc. Natl. Acad. Sci. USA* 71:1838–42
151. Braun, A. G., Radman, M., Grossman, L. 1976. *Biochemistry* 15:4116–20
152. Seeberg, E. 1978. See Ref. 1, pp. 225–28
153. Howard-Flanders, P., Boyce, R. P. 1966. *Radiat. Res. Suppl.* 6:156–84
154. Ikenaga, M., Ichikawa-Ryo, H., Kondo, S. 1975. *J. Mol. Biol.* 92:341–56
155. Boyce, R. P., Howard-Flanders, P. 1964. *Z. Vererbungsl.* 95:345–50
156. Cole, R. S., Levitan, D., Sinden, R. R. 1976. *J. Mol. Biol.* 103:39–59
157. Kelly, R. B., Atkinson, M. R., Huberman, J. A., Kornberg, A. 1969. *Nature* 224:495–501

158. Livingston, D. M., Richardson, C. C. 1975. *J. Biol. Chem.* 250:470–78
159. Chase, J. W., Richardson, C. C. 1974. *J. Biol. Chem.* 249:4553–61
160. Tanaka, J., Sekiguchi, M. 1975. *Biochim. Biophys. Acta* 383:178–87
161. Chase, J. W., Masker, W. E. 1977. *J. Bacteriol.* 130:667–75
162. Shlaes, D. M., Anderson, J. A., Barbour, S. D. 1972. *J. Bacteriol.* 111:723–30
163. Cooper, P. 1977. *Mol. Gen. Genet.* 150:1–12
164. Kushner, S. R. 1974. *J. Bacteriol.* 120:1213–18
165. Cooper, P. K., Hunt, J. G. 1978. See Ref. 1, pp. 255–60
166. Masker, W. E., Chase, J. W. 1978. See Ref. 1, pp. 261–65
167. Friedberg, E. C., Lehman, I. R. 1974. *Biochem. Biophys. Res. Commun.* 58:132–39
168. Rothman, R. H. 1978. *J. Bacteriol.* 136:444–48
169. van Sluis, C. A., Mattern, I. E., Paterson, M. C. 1974. *Mutat. Res.* 25:273–79
170. Masker, W., Hanawalt, P., Shizuya, H. 1973. *Nature New Biol.* 244:242–43
171. Gefter, M. L. 1975. *Ann. Rev. Biochem.* 44:45–78
172. Campbell, J. L., Soll, L., Richardson, C. C. 1972. *Proc. Natl. Acad. Sci. USA* 69:2090–94
173. Ben-Ishai, R., Sharon, R. 1978. *J. Mol. Biol.* 120:423–32
174. Masker, W. E. 1977. *J. Bacteriol.* 129:1415–23
175. Cooper, P. K., Hanawalt, P. C. 1972. *Proc. Natl. Acad. Sci. USA* 69:1156–60
176. Masker, W. E., Simon, T. J., Hanawalt, P. C. 1975. See Ref. 2, pp. 245–54
177. Youngs, D. A., Smith, K. C. 1973. *J. Bacteriol.* 116:175–82
178. Tait, R. C., Harris, A. L., Smith, D. W. 1974. *Proc. Natl. Acad. Sci. USA* 71:675–79
179. Youngs, D. A., van der Schueren, E., Smith, K. C. 1974. *J. Bacteriol.* 117:717–25
180. Hanawalt, P., Burrell, A., Cooper, P., Masker, W. 1975. See Ref. 132, pp. 774–90
181. Ben-Ishai, R., Pugravitsky, E., Sharon, R. 1978. See Ref. 1, pp. 267–70
182. Masker, W. E. 1976. *Biochim. Biophys. Acta* 442:162–73
183. Sharon, R., Miller, C., Ben-Ishai, R. 1975. *J. Bacteriol.* 123:1107–14
184. Youngs, D. A., Smith, K. C. 1977. *Mol. Gen. Genet.* 152:37–41
185. Fornace, A. J. Jr., Kohn, K. W., Kann,

H. E. Jr. 1976. *Proc. Natl. Acad. Sci. USA* 73:39–43
186. Cook, P. R., Brazell, I. A., Pawsey, S. A., Giannelli, F. 1978. *J. Cell Science* 29:117–27
187. Tanaka, K., Sekiguchi, M., Okada, Y. 1975. *Proc. Natl. Acad. Sci. USA* 72:4071–75
188. Tanaka, K., Hayakawa, H., Sekiguchi, M., Okada, Y. 1977. *Proc. Natl. Acad. Sci. USA* 74:2958–62
189. Ciarrocchi, G., Linn, S. 1978. *Proc. Natl. Acad. Sci. USA* 75:1887–91
190. Linn, S., Kuhnlein, U., Deutsch, W. A. 1978. See Ref. 1, pp. 199–203
191. Mortelmans, K., Friedberg, E. C., Slor, H., Thomas, G., Cleaver, J. E. 1976. *Proc. Natl. Acad. Sci. USA* 73:2757–61
192. Friedberg, E. C., Rudé, J. M., Cook, K. H., Ehmann, U. K., Mortelmans, K., Cleaver, J. E., Slor, H. 1977. See Ref. 16, pp. 21–36
193. Nes, I. F., Nissen-Meyer, J. 1978. *Biochim. Biophys. Acta* 520:111–21
194. Waldstein, E., Peller, S., Robel, A., Setlow, R. B. 1978. See Ref. 1, pp. 291–294
195. Ahmed, F. E., Setlow, R. B. 1978. See Ref. 1, pp. 333–36
196. Amacher, D. E., Lieberman, M. W. 1977. *Biochem. Biophys. Res. Commun.* 74:285–90
197. Ahmed, F. E., Setlow, R. B. 1977. *Proc. Natl. Acad. Sci. USA* 74:1548–52
198. Brown, A. J., Fickel, T. H., Cleaver, J. E., Lohman, P. H. M., Wade, M. H., Waters, R. 1979. *Cancer Res.* In press
199. Ahmed, F. E., Setlow, R. B. 1977. *Cancer Res.* 37:3414–19
200. Yamasaki, H., Pulkrabek, P., Grunberger, D., Weinstein, I. B. 1977. *Cancer Res.* 37:3756–60
201. Amacher, D. E., Elliott, J. A., Lieberman, M. W. 1977. *Proc. Natl. Acad. Sci. USA* 74:1553–57
202. Scudiero, D., Norin, A., Karran, P., Strauss, B. 1976. *Cancer Res.* 36:1397–1403
203. Peleg, L., Raz, E., Ben-Ishai, R. 1977. *Exp. Cell Res.* 104:301–07
204. Ley, R. D., Sedita, B. A., Grube, D. D., Fry, R. J. M. 1977. *Cancer Res.* 37:3243–48
205. Weissbach, A. 1977. *Ann. Rev. Biochem.* 46:25–47
206. Bollum, F. J. 1975. *Prog. Nucleic Acid Res. Mol. Biol.* 15:109–44
207. Mosbaugh, D. W., Stalker, D. M., Probst, G. S., Meyer, R. R. 1977. *Biochemistry* 16:1512–18
208. Strauss, B., Bose, K., Altamirano, M.,

Sklar, R., Tatsumi, K. 1978. See Ref. 1, pp. 621–24

209. Brakel, C. L., Blumenthal, A. B. 1977. *Biochemistry* 16:3137–43

210. Bose, K., Karran, P., Strauss, B. 1978. *Proc. Natl. Acad. Sci. USA* 75:794–98

211. Lindahl, T., Gally, J. A., Edelman, G. M. 1969. *Proc. Natl. Acad. Sci. USA* 62:597–603

212. Cook, K. H., Friedberg, E. C. 1978. *Biochemistry* 17:850–57

213. Cook, K. H., Friedberg, E. C. 1978. See Ref. 1, pp. 301–5

214. Doniger, J., Grossman, L. 1976. *J. Biol. Chem.* 251:4579–87

215. Byrnes, J. J., Downey, K. M., Black, V. L., So, A. G. 1976. *Biochemistry* 15: 2817–23

216. Bertazzoni, U., Scovassi, A. I., Brun, G. M. 1977. *Eur. J. Biochem.* 81:237–48

217. Bolden, A., Pedrali Noy, G., Weissbach, A. 1977. *J. Biol. Chem.* 252: 3351–56

218. Hübscher, U., Kuenzle, C. C., Spadari, S. 1977. *Eur. J. Biochem.* 81:249–58

219. Tanaka, S., Koike, K. 1978. *Biochem. Biophys. Res. Commun.* 81:791–97

220. Berger, N. A., Johnson, E. S. 1976. *Biochim. Biophys. Acta* 425:1–17

221. Seki, S., Oda, T. 1977. *Biochim. Biophys. Acta* 476:24–31

222. Umeda, T., Koga, M. 1977. *Biochim. Biophys. Acta* 478:115–27

223. Friedman, D. L., Mueller, G. C. 1968. *Biochim. Biophys. Acta* 161:455–68

224. Edenberg, H. J., Anderson, S., DePamphilis, M. L. 1978. *J. Biol. Chem.* 253:3273–80

225. Su, R. T., DePamphilis, M. L. 1978. *J. Virol.* 28:53–65

226. Stockdale, F. E. 1971. *Science* 171: 1145–47

227. Stockdale, F. E., O'Neill, M. C. 1972. *J. Cell Biol.* 52:589–97

228. Wicha, M., Stockdale, F. E. 1972. *Biochem. Biophys. Res. Commun.* 48: 1079–87

229. Hübscher, U., Kuenzle, C. C., Limacher, W., Schoner, R., Spadari, S. 1978. *Cold Spring Harbor Symp. Quant. Biol.* 43: In press

230. Gautschi, J. R., Young, B. R., Cleaver, J. E. 1973. *Exp. Cell Res.* 76:87–94

231. Pedrali Noy, G., Weissbach, A. 1977. *Biochim. Biophys. Acta* 477:70–83

232. Hahn, G. M., King, D., Yang, S.-J. 1971. *Nature New Biol.* 230:242–44

233. Bertazzoni, U., Stefanini, M., Pedrali Noy, G., Guilotto, E., Nuzzo, F., Falaschi, A., Spadari, S. 1976. *Proc. Natl. Acad. Sci. USA* 73:785–89

234. Otto, B., Baynes, M., Knippers, R. 1977. *Eur. J. Biochem.* 73:17–24

235. Korn, D., Fisher, P. A., Battey, J., Wang, T. S. F. 1978. *Cold Spring Harbor Symp. Quant. Biol.* 43: In press

236. Sheinin, R., Guttman, S. 1977. *Biochim. Biophys. Acta* 479:105–18

237. Basilico, C. 1977. *Adv. Cancer Res.* 24:223–66

238. Regan, J. D., Setlow, R. B. 1974. *Cancer Res.* 34:3318–25

239. Mattern, M. R., Hariharan, P. V., Cerutti, P. A. 1976. *Biochim. Biophys. Acta* 395:48–55

240. Painter, R. B., Young, B. R. 1972. *Mutat. Res.* 14:225–35

241. Setlow, R. B., Faulcon, F. M., Regan, J. D. 1976. *Int. J. Radiat. Biol.* 29:125–36

242. Edenberg, H., Hanawalt, P. 1972. *Biochim. Biophys. Acta* 272:361–72

243. Smith, C. A. 1978. See Ref. 1, pp. 311–14

244. Clarkson, J. M. 1978. *Mutat. Res.* 52:273–84

245. Petinga, R. A., Andrews, A. D., Tarone, R. E., Robbins, J. H. 1977. *Biochim. Biophys. Acta* 479:400–10

246. Paterson, M. C., Lohman, P. H. M., De Weerd-Kastelein, E. A., Westerveld, A. 1974. *Biophys. J.* 14:454–66

247. Ehmann, U. K., Cook, K. H., Friedberg, E. C. 1978. *Biophys. J.* 22:249–64

248. Williams, J. I., Cleaver, J. E. 1978. *Biophys. J.* 22:265–79

249. Carrier, W. L., Smith, D. P., Regan, J. D. 1978. *J. Supramol. Struct.* Suppl. 2, p. 77

250. Smith, C. A., van Zeeland, A. A. 1978. *J. Supramol. Struct.* Suppl. 2, p. 38

251. Ehmann, U. K., Cook, K. H., Friedberg, E. C. 1978. See Ref. 1, pp. 315–18

252. Simons, J. W. I. M. 1978. See Ref. 1, pp. 729–32

253. Edenberg, H. J., Hanawalt, P. C. 1973. *Biochim. Biophys. Acta* 324:206–17

254. Smith, C. A., Hanawalt, P. C. 1976. *Biochim. Biophys. Acta* 447:121–32

255. Lieberman, M. W. 1976. *Int. Rev. Cytol.* 45:1–23

256. Birnboim, H. C., Paterson, M. C. 1978. *J. Mol. Biol.* 121:561–66

257. Felsenfeld, G. 1978. *Nature* 271:115–22

258. McConologue, L. C. 1978. *The distribution of ultraviolet and ionizing radiation induced damage and repair in staphylococcal nuclease sensitive and resistant DNAs of mammalian chromatin.* PhD thesis. Univ. Calif., Los Angeles. 139 pp.

259. Wiesehahn, G. P., Hyde, J. E., Hearst, J. E. 1977. *Biochemistry* 16:925–32

260. Cech, T., Pardue, M. L. 1977. *Cell* 11:631–40
261. Hanson, C. V., Shen, C.-K. J., Hearst, J. E. 1976. *Science* 193:62–64
262. Cech, T., Potter, D., Pardue, M. L. 1977. *Biochemistry* 16:5313–21
263. Ramanathan, R., Rajalakshmi, S., Sarma, D. S. R., Farber, E. 1976. *Cancer Res.* 36:2073–79
264. Metzger, G., Wilhelm, F. X., Wilhelm, M. L. 1977. *Biochem. Biophys. Res. Commun.* 75:703–10
265. Bodell, W. J. 1977. *Nucleic Acids Res.* 4:2619–28
266. Bodell, W. J. 1979. *Nucleic Acids Res.* In press
267. Cleaver, J. E. 1977. *Nature* 270:451–53
268. Smerdon, M. J., Tlsty, T. D., Lieberman, M. W. 1978. *Biochemistry* 17:2377–86
269. Smerdon, M. J., Lieberman, M. W. 1978. *Proc. Natl. Acad. Sci. USA* 75:4238–41
270. Smerdon, M. J., Lieberman, M. W. 1978. See Ref. 1, pp. 327–32
271. Cleaver, J. E., Williams, J. I., Kapp, L., Park, S. D. 1978. See Ref. 1, pp. 85–93
272. Erixon, K., Ahnström, G. 1978. See Ref. 1, pp. 319–322
273. Tlsty, T. D., Lieberman, M. W. 1978. *Nucleic Acids Res.* 5:3261–73
274. Wilkins, R. J., Hart, R. W. 1974. *Nature* 247:35–36
275. Povirk, L. F., Painter, R. B. 1976. *Biochim. Biophys. Acta* 432:267–72
276. Bodell, W. J., Banerjee, M. R. 1976. *Nucleic Acids Res.* 3:1689–701
277. Lieberman, M. W., Poirier, M. C. 1974. *Proc. Natl. Acad. Sci. USA* 71:2461–75
278. Moses, H. L., Webster, R. A., Martin, G. D., Spelsberg, T. C. 1976. *Cancer Res.* 36:2905–10
279. Henson, P. 1978. *J. Mol. Biol.* 119:487–506
280. Waters, R. 1978. See Ref. 1, pp. 323–26
281. Nakayama, H., Pratt, A., Hanawalt, P. 1972. *J. Mol. Biol.* 70:281–89
282. Clark, A. J., Volkert, M. R. 1978. See Ref. 1, pp. 57–72
283. Howard-Flanders, P., Rupp, W. D., Wilkins, B. M., Cole, R. S. 1968. *Cold Spring Harbor Symp. Quant. Biol.* 33:195–205
284. Sedgwick, S. G. 1975. *J. Bacteriol.* 123:154–61
285. Youngs, D. A., Smith, K. C. 1976. *J. Bacteriol.* 125:102–10
286. Smith, K. C., Meun, D. H. C. 1970. *J. Mol. Biol.* 51:459–72
287. Rupp, W. D., Wilde, C. E. III, Reno, D. L., Howard-Flanders, P. 1971. *J. Mol. Biol.* 61:25–44

288. Mosevitsky, M. I. 1976. *J. Mol. Biol.* 100:219–25
289. Ley, R. D. 1973. *Photochem. Photobiol.* 18:87–95
290. Iyer, V. N., Rupp, W. D. 1971. *Biochim. Biophys. Acta* 228:117–26
291. Johnson, R. C., McNeill, W. F. 1978. See Ref. 1, pp. 95–99
292. Lavin, M. F. 1978. *Biophys. J.* 23:247–56
293. Broker, T. R., Lehman, I. R. 1971. *J. Mol. Biol.* 60:131–49
294. Ganesan, A. K. 1974. *J. Mol. Biol.* 87:103–19
295. Ganesan, A. K., Seawell, P. C. 1975. *Mol. Gen. Genet.* 141:189–205
296. Bridges, B. A., Sedgwick, S. G. 1974. *J. Bacteriol.* 117:1077–81
297. Rothman, R. H., Kato, T., Clark, A. J. 1975. See Ref. 2, pp. 283–91
298. Johnson, R. C. 1976. *Biochem. Biophys. Res. Commun.* 70:791–96
299. Sedgwick, S. G., Bridges, B. A. 1974. *Nature* 249:348–49
300. Johnson, R. C. 1978. *J. Bacteriol.* 136:125–30
301. Johnson, R. C. 1977. *Nature* 267:80–81
302. Holloman, W. K., Radding, C. M. 1976. *Proc. Natl. Acad. Sci. USA* 73:3910–14
303. Kobayashi, I., Ikeda, H. 1977. *Mol. Gen. Genet.* 153:237–45
304. Mount, D. W., Low, K. B., Edmiston, S. J. 1972. *J. Bacteriol.* 112:886–93
305. Cleaver, J. E., Thomas, G. H. 1969. *Biochem. Biophys. Res. Commun.* 36:203–8
306. Rupp, W. D., Zipser, E., von Essen, C., Reno, D., Prosnitz, L., Howard-Flanders, P. 1970. In *Time and Dose Relationships as Applied to Radiotherapy*, p. 1. Brookhaven Natl. Lab. Publ. BNL 50203 (C-57)
307. Lehmann, A. R. 1972. *J. Mol. Biol.* 66:319–37
308. Lehmann, A. R., Kirk-Bell, S. 1972. *Eur. J. Biochem.* 31:438–45
309. Fujiwara, Y., Tatsumi, M. 1976. *Mutat. Res.* 37:91–110
310. Lehmann, A. R., Kirk-Bell, S. 1974. *Mutat. Res.* 26:73–82
311. Fujiwara, Y. 1978. See Ref. 1, pp. 519–22
312. Lehmann, A. R., Ormerod, M. G. 1971. *Biochim. Biophys. Acta* 272:191–201
313. Elkind, M. M., Kamper, C. 1970. *Biophys. J.* 10:237–45
314. Lett, J. T., Klucis, E. S., Sun, C. 1970. *Biophys. J.* 10:277–92
315. Huberman, J. A., Riggs, A. D. 1968. *J. Mol. Biol.* 32:327–41

316. Sheinin, R., Humbert, J., Pearlman, R. E. 1978. *Ann. Rev. Biochem.* 47:277–316
317. Edenberg, H. J. 1976. *Biophys. J.* 16:849–60
318. Lehmann, A. R. 1978. See Ref. 1, pp. 485–88
319. Meyn, R. E., Hewitt, R. R., Thomson, L. F., Humphrey, R. M. 1976. *Biophys. J.* 16:517–25
320. Cleaver, J. E. 1965. *Biochim. Biophys. Acta* 108:42–52
321. Klímek, M., Vlasínová, M. 1966. *Int. J. Radiat. Biol.* 11:329–37
322. Doniger, J. 1978. *J. Mol. Biol.* 120:433–46
323. Meneghini, R. 1976. *Biochim. Biophys. Acta* 425:419–27
324. Meneghini, R., Hanawalt, P. 1975. See Ref. 2, pp. 639–42
325. Meneghini, R., Hanawalt, P. 1976. *Biochim. Biophys. Acta* 425:428–37
326. Clarkson, J. M., Hewitt, R. R. 1976. *Biophys. J.* 16:1155–64
327. Fujiwara, Y., Kondo, T. 1974. In *Sunlight and Man-Normal and Abnormal Photobiologic Responses,* ed. T. B. Fitzpaatrick, M. A. Pathek, L. C. Harber, S. Makoto, A. Kukita, pp. 91–106. Tokyo: Tokyo Univ. 870 pp.
328. Painter, R. B. 1974. *Genetics* 78:139–48 (Suppl.)
329. Buhl, S. N., Regan, J. D. 1973. *Nature* 246:484
330. Waters, R., Regan, J. D. 1976. *Biochem. Biophys. Res. Commun.* 72:803–7
331. Waldstein, E. A., Setlow, R. B. 1976. *Int. Congr. Photobiol. 7th,* p. 155. (Abstr.)
332. Fujiwara, Y., Tatsumi, M. 1977. *Mutat. Res.* 43:279–90
333. Lehmann, A. R., Kirk-Bell, S. 1978. *Photochem. Photobiol.* 27:297–307
334. Meneghini, R., Menck, C. F. M. 1978. See Ref. 1, pp. 493–97
335. D'Ambrosio, S. M., Setlow, R. B. 1978. See Ref. 1, pp. 499–503
336. Meyn, R. E., Humphrey, R. M. 1971. *Biophys. J.* 11:295–301
337. Meyn, R. E., Fletcher, S. E. 1978. See Ref. 1, pp. 513–16
338. Ganesan, A. K., Smith, K. C. 1971. *Mol. Gen. Genet.* 113:285–96
339. Villani, G., Boiteux, S., Radman, M. 1978. *Proc. Natl. Acad. Sci. USA* 75:3037–41
340. Higgins, N. P., Kato, K., Strauss, B. 1976. *J. Mol. Biol.* 101:417–25
341. Tatsumi, K., Strauss, B. 1978. See Ref. 1, pp. 523–26
342. Weigle, J. J. 1953. *Proc. Natl. Acad. Sci. USA* 39:628–36
343. Defais, M., Fauquet, P., Radman, M., Errera, M. 1971. *Virology* 43:495–503
344. Witkin, E. M., George, D. L. 1973. *Genetics* 73:91–108 (Suppl.)
345. Witkin, E. M. 1974. *Proc. Natl. Acad. Sci. USA* 71:1930–34
346. Radman, M. 1974. In *Molecular and Environmental Aspects of Mutagenesis,* eds. L. Prakash, F. Sherman, M. W. Miller, C. W. Lawrence, H. W. Taber, pp. 128–42. Springfield, Ill: Thomas. 289 pp.
347. Radman, M. 1975. See Ref. 2, pp. 355–67
348. Castellazzi, M., George, J., Buttin, G. 1972. *Mol. Gen. Genet.* 119:139–52
349. Inouye, M. 1971. *J. Bacteriol.* 106:539–42
350. Little, J. W., Kleid, D. G. 1977. *J. Biol. Chem.* 252:6251–52
351. Roberts, J. W., Roberts, C. W., Craig, N. L. 1978. *Proc. Natl. Acad. Sci. USA* 75:4714–18
352. Meyn, M. S., Rossman, T., Troll, W. 1977. *Proc. Natl. Acad. Sci. USA* 74:1152–56
353. Weinstock, G. M., McEntee, K., Lehman, I. R. 1979. *Proc. Natl. Acad. Sci. USA* 76:126–30
353a. Shibata, T., DasGupta, C., Cunningham, R., Radding, C. 1979. *Proc. Natl. Acad. Sci. USA* 76: In press
354. Little, J. W., Hanawalt, P. C. 1977. *Mol. Gen. Genet.* 150:237–48
355. Gudas, L. J., Pardee, A. B. 1975. *Proc. Natl. Acad. Sci. USA* 72:2330–34
356. Smith, C. L., Oishi, M. 1978. *Proc. Natl. Acad. Sci. USA* 75:1657–61
357. Oishi, M., Smith, C. L. 1978. *Proc. Natl. Acad. Sci. USA* 75:3569–73
358. Gudas, L. J. 1976. *J. Mol. Biol.* 104:567–87
359. Higgins, N. P., Peebles, C. L., Sugino, A., Cozzarelli, N. R. 1978. *Proc. Natl. Acad. Sci. USA* 75:1773–77
360. Gudas, L. J., Pardee, A. B. 1976. *J. Mol. Biol.* 101:459–77
361. Oishi, M. 1978. *Cold Spring Harbor Symp. Quant. Biol.* 43: In press
362. Caillet-Fauquet, P., Defais, M., Radman, M. 1977. *J. Mol. Biol.* 117:95–112
363. Bridges, B. A., Mottershead, R. P., Sedgwick, S. G. 1976. *Mol. Gen. Genet.* 144:53–58
364. Sedgwick, S. G. 1975. *Proc. Natl. Acad. Sci. USA* 72:2753–57
365. Witkin, E. M., Wermundsen, I. E. 1978. *Cold Spring Harbor Symp. Quant. Biol.* 43: In press
366. Green, M. H. L., Bridges, B. A., Eyfjörd, J. E., Muriel, W. J. 1977. In *Col-*

loques Internationaux du CNRS No. 256, pp. 227–36. Paris: CNRS

367. Nishioka, H., Doudney, C. O. 1969. *Mutat. Res.* 8:215–28
368. Nishioka, H., Doudney, C. O. 1970. *Mutat. Res.* 9:349–58
369. Bridges, B. A., Mottershead, R. 1971. *Mutat. Res.* 13:1–8
370. Green, M. H. L., Bridges, B. A., Eyfjörd, J. E., Muriel, W. J. 1977. *Mutat. Res.* 42:33–44
371. Kato, T., Rothman, R. H., Clark, A. J. 1977. *Genetics* 87:1–18
372. Bresler, S. E. 1975. *Mutat. Res.* 29:467–72
373. Sedgwick, S. G. 1976. *Mutat. Res.* 41:185–200
374. Bridges, B. A., Mottershead, R. P. 1978. *Mol. Gen. Genet.* 162:35–41
375. Bridges, B. A., Mottershead, R. P. 1978. *Mutat. Res.* 52:151–59
376. Kato, T. 1977. *Mol. Gen. Genet.* 156:115–20
377. Sedliaková, M., Slezáriková, V., Pirsel, M. 1979. *Mol. Gen. Genet.* 167:209–15
378. Samson, L., Cairns, J. 1977. *Nature* 267:281–83
379. Jeggo, P., Defais, M., Samson, L., Schendel, P. 1977. *Mol. Gen. Genet.* 157:1-9
380. Jeggo, P., Defais, M., Samson, L., Schendel, P. 1978. *Mol. Gen. Genet.* 162:299–305
381. Schendel, P. F., Defais, M., Jeggo, P., Samson, L., Cairns, J. 1978. See Ref. 1, pp. 391–94
382. Schendel, P. F., Robins, P. E. 1978. *Proc. Natl. Acad. Sci. USA* 75:6017–20
383. Pegg, A. E. 1978. *Biochem. Biophys. Res. Commun.* 84:166–73
384. Bockstahler, L. E., Lytle, C. D. 1970. *Biochem. Biophys. Res. Commun.* 41:184–89

385. Bockstahler, L. E., Lytle, C. D. 1971. *J. Virol.* 8:601–2
386. Lytle, C. D. 1979. *Radiation-enhanced virus reactivation in mammalian cells.* Presented at Int. Conf. UV Carcinogenesis, Warrenton, Virginia, 1977. NCI monogr. In press
387. Bockstahler, L. E., Lytle, C. D. 1977. *Photochem. Photobiol.* 25:477–82
388. Sarasin, A. R., Hanawalt, P. C. 1978. *Proc. Natl. Acad. Sci. USA* 75:346–50
389. Lytle, C. D., Coppey, J., Taylor, W. D. 1978. *Nature* 272:60–62
390. Lytle, C. D., Day, R. S. III, Hellman, K. B., Bockstahler, L. E. 1976. *Mutat. Res.* 36:257–64
391. Devoret, R. 1965. *CR Ac. Sci. D* 260:1510–13
392. Day, R. S. III 1978. See Ref. 1, pp. 531–34
393. DasGupta, U. B., Summers, W. C. 1978. *Proc. Natl. Acad. Sci. USA* 75:2378–81
394. Day, R. S. III, Ziolkowski, C. 1978. See Ref. 1, pp. 535–39
395. Sarasin, A. R., Hanawalt, P. C. 1978. See Ref. 1, pp. 547–50
396. Williams, J. I., Cleaver, J. E. 1978. See Ref. 1, pp. 551–54
397. D'Ambrosio, S. M., Setlow, R. B. 1976. *Proc. Natl. Acad. Sci. USA* 73:2396–400
398. D'Ambrosio, S. M., Setlow, R. B. 1978. *Cancer Res.* 38:1147–53
399. Hanawalt, P. C. 1977. See Ref. 16, pp. 1–19
400. Painter, R. B. 1978. *Nature* 275:243–45
401. Montesano, R., Brésil, H., Margison, G. P. 1979. *Cancer Res.* In press
402. Pegg, A. E. 1978. *Nature* 274:182–84
403. Andrews, A. D., Barrett, S. F., Robbins, J. H. 1978. See Ref. 1, pp. 613–16
404. Maher, V. M., Dorney, D. J., Heflich, R. H., Levinson, J. W., Mendrala, A. L., McCormick, J. J. 1978. See Ref. 1, pp. 717–22

Ann. Rev. Biochem. 1979. 48:837–70

HISTONE GENES AND ♦12027
HISTONE MESSENGERS

Laurence H. Kedes[1]

Howard Hughes Medical Institute Labortory at the Department of Medicine,
Stanford University School of Medicine,
and the Veterans Administration Hospital, Palo Alto, California 94304

CONTENTS

[1]This work was supported in part by grants from the National Institutes of Health, the American Cancer Society, and the Veterans Administration. The author is an Investigator of the Howard Hughes Medical Institute.

PERSPECTIVES AND SUMMARY

Form ever follows function

Louis Henri Sullivan
American Architect 1856–1924

If nature, the architect, and evolution, the builder, have followed the advice of Louis Sullivan, then the molecular geneticist must try, like the modern archeologist exploring the architectural remains of an ancient civilization, to infer function from form.[2] Recent advances in the ability to isolate eukaryotic genes, especially with the technology of recombinant DNA cloning, have provided molecular geneticists with numerous shards of the animal genome to examine in minute detail. From these pieces it is hoped that general principals about gene organization will emerge, and that testable concepts of gene regulation and function will be suggested. This review focuses on the details of one of those genome fragments, the genes for histone proteins. The histones, a limited set of small proteins, are the fundamental structural proteins of chromosomal organization and are ubiquitous in Nature. Much is known about the dynamics of their biogenesis. Other protein coding genes that are currently under investigation, such as those for globin, immunoglobulin, and ovalbumin, are expressed only in highly differentiated cells. The histone gene set products, on the other hand, are synthesized by every cell undergoing division.

The genes for histone proteins have now been isolated by recombinant DNA cloning from several species. Many common features have been found both in general structural topology and in terms of detailed DNA sequence analysis. Perhaps the most striking feature is that the genes for the histone

---

[2]Metaphorical apologies to Paul Gross who taught me everything I know about metaphors but unfortunately did not teach me everything *he* knows about metaphors.

protein set are closely clustered. Since eukaryotic cells are perfectly capable of regulating the activity of functionally related genes located at dispersed genomic loci, a major working hypothesis is that the form of histone gene topology follows from as yet undefined functions.

This review deals mainly with the structure and organization of the histone genes and messenger RNAs and provides only a glimpse of the existing information concerning the structure, regulation, and metabolism of the histone proteins. The selection of references is meant to be neither comprehensive nor historical but rather illustrative of our state of knowledge about the histone genes and their messenger RNAs.

INTRODUCTION

Histone proteins are a remarkable set of five small basic polypeptides ubiquitous in eukaryotic organisms. Examination of their interactions with each other and with DNA have provided new insights into the structure of chromatin and chromosomes [see, for example, the review by Elgin & Weintraub (1)]. Recent investigations on the genes and mRNAs that code for histones have demonstrated equally novel features about eukaryotic gene organization. The histone genes of sea urchins have been the most extensively defined of any genera. They serve as a prototype against which the histone genes of other species can be compared. The genes are remarkable in that they are repetitive and tandemly linked, and that the coding regions for five histone proteins are interdigitated with noncoding spacer sequences.

Unfractionated 9S polysomal histone mRNA (2) was used as a probe for hybridization with total sea urchin DNA to show that the genes coding for histones are reiterated several hundred times in the sea urchin genome (3). In addition, each of the separated histone mRNAs hybridizes with the same reiteration frequency as the unfractionated 9S histone mRNA (4–6).

The repetitive histone genes are tandemly linked since high molecular weight histone DNA sequences separate as a satellite from bulk chromosomal DNA after density equilibrium centrifugation on CsCl gradients (3, 7). However, the GC content and buoyant density of DNA coding for histones, predicted from both histone protein amino acid compositions and histone mRNA base compositions, was higher than that of the observed satellite DNA. On the other hand, when the DNA was sheared to fragments of about 1000 base pairs, the histone gene satellite attained a higher buoyant density (3) consistent with the predicted value. This result suggested the existence of AT-rich "spacer" DNA interspersed with approximately equal amounts of the GC rich histone coding sequences. The coding sequences consist of about 0.2% of the total sea urchin DNA. However, the combination of coding plus spacer regions may amount to 0.5% or more of the haploid sea urchin genome (7).

HISTONE GENE ORGANIZATION

General Features

The major breakthroughs in understanding the organization of histone genes were achieved with experiments making use of restriction endonucleases to clone histone DNA in bacterial plasmids (8–12) and to examine both chromosomal and cloned histone DNA (9, 13–19). Restriction endonuclease-treated histone DNA has been hybridized with purified individual histone mRNAs. Such experiments (9, 13, 14, 17, 20) reveal that the genes[3] coding for each of the five major histone proteins are interspersed with one another and that spacer DNA is located between each of the coding regions (9, 14, 15, 17, 18, 20). These multigenic clusters are tandemly linked and identical (or nearly identical). These observations have allowed the construction of maps of histone gene repeat units for several sea urchin species (Figure 1).

In sea urchins all five histone mRNA fractions hybridize within a single repeat unit. Restriction mapping and electron microscopic studies of mRNA gene hybrids or partial denaturation maps of the genes have helped to locate each of the histone coding sequences within the repeat unit (9, 15, 17, 18).

Although the lengths of the repeat units could accommodate more than five coding regions, these studies have shown that each of the mRNAs hybridizes only once per repeat unit. Equally important is the conclusion that proteins, DNA sequences, and mRNAs appear to be entirely colinear by both electron microscope hybrid analysis and DNA sequencing. Thus, there seems to be no intervening sequences within the mRNA coding re-

---

[3]In order to describe the organization of histone genes in a consistent manner we define these genes by their positions relative to the sequences that encode amino acids. In referring to 5' and 3' directions, we only consider the DNA strand that is not transcribed i.e. whose base sequence is synonymous with the mRNA sequence. *Coding region:* The region of the DNA that encodes a histone protein and is located between the ATG start codon and a terminator. *mRNA region:* The region of the DNA that is synonymous with the polysomal mRNA sequence. This region by definition includes the coding region. *mRNA leader:* Any DNA sequence at the 5' end (upstream) of the coding region and present in the polysomal mRNA. *mRNA trailer:* Any DNA sequence at the 3' end (downstream) of the coding region and present in the polysomal mRNA. *Spacer DNA:* Any DNA sequence lying outside of the polysomal mRNA region. Such spacer DNA can be transcribed or nontranscribed. If spacer is transcribed and contiguous with molecules that subsume the mRNA region it must be part of a precursor to polysomal mRNA. Depending on its upstream or downstream position it would be *precursor leader* or *precursor trailer DNA.*

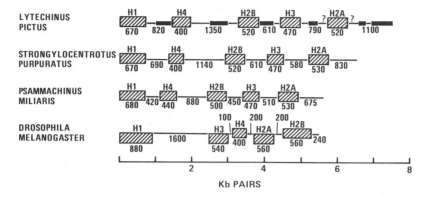

Figure 1 Molecular maps of histone gene repeat units of three sea urchin species and Drosophila Melanogaster. The histone gene organizational maps are calibrated in base pairs. The mRNA coding regions, including their leader and trailer sequences, are depicted by the hatched boxes. Spacer DNA sequences are represented by the lines connecting the boxes. Gene regions on the line are transcribed from left to right (i.e. the 5' end of the anticoding strand is to the left). Gene regions below the line are transcribed right to left. The two nonallelic *L. pictus* repeat units discussed in the text are superimposed on the map since they have essentially identical linear topology. Regions of nonhomology, drawn to scale, are depicted by the thick lines in spacer regions. The "?" symbols around the *L. pictus* H2A gene signify unexamined regions.

The topology, overall length, and strandedness of the *Drosophila* genes are as shown in the Figure. The spacer and coding lengths shown are not accurate. The actual lengths (provided by R. Lifton and M. Goldberg) are, for the coding regions, H1, 950 bases; H3, 520; H4, 365; H2A, 470; and H2B, 420. For the 5 spacers shown in the figure reading from left to right the correct lengths are 1350 bases, 200, 400, 180, and 300.

gions. All the sea urchin histone genes share the same polarity, (i.e. only one of the DNA strands serves as the template for synthesis of all five histone mRNAs) and the same relative 3'–5' order of the five genes on the transcribed strand: H1, H4, H2B, H3 and H2A (see Figure 1). Results of direct sequencing of cloned histone DNAs (21–24) demonstrate correspondence with the amino acid sequences of all five histones. Sequencing data has thus rigorously confirmed these earlier results on the gene organization.

The topology of the genes of several different sea urchin species that are separated by millions of years of evolution, is remarkably conserved. In addition to the features described above, there seems to be a constant rank order of the spacer distances and their relative sizes compared to either the basic repeat length or to the total spacer length (Figure 2).

The Special Case of Lytechinus pictus

A segment of an *L. pictus* histone repeat unit has been cloned by recombinant DNA technology (8). When that segment was used to probe restricted

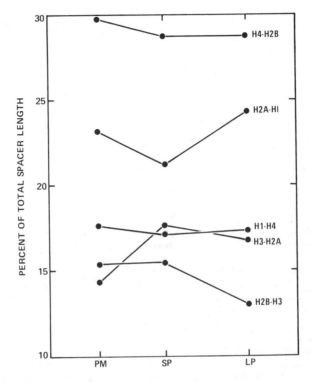

Figure 2 Relative spacer lengths in sea urchin histone gene repeat units. The lengths of 5 spacer regions for the 3 sea urchin species shown in Figure 1 were normalized by dividing by their repeat unit length. The data for the sea urchins *Psammechinus miliaris* (PM), *Strongylocentrotus purpuratus* (SP), and *Lytechinus pictus* (LP) reflect a constancy of relative size and rank order of spacer lengths.

L. pictus chromosomal DNA, it was found that individual sea urchins contained two major classes of nonallelic histone gene repeat units (Figure 1). Both types were subsequently cloned (R. Cohn and L. Kedes, submitted for publication).

TOPOLOGICAL DETAILS   Detailed topological maps of gene and spacer regions have been obtained for the two types of *L. pictus* histone gene repeat units. The major findings of these studies are illustrated in Figure 1. The two repeat units have the following characteristics:

1. They are approximately the same length, 7.3 kb.
2. Their topology and gene order and that of other sea urchin histone gene repeat units is essentially identical.
3. R-loop mapping (25) with histone mRNAs has placed the coding regions of the two clones at analogous loci.

4. Restriction endonuclease mapping, on the other hand, shows extensive differences between the two types of repeats. The restriction site differences cluster mainly in the spacer regions, although a few are found in the coding regions. Those restriction cutting sites that are located at *homologous* topological points are confined to coding regions.

5. Electron microscope heteroduplex analysis of hybrids formed between the two cloned repeat units reveals extensive areas of nonhomology that are confined to spacer DNA. These nonhomologous regions are mapped in Figure 1 as heavy bars on the *L. pictus* gene map. Equally important is the observation that the homology between the repeats is not confined to the coding region but extends in either or both directions into spacer DNA. Thus some of the spacer DNA has been highly conserved and other regions totally divergent. Regions of nonhomology are found in every spacer.

6. The results of experiments that rely on separation of restriction digested genomic DNA, suggest that the two types of repeats are not chromosomally linked and are not intermingled.

7. Thermal elution analyses of histone mRNA synthesized in early development and hybridized with each of the two cloned repeat units reveal essentially identical melting curves.

This data can be interpreted to suggest that both cloned gene units are active in histone mRNA transcription in early embryogenesis and that the *L. pictus* histone mRNAs are an approximately equal mixture of templates transcribed from the two gene variants.

EVOLUTIONARY IMPLICATIONS   The striking feature of histone genes is that two nonallelic clusters of tandemly repeated ones have evolved within the same organism, presumably from a common ancestral gene set. Their topology has been almost precisely conserved, although large regions of sequences in the spacer DNA have totally diverged. Spacer regions immediately flanking the coding sequences are also conserved, although the restriction endonuclease mapping and thermal elution profile suggests that there is some divergence in the coding region analogs. While some restriction site microheterogeneity among the tandem repeats of each class can be detected, each class is essentially homogeneous.

Evolutionary mechanisms that maintain the homogeneity of tandemly repeated genes have been put forward (26). However, it is more difficult to conceive of mechanisms that can maintain the high degree of homology of coding and flanking regions of the two classes of *L. pictus* repeat units while allowing total divergence of the variable spacer regions. One is left then with the notion that the conserved analogous regions, including both coding and flanking spacer sequences, are under tight evolutionary constraint and selective pressure within the species. The evolutionary maintenance of the topology (i.e. location and length) of these variant spacer regions suggests, furthermore, that while their actual DNA sequences are free to diverge their size and location is under selective pressure.

Drosophila Melanogaster

The histone genes of *D. melanogaster* are the only ones other than sea urchin for which a detailed structural map is now available (10, 27).

CHROMOMERIC LOCALIZATION   In situ hybridization of sea urchin (*Psammechinus miliaris, Lytechinus pictus*, or *Strongylocentrotus purpuratus*) histone mRNA (28, 29) or complementary RNA made from cloned sea urchin histone genes (30) has revealed interesting features of the histone gene organization in the polytene salivary gland chromosomes of *Drosophila*. The probe hybridized extensively and exclusively with DNA in region 39 D-E on the left arm of chromosome 2. Probes consisting of individual histone mRNAs or complementary RNAs made from subregions of the histone gene repeat unit of *S. purpuratus* each hybridize throughout the 39 D-E region (30). Such data provided the first evidence that the sequences coding for each of the histone mRNAs may be intermingled in the genome rather than arranged in independent blocks.

CLONED DROSOPHILA HISTONE GENES   Karp & Hogness have isolated a number of recombinant clones of *D. melanogaster* DNA which contain histone mRNA coding sequences (27). They screened a collection of colE1 plasmids containing segments of *Drosophila melanogaster* DNA for their ability to hybridize to labelled sea urchin histone mRNA. The *Drosophila* sequences in these selected clones are homologous to DNA in region 39 D-E of salivary gland chromosomes. Two major types of repeat units were found. In both, the five histone messenger coding regions are intermingled and interdigitate with spacer DNA. The two types of repeat are 4,750 bases long and 5,000 bases long. The only known difference between them is the presence in the longest spacer region of a 250 base pair insertion.

DNA excess hybridization kinetics reveal that there are about 110 copies of the histone gene repeat unit per *Drosophila* haploid genome. Nick-translated, cloned *Drosophila* histone genes were hybridized to restriction-digested genome DNA separated by electrophoresis and blotted onto nitrocellulose membranes by the technique described by Southern [the "Southern blot"; (31)] The results revealed that the longer repeat unit predominates in the genome in a ratio of 3 : 1. The repeats are arranged in tandem. The 4.75 kb and 5.0 kb repeat units can be found next to each other in a nonrandom way whose organization is not yet clear. Thus, the major features of histone gene organization in *D. melanogaster*, namely, tandem repeats containing sequences for each of the five histone proteins, linearly arrayed and interdigitated with spacer DNA, seems to be similar to the organization found in sea urchins. Detailed analysis of *Drosophila* repeat unit, however, has revealed at least one major topological difference.

Lifton et al (10) separated *Drosophila* histone mRNAs by hybridization to subsegments of the cloned *Drosophila* repeat unit. After separating the cloned DNA strands, they found that three of the messengers hybridized to one strand and two to the other (see Figure 1). This result demonstrates that the *Drosophila* histone genes are not all transcribed from the same DNA strand and that, at the least, two sites of transcription initiation must be present. The precise locations of the mRNA coding and spacer regions on the *Drosophila* histone gene repeat unit have been identified by a combination of restriction endonuclease fragment hybridization and DNA sequence analysis (M. Goldberg and D. Hogness, personal communication).

Cloned Drosophila histone gene variants Karp & Hogness [27] have also identified rarer clones of histone DNA sequences in *D. melanogaster*. Two such clones contain, in addition to histone DNA, sequences with no homology with the histone genes. One clone, cDm531, contains 4.6 kb of a 5.0 kb type repeat unit adjacent to a nonhistone 13.4 kb DNA segment. The junction between the histone and nonhistone sequence seems to lie in the H1 coding region (see Figure 1). Karp & Hogness examined restriction endonuclease–digested *D. melanogaster* chromosomal DNA by the Southern blot (31) hybridization technique. They found that the nonhistone sequences of cDm531 not only occur in fragments that corresponded in size to those found in the parent clone but they also occur in many DNA fragments of other sizes. Thus sequences within this 13.4 kb region of nonhistone DNA are scattered throughout the genome as well as being located adjacent to histone DNA sequences.

A second clone, cDm528, contains two histone regions that are separated by a 7 kb nonhistone region. The 7 kb nonhistone DNA contains no homology to the nonhistone region of cDm531. The two flanking histone regions both correspond in type to the 4.75 kb-repeating unit and are oriented in the same direction. The data of Karp & Hogness suggest that the nonhistone DNA of cDm528 is an exact insertion between the H3 and H4 coding regions. The nonhistone DNA sequences of cDm528 are found in many DNA fragments in the total *Drosophila* genome. These fragments do not correspond in size to those present in cDm528. The type of histone gene organization exhibited by cDm528 suggests that, at least in *D. melanogaster*, the tandemly arrayed histone genes may be intermittently interrupted by nonhistone DNA. Topologically, of course, such interruptions are merely insertions in spacer DNA which might contribute to histone repeat unit heterogeneity. The fact that this insertion is not shared by all histone DNA repeat units could be an important clue to a superorder organization of histone genes or, alternatively, an example of a random insertion event that might not prove to be stable in the evolutionary sense.

Histone Genes of Other Species

Several laboratories have been able to obtain preliminary data on histone gene organization in species other than sea urchins and *Drosophila*. The histone gene reiteration frequency has been analyzed by DNA excess hybridization using homologous histone mRNA from *HeLa* cells, (32, 33) mouse (34) and *Xenopus laevis* embryos (35). The reiteration frequency determined in this way is 20–50 fold for *Xenopus,* 10–20 fold for mouse, and 30–40 fold in human placental DNA. Scott & Wells (36) have isolated mRNA for the chicken H5 histone. This protein is a tissue-specific histone and the only one synthesized in avian red blood cells. They determined the reiteration frequency of its genes by DNA excess hybridization and found about ten copies per genome. No data exist on the organization or reiteration of the genes for the other avian histones. Freigan et al (37) have examined the histone gene organization of a number of marine invertebrates. Restriction endonuclease–digested genomic DNA was electrophoretically separated on aqueous gels, blotted onto nitrocellulose by the method of Southern (31), and hybridized with nick-translated sea urchin histone gene clones. By examining the effect of several different restriction enzymes on the hybridization pattern, they were able to determine a histone gene repeat unit length for: *Limulus polyphemus* (horseshoe crab), 4.1 kb; *Echinarachinus parma* (sand dollar), 6.1 kb; two mollusks, *Spisula solidissima* (clam) and *Crassostrea virginica* (oyster), 4.5 and 6.3 kb respectively; *Chaetopterus pergamentaceous* (worm), 5.2 kb; and an additional sea urchin, *Arbacia punctulata,* 7.0 kb. The repeat unit of the sea urchin *M. tripneustes* is 9.5 kb (R. Cohn, unpublished data). In addition, hybridization with subsegments of the *S. purpuratus* cloned fragments also demonstrated linkage of multiple histone protein coding regions within the same sized repeat unit for the other sea urchins, the sand dollar, and the star fish.

In situ hybridization studies with human chromosomes have revealed DNA sequences in chromosome 7 that are homologous to both human histone 4 mRNA and to the cloned sea urchin histone genes [(38, 39); and D. Stephenson, R. Cohn, and L. Kedes, unpublished]. Chandler et al (39) have placed the locus at the distal end of the long arm of this chromosome in the G-negative band, q 34. Since identical results were found using probes from different histone genes of the sea urchin, it is likely that the human gene organization consists of intermingled coding regions arranged in tandem arrays.

Recently, a subset of the histone genes of the yeast *Saccharomyces cerviciae* has been cloned in a bacterial plasmid (L. Hereford and K. Fahner, personal communication). Two clones, about 8 kb in length and from different regions of the genome, contain only H2A and H2B genes. Each clone is a single copy in the yeast genome and the two genes are on opposite strands. Thus, it seems likely that the organization of histone genes in this simple eukaryote is unlike that found to date in sea urchins or *Drosophila*.

Genetic Polymorphism in Sea Urchin Histone Genes

Allelic variants of histone proteins are not common but have been described, for example, in Maize [(40); although allelic variants of histone-modifying proteins are not excluded as a possibility]. Molecular cloning and recombinant DNA technology, however, have allowed detailed analyses of histone gene variations that are independent of and more sensitive than histone protein analyses. K. Gross, P. Little, and E. Southern (personal communication) have found that the sea urchin *E. esculentus* exhibits polymorphism in the size of the major histone DNA repeat unit. The histone genes of *E. esculentus* have been cloned and used as a probe to examine chromosomal DNA of individuals in the species. Repeat lengths of 6 and 7 kb have been observed. Restriction endonuclease mapping indicates that the two kinds of repeats are colinear with respect to recognition sites except for a 1 kb insert in the vicinity of the H4 gene. Examination of the histone DNA of individuals revealed those having only the 6 or 7 kb complement. No interspersion of 6 and 7 kb components in the same tandem array were detected. Somatic and gonadal tissues were found to be identical. This result indicates that the tandem arrays that are alternative alleles at a single chromosomal locus, differ from the kinds of heterogeneity described for *S. purpuratus, D. melanogaster,* and *L. pictus.*

DNA SEQUENCES OF HISTONE GENES

General Features

Extensive sequence analysis of the early histone genes of *Strongylocentrotus purpuratus* (21, 24, 41) and of *Psammechinus miliaris* [22, 23] have recently been published. The general topology of these gene sets is remarkably conserved (see previous sections on gene organization and Figures 1 and 2). Similarly, the general features of the sequences of both the coding and spacer DNA from these two species, members of different families that diverged somewhere between 6×10^7 and 16×10^7 years ago (42), are also remarkably conserved.

The DNA sequences of the sea urchin histone genes demonstrate that the coding sequences are colinear with the amino acid sequences [43–45] for the H4, H3, H2A, and H2B genes, and for those segments of the H1 genes sequenced to date. Thus, inserted DNA sequences in coding regions of structural genes, an increasingly common feature of other eukaryotic coding regions [see for example (46)], are not present in the early histone genes of sea urchins.

AT-rich regions in spacers are not randomly dispersed but tend to cluster, with some long stretches approaching 80% AT. Conversely, long stretches of G + C-rich spacer were not uncommon in *S. purpuratus* (64%). In neither species do the spacer regions show strong evidence of internally repetitive sequences. Although by their very nature AT-rich sequences have fewer degrees of organizational freedom than do random sequences, and thus may tend to appear to be organized into simple sequence repeat units, computer assisted analyses of gene sets of both species failed to conclusively demonstrate any such organization.

Fine Structure

CODON ASSIGNMENTS   Codon usage among each of the mRNA coding regions is far from random. Some amino acids have no third base isocoding bias while others are highly skewed (Table 1). C and G are preferred in third base positions (37% and 24% respectively in *S. purpuratus* and 35% and

Table 1   Codon usage in histone mRNAs[a]

1	2 U Sp[b] Pm			C Sp Pm			A Sp Pm			G Sp Pm			2 3
U	Phe UUU	2	2	Ser UCU	5	8	Tyr UAU	2	1	Cys UGU	2	0	U
	UUC	5	9	UCC	6	5	UAC	11	13	UGC	—	1	C
	Leu UUA	—	—	UCA	3	4	End UAA	—	3	End UGA	—	1	A
	UUG	2	5	UCG	—	2	UAG	3	1	Trp UGG	—	—	G
C	Leu CUU	4	8	Pro CCU	4	9	His CAU	4	7	Arg CGU	10	13	U
	CUC	15	14	CCC	6	6	CAC	4	5	CGC	17	14	C
	CUA	5	4	CCA	5	5	Gln CAA	6	7	CGA	7	8	A
	CUG	9	11	CCG	—	1	CAG	11	14	CGG	2	3	G
A	Ile AUU	5	4	Thr ACU	4	5	Asn AAU	2	1	Ser AGU	6	4	U
	AUC	21	25	ACC	15	17	AAC	10	11	AGC	6	9	C
	AUA	—	1	ACA	6	7	Lys AAA	15	18	Arg AGA	5	5	A
	Met AUG	6	10	ACG	4	5	AAG	38	49	AGG	9	6	G
G	Val GUU	7	5	Ala GCU	17	12	Asp GAU	4	5	Gly GGU	12	13	U
	GUC	14	13	GCC	21	29	GAC	6	6	GGC	12	16	C
	GUA	1	5	GCA	12	13	Glu GAA	8	8	GGA	16	13	A
	GUG	8	9	GCG	1	6	GAG	14	17	GGG	3	8	G

[a]Data derived from Schaffner et al (23), Grunstein & Grunstein (41) and Sures et al (21).
[b]Signify the codon frequencies for *Strongylocentrotus purpuratus* (Sp) and *Psammechinus miliaris* (Pm)

27% in *P. miliaris*). Although C appears as the second position in 109 *S. purpuratus* codons and in 134 *P. miliaris* codons it is followed by a G in only 5 and 14 cases respectively (27 times and 34 times respectively expected by chance). Thus CpG is underrepresented at isocoding sites but was found at the rate expected by random association throughout the sequences as a whole.

Perhaps the most striking aspects of the codon assignment comparisons are the strong parallels in codon preferences between species. Those codons underutilized or overutilized are similarly skewed and to the same degree in both species. Even those codon sets of one species showing only a slight and statistically insignificant preference for one or another triplet, show the parallel preferences in the other species; the leucine and arginine triplets are clear examples of such cases. The interspecific coding sequence conservation accounts for this codon similarity only in part. We must await codon analysis of other nonhistone genes from these species before we can exclude a general species codon preference as the explanation.

AMINO ACID SEQUENCES   Comparisons of the coding region of *P. miliaris* and *S. purpuratus* show no amino acid divergence of the H4 sequence, and limited changes in H2A, H3, and H2B. Direct comparisons of the nucleotide sequences of the coding regions of the H4 mRNAs of two sea urchin species (47) exemplify the concept that, due to degeneracy of the amino acid code, base changes at silent positions can be extensive. Similarly, the DNA sequences of *P. miliaris* and *S. purpuratus* histone coding regions exhibit striking third base changes. Table 2 quantitates the numbers of

Table 2 Comparison of codon divergence of sea urchin histone genes[a]

Gene	No. codons[b] compared	No. silent[c] sites	Predicted[d] random changes	No. observed changes	Conservation[e] ratio (%)
H2A	125	149	105	50	52
H2B	95	120	77	34	56
H3	137	163	102	46	55
H4	52	63	30	19	37

[a] The sequences of *S. purpuratus* (24, 41) were compared to those of *P. miliaris* (23).
[b] The available sequence data are not fully overlapping. Only complete codons were compared.
[c] Silent sites are those codon positions at which mutation could take place without affecting the amino acid sequence.
[d] Corrected for degrees of freedom for silent mutations. The assumption was made that random substitution would lead to 75%, 67%, and 50% divergence at silent sites with 3, 2, and 1 degrees of freedom, respectively. No corrections were made for reversions or convergent changes.
[e] Calculated as $[1 - (\text{observed}/\text{expected})] \times 100$. The sample for the H4 gene is too small to conclude that its homology is significantly different from the other sequences.

changes at silent sites and compares them with the expected number of changes if random substitution had occurred. Since not all substitutions at such codon loci are silent, Table 2 is corrected for the degrees of freedom at each site. It is clear that long segments of the spacer DNAs of these two gene sets of *P. miliaris* and *S. purpuratus* have diverged completely and show no sequence homology. Thus it seems reasonable to conclude that any sequences that have not diverged completely are under selective restraint. Table 2 shows that the observed silent position divergence for each of the genes is about half that expected for random silent substitutions. Thus it is clear that some selective pressure is operative on the nucleotide sequences of the coding regions beyond that maintaining the amino acid sequence. Comparisons with histone genes of a third species should determine whether specific silent sites are maintained or whether there is random silent substitution but at a slow rate. Since many of the isocoding triplets exhibit random utilization, the maintenance of such specific codons at the same locus in two species suggests that this type of selective pressure cannot be related to tRNA availability. It is more likely related to mRNA structure/-function or perhaps to a general restraint on divergence in these genes. The latter possibility seems unlikely in view of the fact that the spacer sequences between the mRNA regions have been able to undergo apparently random divergence and now show essentially no homology.

LEADER AND TRAILER SEQUENCE COMPARISONS Perhaps more surprising is the extensive evolutionary conservation, in *S. purpuratus* and *P. miliaris,* of some of the sequences immediately preceding and following the protein coding regions (Figures 3 and 4). This possibly will help in the location of important regulatory functions. Most important, perhaps, is the fact that these homologies occur in topologically analogous positions relative to the coding region, an observation that immeasurably increases their potential significance far above that calculated from probabilistic considerations. Figure 3 compares the available leader region sequences of *S. purpuratus* and *P. miliaris* histone genes taken from the date of Grunstein & Grunstein (41), Schaffner et al (23), and Sures et al (24).

The trailer regions are not as precisely conserved, base for base, as the leader regions, but they do tend to share the feature that a very AT-rich region immediately follows the end of the mRNA. These features lead us to believe that the leader and trailer sequences of the histone mRNAs are also under stringent selective pressure. These pressures on sequence conservation may not be the same for leader and trailer sequences as for codons. Nor is it clear whether those pressures operate at the transcriptional or translation level or both. In no instance was there a clustering of multiple termination signals following the legitimate stop signal. The sequence

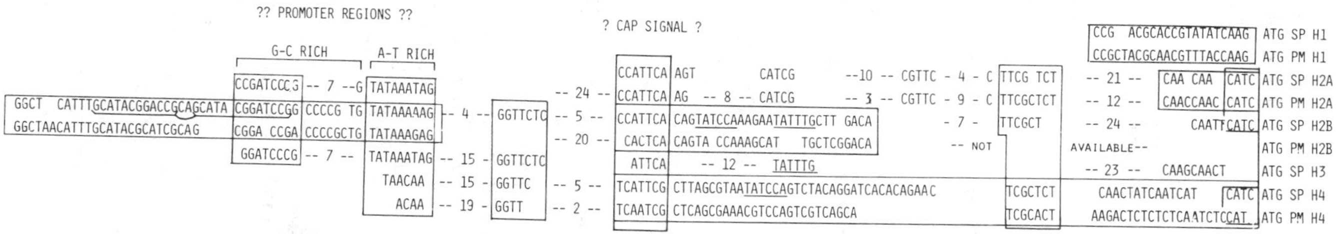

```
?? PROMOTER REGIONS ??                              ? CAP SIGNAL ?

  G-C RICH        A-T RICH

CCGATCCCG -- 7 --G  TATAAATAG                      CCATTCA AGT        CATCG      -10 -- CGTTC - 4 - C  TTCG TCT    CCG ACGCACCGTATATCAAG   ATG SP H1
CGGATCCGG CCCCG TG  TATAAAAG -- 4 --  GGTTCTG       CCATTCA AG -- 8 -- CATCG     -- 3 -- CGTTC - 9 - C  TTCGCTCT   CCGCTACGGCAACGTTTACCAAG ATG PM H1
CGGA CCGA CCCCGCTG  TATAAAGAG                                                                                     -- 21 -- CAA CAA CATC   ATG SP H2A
GGATCCCG -- 7 --    TATAAATAG -- 15 - GGTTCTC       CCATTCA CAGTATCCAAAGAATATTTGCTT GACA           TTCGCT         -- 12 -- CAACCAAC CATC  ATG PM H2A
                    TAACAA -- 15 - GGTTC                                                                          -- 24 -- CAATTCAIC       ATG SP H2B
GGCT CATTTGCATACGGACCGCGAGCATA                      CACTCA CAGTA CCAAAGCAT  TGCTCGGACA            -- NOT --       AVAILABLE--              ATG H2B
GGCTAACATTTGCATACGCATCGCAG                          ATTCA -- 12 -- TATTTG                                         -- 23 -- CAAGCAACT      ATG SP H3
                    ACAA -- 19 - GGTT               TCATTCG CTTAGGTAATATCCAGTCTACAGGATCACACAGAAC   TCGCTCT        CAACTATCAATCAT CATC      ATG SP H4
                                                    TCAATCG CTCAGCGAAAGGTCCAGTCGTCAGCA             TCGCACT        AAGACTCTCTCCAATCTCCAT    ATG PM H4

                                    -- 24 --
                                    -- 5 --
                                    -- 20 --
                                    -- 5 --
                                    -- 2 --
```

Figure 3 Comparison of DNA sequences of leader regions of histone genes of *S. purpuratus* and *P. miliaris.*

DNA sequence data (21, 22, 41) of *S. purpuratus* (SP) and *P. miliaris* (PM) histone genes were compared to computer-assisted search with the help of the MOLGEN Project, which contains a collection of DNA search subroutines. Homologous regions are boxed. In general, vertical boxes surround topologically homologous sequences shared by several genes of both species. Horizontal boxes outline homologous sequences of gene analogs. The sequence identified as a potential Cap Signal is the 5' end of the *S. purpuratus* H4 mRNA and the 5' end of mRNA:DNA heteroduplex homology for the other *S. purpuratus* histone genes. The underlined sequences 5' to the Cap Signal in the SP H2B leader are a short repeat sequence. The underlined sequences 3' to the Cap Signal are short homologies unique to *S. purpuratus.* Numbers in sequences represent known bases that show no obvious homologies. Blank spaces in sequences represent loopouts for maximal alignment. The biological significance of loopouts is discussed in the text. Sequence analysis of *D. melanogaster* histone genes demonstrates a closely related AT-rich "promoter" sequence 5' to each coding region (R. Goldberg and D. Hogness, personal communication). DNA sequence analysis of the *S. purpuratus* H2A gene also reveals a GC-rich and an AT-rich region 30 bases 5' to the "cap signal." The sequence is nearly identical to that for the *S. purpuratus* H2B and H3 genes. In addition, the 5' end of *S. purpuratus* early H2A and H2B mRNAs have been mapped by reverse transcriptase sequencing and both end in or at the cap signal region shown in the figure (S. Levy and I. Sures, unpublished).

```
PM H2A   TAAATTT GTTTGCTACC     TCTTGCAACCTCAACAACGGCC     -- NOT AVAILABLE --
SP H2A   TAGATAGAGTTTGCTCCCGGCAATCTTGAAACCTC  AACGGCCCTTATCAGGGCCACCAATTACTCACGAAAGAATTGTTTCAT
SP H2B   TAGAC      -- 30 --          ACAACGGCCXXA       CAAATAATCAAGAAAGAAT
PM H2B   TAAAC --8--  -- 20 NOT AVAILABLE --      GGCCACCAAACATCCAAGAAAGAATTGTGTCAT
SP H3    TAG        -- 40 --       AACGGCTCTTTTCAGAGCCACCACAACCCCAAGAAAGAXT
PM H1    TGA        -- 32 --       AACGGCTCTTTTCAGAGCCACCACATTTCCACGTAAGACC
```

Figure 4   Comparison of DNA sequences in trailer regions of sea urchin histone genes. DNA sequence data (21, 22) were compared by computer-assisted search with the help of MOL-GEN. Homologous regions are boxed. The H2A analogs are the only pair available for complete comparison. Homologies behind the translation stop signal are also boxed. Numbers represent known bases that show no homology. Blank spaces in sequence represent loopouts.

AAUAAA has been pointed out as a concomitant of eukaryotic mRNA trailer sequences (48). Although this sequence is found in the *Strongylocentrotus* H2B trailer region it is not present in the H2A and H3 genes and, perhaps equally important, it is not present in the *Psammechinus* H2B trailer region. Thus it appears likely that this particular oligonucleotide is not a necessary signal in histone mRNAs and is thus not common to all eukaryotic mRNAs. Since early histone mRNAs are relatively unusual in that they are not polyadenylated, it might be speculated that the AAUAAA sequence is a polyadenylation signal. On the other hand, we must await more sea urchin sequencing data on nonhistone genes before deciding whether the absence of these otherwise ubiquitous signals is species-specific.

LEADER AND TRAILER SEQUENCE HOMOLOGIES   One objective of examining histone DNA sequences and of comparing sequences between species, is the possibility of recognizing regulatory signals in leader and trailer regions. The vertically boxed leader and trailer sequences in Figures 3 and 4 are examples of regions shared by at least two different histone genes that have their topological counterparts in both species. The boxed sequence TCATTCG in the leader of the *S. purpuratus* H4 lies at the 5' end of the H4 mRNA according to Grunstein & Grunstein (41). Since a canonical form of this sequence is shared by all the other genes, it may well mark the start of other mRNA molecules. Further upstream, some 20 bases away, lie other homologous regions designated in Figure 3 as GC- and AT-rich sites. Even if these sites represent remnants of a common evolutionary past for each of the histone genes, the conservation of sequence and topology suggests that their divergence is limited. More likely is the possibility that these conserved regions represent essential regulatory signals. It is clear that the loopouts, often required to appreciate the degenerate homology of the sequences, interfere with the stereochemical identity of the compared segments. Accordingly, only those positions of the segments that are *stereochemically* similar can be assumed to have analogous functions. The high level of *sequence* homology, on the other hand, argues that the complete segments shared a common ancestral sequence and are still under evolutionary restraint.

Regions common to only one class of mRNA are set in the horizontally boxed sequence connecting analog pairs of nucleotides. Of particular note are the strong sequence homologies of leader sequences of the H1, H2A, H2B, and H4 genes of *S. purpuratus* and *P. miliaris*. (Leader and trailer sequences of *P. miliaris* H3 have not been published and are therefore not compared.) The H2B homologies are striking even at a distance greater than 100 bases upstream from the initiation codon. The extensive homologies of some of the leader and trailer regions are even more strongly conserved than the silent mutation sites within the coding regions. If these conserved regions, shared by only one pair of histone gene analogs, represent important regulatory sequences, then it is also likely that the regulation of each histone gene (or mRNA) differs from the others. Whether this difference relates to replication, transcription, or to mRNA structure/function cannot be anticipated.

Homogeneity of Histone Gene Sequences Within Species

Southern blots (31) of restriction-digested chromosomal DNA show limited size heterogeneity of the major histone DNA repeat units. Some microheterogeneity and small differences in unit length of the major repeats are evident (12, 17). In some species the presence of an insert in one or another spacer region leads to the limited types of heterogeneity discussed earlier for *Drosophila melanogaster* and for *Echinus esculentus*. The major conclusion, however, is that the clustered genes have evolved in tandem. The special situation in *Lytechinus pictus* is more complicated but can be reduced to the same argument: although two sets of repeats have arisen, each is independently clustered and the members of any one set are essentially identical.

The limits of this sequence homology can be established by comparisons of independent DNA clones from the same species. Overton & Weinberg (12) have compared restriction maps of two *Strongylocentrotus purpuratus* histone gene repeat units. They found evidence for microheterogeneity, such as restriction enzyme sites being lost or gained. Birnstiel et al (49), among other things, have reported similar kinds of heterogeneity in cloned *Psammechinus miliaris* genes. These types of heterogeneity are likely to be single base changes. Overton & Weinberg (12) have carefully compared restriction fragments from cloned histone genes. They found small differences (10–200 bases) in the lengths of three fragments. These differences seemed to be confined to spacer regions and reflect the chromosomal histone gene microheterogeneity. The extent of this heterogeneity is limited and the essential homogeneity of the overall structure and topology of the histone gene repeat units is far more impressive than their differences (see below).

Sures et al (24) have compared the DNA sequences of several *S. purpuratus* clones. No differences were detected in over 90 bases of spacer and 82 adjacent bases in the histone H3 coding region. Similarly, no differences were found in the coding region of H2A and 370 bases of adjacent trailer region with the exception of one "hot spot" (see below). How do such tandemly repeated genes maintain their overall homogeneity? The histone genes are not unique in this regard since the same seems to hold for ribosomal and 5S genes (50, 51). Smith (26) has postulated that repetitive sequences might maintain homogeneity by undergoing rapid correction or amplification through unequal crossing-over events. Although this hypothesis can explain both the homogeneity of satellite sequences and their rapid divergence, it does not easily explain the homogeneity of highly complex tandem genes such as ribosomal or histone genes. Some mechanism must ensure that if rapid elimination or amplification of mutations within tandemly repetitive genes is carried out by unequal crossing-over events, then these events must occur in register. Conversely some mechanism must prevent frequent or random unequal crossover events within a repeat unit. One particular feature of the histone gene DNA sequences of the sea urchin and *Drosophila* is interesting to consider in this regard. Long, alternating homocopolymer stretches have been found in one of the five spacer segments for each of the three examined gene sets. In *S. purpuratus* the sequence is located between the H2A and H1 genes and has the remarkable form

$$\begin{pmatrix} CT \\ GA \end{pmatrix}_{27}.$$

If recombinational events depend on base pairing, then this simple sequence region will reanneal at a much more rapid rate than any other segment of a 6500 base-pair histone gene repeat unit. Reannealing between $pCpT_{27}$ and $pGpA_{27}$ regions would almost always be the first nucleation event of duplex formation. Branch migration of the duplex would realign the entire repeat unit in precise register. Thus this "zipper" could be responsible for minimizing out-of-register unequal crossover while promoting in-register unequal crossing-over as a gene correction-amplification mechanism. One prediction of such a mechanism is that the CT stretch might unequally align itself and that, as a remnant of prior recombinational events, it would exhibit exaggerated heterogeneity in both length and sequence from one repeat to the next. The CT region does seem to be a mutational hot spot in the histone gene repeat unit of *S. Purpuratus*. The identified DNA sequences of the plasmid pCO2 are essentially identical with those of pSp2 plus pSp17 except in the CT stretch. The CT segment in clone pCO2 is 14 bases shorter and has a single base change (CTC→TTT) in the middle. Of some interest is the observation that the histone DNA sequence of both *P. miliaris* (23) and of *D. melanogaster* (M. Goldberg and D. Hogness, personal communication) both contain such alternating

$$\begin{pmatrix} CT \\ GA \end{pmatrix}_n$$

homocopolymer regions in a spacer region.

Equally interesting, and perhaps strengthening the hypothesis, is the observation that zipper regions may be a feature common to other tandemly repeated genes. The ribosomal repeats of *X. laevis* demonstrate sequence heterogeneity which is confined to a spacer region and is made up of an internally repetitive sequence 50 base pairs long. The number of such repetitive blocks differs from one rDNA repeat unit to the next (50). Similarly, the 5S genes of *Xenopus laevis* demonstrate a confined region of heterogeneity that has as internally repetitive region made up of 15 base pairs [51]. Thus an internally located repetitive sequence may be a common feature of tandem multigene families.

HISTONE mRNAs

Identification

Histone RNAs were first identified by the classical experiments of Borun, Scharff & Robins (52) and of Gallwitz & Mueller (53) who found that during S phase of the mammalian tissue culture cell cycle, radiolabeled RNA with characteristics of messengers and about 9S in size were associated with those polyribosomes that were synthesizing histone-like proteins. The initial identification of the 9S RNAs as histone messengers was based first on the size of the RNA (it was about the size expected to code for the relatively small histone proteins) and second on the appearance of these RNAs at those periods of the cell cycle when histone proteins are synthesized (52, 54–59).

These criteria were used to identify radiolabeled RNA molecules found in great abundance in the polyribosomes of rapidly cleaving sea urchin embryos (60–62). Developing sea urchins undergo 2^{9-10} cell divisions in a 10–12 hr period after fertilization (63). During this period of rapid growth, synthesis of nuclear proteins accounts for 60% of the total protein synthesis (60, 64) and histones represent at least one half of that amount. Thus it is not surprising that mRNAs sufficient to code for such a large amount of histone should have been tentatively identified among the cytoplasmic radio-labeled RNA species (61, 65) as the most prevalent class among the polysome-associated templates.

Three lines of evidence—in vitro template activity, RNA sequence analysis, and homology with histone genes—have been used to demonstrate more rigorously that the polysome-associated 9S RNAs isolated from cleaving sea urchin embryos are the histone mRNAs.

HISTONE TEMPLATE ACTIVITY   The 9S polysomal RNAs contain the templates for five classes of histone proteins (66–68, 70–72). The 9S histone mRNAs can be separated electrophoretically into subfractions specific for individual histone proteins (70, 72). By a combination of techniques characterizing either their size or template activity, histone mRNA templates have been identified in mammalian cells (52, 53, 55–57, 59, 73–77, 79, 80), the toad, *Xenopus laevis* (81–83), the newt, *Triturus viridescens* (84), the surf clam, *Spisula solidissima* (85), rice embryos (86), the brine shrimp, *Artemia salina* (87), and avian red blood cells (88).

mRNA SEQUENCE ANALYSIS   Nucleotide sequences have been obtained for one of the H4 mRNAs of the sea urchins *S. purpuratus* and *L. pictus* (6, 41, 47, 71).

HOMOLOGY WITH CLONED HISTONE GENES   Cloned histone DNA fragments can be used to identify hybridizing mRNAs as histone templates or their putative precursors. Childs et al (89) demonstrated that it is possible to use cloned histone DNA fragments to isolate, from whole cell RNA, radiochemically pure RNAs with monospecific in vitro template activity. In this general way histone mRNAs have been identified in *Drosophila melanogaster* [(90); R. Lifton and D. Hogness, personal communication], and in *X. laevis* (83), HeLa cells (33, 91), and *Triturus viridescens (84).*

Structural Features of Histone mRNAs

THE 3' TERMINUS   Adesnik & Darnell (92) demonstrated that histone mRNAs isolated from synchronized HeLa cells lack poly (A) at the 3' hydroxyl end as assayed by the absence of ribonuclease-resistant poly(U)-binding material. Although it was originally thought that among mammalian messengers only histone mRNAs lacked 3' poly(A), it has become clear that in both sea urchins and mammalian cells a substantial fraction of nonhistone polysomal messenger RNAs also lack poly (A) tails (65, 93). The absence of poly (A) on the sea urchin histone mRNAs that are newly synthesized during early cleavage stages of development was established by the failure to identify ribonuclease-resistant poly(A) containing oligonucleotides in the radiolabeled histone templates (6, 47). Histone mRNA template activity has also been found in the poly(A) minus fraction of cytoplasmic templates in *Xenopus laevis* (35, 81, 83), *Drosophila melanogaster* (90), and *Artemia salina* (87) and *triturus* (84).

Increasing evidence has suggested, however, that at least a portion of some histone mRNAs may be polyadenylated. Such subsets of poly(A) containing histone messenger RNAs have been found in HeLa cells (78), amphibians (81, 84), sea urchins (84), and clams (85). The ability to detect poly(A)-containing histone mRNAs may be related in part to stage-specific switches of histone mRNA sequences. For example, although considerable histone mRNA template activity is found in the poly(A$^+$) RNA of *Xenopus* oocytes, after fertilization most of the activity is found in the poly(A$^-$) fraction [see Ruderman & Pardue (84) *inter alia*.] In another series of experiments Ruderman & Pardue (84) demonstrated the presence of a poly(A)-containing histone template fraction in another amphibian, *Triturus viridescens*, by synthesizing complementary DNA to histone messenger in a reaction dependent on the use of an oligo-dT primer. The product was capable of hybridizing to both *D. melanogaster* and sea urchin histone genes. These experiments eliminated the possibility that histone templates in the poly(A$^+$) fraction were poly(A$^-$) contaminants.

THE 5' TERMINUS   Surrey & Nemer (95) found that sea urchin histone mRNA 5' termini are capped with the structure m^{7}G(5')pppX^{m}pY where the 7-methylguanosine m^{7}G residue is connected from the 5' position through a triphosphate bridge to the 5' position of an adjacent nucleotide (X) methylated in the 2 position. Such structures are characteristic of the 5' mRNA termini of a variety of eukaryotes. [See for example (96) and (97).] The sea urchin cap structure has either unmethylated A or G at the X position in ratio of 4:1.

Two groups agree that the 5' termini of the histone mRNAs of HeLa cells are also capped and the X position is occupied only by methylated purines (96, 97). There is disagreement as to the exact distribution of the various methylated bases found at the X and Y position and as to whether they are dimethylated or monomethylated. These differences may reflect either technical considerations or the *HeLa* cells lines involved.

LEADER AND TRAILER SEQUENCES   The size of histone mRNAs from sea urchins is longer than that required to code for histone proteins. In the absence of any evidence for the post-translational shortening of histone polypeptides, this must mean that histone messengers have nontranslated leader and/or trailer sequences. Some of the oligonucleotides of the *S. purpuratus* H4 mRNA, which could not be construed to code for histone H4 polypeptides, are identical to sequences 5' to the initiator codon found on the *S. purpuratus* H4 gene DNA sequences (41). These oligonucleotides provide direct evidence for the presence of a leader sequence on the H4 mRNA and place the 5' end of the RNA some 78 nucleotides upstream from the initiator codon (see Figure 3). Similarly, heteroduplex analysis mapping of mRNAs with respect to their genes suggests that both leader and trailer sequences exist on all the histone mRNAs of sea urchins (18, 19). Although the size of histone mRNAs as determined by electrophoretic mobility is known, the relative amounts of noncoding sequences in the leader and trailer regions have not yet been established.

INTERNAL MODIFICATIONS   Methylated nucleotides in internal sites are a rare but constant feature of eukaryotic mRNAs (98, 99). Examinations of histone mRNAs of both echinoderms and HeLa cells has failed to detect such bases however (95–97).

Histone mRNA Transcription and Turnover

LINKAGE TO DNA REPLICATION: A NEW CONTROVERSY   Histone protein synthesis is closely coupled to DNA replication in the somatic cells of animals (100), yeast (101), and protozoans (102–104), The molecular mechanisms involved in this coupling have been extensively explored. There is reasonable agreement on the following conclusions:

1. When DNA replication is blocked by agents such as cytosine arabinoside or hydroxyurea, the rate of histone protein synthesis falls abruptly (52, 53, 95, 100).
2. Both polysomal and total cytoplasmic histone mRNA template activity falls abruptly following the completion of S phase or the inhibition of DNA replication (53, 54, 61, 100).
3. In vitro synthesized histone mRNA sequences can only be detected in HeLa cell chromatin when the chromatin is isolated from cells active in DNA replication (105).
4. Taken together, these data strongly suggest that in somatic cells there is a rapid loss of histone mRNA activity from cytoplasm and polyribosomes following the cessation of DNA replication.
5. The situation in echinoderm and amphibian oogenesis and early development is quite different: the synthesis of histone proteins and histone mRNAs is independent of DNA replication. Both the proteins and messengers are stored for later use following fertilization (66, 67, 81, 84, 85, 106, 107).

6. Most workers have speculated that (*a*) the S phase regulation of histone messenger RNA activity in somatic cells is centered on a transcription level control mechanism and (*b*) the abrupt change in template activity following S phase is the result of changes in the rate of histone mRNA turnover or inactivation.

Recent experiments from two laboratories suggest that this popular model may be wrong and that post-transcriptional processing of histone mRNA may be the key feature in histone-DNA coupling. Melli et al (91) have searched for newly synthesized histone gene transcripts in HeLa cell cytoplasm and nucleus in the presence and absence of DNA replication inhibitors. The assay of these workers was the ability of sea urchin histone genes to hybridize to the mammalian histone transcripts. RNA sequences in HeLa cell cytoplasm that hybridized to the heterologous probe were found only during DNA replication, in agreement with the data cited above. Newly labeled nuclear histone RNA, however, could be found throughout the cell cycle.

These groups of investigations conclude that histone RNA transcription is independent of DNA replication whereas polysome accumulation of histone RNA transcripts is tightly coupled to DNA replication. These data suggest that processing or transport of mRNA from nucleus to cytoplasm may be a tightly regulated phenomenon and that regulation of histone gene activity may play a minimal role in the cell cycle of somatic cells.

Two technical aspects of these important experiments require reevaluation. First, the hybridization probe was a heterologous gene set with a highly divergent nucleotide sequence. Second, the method of cell synchronization, double-thymidine block, may allow a backround of non-S phase cells.

Indeed, Stein et al (108) have arrived at the opposite conclusion using a different assay. These workers assessed the presence of histone RNA sequences by hybridization to a radiolabeled complementary DNA probe made from cytoplasmic histone mRNA of HeLa cells. Stein et al isolated HeLa cell RNA from nuclei, polyribosomes, and postribosomal supernatant compartments from cells synchronized by mitotic detachment or incubated in the presence and absence of DNA replication inhibitors. The RNA was unlabeled and was used to drive RNA-DNA hybridization reactions in RNA excess. Stein's data confirm that the concentration of hybridizing histone RNA sequences in polysomal RNA is detectable only when DNA is being replicated. But, more importantly, there were no detectable histone RNA sequences in nuclear RNA samples taken either from cells in G1 phase or cells incubated in the presence or absence of DNA replication inhibitors.

Stein's approach would not have detected short-lived nuclear histone sequences. Resolution of this important controversy must await the development of a pure, homologous DNA probe for mammalian histone RNA.

ARE THERE HIGH-MOLECULAR-WEIGHT PRECURSORS TO HISTONE mRNAs? Intensive searches for the presence of high-molecular-weight or polycistronic precursor molecules, especially in nuclear RNA, have been unsuccessful when cleavage stage embryos were examined (89, 109). Recently, however, Kunkel & Weinberg (109) detected high-molecular-weight nuclear RNA hybridizing to sea urchin histone genes in sea urchin gastrulae. That report suggests that early and late histone genes are not only different in sequence but may be transcribed or processed in different ways. An alternative explanation to account for the apparent difference is that the processing of the early histone gene transcripts may be more rapid than at later stages and processing of a large transcript may be occurring during transcription itself.

Support for the presence of high-molecular-weight histone mRNA in HeLa cells has been provided by Melli et al (110) who hybridized the heterologous sea urchin histone DNA probe to pulse-labeled nuclear RNA. They detected nonpolyadenylated high-molecular-weight RNAs that disappeared from nuclei when the cells were treated with actinomycin D. They interpreted their results to be compatible with the presence of high-molecular-weight histone mRNA precursors. The data do not necessarily mean that these large RNAs are polycistronic, since the organization of the HeLa cell histone gene is not yet known.

Using an entirely different approach, Hackett et al (111) have also suggested that the HeLa cell histone gene transcriptional unit is larger than histone mRNA molecules but that each mRNA is independently transcribed. These authors used the indirect method of UV inactivation mapping to determine that the decay rates of histone mRNA translational activity were similar for each of the five genes. In this method, DNA is damaged by UV irradiation and the assumption is made that sequences further downstream from an RNA transcription initiation site will be more sensitive to the DNA damage than sequences closer to a single initiation site. The data of Hackett et al (111) albeit indirect, support a model of independent transcription of each of five histone messengers of HeLa cells. Target size analysis suggests that although the primary transcripts are one- or twofold larger than the mature cytoplasmic messengers they are much smaller than the size suggested by the data of Melli et al (110). The resolution of these conflicting results must await further experimentation. Preliminary UV inactivation data from sea urchin embryos assayed by hybridization of newly transcribed RNA to histone genes have suggested five independent UV targets at early cleavage stages each of which is slightly larger than cytoplasmic mRNAs (G. Childs and L. Kedes, unpublished). Although the apparent agreement of these indirect UV irradiation assays of histone transcription in HeLa cells and sea urchin embryos is striking, there can be no substitute for a rigorous search for a chemically definable primary gene transcript. In any event, if the high-molecular-weight transcripts prove to be precursors to cytoplasmic mRNAs, then processing of the RNAs is required and its mechanisms should be accessible to experimental examination.

Sea urchin nuclei can transcribe histone mRNA sequences in vitro (112, 113). Although the average size of the total in vitro transcripts is 18–20S, which demonstrates the ability of the system to synthesize high-molecular-weight RNA, and although 2–9% of the transcripts hybridize to histone genes and spacers, no giant polycistronic-sized histone transcripts have been detected [113]. On the other hand, the in vitro histone transcripts in *S. purpuratus* are surprisingly discrete in size and consist of a series of 16 RNAs ranging between 100 and 1100 nucleotides in length. The five largest RNAs made in vitro are each bigger than their cytoplasmic histone mRNA counterparts. Levy et al (113) found that each of the five largest in vitro transcripts hybridized with only one of the five histone genes and adjacent spacer regions. If the in vitro observations are a reflection of in vivo events, then the five large RNAs may represent primary transcripts each longer than a mature cytoplasmic histone mRNA. This implies that post-transcriptional processing at specific intramolecular sites must occur in order to generate the shorter polysomal histone mRNAs.

HETEROGENEITY OF HISTONES—THEIR mRNAS AND THEIR GENES

Histone Proteins: A Brief Overview

The character of the histones and of their genes reflect each other so remarkably that it is appropriate to briefly describe some of the characteristics of the structure and evolution of these polypeptides. A more complete review on histones and histone evolution has appeared in these pages previously (1).

EVOLUTIONARY CONSIDERATIONS   Two of the five histones, H4 and H3, are conserved in the evolutionary sense, in the extreme. H4 of peas and cows for example, differ in only two of 102 amino acid residues. The H2B and H2A proteins, though not as stringently conserved in evolution, demonstrate a high degree of primary sequence conservation in several regions of the polypeptide. These four proteins are bimolecularly represented in the basic nucleosome core particle of eukaryotic chromatin. The restraint on evolutionary divergence of the histone sequences is obviously a reflection of selective pressures brought to bear on the histone genes by the universal nature of the multiple sites of interactions of the polypeptides with each other and with their surrounding DNA. Conversely, those residues within the histone proteins that have been under less stringent evolutionary constraint must define regions of the protein molecules not involved with critical intra- or inter-molecular interactions in the nucleosome. While the fifth histone, H1, is also a participant in chromatin organization, it has undergone more extensive divergence in evolution, with many of the substitutions being nonconservative. Certain regions of it are highly conserved and one may speculate that they are the sites of interaction with the core particle or with DNA.

HISTONES EARLY AND LATE IN DEVELOPMENT   Although the histone proteins fall into five major classes, recent evidence has pointed out that important heterogeneities exist in some of the classes which may reflect functional differences. I am not referring here to the well-recognized alterations in the degree of polypeptide chain phosphorylation, methylation, or acetylation (1, 114). The mechanism of these sorts of variations and their potential importance are beyond the scope of this review. Rather, I am referring to amino acid sequence heterogeneity in a single individual. Amino acid sequence differences of H1 proteins have been found in different tissues in mammalian cells and following response to hormonal stimuli (115, 116). Alterations in the H1 proteins of sea urchins have been described during embryonic development (117–120). More recently, developmental changes in sea urchin histones have been found in H2B and H2A proteins as well (121, 122).

Cohen and his co-workers have extended these studies by examining each of the individual histone proteins during the development of the sea urchin *Strongylocentrotus purpuratus*. These workers have developed a sensitive electrophoretic assay that can distinguish minor variants of histone protein primary structure by differences in affinity of the polypeptides for the non-ionic detergent, Triton X-100 (118, 123–127). Their studies using in vivo pulse-chase labeling experiments have revealed that three electrophoretically distinct cohorts of histone proteins are sequentially synthesized during the early development of the sea urchin. Figure 5 contains a diagrammatic representation of their most recent analysis combined with an analysis of histone mRNA developmental regulation.

It is clear that these varieties of proteins do not exhaust the histone varient repertoire of the sea urchin genome since additional varieties of sperm H2B protein have been described as well (44, 45). Careful analyses

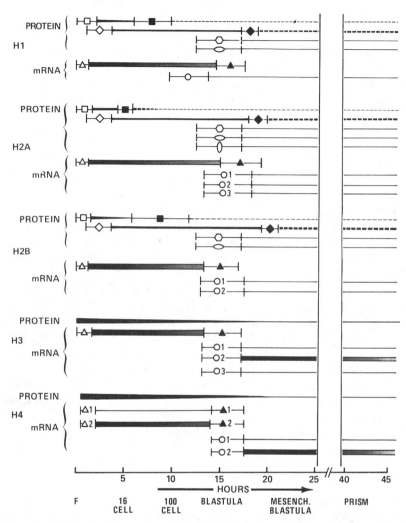

Figure 5 Developmental expression of sea urchin histone genes. The figure is a chronicle of gene expression of histone subtypes during early development of *S. purpuratus,* modified from Newrock et al with additional data as described in the text. A thick line indicates an abundant component and a thin line indicates a relatively minor component. Open symbols represent onset of synthesis and closed symbols represent end of synthesis. Dashed lines indicate continued presence of a product synthesized earlier. Vertical lines delimit the rather imprecise time periods during which production starts or stops. The different symbols represent unique components; i.e. there are two early H2A proteins and three different late subtypes detected while there is only one identified early H2A mRNA subtype and three late subtypes. The early H2A mRNA contains two template activities corresponding to the early proteins but has not yet been separated into two components. The break in the time scale represents the period of gastrulation. F indicates fertilization.

of adult sea urchin histones and oocyte histones have not yet been reported. Although it seems likely that newly synthesized histone proteins do not associate with old nucleosomes, it is not yet clear whether the homologous histone variants that are synthesized simultaneously, intermingle within a given nucleosome particle or even whether they are synthesized in the same cell type. The developmental implications of these histone protein switches remain speculative and have been provocatively reviewed by Newrock et al (122, 128).

Early and Late Histone Genes

The sea urchin histone gene clones examined to date all code for early cleavage stage proteins [the α variants of Newrock et al (128)]. The most compelling evidence for this conclusion comes from direct sequence data of H2A genes of *S. Purpuratus* and *P. Miliaris* in which the codon for methionine occurs at position 50 (23, 24). The early sea urchin H2A protein has a single methionine at residue 50 while the late proteins do not (128). Secondly, although early mRNAs hybridize stringently to the cloned DNA segments, each of the late RNAs, discussed in detail below, shows evidence of sequence divergence (129, 130). Early mRNA persists in cytoplasm, as assayed by in vitro translation experiments, for at least 40 hr after fertilization [(122, 128); G. Childs, R. Maxson, and L. Kedes, unpublished]. Late mRNA can first be detected by the same assay only at about 12 hr. Early mRNA sequences can no longer be pulse-labelled starting at about 12 hr. Thus, although the switch in synthesis is fairly abrupt, the expression of the two gene sets overlaps for at least 20–30 hr. The regulation of histone genes represents a developmentally modulated gene-switching event that can be quantitatively monitored by assay of both mRNA transcription and template activity.

Structural Microheterogeneity

HETEROGENEITY OF HISTONE mRNAS SYNTHESIZED CONTEMPORANEOUSLY   The extensive heterogeneity of histone proteins is paralleled by analogous variations of histone mRNAs. Histone template variants synthesized contemporaneously during a specific stage of sea urchin development were first identified by their heterogeneous electrophoretic mobilities on nondenaturing gels. Grunstein et al (71) detected in this way at least two H4 templates in cleavage stage embryos of both *S. purpuratus* and *L. pictus.* Ribonuclease T1 fingerprint examination of two H4 templates from *L. pictus* demonstrated that most oligonucleotides are held in common but some are unique to each messenger. Thus H4 protein in this sea urchin embryo is translated from at least two mRNA populations which differ in several nucleotides. Since all the H4 proteins synthesized are apparently identical in primary sequence, and do not reflect the template heterogeneity, the nucleotide differences must be confined to isocoding positions and to mRNA leader and trailer sequences.

Gross et al (16) further characterized the heterogeneity of the newly synthesized histone templates from cleavage stage sea urchin embryos by electrophoresing them across a continuous gradient of increasingly denaturing conditions. This approach clearly demonstrated conformational structural differences of essentially all of the histone mRNAs. When such early developmental mRNA preparations are fully denatured however, the template activity for each histone protein behaves as a single entity on electrophoretic separation (89). Thus it seems that the heterogeneity observed among at least the *early* templates is mostly related to sequence differences rather than to length differences. This form of heterogeneity during earliest development could be partly explained if the multiple histone genes active at any given stage of development are not completely homogeneous and minor nucleotide differences have arisen in them during the course of evolution. Such nucleotide changes, especially if predominantly confined to isocoding positions, would result in a minimal degree of polypeptide heterogeneity but could still explain the observed microheterogeneity of mRNAs. However, the simultaneous production of representatives of two histone cohorts at certain developmental states (see Figure 5) suggests that some of the mRNA heterogeneity observed at such developmental time points is further evidence for distinct classes of structural histone genes being differentially expressed.

HETEROGENEITY OF HISTONE mRNAs SYNTHESIZED AT DIFFERENT DEVELOPMENTAL STAGES   The appearance of stage specific histone proteins suggests that different sets of histone genes are uniquely active at different stages of development. This has been confirmed by recent studies on histone mRNAs isolated from embryos at different stages of sea urchin development [(128–130); G. Childs, R. Maxson, and L. Kedes, unpublished]. Total 9S polysomal RNAs are capable of acting as in vitro templates only for those histone proteins being synthesized in vivo at the time of development from which the RNAs were isolated (128). This finding demonstrates that the stage-specific histone protein switches are not related to post-translational modifications of the proteins but are engendered by the appearance of different sets of messengers. Kunkel & Weinberg (129) used ribonuclease sensitivity and thermal elution assays to demonstrate different degrees of homology of the early and the late histone mRNA populations with the cloned sea urchin genes. Each of the five major histone mRNAs synthesized late in development differed from its early counterpart by at least 10% sequence divergence.

Grunstein isolated *Lytechinus pictus* histone mRNAs synthesized at various developmental stages (130) by their ability to hybridize to subsegments of cloned *L. Pictus* histone gene repeat units (8). When autofluorograms of the electrophoretically separated RNAs were compared, marked differences in mobility of the various histone mRNA classes were observed. The major finding was that the transition from blastula stage of development is accompanied by a shift in each and every one of the five major histone mRNA classes. The H1 messenger, late in development, is approximately 40 bases longer than its early counterpart, whereas the H2A, H2B, H3, and H4 messengers are approximately 25–70 bases shorter than their early counterparts. In addition, T1 ribonuclease fingerprints of the early and late H4 mRNAs reveal that while most oligonucleotides are held in common, many are unique to one of the two templates. Thus, the developmental shift in the H4 template is the result of the differential activity of two H4 gene sets. This fingerprint data excludes the possibility that the difference in H4 templates is related to differential processing or shift of the initiation site of H4 transcription. Since no variants of H4 proteins have yet been reported in the development of the sea urchin, it is likely that the variant H4 messengers do not differ at coding positions and that the differences are confined, rather, to isocoding third base loci, and to leader and trailer sequences. As discussed elsewhere, these histone templates are not polyadenylated and their difference in mobility and sequence cannot be explained by differences in poly(A) content.

Childs, Maxson and Kedes (in preparation) have analyzed the individual templates for each of the histones at early and late developmental stages of *S. purpuratus,* by hybridization to subcloned fragments of histone DNA. Their results are also integrated into the diagram of Figure 5. They have examined the in vitro template activity, electrophoretic mobility under denaturing conditions, and thermal elution kinetics of each of the early and late histone mRNAs. The early H2B mRNA is all of one size under denaturing conditions and is template in vitro for H2B (early) protein. Its late counterpart, however, consists of two different RNAs of similar size but about 50 bases shorter than early H2B mRNA. These two RNAs stimulate the synthesis of the two late H2B histones β and γ. The late H3 RNAs consist of three fractions which separate under denaturing conditions. They range from 20–40 nucleotides shorter than their single, homogeneous early H3 mRNA counterpart. When the early H3 template and the three late H3 templates are translated in vitro, they each give rise to an identical H3 protein, indistinguishable in SDS-acrylamide, acid-urea-acrylamide, or Triton X-100-acrylamide gel electrophoresis.

The late H4 mRNAs consist of two species of similar but not identical electrophoretic mobility some 30–40 nucleotides shorter than their single-sized early counterparts. As discussed previously, early H4 templates are heterogeneous and can be separated by electrophoresis in nondenaturing conditions. These early templates differ from each other in their nucleotide sequence but each codes for the identical protein, presumably because the differences are confined to isocoding or noncoding positions. Although the two late forms of H4 mRNA have not been compared in the same way, they code for the same protein and it is reasonable to assume that analogous differences distinguish them.

The late *S. purpuratus* H2A and H1 templates do not anneal with the cloned early genes even under low stringency conditions of hybridization. They have been identified by eluting RNA from various regions of acrylamide gels after electrophoretic separation and then testing their template activity in vitro. The late H2A templates are some 20–50 bases shorter than the early H2A messenger, and the late H1 template activity is 50–60 bases longer. Independent activities of different mobilities could be distinguished which engendered the H2A β, γ, or δ proteins.

It is clear then that there must exist several sets of histone genes. Each set must be differentially regulated and active at different stages of development. In addition, at any stage in development there must be multiple sets of genes operating in parallel and giving rise to variant mRNAs and proteins. The evolutionary conservation, the developmental and stage specific microheterogeneity, and the dynamic aspects of histone protein synthesis and regulation must all, somehow, be able to be explained by the nature of the organization of the histone genes. Many of the newly described aspects of the histone genes presented in this review help explain these characteristics. Those features not yet explained should soon come to light.

ACKNOWLEDGMENTS

I thank Professors M. Birnstiel, L. Cohen, D. Gallwitz, M. Grunstein and E. Weinberg for providing manuscripts before publication. I am grateful to my colleagues S. Levy, I. Sures, R. Cohn, R. Maxson, and G. Childs for critical comments and discussions. I thank Barbara Kartman and Doris Hosmer for invaluable and patient help in the preparation of this manuscript.

Literature Cited

1. Elgin, S. C. R., Weintraub, H. 1975. *Ann. Rev. Biochem.* 22:725–74
2. Kedes, L. H., Gross, P. 1969. *Nature* 223:1335–39
3. Kedes, L. H., Birnstiel, M. L. 1971. *Nature New Biol.* 230:165–69
4. Weinberg, E. S., Birnstiel, M. L., Purdom, I. F., Williamson, R. 1972. *Nature* 240:225–28
5. Grunstein, M., Schedl, P., Kedes, L. 1973. In *Molecular Cytogenetics,* ed. B. Hamkalo, J. Papaconstantinou, pp. 115–23. New York: Plenum
6. Grunstein, M., Schedl, P. 1976. *J. Mol. Biol.* 104:323–49
7. Birnstiel, M. L., Telford, J., Weinberg, E. S., Stafford, D. 1974. *Proc. Natl. Acad. Sci USA* 71:2900–4
8. Kedes, L. H., Chang, A. C. Y., Housman, D., Cohen, S. N. 1975a. *Nature* 255:533–38
9. Cohn, R. H., Lowry, J. C., Kedes, L. H. 1976. *Cell* 9:147–61
10. Lifton, R. P., Goldberg, M. L., Karp, R. W., Hogness, D. S. 1977. *Cold Spring Harbor Symp. Quant. Biol.* 42:1047–51
11. Clarkson, S. G., Smith, H. O., Schaffner, W., Gross, K., Birnstiel, M. L. 1976. *Nucleic Acids Res.* 3:2617–32
12. Overton, G. C., Weinberg, E. S. 1978. *Cell* 14:247–57
13. Weinberg, E. S., Overton, G. C., Shutt, R. H., Reeder, R. H. 1975. *Proc. Nat. Acad. Sci. USA* 72:4815–19
14. Kedes, L. H., Cohn, R. H., Lowry, J. C., Chang, A. C. Y., Cohen, S. N. 1975. *Cell* 6:359–69
15. Portmann, R., Schaffner, W., Birnstiel, M. L. 1976. *Nature* 264:31–34
16. Gross, K. W., Schaffner, W., Telford, J., Birnstiel, M. 1976. *Cell* 8:479–84
17. Schaffner, W., Gross, K., Telford, J., Birnstiel, M. 1976. *Cell* 8:471–78
18. Wu, M., Holmes, D. S., Davidson, N., Cohn, R. H., Kedes, L. H. 1976. *Cell* 9:163–69
19. Holmes, D. S., Cohn, R. H., Kedes, L. H., Davidson, N. 1977. *Biochemistry* 16:1504–12
20. Kedes, L. H. 1976. *Cell* 8:321–31
21. Sures, I., Maxam, A., Cohn, R. H., Kedes, L. H. 1976. *Cell* 9:495–502
22. Birnstiel, M. L., Schaffner, W., Smith, H. O. 1977. *Nature* 266:603–7
23. Schaffner, W., Kunz, G., Daetwyler, H., Telford, J., Smith, H. O., Birnstiel, M. L. 1978. *Cell* 14:655–71
24. Sures, I., Lowry, J., Kedes, L. 1978. *Cell* 15:1033–44
25. White, R. L., Hogness, D. S. 1977. *Cell* 10:177–92
26. Smith, G. P. 1976. *Science* 191:528–35
27. Karp, R. 1979. PhD. thesis. Stanford Univ., Calif.
28. Pardue, M. L., Weinberg, E., Kedes, L., Birnstiel, M. L. 1972 *J. Cell Biol* 55:199a
29. Birnstiel, M. L., Weinberg, E. S., Pardue, M. L. 1973. In *Molecular Cytogenetics,* ed. B. A. Hamkalo, J. Papaconstantinou, pp. 75–93. New York: Plenum
30. Pardue, M. L., Kedes, L. H., Weinberg, E. S., Birnstiel, M. L. 1977. *Chromosoma (Berlin)* 63:135–51
31. Southern, E. M. 1975. *J. Mol. Biol.* 98:503–17
32. Wilson, M. C., Melli, M., Birnstiel, M. L. 1974. *Biochem. Biophys. Res. Commun.* 61:354–58
33. Wilson, M. C., Melli, M. 1977. *J. Mol. Biol.* 110:511–35
34. Jacob, E. 1976. *Eur. J. Biochem.* 65:275–84
35. Jacob, E., Malacinski, G., Birnstiel, M. L. 1976. *Eur. J. Biochem.* 69:45–54
36. Scott, A. C., Wells, J. R. E. 1976. *Nature* 259:635–38
37. Fregien, N., Marchionni, M., Kedes, L. H. 1976. *Bio. Bull.* 151:407(Abstr.)
38. Yu, L. C., Szabo, P., Borun, T. W., Prensky, W. 1978. *Cold Spring Harbor Symp. Quant. Biol.* 42:1101–5
39. Chandler, M., Kedes, L. H., Yunis, J. 1978. *Am. Soc. Hum. Genet P.* 76A-(Abstr.)
40. Stout, J. T., Phillips, R. L. 1973. *Proc. Natl. Acad. Sci. USA* 70:3043–47
41. Grunstein, M., Grunstein, J. 1977. *Cold Spring Harbor Symp. Quant. Biol.* 42:1083–92
42. Durham, J. W. 1966. In *Treatise on Invertebrate Paleontology, Part U, Echinodermata 3,* ed. R. C. Moore, 270–95. Lawrence, Kansas: Univ. Kansas Press
43. Wouters-Tyrou, D., Sautiere, P., Biserte, G. 1976. *FEBS Lett.* 65:225–28
44. Strickland, M., Strickland, W. N., Brandt, W. F., Von Holt, C. 1974. *FEBS Lett.* 40:346–48
45. Strickland, W. N., Schaller, H., Strickland, M., Von Holt, C. 1976. *FEBS Lett.* 66:322–27
46. Tonegawa, S., Maxam, A., Tizard, R., Bernard, O., Gilbert, W. 1978. *Proc. Natl. Acad. Sci. USA* 75:1485–89
47. Grunstein, M., Schedl, P., Kedes, L. H. 1976. *J. Mol. Biol.* 104:351–69
48. Proudfoot, N. J., Brownlee, G. G. 1976. *Nature* 263:211–14

49. Birnstiel, M. L., Kressman, A., Schaffner, W., Portmann, R., Busslinger, M. 1978. *Phil. Trans. R. Soc. Lond. B.* 283:319–24
50. Wellauer, P. K., David, I. B., Brown, D. D., Reeder, R. H. 1976. *J. Mol. Biol.* 105:461–86
51. Federoff, N. V., Brown, D. D. 1978. *Cell* 13:701–16
52. Borun, T. W., Scharff, M. D., Robbins, E. 1967. *Proc. Natl. Acad. Sci. USA* 58:1977–83
53. Gallwitz, D., Mueller, G. C. 1969. *J. Biol. Chem.* 244:5947–52
54. Butler, W. B., Mueller, G. C. 1973. *Biochem. Biophys. Acta* 294:481–96
55. Jacobs-Lorena, M., Gabrielli, F., Borun, T. W., Baglioni, C. 1973. *Biochem. Biophys. Acta* 324:275–81
56. Perry, R. P., Kelley, D. E. 1973. *J. Mol. Biol.* 71:681–96
57. Breindl, M., Gallwitz, D. 1974. *Mol. Biol. Rep.* 1:263–68
58. Breindl, M., Gallwitz, D. 1974. *Eur. J. Biochem.* 45:91–97
59. Borun, T. W., Gabrielli, F., Ajiro, K., Zweidler, A., Baglioni, C. 1975. *Cell* 4:59–67
60. Kedes, L. H., Gross, P. 1969. *J. Mol. Biol.* 42:559–75
61. Kedes, L. H., Gross, P. 1969. *Nature* 223:1335–39
62. Nemer, M., Lindsay, D. T. 1969. *Biochem. Biophys. Res. Commun.* 35:156–60
63. Harvey, E. B. 1956. *The American Arbacia,* Princeton, New Jersey: Princeton Univ. Press. 298 pp.
64. Moav, B., Nemer, M. 1971. *Biochemistry* 10:881–88
65. Nemer, M. 1975. *Cell* 6:559–70
66. Gross, K. W., Jacobs-Lorena, M., Baglioni, C., Gross, P. R. 1973. *Proc. Natl. Soc. Acad. Sci. USA* 70:2614–18
67. Skoultchi, A., Gross, P. R. 1973. *Proc. Natl. Acad. Sci. USA* 70:2840–44
68. Arceci, R. J., Senger, D. R., Gross, P. R. 1976. *Cell* 9:171–78
69. Lifton, R. P., Kedes, L. H. 1976. *Dev. Biol.* 48:47–55
70. Gross, K. W., Probst, E., Schaffner, W., Birnstiel, M. 1976. *Cell* 8:455–69
71. Grunstein, M., Levy, S., Schedl, P., Kedes, L. H. 1973. *Cold Spring Harbor Symp. Quant. Biol* 38:717–24
72. Levy, S., Wood, P., Grunstein, M., Kedes, L. 1975. *Cell* 4:239–48
73. Schochetman, G., Perry, R. P. 1972. *J. Mol. Biol.* 63:591–96
74. Gallwitz, D., Mueller, G. C. 1970. *FEBS Lett.* 6:83–85
75. Gallwitz, D. 1975. *Nature* 257:247–48
76. Gallwitz, D., Traub, U., Traub, P. 1977. *Eur. J. Biochem.* 81:387–94
77. Liberti, P., Fischer-Fantuzzi, L., Vesco, C. 1976 *J. Mol. Biol.* 105:263–73
78. Borun, T. W., Ajiro, K., Zweidler, A., Dolby, T. W., Stephens, R. E. 1977. *J. Biol. Chem.* 252:173–80
79. Stephens, R. E., Pan, C., Ajiro, K., Dolby, T. W., Borun, T. W. 1977. *J. Biol. Chem.* 252:166–72
80. Bos, E., Roskam, W., Gallwitz, D. 1976. *Biochem. Biophys. Res. Comm.* 73:404–10
81. Levenson, R., Marcu, K. 1976. *Cell* 9:311–22
82. Ruderman, J. V., Pardue, M. L. 1977. *Dev. Biol.* 60:48–68
83. Destree, O. H. J., Haenni, A., Birnstiel, M. L. 1977. *Nucleic Acids Res.* 4:801–11
84. Ruderman, J. V., Pardue, M. L. 1978. *J. Biol. Chem.* 253:2018–25
85. Gabrielli, F., Baglioni, C. 1975. *Devel. Biol.* 43:254–63
86. Bhat, S. P., Padayatty, J. D. 1977. *Ind. J. Biochem. Biophys* 14:151–53
87. Amaldi, P., Felicetti, L., Campioni, N. 1978. *Biochem. Biophys. Acta* 518:518–24
88. Knoechel, W. 1975. *Biochem. Biophys. Acta* 395:501–8
89. Childs, G., Levy, S., Kedes, L. H. 1979. *Biochemistry.* 18:208–13
90. Burckhardt, J., Birnstiel, M. L. 1978. *J. Mol. Biol.* 18:61–79
91. Melli, M., Spinelli, G., Arnold, E. 1977. *Cell* 12:167–74
92. Adesnik, M., Darnell, J. E. 1972. *J. Mol. Biol.* 67:397–406
93. Milcarek, C., Price, R., Penman, S. 1974. *Cell* 3:1–10
94. Jacobs-Lorena, M., Baglioni, C., Borun, T. 1972. *Proc. Natl. Acad. Sci.* 69:2095–99
95. Surrey, S., Nemer, M. 1976. *Cell* 9:589–95
96. Moss, B., Gershowitz, A., Weber, L. A., Baglioni, C. 1977. *Cell* 10:113–20
97. Stein, J. L., Stein, G. S., McGuire, P. M. 1977. *Biochemistry* 16:2207–13
98. Wei, C. M., Gershowitz, A., Moss, B. 1975. *Cell* 4:379–86
99. Furuichi, Y., Morgan, M., Shatkin, A. J., Jelinek, W., Salditt-Georgieff, M., Darnell, J. 1975. *Proc. Natl. Acad. Sci.* 72:1904–8
100. Robbins, E., Borun, T. W. 1967. *Proc. Natl. Acad. Sci. USA* 57:409–16
101. Moll, R., Wintersberger, E. 1976. *Proc. Nat. Acad. Sci.* 73:1863–67
102. Prescott, D. M., 1966. *J. Cell Biol.* 31:1–9

103. Alfert, M. 1958. *Exp. Cell Res.* 6:227–35 (Suppl.)
104. Woodard, J., Rasch, E., Swift, H. 1961. *J. Biophys. Biochem. Cytol.* 9:445–62
105. Stein, G., Park, W., Thrall, C., Mans, R., Stein, J. 1975. *Nature* 257:764–67
106. Farquhar, M. N., McCarthy, B. J. 1973. *Biochem. Biophys. Res. Commun.* 53:515–22
107. Adamson, E. D., Woodland, H. R. 1977. *Devel. Biol.* 57:136–49
108. Stein, G., Stein, J., Shephard, E., Park, W., Phillips, I. 1977. *Biochem. Biophys. Res. Commun.* 77:245–52
109. Kunkel, N. S., Hemminiki, K., Weinberg, E. S. 1978. *Biochemistry* 17:2591–98
110. Melli, M., Spinelli, G., Wyssling, H., Arnold, E. 1977. *Cell* 11:651–61
111. Hackett, P. B., Traub, P., Gallwitz, D. 1978. *J. Mol. Biol.* 126:649
112. Shutt, R., Kedes, L. H. 1974. *Cell* 3:283–92
113. Levy, S., Childs, G., Kedes, L. H. 1978. *Cell* 15:151–62
114. Burdick, C. J., Taylor, B. A. 1976. *Exp. Cell Res.* 100:428–33
115. Kinkade, J. M., Cole, R. D. 1966. *J. Biol. Chem.* 241:5798–5805
116. Bustin, M., Cole, R. D. 1968. *J. Biol. Chem.* 243:4500–5
117. Hill, R. J., Poccia, D. L., Doty, P. 1971. *J. Mol. Biol.* 61:445–62
118. Cohen, L. H., Mahowald, A. P., Chalkley, R., Zweidler, A. 1973. *Fed. Proc.* 32:588 (Abstr.)
119. Seale, R., Aronson, A. I. 1973. *J. Mol. Biol.* 75:633–45
120. Ruderman, J. V., Baglioni, C., Gross, P. R. 1974. *Nature* 247:36–38
121. Cohen, L. H., Newrock, K. M., Zweidler, A. 1975. *Science* 190:994–97
122. Newrock, K. M., Alfageme, C. R., Nardi, R. V., Cohen, L. H. 1977. *Cold Spring Harbor Symp. Quant. Biol.* 42:421–31
123. Zweidler, A., Cohen, L. H. 1972. *Fed. Proc.* 31:926(Abstr.)
124. Zweidler, A., Cohen, L. H. 1973. *J. Cell Biol.* 59:378a(Abstr.)
125. Alfageme, C. R., Zweidler, A., Mahowald, A., Cohen, L. H. 1974. *J. Biol. Chem.* 249:3729–36
126. Zweidler, A. 1977. *Meth. Cell Biol.* 17:223–33
127. Franklin, S. G., Zweidler, A. 1977. *Nature* 266:273–75
128. Newrock, K. M., Cohen, L. H. 1978. *Cell* 14:327–36
129. Kunkel, N. S., Weinberg, E. S. 1978. *Cell* 14:313–26
130. Grunstein, M. 1978. *Proc. Natl. Acad. Sci.* 75:4135–43

Ann. Rev. Biochem. 1979. 48:871–922

METABOLITE TRANSPORT IN MITOCHONDRIA

♦12028

Kathryn F. LaNoue and Anton C. Schoolwerth[1]

Departments of Physiology and Medicine, The Milton S. Hershey Medical Center, The Pennsylvania State University, Hershey, Pennsylvania 17033

CONTENTS

[1]This work was supported in part by grant AM-19714 from the National Institutes of Health.

0066-4154/79/0701-0871$01.00

PERSPECTIVES AND SUMMARY

The ubiquitous function of mitochondria from both plants and animals is the conversion of energy obtained from respiration to forms usable by the cell. In most cells, the chief product of mitochondrial metabolism is ATP, but in some instances, it may be heat (1–3) or Ca^{2+} accumulation (4–6). In order for the mitochondria to perform their functions, an enormous traffic of metabolites through the mitochondrial membranes is required. Enzymes that utilize ATP are usually outside mitochondria, whereas those involved in ATP synthesis are inside. In organs that carry out gluconeogenesis and ureogenesis, portions of these synthetic pathways are located within the mitochondrial matrix and other portions outside (7, 8).

Mitochondria are surrounded by two membrane systems that have considerably different biochemical properties. These membranes enclose two compartments which correspond to an intermembrane space and a matrix space contained inside the inner membrane (9–11). The outer membrane has a relatively high lipid to protein ratio and, since there are no invaginations, it has a low surface to volume ratio (12). It is permeable to most small molecules with a molecular weight up to ~5,000. The inner membrane has a low lipid-to-protein ratio, is highly invaginated, and has a high surface-to-volume ratio (12). The permeability characteristics of the inner membrane are highly selective and most hydrophilic substances cannot penetrate its lipid bilayer. Specialized carrier proteins within this membrane catalyze the transport of specific metabolites through the lipid bilayer.

Since the pioneering work of Chappell and co-workers (13, 14) on transport processes in mitochondria, an impressive number of carrier systems have been demonstrated. Basic principles of mitochondrial transport have emerged from these studies. One striking feature of the findings is their essential harmony with important concepts of the chemiosmotic hypothesis of oxidative phosphorylation (15). According to this hypothesis, protons are pumped from the mitochondrial matrix space by enzymes of the electron transport chain. This results in an electrical potential gradient across the inner membrane, negative inside, and a pH gradient, alkaline inside. The total (H^+) electrochemical potential, frequently designated $\Delta\mu_{H^+}$, which can develop across the membrane ranges between 180–220 mvolts (16, 17). The energy of this gradient may be used for ATP synthesis by the mitochondrial ATPase. Transport of metabolite anions against this electrical potential would require the expenditure of a significant amount of cellular energy. However, a number of anions are transported into the mitochondria in symport[2] with protons, which neutralizes the unfavorable charge movement and allows the anions to accumulate in the matrix in proportion to the magnitude of the ΔpH (18). Having entered the matrix, these anions may be exchanged directly for other anions outside the matrix (13). The outward flow of anionic products of mitochondrial metabolism in vivo may be facilitated by the membrane potential (19). The carriers appear to possess molecular mechanisms that make use of the electrochemical potential gradient of protons across the membrane to facilitate transport in either direction. In this way, the carriers determine the direction of the transport processes.

Recent studies have focused on the molecular mechanisms of transport and the interaction of the carriers with the pH and electrical gradients (20, 21). Isolation of some of the carrier proteins has facilitated these studies considerably (22, 23). Some of these mitochondrial carriers may function to control metabolic rates (24), but this is presently a controversial area. For example, there is disagreement as to whether the transporter that catalyzes the exchange of ADP for ATP can normally limit and control rates of cellular respiration (25–27).

INTRODUCTION

This review covers a broad area of sometimes controversial research. Since it is impossible to review the entire field, we have chosen to concentrate on certain aspects. The molecular interaction of carriers with the electrochemical potential gradients of protons will be stressed. Also, the mechanisms that control carriers and how this control may influence flux through certain metabolic pathways are discussed. Several excellent reviews of metabolite transport have been published in the last few years (28–30) and, therefore, the emphasis of this review is on recent work.

[2]A mechanism whereby two molecular species are transported simultaneously in the same direction by a single membrane carrier. Frequently the gradient of one is used to drive the transport of the other against a concentration gradient (e.g. Na^+-dependent glucose transport in the intestinal mucosa).

TECHNIQUES FOR MEASURING METABOLITE TRANSPORT

Although transport across the cell membrane has been actively investigated for years, mitochondrial transport did not develop rapidly because of technical difficulties involved in quantitating rapid metabolite flux through a very small compartment. This was true even though mitochondrial carriers are more active and more abundant than carriers in cell membranes (31, 32). However, technical advances such as described in (33) have permitted significant progress to be made during the last decade. A short discussion of the techniques used to establish the nature of mitochondrial carriers and to measure rates of transport is presented before proceeding to a detailed description of the properties of carriers and the relationship between H^+ ion and metabolite fluxes across the membrane.

Direct Measurements of Transport Rates

MITOCHONDRIAL SWELLING   The existence and specificity of many of the transporters were first demonstrated using techniques that employed mitochondrial swelling (13, 14). Swelling occurs when large amounts of solutes enter the matrix space, causing increased osmotic pressure and volume of the matrix. The extent of swelling can be measured as a decrease in visible light scattering by a suspension of mitochondria. The change in optical density may be used as a qualitative measure of transport. With somewhat more sophisticated analysis, swelling techniques may be used to predict whether an electroneutral or electrogenic movement of ions occurs on a given carrier. When mitochondria are suspended in the salt of a permeant anion and swelling does not occur, it is usually because of electrical or pH imbalances which would result if swelling did occur. K^+ and Na^+ do not normally penetrate the mitochondrial inner membrane (34, 35) and, therefore, mitochondria will not swell in K^+ or Na^+ salts of the permeant anions (Figure 2 A). Thus, massive inward transport of anions as such does not occur because of the resultant electrical imbalance. Similarly, the protonated form of the anion is not massively accumulated because of the proton imbalance (Figure 1) unless there is a sink or mechanism for efflux for the H^+ released on dissociation of the anion in the mitochondrial matrix. However, ammonia can penetrate the membrane and mitochondria swell in isosmotic ammonium salts of anions because the proton transported can be used to form NH_4^+ in the matrix and thus stabilize the matrix pH. In the presence of an ammonium salt, EDTA, and a respiratory inhibitor, swelling indicates that matrix accumulation of the anion has occurred and that the anion is being transported electroneutrally with a proton (Figure 1) or in exchange for a hydroxyl ion. If protons are transported into the matrix with the permeant anion, but are trapped by formation of NH_4^+,

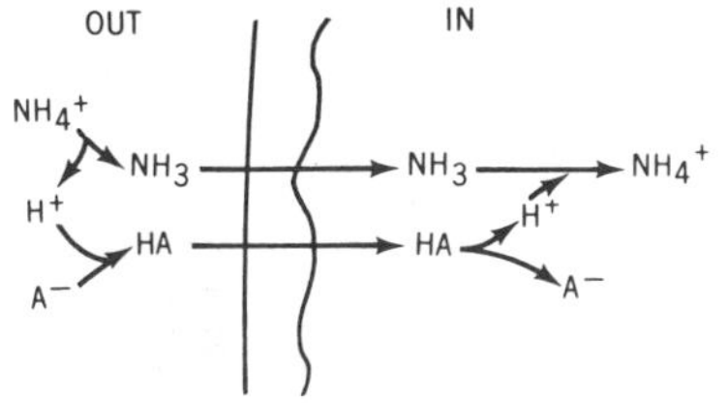

Figure 1.   Ion fluxes across the mitochondrial membranes during swelling in isosmolar ammonium salts. A^- is any permeant anion.

transport can continue indefinitely since there is no resultant charge or proton imbalance and thus swelling results (Figure 1).

Mitochondria also will swell in K^+ salts of permeant anions under certain conditions. It is necessary to make K^+ permeable to the membrane by including the ionophore, valinomycin (36), in the media. In this case, electroneutral transport of the anion with a proton, as illustrated in Figure 2 *A,* results in a proton and charge imbalance, which prevents large anion accumulation and no swelling results. If H^+ can be transported out of the matrix either as the charged ion by active transport during respiration (Figure 2 *C*), or passively with uncoupling agents (Figure 2 *B*) (37), swelling will occur. Swelling in the absence of respiration or uncoupling would indicate that the anion has penetrated in its charged, unprotonated form (38).

Rapid data collection can be achieved by swelling techniques, but they have limited usefulness. This technique is not a quantitative measure of transport, and can be used only to measure uniport[3] (i.e. one-way) mechanisms or exchanges linked to uniport mechanisms, and only when high concentrations of permeant metabolites are employed.

CENTRIFUGAL FILTRATION   A second procedure for studying transport was developed (9, 11, 33) in which the mitochondria are rapidly separated from the incubation media by centrifugation. Changes in matrix metabolite levels due to transport can be assayed either isotopically or enzymatically. The mitochondrial suspension is layered on top of a slightly more dense layer of silicone oil, which is, in turn, layered on top of a solution of 14% perchloric acid. Upon centrifugation, the mitochondria pass through the silicone oil into the perchloric acid which rapidly stops transport and metabolism.

[3]A mechanism whereby one or more molecules are transported across a membrane, not in exchange with another molecule on the opposite side of the membrane. Carriers that operate by this mechanism must be able to reorient themselves across the membrane in the substrate-free form to transport a second molecule of substrate.

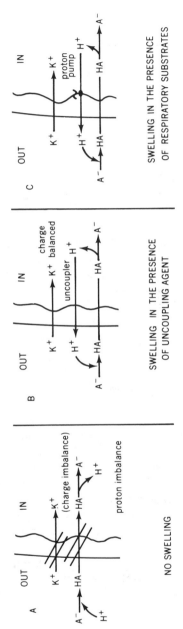

Figure 2. Ion fluxes during mitochondrial swelling in potassium salts of permeant anions. A^- is any permeant anion. *A.* No additions other than valinomycin. *B,* Valinomycin plus an uncoupling agent. *C,* Valinomycin plus a respiratory substrate.

The volumes of the matrix and intermembrane spaces can be assessed by including ^3H_2O and ^{14}C-sucrose, an inner membrane nonpenetrant, in the incubation media and measuring the amount of isotopes in the perchloric acid layer after centrifugation. The ^3H_2O space (total mitochondrial volume) is about 3.3 $\mu l/mg$ and the ^{14}C-sucrose space (intermembrane) is 2.5 $\mu l/mg$, leaving a matrix space of 0.8 $\mu l/mg$ (39). The major disadvantage of the rapid centrifugation method is lack of kinetic resolution. The procedure requires about 20 sec and thus true initial rates of transport cannot be obtained.

Rapid sampling is frequently required for transport studies, particularly in exchange or antiport[4] reactions when the small endogenous matrix pool of substrate does not change size. Thus one must measure rapid turnover of this small pool. Recycling of the substrate makes the uptake almost immediately nonlinear. Under these conditions, the log of the percent equilibration of the internal pool of substrate with externally added labeled substrate is plotted vs time to obtain a pseudo first order rate constant. When this constant is multiplied by the total endogenous pool of the substrate, a measure of the transport rate is obtained (40).

In order to circumvent the problem of slow sampling, another procedure was developed which uses multiple layers of media of different density (41). Centrifuge tubes are prepared by layering silicone oil on 14% $HClO_4$. A reaction mixture of buffer containing the transportable metabolite plus dextran is placed over the heavy oil. The mitochondrial suspension is added as the final layer. The mitochondria are centrifuged into the reaction mixture where they are in contact with the metabolite for only seconds before passing into the perchloric acid layer. The perchloric acid fraction is then assayed to determine the amount of material transported into the matrix during the passage of the mitochondria through the reaction layer. Time required to pass through the reaction layer can be calculated as 2 or 3 sec by including a substrate such as β-OH-butyrate in the reaction layer and measuring product formation.

FILTRATION   Initial rates of transport out of mitochondria have been measured using Millipore filtration techniques to separate mitochondria rapidly from their media (42, 43). Reactions in the filtrate can be rapidly quenched by filtering into perchloric acid, but the mitochondria remain on the filter paper and reactions there cannot be stopped rapidly. Thus this

[4]A transport mechanism whereby one or more molecules are transported across a membrane only in exchange for another molecule on the opposite side. Carriers that operate by this mechanism go through the configurational change which results in transport only when bound to substrate.

procedure is only good for measuring outward transport since the large volume of external media remaining on the filter makes precise measurement of mitochondrial content difficult. Washing the filter prior to extraction of the mitochondria may remove some matrix metabolites. Automatic devices using this procedure have been developed which allow samples to be obtained in the millisecond range (33).

THE INHIBITOR STOP TECHNIQUE   By far the most useful method of measuring transport across the mitochondrial membrane is the so-called inhibitor stop technique (33, 44). However, it is only possible to use this technique if an effective, specific inhibitor of the particular transport process to be studied can be obtained. The transportable metabolite is added to the mitochondrial suspension under appropriate conditions and the inhibitor is added at timed intervals thereafter. The inhibitor must stop transport immediately and completely on addition to the mitochondrial suspension. The mitochondria can then be separated from their media and washed by any convenient procedure and assayed for transported metabolite. The time of transport, using this procedure, is the interval between the additions of metabolite and inhibitor. Time resolution of 1–2 sec is possible. This method was originally developed for the adenine nucleotide carrier and made use of the inhibitor, carboxyatractyloside. The technique has been combined with an automated quench flow apparatus to obtain samples in the millisecond range (45).

Indirect Methods

Indirect methods of measuring transport which depend on an intramitochondrial reaction of the transported species have been developed (13, 46). The rate of production of NADH in the presence of respiratory inhibitors can be used to measure the penetration of substrates of mitochondrial dehydrogenases. In the same way, rates of oxygen utilization can be used if it can be shown by independent means that metabolite transport is rate-limiting for oxygen utilization.

GENERAL CHARACTERISTICS OF THE MITOCHONDRIAL CARRIERS

Nature and Distribution of the Carriers

The activity of specific mitochondrial carriers from different tissues correlates with the function of the mitochondria in situ. The synthesis of ATP is common to all mitochondria and requires entry of oxygen, ADP, phosphate, and electron-rich substrates such as pyruvate and fatty acids. The products of these reactions are water, carbon dioxide, and ATP which must be transported out. Oxygen, water, and carbon dioxide are freely permeable to the mitochondrial membrane. Under some conditions it appears that HCO_3^- can also penetrate the membrane as the negatively charged ion (38). However, specialized carriers are present in most mitochondria to transport phosphate (47), pyruvate (48), and fatty acids (49). Another carrier universally present in mitochondrial membranes is the adenine nucleotide carrier, which catalyzes the exchange of ADP for ATP across the membrane (50, 51). The substrate and inhibitor specificity of the phosphate carrier and the adenine nucleotide carrier from plant, bacterial, and animal sources are very similar. It has been further demonstrated that the adenine nucleotide carriers from a variety of sources are immunologically similar (52).

Two additional carriers appear to have a fairly broad distribution. One of these catalyzes the exchange of malate for α-ketoglutarate and the other catalyzes the exchange of glutamate for aspartate. These are involved in a cyclic transport pathway called the malate-aspartate shuttle (53) and can account for the transport of reducing equivalents from the cytosol to the mitochondrial matrix. NADH as such does not penetrate the mammalian mitochondrial membrane (54). Therefore indirect routes exist for transporting the reducing equivalents formed in the cytosol by glycolysis into the mitochondira. Electrons can be transported into mitochondria as malate on a carrier that exchanges α-ketoglutarate for malate. The reducing equivalent is then passed to NAD in the matrix by the action of malate dehydrogenase. The product of this reaction (oxaloacetate) is transaminated to aspartate by glutamate-oxaloacetate transaminase. The aspartate is then transported out of the mitochondria in exchange for glutamate. The cytosolic aspartate is converted to oxaloacetate and then to malate to complete the cycle. Other pathways have been suggested for the transport of cytosolic reducing equivalents which include the well known α-glycerophosphate pathway (24) and the fatty acid chain elongation pathway (55). However, it appears that in heart, kidney, and liver, at least, the malate-aspartate shuttle is most important (56–58).

The dicarboxylate carrier was originally identified in rat liver mitochondria (14) where it catalyzes a very active electroneutral exchange of citric acid cycle dicarboxylic acids with each other and with phosphate. This carrier is minimally active in heart mitochondria (59), and its presence in the liver probably reflects the need for net transport of citric acid cycle intermediates out of the mitochondria to be used in the biosynthetic pathway for glucose.

A tricarboxylate carrier catalyzes the exchange of citrate, isocitrate, phosphoenolpyruvate, and malate (60). All these metabolites will exchange rapidly with each other across liver and kidney mitochondrial membranes. The tricarboxylate carrier's activity is very low in heart (61) and particularly high in liver. This reflects the cytosolic location of fatty acid biosynthesis in liver. The acyl groups used in hepatic fatty acid synthesis are transported out of the mitochondria as citrate. The activity of liver aconitase is low compared to the citrate carrier and citrate is preferentially transported out into the cytosol. Since the carrier catalyzes the exchange of citrate for malate, depletion of citric acid cycle intermediates does not occur.

A carrier that transports glutamate with a proton has been identified in liver mitochondria (62). The activity of the glutamate transporter is high in liver (63, 64) where it functions to provide substrate for glutamate dehydrogenase as part of the pathway for conversion of amino acids to glucose.

Balance of Proton Movements In Vivo

Protons entering the mitochondria in symport with substrates must be transported back to the cytosol for continuous functioning of metabolic processes. Kinetic studies show that transporters that operate by proton symport mechanisms do not function well unless adequate means are present for neutralizing matrix protons (20, 65, 66). Protons that enter the matrix by electroneutral symport[5] cannot be pumped out by the electron transport chain since this system leaves a negative charge in the matrix and a charge imbalance would result (67). Although the electrogenic outward directed proton pump has dramatic effects on rates of transport and distribution ratios of metabolite anions at equilibrium (67), it cannot be used to balance metabolite-linked electroneutral proton entry. Anion efflux not accompanied by protons could balance the charge but such a process would be very sensitive to uncoupling because the rate of electrogenic anion efflux would be a function of the membrane potential. Many continuous metabolic processes, such as the oxidation of pyruvate and citrate by isolated mitochondria, involve proton-linked substrate entry but are not inhibited by uncoupling agents. Therefore, each proton that enters the mitochondria by electroneutral proton symport in a metabolic pathway not inhibited by uncoupling agents must be balanced by an electroneutral proton efflux. This usually occurs in symport with a metabolic product.

In each case studied to date, when a substrate enters with a proton, a product carries the proton out. The simplest case is illustrated by pyruvate oxidation (Figure 3A). A proton transported (48) into the mitochondria with pyruvate eventually is balanced by efflux of CO_2 which is converted to carbonic acid in the external media. Glutamate oxidation via glutamate dehydrogenase is a more complicated system where glutamate is transported into the mitochondria by proton symport (Figure 3B) (65), but converted to malate by oxidation. Malate then leaves the matrix in exchange

---

[5]A symport mechanism which results in no net movement of charge across the membrane, sometimes because the charges of the two transported species cancel each other.

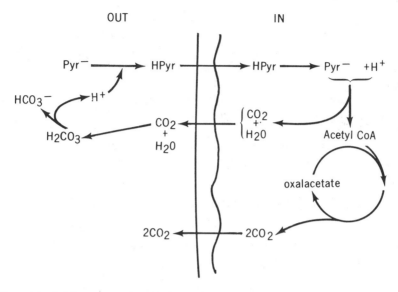

Figure 3A.   Ion fluxes across the mitochondrial membranes during the oxidation of pyruvate. Pyr, pyruvate.

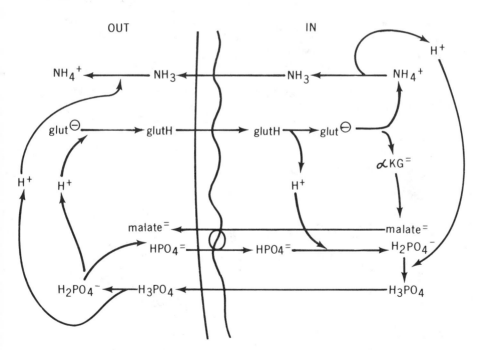

Figure 3B.   Ion fluxes during the oxidation of glutamate. Glut, glutamate; aKG, aKetoglutarate.

for phosphate. Phosphate can be transported out of the matrix by a uniport mechanism which results in an electroneutral loss of two protons. These two protons are balanced by the one proton which entered with glutamate and by efflux of ammonia, the other product of glutamate oxidation, as shown in Figure 3B. Citrate oxidation to malate is another example of proton symport (Figure 3C). Citrate with three negative charges enters the mitochondria in exchange for malate with two charges. A proton enters on the carrier with citrate (68) to maintain the electroneutrality of the exchange. The subsequent oxidation of citrate to malate produces CO_2 which diffuses out of the mitochondria and is hydrated with liberation of a proton.

Coupling Metabolite Transport to Mitochondrial $\Delta\mu_{H^+}$

IONOPHORES AS A TOOL IN MEMBRANE RESEARCH   The availability of ionophores in the study of mitochondrial transport led to a major breakthrough—the demonstration that most metabolites enter mitochondria by electroneutral mechanisms (36, 69). The great usefulness of ionophores lies in the fact that they can be used to measure and manipulate electrical potentials and pH gradients separately (e.g. 70, 71). Ionophores complex with potassium and other metal cations to form a lipophilic "crown" with the inorganic cation bound within the crevice. Thus shielded, the cation is soluble in organic solvents and will pass easily through natural and artificial membranes.

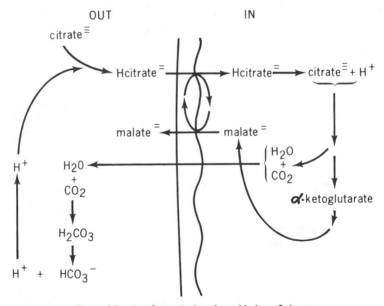

Figure 3C.   Ion fluxes during the oxidation of citrate.

Valinomycin, a cyclic dodecapeptide, is the best known of the neutral ionophores. It binds K^+ and the resultant positively charged complex allows K^+ to equilibrate across membranes in accordance with the electrical potential gradient. At low K^+ concentrations, the K^+ gradient is a measure of the membrane potential (16, 17, 72). However, at higher concentrations of K^+ the potential gradient is dissipated. In mitochondria, K^+ decreases the electrical potential but increases the ΔpH, because the available energy, stored largely as a membrane potential, is converted to ΔpH (73).

Some ionophores are negatively charged due to the presence of a carboxyl group. Nigericin is the best known example of these (74). The neutral K^+ nigericin complex and the protonated but K^+-free ionophore are permeable to membranes. Thus nigericin can act as a catalyst to exchange K^+ for H^+ across membranes and it will rapidly equilibrate the K^+ gradient with the pH gradient. Endogenous matrix levels of K^+ are high, and at high external K^+ levels nigericin has only a small effect on matrix pH. At lower external K^+, however, nigericin catalyzes K^+ efflux and H^+ influx which collapses the ΔpH but does not affect the membrane potential.

Many uncouplers of mitochondrial respiration may be considered as proton ionophores. Dinitrophenol and carbonyl cyanide p-trifluoromethoxyphenylhydrazone (FCCP) both increase the permeability of the mitochondrial membranes to protons (75, 76). These uncoupling agents, therefore, will decrease both electrical and chemical components of the $\Delta\mu_{H^+}$.

CARRIER-MEDIATED ELECTRONEUTRAL TRANSPORT   Accumulation of anionic metabolites in the mitochondrial matrix was initially interpreted to indicate that anions were transported as ions without proton compensation. From the distribution of anions, it appeared that the electrical potential gradient across the mitochondrial membrane was about 20–30 mvolts positive inside (77). However, calculations of the potential from the distribution of K^+ in the presence of valinomycin conflicted with this conclusion. Palmieri & Quagliariello (18) first proposed that anions are accumulated in the mitochondrial matrix by exchange with hydroxyl ions or by proton co-transport. The proposal was based on the observation that the distribution ratio of acetate, malate, succinate, citrate, and malonate across the mitochondrial membrane varies with the ΔpH and not the membrane potential. Effects due to the electrical potential were distinguished from effects due to ΔpH with the use of ionophores. Mitchell & Moyle (37) also published data supporting the conclusion that anionic metabolites were transported electroneutrally with protons.

At equilibrium, the rate of transport in should equal the rate out and the distribution of permeant anions should be proportional to the ΔpH as follows:

$$k(A_e^{n-}) (H_e^+)^n = k(A_i^{n-}) (H_i^+)^n \qquad \text{1.}$$

$$\log (A_i^{n-})/(A_e^{n-}) = n\Delta pH \qquad \text{2.}$$

where k = rate constant of the carrier; n = anionic charge; A_e^{n-} = external concentration of the anion; A_i^{n-} = matrix concentration of the anion; H_e^+ = external hydrogen ion concentration; and H_i^+ = matrix hydrogen ion concentration.

The prediction has been extensively tested (67). The distribution ratios of acetate, phosphate, malate, pyruvate, glutamate, and citrate increase with increasing ΔpH. This is true whether the ΔpH is varied by changing the external pH or whether the internal pH is changed at constant external pH with the use of ionophores. Also, many early studies showed that respiration of mitochondria causes alkalinization of the matrix and also increases the anionic distribution ratios (78, 79).

The predictions of Equation 2 are met only qualitatively. Distribution ratios have been carefully determined for malate, acetate, and phosphate (80). If one plots $\log (A_i^-)/(A_e^-)$ against ΔpH (independently measured by breaking the mitochondrial membrane with detergents) the resultant straight line does not go through the origin. This can be due to the difference in the chemical activity coefficients of the internal and external anion and the suggestion has been made that the intercept at 0 ΔpH is equal to the log of the ratio of the internal and external activity coefficients. In addition, the slope of the line, which should be n, is significantly lower than n and decreases with increasing concentration of (A_e^-). Also, there is a maximal internal anionic capacity (A_{max}^-) which does not vary with the ΔpH or the external buffer pH. The concentration of anion $(A_{1/2}^-)$ which produces half maximal saturation of the internal capacity, however, decreases as the external pH decreases. If the values of (A_{max}^-) and $(A_{1/2}^-)$ are proportional to the rates of entry of the anion, the data obtained by Palmieri et al (80) imply that the V_{max} of entry is independent of pH, but that the binding of (A^-) to the carrier is increased by H^+. Thus, perhaps at infinitely high concentration of A^-, the anion could be transported electrogenically.

The assumption that (A_{max}^-) and $(A_{1/2}^-)$ are proportional to the entry rates, V_{max} and ½ V_{max}, is not necessarily valid, however. Values of (A_{max}^-) and $(A_{1/2})$ for all the anions tested are similar but it has been demonstrated that the three anions tested have very different mechanisms of transport across the membrane.

The initial conclusion that all anions are transported into the mitochondrial matrix as neutral acids was soon shown to be an oversimplification. Phosphate is transported by a neutral proton symport mechanism and this can be demonstrated both directly and indirectly. Mitochondria swell in ammonium phosphate in the absence of respiration (37) (cf Figure 1). Also, an amount of proton disappearance equivalent to the disappearance of phosphate has been measured in lightly buffered mitochondrial suspensions (81). Other metabolite anions such as malate and succinate initially appeared to be transported with protons because they accumulated in proportion to the ΔpH. However, it has now been shown that these metabolites are transported across the membrane in exchange for phosphate and that it is phosphate that is in direct equilibrium with the pH gradient. Thus, phosphate distribution according to the pH gradient indirectly allows the other metabolites to equilibrate according to the pH gradient. If the link of the dicarboxylic acids to the hydrogen ion gradient through phosphate is broken, for example by inhibiting proton-linked phosphate transport with N-ethylmaleimide (82), then the dicarboxylic acids do not accumulate according to the ΔpH. N-ethylmaleimide does not inhibit the dicarboxylate exchange. This conclusion is also supported by the observation that mitochondria do not swell in isosmolar ammonium malate or succinate unless a small amount (5 mM) of phosphate is added to the media (13). The tricarboxylate and the α-ketoglutarate carriers are also indirectly linked to the ΔpH because they catalyze direct exchange of these carboxylic acids with malate, and malate is in turn linked to the phosphate gradient by means of the dicarboxylate carrier. Mitochondria do not swell in ammonium citrate or ammonium α-ketoglutarate unless a small amount of both malate and phosphate have been added to the media (13).

This "cascade" principle linking anion transport to proton gradients through phosphate is also demonstrated in a more direct experiment where the stoichiometry of metabolites and protons transported on the different carrier systems is measured. The earliest stoichiometric measurement of an exchange carrier system was made using the multiple layer centrifugal filtration method and demonstrated a 1:1 exchange of citrate for malate (83). These measurements were later confirmed and expanded by the demonstration that one proton accompanies the movement of trivalent citrate in the exchange for divalent malate (68, 81, 84, 85). In addition, other direct measurements show that a molecule of phosphate leaves the matrix for each molecule of malate that enters (68, 84). These measurements were carried out in the presence of N-ethylmaleimide to prevent phosphate from recycling. Simultaneous recordings of pH changes indicated that phosphate was transported as the divalent ion balancing the charge on divalent malate. Other data indicate that the dicarboxylate carrier catalyzes the 1:1 exchange of different dicarboxylic acids across the membrane.

CARRIER-MEDIATED ELECTROGENIC TRANSPORT   All the carrier systems discussed thus far have catalyzed electroneutral processes though in some cases electroneutrality has been imposed on the transport system by simultaneous proton movements. Certain anions such as thiocyanate, nitrate (37), and to a small extent chloride (86) are transported electrogenically by noncarrier mediated uniports. They tend to be excluded from respiring mitochondria and transport does not respond to the ΔpH across the membrane (70). Two well-characterized carriers have been shown to catalyze exchanges that result in a charge imbalance. These are the adenine nucleotide carrier and the glutamate-aspartate carrier. At neutral pH, ATP is mostly tetravalent and ADP is trivalent. The adenine nucleotide carrier catalyzes exchange of these two nucleotides without simultaneous proton movements (87, 88). Since substrate oxidation in mitochondria results in a negative potential gradient across the mitochondrial membrane, the electrogenic nature of the carrier provides a means of ejecting ATP from the matrix in preference to ADP (89). Even in the absence of ATP synthesis, ADP entry and ATP efflux are stimulated by respiration and inhibited by uncouplers. The fact that ΔpH has little effect on the direction of transport and that valinomycin inhibits ATP efflux, demonstrates the electrical character of the transport (88).

Glutamate and aspartate are both monovalent at neutral pH, but the carrier-mediated exchange becomes electrogenic because a proton is cotransported with the glutamate. Aspartate efflux in exchange for glutamate influx is inhibited by uncouplers and by valinomycin with K^+ (90); both conditions dissipate the electrical potential. Also, an equivalent amount of proton uptake into the matrix was measured during the exchange of glutamate for matrix aspartate (91, 92). Thus aspartate, which is needed in the cytosol for NADH transport and biosynthetic processes is effectively pumped out of the matrix due to the favorable membrane electrical potential. Protons carried into matrix during the exchange are pumped out electrogenically by the electron transport chain. It has been suggested that certain other transport processes are electrogenic; for example, the exchange of glutamine for glutamate (63) and the transport of ornithine (93). However, it now appears that glutamine is transported by a neutral uniport mechanism (94) and that the large accumulation of ornithine in the matrix is due to the accumulation of radioactive products of ornithine transamination and not to ornithine per se (95).

It can be concluded from these studies that the electrical nature of a particular transport process is not governed by the nature or ionic form of the substrates of the carriers, but by proton binding sites on the carriers themselves.

Mechanisms of Carrier Transport

Many lines of evidence indicate that metabolites are transported on protein carriers. The transport systems exhibit saturation kinetics and show a remarkable degree of specificity with respect to their substrates. Many are inhibited by specific toxins isolated from natural sources and by analogues of the natural substrates that have been synthesized (30). A majority of the mitochondrial transport systems are inhibited by sulfhydryl reagents (29, 96, 97). The nature of the particular sulfhydryl reagents that will inhibit a particular transport system may depend on a variety of factors, including hydrophobicity of target sulfhydryl groups, the hydrophobicity of the reagent, and pH (47). Inhibition may be pH-sensitive, which would indicate that the ionic state of the reagent and the sulfhydryl site on the carrier are important.

Kinetic studies of transport as a function of temperature indicate that the activation energy for transporting most metabolites across the mitochondrial membrane is high, usually in the range of 20–30 kcal (29). This implies that the carriers are not designed as simple channels.

In the case of those carriers that transport protons as well as metabolites, a consistent kinetic pattern is emerging which may also provide clues about mechanism. It appears that the proton usually binds to a site on the carrier protein and not on the substrate. The protonated carrier has a lower K_m for the substrate than the unprotonated carrier. Thus, at low concentrations of substrate, protonation of the carrier may be rate-limiting for transport. At high levels of substrate, on the other hand, the rate of transport frequently has an alkaline pH optimum which suggests that deprotonation of the carrier on the opposite side of the membrane may be rate-limiting (e.g. 98, 20). Other data also suggest that deprotonation of the carriers may be relatively slow and occasionally rate-limiting. For example, uptake of anions is frequently stimulated by external H^+ ions and efflux of anions such as phosphate (84), pyruvate (48), and glutamate (70) is inhibited by H^+.

Charge movements which may accompany anionic exchanges of di- and tri-carboxylates may be neutralized by positive charges at the protein-substrate binding site. The possibility that a divalent metal ion is present at the substrate binding site of the α-ketoglutarate, dicarboxylate, and tricarboxlate carriers (99, 100) is suggested by the observation that bathophenanthrolene, a strong metal chelating agent, is an effective competitive inhibitor of transport by these anion exchange carriers.

PROPERTIES OF THE INDIVIDUAL CARRIERS

For the purpose of this review the carriers will be classified according to the nature of their interaction with the proton electrochemical potential gradient. Thus, some carriers catalyze 1. proton-compensated electroneutral transport. Other carriers catalyze 2. electroneutral exchange of anions, 3. neutral metabolite transport in which protons are not involved, and 4. electrogenic exchange of anions.

Proton-Compensated Electroneutral Transporters

THE PHOSPHATE CARRIER   Mitochondrial phosphate transport is the fastest of the mitochondrial transport processes. Other than phosphate, substrates for the carrier include arsenate and monofluoro, but not difluoro, derivatives of phosphate (101, 102). A sulfhydryl group is probably involved in the catalysis of transport since the carrier is inhibited by N-ethylmalei-mide and various mercurials. Evidence that phosphate is transported elec-troneutrally by proton co-transport or in exchange for hydroxyl ions is outlined already above.

Although it is sometimes suggested that the ratio of H^+ to transported phosphate is 1.0, at neutral pH more than one proton must move across the membrane with each phosphate to maintain electroneutrality. When the stoichiometry of protons to phosphate was carefully measured at pH 7.0, it was found to be 1.5 (81). This is close to the anticipated value, because phosphate exists in a nearly 1:1 mixture of HPO_4^{2-} and $H_2PO_4^-$ at pH 7.0 Below pH 7.0 the ratio of H^+/phosphate transported was somewhat lower than that predicted theoretically, possibly due to some small proportion of electrogenic transport.

Recent studies of the transport of fluoro derivatives of phosphate carried out by Freitag & Kadenbach (101) show that monofluorophosphate, but not the difluor derivative, is transported on the carrier. This suggests that divalent phosphate is the ionic species that binds to the carrier and that there are two separate proton binding sites which must be filled before electroneutral transport can occur.

Although phosphate-phosphate exchange kinetics have been measured by the inhibitor stop technique, using an automated timing device in rat liver mitochondria with N-ethylmaleimide as the specific inhibitor (103), attempts to study the precise quantitative kinetics of net uptake have been discouraged by technical problems. Phosphate-phosphate exchange at 0°C is an order of magnitude faster than any of the other mitochondrial trans-port processes studied ($V_{max} = 205$ nmoles/min·mg, and $K_m = 1.6$ mM at pH 7.4) and uptake kinetics would be limited by the ability of the mito-chondria to dispose of the protons accumulated with phosphate.

Two indirect methods, designed especially to reduce the possibility that excess proton accumulation in the matrix would limit transport, have been used to study uptake kinetics. Both use changes of light scattering as a means of detecting transport and thus lack quantitative precision. Results obtained from the first of these methods, ammonium phosphate swelling (37), indicate that the V_{max} of phosphate transport increases with increasing pH. This conclusion was questioned in a later study (104) where mitochon-drial shrinkage, not swelling, was used as a detection technique. Mitochondria were first swollen in calcium acetate in the presence of a respiratory substrate. Subsequent addition of phosphate led to uptake of phosphate, efflux of acetate, and precipitation of calcium phosphate in the matrix. The decrease in mitochondrial volume and the resultant increase in light scattering was monitored by optical density measurements. As in the swelling study, it was possible to measure values for the V_{max} of transport, but not the K_ms. According to this later study, the pH profile was flat between pH 6.2 and 7.2. Possibly efflux of neutral acetic acid was rate-limiting under these conditions.

Recently, Kadenbach and his associates (101, 102) have carried out more detailed studies of the rate of phosphate-dependent swelling in ammonium salts. In isosmotic solutions of ammonium phosphate, the pH optimum for swelling is 7.8 whereas the pH optimum is 6.5 for phosphate dependent swelling in ammonium malate where the phosphate concentration was only 5 mM. Since swelling in ammonium malate is dependent on the activity of the phosphate carrier as well as the dicarboxylate carrier, and the pH 6.5 optimum does not correspond to that of the dicarboxylate carrier, it would appear that the pH optimum for phosphate transport is more acidic at lower than at higher concentrations of phosphate. The data can be explained if the K_m for the carrier is decreased by H^+, and if V_{max} is also decreased. Parenthetically, studies of the effect of pH on distribution suggest that the K_m of the phosphate carrier is lowered by decreasing the pH (79). However, Kadenbach & Freitag (102) conclude that phosphate acts as an allosteric activator of net malate uptake by a malate-proton symport mechanism and that actual phosphate transport across the membrane is not involved in the phosphate-dependent swelling in ammonium malate. This conclusion seems unjustified since numerous studies have directly demonstrated exchange of malate for phosphate in the presence of N-ethylmaleimide (68, 81, 84) which inhibits phosphate transported on the phosphate carrier. On the basis of the data available, one could suggest a tentative model for phosphate transport in which H^+ ion binding to sites on the carrier aid or stimulate phosphate binding. Thus, at low phosphate concentrations, H^+ ion binding to the carrier could be rate-limiting, while at higher concentrations of phosphate, release of H^+ on the other side of the membrane could be rate-limiting. This is possibly a common kinetic pattern that exists among those carriers that catalyze a proton symport mechanism.

Since mitochondria do not swell in valinomycin, K^+, and phosphate unless some means of transporting protons out of matrix is provided (37), it seems unlikely that any electrogenic transport of phosphate occurs. On the other hand, three recent studies (105–107) indicate that under special conditions phosphate can be transported from mitochondria or into submitochondrial particles electrogenically, i.e. with at least one of the proton binding sites unfilled. The studies suggest that when the phosphate concentration is high and transport is not counter to the membrane potential, that proton binding to the carrier may not be an absolute requirement for transport.

Purification of the carrier (107) has been achieved by extracting submitochondrial particles with a neutral detergent, acetyl glucoside, followed by ammonium sulfate fractionation. The use of neutral detergents appears to be very important in obtaining membrane proteins that retain their native activities (108, 109). This protein has not been fully characterized with respect to molecular weight, amino acid sequence, and other characteristics.

Several other laboratories in recent years have isolated a protein that is also presumed to be the phosphate carrier (110–112). The isolation procedure is based on the observation that this carrier contains sulfhydryl group(s) which are sensitive to both N-ethylmaleimide and mercurials. A protein fraction is obtained by treating mitochondria with an unlabeled mercurial followed by unlabeled N-ethylmaleimide (47). The mercurial is then removed from the proteins with a sulfhydryl reagent such as dithioerythritol, which regenerates the phosphate carrier sulfhydryl groups. The mitochondria are then treated with radiolabeled N-ethylmaleimide, which forms a covalent bond to the phosphate carrier that remains during subsequent detergent fractionation. Several proteins are obtained, but the major one appears to have a molecular weight of about 30,000. These studies have yielded little additional information, since the isolated, labeled protein has no activity. Also its identity is somewhat questionable since both the pyruvate carrier (48) and the acyl carnitine carrier (49) are also inhibited by both mersalyl and N-ethylmaleimide.

Studies of the biosynthesis of the phosphate carrier (113, 114) have indicated that two protein components are involved in transport in yeast. Thus, in glucose-grown yeast, no mitochondrial phosphate transport occurs. In chloramphenicol-treated yeast, where mitochondrial protein synthesis is inhibited, transport occurs but the process cannot be saturated like a normal enzymatic reaction. Yeast grown on lactate have mitochondrial phosphate transport kinetics similar to the mammalian system. It was, therefore, concluded that the carrier is composed of a binding component, coded by mitochondrial DNA, and a channel-like component, coded in the nucleus.

THE PYRUVATE CARRIER   The history of the pyruvate carrier has been particularly stormy. An early review of mitochondrial anion transport suggested that no carrier was needed for the transport of this metabolite since it is a monocarboxylic acid (67). Because of its relatively acidic pK this is apparently not the case. Papa et al (48) demonstrated in 1971 that pyruvate transport followed saturation kinetics, was inhibited by sulfhydryl reagents, and was accumulated in proportion to the ΔpH across the membrane. They also showed that pyruvate could exchange across the membrane with certain other monocarboxylic acids, particularly acetoacetate. Not long afterwards this conclusion was questioned by Zahlten et al (115), on the basis that denatured mitochondria also appeared to accumulate pyruvate. However, later studies from Tager's laboratory (116) showed that the interpretation of Zahlten et al (115) was incorrect and that the apparent binding and uptake by denatured protein was simply due to media adhering to the precipitated protein. The conclusion that a specific carrier is involved was further confirmed by the discovery of a specific class of inhibitors, the hydroxycinnamate derivatives (117, 118). The inhibition by the hydroxycinnamate derivatives is based on their ability to react with specific sulfhydryl groups on the protein. Later studies (119) of the transport process show that the pyruvate carrier is inhibited by both N-ethylmaleimide and mersalyl with approximately the same sensitivity as the phosphate carrier.

There is general agreement that pyruvate can be accumulated in the mitochondrial matrix by proton symport. This conclusion is based on the findings that, at equilibrium, the ratio of pyruvate across the membrane is a direct function of ΔpH (119), that the initial rate of pyruvate efflux is stimulated at alkaline pH (48), and that one proton disappears from the media for each pyruvate molecule accumulated in the matrix (48). Numerous kinetic studies have been carried out and there appears to be little general agreement about values for K_m, V_{max}, substrate specificity, and the effect of H^+ on kinetic parameters. Some workers (120, 121) find that dicarboxylic acids, carnitine, and acyl carnitine exchange with pyruvate but others do not (122). The values for the V_{max} of initial pyruvate uptake in liver mitochondria vary from 0.6 nmoles/min·mg at 9°C (123) to 32 nmoles/min·mg at 0°C (119). The K_m values vary from 0.037 (122)–0.84 mM (121). The variability of these results likely reflects differences in the techniques employed. For example, much higher values of V_{max} are obtained when the matrix is made alkaline by providing a respiratory substrate. Thus, in some cases, accumulation of protons causes a rate limitation by preventing the return, or reorientation, of the deprotonated carrier to the outer side of the membrane. In a recent publication by Pande & Parvin (122) particularly low values for the K_m were obtained (5 μM in the case of heart mitochondria and 37 μM in the case of liver). Only the hydroxycinnamate-sensitive portion of pyruvate uptake was considered due to the carrier. It was suggested that hydroxycinnamate-insensitive uptake was due to external binding and noncarrier-mediated diffusion. Samples were taken at a single early time point within 15 sec of the addition of pyruvate, and controls, in which the inhibitor was added before pyruvate, were used to correct the uptake data. Subsequent studies by Halestrap (20) show that hydroxycinnamate does not inhibit immediately but requires at least several seconds or more depending on pH and substrate concentrations.

In Pande and Parvin's study (122) no time courses with and without inhibitor were shown, and thus it is possible that some of the very early hydroxycinnamate-insensitive uptake was due to carrier activity. Because the results obtained are significantly at variance with other studies, they should be interpreted with some caution. Many aspects of the previously reported data cannot be explained solely on the basis that most of the previously reported transport actually represents binding. For example, at low (0.5 mM) pyruvate concentrations, the membrane distribution ratio of pyruvate is increased as buffer pH decreases, but this dependence disappears in the presence of the ionophore nigericin which effects a K^+/H^+ exchange across the membrane and an acidification of the matrix pH (116). According to the binding theory, "bound pyruvate" should respond to the media pH, not the matrix pH. However, valid points made by Pande & Parvin (122) are that above 5 mM, pyruvate transport across the membrane is partially noncarrier-mediated and that respiration from pyruvate above 10 mM is not inhibited by hydroxycinnamate.

Recent data from Halestrap's laboratory (20), and the more recent studies of Paradies & Papa (124) are in agreement concerning the kinetic exchange characteristics of the carrier. Since very different techniques were used, the agreement is reassuring. The V_{max} of exchange of intramitochondrial pyruvate with various carboxylic acids is about 12 nmoles/min·mg at 4.0°C irrespective of the antiporter. In exchange for internal pyruvate, the K_m for external pyruvate at pH 7.0 is 0.11 mM, for acetoacetate it is 0.61 mM, for α-ketobutyrate, 0.17 mM, and for lactate, about 12 mM. Unidirectional transport in or out is slower than exchange and efflux is stimulated at alkaline pH. Several independent studies indicate that the K_m for initial rates of net pyruvate uptake appears to decrease as the buffer pH is decreased (116, 119). V_{max} in these studies was nearly constant. When pH-dependence of swelling in NH_4^+ pyruvate was studied, V_{max} appeared to increase with decreasing pH (116). However, in the presence of 100 mM ammonium pyruvate, carrier-mediated transport is probably insignificant (122). In a more recent paper, Halestrap (20) has used an elegant new method to determine the V_{max} of the pyruvate carrier at different pHs. Pyruvate was actively oxidized in the presence of an uncoupling agent, so that export of the products of metabolism (particularly CO_2) yielded a proton to balance the one brought in on the pyruvate carrier. This prevented artifactual inhibition of transport by matrix acidification. At the same time, transport was inhibited by hydroxycinnamate in a controlled way. Initially, at low hydroxycinnamate levels, no inhibition of respiration occurs because the dehydrogenase, not transport, is rate-limiting. However, when the curve relating the reciprocal of respiration rates to inhibitor concentration is extrapolated to zero inhibitor, it is possible to estimate V_{max}. The data obtained suggest that V_{max} may decrease with decreasing pH. Thus, at very low pyruvate concentration, H^+ may stimulate the rate of transport by increasing the binding of pyruvate to the carrier. At higher pyruvate concentrations, deprotonation of the carrier on the matrix side of the membrane may become rate-limiting, and at even higher concentrations (above physiological) transport is not carrier-mediated.

THE GLUTAMATE CARRIER   Carrier-mediated transport of glutamate in rat liver mitochondria was reported by Azzi et al (62) in 1967. The carrier is dependent on the activity of a sulfhydryl group as demonstrated by its sensitivity to N-ethylmaleimide (125) and other lipophilic sulfhydryl reagents (126). It is not inhibited by mersalyl. The fungicide, avenaciolide, (127) specifically inhibits glutamate/H^+ transport, also because of its interaction with the suflhydryl group. Proton co-transport with glutamate is a generally accepted fact. Thus, there is a positive correlation between the equilibrium uptake of glutamate and the mitochondrial pH gradient (66, 128–130). Efflux is stimulated by matrix acidification, and media alkalinization (70, 126). The ratio of initial glutamate uptake to proton uptake is one (65) and initial rates of glutamate uptake are stimulated by decreasing pH when the glutamate concentration is below the K_m. The K_m for the carrier is about 4 mM at both pH 6.5 and 6.9 (126, 131). V_{max} has been reported as 9.1 nmoles/min·mg at 20°C or 23 nmoles/min·mg at 25°C. Transport is slow and limits glutamate metabolism via glutamate dehydrogenase. Lack of reproducibility in determining V_{max} is related to the fact that matrix pH has not been controlled in the studies of initial uptake kinetics (65). Thus, accumulation of H^+ in the matrix after the first 10–15 sec of transport limits further accumulation.

Studies of mitochondrial swelling in isosmolar solutions of ammonium glutamate (116) showed that uptake was faster at pH 7.9 than at pH 7.1 and intermediate at pH 7.4. This suggests that the V_{max} of glutamate transport increases at increasing pH just as it increases for the other carriers that catalyze a proton symport.

A protein has been isolated from cholate extracts of pig heart submitochondrial particles (132–135) by affinity chromatography on γ-methyl glutamate albumin coreticulated on glass fiber. This protein catalyzes glutamate transport into artificial liposomes and may be the glutamate carrier. The protein binds 9 nmoles of N-ethylmaleimide/mg of protein in the absence of glutamate, but it binds only 6.1 in its presence. N-ethylmaleimide dissociates the protein and reduces its binding for glutamate. The measured K_d for glutamate was 72 μM.

Other mitochondrial proteins bind glutamate, however, and it seems likely that the relatively impure protein fraction isolated contains at least some of these other proteins. Although no glutamate dehydrogenase or glutamate-oxaloacetate transaminase activity is present, inhibited forms of both of these enzymes have been demonstrated in the isolated protein fraction by immunological antibody techniques (135). One third of the protein is transaminase subunits.

Aspartate inhibits glutamate binding to the protein, and the protein catalyzes, to a slight extent, glutamate-aspartate exchange across liposome membranes, which suggests that the glutamate-aspartate carrier may be present in the fraction as well. This does not exclude the presence of the real glutamate-H^+ carrier but suggests the need for further purification.

THE ORNITHINE CARRIER   Since the pK of the δ-amino group of ornithine is 10.8, at neutral pH values this molecule carries one net positive charge. In 1973, Gamble & Lehninger (93) published experiments which showed that mitochondria swell in ornithine phosphate only when succinate is present as a source of energy. Also, the apparent distribution ratio of labeled ornithine was high in the presence of phosphate and succinate, but not in their absence. This led to the conclusion that a carrier existed in the mitochondrial membrane that catalyzed the electrophoretic entry of ornithine$^+$ into the matrix space of rat liver mitochondria. According to the scheme of these workers for the net entry of ornithine, the positive charge was balanced by the electrogenic proton pump of the electron transport chain. The resultant chemical imbalance of protons was balanced by electroneutral proton movement in symport with phosphate or acetate. The scheme works for net uptake of ornithine but continuous inward transport of ornithine in the pathway of urea production would have to be balanced by continuous stoichiometric entry of phosphate in excess of the needs for oxidative phosphorylation. Recent studies by McGivan et al (95) indicate that ornithine is transported by a carrier that exchanges the cation for a proton and thus it fits the classification of an electroneutral proton-compensated transporter. The proton moves in a direction opposite to those considered previously in this review.

It is possible to balance the proton movements across the membrane in the pathway of citrulline formation if ornithine is transported by a proton antiporter mechanism, without the necessity for continuous stoichiometric entry of phosphate or electrogenic proton pumping. A molecule of CO_2 and one of NH_3 enter the mitochondria with ornithine to make citrulline. The two ATP molecules hydrolyzed in this reaction may be generated inside the mitochondria from internal ADP and phosphate. The entering CO_2 molecule hydrates with proton liberation; this proton is neutralized by simultaneous NH_3 entry. Because of the difference in the pKs of the side chains of ornithine vs the product of the reaction, citrulline, a proton appears in the matrix that can be transported out in exchange for the next ornithine coming in. Efflux of the product citrulline results in no gain or loss of protons. McGivan et al (95) have shown that the accumulation of radioactivity reported by Gamble & Lehninger (93) in the presence of ^{3}H-ornithine, succinate, and phosphate is due to products of the reaction of ornithine aminotransferase and that no accumulation occurs in the presence of transaminase inhibitors. When an inhibitor of transamination is present, the radioactive ornithine accumulates in the mitochondrial matrix in proportion to the ΔpH, but in the opposite direction from the permeant anions, acetate and phosphate. Thus, the more acidic the matrix, the more ornithine is found in the matrix at equilibrium. It is difficult to account for the energy requirement for swelling except to assume that the ornithine cation may be permeable without the carrier at very high concentrations. Kinetic parameters of ornithine transport were measured (95) by the centrifugal filtration technique, in which the mitochondria are added directly to the media containing ornithine already layered over silicone oil. The K_m was found to be about 1.0 and the V_{max} 8.3 nmoles/min·mg measured at pH 7.2 and 20°C. This V_{max} seems somewhat low to account for rates of citrulline production. The kinetics did not change on varying the media pH from 6.8 to 7.4, but it is nevertheless possible that faster rates could have been attained at more acidic matrix pH values.

THE TRICARBOXYLATE CARRIER   The operation of a tricarboxylate-malate exchange across the inner mitochondrial membrane was shown in a number of laboratories before 1970 (14, 60, 136–138). The high exchange activity in liver causes citrate to come out of the mitochondria before it can be oxidized. The substrates for the carrier were first identified as citrate, isocitrate, cis-aconitate, and malate. Later phosphoenolpyruvate (139, 140) was also shown to be an effective substrate with high affinity. The carrier operates by a strict antiport mechanism, and unlike the carriers already discussed, it does not catalyze net uptake of substrate, but rather catalyzes the exchange across the membrane of any of its substrates for any other. It is classified in this review as a proton-compensated electroneutral carrier because, in the exhange of citrate for malate, a proton is transported in the same direction as citrate (68) which maintains electroneutrality. Citrate at neutral pH has three negative charges and malate only two. Evidence for the proton co-transport is quite convincing. The ratio of citrate$_{in}$/malate$_{in}$

to citrate$_{out}$/malate$_{out}$ varies in proportion to the membrane ΔpH (141, 142). The accumulation of citrate inside is favored over the accumulation of malate at neutral pH, and the effect is more pronounced when the matrix pH is made more alkaline, with K$^+$ and valinomycin. Efflux of citrate is favored by acidification of the matrix (143) caused by lowering the buffer pH. The stoichiometry of proton uptake has been measured directly in the exchange of citrate for malate (68, 81).

The carrier is inhibited by mersalyl (29), but by higher concentrations than are required for inhibition of the phosphate carrier or the dicarboxylate carriers. Other inhibitors are long chain acyl CoA derivatives (144), bathophenanthroline (100) and α-cetyl citric acid (145). The inhibition by long chain acyl CoA was at one time thought to be physiologically significant, but is now considered too nonspecific to be important. Discovery (146) of the very specific and effective inhibitor, 1,2,3 benzenetricarboxylate, was a major breakthrough in the study of this carrier system because it has allowed the kinetics of the carrier to be studied by the inhibitor stop technique. Two detailed kinetic studies have been reported, one in which mitochondria were loaded with citrate (141) and another in which they were loaded with malate (147). The studies are in substantial agreement, the K_m being measured as 0.12 mM at pH 7.0 in one study (147), and 0.75 mM at pH 7.4 in the other study. V_{max} was not determined accurately in the first study, but was found to be 22 μmoles/min·g protein at 9°C at pH 7.0 when the mitochondria were loaded maximally with malate. The influence of the concentration of the internal substrate was not investigated. Kinetic parameters of the citrate transporter are summarized in Table 1.

Table 1   Characteristics of the tricarboxylate carrier in rat liver mitochondria[a]

Substrates			Inhibitors		
Chemical name	K_m (mM)	V_{max} (nmol/ min · mg protein)	Chemical name	K_i (μM)	Inhibitor binding sites (nmoles/ mg protein)
citrate	0.12–0.25	22.0	1,2,3-benzene-	70	—
cis-aconitate	0.09–0.22	22.0	tricarboxylate	160	0.3–0.4
isocitrate	0.18	22.0	α-cetylcitrate[b]	3	—
phosphoenol-			long chain acyl		
pyruvate	0.11	22.0	CoA	4	—
malate	0.70	22.0	bathophenan-		
succinate	3.40	22.0	throline	25	—

[a] Data from (29, 145, 148).
[b] In addition it has the following kinetic properties: activation energy, 20.1 kcal/mol; turnover site, 430 min^{-1}; and pH dependence optimum at pH70.

Both studies show a pH optimum of transport at pH 7.0. When the influence of pH was studied in detail in the exchange of external citrate for internal malate, it was found that increase in pH above pH 7.0 lowered the K_m for citrate and thus stimulated transport by facilitating substrate binding. The inhibition by H^+ below pH 7 was noncompetitive and possibly due to the rate limitation of carrier deprotonation on the inside of the membrane, in conformity with the kinetic pattern observed for the other proton-translocating carriers.

A citrate-binding protein that may be identified as the carrier, or part of the carrier, has been isolated by affinity chromatography of Triton X-100 extracts of submitochondrial particles (148, 149). The binding specificity of the isolated protein is very similar to the substrate specificity of the carrier. However, the molecular weight is quite low (20,000) so this may be a monomer, or a polypeptide subunit of the functional carrier. As yet, attempts to catalyze transport in liposomes in reconstitution studies have been unsuccessful.

Neutral Anion Antiporters

Two carriers, the dicarboxylate carrier and the α-ketoglutarate carrier, have been identified in the mitochondrial matrix. These carriers catalyze the electroneutral exchange of divalent anions (29). They can be distinguished from each other by their substrate and inhibitor specificity. Both catalyze strict antiport reactions between substrates on one side of the membrane with substrates on the other. The substrates of the dicarboxylate carrier (29, 150) include phosphate, malate, malonate, succinate, sulphate, sulfite, thiosulphate, arsenate, and possibly oxaloacetate.

The other neutral anion antiporter is the α-ketoglutarate carrier whose substrates include α-ketoglutarate, malate, malonate, succinate, and oxalacetate (151). The two carriers appear to have rather similar kinetic characteristics. There is very little pH sensitivity in the presence of K^+ salts (99, 152). There is a characteristic V_{max} independent of the substrate, while K_m varies with the substrate (152, 153). When the concentration of internal substrate is varied over a narrow range, the external K_m remains constant while V_{max} changes. This was true for the α-ketoglutarate carrier in rat heart (154–156) and for the dicarboxylate carrier in rat liver and rat kidney (157). The kinetic pattern suggests that the carriers have two binding sites, one on the inside and one on the outside of the membrane, and that the two substrates bind in random order and that translocation is possible when both are bound. Further work over a wider range of internal substrate concentration is needed. The substrates of the α-ketoglutarate carrier are mutually competitive when present outside the mitochondria which indicates that there is only one external binding site (152). The substrates of the dicarboxylate carrier are also mutually competitive except for phosphate which is a noncompetitive inhibitor of dicarboxylate entry (153); thus, the binding site for phosphate may be separate. Butylmalonate is a fairly specific inhibitor of dicarboxylate transport (150) while the α-ketoglutarate carrier is less sensitive (152). Mersalyl and other mercurials but not N-ethylmaleimide will inhibit both carriers but again the α-ketoglutarate carrier is less sensitive (29, 158, 159). A specific inhibitor of the α-ketoglutarate carrier, phthalonic acid, has recently been identified (160).

A controversy that developed concerning whether oxaloacetate is transported on the dicarboxylate or the α-ketoglutarate carrier (161, 162) may have been resolved by the observation that the exchange of oxaloacetate for malate is sensitive to the α-ketoglutarate specific inhibitor, phthalonic acid (163). However, since physiological levels of oxaloacetate are much lower than the K_m for transport, movement of oxaloacetate across mitochondrial membranes does not occur to any significant extent in the cell. The relatively rapid V_{max} values, low K_ms, and high activation energies (Table 2) suggest that transport of the normal physiological metabolites are not rate-

Table 2 Characteristics of the α-ketogluturate decarboxylate carriers in rat liver mitochondria[a]

Substrates			Inhibitors		
Chemical name	K_m (mM)	V_{max} (9°) (nmol/ min · mg protein)	Chemical name	K_i (μM)	Type of inhibition
Carrier: α-ketoglutarate[b]					
α-ketoglutarate	0.05	43 ± 2	phthalonate[b]	20–30	competitive
malate	0.12	43 ± 2	butylmalonate	1,000	competitive
malonate	1.6	43 ± 2	bathophenan-		
oxalacetate	1.0	43 ± 2	throline	30	competitive
succinate	3.2	43 ± 2	phenylsuccinate	250	competitive
Carrier: dicarboxylate[c]					
malate	0.2	69 ± 6	butylmalonate[c]	350	competitive
malonate	0.4	69 ± 6	phthalonate	2,000	competitive
oxalacetate (?)	—	69 ± 6	bathophenan-		
succinate	1.2	69 ± 6	throline	25	competitive
phosphate	1.5	69 ± 6	phenylsuccinate	710	competitive
arsenate	1.1	69 ± 6	iodobenzyl-		
sulphate	0.3	69 ± 6	malonate	150	competitive
sulphite	—	69 ± 6	organic	100 % inhi-	noncompetitive
thiosulphate	1.0	69 ± 6	mercurial	bition	(before
				30 nmol/mg	substrate)
				protein	

[a] Data from 29, 84, 85, 160.
[b] The α-ketoglutarate carrier has the following kinetic properties: activation energy, 20.1 kcal/ mol; pH-insensitive.
[c] The dicarboxylete carrier has the following kinetic properties: activation energy, 22.0 kcal/ mol; pH-insensitive.

limiting for metabolism. Additional details concerning the properties of these two transporters have been reported in a recent review by Palmieri & Quagliariello (29).

Neutral Metabolite Transporters

THE GLUTAMINE CARRIER   The existence of a neutral uniport mechanism for glutamine in kidney mitochondria was originally proposed by Kovacevic et al in 1970 (164) because kidney mitochondria swell in isotonic glutamine in the absence of other metabolites, ions, or ionophores. This conclusion was questioned in a subsequent publication (63) because the entry of glutamine and efflux of glutamate appeared to be stoichiometrically linked. An active, phosphate-dependent glutaminase is present in the mitochondrial matrix which catalyzes the conversion of glutamine to glutamate plus NH_3, and until recently the glutaminase could not be inhibited without also inhibiting glutamine transport. Therefore, it was impossible to tell whether the apparent stoichiometry was dictated by the enzyme or by the carrier. The suggestion was even made that the two were in some way physically linked (165).

Three reasonable schemes for glutamine entry and glutamate efflux have been proposed (Figure 4). Scheme 1 illustrates the proposal (63) that glutamine is transported by an electrogenic antiporter which exchanges glutamate for glutamine. The electrogenic proton pump is necessary to balance charges. Because the activity of the glutamate/H^+ carrier is very low in kidney (63, 166), this mechanism would limit activity of glutamate dehydrogenase in the kidney because the dehydrogenase can be linked to the oxidation of external glutamate only by the glutamate/H^+ carrier. Data quoted by Crompton & Chappell (63) in support of Scheme 1 are the following: (a) Activation of the electrogenic proton pump by succinate stimulates the formation of glutamate from glutamine in the presence of rotenone. (b) The ratio of glutamine to glutamate in the matrix is increased by succinate, while the sum glutamate plus glutamine remains constant. (c) Only very low flux rates through glutamate dehydrogenase could be detected.

However, aside from the effect of succinate on the matrix ratio of glutamine to glutamate, which may be inaccurate due to technical problems in the rapid separation of the mitochondrial fraction, these data can be equally well accomodated by Schemes 2 or 3 which both allow glutamine to enter by a uniport mechanism and glutamate to leave on a separate carrier. They differ in that, according to Scheme 2, glutamate leaves the matrix electroneutrally with a proton and in Scheme 3 it comes out electrogenically. Scheme 3 requires the activity of the electrogenic proton pump to balance the charges, whereas Scheme 2 does not.

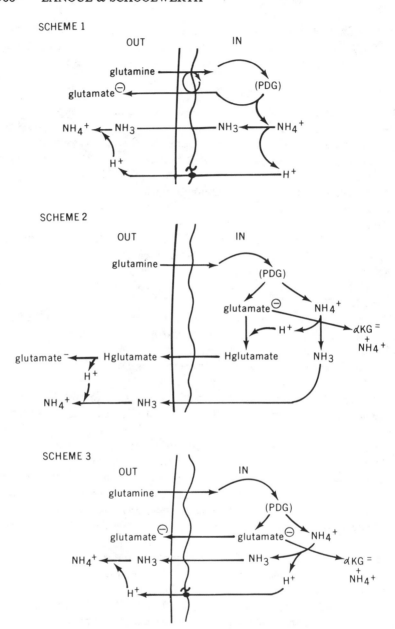

Figure 4. Ion fluxes across the mitochondrial membranes implied in 3 schemes for the metabolic conversion of glutamine to glutamate and ammonia. The schemes are discussed in the text. PDG, phosphate dependent glutaminase; $\sim$, the electrogenic proton pump.

Recent data greatly favor Schemes 2 and 3 over Scheme 1. Very significant flux through glutamate dehydrogenase can be demonstrated under many conditions with glutamine as the substrate in kidney mitochondria (166, 167). Inhibitors of transport of external glutamate into kidney mitochondria do not inhibit the oxidation of glutamine (168), which according to Scheme 1 must proceed through external glutamate. Furthermore, using ^{14}C-glutamine in the presence of unlabeled glutamate, it is possible to show that the $^{14}CO_2$ produced by oxidation of ^{14}C-glutamine does not equilibrate with the cold pool of external glutamate (169). All these data show that the link between glutamine entry and glutamate efflux is not obligatory. It has been stated that the discrepancies may be due to species differences between the pig and rat, but the key experiments have now been done in both species.

The remaining question relates to the electrogenicity of the glutamate efflux. The fact that glutamine enters by an electroneutral uniport is adequately demonstrated by the recent data of Curthoys & Shapiro (94), who were able to separate kidney mitochondria from contaminating brush border membranes with the use of a density gradient and show that glutamine ratios across the membrane are energy-independent. With the use of a recently discovered inhibitor of mitochondrial glutaminase which does not inhibit transport they have confirmed this conclusion in the absence of further metabolism of glutamine (170). Kovacevic (168), the chief proponent of Scheme 3, suggests that glutamate efflux is electrogenic in coupled mitochondria but in uncoupled mitochondria glutamate leaves on the glutamate/H^+ carrier. Scheme 3 would explain the stimulation by energy from succinate oxidation. However, mitochondrial glutaminase has an alkaline pH optimum and glutamate production from glutamine is stimulated at alkaline pH in the presence of energy. Perhaps active proton pumpimg is necessary only to raise the matrix pH to a level sufficiently alkaline to stimulate glutaminase. Recent data (171) show that nigericin, which equilibrates H^+ and K^+ across the mitochondrial membrane, causing matrix acidification, inhibits glutamate production from glutamine. In this system, the effect of uncouplers was similar to the effect of nigericin. This suggests that the effect of uncouplers is not due to a change of the electrical potential across the membrane but only to an effect on the pH in the matrix.

These data do not rule out the possibility that some portion of the glutamate efflux is electrogenic and that accelerated glutamate efflux stimulates glutaminase by lowering product inhibition. Proton-compensated and electrogenic glutamate efflux could be catalyzed by the same carrier and take place simultaneously depending on matrix pH, the proton binding properties of the glutamate carrier, and the membrane potential.

Goldstein & Boylan (171) have recently published kinetic data for the glutamine carrier using the uptake of radioactive glutamine in conjunction with an automated rapid mixing and rapid filtering system to obtain data in the one second range. Thus, they attempted to avoid the problem of confusing effects due to enzyme activity and rate of product efflux with glutamine carrier activity. The results indicate that at pH 7.4 and 23°C, the K_m of the carrier is 2.7 mM and the V_{max} 150–300 nmoles/min·mg. Transport, then, appears to be very rapid compared to rates of glutaminase activity and it seems to the authors of this review that the carrier could not limit metabolic activity. Further results of Goldstein & Boylan (171) show that α-ketoglutarate at 0.3 mM is a competitive inhibitor of the glutamine carrier and will double the apparent K_m for transport.

THE CARNITINE-ACYL CARNITINE EXCHANGE CARRIER   Fatty acids are important substrates for mitochondrial respiration in many tissues, especially mammalian heart. Since these compounds carry a negative charge at neutral pH, transport into the relatively electronegative mitochondrial matrix space would be prevented without the intervention of some compensatory mechanism. The necessary compensatory mechanism used for moving glutamate, pyruvate, and phosphate against the electrical potential gradient is proton co-transport, but a quite different strategy has evolved in the case of fatty acids. It has been shown that fatty acids must be converted to neutral carnitine ester derivatives before entering the mitochondrial matrix space.

The scheme for transporting long chain acyl groups into the mitochondria as originally proposed (172) involved conversion of fatty acids to acyl CoA derivatives, transfer of the acyl groups from CoA to carnitine, and, finally, penetration of acyl carnitine derivatives into the matrix and transfer of the acyl unit to CoA. The observation that no carnitine could be measured in the matrix space of liver mitochondria in the presence of external carnitine led to the conclusion that carnitine derivatives were not transported directly across the membrane. Yates & Garland (173) first suggested that two membrane-bound carnitine acyl CoA transferases exist, one on the inner and one on the outer surface of the membrane. The enzyme was supposed to transfer the acyl unit directly across the membrane without transferring carnitine.

However, in 1975, Pande (174) and shortly thereafter Ramsay & Tubbs (175) reported active carnitine-carnitine and acyl carnitine-carnitine exchange in heart mitochondria, demonstrable because of the presence of a measureable amount of carnitine within the matrix space of heart but not liver mitochondria.

The carrier is classed as an exchange carrier, but it can catalyze net uptake or release of carnitine at a rate about 0.5% of the exchange rate (176). The exchange is very rapid and probably never becomes rate-limiting for the oxidation of long chain acyl groups (177). It is specific for carnitine, acetyl carnitine, and both long and short chain acyl carnitine derivatives. Studies (96, 177) conducted over a very narrow range of endogenous carnitine indicate that the rate of transport may vary directly with the size of the endogenous pool. In isolated heart mitochondria, this pool size depends on isolation conditions (96), the metabolic state of the heart (178), and apparently the age of the animal (177). The K_m for external carnitine is about 5.3 mM and the V_{max} of transport at 20°C is 5 nmoles/min·mg (176). Transport of fatty acyl carnitine in exchange for carnitine is faster and has a lower K_m than carnitine-carnitine exchange (175).

The carrier system, like the phosphate carrier and the pyruvate carrier, is inhibited by both mersalyl and N-ethylmaleimide (96) which suggests the involvement of a sulfhydryl group in catalysis. The competitive nature of the inhibition by mersalyl suggests sulfhydryl group involvement in substrate binding. Derivatives of hydroxycinnamate which inhibit the pyruvate transporter, presumably by reacting with the sulfhydryl group, also inhibit carnitine exchange. A report that pyruvate could exchange for acyl carnitine in the matrix (120) prompted the suggestion that the pyruvate and carnitine carriers were identical, but subsequent studies (179) indicate that this is not the case.

Carnitine and acyl carnitine derivatives do not accumulate in the matrix, and there is no effect of uncoupling agents on carnitine exchange (96), although uncoupling agents have been shown to inhibit the oxidation of acyl carnitine (180, 181). The effect is probably due to loss of citric acid cycle intermediates from the mitochondria, due to swelling and loss of K^+ during preincubation with uncouplers in the absence of malate. Addition of malate prevents the observed effect.

THE NEUTRAL AMINO ACID CARRIER   The existence of a uniport mechanism for neutral amino acid transport is best demonstrated by the observation that mitochondria swell in isosmolar solutions of neutral amino acids (97). There is no proton or cation ionophore requirement for swelling, nor an apparent energy requirement. Carrier mediation is strongly suggested by the observation that L-isomers are transported more rapidly than the D-isomers. Swelling and, therefore, transport is inhibited by mersalyl and other mercurials but not N-ethylmaleimide. Many neutral amino acids are mutually protective against the inhibitory effect of mercurials and, therefore, it is generally concluded that one carrier with rather broad specificity is involved.

There is disagreement about the effect of energy on transport and accumulation. One thorough study (182) of the influence of ATP, uncouplers, and inhibitors of mitochondrial respiration on the accumulation of radioactive L-leucine in mitochondria produced no evidence in support of an energy involvement in the transport process despite the fact that other studies show that leucine accumulates by about twofold in the matrix (183), possibly due to some binding phenomenon. In most studies of amino acid transport, transport activity has been assayed as swelling, or as the accumulation of counts from radioactively labeled amino acids. It would be a mistake to attempt to assay neutral amino acids directly, because the amino acids assayed may be lysosomal, rather than mitochondrial. It has been demonstrated recently (184), that the endogenous pool of amino acids present in liver mitochondrial preparations is derived from liver protein degradation and does not equilibrate with cytoplasmic amino acids. The conclusion is drawn that these amino acids are present in the lysosomal, not the mitochondrial, fraction of the preparation. Probably the pool of neutral amino acids present in the mitochondrial matrix in situ is washed out during isolation.

The metabolism and transport of proline has been studied separately from the other neutral amino acids (185). A separate uniport carrier may exist for proline since it is not a very effective agent in protecting other neutral amino acids against mersalyl inhibition. Like them, however, its transport is inhibited by mersalyl and not by N-ethylmaleimide. Proline is oxidized by liver mitochondria in a two-step process that results in the formation of glutamate. The first enzyme in proline oxidation, proline oxidase, feeds electrons into the mitochondrial electron transport chain beyond the first oxidative phosphorylation site and is thus insensitive to rotenone inhibition. In a recent study (185) of the effect of energy on the accumulation of proline, there was an apparent effect of uncouplers and inhibitors of mitochondrial respiration on proline accumulation.

Proton co-transport on the carrier with proline would produce this energy dependence, but no proton uptake with proline could be detected and thus its mechanism remains a mystery. Since proline oxidation is not significantly inhibited by uncoupling agents, and since no molecular mechanism for the apparent energy effects can be shown, further work in this area is needed to confirm the energy requirment.

Electrogenic Transporters

THE GLUTAMATE-ASPARTATE EXCHANGE CARRIER   Numerous reports (39, 186, 187) appeared in the literature beginning with one by Klingenberg et al (188) which suggested that efflux of aspartate from mitochondria is an energy-dependent process. These reports received little attention, however, until a molecular mechanism for the energy dependence could be demonstrated.

The existence of a separate carrier system for aspartate was first demonstrated by Azzi et al (62) in 1967. These workers found that aspartate would oxidize intramitochondrial pyridine nucleotides (secondary to its conversion to oxaloacetate) only in the presence of added glutamate. The suggestion was made that aspartate could enter the mitochondria only in exchange for glutamate. The existence of a 1 : 1 exchange of glutamate for aspartate was subsequently confirmed, but it was further demonstrated in several laboratories (70, 116, 187) that entry of aspartate in exchange for glutamate is very slow and does not, in fact, occur in energized mitochondria. Other data (39) indicated that exchange in the opposite direction, entry of glutamate, and efflux of aspartate is facilitated by energization, which occurs when glutamate is metabolized by the transamination pathway. In rat heart mitochondria where the activity of the glutamate dehydrogenase pathway is very low, addition of uncouplers to mitochondria-metabolizing glutamate leads to a 90% inhibition of respiration and a 20-fold accumulation of aspartate in the mitochondrial matrix space. It was found that efflux of aspartate is stimulated by coupled respiration and by ATP, but inhibited by uncouplers and valinomycin (90). Efflux is not effected by nigericin and, therefore, one may conclude that transport in the physiological direction of aspartate efflux is stimulated by the presence of an electrical potential gradient across the membrane and that the exchange of glutamate for aspartate is electrical. One would not expect the exchange of monovalent aspartate for monovalent glutamate to involve charge separation unless a proton is transported with glutamate. When a method was devised for loading liver mitochondria with aspartate in the absence of external glutamate, it was possible to measure proton movements across the membrane on addition of glutamate to the aspartate-loaded mitochondria (91). Careful measurements of the stoichiometry of ion movements in a metabolically inhibited system, indicated that for each molecule of aspartate that leaves the matrix, one molecule of glutamate enters together with a proton. This phenomenon clearly demonstrates the principle that the electrical nature of the transporters is not dependent on the ionic state of the transported substrates, but rather on the nature of the carrier and its capacity to bind and transport protons with the metabolites.

The specificity of the carrier is fairly narrow. L-3-hydroxyglutamate will replace glutamate in the exchange for aspartate (189) and cysteine sulfinate (190) can exchange with either glutamate or aspartate. Cysteine sulfinate is transported, like aspartate, without a proton, as the anion. Its exchange with aspartate is electroneutral and energy-independent. On the other hand, the exchange with glutamate is electrogenic and, therefore, dependent on energy. An effective specific inhibitor of glutamate-aspartate exchange has not been identified, but mersalyl appears to be a fairly effective nonspecific inhibitor of transport in rat heart (90).

Because the total activity of the carrier is not exceptionally high, it has been possible to do kinetic studies in rat liver mitochondria using a modified rapid centrifugal filtration technique (191). Both K_m and V_{max} are increased when matrix aspartate levels are raised (19) which indicates that transport occurs by a ping-pong mechanism (192). In addition, use of indirect techniques measuring respiration as a means of quantitating transport, can be justified because of the unidirectional nature of the carrier. Thus, the rate of transport is geared precisely to the rate of transamination. Studies (98) of oxygen uptake at high glutamate levels show that transport is faster (250 nmoles/min·mg) at pH 7.4 than at pH 6.4, (177 nmoles/min·mg). However, since H^+ ions lower the apparent K_m of glutamate from 4.6 mM at pH 7.4 to 1.9 mM at pH 6.4, respiration is faster at pH 6.5 than at pH 7.5 when glutamate levels fall below the K_m. These data are reminiscent of those obtained from the phosphate-H^+ carrier and the other proton translocating carrier systems. At low concentrations of glutamate, proton binding to the carrier may be the rate-limiting step for transport. On the other hand, release of protons on the inner side of the membrane may be rate-limiting when the glutamate level is higher, which establishes a more alkaline pH optimum for transport at the higher metabolite concentrations. The mitochondrial glutamate-oxaloacetate transaminase is pH-insensitive in the pH region 6–8 (193) and the change in pH sensitivity of glutamate oxidation with glutamate concentration has been observed in heart (98), liver (K. F. LaNoue, unpublished data), and kidney mitochondria (194). The proposed sequence of events to account for the data has been partially confirmed in rat liver (191). When the pH of the intra- and extra-mitochondrial compartments were varied independently with the use of nigericin at different media pHs, it was found that aspartate efflux from the mitochondria at high glutamate levels had an alkaline pH optimum, but transport was sensitive only to the internal, not to the external pH.

THE ADENINE NUCLEOTIDE CARRIER   A separate system for the transport of adenine nucleotides was identified in 1965 (195). Since then, this important system has been studied in a wide variety of cells and two recent reviews have been published (50, 51).

The properties of the adenine nucleotide carrier are summarized in Table 3. The carrier catalyzes a 1:1 exchange between intramitochondrial and extramitochondrial adenine nucleotides, but of the natural nucleotides, only ADP and ATP are transported (196). Some synthetic analogues of ADP and ATP can also be transported, however, and these include methylene phosphonic analogues (197, 198), deoxyribose analogues (197), and thio-

Table 3 Characteristics of the adenine nucleotide carrier in rat liver mitochondria[ab]

Substrates			Inhibitors			
Chemical name	K_m (μM)	V_{max} (nmol/ min · mg protein)	Chemical name	K_i (μM)	Type	Number of inhibitor sites (nmoles/ mg protein)
ADP	1–10	150–200	atractyloside	0.02–1	competitive nonpenetrant	0.18–.024
ATP	150 (ener- gized)		carboxyatractyl oside	—	noncompetitive nonpenetrant	0.18–.024
	1 (ener- gized)		bongkrekate	—	uncompetitive nonpenetrant	
			isobongkrekate	—	uncompetitive nonpenetrant	
			long chain acyl CoA	0.2	competitive nonpenetrant	
			α-cetylcitrate	—	competitive	

[a] Data from (50, 51).
[b] In addition it has the following kinetic properties: activation energy, at 0–18°C, 29 kcal/ mol; at 18–30°C, 11 kcal/mol; turnover/site, per carboxyatractyloside binding sites or per 60,000 Mw, 500 min⁻¹; pH-insensitive.

phosphate analogues (199). Although adenine nucleotide transport is rapid, there is evidence that it may be the rate-limiting step in cellular respiration (25,200).

The properties of several highly specific and effective transport inhibitors are discussed in detail in a recent review (201). These inhibitors include two plant toxins, atractyloside and carboxyatractyloside (202) derived from the thistle, *Atractyles gummifera,* and a bacterial toxin, bongkrekic acid (203, 204). Long chain acyl CoA derivatives are nonspecific inhibitors of the carrier (205). N-ethylmaleimide will also inhibit the adenine nucleotide carrier (206) but only in the presence of ADP.

The influence of energy   It is generally accepted that the adenine nucleotide carrier is electrogenic. Charge separation across the membranes occurs during the exchange of ATP, which is largely tetravalent, for trivalent ADP. Evidence that the exchange is electrogenic comes from many differ- ent types of experiments. During respiration when mitochondria synthesize ATP from ADP and phosphate, transport of ADP into the mitochondria is much faster than inward transport of ATP (89). This occurs even when the concentration of ATP is many fold higher than the ADP (207). Because of the preference for inward ADP transport in the presence of respiratory substrates, the ratio of ATP/ADP is much higher outside than inside the mitochondria. In the presence of uncouplers (208), this difference disap- pears as does the kinetic preference for ADP. Recent data (88) show that the ratio of the internal ATP/ADP ratio to the external ATP/ADP ratio is a linear function of the membrane potential with a slope of 0.85 when the potential is varied with K^+ in the presence of valinomycin. 3,3'-Dipropylthiodicarboxyamine iodide is a dye that binds to the mitochondrial membrane and undergoes fluorescence changes coincident with membrane potential changes. ATP/ADP exchange across the mitochondrial membrane causes a transient fluorescence change indicative of an increase in the membrane potential (209).

Early studies concluded that the carrier was only partially electrogenic since the exchange of ADP for ATP was about 50% proton-compensated (67, 210). More recent studies show that the carrier catalyzes a completely electrogenic exchange of ATP and ADP with no proton cotransport when valinomycin and K^+ are included in the medium to minimize artifactual proton movements (87). Since the ATP/ADP exchange is fully electrogenic, some problems arise in balancing charge and hydrogen ion movements during the synthesis of ATP. According to the original proposal of Mitchell (15), two protons are pumped out of the mitochondria at each coupling site of the electron transport chain. Synthesis of ATP results in the return of these two protons. Utilization of half of the available energy for a transport process would be thermodynamically wasteful. In addition, the measured membrane potential is not sufficient to drive the intramitochondrial synthesis of ATP (16). The energy available from the membrane potential depends on the number of charges that move across the membrane during the chemical reaction. However, numerous recent studies (16, 207, 211–214) suggest that the original proton stoichiometry of the chemiosmotic hypothesis is incorrect. Direct measurements of proton movements indicate that four protons are transported by the electron transport chain for each site of ATP synthesis. Three of these protons are thought to be transported back during ATP synthesis, thus providing the energy needed to overcome the large free energy difference between ADP plus phosphate and ATP. One charge is left for ATP transport which leaves the membrane electrically balanced.

Balancing proton movements is somewhat more complicated. Phosphate is a monovalent anion at pH 6 and the electroneutral transport of phosphate at this low pH is accompanied by one free proton per phosphate molecule. Under these conditions, four protons are transported out of the matrix by the electron transport chain, three return during ATP synthesis (214), and the difference is made up by electroneutral transport of phosphate with release of one free proton in the matrix per site of ATP synthesis. However, since phosphate is a mixture of monovalent and divalent anions at pH 7.0, more than one free proton is released per molecule of phosphate. In practice, no proton imbalance occurs since the extra free protons that appear in the matrix are consumed in the synthesis of ATP. The free protons that disappear from the cytoplasm, reappear when the ATP is hydrolyzed by the energy requiring reactions of the cell.

A recent report (215) suggests that under certain conditions when ATP enters the mitochondria to drive metabolic processes, transport may be electroneutral due to cotransported anionic phosphate. This report requires further confirmation since the activity of the electroneutral process for ATP entry must be very low under most circumstances to prevent futile cycling and uncoupling.

Kinetic studies   There have been numerous kinetic studies of the adenine nucleotide carrier (45, 50, 51). Excellent time resolution is possible in these studies because the inhibition by atractyloside and carboxyatractyloside is both specific and fast. The internal nucleotides are labeled with isotope and then the equilibration of the labeled internal nucleotide pool is measured as a function of time. The reaction is first order because a plot of log percent equilibration vs time is initially linear. The entire nucleotide pool does not equilibrate rapidly, however, because AMP is not a substrate for the carrier.

The nature of both the internal and external nucleotides, as well as the presence or absence of a membrane potential, profoundly affects the kinetic parameters of transport. One recent study (216) shows that variations in the composition of the internal nucleotides cause parallel changes in the K_m and V_{max} for external ADP transport. This suggests a ping-pong mechanism of transport. The K_m for the transport of external ADP (1–10 μM) (217) is not strongly influenced by the energy state of the mitochondria. The K_m of ATP, however, is dramatically affected—similar to that of ADP in the uncoupled state, but 50–100-fold higher when a respiratory substrate is present (207). The V_{max} of ATP transport into mitochondria is also lower in the energized state but this effect is smaller, two- to threefold.

Little is known about the effect of the internal nucleotides on transport. It was stated at one time (89) that ATP and ADP on the inside of the membrane are kinetically indistinguishable even in the presence of a respiratory substrate, but it is now known that ATP is the preferred substrate when the membrane potential is physiologically high (50, 218, 219).

Another experimental approach to measuring the kinetic parameters from the internal face of the membrane is to study submitochondrial particles where the membrane is turned inside out. The difficulty with this approach is that the carrier activity (V_{max}) is low in these particles, while mechanisms and substrate and inhibitor specificity are preserved. The addition of SCN$^-$ lowers the membrane potential of the particles and causes a twofold decrease in the V_{max} of ATP and an eightfold increase in the K_m of transport (21). Thus it is interesting to note the somewhat perplexing fact that on both sides of the membrane, the potential across the membrane appears to have a larger effect on the binding of substrate to the carrier than on the V_{max} of transport.

Isolation of the carrier protein   The carrier protein for adenine nucleotide transport has been isolated in water soluble form in a number of laboratories (22, 23, 31, 220). The most successful isolation procedures utilize carboxyatractyloside binding. Up to 10% of the protein of the mitochondrial inner membrane can be isolated, and bound to carboxyatractyloside in a chromatographically pure form (221). Attempts to isolate the free carrier have been hampered by its instability and sensitivity to proteolytic digestion. However, using a freshly isolated protein preparation of free carrier, it has been possible to reconstitute some adenine nucleotide transport in artificial liposomes (222, 223). Comparison of data from different laboratories (22, 23, 220, 221) suggests that the carboxyatractyloside-binding protein is a dimer of subunits with a molecular weight of 30,000. Whether these subunits are absolutely identical is still unknown.

Bongkrekic acid has also been used to protect the protein from denaturation during the isolation procedure (224). Comparison of the properties of the isolated carboxyatractyloside protein with bongkrekic acid protein provides some interesting clues about the mechanism of transport. For example, it has been shown that carboxyatractyloside binds to the carrier from the outer surface of the membrane while bongkrekic acid binding is dependent upon the presence of a small amount of ADP outside the mitochondria. During bongkrekate inhibition, the amount of ADP (or ATP) carried into the matrix is equivalent to the amount of carrier present. A number of years ago, Klingenberg & Buckholz (225) interpreted these findings to mean that the carrier could exist in two forms, the *m* form with the substrate binding site exposed to the matrix side of the membrane, which could bind bongkrekic acid, and the *c* form with the substrate binding side exposed to the cytoplasmic surface, which could bind carboxyatractyloside but not bonkrekic acid. The *c* form could be converted to the *m* form, thus catalyzing the transport of ADP across the membrane, but this conformational change could occur only when substrate was bound to the carrier. Isolation of the two forms of the adenine nucleotide carrier confirms this interpretation. While the carboxyatractyloside protein and the bongkrekate proteins have the same molecular weight and same amino acid composition (224) they are immunologically different (226). In addition, they differ in their sensitivity to sulfhydryl reagents and to proteolytic digestion. However, the carboxyatractyloside protein can be converted to a protein that binds bongkrekate in solution, when ADP and bongkrekate are both present but not when either one is present alone (224). This may provide a unique opportunity to study a transport process carried out by a solubilized carrier protein.

THE INFLUENCE OF MITOCHONDRIAL CARRIERS ON METABOLIC FLUXES

Experimental Approaches

The role of mitochondrial metabolite carriers in the regulation of intermediary metabolism has only recently been considered. Although the interrelationship of the many translocators makes determination of a possible regulatory role for mitochondrial carriers difficult, the same principles used to identify controlling enzyme steps may be employed. As discussed by Rolleston (227), the best way to define whether or not a reaction is regulatory (displaced from thermodynamic equilibrium) is to compare the mass action ratio of the reactants and products with the known equilibrium constant. In the case of the mitochondrial metabolite translocators, the kinetic characteristics of the translocator must be taken into account with respect to the substrate and product concentrations on either side of the mitochondrial membrane and to the characteristics of the interrelated enzymatic steps.

An obvious drawback in assessing the role of transport in the control of metabolism was the lack of techniques for measuring metabolite concentrations in the cytoplasmic and mitochondrial compartments separately as opposed to total tissue measurements. An indirect approach is the metabolite indicator method described by Krebs & Veech (228). In this method, total tissue concentrations of reactants and products are measured. This information is combined with calculations of their contents in the separate mitochondrial and cytosolic spaces, assuming that certain reactions in which they participate are held close to equilibrium ("near-equilibrium"). These indirect methods are based on the calculation of the cytosolic free NAD/NADH from the lactate dehydrogenase equilibrium and the mitochondrial-free NAD/NADH from the β-hydroxybutyrate dehydrogenase equilibrium. The disadvantages of this indirect method are discussed by Williamson (229).

More recently, three direct methods have been developed to measure the concentration of various metabolites in the mitochondrial and cytosolic compartments (230, 231, 232). These methods are designed to effect physical separation of the two compartments by procedures that avoid metabolic activity during the fractionation and hence provide an estimate of compartment composition in situ. The three methods are: 1. fractionation of freeze-clamped tissue utilizing density gradient centrifugation in organic solvents (230); 2. treatment of isolated cells with digitonin to lyse the plasma membrane, after which the two cell compartments are separated by silicone oil filtration (231); and 3. the cavitation method, based on disruption of the plasma membrane of isolated cells by mechanical shearing forces, followed by silicone oil filtration (232). Each of these methods has drawbacks, but each offers considerable advantages over the previously used indirect method (233). Recent data from the direct methods suggest an important role for carriers in the regulation of metabolism. On theoretical grounds, the carriers most likely to have a regulatory role would be those that are functionally unidirectional and/or slow. These include the adenine nucleotide translocator (and respiration) (25), the glutamate-aspartate carrier (and reducing equivalent transport) (53), the dicarboxylate carrier (and ureogenesis and gluconeogenesis) (8), the pyruvate carrier (and gluconeogenesis) (121, 234), and glutamine and glutamate transport (and renal ammoniagenesis) (235). The clearest example of the great importance of mitochondrial metabolite transport in cell metabolism is the transport of reducing equivalents into mitochondria by the malate-aspartate cycle (24, 28). The roles of the adenine nucleotide carrier, the glutamine carrier, and the pyruvate carrier are also being actively studied.

Control of Respiration by the Adenine Nucleotide Carrier

An important question of mitochondrial bioenergetics is the mechanism of control of mitochondrial respiration. It was initially proposed that respiration was kinetically controlled by ADP availability (236, 237), by the so-called phosphorylation potential, $[ATP]/[ADP][P_i]$ (238–240), and by ADP translocation (241–244). The important question of whether the adenine nucleotide translocator regulates mitochondrial respiration remains an unresolved and highly controversial issue.

Wilson et al (245, 246) and Stubbs et al (247) have argued on theoretical grounds that since near equilibrium exists between the electron transport chain (up to, but not including cytochrome aa_3) and the cytosolic phosphorylation state, that adenine nucleotide translocation, which is an obligatory intermediate step in the reaction sequence, cannot be rate-limiting. That is, if the overall system of oxidative phosphorylation is at near equilibrium, all intermediary steps must also be at equilibrium. Additionally, if the intramitochondrial and extramitochondrial phosphorylation states are truly different, it must follow that any differences in activities of ATP, ADP, and P_i must be coupled to ion gradients and/or potentials across the inner mitochondrial membrane in such a way that translocation occurs without loss of free energy. These workers note (247) that since the K_m of the translocator for ADP is 1–10 μM (50, 51, 207) and the cytosolic ADP content is 300–600 μM (248), the translocator should operate at full capacity. Finally, while confirming that atractyloside and carboxyatractyloside inhibit gluconeogenesis in hepatocytes as shown by Akerboom et al [see below and (25)], Stubbs et al (247) suggest that this is not proof that the carrier is normally rate-limiting, since any step in a reaction sequence may become rate-limiting if sufficiently inhibited.

More recently, Ericinska et al (249) demonstrated that the rate of respiration of liver cells paralleled the [ATP]/[ADP][P$_i$], not the [ATP]/[ADP] ratio of the cells. It should be noted that in these experiments, total cell metabolite concentrations were measured, the phosphate being varied by incubation with 10 mM fructose or glycerol. This resulted in decreased phosphate concentrations from 4 μmol/gm to < 1 μmol/gm. This point is of some importance since other workers [see below and (250–252)] have demonstrated that phosphate becomes important at unphysiologically low phosphate levels (below 5 μmol/gm), and may obscure the regulatory role of the adenine nucleotide translocator. A recent paper by Holian et al (253) has countered this criticism by studying isolated mitochondria at varied ATP and phosphate concentrations. With glutamate-malate as substrates, ATP was altered from 1 to 8 mM at a constant phosphate concentration of 4 mM. In addition, phosphate was varied (1 to 8 mM) at a constant ATP concentration of 8 mM. The respiratory rate was shown to be inversely proportional to log [ATP]/[ADP][P$_i$]. Since the respiratory rate correlated more closely with this logarithmic function than with log [ATP]/[ADP], it was concluded that adenine nucleotide translocation could not control the rate of respiration and was not a rate-limiting reaction in oxidative phosphorylation.

A recent paper by Davis & Davis-van Thienen (254) of rather similar experimental design came to opposite conclusions. The chief experimental difference lay in the use of purified ATPase, which was used to obtain steady-state rates of respiration and [ATP]/[ADP] ratios. As long as phosphate was not limiting for phosphorylation (>3 mM), the rates of ATPase-stimulated respiration were highly correlated with [ATP]/[ADP] ratios, not with the [ATP]/[ADP][P$_i$] ratio. The data, therefore, are consistent with a rate-limiting role for the translocator when the free [ATP]/[ADP] and [P$_i$] are in the range thought to exist in the intact cell.

Numerous other experiments have supported a rate-limiting role for the adenine nucleotide carrier. Davis & Lumeng (200) utilized purified ATPase to maintain constant [ATP]/[ADP] and [P$_i$] levels. They observed that the steady-state [ATP]/[ADP] ratio generated outside the mitochondria decreased as the respiration rate increased toward state 3; matrix [ATP]/[ADP] ratios were not altered. Küster et al (255) used a hexokinase-glucose trap as an ADP-regenerating system to provide a stationary state of partially limited respiration in rat liver mitochondria. At a constant phosphate concentration of 10 mM, these workers found that mitochondrial respiration was controlled by the extramitochondrial [ATP]/[ADP] ratio, independent of the total concentration of adenine nucleotides and the nature of the substrates utilized. Control was observed when the ratio was between 5 and 10 at 10 mM phosphate. At lower ratios of [ATP]/[ADP], the mitochondria were in their maximum phosphorylating state and the rate of oxidative phosphorylation became independent of the adenine nucleotide pattern. Similar findings were observed by Soboll et al (256) in perfused rat livers utilizing techniques of fractionation of freeze-fixated tissue in nonaqueous solvents. Although the correlation was not tight, O_2 consumption correlated best with external (cytosolic) rather than internal (mitochondrial) [ATP]/[ADP] ratio or the phosphorylation potential.

Further evidence for the regulatory role of the adenine nucleotide translocator on mitochondrial respiration was reported by Akerboom et al (25). Atractyloside was shown to inhibit glucose production, and, to a lesser extent, urea synthesis from lactate plus pyruvate or from alanine, which suggests that the translocase was rate-limiting. More recently, Akerboom et al (233) and van der Meer et al (257) interpreted their data to indicate that the whole system of oxidative phosphorylation is far from thermodynamic equilibrium. Therefore, O_2 uptake can best be described in terms of irreversible thermodynamics, a framework in which the adenine nucleotide translocator may be the limiting step.

At the present time, although the bulk of the extant data are highly suggestive, more definitive data are required to determine the potential rate-limiting role of the adenine nucleotide carrier in mitochondrial respiration.

Ammonia Formation by the Kidney

During metabolic acidosis in some mammals (including man), homeostasis is maintained by a several-fold increase in the renal excretion of hydrogen ions in the form of ammonium. The increased ammonium excretion is derived mainly from increased renal ammonia production resulting from glutamine by the glutaminase I pathway. This pathway is intramitochondrial and involves phosphate-dependent glutaminase and glutamate dehydrogenase. Several factors have been postulated to be rate-limiting for the augmented renal ammonia synthesis during metabolic acidosis (235). The more attractive theories suggest that ammoniagenesis is regulated by activation of either glutamine transport or of the enzymes of glutamine metabolism. It should be pointed out that significant species differences may exist, particularly between the rat and dog (the two most extensively studied species) with respect to renal ammonia production.

Since both phosphate-dependent glutaminase and glutamate dehydrogenase are located within the mitochondria, it was suggested that transport of glutamine into mitochondria might be a factor limiting its metabolism (235). Recent studies by Adam & Simpson (165, 258, 259) demonstrated that uptake of [^{14}C]glutamine by rat and dog kidney mitochondria was increased in acute and chronic metabolic acidosis. These findings were confirmed in rat kidney mitochondria by Goldstein (260), Goldstein & Boylan (171), and Tannen & Kunin (261). Although these experiments were performed with rotenone-inhibited mitochondria to prevent glutamate metabolism, they could not distinguish the activation of a glutamine transporter from the activation of glutaminase. Activation of glutamine transport was postulated to be the regulatory step since, despite the increase in matrix counts from glutamine, glutamine itself could not be detected in the matrix space (165, 258, 259).

Recent studies reported by Curthoys & Shapiro (94) have suggested that glutaminase activity, and not glutamine transport, is the rate-limiting step in rat renal mitochondrial metabolism. Using Ficoll density gradients to purify the mitochondria of contaminating γ-glutamyl-transpeptidase in vesicular brush borders, these workers were able to detect glutamine in the matrix space. Moreover, they found that the concentration of glutamine was always lower in the matrix than in the incubation medium and that matrix glutamine did not change in acidosis despite an increase in the amount of radioactivity from the product of glutamine metabolism, glutamate. Since Simpson & Adam (258, 259) could detect no glutamine in the matrix space of dog kidney mitochondria, which presumably would not be contaminated by γ-glutamyl-transpeptidase, these results cannot be extrapolated to this species without additional studies.

More recently, Curthoys, Shapiro & Clark (170, 262) demonstrated that in the presence of a specific inhibitor of glutaminase (262), the increased uptake of radioactivity into the matrix space in acidosis was prevented. These findings indicate that, at least in the rat, activation of glutaminase is an important process in the augmented renal ammoniagenesis in metabolic acidosis (263). Additional studies recently reported by Schoolwerth et al (166) have indicated that activation of glutamate dehydrogenase also contributes substantially to the accelerated ammonia synthesis in acidosis.

Further studies are needed to determine the regulatory factors of ammoniagenesis in acidosis. Whether this process is tightly regulated or regulated at all by mitochondrial transport processes, will require further study. As mentioned previously, glutamate/H$^+$ transport is very slow in kidney mitochondria. However, the mechanism of glutamate efflux has not been clearly elucidated. Since glutamate is a potent end-product inhibitor of glutaminase, it is possible that efflux of glutamate from mitochondria plays an important role. Additionally, the studies of Goldstein & Boylan (171,

264) have indicated that α-ketoglutarate at physiological concentrations inhibits uptake and deamidation of glutamine. Whether this effect is exerted at the level of glutamine transport or at another step is not clearly established at this time.

Possible Hormonal Controls of Metabolite Transport

Hormones that stimulate or inhibit fluxes in metabolic pathways have not been shown to affect specific mitochondrial carriers or even specific mitochondrial enzymes, despite the fact that mitochondrial proteins are phosphorylated in response to hormonal activity (265). Several studies recently reported have suggested that glucagon may exert a role in the control of mitochondrial metabolism.

Glucagon stimulates gluconeogenesis from pyruvate and lactate at an intermediate point between pyruvate and phosphoenolpyruvate. This effect of glucagon was suggested to be due to an enhanced rate of pyruvate transport (266), a proposal for which data have recently been reported (121, 234). The effect of glucagon has been clarified by Halestrap (267, 268), who previously (123) had been unable to demonstrate an effect of glucagon on the pyruvate transporter. He now indicates that glucagon stimulates the respiratory chain between cytochromes c_1 and c. Under metabolizing conditions, this leads to an increase in $\Delta\mu_H^+$ and ATP concentrations as well as an increase in ΔpH. The latter increase in matrix pH under metabolizing conditions explains the stimulation of pyruvate uptake caused by glucagon. This important stimulation of a metabolic pathway by matrix alkalinization emphasizes the role played by proton translocation in regulating the fluxes of metabolites across the membrane.

SUMMARY

It can be concluded from the studies discussed in this review that an important aspect of the specificity of the mitochondrial carriers is their ability to bind and transport protons with certain metabolites, but not with others. Proton binding provides a means by which the carriers can utilize the large electrochemical potential gradient of protons as an energy source and to give directionality to the transport processes.

Rapid progress in this field of research has been made in the last decade due to technical advances that make it possible to measure transport of protons and metabolites on a nanomolar scale through a very small cellular compartment, and to theoretical advances following the suggestions of the chemiosmotic hypothesis of oxidative phosphorylation. Technical problems, however, still present real barriers to progress, especially with regard to detailed kinetic analysis of the different carriers. Research is needed to improve the time resolution of measurements and to provide more accurate means of controlling the intramitochondrial pH and metabolite levels. Probably the most important area for future research lies in the investigation of the role the carriers play in metabolic regulation. Here also, technical advances in rapid fractionation must precede the performance of definitive experiments.

Literature Cited

1. Nicholls, D. 1977. *Eur. J. Biochem.* 77:349–56
2. Himms-Hagen, J. 1976. *Ann. Rev. Physiol.* 38:315–51
3. Flatmark, T., Pedersen, J. I. 1975. *Biochim. Biophys. Acta* 416:53–103
4. Lee, N. H., Shapiro, I. M. 1978. *J. Membr. Biol.* 41:349–60
5. Malmstrom, K., Carafoli, E. 1977. *Arch. Biochem. Biophys.* 182:657–66
6. Lehninger, A. L. 1970. *Biochem. J.* 119:129–38
7. Meijer, A. J., Gimpel, J. A., Deleeuw, G. A., Tager, J. M., Williamson, J. R. 1975. *J. Biol. Chem.* 250:7728–38
8. Williamson, J. R. 1976. In *Use of Isolated Liver Cells and Kidney Tubules in Metabolic Studies*, ed. J. M. Tager, H. D. Soling, J. R. Williamson, pp. 79–95. Amsterdam: North Holland
9. Werkeiser, W. C., Bartley, W. 1957. *Biochem. J.* 66:79–91
10. Pfaff, E. 1965. *Unspezifiche Permeabilitat und spezificher Austauche der adeninnukleotidesals Beispiel mitochondrialer Kompartmentierung* PhD thesis. Univ. of Marburg, Marburg, Germany
11. Pfaff, E., Klingenberg, M., Ritt, E., Vogell, W. 1968. *Eur. J. Biochem.* 5: 222–32
12. Parsons, D. F., Yano, Y. 1967. *Biochim. Biophys. Acta* 135:362–64
13. Chappell, J. B. 1968. *Br. Med. Bull.* 24:150–57
14. Chappell, J. B., Haarhoff, K. M. 1966. In *Biochemistry of Mitochondria*, ed. E. C. Slater, Z. Kaniuga, L. Wojtczak, pp. 75–92. New York: Academic
15. Mitchell, P. 1966. *Chemiosmotic Coupling in Oxidative and Photosynthetic Phosphorylation*, Res. Rep. No. 66/1, May 1966, Bodwin, Cornwall, England: Glynn Res. Ltd.
16. Nicholls, D. G. 1974. *Eur. J. Biochem.* 50:305–15
17. Padan, E., Rottenberg, H. 1973. *Eur. J. Biochem.* 40:431–37
18. Palmieri, F., Quagliariello, E. 1969. *Eur. J. Biochem.* 8:473–81
19. LaNoue, K. F., Tischler, M. E. 1976. In *Mitochondria*, ed. L. Packer, A. Gomez-Puyou, pp. 61–68. New York: Academic
20. Halestrap, A. P. 1978. *Biochem. J.* 172:377–87
21. Lauquin, G. J. M., Villiers, C., Michejda, J., Brandolin, G., Boulay, F., Cesarini, R., Vignais, P. V. 1978. In *The Proton and Calcium Pumps.* ed. G. F. Azzoni, M. Avron, J. C. Metcalfe, E. Quagliariello, N. Siliprandi, pp. 251–62. Amsterdam: Elsevier/North-Holland Biomed. Press
22. Klingenberg, M., Riccio, P., Aquila, H., Schmiedt, B., Grebe, K., Topitsch, P. 1974. In *Membrane Proteins in Transport and Phosphorylation.* ed. G. F. Azzoni, M. Avron, J. C. Metcalfe, E. Quagliariello, N. Silliprandi, pp. 229–43. Amsterdam: Elsevier/North-Holland Biomed. Press
23. Brandolin, G., Meyer, C., DeFaye, G., Vignais, P. M., Vignais, P. V. 1974. *FEBS Lett.* 46:149–53
24. Williamson, J. R. 1976. In *Gluconeogenesis*, ed. R. W. Hanson, M. S. Mehlman, pp. 165–220. New York: Wiley-Interscience
25. Akerboom, T. P. M., Bookelman, H., Tager, J. M. 1977. *FEBS Lett.* 74:50–54
26. Stubbs, M., Vignais, P. V., Krebs, H. A. 1978. *Biochem. J.* 172:333–42
27. Soboll, S., Scholz, R., Heldt, H. W. 1978. *Eur. J. Biochem.* 87:377–90
28. Meijer, A. J., Van Dam, K. 1974. *Biochim. Biophys. Acta* 346:213–44
29. Palmieri, F., Quagliariello, E. 1978. In *Bioenergetics at Mitochondrial and Cellular Levels*, ed. L. Wojtczak, E. Lenartowicz, J. Zborowski, pp. 5–38. Warsaw: Nencki Inst. Exp. Biol.
30. Fonyo, A., Palmieri, F., Quagliariello, E. 1976. In *Horizons in Biochemistry and Biophysics*, eds. E. Quagliariello, F. Palmieri, T. P. Singer, 2:60–105. Addisons-Wesley, Reading, Mass.
31. Riccio, P., Aquila, H., Klingenberg, M. 1975. *FEBS Lett.* 56:129–32

32. Van Steveninck, J., Weed, R. I., Rothstein, A. 1965. *J. Gen. Physiol.* 48:617–32
33. Palmieri, F., Klingenberg, M. 1979. *Methods Enzymol.* 56: In press
34. Bartley, W., Davies, R. E. 1954. *Biochem. J.* 57:37–49
35. Price, C. A., Fonnesu, A., Davies, R. E. 1957. *Biochem. J.* 64:754–68
36. Pressman, B. C. 1968. *Fed. Proc.* 27:1283–88
37. Mitchell, P., Moyle, J. 1969. *Eur. J. Biochem.* 9:149–55
38. Selwyn, M. J., Walker, H. A. 1977. *Biochem. J.* 166:137–39
39. LaNoue, K. F., Walajtys, E. I., Williamson, J. R. 1973. *J. Biol. Chem.* 248:7171–83
40. Quagliariello, E., Palmieri, F., Prezioso, G., Klingenberg, M. 1969. *FEBS Lett.* 4:251–54
41. Klingenberg, M., Pfaff, E., Kroger, A. 1964. In *Rapid Mixing and Sampling Techniques in Biochemistry*, ed. B. Chance. New York: Academic p. 333
42. LaNoue, K. F., Bryla, J., Williamson, J. R. 1972. *J. Biol. Chem.* 247:667–79
43. Stucki, J. W., Brawand, F., Walter, P. 1972. *Eur. J. Biochem.* 27:181–91
44. Pfaff, E., Klingenberg, M. 1968. *Eur. J. Biochem.* 6:66–79
45. Nohl, H., Klingenberg, M. 1978. *Biochim. Biophys. Acta.* 503:155–69
46. Kovacevic, Z., Morris, H. P. 1972. *Cancer Res.* 32:326–33
47. Fonyo, A., Ligeti, E., Palmieri, F., Quagliariello, E. 1975. In *Biomembranes, Structure and Function*, ed. G. Gardos, I. Szasz, pp. 287–306. Hungary: Akad. Kiado and Amsterdam: North-Holland
48. Papa, S., Francavilla, A., Paradies, G., Meduri, B. 1971. *FEBS Lett.* 12:285–88
49. Pande, S. 1975. *Proc. Natl. Acad. Sci. USA* 72:883–87
50. Klingenberg, M. 1976. In *The Enzymes of Biological Membranes: Membrane Transport*, ed. A. N. Martonosi, 3:383–438. New York: Plenum
51. Vignais, P. V. 1976. *Biochim. Biophys. Acta* 456:1–38
52. Eiermann, W., Aquila, H., Klingenberg, M. 1976. *FEBS Lett.* 74:209–14
53. LaNoue, K. F., Williamson, J. R. 1971. *Metabolism* 20:119–40
54. Purvis, J. L., Lowenstein, J. M. 1961. *J. Biol. Chem.* 236:2794–803
55. Whereat, A. F., Orishimo, M. W., Nelson, J., Phillips, S. J. 1969. *J. Biol. Chem.* 244:6498–506
56. Safer, B., Smith, C. M., Williamson, J. R. 1971. *J. Mol. Cell. Cardiol.* 2:111–24
57. Rognstad, R., Katz, J. 1970. *Biochem. J.* 116:483–89
58. Williamson, J. R., Jakob, A., Refino, C. 1971. *J. Biol. Chem.* 246:7632–41
59. Sluse, F. E., Meijer, A. J., Tager, J. M. 1971. *FEBS Lett.* 18:149–53
60. Robinson, B. H., Chappell, J. B. 1967. *Biochem. Biophys. Res. Commun.* 28:249–55
61. Robinson, B. H., Oei, J. 1975. *Can J. Biochem.* 53:643–47
62. Azzi, A., Chappell, J. B., Robinson, B. H. 1967. *Biochem. Biophys. Res. Commun.* 29:148–52
63. Crompton, M., Chappell, J. B. 1973. *Biochem. J.* 132:35–46
64. Brand, M. D., Chappell, J. B. 1974. *Biochem. J.* 140:205–10
65. Hoek, J. B., Njogu, R. M. 1976. *FEBS Lett.* 71:341–46
66. Debise, R., Briand, Y., Durand, R., Gachon, P., Jeminet, G. 1977. *Biochimie.* 59:497–508
67. Klingenberg, M. 1970. *Essays Biochem.* 6:119–59
68. Papa, S., Lofrumento, N. E., Kanduc, D., Paradies, G., Quagliariello, E. 1971. *Eur. J. Biochem.* 22:134–43
69. Henderson, P. J. F., McGivan, J. D., Chappell, J. B. 1969. *Biochem. J.* 111:521–35
70. Palmieri, F., Genchi, G., Quagliariello, E. 1971. *Experientia Suppl.* 18:505–12
71. Moore, C., Pressman, B. C. 1964. *Biochem. Biophys. Res. Commun.* 15:562–67
72. Mitchell, P., Moyle, J. 1969. *Eur. J. Biochem.* 7:471–84
73. Quagliariello, E., Palmieri, F. 1970. *FEBS Lett.* 8:105–08
74. Cockrell, R. S., Harris, E. J., Pressman, B. C. 1967. *Nature* 215:1487–88
75. Racker, E. 1976. In *A New Look At Mechanisms in Bioenergetics*, pp. 60–61 New York: Academic
76. Skulachev, V. P. 1971. *Curr. Top. Bioenerg.* 4:127–90
77. Harris, E. J., Hofer, M. P., Pressman, B. C. 1967. *Biochem.* 6:1348–60
78. Harris, E. J., Manger, J. R. 1968. *Biochem. J.* 109:239–46
79. Quagliariello, E., Palmieri, F. 1967. *Eur. J. Biochem.* 4:20–27
80. Palmieri, F., Quagliariello, E., Klingenberg, M. 1970. *Eur. J. Biochem.* 17:230–38
81. McGivan, J. D., Klingenberg, M. 1971. *Eur. J. Biochem.* 20:392–99
82. Meijer, A. J., Tager, J. M. 1969. *Biochim. Biophys. Acta* 189:136–39
83. Palmieri, F., Quagliariello, E. 1968. In *Mitochondria Structure and Function.*

FEBS Meet. 5th Prague. Czechoslovak Biochem. Soc. (Abstr. 532)
84. Papa, S., Lofrumento, N. E., Quagliariello, E., Meijer, A. J., Tager, J. M. 1970. *J. Bioenerg.* 1:287–307
85. Robinson, B. H., Williams, G. R., Halperin, M. L., Leznoff, C. C. 1971. *J. Biol. Chem.* 246:5280–86
86. Weiner, M. W. 1975. *Am. J. Physiol.* 228:122–26
87. LaNoue, K., Mizani, S. M., Klingenberg, M. 1978. *J. Biol. Chem.* 253: 191–98
88. Klingenberg, M., Rottenberg, H. 1977. *Eur. J. Biochem.* 73:125–30
89. Heldt, H. W., Klingenberg, M., Milovancev, M. 1972. *Eur. J. Biochem.* 30:434–40
90. LaNoue, K. F., Bryla, J., Bassett, D. J. P. 1974. *J. Biol. Chem.* 249:7514–21
91. LaNoue, K. F., Tischler, M. E. 1974. *J. Biol. Chem.* 249:7522–28
92. LaNoue, K. F., Meijer, A. J., Brouwer, A. 1974. *Arch. Biochem. Biophys.* 161: 544–50
93. Gamble, J. G., Lehninger, A. L. 1973. *J. Biol. Chem.* 248:610–18
94. Curthoys, N. P., Shapiro, R. A. 1978. *J. Biol. Chem.* 253:63–68
95. McGivan, J. D., Bradford, N. M., Beavis, A. D. 1977. *Biochem. J.* 162:147–56
96. Pande, S. V., Parvin, R. 1976. *J. Biol. Chem.* 251:6683–91
97. Cybulski, R. L., Fisher, R. R. 1977. *Biochemistry* 16:5116–20
98. LaNoue, K., Duszynski, J. 1978. See Ref. 21, pp. 297–307
99. Meisner, H., Palmieri, F., Quagliariello, E. 1972. *Biochemistry* 11:949–55
100. Passarella, S., Palmieri, F., Quagliariello, E. 1973. *FEBS Lett.* 38:91–5
101. Freitag, H., Kadenbach, B. 1978. *Eur. J. Biochem.* 83:53–7
102. Kadenbach, B., Freitag, H., Kolbe, H. 1978. *FEBS Lett.* 89:161–64
103. Coty, W. A., Pedersen, P. L. 1974. *J. Biol. Chem.* 249:2593–98
104. Fonyo, A., Palmieri, F., Ritvay, J., Quagliariello, E. 1974. See Ref. 22, pp. 283–86
105. Azzoni, G. F., Massari, S., Pozzan, T. 1976. *Biochim. Biophys. Acta* 423:15–26
106. Wehrle, J. P., Antron, N. M., Pedersen, P. L. 1979. *J. Biol. Chem.* 253:8598–603
107. Banerjee, R. K., Shertzer, H. G., Kanner, B. I., Racker, E. 1977. *Biochem. Biophys. Res. Commun.* 75:772–78
108. Klingenberg, M., Aquila, H., Kramer, R., Babel, W., Feckl, J. 1977. In *Biochemistry of Membrane Transport,*

FEBS Symp. 42nd, pp. 567–79. Germany: Springer (Berlin)
109. Helenius, A., Simons, K. 1975. *Biochim. Biophys. Acta* 415:29–79
110. Coty, W. A. Pedersen, P. L. 1975. *J. Biol. Chem.* 250:3515–21
111. Hadvary, P., Kadenbach, B. 1976. *Eur. J. Biochem.* 67:573–81
112. Touraille, S., Briand, Y., Durand, R. 1977. *FEBS Lett.* 84:119–23
113. Guerin, B., Guerin, M., Napias, C., Rigoulet, M. 1977. *Biochem. Soc. Trans.* 5:504–6
114. Rigoulet, M., Guerin, M., Guerin, B. 1977. *Biochim. Biophys. Acta.* 471: 280–95
115. Zahlten, R. N., Hochberg, A. A., Stratman, F. W., Lardy, H. A. 1972. *FEBS Lett.* 21:11–13
116. Brouwer, A., Smits, G. G., Tas, J., Meijer, A. J., Tager, J. M. 1973. *Biochemie* 55:717–25
117. Halestrap, A. P., Denton, R. M. 1974. *Biochem. J.* 138:313–16
118. Halestrap, A. P. 1976. *Biochem. J.* 156:181–83
119. Papa, S., Paradies, G. 1974. *Eur. J. Biochem.* 49:265–74
120. Mowbry, J. 1975. *Biochem. J.* 148: 41–47
121. Titherage, M. A., Coore, H. G. 1976. *FEBS Lett.* 63:45–50
122. Pande, S. V., Parvin, R. 1978. *J. Biol. Chem.* 253:1563–73
123. Halestrap, A. P. 1975. *Biochem. J.* 148:85–96
124. Paradies, G. Papa, S. 1975. *FEBS Lett.* 52:149–52
125. Meijer, A. J., Brouwer, A., Reijngoud, D. J., Hoek, J. B., Tager, J. M. 1972. *Biochim. Biophys. Acta* 283:421–29
126. Meyer, J., Vignais, P. M. 1973. *Biochim. Biophys. Acta* 325:375–84
127. McGivan, J. D., Chappell, J. B. 1970. *Biochem. J.* 116:37P–38P
128. Palmieri, F., Genchi, G., Quagliariello, E. 1973. *Boll. Soc. Biol. Sper.* 49:270–76
129. Debise, R., Gachon, P., Durand, R. 1978. *FEBS Lett.* 85:25–29
130. Meijer, A. J. 1971. *Anion translocation in mitochondria.* PhD thesis. Univ. Amsterdam, Amsterdam, The Netherlands
131. Bradford, N. M., McGivan, J. D. 1973. *Biochem. J.* 134:1023–29
132. Gautheron, D. C., Julliard, J. H., Godenot, C. 1974. See Ref. 22, pp. 91–96
133. Julliard, J. H., Gautheron, D. C. 1976. In *Use of Isolated Liver Cells and Kidney Tubules in Metabolic Studies,* ed. J. M. Tager, H. D. Soling, J. R. William-

son, pp. 98–101. Amsterdam: North-Holland
134. Julliard, J. H., Gautheron, D. C. 1973. *FEBS Lett.* 37:10–16
135. Julliard, J. H., Gautheron, D. C. 1978. *Biochim. Biophys. Acta* 503:223–37
136. Chappell, J. B., Robinson, B. H. 1968. *Biochem. Soc. Symp.* 27:123
137. Harris, E. J. 1969. In *The Energy Level and Metabolic Control in Mitochondria,* ed. S. Papa, J. M. Tager, E. Quagliariello, E. C. Slater, pp. 31. Bari:Adriatica Editrice
138. Meijer, A. J., Tager, J. M., Van Dam, K. 1969. In *The Energy Level and Metabolic Control in Mitochondria.* ed. S. Papa, J. M. Tager, E. Quagliariello, E. C. Slater. Bari: Adriatica Editrice p. 147
139. Robinson, B. H. 1971. *FEBS Lett.* 16:267–71
140. Robinson, B. H. 1971. *FEBS Lett.* 14:309–12
141. Robinson, B. H., Williams, G. R., Halperin, M. L., Leznoff, C. C. 1971. *J. Biol. Chem.* 246:5280–86
142. Stucki, J. W. 1976. *FEBS Lett.* 61: 171–75
143. Harris, E. J., Berent, C. 1969. *Biochem. J.* 115:645–52
144. Halperin, M. L., Cheema-Dhadli, S., Taylor, W. M., Fritz, I. 1975. *Adv. Enz. Regul.* 13:435–45
145. Stipani, I., Francia, F., Palmieri, F., Quagliariello, E. 1977. *Bull. Mol. Biol. Med.* 2:72–79
146. Robinson, B. H., Williams, G. R., Halperin, M. L., Leznoff, C. C. 1971. *Eur. J. Biochem.* 20:65–71
147. Palmieri, F., Stipani, I., Quagliariello, E., Klingenberg, M. 1972. *Eur. J. Biochem.* 26:587–94
148. Palmieri, F., Genchi, G., Stipani, I., Quagliariello, E. 1977. In *Structure and Function of Energy Transducing Membranes,* ed. K. VanDam, B. F. van Gelder, pp. 251–60. Amsterdam: North-Holland Biomed. Press
149. Palmieri, F., Genchi, G., Stipani, I., Riccio, P., Quagliariello, E. 1977. *Biochem. Soc. Trans.* 5:527–31
150. Johnson, R. N., Chappell, J. B. 1973. *Biochem. J.* 134:769–74
151. DeHaan, E. J., Tager, J. M. 1968. *Biochim. Biophys. Acta.* 153:98–112
152. Palmieri, F., Quagliariello, E., Klingenberg, M. 1972. *Eur. J. Biochem.* 29: 408–16
153. Palmieri, F., Prezioso, G., Quagliariello, E., Klingenberg, M. 1971. *Eur. J. Biochem.* 22:66–74
154. Sluse, F. E., Meijer, A. J., Tager, J. M. 1971. *FEBS Lett.* 18:149–53

155. Sluse, F. E., Ranson, M., Liebecq, C. 1972. *Eur. J. Biochem.* 25:207–17
156. Sluse, F. E., Goffart, G., Liebecq, C. 1973. *Eur. J. Biochem.* 32:283–91
157. Cheema-Dhadli, S., Halperin, M. L. 1978. *Can. J. Biochem.* 56:23–28
158. Quagliariello, E., Palmieri, F. 1972. In *Biochemistry and Biophysics of Mitochondrial Membranes,* pp. 659–80. New York: Academic
159. Palmieri, F., Passarella, S., Stipani, I., Quagliariello, E. 1974. *Biochim. Biophys. Acta* 333:195–208
160. Meijer, A. J., Van Woerkom, G. M., Eggelte, T. A. 1976. *Biochim. Biophys. Acta* 430:53–61
161. Gimpel, J. A., DeHaan, E. J., Tager, J. M. 1973. *Biochim. Biophys. Acta* 292: 582–91
162. Passarella, S., Palmieri, F., Quagliariello, E. 1977. *Arch. Biochem. Biophys.* 180:160–68
163. Passarella, S., Palmieri, F., Quagliariello, E. 1977. In *Bioenergetics of Membranes,* ed. L. Packer, G. C. Papageorgiou, A. Trebst, pp. 425–34. Netherlands: Elsevier (Amsterdam) and Netherlands: North-Holland Biomed. Press
164. Kovacevic, Z., McGivan, J. D., Chappell, J. B. 1970. *Biochem. J.* 118:265–74
165. Adam, W., Simpson, D. P. 1974. *J. Clin. Invest.* 54:165–74
166. Schoolwerth, A. C., Nazar, B. L., LaNoue, K. F. 1978. *J. Biol. Chem.* 253:6177–83
167. Kovacevic, Z. 1971. *Biochem. J.* 125: 757–63
168. Kovacevic, Z. 1975. *Biochim. Biophys. Acta.* 396:325–34
169. Brosnan, J. T., Hall, B. 1977. *Biochem. J.* 164:331–37
170. Curthoys, N. P., Shapiro, R. A. 1978. *Int. Workshop Ammoniagenesis,* 1st, *Mont Gabriel, Quebec, Canada*
171. Goldstein, L., Boylan, J. M. 1978. *Am. J. Physiol.* 234:F514–21
172. Fritz, I. B., Yue, K. T. N. 1963. *J. Lipid Res.* 4:279–88
173. Yates, D. W., Garland, P. B. 1970. *Biochem. J.* 119:547–52
174. Pande, S. V. 1975. *Proc. Natl. Acad. Sci. USA* 72:883–87
175. Ramsay, R. R., Tubbs, P. K. 1975. *FEBS Lett.* 54:21–25
176. Ramsay, R. R., Tubbs, P. K. 1976. *Eur. J. Biochem.* 69:299–303
177. Hansford, R. G. 1978. *Biochem. J.* 170:285–95
178. Idell-Wenger, J. A., Grotyohann, L. W., Neely, J. R. 1978. *J. Biol. Chem.* 253:4310–18

179. Parvin, R., Pande, S. V. 1978. *J. Biol. Chem.* 253:1944–46
180. Levitsky, D. O., Skulachev, V. P. 1972. *Biochim. Biophys. Acta* 275:33–50
181. Osmundsen, H., Bremer, J. 1976. *FEBS Lett.* 69:221–24
182. King, M. J., Diwan, J. J. 1973. *Arch. Biochem. Biophys.* 159:166–73
183. McGivan, J. D., Bradford, N. M., Crompton, M., Chappell, J. B. 1973 *Biochem. J.* 134:209–15
184. Ward, W., Mortimore, G. E. 1978. *J. Biol. Chem.* 253:3581–87
185. Meyer, J. 1977. *Arch. Biochem. Biophys.* 178:387–95
186. DeHaan, E. J., Tager, J. M., Slater, E. C. 1967. *Biochim. Biophys. Acta.* 131:1–13
187. LaNoue, K. F., Williamson, J. R. 1971. *Metabolism* 20:119–40
188. Klingenberg, M., von Haefen, H., Wenske, G. 1965. *Biochem. Z.* 343:452–78
189. Tischler, M. E. 1977. PhD thesis. Univ. Pennsylvania, Philadelphia, Pa. p. 90
190. Palmieri, F., Stipani, I., Iacobazi, I. 1978. In *Frontiers of Biological Energetics,* Vol II, ed. P. L. Dutton, J. S. Leigh, A. Scarpa. pp. 1161–69, New York: Academic. In press
191. Tischler, M. E., Pachence, J., Williamson, J. R., LaNoue, K. F. Rottenberg, H. 1976. *Arch. Biochem. Biophys.* 173:448–62
192. Cleland, W. W. 1967. *Ann. Rev. Biochem.* 36:77–112
193. LaNoue, K. F., Zimmerman, U. J. P. 1974. *Fed. Proc.* 33:1360
194. Schoolwerth, A. C., Nazar, B. L., La-Noue, K. F. 1977. *Kidney Int.* 12:466
195. Pfaff, E., Klingenberg, M., Heldt, H. W. 1965. *Biochim. Biophys. Acta* 104:314–15
196. Pfaff, E., Klingenberg, M. 1968. *Eur. J. Biochem.* 6:66–79
197. Duee, E. D., Vignais, P. V. 1969. *J. Biol. Chem.* 244:3920–31
198. Duee, E. D., Vignais, P. V. 1968. *Biochem. Biophys. Res. Commun.* 30:420–27
199. Schlimme, E., Lamprecht, W., Eckstein, F., Goody, R. S. 1973. *Eur. J. Biochem.* 40:485–91
200. Davis, E. J., Lumeng, L. 1975. *J. Biol. Chem.* 250:2275–82
201. Vignais, P. V., Vignais, P. M., Lauquin, G., Morel, F. 1973. *Biochemie* 55:763–78
202. Vignais, P. V., Vignais, P. M., DeFaye, G. 1973. *Biochemistry* 12:1508–19
203. Henderson, P. J. F., Lardy, H. A. 1970. *J. Biol. Chem.* 245:1319–26
204. Klingenberg, M., Grebe, K., Heldt, H. W. 1970. *Biochem. Biophys. Res. Commun.* 39:344–51
205. Pande, S. V., Blanchaer, M. C. 1971. *J. Biol. Chem.* 246:402–11
206. Vignais, P. V., Vignais, P. M. 1972. *FEBS Lett.* 26:27–31
207. Souverijn, J. H. M., Huisman, L. A., Rosing, J., Kemp, A. Jr., 1973. *Biochim. Biophys. Acta* 305:185–98
208. Slater, E. C., Rosing, J., Mol, A. 1973. *Biochim. Biophys. Acta* 292:534–53
209. Laris, P. C. 1977. *Biochim. Biophys. Acta* 459:110–18
210. Wulf, R., Kaltstein, A., Klingenberg, M. 1978. *Eur. J. Biochem.* 82:585–92
211. Brand, M. D., Reynafarje, B., Lehninger, A. L. 1976. *Proc. Natl. Acad. Sci. USA* 73:437–41
212. Brand, M. D., Chen, C. H., Lehninger, A. L. 1976. *J. Biol. Chem.* 251:968–74
213. Brand, M. D., Reynafarje, B., Lehninger, A. L. 1976. *J. Biol. Chem.* 251:5670–79
214. Reynafarje, B., Brand, M. D., Lehninger, A. L. 1976. *J. Biol. Chem.* 251:7442–51
215. Lehninger, A. L., Reynafarje, B. 1978. *Proc. Natl. Acad. Sci. USA.* 75:4788–92
216. Vignais, P. V., Lauquin, G. J. M., Vignais, P. M. 1976. In *Mitochondria, Bioenergetics, Biogenesis, and Membrane Structure,* ed. L. Packer, A. Gomez-Puyou, pp. 109–25. New York: Academic
217. Pfaff, E., Heldt, H. W., Klingenberg, M. 1969. *Eur. J. Biochem.* 10:484–93
218. Duszynski, J., Savina, H. Z., Wojtczak, L. 1978. *FEBS Lett.* 86:9–13
219. Klingenberg, M. 1976. In *Myocardial Failure International Boehringer Symposium,* ed. G. Precker, A. Weber, J. Goodwin, pp. 153–61. Berlin/Heidelberg/New York: Springer
220. Bojanovski, D., Schlimme, E., Wang, C. S., Alaupovic, P. 1976. *Eur. J. Biochem.* 71:539–48
221. Klingenberg, M., Riccio, P., Aquila, H. 1978. *Biochim. Biophys. Acta* 503:193–210
222. Kramer, R., Aquila, H., Klingenberg, M. 1977. *Biochemistry* 16:4949–53
223. Kramer, R., Klingenberg, M. 1977. *FEBS Lett.* 82:363–67
224. Aquila, H., Eiermann, W., Babel, W., Klingenberg, M. 1978. *Eur. J. Biochem.* 85:549–60
225. Klingenberg, M., Buckholz, M. 1973. *Eur. J. Biochem.* 38:346–58
226. Buchanan, B. B., Eiermann, W., Riccio, P., Aquila, H., Klingenberg, M. 1976. *Proc. Natl. Acad. Sci. USA* 73:2280–84

227. Rolleston, F. S. 1972. *Curr. Top. Cell. Regul.* 5:47–75
228. Krebs, H. A., Veech, R. L. 1969. *Adv. Enzymol. Regul.* 7:397–413
229. Williamson, R. R. 1976. In *Gluconeogenesis,* ed. R. W. Hanson, M. A. Mehlman, pp. 194–204. New York: Wiley
230. Elbers, R., Heldt, H. W., Schmucker, P., Soboll, S., Wiese, H. 1974. *Hoppe-Seylers Z. Physiol. Chem.* 355:378–93
231. Zuurendonk, P. F., Tager, J. M. 1974. *Biochim. Biophys. Acta.* 333:393–99
232. Tischler, M. E., Hecht, P., Williamson, J. R. 1977. *Arch. Biochem. Biophys.* 181:273–92
233. Akerboom, T. P. M., Bookelman, H., Zuurendonk, P. F., van der Meer, R., Tager, J. M. 1978. *Eur. J. Biochem.* 84:413–20
234. Titherage, M. A., Coore, H. G. 1976. *FEBS Lett.* 71:73–78
235. Pitts, R. F. 1972. *Kidney Int.* 1:297–305
236. Lardy, H. A. 1956. In *Proc. Int. Congr. Biochem., 3rd, Brussels, 1955,* pp. 287–94. New York: Academic
237. Chance, B., Williams, G. R. 1956. *Adv. Enzymol.* 17:65–134
238. Klingenberg, M. 1969. See Ref. 138, pp. 189–93
239. Klingenberg, M. 1968. In *Biological Oxidations,* ed. T. P. Singer, pp. 3–54. New York: Wiley
240. Slater, E. C. 1969. See Ref. 138, pp. 255–59
241. Heldt, H. W. 1966. In *Regulation of Metabolic Processes in Mitochondria,* ed. J. M. Tager, S. Papa, E. Quagliariello, E. C. Slater, p. 51. Netherlands: Elsevier (Amsterdam)
242. Heldt, H. W., 1967. In *Mitochondrial Structure and Compartmentation,* ed. E. Quagliariello, S. Papa, E. C. Slater, J. M. Tager, pp. 60–267. Bari, Italy: Adriatica Editrice
243. Heldt, H. W., Klingenberg, M. 1968. *Eur. J. Biochem.* 4:1–8
244. Kemp, A. Jr., Groot, G. S. P., Reitsma, H. J. 1969. *Biochim. Biophys. Acta* 180:28–34
245. Wilson, D. F., Stubbs, M., Veech, R. L., Erecinska, M., Krebs, H. A. 1974. *Biochem. J.* 140:57–64
246. Wilson, D. F., Stubbs, M., Oshino, N., Erecinska, M. 1974. *Biochemistry* 13:5305–11
247. Stubbs, M., Vignais, P. V., Krebs, H. A. 1978. *Biochem. J.* 172:333–42
248. Siess, E. A., Wieland, O. H. 1976. *Biochem. J.* 156:91–92
249. Erecinska, M., Stubbs, M., Miyata, Y., Ditre, C. M., Wilson, D. F. 1977. *Biochim. Biophys. Acta.* 462:20–35
250. Chance, B. 1958. *Ciba Found. Symp.* pp. 91–121
251. Lee, C. P., Ernster, L. 1968. *Eur. J. Biochem.* 3:385–390
252. Kayalar, C., Rosing, J., Boyer, P. D. 1976. *Biochem. Biophys. Res. Commun.* 72:1153–59
253. Holian, A., Owen, C. S., Wilson, D. F. 1977. *Arch. Biochem. Biophys.* 181:164–71
254. Davis, E. J., Davis-van Thienen, W. I. A. 1978. *Biochem. Biophys. Res. Commun.* 83:1260–66
255. Küster, U., Bohnensack, R., Kunz, W. 1976. *Biochim. Biophys. Acta.* 440:391–402
256. Soboll, S., Scholz, R., Heldt, H. W. 1978. *Eur. J. Biochem.* 87:377–90
257. van der Meer, R., Akerboom, T. P. M., Groen, A. K., Tager, J. M. 1978. *Eur. J. Biochem.* 84:421–28
258. Simpson, D. P., Adam, W. 1975. *J. Biol. Chem.* 250:8148–58
259. Simpson, D. P. 1975. *Med. Clin. North Am.* 59:555–67
260. Goldstein, L. 1975. *Am. J. Physiol.* 229:1027–33
261. Tannen, R. L., Kunin, A. S. 1976. *Am. J. Physiol.* 231:1631–37
262. Shapiro, R. A., Clark, V. M., Curthoys, N. P. 1978. *J. Biol. Chem.* 253:7086–90
263. Shapiro, R. A., Curthoys, N. P. 1978. *FEBS Lett.* 91:49–52
264. Goldstein, L. 1976. *Biochem. Biophys. Res. Commun.* 70:1136–41
265. Zahlten, R. N., Hochberg, A. A., Stratman, F. W., Lardy, H. A. 1972. *Proc. Natl. Acad. Sci. USA* 69:800–4
266. Adam, P. A. J., Haynes, R. C. 1969. *J. Biol. Chem.* 244:6444–50
267. Halestrap, A. P. 1978. *Biochem. J.* 172:389–98
268. Halestrap, A. P. 1978. *Biochem. J.* 172:399–405

Ann. Rev. Biochem. 1979. 48:923–59

PHOSPHORYLATION-DEPHOSPHORYLATION OF ENZYMES

◆12029

Edwin G. Krebs and Joseph A. Beavo

Laboratory of Molecular Pharmacology, Howard Hughes Medical Institute, and the Department of Pharmacology, SJ-30, University of Washington, Seattle, Washington 98195

CONTENTS

923

0066-4154/79/0701-0923$01.00

PERSPECTIVES AND SUMMARY

Recognition of the reversible covalent modification of proteins as a major regulatory process was a result of work on glycogen metabolism. Glycogen phosphorylase was shown to exist in two interconvertible species, now known to be nonphosphorylated and phosphorylated forms of the enzyme. Since elucidation of this first example, other reversible covalent protein modifications have been described including acetylation-deacetylation, adenylylation-deadenylylation, uridylylation-deuridylylation, and methylation-demethylation. The number of enzymes known to undergo phosphorylation-dephosphorylation has risen to more than twenty, and the scope of this particular type of regulatory mechanism has broadened to include many nonenzymic proteins. The present review is restricted to the phosphorylation and dephosphorylation of enzymes.

For the phosphorylation and dephosphorylation of enzymes to serve a regulatory function, these processes must in turn be regulated. This requires that the protein kinases and/or the phosphoprotein phosphatases be controlled. The kinases are often regulated by specific effector compounds or messengers. Thus there exist cAMP-dependent protein kinases, calcium-dependent protein kinases, a cGMP-dependent protein kinase, and a double-stranded RNA-dependent protein kinase. The phosphoprotein phosphatases are apparently regulated, not through their direct interaction with specific effectors, but instead are often controlled by "substrate-directed effects" in which a given metabolite or metal ion combines with the phosphoenzyme substrate causing it to become a "better" or "poorer" substrate for the phosphatase. The phosphoprotein phosphatases are also regulated as a result of protein-protein interactions. Finally, with respect to phosphoprotein phosphatase reactions involving substrates having multiple phosphorylation sites, it is possible that phosphorylation at one site may regulate the dephosphorylation of another.

INTRODUCTION

It has been estimated that proteins undergo at least 125 different kinds of post-translational covalent modification involving derivatization of their individual amino acid residues (1). However, within this broad spectrum of modification reactions, there are only a few processes that are recognized as being readily reversible. Excluding the formation of transient enzyme-substrate complexes involving covalent bonding or other covalent modifications that occur as a part of reaction mechanisms, one can recognize no more than five or six types of reversible covalent protein modifications: (a) phosphorylation-dephosphorylation, as first elucidated for glycogen phosphorylase (2, 3), (b) acetylation-deacetylation (4), (c) adenylylation-deadenylylation (5, 6), (d) uridylylation-deuridylylation (7), (e) methylation-demethylation (8), and (f) S-S/SH interconversions. The reversible covalent modifications are viewed as regulatory in nature and share certain properties. The present review, however, is concerned only with the phosphorylation-dephosphorylation of proteins; and within this group emphasis will be placed on the phosphorylation and dephosphorylation of enzymes.

The reactions involved in the enzyme-catalyzed phosphorylation and dephosphorylation of proteins are shown in Equations 1 and 2.

$$\text{Protein} + n\text{NTP} \xrightleftharpoons[\quad]{\substack{\text{protein} \\ \text{kinase}}} \text{Protein--P}_n + n\text{NDP} \qquad 1.$$

$$\text{Protein--P}_n + n\text{H}_2\text{O} \xrightleftharpoons[\quad]{\substack{\text{phosphoprotein} \\ \text{phosphatase}}} \text{Protein} + n\text{P}_i \qquad 2.$$

In general, NTP is ATP, but at least one protein kinase is known in which GTP is almost as effective as ATP. This latter enzyme was first described by Rodnight & Lavin (9) as a brain phosvitin kinase and later emerged as a reticulocyte casein kinase (10) and as a muscle glycogen synthase kinase (11). The amino acid residue(s) to which the phosphoryl group is transferred in reactions of the type shown in Equation 1 is usually a serine or a threonine, but instances have been reported in which protein kinases catalyzed the transfer of the phosphoryl group from ATP to histidine or lysine residues in proteins (12).

In addition to the dephosphorylation of phosphoproteins catalyzed by phosphoprotein phosphatases (Equation 2), it has also been shown that protein kinase reactions themselves (Equation 1) are reversible. This was first seen by Rabinowitz & Lipmann (13) and also by Lerch et al (14) who were studying a yeast protein kinase that catalyzes the phosphorylation of phosvitin and casein. Shizuta et al (15) measured the equilibrium position for the cyclic AMP-dependent protein kinase–catalyzed phosphorylation of casein by ATP and obtained a k_{eq} value of 24. The standard free energy of hydrolysis of protein-bound phosphate in casein was calculated to be –6.5 kcal mol^{-1}. Shizuta et al (16) reexamined the question of the reversibility of the phosphorylase kinase reaction and showed that this reaction is also reversible but only when measured in the presence of glucose, an allosteric effector of phosphorylase (see section on glycogen phosphorylase). Rosen & Erlichman (17) found that autophosphorylation of type II cyclic AMP-dependent protein kinase, i.e. phosphorylation of the regulatory subunit, is reversible. The possible physiological significance of the dephosphorylation of phosphoproteins by direct reversal of protein kinase–catalyzed reactions is unknown.

Reactions of the type shown in Equations 1 and 2 occur in all types of eukaryotic cells that have been examined but until very recently were not thought to occur in prokaryotes except after phage infection (18). Wang & Koshland have recently reported, however, that *Salmonella typhimurium* contains one or more protein kinases capable of catalyzing the phosphorylation of endogenous proteins (19). Bacterial proteins can often be phosphorylated by animal tissue protein kinases in vitro (20). Such protein phosphorylation reactions are probably of little physiological significance but do provide a way to mark proteins with radioactive phosphorus (21).

For protein phosphorylation-dephosphorylation reactions to function in regulation, it is apparent that appropriate signals should bring about changes in the relative concentrations of the phosphorylated and nonphosphorylated forms of the protein substrates. From Equations 1 and 2 it can be seen that this could occur through control of the protein kinase step, the phosphoprotein phosphatase step, or through the simultaneous regulation of both reactions. These controls could involve rapid or immediate responses, as might occur due to fluctuations in the levels of effector molecules, or they might be mediated by adaptive changes that alter the ratio of protein kinase to phosphoprotein phosphatase within the cell. The present review is restricted to the first type of regulation, but the reader's attention is called to significant work that has been done on the regulation of protein kinase and phosphoprotein phosphatase levels (22–28).

PROTEIN KINASES

The protein kinases (and the phosphoprotein phosphatases) are reviewed primarily from the standpoint of their regulation and specificity. The reader is referred to earlier reviews (29–34) for a coverage of their isolation, physicochemical properties, and for kinetic studies. Investigators have paid more attention to the protein kinases and their control than to the phosphatases. This may be due to the fact that the kinases are easier to purify than the phosphatases. Alternatively, it is possible that dynamic control is exerted primarily at the kinase step and investigators may have bypassed the phosphatases in favor of studying the more dynamic kinases.

Classification

A number of protein kinases are regulated through their direct interaction with specific regulatory agents, which serve as messengers in relaying signals from outside the cell, and a classification scheme based on these agents has been developed (Table 1). The cAMP-dependent protein kinase, recognized as a distinct multifunctional enzyme in 1968 (37), was the first protein kinase named in this manner. This was followed by the cGMP-dependent protein kinase originally detected in lobster muscle (38) but now studied in several mammalian tissues (39–44). The third set of enzymes classified in this manner were the Ca^{2+}-dependent protein kinases (45). The most recent protein kinase to join the family of enzymes regulated by specific agents is the double-stranded RNA-dependent protein kinase, found in cells treated with interferon and also detected in erythrocyte lysates (46–50). The nonspecified or "messenger-independent" protein kinases (Category 5 in Table 1) constitute a group of enzymes for which no direct acting effectors are known. These enzymes are usually named according to the substrates on which they act. It is probable that many of the protein kinases in the nonspecified group will be found to be the targets for specific regulators in the future.

The Ca^{2+}-dependent protein kinases deserve special comment because this category contains enzymes having different specificities and because an alternative, albeit related, name can be used for them. Phosphorylase kinase was recognized as having a Ca^{2+} requirement in 1964 (51). Later, this

Table 1 Classification of protein kinases based on regulation by specific agents

Category	Designation	Recognized entities
1	cAMP-dependent protein kinases	Type I and Type II[a]
2	cGMP-dependent protein kinases	one known entity
3	Ca^{2+}-dependent protein kinases	phosphorylase kinase and myosin light chain kinase
4	double-stranded RNA-dependent	one known entity
5	nonspecified or messenger-independent protein kinases	many known examples

[a] The two types of cAMP-dependent protein kinase differ in the nature of their regulatory subunits but have identical catalytic subunits (35, 36).

requirement was quantified, and it was shown that stimulation of the enzyme by Ca^{2+} occurs within the physiological range of metal ion concentrations (52). The designation of phosphorylase kinase as a Ca^{2+}-dependent protein kinase occurred after it had been recognized as a potentially multifunctional enzyme (45). Recently it was shown (53, 54) that myosin light chain kinase, a Ca^{2+}-dependent protein kinase with a different specificity (55), acquires its Ca^{2+} dependency by virtue of its interaction with the Ca^{2+}-dependent modulator of cyclic nucleotide phosphodiesterase (56). Waisman et al (57) have shown that the light chain kinase is multifunctional and also catalyzes the phosphorylation of phosphorylase kinase and histone. They have proposed that it be referred to as modulator-dependent protein kinase. The same designation would also be applicable to phosphorylase kinase in that this enzyme may owe its Ca^{2+}-dependency to the presence of the modulator protein (58). It is probable that additional Ca^{2+} (modulator)-dependent protein kinases will be described; e.g. a brain membrane protein kinase studied by Schulman & Greengard appears to be in this category (58a).

Regulation

cAMP-DEPENDENT PROTEIN KINASE

Regulation by cAMP   It is generally accepted (29–34) that both Types I and II cAMP-dependent protein kinases are activated as follows:

$$R_2C_2 \; + \; 2cAMP \; \rightleftharpoons \; R_2(cAMP)_2 \; + \; 2C \qquad\qquad 3.$$
$$\begin{array}{lll} \text{(inactive} & & \text{(active} \\ \text{holoenzyme} & & \text{catalytic} \\ \text{form)} & & \text{subunit)} \end{array}$$

R represents regulatory and C catalytic subunits respectively. Ogez & Segel (59) and Swillens & Dumont (60) have discussed the potential pathways for this overall reaction. The first authors favored models in which C dissociates from the holoenzyme prior to the binding of cAMP to R, whereas the second group thought it more likely that dissociation of the subunits occurs subsequent to the binding of cAMP to R_2C_2. Unpublished data from this laboratory (S. E. Builder, J. A. Beavo, and E. G. Krebs) and the work of Chau et al (61) support the latter concept. A ternary complex of R, C, and cAMP may be enzymatically active. The *raison d'être* for Types I and II cAMP-dependent protein kinase has not been elucidated. Since their substrate specificities are the same, it is probable that their significance may lie in their respective locations within the cell and/or their differing responses to cAMP. In cardiac muscle, it is known that a significant fraction of the Type II enzyme is membrane-bound (62, 63). The responsiveness of Type

II protein kinase to cAMP is altered by variation in the state of phosphorylation of its regulatory subunit (64), a property not shared by the Type I enzyme. The direction of this effect is such that phosphorylation of the Type II enzyme enhances its response to cAMP with respect to binding (36) and dissociation (64). It is of interest that the interaction of the cAMP-dependent protein kinase with heme alters its response to cAMP (65).

Regulation by protein inhibitor(s)   Tissues contain heat-stable protein inhibitors of the cAMP-dependent protein kinase. The most extensively studied inhibitor is that obtained from rabbit skeletal muscle (66), which has been purified to homogeneity (67). The homogeneous protein was found to have a molecular weight of 11,300 (67), less than half the value reported originally by Walsh et al (66) and approximately half the value determined for the inhibitor from bovine heart (68). The reason for this discrepancy is unknown but could reflect the state of aggregation. The inhibitor binds to the catalytic subunit of the cAMP-dependent protein kinase with a $K_i \simeq 2 \times 10^{-9}$ M and is competitive with respect to protein substrates for the kinase (67). The function of the inhibitor remains uncertain. Even in a tissue such as muscle in which it is relatively abundant, its concentration is such that only 20% of the cAMP-dependent protein kinase could be blocked (69). Thus, although the inhibitor may act to inhibit any free catalytic subunit present under basal conditions, and thereby nullify the effects of low concentrations of cAMP (70), it could not be much of a factor at high levels of the cyclic nucleotide. Several reports indicate that the concentration of the inhibitor in tissues may change under different physiological conditions (69, 71, 72). The inhibitor continues to be a useful tool for distinguishing the cAMP-dependent protein kinase from other protein kinases (73). A second inhibitor of cAMP-dependent protein kinase has been isolated and characterized from the rat testis (74). This protein, with a molecular weight of 26,100, is acidic and resembles the muscle protein in this respect, but it has a different amino acid composition. Moreover, the testis protein also inhibits the cAMP phosphodiesterase. A third inhibitor of cAMP-dependent kinases that has been reported is the "Type II inhibitor" from brain, which inhibits protein phosphorylation catalyzed by several protein kinases including the cAMP-dependent protein kinase (75).

cGMP-DEPENDENT PROTEIN KINASE   The most definitive data available with respect to activation of the cGMP-dependent protein kinase are those obtained with the homogeneous enzyme obtained from beef lung (39–43) and the highly purified enzyme from the silk worm (76, 77). The beef lung enzyme has a molecular weight of 150,000 (39) or 165,000 (40) and is made up of two apparently identical subunits with molecular weights determined to be 74,000 (39) or 81,000 (40). Unlike the cAMP-dependent protein kinase, the cGMP-dependent enzyme does not contain separate regulatory subunits and is not dissociated by its cyclic nucleotide activator, i.e. cGMP-binding and catalytic activity reside on a single polypeptide chain (39, 41, 42, 76, 77). The enzyme can be split by trypsin into catalytic and cyclic GMP-binding fragments of approximately equal size which indicates that two distinct functional sites are present (77). The possibility exists that limited proteolysis may have been a factor in studies which suggest the presence of separate regulatory and catalytic subunits in cGMP-dependent protein kinases (78–81). A factor(s) of unknown physiological significance with respect to control of cGMP-dependent protein kinase activity is the modulator protein(s) that stimulates its activity under certain conditions in vitro (82, 83). The cGMP-dependent protein kinase is not inhibited by the heat-stable protein inhibitor of cAMP-dependent protein kinase (39, 76). The cGMP-dependent protein kinase, like its cAMP-dependent counterpart, undergoes self-phosphorylation (84).

Ca^{2+}-DEPENDENT PROTEIN KINASES   Until recently only one example of a Ca^{2+}-dependent protein kinase, i.e. purified rabbit skeletal muscle phosphorylase kinase, was available for studies directed toward elucidating the mechanism of activation of this type of enzyme by the metal ion. With the finding of a second Ca^{2+}-dependent protein kinase, the myosin light chain kinase (55), further insight into the problem has been gained. As indicated above this enzyme was shown to depend on the presence of the Ca^{2+}-dependent modulator of cyclic nucleotide phosphodiesterase (56) for its sensitivity to Ca^{2+} (53, 54, 57). Recent work suggests that the activation of muscle phosphorylase kinase by Ca^{2+} may also be due to the presence of the modulator (58). In this instance the modulator is tightly bound and remains with the enzyme even though all of the steps in the classical muscle phosphorylase kinase preparative procedure are carried out in the presence of millimolar concentrations of EDTA (85).

Specificity

cAMP-DEPENDENT PROTEIN KINASE   Despite the existence of early reports showing that peptides derived from protein substrates could be phosphorylated by phosvitin kinase (13) and phosphorylase kinase (86), it was initially assumed that substrates for the cAMP-dependent protein kinase had to contain some specific three-dimensional configuration. This conclusion was reached (31, 87) when it was found that the phosphorylation site sequences for several substrates of this enzyme displayed no apparent similarity. The concept broke down, however, when it was found that the cAMP-dependent protein kinase would also phosphorylate small peptides (88). Furthermore, it was found that the denaturation of proteins could enhance their susceptibility to phosphorylation (89, 90). Thus, it was clear that the primary structure of protein substrates played a major role in determining specificity. The importance of a basic amino acid residue on the amino terminal side of phosphorylated serines or threonines became apparent as a result of (a) studies using genetic variants of β-casein (91), (b) a consideration of phosphorylated site sequences in lysine-rich histones determined from peptides derived using proteases other than trypsin (92), and (c) studies on synthetic peptides as potential substrates for the cAMP-dependent protein kinase (93, 94). The importance of two adjacent basic amino acid sequence at the phosphorylated site in pig liver (type L) pyruvate kinase (95) and confirmed by studies with synthetic peptides (96, 97). Additional information on phosphorylated site sequences in natural substrates for the enzyme (98–102) and further synthetic peptide work (103, 104) helped to extend these concepts. Two phosphorylated site sequences, not heretofore reported (T. S. Huang, and E. G. Krebs, in preparation), are the extended sequence for one of the sites phosphorylated by the cAMP-dependent protein kinase in rabbit skeletal muscle glycogen synthase:

Ser-Ser-Gly-Gly-Ser-Lys-Arg-Ser-Asn-Ser(P)-Val-Asp-Thr-Ser-Ser-Leu-Ser-Pro-Pro-Thr-Gly-Ser-Leu-Ser-Ser-Ala-Pro-Leu-Gly-Glu-Gln-Asp-Arg;

and the sequence at the phosphorylated site in bovine Type II regulatory subunit of cAMP-dependent protein kinase:

Asp-Arg-Arg-Val-Ser(P)-Val.

These last sequences illustrate the two main patterns that have been seen in natural enzyme substrates for the kinase; namely, Lys-Arg-X-X-Ser(P) and Arg-Arg-X-Ser(P). In the first instance, the phosphorylated serine is three residues C-terminal to the pair of basic amino acids "position C," according to Shenolikar & Cohen (102) whereas in the second pattern the phosphorylated serine is two residues C-terminal to the basic pair, or in "position B." Amino acids with hydrophobic side chains usually occur on both sides of the phosphorylated serine. Although appropriate peptides can serve as good substrate for the cAMP-dependent protein kinase with K_m values in the μ molar range and V_{max} values as high or higher than those seen with intact protein substrates (96, 97), it is probable that the enzyme recognizes some preferred configuration that these substrates can assume. Small et al (105) have predicted on the basis of the sequences present in phosphorylated sites of a number of proteins, that these regions will often be found in β-bends.

cGMP-DEPENDENT PROTEIN KINASE   The protein or peptide substrate specificity for the cGMP-dependent protein kinase is similar but not identical to that of the cAMP-dependent protein kinase. The cGMP-dependent enzyme catalyzes the phosphorylation of all five of the histone fractions (83). Furthermore, both kinases will phosphorylate pyruvate kinase (41), glycogen synthase (41), phosphorylase b kinase (41, 106), the hormone-sensitive lipase (106), cholesterol esterase (106), fructose-1,6-diphosphatase (107), cardiac troponin (108, 109), and certain ribosomal proteins (110). Work with synthetic peptides (41, 111, 112) has confirmed that the cGMP and cAMP-dependent protein kinases have similar but not identical structural determinants of specificity that can be identified in the linear amino acid sequence involving the phosphorylation site. In the best analyzed example, it has been shown that peptides corresponding to the phosphorylation sites around serine 32 and serine 36 in H2B histone show different relative "preferences" for the cGMP- and cAMP-dependent protein kinase (112) in keeping with the earlier observation of Hashimoto et al (113) with the intact protein. An early finding of Casnellie & Greengard (114) shows that mammalian smooth muscle membranes appear to contain proteins whose phosphorylation is stimulated specifically by cGMP.

Ca^{2+}-DEPENDENT PROTEIN KINASES

Phosphorylase kinase   Glycogen phosphorylase b is the only clearly defined physiologically significant substrate for phosphorylase kinase. However, this enzyme also catalyzes the phosphorylation of casein (115) and free troponin I (116, 117) in vitro. Moreover, it can undergo autophosphorylation (115, 118). There is some indication that phosphorylase kinase may catalyze the phosphorylation of a substrate in the sarcoplasmic reticulum but this has not been clearly defined (119, 120). Most recently, it has been reported that phosphorylase kinase catalyzes the phosphorylation of glycogen synthase I in vitro (121).

The sequence of amino acids found at the phosphorylated site in rabbit muscle phosphorylase a is Ser-Asp-Glu-Glu-Lys-Arg-Lys-Glu-Ile-Ser(P)-Val-ARg-Gly-Leu (86). Tessmer et al (122), using synthetic peptides, showed at the first six amino acid residues are not essential determinants of specificity, but substitutions in the next six positions have important effects on kinetic parameters. In particular, the arginine residue in the second position C-terminal to the phosphorylatable serine appears to be of critical importance. It is of interest that an arginine in this position acts as a negative determinant for the cAMP-dependent protein kinase (97). Synthetic peptides are poorer substrates for phosphorylase kinase than phosphorylase b which suggests that orders of structure above the primary level play a role in determining the substrate specificity for this enzyme.

Myosin light chain kinase   The Ca^{2+}-requiring myosin light chain kinase (55, 123) also catalyzes the phosphorylation of histones and phosphorylase kinase, apparently at rates comparable to those found for the phosphorylation of myosin light chains (57). The amino acid sequence (123, 123a) at the phosphorylated site in the phosphorylatable light chain or rabbit skeletal muscle, Arg-Ala-Ala-Ala-Glu-Gly-Gly-(Ser,SerP)-Asn-VAl-Phe, does not by itself reveal what particular residues may be important in determining specificity.

PYRUVIC DEHYDROGENASE KINASE   Of the protein kinases in the nonspecified group (Table I) that are known to have a role in the regulation of enzyme activity, the only one for which much information is available concerning specificity is pyruvic dehydrogenase kinase, which catalyzes the phosphorylation of the α subunit of the enzyme (124). Three serine residues can be phosphorylated but regulation (inactivation) of the enzyme appears to be associated with only one of the three sites (125), i.e. Site 1. The sites, as defined by their amino acid sequences, are shown for the bovine heart and kidney dehydrogenase in the following peptides that have been isolated:

Tyr-His-Gly-His-Ser(P)-Met-Ser-Asn-Pro-Gly-Val-Ser(P)-Tyr-Arg
              ↑   ↑
           Site 1   Site 2
Tyr-Gly-Met-Gly-Thr-Ser(P)-Val-Glu-Arg
                    ↑
                 Site 3

Scheme 1

It has been determined that peptides derived from pyruvic dehydrogenase by tryptic digestion can serve as substrates for pyruvic dehydrogenase kinase (126).

PHOSPHOPROTEIN PHOSPHATASES

Multifunctional Phosphoprotein Phosphatase

Early evidence indicated that a single phosphoprotein phosphatase catalyzing the dephosphorylation of a variety of protein substrates including phosphorylase a, the inhibitory subunit of troponin (troponin-I), histones, glycogen synthase D, phosphorylase kinase, and others could be obtained from various animal tissues (127–131). Such an enzyme, having a molecular weight of ∼35,000, was purified to homogeneity from liver (132, 133) and heart (134, 135). Similarly, Antoniw et al (136), and Ray & England (137) obtained evidence for a multifunctional phosphatase that appears to fractionate as a single enzyme. In their purification of the multifunctional phosphoprotein phosphatase, Brandt et al (138) noted that an alcohol precipitation step, which they employed, caused a marked activation of the enzyme and resulted in its conversion from a higher molecular weight form(s) to a form of M_r 35,000. These authors raised the possibility that the isolated enzyme might represent a catalytic subunit derived from a larger complex. A number of other reports support this concept (139–141). Lee and his group have strengthened their evidence for the existence of a native complex or holoenzyme form of the phosphatase, which they refer to as protein phosphatase-H, and a catalytic subunit, protein phosphatase-C (142). Protein phosphatase-H has been purified and preliminary evidence shows that it consists of two catalytic subunits (M_4 35,000) complexed with one inhibitor subunit (M_4 65,000). The form of phosphorylase phosphatase purified by Gratecos et al (143) has a molecular weight of approximately 33,000 and thus resembles protein phosphatase-C discussed above but it appears to be more specific than the latter enzyme.

The multifunctional phosphoprotein phosphatase is regulated by (a) competition between various phosphoprotein substrates for the enzyme, (b) the noncompetitive interaction of a protein modifier(s) with phosphatase, and (c) the specific interaction of ligands with the phosphoprotein substrates of the enzyme, i.e. by substrate-directed effects. The first category of regulation can be illustrated by the inhibitory effect of phosphorylase a on the glycogen synthetase phosphatase reaction in liver, as first elucidated by Hers and his collaborators (144–146), although it is not universally accepted that this represents a competitive phenomenon. Phosphorylase kinase is also known to interact with the phosphatase and inhibit its activity relative to the dephosphorylation of phosphorylase a (147, 148). The second category of regulation is illustrated by the roles of the heat stable muscle inhibitors (Inhibitors 1 and 2) discovered by Huang & Glinsmann (149–151) and that of a heat-stable inhibitor of the phosphatase present in liver (152). Inhibitor 1 is of particular interest because it depends upon phosphorylation by the cAMP-dependent protein kinase in order to serve as an inhibitor (149, 153–155). Nimmo & Cohen (156) have purified Inhibitor 1 to a state approaching homogeneity and have estimated that its concentration in muscle is at least as high as the concentration of the multifunctional phosphatase. Inhibition of the phosphatase by Inhibitor 1 is reported to be noncompetitive with respect to the phosphoprotein substrate (154, 157). It is unclear whether or not heat-stable protein inhibitors of the multifunctional phosphatase are bound to the enzyme when it is present in crude extracts (see discussion of the native complex or "holoenzyme" above). Regulation of the multifunctional phosphatase by substrate-directed effects (the third category) is discussed under specific enzymes.

Specific Phosphoprotein Phosphatases

The existence of specific phosphoprotein phosphatases cannot be ruled out. Antoniw et al (136) found that their protein phosphatase-III (the multifunctional enzyme) accounted for only 75% of the total glycogen synthase phosphatase activity at a given stage of their fractionation. Moreover, their protein phosphatases-I and -II each possessed 10–15% of the total synthase phosphatase. In addition, the titration of tissue extracts with Inhibitors 1 and 2, which act on the multifunctional protein phosphatase, showed that no more than 50–80% of the phosphorylase phosphatase activity could be inhibited depending on the tissue examined (155). These reports and others certainly allow for the presence of enzymes that may be more or less specific for glycogen synthetase or phosphorylase (143, 158–160). It should be noted that a phosphatase specific for the dephosphorylation of the α subunit of phosphorylase kinase has been separated from the multifunctional enzyme. There is a possibility that substrate-specific phosphatases may be composed of a nonspecific catalytic subunit (multifunctional phosphatase?) that exists in different combinations with a selection of regulatory subunits that in some manner dictate specificity (161). A number of phosphoprotein phosphatase activities have been described in relation to the dephosphorylation of proteins outside of the area of glycogen metabolism. For example, pyruvic dehydrogenase phosphatase represents an intramitochondrial enzyme whose only known substrate is the dehydrogenase (162) or peptides derived from it (126).

THE PHOSPHORYLATION-DEPHOSPHORYLATION OF SPECIFIC ENZYMES

Twenty-one enzymes have been reported to undergo phosphorylation-dephosphorylation reactions (Table 2) and additional enzymes may belong in this group (184–186). Because of the heightened interest in these reactions it becomes important to consider criteria that should be satisfied to establish that a given enzyme undergoes physiologically significant phosphorylation-dephosphorylation. An attempt to formulate such a set of criteria with reference to proteins undergoing phosphorylation reactions catalyzed by the cAMP-dependent protein kinase was made several years ago (34, 187). Drawing from these earlier efforts but extending the concepts to include phosphorylation reactions catalyzed by enzymes other than the cAMP-dependent protein kinase, we have now drawn up a new set of criteria for establishing that an enzyme undergoes physiologically significant phosphorylation-dephosphorylation:

1. Demonstration in vitro that the enzyme can be phosphorylated stoichiometrically at a significant rate in a reaction(s) catalyzed by an appropriate protein kinase(s) and dephosphorylated by a phosphoprotein phosphatase(s).

2. Demonstration that functional properties of the enzyme undergo meaningful changes that correlate with the degree of phosphorylation.

3. Demonstration that the enzyme can be phosphorylated and dephosphorylated in vivo or in an intact cell system with accompanying functional changes.

4. Correlation of cellular levels of protein kinase and/or phosphoprotein phosphatase effectors and the extent of phosphorylation of the enzyme.

For enzymes that are phosphorylated at multiple sites the same set of criteria can be applied, but under these circumstances it is necessary to examine the significance of each phosphorylation event separately (34).

It is apparent that considerable judgment must be applied in using the criteria listed above. For example, in considering whether or not a phosphorylation reaction occurs at a "significant rate" an investigator would want to determine whether the rate is commensurate with the postulated role for that reaction within the cell. It is obviously important that the protein kinase(s) and phosphoprotein phosphatase(s) shown to catalyze the phosphorylation of a given enzyme be located in the cell-type and/or subcellular compartment from which the enzyme was derived. Changes in the func-

Table 2 Initial reports on enzymes undergoing phosphorylation-dephosphorylation

Enzyme	Year reported	References
Glycogen phosphorylase	1955	2, 3
Phosphorylase kinase	1959	163
Glycogen synthase	1963	164
Hormone-sensitive lipase	1964, 1970[a]	165–167
Fructone-1,6-biphosphatase	1966, 1977[a]	107, 168
Pyruvic dehydrogenase	1969	169
Hydroxymethylglutaryl CoA reductase	1973	170
Acetyl CoA carboxylase	1973	171
DNA-dependent RNA polymerase	1973	172
Pyruvate kinase (liver)	1974	173
Cholesterol ester hydrolase	1974	174
R subunit of Type II cAMP-dependent protein kinase	1974	64
Reverse transcriptase	1975	175
Phosphofructokinase (liver)	1975	176
Tyrosine hydroxylase	1975	177
Phosphorylase phosphatase inhibitor	1975	149
Phenylalanine hydroxylase	1976	178
eIF2-kinase	1977	179
cGMP-dependent protein kinase	1977	84
Tryptophan hydroxylase	1977	180, 181
NAD-dependent glutamate dehydrogenase (yeast)	1978	182
Glycerophosphate acyltransferase	1978	183

[a] In those instances in which a substantial period of time elapsed between initial and subsequent reports two dates are listed.

tional properties of a given enzyme, such as alteration in K_m, V_{max}, etc, should be in the appropriate direction. It should be noted, however, that in some instances physiologically significant functional changes in an enzyme may not be readily demonstrable in a simple in vitro assay. This could occur, for example, if the purpose served by phosphorylation of an enzyme was its translocation from one cellular compartment to another. Functional changes could also be missed in those instances that involve enzymes catalyzing complex reactions in which it is not currently possible to duplicate the intracellular environment in vitro. The third criterion listed above is without question the most difficult one of the set for investigators to satisfy, and an impressive backlog of candidates for examination with respect to in vivo phosphorylation has arisen. This is particularly true for some of the enzyme substrates that have been shown to be phosphorylated by a number of different protein kinases. With respect to the fourth criterion, it is clear that a direct correlation would exist between the level of a specific kinase effector and the extent of enzyme phosphorylation only in a limited range of effector concentration. In those instances in which the phosphoprotein phosphatase step is also regulated actively a very complicated picture will result.

No attempt is made to review in depth work on each of the individual phosphorylatable enzymes. Instead, the enzymes listed in Table 2 will be discussed largely with respect to the criteria discussed above. No further comments concerning phosphorylation of the R subunit of Type II cAMP-dependent protein kinase or the cGMP-dependent protein kinase will be made. The phosphorylation of eIF2 kinase, covered in the chapter by Ochoa and de Haro in this volume, is not discussed.

Phosphorylase Kinase

MUSCLE PHOSPHORYLASE KINASE

Phosphorylation by cAMP-dependent protein kinase Phosphorylase kinase was the first substrate recognized for the cAMP-dependent protein kinase, and its phosphorylation and dephosphorylation have been studied thoroughly. Thus, the reader is referred to other reviews (34, 188, 189) for information with respect to extensive work on its phosphorylation in vitro. Only a limited amount of work has been carried out on the phosphorylation of this enzyme in vivo. Kinetic data consistent with phosphorylation of phosphorylase kinase in intact muscle as a result of epinephrine stimulation were obtained as early as 1964 (190) but later attempts to demonstrate the actual phosphorylation of the enzyme were unsuccessful (191). Yeaman & Cohen (192) did succeed in showing that epinephrine stimulates phosphorylation of the kinase in vivo; moreover, they showed that the phosphorylated sites are identical to those obtained in vitro. In considering their work, however, the question could arise as to whether the phosphorylation that they observed did, in fact, occur in vivo or whether it occurred during the grinding and extraction of muscle. These workers did not employ quick-freezing techniques of the type commonly used in studying phosphorylation reactions in vivo. For studies relating to the correlation between cyclic AMP levels and phosphorylase kinase activation, i.e. application of the fourth criterion listed on page 936, the reader is again referred to other reviews (34, 188, 189).

Phosphorylation by Ca²⁺-dependent protein kinases   No evidence has been offered that the autophosphorylation (115, 118) and the phosphorylation of muscle phosphorylase kinase catalyzed by myosin light-chain kinase (modulator-dependent) protein kinase (57) that have been observed in vitro are of physiological significance. Electrical stimulation of muscle, which would make Ca^{2+} available to the kinases, was found to cause a very slight increase in the pH 6.8/8.2 activity ratio of the enzyme (190); however this finding was not substantiated by later work in two different laboratories (193, 194). Furthermore, in experiments involving cardiac tissue Dobson & Mayer (195) found that anoxia caused conversion of phosphorylase *b* to *a*, without phosphorylase kinase activation, by a mechanism that probably involved an influx of Ca^{2+}.

Phosphorylation by other protein kinases   In addition to the reactions involving muscle phosphorylase kinase cited above, this enzyme has been reported to undergo in vitro phosphorylation reactions catalyzed by beef lung cGMP-dependent protein kinase (14, 106) and by a nonspecified protein kinase (protein kinase M) from brain that is activated by proteases (196). The reaction catalyzed by the cGMP-dependent protein kinase is very slow compared to that of the cAMP-dependent kinase, occurring at a rate less than 5% of that obtained with the latter enzyme (197). In view of the low concentration of the cGMP-dependent protein kinase in muscle this reaction is probably of no physiological significance. No information bearing on the possible significance of phosphorylase kinase phosphorylation by protein kinase M is available.

LIVER PHOSPHORYLASE KINASE   There is renewed interest in determining how cAMP causes activation of phosphorylase in liver (198–202). This has been a difficult problem to resolve because of the inability of investigators to isolate liver phosphorylase kinase in a form that is still responsive to regulation by phosphorylation-dephosphorylation. Vandenheede et al (199), using the heat-stable inhibitor of cAMP-dependent protein kinase to prevent in vitro changes, showed that glucagon activates phosphorylase kinase in isolated liver cells. The same authors (200) showed that activated liver phosphorylase kinase can be reversibly inactivated in a Mg^{2+}-requiring reaction in extracts. Reactivation of the kinase, which was more rapid in the presence of cAMP, was blocked by the heat-stable inhibitor of cAMP-dependent protein kinase. These observations and others (201, 202) are all in keeping with a regulatory cascade scheme analogous to that of muscle. Vandenheede et al (201) have shown that purified catalytic subunit of the cAMP-dependent protein kinase can be substituted for cAMP in promoting the formation of phosphorylase a in liver extracts. As will be discussed in the next section, liver phosphorylase kinase has been shown to be stimulated by calcium (203, 204).

Glycogen Phosphorylase

Glycogen phosphorylase, the first cellular protein recognized as undergoing enzymic phosphorylation-dephosphorylation (2, 3), has been reviewed extensively (205–207). Accordingly, the present treatment is restricted to the next two topics; the first represents an area in which there has been significant new developments, the second serves as a good prototype for other phosphorylation-dephosphorylation systems (see the next two subsections).

α-ADRENERGIC STIMULATION OF PHOSPHORYLASE a FORMATION It has been shown that α-adrenergic agents (202, 208–212), vasopressin (213) and angiotensin, (214) cause phosphorylase a formation in liver or fat cells (215) without an increase in cAMP levels, cAMP-dependent protein kinase activation, or phosphorylase kinase activation. A possible mechanism for the action of these agents is the direct stimulation of phosphorylase kinase by Ca^{2+} (203, 204), and strong support for this mechanism has been obtained (201, 211, 212, 216, 217). In general, such support is based on the fact that the agents work poorly, if at all, when the cells are present in a calcium-free medium. In a recent study, however, Blackmore et al (217) have shown that the source of calcium for stimulation of phosphorylase kinase may be due to mobilization of intracellular calcium rather than an enhanced influx from outside the cell.

REGULATION OF PHOSPHORYLASE INACTIVATION BY SUBSTRATE-DIRECTED EFFECTS   Inhibition of the phosphorylase phosphatase reaction by AMP and its stimulation by caffeine were shown to be due to the interaction of these substances with the substrate, since neither compound had any effect on the dehosphorylation of a small phosphopeptide derived from phosphorylase a (86). Subsequently it was shown that the stimulatory effect of glucose-6-P (218), glucose (219), and glycogen, as well as the inhibitory effect of glucose-1-P (220), on the phosphorylase phosphatase reaction were also substrate-directed (221–224). It was shown that glucose and glucose-6-P act in opposition to AMP, i.e. the configuration of phosphorylase a that results from the binding of glucose or glucose-6-P provides a better substrate for the phosphatase than that which is favored by the binding of AMP and vice versa (222, 223). Martensen et al (225) found that glucose and glucose-6-P stimulate the reaction by causing an increase in V_{max}, whereas the effect of glycogen is due to a lowering of the K_m. Recent X-ray crystallographic work reported by Kasvinsky et al (226) provides evidence suggesting that the allosteric transition undergone by phosphorylase a, either when it binds AMP at the high affinity "nucleotide binding site" or when it binds glucose 1-phosphate at the active site, may be such that the serine-bound phosphate (Serine 14 near the NH_2 terminus) becomes unavailable to the phosphatase. Conversely, the binding of glucose or glucose-6-P at the active site appears to favor a conformation in which that region is exposed. Caffeine or other effectors binding at what has been identified as the "nucleoside site" also favor a conformation that renders phosphorylase a susceptible to phosphatase attack. It should be emphasized that the regulation of the phosphorylase phosphatase reaction by metabolites, particularly the regulation of this reaction by glucose in the liver (144–146), is of physiological significance.

Glycogen Synthase

Detailed reviews on the regulation of glycogen synthetase by phosphorylation-dephosphorylation have appeared recently (34, 227, 228). No enzyme better illustrates the complexity of determining the physiological relevance of multiple phosphorylation-dephosphorylation reactions. In 1971 Smith et al (229) reported that the 85,000–90,000 dalton subunit of glycogen synthase D contains 6 moles of covalently bound phosphate, and the suggestion was made (230) that six small subunits of ~15,000 daltons, each containing an identical phosphate-binding site, might exist. An entirely different explanation for a stoichiometry greater than one was introduced, however, by the finding of Nimmo & Cohen (11) that more than one protein kinase is involved in catalyzing the phosphorylation of the synthase. These workers showed that a phosvitin kinase present in skeletal muscle can also serve as a synthase kinase, although its action did not result in profound change in the activity of the synthase (11). A clear example of a cAMP-independent protein kinase that catalyzes the phosphorylation as well as the synthase I to D conversion was discovered by Schlender & Reimann in the renal medulla and other tissues (231, 232). Work in several laboratories amply confirmed the existence of at least one such enzyme (233–237). In addition to the phosphorylation and synthase I to D conversion catalyzed by the cAMP-independent synthase kinase and the cAMP-dependent protein kinase, it has been shown that purified cGMP-dependent protein kinase catalyzes this reaction in vitro (41), albeit at a relatively slow rate (197). It has also been reported that the nonspecified protein kinase of Nishizuka and co-workers, protein kinase M, catalyzes this reaction (238). Finally, Roach et al (121) have reported that phosphorylase kinase may have a functional role in this regard in that it catalyzes the phosphorylation of synthase I in a Ca^{2+}-dependent reaction. The major challenge now facing investigators is to establish whether there are physiological roles for the multiple phosphorylations catalyzed by the various protein kinases that act on glycogen synthase. This will require a clear definition of the specific sites in the synthase that are phosphorylated in vitro by each kinase and the development of methods for studying site-specific phosphorylation reactions in vivo. Progress in this direction is being made in several laboratories (227, 233, 239, 240).

Regulation of the relative concentrations of glycogen synthase I and D may also occur through control of one or more phosphoprotein phosphatases. Hutson et al (241) have obtained evidence that the multifunctional phosphatase catalyzes the dephosphorylation of synthase D in a reaction that is regulated by the extent of phosphorylation of the substrate. This type of control, orginally described with respect to the dephosphorylation of phosphorylase kinase by Cohen and co-workers (34), may constitute a general process that should be added to the list of mechanisms described earlier for the regulation of phosphoprotein phosphatases. Control of the glycogen synthase I to D reaction by the substrate-directed effects of metabolites (reviewed in 146) is not discussed here. With respect to this last type of regulation, however, it can be anticipated that work on the structure of glycogen synthase, comparable to that being carried out with phosphorylase, will eventually provide a clearer understanding of the various effects that have been described. The concept (144, 145) that the glycogen synthase phosphatase reaction is inhibited by phosphorylase *a,* possibly by competition between phosphorylase *a* and synthase D for the multifunctional phosphatase, continues to gain support (208, 242).

Phosphofructokinase

As phosphofructokinase (PFK) is one of the most highly regulated enzymes in carbohydrate metabolism, it has been the subject of intense study, and several reviews of its properties and regulation have appeared (243–145). Most of the work concerning its possible regulation by phosphorylation-dephosphorylation has dealt with the major liver and muscle isozymes of PFK. Brand & Söling and their co-workers (176, 246) reported that the activity of the liver enzyme is stimulated by MgATP and reversibly inhibited by Mg^{2+}. [^{32}P]phosphoserine can be recovered in specific immunoprecipitates of the enzyme after preincubation with [γ-^{32}P]ATP, and the amounts vary with the metabolic state of the animal (i.e. starved or fed) and the extraction conditions used in the isolation. For example, when they are extracted under conditions where the kinases are active in the homogenate, the activity is uniformly high, but when the kinase and phosphatase activity is inhibited, differences are seen between fed and starved states.

Several investigators have reported that the purified isozyme from muscle contains covalently bound phosphate. Hofer & Fürst (247) and Riquelme et al (248) reported that ^{32}P was found in muscle PFK after injection of ^{32}P$_i$; however the data of Hussey et al (249), who used inorganic phosphorus measurements, indicate that much less than stoichiometric amounts are found. No change in PFK activity due to phosphorylation has been demonstrated in the muscle system.

In summary, although there is suggestive data, particularly in liver, that PFK can be phosphorylated and that phosphorylation may regulate the enzyme, the data are far from clear. It will probably remain so until good correlations between phosphorylation of a specific site can be correlated with an activity change. Furthermore, the nature of the kinase(s) and phosphatase(s) and their regulation are unknown for each of these systems.

Pyruvate Kinase

The general properties of pyruvate kinase have been discussed (250, 251) and the regulation of this enzyme by phosphorylation-dephosphorylation has been reviewed recently by Engström (252, 253). At present there is strong evidence that the L-type isozyme of pyruvate kinase is modified and regulated by a cyclic AMP-dependent phosphorylation in vitro and in vivo. Furthermore, of the several isozymes of pyruvate kinase isolated (254–258) only the L form has been shown to be a good substrate for cAMP-dependent protein kinase (173, 259). The rate of phosphorylation is rapid and the extent of phosphorylation correlates well with a decrease in the activity of the enzyme (260, 261). Mechanistically, phosphorylation causes a decrease in activity by altering the affinity of one of the substrates, phosphoenolpyruvate (260, 261). This effect can be overcome by increases in the concentration of phosphoenolpyruvate, or the allosteric activator, fructose 1:6 diphosphate. Similarly, the effect of phosphorylation is decreased at lower pH (e.g. pH 6.8) but increased in the presence of ATP and alanine which are thought to be important allosteric inhibitors of the enzyme (253, 260, 261). Pyruvate kinase nicely illustrates the generalization that positive correlations between covalent modifications and enzyme activity are most likely to be detected at suboptimal reaction conditions. Moreover, in this case, these nonsaturating substrate and effector concentrations approximate those existing in the cell and therefore make it likely that the phosphorylation is physiologically relevant. Since to date only one phosphorylated site has been found in pyruvate kinase (262, 263), correlations between activity and the state of phosphorylation are easily made (264). Furthermore, only the cyclic nucleotide–dependent protein kinases have been found to catalyze this phosphorylation (197, 252). As the levels of cGMP and cGMP-dependent protein kinase are quite low in liver and kidney, the location of the L-type enzyme, it is presumed that cAMP and the cAMP-dependent protein kinases are the physiologically important regulators.

Considerable evidence also has been obtained that indicates that phosphorylation-dephosphorylation is an important regulatory mechanism in vivo. In the early studies, changes in the kinetic properties of pyruvate kinase similar to those shown to be mediated by phosphorylation in vitro were shown to occur after various hormone treatments (265–268). More recently the extent of phosphorylation in vivo has been directly correlated with activity changes and cAMP levels (269–272). One potentially important observation is that several of the allosteric effectors of the enzyme, which directly affect activity, also affect the state of phosphorylation (272a).

Data on dephosphorylation is less well developed. It has been shown that the rate of dephosphorylation by partially purified phosphatase preparations are comparable to those of phosphorylated histone or protamine (264, 273). Interestingly, dephosphorylation is inhibited by fructose 1, 6-diphosphate, ATP, and ADP which may imply that allosteric effects on the substrate are important. However, most of these effects are reversed by mM concentration of Mn^{2+} or Mg^{2+} which implies that they may be secondary to metal chelation effects.

Fructose-1,6-Diphosphatase (FDPase)

Although changes in FDPase activity after hormone treatment have been reported (266, 268, 274), to date only one group has shown that it is phosphorylated (107). They report that stoichiometric amounts of phosphate are incorporated by high levels of pure cAMP-dependent protein kinase catalytic subunit. Furthermore, the phosphorylation is accompanied by a small (40%) but reproducible increase in the activity. Finally, ^{32}P was found in the enzyme after in vivo injection of ^{32}P and subsequent isolation by immunoprecipitation techniques. In short, the evidence to date is very suggestive that FDPase activity may be modulated by phosphorylation-dephosphorylation.

Pyruvate Dehydrogenase (PDH)

The general enzymatic properties and regulation of PDH activity have been the subject of several recent reviews (275–280). It was originally shown by Linn et al (162) that the PDH complex contained a kinase and phosphatase which altered PDH activity by stoichiometric phosphorylation-dephosphorylation at rates that were fast enough to be physiologically relevant. The phosphorylation occurs at three sites on the α subunit of the decarboxylase (281, 282) and the activity of the enzyme has been shown to correlate with the state of phosphorylation of the first site both in vitro and in vivo (282–287). Several peptide fragments of PDH will serve as substrates for both the kinase and phosphatase which indicates that primary sequence determinants are of importance in determining specificity (288). However, basic residues are not found in close proximity to the phosphorylation site as in substrates for the cyclic nucleotide–dependent protein kinases. The function of the multiple phosphorylations of PDH is not yet clear. There is some evidence (281) that the phosphorylation of the second and possibly third sites is dependent on the rapid phosphorylation of first site. By analogy to the regulation of the glycogen synthase and phosphorylase kinase systems discussed previously, the multiple phosphorylations may regulate the rates of dephosphorylation and reactivation of the enzyme. Very recently some direct evidence for this concept has been presented (289).

In addition to this site/site type of regulation, the same allosteric effectors which *directly* affect the PDH activity (e.g. high NADH/NAD, ATP/ADP, and AcCoA/CoA ratios) also increase the rate of phosphorylation of the complex (290–293). On the other hand, high levels of pyruvate, dichloroacetate, Ca^{2+}, K^+, and thiamine pyrophosphate derivatives have been shown to inhibit phosphorylation (291). It seems likely that some of these agents influence phosphorylation by affecting the substrates. However, until good alternative substrates of the kinase, which do not bind these agents, are tested, definitive data on the mechanism by which they act will be limited. In this regard, small peptides that include the phosphorylated site have recently been shown to be reasonable substrates for the kinase and phosphatase and may prove useful (288).

Since PDH is located in the mitochondrial matrix, studies on phosphorylation-dephosphorylation in this intact organelle could be carried out (294–295). An advantage of this system is that all of the ATP pools can be highly and equally labeled with ^{32}P. This, and the fact that few other mitochondrial proteins are phosphorylated, has allowed good correlations between activity and the degree of phosphorylation to be demonstrated. The availability of specific antibody has also been helpful in showing such changes in whole tissue extracts (295).

The majority of studies on hormone effects on PDH activity have centered around insulin (280, 296) which causes an increase in PDH activity. Recent studies have been largely concerned with the mechanism by which this increase occurs. Since insulin causes little if any change in the levels of the metabolites which affect PDH activity (294) other mechanisms must be involved (294). Although in most cases the apparent rate of ^{32}P incorporation into PDH appears to be increased (295), the net phosphate incorporation into mitochondrial PDH is decreased by insulin. Since the specific activity of the ATP does not change with insulin treatment, insulin may be acting by stimulating the PDH phosphatase activity; the increased dephosphorylation would favor phosphate turnover. Other studies indicate that the state of phosphorylation of the second and third sites is important in controlling phosphatase activity (296, 297).

Hormone-Sensitive Lipase

In 1964 (165) Rizak showed that the hormone-sensitive lipase in fat cell homogenates could be activated under phosphorylating conditions in a reaction that was stimulated by cAMP, and definitive work demonstrating that this activation involved the cAMP-dependent protein kinase was reported in 1970 (166, 167). This lipase has recently been the subject of several reviews (298, 299).

Phosphorylation and dephosphorylation of the hormone-sensitive lipase have been difficult to study because of difficulties in obtaining a homogeneous enzyme. Nonetheless, it has been shown that in a partially purified preparation the enzyme undergoes phosphorylation with a concomitant increase in activity (300, 301). Activation is best seen as low salt concentrations (299) and results in a lowering of the apparent K_m for the triglyceride substrate without a change in V_{max} (302). Dephosphorylation and inactivation of phosphorylated hormone-sensitive lipase catalyzed by a phosphoprotein phosphatase has been demonstrated (303, 304). With respect to in vivo data, only limited information is available for the hormone-sensitive lipase. The enzyme as obtained from fat cells treated with lipolytic hormones manifests less protein kinase–dependent activation than enzyme prepared from control cells (305), consistent with its having been phosphorylated and activated in the intact cell. An abundant literature, which is not reviewed here, exists with respect to the effect of lipolytic hormones on cAMP levels in fat cells and the correlation of cAMP with the release of free fatty acids. Corbin et al (306) have also shown that lipolytic hormones cause dissociation of the cAMP-dependent protein kinase in fat cells. The hormone-sensitive lipase can be activated by cGMP-dependent protein kinase in vitro (106, 299), but the functional significance of this reaction is unknown.

Cholesterol Ester Hydrolase

Cholesterol ester hydrolase is treated as a distinct entity in Table 2, but it is possible that this enzyme is identical to the hormone-sensitive lipase discussed above (302, 307).

Early work showed that cAMP mediates the action(s) of ACTH in the adrenals, and that at least one of its functions is to cause the breakdown of cholesterol esters (reviewed in 308) presumably due to activation of cholesterol esterase (309, 310). It was then found that the esterase is activated reversibly in crude bovine adrenal cortical fractions under appropriate phosphorylating conditions in the presence of cAMP-dependent protein kinase (174, 311). These observations were later extended using more purified enzyme preparations from the adrenal (312, 313). Cholesterol ester hydrolase activity stimulated by cAMP-dependent protein kinase has also been studied in adipose tissue preparations (302. 314). The study of Khoo et al (302) is of particular significance relative to the question of the possible identity of the hormone-sensitive lipase and cholesterol ester hydrolase. The maximal degree of activation obtainable in vitro decreases when the tissue has been previously exposed to glucagon, which indicates that the hormone-induced activation is probably similar or identical to activation catalyzed by the protein kinase in vitro (302).

Acetyl CoA Carboxylase

Evidence that acetyl CoA carboxylase is regulated by phosphorylation-dephosphorylation has been accumulating steadily. Inoue & Lowenstein (315) found that the enzyme from rat liver contains protein-bound phosphate, and shortly thereafter Carlson & Kim (171, 316, 317) reported that the partially purified carboxylase could be phosphorylated and inactivated by preincubation with $[\gamma\text{-}^{32}P]ATP$ in the presence of Mg^{2+}. Addition of cAMP did not increase the rate of the reaction. Carlson & Kim also showed (171, 361) that the carboxylase could be dephosphorylated and reactivated presumably due to the action of a phosphoprotein phosphatase. They reported that phosphorylation of the carboxylase decreased the ability of low concentrations of citrate to activate the enzyme and increased its sensitivity to inhibition by palmityl CoA (317). Recently Lee & Kim (318) repeated the earlier experiments from that laboratory using more highly purified components. Hardie & Cohen (319) purified acetyl CoA carboxylase from lactating mammary glands of the rabbit and found that it can be phosphorylated by the cAMP-dependent protein kinase, and that, in addition, it undergoes a slow phosphorylation reaction catalyzed by a nonspecified protein kinase present as a contaminant in their preparation; they did not report what effect phosphorylation by either enzyme had on the activity of the carboxylase. The indication that more than one protein kinase may be involved in catalyzing the phosphorylation of the carboxylase is in keeping with data showing that the isolated enzyme from various sources contains anywhere from 2–6 mol of bound phosphate per subunit (315, 320). It has been demonstrated that acetyl CoA carboxylase can be phosphorylated in intact fat cells incubated with $^{32}P_i$ (321); the extent of phosphorylation observed was not affected by insulin, however, which is known to increase the activity of the enzyme (322, 323). Glucagon has been reported to decrease the activity of liver acetyl CoA carboxylase in vivo (324) and epinephrine that of the enzyme from epidyimal fat cells (323). Moreover, it has been possible to demonstrate inhibition of the enzyme by dibutyryl cAMP in liver slices (325). In cultured liver cells Pekala et al (325a) found no effect of dibutyryl cAMP on the extent of phosphorylation of acetyl CoA carboxylase.

Glycerophosphate Acyltransferase

It has been shown that exposure of fat cells to epinephrine causes a decrease in the activity of glycerophosphate acyltransferase that persists through freeze-thaw steps and extraction procedures (326). This effect is blocked by propanolol and also by insulin (327). Insulin alone has variable effects depending upon the presence or absence of carbohydrate substrates in the medium (326). Evidence has been obtained recently, using a cell-free preparation, that the mechanism of the epinephrine effect in all probability involves phosphorylation of glycerophosphate acyltransferase by the cAMP-dependent protein kinase (183).

Hydroxymethylglutaryl CoA Reductase

The general properties and regulation of the reductase have been the subject of recent reviews (328, 329), and initial data, which suggest that it may be regulated by phosphorylation-dephosphorylation, have been presented (330). Several lines of investigation indicate that the enzyme may be regulated by reversible covalent modification. The activity of the enzyme has been shown to be increased by insulin, decreased by glucagon (331–334), and decreased by exogenously added cAMP (335). In addition, in vitro studies with crude enzyme preparations indicated that cAMP and MgATP causes a decrease in enzyme activity, and, conversely, incubation without ATP causes reversal of this inhibition (170, 336). Furthermore, the increase in activity due to preincubation without ATP can be prevented by NaF, a known phosphoprotein phosphatase inhibitor (336–339) and mimicked by incubation with purified phosphatase (338).

More recently, it has been reported that after incubation with $[\gamma\text{-}^{32}P]$-ATP (340), specific immunoprecipitates of HMG CoA reductase contained ^{32}P. The same authors also report that the incorporation of alkali-labile ^{32}P into microsomal protein correlated well with the inactivation of enzyme activity. Another recent report (338) indicates that the system may be even more complex in that the inactivating system (i.e. a protein kinase) is itself also inactivated by phosphatase treatment and reactivated by MgATP. Although the nature of both of these kinases are not well defined, it appears from these studies that both cAMP-dependent and cAMP-independent are involved.

In summary, there is now reasonably good evidence that HMG CoA reductase may be modulated by phosphorylation-dephosphorylation. However, the details of the mechanism involved, the regulation and specificites of the kinases and phosphatases, and the connection with hormonal and other metabolically induced changes have yet to be clearly demonstrated. It is tempting to speculate that a regulatory system much like that described for glycogen synthase may eventually be shown to operate.

Tyrosine Hydroxylase

In 1975 Morgenroth et al (177) showed that crude tyrosine hydroxylase could apparently be activated by the cAMP-dependent protein kinase. Lovenberg et al (341) carried out similar studies and in addition sought evidence that the hydroxylase was actually being phosphorylated in the activation reaction. Surprisingly, neither they nor Lloyd & Kaufman (342) were able to show any phosphate uptake even though the enzyme was activated extensively. It was postulated that the component undergoing phosphorylation might be an activator protein rather than tyrosine hydroxylase itself (342, 343). In both of these negative studies the technique used for isolating the putative phosphorylated form of the hydroxylase involved its precipitation with specific antibody, a procedure requiring a period of incubation during which phosphoprotein phosphatases could have been acting. More recently, evidence has been obtained demonstrating that tyrosine hydroxylase is indeed being phosphorylated in the activation reaction (344, 345). The above studies and others (346–348) show that regulation of the hydroxylase by phosphorylation occurs in various parts of the brain and also in the adrenal medulla. Phosphorylation of the hydroxylase lowers the K_m for the pteridine cofactor and decreases the affinity for end-product inhibitors (349). The effect on V_{max} is variable depending on the source and/or the extent of purification of the enzyme (350). Tyrosine hydroxylase can be phosphorylated in intact cells (351) but the extent of phosphorylation with various physiological stimuli has not been examined. It is known, however, that an enzyme undergoes changes in kinetic properties in vivo that are similar to those caused by phosphorylation in vitro (347). Correlations have been shown between physiological stimuli affecting cAMP levels, and/or Ca^{2+} influx into the cell, and tyrosine hydroxylase activation (352, 353).

Phenylalanine Hydroxylase

Liver phenylalanine hydroxylase contains approximately 0.3 mol of protein-bound phosphate per 50,000 molecular weight subunit (354). The purified enzyme can be further phosphorylated in the presence of Mg^{2+}, $[\gamma^{32}\text{-P}]ATP$, cAMP, and cAMP-dependent protein kinase to a final level of 1.0 mol of total protein-bound phosphate (labeled and unlabeled) per subunit. Phosphorylation of the hydroxylase is accompanied by a 2.5- to 3.0-fold increase in its activity as measured using the naturally occurring cofactor, tetrahydrobiopterin. The kinetic properties of fully phosphorylated phenylalanine hydroxylase are essentially the same as those of the "native" enzyme as isolated, i.e. the partially phosphorylated form, except for a higher V_{max}. It has not been possible thus far to obtain fully dephosphorylated phenylalanine hydroxylase using a partially purified liver phosphoprotein phosphatase (355). Phosphatase treatment does cause reversal of the activation that had been achieved by subjecting the enzyme to phosphorylating conditions. In a recent study (356) Donlon & Kaufman have shown that liver phenylalanine hydroxylase can be phosphorylated and activated in vivo in response to glucagon.

Tryptophan Hydroxylase

The most recent hydroxylase found to be activated by enzymic phosphorylation is tryptophan hydroxylase (180, 357, 358). Although only preliminary data are available, the reports suggest that the protein kinase involved may be a Ca^{2+}-requiring enzyme, although in one of the studies (180) the phosphorylation reaction was also stimulated by cAMP. Phosphorylation of the enzyme leads to a reduction in the apparent K_m for the pteridine cofactor (180, 358). In one set of experiments (180) it was found that the K_m for tryptophan is also reduced by phosphorylation.

NAD-Dependent Glutamate Dehydrogenase (Yeast)

In a recent report it was shown that yeast NAD-dependent glutamate dehydrogenase exists in two forms that were designated as a and b (359). NAD-dependent glutamate dehydrogenase b is found in starved yeast cells and manifests reduced affinity for L glutamate as compared to the a form. When the starvation experiment was carried out in the presence of ^{32}P it was found that the enzyme had incorporated phosphate (182). Fully active enzyme contained 0.09 mol of P per mol of enzyme subunit, whereas the b form contained 1.25 mol of P per mol of enzyme subunit. The dephosphorylation of NAD-dependent glutamate dehydrogenase b could be achieved with E. coli alkaline phosphatase or with crude yeast extract as a source of enzyme.

DNA-Dependent RNA Polymerase

Since hormones that affect other systems via phosphorylation mechanisms are known to affect RNA and protein synthesis, considerable work has been carried out on the possible effects of reversible phosphorylation on the proteins involved in these events (for review see 360–362). Other more general reviews of the regulation of transcription in response to physiological stimuli have also been presented (363, 364).

Early reports by Jungmann et al (365, 366), Martelo et al (367, 368), and Dahmus (369) all indicated that in vitro treatment of RNA polymerase I and II with partially purified protein kinase preparations and ATP caused an increase in polymerase activity. These studies have been extended to show that both cAMP-dependent and -independent protein kinases can catalyze phosphorylation and activation of calf thymus polymerase II (370, 371). In addition the activity can be decreased and the serine phosphate content reduced by treatment with alkaline phosphatase (370). Unfortunately, the changes in activity have not generally been large (usually less than three-fold), and more importantly, it has not yet been possible to demonstrate a good stoichiometric correlation between phosphorylation and activity.

A major reason for these latter difficulties may be explained by the observation that a rather large number of the subunits and/or tightly associated proteins in the polymerase complex are phosphorylated. In particular, experiments with yeast that have been grown in medium containing $^{32}P_i$ (372) indicate that the 24,000 mol wt subunit of all the polymerases, and a large number of other subunits in the polymerases I and III are phosphorylated. One might assume that similar multiple phosphorylations occur in mammalian RNA polymerases, particularly in polymerase I and III. Furthermore, it has been shown that multiple kinases catalyze the phosphorylation of polymerase II in calf thymus (40) although it not known if different sites are involved. Therefore, by analogy to some of the other enzyme systems described in this review, where multiple kinases and multiple phosphorylations can all affect enzyme activity, it is not surprising that there might be some difficulty in trying to correlate enzyme activity with total phosphate content.

It also has been reported (373) that E. coli RNA polymerase is phosphorylated on a threonine residue(s) by a bacteriophage T_4–induced protein kinase. Analysis of immunoprecipitates indicated that both the B and B' subunits are phosphorylated. Little is known about the specificity and regulation of the kinase involved, or if any activity changes occur in response to phosphorylation. Similarly, (374) crude preparations containing cAMP-dependent and -independent enzymes and added ATP can increase the activity of a Rous sarcoma virus RNA–dependent DNA nucleotidyl transferase (reverse transcriptase). Little is known about the possible function of this process.

CONCLUDING REMARKS

Table 2 shows that we are in a period in which new examples of enzymes undergoing phosphorylation-dephosphorylation are being revealed at a rapid, albeit a fairly steady, rate. Thus, it can be anticipated that if the present trend continues, another 10–15 enzymes could be added to the group within the next 5 years. Good candidates for the list include enzymes known to be subject to hormonal control and, interestingly, those enzymes already known to be regulated by allosteric effectors. The existence of what might appear to be a surfeit of regulatory mechanisms does not in any way negate the likelihood of an enzyme also being regulated by covalent modification. Indeed, consideration of the examples in Table 2 would imply that the opposite may be true. Just as the number of substrates known to undergo phosphorylation-dephosphorylation are increasing rapidly, so also are the number of kinases that catalyze these reactions. The recent increase in knowledge concerning these enzymes and substrates is leading to a more detailed understanding of the regulation of many physiologically important processes.

ACKNOWLEDGMENTS

The authors would like to thank Ms. Sally Hopkins for her efforts during the preparation of this manuscript. We also would like to thank the investigators who provided us with copies of their work in press. Work in J. A. Beavo's laboratory is supported in part by the American Heart Association and USPHS Grant AM21723, and he is an Established Investigator of the American Heart Association.

Literature Cited

1. Uy, R., Wold, F. 1977. *Science* 198: 890–96
2. Fischer, E. H., Krebs, E. G. 1955. *J. Biol. Chem.* 216:121–32
3. Sutherland, E. W. Jr., Wosilait, W. D. 1955. *Nature* 175:169–71
4. Gershey, E. L., Vidali, G., Allfrey, V. G. 1968. *J. Biol. Chem.* 243:5018–22
5. Holzer, H., Mecke, D., Wulff, K., Liess, K., Heilmeyer, L. Jr. 1967. *Adv. Enzyme Regul.* 5:211–25
6. Stadtman, E. R., Shapiro, B. M., Ginsburg, A., Kingdon, H. S., Denton, M. D. 1968. *Brookhaven Symp. Biol.* 21:378–96
7. Adler, S. P., Mangum, J. H., Magni, G., Stadtman, E. R. 1974. In *Metabolic Interconversion of Enzymes,* 1973, ed. E. H. Fischer, E. G. Krebs, H. Neurah, E. R. Stadtman, 221–33. Berlin: Springer. 399 pp.
8. Kort, E. N., Goy, M. F., Larsen, S. H., Adler, J. 1975. *Proc. Natl. Acad. Sci. USA* 72:3939–43
9. Rodnight, R., Lavin, B. E. 1964. *Biochem. J.* 93:84–91
10. Traugh, J. A., Mumby, M., Traut, R. R. 1973. *Proc. Natl. Acad. Sci. USA* 70:373–76
11. Nimmo, H. G., Cohen, P. 1974. *FEBS Lett.* 47:162–66

12. Smith, D. L., Chen, C. C., Bruegger, B. B., Holtz, S. L., Halpern, R. M., Smith, R. A. 1974. *Biochemistry* 13: 3780–85
13. Rabinowitz, M., Lipmann, F. 1960. *J. Biol. Chem.* 235:1043–50
14. Lerch, K., Muir, L. W., Fischer, E. H. 1975. *Biochemistry* 14:2015–23
15. Shizuta, Y., Beavo, J. A., Bechtel, P. J., Hofmann, F., Krebs, E. G. 1975. *J. Biol. Chem.* 250:6891–96
16. Shizuta, Y., Khandelwal, R. L., Maller, J. L., Vandenheede, J. R., Krebs, E. G. 1977. *J. Biol. Chem.* 252:3408–13
17. Rosen, O. M., Erlichman, J. 1975. *J. Biol. Chem.* 250:7788–94
18. Rahmsdorf, H. J., Pai, S. H., Ponta, H., Herrlich, P., Roskoski, R. Jr., Schweiger, M., Studier, F. W. 1974. *Proc. Natl. Acad. Sci. USA* 71:586–89
19. Wang, J. Y. J., Koshland, D. E. Jr. 1978. *J. Biol. Chem.* 253:7605–8
20. Martelo, O. J., Woo, S. L. C., Reimann, E. M., Davie, E. W. 1970. *Biochemistry* 9:4807–13
21. Fakunding, J. L., Traugh, J. A., Traut, R. R., Hershey, J. W. B. 1972. *J. Biol. Chem.* 247:6365–67
22. Mersmann, H. J., Segal, H. L. 1969. *J. Biol. Chem.* 244:1701–4
23. Jennissen, H. P., Hörl, W. H., Gröschel-Stewart, U., Velick, S. F., Heilmeyer, L. M. G. Jr. 1976. In *Metabolic Interconversion of Enzymes 1975*, ed. S. Shaltiel, 19–26. New York: Springer. 234 pp.
24. Costa, M., Gerner, E. W., Russell, D. H. 1976. *J. Biol. Chem.* 251:3313–19
25. Lee, P. C., Radloff, D., Schwepe, J. S., Jungmann, R. A. 1976. *J. Biol. Chem.* 251:914–21
26. Gharrett, A. J., Malkinson, A. M., Sheppard, J. R. 1976. *Nature* 264: 673–75
27. Fuller, D. J. M., Byus, C. V., Russell, D. H. 1978. *Proc. Natl. Acad. Sci. USA* 75:223–27
28. Vanstapel, F., Stalmans, W. 1978. *Arch. Int. Physiol. Biochim.* 86:966–7
29. Krebs, E. G. 1972. *Curr. Top. Cell. Regul.* 5:99–133
30. Walsh, D. A., Krebs, E. G. 1973. *Enzymes* 8:555–81
31. Langan, T. A. 1973. *Adv. Cyclic Nucleotide Res.* 3:99–153
32. Taborsky, G. 1974. *Adv. Protein Chem.* 28:1–210
33. Rubin, C. S., Rosen, O. M. 1975. *Ann. Rev. Biochem.* 44:831–87
34. Nimmo, H. G., Cohen, P. 1977. *Adv. Cyclic Nucleotide Res.* 8:145–266
34a. Carlson, G. M., Bechtel, P. J., Graves, D. J. 1979. *Adv. Enzymol.* 50: In press

35. Corbin, J. D., Keely, S. L., Park, C. R. 1975. *J. Biol. Chem.* 250:218–25
36. Hofmann, F., Beavo, J. A., Bechtel, P. J., Krebs, E. G. 1975. *J. Biol. Chem.* 250:7795–801
37. Walsh, D. A., Perkins, J. P., Krebs, E. G. 1968. *J. Biol. Chem.* 243:3763–65
38. Kuo, J. F., Greengard, P. 1970. *J. Biol. Chem.* 245:2493–98
39. Gill, G. N., Holdy, K. E., Walton, G. M., Kanstein, C. B. 1976. *Proc. Natl. Acad. Sci. USA* 73:3918–22
40. Lincoln, T. M., Dills, W. L. Jr., Corbin, J. D. 1977. *J. Biol. Chem.* 252:4269–75
41. Lincoln, T. M., Corbin, J. D. 1977. *Proc. Natl. Acad. Sci. USA* 74:3239–43
42. Gill, G. N., Walton, G. M., Sperry, P. J. 1977. *J. Biol. Chem.* 252:6443–49
43. Lincoln, T. M., Flockhart, D. A., Corbin, J. D. 1978. *J. Biol. Chem.* 253:6002–9
44. Corbin, J. D., Lincoln, T. M. 1978. *Adv. Cyclic Nucleotide Res.* 9:159–70
45. Krebs, E. G., Stull, J. T., England, P. J., Huang, T. S., Brostrom, C. O., Vandenheede, J. R. 1973. In *Protein Phosphorylation in Control Mechanisms, Miami Winter Symp.* 5:31–45
46. Lebleu, B., Sen, G. C., Shaila, S., Cabrer, B., Lengyel, P. 1976. *Proc. Natl. Acad. Sci. USA* 73:3107–111
47. Zilberstein, A., Federman, P., Shulman, L., Revel, M. 1976. *FEBS Lett.* 68: 119–24
48. Roberts, W. K., Hovanessian, A., Brown, R. E., Clemens, M. J., Kerr, I. M. 1976. *Nature* 264:477–80
49. Ernst, V., Levin, D. H., Ranu, R. S., London, I. M. 1976. *Proc. Natl. Acad. Sci USA* 73:1112–16
50. Sen, G. C., Taira, H., Lengyel, P. 1978. *J. Biol. Chem.* 253:5915–21
51. Meyer, W. L., Fischer, E. H., Krebs, E. G. 1964. *Biochemistry* 3:1033–39
52. Ozawa, E., Hosoi, K., Ebashi, S. 1967. *J. Biochem.* 61:531–33
53. Dabrowska, R., Sherry J. M. F., Aromatorio, D. K., Hartshorne, D. J. 1978. *Biochemistry* 17:253–58
54. Yagi, K., Yazawa, M., Kakiuchi, S., Ohshima, M., Uenishi, K. 1978. *J. Biol. Chem.* 253:1338–40
55. Pires, E., Perry, S. V., Thomas, M. A. W. 1974. *FEBS Lett.* 41:292–96
56. Cheung, W. Y. 1967. *Biochem. Biophys. Res. Commun.* 29:478–82
57. Waisman, D. M., Singh, T. J., Wang, J. H. 1978. *J. Biol. Chem.* 253:3387–90
58. Cohen, P., Burchell, A., Foulkes, J. G., Cohen, P. T. W., Vanaman, T. C., Nairn, A. C. 1978. *FEBS Lett.* 92: 287–93

58a. Schulman, H., Greengard, P. 1978. *Proc. Natl. Acad. Sci. USA* 75:5432–36
59. Ogez, J. R., Segel, I. H. 1976. *J. Biol. Chem.* 251:4551–56
60. Swillens, S., Dumont, J. E. 1976. *J. Mol. Med.* 1:273–88
61. Chau, V., Anderson, C., Huang, C., Huang, L. C. 1978. *Fed. Proc.* 37(6): 1329 (Abstr.)
62. Corbin, J. D., Keely, S. L. 1977. *J. Biol. Chem.* 252:910–18
63. Corbin, J. D., Sugden, P. H., Lincoln, T. M., Keely, S. L. 1977. *J. Biol. Chem.* 252:3854–61
64. Erlichman, J., Rosenfeld, R., Rosen, O. M. 1974. *J. Biol. Chem.* 249:5000–3
65. Datta, A., DeHaro, C., Ochoa, S. 1978. *Proc. Natl. Acad. Sci. USA* 75:1148–52
66. Walsh, D. A., Ashby, C. D., Gonzalez, C., Calkins, D., Fischer, E. H., Krebs, E. G. 1971. *J. Biol. Chem.* 246:1977–85
67. Demaille, J. G., Peters, K. A., Fischer, E. H. 1977. *Biochemistry* 16:3080–86
68. Weber, H., Rosen, O. M. 1977. *J. Cyclic Nucleotide Res.* 3:415–24
69. Walsh, D. A., Ashby, C. D. 1973. *Recent Progr. Horm. Res.* 29:329–59
70. Beavo, J. A., Bechtel, P. J., Krebs, E. G. 1974. *Proc. Natl. Acad. Sci. USA* 71:3580–83
71. Kuo, J. F. 1975. *Biochem. Biophys. Res. Commun.* 65:1214–20
72. Costa, M. 1977. *Biochem. Biophys. Res. Commun.* 78:1311–18
73. Traugh, J. A., Ashby, C. D., Walsh, D. A. 1974. *Methods Enzymol.* 38: 290–99
74. Beale, E. G., Dedman, J. R., Means, A. R. 1977. *J. Biol. Chem.* 252:6322–27
75. Szmigielski, A., Guidotti, A., Costa, E. 1977. *J. Biol. Chem.* 252:3848–53
76. Takai, Y., Nakaya, S., Inoue, M., Kishimoto, A., Nishiyama, K., Yamamura, H., Nishizuka, Y. 1976. *J. Biol. Chem.* 251:1481–87
77. Inoue, M., Kishimoto, A., Takai, Y., Nishizuka, Y. 1976. *J. Biol. Chem.* 251:4476–78
78. Miyamoto, E., Petzold, G. L., Kuo, J. F., Greengard, P. 1973. *J. Biol. Chem.* 248:179–89
79. Van Leemput-Coutrez, M., Camus, J., Christophe, J. 1973. *Biochem. Biophys. Res. Commun.* 54:182–90
80. Kobayashi, R., Fang, V. S. 1976. *Biochem. Biophys. Res. Commun.* 69: 1080–87
81. Shoji, M., Patrick, J. G., Tse, J., Kuo, J. F. 1977. *J. Biol. Chem.* 252:4347–53
82. Shoji, M., Brackett, N. L., Tse, J., Shapira, R., Kuo, J. F. 1978. *J. Biol. Chem.* 253:3427–34

83. Yamamoto, M., Takai, Y., Hashimoto, E., Nishizuka, Y. 1977. *J. Biochem.* 81:1857–62
84. deJonge, H. R., Rosen, O. M. 1977. *J. Biol. Chem.* 252:2780–83
85. Brostrom, C. O., Hunkeler, F. L., Krebs, E. G. 1971. *J. Biol. Chem.* 246:1961–67
86. Nolan, C., Novoa, W. B., Krebs, E. G., Fischer, E. H. 1964. *Biochemistry* 3:542–51
87. Cohen, P., Watson, D. C., Dixon, G. H. 1975. *Eur. J. Biochem.* 51:79–92
88. Daile, P., Carnegie, P. R. 1974. *Biochem. Biophys. Res. Commun.* 61: 852–58
89. Bylund, D. B., Krebs, E. G. 1975. *J. Biol. Chem.* 250:6355–61
90. Humble, E., Berglund, L., Titanji, V., Ljungström, O., Edlund, B., Zetterqvist, Ö., Engström, L. 1975. *Biochem. Biophys. Res. Commun.* 66:614–21
91. Kemp, B. E., Bylund, D. B., Huang, T. S., Krebs, E. G. 1975. *Proc. Natl. Acad. Sci. USA* 72:3448–52
92. Shlyapnikov, S. V., Arutyunyan, A. A., Kurochkin, S. N., Memelova, L. V., Nesterova, M. V., Sashchenko, L. P., Severin, E. S. 1975. *FEBS Lett.* 53: 316–19
93. Daile, P., Carnegie, P. R., Young, J. D. 1975. *Nature* 257:416–18
94. Kemp, B. E., Benjamini, E., Krebs, E. G. 1976. *Proc. Natl. Acad. Sci. USA* 73:1038–42
95. Hjelmquist, G., Andersson, J., Edlund, B., Engström, L. 1974. *Biochem. Biophys. Res. Commun.* 61:559–63
96. Zetterqvist, Ö., Ragnarsson, U., Humble, E., Berglund, L., Engström, L. 1976. *Biochem. Biophys. Res. Commun.* 70:696–703
97. Kemp, B. E., Graves, D. J., Benjamini, E., Krebs, E. G. 1977. *J. Biol. Chem.* 252:4888–94
98. Yeaman, S. J., Cohen, P., Watson, D. C., Dixon, G. H. 1977. *Biochem. J.* 162:411–21
99. Huang, T. S., Krebs, E. G. 1977. *Biochem. Biophys. Res. Commun.* 75: 643–50
100. Proud, C. G., Rylatt, D. B., Yeaman, S. J., Cohen, P. 1977. *FEBS Lett.* 80:435–42
101. Cohen, P., Rylatt, D. B., Nimmo, G. A. 1977. *FEBS Lett.* 76:182–86
102. Shenolikar, S., Cohen, P. 1978. *FEBS Lett.* 86:92–98
103. Pomerantz, A. H., Allfrey, V. G., Merrifield, R. B., Johnson, E. M. 1977. *Proc. Natl. Acad. Sci. USA* 74:4261–65
104. Turaev, O. D., Burichenko, V. K. 1978. *FEBS Lett.* 88:59–61

105. Small, D., Chou, P. Y., Fasman, G. D. 1977. *Biochem. Biophys. Res. Commun.* 79:341–46
106. Khoo, J. C., Sperry, P. J., Gill, G. N., Steinberg, D. 1977. *Proc. Natl. Acad. Sci. USA* 74:4843–47
107. Riou, J. P., Claus, T. H., Flockhart, D. A., Corbin, J. D., Pilkis, S. J. 1977. *Proc. Natl. Acad. Sci. USA* 74:4615–19
108. Blumenthal, D. K., Stull, J. T., Gill, G. N. 1978. *J. Biol. Chem.* 253:334–36
109. Lincoln, T. M., Corbin, J. D. 1978. *J. Biol. Chem.* 253:337–39
110. Chihara-Nakashima, M., Hashimoto, E., Nishizuka, Y. 1977. *J. Biochem.* 81:1863–67
111. Edlund, B., Zetterqvist, Ö., Ragnarsson, U., Engström, L. 1977. *Biochem. Biophys. Res. Commun.* 79:139–44
112. Glass, D. B., Krebs, E. G. 1978. *Fed. Proc.* 37(6):1329 (Abstr.)
113. Hashimoto, E., Takeda, M., Nishizuka, Y., Hamana, K., Iwai, K. 1976. *J. Biol. Chem.* 251:6287–93
114. Casnellie, J. E., Greengard, P. 1974. *Proc. Natl. Acad. Sci. USA* 71:1891–95
115. DeLange, R. J., Kemp, R. G., Riley, W. D., Cooper, R. A., Krebs, E. G. 1968. *J. Biol. Chem.* 243:2200–8
116. Moir, A. J. G., Wilkinson, J. M., Perry, S. V. 1974. *FEBS Lett.* 42:253–56
117. Huang, T. S., Bylund, D. B., Stull, J. T., Krebs, E. G. 1974. *FEBS Lett.* 42:249–52
118. Wang, J. H., Stull, J. T., Huang, T. S., Krebs, E. G. 1976. *J. Biol. Chem.* 251:4521–27
119. Schwartz, A., Entman, M. L., Kaniike, K., Lane, L. K., VanWinkle, W. B., Bornet, E. P. 1976. *Biochim. Biophys. Acta* 426:57–72
120. Hörl, W. H., Heilmeyer, L. M. G. Jr. 1978. *Biochemistry* 17:766–72
121. Roach, P. J., DePaoli-Roach, A. A., Larner, J. 1978. *J. Cyclic Nucleotide Res.* In press
122. Tessmer, G. W., Skuster, J. R., Tabatabai, L. B., Graves, D. J. 1977. *J. Biol. Chem.* 252:5666–71
123. Perrie, W. T., Smillie, L. B., Perry, S. V. 1973. *Biochem. J.* 135:151–64
123a. Jakes, R., Northrop, F., Kendrick-Jones, J. 1976. *FEBS Lett.* 70:229–34
124. Barrera, C. R., Namihira, G., Hamilton, L., Munk, P., Eley, M. H., Linn, T. C., Reed, L. J. 1972. *Arch. Biochem. Biophys.* 148:343–58
125. Yeaman, S. J., Hutcheson, E. T., Roche, T. E., Pettit, F. H., Brown, J. R., Reed, L. J.. Watson, D. C., Dixon, G. H. 1978. *Biochemistry* 17:2364–70
126. Davis, P. F., Pettit, F. H., Reed, L. J. 1977. *Biochem. Biophys. Res. Commun.* 75:541–49
127. England, P. J., Stull, J. T., Krebs, E. G. 1972. *J. Biol. Chem.* 247:5275–77
128. Kato, K., Bishop, J. S. 1972. *J. Biol. Chem.* 247:7420–29
129. Zieve, F. J., Glinsmann, W. H. 1973. *Biochem. Biophys. Res. Commun.* 50:872–78
130. Kato, K., Sato, S. 1974. *Biochim. Biophys. Acta* 358:299–307
131. Kanai, C., Thomas, J. A. 1974. *J. Biol. Chem.* 249:6459–67
132. Killilea, S. D., Brandt, H., Lee, E. Y. C., Whelan, W. J. 1976. *J. Biol. Chem.* 251:2363–68
133. Khandelwal, R. L., Vandenheede, J. R., Krebs, E. G. 1976. *J. Biol. Chem.* 251:4850–58
134. Chou, C. K., Alfano, J., Rosen, O. M. 1977. *J. Biol. Chem.* 252:2855–59
135. Li, H. C., Hsiao, K. J., Chan, W. W. S. 1978. *Eur. J. Biochem.* 84:215–25
136. Antoniw, J. F., Nimmo, H. G., Yeaman, S. J., Cohen, P. 1977. *Biochem. J.* 162:423–33
137. Ray, K. P., England, P. J. 1976. *Biochem. J.* 157:369–80
138. Brandt, H., Capulong, Z. L., Lee, E. Y. C. 1975. *J. Biol. Chem.* 250:8038–44
139. Kato, K., Kobayashi, M., Sato, S. 1974. *Biochim. Biophys. Acta* 371:89–101
140. Kobayashi, M., Kato, K. 1977. *J. Biochem.* 81:93–97
141. Merlevede, W., Defreyn, G., Goris, J., Kalala, L. R., Roosemont, J. 1976. *Arch. Int. Physiol. Biochem.* 84:359–423
142. Lee, E. Y. C., Mellgren, R. L., Killilea, S. D., Aylward, J. H. 1978. In *Regulatory Mechanisms of Carbohydrate Metabolism*, ed. V. Esmann, 42:327–46. Oxford/New York: Pergamon
143. Gratecos, D., Detwiler, T. C., Hurd, S., Fischer, E. H. 1977. *Biochemistry* 16:4812–17
144. Stalmans, W., DeWulf, H., Hers, H. G. 1971. *Eur. J. Biochem.* 18:582–87
145. Stalmans, W., DeWulf, H., Hue, L., Hers, H. G. 1974. *Eur. J. Biochem.* 41:127–34
146. Hers, H. G. 1976. *Ann. Rev. Biochem.* 45:167–89
147. Bot, G., Varsányi, M., Gergely, P. 1975. *FEBS Lett.* 50:351–54
148. Gergely, P., Dombrádi, V., Bot, G. 1978. *FEBS Lett.* 93:239–41
149. Huang, F. L., Glinsmann, W. H. 1975. *Proc. Natl. Acad. Sci. USA* 72:3004–8
150. Huang, F. L., Glinsmann, W. H. 1976. *FEBS Lett.* 62:326–29

151. Huang, F. L., Glinsmann, W. H. 1976. *Eur. J. Biochem.* 70:419–26
152. Khandelwal, R. L., Zinman, S. M. 1978. *J. Biol. Chem.* 253:560–65
153. Tóth, G., Gergely, P., Parsadanian, H. K., Bot, G. 1977. *Acta Biochim. Biophys. Acad. Sci. Hung.* 12:389–98
154. Nimmo, G. A., Cohen, P. 1978. *Eur. J. Biochem.* 87:353–65
155. Burchell, A., Foulkes, J. G., Cohen, P. T. W., Condon, G. D., Cohen, P. 1978. *FEBS Lett.* 92:68–72
156. Nimmo, G. A., Cohen, P. 1978. *Eur. J. Biochem.* 87:341–51
157. Nakai, C., Glinsmann, W. 1977. *Mol. Cell. Biochem.* 15:133–39
158. Kikuchi, K., Tamura, S., Hiraga, A., Tsuiki, S. 1977. *Biochem. Biophys. Res. Commun.* 75:29–37
159. Tan, A. W. H., Nuttall, F. Q. 1978. *Biochim. Biophys. Acta* 522:139–50
160. Laloux, M., Stalmans, W., Hers, H. G. 1978. *Eur. J. Biochem.* 92:15–24
161. Imaoka, T., Imazu, M., Ishida, N., Takeda, M. 1978. *Biochim. Biophys. Acta* 523:109–20
162. Linn, T. C., Pettit, F. H., Hucho, F., Reed, L. J. 1969. *Proc. Natl. Acad. Sci. USA* 64:227–34
163. Krebs, E. G., Graves, D. J., Fischer, E. H. 1959. *J. Biol. Chem.* 234:2867–73
164. Friedman, D. L., Larner, J. 1963. *Biochemistry* 2:669–75
165. Rizack, M. A. 1964. *J. Biol. Chem.* 239:392–95
166. Corbin, J. D., Reimann, E. M., Walsh, D. A., Krebs, E. G. 1970. *J. Biol. Chem.* 245:4849–51
167. Huttunen, J. K., Steinberg, D., Mayer, S. E. 1970. *Proc. Natl. Acad. Sci. USA* 67:290–95
168. Mendicino, J., Beaudreau, C., Bhattacharyya, R. N. 1966. *Arch. Biochem. Biophys.* 116:436–45
169. Linn, T. C., Pettit, F. H., Reed, L. J. 1969. *Proc. Natl. Acad. Sci. USA* 62:234–41
170. Beg, Z. H., Allmann, D. W., Gibson, D. M. 1973. *Biochem. Biophys. Res. Commun.* 54:1362–69
171. Carlson, C. A., Kim, K. H. 1973. *J. Biol. Chem.* 248:378–80
172. Martelo, O. J. 1973. In *Protein Phosphorylation in Control Mechanisms, Miami Winter Symp.* eds. F. Huijing, E. Y. C. Lee, 5:199–216. New York: Academic
173. Ljungström, O., Hjelmquist, G., Engström, L. 1974. *Biochim. Biophys. Acta* 358:289–98
174. Trzeciak, W. H., Boyd, G. S. 1974. *Eur. J. Biochem.* 46:201–7
175. Lee, S. G., Miceli, M. V., Jungmann, R. A., Hung, P. P. 1975. *Proc. Natl. Acad. Sci. USA* 72:2945–49
176. Brand, I. A., Söling, H. D. 1975. *FEBS Lett.* 57:163–68
177. Morgenroth, V. H. III, Hegstrand, L. R., Roth, R. H., Greengard, P. 1975. *J. Biol. Chem.* 250:1946–48
178. Milstien, S., Abita, J. P., Chang, N., Kaufman, S. 1976. *Proc. Natl. Acad. Sci. USA* 73:1591–93
179. Datta, A., DeHaro, C., Sierra, J. M., Ochoa, S. 1977. *Proc. Natl. Acad. Sci. USA* 74:1463–67
180. Hamon, M., Bourgoin, S., Héry, F., Simonnet, G. 1978. *Mol. Pharmacol.* 14:99–110
181. Hamon, M., Bourgoin, S., Artaud, F., Héry, F. 1977. *J. Neurochem.* 28:811–18
182. Hemmings, B. A. 1978. *J. Biol. Chem.* 253:5255–58
183. Nimmo, H. G., Houston, B. 1978. *Biochem. J.* 176:607–10
184. Narumi, S., Miyamoto, E. 1974. *Biochim. Biophys. Acta* 350:215–24
185. Qureshi, A. A., Jenik, R. A., Kim, M., Larnitzo, F. A., Porter, J. W. 1975. *Biochem. Biophys. Res. Commun.* 66:344–51
186. Lee, K. L., Nickol, J. M. 1974. *J. Biol. Chem.* 249:6024–26
187. Krebs, E. G. 1972. *Proc. IV Intern. Cong. Endocrin., Intern. Congr. Ser. No. 273, Amsterdam,* pp. 17–29
188. Walsh, D. A., Cooper, R. H. 1979. *Biochem. Actions Horm.* 6:In press
189. Krebs, E. G., Preiss, J. 1976. In *MTP International Review of Science. Biochemistry of Carbohydrates,* ed. W. J. Whelan, 5:337–89. Baltimore: Univ. Park Press. 441 pp.
190. Posner, J. B., Stern, R., Krebs, E. G. 1965. *J. Biol. Chem.* 240:982–85
191. Mayer, S. E., Krebs, E. G. 1970. *J. Biol. Chem.* 245:3153–60
192. Yeaman, S. J., Cohen, P. 1975. *Eur. J. Biochem.* 51:93–104
193. Drummond, G. I., Harwood, J. P., Powell, C. A. 1969. *J. Biol. Chem.* 244:4235–40
194. Stull, J. T., Mayer, S. E. 1971. *J. Biol. Chem.* 246:5716–23
195. Dobson, J. G. Jr., Mayer, S. E. 1973. *Circul. Res.* 33:412–20
196. Kishimoto, A., Takai, Y., Nishizuka, Y. 1977. *J. Biol. Chem.* 252:7449–52
197. Lincoln, T. M., Corbin, J. D. 1978. *J. Cyclic Nucleotide Res.* 4:3–14
198. VanDeWerve, G., VanDenBerghe, G., Hers, H. G. 1974. *Eur. J. Biochem.* 41:97–102

199. Vandenheede, J. R., Keppens, S., De-Wulf, H. 1976. *FEBS Lett.* 61:213–17
200. Vandenheede, J. R., Keppens, S., De-Wulf, H. 1977. *Biochim. Biophys. Acta* 481:463–70
201. Vandenheede, J. R., Khandelwal, R. L., Krebs, E. G. 1977. *J. Biol. Chem.* 252:7488–94
202. VanDeWerve, G., Hue, L., Hers, H. G. 1977. *Biochem. J.* 162:135–42
203. Khoo, J. C., Steinberg, D. L. 1975. *FEBS Lett.* 57:68–72
204. Shimazu, T., Amakawa, A. 1975. *Biochim. Biophys. Acta* 385:242–56
205. Fischer, E. H., Heilmeyer, L. M. G. Jr., Haschke, R. H. 1971. *Curr. Top. Cell. Regul.* 4:211–51
206. Graves, D. J., Wang, J. H. 1972. *Enzymes* 7:435–82
207. Busby, S. J. W., Radda, G. K. 1976. *Curr. Top. Cell. Regul.* 10:89–160
208. Hutson, N. J., Brumley, F. T., Assimacopoulos, F. D., Harper, S. C., Exton, J. H. 1976. *J. Biol. Chem.* 251:5200–8
209. Cherrington, A. D., Assimacopoulos, F. D., Harper, S. C., Corbin, J. D., Park, C. R., Exton, J. H. 1976. *J. Biol. Chem.* 251:5209–18
210. Birnbaum, M. J., Fain, J. N. 1977. *J. Biol. Chem.* 252:528–35
211. Keppens, S., Vandenheede, J. R., De-Wulf, H. 1977. *Biochim. Biophys. Acta* 496:448–57
212. Assimacopoulos-Jeannet, F. D., Blackmore, P. F., Exton, J. H. 1977. *J. Biol. Chem.* 252:2662–69
213. Keppens, S., DeWulf, H. 1975. *FEBS Lett.* 51:29–32
214. Keppens, S., DeWulf, H. 1976. *FEBS Lett.* 68:279–82
215. Lawrence, J. C. Jr., Larner, J. 1977. *Mol. Pharmacol.* 13:1060–75
216. Chan, T. M., Exton, J. H. 1977. *J. Biol. Chem.* 252:8645–51
217. Blackmore, P. F., Brumley, F. T., Marks, J. L., Exton, J. H. 1978. *J. Biol. Chem.* 253:4851–58
218. Hurd, S. S., Teller, D., Fischer, E. H. 1966. *Biochem. Biophys. Res. Commun.* 24:79–84
219. Holmes, P. A., Mansour, T. E. 1968. *Biochim. Biophys. Acta* 156:275–84
220. Cori, G. T., Cori, C. F. 1945. *J. Biol. Chem.* 158:321–32
221. Stalmans, W., DeWulf, H., Lederer, B., Hers, H. G. 1970. *Eur. J. Biochem.* 15:9–12
222. Bot, G., Dósa, I. 1971. *Acta Biochim. Biophys. Acad. Sci. Hung.* 6:73–87
223. Stalmans, W., Laloux, M., Hers, H. G. 1974. *Eur. J. Biochem.* 49:415–27

224. Bailey, J. M., Whelan, W. J. 1972. *Biochem. Biophys. Res. Commun.* 46:191–97
225. Martensen, T. M., Brotherton, J. E., Graves, D. J. 1973. *J. Biol. Chem.* 248:8329–36
226. Kasvinsky, P. J., Madsen, N. B., Sygusch, J., Fletterick, R. J. 1978. *J. Biol. Chem.* 253:3343–51
227. Roach, P. J., Larner, J. 1977. *Mol. Cell. Biochem.* 15:179–200
228. Soderling, T. R., Park, C. R. 1974. *Rec. Adv. Cyclic Nucleotide Res.* 4:283–333
229. Smith, C. H., Brown, N. E., Larner, J. 1971. *Biochim. Biophys. Acta* 242:81–88
230. Rebhun, L. I., Smith, C., Larner, J. 1973. *Mol. Cell. Biochem.* 1:55–61
231. Schlender, K. K., Reimann, E. M. 1975. *Proc. Natl. Acad. Sci. USA* 72:2197–2201
232. Schlender, K. K., Reimann, E. M. 1977. *J. Biol. Chem.* 252:2384–89
233. Nimmo, H. G., Proud, C. G., Cohen, P. 1976. *Eur. J. Biochem.* 68:31–44
234. Huang, K. P., Huang, F. L., Glinsmann, W. H., Robinson, J. C. 1975. *Biochem. Biophys. Res. Commun.* 65:1163–69
235. Itarte, E., Robinson, J. C., Huang, K. P. 1977. *J. Biol. Chem.* 252:1231–34
236. Soderling, T. R., Jett, M. F., Hutson, N. J., Khatra, B. S. 1977. *J. Biol. Chem.* 252:7517–24
237. Brown, J. H., Thompson, B., Mayer, S. E. 1977. *Biochemistry* 16:5501–8
238. Kishimoto, A., Mori, T., Takai, Y., Nishizuka, Y. 1978. *J. Biochem.* 84:47–53
239. Huang, T. S., Krebs, E. G. 1977. *Biochem. Biophys. Res. Commun.* 75:643–650
240. Huang, K. P., Huang, F. L., Glinsmann, W. H., Robinson, J. C. 1976. *Arch. Biochem. Biophys.* 173:162–70
241. Hutson, N. J., Khatra, B. S., Soderling, T. R. 1978. *J. Biol. Chem.* 253:2540–45
242. Witters, L. A., Avruch, J. 1978. *Biochemistry* 17:406–10
243. Bloxham, D. P., Lardy, H. A. 1973. *Enzymes* 8:238–78
244. Wood, H. G. 1977. *Adv. Enzymol.* 45:85–155
245. Hofmann, E. 1976. *Rev. Physiol. Biochem. Pharmacol.* 75:1–68
246. Brand, I. A., Müller, M. K., Unger, C., Söling, H. D. 1976. *FEBS Lett.* 68:271–74
247. Hofer, H. W., Fürst, M. 1976. *FEBS Lett.* 62:118–22
248. Riquelme, P. T., Fox, R. W., Kemp, R. G. 1978. *Biochem. Biophys. Res. Commun.* 81:864–70

249. Hussey, C. R., Liddle, P. F., Ardron, D., Kellett, G. L. 1977. *Eur. J. Biochem.* 80:497–506
250. Kayne, F. J. 1973. *Enzymes* 8:353–82
251. Seubert, W., Schoner, W. 1971. *Curr. Top. Cell. Regul.* 3:237–67
252. Engström, L. 1978. *Curr. Top. Cell. Regul.* 13:29–51
253. Engström, L. 1978. *Proc. FEBS Meet. 11th, Copenhagen.* New York: Pergamon. pp. 53–60
254. Carbonnel, J., Felíu, J. E., Marco, R., Sols, A. 1973. *Eur. J. Biochem.* 37: 148–56
255. Ibsen, K. H., Trippet, P. 1972. *Biochemistry* 11:4442–50
256. Imamura, K., Tanaka, T. 1972. *J. Biochem.* 71:1043–51
257. Imamura, K., Taniuchi, K., Tanaka, T. 1972. 1972. *J. Biochem.* 72:1001–15
258. Harkins, R. N., Black, J. A., Rittenberg, M. B. 1977. *Biochemistry* 16: 3831–37
259. Engström, L., Berglund, L., Bergström, G., Hjelmquist, G., Ljungström, O. 1974. In *Lipmann Symposium: Energy, Regulation and Biosynthesis in Molecular Biology,* ed. D. Richter, pp. 192–204. Berlin: deGruyter. 701 pp.
260. Ekman, P., Dahlqvist, U., Humble, E., Engström, L. 1976. *Biochim. Biophys. Acta* 429:374–82
261. Ljungström, O., Berglund, L., Engström, L. 1976. *Eur. J. Biochem.* 68:497–506
262. Hjelmquist, G., Andersson, J., Edlund, B., Engström, L., 1974. *Biochem. Biophys. Res. Commun.* 61:559–63
236. Edlund, B., Andersson, J., Titanji, V., Dahlqvist, U., Ekman, P., Zetterqvist, Ö., Engström, L. 1975. *Biochem. Biophys. Res. Commun.* 67:1516–21
264. Titanji, V. P. K., Zetterqvist, Ö., Engström, L. 1976. *Biochim. Biophys. Acta* 422:98–108
265. Taunton, O. D., Stifel, F. B., Greene, H. L., Herman, R. H. 1974. *J. Biol. Chem.* 249:7228–39
266. Riou, J. P., Claus, T. H., Pilkis, S. J. 1976. *Biochem. Biophys. Res. Commun.* 73:591–99
267. Blair, J. B., Cimbala, M. A., Foster, J. L., Morgan, R. A. 1976. *J. Biol. Chem.* 251:3756–62
268. Felíu, J. E., Hue, L., Hers, H. G. 1976. *Proc. Natl. Acad. Sci. USA* 73:2762–66
269. Riou, J. P., Claus, T. H., Pilkis, S. J. 1978. *J. Biol. Chem.* 253:656–59
270. Ishibashi, H., Cottam, G. L. 1978. *J. Biol. Chem.* 253:8767–77
271. Foster, J. L., Blair, J. B. 1979. *J. Biol. Chem.* In press

272. Felíu, J. E., Hue, L., Hers, H. G. 1977. *Eur. J. Biochem.* 81:609–17
272a. Pilkis, S. J., Pilkis, J., Claus, T. H. 1978. *Biochem. Biophys. Res. Commun.* 81:139–46
273. Titanji, V. P. K. 1977. *Biochim. Biophys. Acta* 481:140–51
274. Pontremoli, S., DeFlora, A., Salamino, F., Melloni, E., Horecker, B. L. 1975. *Proc. Natl. Acad. Sci. USA* 72:2969–73
275. Reed, L. J. 1974. *Acc. Chem. Res.* 7:40–46
276. Reed, L. J., Pettit, F. H., Roche, T. E., Pelley, J. W., Butterworth, P. J. 1976. In *Metabolic Interconversion of Enzymes 1975,* ed. S. Shaltiel, pp. 121–24. Berlin: Springer
277. Denton, R. M., Randle, P. J., Bridges, B. J., Cooper, R. H., Kerbey, A. L., Pask, H. T., Severson, D. L., Stansbie, D., Whitehouse, S. 1975. *Mol. Cell. Biochem.* 9:27–53
278. Denton, R. M., Hughes, W. A. 1978. *Int. J. Biochem.* 9:545–57
279. Reed, L., Pettit, F., Yeaman, S. 1978. In *Microenvironments and Metabolic Compartmentation* ed. P. A. Srere, R. W. Estabrook, pp. 305–21. New York: Academic.
280. Denton, R. M., Hughes, W. A., Bridges, B. J., Brownsey, R. W., McCormack, J. G., Stansbie, D. 1978. *Hormones Cell Regul.* 2:191–208
281. Yeaman, S. J., Hutcheson, E. T., Roche, T. E., Pettit, F. H., Brown, J. R., Reed, L. J., Watson, D. C., Dixon, G. H. 1978. *Biochemistry* 17:2364–70
282. Hughes, W. A., Denton, R. M. 1979. *Biochem. Soc. Trans.* In press
283. Denton, R. M., Randle, P. J., Bridges, B. J., Cooper, R. H., Kerbey, A. L., Pask, H. T., Severson, D. L., Stansbie, D., Whitehouse, S. 1975. *Mol. Cell. Biochem.* 9:27–53
284. Taylor, S. I., Mukherjee, C., Jungas, R. L. 1975. *J. Biol. Chem.* 250:2028–35
285. Wieland, O. H., Portenhauser, R. 1974. *Eur. J. Biochem.* 45:577–88
286. Batenburg, J. J., Olson, M. S. 1976. *J. Biol. Chem.* 251:1364–70
287. Hansford, R. G. 1976. *J. Biol. Chem.* 251:5483–89
288. Davis, P. F., Pettit, F. H., Reed, L. J. 1977. *Biochem. Biophys. Res. Commun.* 75:541–49
289. Sugden, P. H., Hutson, N. J., Kerbey, A. L., Randle, P. J. 1978. *Biochem. J.* 169:433–35
290. Pettit, F. H., Pelley, J. W., Reed, L. J. 1975. *Biochem. Biophys. Res. Commun.* 65:575–82

291. Butler, J. R., Pettit, F. H., Davis, P. F., Reed, L. J. 1977. *Biochem. Biophys. Res. Commun.* 74:1667–74
292. Hansford, R. G. 1976. *J. Biol. Chem.* 251:5483–89
293. Cate, R. L., Roche, T. E. 1978. *J. Biol. Chem.* 253:496–503
294. Hughes, W. A., Denton, R. M. 1976. *Nature* 264:471–73
295. Hughes, W. A., Brownsey, R., Denton, R. M. 1977. *Phosphorylated Proteins and Their Related Enzymes* pp. 17–31. Inf. Retr. Ltd. 1 Faconberg Ct., London W1V 5FG
296. Hutson, N. J., Kerbey, A. L., Randle, P. J., Sugden, P. H. 1979. *J. Supramol. Struct.* In press
297. Hutson, N. J., Kerbey, A. L., Randle, P. J., Sugden, P. H. 1978. *Biochem. J.* 173:669–80
298. Steinberg, D. 1976. *Adv. Cyclic Nucleotide Res.* 7:157–98
299. Steinberg, D. 1978. In *Molecular Biology and Pharmacology of Cyclic Nucleotides,* ed. G. Folco, R. Paoleti, pp. 95–108. Amsterdam: North Holland Elsevier
300. Huttunen, J. K., Steinberg, D. 1971. *Biochim. Biophys. Acta* 293:411–27
301. Kissebah, A. H., Vydelingum, N., Tulloch, B. R., Hope-Gill, H., Fraser, T. R. 1974. *Horm. Metab. Res.* 6:247–55
302. Khoo, J. C., Steinberg, D., Huang, J. J., Vagelos, P. R. 1976. *J. Biol. Chem.* 251:2882–90
303. Khoo, J. C., Steinberg, D. 1974. *J. Lipid Res.* 15:602–10
304. Severson, D. L., Khoo, J. C., Steinberg, D. 1977. *J. Biol. Chem.* 252:1484–89
305. Huttunen, J. K., Heller, R., Steinberg, D. 1971. *Trans. Assoc. Am. Phys.* 84:162–71
306. Soderling, T. R., Corbin, J. D., Park, C. R. 1973. *J. Biol. Chem.* 248:1822–29
307. Gorban, A. M. S., Boyd, G. S. 1977. *FEBS Lett.* 79:54–58
308. Garren, L. D., Gill, G. N., Masui, H., Walton, G. M. 1971. *Recent Prog. Horm. Res.* 27:433–78
309. Shima, S., Mitsunaga, M., Nakao, T. 1972. *Endocrinology* 90:808–14
310. Trzeciak, W. H., Boyd, G. S. 1973. *Eur. J. Biochem.* 37:327–33
311. Naghshineh, S., Treadwell, C. R., Gallo, L., Vahouny, G. V. 1974. *Biochem. Biophys. Res. Commun.* 61:1076–82
312. Beckett, G. J., Boyd, G. S. 1977. *Eur. J. Biochem.* 72:223–33
313. Wallat, S., Kunau, W. H. 1976. *Hoppe-Seylers Z. Physiol. Chem.* 357:949–60
314. Pittman, R. C., Khoo, J. C., Steinberg, D. 1975. *J. Biol. Chem.* 250:4505–11
315. Inoue, H., Lowenstein, J. M. 1972. *J. Biol. Chem.* 247:4825–32
316. Carlson, C. A., Kim, K. H. 1974. *Arch. Biochem. Biophys.* 164:478–89
317. Carlson, C. A., Kim, K. H. 1974. *Arch. Biochem. Biophys.* 164:490–501
318. Lee, K. H., Kim, K. H. 1977. *J. Biol. Chem.* 252:1748–51
319. Hardie, D. G., Cohen, P. 1979. *FEBS Lett.* 91:1–70
320. Ahmad, F., Ahmad, P. M., Pieretti, L., Watters, G. T. 1978. *J. Biol. Chem.* 253:1733–37
321. Brownsey, R. W., Hughes, W. A., Denton, R. M., Mayer, R. J. 1977. *Biochem. J.* 168:441–45
322. Halestrap, A. P., Denton, R. M. 1973. *Biochem. J.* 132:509–17
323. Lee, K. H., Thrall, T., Kim, K. H. 1973. *Biochem. Biophys. Res. Commun.* 54:1133–40
324. Klain, G. J., Weiser, P. C. 1973. *Biochem. Biophys. Res. Commun.* 55:76–83
325. Allred, J. B., Roehrig, K. L. 1973. *J. Biol. Chem.* 248:4131–33
325a. Pekala, P. H., Meredith, M. J., Tarlow, D. M., Lane, M. D. 1978. *J. Biol. Chem.* 253:5267–69
326. Sooranna, S. R., Saggerson, E. D. 1976. *FEBS Lett.* 64:36–39
327. Sooranna, S. R., Saggerson, E. D. 1978. *FEBS Lett.* 90:141–44
328. Rodwell, V. W., Nordstrom, J. L., Mitschelen, J. J. 1976. *Adv. Lipid Res.* 14:1–74
329. Beytia, R. D., Porter, J. W. 1976. *Ann. Rev. Biochem.* 45:113–42
330. Gibson, D. M., Ingebritsen, T. S. 1978. *Life Sci.* 23:2649–64
331. Lakshmanan, M. R., Nepokroeff, C. M., Ness, G. C., Dugan, R. E., Porter, J. W. 1973. *Biochem. Biophys. Res. Commun.* 50:704–10
332. Nepokroeff, C. M., Lakshmanan, M. R., Ness, G. C., Dugan, R. E., Porter, J. W. 1974. *Arch. Biochem. Biophys.* 160:387–93
333. Dugan, R. E., Ness, G. C., Lakshmanan, M. R., Nepokroeff, C. M., Porter, J. W. 1974. *Arch. Biochem. Biophys.* 161:499–504
334. Lakshmanan, M. R., Dugan, R. E., Nepokroeff, C. M., Ness, G. C., Porter, J. W. 1975. *Arch. Biochem. Biophys.* 168:89–95
335. Bricker, L. A., Levey, G. S. 1972. *J. Biol. Chem.* 247:4914–15
336. Goodwin, C. D., Margolis, S. 1973. *J. Biol. Chem.* 248:7610–13

337. Berndt, J., Hegardt, F. G., Bove, J., Gaumert, R., Still, J., Cardo, M. T. 1976. *Hoppe-Seylers Z. Physiol. Chem.* 357:1277–82
338. Ingebritsen, T. S., Lee, H. S., Parker, R. A., Gibson, D. M. 1978. *Biochem. Biophys. Res. Commun.* 81:1268–77
339. Goodwin, C. D., Margolis, S. 1978. *J. Lipid Res.* 19:81–90
340. Beg, Z. H., Stonik, J. A., Brewer, H. B. Jr. 1978. *Proc. Natl. Acad. Sci. USA* 75:3678–82
341. Lovenberg, W., Bruckwick, E. A., Hanbauer, I. 1975. *Proc. Natl. Acad. Sci. USA* 72:2955–58
342. Lloyd, T., Kaufman, S. 1975. *Biochem. Biophys. Res. Commun.* 66:907–13
343. Roth, R. H., Walters, J. R., Murrin, L. C., Morgenroth, V. H. III, 1975. In *Preand Postsynaptic Receptors,* ed. E. Usdin, W. E. Bunney Jr., pp. 5–48. New York: Dekker
344. Raese, J. D., Edelman, A. M., Lazar, M. A., Barchas, J. D. 1977. In *Structure and Function of Monoamine Enzymes,* ed. E. Usdin, N. Weiner, M. B. H. Youdim, pp. 383–400. New York: Dekker
345. Yamauchi, T., Fujisawa, H. 1978. *Biochem. Biophys. Res. Commun.* 82: 514–17
346. Goldstein, M., Bronaugh, R. L., Ebstein, B., Roberge, C. 1976. *Brain Res.* 109:563–74
347. Zivkovic, B., Guidotti, A., Costa, E. 1976. *J. Cyclic Nucleotide Res.* 2:1–10
348. Morita, K., Oka, M., Izumi, F. 1977. *FEBS Lett.* 76:148–50
349. Ames, M. M., Lerner, P., Lovenberg, W. 1978. *J. Biol. Chem.* 253:27–31
350. Hoeldtke, R., Kaufman, S. 1977. *J. Biol. Chem.* 252:3160–69
351. Letendre, C. H., MacDonnell, P. C., Guroff, G. 1977. *Biochem. Biophys. Res. Commun.* 74:891–97
352. Roth, R. H., Morgenroth, V. H. III, Salzman, P. M. 1975. *Naunyn-Schmiedeberg's Arch. Pharmacol.* 289: 327–43
353. Roth, R. H., Salzman, P. M., Nowycky, M. C. 1978. In *Psychopharmacology: A Generation of Progress,* ed. M. A. Lipton, A. DiMascio, K. F. Killam, pp. 185–98. New York: Raven
354. Abita, J. P., Milstien, S., Chang, N., Kaufman, S. 1976. *J. Biol. Chem.* 251:5310–14
355. Jedlicki, E., Kaufman, S., Milstien, S. 1977. *J. Biol. Chem.* 252:7711–14
356. Donlon, J., Kaufman, S. 1978. *J. Biol. Chem.* 253:6657–59
357. Lysz, T. W., Sze, P. Y. 1978. *Trans. Am. Soc. Neurochem.* 9:128
358. Kuhn, D. M., Vogel, R. L., Lovenberg, W. 1978. *Biochem. Biophys. Res. Commun.* 82:759–66
359. Hemmings, B. A., Sims, A. P. 1977. *Eur. J. Biochem.* 80:143–51
360. Kleinsmith, L. J. 1975. *J. Cell. Physiol.* 85:459–76
361. Kleinsmith, L. J., Stein, J., Stein, G. 1975. In *Chromosomal Proteins and Their Role in the Regulation of Gene Expression,* ed. G. S. Stein, L. J. Kleinsmith, pp. 59–66. New York: Academic
362. Jungmann, R. A., Kranias, E. G. 1977. *Int. J. Biochem.* 8:819–30
363. Rutter, W. J., Goldberg, M. L., Perriard, J. C. 1974. In *Biochemistry of Cell Differentiation,* ed. J. Paul, pp. 161–80. Baltimore: Univ. Park Press
364. Roeder, R. G. 1976. In *RNA Polymerase,* ed. R. Losick, M. Chamberlin, pp. 285–329. Cold Spring Harbor, NY: Cold Spring Harbor Lab.
365. Jungmann, R. A., Hiestand, P. C., Schweppe, J. S. 1974. *J. Biol. Chem.* 249:5444–51
366. Jungmann, R. A., Kranias, E. G. 1976. In *Advances in Biochemical Psychopharmacology,* ed. E. Costa, G. Giacobini, R. Paoletti, pp. 413–28, New York: Raven
367. Martelo, O. J., Hirsch, J. 1974. *Biochem. Biophys. Res. Commun.* 58: 1008–15
368. Hirsch, J., Martelo, O. J. 1976. *J. Biol. Chem.* 251:5408–13
369. Dahmus, M. E. 1976. *Biochemistry* 15:1821–29
370. Kranias, E. G., Schweppe, J. S., Jungmann, R. A. 1977. *J. Biol. Chem.* 252:6750–58
371. Kranias, E. G., Jungmann, R. A. 1978. *Biochim. Biophys. Acta* 517:439–46
372. Bell, G. I., Valenzuela, P., Rutter, W. J. 1977. *J. Biol. Chem.* 252:3082–91
373. Zillig, W., Fujiki, H., Blum, W., Janeković, D., Schweiger, M., Rahmsdorf, H. J., Ponta, H., Hirsch-Kauffmann, M. 1975. *Proc. Natl. Acad. Sci. USA* 72:2506–10
374. Lee, S. G., Miceli, M. V., Jungmann, R. A., Hung, P. P. 1975. *Proc. Natl. Acad. Sci. USA* 72:2945–49

Ann. Rev. Biochem. 1979. 48:961–97

THREE-DIMENSIONAL STRUCTURE OF IMMUNOGLOBULINS

♦12030

L. Mario Amzel and Roberto J. Poljak

Department of Biophysics, Johns Hopkins University School of Medicine,
725 N. Wolfe Street, Baltimore, Maryland 21205

CONTENTS

PERSPECTIVES AND SUMMARY

Immunoglobulins are serum glycoproteins synthesized by vertebrates as antibodies against different antigens. They generally consist of two polypeptide chains called the light (L) chain and the heavy (H) chain (mol wts

961

0066-4154/79/0701-0961$01.00

25,000 and 50,000, respectively). The three-dimensional structure of several immunoglobulins and their fragments have been determined by single crystal x-ray diffraction methods. These studies have shown that immunoglobulins are multimeric proteins consisting of globular subunits arranged in pairs. These subunits of the L and H chains contain homologous amino acid sequences ("homology regions") about 110 amino acids long, and share a common pattern of three-dimensional chain folding ("immunoglobulin-fold").

The predominant secondary structure of the subunits is antiparallel β-pleated sheet. Each subunit contains two twisted, roughly parallel sheets formed by three and four antiparallel strands. The internal volume of the subunits is tightly packed with hydrophobic side chains enveloped by the two β-sheets. About 50% of the amino acid residues of the subunits form part of the β-sheets and have highly conserved sequences in different immunoglobulins.

The antigen-combining sites of immunoglobulins are formed by the two N-terminal subunits of the L and H chains and occur at the ends of the molecules, fully exposed to solvent. The conformation of the combining sites is determined by the amino acid sequence of segments of the H and L polypeptide chains ("complementarity-determining residues"). The amino acid sequences of these segments, unique to each different immunoglobulin, determine the specificities of antibody molecules. The structures of several ligand-immunoglobulin complexes determined by X-ray diffraction have provided useful models of antigen-antibody interactions.

The segments of polypeptide chain connecting the globular immonoglobulin domains show different degrees of flexibility. This flexibility of antibodies could be important for the mechanism of antigen binding and the ensuing activation of secondary or effector functions involving domains of the structure removed from the antigen-combining site.

Studies on the primary and tertiary structure of immunoglobulins substantiate the postulate that the different homology subunits arose during evolution by a mechanism of gene duplication and diversification. This mechanism provided the structural basis for the different functions of antibody molecules. The evolutionary mechanism did not alter the overall folding pattern of the subunits. However, within an immunoglobulin molecule the subunits pair in three different types of geometrical arrangement. These different pairing schemes are determined by amino acid substitutions in the β-sheets of the subunits. The major amino acid differences between immunoglobulin of the same class and of the same animal species occur in the loops connecting the β-pleated sheet strands.

Starting from the known structure of immunoglobulin fragments some attempts have been made to predict the conformation of the combining site of other immunoglobulins based solely on their amino acid sequences. Trial models have been derived by this general procedure but there is not yet enough information to evaluate their accuracy. The determination of three-dimensional structures of additional immunoglobulin-combining sites will be necessary to ascertain the ultimate potential of this approach.

The studies on the three-dimensional structures of immunoglobulins and their fragments, reviewed below, have provided invaluable information for the understanding of the structural basis of antibody function. Further studies will be needed, however, to complete our knowledge of the chemical basis of antibody specificity and to understand the triggering and the structural localization of effector functions.

INTRODUCTION

Several reviews on the function and the three-dimensional structure of immunoglobulins have been published (1–8) including one in this series (2). Consequently, we limit ourselves to a brief outline of the most recent results in the study of the structure of immunoglobulins by X-ray diffraction and a brief discussion of related topics.

Antibodies belong to the class of serum proteins called γ-globulins or immunoglobulins (Ig). Myeloma proteins, associated with the spontaneous occurrence of multiple myelomatosis and other pathological lympho-proliferative disorders in man, and with experimentally induced tumors in mice, are closely related to serum immunoglobulins and antibodies by a number of structural and functional properties. Human or murine myeloma proteins can be obtained in large, easily purified quantities that provide suitable material for structural studies. Immunoglobulins (Ig) are divided into major classes or isotypes characterized by their H-chain type. They all contain carbohydrates, largely hexose and hexosamine but also sialic acid and fucose, covalently attached to the protein moiety. The IgG class is the most abundant in normal serum and the most commonly observed class in human myeloma immunoglobulins. Its diagrammatic structure, including homology regions, is shown in Figure 1. The C_H1, C_H2, C_H3, and C_L regions are highly homologous to each other and less homologous to V_H and V_L. The N-terminal variable regions, V_H and V_L, are highly homologous to each other.

The L chains of IgG can be antigenically classified into two isotypes (or classes) called κ and λ, each characterized by a unique (or nearly unique) sequence in their C-terminal regions. IgM, IgA, IgD, and IgE possess similar κ and λ light chains but their H chains (called μ, α, δ, and ϵ, respectively) are different and are specific to each class. Amino acid sequence studies of human myeloma L chains have shown that in those of the same class (κ or λ) C_L's have constant amino acid sequences while V_L's have variable sequences. Because of the genetic and possible functional implications, the patterns of variability of L-chain sequences have been extensively analyzed. It is observed that within a given class of L chains there are sequences that are very similar to each other which define a "subgroup." Three or four such subgroups have been proposed for human κ-chains and five for human λ-chains. All chains within a subgroup are very similar in sequence except at certain positions within V_L, where extreme variability is observed. Kabat & Wu (9, 10) proposed that these hypervariable sequences constitute the regions of the L-chain structure that come in contact with antigen, so that the presence of different sequences in these regions will

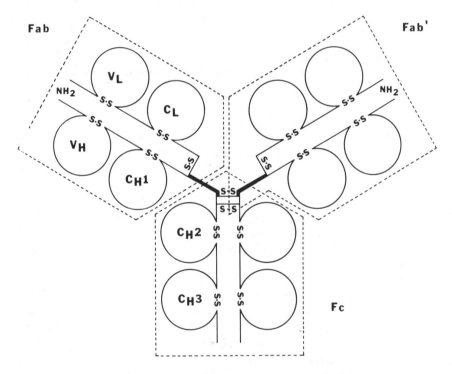

Figure 1 Diagram of a human immunoglobulin (IgG1) molecule. The light (L) chains (mol wt ~25,000) are divided into two homology regions, V_L and C_L. The heavy (H) chains (mol wt ~50,000) are divided into four homology regions, V_H, C_H1, C_H2, and C_H3. The C_H1 and C_H2 are joined by a "hinge" region indicated by a thicker line. Cleavage of the IgG1 molecule by papain generates Fab fragments (mol wt ~50,000) consisting of an L and an Fd polypeptide chain, and Fc fragments (mol wt ~50,000). Cleavage by pepsin followed by reduction of inter-H-chain disulfide bonds generates an Fab' fragment consisting of an L and an Fd' polypeptide chain. Interchain and intrachain disulfide bonds and the N termini of the L and H chains are indicated. [Reproduced from (52) with permission.]

result in different antibody specificities. Comparative studies on H chains of the same class have shown that the sequences of C_H1, C_H2, and C_H3 remain constant whereas those of V_H display variability. Just as in the L chain, the variable region of the H chain occurs at the N-terminal end of the molecule, is approximately 110 amino acid residues long, and also contains hypervariable regions (11).

Pepsin, papain, trypsin, and other enzymes split bonds in the "hinge" region which links Fab to Fc (see Figure 1) and which, by this criterion, appears openly accessible to solvent. In this review we use the term "region" (Fab region, Fc region) to denote that part of the immunoglobulin structure that corresponds to fragments such as Fab and Fc.

BASIC STRUCTURAL PATTERNS
OF IMMUNOGLOBULINS

Homology Subunits

The homology regions of immunoglobulin chains fold into independent, compact units of three-dimensional structure (homology subunits). X-ray diffraction studies (12, 13) have shown that all homology subunits of immunoglobulins share a common pattern of three-dimensional chain folding ("immunoglobulin fold") shown in Figures 2 and 3. The immunoglobulin fold consists of two twisted, stacked β-pleated sheets that surround an internal volume tightly packed with hydrophobic side chains. These two β-sheets are covalently linked by an intrachain disulfide bridge in the inner volume, in a direction approximately perpendicular to the plane of the sheets. The V subunits have an extra length of polypeptide chain that forms the two-stranded loop represented by dotted lines in Figure 2. Stereo diagrams of the four homology subunits of the Fab fragment of human IgG New are shown in Figure 4. Comparison of these diagrams shows the structural homology of the subunits. The extra loop characteristic of the V subunits is clearly seen in V_H. The V_L subunit of Fab New has a deletion of seven amino acids in this region and therefore the extra loop is appreciably shorter. The structures of the different homology subunits of Fab New are also represented schematically in Figure 5 where the two β-pleated sheets are shown as two clusters of polypeptide chain, with hydrogen bonds between main chain atoms represented by dotted lines.

Several labelling systems (2, 3, 14) have been used to name the different segments of polypeptide chain according to their position in the immunoglobulin fold. For the purpose of this discussion we introduce the modified labelling represented in Figure 6 that includes many of the features of the previously proposed nomenclatures. The four-stranded β-pleated sheet, called 4β, contains strands $4\beta1$, $4\beta2$, $4\beta3$, and $4\beta4$. The three-stranded

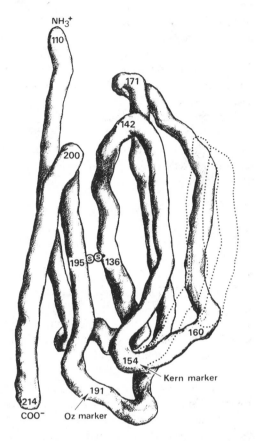

Figure 2 Diagram of the basic "immunoglobulin-fold." Solid trace shows the folding of the polypeptide chain in the constant subunits (C_L and C_H). Numbers designate L (λ) chain residues, beginning at NH_3^+ which corresponds to residue 110 for the L chain. Broken lines indicate the additional loop of polypeptide chain characteristic of the V_L and V_H subunits.

β-pleated sheet, called *3β*, contains strands *3β1*, *3β2*, and *3β3*. The two strands of the extra loop characteristic of the V homology subunits also occur in the *3β* sheet and are called *3β4* and *3β5* (dotted lines in Figure 6). The modified *3β* sheets of the V subunits are called *3β'*. The regions containing the segments (loops) that connect the β-sheet strands are labelled according to whether they occur at the same side as the N-terminal front loops, *fl'*, or the C-terminal back loops, *bl*, of the immunoglobulin fold. The front end contains the connecting segments *fl1*, *fl2*, and *fl3* while the back end contains segments *bl1*, *bl2*, and *bl3*. The connecting segments *fl4* and *bl4* are associated with the extra loop and occur only in V subunits (dotted lines in Figure 6). In some cases (e.g. some *fl3*'s) these segments

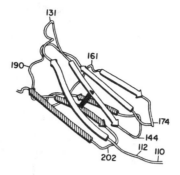

Figure 3 Schematic Diagram of the α-carbon backbone of the C$_L$ homology subunit showing two planes of β-pleated sheet (one containing four hydrogen-bonded antiparallel chains shown by white arrows and another containing three hydrogen-bonded antiparallel chains shown by striated arrows). These two twisted and roughly parallel sheets surround the intrachain disulfide bond (shown in black) which links the two sheets in a direction approximately perpendicular to their planes. [Reproduced from Edmundson et al (32) with permission.]

form regular reverse turns (β-bends, etc) while in most cases they show no secondary structure. Several connecting segments sometimes contain short α-helical stretches (approximately one helical turn).

About 50% of the residues in each homology subunit are contained in the two β-pleated sheets (strands *4β*1–*4β*4 and *3β*1–*3β*3 or *3β*5). The expected pattern of alternating polar-apolar residues is observed in the β-sheets, especially in the sheets that are not involved in the intersubunit interactions. In this alternating pattern of the V subunits, serine and threonine are the most frequently observed amino acids at the surface, exposed to solvent. In contrast, there is a greater diversity of residues at the surface of C-domains (14). The three-dimensional structure of the β-sheet strands is highly conserved between the different homology subunits (15, 16), while most of the structural differences occur in the connecting segments. The hypervariable regions of the V subunits (V$_L$, V$_H$) form the connecting segments *fl*1, *fl*2, and *fl*4 which occur in spatial proximity in the three-dimensional structure. These segments contain the largest structural differences between different V homology subunits (17) and determine the conformation of the antigen-combining site.

Subunit Interactions

In immunoglobulin molecules a large number of noncovalent interactions stabilize the arrangement of the different homology subunits. These interactions occur between adjacent subunits of the same chain (*cis* interactions) and between subunits in different chains (*trans* interactions). *Trans* interactions are in general quite extensive and stabilize the structural domains

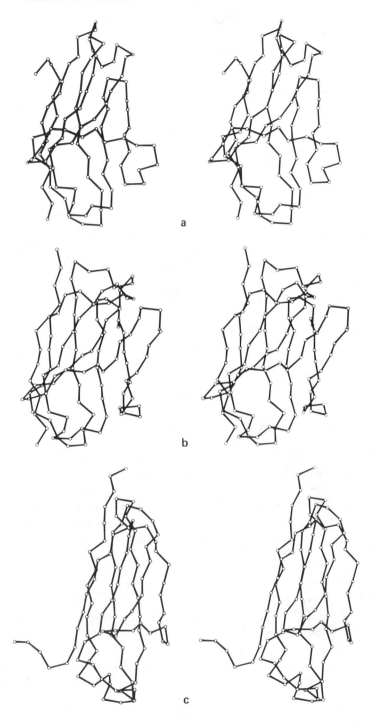

a

b

c

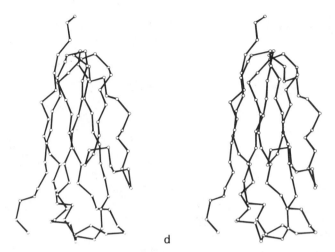

d

Figure 4   Stereo pair drawings of the α-carbon backbones of the four homology subunits of Fab New (V_L, V_H, C_L and C_H1) viewed in approximately the same orientation. The structural similarity is clearly shown in this drawing. The V_H homology subunit, *b*, shows the 'extra loop' characteristic of the V subunits. The V_L subunit of Fab New, *a*, has a much shorter 'extra loop.'

formed by pairs of homology subunits. Conversely, *cis* interactions involve a small number of contact residues and the subunit arrangements stabilized by these interactions are generally flexible (18, 19). These interactions can be seen for example in the structure of the Fab fragment of human IgG New (Figure 7).

SUBUNIT PAIRS AND STRUCTURAL DOMAINS: *TRANS* INTERAC-
TIONS   In immunoglobulins the homology subunits are found associated in pairs (20, 21) through *trans* interactions. With the exception of the C_H2 region the structural association between subunits is very close and involves a large area of contact. Two types of close associations are observed in subunit pairs. The first (V-type pair) has been observed in the structures of V_κ dimers (22–24), and in the V regions of Fab fragments (13, 25, 26), whole immunoglobulins (19, 27), and an L chain (λ) dimer (12). In all these cases the contact area is formed by the modified $3\beta'$ sheets of both subunits (strands $3\beta1$–$3\beta5$). The geometry of this type of pairs is such that the associated subunits are related by approximate (V_L-V_H pairs) or exact (V_L-V_L pairs) twofold axes of symmetry (Figure 7). The interactions between subunits are very extensive and in the case of V_L-V_H pairs the association constant was estimated to be larger than $10^8 M^{-1}$ (28). The arrangement of subunits in all V-type pairs studied is similar even when a V_L-V_L pair and a V_L-V_H pair are compared. Furthermore, this arrangement

seems to be independent of L- and H-chain classes since κ- and λ-L chains and γ- and α-H chains associate in similar patterns in the structures that have been studied thus far. The expected alternating pattern of polar-apolar residues is clearly observed in the 4β sheets of the V subunits while the residues at the outer face of the $3\beta'$ sheets are less polar and participate in

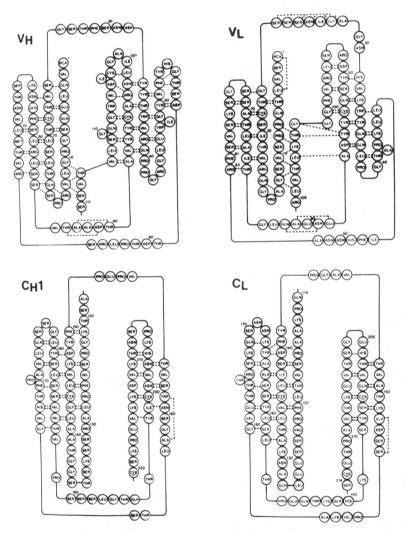

Figure 5 Diagram of hydrogen bonding (broken line) between the main chain atoms of the V_H, V_L, C_H1, and C_L homology region of Fab New. The hydrogen bonded clusters correspond to the two β-sheet structures of each subunit. Cysteine residues that participate in intrachain disulfide bonds are underlined.

the *trans* interactions. The *fl* segments containing the hypervariable regions occur in close spatial proximity defining a large cavity in V_L-V_L pairs and the antibody-combining site in V_L-V_H pairs.

The second type of close association (C-type pairs) is observed in the structures of an L-chain dimer (12), the C regions of three Fab fragments (13, 25, 26), and two immunoglobulins (19, 27) and an Fc fragment (21). In this type of association the contact area is formed by the 4β sheets of the homology subunits (strands $4\beta1$–$4\beta4$). To this type belong the C_L-C_L, C_L-C_H1, and C_H3-C_H3 pairs. Again, the geometrical arrangement of subunits in these pairs is highly conserved and involves an exact or an approximate two fold axis of symmetry (Figure 7). In these C-type pairs the conservation of the structure extends to the geometrical arrangement of subunits in different structural domains of the same immunoglobulin (i.e. C_L-C_H1 and C_H3-C_H3).

A third type of association of homology subunits is that observed in C_H2 pairs. This association is observed in the structures of an Fc fragment (21) and a whole immunoglobulin (27) and is much weaker than those of the V- and C-type pairs, in agreement with the available physical chemical information (28, 29).

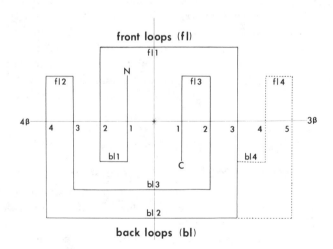

Figure 6 Diagrammatic representation of the β-sheet structure of the homology subunits. The antiparallel strands *1, 2, 3,* and *4* of the 4β sheet begin at N (the "N terminus" of each homology subunit). The 3β sheet consists of strands *1, 2,* and *3,* and in the V_H and V_L subunits only, of additional strands (dotted line) *4* and *5*. The strands are connected by "front" loops (*fl*) which in the case of V_H and V_L are exposed to solvent, and by "back" loops (*bl*). C denotes the "C terminus" of each homology subunit. See Figure 5 to correlate the features of this diagram with actual sequences.

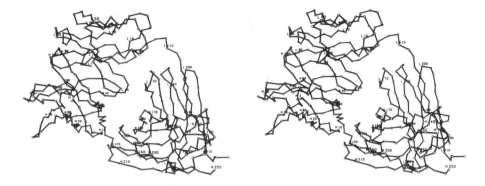

Figure 7 Stereo pair drawing of the β-carbon backbone of Fab New. The interactions between subunits can be clearly seen in this diagram. The interactions between subunits of different chains (*trans*) are much more extensive than those between subunits of the same chain (*cis*).

CIS INTERACTIONS   The interactions between consecutive subunits of the same chain in immunoglobulin molecules are very limited and involve a small number of contacts. These contacts are made between some of the *bl* segments of one subunit and *fl* segments of the next subunit of the same chain. At least three different arrangements involving C_H1-V_H and three involving C_L-V_L were observed in the structures studied so far. Apart from indicating the presence of flexibility in the "switch regions" connecting V to C, no other common characteristic of these V-C *cis* interactions has emerged from these studies. We discuss *cis* interactions in more detail in our review of the structures studied by X-ray diffraction methods.

X-RAY DIFFRACTION STUDIES OF IMMUNOGLOBULIN L CHAINS

Three-Dimensional Structures of V_κ Fragments

The three-dimensional structures of three dimeric V_κ fragments (Rei, Au, and Roy) have been reported (22–24, 30). These fragments show the same overall three-dimensional structure of the V-type dimers described in the previous section. An interesting observation was made concerning Tyr 49, a constant residue in the three V_κ dimers which could therefore be expected to adopt the same conformation in the three structures. However, Tyr 49 occurs in close spatial proximity to residue 96 which is a Leu in Roy, a Tyr in Rei, and a Trp in Au. The proximity of these residues at position 96 imposes different conformations on the side chain of Tyr 49 in the three dimers. Thus, amino acid substitutions in the hypervariable regions are shown to have an effect not only in the character of the position at which they occur but also at neighboring positions of the antigen-combining site.

Although there is now a general understanding about the relationship between molecular structure and antigen binding, and in particular about hypervariability and the antigen-combining site, the specific details about the effect of amino acid substitutions, and the extent of the antigen-antibody interactions are not known. Studies such as those reported above for closely related κ-chains are of great interest to our understanding of this topic.

The Mcg Dimer

The crystal structure of the Bence-Jones protein (L-chain dimer) Mcg has been analyzed to 2.3 Å resolution and partially refined (14). The complete amino acid sequence of the λ-chain Mcg has also been determined (31) and correlated with the electron density map. The structure of the Mcg L-chain dimer is similar to that of Fab fragments. The cavity determined by the hypervariable regions is also similar to that observed in Fabs but contains a solvent channel not observed in V_L-V_H pairs. The fact that the contact residues between V_L-V_L and those between C_L-C_L are nonhomologous has been described as arising from "rotational allomerism" (14). Edmundson et al (14) suggest that the genetic changes necessary to bring about suitable amino acid contacts leading to rotational allomerism of V and C domains are key steps in the evolution of immunoglobulins.

The cavity and also the solvent channel of the Mcg dimer were found to provide binding sites (located by difference-Fourier maps) for a number of ligands (32). These included an iodinated derivative of 1-fluoro-2,4 dinitro-benzene, DNP derivatives of lysine and leucine, 5-acetyluracil, menadione, caffeine, theophylline, ϵ-dansyl lysine, phenantroline, and colchicine. Edmundson et al (32) suggest that the L-chain dimer behaves as a primitive antibody. However it is difficult to ascertain the physiological significance of this binding activity since the complete Mcg myeloma IgG protein (which occurs in the Mcg serum) does not appear to bind the same ligands.

A most interesting aspect of the Mcg structure is that although the L chains of the dimer have identical amino acid sequence their quaternary structures are different. This difference can be described as one in the angle made by the major axes of V_L and C_L in one chain (70°) and the corresponding angle in the second L chain of the dimer (110°). In this respect the L-chain dimer resembles the Fab M603 and New structures (described below).

STRUCTURE OF FAB FRAGMENTS

The Three-Dimensional Structure of Fab M603

The structure of Fab-M603 [(mouse IgA, κ; (25)] is very similar to that of the previously reported Fab New (human IgG, λ) which shows that the structures of immunoglobulins are conserved across species and H- and L-chain classes. The pseudo twofold axes relating V_H to V_L and C_H1 to C_L make an angle of 135° (Figure 8). The long axes of the subunits of the L chain make an angle of approximately 100° while the corresponding angle for the H chain is approximately 80°. Segal et al (25) postulate that the close approach of the two H-chain subunits is facilitated by the presence at the interface of amino acids with small side chains, such as Gly 8-Gly 9-Gly 10 of strand $4\beta1$ and Gly 109-Ala 110-Gly 111 of strand $3\beta1$. The equivalent residues in Fab New are Gly 8-Pro 9-Gly 10 and Gly 108-Gln 109-Gly 110. In Fab New, however, only Gly 10 is in a position such that a bulkier side chain could interfere with the observed contacts between V_H and C_H1. Furthermore the sequence Gly-Gln-Gly (positions 108–110) of V_H M603 is homologous to Gly-Gly-Gly (positions 100–102) of the λ-chain of IgG New which assumes, however, an open structure with fewer V_L-C_L

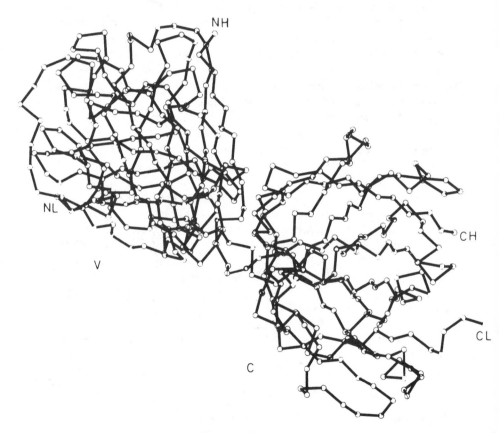

Figure 8 Drawing of the α-carbon positions of the Fab fragment of mouse myeloma M603, projected in a direction approximately perpendicular to the "switch regions." [Reproduced from Huber et al (53) with permission.]

contacts. This sequence was previously shown to participate in intrasubunit interactions (18).

The site of hapten binding of M603 was identified by diffusing phosphorylcholine into Fab crystals and calculating a difference-Fourier map. It is located in a large wedge-shaped cavity approximately 12 Å deep, 15 Å wide, and 20 Å long which is lined with hypervariable residues. Only five hypervariable regions, designated as L1, L3, H1, H2, and H3 contribute to the formation of the cavity. L2 does not contribute to the lining of this cavity because it is screened by the large *f*l1 loop of the first hypervariable region. Phosphorylcholine occupies only a small part of the cavity and it is bound mainly to the H chain. The choline group is buried in the interior of the cavity while the phosphate group remains closer to the exterior. This mode of binding is consistent with the occurrence of the choline moiety as the most exposed determinant of natural phosphorylcholine antigens (33). The phosphate group is hydrogen bonded through two of its oxygens to the hydroxyl group of Tyr 33 (H chain) and to an amino group of the side chain of Arg 52 (H chain) (Figure 9). The positive charge of the guanidinium

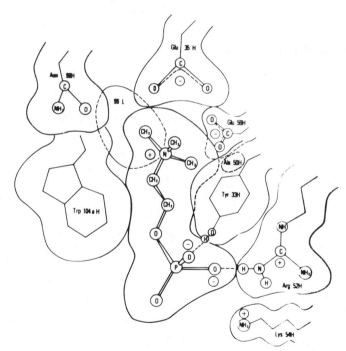

Figure 9 Schematic representation of the specific interactions between phosphorylcholine and the protein side groups in M603. The binding cavity is located in the cleft between the L and H chains. Choline binds in the interior while the phosphate group is towards the exterior of the cavity. [Reproduced from Padlan et al (34) with permission.]

group also contributes to the binding of the negatively charged phosphate. In addition, the amino group of Lys 54 (H) is in close proximity with the phosphate and probably contributes to stabilize the complex. The positively charged trimethylammonium group interacts with the negatively charged side chains of Glu 35(H) and Glu 59(H). Furthermore, there are extensive van der Waals interactions between atoms of the hapten and main chain atoms of residues 102 and 103 of the H chain and residues 91–94 of the L chain, and with the ring atoms of H-chain Tyr 33 and Trp 104a. In a recent paper Padlan et al (34) compare the sequences of the hypervariable regions of the H chain of the seven phosphorylcholine-binding mouse myeloma proteins (Table 1). They find that all residues identified as participating in phosphorylcholine binding in M603 (Tyr 33, Glu 35, Arg 52, Lys 54, and Glu 59) are conserved in all seven H chains. In addition, position 50 is Ala in all the chains and position 104a is Trp in all except M167. Based on this evidence the authors suggest that the binding site of these phosphorylcholine binding proteins is very similar. Moreover, they predict that the hapten will bind to them in essentially the same manner. The differences in affinity for phosphorylcholine (35) are attributed to amino acid substitutions in other residues.

The Structure of Fab New

The description of the immunoglobulin fold, and the characterization of the homology subunits and of subunit pairs (structural domains), were made on the bases of the structure of Fab New determined at 6.0 Å (20) and 2.8 Å (13) resolution. These studies showed that the Fab fragment can be divided into two structural domains, V and C, of approximate dimensions 40 × 50 × 40 Å. Each structural domain was shown to contain two homology subunits (V_H and V_L; C_H1 and C_L) related by an approximate twofold axis of symmetry (Figure 7). The V and C structural domains of Fab New are not colinear and their pseudo twofold axes of symmetry form an angle of approximately 137°. The L chain is more extended than the H chain so that the angle formed by the major axes of the V_L and C_L subunits is approximately 110° while that of V_H and C_H1 is approximately 80°. From the extensive sequence homologies of different classes of immunoglobulins, the sharing of a peptide chain folding pattern in V_H, V_L, C_H1, and C_L and the fact that different intrachain and interchain disulfide bonds of immunoglobulins could be explained on the basis of the Fab New model, it was concluded that the immunoglobulin fold and the association of subunits in globular domains are general features of all classes of immunoglobulins in all animal species (13).

Hypervariable regions of the L and H chain occur in close spatial proximity, surrounding a shallow groove that was identified as the binding site of

Table 1 Heavy chain complementarity region sequences of mouse phosphorylcholine-binding immunoglobulins[a]

	H1		H2						H3					
	31 … 35		50	55	58 a	b	60	65	100 a		(D) S Y	104	a	b
M167	D F Y M E	A A S R S		K A H	D Y R		T E Y S A S V K G		D A	D Y G	(D) S Y	F	G	Y
M511	B			D		T				G	S		W	
T15				N		T				Y–	S		W	
S107	B		B	N		T				Y–	S		W	
H8	B		B	N		T				Y–	B		W	
M603			N – G N K			T			N	Y–	S –		T	W
W3207	B		B	N		T				Y				

[a] Reproduced from Padlan et al (34) with permission. The one letter code for amino acids is as used in Figure 11.

the Fab fragment (36). Fab New binds several haptens at this site with affinity constants ranging from $10^3 M^{-1}$–$10^5 M^{-1}$. A schematic diagram of the binding of a γ-hydroxyl derivative of vitamin K_1 ($K_1 OH$) to Fab New is shown in Figure 10.

The structure of Fab New has been partially refined and the improved model used to quantitatively compare the homology subunits (15) V_H, V_L, $C_H 1$, and C_L using the method of Rao & Rossman (37). All the possible pairs of homology subunits were aligned and superimposed and the matching was optimized by minimizing the sum of the squares of the distances between equivalent $C\alpha$'s. Table 2 shows that the largest numbers of equivalenced $C\alpha$'s occurring at distances shorter than 1.5 Å and 3.0 Å are found when comparing V_H to V_L and $C_H 1$ to C_L. The average minimum base change per codon for equivalenced amino acids is smallest for these pairs and, in general, seems to be in inverse relation to the number of $C\alpha$'s equivalenced. Moreover, the average minimum base change per codon is,

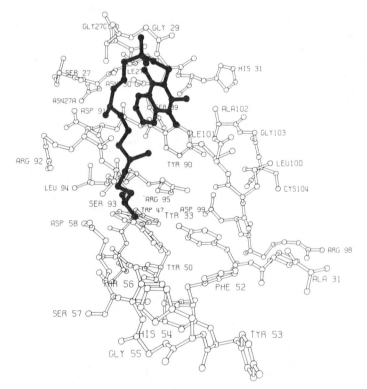

Figure 10   Schematic representation of vitamin $K_1 OH$ bound to the combining region of Fab New. Some of the amino acid residues in this figure correspond to a tentative sequence and are different from those of Figures 11 and 13 that contain the final sequence.

in all cases, smaller when the more stringent condition ($d_{C\alpha\text{-}C\alpha} \leqslant 1.5$ Å) is used for the scoring of equivalences. Thus, the conservation of fine structural details seems to be mainly a reflection of the conservation of sequence. The quantitative three-dimensional matching procedure leads to amino acid sequence alignments that clearly reflect the well-established homologies between V_H and V_L and between C_H1 and C_L (Figure 11). The closest sequence similarities are found between C_H1 and C_L and between V_H and V_L. In addition Table 2 and Figure 11 show that there is considerable homology between V and C regions. These results indicate that all homology subunits contain a basic core of amino acid residues with highly preserved three-dimensional structure. The chemical nature of these residues is also preserved as shown by the low values of the average minimum base change per codon. This basic core of residues forms part of the two β-sheets (*3β* and *4β*) and includes the two cysteines of the intrachain disulfide bonds and a constant Trp occurring at 14 or 15 positions after the first Cys of the disulfide bond. These findings reinforce the postulate that the different homology regions of immunoglobulins appeared during evolution through a gene duplication mechanism.

The refined model of Fab New has been used to evaluate the chemical contacts between the different homology subunits (15). The results of this analysis are diagrammatically presented in Figure 12, and show that the *trans*-interactions between V_H and V_L and between C_H1 and C_L are far more extensive than the *cis*-interactions between V_L and C_L and between C_H1 and V_H. Also, most of the residues involved in *trans*-interactions in C subunits occur in the *4β* sheets while in the V subunits they occur in the *3β'* sheets. The *cis*-interactions occur in *bl* segments in the V subunits and in *fl* segments in C subunits. The smaller number of contacts between V_L and C_L than between V_H and C_H1 are a consequence of the more extended structure of the L chain. A more detailed description of the intersubunit contacts is given in Table 3. The contacts between V_H and V_L are of

Table 2 Alignment of α-carbon coordinates of four homology subunits of Fab (New)

Subunits	Number of C pairs equivalenced with $d_{C\alpha-C\alpha}$(Å) $\leqslant 1.5$	Average minimum base change per codon for $d_{C\alpha-C\alpha}$(Å) $\leqslant 1.5$	Number of C pairs equivalenced with $d_{C\alpha-C\alpha}$(Å) $\leqslant 3.0$	Average minimum base change per codon for $d_{C\alpha-C\alpha}$(Å) $\leqslant 3.0$
V_H-V_L	56	0.98	81	0.97
C_H1-C_L	60	0.71	82	0.80
C_L-V_L	40	1.03	66	1.23
C_L-V_H	29	1.04	59	1.28
C_H1-V_L	27	1.04	58	1.24
C_H1-V_H	25	1.29	49	1.40

```
                    1              10        20      27 a b c  30              40               50
VL      - - - - Z S V L T Q P P S V S G A P - G Q R V T I ⱡ C T G S S N I G A G N H V K W Y Q Q L P G T A P K - L L I P H N N A R * * * *
                    1              10        20            30              40               50            60
VH      - - - - Z V Q L E Q S G P G L V R P - S Q T L S L T C T V S G S T F S N D - Y Y T W V R Q P P G R G L E W I G Y V F Y H G T S D T D
        110            120       130       140             150
CL      Q P K A A P S V T L F P P S S E E L Q A N K A T L V C L I S D F Y P G A V - T V A W K - - A D S S - - - - - - - - - - - -
        120            130       140             150             160
CH1     A S T K G P S V F P L A P S S K S T S G G T A A L G C L V K D Y F P E P V - T V S W N - - - S G - - - - - - - - - - - - - -

        61          70              80              90              100           109
VL      * * * - - F S V S K S G - - - - - - - - - - S S A T L A I T G L Q A E D E A D Y Y C Q S Y D R S L R * * - V F G G G T K L T V L R
                    70            80              90              100       110             117
VH      T P L R S R V T M L V N T - S - - - - - - K N Q F S L R L S S V T A A D T A V Y Y C A R N L I A G - C I D V W G Q G S L V T V S S
        160          170           180             190           200       210             214
CL      - P V K A - - G V E T T P S K Q S N N K Y A A S Y L S L T P E Q W K S H K S Y S C Q V T H - - E G S T - V E K T - V A P T E C S
        170          180           190           200           210           220
CH1     - A L T S - - G V H T F P A V L Q S S G L Y S L S S V V T V P S S S L G T - Q T Y I C N V N H K P S N T K - V D K K - V E P K S C
```

Figure 11 Amino acid sequences of the V_L, V_H, C_L, and C_H1 homology region of Fab New aligned by comparison of their three-dimensional structures. Dashes indicate gaps introduced to maximize alignment of the three-dimensional structures. Asterisks indicate a deletion in the V_L sequence. Abbreviations for amino acids are as given in *Atlas of Protein Sequence and Structure*, Vol. 5. 1972. ed. M. O. Dayhoff. Washington, DC: Natl. Biomed. Res. Found. [Reproduced from (15) with permission.]

particular interest in relation to the structural viability of different V_H- and V_L-chain pairs. Unrestricted pairing of different H and L chains would provide a simple mechanism for increasing antibody diversity from a given number of H- and L-chain genes. Three types of V_H-V_L contacts were discussed: 1. those at the core of the contacting region, which involves residues that are invariant or semi-invariant in V_H and V_L sequences; 2. those made by conserved residues with hypervariable residues; and 3. those made between hypervariable residues.

The first type of contact occurs at the core of the contacting region between V_H and V_L and involves equivalent residues from the two chains. These contacts include residues Val 37, Gln 39, Leu 45, Tyr 94, and Trp 107 in V_H and residues Tyr 35, Gln 37, Ala 42, Pro 43, Tyr 86, and Phe 99 in V_L. These residues are structurally homologous between both chains with the exception of Ala 42 of the L chain. The contacts between these homologous residues of V_H and V_L account for about 50% of all contacts listed in Table 3. All these residues are invariant or are replaced by homologous residues in V_L and V_H sequences from different classes and different animal species. The presence of these invariant or nearly invariant residues at the main V_H-V_L contact area together with the constant C_H1-C_L contacts, explains the property that different H and L chains can recombine to form new immunoglobulin molecules (38, 15).

The second and third types of contacts are more difficult to evaluate in general terms, since they are different for different immunoglobulins. These kinds of contacts could perhaps provide a structural explanation of the preferred reassociation sometimes observed between complementary H and L chains derived from a single immunoglobulin molecule (38).

The interactions between the constant domains C_H1 and C_L are very extensive (Figure 12 and Table 3). The core of the contact region is formed by residues which appear to be invariant or nearly invariant in the H- and L-chain sequences of different classes and of different animal species (15).

The refined structure of Fab New provided more precise information about the structural homology of the subunits, the interactions between L and H chains, and the conformation of the residues in the hypervariable regions (Figure 13). Further work in this system is being directed toward detailed studies of hapten-combining site complexes.

The Structure of Fab Kol

The study of Fab Kol (26) is of particular interest since the crystal structure of the parent IgG molecule has also been determined (see below). IgG Kol and Fab Kol can thus be used to examine the possible influences that the Fc region may have on the tertiary and quaternary structure of the Fab region.

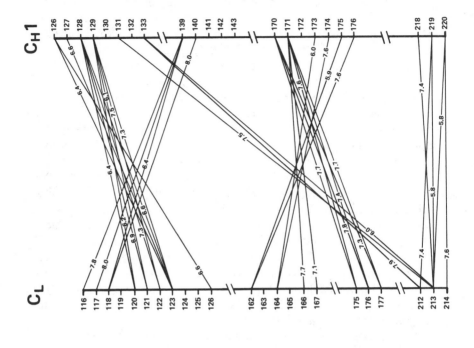

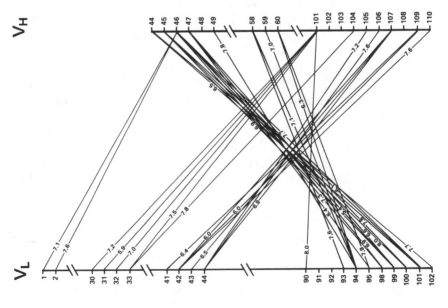

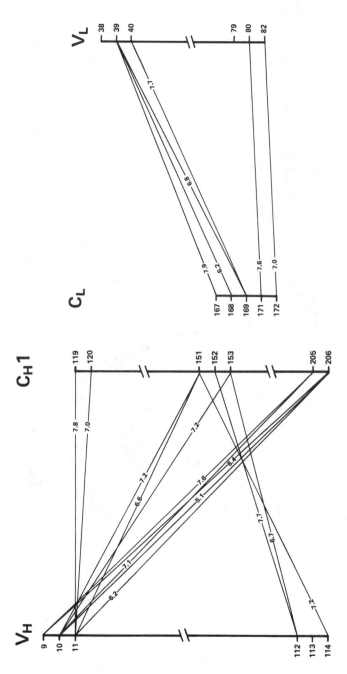

Figure 12 Intersubunit α-carbon contacts at distances of 8 Å or less. Contacts are indicated by lines joining the corresponding amino acid residue numbers. Numbers on the lines indicate the Cα-Cα distance in Angstroms. [Reproduced from (15) with permission.]

Table 3 Intersubunit contacts[a]

V_H	V_L	No. of contacts	C_H1	C_L	No. of contacts
Val 37	Phe 99	1	Phe 126	Glu 126	14
Gln 39	Gln 37	5	Phe 126	Glu 125	1
Arg 43	Asp 84	14	Phe 126	Ser 123	3
Arg 43	Tyr 86	7	Leu 128	Phe 120	20
Arg 43	Gln 37	3	Leu 128	Val 135	2
Leu 45	Tyr 86	4	Leu 128	Pro 121	1
Leu 45	Phe 99	2	Ala 129	Phe 120	8
Glu 46	Phe 99	3	Ala 129	Pro 121	2
Trp 47	Arg 95	19	Lys 133	Glu 212	1
Trp 47	Leu 94	8	Thr 139	Thr 118	3
Trp 47	Ser 93	1	Thr 139	Lys 206	2
Asp 58	Ser 93	2	Ala 141	Phe 120	6
Asp 60	Leu 94	3	Leu 142	Phe 120	4
Thr 61	Leu 94	1	Gly 143	Phe 120	5
Tyr 94	Ala 42	6	Leu 145	Tyr 179	1
Asn 98	Arg 95	8	Leu 145	Val 135	1
Leu 99	Arg 95	3	Lys 147	Glu 126	1
Ala 101	Tyr 90	6	Lys 147	Lys 131	3
Ala 101	His 31	7	Lys 147	Thr 133	2
Ala 101	Lys 33	2	Phe 170	Leu 137	10
Gly 102	Lys 33	6	Phe 170	Ile 138	4
Ile 104	Tyr 35	3	Phe 170	Ser 177	4
Ile 104	Gln 88	3	Pro 171	Ser 167	2
Ile 104	Leu 45	2	Pro 171	Ala 175	1
Trp 107	Pro 43	21	Val 173	Tyr 179	6
Trp 107	Ala 42	2	Gln 175	Glu 162	7
Trp 107	Phe 99	4	Ser 176	Glu 162	8
Trp 107	Tyr 35	2	Leu 182	Tyr 179	2
			Ser 183	Tyr 179	6

C_H1	V_H				
			Ser 183	Val 135	1
			Ser 183	Leu 137	1
Ala 118	Leu 11	2	Val 185	Leu 137	3
Ser 119	Leu 11	2	Val 185	Phe 120	3
Thr 120	Leu 11	3	Lys 218	Cys 213	2
Phe 150	Leu 11	3	Ser 219	Glu 212	2
Phe 150	Thr 114	1	Ser 219	Cys 213	6
Pro 151	Leu 11	2	Cys 220	Cys 213	7
Pro 151	Thr 114	2			
Glu 152	Leu 112	4			
Pro 153	Leu 112	7			

C_L	V_L				
Gln 110	Glu 82	1
Lys 168	Pro 39	5
Asn 172	Glu 82	7

[a] The number of interatomic distances not larger than 1.2 times the sum of the Van der Waals radii (C–C = 4.32 Å; O–O ≤ 3.65 Å; N–N ≤ 3.72 Å; C–O ≤ 3.98 Å; and C–N ≤ 4.02 Å) are listed. [Reproduced from (15) with permission.]

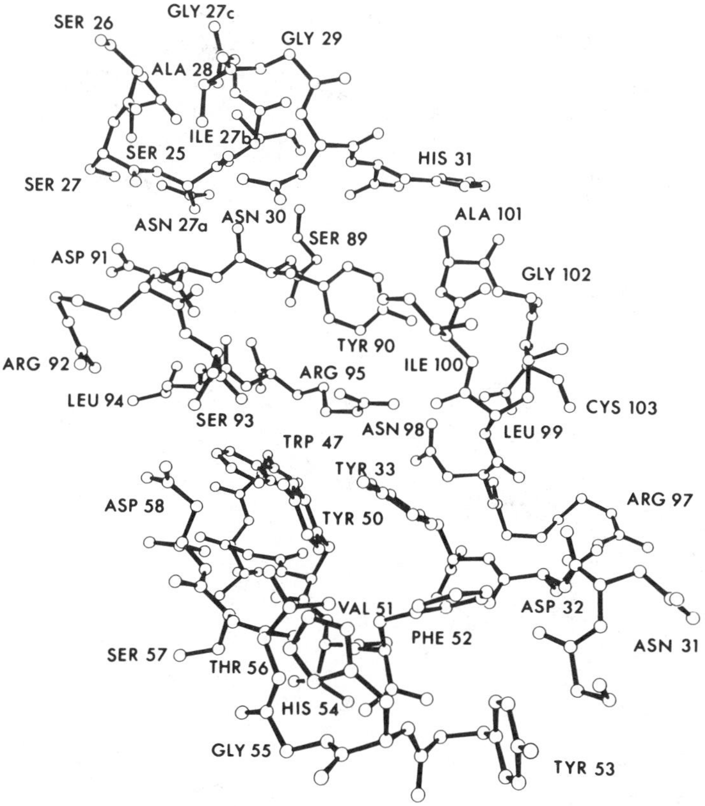

Figure 13 View of some of the amino acid residues at the combining site of IgG New.

The crystal structure of Fab Kol was determined by the multiple isomorphous heavy atom substitution method to a resolution of 3.0 Å. The Cα coordinates of the V_κ Rei dimer (22) were fitted to the electron density map of the V domain. The Cα coordinates of C_H1 and C_L from M603 and from Fab New were fitted to the electron density of the C domains. The fitting of these different coordinates to the electron density map of Fab Kol helped distinguish its H and L chains. Since the amino acid sequences of V_H and V_L of Kol have not yet been obtained, the model is based on tentative assignments of amino acid side chains, in particular in the segments corresponding to the hypervariable regions. From the crystallographic model the L chain was identified as a λ-chain. The tentative model was subjected to constrained crystallographic refinement.

As expected, the tertiary structure of the Fab Kol homology subunits is essentially the same as that reported before for other immunoglobulins. The quaternary structure of Fab Kol (Figure 14) differs, however, from that found in Fab M603 (Figure 9) and Fab New (Figure 7). This difference can be described as a change in the relative orientations of the V and C homology subunits or as a change in the angle made by the approximate twofold axes of symmetry relating V_H to V_L and C_H1 to C_L. The changed angular relationship between the V and C domains is due to a change in

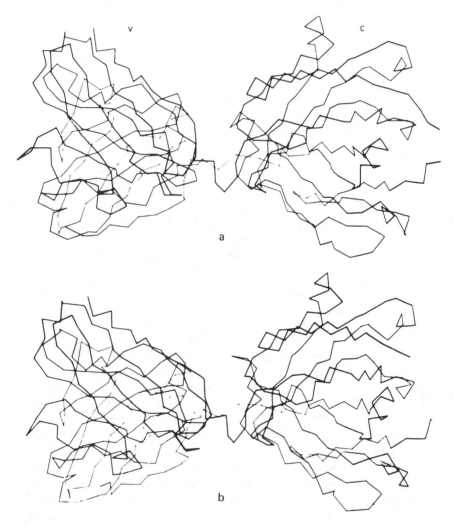

Figure 14 Cα drawing of *a* Fab Kol and of *b* the Fab portion of IgG Kol viewed along an axis through the switch peptides. [Reproduced from Matsushima et al (26) with permission.]

conformation at the switch regions ("elbow bending") such that the V and C domains are more nearly colinear in Fab Kol. As a result of the change in quaternary structure there are fewer V-C contacts in Fab Kol than in Fab New. Matsushima et al postulate (26) that the intermolecular contacts (crystal packing) rather than the V-C *cis*-contacts may be the factor influencing the bending of the switch regions.

In Fab Kol the crystal packing is a result of contacts between hypervariable regions and C_H1-C_L residues. The same contacts are observed in the crystalline IgG Kol, which is a cryoglobulin. The authors suggest that these contacts may explain the phenomenon of cold precipitation. This suggestion is in line with that presented in a recent biochemical study on cryoglobulins (39). The structure of Fab Kol will be discussed further in connection with that of the parent protein, IgG Kol.

STRUCTURE OF Fc FRAGMENTS

The Structure of the Human Fc Fragment

The structure of a human Fc fragment has been determined to 3.5 Å resolution using a combination of multiple isomorphous heavy atom replacement and molecular replacement techniques (21). The Fc structure has been described as having the approximate shape of a "mickey mouse" with the C_H2 domains forming the ears and the C_H3 subunits forming a globular head (see Figure 15). The overall tertiary structure of C_H2 and C_H3 is as described for the immunoglobulin fold. A loosely folded segment of polypeptide chain extending from Ser 337 to Gln 342 connects the two domains. This segment of polypeptide chain is exposed to solvent, a feature that explains its susceptibility to proteolytic attack by enzymes. The C_H3 subunits interact very closely in a pattern which is similar to the C_H1-C_L interactions described above for Fab fragments. The C_H2 subunits show no close interaction with each other. The N terminus, including the sequence -Cys-Pro-Pro-Cys- appears to be disordered since it cannot be traced in the electron density map; the segment that follows it in the N-terminal sequence (to Pro 238) may be in close contact with part of the carbohydrate chains attached to Asn 297. This sequence and the carbohydrate chains lie in between the C_H2 subunits (see Figure 15). The structure of the carbohydrate chains is somewhat tentative but their general location and overall conformation are clear. The electron density assigned to carbohydrate is compatible with a branched chain of the general type:

$$\begin{array}{c} \quad\quad\quad\;\text{H} \quad\quad\quad\quad \text{H–H} \\ \quad\quad\quad\;| \quad\quad\quad\quad\;\; \nearrow \\ \text{Asn 297–H–H–H–H–H} \\ \quad\quad\quad\quad\quad\quad\quad\; \searrow \\ \quad\quad\quad\quad\quad\quad\quad\quad \text{H–H} \end{array}$$

R

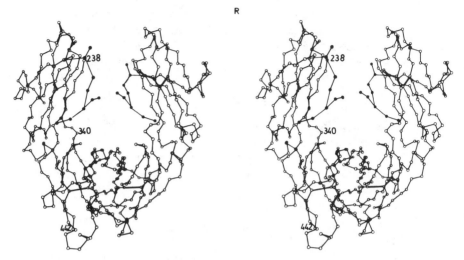

Figure 15   Stereo drawing of the Cα carbon positions and the centers of the carbohydrate hexose units of the Fc fragment. The black dots represent the approximate centers of carbohydrate hexose units. The carbohydrate attachment site is Asn 297. The positions of the α carbons are indicated by the open circles. The disulfide bonds are indicated. [Reproduced from Huber et al (53) with permission.]

where H is a hexose unit. This chain appears to be longer than those that have been observed in myeloma proteins. The attachment site (Asn 297) is at an accessible turn of the polypeptide chain, in agreement with a post-translational attachment of the sugar moiety by specific transferase enzymes.

STRUCTURE OF IgG MOLECULES

The crystalline human myeloma IgG1 Kol has been studied by X-ray diffraction techniques to a resolution of 5 Å (19), 4.0 Å, and 3.5 Å (26), and subsequently refined by constrained crystallographic procedures (40). This study is of particular interest due to the fact that the crystals diffract to a resolution of 3.5 Å, and also, that IgG Kol appears to contain a normal hinge region, unlike other crystalline immunoglobulins that have been studied (41–43). However, the structural analysis of IgG Kol revealed that no electron density could be assigned to the Fc region. This part of the molecule seems to be disordered in the crystal, probably due to the presence of an intact hinge region which is capable of motion even in the crystal lattice. The electron density corresponding to the Fab regions of the molecule could be traced to residues 213 (H chain) and 209 (L chain) and beyond these points, in a tentative way, down to the hinge region sequence -Cys-Pro-Pro-

R

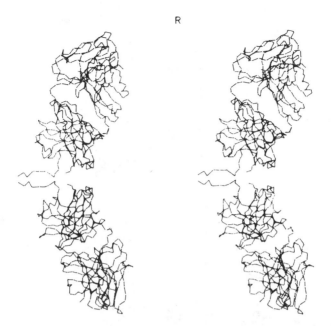

Figure 16   Stereo pair of the Cα positions of IgG Ko*l*. The Fc portion of the molecule was disordered in these crystals and is not represented in the diagram. [Reproduced from Matsushima et al (26) with permission.]

Cys-Pro- (residues 226–230, see Figure 16). The Fc part cannot be traced or even be assigned to a general area in the unit cell of the crystal without overlap problems. However, the tight packing around the hinge peptide in the crystal structure requires that this peptide be rather extended. From this interpretation the authors conclude that the C_H2 domains do not come close to the Fab region, that is to say, that there are no contacts between C_H1 and C_H2 except for those resulting from the continuity of the peptide chain.

An interesting feature of the IgG Kol is the quaternary structure of its Fab regions. As described above for Fab Kol, the V_H-V_L domain is more linear with the C_H1-C_L domain than in the crystal structures of the Fab fragments M603 and New (Figures 7, 9, and 17). Although an 8° difference in the relative orientations of the V and C domains was detected, the Fab model still provided a suitable description of the IgG structure.

The three-dimensional structure of human IgG1 (κ) Dob, a cryoglobulin, has been determined to a nominal resolution of 6 Å (41) using the multiple isomorphous heavy atom replacement technique. Recently, this structure has been reinvestigated (27) by fitting the coordinates of the Fc fragment (21) and those of the Fab M603 fragment (25) to the electron density map. This analysis provides a reasonably accurate description of the relative

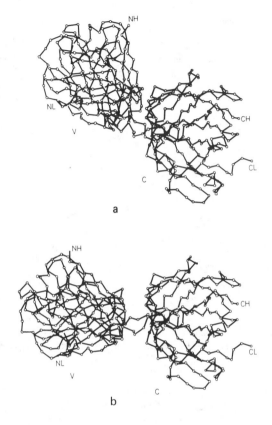

Figure 17 Drawing of α-carbon atoms of the structures of *a* Fab M603 (McPC 603) and of *b* the Fab part of IgG Ko*l*. Both molecules are shown in the same orientation to facilitate their comparison. The N and C termini of all chains are labelled. The constant regions of both molecules on exactly the same orientation showing a different position for the variable regions. [Reproduced from Huber et al (53) with permission.]

positions and orientations of the domains of the molecule although it does not extend to finer details of secondary or tertiary structure (Figure 18). The resulting model shows several interesting features. The structure of the Dob Fc part corresponds closely to that of the Fc fragment within the limits of the resolution and the fitting procedure mentioned above. The C_H2 domains are separated by the carbohydrate chains and make no contact with each other (see Figure 18). The carbohydrate moiety is wedged between C_H1 and C_H2 which prevents close contacts between them (see Figure 19). In the Fab part, a distinction could be made between the H and L chains based on the fact that the L chains are linked to each other by a disulfide bond at their C termini. The angle made by the pseudo twofold axes relating V_H to V_L and C_H1 to C_L is 147°.

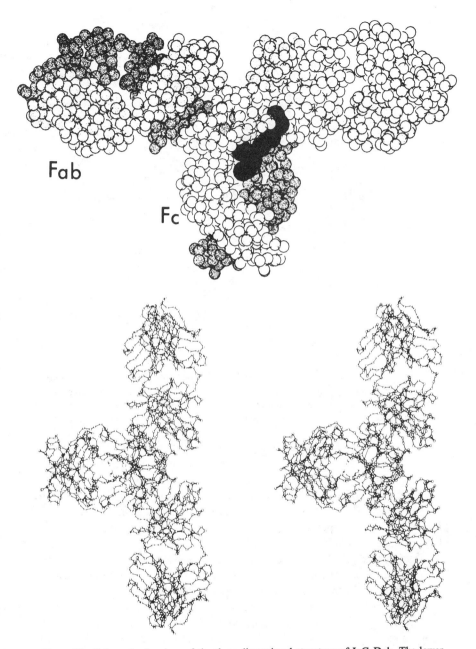

Figure 18 Schematic drawings of the three-dimensional structure of IgG Dob. The lower panel contains stereo pairs of the α-carbon position (small circles) and the positions of the carbohydrate hexose unites (large circles). The twofold axis of symmetry relating the two halves of the molecule is horizontal. One complete H chain is white while the other is dark gray. The two L chains are lightly shaded and the hexose units are represented as large black spheres. The upper panel contains a space-filling view of the same molecule rotated 90° in the plane of the drawing. [Reproduced from Silverton et al (27) with permission.]

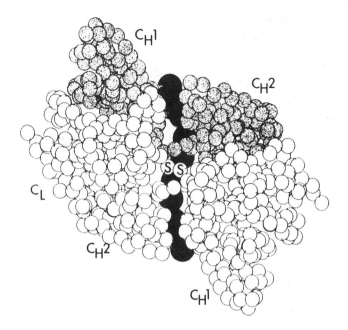

Figure 19 Space-filling view of the constant regions of IgG Dob looking down the twofold axis of symmetry. The shading is as described for Figure 18. The carbohydrate chains seem to provide most of the contacts between C_H2 subunits. [Reproduced from Silverton et al (27) with permission.]

A third crystalline human IgG protein, Mcg, is under crystallographic study. Its hinge region is also affected by a deletion of fifteen residues (42). Only preliminary results have so far been reported (43).

Based mainly on the results of the crystallographic studies of the Fab fragments M603 and New, and those of the human Fc and of IgG Kol, Huber et al (44) proposed an attractive structural model to explain a putative conformational change that could be transmitted from the antigen-combining site. The transmission of such a conformational signal to the $C\gamma2$ domain could be essential for a secondary function such as complement fixation. This model was partly based on the observation that isolated Fab fragments have different quaternary structures from that of the complete immunoglobulin IgG Kol. The determination of the structure of Fab Kol however showed the same V-C arrangement as that of IgG Kol. In addition, in IgG Dob the angle between the pseudo twofold axes in the V_H, V_L, and C_H1, C_L domains is intermediate between those observed in the Fabs M603 and New and that observed in IgG Kol. Thus, the model is not sustained by the more recent structural analyses.

Perhaps the most striking feature that the crystallographic studies of immunoglobulins have revealed is that of molecular flexibility. Segmental flexibility had been well-established before the crystallographic studies; the major structural site of this property is the hinge region. Taken together, the crystallographic studies of immunoglobulins and their fragments indicate that flexibility is also present in the switch regions connecting V_H to C_H1, V_L to C_L, and, possibly, the segment connecting C_H2 to C_H3. Segmental flexibility (hinge region) and intrasegmental flexibility (switch regions) may facilitate antigen-binding by allowing an optimal fit of the combining site to antigenic determinants occurring at varying distances and angles.

MODEL BUILDING OF ANTIGEN-BINDING SITES

The specificity and the affinity of antigen-combining sites of immunoglobulins are determined by the sequence of their hypervariable regions. These regions can be considered as attached to the highly conserved rigid framework of the V homology subunits formed by the 4β and $3\beta'$ sheets and by the bl-connecting segments. The hypervariable regions, on the other hand, are contained in the connecting loops $fl1$, $fl3$, and $fl4$ and are highly variable in both amino acid sequence and structure. The larger structural differences are those arising from the existence of deletions and insertions in these regions. However, the number of amino acids contained in these loops is not very large and in many cases, different immunoglobulins have loops of the same length. Therefore, starting from the known three-dimensional structures and the amino acid sequences of different immunoglobulins, attempts could be made at predicting the conformation of their antigen-combining sites. The simplest approach consists of building a model with the desired sequence following the framework and loop structures of other immunoglobulins that have been determined by crystallographic studies. The description of the phosphorylcholine-binding site of several mouse myeloma proteins based on the structure of M603 (discussed in a previous section) is a good example of model building using highly homologous sequences (34). Another example is that of the analysis of the effects of amino acid substitutions at specified positions in mouse λ chains, based on the structure of Fab (λ) New (45). The conformation of the combining site of mouse myeloma protein MOPC315 has also been described on the basis of the structures of other immunoglobulins (18, 46).

The future availability of a larger number of known three-dimensional structures of immunoglobulins and other proteins will help make predictive schemes more accurate. The use of more elaborate techniques as described for other systems (energy minimization, resonance methods, etc) will be necessary for constructing detailed models of antigen-combining sites based solely on sequence data. Any predictive scheme should be tested by its ability to predict a known structure that has not been included in the data base of the scheme. Clearly, this condition is not met by the attempts described above. Unfortunately the number of immunoglobulin-combining sites with known three-dimensional structure available so far is insufficient to propose and test more elaborate predictive schemes.

SIMILARITIES IN THE POLYPEPTIDE CHAIN FOLDING OF IMMUNOGLOBULINS AND OTHER PROTEINS

Although no amino acids sequence homology can be detected between the bovine Cu,Zn superoxide dismutase enzyme and immunoglobulins, there is a striking similarity in their three-dimensional structures (16). Superoxide dismutases are not related to immunoglobulins by their function since they are intracellular metalloenzymes that process superoxide radicals into O_2 and H_2O. Figure 20 gives a diagrammatic representation of the polypeptide chain folding in both proteins. As can be seen from the figure, superoxide dismutase contains an additional N-terminal strand that contributes to the β-pleated sheet structure. The remaining strands of superoxide dismutase are topologically equivalent to those of an immunoglobulin C region. The packing of subunits is different in the two proteins, since the two superoxide dismutase subunits that constitute a dimer make contacts that are not similar to those made by the V or C domains in immunoglobulins.

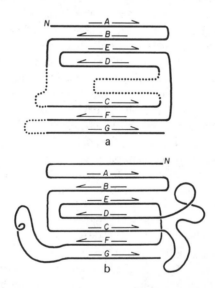

Figure 20   Schematic representation of the folding of the polypeptide chains of *a* the variable region of an immunoglobulin and *b* of bovine Cu,Zn superoxide dismutase. Strands common to both structures are labeled A to G. Dotted lines indicate the hypervariable regions of the immunoglobulin V subunit. [Reproduced from Richardson et al (16) with permission.]

The folding of the polypeptide chain in the blue Cu protein azurin from *Pseudomonas aeruginosa* (47) and in the Cu-containing protein plastocyanin from poplar leaves (48) is topologically equivalent to that of superoxide dismutase, and consequently, to the immunoglobulin fold. Adman et al (47) postulate that similar chain folding in proteins of widely different function may arise from a folding process by which strands having short connecting loops (A and the N-terminal strand; E and D; F and G, Figure 18) assume their secondary structure first. The process is then completed by a "super folding" into the final structure. However, as pointed out by Richardson et al (16) the process of protein folding is not yet sufficiently understood to judge the merits of this proposal. A second alternative, a common evolutionary origin followed by gene duplication and amino acid sequence divergence favoring new functional requirements remains an appealing explanation of the observed similarity in three-dimensional structure.

The amino acid sequence of β_2 microglobulin, a polypeptide chain associated with the heavy chains of histocompatibility antigens, is highly homologous to those of the C regions of immunoglobulins (49). This and other facts have led to speculation that the heavier chain (mol wt 45,000) of histocompatibility antigens may share sequence homologies with immunoglobulins, which would indicate a common evolutionary origin. It has recently been shown (50) that indeed the heavy chain of a human histocompatibility antigen (HLA-B7) has a segment of polypeptide chain with an amino acid sequence highly homologous to that of a human V_H sequence. More complete sequence data is awaited to ascertain the extent of homology, the possible existence of similar domains in both structures, etc. Haptoglobins (50) and the C-reactive protein (51) also show sequence homology with immunoglobulins which leads to the expectation of topological similarity in their three-dimensional folding.

Acknowledgments

The authors are grateful to Mrs. Arleen Skaist for her devoted and patient secretarial work in preparing this review. The research work in this laboratory has been supported by Research Grant AI 08202 from the National Institutes of Health. During the preparation of this paper R. J. Poljak was a Faculty Scholar of the Josiah Macy Jr. Foundation.

Literature Cited

1. Nisonoff, A., Hopper, J. E., Spring, S. B. 1975. *The Antibody Molecule*, New York: Academic pp. 542
2. Davies, D. R., Padlan, E. A., Segal, D. M. 1975. *Ann. Rev. Biochem.* 44: 639–67
3. Beale, D., Feinstein, A. 1976. *Q. Rev. Biophys.* 9:135–71
4. Poljak, R. J. 1975. *Adv. Immunol.* 21:1–33
5. Poljak, R. J. 1975. *Nature* 256:373–76
6. Poljak, R. J., Amzel, L. M., Phizackerley, R. P. 1976. *Prog. Biophys. Mol. Biol.* 31:67–93
7. Padlan, E. A. 1977. *Q. Rev. Biophys.* 10:35–65
8. Poljak, R. J. 1978. *CRC Crit. Rev. Biochem.* 5:45–84
9. Wu, T. T., Kabat, E. A. 1970. *J. Exp. Med.* 132:211–50
10. Kabat, E. A., Wu, T. T. 1971. *Ann. NY Acad. Sci.* 190:382–91
11. Kehoe, J. M., Capra, J. D. 1971. *Proc. Natl. Acad. Sci. USA* 68:2019–21
12. Schiffer, M., Girling, R. L., Ely, K. R., Edmundson, A. B. 1973. *Biochemistry* 12:4620–31
13. Poljak, R. J., Amzel, L. M., Avey, H. P., Chen, B. L., Phizackerley, R. P., Saul, F. 1973. *Proc. Natl. Acad. Sci. USA* 70:3305–10
14. Edmundson, A. B., Ely, K. R., Abola, E. E., Schiffer, M., Panagiotopoulos, N. 1975. *Biochemistry* 18:3953–61
15. Saul, F., Amzel, L. M., Poljak, R. J. 1978. *J. Biol. Chem.* 253:585–97
16. Richardson, J. S., Richardson, D. C., Thomas, K. A., Silverton, E. W., Davies, D. R. 1976. *J. Mol. Biol.* 102: 221–35
17. Segal, O. M., Padlan, E. A., Cohen, G. H., Silverton, E. W., Davies, D. R., Rudikoff, S., Potter, M. 1974. *Progress in Immunology-II. Immunochemical Aspects*, ed. L. Breat, J. Holborow, 1:93–101. Amsterdam: North-Holland. 337 pp.
18. Poljak, R. J., Amzel, L. M., Chen, B. L., Phizackerley, R. P., Saul, F. 1974. *Proc. Natl. Acad. Sci. USA* 71:3440–44
19. Colman, P. M., Deisenhofer, J., Huber, R., Palm, W. 1976. *J. Mol. Biol.* 100:257–82
20. Poljak, R. J., Amzel, L. M., Avey, H. P., Becka, L. N., Nisonoff, A. 1972. *Nature New Biol.* 235:137–40
21. Deisenhofer, J., Colman, P. M., Epp, O., Huber, R. 1976. *Hoppe-Seylers Z. Physiol. Chem.* 357:1421–34
22. Epp, O., Coleman, P. M., Fehlhammer, H., Bode, W., Schiffer, M., Huber, R., Palm, W. 1974. *Eur. J. Biochem.* 45: 513–24
23. Fehlhammer, H., Schiffer, M., Epp, O., Colman, P. M., Lattman, E. E., Schwager, P., Steigemann, W., Schramm, H. J. 1975. *Biophys. Struct. Mech.* 1:139–46
24. Colman, P. M., Schramm, H. J., Guss, J. M. 1977. *J. Mol. Biol.* 116:73–79
25. Segal, D. M., Padlan, E. A., Cohen, G. H., Rudikoff, S., Potter, M., Davies, D. R. 1974. *Proc. Natl. Acad. Sci. USA* 71:4298–302
26. Matsushima, M., Marquart, M., Jones, T. A., Colman, P. M., Bartels, K., Huber, R., Palm, W. 1977. *J. Mol. Biol.* 121:441–59
27. Silverton, E. W., Navia, M. A., Davies, D. R. 1977. *Proc. Natl. Acad. Sci. USA* 74:5140–44
28. Hochman, J., Garish, M., Inbar, D., Givol, D. 1976. *Biochemistry* 15: 2706–10
29. Ellerson, J. R., Yasmeen, D., Painter, R. H., Dorrington, K. J. 1976. *J. Immunol.* 116:510–17
30. Hilschmann, N. 1967. *Hoppe-Seylers Z. Physiol. Chem.* 348:1077–80
31. Fett, J. W., Deutsch, H. F. 1974. *Biochemistry* 13:4102–14
32. Edmundson, A. B., Ely, K. R., Girling, R. L., Abola, E. E., Schiffer, M., Westholm, F. A., Fausch, M. D., Deutsch, H. F. 1974. *Biochemistry* 13:3816–27
33. Watson, M. J., Baddiley, J. 1974. *Biochem. J.* 137:399–404
34. Padlan, E. A., Davies, D. R., Rudikoff, S., Potter, M. 1976. *Immunochemistry* 13:945–49
35. Leon, M. A., Young, N. M. 1971. *Biochemistry* 10:1424–29
36. Amzel, L. M., Poljak, R. J., Saul, F., Varga, J. M., Richards, F. F. 1974. *Proc. Natl. Acad. Sci. USA* 71:1427–30
37. Rao, S. T., Rossmann, M. G. 1973. *J. Mol. Biol.* 76:241–56
38. DePreval, C., Fougereau, M. 1976. *J. Mol. Biol.* 102:657–78
39. Middaugh, C. R., Gerber-Jenson, B., Hurvitz, A., Paluszek, A., Scheffel, C., Litman, G. W. 1978. *Proc. Natl. Acad. Sci. USA* 75:3440–44
40. Deisenhofer, J., Steigemann, W. 1975. *Acta Crystallogr. Sect. B* 31:238–50
41. Sarma, V. R., Silverton, E. W., Davies, D. R., Terry, W. D. 1971. *J. Biol. Chem.* 246:3753–59
42. Deutsch, H. F., Susuki, T. 1971. *Ann. NY Acad. Sci.* 190:472–85
43. Edmundson, A. B., Schiffer, M., Wood, M. K., Hardman, K. D., Ely, K. R.,

Ainsworth, C. F. 1971. *Cold Spring Harbor Symp. Quant. Biol.* 36:427–32

44. Huber, R., Deisenhofer, J., Colman, P. M., Matsushima, M., Palm, W. 1976. *Nature* 264:415–20

45. Poljak, R. J., Amzel, L. M., Chen, B. L., Chiu, Y. Y., Phizackerley, R. P., Saul, F., Ysern, X. 1977. *Cold Spring Harbor Symp. Quant. Biol.* 41:639–45

46. Padlan, E. A., Davies, D. R., Pecht, I., Givol, D., Wright, C. 1977. *Cold Spring Harbor Symp. Quant. Biol.* 41:627–37

47. Adman, E. T., Stenkamp, R. E., Sieker, L. C., Jensen, L. H. 1978. *J. Mol. Biol.* 123:35–47

48. Colman, P. M., Freeman, H. C., Guss, J. M., Murata, M., Norris, V. A., Ramshaw, J. A. M., Venkatappa, M. P. 1978. *Nature* 272:319–24

49. Peterson, P. A., Cunningham, B. A., Berggard, I., Edelman, G. M. 1972. *Proc. Natl. Acad. Sci. USA* 72:1612–16

50. Terhorst, C., Robb, R., Jones, C., Strominger, J. L. 1977. *Proc. Natl. Acad. Sci. USA* 74:4002–6

51. Osmand, A. P., Gewurz, H., Friedenson, B. 1977. *Proc. Natl. Acad. Sci. USA* 74:1214–18

52. Poljak, R. J. 1973. In *Contemporary Topics in Molecular Immunology,* ed. R. A. Reisfeld, W. J. Mandy, 2:1–26. New York: Plenum

53. Huber, R., Deisenhofer, J., Colman, P. M., Matsushima, M., Palm, W. 1976. *The Immune System,* 27th Mosbach Colloquium, pp. 26–40. Berlin: Springer

Ann. Rev. Biochem. 1979. 48:999–1034

INITIATION OF DNA SYNTHESIS IN *ESCHERICHIA COLI*[1]

❖12031

Jun-ichi Tomizawa and Gerald Selzer

Laboratory of Molecular Biology, National Institute of Arthritis,
Metabolism, and Digestive Diseases, National Institutes of Health,
Bethesda, Maryland 20014

CONTENTS

PERSPECTIVES AND SUMMARY

Because DNA contains the genetic information of the cell, its replication is one of the most important of the events that occur as the cell grows. The process that carries out this replication can be conveniently subdivided into three stages: initiation, continuation, and termination. In this review, we describe recent progress in understanding the mechanisms by which DNA synthesis can initiate in *Escherichia coli*. The stages that follow initiation

[1]The US Government has the right to retain a nonexclusive, royalty-free license in and to any copyright covering this paper.

are not discussed except when they provide information about initiation itself.

Many approaches have been taken to the study of DNA synthesis. These include physical and chemical studies on the structure of DNA, biochemical studies on enzymatic mechanisms of DNA synthesis (1–9), and genetic studies of the functions required for DNA replication and its regulation (10). For a long time these studies were performed more or less separately, and it is only recently that a comprehensive understanding of DNA replication has begun to emerge.

Both DNA and RNA are synthesized by the ordered polymerization of nucleotides in a sequence complementary to that of the template strand. While RNA polymerases can initiate polymerization at specific nucleotide sequences, no known DNA polymerase can copy a template unless presented with the 3'-hydroxyl terminus of a polynucleotide primer that is based-paired to a single strand of the DNA template. Thus it is the process of primer formation that determines the site where the initiation of DNA synthesis occurs.

Considerable information about primer formation has been gained through the study of in vitro DNA synthesis using the single-stranded DNAs of small phages like fd and ϕX174 as templates. These studies have shown that both RNA polymerase and the *dnaG* protein can make primers for DNA synthesis. Depending on the kind of template DNA, one or the other enzyme is used.

Initiation of DNA synthesis on double-stranded DNA requires separation of the DNA strands in addition to the formation of primer. One way to accomplish this is by making use of the local strand separation that occurs during transcription by RNA polymerase. In this case, the transcript may itself serve as primer. Another mechanism involves the nicking of one strand to expose a 3'-hydroxyl group. The strands can be separated at the nick by an enzyme, and the 3'-hydroxyl group can then serve as primer.

As double-stranded DNA replicates, growth of one of the two progeny strands is usually occurring at the 5' end of the strand, even though polymerases are only known to add nucleotides to 3' ends. It is now clear that this growth is accomplished through the addition of small DNA fragments to the 5' end. Each of these fragments is synthesized shortly before it joins to the strand.

Thus there are two roles for primer synthesis during replication. One of these is the initiation of replication of the genome, while the other is the initiation of synthesis of each of the small fragments made during elongation of the progeny strands. Despite this difference in function, the two kinds of primer synthesis could be carried out by the same mechanism.

The elucidation of the nucleotide sequences of regions where DNA synthesis initiates is an area in which major progress has recently been made.

Most of the known sequences can be arranged into extensive secondary structures. As yet, however, there is practically nothing known about the significance of these structures. Their role in the protein-nucleic acid interactions that occur during the initiation of DNA synthesis is an interesting problem that is now accessible.

Characterization of the proteins that participate in DNA replication and the precise determination of the location where DNA replication initiates became possible only after the development of in vitro systems that carry out DNA replication. For this reason, our discussion of the basic mechanism of initiation depends heavily on information obtained from in vitro experiments. Since it is difficult to perform experiments with a large genetic element like the chromosome of *E. coli,* much of this information has been gained by use of the DNAs of small viruses and plasmids. Recently, small plasmids that appear to contain the origins of replication of large bacterial and viral genetic elements have been isolated. Future in vitro studies of the replication of such plasmids should clarify the biochemical mechanisms involved in initiation of replication at these origins.

DNA SYNTHESIS ON SINGLE-STRANDED CIRCULAR DNA

A number of small phages of *E. coli* are known to have circular single-stranded DNA genomes of similar size (about 6000 bases). Although these phages can be divided into two groups (filamentous and isometric) on the basis of morphological differences, their genomes replicate by similar mechanisms. In the first stage of this replication, the viral DNA [(+) strand] is converted to a circular duplex called an RF molecule. The process, which is sometimes called SS → RF synthesis, involves synthesis of the complementary strand [(−) strand] in a reaction that requires no phage-specified protein. The next stage of DNA synthesis (RF → RF synthesis) begins when a phage-encoded protein breaks the RF molecule at a specific position on its (+) strand. The new (+) strand is then synthesized by extension of the 3' end at the break by a mechanism involving strand displacement. Completion of synthesis yields a new circular (+) strand. The displaced (+) strand may serve as template for synthesis of the (−) strand of a new RF molecule which could begin while (+) strand synthesis is still in progress. The third stage of replication (RF → SS synthesis) occurs at later times in infection when synthesis of (−) strands on the displaced (+) strand is blocked by viral proteins that encapsidate the (+) strand.

Synthesis of (−) strands on single-stranded viral DNA initiates with the formation of a primer synthesized either by RNA polymerase (fd and other filamentous phage) or by the *dnaG* protein (ϕX174 and other isometric phage). Subsequent DNA synthesis is carried out by the DNA elongation components. In most in vitro experiments, elongation of the primer has been carried out by DNA polymerase III in combination with *dnaZ* protein, and the DNA elongation factors I and III (8, 11), or else by a protein complex that contains the polymerase and its accessory factors (DNA polymerase III holoenzyme) (12, 13). However, elongation can also be performed by DNA polymerase I or DNA polymerase II. Since recent reviews of the subject are available (7, 8), the DNA elongation reaction is not discussed here.

In the section that follows, the in vitro synthesis of complementary strands is examined in some detail. Following this, the mechanisms of RF → RF and RF → SS synthesis are described. [Reviews: in vivo, (14–18); in vitro, (6, 8).]

Priming by RNA Polymerase

The synthesis of DNA on (+) strands of filamentous phage M13 (or fd) in cell extracts initiates at a specific region of the template (19). In addition to dNTPs, the synthesis requires all four rNTPs (20, 21), and is inhibited by rifampicin (21, 22), a specific inhibitor of *E. coli* DNA-dependent RNA polymerase. The products of synthesis are (−) strands of nearly full length.

The synthesis of (−) strands can also be carried out with purified proteins, namely DNA-dependent RNA polymerase, DNA binding protein I [DBP, (23)], and the DNA elongation components (24). It has been proposed that the primer is synthesized at a unique position that is not covered with DBP and is therefore accessible to the RNA polymerase (6, 24). According to this model, a duplex hairpin structure in the (+) strand blocks binding of DBP to the region where the primer is formed.

The structure of fd DNA in the region where RNA polymerase binds has been characterized by the following experiments. Exhaustive DNase digestion of fd DNA in the presence of RNA polymerase and DBP yielded protected fragments of about 120 nucleotides in length (25). Pyrimidine tract analysis of these RNA polymerase-protected fragments showed that they come from a region known to contain the origin of synthesis of the (−) strand (26). Examination of the nucleotide sequence of this region further showed that the protected fragment could form two separate hairpin structures. As described below, one of the hairpin structures includes the DNA that specifies the primer of (−) strand synthesis. The other hairpin has been considered unimportant for replication because its alteration by the insertion of foreign DNA does not abolish the infectivity of the DNA (27). However, the site of this insertion is near the loop of the hairpin; thus the importance of the stem region has not been determined.

When fd DNA is incubated with RNA polymerase, DBP, and rNTPs, a transcript of about 30 nucleotides in length is made. The results of restriction enzyme analysis of DNA fragments formed by extension of the RNA transcript by DNA polymerase I have been combined with RNA fingerprint analysis of the primer itself to provide the exact location of the RNA on the nucleotide sequence of fd DNA (28). The transcription initiates with ATP at a dTMP that is not base-paired, but is located near the stem of the hairpin structure containing the primer sequence. There is no nearby heptanucleotide sequence of the type usually found in promoter sequences (29, 30). The transcription proceeds toward the loop of the hairpin structure. With the progress of transcription, DBP may complex the opposition strand of the hairpin as the structure melts. Since transcription terminates anywhere within a region of several nucleotides, it appears possible that termination results when RNA polymerase encounters DBP in the region of the melted hairpin structure (28).

The nucleotide sequences of the region containing the origin of (–) strand synthesis of M13 (31) and f1 (32) DNAs have also been determined. These have extensive homology with the fd sequence and, in particular, have almost completely conserved hairpin structures.

As discussed in the next section, ϕX (–) strand synthesis in cell extracts does not use RNA polymerase for its primer synthesis. Nonetheless, ϕX (–) strands can be synthesized by the same mixture of RNA polymerase, DBP, and DNA elongation proteins used for fd (–) strand synthesis (33, 34). If the purified RNA polymerase is replaced by a crude enzyme preparation with RNA polymerase activity, ϕX (–) strand DNA can no longer be made, although M13 DNA is still synthesized (33). Starting with a cell extract, three proteins have been obtained that specifically inhibit ϕX DNA synthesis when added together to the enzyme mixture containing purified RNA polymerase (34, 35). One of these has been identified as ribonuclease H (RNase H); the other two are referred to as discriminatory factors α and β. Although the exact role of these proteins in the discriminatory effect is unknown, the involvement of RNase H suggests that the effect may result from the degradation of ϕX transcripts which have potential priming function.

Priming by dnaG Protein

With DNA of the isometric phages (ϕX174, G4, ST-1, etc), the first step in (–) strand synthesis involves formation of a primer by the *dnaG* protein. Remarkably, the various isometric phages differ with respect to the number of factors, besides the *dnaG* protein, that they use for primer formation.

Synthesis of (–) strands of G4 (or ST-1) can be demonstrated with purified *dnaG* protein, DBP, and the DNA elongation components (36–41). The process of DNA synthesis by these proteins consists of several discrete steps (8, 39–41). In the first step, DBP covers the template DNA and the *dnaG* protein then binds. No nucleotide cofactor is required in the process. Once formed, the DNA-protein complex can be isolated by gel filtration (41–43). The complex contains approximately one molecule of *dnaG* protein for each molecule of template DNA even when excess protein is present in the incubation. Binding of the *dnaG* protein in the presence of DBP is specific for G4 (or ST-1) DNA since no stable complex is formed with fd or ϕX DNA (42, 43).

The exact NTP requirement for primer synthesis by *dnaG* protein has been a subject of some controversy (36, 40). A partial resolution was provided by the discovery that the *dnaG* protein can synthesize oligonucleotides that consist of ribo-, deoxyribo-, or mixed ribo- and deoxyribonucleotides (41, 44, 45). To some extent, ribonucleotides and deoxyribonucleotides are used interchangeably for primer formation. Efficient G4 primer synthesis can occur in the absence of all rNTPs but only if ADP is present (42, 43). For ST-1 primer synthesis, even ADP is not required. In its absence, the ST-1 primer initiates with dATP at the 5' end (42, 43). Irrespective of its chemical composition, the oligonucleotide that can be elongated by the DNA elongation components is called a primer, and the *dnaG* protein is called primase (46).

When all four rNTPs are present, an oligoribonucleotide of 25–28 residues is synthesized on G4 DNA by the *dnaG* protein and DBP. The base sequence of the oligonucleotide has been determined (47). Regardless of its exact length, the product can serve as a primer for DNA synthesis (47). Recently it has been shown that the size of the oligonucleotide products is affected by the conditions of the reaction (48). When a low salt concentration (50 mM NaCl) and a high temperature (39°) are used, most of the oligonucleotides made on G4, ST-1, ϕK, and α3 DNAs have a length of 28 bases. When the salt concentration is raised (250 mM) and the temperature lowered (15°), the length decreases to 14 bases with G4 and 11 bases with the other DNAs.

The nucleotide sequence of the region of DNA where the synthesis of (−) strands initiates has been determined for G4 (49, 50) and for ST-1, ϕK, and α3 (51). Within this region (about 300 nucleotides), the sequences of ST-1, ϕK, and α3 are similar to each other, while differing significantly from that of G4. There are two stretches, one 42, the other 45 bases in length, that are quite similar in all four phages. These two stretches are separated by 13 bases whose sequence differs among the various DNAs (51).

In each case, the region that specifies the primer lies within one of the conserved stretches. The sequence of this region in ST-1 is identical to that of ϕK and α3, but different at 5 positions from that of G4. Despite the difference, either sequence can be arranged in a similar hairpin structure with a stem of 8 base pairs and a loop of 5 bases.

Because of the effect of salt and temperature noted above, it appears that the exact size of the product made by the *dnaG* protein may be determined by the stability of this hairpin structure. Comparison of the various sequences does not suggest a sequence or structure, other than that of the hairpin itself, which could play a specific role in the interaction of the DNA with the *dnaG* protein.

Priming by dnaG Protein in Conjunction with Other Proteins

Synthesis of ϕX (–) strand requires at least 11 proteins, only 6 of which are needed for G4 (–) strand synthesis. All of the additional proteins act before or during synthesis of the primer which is again carried out by the *dnaG* protein. The additional proteins include the *dnaB* and *dnaC* proteins, factors X, Y, and Z (8, 43, 52). These three factors probably correspond to proteins i, n', and n, respectively, which have been independently isolated and named (13, 53). In the absence of any of these components, the *dnaG* protein cannot catalyze primer formation with ϕX DNA as template. However, primer synthesis can occur in the absence of the proteins required for DNA elongation (43, 54).

The sequence of events during priming can be summarized briefly as follows. First, DBP covers the DNA and one or several molecules of factor Y (protein n') and Z (protein n) bind as well. Then the *dnaB* protein is transferred to the protein-covered DNA in a reaction that requires the *dnaC* protein, factor X (protein i), and ATP. The net result of these reactions is formation of a DNA-protein complex containing the *dnaB* protein, factors Y and Z (proteins n' and n), and DBP (8, 43, 54). The *dnaG* protein now binds and begins synthesis.

The reactions that occur after binding of the *dnaB* protein are as yet unclear. However, some insight into the mechanism may be provided by the following results obtained through the use of antibodies (55, 56). Addition of antibody directed against the i, n, or *dnaB* proteins inhibits *dnaG*-dependent transcription and also DNA synthesis by the reconstituted system for ϕX (–) strand synthesis. Once the DNA-protein complexes containing the *dnaB* protein are formed, transcription is inhibited by the anti-*dnaB* antibody, but not by the others. If primer has been made, none of these antibodies inhibits DNA synthesis. These results suggest that proteins i and n function before binding of the *dnaB* protein to the DNA, and that once primer is made, even the *dnaB* protein is not needed for subsequent DNA synthesis.

One drawback to this kind of experiment is the possibility that a particular antibody may be ineffective because the site it recognizes on a protein is masked in a complex. Another is the difficulty in performing precise kinetic analysis. Firmer conclusions about the roles of the various components probably require a more direct analysis of the proteins present in the intermediate complexes.

Unlike transcription of G4 DNA, where a single unique primer is synthesized, a yield of six to eight oligonucleotides per template can be obtained when ϕX DNA is used (57). These oligonucleotides vary in length from 16–50 residues and appear to be of heterogeneous sequence. Even under conditions where a yield of one or less oligonucleotide per template is obtained, no strongly preferred site of transcription is evident (57). From the available data, no conclusion can be drawn about the regions transcribed or about which transcripts actually serve as primers.

Recently it has been shown that protein n' (factor Y) binds preferentially to the single-stranded *Hae*III-Z1 fragment, which contains 25% of the ϕX genome (13). Thus it appears possible that the *dnaB* protein, guided by protein n', binds at a unique site on ϕX DNA. To reconcile the heterogeneity of the transcripts with the assumption that the *dnaG* protein binds and initiates primer synthesis only where the *dnaB* protein is present, it has been proposed that the *dnaB* protein, once bound, migrates on the template (56, 57).

As is the case for G4 (or ST-1) primer synthesis, the *dnaG* protein can synthesize ribo-, deoxyribo-, and mixed ribo- and deoxyribooligonucleotides on ϕX DNA (43, 57). It appears that similar numbers of products are made with either rNTPs or dNTPs as substrate, but that the probability of termination is increased by the presence of dNTPs. Possibly because of the variety of products, the nucleotide requirements for effective ϕX primer synthesis (57, 58) have been far less simple to establish. Except for an absolute requirement for ATP (21, 43), the efficiency with which dNTPs substitute for the corresponding rNTPs is still an open question.

With the obvious exception of ϕX, the nucleotide sequences of the isometric phages are very similar in the region used as the origin of (–) strand synthesis. This also appears to be true for the analogous region of the filamentous phages. Since in all these cases the regions are intercistronic, it seems likely that the sequence is conserved primarily because of the necessity of a specific site for initiation of DNA synthesis. Larger genomes, such as the *E. coli* chromosome, need to repeat the initiation events required for discontinuous synthesis many times during a round of replication. Use of such a site-specific mechanism, which would require many copies of the same long sequence to be present in the genome, could lead to disintegration of the chromosome through recombination. From this viewpoint, the mechanism employed for ϕX (–) strand synthesis is a more suitable method for replication of large genomes. For the moment, the existence of the *dnaG* reaction observed with G4 remains a puzzle.

DNA SYNTHESIS ON DOUBLE-STRANDED DNA

DNA Synthesis from a 3'-hydroxyl End at a Strand Break

In some cellular processes, DNA synthesis initiates at a pre-existing 3' hydroxyl end of a DNA strand. Such processes include the DNA synthesis which seals gaps formed during replication, genetic recombination, and excision repair. In other cases, a 3'-hydroxyl end is created by specific nicking of duplex DNA and then used as a primer for initiation of DNA synthesis. The synthesis of (+) strands on RF molecules of the small single-stranded DNA phages provides the best-studied example of this kind of initiation (59). In the discussion that follows, DNA synthesis on ϕX RF molecules is described in some detail, and then a briefer description of other systems is given.

In vivo studies have shown that DNA synthesis on ϕX RF molecules requires the phage gene A protein (60). This protein breaks the (+) strand of an RF molecule (61, 62) in the $Hind$II-R3 segment that includes a portion of the gene A itself (63). The protein remains attached covalently to the 5' end of the broken strand (64, 65). Both supercoiled DNA and circular (+) strand DNA are susceptible to the action of the protein, but relaxed closed-circular, open-circular, and linear duplex DNAs are not (64, 65). The nucleotide sequence at the 3' terminus of the cleaved (+) strand has been determined (66). The terminus is located at a unique position in the known nucleotide sequence of ϕX DNA (67). That the action of the gene A protein is actually the first step in initiation of DNA synthesis has been suggested by an in vivo experiment which shows that the $Hind$II-R3 segment contains the region synthesized first during RF replication (68).

DNA synthesis on RF molecules is observed in extracts from uninfected bacteria that are supplemented with the gene A protein (69–71). These extracts are inactive when prepared from a mutant strain (rep^-) unable to support ϕX replication in vivo (72). A set of proteins that can carry out DNA synthesis on RF molecules has been purified (69–71). When supplied with RF molecules, the mixture of gene A protein, rep protein, DBP, and DNA polymerase III holoenzyme produces (+) strand circles in numbers that far exceed the number of input template molecules (69). Following fixation with formaldehyde and glutaraldehyde, replicating molecules from this purified enzyme system have been examined in the electron microscope and found to have a single-stranded loop attached to a double-stranded circle at a single point ("looped" rolling circle) (73, 74). This observation suggests that the 5' end moves along the circular molecule with the replication fork. When a round of synthesis is completed, a circular (+) strand could be produced through a transfer reaction catalyzed by the gene A protein (73).

Although the strands of an RF molecule separate as DNA synthesis proceeds, DNA synthesis is not obligatory for strand separation. In the absence of the DNA polymerase III holoenzyme, the RF molecule is separated into circular and linear strands by a mixture of the gene *A* protein, *rep* protein, and DBP (75–77). Recently, it has been shown that even the gene *A* protein is not required for strand separation (78). Together with DBP, the *rep* protein can separate the strands of DNA that contains gaps, but not the strands of nicked DNA. Since the addition of DBP before the *rep* protein prevents the reaction, the single-stranded region is probably required for binding of the *rep* protein (78). Two molecules of ATP are hydrolyzed for every base pair melted, whether or not the strand separation is coupled to DNA synthesis (77, 78).

Although the in vitro system which contains only the gene *A* protein, *rep* protein, DBP, and the DNA polymerase III holoenzyme produces complete circular (+) strands, other results suggest that additional *E. coli* proteins may participate in (+) strand synthesis in vivo. For example, a partially purified bacterial extract supplemented with the gene *A* protein cannot support synthesis of (+) strand from RF molecules if its *dnaB, C,* or *G* protein has been inactivated (79). The synthesis of (+) strands in vivo may also require these same gene products (80). Other experiments suggest that both strands of the ϕX RF molecule are replicated in vivo by a discontinuous mechanism (81), a finding hard to reconcile with the simple model presented above. These various results indicate to us that there may be factors or reaction conditions that determine the exact reaction mechanism to be used for in vivo (+) strand synthesis. Changes in these conditions or loss of the factors during purification of the proteins used for in vitro synthesis may permit the occurrence of reactions that normally play a minor role in synthesis in the cell.

The site of the origin used for synthesis of the G4 (+) strand has been deduced from the location of the gaps in open-circular molecules that accumulate at late times in infection (82, 83), as well as from the position of the end of the tail in rolling circle molecules (84). This site is located about 40% of unit length away from the origin of (–) strand synthesis. When the nucleotide sequence of the region containing the origin of G4 (+) strand synthesis is compared with that around the site cleaved by the gene *A* protein in ϕX DNA, an identical 30 nucleotide sequence can be seen in both DNAs (48). This is the longest stretch of sequence homology in these two DNAs (85). These results, coupled with the observation that G4 is unable to replicate in *rep*⁻ bacteria (86), suggest that G4 also makes use of a site-specific endonuclease to initiate RF → RF replication. There is, as yet, no direct evidence for such a nuclease.

Despite the apparent similarity of the φX and G4 initiation mechanisms, RF molecules of G4 are not found in rolling-circle structures during the early phase of infection. Instead, displacement loop (D- loop) structures similar to those seen in replicating mitochondrial DNA (87) are observed (88). As mentioned above, circles with a single-stranded tail (rolling circles) are found at later times (85).

Replication of RF molecules of the filamentous phages (M13, fd, and f1) depends on the phage gene *II* protein (89, 90) and the host *rep* protein (91). The gene *II* protein breaks the (+) strand at a specific site (92). In contrast to the φX gene *A* protein, the gene *II* protein does not attach covalently to the cleaved strand (93). The cleavage site is separated from the site where synthesis of the primer for (–) strand synthesis initiates by 24 nucleotides (92). The inference that the cleavage site is the origin of (+) strand synthesis is consistent with the observation that, in vivo, the origins of synthesis of the (+) and (–) strands of f1 DNA are located close to one another (94).

As mentioned above, the nucleotide sequence in the origin region is very similar in these filamentous phages (27, 31, 32). In this region there is a six-base homology with the sequence near the origin used for φX (+) strand synthesis. Remarkably, the gene *II* protein cleaves fd DNA within this short homology while the φX gene *A* protein cuts φX DNA just 3 bases outside it.

In vivo RF → RF replication of M13 requires the *dnaG* (95, 96) and *dnaA* (97) proteins. However, these proteins are not required during either SS → RF or RF → SS synthesis in vivo. Thus there appears to be a requirement for certain proteins that is unique to the mechanism that produces progeny RF. The availability of DBP or of phage capsid proteins may determine the exact mechanism to be employed and thus the protein requirement.

Synthesis of (+) strands on fd RF molecules by the gene *II* protein, DBP and DNA polymerase III holoenzyme has been reported (93). It is possible that the *rep* protein, which is required for in vivo replication of the RF molecules, was present in the preparation of gene *II* protein that was used.

Initiation of DNA synthesis by the breakage of a strand is not limited to these small DNA phages. In vivo experiments show that replication of phage P2 DNA requires the P2 gene *A* protein (98) which helps catalyze a strand-specific break in the region where replication initiates (99, 100). As with that of the small phage, replication of P2 requires the host *rep* protein (101).

Nicking and extension of DNA strands may also occur during phage T7 DNA synthesis. In the absence of rNTPs, the T7 DNA polymerase and T7 gene 4 protein can catalyze DNA synthesis from preexisting nicks (102–

104). Normally, no synthesis occurs when these proteins are incubated with intact T7 DNA. However, if an *E. coli* protein, named initiation protein, is also included in the incubation, synthesis initiates from nicks introduced at three specific sites on the light strand (104). Alone, none of these proteins exhibit any nucleolytic activity. Approximately 70% of the synthesis initiates at a site about 18% from the left end of the linear DNA and proceeds rightward to give a product in which the newly synthesized DNA is attached covalently to the parental light strand. Although this site of initiation is close to the site of initiation of in vivo replication, if not identical to it, the synthesis does not give rise to the replication loop intermediate observed in vivo (2).

Recently a hybrid plasmid containing the in vivo origin of T7 has been constructed and shown to replicate in extracts of T7-infected bacteria (105). As yet, little is known either about the proteins required for this replication, or about the DNA produced. Future study of the system should prove useful in understanding the T7 initiation process.

Processing of Primer RNA and Use of Two Initiation Mechanisms

These two phenomena are observed during in vitro replication of ColE1 DNA. Although unrelated, they are presented together in this section. Recent reviews of ColE1 DNA replication are available (8, 106, 107, 107a).

Synthesis of ColE1 DNA can be carried out in vitro using extracts of *E. coli* cells (108). Extracts of bacteria that do not carry ColE1 are effective, even when de novo protein synthesis is blocked during the incubation (109). This result indicates that proteins encoded by the plasmid are not required for its replication in vitro. That this is also the case in vivo has been shown by the replication of ColE1-phage hybrids following infection of chloramphenicol-treated cells (110, 111).

The major products of the in vitro reaction are completely replicated molecules and other molecules (early replicative intermediates) having a small replication loop in a unique region. Such loops contain newly synthesized small DNA fragments (6S fragments) (108, 112, 113). A specific fragment of L-strand DNA (6S L fragment) is the first product of synthesis (114).

Synthesis of the 6S fragments in cell extracts can be inhibited by rifampicin (108), novobiocin (115), or nalidixic acid (116), which indicates the involvement of RNA polymerase and DNA gyrase. Synthesis does not occur in extracts made from *polA*⁻ bacteria, which indicates the participation of DNA polymerase I (107, 117). In vivo ColE1 replication also requires DNA polymerase I (118) and RNA polymerase (119).

The in vitro origin of replication has been located by analysis of the 6S L fragments (120). The precise location where synthesis of the DNA in the 6S L fragments begins was determined by treating the DNA with alkali to remove a possible primer RNA, labeling the 5' ends exposed by the alkali with ^{32}P, and then digesting the DNA with restriction enzymes. The sizes of the labeled fragments indicated the position of sites at which DNA synthesis begins. These sites, which are defined as the origin of ColE1 DNA replication, are located almost exclusively at any of three neighboring nucleotides. The presence of specific RNA-DNA linkages has been confirmed by nearest neighbor analysis of fragments synthesized with [α-^{32}P]dNTP (121). The origin of in vivo replication appears to be the same site used as the origin for in vitro replication (122).

Some DNA is synthesized on supercoiled ColE1 DNA by a mixture of RNA polymerase and DNA polymerase I (123). A protein, later identified as RNase H, has been isolated from cell extracts by virtue of its ability to stimulate DNA synthesis at the specific origin when added to the enzyme mixture. In the presence of RNase H, a majority of the products of synthesis (about 300 nucleotides in length) initiate at the specific origin described above. Without RNase H, only 20% as many fragments are made, none of which begin at the specific origin (123). RNase H appears to process the RNA, thus creating a specific end at which DNA synthesis can begin. This conclusion is based on the observation that the distribution of the 5' ends of the DNA molecules among the several nucleotides at the origin is altered by the exact concentration of RNase H in the reaction mixture (123). The 5' end of the primer for DNA synthesis has not yet been identified.

As described above, the synthesis of ColE1 by a mixture of DNA polymerase I, RNA polymerase, and RNase H occurs in the absence of DNA gyrase even though this enzyme is required for synthesis in extracts. Addition of gyrase to the mixture results in the formation of long L-strand DNA although H-strand DNA is still not made (123). However, the addition of DNA gyrase does not alter the number of molecules on which synthesis initiates as long as a supercoiled template is used. It has been inferred from this that a role of DNA gyrase in synthesis of 6S L fragments by extracts is to counteract the action of topoisomerases that reduce the superhelical density of the template (123).

The region of ColE1 DNA that is essential for its replication has been deduced from the properties of plasmids obtained by ligating various restriction fragments of ColE1-type plasmids to other replicons or to a fragment that confers a selectable property (124–126). The results show that only a small portion of the plasmid genome is required for its replication and maintenance in bacteria. At this time, the essential region is estimated to contain no more than 580 base pairs of which only 13 base pairs are downstream from the origin of replication (125). The essential region must specify the primer RNA and, in addition, may carry a recognition site for a hypothetical enzyme that ensures separation of daughter molecules (107, 127). The possibility that the region also encodes a protein required for plasmid maintenance, but not for replication per se, has not been excluded. An RNA about 110 nucleotides in length (128) is transcribed from the part of this region which is most distant from the origin of replication. It has been proposed that this RNA is processed and then hybridizes to the origin region where it serves as the primer for DNA synthesis (125). So far, no supporting data are available.

ColE1 DNA replicates unidirectionally, and synthesis of both strands is usually coordinated (113, 129, 130). Once synthesis of the 6S L-DNA starts, RNA polymerase is no longer needed (114). On the other hand, some functions that do not participate in the synthesis of 6S L-DNA are involved in the subsequent DNA synthesis that yields complete molecules. These functions include *dnaB, dnaC, dnaZ,* and DNA polymerase III [(107, 107a, 117, 131, 132); Y. Sakakibara, personal communication; T. Itoh and J. Tomizawa, unpublished results)]. Participation of *dnaG* protein in ColE1 DNA replication in vivo has been suggested (133).

Two possible mechanisms for the synthesis of ColE1 DNA are schematically presented in Figure 1 and can be summarized as follows: (*a*) Synthesis of the L fragment from the origin requires a unique set of proteins. A different set of proteins is required for synthesis of the rest of the molecule; (*b*) The L strand, as a leading strand, is extended continuously (Model *A*) or else discontinuously (Model *B*); (c) The H strand, as a lagging strand, is synthesized discontinuously.

In either model, the use of a unique origin for ColE1 DNA replication results from some site-specific property of the mechanism used to synthesize the first portion of the L strand. Since this synthesis separates the strands of the duplex molecule, synthesis of the first H strand fragment can now

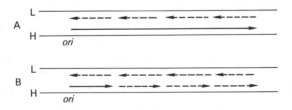

Figure 1   Schematic representation of DNA fragments synthesized on the H and L strands. The continuous and discontinuous arrows represent DNA fragments synthesized by different mechanisms. The two models, *A* and *B*, are described in the text.

initiate on the exposed L strand by use of a second mechanism that requires a single-stranded template and other gene products. It is with respect to the synthesis of the rest of the molecule that the models diverge.

Model *A* supposes continuous synthesis of the L strand. This is suggested by results obtained with purified enzymes. These show that extensive elongation of the L strand can occur when purified RNA polymerase, RNase H, DNA polymerase I, and DNA gyrase are used to synthesize ColE1 DNA (123). If this model is correct, the requirement for certain gene products (e.g. the *dnaB* protein) during extension of the L strand in extracts, may simply result from a necessity to coordinate the synthesis of H and L strands in extracts and in vivo.

Model *B* supposes discontinuous synthesis of the L strand. This possibility has been suggested by experiments that used extracts made from bacteria containing a temperature-sensitive DNA ligase (134). In these extracts, DNA fragments of about 7S are synthesized on various regions of both strands of ColE1 DNA at high temperature. Control experiments indicate that the small fragments are not generated by a repair process that acts on misincorporated nucleotides (135). Thus there is some reason to think that the fragments are formed as intermediates in the elongation of both strands.

Activation of the Origin by Transcription

Transcriptional activation is a concept that arose from studies of the replication of phage λ (136, 137). Although the mechanism that activates λ DNA replication is not fully understood, the concept itself is important in considering the initiation of DNA synthesis on double-stranded DNA.

Replication of λ DNA initiates at a fixed origin and proceeds bidirectionally (138). The characterization of stable plasmids composed entirely of small portions of λ DNA has shown that all the information needed for the replication of λ DNA lies between 78% and 83% on the physical map (139) (Figure 2). A more precise location of the origin itself is given by mutants (now called *ori⁻*) which have a *cis*-dominant defect in replication and are, therefore, believed to have an altered origin of replication (137, 140). The mutations responsible for the defect map in gene *O* (140) and, as will be discussed below, are small deletions or, in one case, a single base change (141, 142).

In vivo, the replication of λ DNA depends on the phage *O* and *P* proteins (143) and on several bacterial proteins including the *dnaB, dnaG, dnaJ* (144, 145), and *dnaK* (144, 145) products, DNA polymerase III, and RNA polymerase (146, 147). Genetic complementation tests show that the *O*-like proteins of various lambdoid phages cannot substitute for each other, although in some cases, their *P*-like functions can (147–149). Measurements of the ability of hybrids formed by recombination between λ and φ80 or

82 to make use of the *O* function of these various phages show that the inability of λ to use other *O*-like proteins is due to sequences present in the N-terminal region of its own gene *O* (150). A similar analysis suggests that the ability of λ to use the *P*-like functions of other phage may be governed by the C-terminal region of the *O* protein (150). The interaction of the *O* and *P* proteins has also been inferred from the ability of some gene *P* mutations to suppress a gene *O ts* mutation (151). Since the *P* protein has been shown to interact with the *dnaB* protein (152, 153), it appears possible that all three proteins may act as a complex.

Even when all necessary gene products are supplied in *trans*, λ DNA cannot replicate in the presence of repressor, an effect which is termed replication inhibition (154). Genetic studies of replication inhibition suggest that it is due to inhibition of rightward transcription from the P_R promoter, and that either the transcript need not be translated, or else a required translation product is *cis*-acting (136).

The possibility that a *cis*-acting protein is required is unlikely because of the properties of some λ mutants (*ri^c*) that overcome replication inhibition (137). These were obtained by mixed infection of a λ lysogen with λ and a heteroimmune helper phage. After several cycles of selection, two types of mutants able to replicate in the presence of λ repressor were obtained; those in one class express the *O* and *P* functions constitutively, while mutants of the second type (*ri^cD, ri^c5b*) express the *P* function constitutively as the result of mutations mapping within the *O* gene itself (155). Even in the presence of repressor, the *ri^cD* mutation allows transcription that starts at the site of the mutation and proceeds to the right (156). Since the *ri^cD* mutation enables λ to replicate when transcription from other promoters is repressed, any *cis*-acting protein should be encoded by the

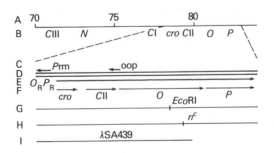

Figure 2 Maps of a portion of phage λ genome that controls transcription and DNA synthesis. The information is taken from references cited in the text. *A,* distance from the left end of the DNA molecule expressed as a percentage of its total length; *B,* approximate position of genes; *C* and *E,* sites of promoters and direction of transcription; *D,* the region contained in λ*dvh*93; *F,* extent and direction of translation; *G, Eco*RI cleavage site; *H,* location of the *ri^c* mutations; *I,* the region retained in λSA439.

region to the right of the mutation. However, most of this region is deleted in a mutant (λSA439) (Figure 2) which can, nonetheless, synthesize DNA when the O and P proteins are supplied in *trans* (157). Since the only region present in λSA439 that can be transcribed from ri^cD is an internal portion of the O gene, transcription from ri^cD should not be required for synthesis of a *cis*-acting protein (137).

It has been proposed that the transcription could act by altering the structure of the origin region and/or by moving the origin to a site in the cell where it can be replicated (137). These explanations were proposed before the concept of RNA priming of DNA synthesis was established, and the first possibility may now be divided into two different ones, depending upon whether the transcript is to be used as a primer for DNA synthesis or not. While it is not yet possible to completely rule out any of these proposals, recent results obtained through the cloning and sequencing of various regions around the origin of replication have provided some additional insight into the problem.

Digestion of λ DNA by *Eco*RI produces a fragment containing a region that extends from the immunity region to a point inside O gene (66%–81% on the physical map). When this fragment is inserted into a λ-ϕ80 hybrid vector lacking the λ origin region, the composite phage now shows λ-specific replication (158). This result suggests that the region covered by the fragment contains the λ origin of replication. Furthermore, a combination of genetic analysis and DNA sequence determination shows that at least four of the ori^- mutations described above cluster in a small region just to the left of the *Eco*RI site within gene O (140, 141, 155) (Figure 2).

In an attempt to delimit the required region to the left of the *Eco*RI site, portions of the λ DNA in the composite phage have been deleted by in vitro techniques (159). The results show that the phage can still grow when all λ DNA to the left of the middle of the cI gene has been removed. Furthermore, the additional deletion of nearly all the DNA lying between the P_R promoter and a site well within gene O still gives a phage that can replicate.

These results show that, at most, the 190 base-pair segment of DNA just to the left of the *Eco*RI site is required in conjunction with the P_R promoter to give O protein-dependent replication. Remarkably, the two ri^c mutations in gene O are located to the right of this *Eco*RI site (150). Presence of the ri^c5b mutation causes transcription which begins 135 ± 5 base pairs to the right of the *Eco*RI site and proceeds to the right (150). Thus there is no apparent requirement for transcription in the region believed to contain the origin. This is true even if the region extends as far as 20 base pairs to the right of the *Eco*RI site, as has been reported (142).

An important objection to such a conclusion is raised by the possibility that the ri^c mutation creates a new origin for λ replication. This is excluded by the following experiment (150). Phage 82 and λ are related phages but their *O* proteins are phage-specific. A hybrid (*rep*82 : λ) was constructed in which the origin region to the left of the *Eco*RI site is derived from 82 and the region to the right from λ. This phage requires the 82-type *O* protein. If the *ri*^{c} mutation creates a new origin, *rep*82 : λ*ri*^{c} phage should not require the 82-type *O* protein for replication under the repressed condition. In fact, the 82-type *O* protein, but not the λ-type *O* protein, supports replication of the phage. Therefore, it appears likely that the *ri*^{c} mutation does not create a new origin. Rather, these results argue that transcription from the *ri*^{c}*5b* mutation somehow activates the normal origin.

The results described above indicate that a small region to the left of the *Eco*RI site is sufficient for λ-specific initiation of replication when the *O* and *P* proteins are supplied in *trans*. A similar conclusion has been drawn from independent experiments in which composite plasmids were created by insertion of restriction fragments from λ*dv* into a ColE1-type vector, and the maintenance of the plasmids under *polA*⁻ conditions was determined (142). This time, the *O* and *P* proteins were supplied by a second hybrid plasmid which contains all of λ*dv*. Since the vector itself is not maintained in the absence of *polA* function, a positive result indicated the presence of a functional λ origin.

The results show two regions of importance; one is a 199 base-pair segment (the A region) directly to the left of the *Eco*RI site, the other an 89 base-pair segment (the B region) of the *cII* gene. When a plasmid contains only the B region, it can replicate under *polA*⁻ conditions, but only if the segment containing the B region is transcribed. Results obtained through use of a somewhat different type of complementing plasmid than that described above suggest that this replication requires the *P* protein and not the *O* protein. On the other hand, when both the A and B regions are present in the same relative orientation found in phage λ, plasmid maintenance is now dependent on both the *O* and *P* proteins as well as transcription (142).

The cause of disparity between the results obtained with phage and plasmid vectors is not known. It could simply be due to different contexts around the cloning sites. More likely, it may arise from differences in the kind and amount of replication required by the two vector systems. In particular, the analysis of the origin with the phage vector has been done in a way that requires production of phage particles in addition to DNA replication. In the plasmid system, where bacterial growth was measured, there was a requirement that both daughter cells receive some plasmid DNA at the time of cell division.

Some characteristics of the sequence of the DNA (about 160 base pairs) directly to the left of the *Eco*RI site may be worthy of mention (142, 159). The region contains a series of tandem repeats of a 19 base-pair sequence that has internal dyad symmetry. Upon strand separation, the region can be folded into remarkable secondary structures which are assumed to play a role in recognition of the DNA by the proteins required for DNA synthesis (142, 160). One *ori*⁻ deletion mutation is located in this region. In addition, there is an inverted repeat of about 30 base pairs that includes the *Eco*RI site near its center. Between these structures, a highly A-T rich sequence (about 40 base pairs), with an asymmetric arrangement of pyrimidines and purines, exists. Two *ori*⁻ deletions occur in this region. The analogous region of ϕ80 has four nearly complete tandem repeats of a 12 base-pair sequence, a somewhat similar middle region, and an almost identical inverted repeat. The tandemly repeated sequence differs significantly from that of λ.

It has been proposed that the required transcription separates the two strands of DNA in the origin region, thereby permitting the strands to fold into the secondary structure described above. DNA synthesis could then begin on the exposed single-stranded DNA (142). If this explanation of transcriptional activation is correct, it is still not clear how the *ri*ᶜ mutations activate the origin.

We now have a fairly thorough knowledge of the genetics and chemical structure of the origin of replication of λ. Nonetheless, we know very little about the biochemical mechanism of initiation of λ DNA synthesis. It is vital to determine the actual site at which DNA synthesis initiates and the biochemical mechanisms that carry out this reaction.

Formation of Nascent Fragments

The process of DNA chain elongation by sequential synthesis and joining of small nascent fragments (Okazaki fragments) of DNA is called discontinuous synthesis (161). Taking into consideration the polarity of chain growth catalyzed by known DNA polymerases, the strand that elongates in an overall direction of 3' → 5' (the "lagging" strand) must be synthesized discontinuously. In principle, the other strand (the "leading" strand), which grows in an overall direction of 5' → 3', can elongate continuously. Nonetheless, it could conceivably be synthesized by a discontinuous mechanism.

Originally, the presence of labeled fragments following a short pulse of radioactive precursor was taken to be evidence for the occurrence of discontinuous synthesis. However, fragments can be formed by post-replication degradation of DNA as, for example, is known to occur during removal of misincorporated nucleotides (135). Therefore, the mere existence of fragments does not itself constitute sufficient evidence of discontinuous synthesis. Rather, one has to know whether or not the fragments are actually formed by synthesis.

One approach to the problem is to examine the fragments for special structures at their 5' end which could specify the mechanism of formation.

Thus fragments labeled during brief pulses have been tested for the presence of a covalently bound RNA primer. Various methods have been applied to detect the DNA-bound RNA and, in a number of cases, its presence has been claimed (162). However, this interpretation of the data has been disputed (7). Even the use of alkali as a specific agent for the removal of RNA bound covalently to the 5' end of the DNA has been criticized on the grounds that 5'-OH termini can also be created by such treatment of the fragments formed during elimination of misincorporated nucleotides (163). Because of these complications, only a direct chemical demonstration of RNA-DNA covalent linkage at the 5' ends of DNA fragments can provide conclusive evidence for RNA priming in the discontinuous mechanism. The results discussed in this section demonstrate the formation of RNA-primed DNA fragments during replication of phage T7 DNA in vitro and in vivo. Similar results obtained with phage T4 and *E. coli* are also described.

DNA is synthesized on both native and denatured T7 DNA in vitro by a mixture of T7 DNA polymerase, and the T7 gene 4 protein (164, 165). When single-stranded DNA is used as template, synthesis is dependent on the presence of rATP and rCTP (166). Sequence analysis of small fragments formed during the reaction has shown that the primer(s) used are pppApCpCpA (166, 167) or pppApCpCpC (104). Although the RNA-primed DNA fragments are formed on native T7 DNA, this DNA is not as effective a template as is denatured T7 DNA. On native DNA, leading strand synthesis may initiate at a nick and the RNA-primed fragments then form on the displaced single strand.

Primer RNA can be synthesized on single-stranded DNA by gene 4 protein in the absence of DNA polymerase and dNTPs (104, 166). In the presence of single-stranded DNA, the gene 4 protein also has a nucleoside 5'-triphosphatase activity (168, 169). When double-stranded DNA is used as template, hydrolysis of NTPs is interdependent with DNA synthesis.

The study of in vivo synthesis of T7 DNA has also led to the identification of RNA-primed fragments (162). A small fraction of the fragments retain a triphosphate at their 5' end and thus must have an intact primer. Most of these have the structure $pppApC(p-rN)_2$ which is consistent with either of the sequences established in vitro. Shorter segments of RNA are found on the majority of the fragments. Since these lack the 5' triphosphate, they are presumed to arise through partial degradation of the tetramer. Their base composition is mainly A and C, again in good agreement with the in vitro results.

An attempt to determine the number of sites at which initiation can occur in vivo has also been made (162). The results suggest that even within a short segment of the T7 chromosome (a 350 nucleotide *Hpa*II fragment), there are several sites at which synthesis of the fragments can initiate. If so, this suggests that initiation does not require a special sequence other than that of the primer itself. It is not known if the primer is made at the site where it is used.

When the products of phage T4 genes 41, 43, 44, 45, and 62 are incubated with single-stranded DNA, rNTP-dependent DNA synthesis is observed, but only if another protein, called protein X, is also present (170–173). Protein X activity is absent from extracts of cells infected with phage mutant in gene 61, but it has not yet been shown that gene 61 actually encodes protein X (172, 173). DNA synthesis can also occur if nicked ColE1 DNA is used as a template. In this case, the gene 41 and X proteins, as well as rNTPs, can all be omitted, but another T4 protein, the gene 32 product, must be added (172). No synthesis is observed when intact duplex DNAs are supplied as template, and it is possible that additional proteins are required for initiation with such templates.

The T4 chain initiation reaction that occurs on single-stranded DNA requires ATP and CTP, while GTP and UTP stimulate the reaction (172, 173). When the gene 41 and X proteins are incubated with ATP, CTP, and fd (+) strand DNA, short oligonucleotides are synthesized in large yield (173). Most of the products contain six to eight nucleotides. When GTP and UTP are also present, they are incorporated into oligonucleotides of similar lengths. At least two kinds of products are made from ATP and CTP, and at least several kinds when the four rNTPs are used. Some of these oligonucleotides can be used as primer for DNA synthesis when the necessary components are supplied (173).

The results of other experiments partially complement these observations by providing evidence of the in vivo formation of RNA-primed DNA fragments (162). More than 40% of the primers appear to be pentanucleotides with a sequence at their 5' end mainly of pApC. A primer with a 5'-nucleoside triphosphate has not been isolated.

The presence of RNA-primed DNA fragments in *E. coli* has also been demonstrated (174, 175). Small fragments were isolated from a strain carrying a *polAex⁻* mutation. The 5' ends of the fragments were labeled with ^{32}P and the fragments then digested with pancreatic DNase or with 3' → 5' exonuclease of T4 DNA polymerase. The digestion products included [5'-^{32}P] ribo-deoxyribo-oligonucleotides containing two to six nucleotides (174). In a similarly designed experiment, the fragments labeled with ^{32}P at the 5' ends were hydrolyzed by alkali. Some label was found in 2'(3'),5'-ribonucleoside diphosphates, predominantly pAp and pGp (175).

In these three cases, where the presence of RNA-primed DNA fragments has been clearly shown, it has not been demonstrated that such fragments are synthesized on both strands in the same region of the genome.

INITIATION OF REPLICATION OF THE *E. COLI* CHROMOSOME

The chromosome of *E. coli* is a circular duplex DNA that is folded into a series of approximately one hundred supercoiled loops. Replication initiates within a defined region on the genome and then proceeds bidirectionally by a semi-conservative mechanism. During the elongation process, one or perhaps both strands are synthesized discontinuously. Genetic studies show that the *dnaB, dnaC,* and *dnaG* functions participate in this process together with functions needed for unwinding, polymerization, ligation, and supercoiling of DNA (8, 10). While information about the elongation reaction has been provided by studies that employ cell lysates or toluene-treated cells (3, 4), as yet no information about the initiation reaction is available from such in vitro studies. Thus the following description of events occurring during initiation is based on analyses of replication in vivo. The region of the DNA where these events are thought to occur will be described in the subsequent section.

Function of Required Gene Products

A number of genes whose products participate in replication of the *E. coli* chromosome have been identified by the study of conditionally lethal mutants (10). Whereas many mutants stop DNA synthesis immediately when shifted to a nonpermissive temperature (elongation-defective mutants), others appear to be able to finish rounds of replication in progress at the time of the shift. This second class of mutants, termed initiation-defective mutants, includes both *dnaA* and *dnaC* mutants. As described below, at least one *dnaB* mutant also falls into this class (176). Mutations in several other genes, (*dnaI, J, K,* and *P*) give this phenotype but are not well characterized (9, 144, 145).

The addition of rifampicin and chloramphenicol to inhibit RNA and/or protein synthesis in *E. coli* also inhibits chromosome replication, but not until replication in progress at the time of addition is completed. Thus some new RNA and/or protein must be synthesized to initiate a new round of replication (10, 177). However, the arrest of RNA synthesis is still inhibitory at a time in the cell cycle when arrest of protein synthesis no longer prevents replication (178, 179). Mutants (*sdr^c*) capable of continuing several rounds of chromosomal DNA synthesis in the presence of chloramphenicol have now been isolated (180, 181). Each new round of replication requires the *dnaA* and *C* gene products and is inhibited by rifampicin. Because the *sdr^c* mutation is recessive, it has been suggested that the chloramphenicol-sensitive step in wild-type bacteria is the synthesis of a protein that counteracts a negative regulatory function. This regulatory function is believed to be defective in the mutants.

Experiments that examine the interaction of various initiation-defective mutants and the two drugs mentioned above suggest a particular order of events at the time of initiation. For example, if temperature-sensitive *dnaA* or *dnaC* mutants are exposed to a nonpermissive temperature for a time sufficient for ongoing rounds of replication to finish, a new round of replication initiates when the cells are returned to the permissive temperature whether or not chloramphenicol is present (182, 183). This suggests that the necessary protein synthesis can occur in the absence of either gene product. On the other hand, when rifampicin is present at the time of return to permissive temperature, the *dnaA* mutant cannot resume replication although the *dnaC* mutant can (183–185). This suggests that the action of the *dnaA* product precedes or is concurrent with the rifampicin-sensitive step, but that in either case, both precede the reaction that requires the *dnaC* product.

One biochemical function of *dnaA* protein has been suggested by the observation that some *dnaA* mutations are suppressed by mutation to rifampicin resistance (186, 187). These suppressing mutations occur in the structural gene for the β subunit of RNA polymerase which suggests, among other possibilities, a direct interaction of the *dnaA* protein and RNA polymerase. Obviously, such a proposal is consistent with a requirement for active *dnaA* protein during the rifampicin-sensitive step of initiation.

A second kind of suppression of *dnaAts* mutations is observed when the DNAs of any of a number of plasmids or phages are present in the bacterial chromosome. For example, integration of the F or R factors into the chromosomes allows the growth of *dnaAts* bacteria at nonpermissive temperatures (188–191). The integration of certain phages, such as P1 (192) and P2*sig5* (98), can give a similar result. When the F factor (M. G. Chandler and L. Caro, personal communication), R100-1 factor (193), or the phage P2*sig5* (194, 195) is used for this integrative suppression, the origin of chromosome replication in cells growing at high temperature is in the region where the integrated DNA resides. However, when such cells are grown at low temperature, replication of the chromosome starts frequently at the *E. coli* origin (196). The simplest interpretation of these experiments is that the *dnaA* protein is required for initiation of replication of the *E. coli* chromosome only at its own origin.

For some time, all *dnaB* mutants have been classified as elongation-defectives and the possibility of a role for the *dnaB* protein in initiation was speculative. Now, the characterization of *dnaB252,* an exceptional mutant that has an initiation-defective phenotype (176), provides evidence for direct participation of the protein in initiation. This temperature-sensitive mutant cannot begin, but can continue chromosome replication at high temperature. Furthermore, unlike other *dnaB* mutants, it can support replication of phage λ at high temperature (197). The purified *dnaB252* protein is unable to aid in the synthesis in vitro of ϕX (–) strands, but does retain a DNA-dependent ATPase activity (197) characteristic of the wild-type *dnaB* protein (198, 199). In contrast to this result, the *dnaB* protein of an elongation-defective *dnaBts* mutant has been found to have a temperature-sensitive ATPase activity (199). These observations suggest that the requirement for *dnaB* protein during both elongation of the *E. coli* chromosome and replication of λ DNA can be satisfied by the ATP-consuming activity of the protein, while the initiation of chromosome replication and the synthesis of ϕX DNA may require another activity, or possibly both. Such a difference in the mechanisms of action of the *dnaB* protein during initiation and elongation of the chromosome may explain the properties of another *dnaBts* mutant, BT165/70. A shift to high temperature reversibly inactivates the ability of this mutant to synthesize DNA, but irreversibly damages its ability to initiate new rounds of synthesis (200).

As described above, results obtained with the study of ϕX DNA replication in vitro show that the *dnaB* and *C* products act in concert. However, in contrast to the initiation-defective class of *dnaC* mutants, the mutant *dnaB252* cannot initiate new rounds of replication when returned to low temperature in the presence of rifampicin. If the *dnaB* and *C* proteins also act together during initiation of chromosome replication, the *dnaB252* mutant must somehow be less efficient in making use of the *dnaA*-dependent transcript than are the *dnaC* mutants so far tested. This might occur if the *dnaB252* product is relatively slow to renature and the transcript is unstable.

In concluding this section, we would like to present three models for how initiation of replication of the *E. coli* chromosome occurs. Because of the lack of sufficient information about the biochemical process that initiates replication of *E. coli* chromosome, the models are based on knowledge obtained with small genomes as described in previous sections, and on fragmentary information about chromosome replication obtained with *E. coli* in vivo. Regulatory mechanisms have not been taken into consideration.

1. RNA polymerase "activates" the origin, but the *dnaG* protein primes initiation of chromosome replication. In this model, RNA polymerase somehow activates the origin, but does not synthesize the primer. For example, transcription could separate DNA strands in the region of the origin and thus provide a single-stranded DNA template for primer synthesis performed by the *dnaG* protein. To ensure that this only occurs at the origin, one assumes that a special structure in the origin region stabilizes the separated strands, and/or that the region contains a specific attachment site either for the *dnaG* protein or for other proteins that mediate its binding to the DNA.

2. RNA polymerase primes initiation of chromosome replication. In this model, the first primer for DNA synthesis is made by RNA polymerase. Thus DNA synthesis could stabilize the separated strands and allow transcription by the *dnaG* protein to initiate lagging strand synthesis. In either this or the previous model, the lagging strand may extend through the region where the leading strand initiates, thus becoming the leading strand for replication of the chromosome in the opposite direction. In this way, initiation of bidirectional replication could be accomplished.

3. A nick at the origin triggers initiation of chromosome replication. Following the nick at a specific site, the 3' end at the break would serve as primer. Once synthesis begins, the 5' end would be transferred to reform an intact parental strand. Replication in the other direction could be initiated by the *dnaG* protein as described above. Transcription by RNA polymerase would somehow be required in a preparatory stage before the nicking occurs.

So far, no evidence points to one of these as the mechanism operating at the origin. Obviously, identification of the nature of the primer for initial synthesis at the origin is crucial. An RNA species that seems to be synthesized under *dnaC⁻*, but not *dnaA⁻*, conditions has been described (201). The RNA has been termed "origin-RNA" though neither synthesis nor utilization of the RNA in the origin region has been demonstrated. There is no evidence that this RNA has a role in initiation of replication from the normal origin.

The third model appears the least satisfactory because it provides no clear role for RNA polymerase. Nonetheless, there are two known situations where initiation probably occurs in the manner described by this model. One occurs when phage P2*sig5* is integrated in a *dnaAts* strain, thus suppressing the temperature-sensitive phenotype. For replication to occur, the phage gene *A* protein, a specific endonuclease (202), has to be active (98). Following initiation, replication appears to proceed unidirectionally from the site of phage integration and only later to go in both directions (194, 195). The other situation occurs during bacterial mating. The replication that occurs during transfer of the chromosome is probably carried out by the third mechanism except that here the 5' end at the nick is transferred to the recipient bacteria (203, 204).

Detection of the Origin of Replication

Genetic studies have located the origin of replication within the vicinity of the *ilv* gene (205, 206). A physical map of DNA in the origin region has been obtained by labeling the chromosome of *dnaA* and *dnaC* temperature-sensitive initiation mutants just at the start of a synchronous round of replication (207). The distribution of radioactivity among restriction fragments of the pulse-labeled DNA showed that the origin of replication was located within or very near a *Hin*dIII segment of 1.3 kb (kilobase) in length. This segment is itself situated within an 8.6 kb *Eco*RI segment. Studies of various plasmids that carry the *ilv* region have now allowed the refinement and correlation of the genetic and physical maps. This line of work began with the isolation of an F' factor carrying the putative origin of replication.

While most F' factors are maintained in F⁻ bacteria, only a few spontaneously arising F' factors (F' *poh*⁺) are readily carried by *Hfr* bacteria (208). The *poh*⁺ phenotype is invariably associated with the presence in the F' factor of the *bglB-rbsK* region (209) (Figure 3). These observations are thought to indicate that the site where *E. coli* chromosome replication initiates lies somewhere within the region. The presence of the bacterial origin on F' factors that contain the *bglB-rbsK* region has also been used to explain the finding that strains that carry the factors exhibit slow growth and disturbed cell division (210).

The physical and genetic maps were correlated by restriction enzyme analysis of the DNAs of various F' factors containing portions of the *bglB-rbsK* region (211) (Fig. 3). This correlation was also established by experiments in which restriction fragments formed by *Eco*RI digestion of the entire *E. coli* genome were ligated to a DNA fragment containing a gene responsible for ampicillin resistance. By using this DNA to transform bacteria to ampicillin resistance, three self-replicating plasmids were isolated (212, 213). These were all found to contain the 8.6 kb *Eco*RI segment. One of the plasmids (pSY211) can be integrated into the *E. coli* chromosome at sites between the *uncA* and *rbsK* genes by *recA*-mediated recombination, which indicates that the plasmid contains *E. coli* DNA coming from the region close to these genes (Figure 3).

In an independent set of experiments, a series of λ-transducing phages containing the region around the *asn* gene were isolated (214, 215). The *asn* gene has been shown to lie in the 8.6 kb *Eco*RI segment (211). Some

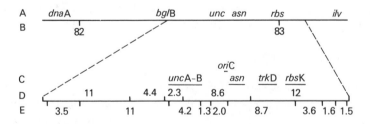

Figure 3 Genetic and physical maps of the region around the *E. coli* origin of replication. *A* and *C*, approximate position of genes; *B*, map positions in terms of minutes; *D* and *E*, sites of cleavage by *Eco*RI (*D*) and *Hin*dIII (*E*), and the sizes of segments in kilobases.

of the λ*asn*-transducing phages were found to establish themselves as plasmids following infection of a λ lysogen. Genetic and physical analyses of various deletion derivatives obtained from the phages led to the conclusion that the origin of replication used by these λ*asn* plasmids is located between the *unc* and *asn* genes (214, 215).

The functional origin present in the λ*asn* plasmids and in pSY211 has been termed *ori*C, even though this name was originally proposed as a designation for the origin of replication of the bacterial chromosome (208). To avoid confusion, we also use *ori*C to indicate the origin of replication of these plasmids, but will return below to the question of the relationship of *ori*C and the chromosome origin.

Using DNAs of deletion mutants of plasmids that have the *ori*C of pSY211 or of λ*asn,* the nucleotide sequence of the region containing *ori*C (about 500 base pairs) has been determined (213, 216–218). The sequence shows a high degree of repetitiveness and a remarkable degree of homology with the sequence of the putative replication origin of phage λ and with the sequence of a stretch of the G4 genome upstream from the site where the primer for synthesis of the (–) strand initiates. This homology is represented twice in the *ori*C region, once in each orientation with respect to the rest of the chromosome. By analogy to the G4 initiation mechanism, the homologous sequence is presumed to be a recognition site for the *dna*G protein (213, 217). However, much of the sequence that is common to the *ori*C, λ, and G4 DNAs is not found in the corresponding region of the ST-1, φK, and α3 genomes (51).

In concluding this section, we would like to examine the criteria that establish *ori*C as the "real" *E. coli* origin of replication. First, only one to three copies of the plasmid carrying an intact 8.6 kb *Eco*RI segment are present in the cell, and the plasmid is readily lost (216, 219). Cells carrying the plasmid multiply slowly and it appears possible that the *ori*C of the plasmid and the origin of the chromosome compete for a limiting substance in the cell. However, deletion of a portion of the plasmid DNA allows more rapid multiplication by its host and increases the number of plasmid copies per cell. The portion of the plasmid that is lost in such deletion derivatives includes part of the *unc* genes whose products participate in electron transport (219). Further analysis is required to establish the cause of this effect and, in particular, to determine whether or not overproduction of the *unc* gene products causes poor growth.

Replication of an *unc* deletion derivative has been shown to require the *dnaA* and *dnaC* proteins, and to be quickly inhibited by the addition of rifampicin (219). However, the addition of chloramphenicol gives only gradual inhibition (219). Thus, at least in part, replication of the plasmid resembles that of the *E. coli* chromosome. Unlike the λ*asn,* plasmid pSY211 and the two other plasmids obtained simultaneously were selected without the use of any chromosomal markers. Nonetheless, all these isolates contain the same 8.6 kb *Eco*RI segment. As described above, independent experiments indicate that this segment is the first region of the chromosome to replicate following the shift of a *dnaA* or *dnaC* mutant to a permissive temperature (207). These results suggest that the *oriC* may represent a unique origin for chromosome replication and that no other region of the DNA may be able to endow the ability to self-replicate.

In the assessment of this information, however, two observations appear important. First, sequences that can serve as functional initiation sites, but normally do not, may be present in a chromosome. As discussed above, the analysis of plasmids containing the phage λ origin (142) provides an example of this and, in addition, points to the possibility that such sequences may come from a region close to the "real" initiation site itself. This means that chromosomal location, while useful, is not a sufficient criterion for the identification of the normal origin.

Second, the details of the strong selection used to isolate plasmids like pSY211 may influence the uniqueness of the products that are obtained. In fact, when the procedure used to isolate pSY211 was repeated, this time using an F⁻ rather than Hfr strain as host, a self-replicating plasmid that does not contain the 8.6 kb *Eco*RI segment was isolated (220). The origin present on this plasmid is not homologous to *oriC,* and so has been named *oriJ.* Neither the location of *oriJ* on the *E. coli* chromosome nor its role, if any, in the normal mechanism of initiation of chromosome replication has been established.

REGULATION OF DNA REPLICATION

A major question in understanding the regulation of DNA replication is how the number of copies of a genetic element is maintained in the cell at a certain characteristic value, the copy number. It is likely that this maintenance is achieved through regulation of the frequency and timing of initiation of replication with respect to growth of the cell. To illustrate the problems involved, we briefly discuss several proposed mechanisms of regulation. No attempt is made to cover the literature.

The replicon model (221) was an early attempt to provide a framework for understanding regulation and, in particular, to show that the same kinds of elements that regulated gene expression could serve to regulate replication. The model states that (*a*) a genetic element such as a chromosome or episome constitutes a unit of replication, or *replicon,* and that (*b*) the replicon itself carries two specific determinants. One determinant is a structural gene that controls the synthesis of a specific *initiator,* while the other, the *replicator,* is a specific recognition element (analogous to the operator) upon which the initiator acts to permit replication. Although presented as a positive regulatory mechanism (221), the model itself does not specify how the amount of active initiator in the cell is to be regulated. Therefore, in its basic structure, the model does not predict the actual mode of regulation.

In their discussion of the replicon model, its authors provided several examples to illustrate how the level of active initiator might be determined by negative control of initiator synthesis or by positive control of the activity of the initiator once made. As an example of the former, they proposed that the phage repressor blocks synthesis of an initiator in λ lysogens. As an example of the latter, it was proposed that the bacterial chromosome is attached through its initiator to the bacterial membrane, and that initiation is positively regulated by growth of the membrane (222). Though meant as an example, the latter proposal has been frequently included as part of the replicon model itself.

Recently it has been suggested that the phenomenon of autoregulation (223) provides an attractive mechanism for the regulation of initiator production (224). The replication of phage λ as a plasmid provides an instructive example. Under certain conditions, λ uses its own specific replication functions to replicate as a plasmid (225, 226). Study of deletion variants called λdv (225) has shown that three phage products are invariably required for plasmid formation (227, 228). These are the *cro, O* and *P* proteins which are produced as a result of transcription from promoter P_R (Figure 2). The *O* and *P* proteins act as initiator, while the *cro* protein represses transcription and thereby regulates the supply of all three proteins (229). Through this simple feedback mechanism, growth of the cell can lower the intracellular concentration of the *cro* protein, thus triggering transcription and replication.

So far, all bacterial genetic elements examined have one or possibly a few specific sites where replication initiates. Thus the existence of replicators appears to be universal. The occurrence of initiator-like proteins is widespread, but there is, as yet, little direct evidence that these proteins actually regulate replication. Furthermore, it is now clear that an initiator is not produced by all genomes. An example is provided by the plasmid ColE1 whose replication does not require any plasmid-encoded proteins. In this case, replication could be limited simply by the synthetic capacity of the host cell, or possibly by a negatively acting plasmid function.

A third type of model proposes that a negatively acting element, rather than an initiator, regulates the rate of initiation (230, 231). In this model, regulation is accomplished by coupling synthesis of the negative element, or inhibitor, to replication of the gene that produces it. Thus, each initiation event leads to replication of the gene and production of the inhibitor which then represses further initiation. Growth of the cell reduces the concentration of the inhibitor and thereby increases the probability of initiation. Formally, this model can explain both the maintenance of replicons at a unique copy number and the timing of initiation with respect to the cell cycle. As does the autoregulatory model described above, it employs a feedback mechanism that links gene expression and replication.

This model has been used to explain the replication properties of several plasmids, one of which is the sex factor F. In exponentially growing Hfr cells, where F is found as part of the chromosome, replication usually initiates at the chromosomal origin and not at that of the integrated F factor (232). To explain this by means of the negative regulation model, it is necessary that the effective copy number of F in the Hfr cells be higher than it would be in F^+ cells. This would happen if, on the average, the F DNA replicates sooner in the cell cycle when integrated than it does when present as a plasmid. Thus repression can be maintained in the absence of initiation at the F origin. There is now some experimental support for such an increase in the effective copy number of F in Hfr strains (233, 234).

Studies of the composite plasmid pSC134 have yielded results similar to those obtained with the F factor, and have been used to support the existence of replicon-specific negative regulation of initiation (235). This plasmid contains the entire genomes of two compatible plasmids, pSC101 and ColE1. In wild-type bacteria, the composite plasmid uses its ColE1 replication origin exclusively, and exists at a copy number of 15 per cell, or about three times higher than that of pSC101 in the same strain. In their discussion of this result, the authors argue that the turn-off of the pSC101 origin can be most easily explained by the existence of a specific inhibitor of the pSC101 origin. According to the model presented above, the frequent replication of pSC101 DNA present in the composite plasmid should lead to overproduction of a specific repressor of the pSC101 origin (235).

A regulatory model that makes use of an initiator can also explain the results obtained with pSC134. This model requires that (a) the rate of initiator production be independent of the number of copies of the replicon present, (b) all or nearly all initiators be bound to the DNA at any time, and (c) multiple initiator molecules be bound to the same replicator for initiation to occur. If binding of the initiators is random, then an increase in the number of copies of the replicon will decrease the chance that any replicator has the required number of bound initiators, and fewer initiation events will occur. The required control of synthesis of the initiator could be readily obtained with an autoregulatory mechanism of the type described above. As has been observed elsewhere, a similar effect might be observed if a host function limits initiation at the pSC101 origin (231).

This model can also describe much of what is known about the control of F factor replication. However, by introducing it here, we do not mean to imply that it provides a better explanation of the pSC134 or F factor data. Rather, our intention is only to point out that the response of an initiator-type mechanism to a change in DNA concentration can be dictated by the exact details of initiator supply and action.

To date, the only genome whose regulatory mechanism is clearly established is λdv. While each of the models described above provides some insight into how the initiation of replication can be regulated, much remains to be learned.

CONCLUSIONS

Because primers for DNA synthesis play a central role in the initiation of DNA replication, much of the work reviewed here has concerned the reactions involved in primer formation. The known initiation mechanisms require synthesis of an RNA primer, or else the nicking of one strand of a duplex DNA template. Synthesis of the RNA primers can occur on single- or double-stranded templates, and, in at least one case, post-transcriptional modification plays an important role in forming the primer. In some cases, the primer is made at a unique site on the template, whereas in others, a primer may be produced at a large number of sites.

Knowledge of these mechanisms has come primarily from the study of in vitro DNA synthesis on small single- or double-stranded DNAs. The initiation mechanisms employed by large genomes are poorly understood, in part because of the difficulty of in vitro experimentation with large DNA molecules. The use of modern genetic techniques to construct small plasmids that contain portions of the large genomes holds promise of rapid progress in the understanding of how such genomes initiate and regulate their replication.

ACKNOWLEDGMENTS

We wish to thank R. E. Bird, M. Gellert, T. Itoh, J. L. Rosner, Y. Sakakibara, L. Silver, S. Wickner, and K. Yamaguchi for advice, and J. Smallwood for secretarial help. We also thank all of those investigators who sent us preprints before publication.

Literature Cited

1. Kornberg, A. 1974. *DNA Synthesis.* San Francisco: Freeman. 399 pp.
2. Dressler, D. 1975. *Ann. Rev. Microbiol.* 29:525–59
3. Gefter, M. L. 1975. *Ann. Rev. Biochem.* 44:45–78
4. Geider, K. 1976. *Curr. Top. Microbiol. Immunol.* 74:55–112
5. Jovin, T. M. 1976. *Ann. Rev. Biochem.* 45:889–920
6. Kornberg, A. 1976. In *RNA Polymerase,* ed. R. Losick, M. Chamberlin, pp. 331–52. Cold Spring Harbor, NY: Cold Spring Harbor Lab. 899 pp.
7. Alberts, B., Sternglanz, R. 1977. *Nature* 269:655–61
8. Wickner, S. H. 1978. *Ann. Rev. Biochem.* 47:1163–91
9. Champoux, J. J. 1978. *Ann. Rev. Biochem.* 47:449–79
10. Wechsler, J. A. 1977. In *DNA Synthesis, Present and Future,* ed. I. Molineux, M. Kohiyama, pp. 49–70. New York: Plenum. 1161 pp.
11. Wickner, S. 1977. *Proc. Natl. Acad. Sci. USA* 73:3511–15
12. McHenry, C., Kornberg, A. 1977. *J. Biol. Chem.* 252:6478–84
13. Meyer, R. R., Shlomai, J., Kobori, J., Bates, D. L., Rowen, L., McMacken, R., Ueda, K., Kornberg, A. 1978. *Cold Spring Harbor Symp. Quant. Biol.* 43: In press
14. Denhardt, D. T. 1975. *Crit. Rev. Microbiol.* 4:161–223
15. Denhardt, D. T. 1977. In *Comprehensive Virology,* ed. H. Fraenkel-Conrat, R. R. Wagner, pp. 1–104. New York/London: Plenum. 300 pp.
16. Dressler, D., Hourcade, D., Koth, K., Sims, J. 1979. In *The Single-Stranded DNA Phages,* ed. D. Dressler, D. Denhardt, D. Ray. Cold Spring Harbor, NY: Cold Spring Harbor Lab. In press
17. Ray, D. S. 1977. See Ref. 15, pp. 105–78
18. Ray, D. S. 1979. See Ref. 16. In press
19. Tabak, H. F., Griffith, J., Geider, K., Schaller, H., Kornberg, A. 1974. *J. Biol. Chem.* 249:3049–54
20. Wickner, W. T., Brutlag, D., Schekman, R., Kornberg, A. 1972. *Proc. Natl. Acad. Sci. USA* 69:965–69
21. Wickner, R. B., Wright, M., Wickner, S., Hurwitz, J. 1972. *Proc. Natl. Acad. Sci. USA* 69:3233–37
22. Brutlag, D., Schekman, R., Kornberg, A. 1971. *Proc. Natl. Acad. Sci. USA* 68:2826–29
23. Sigal, N., Delius, H., Kornberg, T., Gefter, M. L., Alberts, B. 1972. *Proc. Natl. Acad. Sci. USA* 69:3537–41
24. Geider, K., Kornberg, A. 1974. *J. Biol. Chem.* 249:3999–4005
25. Schaller, H., Uhlmann, A., Geider, K. 1976. *Proc. Natl. Acad. Sci. USA* 73:49–53
26. Gray, C. P., Sommer, R., Polke, C., Beck, E., Schaller, H. 1978. *Proc. Natl. Acad. Sci. USA* 75:50–53
27. Schaller, H. 1978. *Cold Spring Harbor Symp. Quant. Biol.* 43: In press
28. Geider, K., Beck, E., Schaller, H. 1978. *Proc. Natl. Acad. Sci. USA* 75:645–49
29. Schaller, H., Gray, C., Herrmann, K. 1975. *Proc. Natl. Acad. Sci. USA* 72:737–41
30. Pribnow, D. 1975. *J. Mol. Biol.* 99:419–43
31. Suggs, S. V., Ray, D. S. 1978. *Cold Spring Harbor Symp. Quant. Biol.* 43: In press
32. Horiuchi, K., Ravetch, J. V., Zinder, N. D. 1978. *Cold Spring Harbor Symp. Quant. Biol.* 43: In press
33. Wickner, W., Kornberg, A. 1974. *Proc. Natl. Acad. Sci. USA* 71:4425–28
34. Vicuna, R., Hurwitz, J., Wallace, S., Girard, M. 1977. *J. Biol. Chem.* 252:2524–33
35. Vicuna, R., Ikeda, J., Hurwitz, J. 1977. *J. Biol. Chem.* 252:2534–44
36. Schekman, R., Weiner, A., Kornberg, A. 1974. *Science* 186:987–93
37. Zechel, K., Bouché, J.-P., Kornberg, A. 1975. *J. Biol. Chem.* 250:4684–89
38. Bouché, J.-P., Zechel, K., Kornberg, A. 1975. *J. Biol. Chem.* 250:5995–6001
39. Rowen, L., Kornberg, A. 1978. *J. Biol. Chem.* 253:758–64
40. Wickner, S., Hurwitz, J. 1975. In *DNA Synthesis and Its Regulation,* ed. M. Goulian, P. Hanawalt, C. F. Fox, pp. 227–38. Menlo Park, Calif: Benjamin. 880 pp.
41. Wickner, S. 1977. *Proc. Natl. Acad. Sci. USA* 74:2815–19
42. Wickner, S. 1978. *Nato Adv. Stud.* 17:113–20
43. Wickner, S. 1979. See Ref. 16. In press
44. McMacken, R., Bouché, J.-P., Rowen, S. L., Weiner, J. H., Ueda, K., Thelander, L., McHenry, C., Kornberg, A. 1977. In *Nucleic Acid-Protein Recognition,* ed. H. J. Vogel, pp. 15–29. New York: Academic. 587 pp.
45. Kornberg, A. 1977. *Biochem. Soc. Trans.* 5:359–74
46. Rowen, L., Kornberg, A. 1978. *J. Biol. Chem.* 253:770–74
47. Bouché, J.-P., Rowen, L., Kornberg, A. 1978. *J. Biol. Chem.* 253:765–69

48. Capon, D., Gefter, M. 1978. *Cold Spring Harbor Symp. Quant. Biol.* 43: In press
49. Fiddes, J. C., Barrell, B. G., Godson, G. N. 1978. *Proc. Natl. Acad. Sci. USA* 75:1081–85
50. Sims, J., Dressler, D. 1978. *Proc. Natl. Acad. Sci. USA* 75:3094–98
51. Sims, J., Koths, K., Dressler, D. 1978. *Cold Spring Harbor Symp. Quant. Biol.* 43: In press
52. Wickner, S., Hurwitz, J. 1974. *Proc. Natl. Acad. Sci. USA* 71:4120–24
53. Schekman, R., Weiner, J. H., Weiner, A., Kornberg, A. 1975. *J. Biol. Chem.* 250:5859–65
54. Weiner, J. H., McMacken, R., Kornberg, A. 1976. *Proc. Natl. Acad. Sci. USA* 73:752–56
55. McMacken, R., Ueda, K., Kornberg, A. 1977. *Proc. Natl. Acad. Sci. USA* 74:4190–94
56. Ueda, K., McMacken, R., Kornberg, A. 1978. *J. Biol. Chem.* 253:261–69
57. McMacken, R., Kornberg, A. 1978. *J. Biol. Chem.* 253:3313–19
58. Ray, R., Capon, D., Gefter, M. 1976. *Biochem. Biophys. Res. Commun.* 70:506–12
59. Gilbert, W., Dressler, D. H. 1968. *Cold Spring Harbor Symp. Quant. Biol.* 33:473–84
60. Tessman, E. S. 1966. *J. Mol. Biol.* 17:218–36
61. Francke, B., Ray, D. S. 1971. *J. Mol. Biol.* 61:565–86
62. Henry, T. J., Knippers, R. 1974. *Proc. Natl. Acad. Sci. USA* 71:1549–53
63. Ikeda, J., Yudelevich, A., Hurwitz, J. 1976. *Proc. Natl. Acad. Sci. USA* 73:2669–73
64. Eisenberg, S., Scott, J. F., Kornberg, A. 1978. *Cold Spring Harbor Symp. Quant. Biol.* 43: In press
65. Sumida-Yasumoto, C., Ikeda, J., Benz, E., Marians, K. T., Vicuna, R., Sugrue, S., Zipursky, S. L., Hurwitz, J. 1978. *Cold Spring Harbor Symp. Quant. Biol.* 43: In press
66. Langeveld, S. A., van Mansfeld, A. D. M., Baas, P. D., Jansz, H. S., van Arkel, G. A., Weisbeek, P. J. 1978. *Nature* 271:417–20
67. Sanger, F., Air, G. M., Barrell, B. G., Brown, N. L., Coulson, A. R., Fiddes, J. C. Hutchison, C. A. III, Slocombe, P. M., Smith, M. 1977. *Nature* 265:687–95
68. Godson, G. N. 1974. *J. Mol. Biol.* 90:127–41
69. Eisenberg, S., Scott, J. F., Kornberg, A.

1976. *Proc. Natl. Acad. Sci. USA* 73: 1594–97
70. Eisenberg, S., Scott, J. F., Kornberg, A. 1976. *Proc. Natl. Acad. Sci. USA* 73:3151–55
71. Sumida-Yasumoto, C., Yudelevich, A., Hurwitz, J. 1976. *Proc. Natl. Acad. Sci. USA* 73:1887–91
72. Denhardt, D. T., Dressler, D. H., Hathaway, A. 1967. *Proc. Natl. Acad. Sci. USA* 57:813–20
73. Eisenberg, S., Griffith, J., Kornberg, A. 1977. *Proc. Natl. Acad. Sci. USA* 74:3198–3202
74. Kornberg, A. 1978. *Nato Adv. Stud.* 17: 705–28
75. Scott, J. F., Eisenberg, S., Bertsch, L. L., Kornberg, A. 1977. *Proc. Natl. Acad. Sci. USA* 74:193–97
76. Scott, J. F., Kornberg, A. 1978. *J. Biol. Chem.* 253:3292–97
77. Kornberg, A., Scott, J. F., Bertsch, L. L. 1978. *J. Biol. Chem.* 253:3298–3304
78. Yarranton, G., Gefter, M. 1979. *Proc. Natl. Acad. Sci. USA.* In press
79. Sumida-Yasumoto, C., Hurwitz, J. 1977. *Proc. Natl. Acad. Sci. USA* 74:4195–99
80. Denhardt, D. T., Eisenberg, S., Harbers, B., Lane, H. E. D., McFadden, G. 1975. See Ref. 40, pp. 398–422
81. Machida, Y., Okazaki, T., Okazaki, R. 1977. *Proc. Natl. Acad. Sci. USA* 74:2776–79
82. Ray, D. S., Dueber, J. 1977. *J. Mol. Biol.* 113:651–61
83. Martin, D. M., Godson, G. N. 1977. *J. Mol. Biol.* 117:321–35
84. Godson, G. N. 1977. *J. Mol. Biol.* 117:337–51
85. Godson, G. N., Barrell, B. G., Staden, R., Fiddes, J. C. 1978. *Nature* 276: 236–47
86. Taketo, A. 1975. *Proc. Mol. Biol. Meet. Jpn., 1975*, pp. 118–20. Tokyo: Kyoritsu Shuppan Co. 124 pp.
87. Kasamatsu, H., Vinograd, J. 1974. *Ann. Biochem.* 43:695–719
88. Godson, G. N. 1977. *J. Mol. Biol.* 117:353–67
89. Pratt, D., Erdahl, W. S. 1968. *J. Mol. Biol.* 37:181–200
90. Ray, D. S. 1969. *J. Mol. Biol.* 43:631–43
91. Fidanián, H. M., Ray, D. S. 1972. *J. Mol. Biol.* 72:51–63
92. Meyer, T. F., Geider, K., Kurz, C., Schaller, H. 1979. *Nature.* In press
93. Geider, K., Meyer, T. F. 1978. *Cold Spring Harbor Symp. Quant. Biol.* 43: In press
94. Horiuchi, K., Zinder, N. D. 1976. *Proc. Natl. Acad. Sci. USA* 73:2341–45

95. Ray, D. S., Dueber, J., Suggs, S. 1975. *J. Virol.* 16:348–55
96. Dasgupta, S., Mitra, S. 1976. *Eur. J. Biochem.* 67:47–51
97. Mitra, S., Stallions, D. R. 1976. *Eur. J. Biochem.* 67:37–45
98. Lindahl, G., Hirota, Y., Jacob, F. 1971. *Proc. Natl. Acad. Sci. USA* 68:2407–11
99. Geisselsoder, J. 1976. *J. Mol. Biol.* 100:13–22
100. Chattoraj, D. K. 1978. *Proc. Natl. Acad. Sci. USA* 75:1685–89
101. Calendar, R., Lindquist, B. H., Sironi, G., Clark, A. J. 1970. *Virology* 40: 72–83
102. Kolodner, R., Masamune, Y., LeClerc, J. E., Richardson, C. C. 1978. *J. Biol. Chem.* 253:566–73
103. Kolodner, R., Richardson, C. C. 1978. *J. Biol. Chem.* 253:574–84
104. Richardson, C. C., Romano, L. J., Kolodner, R., LeClerc, J. E., Tamanoi, F., Engler, M. J., Dean, F. B., Richardson, D. S. 1978. *Cold Spring Harbor Symp. Quant. Biol.* 43: In press
105. Campbell, J. L., Tamanoi, F., Richardson, C. C., Studier, F. W. 1978. *Cold Spring Harbor Symp. Quant. Biol.* 43: In press
106. Helinski, D. R. 1976. *Fed. Proc. Fed. Amer. Acad. Sci. Exp. Biol.* 35:2026–30
107. Tomizawa, J. 1978. *Nato Adv. Stud.* 17: 797–826
107a. Staudenbauer, W. L. 1978. *Curr. Top. Microbiol. Immunol.* 83:94–156
108. Sakakibara, Y., Tomizawa, J. 1974. *Proc. Natl. Acad. Sci. USA* 71:802–6
109. Tomizawa, J., Sakakibara, Y., Kakefuda, T. 1975. *Proc. Natl. Acad. Sci. USA* 72:1050–54
110. Donoghue, D. J., Sharp, P. A. 1978. *J. Bacteriol.* 133:1287–94
111. Kahn, M., Helinski, D. R. 1978. *Proc. Natl. Acad. Sci. USA* 75:2200–4
112. Sakakibara, Y., Tomizawa, J. 1974. *Proc. Natl. Acad. Sci. USA* 71:1403–7
113. Tomizawa, J., Sakakibara, Y., Kakefuda, T. 1974. *Proc. Natl. Acad. Sci. USA* 71:2260–64
114. Tomizawa, J. 1975. *Nature* 257:253–54
115. Gellert, M., O'Dea, M. H., Itoh, T., Tomizawa, J. 1976. *Proc. Natl. Acad. Sci. USA* 73:4474–78
116. Gellert, M., Mizuuchi, K., O'Dea, M. H., Itoh, T., Tomizawa, J. 1977. *Proc. Natl. Acad. Sci. USA* 74:4772–76
117. Staudenbauer, W. L. 1976. *Molec. Gen. Genet.* 149:151–58
118. Kingsbury, D. T., Helinski, D. R. 1970. *Biochem. Biophys. Res. Commun.* 41: 1534–44
119. Clewell, D. B., Evenchik, B., Cranston, J. W. 1972. *Nature New Biol.* 237:29–31
120. Tomizawa, J., Ohmori, H., Bird, R. E. 1977. *Proc. Natl. Acad. Sci. USA* 74:1865–69
121. Bird, R. E., Tomizawa, J. 1978. *J. Mol. Biol.* 120:137–43
122. Bolivar, F., Betlach, M. C., Heyneker, H. L., Shine, J., Rodriguez, R. L., Boyer, H. W. 1977. *Proc. Natl. Acad. Sci. USA* 74:5265–69
123. Itoh, T., Tomizawa, J. 1978. *Cold Spring Harbor Symp. Quant. Biol.* 43: In press
124. Ohmori, H., Tomizawa, J. 1979. *Molec. Gen. Genet.* In press
125. Backman, K., Betlach, M., Boyer, H. W., Yanofsky, S. 1978. *Cold Spring Harbor Symp. Quant. Biol.* 43: In press
126. Kahn, M. L., Figurski, D., Ito, L., Helinski, D. R. 1978. *Cold Spring Harbor Symp. Quant. Biol.* 43: In press
127. Sakakibara, Y., Suzuki, K., Tomizawa, J. 1976. *J. Mol. Biol.* 108:569–82
128. Levine, A. D., Rupp, W. D. 1978. In *Microbiology 1978,* ed. D. Schlessinger, pp. 163–66. Washington, DC: Am. Soc. Microbiol.
129. Inselburg, J. 1974. *Proc. Natl. Acad. Sci. USA* 71:2256–59
130. Lovett, M. A., Katz, L., Helinski, D. R. 1974. *Nature* 251:337–40
131. Staudenbauer, W. L. 1977. *Molec. Gen. Genet.* 156:27–34
132. Staudenbauer, W. L., Lanka, E., Schuster, H. 1978. *Molec. Gen. Genet.* 162:242–49
133. Collins, J., Williams, P., Helinski, D. R. 1975. *Molec. Gen. Genet.* 136:273–89
134. Sakakibara, Y. 1978. *J. Mol. Biol.* 124:373–89
135. Tye, B.-K., Nyman, P.-O., Lehman, I. R., Hochhauser, S., Weiss, B. 1977. *Proc. Natl. Acad. Sci. USA* 74:154–57
136. Dove, W. F., Hargrove, E., Ohashi, M., Haugli, F., Guha, A. 1969. *Japan J. Genet.* 44: Suppl. 1, pp. 11–22
137. Dove, W. F., Inokuchi, H., Stevens, W. F. 1971. In *The Bacteriophage Lambda,* ed. A. D. Hershey, pp. 747–71. Cold Spring Harbor, NY: Cold Spring Harbor Lab. 792 pp.
138. Schnös, M., Inman, R. B. 1970. *J. Mol. Biol.* 51:61–73
139. Streeck, R. E., Hobom, G. 1975. *Eur. J. Biochem.* 57:595–606
140. Rambach, A. 1973. *Virology* 54:270–77
141. Denniston-Thompson, K., Moore, D. D., Kruger, K. E., Furth, M. E., Blattner, F. R. 1977. *Science* 198:1051–56
142. Hobom, G., Lusky, M., Grosschedl, R.,

Scherer, G. 1978. *Cold Spring Harbor Symp. Quant. Biol.* 43: In press
143. Ogawa, T., Tomizawa, J. 1968. *J. Mol. Biol.* 38:217–25
144. Saito, H., Uchida, H. 1978. *Molec. Gen. Genet.* 164:1–8
145. Yochem, J., Uchida, H., Sunshine, M., Saito, H., Georgopoulos, C. P., Feiss, M. 1978. *Molec. Gen. Genet.* 164:9–14
146. Herskowitz, I. 1973. *Ann. Rev. Genet.* 7:289–324
147. Skalka, A. M. 1978. *Curr. Top. Mol. Biol. Immunol.* 78:201–37
148. Dove, W. 1968. *Ann. Rev. Genet.* 2:305–40
149. Szpirer, J., Brachet, P. 1970. *Molec. Gen. Genet.* 108:78–92
150. Furth, M. E., Yates, J. L., Dove, W. F. 1978. *Cold Spring Harbor Symp. Quant. Biol.* 43: In press
151. Tomizawa, J. 1971. See Ref. 137, pp. 549–52
152. Georgopoulos, C. P., Herskowitz, I. 1971. See Ref. 137, pp. 553–64
153. Wickner, S. H. 1978. *Cold Spring Harbor Symp. Quant. Biol.* 43: In press
154. Thomas, R., Bertani, L. E. 1964. *Virology* 24:241–53
155. Furth, M. E., Blattner, F. R., McLeester, C., Dove, W. F. 1977. *Science* 198:1046–51
156. Nijkamp, H. J. J., Szybalski, W., Ohashi, M., Dove, W. F. 1971. *Molec. Gen. Genet.* 114:80–88
157. Stevens, W. F., Adhya, S., Szybalski, W. 1971. See Ref. 137, pp. 515–33
158. Moore, D. D., Denniston-Thompson, K., Furth, M. E., Williams, B. G., Blattner, F. R. 1977. *Science* 198:1041–46
159. Moore, D. D., Denniston-Thompson, K., Kruger, K. E., Furth, M. E., Williams, B. G., Daniels, D. L., Blattner, F. R. 1978. *Cold Spring Harbor Symp. Quant. Biol.* 43: In press
160. Schwarz, E., Scherer, G., Hobom, G., Kössel, H. 1978. *Nature* 272:410–14
161. Okazaki, R., Okazaki, T., Sakabe, K., Sugimoto, R., Kainuma, A., Sugino, A., Iwatsuki, N. 1968. *Cold Spring Harbor Symp. Quant. Biol.* 33:129–43
162. Okazaki, T., Kurosawa, Y., Ogawa, T., Seki, T., Shinozaki, K., Hirose, S., Fujiyama, A., Kohara, Y., Machida, Y., Tamanoi, F., Hozumi, T. 1978. *Cold Spring Harbor Symp. Quant. Biol.* 43: In press
163. Thomas, K. R., Ramos, P. M., Lundquist, R., Olivera, B. M. 1978. *Cold Spring Harbor Symp. Quant. Biol.* 43: In press
164. Hinkle, D. C., Richardson, C. C. 1975. *J. Biol. Chem.* 250:5523–29
165. Scherzinger, E., Klotz, G. 1975. *Molec. Gen. Genet.* 141:233–49
166. Scherzinger, E., Lanka, E., Morelli, G., Seiffert, D., Yuki, A. 1977. *Eur. J. Biochem.* 72:543–58
167. Scherzinger, E., Lanka, E., Hillenbrand, G. 1977. *Nucleic Acids Res.* 4:4151–63
168. Kolodner, R., Richardson, C. C. 1977. *Proc. Natl. Acad. Sci. USA* 74:1525–29
169. Hillenbrand, G., Morelli, G., Lanka, E., Scherzinger, E. 1978. *Cold Spring Harbor Symp. Quant. Biol.* 43: In press
170. Alberts, B. M., Morris, C. F., Mace, D., Sinha, N., Bittner, M., Moran, L. 1975. See Ref. 40, pp. 241–69
171. Alberts, B. M., Barry, J., Bittner, M., Davies, M., Hama-Inaba, H., Liu, C. C., Mace, D., Moran, L., Morris, C. F., Piperno, J., Sinha, N. K. 1977. See Ref. 44, pp. 31–63
172. Silver, L. L., Nossal, N. G. 1978. *Cold Spring Harbor Symp. Quant. Biol.* 43: In press
173. Liu, C. C., Burke, R. L., Hibner, U., Barry, J., Alberts, B. M. 1978. *Cold Spring Harbor Symp. Quant. Biol.* 43: In press
174. Ogawa, T., Hirose, S., Okazaki, T., Okazaki, R. 1977. *J. Mol. Biol.* 112:121–40
175. Miyamoto, C., Denhardt, D. T. 1977. *J. Mol. Biol.* 116:681–707
176. Zyskind, J. W., Smith, D. W. 1977. *J. Bacteriol.* 129:1476–86
177. Lark, K. G. 1978. In *Biological Regulation and Development,* ed. R. Goldberger, pp. 201–17. New York: Plenum. In press
178. Lark, K. G. 1972. *J. Mol. Biol.* 64:47–60
179. Messer, W. 1972. *J. Bacteriol.* 112:7–12
180. Kogoma, T. 1978. *J. Mol. Biol.* 121:55–69
181. Kogoma, T., Cunnaughton, M. J., Alizadeh, B. A. 1978. *Ann. Rev. Biochem.* 47:113–20
182. Abe, M., Tomizawa, J. 1971. *Genetics* 69:1–15
183. Hiraga, S., Saitoh, T. 1974. *Mol. Gen. Genet.* 132:49–62
184. Saitoh, T., Hiraga, S. 1975. *Mol. Gen. Genet.* 137:249–61
185. Zyskind, J. W., Deen, L. T., Smith, D. W. 1977. *J. Bacteriol.* 129:1466–75
186. Bagdasarian, M. M., Izakowska, M., Bagdasarian, M. 1977. *J. Bacteriol.* 130:577–82
187. Bagdasarian, M. M., Izakowska, M., Natorff, R., Bagdasarian, M. 1978. *Ann. Rev. Biochem.* 47:101–12

188. Nishimura, Y., Caro, L., Berg, C. M., Hirota, Y. 1971. *J. Mol. Biol.* 55:441–56

189. Tresguerres, E. F., Nandadasa, H. G., Pritchard, R. H. 1975. *J. Bacteriol.* 121:554–61

190. Nishimura, A., Nishimura, Y., Caro, L. 1973. *J. Bacteriol.* 116:1035–42

191. Sotomura, M., Yoshikawa, M. 1975. *J. Bacteriol.* 122:623–28

192. Chesney, R. H., Scott, J. R. 1978. *Plasmid* 1:145–63

193. Bird, R. E., Chandler, M., Caro, L. 1976. *J. Bacteriol.* 126:1215–23

194. Kuempel, P. L., Duerr, S. A., Seeley, N. R. 1977. *Proc. Natl. Acad. Sci. USA* 74:3927–31

195. Kuempel, P. L., Duerr, S. A., Maglothin, P. D. 1978. *J. Bacteriol.* 134:902–12

196. Chandler, M., Silver, L., Caro, L. 1977. *J. Bacteriol.* 131:421–30

197. Lanka, E., Geschke, B., Schuster, H. 1978. *Proc. Natl. Acad. Sci. USA* 75:799–803

198. Wickner, S., Wright, M., Hurwitz, J. 1974. *Proc. Natl. Acad. Sci. USA* 71:783–87

199. Lanka, E., Edelbluth, C., Schlicht, M., Schuster, H. 1978. *J. Biol. Chem.* 253:5847–51

200. Kogoma, T. 1976. *J. Mol. Biol.* 103:191–97

201. Messer, W., Dankwarth, L., Tippe-Schindler, R., Womack, J. E., Zahn, G. 1975. See Ref. 40, pp. 602–17

202. Chattoraj, D. K. 1978. *Proc. Natl. Acad. Sci. USA* 75:1685–89

203. Rupp, W. D., Ihler, G. 1968. *Cold Spring Harbor Symp. Quant. Biol.* 33:647–50

204. Ohki, M., Tomizawa, J. 1968. *Cold Spring Harbor Symp. Quant. Biol.* 33:651–58

205. Bird, R. E., Louarn, J. M., Martuscelli, J., Caro, L. 1972. *J. Mol. Biol.* 70:549–66

206. Louarn, J., Funderburgh, M., Bird, R. E. 1974. *J. Bacteriol.* 120:1–5

207. Marsh, R. C., Worcel, A. 1977. *Proc. Natl. Acad. Sci. USA* 74:2720–24

208. Hiraga, S. 1976. *Proc. Natl. Acad. Sci. USA* 73:198–202

209. Joh, K., Ogura, T., Hiraga, S. 1977. *Proc. Mol. Biol. Meet. Jpn.*, 1977, pp. 114–16. Tokyo: Kyoritsu Shuppan Co. 147 pp.

210. Masters, M. 1975. *Mol. Gen. Genet.* 143:105–11

211. von Meyenburg, K., Hansen, F. G., Nielsen, L. D., Jorgensen, P. 1977. *Mol. Gen. Genet.* 158:101–9

212. Yasuda, S., Hirota, Y. 1977. *Proc. Natl. Acad. Sci. USA* 74:5458–62

213. Sugimoto, K., Oka, A., Sugisaki, H., Takanami, M., Nishimura, A., Yasuda, S., Hirota, Y. 1979. *Proc. Natl. Acad. Sci. USA.* 76:575–79

214. von Meyenburg, K., Hansen, F. G., Nielsen, L. D., Riise, E. 1978. *Mol. Gen. Genet.* 160:287–95

215. Miki, T., Hiraga, S., Nagata, T., Yura, T. 1978. *Proc. Natl. Acad. Sci. USA* 75:5099–103

216. Hirota, Y., Yasuda, A., Nishimura, A., Takeda, Y., Yamada, M., Sugimoto, K., Sugisaki, H., Oka, A., Takanami, M. 1978. *Cold Spring Harbor Symp. Quant. Biol.* 43: In press

217. Meijer, M., Beck, E., Hansen, F. G., Bergmans, H. E. N., Messer, W., von Meyenburg, K., Schaller, H. 1979. *Proc. Natl. Acad. Sci. USA.* 76:580–84

218. Messer, W., Meijer, M., Bergmans, H. E. N., Hansen, F. G., von Meyenburg, K. 1978. *Cold Spring Harbor Symp. Quant. Biol.* 43: In press

219. von Meyenburg, K., Hansen, F. G., Riise, E., Bergmans, H. E. N., Meijer, M., Messer, W. 1978. *Cold Spring Harbor Symp. Quant. Biol.* 43: In press

220. Diaz, R., Pritchard, R. H. 1978. *Nature* 275:561–64

221. Jacob, F., Brenner, S. 1963. *CR Acad. Sci.* 256:298–300

222. Jacob, F., Brenner, S., Cuzin, F. 1963. *Cold Spring Harbor Symp. Quant. Biol.* 28:329–48

223. Goldberger, E. 1974. *Science* 183:810–16

224. Sompayrac, L., Maaloe, O. 1973. *Nature New Biol.* 241:133–35

225. Matsubara, K., Kaiser, A. D. 1968. *Cold Spring Harbor Symp. Quant. Biol.* 33:769–75

226. Signer, E. 1969. *Nature* 223:158–60

227. Berg, D. E. 1974. *Virology* 62:224–33

228. Matsubara, K. 1976. *J. Mol. Biol.* 102:427–39

229. Matsubara, K., Takeda, Y. 1975. *Mol. Gen. Genet.* 142:225–30

230. Pritchard, R. H., Barth, P. T., Collins, J. 1967. *Symp. Soc. Gen. Microbiol.* 19:263–97

231. Pritchard, R. H. 1978. See Ref. 10, pp. 1–26

232. Chandler, M., Silver, L., Roth, Y., Caro, L. 1976. *J. Mol. Biol.* 104:517–23

233. Collins, J., Pritchard, R. H. 1973. *J. Mol. Biol.* 78:143–55

234. Pritchard, R. H., Chandler, M. G., Collins, J. 1975. *Mol. Gen. Genet.* 138:143–55

235. Cabello, F., Timmis, K., Cohen, S. N. 1976. *Nature* 259:285–90

Ann. Rev. Biochem. 1979. 48:1035–69

RNA PROCESSING AND THE ♦12032
INTERVENING SEQUENCE PROBLEM

John Abelson

Department of Chemistry, University of California, San Diego,
La Jolla, California 92093

CONTENTS

PERSPECTIVES AND SUMMARY

RNA processing is the collection of enzymatic reactions that transform a primary transcription product into a mature functioning molecule. These enzymatic reactions are of several kinds. There are the endo- and exonucleases which alter the size of the transcript. There is the addition of nucleotides either to the 5' end of the molecule—(the "cap") or to the 3' end (the PolyA or CCA addition). Also there are a myriad of enzymes that modify the canonical nucleosides.

1035

0066-4154/79/0701-1035$01.00

In order to understand the complete set of reactions in the processing pathway which lead to the final product it is necessary to know the structures of both the initial transcription product and the mature RNA molecule. One can then search for intermediates or precursors that will serve as substrates for the detection and purification of processing enzymes.

There have been tremendous advances in this field in the three years since Perry reviewed the subject (1). They have been mostly due to the rapid acquisition of information about the structure of genes and their initial transcription products, and have come mainly through the application of recombinant DNA technology and the development of rapid DNA sequencing techniques (2, 3). The most important result is that the genes of eukaryotes are not necessarily colinear with their products—instead they contain interruptions in the coding sequence which have been called intervening sequences or introns (4). (I shall call them intervening sequences.) The phenomenon was first discovered during an investigation of the structure of adenovirus mRNA, and in succeeding months reports appeared of the split gene phenomenon in other viral genomes, chromosomal genes, rRNA genes, and tRNA genes. A wider range of English usage was required to express the surprise. Thus words like amazing, astounding, and baroque have uncharacteristically made their way into the scientific literature.

It was immediately realized that the split gene phenomenon implied a new RNA processing reaction. Although there were other possibilities the favored mechanism for the removal of the intervening sequences was a reaction that has come to be called RNA splicing. In this mechanism the entire split gene, including the intervening sequences, is transcribed to form a precursor. The intervening sequences are then removed at the RNA level by the RNA splicing enzyme which clips out the intervening sequences and then rejoins the ends to form the mature RNA molecule. There is now substantial evidence that this is the primary mechanism for the removal of intervening sequences.

In the field of prokaryotic RNA processing new results have supplied detailed information concerning the pathway of processing and the enzymatic mechanisms by which they are accomplished. The rationale of this review is that the more detailed information available in several prokaryotic systems can shed light on the general mechanisms of RNA processing. Thus the RNA processing steps in several systems are described in some depth. In the eukaryotic field the focus is on the existence of intervening sequences and the RNA splicing mechanism. The possible mechanisms of substrate recognition are discussed in light of the prokaryotic experience. I have limited myself to discussing the steps that alter the size of the transcript— nucleases and splicing. Excellent reviews have appeared on modified nucleosides in tRNA (5), on tRNA biosynthesis (6), and on capping (7).

PROCESSING OF RNA IN PROKARYOTES

Escherichia coli tRNA

Much of the early knowledge of tRNA processing came from the *Escherichia coli* tRNA$_I$^{Tyr} system (6). Work on this system has continued and now the complete DNA sequence of the region is known (8–12) and the transcription unit has been fully defined (13). The gene has been completely synthesized (14) de novo and has been shown to function correctly when reinserted into *E. coli* (15). Thus this tRNA gene is one of the most intensively studied although it cannot yet be said that all of the steps in the synthesis of tRNA$_I$^{Tyr} are understood.

The tRNA$_I$^{Tyr} transcription unit contains two identical structural genes separated by 200 nucleotides. One of these genes, together with the interstitial sequences, can be deleted in an unequal recombination event (16). Transducing phages carrying either the doublet or the singlet genes have been isolated. Most experiments have been carried out with the singlet transcription unit which contains approximately 350 nucleotides. The transcription start site is located 41 nucleotides upstream from the 5' end of the mature RNA. In vitro transcription studies with purified restriction fragments derived from ϕ80 pSu$_3$$^+$ DNA strongly indicate that transcription continues 224–226 nucleotides beyond the CCA 3' OH end of the tRNA (the CCA end is encoded in the DNA) and terminates at a rho-dependent termination site (13). A bizarre feature of the DNA sequence following the 3' of the tRNA structural gene end is that it consists of a series of three 178 base-pair repeating units. The last 16 bases of the tRNA are included in the basic repeat unit (12).

Thus at least two nucleolytic reactions are required to mature the tRNA from the larger de novo transcript. One of these, the reaction that matures the 5' end has been studied in detail. This was possible because a precursor to tRNA$_I$^{Tyr} was isolated that contains the unprocessed 41 nucleotides at the 5' end (including the 5' triphosphate) and an extra three nucleotides at the 3' end (9). This precursor can be matured at the 5' end with the enzyme RNase P (17). As we shall see, this enzyme has a broad specificity and in fact must be involved in the synthesis of most or all *E. coli* tRNAs. This fact places considerable restraint on the substrate specificity of RNase P. The specificity of the enzyme has been further studied through mutants of the Su III gene (an amber suppressor allele of the tRNA$_I$^{Tyr} gene) (6). For example, a transition of G to T at position 15 loosens the structure of the tRNA (18, 19) (as judged by the chemical reactivity of the bases) and also causes an accumulation of the precursor. A considerable body of evidence now indicates that RNase P must recognize general features of tRNA structure in order to make the precise maturation at the 5' end. Sequence recognition per se does not seem to be involved.

Despite a good deal of effort, RNase P has not yet been completely purified. The enzyme has unusual purification properties that suggest that it is tightly bound to RNA [(20), C. Guthrie, personal communication]. For example, the activity is eluted from DEAE Sephadex at 0.4 M NaCl. The buoyant density of the enzyme is 1.71 g/cc, characteristic of a ribonucleoprotein complex composed of 80% RNA and 20% protein. The activity of the enzyme is destroyed by treatment with micrococcal nuclease or RNase A digestion under controlled conditions. The major RNA component in the most highly purified preparations is about 300 nucleotides in length (20). The intriguing possibility is that an RNA molecule is an essential component of the enzyme, but this remains to be rigorously proved. The purification and characterization of this enigmatic enzyme remains as one of the most important problems in prokaryotic RNA processing.

The processing events at the 3' end of $tRNA_I^{Tyr}$ are less well characterized. Bikoff, Gefter & LaRue (21, 22) have studied the in vitro synthesis of $tRNA_I^{Tyr}$ using the total $\phi 80\ pSu_I^+$ DNA. By separation of and reconstitution of fractions from a crude extract they implicated both an endonuclease and an exonuclease in the processing of the 3' end. Sekiya et al (23) have used the more defined synthetic DNAs as template. The synthetic DNA consists of the promoter, the 5'-terminal precursor sequence, the mature tRNA gene, and 16 base-pairs distal to the 3' portion of the structural gene. This latter region would be expected to form a hairpin consisting of five base pairs. Using this DNA as template, $tRNA_I^{Tyr}$ is correctly processed in a crude in vitro system. It is postulated that an endonuclease recognizes and cleaves the hairpin loop and the remaining seven bases of the precursor RNA are removed by an exonuclease. Exonuclease maturation of the 3' end of tRNA precursors have been suggested by several other groups (21, 24, 25), but until recently the activity had not been purified. A plausible candidate for this function, RNase D, has now been purified approximately 250-fold by Ghosh & Deutscher (26). The synthetic substrate used in this purification was prepared by adding two ^{14}C– labeled cytidine residues to the end of tRNA to form tRNA CCA $C^*\ C^*_{OH}$ using an aberrant activity of the rabbit tRNA-nucleotidyl transferase. RNase D (a) removes the two terminal residues, but does not degrade the tRNA; (b) matures the tyrosine precursor, but only if the 5' end is first matured with RNase P; and (c) degrades tRNAs that have been partially degraded by snake venom phosphodiesterase (27). The activity is probably the same as the RNase P III of Bikoff et al (21, 22) and the RNase Q of Shimura and colleagues (24). The enzyme must recognize the tertiary structure of tRNA and the CCA residues in order to perform this function.

The fact that the enzyme RNase P is required for the synthesis of suppressor tRNAs allowed for the isolation of temperature-sensitive mutants of

E. coli defective in RNase P at the nonpermissive temperature. Two groups have isolated RNase P^{ts} mutants—Schedl & Primakoff (28) who isolated a single mutant A49, and Ozeki et al who isolated two mutants both of which are defective in RNase P at the nonpermissive temperature. These mutants, ts 241 and ts 709, map at different locations on the *E. coli* chromosome (29, 30). The RNase P^{ts} mutants have allowed the characterization of a large number of tRNA precursors. When *E. coli* carrying an RNase P^{ts} mutation is labeled with $^{32}PO_4$ at the nonpermissive temperature more than 30 tRNA precursors accumulate (31–34). These can be separated by polyacrylamide gel electrophoresis and further characterized. Under these conditions of labeling all tRNAs are accumulated as precursors, which indicates that RNase P is required for the synthesis of all tRNA in *E. coli.* Some of the precursors are monomeric in form, analogous to the tRNA$_I^{Tyr}$ precursor, but others contain two, three, or perhaps as many as five tRNA sequences. These can be different tRNAs, e.g. tRNA$_m^{Met}$, tRNA$_I^{Gln}$, and a third, unidentified, tRNA are found together in a precursor (34); or the precursors can consist of tandem arrays of a single tRNA, e.g. as many as four molecules of tRNA^{Leu} may be linked in a single precursor (31).

Shimura and his colleagues have studied the processing of these tRNA precursors in vitro. The processing of the monomeric precursors is similar to the mechanisms discussed for pre-tRNA^{Tyr}. First the 5' end of the precursor is matured by RNase P and subsequently the 3' end is matured by an exonuclease which Sakano & Shimura have designated RNase Q. All of the monomeric precursors are stable in heat-inactivated extracts of *E. coli* ts 241 (the RNase P^{ts} mutant). Most of the polymeric precursors are cleaved under these conditions. Shimura & Sakano designated this activity RNase O (24).

Beyond any other benefit, these results have given us some idea of the organization of the tRNA genes. It is known (from the genetic mapping of suppressor genes) that tRNA genes in *E. coli* are not completely clustered, but structural analyses of several precursors show that there is some clustering.

Bacteriophage T4 tRNA

Bacteriophage T4 carries the genes for eight tRNAs and two stable RNAs of unknown function (35, 36). All but one of these RNAs has been sequenced.

This system has been an important one in studying the biosynthesis of tRNA. Perhaps the most important advantage of the T4 system is that a comprehensive genetic approach has been made and as a result there are mutants in six of the ten RNA genes (37–39) and a set of deletions which have helped in the mapping of the tRNA genes (37). The genetic results show the tRNA genes are linked in two clusters between the genes *e* (lysozyme) and 57. The small cluster contains the tRNA^{Arg} gene and the genes for the two stable RNAs Band 1 and Band 2. This cluster is 500 base pairs away from a larger cluster containing seven tRNA genes (see Figure 1). Studies of the pathway of synthesis of the tRNAs support this notion of close linkage.

Six of the tRNAs are synthesized via dimeric tRNA precursors. These are the tRNA^{Gln}-tRNA^{Leu}, the tRNA^{Pro}-tRNA^{Ser}, and the tRNA^{Thr}-tRNA^{Ileu} precursors (38) and all of them have been sequenced. The structures of each of these precursors are different in an interesting way involving the CCA-OH end of the tRNA. In the tRNA^{Pro}-tRNA^{Ser} precursor, the CCA ends are not present (38). In the tRNA^{Gln}-tRNA^{Leu} precursor the tRNA^{Leu} CCA end is present, but the tRNA^{Gln} CCA is not (40). In the tRNA^{Thr}-tRNA^{Ileu} precursor the tRNA^{Thr} CCA is present, but the tRNA^{Ileu} is not (C. Guthrie, personal communication). McClain and his colleagues have shown that the disposition of the 3' end is crucial in the processing of each of the precursors (reviewed in 41). In order for RNase P to optimally process the Pro-Ser precursor the CCA end must be added to the 3' end of tRNA^{Ser}. This would require exonuclease removal of three bases followed by CCA end addition. These requirements have been demonstrated genetically. A mutant strain of *E. coli,* BN, lacks the exonuclease (42, 43), [which is different from the RNase D/RNase Q activity (26)]. Deutscher has isolated mutants of *E. coli* that are very low in tRNA nucleotidyl transferase (44). In both classes of mutants a suppressor allele of the T4 tRNA^{Ser} is nonfunctional (42, 45).

All of the T4 tRNAs require RNase P for their synthesis (30, 37). Two dimeric precursors, tRNA^{Pro}-tRNA^{Ser} and tRNA^{Gln}-tRNA^{Leu}, accumulate in the absence of RNase P. The third dimeric precursor, tRNA^{Thr}-

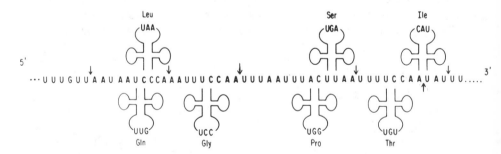

Figure 1 Schematic representation of the transcript of the T4 tRNA gene cluster (K. Fukada and J. Abelson, unpublished). The arrows represent sites of cleavage that must occur in order to give the tRNA precursors that accumulate in the absence of RNase P.

tRNA^{Ileu} does not. Monomeric precursors for tRNAs Gly and Thr accumulate and, surprisingly, the 140 nucleotide Band 1 RNA also contains additional nucleotides at its 5' end in the absence of RNase P.

Recently the T4 tRNA gene cluster has been isolated by recombinant DNA techniques (46). This work has resulted in two important pieces of information relevant to the transcription and processing of the T4 tRNAs. First, the promoter has been located 1000 base pairs upstream from the tRNA^{Gln} gene. This confirms what has been long suspected (47, 48). The tRNAs are transcribed as part of a long, probably multifunctional transcript. Second, the DNA sequence of the cluster of seven tRNA genes has been determined (K. Fukada and J. Abelson, unpublished). This sequence, presented schematically in Figure 1, helps to answer the question of how the precursors which accumulate in the absence of RNase P are generated. The RNase P precursors are simply joined together in the DNA sequence which suggests that single endonuclease cleavages separate each precursor. The positions of these cleavages (Figure 1) imply an endonuclease that recognizes tRNA and hydrolyzes a phosphodiester bond producing the CCA end where it is coded (tRNAs Leu and Gly) or producing a 3' end of equivalent length (e.g.-UAA$_{OH}$ in tRNA^{Ser}). The two dimers, tRNA^{Pro}-tRNA^{Ser} and tRNA^{Gln}-tRNA^{Leu}, are not cleaved by this enzyme, but it is possible that the tRNA^{Thr}-tRNA^{Ileu} dimer with its larger interstitial sequence, including CCA, is cleaved. We believe that this endonuclease is a host function, but such an activity has not yet been described.

All of the T4 tRNAs are synthesized in a single transcript and yet the amounts of different mature tRNA species can vary as much as fourfold. It appears that the precise arrangement of sequences surrounding the T4 tRNA genes serves to modulate the efficiency of RNA processing thus controlling the level of the various tRNAs.

Escherichia coli Ribosomal RNA

In *E. coli* there are at least seven ribosomal RNA transcription units. These are dispersed in the genome and contain, in single transcription units, 16S RNA, 23S RNA, 5S RNA, and one or more tRNAs (reviewed by Nomura, Morgan & Jaskunas, 49). The tRNA genes are located either in the spacer region between 16S and 23S rRNA genes (50–52) or at the 3' end beyond the 5S RNA (53). Three of the rRNA operons contain both tRNA$_{1B}$^{Ala} and tRNA^{Ile} genes in the spacer region while the remaining four operons contain only the gene for tRNA$_2$^{Glu}. In one rRNA operon the tRNA^{Trp} and tRNA$_1$^{Asp} genes are at the 3' end of the transcription unit (53).

The rRNA operons have been isolated either by use of transducing phages or by recombinant DNA methods. DNA sequencing of the rRNA cistrons is now well under way. The 16S rRNA genes (54), the tRNA spacer region (55), and interesting spacer or leader regions (56) involved in RNA processing have now been sequenced and no doubt the entire sequence—greater than 5000 base pairs—will soon be completed. The sequence information considerably sharpens our picture of the structure and hence the substrate specificities of the various enzymes that process the rRNA transcript.

Ordinarily, processing of rRNA is coupled to transcription, but in a mutant of *E. coli* deficient in RNase III the entire 30S RNA transcript accumulates (57, 58). This RNA can be used as a substrate in RNA processing reactions. For example, treatment of 30S RNA with purified RNase III generates precursors to 16S, 23S, and 5S RNA (57–60) called p16S$_{III}$, p23S$_{III}$ and p5S$_{III}$) as well as fragments containing the tRNAs (51, 61). p16S$_{III}$ is identical to a 17S precursor that can be isolated (in ribosomes) from pulse-labelled (62–64) or chloramphenicol-treated (65) cells. p23S$_{III}$ is less well-characterized, but is also probably an immediate precursor to 23S RNA. p5S$_{III}$ contains approximately 300 nucleotides (59)—twice as many as the mature 5S RNA.

The DNA sequences of the regions surrounding the 16S RNA gene have been determined (56). This sequence contains the 5' and 3' ends of p16S$_{III}$. Examination of the sequence leads to a very interesting conclusion. The 5' and 3' ends of p16S$_{III}$ can form 26 continuous base pairs (Figure 2). The RNase III cleavage sites which generate the 5' and 3' ends of p16S$_{III}$ are found to be within two base pairs of each other on opposite strands of the double helical RNA. This substrate can be isolated by treating the RNA isolated from an RNase III, RNase P double mutant of *E. coli* with single-strand specific nucleases (66). One of the small double-stranded RNAs isolated in this way is homologous with the stem of the 16S precursor structure shown in Figure 2 and can be cleaved with RNase III in vitro (59). The sequence complementarity in this region was also revealed by observation of looped structures in electron micrographs of the single-stranded ribosomal DNA (67). Thus we are led to the remarkable conclusion that the substrate for the RNase III production of p16S$_{III}$ is a hairpin with a loop consisting of the entire sequence (1600 nucleotides) of 16S rRNA.

p16S$_{III}$ RNA is acted on by several other ribonucleases, one of which is called RNase M16. This enzyme generates the mature 5' end of 16S rRNA and requires the assembly of ribosomal protein S4 to achieve correct cleavage (68). Another activity has been isolated which produces the mature 3' end of 16S RNA (69).

A similar double-stranded structure can be constructed by pairing the 5' and 3' sequences surrounding 23S RNA [(70; Figure 2]. In this case the RNase III cleavage sites are not yet known, but it is likely that this structure forms the substrate for the generation of p23S$_{III}$. The enzymes involved in maturation of p23S$_{III}$ RNA have been identified.

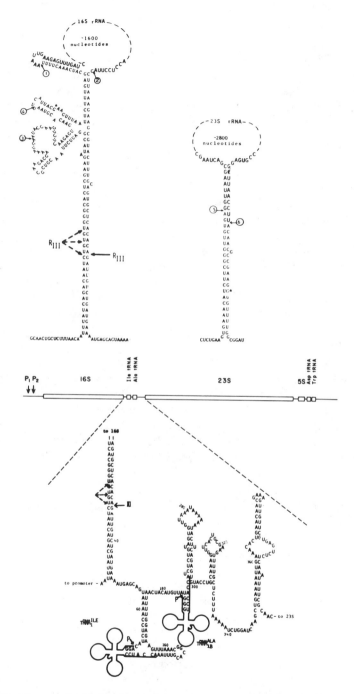

Figure 2 The map of the *rrnX* operon is shown in the center. Nucleotide sequences surrounding the 16S RNA (56) and 23S RNA (70) are shown above. Sites 1 and 2 are the mature ends of 16S RNA. The nucleotide sequence of the tRNA spacer region is shown below (55).

The 437 base-pair spacer region containing the genes for $tRNA_I^{Ile}$ and $tRNA_{1B}^{Ala}$ has been sequenced [(55); Figure 2]. In vitro studies of RNA processing of the 30S precursor have been performed. When 30S RNA is treated with crude extracts the tRNAs are completely processed (51). In vivo, under conditions in which RNase III and RNase P are not present, a 19S RNA accumulates that contains 16S RNA and $tRNA_2^{Glu}$ (71). Thus these two enzymes are responsible for all of the cuts occurring between the 3' end of 16S RNA and the 3' end of $tRNA_2^{Glu}$. RNase III generates the 3' end of $p16S_{III}$, the 5' end of $p23S_{III}$, and, in addition, cleaves the spacer region in vitro at position 290 [(61, 70) and Figure 2] to produce a dimeric tRNA precursor. This precursor can be cleaved in vitro by pure RNase P to produce the 5' end of $tRNA_{1B}^{Ala}$ (61). The reaction producing the 5' end of $tRNA_I^{Ile}$ does not occur in vitro although it must be true that RNase P accomplishes this function in vivo. Perhaps the secondary structure in this region inhibits the RNase P reaction in vitro. When the processing is coupled to transcription in vivo this may not be a problem. At least one more enzyme must act to generate the 3' ends of the tRNAs. (Could this activity be the same as the one invoked for the processing of the 3' ends of the T4 tRNA?)

A temperature-sensitive mutant, rne-3071, has been isolated that accumulates a 9S precursor containing the 5S RNA (72, 73). This precursor does not accumulate in the rnc (RNase III⁻) rne double mutant which indicates that the 5' end of this RNA is created by the RNase III cleavage at the 3' end of the 23S RNA. Instead, the 5S RNA is linked to 23S RNA in a 25S RNA molecule. A p5S RNA (74) can be matured from the 9S precursor by a partially purified enzyme called RNase E (75). This activity can also release p5S RNA from the 25S RNA. Although precursors to 16S and 23S rRNA accumulate in the absence of protein synthesis, only a very limited precursor to 5S, p5S RNA is seen. This precursor contains three extra nucleotides at the 5' end and these may be removed exonucleolytically (74).

In *Bacillus subtilis*, however, two large presursors to 5S RNA accumulate in the absence of protein synthesis. One of these, $p5_A$, has been studied extensively by Pace and his colleagues (76). The $p5_A$ precursor shown in Figure 3 is 179 nucleotides in length and contains 21 extra nucleotides at its 5' end and 42 extra nucleotides at the 3' end (77). The precursor is matured accurately in a ribosomal wash fraction of a *B. subtilis* extract. A single enzymatic activity termed RNase M5 is responsible for both cleavages that generate the 5S RNA. During purification on DEAE the enzyme separates into two components termed α and β, neither of which alone has endonuclease activity (78). The β component (but not the α component) forms a specific complex with $p5_A$ which is retained on nitrocellulose filters. Thus the β component may be a recognition subunit and the α component the enzymatic subunit. In the probable secondary structure of p5$_A$ (Figure 3) a helical section of 13 base pairs places the two cleavage sites in close proximity to each other. Extensive manipulations of the p5$_A$ substrate have shown that the base-pairing of the molecule in this region very likely is an important feature of the substrate but that the precursor-specific 5' and 3' fragments are not involved in the recognition. The most elegant of these involved the preparation of artificial substrates by T4 RNA ligase-catalyzed addition of oligonucleotides to the 5' end of the mature 5S RNA (79). This is done by preparing half molecules of either mature 5S RNA or p5$_A$ by partial digestion with RNase T2 [Figure 3; (80)]. Under selective conditions there is a single cleavage by T2 at residue 60. The purified 5' end of mature 5S RNA is used as the donor in an RNA ligase reaction with e.g. (U$_p$)$_3$ G$_{OH}$. The elongated molecule is then annealed with either the 3' end of p5$_A$ or the 3' end of mature 5S RNA to form a nicked substrate. In the former case the 3' precursor-specific fragment and the oligonucleotide are both efficiently released. In the latter substrate the oligonucleotide is released. Other oligonucleotides were tested as substrates. U$_3$G could be substituted by C$_6$G, but A$_3$C or A$_6$C are not as effective in potentiating the release of the 3' precursor fragment. Further studies with hybrid structures constructed by annealing of partial cleavage products of p5$_A$ clearly established that RNase M$_5$ must be recognizing some feature of the tertiary structure of the mature 5S RNA. The sequence and secondary structure of the distal portions of the precursor are not important provided that they allow base-pairing at the site of enzyme action.

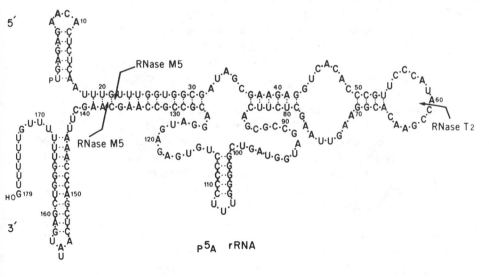

Figure 3 The nucleotide sequence of the *B. subtilis* 5S RNA precursor (77).

Bacteriophage T7 Early mRNA

The processing of early T7 mRNA remains the single well-documented case of endonucleolytic processing of mRNA in prokaryotes (although it may be that endonucleases are involved in the breakdown of mRNA). The early mRNAs in bacteriophage T7 are transcribed from three closely linked early promoters, near one end of the phage DNA. Termination of transcription occurs at a natural termination site 7500 bases from the promoter (81). The transcript is cleaved either in vivo or in vitro by RNase III to yield five mRNAs, and an interstitial fragment which arises from cleavage by RNase III at two closely spaced sites between genes 1.1 and 1.3 (81).

RNase III was originally detected as a nuclease capable of degrading double-stranded RNA (82). Subsequent investigations have shown that the products of this reaction are a random collection of 5'-phosphorylated oligonucleotides about 15 nucleotides long (83, 84). RNase III has been purified by Crouch (83) and by Dunn (85). Dunn has purified the enzyme to homogeneity by use of poly I-C affinity chromatography. The enzyme has a monomer molecular weight of 25,000. Estimation of the molecular weight by gel filtration indicates that it is a dimer.

By contrast to the random cleavage of double-strand RNA, RNase III is exceedingly specific in its processing of natural substrates in vitro. Four of the RNase III processing sites in T7 RNA have been sequenced [(86–88); J. J. Dunn, personal communication]. The comparison of these sequences does not reveal extensive sequence homology but there are similarities in the secondary structures. Each of the substrates consist of two helical segments of 9–11 base pairs separated by an internal bubble. Chain scission occurs within the bubble on one or both sides. For a more extensive discussion of these structures see Robertson & Barany (66).

Examination of the p16S$_{III}$ rRNA substrate (Figure 2) reveals that it does not contain a bubble. All of the sites have double-stranded character and this must certainly be an important feature of the recognition since all natural cleavages are strongly inhibited by double-stranded RNA. And yet double-stranded sequences per se are not necessarily substrates. MS2 RNA has a number of extensive hairpin regions but none of the phage RNAs are cleaved by RNase III. The concentration of monovalent cation is important in the specificity. Primary sites such as the T7 processing sites are recognized exclusively at salt concentrations higher than 0.1 M (85). At lower salt concentrations secondary sites in both T7 mRNA and ribosomal RNA are cleaved (85). These secondary cleavages are also quite specific. The cleavage of T4 species I RNA is an example of a specific secondary site cleavage. This RNA is not cleaved in vivo but in vitro at 0.03 M [NH$_4^+$]; it is specifically cleaved at two sites on opposite sides of a hairpin structure (89).

RNA Processing in Prokaryotes: The Take Home Lessons

The outlines in this field were clear four years ago and can be seen in the Brookhaven Symposium volume (31). Since then our perception of the problems has sharpened considerably. The nucleotide sequences of a number of substrates have been determined so that the entire processing pathway of several RNA transcripts is close to being understood. Several of the enzymes have been purified but studies on the mechanism of substrate recognition have only just begun. We are beginning, however, to reach some conclusions and it is perhaps worthwhile to list them:

1. All stable RNA molecules in *E. coli* require processing steps for their synthesis. By contrast, T7 mRNA remains the only well-documented case of processing of mRNA.

2. A relatively small number of nucleases carry out a large number of RNA processing reactions. I can count only about ten nucleases that have been implicated in the RNA processing reactions in *E. coli*. I doubt that the true number would be as much as twice that.

3. In the case of endonucleases, all processing enzymes that have been studied create polynucleotide chains with 5' phosphates and 3'-hydroxyl terminii. In general, all exonucleases mature 3' ends although the maturation of p5S RNA is possibly an exception to this rule.

4. The recognition of substrate by enzyme always involves a large structural component. In the case of RNase III, double helical RNA plays an important role and it is clear that both product and precursor-specific sequences contribute to the specificity of the reaction. In the case of RNase P and RNase M5, recognition of substrate seems to lie entirely in the structure and/or sequence of the product. Synthetic sequences unrelated to the natural precursor can be joined to the 5' end of mature tRNA (C. Guthrie and O. Uhlenbeck, personal communication) or to the 5' end of *B. subtilis* 5S RNA (79) and these molecules are seen as substrates by their respective processing enzymes. An RNA processing enzyme that recognizes sequence per se, as in the case of DNA restriction endonucleases, has never been found. This is true despite the fact that ribonucleases can recognize sequence (cf RNase T1).

5. In general RNA processing enzymes can recognize RNA directly but the important exception to this is in the processing of $p16S_{III}$ which occurs only when at least some ribosomal proteins are bound to the RNA (68, 69).

6. There appear to be multiple pathways of RNA processing. Thus RNase III plays a primary role in the processing of ribosomal RNA but is not absolutely required for the synthesis of ribosomal RNA. The necessary cuts in the spacer region can be made by the tRNA processing enzymes and these are apparently sufficient to allow further maturation of 16 and 23S RNA.

PROCESSING OF RNA IN EUKARYOTES

This half of the review focusses on the recent developments in the knowledge of gene structure in eukaryotes. The fact that many genes are split—that is, they contain intervening sequences interrupting the coding regions—implicates a crucial RNA processing step in the expression of the gene. In principal there are a number of possible mechanisms that the cell could employ to express a gene containing intervening sequences. In special cases it is clear that several mechanisms are used. For example, recombination at both the DNA and RNA level are involved in the expression of the immunoglobulin gene. It is expected, however, that the predominant mechanism is a new reaction of RNA processing that has come to be called splicing. In the RNA splicing reaction an internal section of a precursor RNA is excised leaving a covalently closed RNA product.

The structures of genes containing intervening sequences determine exactly what the RNA splicing machinery must be able to do. Accordingly, I first review what is known about the structures of genes containing intervening sequences and then discuss what is known about the mechanisms of RNA splicing.

Intervening Sequences in Eukaryotic Genes

HEMOGLOBIN   In the BALB/C mouse the 144 amino acid β Hb polypeptide is coded for by two nonallelic but closely linked genes (90). Leder and his colleagues have cloned both of these genes in λ vectors (91, 92). Analysis of the DNA by R loop mapping with 9S β Hb mRNA revealed the surprising result that the DNA is interrupted by two intervening sequences (93). The entire β Hb gene has now been sequenced (94). The smaller intervening sequence interrupts codon 30 and 31 and is 116 base pairs in length (94, 95). The larger intervening sequence interrupts codons 104 and 105 and is 646 base pairs in length. There are 52 base pairs in the 5' untranslated portion of the gene and 110 base pairs in the 3' untranslated region (Figure 4).

With regards to the natural history of the intervening sequences it is interesting to review what is known about the persistence of the intervening sequences in the β Hb genes.

1. Analysis of the genomic β Hb genes from rabbit (95) and human (96) reveals that two intervening sequences are also present in these genes and that they are to be found in exactly the same positions—i.e. between codons 30 and 31 and codons 104 and 105. Lawn et al (96) have also cloned the closely linked δ Hb gene; this also has a large intervening sequence, probably between codons 104 and 105. Analysis of the δ Hb gene is not complete so information is not yet available concerning the small intervening sequence.

A detailed comparison between the intervening sequences in the mouse and rabbit Hb genes has been performed (95). The small intervening sequence in the rabbit Hb gene has been sequenced. The rabbit sequence contains 126 base pairs and the mouse 115 or 116 base pairs. Aside from homologies in the flanking sequences to be discussed later these sequences are strikingly *dissimilar* especially in comparison to the conservation of sequence in the coding region. The large intervening sequences are also divergent between mouse and rabbit but are about the same size, 646 and ~580 base pairs respectively.

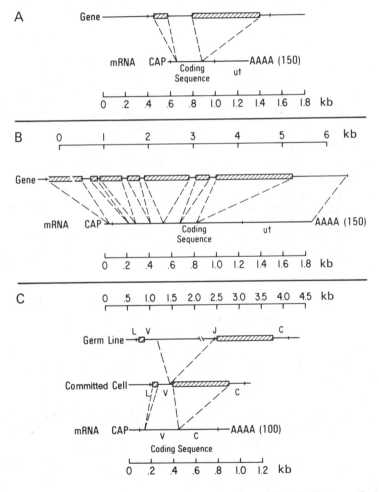

Figure 4 Maps of the genes and mRNAs for *A,* mouse β hemoglobin, *B,* chicken ovalbumin, and *C,* mouse λ light chain. For mouse immunoglobulin the genes from both embryonic and committed cells are shown. The intervening sequences are shown as thicker hatched lines.

The human β Hb gene contains 900 base pairs and the δ Hb gene 950 base pairs in the large intervening sequence (96) so that these two sequences, though found in the same positions, are longer than the corresponding sequences in mouse and rabbit. The β and δ large intervening sequences show little homology.

2. Within the mouse the two closely linked β Hb genes have been compared. Both genes contain the large intervening sequence at the same position, but heteroduplex studies between the genes show that only the coding regions are homologous (92). Taken together these studies show that the intervening sequences in the β Hb gene diverge much more rapidly than the coding sequences, but over a fairly long period of evolutionary time the intervening sequences are retained and their sizes have remained relatively constant.

3. Leder and his colleagues have cloned an α Hb gene from mouse (97). Analysis of the DNA by R-loop mapping and DNA sequencing shows that it too has two intervening sequences located in exactly analogous positions to the interruptions in the β Hb gene. The intervening sequences in this case are smaller; ~100 base pairs and ~150 base pairs for the small and large sequences respectively. It is thought that the vertebrate α and β Hb genes diverged from a common ancestor 500 million years ago. Thus the intervening sequences have remained during that period of time, but their sizes and sequences have diverged more than the coding sequences.

4. Jeffreys & Flavell (98) using the Southern-blot procedure (99) have shown that the large intervening sequence is present in rabbit genomic DNA regardless of the tissue from which the DNA was isolated. The sources included erythroid cells, so that whether the gene is being transcribed or not the intervening sequence is present.

OVALBUMIN   The most baroque example of the IVS phenomenon is to be found in the ovalbumin gene of chicken which has now been shown to contain a minimum of seven intervening sequences [(100–108); Figure 4].

The mRNA for ovalbumin has a 76 base leader, the 1149 base coding sequence and a 634 base untranslated 3' sequence preceding the polyA end. The entire cDNA copy, save 12 base pairs at the 5' end, has been cloned and sequenced (109).

Sorting out the situation in the genomic DNA has been a *tour de force* of the recombinant DNA technology because three of the intervening sequences contain *Eco RI* sites so that the gene is contained in four contiguous *Eco RI* fragments. All of those have now been cloned (104–107) and the analysis, including extensive DNA sequencing (110, 111), shows that at least one intervening sequence is located in the 5' leader region of the mRNA at a position 19 bases before the initiation codon (110). Six intervening sequences are found in the coding region and none are found in the large 3' untranslated region. Altogether the intervening sequences increase the genomic occupation of the 1859 base mRNA by more than a factor of three.

An interesting phenomenon has been observed in the analysis of total genome digests of chicken DNA. There is an apparently allelic variation in one of the fragments (102). The 1.8 kb *Eco RI* fragment containing two intervening sequences is sometimes substituted by a 1.3 kb fragment. This variation may represent a change in the intervening sequence which has no effect on the structure of ovalbumin and no detectable phenotype.

The study of ovalbumin gene expression in the chicken has been one of long-standing interest because the expression of the gene in the oviduct is hormonally controlled. New information on the structure of the gene should present the opportunity to study the role of intervening sequences in the control of expression of this gene.

IMMUNOGLOBULIN LIGHT CHAIN   Each immunoglobulin molecule (IgG) consists of four polypeptide chains—two identical light chains and two identical heavy chains. Normally mammals produce a large number of antibodies with potentially, perhaps, a million different specificities. It has been possible, however, to study individual antibodies through the use of plasma cell tumors (myelomas) which produce single light or single heavy chains. The light chain is composed of two distinct domains—the N-terminal variable region and the C-terminal constant region. In the mouse there are three classes of constant regions λ_1, λ_2 and κ. The variable region is so named because the differences from antibody to antibody occur here, clustered in three hypervariable regions around positions 30, 55, and 90. [For a review of this subject see (112)].

In the case of the mouse light chain it is now clear that the coding regions for the constant and variable region are some distance from each other in the germ line DNA, but are brought closer together in the committed cell. In the experiments of Hozumi & Tonegawa (113), restriction fragments of total genomic DNA from germ line or a κ-producing myeloma cell were separated by electrophoresis. Hybridization with radioactive κ mRNA indicated that the DNA sequences complementary to κ mRNA had been rearranged in the process of differentiation of the committed cell. These experiments have also been performed by Rabbitts & Forster (114) with different κ chain tumors.

More recently genomic DNA fragments containing the $V_{\lambda 1}$, $V_{\lambda 2}$, and $C_{\lambda 1}$ regions have been cloned from embryonic mouse DNA. A clone containing both $V_{\lambda 1}$ and $C_{\lambda 1}$ regions from the λ chain-producing tumor H2020 was also isolated (115–118). Several interesting features were discovered in the analysis of these clones (Figure 4).

1. In the germ line V gene the coding sequence is not intact. An intervening sequence of 93 bases is found in the coding region of the hydrophobic signal peptide four codons after the initiation codon (119).
2. Extensive comparisons of the protein sequences of light chains have led to the conclusion that the V region extends to position 112. The coding sequence in the germ line V gene, however, extends only to position 98 (119).
3. In the germ line C gene the coding sequence for the remaining 14 amino acids of the gene are found. This segment, now called the J region, is located 1250 base pairs from the C gene (118). As Brack et al (118) point out, this small J sequence has the interesting property of having a DNA recombination site in its left half and an RNA recombination site in its right half.
4. In the committed cell the recombination that generates the λ light chain gene takes place between J and V to produce a gene with two intervening sequences, one in the signal peptide and the other a sequence of 1250 base pairs separating the classical V and C regions (118).

It is clear from these results that both DNA and RNA recombination are required for the expression of an immunoglobulin gene.

VIRUSES   As noted earlier, the RNA splicing phenomena was discovered first in the DNA tumor viruses. It is now clear that the pattern of expression of the viral genes is extremely complex, with a large number of splicing events taking place. The complexity, and the rapid rate at which progress is being made in this area, preclude a comprehensive review of viral RNA processing here. (Several authors are preparing reviews and it is clear now that a lengthy one will be required to explore, e.g., the adenovirus story alone.)

It is, however, important to outline some of the major points that have been learned in this field, because the results show different and very interesting uses of RNA processing that have not so far been observed elsewhere.

Adenovirus   Adenovirus 2 (Ad 2) is one of a class of DNA viruses that causes respiratory infection in humans and can transform rat primary cell lines. The Ad 2 genome is a linear DNA molecule of 35,000 base pairs. Following infection the genome is transcribed in two general programs—the early and late phases. It was through the study of the late mRNAs that the phenomenon of splicing was discovered (120–123).

In the late phase of Ad 2 infection there is a single transcription start site located at map position *16* [(124); Figure 5]. Even though some of the late genes are many thousands of nucleotides away, each late mRNA possesses the same 200 nucleotide tripartite leader consisting of noncontiguous se-

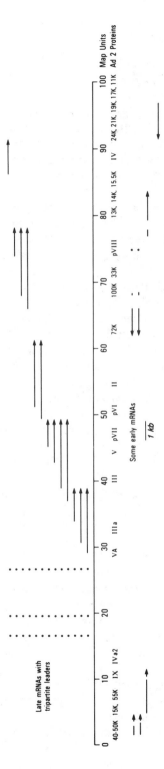

Figure 5 Transcription map of Adenovirus (Ad 2). The late transcripts are as described by Chow & Broker (125). Only the major early transcripts are shown [(139), Chow and Broker, personal communication]. The arrows represent 5' → 3' direction and the sites of poly(A) addition.

quences encoded at map positions 16, 20, and 27. Altogether there are 13 different mRNAs with this structure (120, 121, 125, 126). The precursor to these RNAs is a long RNA molecule reminiscent of heterogeneous nuclear RNA (Hn RNA) (127). The RNA is capped (128, 129) and polyadenylated (130) and in its longest form must extend from coordinate 16 to 91.5, at the far end of the genome. This precursor RNA is not found in the cytoplasm, thus the processing events forming the tripartite leader and joining it to the distant coding sequence must take place within the nucleus or on the way out.

An important clue in deciphering this surprising mode of gene expression was that all of the late mRNAs have the same capped oligonucleotide at the 5' end (131, 132). This sequence has now been located in the Ad 2 genome at coordinate 16.4 by sequencing the DNA (133). Analysis of nuclear RNA shows that no late transcripts containing sequences to the left of this point can be found. Thus it is reasonable to conclude that this must be the start site for late RNA transcription, and the region proceeding it a "promoter" for RNA polymerase II (134, 135). Several important points arise from this conclusion. First, in this case, the de novo 5' end of the transcript must be immediately capped, and this is the sequence that appears in all late mRNAs. (It is important to caution that the putative 5'-triphosphate precursor to the cap has not yet been detected.) Second, if this is in fact the start site, then it defines for the first time with any certainty the sequence of a putative RNA polymerase II promoter. It is remarkable that the sequence of this "promoter" bears a strong resemblance to analogous genomic sequences located upstream from the capped sequence of βHb mRNA from mouse (94) and rabbit (T. Maniatis, personal communication). These results hold the promise that it will be possible to define eukaryotic transcription units and thus define the pathway of mRNA synthesis.

The early Ad 2 mRNAs are transcribed from four regions of the viral genome [(136, 137); Figure 5]. Most of the early cytoplasmic mRNAs also have spliced structures [(138, 139); L. Chow and T. Broker, personal communication; also see Figure 5]. The significance of the multiplicity of mRNAs produced in each of these regions is not yet clear. In general, more mRNAs than protein products have been identified in each of the regions. Of particular interest is the early region at the left of the map. This region codes for protein(s) required for transformation. We discuss below possible similarities to this region and the analogous regions in SV40 and polyoma.

SV40 and polyoma   These small DNA viruses (members of the Papova family) are capable of transforming cells from several species but only grow lytically on other hosts. Each virus particle contains a covalently closed circular DNA molecule of approximately 5000 base pairs. Both of these

DNA molecules have been completely sequenced [(140, 141); B Griffin, personal communication; T. Friedman, personal communication], and the details of their transcription are being rapidly worked out at the sequence level.

As in the case of adenovirus, there are early and late phases of transcription, but in these smaller viruses the regions are distinct with roughly half of the genome devoted to each phase (Figure 6).

In the early phase of infection a related group of proteins, called the T (for transformation) antigens, are synthesized. In SV40, two T antigens have been detected. Large T is a 90,000 molecular weight (142) phosphorylated protein (143). It is required for DNA replication and for the initiation and maintenance of transformation. A second smaller protein— small t, molecular weight ~20,000, has peptide sequences in common with large T (144–146) and is also required for transformation (149) but not for lytic growth. In polyoma, the picture is more complex. In addition to small t and large T, there is also a middle T (molecular weight ~60,000 (147–149).

The synthesis of the mRNAs for the T antigens is a fascinating use of RNA processing for the purpose of generating more than one protein from the same DNA sequence.

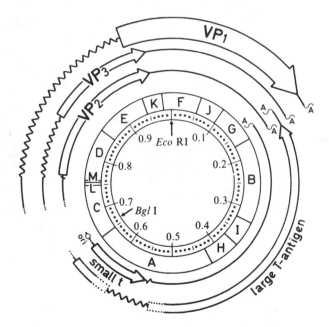

Figure 6   Transcription map of SV40 [taken from Fiers et al (141)]. Only the major transcripts are shown.

In SV40, large T is synthesized from a 19S mRNA. Small t is synthesized by a slightly larger mRNA (144, 145). Both mRNAs are spliced (150). Both mRNAs have the same 5' end, but a larger segment (comprising sequences between .54 and .60) is removed from the large T mRNA (see Figure 5). The DNA sequence of the SV40 DNA (and of polyoma DNA) reveals that there are translation stop codons in all three reading frames in the region .54–.60. Termination of translation in this region results in the synthesis of the smaller t antigen. They are removed by splicing in the large T mRNA which enables the synthesis of a much larger product having the same sequence at the amino terminus.

The RNA splicing pattern observed in the large T mRNA explained an earlier puzzling observation (151). Deletion mutants of SV40 in the region .55–.59 produce a normal large T. The observation was paradoxical because the region from .59–.17 (the end of the early region) is too small to code for the large T polypeptide. The paradox is explained by the splicing of the mRNA—apparently the splicing takes place normally even when much of the intervening sequence is deleted. As one would now predict, the .55–.59 deletion mutants do not produce small t. This class of mutants were first observed in polyoma (152). They were isolated as host-dependent mutants that have lost the ability to transform permissive hosts. These mutants also fail to produce the polyoma small t (148).

The role of these proteins in generating the oncogenic response is not clear. The phenomenon, however, may be one that extends beyond the papova viruses. The pattern of splicing of the early Ad 2 mRNAs in the 1.5–11.5 region, which codes for the transformation functions (Figure 5), is remarkably similar to that seen in the early region of SV40 and polyoma (139). Could there be a family of T antigens in Ad 2 as well?

In SV40, there are three proteins synthesized late in infection (reviewed in 140, 141), the major capsid protein VP 1 and two minor proteins, VP 2 and VP 3. VP 1 is coded for by a region between coordinates 0.95 and .16, but the VP 1 mRNA contains a 203 nucleotide leader mRNA that is coded for ~1000 nucleotides upstream at coordinate .72 (153–158). The coding regions for VP 2 and VP 3 overlap with each other and partially overlap with the VP 1 gene. The mRNAs for VP 2 and VP 3 are also spliced and contain at least some of the leader sequences in common with VP 1 mRNA (153–158). The function of splicing in the case of these mRNAs is not yet clear, but current models postulate that the different leader sequences regulate the level of mRNA utilization.

Retroviruses The avian sarcoma virus (ASV) genome contains genes for four functions. These map on the genome in the order 5'-*gag* (structural proteins of the virus core), *pol* (reverse transcriptase), *env* (the glycoprotein of the viral envelope), and *src* (a protein responsible for transformation). The 38S genomic RNA serves as mRNA for *gag* and *pol*, a 28S mRNA produces *env*, and a 21S mRNA encodes *src*. Several laboratories have shown that the 28S and 21S mRNAs contain sequences identical to a segment at the 5' end of the 38S mRNA (159–161). These sequences are unique to the mRNAs and are not found reiterated in the 38S genomic RNA. The leader sequences are at least 104 nucleotides in length (161) and are presumably produced by RNA splicing. Since the substrate for this splicing must be the intact 38S RNA, this virus system offers a readily available substrate for the study of RNA splicing in vitro. An analogous leader for the murine leukemia virus, 21S *env* mRNA, has also been observed (162, 163).

RIBOSOMAL RNA   Intervening sequences have also been discovered in rRNA genes. An intervening sequence of variable length is found in approximately two thirds of the 28S rRNA genes of *Drosophila*, located mainly in the X chromosome (164–167). It appears, however, that the genes that contain intervening sequences are not transcribed, (I. Dawid, personal communication).

Yeast (*Saccharomyces cerevisiae*) mitochondrial DNA contains a single gene for the 21S rRNA gene (the large RNA subunit in mitochondrial ribosomes). This gene contains an intervening sequence of approximately 1000 base pairs (168). It is interesting to note that the position of this intervening sequence coincides with a genetic locus called "omega" which affects the reciprocity of genetic recombination in mitochondria (169). In omega⁻ strains, the region corresponding to the intervening sequence is deleted (170). (It is not known, however, whether this deletion removes the intervening sequence exactly.) A similar intervening sequence has been discovered in the two genes for the 23S rRNA in the chloroplast DNA of *Chlamydomonas reinhardii* (171). In this case, the intervening sequence is also about 1000 base pairs and interrupts the coding sequence at a site 270 base pairs from the 5' end of the rRNA.

The discovery of intervening sequence in organelle DNA raises a number of fascinating questions. Do the organelles produce their own RNA splicing enzyme or do they use the machinery of the host? What significance does this development have for the long-held notion that organelles have a prokaryotic origin?

tRNA   The first investigations of eukaryotic tRNA genes revealed that they too contain intervening sequences. The yeast tRNA^{Tyr} genes contain an interruption of 14 base pairs just following the anticodon (172). In yeast there are eight tRNA^{Tyr} genes. Three of the genes have been sequenced and the intervening sequences are identical save a one-base pair substitution. It is likely that all eight genes contain intervening sequences. The yeast tRNA^{Phe} genes contain an intervening sequence of 18 or 19 base pairs in an equivalent position just following the anticodon (173). Again the intervening sequences are nearly identical in the three tRNA^{Phe} genes that have been sequenced, but show no relationship to the tyrosine sequences.

These results suggested that perhaps all tRNA genes in eukaryotes contain intervening sequences. This turned out not to be the case. So far, the inventory in yeast shows that tRNA^{Trp}, tRNA$_3$^{Leu} (S. Kang, R. Ogden, and J. Abelson, unpublished), and tRNA$_{UCG}$^{Ser} (C. Guthrie, personal communication) genes *do* contain intervening sequences. tRNA$_1$^{Ser} (G. Page and B. Hall, personal communication), tRNA^{Asp}, and tRNA$_3$^{Arg} genes (H. Sakano, J. Beckmann, and J. Abelson, unpublished) *do not* contain intervening sequences. It is interesting that the latter two genes are sometimes found clustered in the yeast genome (174). Sequence analysis of one of these clones showed that the genes are separated by nine base pairs which indicates that these genes are cotranscribed.

Mechanisms of RNA Splicing

The presence of intervening sequences in eukaryotic genes dictates an extraordinary mechanism for the expression of the gene. Of the possible mechanisms, RNA splicing seemed the most likely, but to detect the enzyme(s) responsible for the specific cleavage and ligation, one needed a substrate. The yeast tRNA system provided a substrate to study this reaction.

A temperature-sensitive mutant of yeast (ts 136) thought to be defective in transport of RNA from nucleus to cytoplasm (175) accumulates 4.5S tRNA precursors at the nonpermissive temperature (176). An analysis of these precursors revealed that they are a restricted set of the possible tRNA precursors: precursors of tRNA^{Phe}, tRNA^{Tyr}, tRNA$_3$^{Leu}, tRNA$_{UCG}$^{Ser}, tRNA^{Trp}, and three other unidentified tRNA precursors accumulate (177). Other tRNA precursors are notably absent (although other evidence suggests that most yeast tRNAs are synthesized via 4.5S precursors (178). Sequence analysis of the pre-tRNA^{Tyr} and pre-tRNA^{Phe} revealed that they contain the intervening sequences predicted by the DNA sequences of the genes (177). Further, these precursors are partially matured. They contain the mature 5' and 3' CCA ends and some, but not all, of the modified nucleosides.

The tRNA precursors have been used as substrates to search for an RNA splicing activity in yeast extracts and this activity can readily be detected in yeast ribosomal wash fractions (177). In the in vitro reaction, the intervening sequence is accurately excised and the polynucleotide chains covalently rejoined converting the precursor to the mature tRNA. The activity requires magnesium, monovalent cation, and ATP (177, 179).

Preliminary experiments indicate that the nuclease and ligase activities can act independently. In the absence of ATP or in the presence of mature tRNA, the intervening sequence is excised but the ends are not resealed by the ligase activity (C. Peebles, R. Ogden, G. Knapp, and J. Abelson, unpublished). Thus, it appears that each half of the reaction can proceed independently. Purification of the enzymes is clearly needed and, with the present assay, it should be possible to accomplish this fairly soon.

It is proving more difficult to detect RNA splicing activities in higher organisms, but there are two indications that they exist (as certainly they must).

1. Blanchard et al (180) have detected in vitro splicing of the early mRNA for the 72K protein in adenovirus (see Figure 4). In their experiments the precursor to 72K mRNA was shown to be present in nuclei isolated from pulse-labeled Ad 2-infected cells. Treatment of these nuclei with cytoplasmic extract resulted in conversion of precursor to mature spliced mRNA.
2. In collaboration with Söll's group we (R. Ogden, J. Beckmann, J. Abelson, D. Söll, and O. Schmidt) have been studying the expression of yeast tRNA genes in an in vitro system prepared from the germinal vesicles of *Xenopus laevis* oocytes (181). We have shown that the cloned yeast tRNA^{Trp} gene which contains an intervening sequence is expressed in this system. We can detect both precursor and mature tRNA, which has been correctly spliced. Thus *Xenopus* must have an RNA splicing system.

Neither of these systems is very promising as an assay for enzyme purification. It is hoped that further characterization of yeast RNA splicing enzyme may reveal special activities of this class of enzyme (e.g. RNA ligase with synthetic substrates) which will make it possible to purify the enzyme without the necessity of using substrates which are difficult or impossible to obtain.

How many RNA splicing enzymes are there? This is a crucial question, and, at the moment, we can only guess. The RNA processing reactions in *E. coli* are probably accomplished by around ten enzymes, each of which has multiple functions. Francis Crick estimates (182) that there are more than ten but less than one hundred RNA splicing enzymes, which seems a good guess.

Without knowing how many enzymes there are, it is difficult to speculate about substrate recognition, but the availability of sequence information at splice points does place some limits on what the recognition could be. Table

1 is a compendium of known sequences at the junctions of a number of splice points. Two features are notable in this Table. First, in most cases, the splice occurs between repeated sequences so that it is not possible to decide exactly where the splice has occurred. (Eventually it may be possible to decide by sequencing of the spliced-out product.) Second, a rule (110) (called Chambon's rule because he first noticed it) seems to emerge, that all splice points *could* have a common sequence at the extremes of the intervening sequence such that these sequences are always-(GU ..... AG)-. This rule does not apply, however, to the splice points in the tRNA genes, which implies a different recognition mechanism for those substrates. Sequence

Table 1 Sequences surrounding the splice points in split genes

Ovalbumin		Intervening Sequence		Reference
A A A T A A G|G T G A G C C	IVS 2	A T T A C A G|G T T G T T		
A G C T C A G|G T A C A G A	IVS 3	T A T T C A G|T G T G G C		
C C T G C C A|G T A A G T T	IVS 4	T T T A C A G|G A A T A C		110, 111
A G A A A T G|G T A A G G T	IVS 5	C T T A A A G|G A A T T A		
G A C T G A G|G T A T A T G	IVS 6	G C T C C A G|C A A G A A		
T G A G C A G|G T A T G G C	IVS 7	C T T G C A G|C T T G A G		
SV40				
Late 19S T C A G A A G|G T A C C T A		T T T C C A G|G T C C A T		
Late 19S C G T T A A G|G T T C G T A		T T T C C A G|G T C C A T		
Late 19S T T A A C T G|G T A A G T T		T T T C C A G|G T C C A T		S. Weissman,
Late 16S T T A A C T G|G T A A G T T		C T T C T A G|G C C T G T		personal communication
Early T A A C T G A G|G T A T T T G		A T T T T A G|A T T C C A		
Early t C T A T A A G|G T A A A T G		A T T T T A G|A T T C C A		
Hemoglobin				
Mouse β^maj T G G G C A G|G T T G G T A	IVS I (Small)	T T T T T A G|G C T G C T	94, 95	
Mouse β^maj C T T C A G G|G T G A G T C	IVS 2	C C C A C A G|C T C C T G	94	
Rabbit β T G G G C A G|G T T G G T A	IVS I	T T C T C A G|G C T G C T	95	
Rabbit β C T T C A G G|G T G A G T T	IVS 2	C C T A C A G|T C T C C T	C. Weissman, personal communication	
Immunoglobin				
Mouse λI A G C T C A G|G T C A G C A	IVS I	T T T G C A G|G G G C C A	S. Tonegawa, personal communication	
Mouse λI G T C C T A G|G T G A G T G	IVS 2	C C T G C G G|C C A G C		
Mouse λ2 T G C T C A G|G T C A G C A	IVS I	T T T G C A G|G A G C C A	119	
Yeast tRNAs				
tRNA^Tyr C U G U A A|U U U A ᵁ_C		A C G A A|A U C U U	172	
tRNA^Phe C U G A A G|A A A ᵁ_A A		A A G U U|A U C U G	173	
tRNA^Ser_UCG C U C G A A|U G G A A		U C G C U|A U C U C	C. Guthrie, personal communication	
tRNA^Trp C U C C A A|U U A A A		U U G C A|A U C G A	Kang, Ogden, Abelson unpublished	
‾‾‾‾ anti- codon				

correlations can also be found in regions adjoining the splice point. Comparison of the various α and β Hb genes suggest that the sequences adjoining splice points have been most conserved in evolution. In the β Hb genes, for example, both splice points take place at sequences in the RNA coding for Arg, Leu, Leu, or Lys, Leu, Leu. By contrast, as pointed out earlier, there is almost no conservation of sequence in the inserts.

Provocative as the sequence homologies are, they obviously do not give sufficient specificity to provide the exact pattern of splicing events that must take place. We have seen that the specificity of RNA processing in prokaryotes is entirely determined by some feature of RNA tertiary structure. It would be surprising if tertiary structure did not also play a role in substrate recognition in RNA splicing, and the limited sequences homologies would seem to dictate an important role for structure. Certainly the preliminary evidence for the recognition of yeast tRNA precursors would force us to invoke tertiary structure. The problem with this is that we do not know enough about RNA structure to be able to predict tertiary from secondary structure. It cannot be doubted, however, that RNA molecules do have definite structures. The sequence of the *E. coli* rRNA, Figure 2, demonstrates that the 30S precursor reliably folds so that processing sites are formed by the conjunction of sequences as far away from each other as 3000 nucleotides in the linear sequence. Unfortuanately, examination of a number of precursor sequences have not revealed obvious secondary structures, e.g. hairpins, that would guarantee conjunction of splice points. There are some hints, however, that would suggest that the important sequence determinants are in the coding sequences, the products, and not in the intervening sequences:

1. In the Hb genes there is a striking divergence of sequence in the intervening sequences while sequences in flanking regions to splice points are conserved.
2. In SV40 most of the intervening sequence removed in formation of the large T mRNA can be genetically deleted without affecting the specificity of splicing.
3. Deletions in the coding sequence of SV40 affect splicing when they are near splice points. No stable mRNA is produced when splice points are deleted (182a).

For some time, it has been known that the nuclei of higher eukaryotic cells contain a disperse collection of large RNA molecules (Hn RNA), [see the review by Darnell (183)]. Most of the RNA sequences do not reach the cytoplasm. It has long been believed that Hn RNA is precursor to mRNA. With the present capability to obtain pure hybridization probes, and with improved methods for the separation and detection of RNA precursors

(184), it has been possible to study the pathway of synthesis of several mRNAs. The results of these studies support the idea that mRNA is synthesized via larger precursors. It is clear also that RNA splicing plays an important role in the maturation of these precursors.

The β Hb mRNA of mouse is synthesized via a 1.5 kb 15S precursor (185–188). The precursor is capped at its 5' end (189) and has a 150 nucleotide poly(A) tract (188–190) at its 3' end (191). The sequences at the end are the same as in the mRNA (192). It has been demonstrated that the 15S precursor has two intervening sequences (191, 193) and that these can hybridize to the intervening sequences in the cloned genomic DNA (192). Thus all of the extra sequences in this precursor are internal. The existence of this precursor would indicate that splicing is the last of the RNA processing steps in the synthesis of the Hb mRNA. A precursor to mouse α Hb mRNA has also been observed (185, 194) and it is smaller (850 nucleotides) which presumably reflects the fact that the intervening sequences in the mouse α Hb gene are smaller than those in the β Hb gene.

The mRNA for the κ (MOPC 21) light chain of mouse immunoglobulin is synthesized via a 10 kb, 40S precursor (195). In this case, the precursor must contain a very long segment prior to the coding sequence (196). Preliminary evidence suggests that this region is spliced out (R. Wall, personal communication). Two further splicing reactions, the first joining the N-terminal leader to the V region, and the second joining the J region to the C region, must be made. A 4.6 kb intermediate was observed in which all but the last of these have been made (R. Wall, personal communication).

Precursors to the ovalbumin mRNA have also been detected (197). The longest of these is 7.8 kb and is long enough to contain the entire transcript. Six other smaller precursors were detected. This pretty result could be interpreted to mean that the six intervening sequences are removed one at a time and each intermediate was detected. Unfortunately one cannot yet draw that conclusion (and Roop et al do not), but it is shown that the intervening sequences nearer the 5' end disappear first.

The average size of these three precursors is three to four times the size of their mature products. This agrees with the estimates for the difference in size between Hn RNA and cytoplasmic mRNA (183). Thus it is possible that the extra sequences in Hn RNA are internal.

On the basis of these studies, a cohesive picture of mRNA synthesis in eukaryotes is beginning to emerge. RNA polymerase II initiates transcription at a defined site (cf the adenovirus late mRNA transcription unit). Capping occurs very shortly after initiation of transcription of nascent RNA (133). From the very limited data available, it may be that the de novo 5' end of the mRNA is capped directly (133, 197a). In any case, most nuclear caps are destined for transport to the cytoplasm as mRNA, so capping does not occur indiscriminately (198). Transcription is continued until the chain is released.

Chain release could take place at a unique site or at any of a number of sites, as in the late mRNA transcription unit in Ad 2. The mechanism of chain release is unclear at this time. It could be by transcription termination or the 3' ends could be generated by endonucleolytic cleavage as in the case of the bacteriophage T7 mRNAs. (In Ad 2 late transcription, it appears that transcription can continue beyond a polyA addition site, but this distal RNA probably does not reach the cytoplasm [(198a), M. Gefter, personal communication]. PolyA is polymerized onto the 3' end of the mRNA. RNA splicing occurs, probably as a terminal step before transport. This picture is undoubtedly too simplistic to be universally true. We already know that there are transcription units in which there are no intervening sequences and/or no polyA addition (it happens that both are true of the histone mRNAs).

Absent from this picture is any knowledge of the process of RNA transport. It is reported that RNA is transported as a protein bound "particle" [reviewed by Spirin, (199)]. When do these proteins become associated with the RNA? Is the actual substrate for processing an RNA protein complex? More work is needed on this aspect of the process.

RNA Processing Involving Endonucleases

In focussing on the role of RNA splicing in the maturation of eukaryotic RNA molecules, I do not wish to give the impression that endonucleases are not involved. (But as H. Robertson has pointed out to me, it may be artificial to distinguish between the cleavages made during splicing and those made in the absence of splices. The same enzymes may be involved.) It is clear that endonucleases must be involved in the maturation of tRNA, rRNA, and probably some mRNAs.

A ribonuclease III–like activity has been detected in mammalian and avian cells (200–205). It is possible that this enzyme could be involved in the processing of rRNA. The pathway of rRNA maturation has been worked out (see 1). Most eukaryotic rRNA transcription units do not contain intervening sequences. A number of endonucleolytic cleavage steps are therefore required and the RNase III–like activities may be one of them.

A ribonuclease P–like activity has been partially purified from human KB cells (206).

The clear involvement of an endonuclease in the processing of mRNA from eukaryotes has not been demonstrated. It is not yet known whether caps or poly(A) can be added to mRNA following endonuclease action. As more transcription units are characterized this question should be answered.

EPILOGUE

We are led to the conclusion that RNA processing is a major mechanism in the expression of the eukaryotic gene. It is such a reliable mechanism that the cell can happily endure seven intervening sequences in the ovalbumin gene and remove all of them accurately to the base pair.

The obvious question is, how did it come about? Wouldn't it be easier just to generate transcripts that would function de novo. The answer to this question may lie in the distant past and fascinating as this subject is I do not intend to discuss the evolutionary implications of intervening sequences. The reader is referred to the imaginative review by F. Crick (182). A related question is, what is it good for? We have seen that RNA splicing is used for diverse purposes. SV40 uses RNA splicing for gene compaction—to generate two proteins from the same DNA sequence. The function of RNA splicing in tRNA synthesis is unclear, but it is obviously different from the function in SV40. The RNA splicing mechanism must have a number of regulatory roles. It is tempting to speculate that one of these is the control of gene expression. The expression of the gene could be regulated at a number of points and RNA processing is one of them.

There is a tantalizing result that suggests that RNA splicing could play a role in differentiation. Segal and Khoury (personal communication) have been investigating the growth of SV40 in the F9 cell line derived from a mouse terratocarcinoma. This cell line resembles totipotent stem cells in that it can differentiate to form a number of different cell types. SV40 can productively infect differentiated cells, but *not* the F9 cells. Segal and Khoury have shown that SV40 DNA reaches the nucleus and is transcribed in SV40-infected F9 cells, but the RNA is not spliced and thus is nonfunctional. This result suggests that the ability to splice SV40 mRNA appears in differentiated cells, but is absent in stem cells.

This field is obviously in a state of rapid flux and it will be fascinating to see where it goes. I believe that a number of other uses of RNA processing will be uncovered. It is probable that the mechanisms of RNA splicing will be worked out in the near future. We can hope that some understanding of the origins of intervening sequences will finally become apparent.

ACKNOWLEDGMENTS

It is a pleasure to acknowledge valuable conversations with Christine Guthrie, Francis Crick, Peter Geiduschek, and Philip Leder, which stimulated my thinking about RNA processing. Thanks to Francis Crick, Peter Geiduschek, Robert Perry, Elsebet Lund and Babette Coté for critical readings of the manuscript. I thank the many investigators who made their results available prior to publication. My research during this period has been supported by a grant from the National Institutes of Health, CA-10984.

Literature Cited

1. Perry, R. P. 1976. *Ann. Rev. Biochem.* 45:605–29
2. Maxam, A. M., Gilbert, W. 1977. *Proc. Natl. Acad. Sci. USA* 74:560–64
3. Sanger, F., Nicklen, S., Coulson, A. R. 1977. *Proc. Natl. Acad. Sci. USA* 74:5463–67
4. Gilbert, W. 1978. *Nature* 271:501
5. McCloskey, J. A., Nishimura, S. 1977. *Acc. Chem. Res.* 10:403–10
6. Smith, J. D. 1976. *Prog. Nucl. Acid Res. Mol. Biol.* 16:25–73
7. Shatkin, A. J. 1976. *Cell* 9:645–53
8. Goodman, H. M., Abelson, J. N., Landy, A., Zadrazil, S., Smith, J. D. 1970. *Eur. J. Biochem.* 13:461–83
9. Altman, S., Smith, J. D. 1971. *Nature New Biol.* 233:35–39
10. Sekiya, T., Khorana, H. G. 1974. *Proc. Natl. Acad. Sci. USA* 71:2978–82
11. Sekiya, T., Contreras, R., Küpper, H., Landy, A., Khorana, H. G. 1976. *J. Biol. Chem.* 251:5124–40
12. Egan, J., Landy, A. 1978. *J. Biol. Chem.* 253:3607–22
13. Küpper, H., Sekiya, T., Rosenberg, M., Egan, J., Landy, A. 1978. *Nature* 272:423–28
14. Sekiya, T., Takeya, T., Brown, E. L., Belagaje, R., Contreras, R., Fritz, H. J., Gait, M. J., Lees, R. G., Norris, K., Ryan, M. J., Khorana, H. G. 1979. *J. Biol. Chem.* 254: In press
15. Ryan, M. J., Belagaje, R., Brown, E. L., Fritz, H. J., Fritz, R. H., Gait, M. J., Küpper, H., Lees, R. G., Norris, K., Sekiya, T., Takeya, T. 1977. *Fed. Proc.* 36:732
16. Russell, R. L., Abelson, J. N., Landy, A., Gefter, M. L., Brenner, S., Smith, J. D. 1970. *J. Mol. Biol.* 47:1–13
17. Robertson, H. D., Altman, S., Smith, J. D. 1972. *J. Biol. Chem.* 247:5243–51
18. Cashmore, T. 1971. *Nature New Biol.* 230:236–39
19. Leon, V., Altman, S., Crothers, D. M. 1977. *J. Mol. Biol.* 113:253–65
20. Stark, B., Kole, R., Bowman, E., Altman, S. 1978. *Proc. Natl. Acad. Sci. USA* 75:3717–21
21. Bikoff, E. K., Gefter, M. L. 1975. *J. Biol. Chem.* 250:6240–47
22. Bikoff, E. K., LaRue, B. F., Gefter, M. L. 1975. *J. Biol. Chem.* 250: 6248–55
23. Sekiya, T., Contreras, R., Takeya, T., Khorana, H. G. 1979. *J. Biol. Chem.* 254: In press
24. Shimura, Y., Sakano, H., Nagawa, F. 1978. *Eur. J. Biochem.* 86:267–81
25. Schedl, P., Roberts, J., Primakoff, P. 1976. *Cell* 8:581–94
26. Ghosh, R. K., Deutscher, M. P. 1978. *Nucleic Acids Res.* 5:3831–42
27. Ghosh, R. K., Deutscher, M. P. 1978. *J. Biol. Chem.* 251:6646–52
28. Schedl, P., Primakoff, P. 1973. *Proc. Natl. Acad. Sci. USA* 70:2091–95
29. Ozeki, H., Sakano, H., Yamada, S., Ikemura, T., Shimura, Y. 1975. *Brookhaven Symp. Biol.* 26:89–105
30. Sakano, H., Yamada, S., Ikemura, T., Shimura, Y., Ozeki, H. 1974. *Nucleic Acids Res.* 1:355–71
31. Schedl, P., Primakoff, P., Roberts, J. 1975. *Brookhaven Symp. Biol.* 26:53–76
32. Sakano, H., Shimura, Y. 1975. *Proc. Natl. Acad. Sci. USA* 72:3369–73
33. Ilgen, C., Kirk, L. L., Carbon, J. 1976. *J. Biol. Chem.* 251:922–29
34. Ikemura, T., Shimura, Y., Sakano, H., Ozeki, H. 1975. *J. Mol. Biol.* 96:69–86
35. Hsu, W. T., Foft, J. W., Weiss, S. B. 1967. *Proc. Natl. Acad. Sci. USA* 58:2028–35
36. Daniel, V., Littauer, U. Z. 1970. *Science* 167:1682–88
37. Abelson, J., Fukada, K., Johnson, P., Lamfrom, H., Nierlich, D. P., Otsuka, A., Paddock, G. V., Pinkerton, T. C., Sarabhai, A., Stahl, S., Wilson, J. H., Yesian, H. 1975. *Brookhaven Symp. Biol.* 26:77–88
38. Guthrie, C., Seidman, J. G., Comer, M. M., Bock, R. M., Schmidt, F. J., Barrell, B. G., McClain, W. H. 1975. *Brookhaven Symp. Biol.* 26:106–23
39. McClain, W. H., Seidman, J. G. 1975. *Nature* 257:106–10
40. Guthrie, C. 1975. *J. Mol. Biol.* 95: 529–47
41. McClain, W. H. 1977. *Acc. Chem. Res.* 10:418–25
42. Maisurian, A. N., Buyanovskaya, E. A. 1973. *Mol. Gen. Genet.* 120:227–29
43. Seidman, J. G., Schmidt, F. J., Foss, K., McClain, W. H. 1975. *Cell* 5:389–400
44. Deutscher, M. P., Hilderman, R. H. 1974. *J. Bacteriol.* 118:621–27
45. Deutscher, M. P., Foulds, J., McClain, W. H. 1974. *J. Biol. Chem.* 249: 6696–99
46. Fukada, K., Gossens, L., Abelson, J. 1979. *J. Mol. Biol.* Manuscript submitted
47. Kaplan, D. A., Nierlich, D. P. 1975. *J. Biol. Chem.* 250:934–38
48. Goldfarb, A., Seaman, E., Daniel, V. 1978. *Nature* 273:562–64
49. Nomura, M., Morgan, E. A., Jaskunas, S. R. 1977. *Ann. Rev. Genet.* 11:297–347

50. Lund, E., Dahlberg, J. E., Lindahl, L., Jaskunas, S. R., Dennis, P. P., Nomura, M. 1976. *Cell* 7:165–77

51. Lund, E., Dahlberg, J. E. 1977. *Cell* 11:247–62

52. Ikemura, T., Nomura, M. 1977. *Cell* 11:247–62

53. Morgan, E. A., Ikemura, T., Lindahl, L., Fallon, A. M., Nomura, M. 1978. *Cell* 13:335–44

54. Brosius, J., Palmer, M. L., Kennedy, P. J., Noller, H. F. 1978. *Proc. Natl. Acad. Sci. USA* 75:4801–5

55. Young, R. A., Macklis, R., Steitz, J. A. 1979. *J. Biol. Chem.* In press

56. Young, R. A., Steitz, J. A. 1978. *Proc. Natl. Acad. Sci. USA* 75:3593–97

57. Dunn, J. J., Studier, F. W. 1973. *Proc. Natl. Acad. Sci. USA* 70:3296–3300

58. Nikolaev, N., Silengo, L., Schlessinger, D. 1973. *Proc. Natl. Acad. Sci. USA* 70:3361–65

59. Ginsburg, D., Steitz, J. A. 1975. *J. Biol. Chem.* 250:5647–54

60. Hayes, F., Vasseur, M., Nikolaev, N., Schlessinger, D., Sriwidada, J., Krol, A., Branlant, C. 1975. *FEBS Lett.* 56:85–91

61. Lund, E., Dahlberg, J. E., Guthrie, C. 1979. *tRNA* eds. Abelson, J., Söll, D., Schimmel, P. Cold Spring Harbor, New York: Cold Spring Harbor Lab. In press

62. Sogin, M., Pace, B., Pace, N. R., Woese, C. R. 1971. *Nature New Biol.* 232:48–49

63. Brownlee, G. G., Cartwright, E. 1971. *Nature New Biol.* 232:50–52

64. Hayes, F., Hayes, D., Fellner, P., Ehresmann, C. 1971. *Nature New Biol.* 232:54–55

65. Lowry, C. V., Dahlberg, J. E. 1971. *Nature New Biol.* 232:52–54

66. Robertson, H. D., Barany, F. 1979. In *Proc. FEBS Congr.* 12th, Oxford New York Frankfurt: Pergamon. In press

67. Wu, M., Davidson, N. 1975. *Proc. Natl. Acad. Sci. USA* 72:4506–10

68. Dahlberg, A. E., Tokimatsu, H., Zahalak, M., Reynolds, F., Calvert, P. C., Rabson, A. B., Lund, E., Dahlberg, J. E. 1978. *Proc. Natl. Acad. Sci. USA* 75:3598–3602

69. Hayes, F., Vasseur, M. 1976. *Eur. J. Biochem.* 61:433–42

70. Young, R. A., Bram, R. J., Steitz, J. A. 1979. See Ref. 61, pp.

71. Gegenheimer, P., Apirion, D. 1978. *Cell* 15:527–39

72. Apirion, D. 1979. *Genetics.* In press

73. Apirion, D., Lassar, A. B. 1978. *J. Biol. Chem.* 253:1737–42

74. Jordan, B. R., Forget, B. G., Monier, R. 1971. *J. Mol. Biol.* 55:407–21

75. Ghora, B. K., Apirion, D. 1979. *Cell.* In press

76. Pace, N. R., Pato, M. L., McKibbin, J., Radcliffe, C. W. 1973. *J. Mol. Biol.* 75:619–31

77. Sogin, M. L., Pace, N. R., Rosenberg, M., Weissman, S. M. 1976. *J. Biol. Chem.* 251:3480–88

78. Sogin, M. L., Pace, B., Pace, N. R. 1977. *J. Biol. Chem.* 252:1350–57

79. Meyhack, B., Pace, B., Uhlenbeck, O. C., Pace, N. R. 1978. *Proc. Natl. Acad. Sci. USA* 75:3045–49

80. Meyhack, B., Pace, B., Pace, N. R. 1977. *Biochemistry* 16:5009–15

81. Dunn, J. J., Studier, F. W. 1975. *Brookhaven Symp. Biol.* 26:267–76

82. Robertson, H. D., Webster, R. E., Zinder, N. D. 1968. *J. Biol. Chem.* 243:82–91

83. Crouch, R. J. 1974. *J. Biol. Chem.* 249:1314–16

84. Robertson, H. D., Dunn, J. J. 1975. *J. Biol. Chem.* 250:3050–56

85. Dunn, J. J. 1976. *J. Biol. Chem.* 251:3807–14

86. Robertson, H. D., Dickson, E., Dunn, J. J. 1977. *Proc. Natl. Acad. Sci. USA* 74:822–26

87. Rosenberg, M., Kramer, R. A. 1977. *Proc. Natl. Acad. Sci. USA* 74:984–88

88. Oakley, J. L., Coleman, J. E. 1977. *Proc. Natl. Acad. Sci. USA* 74:4266–70

89. Paddock, G., Abelson, J. 1973. *Nature New Biol.* 246:2–6

90. Popp, R. A., Bailiff, E. G. 1973. *Biochem. Biophys. Acta* 303:52–60

91. Tilghman, S. M., Tiemeier, D. C., Polsky, F., Edgell, M. H., Seidman, J. G., Leder, A., Enquist, L. W., Norman, B., Leder, P. 1977. *Proc. Natl. Acad. Sci. USA* 74:4406–10

92. Tiemeier, D. C., Tilghman, S. M., Polsky, F. I., Seidman, J. G., Leder, A., Edgell, M. H., Leder, P. 1978. *Cell* 14:237–45

93. Tilghman, S. M., Tiemeier, D. C., Seidman, J. G., Peterlin, B. M., Sullivan, M., Maizel, J. V., Leder, P. 1978. *Proc. Natl. Acad. Sci. USA* 75:725–29

94. Konkel, D. A., Tilghman, S. M., Leder, P. 1978. *Cell.* In press

95. van den Berg, J., van Ooyen, A., Mantei, N., Schamböck, A., Grosveld, G., Flavell, R. A., Weissmann, C. 1978. *Nature* 276:37–44

96. Lawn, R. M., Fritsch, E. F., Parker, R. C., Blake, G., Maniatis, T. 1979. *Cell.* In press

97. Leder, A., Miller, H., Hamer, D., Seidman, J. G., Norman, B., Sullivan, M.,

Leder, P. 1978. *Proc. Natl. Acad. Sci. USA* 75:6187–191
98. Jeffreys, A. J., Flavell, R. A. 1977. *Cell* 12:1097–1108
99. Southern, E. M. 1975. *J. Mol. Biol.* 98:503–17
100. Breathnach, R., Mandel, J. L., Chambon, P. 1977. *Nature* 270:314–19
101. Doel, M. T., Houghton, M., Cook, E. A., Carey, N. H. 1977. *Nucleic Acids Res.* 4:3701–13
102. Weinstock, R., Sweet, R., Weiss, M., Cedar, H., Axel, R. 1978. *Proc. Natl. Acad. Sci. USA* 75:1299–1303
103. Lai, E. C., Woo, S. L. C., Dugaiczyk, A., Catterall, J. F., O'Malley, B. W. 1978. *Proc. Natl. Acad. Sci. USA* 75:2205–9
104. Garapin, A. C., Lepennec, J. P., Roskam, W., Perrin, F., Cami, B., Krust, A., Breathnach, R., Chambon, P., Kourilsky, P. 1978. *Nature* 273:349–54
105. Woo, S. L. C., Dugaiczyk, A., Tsai, M. J., Lai, E. C., Catterall, J. F., O'Malley, B. W. 1978. *Proc. Natl. Acad. Sci. USA* 75:3688–92
106. Garapin, A. C., Cami, B., Roskam, W., Kourilsky, P., Lepennec, J. P., Perrin, F., Gerlinger, P., Cochet, M., Chambon, P. 1978. *Cell* 14:629–39
107. Dugaiczyk, A., Woo, S. L. C., Lai, E. C., Mace, M. L., McReynolds, L., O'Malley, B. W. 1978. *Nature* 274:328–33
108. Mandel, J. L., Breathnach, R., Gerlinger, P., LeMeur, M., Gannon, F., Chambon, P. 1978. *Cell* 14:641–53
109. McReynolds, L., O'Malley, B. W., Nisbet, A. D., Fothergill, J. E., Givol, D., Fields, S., Robertson, M., Brownlee, G. G. 1978. *Nature* 273:723–28
110. Breathnach, R., Benoist, C., O'Hare, K., Gannon, F., Chambon, P. 1978. *Proc. Natl. Acad. Sci. USA* 75:4853–57
111. Catterall, J. F., O'Malley, B. W., Robertson, M. A., Staden, R., Tanaka, Y., Brownlee, G. G. 1978. *Nature* 275:510–13
112. Williamson, A. R. 1976. *Ann. Rev. Biochem.* 45:467–500
113. Hozumi, N., Tonegawa, S. 1976. *Proc. Natl. Acad. Sci. USA* 73:3628–32
114. Rabbitts, T. H., Forster, A. 1978. *Cell* 13:319–27
115. Brack, C., Tonegawa, S. 1977. *Proc. Natl. Acad. Sci. USA* 74:5652–56
116. Tonegawa, S., Brack, C., Hozumi, N., Schuller, R. 1977. *Proc. Natl. Acad. Sci. USA* 74:3518–22
117. Hozumi, N., Brack, C., Pirrotta, V., Lenhard-Schuller, R., Tonegawa, S. 1978. *Nucleic Acids Res.* 5:1779–99
118. Brack, C., Hirama, M., Lenhard-Schuller, R., Tonegawa, S. 1978. *Cell* 15:1–14
119. Tonegawa, S., Maxam, A. M., Tizard, R., Bernard, O., Gilbert, W. 1978. *Proc. Natl. Acad. Sci. USA* 75:1485–89
120. Berget, S. M., Moore, C., Sharp, P. A. 1977. *Proc. Natl. Acad. Sci. USA* 74:3171–75
121. Chow, L. T., Gelinas, R. E., Broker, T. R., Roberts, R. J. 1977. *Cell* 12:1–8
122. Broker, T. R., Chow, L. T., Dunn, A. R., Gelinas, R. E., Hassell, J. A., Klessig, D. F., Lewis, J. B., Roberts, R. J., Zain, B. S. 1977. *Cold Spring Harbor Symp. Quant. Biol.* 42:531–53
123. Berget, S. M., Berk, A. J., Harrison, T., Sharp, P. A. 1977. *Cold Spring Harbor Symp. Quant. Biol.* 42:523–30
124. Evans, R. M., Fraser, N., Ziff, E., Weber, J., Wilson, M., Darnell, J. E. 1977. *Cell* 12:733–39
125. Chow, L. T., Broker, T. R. 1978. *Cell* 15:497–510
126. Meyer, J., Neuwald, P. D., Lai, S. P., Maizel, J. V., Westphal, H. 1977. *J. Virol.* 21:1010–18
127. Bachenheimer, S., Darnell, J. E. 1975. *Proc. Natl. Acad. Sci. USA* 72:4445–49
128. Moss, B., Koczot, F. 1976. *J. Virol.* 17:385–92
129. Somer, S., Salditt-Georgieff, M., Bachenheimer, S., Darnell, J. E., Furuichi, Y., Morgan, M., Shatkin, A. J. 1976. *Nucleic Acid Res.* 3:749–65
130. Philipson, L., Wall, R., Glickman, G., Darnell, J. E. 1971. *Proc. Natl. Acad. Sci. USA* 68:2806–9
131. Gelinas, R. E., Roberts, R. J. 1977. *Cell* 11:533–44
132. Klessig, D. F. 1977. *Cell* 12:9–21
133. Ziff, E. B., Evans, R. M. 1978. *Cell.* 15:1463–75
134. Price, R., Penman, S. 1972. *J. Virol.* 9:621–26
135. Wallace, R. D., Kates, J. 1972. *J. Virol.* 9:627–35
136. Sharp, P. A., Gallimore, P. H., Flint, S. J. 1974. *Cold Spring Harbor Symp. Quant. Biol.* 39:457–74
137. Petterson, U., Tibbetts, C., Philipson, L. 1976. *J. Mol. Biol.* 101:479–501
138. Kitchingham, G. R., Lai, S. P., Westphal, H. 1977. *Proc. Natl. Acad. Sci. USA* 74:4392–95
139. Berk, A. J., Sharp, P. A. 1978. *Cell* 14:695–711
140. Reddy, V. B., Thimmappaya, B., Dhar, R., Subramaniam, K. N., Zain, B. S., Pan, J., Ghosh, P. K., Celma, M. L., Weissman, S. M. 1978. *Science* 200:494–502

141. Fiers, W., Contreras, R., Haegeman, G., Rogiers, R., Van de Voorde, A., Van Heuverswyn, H., Van Herreweghe, J., Volckaert, G., Ysebaert, M. 1978. *Nature* 273:113–20
142. Rundell, K., Collins, J. K., Tegtmeyer, P., Ozer, H. L., Lai, C. J., Nathans, D. 1977. *J. Virol.* 21:636–46
143. Tegtmeyer, P., Rundell, K., Collins, J. K. 1977. *J. Virol.* 21:647–57
144. Prives, C., Gilboa, E., Revel, M., Winocour, E. 1977. *Proc. Natl. Acad. Sci. USA* 74:457–61
145. Paucha, E., Harvey, R., Smith, R., Smith, A. E. 1977. *INSERM Colloq.* 69:189–98
146. Crawford, L. V., Cole, C. N., Smith, A. E., Paucha, E., Tegtmeyer, P., Rundell, K., Berg, P. 1978. *Proc. Natl. Acad. Sci. USA* 75:117–21
147. Ito, Y., Spurr, N., Dulbecco, R. 1977. *Proc. Natl. Acad. Sci. USA* 74:1259–63
148. Schaffhausen, B. S., Silver, J. E., Benjamin, T. L. 1978. *Proc. Natl. Acad. Sci. USA* 75:79–83
149. Hutchinson, M. A., Hunter, T., Eckhart, W. 1978. *Cell* 15:65–77
150. Berk, A. J., Sharp, P. A. 1978. *Proc. Natl. Acad. Sci. USA* 75:1274–78
151. Shenk, T. E., Carbon, J., Berg, P. 1976. *J. Virol.* 18:664–71
152. Benjamin, T. L. 1970. *Proc. Natl. Acad. Sci. USA* 67:394–99
153. Aloni, Y., Dhar, R., Laub, O., Horowitz, M., Khoury, G. 1977. *Proc. Natl. Acad. Sci. USA* 74:3686–90
154. Hsu, M.-T., Ford, J. 1977. *Proc. Natl. Acad. Sci. USA* 74:4982–85
155. Celma, M. L., Dhar, R., Pan, J., Weissman, S. M. 1977. *Nucleic Acids Res.* 4:2549–59
156. Haegeman, G., Fiers, W. 1978. *Nature* 273:70–73
157. Ghosh, P. K., Reddy, V. B., Swinscoe, J., Choudary, P. V., Lebowitz, P., Weissman, S. M. 1978. *J. Biol. Chem.* In press
158. Reddy, V. B., Ghosh, P. K., Swinscoe, J., Lebowitz, P., Weissman, S. 1979. *Proc. Natl. Acad. Sci. USA* 76: In press
159. Mellon, P., Duesberg, P. H. 1977. *Nature* 270:631–34
160. Kryzek, R. A., Collett, M. S., Lau, A. F., Perdue, M. L., Leis, J. P., Faras, A. J. 1978. *Proc. Natl. Acad. Sci. USA* 75:1284–88
161. Cordell, B., Weiss, S. R., Varmus, H. E., Bishop, J. M. 1978. *Cell* 15:79–91
162. Rothenberg, E., Donoghue, D. J., Baltimore, D. 1978. *Cell* 13:435–51
163. Fan, H., Verma, I. 1978. *J. Virol.* 26:468–78

164. White, R. L., Hogness, D. S. 1977. *Cell* 10:177–92
165. Glover, D. M., Hogness, D. S. 1977. *Cell* 10:167–76
166. Wellauer, P. K., David, I. B. 1977. *Cell* 10:193–212
167. Pellegrini, M., Manning, J., Davidson, N. 1977. *Cell* 10:213–24
168. Bos, J. L., Heyting, C., Borst, P. 1978. *Nature* 275:336–38
169. Dujon, B., Bolotin-Fukuhara, M., Coen, D., Deutsch, J., Netter, P., Slonimski, P. P., Weill, L. 1976. *Mol. Gen. Genet.* 143:131–65
170. Sanders, J. P. M., Heyting, C., Verbeet, M. P., Meijlink, F. C. P. W., Borst, P. 1977. *Mol. Gen. Genet.* 157:239–61
171. Rochaix, J. D., Malnoe, P. 1978. *Cell* 15:661–70
172. Goodman, H. M., Olson, M. V., Hall, B. D. 1977. *Proc. Natl. Acad. Sci. USA* 74:5453–57
173. Valenzuela, P., Venegas, A., Weinberg, F., Bishop, R., Rutter, W. J. 1978. *Proc. Natl. Acad. Aci. USA* 75:190–94
174. Beckmann, J. S., Johnson, P. F., Abelson, J. 1977. *Science* 196:205–8
175. Hutchinson, H. T., Hartwell, L. H., McLaughlin, C. S. 1969. *J. Bacteriol.* 99:807–14
176. Hopper, A. K., Banks, F., Evangelidis, V. 1978. *Cell* 14:211–20
177. Knapp, G., Beckmann, J. S., Johnson, P. F., Fuhrman, S. A., Abelson, J. 1978. *Cell* 14:221–36
178. Blatt, B., Feldmann, H. 1973. *FEBS Lett.* 37:129–33
179. O'Farrell, P. Z., Cordell, B., Valenzuela, P., Rutter, W. J., Goodman, H. M. 1978. *Nature* 274:438–45
180. Blanchard, J.-M., Weber, J., Jelenek, W., Darnell, J. E. 1978. *Proc. Natl. Acad. Sci. USA* 75:5344–48
181. Birkenmeier, E. H., Brown, D. D., Jordan, E. 1978. *Cell* 15:1077–86
182. Crick, F. H. C. 1979. *Science.* In press
182a. Lai, C.-J., Khoury, G. 1979. *Proc. Natl. Acad. Sci. USA* 76:71–75
183. Darnell, J. E. 1979. *Prog. Nucleic Acids Res. Mol. Biol.* In press
184. Alwine, J. C., Kemp, D. J., Stark, G. R. 1977. *Proc. Natl. Acad. Sci. USA* 74:5350–54
185. Ross, J. 1976. *J. Mol. Biol.* 106:403–20
186. Curtis, P. J., Weissmann, C. 1976. *J. Mol. Biol.* 106:1061–75
187. Kwan, S.-P., Wood, T. G., Lingrel, J. B. 1977. *Proc. Natl. Acad. Sci. USA* 74:178–82
188. Bastos, R. N., Aviv, H. 1977. *Cell* 11:641–50

189. Curtis, P. J., Mantei, N., Weissmann, C. 1977. *Cold Spring Harbor Symp. Quant. Biol.* 42:971–84
190. Ross, J., Knecht, D. A. 1978. *J. Mol. Biol.* 119:1–20
191. Kinniburgh, A. J., Mertz, J. E., Ross, J. 1978. *Cell* 14:681–93
192. Tilghman, S. M., Curtis, P. J., Tiemeier, D. C., Leder, P., Weissmann, C. 1978. *Proc. Natl. Acad. Sci. USA* 75:1309–13
193. Smith, K., Lingrel, J. B. 1978. *Nucleic Acids Res.* 5:3295–302
194. Curtis, P. J., Mantei, N., van den Berg, J., Weissmann, C. 1977. *Proc. Natl. Acad. Sci. USA* 74:3184–88
195. Gilmore-Hebert, M., Wall, R. 1978. *Proc. Natl. Acad. Sci. USA* 75:342–45
196. Gilmore-Hebert, M., Hercules, K., Komaromy, M., Wall, R. 1978. *Proc. Natl. Acad. Sci. USA* 75:6044–48
197. Roop, D. R., Nordstrom, J. L., Tsai, S. Y., Tsai, M.-J., O'Malley, B. W. 1978. *Cell* 15:671–85
197a. Schibler, U., Perry, R. P. 1977. *Nucleic Acids Res.* 12:4133–49
198. Perry, R. P., Kelley, D. E. 1977. *Cell* 8:433–42
198a. Nevins, J. R., Darnell, J. E., 1978. *Cell* 15:1477–93
199. Spirin, A. 1979. *Adv. Nucleic Acid Res. Mol. Biol.* In press
200. Hall, S. H., Crouch, R. J. 1977. *J. Biol. Chem.* 252:4092–97
201. Büsen, W., Hausen, P. 1975. *Eur. J. Biochem.* 52:179–90
202. Rech, J., Cathala, G., Jeanteur, P. 1976. *Nucleic Acids Res.* 3:2055–65
203. Ohtsuki, K., Groner, Y., Hurwitz, J. 1977. *J. Biol. Chem.* 252:483–91
204. Robertson, H. D., Mathews, M. B. 1973. *Proc. Natl. Acad. Sci. USA* 70:225–29
205. Nikolaev, N., Birge, C. H., Gotoh, S., Glazier, K., Schlessinger, D. 1975. *Brookhaven Symp. Biol.* 26:175–93
206. Koski, R. A., Bothwell, A. L. M., Altman, S. 1976. *Cell* 9:101–16

AUTHOR INDEX

1112 AUTHOR INDEX

1116   AUTHOR INDEX

SUBJECT INDEX

CUMULATIVE INDEXES

CONTRIBUTING AUTHORS, VOLUMES 44–48

A

Abelson, J., 48:1035–69
Acs, G., 45:375–408
Adhya, S., 47:967–96
Adler, J., 44:341–56
Alberts, B. M., 45:721–46
Amzel, L. M., 48:961–97
Andersen, H. C., 47:359–83
Anderson, W. B., 44:491–522
Avron, M., 46:143–55

B

Baldwin, R. L., 44:453–75
Baltz, R. H., 45:11–37
Barden, R. E., 46:385–413
Barker, H. A., 47:1–33
Beavo, J. A., 48:923–59
Bernhardt, J., 46:117–41
Beychok, S., 48:217–50
Beytía, E. D., 45:113–42
Bishop, J. M., 47:35–88
Bisswanger, H., 45:143–66
Blankenship, R. E., 47:635–53
Bleich, H., 44:477–90
Bloch, K., 46:263–98
Blumenthal, T., 48:525–48
Boyer, P. D., 46:957–66
Bradshaw, R. A., 47:191–216
Brady, R. O., 47:687–713
Braun, A., 44:19–43
Breslow, E., 48:251–74
Brimacombe, R., 47:217–49
Brown, M. S., 46:897–930
Bruice, T. C., 45:331–73
Burris, R. H., 45:409–25

C

Cantoni, G. L., 44:435–51
Carlson, S. S., 46:573–639
Carmichael, G. G., 48:525–48
Carpenter, G., 48:193–216
Casjens, S., 44:555–611
Chambon, P., 44:613–38
Champoux, J. J., 47:449–79
Chance, B., 46:967–80
Changeux, J.-P.,

47:317–57
Chargaff, E., 44:1–18
Chou, P. Y., 47:251–76
Chowdhry, V., 48:293–325
Christman, J. K., 45:375–408
Clarke, M., 46:797–822
Cohen, S., 48:193–216
Cohn, Z. A., 46:669–722
Cooper, P. K., 48:783–836
Cormier, M. J., 44:255–72
Cowburn, D., 44:477–90
Cox, G. B., 48:103–31
Coy, D. H., 47:89–128
Coyette, J., 48:73–101
Cozzarelli, N. R., 46:641–68
Craig, L. C., 44:477–90
Crane, F. L., 46:439–69
Cronan, J. E. Jr., 47:163–89
Crowther, R. A., 44:161–82
Czech, M. P., 46:359–84

D

Danielsson, H., 44:233–53
Davie, E. W., 44:799–829
Davies, D. R., 44:639–67
Davoli, D., 46:471–522
de Haro, C., 48:549–80
DeLuca, H. F., 45:631–66
de Meis, L., 48:275–92
DePierre, J. W., 46:201–62
DiRienzo, J. M., 47:481–532
Doisy, E. A., 45:1–9
Downie, J. A., 48:103–31
Drake, J. W., 45:11–37
Dubrow, R., 47:715–50
Dusart, J., 48:73–101

E

Edelstein, S. J., 44:209–32
Elgin, S. C. R., 44:725–74
Erlanger, B. F., 45:267–83
Ernster, L., 46:201–62; 981–95

F

Fareed, G. C., 46:471–522
Fasman, G. D., 47:251–76
Feldberg, R.,

A (col 3)

Felig, P., 44:933–55
Fessler, J. H., 47:129–62
Fessler, L. I., 47:129–62
Fisher, H. W., 48:649–79
Fox, I. H., 47:655–86
Frazier, W., 48:491–523
Frère, J.-M., 48:73–101
Fridovich, I., 44:147–59
Frieden, C., 48:471–89
Friedman, F. K., 48:217–50
Fry, M., 44:775–97
Fujikawa, K., 44:799–829
Furthmayr, H., 45:667–98

G

Gaber, B. P. 46:553–72
Ganesan, A. K., 48:783–836
Gefter, M. L., 44:45–78
Ghuysen, J.-M., 48:73–101
Gibson, F., 48:103–31
Glaser, L., 48:491–523
Goldberg, A. L., 45:747–803
Goldberg, N. D., 46:823–96
Goldstein, J. L., 46:897–930
Goldyne, M., 47:997–1029
Goody, R. S., 45:427–65
Gospodarowicz, D., 45:531–58
Gottesman, M., 47:967–96
Gotto, A. M. Jr., 44:183–207;
47:751–77
Granström, E., 44:669–95;
47:997–1029
Green, K., 44:669–95
Groner, Y., 47:1079–1126
Grossman, L., 44:19–43
Gunsalus, I. C., 44:377–407

H

Haddox, M. K., 46:823–96
Hamberg, M., 44:669–95;
47:997–1029
Hamlin, J. L., 47:715–50
Hammarström, S., 44:669–95;
47:997–1029
Hanawalt, P. C., 48:783–836
Hanson, K. R., 45:307–30
Hayaishi, O., 46:95–116

1160

CHAPTER TITLES, VOLUMES 44–48

ORDER FORM   ANNUAL REVIEWS INC.

Please list the volumes you wish to order. If you wish a standing order (the latest volume sent to you automatically each year), indicate volume number to begin order. Volumes not yet published will be shipped in month and year indicated. Prices subject to change without notice.

		Regular Order Please send:	Standing Order Begin with:
ANNUAL REVIEW SERIES Prices Postpaid, per volume			

Annual Review of ANTHROPOLOGY
    Vols. 1–8 (1972–79): $17.00 USA; $17.50 elsewhere
    Vols. 9–10 (1980–81): $20.00 USA; $21.00 elsewhere
    Vol. 11 (avail. Oct. 1982): $22.00 USA; $25.00 elsewhere   Vol(s). _____ Vol. _____

Annual Review of ASTRONOMY AND ASTROPHYSICS
    Vols. 1–17 (1963–79): $17.00 USA; $17.50 elsewhere
    Vols. 18–19 (1980–81): $20.00 USA; $21.00 elsewhere
    Vol. 20 (avail. Sept. 1982): $22.00 USA; $25.00 elsewhere   Vol(s). _____ Vol. _____

Annual Review of BIOCHEMISTRY
    Vols. 28–48 (1959–79): $18.00 USA; $18.50 elsewhere
    Vols. 49–50 (1980–81): $21.00 USA; $22.00 elsewhere
    Vol. 51 (avail. July 1982): $23.00 USA; $26.00 elsewhere   Vol(s). _____ Vol. _____

Annual Review of BIOPHYSICS AND BIOENGINEERING
    Vols. 1–9 (1972–80): $17.00 USA; $17.50 elsewhere
    Vol. 10 (1981): $20.00 USA; $21.00 elsewhere
    Vol. 11 (avail. June 1982): $22.00 USA; $25.00 elsewhere   Vol(s). _____ Vol. _____

Annual Review of EARTH AND PLANETARY SCIENCES
    Vols. 1–8 (1973–80): $17.00 USA; $17.50 elsewhere
    Vol. 9 (1981): $20.00 USA; $21.00 elsewhere
    Vol. 10 (avail. May 1982): $22.00 USA; $25.00 elsewhere   Vol(s). _____ Vol. _____

Annual Review of ECOLOGY AND SYSTEMATICS
    Vols. 1–10 (1970–79): $17.00 USA; $17.50 elsewhere
    Vols. 11–12 (1980–81): $20.00 USA; $21.00 elsewhere
    Vol. 13 (avail. Nov. 1982): $22.00 USA; $25.00 elsewhere   Vol(s). _____ Vol. _____

Annual Review of ENERGY
    Vols. 1–4 (1976–79): $17.00 USA; $17.50 elsewhere
    Vols. 5–6 (1980–81): $20.00 USA; $21.00 elsewhere
    Vol. 7 (avail. Oct. 1982): $22.00 USA; $25.00 elsewhere   Vol(s). _____ Vol. _____

Annual Review of ENTOMOLOGY
    Vols. 7–25 (1962–80): $17.00 USA; $17.50 elsewhere
    Vol. 26 (1981): $20.00 USA; $21.00 elsewhere
    Vol. 27 (avail. Jan. 1982): $22.00 USA; $25.00 elsewhere   Vol(s). _____ Vol. _____

Annual Review of FLUID MECHANICS
    Vols. 1–12 (1969–80): $17.00 USA; $17.50 elsewhere
    Vol. 13 (1981): $20.00 USA; $21.00 elsewhere
    Vol. 14 (avail. Jan 1982): $22.00 USA; $25.00 elsewhere   Vol(s). _____ Vol. _____

Annual Review of GENETICS
    Vols. 1–13 (1967–79): $17.00 USA; $17.50 elsewhere
    Vols. 14–15 (1980–81): $20.00 USA; $21.00 elsewhere
    Vol. 16 (avail. Dec. 1982): $22.00 USA; $25.00 elsewhere   Vol(s). _____ Vol. _____

Annual Review of MATERIALS SCIENCE
    Vols. 1–9 (1971–79): $17.00 USA; $17.50 elsewhere
    Vols. 10–11 (1980–81): $20.00 USA; $21.00 elsewhere
    Vol. 12 (avail. Aug. 1982): $22.00 USA; $25.00 elsewhere   Vol(s). _____ Vol. _____

Annual Review of MEDICINE: Selected Topics in the Clinical Sciences
    Vols. 1–3, 5–15, 17–31 (1950–52, 1954–64, 1966–80): $17.00 USA; $17.50 elsewhere
    Vol. 32 (1981): $20.00 USA; $21.00 elsewhere
    Vol. 33 (avail. Apr. 1982): $22.00 USA; $25.00 elsewhere   Vol(s). _____ Vol. _____

Annual Review of MICROBIOLOGY
    Vols. 15–33 (1961–79): $17.00 USA; $17.50 elsewhere
    Vols. 34–35 (1980–81): $20.00 USA; $21.00 elsewhere
    Vol. 36 (avail. Oct. 1982): $22.00 USA; $25.00 elsewhere   Vol(s). _____ Vol. _____

Annual Review of NEUROSCIENCE
    Vols. 1–3 (1978–80): $17.00 USA; $17.50 elsewhere
    Vol. 4 (1981): $20.00 USA; $21.00 elsewhere
    Vol. 5 (avail. Mar. 1982): $22.00 USA; $25.00 elsewhere   Vol(s). _____ Vol. _____

Annual Review of NUCLEAR AND PARTICLE SCIENCE
    Vols. 9–29 (1959–79): $19.50 USA; $20.00 elsewhere
    Vols. 30–31 (1980–81): $22.50 USA; $23.50 elsewhere
    Vol. 32 (avail. Dec. 1982): $25.00 USA; $28.00 elsewhere   Vol(s). _____ Vol. _____

Annual Review of NUTRITION
    Vol. 1 (1981): $20.00 USA; $21.00 elsewhere
    Vol. 2 (avail. July 1982): $22.00 USA; $25.00 elsewhere   Vol(s). _____ Vol. _____

(continued on reverse)

Annual Review of PHARMACOLOGY AND TOXICOLOGY
Vols. 1–3, 5–20 (1961–63, 1965–80): $17.00 USA; $17.50 elsewhere
Vol. 21 (1981): $20.00 USA; $21.00 elsewhere
Vol. 22 (avail. Apr. 1982): $22.00 USA; $25.00 elsewhere   Vol(s). _____ Vol. _____

Annual Review of PHYSICAL CHEMISTRY
Vols. 10–21, 23–30 (1959–70, 1972–79): $17.00 USA; $17.50 elsewhere
Vols. 31–32 (1980–81): $20.00 USA; $21.00 elsewhere
Vol. 33 (avail. Nov. 1982): $22.00 USA; $25.00 elsewhere   Vol(s). _____ Vol. _____

Annual Review of PHYSIOLOGY
Vols. 18–42 (1956–80): $17.00 USA; $17.50 elsewhere
Vol. 43 (1981): $20.00 USA; $21.00 elsewhere
Vol. 44 (avail. Mar. 1982): $22.00 USA; $25.00 elsewhere   Vol(s). _____ Vol. _____

Annual Review of PHYTOPATHOLOGY
Vols. 1–17 (1963–79): $17.00 USA; $17.50 elsewhere
Vols. 18–19 (1980–81): $20.00 USA; $21.00 elsewhere
Vol. 20 (avail. Sept. 1982): $22.00 USA; $25.00 elsewhere   Vol(s). _____ Vol. _____

Annual Review of PLANT PHYSIOLOGY
Vols. 10–31 (1959–80): $17.00 USA; $17.50 elsewhere
Vol. 32 (1981): $20.00 USA; $21.00 elsewhere
Vol. 33 (avail. June 1982): $22.00 USA; $25.00 elsewhere   Vol(s). _____ Vol. _____

Annual Review of PSYCHOLOGY
Vols. 4, 5, 8, 10–31 (1953, 1957, 1959–80): $17.00 USA; $17.50 elsewhere
Vol. 32 (1981): $20.00 USA; $21.00 elsewhere
Vol. 33 (avail. Feb. 1982): $22.00 USA; $25.00 elsewhere   Vol(s). _____ Vol. _____

Annual Review of PUBLIC HEALTH
Vol. 1 (1980): $17.00 USA; $17.50 elsewhere
Vol. 2 (1981): $20.00 USA; $21.00 elsewhere
Vol. 3 (avail. May 1982): $22.00 USA; $25.00 elsewhere   Vol(s). _____ Vol. _____

Annual Review of SOCIOLOGY
Vols. 1–5 (1975–79): $17.00 USA; $17.50 elsewhere
Vols. 6–7 (1980–81): $20.00 USA; $21.00 elsewhere
Vol. 8 (avail. Aug. 1982): $22.00 USA; $25.00 elsewhere   Vol(s). _____ Vol. _____

SPECIAL PUBLICATIONS	Prices Postpaid, per volume	Regular Order Please send:

Annual Reviews Reprints: Cell Membranes, 1975–1977
(published 1978)   Soft cover: $12.00 USA; $12.50 elsewhere   _____ copy(ies)

Annual Reviews Reprints: Cell Membranes, 1978–1980
(published 1981)   Hardcover $28.00 USA; $29.00 elsewhere   _____ copy(ies)

Annual Reviews Reprints: Immunology, 1977–1979
(published 1980)   Softcover $12.00 USA; $12.50 elsewhere   _____ copy(ies)

History of Entomology
(published 1973)   Clothbound $10.00 USA; $10.50 elsewhere   _____ copy(ies)

Intelligence & Affectivity: Their Relationship During Child Development, by Jean Piaget
(published 1981)   Hardcover $8.00 USA; $9.00 elsewhere   _____ copy(ies)

Telescopes for the 1980s
(avail. Aug. 1981)   Hardcover $27.00 USA; $28.00 elsewhere   _____ copy(ies)

The Excitement & Fascination of Science, Volume 1
(published 1965)   Clothbound $6.50 USA; $7.00 elsewhere   _____ copy(ies)

The Excitement & Fascination of Science, Volume 2
(published 1978)   Hardcover $12.00 USA; $12.50 elsewhere   _____ copy(ies)
                      Soft cover $10.00 USA; $10.50 elsewhere   _____ copy(ies)

To: ANNUAL REVIEWS INC, 4139 El Camino Way, Palo Alto, CA 94306 USA   (Tel. 415-493-4400)

Please enter my order for the publications checked above.

Amount of remittance enclosed $ _____   California residents, please add applicable sales tax.
Please bill me ☐   Prices subject to change without notice.
Institutional purchase order # _____

Name _____

Address _____

_____ Zip Code _____

Signed _____ Date _____

☐ Please send free copy of the current *Prospectus* each year.
☐ Send free brochure listing contents of recent back volumes for Annual Review(s) of _____